T0199401

Generalized Calculus with Applications to Matter and Forces

MATHEMATICS AND PHYSICS FOR SCIENCE AND TECHNOLOGY

Series Editor

L.M.B.C. Campos

Director of the Center for Aeronautical
and Space Science and Technology

Lisbon Technical University

MATHEMATICS AND PHYSICS FOR SCIENCE AND TECHNOLOGY

Generalized Calculus with Applications to Matter and Forces

L.M.B.C. Campos

Director of the Center for Aeronautical
and Space Science and Technology
Lisbon Technical University

CRC Press
Taylor & Francis Group
Boca Raton London New York

CRC Press is an imprint of the
Taylor & Francis Group, an **informa** business

CRC Press
Taylor & Francis Group
6000 Broken Sound Parkway NW, Suite 300
Boca Raton, FL 33487-2742

First issued in paperback 2019

© 2014 by Taylor & Francis Group, LLC
CRC Press is an imprint of Taylor & Francis Group, an Informa business

No claim to original U.S. Government works

ISBN-13: 978-1-4200-7115-3 (hbk)
ISBN-13: 978-0-367-37872-1 (pbk)

This book contains information obtained from authentic and highly regarded sources. Reasonable efforts have been made to publish reliable data and information, but the author and publisher cannot assume responsibility for the validity of all materials or the consequences of their use. The authors and publishers have attempted to trace the copyright holders of all material reproduced in this publication and apologize to copyright holders if permission to publish in this form has not been obtained. If any copyright material has not been acknowledged please write and let us know so we may rectify in any future reprint.

Except as permitted under U.S. Copyright Law, no part of this book may be reprinted, reproduced, transmitted, or utilized in any form by any electronic, mechanical, or other means, now known or hereafter invented, including photocopying, microfilming, and recording, or in any information storage or retrieval system, without written permission from the publishers.

For permission to photocopy or use material electronically from this work, please access www.copyright.com (http://www.copyright.com/) or contact the Copyright Clearance Center, Inc. (CCC), 222 Rosewood Drive, Danvers, MA 01923, 978-750-8400. CCC is a not-for-profit organization that provides licenses and registration for a variety of users. For organizations that have been granted a photocopy license by the CCC, a separate system of payment has been arranged.

Trademark Notice: Product or corporate names may be trademarks or registered trademarks, and are used only for identification and explanation without intent to infringe.

Library of Congress Cataloging-in-Publication Data

Campos, Luis Manuel Braga da Costa.
   Generalized calculus with applications to matter and forces / Luis Manuel Braga da Costa Campos.
      pages cm. -- (Mathematics and physics for science and technology ; 3)
   Includes bibliographical references and index.
   ISBN 978-1-4200-7115-3
   1. Calculus, Operational. 2. Integral transforms. 3. Theory of distributions (Functional analysis) I. Title.

   QA432.C28 2014
   530.15'5--dc23
                                                      2013036877

Visit the Taylor & Francis Web site at
http://www.taylorandfrancis.com

and the CRC Press Web site at
http://www.crcpress.com

Contents

List of Tables, Notes, Diagrams, Classifications, and Lists

Tables

Notes

Diagrams

Classifications

Lists

Series Preface

The aim of the Mathematics and Physics for Science and Technology series is to describe mathematical methods as they are applied to model natural physical phenomena and to solve scientific and technological problems. The emphasis is on the application, including formulation of the problem, detailed solution, and interpretation of results. Mathematical methods are presented in detail to justify every step of the solution and to avoid superfluous assumptions.

The main areas of physics are covered, as follows:

- Mechanics of particles, rigid bodies, deformable solids, and fluids
- Electromagnetism, thermodynamics, and statistical physics as well as their classical, relativistic, and quantum formulations
- Interactions and combined effects (e.g., thermal stresses, magnetohydrodynamics, plasmas, piezoelectricity, and chemically reacting and radiating flows)

The examples and problems covered in this book include natural phenomena in our environment, geophysics, and astrophysics; the technological implications in various branches of engineering; and other mathematical models in the biological, economic, and social sciences.

The coverage of the areas of mathematics and branches of physics is sufficient to lay the foundations of all branches of engineering, as follows:

- *Mechanical engineering*, including machines, engines, structures, and vehicles
- *Civil engineering*, including structures and hydraulics
- *Electrical engineering*, including circuits, waves, and quantum effects
- *Chemical engineering*, including transport phenomena and multiphase media
- *Computer engineering*, including analytical and computational methods and associated algorithms

Interdisciplinary areas such as electromechanics and aerospace engineering are focused on. These require combined knowledge of several areas and have an increasing importance in modern technology.

Analogies are applied in an efficient and concise way, across distinct disciplines, while also stressing the differences and aspects specific to each area, as follows:

- Potential flow, electrostatics, magnetostatics, gravity field, steady heat conduction, and plane elasticity and viscous flow
- Acoustic, elastic, electromagnetic, internal, and surface waves
- Diffusion of mass, electricity, and momentum

The acoustic, internal, inertial, and magnetic waves appear combined as magneto-acoustic–gravity–inertial (MAGI) waves in a compressible, ionized, stratified, and rotating fluid. The simplest exact solutions of the MAGI wave equation require special functions. Thus, the topic of MAGI waves combines three subjects—gravitational field, fluid mechanics, and electromagnetism—and uses a complex analysis, differential equations, and special functions. This is not such a remote subject, since many astrophysical phenomena involve a combination of several of these effects, as does the technology of controlled nuclear fusion. The latter is the main source of energy in stars and in the universe; if harnessed, it would provide a clear and inexhaustible source of energy on Earth. Closer to our everyday experience there is a variety of electromechanical and control systems that use modern interdisciplinary technology. The aim of this series is to build up knowledge seamlessly from undergraduate to research level, across a range of subjects, to cover contemporary or likely interdisciplinary needs. This requires a consistent treatment of all subjects so that their combination fits together as a whole.

The approach followed in this series is a combined study of mathematics, physics, and engineering so that the theoretical concepts are presented along with practical examples. The mathematical methods are applied without delay to "real" problems, not just to exercises. The electromechanical and other analogies combine different disciplines, which is the basis of much of modern interdisciplinary science and technology. Starting with the simpler mathematical methods, and consolidating them with the detailed solutions of physical and engineering problems, gradually widens the range of topics that can be covered. The traditional method of treating individual disciplines separately remains a possibility, by focusing on mathematical disciplines (such as complex functions) or sets of applications (such as fluid mechanics). Combined multidisciplinary study has the advantage of connecting mathematics, physics, and technology at an earlier stage. Moreover, preserving that link provides a broader view of the subject and the ability to innovate. Innovation requires an understanding of the aims, the physical phenomena that can implement them, and the mathematical methods that quantify the expected results. The combined interdisciplinary approach to the study of mathematics, physics, and engineering is thus a direct introduction to scientific discovery and technological innovation.

Preface

Complex Analysis with Applications to Flows and Fields (Book 1) dealt with the theory of complex functions that is used in *Transcendental Representations with Applications to Solids and Fluids* (Book 2) to address elementary and higher transcendental functions. This book (Book 3) completes the trilogy on the theory of functions by considering generalized functions, such as the Heaviside unit jump, Dirac unit impulse, and its derivatives of all orders. The latter exist because of the fundamental property of generalized functions of being infinitely differentiable. The two main approaches to generalized functions are presented (1) as a nonuniform limit of a family of ordinary functions and (2) as a functional over a set of test functions from which properties are inherited. The second approach is developed more extensively to include multidimensional generalized functions whose arguments are ordinary functions of several variables. This allows integration along curves, surfaces, and hypersurfaces and relates to the invariant differential operator's gradient, curl, and divergence.

The theory of complex (Book 1) and transcendental (Book 2) functions was applied to plane potential fields, and the theory of generalized functions extends the applications to potential fields in three (and higher) dimensions, including axisymmetric cases such as (1) potential, irrotational, and solenoidal flows; (2) electro- and magnetostatics; (3) gravity field; (4) the mechanics of deformable bodies, like strings, membranes, plates, or bars; and (5) steady heat conduction and other analogous problems. The complex and transcendental functions can be used to represent singularities in the plane like multipoles, and the generalized functions (Book 3) extend this to three and higher dimensions. The calculation of the influence or Green function, defined as the fundamental solution of a differential equation forced by a unit impulse, is one of the major applications of generalized functions; it applies to the five classes (1) to (5) of problems mentioned before, as well as to others.

The generalized functions may be used to represent (1) singularities like point masses, electric charges or currents, and concentrated forces or torques as well as (2) their continuous or discontinuous distribution along curves, surfaces, and other subspaces. The influence or Green function specifies the field or response associated with this forcing. For linear differential equations with linear boundary and/or initial conditions, the principle of superposition specifies the fields or responses for arbitrary forcing. This may take the form of convolution integrals whose existence is proved in the context of functional analysis. For symmetric influence functions, the reciprocity principle allows the interchange of sources (forcing) and fields (responses). The reciprocity principle applies to self-adjoint differential operators with suitable boundary and/or initial conditions. Thus, Book 3, *Generalized Calculus with Applications to Matter and Forces*, covers generalized functions to complete the theory of functions and its applications to potential fields, relating them to broader follow-on topics like differential equations.

Organization of the Book

Book 3 is organized in a similar way to Book 2, which follows the organization in Book 1 and the general pattern of the series (Mathematics and Physics for Science and Technology). It consists of 10 chapters. Chapters 1, 3, 5, and 7 present mathematical developments applied to physical and/or technological problems covered in Chapters 2, 4, 6, and 8. Chapter 9 deals with fundamental mathematical aspects. Finally, Chapter 10 consists of 20 examples based on the contents of the preceding chapters as test cases or generalizations.

Each chapter consists of nine sections, with a variable number of subsections. The contents of each chapter and section are previewed by a short introduction. A conclusion at the end of each chapter provides a summary, captured succinctly in classifications, diagrams, figures, panels, lists, and tables. The Notes section briefly addresses related subjects, which may receive fuller treatment in subsequent books in this series. The chapters are preceded by lists of contents and mathematical and physical symbols and followed by references, bibliography, and an index of subjects.

The parts of this book are indicated by a "chapter-section-subsection" format—for example, Chapter 3, Section 3.6, Subsection 3.6.2. The equations are numbered sequentially in each chapter and are indicated in parentheses—for example, (4.62) refers to equation 62 in Chapter 4 and (4.62c) refers to the third equation of that set. When referring to earlier books, roman numerals appear at the beginning—for example, II.8.6.1 refers to Book 2, Chapter 8, Section 6, Subsection 1; (I.30.24) refers to Book 1, Chapter 30, Equation 24; and C.I.29.1 indicates Classification 1 of Chapter 29 in Book 1.

Acknowledgments

The third book of the series justifies renewing some of the acknowledgments made in the first two books, especially to those who contributed more directly to give this book its final form, namely, L. Sousa and S. Pernadas for help with the manuscripts and J. Coelho for all the drawings. Last but not least, I would like to acknowledge my wife for helping in the preparation of this work.

Author

Luis Manuel Braga da Costa Campos graduated in 1972 as a mechanical engineer from the Instituto Superior Tecnico (IST) of Lisbon Technical University. The tutorials as student (1970) were followed by a career in the same institution through all levels: assistant (1972), assistant with tenure (1974), auxiliary professor (1978), assistant professor (1982), and chair of applied mathematics and mechanics (1985). He has been serving as the coordinator of undergraduate and postgraduate degree courses on aerospace engineering since their creation in 1991. He also serves as the coordinator of the Applied and Aerospace Mechanics Group in the Department of Mechanical Engineering and is the founder and director of the Center for Aeronautical and Space Science and Technology.

Campos received his doctorate on "waves in fluids" from the Engineering Department of Cambridge University, England (1977). This was followed by a Senior Rouse Ball Scholarship at Trinity College, also on leave from IST. His first sabbatical (1984) was as a senior visitor in the Department of Applied Mathematics and Theoretical Physics at Cambridge University, England. His second sabbatical (1991) was as an Alexander von Humboldt Scholar at the Max-Planck Institut fur Aeronomic in Katlenburg-Lindau, Germany. Further sabbaticals abroad were not pursued due to major commitments at his home institution, in addition to his participation in scientific meetings, individual or national representation in international institutions, and collaborative research projects.

Campos received the von Karman Medal in 2002 from the Advisory Group for Aerospace Research and Development (AGARD) and Research and Technology Organization (RTO). He served as a vice chairman of the System Concepts and Integration Panel and as a chairman of the Flight Mechanics Panel and of the Flight Vehicle Integration Panels at AGARD/RTO. He was also a member of the Flight Test Techniques Working Group, collaborating on the creation of an independent flight test capability active in Portugal since 1986 and used in national and international projects, including Eurocontrol and the European Space Agency (ESA). His role in the ESA has included serving in various committees as national representative up to the level of the council of ministers.

Campos has participated in activities sponsored by the European Union, including 27 research projects with industry, research, and academic institutions. He has been a member of various committees and has served as vice chairman of the Aeronautical Science and Technology Advisory Committee and on the Space Advisory Panel on the future role of the EU in space. He has also been a member of the Space Science Committee of the European Science Foundation, which liaises with the Space Science Board of the National Science Foundation of the United States. Campos has been a member of the Committee for Peaceful Uses of Outer Space (COPUOS) of the United Nations and has represented the committee in various capacities at international events. He has served as a member and vice chairman of the Portuguese Academy of Engineering; a fellow of the Royal Aeronautical Society, Royal Astronomical Society, and Cambridge Philosophical Society; and associate fellow of the American Institute of Aeronautics and Astronautics.

Campos has several publications to his credit, including 9 books, 132 papers in 56 journals, and 213 communications to symposia. Also acting as reviewer for 29 different journals, including

Mathematics Reviews. His research interests focus on acoustics, magnetohydrodynamics, special functions, and flight dynamics. His work on acoustics deals with the generation, propagation, and refraction of sound in flows with mostly aeronautical applications. His work on magnetohydrodynamics is concerned with magneto-acoustic–gravity–inertial waves in solar–terrestrial and stellar physics. Developments on special functions have been mostly based on differintegration operators, generalizing the ordinary derivative and primitive to complex order. His work on flight dynamics deals with aircraft and rockets, including trajectory optimization, performance, stability, control, and atmospheric disturbances.

The range of topics from mathematics to physics to engineering fits with the aims and contents of this series. Most of the contents of the series have been the subject of university lectures, which test its accessibility to students; the choice of topics and examples relates partially to the scientific and research interests of the author. Campos' professional activities are balanced by other cultural and humanistic interests, which are reflected in a literary work. Complementary nontechnical interests include classical music (mostly orchestral and choral), plastic arts (painting, sculpture, architecture), social sciences (psychology and biography), history (classical, renaissance, and European overseas expansion in the fifteenth to seventeenth centuries), and technology (automotive, photo, audio). He is listed in various biographical publications, including *Who's Who in the World* since 1986, *Who's Who in Science and Technology* since 1994, and *Who's Who in America* since 2011.

Mathematical Symbols

The mathematical symbols presented in the following are those of more common use in the context of (1) sets, quantifiers, and logic; (2) numbers, ordering, and bounds; (3) operations, limits, and convergence; (4) vectors, matrices, and tensors; (5) derivatives and differential operators; (6) functional spaces; (7) geometries; and (8) ordinary, generalized, and special functions. It concludes with a list of functional spaces, most but not all of which appear in this book. The section where the symbol first appears may be indicated after a colon—for example, "III.1.1.4" means Subsection 1.1.4 of Book 3 and "E.I.30.20" means Example 30.20 of Book 1 (see Organization of the Book at the end of the Preface).

Sets, Quantifiers, and Logic

Sets

$A \equiv \{x:...\}$—set whose elements x have the property...
$A \cup B$—union of sets A and B
$A \cap B$—intersection of sets A and B
$A \supset B$—set A contains set B
$A \subset B$—set A is contained in set B

Quantifiers

$\forall_{x \in A}$—for all x belonging to A holds...
$\exists_{x \in A}$—there exists at least one x belonging to A such that...
$\exists^1_{x \in A}$—there exists one and only one x belonging to A such that...
$\exists^\infty_{x \in A}$—there exist infinitely many x belonging to A such that...

Logic

$a \wedge b$—a and b
$a \vee b$—or (inclusive): a or b or both
$a \veebar b$—or (exclusive): a or b but not both
$a \Rightarrow b$—implication: a implies b
$a \Leftrightarrow b$—equivalence: a implies b and b implies a
$a \not\Rightarrow b$—nonimplication: a may not imply b

Constants

$e = 2.7182\ 81828\ 45904\ 52353\ 60287$
$\pi = 3.1415\ 92653\ 58979\ 32384\ 62643$
$\gamma = 0.5772\ 15664\ 90153\ 28606\ 06512$
$\log 10 = 2.3025\ 85092\ 99404\ 56840\ 179915$

Numbers, Ordering, and Operations

Types of Numbers

|C—complex numbers: I.1.3
|C^n—ordered sets of n complex numbers
|F—transfinite numbers: II.9.7–II.9.9
|H—hypercomplex numbers
|I—irrational numbers: real nonrational numbers: I.1.2, II.9.7.2
|L—rational numbers: ratios of integers: I.1.1
|N—natural numbers: positive integers: I.1.1
|N_0—nonnegative integers: zero plus natural numbers: I.1.1
|P—prime numbers: numbers without divisors
|Q—quaternions: I.1.9
|R—real numbers: I.1.2, II.9.7.3
|R^n—ordered sets of n real numbers
|Z—integer numbers: I.1.1, II.9.7.1

Complex Numbers

|...|—modulus of complex number: I.1.4
arg (...)—argument of complex number: I.1.4
Re (...)—real part of complex number: I.1.3
Im (...)—imaginary part of complex number: I.1.3
... *—conjugate of complex number: I.1.6

Relations and Ordering

$a > b$—a greater than b
$a \geq b$—a greater or equal to b
$a = b$—a equal to b
$a \leq b$—a smaller or equal to b
$a < b$—a smaller than b
sup (...)—supremum: smallest number larger or equal than all numbers in the set...
max (...)—maximum: largest number in set...
min (...)—minimum: smallest number in set...
inf (...)—infimum: largest number smaller or equal than all numbers in set...

Operations between Numbers

$a + b$—sum: a plus b
$a - b$—difference: a minus b
$a \times b$—product: a times b
a/b—ratio: a divided by b (alternative $a{:}b$)
a^b—power: a to the power b
$\sqrt[b]{a}$—root: root b of a

Functions, Limits, and Convergence

Limits and Values

lim—limit when x tends to a: $x \to a$: I.11.1

l.i.m.—limit in the mean

$a \sim O(b)$—a is of order b: $\lim b/a \neq 0, \infty$: I.19.7

$a \sim o(b)$—b is of lower order than a: $\lim b/a = 0$: I.19.7

$f(a)$—value of function f at point a

$f(a + 0)$—right-hand limit at a: III.1.8.1

$f(a - 0)$—left-hand limit at a: III.1.8.1

$f_{(n)}(a)$—residue at pole of order n at a: I.15.8, II.1.1.4

$\bar{B}$ or M—upper bound: $|f(z)| \leq \bar{B}$ for z in … : I.39.2

$\underline{B}$ or m—lower bound: $|f(z)| \geq \underline{B}$ for z in … : I.39.2

$f \circ g$—composition of functions f and g: N.I.38.2

$f * g$—convolution of functions f and g: III.7.4

Iterated Sums and Products

$\sum\limits_{a}$—sum over a set

$\sum\limits_{n=a}^{b}$—sum from $n = a$ to $n = b$

$\sum\limits_{n,m=a}^{b}$—double sum over $n, m = a,…,b$

$\prod\limits_{a}$—product over a set

$\prod\limits_{n=a}^{b}$—product from $n = a$ to $n = b$

Convergence

A.C.—absolutely convergent: I.21.2

A.D.—absolutely divergent: I.21.2

C.—convergent: I.21.2

C.C.—conditionally convergent: I.21.2, II.9.4

Cn—converges to class n: (C0 ≡ C): II.9.6

D.—divergent: 21.1, II.9.2.1

N.C.—nonconvergent: divergent or oscillatory: I.21.1

O.—oscillatory: I.21.1, II.9.2.1

T.C.—totally convergent: I.21.7

U.C.—uniformly convergent: I.21.5

applies to

   —power series: I.21.1, II.1

   —series of fractions: I.36.6, II.1.2

   —infinite products: I.36.6, II.1.4

   —continued fractions: II.1.6

Integrals

$\int \ldots dx$—primitive of …with regard to x: I.13.1

$\int_y \ldots dx$—indefinite integral of… at y: I.13.2

$\int_a^b \cdots dx$—definite integral of… between a and b: I.13.2

$\fint \cdots dx$—Cauchy principal value of integral: I.17.8

$\fint \cdots dx$—Hadamard finite value of integral: III.9.2

$\int^{(z+)}$—integral along a loop around z in the positive (counterclockwise) direction: I.13.5

$\int^{(z-)}$—idem in the negative (clockwise) direction: 13.5

$\int_L$—integral along a path L: I.13.2

$\oint_C^{(+)}$—integral along a closed path or loop C in the positive direction: I.13.5

$\oint_C^{(-)}$—integral along a closed path or loop C in the negative direction: I.13.5

Vectors, Matrices, and Tensors

Vectors

$\vec{A} \cdot \vec{B}$—inner product
$\vec{A} \wedge \vec{B}$—outer product
$\vec{A} \cdot (\vec{B} \wedge \vec{C})$—mixed product
$\vec{A} \wedge (\vec{B} \wedge \vec{C})$—double outer product
$\left|\vec{A}\right|$—modulus
ang $(\vec{A}, \vec{B})$—angle of vector $\vec{B}$ with vector $\vec{A}$

Matrices

δ_j^i—identity matrix: III.5.8.6
$\overset{c}{A}_j^i$—matrix of cofactors: N.III.9.10
$\overset{I}{A}_j^i$—inverse matrix: N.III.9.10
$\mathrm{Det}\left(A_j^i\right)$—determinant of matrix: N.III.9.11
$\mathrm{Ra}\left(A_j^i\right)$—rank of matrix: N.III.5.5.1

Tensors

$(\vartheta)T^{i_1\ldots i_q}_{j_1\ldots j_q}$—tensor with weight ϑ, contravariance p and covariance q: N.III.9.14

$\overline{T}^{i_1\ldots i_q}_{j_1\ldots j_q}$—tensor capacity (with weight $\vartheta = 1$): N.III.9.14

$\widetilde{T}^{i_1\ldots i_q}_{j_1\ldots j_q}$—tensor density (with weight $\vartheta = -1$): N.III.9.14

$\delta^{i_1\ldots i_N}_{j_1\ldots j_N}$—identity symbol: N.III.9.9

$e_{i_1\ldots i_N}$—covariant permutation symbol: N.III.9.9

$e_{\ell_1\ldots i_N}$—contravariant permutation symbol: N.III.9.9

Operations

$T_{(i_1\ldots i_p)}$—mixing: N.III.9.12

$T_{[i_1\ldots i_p]}$—alternation: N.III.9.12

T^{ij}_{ik}—contraction: N.III.9.12

$T^i_{jk}\,\delta^j_{i\ell}$—transvection: N.III.9.12

Derivatives and Differential Operators

Differentials and Derivatives

$d\Phi$—differential of Φ

$d\Phi/dt$—derivative of Φ with regard to t

$\partial\Phi/\partial t \equiv \partial_t\Phi$—partial derivative of Φ with regard to t

$\partial\Phi/\partial x_i \equiv \partial_i\Phi \equiv \Phi_{,i}$—partial derivative of Φ with regard to x_i

$\partial^n\Phi/\partial x_{i_1}\ldots\partial x_{i_n} \equiv \partial_{i_1\ldots i_1}\Phi \equiv \Phi_{i_1\ldots i_n}$—$n$-order partial derivative of Φ with regard to $x_{i_1}, \ldots, x_{i_n}$

Vector Operators

$\nabla\Phi = \partial_i\Phi$—gradient of a scalar: III.6.1.3

$\nabla\cdot\vec{A} = \partial_i A_i$—divergence of a vector: III.6.1.4

$\nabla\wedge\vec{A} = e_{ijk}\partial_j A_k$—curl of a vector: III.6.1.5

$\nabla^2 = \partial_{ii} = \partial_i\partial_i$—scalar Laplacian: III.6.1.6

∇^2—vector Laplacian: III.6.1.7

$\bar{\nabla}^2$—modified Laplacian: III.6.2.6

Tensor Operators

$\partial_{[i_{M+1}}U_{i_1\ldots i_n]}$—curl of a covariant M-vector: N.III.9.20

$\partial_{i_{M+1}}U^{i_1\ldots i_{M+1}}$—divergence of a contravariant $(M + 1)$ – vector density: N.III.9.20

Adjointness

$\left\{L\left(\partial/\partial x_i\right)\right\}\Phi$—linear differential operator: III.7.6

$\left\{\left[\left(\partial/\partial x_i\right)\right]\right\}\Psi$—adjoint differential operator: III.7.6

$W\left(\Phi, \Psi\right)$—bilinear concomitant: III.7.6

Functional Spaces

The spaces of functions are denoted by calligraphic letters, in alphabetical order:

...(a, b)—set of functions over interval from a to b

omission of interval: set of functions over real line $)-\infty, +\infty($

$\mathcal{A}$ (...)—analytic functions in ...: I.27.1: III.3.3.5

$\bar{\mathcal{A}}$ (...)—monogenic functions in ...: I.31.1

$\mathcal{B}$ (...)—bounded functions in ...: $\mathcal{B} \equiv \mathcal{B}^0$: I.13.3

$\mathcal{B}^n$ (...)—functions with bounded nth derivative in ...

$\mathcal{C}$ (...)—continuous functions in ...: $\mathcal{C} \equiv \mathcal{C}^0$: I.11.2; III.3.1.3

$\mathcal{C}^n$ (...)—functions with continuous nth derivative in ... III.3.2.1

$\bar{\mathcal{C}}$ (...)—piecewise continuous functions in ...: $\bar{\mathcal{C}} \equiv \bar{\mathcal{C}}^0$

$\bar{\mathcal{C}}^n$ (...)—functions with piecewise continuous nth derivative in ...

$\tilde{\mathcal{C}}$ (...)—uniformly continuous functions in...: I.13.4

$\tilde{\mathcal{C}}^n$ (...)—functions with uniformly continuous nth derivative in...

$\mathcal{D}$ (...)—differentiable functions in ...: $\mathcal{D} \equiv \mathcal{D}^0$: I.11.2; III.3.1.3

$\mathcal{D}^n$ (...)—n-times differentiable functions in ... III.3.2.1

$\mathcal{D}^\infty$ (...)—infinitely differentiable functions or smooth in ...: I.27.1; III.3.3.4

$\bar{\mathcal{D}}$ (...)—piecewise differentiable functions in ...: $\bar{\mathcal{D}} \equiv \bar{\mathcal{D}}^0$

$\bar{\mathcal{D}}^n$ (...)—functions with piecewise continuous nth derivative in ...

$\mathcal{E}$ (...)—Riemann integrable functions in ...: I.13.2; III.3.1.1

$\bar{\mathcal{E}}$ (...)—Lebesgue integrable functions in ...

$\mathcal{F}$ (...)—functions of bounded oscillation (or bounded fluctuation or bounded variation) in ...; $\mathcal{F} \equiv \mathcal{E} \equiv \mathcal{F}^0$: I.27.9.5

$\mathcal{F}^n$ (...)—functions with nth derivative of bounded oscillation (or fluctuation or variation) in ...

$\mathcal{G}$ (...)—generalized functions (or distributions) in ...: III.3.4.1

$\mathcal{H}$ (...)—harmonic functions in ...: I.11.4, II.4.6.4, III.9.3

$\mathcal{H}_2$ (...)—biharmonic functions in ...: II.4.6.4

$\mathcal{H}_n$ (...)—multiharmonic functions of order n in ...: II.4.6.6

$\mathcal{I}$(...)—integral functions in ...: I.27.9, II.1.1.7

$\mathcal{I}_m$ (...)—rational–integral functions of degree m in... $\mathcal{I} \equiv \mathcal{I}_0$: I.27.9, II.1.1.9

$\mathcal{J}$ (...)—square integrable functions with a complete orthogonal set of functions—*Hilbert space*

$\bar{\mathcal{K}}$ (...)—Lipshitz functions in ...

$\mathcal{K}^n$ (...)—homogeneous functions of degree n in ...

$\mathcal{L}^1$ (...)—absolutely integrable functions in ...: N.III.3.7

$\mathcal{L}^2$ (...)—square integrable functions in ...: III.7.1.1

$\mathcal{L}^p$ (...)—functions with power p of modulus integrable in ... —*normed* space: III.7.1.1

$\mathcal{M}^+$ (...)—monotonic increasing functions in ...: I.9.1.1

$\mathcal{M}_0^+$ (...)—monotonic nondecreasing functions in ...: I.9.1.1

$\mathcal{M}_0^-$ (...)—monotonic nonincreasing functions in...: I 9.1.2

$\mathcal{M}^-$ (...)—monotonic decreasing functions in ...: I.9.1.2

$\mathcal{N}$(...)—null functions in ...

$\mathcal{O}$ (...)—orthogonal systems of functions in ...: II.5.7.2

$\bar{\mathcal{O}}$ (...)—orthonormal systems of functions in ...: II.5.7.2

$\tilde{\mathcal{O}}$ (...)—complete orthogonal systems of functions in ...: II.5.7.2

$\mathcal{P}$ (...)—polynomials in ...: I.27.7, II.1.1.6

$\mathcal{P}_n$ (...)—polynomials of degree n in ...: I.27.7, II.1.1.6

$\mathcal{Q}$ (...)—rational functions in...: I.27.7, II.1.1.6

Q_n^m (...)—rational functions of degrees n, m in...

$\mathcal{R}$ (...)—real functions, that is, with the real line as the range

S (...)—complex functions, that is, with the complex plane as the range

$\mathcal{T}^0$ (...)—functions with compact support, that is, which vanish outside a finite interval: III.3.3.2

$\mathcal{T}^n$ (...)—excellent functions of order n: n-times differentiable functions with compact support: III.3.3.2

$\mathcal{T}^\infty$ (...)—excellent functions: smooth or infinitely differentiable functions with compact support: II.3.3.4

$\mathcal{U}$ (...)—single-valued functions in...: I.9.1

$\tilde{\mathcal{U}}$ (...)—injective functions in...: I.9.1

$\bar{\mathcal{U}}$ (...)—surjective functions in...: I.9.1

$\tilde{\bar{\mathcal{U}}}$ (...)—bijective functions in...: I.9.1

$\mathcal{U}_n$ (...)—multivalued functions with n branches in...: I.7.1

$\mathcal{U}_\infty$ (...)—many-valued functions in...: I.7.2

$\mathcal{U}^1$ (...)—univalent functions, in ...: I.37.4

$\mathcal{U}^m$ (...)—multivalent functions taking m values in...: I.37.4

$\mathcal{U}^\infty$ (...)—manyvalent functions in ...: N.I.37.4

$\mathcal{U}_n^m$ (...)—multivalued multivalent functions with n branches and m values in ...: N.I.37.4

$\mathcal{V}$(...)—good functions, that is, with slow decay at infinity faster than some power: III.3.3.1

$\mathcal{V}_n$ (...)—good or slow decay functions of degree n, that is, with decay at infinity faster than the inverse of a polynomial of degree n: III.3.3.1

$\bar{\mathcal{V}}$ (...)—fairly good or slow growth functions, that is, with growth at infinity slower than some power: III.3.3.1

$\mathcal{V}^n$ (...)—fairly good or slow growth functions of degree n, that is, with growth at infinity slower than a polynomial of degree n: III.3.3.1

$\mathcal{V}_\infty$ (...)—very good or fast decay functions, that is, with faster decay at infinity than any power: III.3.3.1

$\mathcal{V}_\infty^\infty$ (...)—superlative functions, that is, smooth functions of fast decay: III.3.3.6

$\mathcal{W}_q^p$ (...)—functions with generalized derivatives of orders up to q such that the power p of the modulus is integrable ... —*Sobolev* space

X_0 (...)—self-inverse linear functions in ....: I.38.1

X_1 (...)—linear functions in ...: I.35.2

X_2 (...)—bilinear, homographic, or Mobius functions in ...: I.35.4

X_3 (...)—self-inverse bilinear functions in ...: I.35.5

X_a (...)—automorphic functions in ...: I.37.6

X_m (...)—isometric mappings in ...: I.35.1

X_r (...)—rotation mappings in ...: II.35.1

X_t (...)—translation mappings in ...: II.35.1

$\mathcal{Y}$ (...)—meromorphic functions in...: I.27.9, II.1.1.10

$\mathcal{Z}1$ (...)—polymorphic functions in...: I.27.9, II.1.1.10

Geometries

$\mathbb{A}_N$—rectilinear: N.III.9.6

$\mathbb{M}_N$—metric: N.III.9.35

$\mathbb{N}_N$—orthogonal: N.III.9.40

$\mathbb{O}_N$—orthonormal: N.III.9.40

$\mathbb{X}_N$—curvilinear: N.III.9.6

$\mathbb{X}_N^2$—curvilinear with curvature: N.III.9.17

Ordinary, Generalized, and Special Functions

$\cos(z)$—circular cosine: II.5.1.1
$\cosh(z)$—hyperbolic cosine: II.5.1.1
$\cot(z)$—circular cotangent: II.5.2.1
$\coth(z)$—circular hyperbolic cotangent: II.5.2.1
$\csc(z)$—circular cosecant: II.5.2.1
$\operatorname{csch}(z)$—hyperbolic cosecant: II.5.2.1
$\operatorname{erf}(z)$—error function: III.1.2.1
$\exp(z)$—exponential: II.3.1
$\log(z)$—logarithm of base e: II.3.5
$\log_a(z)$—logarithm of base a: II.3.7
$\sec(z)$—circular secant : II.5.2.1
$\operatorname{sech}(z)$—hyperbolic secant: II.5.2.1
$\operatorname{sgn}(z)$—sign function: I.36.4.1, III.1.7.2
$\sin(z)$—circular sine: II.5.1.1
$\sinh(z)$—hyperbolic sine: II.5.1.1
$\tan(z)$—circular tangent: II.5.2
$\tanh(z)$—hyperbolic tangent: II.5.2
B_n—Bernoulli number: II.7.1.8
E_n—Euler number: II.7.1.7
$F(c;z)$—hypogeometric function: II.1.9.1
$F(a;c;z)$—confluent hypergeometric function: II.3.9.8
$F(a,b;c;z)$—Gaussian hypergeometric function: II.3.9.7
$_pF_q(a_1,...,a_p;c_1...,c_q;z)$—generalized hypergeometric function: E.I.30.20, N.II.3.3
$H(x)$—Heaviside unit jump: III.1.2
$H_n(x)$—Hermite polynomial of degree n: III.1.1.6
$\delta(x)$—Dirac unit impulse: III.1.3
$\Gamma(x)$—Gamma function: N.III.1.8

Miscellaneous Functions

$p(x)$—probability density function: N.III.1.3
$h_\sigma(x)$—family of functions tending to the unit jump: III.1.2.2
$H_\sigma(x)$—family of functions tending to the unit jump. III.1.2.1
$C(x)$—central characteristic function of a random process: N.III.1.5
$D(x)$—noncentral characteristic function: N.III.1.6
$C(\vec{x})$—generalized function line integral: III.5.6.1
$G(x;\xi)$—influence or Green function: III.2.3.1
$\bar{G}(x;\xi)$—reciprocal influence function: III.2.3.6
$P(x)$—cumulative probability function: N.III.1.3
$S(\vec{x})$—generalized function hypersurface integration: III.5.6.2
$\gamma_\sigma(x)$—family of functions tending to the unit impulse with cylindrical symmetry: E.III.10.12.2
$\delta_\sigma(x)$—family of functions tending to the unit impulse: III.1.3.2
$\Delta_\sigma(x)$—family of functions tending to the unit impulse: III.1.3.2

Physical Symbols

The physical symbols are divided into lower- and uppercase Latin and Greek letters, with each set listed in alphabetical order. The same letter may have different meanings in contexts where no ambiguity can arise. The section where the symbol first appears may be indicated by a colon, for example, "5.8.3" refers to Section 5.8.3 and "N.9.35" to Note 9.35 (always of Book 3).

Lowercase Arabic Letters

a—acceleration: N.2.7;
   —half length of a straight segment: N.5.7.
   —radius of a circle: N.5.12.
   —radius of a cylinder: N.5.17.
   —radius of sphere: 6.5.1.
a_n—coefficients: 2.6.12
b—radius of axisymmetric fairing or body in a flow: 6.3.4
b_n—coefficients: 2.2.1
c—half distance between monopoles: 6.4.1
   —distance between centers of spheres: 6.9.1
   —speed of light in vacuo: N.5.6
c_{em}—speed of electromagnetic waves: N.7.14
c_0—speed of sound: N.7.14
c_n—coefficients: 2.8.3
d_n—coefficients: N.4.6
e—electric charge: N.5.4
$\vec{e}_i$—unit base vector: 2.1.4
e_{ij}—two-dimensional permutation symbol: 5.8.3
e_{ijk}—three-dimensional permutation symbol: 5.7.4
$e_{i_1 \cdots i_N}$—N-dimensional covariant permutation symbol: N.9.9
$e^{i_1 \cdots i_N}$—N-dimensional contravariant permutation symbol: N.9.9
e^i_{j}—contravariant base vectors: N.9.38
e^i_j—covariant base vectors: N.9.38
f—shear stress: 2.1.2
$\vec{f}$—force density per unit volume: N.6.6
g—acceleration of gravity: 2.4.1
   —number of degrees of freedom: 4.2.8
   —function in boundary condition: 9.1.2
   —determinant of covariant metric tensor: N.9.35

g_{ij}—covariant metric tensor: N.9.35

g^{ij}—contravariant metric tensor: N.9.35

h—height of support of an elastic string: 2.3.1

  —height of cross section of a bar: 4.1.7

  —aiming distance: 6.5.2

  —surface thermal conductivity: 9.1.2

h_i—scale factors: 6.1.1

$\vec{j}$—density of electric current: N.5.6

k—curvature: 2.1.2

  —square root of the ratio of tension to bending stiffness: 4.6.1

  —step of a helix: N.5.17

  —resilience of spring: 7.7.3

  —wavenumber: N.7.2

  —thermal conductivity: N.8.5

k_{ij}—thermal conductivity tensor: N.8.5

ℓ—length of a deflected string: 2.6.12

$\vec{\ell}$—relative position vector of observer x and source $\vec{y}$: 8.3.1

m—mass of a body: 6.5.5

m_0—added mass of entrained fluid: 6.5.5

$\bar{m}$—total mass of body plus entrained fluid : 6.5.5

p—probability density: N.1.3

  —linear momentum: N.2.7

  —resilience of straight spring: 4.2.7

  —pressure: 6.2.9

p_0—stagnation pressure: 6.2.10

P_*—dynamic pressure: 6.3.5

p_∞—pressure in free stream: 6.3.5

$\vec{p}$—linear momentum: 8.5.2

q—nonlinearity parameter: 2.6.14

  —density of electric charge: N.5.4

  —density of flow sources/sinks: 8.9.5

  —mass flux: N.5.29

  —resilience of rotary spring: 4.2.7

r—polar coordinate: 1.1.1

  —cylindrical coordinate: distance from axis: 6.1.2

s—arc length: 2.1.1

$\vec{s}$—unit tangent vector: 5.7.6

t—time: N.2.7

u—longitudinal displacement: N.2.5

$\vec{u}$—displacement vector: 4.2.1

v—velocity: N.2.7

$\vec{v}$—velocity vector: 6.2.1

w—heat rate of source/sink: N.8.5

x—Cartesian coordinate: 1.1.1

$\vec{x}$—position vector of observer: N.5.4

$\vec{x}_\pm$—position vector of observer relative to multipole images in a plane: 8.7.1

$\vec{x}_{\pm\pm}$—position vector of observer relative to multipole images on orthogonal planes: 8.8.1

y—Cartesian coordinate: 2.1

y_G—coordinate of center of mass: 4.1.2

$\vec{y}$—position vector of source: N.5.4

z—Cartesian coordinate: 6.1.1

$\vec{z}$—normalized relative position of source and observer: N.8.13

Uppercase Arabic Letters

A—area: 6.3.5.

  —Jacobian of a rectilinear coordinate transformation: N.9.6

A_n—coefficients in decomposition of the unit impulse of a function: 5.1.2

$\vec{A}$—vector potential: 5.8.2

$A_i^{i'}, A_{i'}^{i}$—direct and inverse matrices of a linear coordinate transformation: N.9.6

B—bending stiffness of elastic bar: 4.1.3

B_n—coefficients in the decomposition of the unit derivative impulse of a function: 5.1.3

$\vec{B}$—magnetic induction vector: N.5.6

C—torsional stiffness: N.2.4

C_D—drag coefficient: 6.3.5

C_i—eigenvalues of constitutive tensor : N.8.12

C_n—central moment: N.1.3

    —coefficient of Fourier series: N.3.3

C_p—pressure coefficient: 6.3.5

C_{ij}—constitutive tensor: N.8.11

$\bar{C}_{ij}$—symmetric part of constitutive tensor: N.8.11

D—distance function: N.5.7

  —drag force: 6.3.5

  —bending stiffness of elastic plate: 7.8.12

D_n—coefficients in the generalized function line integral: 5.6.1

$\dot{D}$ —dilatation: N.7.14

$\vec{D}$—electric displacement vector: N.7.12

E—Young's modulus of elasticity: N.2.5

$\bar{E}$—nonlinear modulus of elasticity: N.2.6

E_d—elastic energy: 2.2.1

E_v—kinetic energy: 6.5.4

$\vec{E}$—electric field vector: N.5.4

F—transverse force: 2.1.2

    —inertia force: N.2.7

$\vec{F}_+$—electric force of repulsion: 8.9.16

$\vec{F}_-$—electric force attraction: 8.9.17

$\vec{F}_a$—external applied force: N.8.7

$\vec{F}_b$—compression force: N.8.10

$\vec{F}_e$—electric force: N.8.4

$\vec{F}_g$—gravity force vector: N.8.3

$\vec{F}_\ell$—vortical force: N.8.10

$\vec{F}_m$—magnetic force vector: N.8.10

$\vec{F}_v$—hydrodynamic force: 8.9.15

G—gravitational constant: N.8.3

G_N—influence function of Laplace operator in N-dimensions: N.9.2.2

$\vec{G}$—heat flux: N.8.5

H—enthalpy: 6.2.9
H_0—stagnation enthalpy: 6.2.9
$\vec{H}$—magnetic field vector: N.7.13
I—moment of inertia: 4.1.2
J—Jacobian: 5.5.4
$\vec{J}$—electric current: N.5.6
J^i—mass flux: N.9.14
L—horizontal distance between supports: 2.3.1
$\vec{L}$—angular momentum: 8.5.2
M—bending moment: 4.1.2
M_n—central moment: N.1.4
M_z—axial moment: N.2.4
$\vec{M}$—magnetic moment: 8.5.2
      —nonunit normal to a surface: 5.6.2
N—number of turns of helix: N.5.19
      —dimension of a space: N.9.6
$N_\pm$—bending moment at the supports: 4.3.1
$\vec{N}$—unit normal to a surface: 5.6.2
P—concentrated load: 2.3.1
P_b—critical load for buckling: 4.9.2
P_c—critical load for collapse instability: 4.9.7
P_d—critical load for the displacement instability: 4.9.9
P_e—critical load for twist instability: 4.9.9
P_0—moment of monopole: 6.4.3
P_1—moment of axial dipole: 6.4.3
P_2—moment of axial quadrupole: 6.4.4
P_n—moment of axial multipole of order n: 6.4.5
$\vec{P}_1$—vector moment of multiaxial dipole: 6.4.7–6.4.8
P_{ij}—matrix moment of multiaxial quadrupole: 6.4.7–6.4.8
$P_{i_1\ldots i_n}$—multiplicity moment of multiaxial multipole of order n: 6.4.7
P_n—moment of 2^n-multipole 3.2, e.g., monopole P_0, dipole P_1, quadrupole P_2: 6.4.7
Q—concentrated torque: 4.5.1
      —volume flow rate of point source/sink: 6.3.3.
R—radius of curvature of a curve: 4.1.1
      —spherical coordinate: radial distance from origin: 6.1.2
$R_\pm$—transverse reaction forces at the supports: 2.3.1
      —distance of observer from two ends of a straight segment: N.5.7
      —distance of a point on a sphere from the north/south pole: 6.8.5
      —distance of an observer from multipole images on a plane: 8.7.1
$R_{\pm\pm}$—distance of an observer from multipole images on two orthogonal planes: 8.8.1
R_0—distance of an observer from multipole images: 8.7.1
$R_{1,2}$—distance of an observer from multipoles and their images on a sphere: 8.9.1
S—area element of a surface: 1.1.1
      —strain: N.2.5
      —distance from the center of second sphere: 6.9.2
      —entropy: 6.2.9
$S_\pm$—longitudinal reaction forces at the supports: 2.6.8
S_N—area of the hypersphere in N-dimensional space: 5.9.1
S_{ij}—strain tensor: 4.1.1

T—tangential tension along an elastic string or bar: 2.1.1
    —momentum flux: N.5.29
    —temperature: 6.2.9
T_{ij}—stress tensor: 4.2.1
T^{ij}—momentum flux tensor: N.9.14
V—volume: N.9.13
V_N—volume of the hypersphere in N-dimensional space: 5.9.1
$V^{i_1 \ldots i_M}$—volume in an M-dimensional subspace of an N-dimensional space: N.9.13
$V_{i_{M+1} \ldots i_N}$—dual volume of an M-dimensional subspace of an N-dimensional space: N.9.13
W—work: 2.2.2
    —bilinear comitant: 7.6.3
X—longitudinal apex coordinate: 2.8.1
    —Jacobian of coordinate transformation: N.9.6
$\bar{X}$—dimensionless apex coordinates: 2.9.8
$X_i^{i}, X_{i'}^{i}$—direct and inverse matrices of a curvilinear coordinate transformation: N.9.6
Y—transverse apex coordinate: 2.8.1
$\bar{Y}$—dimensionless apex coordinate: 2.9.8
$\vec{Y}$—complex bilinear concomitant: N.7.17
Z—shape of string in apex coordinates: 2.8.1
$\bar{Z}$—shape of string in dimensionless apex coordinates: 2.9.8

Small Greek Letters

α—starting angle of a helix: N.5.17
    —diffusivity: N.7.2
α_e—ohmic electrical diffusivity: N.7.14
β—ending angle of a helix: N.5.17
χ—gauge potential: 5.8.2
δ—maximum deflection: 2.4.2
δ_j^{i}—identity matrix: 5.8.6
$\delta_{j_1 \ldots j_n}^{i_1 \ldots i_n}$ —Kronecker delta or identity symbol: N.9.9
ε—dielectric permittivity: N.5.4
ε_{ij}—dielectric permittivity tensor: 7.8.4
$\varepsilon_{i_1 \ldots i_N}$—$N$-dimensional covariant permutation tensor: N.9.36
$\varepsilon^{i_1 \ldots i_N}$—$N$-dimensional contravariant permutation tensor: N.9.36
ϕ—angle relative to the center of the second sphere: 6.9.2
γ—angle of inclination of a helix: N.5.17
η—coordinate of catenary: 2.9.1
    —static shear viscosity: N.6.6
$\bar{\eta}$—kinematic shear viscosity: N.6.6
φ—polar angle: 1.1.1
    —cylindrical coordinate: azimuth: 6.1.2
    —spherical coordinate: longitude: 6.1.2
κ—thermal conductivity scalar in isotropic medium: N.8.5
κ_{ij}—thermal conductivity tensor in anisotropic medium: N.8.5
λ—damping: 7.7.5
μ—mean value: N.1.3
    —magnetic permeability: N.5.6
μ_{ij}—magnetic permeability tensor: N.8.9

ν—mass density per unit volume: 2.7.1
　　—ratio of nonlinear to linear elastic moduli: N.4.4
ω—angular frequency: N.7.2
θ—maximum angle of inclination: 2.4.2
　　—spherical coordinate: colatitude measured from polar axis: 6.1.2
θ_n—colatitudes in hyperspherical coordinates ($n = 1, ..., N-2$): 9.7.2
　　—colatitudes in hypercylindrical coordinates ($n = 1, ..., N-3$): 9.7.3
$\theta_\pm$—angles of observer from two ends of a straight segment: N.5.7
$\theta_{1,2}$—angles of observer from multipole and its image on a sphere: 8.9.3
$\bar{\theta}$—transformation of θ_1 to reciprocal point on the sphere: 8.9.3
ρ—mass density per unit length or volume: 2.4.1
σ—root mean square (r.m.s.) value: N.1.3
　　—ohmic electric conductivity of isotropic material: 6.8.3
　　—electric charge per unit area: 8.7.1
σ_N—area of the unit hypersphere in N-dimensional space: 5.9.1
τ—torsion rate: N.2.4
$\bar{\varpi}$—vorticity: 6.2.7
ψ—dimensionless parameter: 2.9.1
ζ—transverse string or bar: 2.1.1
　　—bulk viscosity: N.6.6

Uppercase Greek Letters and Others

Φ—scalar potential: 5.8.1
$\bar{\Phi}$—reciprocal potential: 6.7.2
Φ_-—potential for second sphere theorem: 6.7.2
$\tilde{\Phi}$—potential for third sphere theorem: 6.7.3
$\hat{\Phi}$—potential for fourth sphere theorem: 6.7.4
Φ_e—electric potential: 6.8.4
Φ_g—gravity potential: N.8.3
Φ_j—potential of electric currents: 6.8.4
Φ_n—potential of multipole of order n: 8.3.7
Φ_v—velocity potential: 6.8.4
$\Phi^{(1,2)}$—potential of first and second sphere in a flow: 6.9.1
Λ—source or forcing function: N.8.11
$\vec{\Omega}$—angular velocity of rotation: 8.9.12
Ψ—field or stream function: 5.8.3
$\bar{\Psi}$—reciprocal field function: 6.7.1
Ψ_+—field function for first sphere theorem: 6.7.1
Ψ_n—field or stream function axisymmetric multipole of order n: 8.3.7

Introduction

An isolated singularity of a complex function is a point- (line-) singularity in a two (three)-dimensional space; in order to represent point-singularities in three-dimensional space, generalized functions (also called distributions) may be used. The concept of generalized function extends beyond that of ordinary functions and applies to any number of variables, for example, one, two, three, or more respectively for real, complex, spatial, or higher-dimensional generalized functions. The replacement of an ordinary function by the corresponding generalized function allows derivation to any order (Chapter 1) but is restricted to linear operations (Chapter 3) in generalized calculus. Thus, generalized functions are most useful in linear problems involving singularities, for example, finding the shape of an elastic string (Chapter 2) (bar [Chapter 4]) subject to concentrated forces (or torques); these problems have nonlinear extensions for large slopes and hence also for large deflections. The concentrated forces and moments are represented by one- (Chapters 1 and 3), two-, or three-dimensional (Chapters 5 and 7) generalized functions; other examples include line- (point-) monopoles and dipoles (Chapter I.12 and Chapter 8) that can be extended from two (three) dimensions to N dimensions (Chapter 9). The latter can be used to specify the potential field around a cylinder (Chapters I.24, I.26, and I.28) (sphere [Chapter 6]), and higher-order multipoles represent other two- (three-) dimensional body shapes; the method of images of singularities can be used in two (three) dimensions for straight lines (Chapter I.16) (plane walls [Chapter 8]) and cylindrical (Chapter I.28) (spherical [Chapter 6]) obstacles. All these examples concern solutions of the Laplace or Poisson equation and hence specify harmonic functions (Chapter 9) that have a number of properties analogous in two and three or more dimensions. The use of generalized functions to represent singularities of linear differential equations extends to other types (Chapter 7), for example, the heat and wave equations.

1

Limit of a Sequence of Functions

Although the generalized function is a generalization of the concept of ordinary function, the rigorous definition of the former is based on the latter. One way to approach the theory of generalized functions is to consider the limit of a sequence of ordinary functions: (1) if all functions of the sequence take a fixed value at a given point or have a given integral, then the limit function will inherit the same property, regardless of whether the convergence is uniform or not; and (2) if the convergence is nonuniform, then, even if the functions of the sequence are continuous or differentiable or analytic, the limit function that is the generalized function may not inherit these properties (Section I.23.4), that is, it could be discontinuous or even singular. For example, the Gaussian functions with unit integral and variance σ are all analytic; the limit σ → 0 of zero variance is the Dirac impulse (Section 1.3) that is a generalized function, which is zero everywhere except at the origin, where it is infinite. The nature of the singularity at the origin is specified by the integral across it being unity. This suggests that the primitive of the Dirac impulse is the Heaviside unit step that is a generalized function that is zero before the origin and unity after (Section 1.2). It can be verified rigorously that this unit step function is the limit as σ → 0 of the integrals of Gaussian functions that specify error functions. If instead the derivatives of the Gaussian functions are considered, the limit as σ → 0 specifies (Section 1.4) the derivatives of the Dirac impulse; since the Gaussian functions are analytic, they have derivatives of all orders, and thereby derivatives of all orders of the Dirac impulse and Heaviside unit step can be obtained (Section 1.5). The Heaviside unit step specifies a (Section 1.8) discontinuity with jump unity and can be used to represent any function with isolated, finite discontinuities (Section 1.8); taking the corresponding generalized function, the derivatives of all orders can be obtained (Section 1.9), including at the points of discontinuity, in terms of Dirac impulse and its derivatives. This result relies on the properties of the unit step and impulse as generalized functions (Section 1.6); these can be used to construct other generalized functions, for example, the sign and the modulus (Section 1.7).

1.1 Evaluation of Integrals of Gaussian Functions

The first approach to the theory of generalized functions (Sections 1.1 through 1.4) will rely on sequences of Gaussian and related functions; a second approach is presented subsequently (in Chapters 3 and 5). As a preliminary result is performed the evaluation of the Gaussian integral (Subsection 1.1.1) in two ways (Subsection 1.1.2), followed by some extensions (Subsections 1.1.3 through 1.1.7).

1.1.1 Evaluation of the Basic Gaussian Integral

The evaluation of the integral of a Gaussian function is a nonelementary result, to be used subsequently: *the Gaussian function* $\exp(-x^2)$ *is integrable along the real axis* $-\infty < x < +\infty$, *and its bilateral (1.1a) and unilateral (1.1b) improper integrals of the first kind are given respectively by*

$$\int_{-\infty}^{+\infty}\exp\left(-x^2\right)dx = \sqrt{\pi}, \quad \int_0^\infty \exp\left(-x^2\right)dx = \frac{\sqrt{\pi}}{2}, \quad \sqrt{\pi} = 1.772453851. \tag{1.1a–c}$$

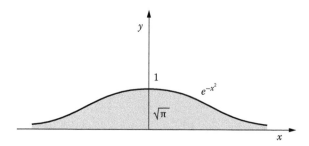

FIGURE 1.1 The area under a Gaussian leads to the basic Gaussian integral that is the starting point for several generalizations (Section 1.1, Notes 1.9 through 1.12).

The result (1.1a) follows from (1.1b) because the Gaussian function is even; it implies that the area between the function and the real axis (Figure 1.1) is (1.1c). The definition (I.17.2b) of improper integral of the first kind implies that (1.1b) can be calculated as (1.2a):

$$\lim_{A\to\infty} I(A) = \frac{\pi}{4}, \quad I(A) = \left\{ \int_0^A \exp(-x^2)\,dx \right\}^2 = \int_0^A dx \int_0^A dy\, \exp(-x^2 - y^2), \tag{1.2a,b}$$

where the square (1.2b) was used. Thus, (1.2a,b) is the result to be proved next. The Gaussian function that appears as the integrand in (1.2b) is positive everywhere, and thus its integral over the square of vertices at $(0, 0)$, $(0, A)$, (A, A), and $(A, 0)$ is larger (smaller) than (Figure 1.2) the integral over a quarter-circle $x, y > 0$ of radius equal to the length of one side A (diagonal $A\sqrt{2}$):

$$J(A) < I(A) < J(A\sqrt{2}), \quad J(A) \equiv \int\int_{x^2+y^2\le A^2;\, x,y>0} \exp(-x^2 - y^2)\,dx\,dy. \tag{1.3a,b}$$

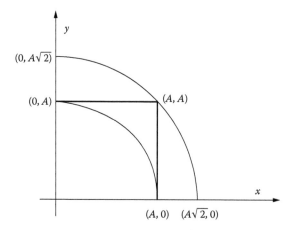

FIGURE 1.2 Since the primitive of the Gaussian is not an elementary function, the Gaussian integral (Figure 1.1) is evaluated taking the square (Figure 1.2); since the integrand is positive a lower (upper) bound is obtained replacing the square by an interior (exterior) quarter circle with radius equal to the side (diagonal). As the radius tends to infinity the fast decay of the Gaussian integrand implies that the lower and upper bounds coincide, determining uniquely the value of the Gaussian integral.

The change is made from Cartesian (x, y) to polar (r, φ) coordinates (I.1.11) $\equiv$ (1.4a):

$$x^2 + y^2 = r^2; \quad h_r = 1, \quad h_\varphi = r: \quad dx\, dy = dS = r\, dr\, d\varphi, \tag{1.4a–d}$$

and dS is the area element (1.4d) using the scale factors (I.11.44a,b) $\equiv$ (1.4b,c) for polar coordinates. The evaluation of the integral (1.3b) over the quarter-circle becomes elementary:

$$J(A) = \int_0^{\pi/2} d\varphi \int_0^A \exp(-r^2) r\, dr = \frac{\pi}{2}\left[-\frac{1}{2}\exp(-r^2) \right]_0^A = \frac{\pi}{4}\left[1 - \exp(-A^2) \right]. \tag{1.5}$$

The integral over the square (1.2b) is bounded (1.3a) by the integrals over two quarter-circles:

$$\frac{\pi}{4} = \lim_{A\to\infty} \frac{\pi}{4}\left[1 - \exp(-A^2) \right] \leq \lim_{A\to\infty} I(A) \leq \lim_{A\to\infty} \frac{\pi}{4}\left[1 - \exp(-2A^2) \right] = \frac{\pi}{4}, \tag{1.6}$$

and since the latter tend to the same value as $A \to \infty$, the result (1.2a) follows. Note that the preceding proof concerns both (1) the existence of the Gaussian integral, implied by the upper bound in (1.6); and (2) its value, determined by the coincidence of upper and lower bounds in (1.6).

1.1.2 Alternative Evaluation of the Gaussian Integral

An alternate proof, separating these two aspects, follows. The Gaussian function has upper bounds:

$$\exp(-x^2) \leq \begin{cases} 1 & \text{if } |x| \leq 1, \tag{1.7a} \\ \exp(-|x|) & \text{if } |x| \geq 1, \tag{1.7b} \end{cases}$$

implying that the unilateral improper integral of the first kind has an upper bound:

$$\int_0^\infty \exp(-x^2)\, dx \leq \int_0^1 dx + \int_1^\infty e^{-x} dx = 1 + \frac{1}{e}, \tag{1.8}$$

and hence exists. Thus, the square of the integral (1.1b) may be evaluated in the (x, y)-plane over any region whose boundary points all tend to infinity; choosing a circle of radius $R \to \infty$, polar coordinates (1.4a–d) can be used in

$$\left\{ \int_0^\infty \exp(-x^2)\, dx \right\}^2 = \int_0^\infty dx \int_0^\infty dy\, \exp(-x^2 - y^2) = \int_0^{\pi/2} d\varphi \int_0^\infty r\, \exp(-r^2)\, dr = \frac{\pi}{2}\left[\frac{-\exp(-r^2)}{2} \right]_0^\infty = \frac{\pi}{4}. \tag{1.9}$$

The proof (1.9) of (1.2b) is essentially equivalent to the preceding (1.6). The two basic Gaussian integrals (1.1a,b) are sufficient for the introduction to distributions or generalized functions in the present Chapter 1; thus, it is possible to proceed directly to the next Section 1.2. The remaining Subsections 1.1.3 through 1.1.7 of the present Section 1.1 concern extensions of the Gaussian integrals (1.1a,b) for future use; further extensions of these Gaussian integrals (Section 1.2) appear in Notes 1.9 through 1.12 and are related to the Gamma function (Note 1.8).

1.1.3 Gaussian Integrals with Trigonometric Factors

There are a number of simple extensions of the Gaussian integral, which will be of use in several subsequent occasions. The change of variable (1.10a) leads from (1.1a) to (1.10b):

$$x = \frac{y}{\sqrt{\alpha}}: \quad \int_{-\infty}^{+\infty} \exp\left(-\alpha x^2\right) dx = \frac{1}{\sqrt{\alpha}} \int_{-\infty}^{+\infty} \exp\left(-y^2\right) dy = \sqrt{\frac{\pi}{\alpha}}. \tag{1.10a,b}$$

If $\alpha = 1$, then (1.10b) reduces to (1.1a). As α increases, the Gaussian decays faster with increasing x, and thus the area under the curve (Figure 1.3a) decreases (1.10b). The integral can be estimated numerically with good accuracy because the integrand is a monotonic decreasing function. A further change of independent variable (1.11a),

$$y = x\sqrt{\alpha} - \frac{i\beta}{2\sqrt{\alpha}}: \quad -y^2 = -x^2\alpha + i\beta x + \frac{\beta^2}{4\alpha}, \quad dy = \sqrt{\alpha}\, dx, \tag{1.11a-c}$$

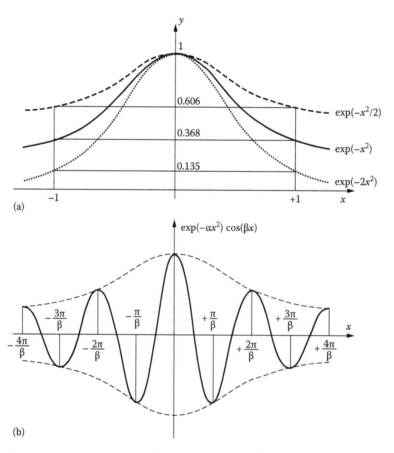

(a)

(b)

FIGURE 1.3 The Gaussian (a) may be used as the envelope of a cosine (b), and its integral generalizes the Gaussian integral (Figures 1.1 and 1.2).

leads by (1.11b,c) to (1.12):

$$\int_{-\infty}^{+\infty} \exp\left(-\alpha x^2 + i\beta x\right)dx = \exp\left(-\frac{\beta^2}{4\alpha}\right)\frac{1}{\sqrt{\alpha}}\int_{-\infty}^{+\infty} \exp\left(-y^2\right)dy. \tag{1.12}$$

This can be evaluated using (1.2a):

$$\int_{-\infty}^{+\infty} \exp\left(-\alpha x^2 + i\beta x\right)dx = \int_{-\infty}^{+\infty} \exp\left(-\alpha x^2\right)\cos(\beta x)dx = \sqrt{\frac{\pi}{\alpha}}\exp\left(-\frac{\beta^2}{4\alpha}\right). \tag{1.13}$$

The real part may be taken in (1.13) since the imaginary part is zero because the integrand then involves $\sin(\beta x)$, which is an odd function of x. If $\beta = 0$, then (1.13) reduces to (1.10b). The integrand in (1.13) is a cosine (Figure 1.3b) with a Gaussian envelope (Figure 1.3a). The integral (1.13) decreases for (1) larger α since the area under the envelope reduces; (2) larger β since a shorter period of oscillation leads to a closer cancellation between successive intervals of the real axis where the integrand is alternatively positive and negative. The integral is difficult to estimate numerically with good accuracy for large β because (1) the integrand oscillates rapidly; (2) the sum of terms with opposite signs increases the absolute error; (3) the final value is small; (4) the relative error, that is, the ratio of (3) to (4), may not be small.

1.1.4 Gaussian Integrals with Products of Power and Trigonometric Factors

The integrand in (1.13) is uniformly continuous (Section I.21.4) in β, and thus the integral (1.13) can be differentiated (I.13.14) with regard to the parameter β under integral sign n times:

$$\int_{-\infty}^{+\infty} (ix)^n \exp\left(-\alpha x^2 + i\beta x\right)dx = \frac{\partial^n}{\partial\beta^n}\left\{\int_{-\infty}^{+\infty} \exp\left(-\alpha x^2 + i\beta x\right)dx\right\}$$

$$= \frac{\partial^n}{\partial\beta^n}\left[\sqrt{\frac{\pi}{\alpha}}\exp\left(-\frac{\beta^2}{4\alpha}\right)\right]. \tag{1.14}$$

The integral (1.14) is not zero for $n = 2m$ even ($n = 2m + 1$ odd) if the real (1.15a) [imaginary (1.15b)] part is taken:

$$\int_{-\infty}^{+\infty} x^{2m} \exp\left(-\alpha x^2\right)\cos(\beta x)dx = (-)^m \frac{\partial^{2m}}{\partial\beta^{2m}}\left[\sqrt{\frac{\pi}{\alpha}}\exp\left(-\frac{\beta^2}{4\alpha}\right)\right], \tag{1.15a}$$

$$\int_{-\infty}^{+\infty} x^{2m+1} \exp\left(-\alpha x^2\right)\sin(\beta x)dx = (-)^{m+1} \frac{\partial^{2m+1}}{\partial\beta^{2m+1}}\left[\sqrt{\frac{\pi}{\alpha}}\exp\left(-\frac{\beta^2}{4\alpha}\right)\right]. \tag{1.15b}$$

For $m = 0$, (1.15a) reduces to (1.13). The integrand in (1.14) is a sinusoid (Figure 1.4b) whose envelope is the product of a power by a Gaussian; thus, the envelope (Figure 1.4a) is dominated by the power for small x and the Gaussian for larger x.

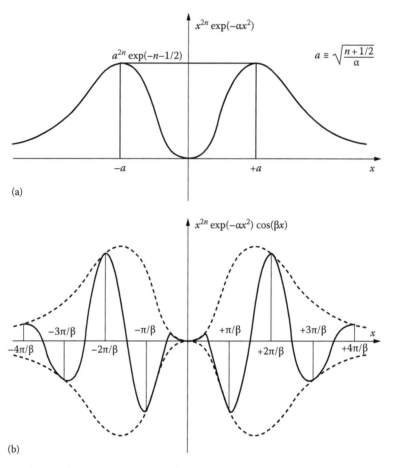

FIGURE 1.4 Another envelope (a) of a sinusoidal function (b) is the product of a power by a Gaussian, that is dominated by the former (latter) for small (large) values of the independent variable, and leads to a further generalization of the Gaussian integral.

1.1.5 Gaussian Integrals with Power Factors

Since (1.15b) does not apply in the particular case $\beta = 0$, it is necessary to start again with (1.13), using the uniform convergence with regard to α, to differentiate m times under the integral sign:

$$\int_{-\infty}^{+\infty}\left(-x^2\right)^m \exp\left(-\alpha x^2 + i\beta x\right)dx = \frac{\partial^m}{\partial\alpha^m}\left[\int_{-\infty}^{+\infty}\exp\left(-\alpha x^2 + i\beta x\right)dx\right]$$

$$= \frac{\partial^m}{\partial\alpha^m}\left[\sqrt{\frac{\pi}{\alpha}}\exp\left(-\frac{\beta^2}{4\alpha}\right)\right]. \tag{1.16a}$$

The real part is nonzero, leading to

$$\int_{-\infty}^{+\infty}x^{2m}\exp\left(-\alpha x^2\right)\cos(\beta x)dx = (-)^m\frac{\partial^m}{\partial\alpha^m}\left[\sqrt{\frac{\pi}{\alpha}}\exp\left(-\frac{\beta^2}{4\alpha}\right)\right]. \tag{1.16b}$$

This is an alternative to (1.15a) ≡ (1.16b) for β ≠ 0 and (1) does not include odd powers like (1.15b), in which in the case (1.16b) would lead to zero; and (2) unlike (1.15a,b), it applies if β = 0, namely,

$$\int_{-\infty}^{+\infty} x^{2m} \exp\left(-\alpha x^2\right) dx = (-)^m \frac{\partial^m}{\partial \alpha^m} \sqrt{\frac{\pi}{\alpha}} = \sqrt{\pi} \alpha^{-1/2-m} \frac{1}{2} \cdot \frac{3}{2} \cdots \left(\frac{1}{2} + m - 1\right)$$

$$= \sqrt{\pi} \alpha^{-1/2-m} 2^{-m} \, 1.3.5 \ldots (2m-1). \tag{1.17a}$$

This simplifies to (1.17d):

$$m \ge 1; \ (2m-1)!! \equiv (2m-1)(2m-3)\ldots5.3.1: \ \int_{-\infty}^{+\infty} x^{2m} \exp\left(-\alpha x^2\right) dx = (2m-1)!!(2\alpha)^{-m} \sqrt{\frac{\pi}{\alpha}}, \tag{1.17b–d}$$

where (1.17b) denotes the double factorial, and the expression holds for (1.17a). The integral is the area under the product of the power and the Gaussian (Figure 1.4a) and decreases for increasing α since then the Gaussian decays faster. The dependence on m is less clear since increasing m increases both the numerator and denominator of (1.17d), because (1) larger m leads to smaller x^m for $x < 1$ and larger x^m for $x > 1$; (2) for $x > 1$, the power x^m slows down the decay of the Gaussian, though the latter always dominates as $x \to \infty$.

1.1.6 Evaluation of Gaussian Integrals in terms of Hermite Polynomials

The definition of **Hermite (1864) polynomials**,

$$H_n(z) \equiv (-)^n \exp\left(z^2\right) \frac{d^n}{dz^n}\left[\exp\left(-z^2\right)\right], \tag{1.18}$$

yields for the first five (1.19a) degrees of (1.19b–f):

$$n = 0, 1, \ldots, 5: \quad H_n(z) = \{1, 2z, 4z^2 - 2, 8z^3 - 12z, 16z^4 - 48z^2 + 12\}. \tag{1.19a–f}$$

The Hermite polynomials replace the derivatives in the Gaussian integrals (1.15a) ≡ (1.20a) [(1.15b) ≡ (1.20b)]:

$$\int_{-\infty}^{+\infty} x^{2m} \exp\left(-\alpha x^2\right) \cos\left(\beta x\right) dx = (-)^m \sqrt{\frac{\pi}{\alpha}} (4\alpha)^{-m} H_{2m}\left(\frac{\beta}{2\sqrt{\alpha}}\right) \exp\left(-\frac{\beta^2}{4\alpha}\right), \tag{1.20a}$$

$$\int_{-\infty}^{+\infty} x^{2m+1} \exp\left(-\alpha x^2\right) \sin\left(\beta x\right) dx = (-)^m \frac{\sqrt{\pi}}{2\alpha} (4\alpha)^{-m} H_{2m+1}\left(\frac{\beta}{2\sqrt{\alpha}}\right) \exp\left(-\frac{\beta^2}{4\alpha}\right), \tag{1.20b}$$

as follows from (1.15a) ≡ (1.21b) [(1.15b) ≡ (1.21c)] with the change of variable (1.21a):

$$z \equiv \frac{\beta}{2\sqrt{\alpha}}: \quad \int_{-\infty}^{+\infty} x^{2m} \exp\left(-\alpha x^2\right)\cos(\beta x)\,dx = (-)^m \sqrt{\frac{\pi}{\alpha}} \frac{\partial^{2m}}{\partial \left(2z\sqrt{\alpha}\right)^{2m}} \left[\exp\left(-z^2\right)\right]$$

$$= (-)^m \sqrt{\frac{\pi}{\alpha}} \left(2\sqrt{\alpha}\right)^{-2m} H_{2m}(z)\exp\left(-z^2\right), \quad (1.21a,b)$$

$$\int_{-\infty}^{+\infty} x^{2m+1} \exp\left(-\alpha x^2\right)\sin(\beta x)\,dx = (-)^{m+1} \sqrt{\frac{\pi}{\alpha}} \frac{\partial^{2m+1}}{\partial \left(2z\sqrt{\alpha}\right)^{2m+1}} \left[\exp\left(-z^2\right)\right]$$

$$= (-)^{m+1} \sqrt{\frac{\pi}{\alpha}} \left(2\sqrt{\alpha}\right)^{-2m-1} (-)^{2m+1} H_{2m+1}(z)\exp\left(-z^2\right). \quad (1.21c)$$

The simplification of (1.21b) [(1.21c)] leads to (1.20a) [(1.20b)].

1.1.7 Summary and Examples of Gaussian Integrals

Thus have been obtained *(1) the basic Gaussian integrals (1.1a,b) by two alternate methods in Subsections 1.1.1 and 1.1.2; (2/3) the first pair of extensions involving Gaussian (1.10b) [trigonometric (1.13)] functions in Subsection 1.1.3; (4/5) there is a second pair of extensions involving powers (1.17b–d) [products of powers and trigonometric functions (1.15a,b)] in Subsection 1.1.4 (Subsection 1.1.5); (6/7) the Gaussian integrals with cosine (sine) times an even (odd) power can be evaluated using derivatives (1.15a) [(1.15b)] in Subsection 1.1.4 or Hermite polynomials (1.20a) [(1.20b)] in Subsection 1.1.6.* The simplest examples beyond (1.1a,b, 1.10b, 1.13) are

$$\int_{-\infty}^{+\infty} x^2 \exp\left(-\alpha x^2\right)dx = -\frac{\partial}{\partial \alpha}\left(\sqrt{\frac{\pi}{\alpha}}\right) = \frac{\sqrt{\pi}}{2}\alpha^{-3/2}; \quad (1.22a)$$

$$\int_{-\infty}^{+\infty} x^4 \exp\left(-\alpha x^2\right)dx = \frac{\partial^2}{\partial \alpha^2}\left(\sqrt{\frac{\pi}{\alpha}}\right) = \sqrt{\pi}\frac{3}{4}\alpha^{-5/2}; \quad (1.22b)$$

$$\int_{-\infty}^{+\infty} x \exp\left(-\alpha x^2\right)\sin(\beta x)\,dx = -\sqrt{\frac{\pi}{\alpha}}\frac{\partial}{\partial \beta}\left[\exp\left(-\frac{\beta^2}{4\alpha}\right)\right] = \sqrt{\frac{\pi}{\alpha}}\frac{\beta}{2\alpha}\exp\left(-\frac{\beta^2}{4\alpha}\right)$$

$$= \frac{\sqrt{\pi}}{2\alpha} H_1\left(\frac{\beta}{2\sqrt{\alpha}}\right)\exp\left(-\frac{\beta^2}{4\alpha}\right); \quad (1.22c)$$

$$\int_{-\infty}^{+\infty} x^2 \exp\left(-\alpha x^2\right)\cos(\beta x)\,dx = -\sqrt{\frac{\pi}{\alpha}}\frac{\partial^2}{\partial \beta^2}\left[\exp\left(-\frac{\beta^2}{4\alpha}\right)\right] = \sqrt{\frac{\pi}{\alpha}}\left(\frac{1}{2\alpha} - \frac{\beta^2}{4\alpha^2}\right)\exp\left(-\frac{\beta^2}{4\alpha}\right)$$

$$= -\sqrt{\frac{\pi}{\alpha}}\frac{1}{4\alpha} H_2\left(\frac{\beta}{2\sqrt{\alpha}}\right)\exp\left(-\frac{\beta^2}{4\alpha}\right); \quad (1.22d)$$

$$\int_{-\infty}^{+\infty} x^3 \exp\left(-\alpha x^2\right)\sin\left(\beta x\right)dx = \sqrt{\frac{\pi}{\alpha}}\,\frac{\partial^3}{\partial\beta^3}\left[\exp\left(-\frac{\beta^2}{4\alpha}\right)\right] = \sqrt{\frac{\pi}{\alpha}}\,\frac{\beta}{4\alpha^2}\left(3-\frac{\beta^2}{2\alpha}\right)\exp\left(-\frac{\beta^2}{4\alpha}\right)$$

$$= -\frac{\sqrt{\pi}}{2\alpha}\frac{1}{4\alpha}H_3\left(\frac{\beta}{2\sqrt{\alpha}}\right)\exp\left(-\frac{\beta^2}{4\alpha}\right); \tag{1.22e}$$

$$\int_{-\infty}^{+\infty} x^4 \exp\left(-\alpha x^2\right)\cos\left(\beta x\right)dx = \sqrt{\frac{\pi}{\alpha}}\,\frac{\partial^4}{\partial\beta^4}\left[\exp\left(-\frac{\beta^2}{4\alpha}\right)\right]$$

$$= \sqrt{\frac{\pi}{\alpha}}\,\frac{1}{4\alpha^2}\left(3-\frac{3\beta^2}{\alpha}+\frac{\beta^4}{4\alpha^2}\right)\exp\left(-\frac{\beta^2}{4\alpha}\right)$$

$$= \sqrt{\frac{\pi}{\alpha}}\,\frac{1}{16\alpha^2}H_4\left(\frac{\beta}{2\sqrt{\alpha}}\right)\exp\left(-\frac{\beta^2}{4\alpha}\right); \tag{1.22f}$$

as follows: (1.22a,b) from (1.17a) $\equiv$ (1.17d) with $m = 1, 2$; (1.22c) from (1.15b) $\equiv$ (1.20b, 1.19c) with $m = 0$, $n = 1$; (1.22d) from (1.15a) $\equiv$ (1.20a, 1.19d) with $m = 1, n = 2$; (1.22e) from (1.15b) $\equiv$ (1.20b, 1.19e) with $m = 1, n = 3$; (1.22f) from (1.15a) $\equiv$ (1.20a, 1.19f) with $m = 2, n = 4$. The Gaussian integrals (Examples 10.14 and 10.17) are related with the normal or Gaussian probability distribution (Notes 1.3 through 1.7). The Gamma function (Note 1.8) can be used to evaluate not only the preceding Gaussian integrals (Section 1.1) but also their generalizations involving powers with nonintegral exponents (Notes 1.9 through 1.12) and further relations concerning Hermite polynomials (Notes 1.13 and 1.14).

1.2 Unit Jump as a Generalized Function

The Heaviside (1876) unit step function can be defined as the nonuniform limit of more than one family of smooth functions, for example, the error functions (Subsection 1.2.1) or hyperbolic tangents (Subsection 1.2.2).

1.2.1 Unit Jump as the Limit of Error Functions

The Gaussian integral (1.10b) can be written in the equivalent (1.23a) forms (1.23b,c):

$$\sigma \equiv \frac{1}{\alpha}: \quad \frac{1}{\sqrt{\pi\sigma}}\int_{-\infty}^{+\infty}\exp\left(-\frac{x^2}{\sigma}\right)dx = 1, \quad \frac{1}{\sqrt{\pi\sigma}}\int_{0}^{\infty}\exp\left(-\frac{x^2}{\sigma}\right)dx = \frac{1}{2}, \tag{1.23a–c}$$

where σ denotes the r.m.s. (root mean square) value of a Gaussian probability distribution with zero mean (Note 1.3). This suggests the introduction of the sequence of functions

$$H_\sigma(x) = \frac{1}{\sqrt{\pi\sigma}}\int_{-\infty}^{x}\exp\left(-\frac{y^2}{\sigma}\right)dy, \tag{1.24}$$

which is specified by incomplete (1.24) Gaussian integrals (1.23b); it is equivalent to (1.25b)

$$z \equiv \frac{y}{\sqrt{\sigma}}: \quad H_\sigma(x) \equiv \frac{1}{\sqrt{\pi}}\int_{-\infty}^{x/\sqrt{\sigma}}\exp\left(-z^2\right)dz \equiv \mathrm{erf}\left(\frac{x}{\sqrt{\sigma}}\right), \tag{1.25a,b}$$

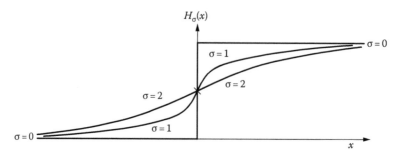

FIGURE 1.5   A family of functions: (1) starting with zero value at $-\infty$, (2) tending to unity at $+\infty$; (3) passing at the origin with value 1/2; and (4) with unbounded slope at the origin as $\sigma \to 0$, specifies the unit jump; the corresponding generalized function (Heaviside, 1876) represents a unit jump or step, like the switching on of an electric circuit.

an **error function** erf of argument (1.25a). The functions $H_\sigma(x)$ in (1.24) are analytic in the variable $-\infty < x < +\infty$, and the parameter σ identifies the element of the sequence. All functions of the sequence (1) coincide at the three points $x = -\infty, 0, +\infty$:

$$H_\sigma(-\infty) = 0, \quad H_\sigma(0) = \frac{1}{2}, \quad H_\sigma(\infty) = 1; \tag{1.26a–c}$$

and (2) can be distinguished by the slope at the origin:

$$H'_\sigma(0) \equiv \lim_{x \to 0} \frac{dH_\sigma}{dx} = \lim_{x \to 0} \frac{1}{\sqrt{\pi\sigma}} \exp\left(-\frac{x^2}{\sigma}\right) = \frac{1}{\sqrt{\pi\sigma}}, \tag{1.27}$$

which increases as σ reduces. Taking $\sigma \to 0$, the **limit function** of the sequence (1.24) is found to be (Figure 1.5) the **Heaviside (1876) unit step function**

$$H(x) \equiv \lim_{\sigma \to \infty} H_\sigma(x) = \begin{cases} 0 & \text{if } x < 0 & (1.28a) \\[2mm] \dfrac{1}{2} & \text{if } x = 0 & (1.28b) \\[2mm] 1 & \text{if } x > 0, & (1.28c) \end{cases}$$

which is zero (1.28a) [unity (1.28c)] before (after) the origin, where it has a unit jump; its value at the origin (1.28b) is the arithmetic mean of the right- and left-hand limits. Although the functions of the sequence (1.24) are continuous, the limit generalized function (1.28a–c) is discontinuous, because the convergence is nonuniform (Section I.4.4) in the parameter σ at the origin $\sigma \to 0$.

1.2.2 Unit Jump as the Limit of Hyperbolic Tangents

Much as a given limit may be approached by different functions, for example,

$$\lim_{x \to 0} \frac{\sin x}{x} = 1 = \lim_{x \to 0} \frac{1}{1+x}, \tag{1.29}$$

a given generalized function may be the limit of distinct sequences of functions. For example, the sequence of hyperbolic tangents,

$$h_\sigma(x) = \frac{1}{2}\left[1 + \tanh\left(\frac{x}{\sigma}\right)\right], \tag{1.30}$$

has the properties (1.26a–c), but the slope at the origin,

$$h_\sigma'(0) = \frac{1}{2\sigma}\lim_{x\to 0}\sec h^2\left(\frac{x}{\sigma}\right) = \frac{1}{2\sigma}, \tag{1.31}$$

is distinct from (1.27), although it also diverges as $\sigma \to 0$. Thus, although the sequences of functions (1.24) and (1.30) are distinct (1.32a),

$$H_\sigma(x) \ne h_\sigma(x), \qquad \lim_{\sigma\to 0} H_\sigma(x) = H(x) = \lim_{\sigma\to 0} h_\sigma(x), \tag{1.32a,b}$$

they tend as $\sigma \to 0$ to the same limit generalized function, namely, (1.28a–c). Thus, *the **unit step** or Heaviside (1876) generalized function (1.28a–c) is a discontinuous function that is the nonuniform limit (1.32b) as $\sigma \to 0$ of distinct (1.32a) sequences of ordinary functions:*

$$\lim_{\sigma\to 0}\mathrm{erf}\left(\frac{x}{\sqrt{\sigma}}\right) = H(x) = \lim_{\sigma\to 0}\frac{1 + \tanh(x/\sigma)}{2}, \tag{1.33a,b}$$

namely, the error functions (1.33a) ≡ (1.24) and the hyperbolic tangents (1.33b) ≡ (1.30).

1.3 Unit Impulse as a Generalized Function

The definition of derivative of a generalized function (Subsection 1.3.1) applied to the Heaviside unit jump (Section 1.2) leads to the Dirac unit impulse (Subsection 1.3.2).

1.3.1 Definition of Derivative of a Generalized Function

The Dirac (1930) unit impulse is the generalized function defined as the derivative of the Heaviside unit step (1.24). If the differentiation is attempted as for ordinary functions, it follows that (1) the Dirac impulse is zero (1.34a) for $x \ne 0$, because the Heaviside unit step is a constant for $x \ne 0$; and (2) at $x = 0$, the Heaviside unit step jumps from 0 at $x = 0-$ to 1 at $x = 0+$, so the **Dirac unit impulse or delta function** is plus infinity (1.34b):

$$\delta(x) \equiv \frac{\mathrm{d}}{\mathrm{d}x}\left[H(x)\right] = \begin{cases} 0 & \text{if } x \ne 0 \\ +\infty & \text{if } x = 0. \end{cases} \tag{1.34a} \\ \tag{1.34b}$$

The nature of the singularity at $x = 0$ is not well specified by (1.34a,b); for example, a nonunit jump would give the same results. In order to specify the type of singularity, the unit impulse is integrated assuming the ordinary rule of primitivation to hold:

$$b > a: \qquad \int_a^b \delta(x)\,\mathrm{d}x = \int_a^b H'(x)\,\mathrm{d}x = H(b) - H(a) = \begin{cases} 0 & \text{if } ab > 0 \\ 1 & \text{if } ab < 0, \end{cases} \tag{1.35a} \\ \tag{1.35b}$$

showing that the integral is zero (1.35a) [unity (1.35b)] if the origin does not $b > a > 0$ (does $b > 0 > a$) lie in the range of integration. None of these reasonings is rigorous, since it was assumed without proof that differentiation and integration properties extend from ordinary to generalized functions. In order to justify the preceding results, the start must be the definition of **derivative of a generalized function**: *if the sequence of differentiable ordinary functions $H_\sigma(x)$ tends to the generalized function $H(x)$ as $\sigma \to 0$, then the derivative of the generalized function $H(x)$ is defined as the limit as $\sigma \to 0$ of the sequence of derivatives of the ordinary functions if the limit exists*:

$$H(x) \equiv \lim_{\sigma \to 0} H_\sigma(x) \Rightarrow H'(x) \equiv \lim_{\sigma \to 0} H'_\sigma(x). \tag{1.36}$$

Since the limit $\sigma \to 0$ is generally nonuniform, it cannot be exchanged with the derivation (Section I.21.5).

1.3.2 Unit Impulse as the Derivative of the Unit Step

In order to apply the preceding definition to the Dirac impulse, consider the sequence of derivatives of the error functions (1.24) $\equiv$ (1.37) that are (Figure 1.6) the Gaussian functions:

$$\delta_\sigma(x) = \frac{d}{dx}\left[H_\sigma(x)\right] = \frac{1}{\sqrt{\pi\sigma}} \exp\left(-\frac{x^2}{\sigma}\right). \tag{1.37}$$

These (1) vanish (1.38a) at $\pm\infty$:

$$\delta_\sigma(\pm\infty) = 0, \quad \delta_{\sigma\,\text{max}} = \delta_\sigma(0) = \frac{1}{\sqrt{\pi\sigma}}, \tag{1.38a,b}$$

and (2) have a maximum at the origin (1.38b) that diverges as $\sigma \to 0$. Alternatively, the sequence of rational functions (1.39a) with a parameter σ can be considered:

$$\Delta_\sigma(x) = \frac{1}{\pi} \frac{\sigma}{\sigma^2 + x^2} \quad \Delta_\sigma(x) \neq \delta_\sigma(x). \tag{1.39a,b}$$

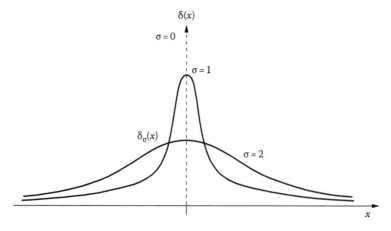

FIGURE 1.6 Derivatives of the family of functions in Figure 1.5 have (1/2) zero slope at $\pm\infty$, and (2) slope increasing monotonically toward the origin, with (3) peak value diverging as $\sigma \to 0$ while (4) keeping the integral equal to unity. The corresponding generalized function (Dirac, 1930) represents a unit impulse such as an instantaneous force imparting a unit momentum.

These are (1.39b) distinct from (1.37), but also (1) vanish (1.40a) at $\pm\infty$:

$$\Delta_\sigma(\pm\infty)=0, \qquad \Delta_{\sigma\max}=\Delta_\sigma(0)=\frac{1}{\pi\sigma}, \tag{1.40a,b}$$

and (2) have a maximum at the origin (1.40b) distinct from (1.38b), but also diverging as $\sigma \to 0$. *The distinct (1.39b) families of Gaussian (1.37) $\equiv$ (1.41a) and rational (1.39a) $\equiv$ (1.41b) functions in the limit $\sigma \to 0$ specify the generalized function* **unit impulse** *or Dirac (1930) delta:*

$$\lim_{\sigma\to0}\frac{1}{\sqrt{\pi\sigma}}\exp\left(-\frac{x^2}{\sigma}\right)=\delta(x)=\lim_{\sigma\to0}\frac{1}{\pi}\frac{\sigma}{\sigma^2+x^2}. \tag{1.41a,b}$$

It has the **integral property** *(1.42a) that is shared (1.42b,c) with all members of the two families:*

$$\int_{-\infty}^{+\infty}\delta(x)\mathrm{d}x=1, \qquad \int_{-\infty}^{+\infty}\delta_\sigma(x)\mathrm{d}x=1=\int_{-\infty}^{+\infty}\Delta_\sigma(x)\mathrm{d}x. \tag{1.42a–c}$$

Its interpretation is that the Dirac unit impulse represents a **point monopole**. The property (1.42b) results from (1.37, 1.23b), and (1.42c) follows from (1.39a) using the change of variable (1.43a) in (1.43b):

$$y\equiv\frac{x}{\sigma}: \qquad \int_{-\infty}^{+\infty}\Delta_\sigma(x)\mathrm{d}x=\frac{1}{\pi}\int_{-\infty}^{+\infty}\frac{\mathrm{d}y}{1+y^2}=\frac{1}{\pi}\big[\arctan(y)\big]_{-\infty}^{+\infty}\equiv1. \tag{1.43a,b}$$

Geometrically (Figure 1.6), since the area under the Gaussian (1.37) [rational (1.39b)] functions is always unity (1.42b) $\equiv$ (1.42c) = (I.1.42) $\equiv$ (1.35b), as $\sigma \to 0$ and the limit function vanishes (1.34a) for all $x \ne 0$, it must diverge to $+\infty$ at the origin (1.34b); also, the integral not including the origin must be zero (1.35a).

1.4 First Derivative of the Unit Impulse

The Dirac unit impulse (its derivative) is [Subsection 1.3.1 (Subsection 1.4.1)] the generalized function that represents [Subsection 1.3.2 (Subsection 1.4.2)] a unit monopole (dipole).

1.4.1 Unit Impulse (Derivative of) as a Monopole (Dipole)

The derivative of the unit impulse (1.34a,b) interpreted (Figure 1.7) as an ordinary function is (1) zero outside the origin, because the unit impulse is zero for $x \ne 0$; (2) as x tends to the origin from below $x \to 0-$ (above $x \to 0+$), the unit impulse changes from 0 to $+\infty$ ($+\infty$ to 0), so that its derivative (1.45b) is $+\infty(-\infty)$ at $x = 0 - (x = 0+)$; (3) the unit impulse is even (1.44a) like the families of functions (1.37, 1.39a):

$$\delta(x)=\delta(-x), \qquad \delta'(x)=-\delta'(-x), \qquad \delta(0)=-\delta'(0)=0, \tag{1.44a–c}$$

and hence its derivative is odd (1.44b) and vanishes at the origin (1.44c) $\equiv$ (1.45c). Thus, it follows that the derivate of the unit impulse satisfies

$$\delta'(x)\equiv\frac{\mathrm{d}}{\mathrm{d}x}\big[\delta(x)\big]=\frac{\mathrm{d}^2}{\mathrm{d}x^2}\big[H(x)\big]=\begin{cases}0 & \text{if } x\ne0, & (1.45a)\\ \pm\infty & \text{if } x=\mp0, & (1.45b)\\ 0 & \text{if } x=0; & (1.45c)\end{cases}$$

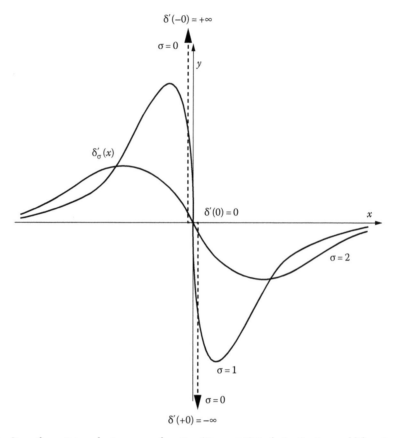

FIGURE 1.7   Since the unit impulse is an even function (Figure 1.6) its derivative is an odd function (Figure 1.7) that vanishes everywhere including at the origin, except just before −0 and after +0 where it takes the values respectively +∞ and −∞. Its integral is zero because it is an odd function, and multiplication by x leads to an even function whose integral is negative and equal to −1. Thus whereas the unit impulse represents a monopole with moment unity, its first-order derivative represents a dipole with moment −1, and the Nth order derivative of the unit impulse corresponds to a multipole of order 2^n with moment $n!(-)^n$.

the derivative Dirac impulse is thus zero everywhere (1.45a) including at the origin (1.45c), where it has a singularity (1.45b) going from +∞ to −∞ through zero. In order to specify the nature of the singularity, an integration may be used; the derivate of the Dirac impulse is odd, and thus its integral is zero (1.46a):

$$\int_{-\infty}^{+\infty}\delta'(x)\,dx = 0; \quad \int_{-\infty}^{+\infty}x\,\delta'(x)\,dx = -1. \tag{1.46a,b}$$

In order to have a nonzero integral, an even integrand is taken, for example, the product of $\delta'(x)$ by x, leading to the integral property (1.46b), stating that *the first moment is a constant, that is, the derivative unit impulse $\delta'(x)$ is a generalized function that represents a unit* **dipole**. The results (1.46a,b) can be deduced, on the assumption that the ordinary rules of derivation hold:

$$\int_{-\infty}^{+\infty}\delta'(x)\,dx = \delta(+\infty)-\delta(-\infty)=0, \quad \int_{-\infty}^{+\infty}x\,\delta'(x)\,dx = \left[x\,\delta(x)\right]_{-\infty}^{+\infty}-\int_{-\infty}^{+\infty}\delta(x)\,dx = -1, \tag{1.47a,b}$$

that is, in (1.47b) was performed an integration by parts, and (1.42a) and (1.79d) were used.

1.4.2 Derivative of Unit Impulse as a Limit of Families of Derivatives

In order to prove rigorously (1.45a–c, 1.46a,b), derivatives (1.48a) [(1.48b)] of the Gaussian (1.37) [rational (1.39a)] functions are considered:

$$\delta'_\sigma(x) \equiv \frac{d}{dx}\big[\delta_\sigma(x)\big] = -\frac{2x}{\sigma\sqrt{\sigma\pi}}e^{-x^2/\sigma}, \quad \Delta'_\sigma(x) \equiv \frac{d}{dx}\big[\Delta_\sigma(x)\big] = -\frac{1}{\pi}\frac{2x\sigma}{\left(x^2+\sigma^2\right)^2}; \quad (1.48a,b)$$

(1) both vanish at $\pm\infty$ and the origin:

$$\delta'_\sigma(\pm\infty) = 0 = \delta'_\sigma(0), \quad \Delta'_\sigma(\pm\infty) = 0 = \Delta'_\sigma(0); \quad (1.49a,b)$$

(2) the functions (1.48a) [(1.48b)] have a distinct (1.50a) [(1.50b)] maximum (minimum) for different negative (positive) values of the variable:

$$\delta'_{\sigma\,\text{max,min}} = \delta'_\sigma\left(\mp\sqrt{\frac{\sigma}{2}}\right) = \pm\frac{1}{\sigma}\sqrt{\frac{2}{\pi e}}, \quad \Delta'_{\sigma\,\text{max,min}} = \Delta'_\sigma\left(\mp\frac{\sigma}{\sqrt{3}}\right) = \pm\frac{3\sqrt{3}}{8\pi\sigma^2}; \quad (1.50a,b)$$

and (3) as $\sigma \to 0$, both (1.48a) [(1.48b)] vanish for all x except the extrema, and the maximum (minimum) becomes $+\infty(-\infty)$ in (1.50a) [(1.50b)] as σ tends to $-0(+0)$, confirming (1.45a–c). The extrema, that is, maxima and minima, of the functions (1.48a) [(1.48b)] correspond to zeros of (1.51a) [(1.52a)], that is, (1.51b) [(1.52b)]:

$$0 = \frac{d}{dx}\left[x\exp\left(-\frac{x^2}{\sigma}\right)\right] = \left(1 - \frac{2x^2}{\sigma}\right)\exp\left(-\frac{x^2}{\sigma}\right), \quad x = \pm\sqrt{\frac{\sigma}{2}}, \quad (1.51a,b)$$

$$0 = \frac{d}{dx}\left[\frac{x}{\left(x^2+\sigma^2\right)^2}\right] = \frac{\sigma^2 - 3x^2}{\left(x^2+\sigma^2\right)^3}, \quad x = \pm\frac{\sigma}{\sqrt{3}}; \quad (1.52a,b)$$

in both cases, the negative (positive) value of the argument (1.51b) [(1.52b)] leads to a positive (negative) value of the function (1.48a) [(1.48b)] and hence to the maximum (minimum) in (1.50a) [(1.50b)]. Graphically, the derivative of Gaussian functions (Figure 1.7) are odd functions whose extrema diverge as they approach the origin for $\sigma \to 0$, so as to (1) lead to zero integrals (1.53a) [(1.53b)] by (1.49a) [(1.49b)]:

$$\int_{-\infty}^{\infty}\delta'_\sigma(x)dx = \delta_\sigma(+\infty) - \delta_\sigma(-\infty) = 0; \quad \int_{-\infty}^{\infty}\Delta'_\sigma(x)dx = \Delta_\sigma(-\infty) - \Delta_\sigma(+\infty) = 0, \quad (1.53a,b)$$

corresponding to (1.46a); and (2) corresponding to (1.46b), the same value holds (1.54a) [(1.54b)] for the first moment of all the functions (1.48a) [(1.48b)]:

$$\int_{-\infty}^{+\infty}\delta'_\sigma(x)x\,dx = -2\sqrt{\frac{\sigma}{\pi}}\int_{-\infty}^{+\infty}\frac{x^2}{\sigma^2}\exp\left(-\frac{x^2}{\sigma}\right)dx$$

$$= -2\sqrt{\frac{\sigma}{\pi}}\frac{d}{d\sigma}\int_{-\infty}^{+\infty}\exp\left(-\frac{x^2}{\sigma}\right)dx = -2\sqrt{\frac{\sigma}{\pi}}\frac{d}{d\sigma}\sqrt{\pi\sigma} = -1, \quad (1.54a)$$

$$\int\limits_{-\infty}^{+\infty}\Delta_\sigma'(x)x\,dx = -2\frac{\sigma}{\pi}\int\limits_{-\infty}^{+\infty}\frac{x^2}{\left(x^2+\sigma^2\right)^2}\,dx = \frac{\sigma}{\pi}\int\limits_{-\infty}^{+\infty}x\,d\left(\frac{1}{x^2+\sigma^2}\right)$$

$$= \frac{\sigma}{\pi}\left[\frac{x}{x^2+\sigma^2}\right]_{-\infty}^{+\infty} - \frac{\sigma}{\pi}\int\limits_{-\infty}^{+\infty}\frac{dx}{x^2+\sigma^2} = -\frac{1}{\pi}\left[\arctan\left(\frac{x}{\sigma}\right)\right]_{-\infty}^{+\infty} = -1. \qquad (1.54b)$$

The parametric differentiation outside the integral sign in (1.54a) is allowed, because the integral is uniformly convergent (I.13.40) and is similar to (1.17a). Note that the product of (1.48a,b) by x is negative everywhere (Figure 1.7), so that the integrals (1.54a,b) $\equiv$ (1.46b) must also be negative. *The distinct families of functions (1.48a) [(1.48b)] specify in the limit $\sigma \to 0$ the generalized function (1.45a–c)* **first derivative of the unit impulse:**

$$\lim_{\sigma\to0}\frac{2x}{\sigma\sqrt{\sigma\pi}}e^{-x^2/\sigma} = -\delta'(x) = \lim_{\sigma\to0}\frac{1}{\pi}\frac{2x\sigma}{\left(x^2+\sigma^2\right)^2}. \qquad (1.55a,b)$$

All the functions have zero integral (1.46a) $\equiv$ (1.53a) $\equiv$ (1.53b) and first moment equal to minus unity (1.46b) $\equiv$ (1.54a) $\equiv$ (1.54b).

1.5 Derivative of Order *N* of the Unit Impulse

The nth derivative of the unit impulse is a generalized function that represents (Subsection 1.5.1) a multipole of order n, since the first nonvanishing moment (Subsection 1.5.2) is of order n.

1.5.1 Multipole of Order *N* as the *N*th Derivative of the Unit Impulse

The preceding process (1.36) of derivation of the Heaviside unit step (Dirac unit impulse) as generalized functions can proceed indefinitely because (1) the functions in the families (1.24, 1.30) [(1.37, 1.39a)] are all smooth, that is, have derivatives of all orders; and (2) the sequence of derivatives of order $n + 1(n)$ tends to a limit as $\sigma \to 0$, thus specifying the nth derivative of the unit impulse as the generalized function:

$$\lim_{\sigma\to0}H_\sigma^{(n+1)}(x) = \lim_{\sigma\to0}\delta_\sigma^{(n)}(x) \equiv \delta^{(n)}(x) \equiv \lim_{\sigma\to0}\Delta_\sigma^{(n)}(x) = \lim_{\sigma\to0}h_\sigma^{(n+1)}(x). \qquad (1.56)$$

The nth derivative of the unit impulse is zero everywhere except at the origin where it changes between $\pm\infty$ a number n of times:

$$\delta^{(n)}(x) \equiv \frac{d^n}{dx^n}[\delta(x)] = \frac{d^{n+1}}{dx^{n+1}}\left[H(x)\right] = \begin{cases} 0 & \text{if } x\neq0, & (1.57a) \\ \pm\infty & \text{if } x=0, & (1.57b) \end{cases}$$

with the first value as $x \to 0-$ being always $+\infty$, and the last $-\infty(+\infty)$ for n odd (even). The value (1.57c) [(1.57d)] at $x = 0$ central is ∞ (zero) for n even (odd):

$$\delta^{(2n)}(0) = \infty, \quad \delta^{(2n+1)}(0) = 0. \qquad (1.57c,d)$$

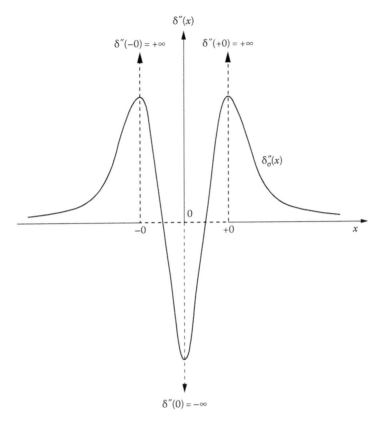

FIGURE 1.8 Proceeding from the generalized function unit impulse (Figure 1.6) to its first (second) derivatives [Figure 1.7 (Figure 1.8)] each additional differentiation: (1) keeps the zero value away from the origin; (2) adds one more jump between $\pm\infty$ across the origin, that is n jumps for the nth derivative; (3) the value at the origin is zero for the derivatives of add order that are odd functions; and (4) the value at the origin is $(-)^m \infty$ for the derivatives of even order $n = 2m$, that $+\infty$ $(-\infty)$ for order $4m$ $(4m - 2)$, with m positive integer.

For example, (1.34b) [(1.45c)] correspond to (1.57c) [(1.57d)] with $n = 0$. The next example is (Figure 1.8) the second-order derivative of the unit impulse:

$$\delta''(x) \equiv \frac{d^2}{dx^2}\big[\delta(x)\big] = \frac{d^3}{dx^3}\big[H(x)\big] = \begin{cases} 0 & \text{if } x \neq 0, & \text{(1.58a)} \\ +\infty & \text{if } x \to \pm 0, & \text{(1.58b)} \\ -\infty & \text{if } x = 0, & \text{(1.58c)} \end{cases}$$

which (1) is zero everywhere (1.58a) except when the origin is approached; (2) if the origin is approached from below -0 (above $+0$), the first derivative (Figure 1.7) of the unit impulse $\delta'(x)$ changes from 0 to $+\infty$ $(-\infty)$, leading to the slope $+\infty$ in (1.58b) that is the value of $\delta''(x)$; and (3) since $\delta'(x)$ changes from $+\infty$ at -0 to $-\infty$ to $+0$, its slope is $-\infty$, specifying the value of $\delta''(x)$ at the origin (1.59c). *The nature of the singularity of the **n**th deriva-* **tive unit impulse** *or Dirac generalized function (1.57a,b) at the origin is specified by the set of integral properties*

$$\int_{-\infty}^{+\infty} x^m \delta^{(n)}(x)\,dx = \begin{cases} 0 & \text{if } m = 0,...., n-1, & \text{(1.59a)} \\ n!\,(-)^n & \text{if } m = n, & \text{(1.59b)} \end{cases}$$

which generalize (1.46a,b) and show that $\delta^{(n)}(x)$ represents a multipole of order n, with nth moment (1.59b).

1.5.2 Derivatives of All Orders of the Unit Impulse

The last result in (1.59b) can be obtained using integration by parts as in (1.47a,b), that was shown to yield the same results as the rigorous process (1.54a,b). The proof is done by induction, starting with $n = 1 = m$:

$$\int_a^b x\delta'(x)\,dx = \int_a^b x\,d\big[\delta(x)\big] = \big[x\delta(x)\big]_a^b - \int_a^b \delta(x)\,dx. \tag{1.60}$$

The first term is zero by (1.79d), and the second is given by (1.35a,b):

$$\int_a^b x\delta'(x)\,dx = \begin{cases} -1 & \text{if } a < 0 < b, & (1.61a)\\ 0 & \text{if } ab > 0, & (1.61b)\\ +1 & \text{if } a > 0 > b. & (1.61c) \end{cases}$$

If the formula holds (1.59b) for $(n-1)$, it can be proved that it holds for n as follows:

$$\int_a^b x^n\delta^{(n)}(x)\,dx = \int_a^b x^n\,d\big[\delta^{(n-1)}(x)\big] = \big[x^n\delta^{(n-1)}(x)\big]_a^b - \int_a^b \delta^{(n-1)}(x)\,d(x^n). \tag{1.62}$$

The first term on the r.h.s. (right-hand side) of (1.62) vanishes by (1.80a), and the second leads to

$$\int_a^b x^n\delta^{(n)}(x)\,dx = -n\int_a^b x^{n-1}\delta^{(n-1)}(x)\,dx = -(-)^{n-2}\,n!\int_a^b x\delta'(x)\,dx. \tag{1.63}$$

Use of (1.61a–c) leads to

$$\int_a^b x^{(n)}\delta^{(n)}(x)\,dx = \begin{cases} n!(-)^n & \text{if } a < 0 < b, & (1.64a)\\ 0 & \text{if } ab > 0, & (1.64b)\\ n!(-)^{n+1} & \text{if } a > 0 > b. & (1.64c) \end{cases}$$

Setting $a = -\infty$, $b = +\infty$ in (1.64a) yields (1.59b).

1.6 Integration, Substitution, and Product Properties

Some of the preceding proofs, namely (1.47b, 1.60, 1.62), have anticipated the properties (1.79d, 1.80a) concerning the product of ordinary and generalized functions. The latter are one instance of a set of properties of the unit impulse and its derivatives, namely the integration (Subsection 1.6.1) [substitution and product (Subsection 1.6.2)] properties.

1.6.1 Integration Property for the Derivatives of the Unit Impulse

The integral of the product of an ordinary n-times differentiable function (1.65a) by the nth derivate of the generalized function unit impulse is defined by the limit (1.65b):

$$f \in \mathcal{D}^n(a,b): \quad I \equiv \int_a^b f(x)\delta^{(n)}(x)dx \equiv \lim_{\sigma \to 0} \int_a^b f(x)\delta_\sigma^{(n)}(x)dx. \tag{1.65a,b}$$

The nth derivative of the Gaussian function (1.37) involves the Hermite polynomials (1.18) as coefficients:

$$\delta_\sigma^{(n)}(x) \equiv \frac{d^n}{dx^n}\left[\delta_\sigma(x)\right] = \frac{1}{\sqrt{\pi}\sigma}\frac{d^n}{dx^n}\left[\exp\left(-\frac{x^2}{\sigma}\right)\right] = \frac{\sigma^{-(n+1)/2}}{\sqrt{\pi}}\frac{d^n}{d\left(x/\sqrt{\sigma}\right)^n}\left[\exp\left(-\frac{x^2}{\sigma}\right)\right]$$

$$= \frac{(-)^n}{\sqrt{\pi}}\sigma^{-(n+1)/2}\exp\left(-\frac{x^2}{\sigma}\right)H_n\left(\frac{x}{\sqrt{\sigma}}\right). \tag{1.66}$$

Outside the origin (1.67a), the exponential dominates all polynomials in the limit (1.67b):

$$x \neq 0: \quad \lim_{\sigma \to 0}\delta_\sigma^{(n)}(x) = \frac{(-)^n}{\sqrt{\pi}}\lim_{\sigma \to 0}\sigma^{-(n+1)/2}\exp\left(-\frac{x^2}{\sigma}\right) = 0. \tag{1.67a,b}$$

Thus integrating (1.65b) by parts n times, all terms at the ends of the interval of integration (a,b) vanish by (1.67b); what remains is the integral multiplied by $(-)^n$ with the derivatives passed to the function that was assumed (1.65a) to be n-times differentiable in the interval of integration:

$$I \equiv (-)^n\lim_{\sigma \to 0}\int_a^b f^{(n)}(x)\delta_\sigma(x)dx = \frac{(-)^n}{\sqrt{\pi}}\int_a^b f^{(n)}(x)\exp\left(-\frac{x^2}{\sigma}\right)\frac{dx}{\sqrt{\sigma}}; \tag{1.68}$$

in (1.68) $\equiv$ (1.69b) a change of variable (1.69a) may be performed:

$$y = \frac{x}{\sqrt{\sigma}}: \quad I = \frac{(-)^n}{\sqrt{\pi}}\lim_{\sigma \to 0}\int_{a/\sqrt{\sigma}}^{b/\sqrt{\sigma}} f^{(n)}\left(y\sqrt{\sigma}\right)\exp\left(-y^2\right)dy$$

$$= \frac{(-)^n}{\sqrt{\pi}}f^{(n)}(0)\int_{-\infty}^{+\infty}\exp\left(-y^2\right)dy = (-)^n f^{(n)}(0), \tag{1.69a,b}$$

where it was assumed that $f^{(n)}(x)$ is uniformly continuous at the origin. The property (1.65b, 1.69b) $\equiv$ (1.70b) applied to a real n-times differentiable function (1.70a),

$$f(x) \in \mathcal{D}^n(|R): \quad \int_{-\infty}^{+\infty}f(x)\delta^{(n)}(x)dx = (-)^n f^{(n)}(0), \tag{1.70a,b}$$

can be used in particular to prove (1.71a,b, 1.72a,b, 1.73a,b):

$$f(x) = 1: \quad \int_{-\infty}^{+\infty}\delta(x)dx = 1, \tag{1.71a,b}$$

$$f(x) = x: \quad \int_{-\infty}^{+\infty} x\delta'(x)\,dx = -\lim_{x\to 0}\frac{d(x)}{dx} = -1, \tag{1.72a,b}$$

$$f(x) = x^n: \quad \int_{-\infty}^{+\infty} x^n \delta^{(n)}(x)\,dx = (-)^n \lim_{x\to 0}\frac{d^n}{dx^n}(x^n) = n!(-)^n, \tag{1.73a,b}$$

in agreement with respectively (1.42a, 1.46b, 1.59b). The translation (1.74a) allows the result (1.70b) to be restated (1.74b):

$$y = x - \xi: \quad \int_{-\infty}^{+\infty} f(x)\delta^{(n)}(x-\xi)\,dx = \int_{-\infty}^{+\infty} f(y+\xi)\delta^{(n)}(y)\,dy = (-)^n f^{(n)}(\xi), \tag{1.74a,b}$$

where the point $x = \xi$ lies in the range $a < x < b$ of integration, otherwise the expression would vanish:

$$b > a: \quad \int_a^b f(x)\delta^{(n)}(x-\xi)\,dx = \begin{cases} 0 & \text{if } \xi < a \text{ or } \xi > b, \tag{1.75a} \\ (-)^n f^{(n)}(\xi) & \text{if } a < \xi < b. \tag{1.75b} \end{cases}$$

This proves the **integration property**: *if the function (1.65a) is differentiable n-times in $a \leq x \leq b$, and its nth derivative is uniformly continuous at $x = \xi$, then it has the integration property (1.75a,b), with regard to the generalized function nth derivative of the unit impulse.* The result (1.75a,b) contains (1.59a,b), as the particular case $a = -\infty$, $b = +\infty$, $f(x) = x^n$. The fundamental property (1.75a,b) can be used to define the *n*th derivative of the unit impulse as a functional (Chapter 3), as an alternative approach to the theory of generalized functions based on the singular limit of a family of regular functionals with regard to a parameter (Chapter 1).

1.6.2 Substitution and Product Properties of the Derivatives of the Unit Impulse

In the case $n = 0$, the function $f(x)$ need only be integrable in (a, b) and uniformly continuous at $x = \xi$ and (1.75b) yields

$$a < \xi < b: \quad \int_a^b f(x)\delta(x-\xi)\,dx = f(\xi) = \int_a^b f(\xi)\delta(x-\xi)\,dx, \tag{1.76}$$

where (1.35b) was used; the validity of the substitution of x for ξ in $f(z)$ can be confirmed from (1.37):

$$\lim_{\sigma\to 0} f(x)\delta_\sigma(x-\xi) = \lim_{\sigma\to 0} f(\xi)\delta(x-\xi), \tag{1.77}$$

because both sides vanish for $x \neq \xi$, and they coincide for $x = \xi$. Thus has been proved the **substitution property** of *the Dirac distribution $\delta(x)$, relative to an integrable function $f(x)$, uniformly continuous at $x = \xi$*:

$$f \in \mathcal{E}(|R) \cap \overline{C}(\xi): \quad f(x)\delta(x-\xi) = f(\xi)\delta(x-\xi). \tag{1.78a,b}$$

There could be the temptation to justify the result as follows: (1) for $x \neq \xi$, both sides of (1.78b) are zero, hence it holds; and (2) for $x = \xi$ holds the replacement of x by ξ in the uniformly continuous function $f(x)$. While the argument (1) is correct, the statement (2) is not conclusive: Both sides of (1.78b) are singular at $x = \xi$, and it must be proved that the singularity is of the same nature. This was proved by considering (1.76) $\equiv$ (1.77). If unit impulse is replaced by a derivative, neither (1.76) $\equiv$ (1.77) nor (1.78b) hold, showing that the argument (2) would be fallacious in this case. Choosing in (1.78b) the origin (1.79a) leads to (1.79b):

$$\xi = 0: \quad f(x)\delta(x) = f(0)\delta(x); \quad f(x) = x: \quad x\delta(x) = 0 \Leftrightarrow \delta(x) \sim o\left(\frac{1}{x}\right), \quad (1.79\text{a–e})$$

and for the identity function (1.79c) follows (1.79d). The latter (1.79d) $\equiv$ (1.79e) shows that the singularity of the unit impulse at the origin (1.34b) is weaker (Section I.19.7) than $1/x$. The result (1.79d,e) is a particular case of

$$x^n \delta^{(n-1)}(x) = 0 \Leftrightarrow \delta^{(n-1)}(x) \sim o\left(\frac{1}{x^n}\right), \quad (1.80\text{a,b})$$

which shows that the singularity at the origin of the $(n-1)$th derivative of the unit impulse (1.57b) is weaker (Section I.19.7) than $1/x^n$; this result will be proved subsequently (3.49a–c) by a different method (Chapter 3). The case (1.79d) is considered next. It is obviously true for $x \neq 0$, since x is finite and $\delta(x) = 0$; it is less obvious for $x = 0$, since $\delta(0) = \infty$, leading to an indetermination of the type $0 \times \infty$. Noting that $x\delta_o(x)$ is an odd function, it follows that it must vanish at the origin, thus completing the proof of (1.79d).

1.7 Sign and Modulus and Related Generalized Functions

The unit jump (Section 1.2) and unit impulse (Section 1.3) and their derivatives (Sections 1.4 and 1.5) and properties (Section 1.6) can be used to construct other generalized functions, for example, the sign (Subsection 1.7.2) and modulus (Subsection 1.7.3) generalized functions; these have derivatives of all orders as generalized functions; the concept of symmetric (skew-symmetric) ordinary function also extends (Subsection 1.7.1) to the generalized functions.

1.7.1 Symmetric and Skew-Symmetric Generalized Functions

The Heaviside unit step (1.28a–c) has the property

$$H(x) + H(-x) = 1, \quad (1.81)$$

which holds because the two terms on the l.h.s. are, respectively, (0, 1) for $x < 0$, (1/2, 1/2) for $x = 0$ and (1, 0) for $x > 0$, so that their sum is unity in all cases. The result (1.81) implies that *the generalized function unit jump is an odd function about the horizontal line* $y = 1/2$:

$$h(x) \equiv H(x) - \frac{1}{2} = \frac{1}{2} - H(-x) = -\left(H(-x) - \frac{1}{2}\right) = -h(-x). \quad (1.82)$$

Differentiating $(n + 1)$-times (1.81) leads to

$$\delta^{(n)}(x) \equiv \frac{d^{n+1}}{dx^{n+1}}[H(x)] = \frac{d^{n+1}}{dx^{n+1}}\left[1 - H(-x)\right]$$

$$= -(-)^{n+1}\frac{d^{n+1}}{d(-x)^{n+1}}\left[H(-x)\right] = (-)^{(n)}\delta^{(n)}(-x). \quad (1.83)$$

It follows that the *unit impulse (1.34a,b)* ≡ *(1.41a,b) is an even function (1.44a), and its nth derivative*

$$\delta^{(n)}(x) = (-)^n \delta^{(n)}(-x),$$

(1.84)

is even (odd) for even (odd) $n = 2m(n = 2m + 1)$:

$$\delta^{(2m)}(x) = \delta^{(2m)}(-x), \quad \delta^{(2m+1)}(x) = -\delta^{(2m+1)}(-x);$$

(1.85a,b)

for example, (1.44a,b) for $m = 0$.

1.7.2 Sign and Its Derivatives as Generalized Functions

The unit jump and unit impulse can be used to construct other generalized functions; for example (Figure 1.9), the **sign function** sgn (x) is $+1(-1)$ for positive $x > 0$ $(x < 0)$ and zero at the origin:

$$\text{sgn}(x) = \begin{cases} -1 & \text{if } x < 0, \\ 0 & \text{if } x = 0, \\ +1 & \text{if } x > 0. \end{cases}$$

(1.86a)

(1.86b)

(1.86c)

It coincides (1.81) with twice the odd part of the unit jump (1.28a–c):

$$\text{sgn}(x) = H(x) - H(-x) = 2H(x) - 1 = 1 - 2H(-x) = -\text{sgn}(-x),$$

(1.87)

Thus, the sign function is constant everywhere except at the origin, where it has a jump equal to 2, implying that its derivate is (1.88a) twice the unit impulse:

$$\frac{d}{dx}\left[\text{sgn}(x)\right] = 2\delta(x), \quad \frac{d^n}{dx^n}\left[\text{sgn}(x)\right] = 2\delta^{(n-1)}(x),$$

(1.88a,b)

and the result extends to higher-order derivates (1.88a,b).

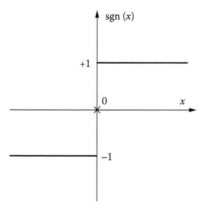

FIGURE 1.9 The sign function (Figure 1.8) is +1(−1) for positive (negative) argument and zero at the origin. Thus it equals the unit jump (Figure 1.5) multiplied by 2 and shifted upward by 1.

1.7.3 Modulus and Its Derivates as Generalized Functions

The sign function can be used to define another generalized function, namely, (Figure 1.10) the **modulus function** $|x|$, that equals $+x(-x)$ for positive (negative) x and thus is always positive, except at the origin where it vanishes:

$$|x| \equiv x\,\mathrm{sgn}(x) = \begin{cases} -x & \text{if } x \le 0, \\ \\ +x & \text{if } x \ge 0. \end{cases}$$

(1.89a)

(1.89b)

The modulus distribution is continuous at the origin, where it has discontinuous derivates,

$$\frac{\mathrm{d}|x|}{\mathrm{d}x} = \mathrm{sgn}(x) = \begin{cases} -1 & \text{if } x < 0, \\ 0 & \text{if } x = 0, \\ +1 & \text{if } x > 0, \end{cases}$$

(1.90a)

(1.90b)

(1.90c)

of the argument equal to the sign; this can be verified from

$$\frac{\mathrm{d}|x|}{\mathrm{d}x} = \frac{\mathrm{d}}{\mathrm{d}x}\big[x\,\mathrm{sgn}(x)\big] = \mathrm{sgn}(x) + 2x\,\delta(x),$$

(1.91)

where the last term vanishes by (1.79d). From (1.91), it follows that *the nth derivative of the modulus function (1.89a,b) as a generalized function equals (1.92) the (n − 1) the derivative of the sign that is twice the (n − 2)th derivate of the unit impulse:*

$$\frac{\mathrm{d}^{n}|x|}{\mathrm{d}x^{n}} = \frac{\mathrm{d}^{n-1}}{\mathrm{d}x^{n-1}}\big[\mathrm{sgn}(x)\big] = 2\delta^{(n-2)}(x).$$

(1.92)

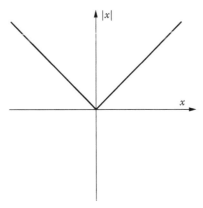

FIGURE 1.10 The modulus function (Figure 1.9) equals the argument for positive values, and reverses the sign for negative values, so that it vanishes at the origin. It equals its argument multiplied by the sign function (Figure 1.9) and thus its slope has a jump of 2 at the origin.

1.8 Nonremovable, Isolated, and Finite Discontinuities

The classification of discontinuities of an ordinary function (Subsection 1.8.1) shows that using the unit jump, it can be replaced by a generalized function (Subsection 1.8.2) if it has a finite number of finite discontinuities.

1.8.1 Classification of Discontinuities of an Ordinary Function

If the function $f(x)$ is continuous at $x = a$, then (Section I.13.2) both the **right** $f(a + 0)$ and **left** $f(a − 0)$ **limits** exist and equal the value of the function at a:

$$f \in C; \varepsilon, \delta > 0: \quad \lim_{\varepsilon \to 0} f(a + \varepsilon) \equiv f(a + 0) = f(a) = f(a - 0) \equiv \lim_{\delta \to 0} f(a - \delta). \tag{1.93}$$

It the right and left hand limits exist, then discontinuity of a function at point $x = a$ may be classified as follows:

$$\text{discontinuity} \begin{cases} \text{removable}: & f(a+0) = f(a-0) \neq f(a), & \text{(1.94a)} \\ \text{nonremovable}: & f(a+0) \neq f(a-0), & \text{(1.94b)} \end{cases}$$

(1) **removable** (1.94a) if the left and right limits coincide but the value of the function is different (Figure 1.11a), that is, continuity may be restored by changing the value of the function at one point only; and (2) **nonremovable** (1.94b) if the left and right limits are different (Figure 1.11b), so that in order that the function becomes continuous, it must be changed over a neighborhood of a. A nonremovable discontinuity is associated with a **jump** of the function:

$$\Delta f(a) \equiv f(a+0) - f(a-0) \begin{cases} \leq M < \infty & \text{finite,} & \text{(1.95a)} \\ = \pm\infty & \text{infinite,} & \text{(1.95b)} \end{cases}$$

and it is a **finite (infinite) discontinuity** (1.95a) [(1.95b)] if the jump is bounded (unbounded) as in Figure 1.11b (Figure 1.11c). For example, the sign function (1.86a–c) has a finite discontinuity with jump two at the origin, and the function $1/x$ has an infinite discontinuity at the origin, since $1/x \to \pm\infty$ as $x \to \pm0$. A removable discontinuity (1.94a) has zero jump. The discontinuity of a function $f(x)$ at a point $x = a$ is **isolated** if there exists a neighborhood $V_\varepsilon(a)$ for some $\varepsilon > 0$ not containing any other discontinuity:

$$\exists_{\varepsilon > 0} \forall_{b \in |R} : \quad 0 < |b - a| < \varepsilon \Rightarrow f(b+0) = f(b) = f(b-0). \tag{1.96}$$

The discontinuity is **nonisolated** if every neighborhood $V_\delta(a)$, for any $\delta > 0$, contains other discontinuities. For example, the **Dirichlet function** $D(x)$, equal to zero (1.97b) [unity (1.97a)] for irrational (rational) variables,

$$D(x) = \begin{cases} 1 & \text{if } x \in |R - |I \equiv |L, & \text{(1.97a)} \\ 0 & \text{if } x \in |I \equiv |R - |I, & \text{(1.97b)} \end{cases}$$

has finite discontinuities with unit jump that are everywhere dense on the real axis, because every real interval contains an infinite number of rational and irrational points (Subsection II.9.7.2).

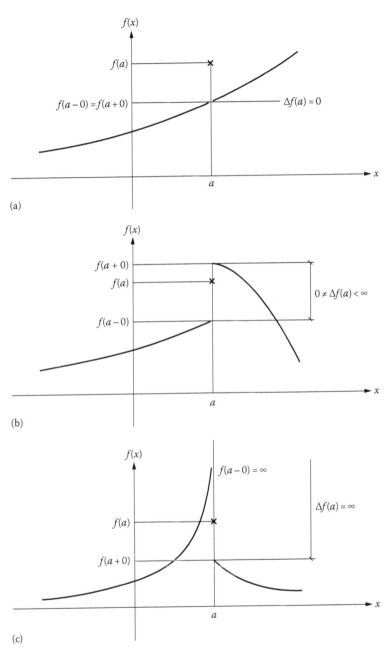

FIGURE 1.11 The simplest type of discontinuity is removable, when the right-hand limits are the same (a) but the value of the function is distinct; changing the value of the function at one point, to the same as the left- and right-hand limits restores continuity. If the left- and right-hand limits are distinct the discontinuity is nonremovable, since continuity could only be restored by changing the function over a neighborhood of the discontinuity. In this case, there is a jump of the value of the function across the discontinuity equal to the difference of right- and left-hand limits. The discontinuity is finite (infinite) according to whether the value of the jump is finite (infinite) in (b) [(c)].

1.8.2 Function with a Finite Number of Finite Discontinuities

If the function f(x) has a finite number of discontinuities in the interval (x_0, x_n), then these are isolated, and a partition $x_0 < x_1 < \cdots < x_{n-1} < x_n$ can be made, such that each interval (x_k, x_{k+1}) with $k = 0, \ldots, n-1$ contains only one discontinuity. Thus, it is sufficient to consider one finite, nonremovable, isolated discontinuity to show that an ordinary function that is piecewise differentiable except at a finite number of points of finite discontinuity can be replaced by a generalized function that is differentiable everywhere including at the points of discontinuity. Suppose (Figure 1.12a) that the function f(x) coincides with $f_1(f_2)$ for $x < a$ ($x > a$):

$$F(x) = \begin{cases} f_1(x) & \text{if } x < a, \\ f_2(x) & \text{if } x > a, \end{cases}$$

(1.98a)

(1.98b)

but is not defined at the point of discontinuity $x = a$, where it has a finite jump (1.95a). The associated generalized function is

$$F(x) = f_1(x)H(a-x) + f_2(x)H(x-a)$$

$$= f_1(x) + [f_2(x) - f_1(x)]H(x-a)$$

$$= f_2(x) - [f_2(x) - f_1(x)]H(a-x),$$

(1.99a–c)

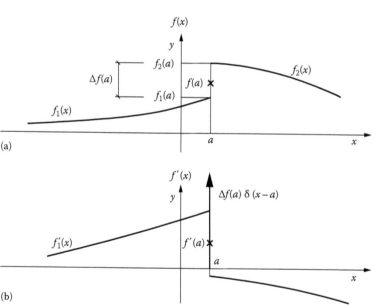

(a)

(b)

FIGURE 1.12 An ordinary function with a finite number of finite discontinuities can be replaced by a generalized function that: (1) coincides with the ordinary function at the points of continuity; and (2) uses the unit jump to represent the finite discontinuities, that are isolated (a) because they are finite in number. If the ordinary function has piecewise Nth order derivative at all points except those of discontinuity, then the corresponding generalized function has derivatives of orders up N including at the points of discontinuity; the Nth order derivative of the generalized function involves the Nth order derivative of the ordinary function at the points where this Nth order derivative is continuous; the discontinuities of the ordinary function or its derivatives of order less than N appear as jumps multiplied by derivatives of order $N-n$ of the unit impulse at the points of discontinuity. For example, the first-order derivative (b) of the ordinary function (a) involves the product of the jump by the unit impulse at the point of discontinuity.

where (1.28a–c, 1.81) are used. The generalized function (1.99) does (1/2) coincide with the ordinary function $f_1(f_2)$ before (1.98a) [after (1.98b)] the point $x = a$; and (3) at $x = a$, it takes the arithmetic mean value between the right- and left-hand limits:

$$2F(a) = f_1(a) + f_2(a) = f(a+0) + f(a-0). \tag{1.100}$$

The main advantage of the generalized function (1.99) over the original ordinary function (1.98a,b) is that if the ordinary function is piecewise differentiable, the generalized function is differentiable including at the point of discontinuity, where it has a finite jump (1.95a) ≡ (1.101):

$$\Delta f(a) \equiv f(a+0) - f(a-0) = f_2(a) - f_1(a). \tag{1.101}$$

This will be proved next (Section 1.9).

1.9 Jump of a Function and of Its Successive Derivates

Since an ordinary function with a finite number of finite discontinuities (Subsection 2.8.1) can be replaced by a generalized function (Subsection 1.8.2), the latter can be used to obtain the first derivative (Subsection 1.9.1) as well as derivatives of any order (Subsection 1.9.2).

1.9.1 Derivative of a Discontinuous Function

Assuming that the ordinary function (1.98a,b) is piecewise differentiable (1.102a) except at the point of discontinuity (1.101), the associated generalized function (1.99a) may be differentiated (1.102b):

$$f \in \bar{\mathcal{D}}: \quad F'(x) - f_1'(x) H(a-x) - f_2'(x) H(x-a)$$
$$= -f_1(x)\delta(a-x) + f_2(x)\delta(x-a)$$
$$= \left[f_2(x) - f_1(x) \right]\delta(x-a) = \Delta f(a)\delta(x-a), \tag{1.102a,b}$$

where (1) on the l.h.s. is the piecewise ordinary differentiation before and after the point of discontinuity; and (2) the terms on the r.h.s. are concentrated at the point of discontinuity. The latter (2) involve the unit impulse (1) as the derivative of the unit jump (1.34a,b); and (2/3) its symmetry (1.44a) and substitution (1.78a,b) properties. It follows (1.102a,b) ≡ (1.103a,b) that,

$$f \in \bar{\mathcal{D}}: \quad F'(x) = f_1'(x) H(a-x) + f_2'(x) H(x-a) + \Delta f(a)\delta(x-a), \tag{1.103a,b}$$

the first-order derivate of the generalized function (1.99a–c) associated with the ordinary function (1.98a,b), piecewise differentiable (1.103a) except at a finite isolated discontinuity, consists (Figure 1.12b) of (1.103b) (1) the derivate of the ordinary function at the regular points $x \neq a$; (2) plus the jump of the function (1.101) multiplied by a unit impulse concentrated at the point of discontinuity $x = a$. For example, in the case of the sign function (1.90a–c) ≡ (1.104a,b), the discontinuity is at the origin (1.104c) with jump 2 in (1.104d), and substituting

$$f_2(x) = -1, \quad f_2(x) = +1, \quad a = 0, \quad \Delta f(0) = 2: \quad \text{sgn}'(x) = 2\delta(x) \tag{1.104a–e}$$

into (1.103b) leads to (1.104e) in agreement with (1.88a).

1.9.2 Derivatives and Jumps of Higher Order

If the function, like the modulus (1.89a,b), has discontinuous derivatives, with jumps,

$$n = 0,\ldots,N: \quad \Delta f^{(n)}(a) \equiv f^{(n)}(a+0) - f^{(n)}(a-0), \tag{1.105}$$

these appear in higher-order differentiations. For example, if the ordinary function (1.98a,b) is piecewise twice differentiable (1.106a), the second-order derivative of the associated generalized function (1.99a) may be calculated differentiating once more (1.103b):

$$f \in \bar{D}^2: \quad F''(x) - f_1''(x)H(a-x) - f_2''(x)H(x-a) - \Delta f(a)\delta'(x-a)$$

$$= -f_1'(x)\delta(a-x) + f_2'(x)\delta(x-a)$$

$$= \left[f_2'(x) - f_1'(x)\right]\delta(x-a) = \Delta f'(a)\delta(x-a), \tag{1.106a,b}$$

it involves the jump of the first derivate of the function. For example, in the case (1.89a,b) ≡ (1.107a,b) of the modulus function, it is continuous at the origin (1.107c), but there is a discontinuity with a jump (1.107d) at the origin in the first derivate:

$$f_1(x) = -x, \quad f_2(x) = +x, \quad \Delta f(0) = 0, \quad \Delta f'(0) = 2: \quad \frac{d^2|x|}{dx^2} = 2\delta(x), \tag{1.107a-e}$$

and substituting (1.107a–d) in (1.106b) leads to (1.107e) that coincides with (1.92), for $n = 2$.

The formula (1.106a,b) can be generalized to the Nth derivate:

$$f(x) \in \bar{D}^N: \quad F^{(N)}(x) = f_1^{(N)}(x)H(a-x) + f_2^{(N)}H(x-a) + \sum_{n=0}^{N-1} \Delta f^{(N-n-1)}(a)\delta^{(n)}(x-a). \tag{1.108a,b}$$

This can be proved by induction since (1) it holds for $N = 1$(2), when it reduces to (1.103a,b) [(1.106a,b)]; and (2) if it holds for N, then it also applies for $N + 1$, because from (1.108a,b) follows (1.109a,c):

$$f(x) \in \bar{D}^{N+1}; \quad m = n+1: \quad F^{(N+1)}(x) - f_1^{(N+1)}(x)H(a-x) - f_2^{(N+1)}(x)H(x-a)$$

$$= \left[f_2^{(N)}(x) - f_1^{(N)}(x)\right]\delta(x-a) + \sum_{n=0}^{N-1} \Delta f^{(N-n-1)}(a)\delta^{(n+1)}(x-a)$$

$$= \Delta f^{(N)}(a)\delta(x-a) + \sum_{m=1}^{N} \Delta f^{(N-m)}(a)\delta^{(m)}(x-a) = \sum_{m=0}^{N} \Delta f^{(N-m)}(a)\delta^{(m)}(x-a), \tag{1.109a-c}$$

where (1) it is assumed that the function is $(N + 1)$-times piecewise differentiable (1.190a); (2) the change of index of summation (1.109b) was made; and (3) the proof is completed noting that (1.109c) coincides with (1.108a,b) replacing N by $N + 1$. Thus has been obtained the **theorem of differentiation of a discontinuous function**: *if the ordinary function (1.98a,b) is piecewise differentiable (1.108a) up to and including order N, except at a point x = a, where its derivatives of lower orders 0, 1, …, N – 1 have jumps (1.105), then the associated generalized function (1.99a–c) has derivates of all orders up to N in (1.108b), consisting of (1) the Nth derivate of the function at all points x ≠ a; and (2) at the point x = a, the sum of the jumps (1.105)* *of the derivates of order n = 0,..., N − 1 multiplied by the derivate of order N − 1 − n of the unit impulse distribution concentrated at x = a.* Since the $(N − n − 1)$th derivative of the unit impulse is (1.34a,b) the $(N–n)$-th derivative of the unit jump and is multiplied by the jump of the nth derivative of the function in (1.108b), the orders of the derivatives of all terms add to N. The differentiation rule is applied to several generalized functions corresponding to discontinuous ordinary functions in Subsections 2.5.1, 2.5.4, and 2.5.7 and in Example 10.1.

NOTE 1.1: Generalized Functions as Nonuniform Limits of Ordinary Functions

The approach based on the limit of a family of functions $f_\sigma(x)$ with regard to a parameter leads to ordinary (generalized) functions if the limit is uniform (1.110a) [nonuniform (1.110b)]:

$$f_\sigma(x) \in \mathcal{U}(a,b): \quad \lim_{\sigma \to 0} f_\sigma(x) = f(x) = \begin{cases} \in \mathcal{U} & \text{uniform limit,} \hspace{2cm} (1.110a) \\ \in \mathcal{G} & \text{nonuniform limit.} \hspace{1.6cm} (1.110b) \end{cases}$$

This approach to generalized functions thus relies on the properties of nonuniform limits (Section I.24.5); for example, the nonuniform limit of a sequence of continuous/differentiable/integrable functions may be discontinuous/nondifferentiable/nonintegrable, and may fail to define an ordinary or injective function, such as the derivatives of the unit impulse that have no single well-defined value at the origin. The limit of a sequence of functions can help visualize how the peculiar properties of some generalized functions arise; for example, (1) the Gaussians with unit integral and decreasing variance (Figure 1.6) become zero everywhere except at the origin where they diverge to infinity leading to the unit impulse; (2) the nth derivative ($n = 1$ in Figure 1.7 and $n = 2$ in Figure 1.8) is still zero everywhere except at the origin where it has two $(n + 1)$ infinite jumps leading to the nth derivative of the unit impulse; (3) the first primitive or error function (Figure 1.5) is monotonic increasing from 0 at $−\infty$ to at $+\infty$ and reduces to a unit jump at the origin for zero variance; and (4) further primitives are continuous (Section 3.5). The approach to generalized functions via the nonuniform parametric limit of a family of ordinary functions implies that the proof of any property requires a limit of a "similar" property of the family; since the same generalized function can arise from distinct families of ordinary functions, the choice of a particular family is circumstantial rather than essential. In this respect, it is preferable to base the theory of generalized functions on the general properties of certain classes of test functions (Chapter 3); this leads to deductions of the properties of generalized functions that are often simpler than dealing with nonuniform limits of families of ordinary functions. The latter have already served (Chapter 1) the useful purpose of showing intuitively how the generalized arise out of the ordinary functions; the alternative approach based on functionals (Chapter 3) will be the basis of further developments; for example, it extends readily to (1) multiple-dimensional functions whose arguments are ordinary functions of several variables; (Chapter 5); (2) their applications such as (2-a) concentrated forces and moments (Chapters 2 and 4); (2-b) monopoles, dipoles, quadrupoles, and multipoles of any order (Chapters 6 and 8) and any dimension (Chapter 9); and (3) they also relate to the solution of linear ordinary and partial differential equations with forcing (Chapter 7) via the Green or influence function defined as the response to the Dirac unit impulse.

NOTE 1.2: Gaussian Integrals and the Generalized Calculus

One of the families of functions whose nonuniform limit is the unit impulse is the set of Gaussians with decreasing r.m.s. value tending to zero. The Gaussian appears in several contexts including probability distributions (Notes 1.3 through 1.7), unsteady heat conduction, and statistical mechanics. The related functions include the error function (Section 1.2, Note 1.3) and the Fresnel and exponential integrals. Several of these functions involve Gaussian integrals, of which those in Section 1.1 are a first limited sample, sufficient for the present Chapter 1. The Gaussian integrals involving integral powers may be extended to nonintegral or complex exponents, like the other parameters, that may also take complex values. These Gaussian integrals involving powers with complex exponents (Notes 1.9 through 1.12) may be evaluated in terms of the Gamma function (Note 1.8) and are related to the Hermite polynomials (Subsections 1.1.6 and 1.1.7, Notes 1.13 and 1.14). The Gaussian integrals owe their name to the Gaussian or normal probability distribution considered next (Notes 1.3 through 1.7). The designation of "generalized function" is preferred to "distribution" precisely to avoid confusion with probability distributions or other distributions: of mass, electric charge, or current, etc.... On the other hand, the generalized functions can also be defined as functionals (Chapter 3). The most important generalized functions also have multiple names; for example, (1) the unit jump was introduced by Heaviside (1876) in electrical theory and is known as Heaviside function or Heaviside unit step function; and (2) the unit impulse was introduced by Dirac (1930) in quantum mechanics and is known as the Dirac delta function. For brevity will be used henceforth the designations unit jump function for (1), unit impulse function for (2), and derivative unit impulse functions for the derivatives of (2). The designation generalized function is preferred to distribution to distinguish from ordinary functions, and generalized calculus means the extension of the calculus from ordinary to generalized functions.

NOTE 1.3: Gaussian or Normal Probability Distribution

The Gaussian probability density is defined on the real line (1.111a) by (1.111b):

$$-\infty < x < +\infty: \quad p(x) = a \exp[-b(x-\mu)^2], \tag{1.111a,b}$$

involving three parameters: (1) the **mean value** μ where the probability density has a maximum $p(\mu) = a$; (2) the parameter a is related to b by the **normalization condition** (1.112b) of unit total probability:

$$y \equiv x - \mu: \quad 1 = \int_{-\infty}^{+\infty} p(x)\,dx = a \int_{-\infty}^{+\infty} \exp(-by^2)\,dy = a\sqrt{\frac{\pi}{b}}, \tag{1.112a,b}$$

where the change of variable (1.112a) and the Gaussian integral (1.10a,b) were used; and (3) a further relation between (a,b) is provided by a **central moment** of any order n:

$$C_n \equiv \left\langle (x-\mu)^n \right\rangle \equiv \int_{-\infty}^{+\infty} (x-\mu)^n p(x)\,dx. \tag{1.113}$$

For example, in the case of the Gaussian probability distribution (1.111b, 1.112a),

$$C_n = a \int_{-\infty}^{+\infty} y^n \exp(-by^2)\,dy, \tag{1.114}$$

all moments (1) of odd order vanish (1.115b), as follows from (1.114):

$$C_{2n+1} = 0; \quad C_{2n} = (2n-1)!!(2b)^{-n}; \tag{1.115a,b}$$

and (2) of even order are given by (1.115b), as follows from (1.114):

$$C_{2n} = a \int_{-\infty}^{+\infty} y^{2n} \exp(-by^2)\,dy = (2n-1)!!\, a\sqrt{\frac{\pi}{b}}(2b)^{-n}, \tag{1.116}$$

where (1.17d) was used; substitution of (1.112b) in (1.116) leads to (1.115b).

The vanishing of the central moment of first order (1.115a) with $n = 0$ proves (1.117a) that μ is the **mean value** (1.17b):

$$0 = C_1 = \langle x \rangle - \mu : \quad \mu = \langle x \rangle = \int_{-\infty}^{+\infty} x \, p(x) \mathrm{d}x. \tag{1.117a,b}$$

The nonvanishing central moment of lowest order is the **variance**, whose square root is the **r.m.s. value:**

$$\sigma^2 = C_2 = \left\langle (x - \mu)^2 \right\rangle = \frac{a}{2b} \sqrt{\frac{\pi}{b}} = \frac{1}{2b}. \tag{1.118}$$

Using the normalization condition (1.112b) simplifies (1.118) to (1.119a):

$$b = \frac{1}{2\sigma^2}, \quad a = \sqrt{\frac{b}{\pi}} = \frac{1}{\sigma\sqrt{2\pi}}, \tag{1.119a,b}$$

and thus determines both parameters (a,b) in (1.111b) $\equiv$ (1.120a):

$$p(x) = \frac{1}{\sigma\sqrt{2\pi}} \exp\left[-\frac{(x - \mu)^2}{2\sigma^2} \right]. \tag{1.120a}$$

Thus, *the Gaussian probability distribution with mean value μ and r.m.s. value σ is specified by the* **probability density function** *(1.120a); the* **cumulative probability function** *is specified by (1.120b):*

$$P(x) \equiv \int_{-\infty}^{x} p(\xi) \mathrm{d}\xi = \mathrm{erf}\left(\frac{x - \mu}{\sigma\sqrt{2}} \right), \tag{1.120b}$$

in terms of the error functions (1.25b), as follows from (1.121b):

$$z \equiv \frac{\xi - \mu}{\sigma\sqrt{2}} : \quad P(x) = \frac{1}{\sigma\sqrt{2\pi}} \int_{-\infty}^{x} \exp\left[-\left(\frac{\xi - \mu}{\sigma\sqrt{2}} \right)^2 \right] \mathrm{d}\xi$$

$$= \frac{1}{\sqrt{\pi}} \int_{-\infty}^{(x-\mu)/\sigma\sqrt{2}} \exp\left(-z^2 \right) \mathrm{d}z = \mathrm{erf}\left(\frac{x - \mu}{\sigma\sqrt{2}} \right), \tag{1.121a,b}$$

using the change of variable (1.121a). The two parameters (1.120) of Gaussian probability distribution (Figure 1.13b) are (1) the mean value that indicates the location of the peak (Figure 1.13a) or maximum probability (1.122a):

$$p_{\max} = p(\mu) = \frac{1}{\sigma\sqrt{2\pi}}; \quad p(k\mu) = p(\mu) \exp\left[-\left(\frac{k-1}{2} \right)^2 \frac{\mu^2}{\sigma^2} \right], \tag{1.122a,b}$$

and (2) the variance σ^2 or r.m.s value σ that is smaller (Figure 1.13c) for a higher maximum probability (1.122a) and a smaller spread around (or faster decay away from) the maximum (1.122b).

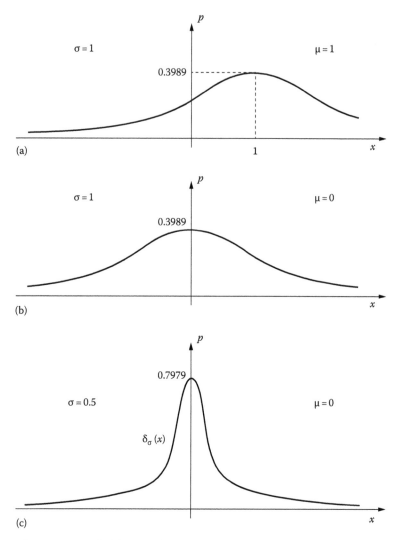

FIGURE 1.13 The Gaussian probability distribution is specified by two parameters in the probability density function (b). The mean value μ indicates the peak or location of maximum probability. A larger (smaller) r.m.s. value σ implies a lower (higher) peak and slower (faster) decay away from it. The mean value is zero in (b,c) and unity in (a); the r.m.s. value is unity in (a,b) and one-half in (c).

NOTE 1.4: Central and Noncentral Moments of All Orders

For any normalized (1.123a) probability distribution, the central moments (1.113) and noncentral moments relative to the origin (1.123b)

$$1 = \int_{-\infty}^{+\infty} p(x)\,dx: \quad M_n \equiv \langle x^n \rangle \equiv \int_{-\infty}^{+\infty} x^n p(x)\,dx, \tag{1.123a,b}$$

are related by

$$C_n = \sum_{m=0}^{n} \frac{n!(-)^{n-m}}{m!(n-m)!}\mu^{n-m}M_m, \quad M_n = \sum_{m=0}^{n} \frac{n!}{m!(n-m)!}\mu^{n-m}C_m, \tag{1.124a,b}$$

so that the central (noncentral) moments of order n involve all central (noncentral) moments of equal or lower order m = 0, 1, ..., n. The proof of (1.124a,b) follows from the binomial theorem (I.25.38):

$$C_n = \left\langle \left(x-\mu\right)^n \right\rangle = \sum_{m=0}^{n} \binom{n}{m}\left(-\mu\right)^{n-m}\left\langle x^m \right\rangle, \tag{1.125a}$$

$$M_n = \left\langle \left(x-\mu+\mu\right)^n \right\rangle = \sum_{m=0}^{n} \binom{n}{m}\mu^{n-m}\left\langle \left(x-\mu\right)^m \right\rangle. \tag{1.125b}$$

In the case of the Gaussian probability distribution (1.120a) ≡ (1.120b), the central moments (1.115a,b, 1.119a,b) are specified by the r.m.s. value alone:

$$C_{2n+1} = 0, \quad C_{2n} = \left(2n-1\right)!!\sigma^{2n}. \tag{1.126a,b}$$

The noncentral moments relative to the origin (1.123b) involve (1.124b):

$$M_n = \sum_{m=0}^{\leq n/2} \frac{n!\left(2m-1\right)!!}{\left(2m\right)!\left(n-2m\right)!}\mu^{n-2m}\sigma^{2m}, \tag{1.127}$$

and also the mean value in (1.127).

The main moments are by increasing order (1) the mean value (1.117b) or moment of first order M_1 relative to the origin (1.123b); (2) the variance (1.118) or second central moment that defines the r.m.s. value as its square root; and (3/4) the next independent moments of lower orders are the central moment of third (1.128a) [fourth (1.128b)] order:

$$C_3 \equiv \left\langle \left(x-\mu\right)^3 \right\rangle, \quad C_4 = \left\langle \left(x-\mu\right)^4 \right\rangle. \tag{1.128a,b}$$

The noncentral moment of order 2 relative to the origin is given by (1.129b) because the first central moment is zero (1.129a):

$$C_1 = \left\langle x \right\rangle - \mu = 0: \quad M_2 \equiv \left\langle x^2 \right\rangle = \left\langle \left(x-\mu+\mu\right)^2 \right\rangle = \left\langle \left(x-\mu\right)^2 \right\rangle + \mu^2 = \sigma^2 + \mu^2. \tag{1.129a,b}$$

The noncentral moments of third (fourth) order relative to the origin are given for any probability distribution by (1.130a) [(1.130b)]:

$$M_3 \equiv \left\langle x^3 \right\rangle = \left\langle \left(x-\mu+\mu\right)^3 \right\rangle = C_3 + 3\mu C_2 + \mu^3, \tag{1.130a}$$

$$M_4 \equiv \left\langle x^4 \right\rangle = \left\langle \left(x-\mu+\mu\right)^4 \right\rangle = C_4 + 4\mu C_3 + 6\mu^2 C_2 + \mu^4. \tag{1.130b}$$

In the case of the Gaussian probability distribution, (1) the first four nonzero central moments (1.126b) are

$$\left\{C_2, C_4, C_8, C_{10}\right\} = \left\{\sigma^2, 3\sigma^4, 15\sigma^6, 105\sigma^8, 945\sigma^{10}\right\}; \tag{1.131a-e}$$

and (2) only the first two appear (1.117b, 1.129b, 1.130a,b) in the first four noncentral moments relative to the origin:

$$\{M_1, M_2, M_3, M_4\} = \left\{\mu, \sigma^2 + \mu^2, \mu(\mu^2 + 3\sigma^2), 3\sigma^4 + 6\sigma^2\mu^2 + \mu^4\right\}. \tag{1.132a--d}$$

NOTE 1.5: Central Characteristic Function of a Probability Distribution

The central characteristic function is defined as the Fourier transform of the probability density function with the deviation from the mean $x - \mu$ as the variable:

$$C(a) \equiv \left\langle \exp\left[ia(x-\mu)\right]\right\rangle = \int_{-\infty}^{+\infty} p(x)\exp\left[ia(x-\mu)\right]dx. \tag{1.133}$$

The power series for the exponential (I.23.60) $\equiv$ (II.1.26b) leads to

$$C(a) = \sum_{n=0}^{\infty} \frac{(ia)^n}{n!}\left\langle(x-\mu)^n\right\rangle = \sum_{n=0}^{\infty} \frac{(ia)^n}{n!}C_n, \tag{1.134a,b}$$

through the coefficients of powers of a, all the central moments (1.113) of the probability distribution. The series (1.134a,b) converges if the characteristic function is analytic (1.135a) in a neighborhood of the origin, in which case it coincides with the Maclaurin series (I.23.34b) $\equiv$ (II.1.24a,b) $\equiv$ (1.135b):

$$C \in \mathcal{A}: \quad C(a) = \sum_{n=0}^{\infty} \frac{a^n}{n!}C^{(n)}(0); \quad C_n = i^{-n}\lim_{a\to 0}\frac{d^n C}{da^n}. \tag{1.135a--c}$$

Comparing (1.134b) $\equiv$ (1.135b), it follows that all the central moments are specified (1.135c) by the derivatives of the characteristic function at the origin. In the case of the Gaussian probability distribution (1.120a), the central characteristic function (1.133) is given by (1.136b):

$$y = x - \mu: \quad C(a) = \frac{1}{\sigma\sqrt{2\pi}}\int_{-\infty}^{+\infty}\exp\left[ia(x-\mu) - \frac{(x-\mu)^2}{2\sigma^2}\right]dx$$

$$= \frac{1}{\sigma\sqrt{2\pi}}\int_{-\infty}^{+\infty}\exp\left(iay - \frac{y^2}{2\sigma^2}\right)dy = \exp\left(-\frac{a^2\sigma^2}{2}\right), \tag{1.136a,b}$$

using the change of variable (1.136a) and the integral (1.13). Expanding the characteristic function (1.136b) $\equiv$ (1.137a) in powers

$$C(a) = \exp\left(-\frac{a^2\sigma^2}{2}\right) = \sum_{n=0}^{\infty}\frac{(-)^n}{n!}a^{2n}\left(\frac{\sigma^2}{2}\right)^n, \quad C_{2n+1} = 0, \tag{1.137a,b}$$

and comparing with (1.134b) shows that the central moments of (1) odd order are zero (1.137b) $\equiv$ (1.115a); and (2) even order are given by

$$C_{2n} = \frac{(2n)!}{n!}\left(\frac{\sigma^2}{2}\right)^n = \sigma^{2n}\frac{(2n)!}{(2n)!!} = (2n-1)!!\sigma^{2n}, \tag{1.138}$$

in agreement with (1.138) $\equiv$ (1.115b).

NOTE 1.6: Noncentral Characteristic Function as a Fourier Transform

The **noncentral characteristic function** is defined by (1.139a):

$$D(a) \equiv \langle \exp(iax) \rangle = \exp(ia\mu)C(a), \qquad (1.139\text{a,b})$$

and is related to the central characteristic function (1.133) by (1.139b). The noncentral characteristic function (1.139a) is the Fourier transform (1.140a) of the probability density function:

$$D(a) = \int_{-\infty}^{+\infty} p(x)e^{iax}\mathrm{d}x; \quad p(x) = \frac{1}{2\pi}\int_{-\infty}^{+\infty} D(a)e^{-iax}\mathrm{d}a. \qquad (1.140\text{a,b})$$

The characteristic function (1.140a) and probability density function (1.140b) are equivalent descriptions of the probability distribution, since the latter can be obtained from the former by an inverse Fourier transform (1.140b). The exponential series (I.23.60) ≡ (II.1.26b) leads (1.139a) to (1.141a):

$$D(a) \equiv \sum_{n=0}^{\infty} \frac{(ia)^n}{n!}\langle x^n \rangle = \sum_{n=0}^{\infty} \frac{(ia)^n}{n!} M_n, \qquad (1.141\text{a,b})$$

which specifies all noncentral moments (1.123b) as the coefficients of powers of a. The series (1.141a,b) converges if the noncentral characteristic function is analytic in a neighborhood of the origin (1.142a), in which case (1.141b) coincides with its Maclaurin series (I.23.34b) ≡ (I.1.24a,b) ≡ (1.142b):

$$D \in \mathcal{A}: \quad D(a) = \sum_{n=0}^{\infty} \frac{a^n}{n!} D^{(n)}(0); \quad M_n = i^{-n}\lim_{a\to 0}\frac{\mathrm{d}^n D}{\mathrm{d}a^n}. \qquad (1.142\text{a--c})$$

Comparing (1.141b) ≡ (1.142b), it follows that the noncentral moments are specified (1.142c) by the derivatives of the noncentral characteristic function at the origin.

Using (1.134b) and (1.141b) in (1.139b) leads to (1.143b):

$$n = m + k: \quad \sum_{n=0}^{\infty}\frac{(ia)^n}{n!}M_n = e^{ia\mu}\sum_{m=0}^{\infty}\frac{(ia)^m}{m!}C_m = \sum_{m,k=0}^{\infty}\frac{\mu^k(ia)^{m+k}}{k!m!}C_m = \sum_{n=0}^{\infty}\sum_{m=0}^{n}\frac{\mu^{n-m}(ia)^n}{(n-m)!m!}C_m, \qquad (1.143\text{a,b})$$

where (1.143a) was used, so that the inner sum is over $m = 0, ..., n$ because (1) there are no powers μ^{n-m} with negative exponents; and (2) equivalently for $n - m$ negative $(n - m)! = \infty$ so the corresponding terms do not appear in (1.143b). Equating powers of $(ia)^n$ in (1.143b) proves the relation (1.124b) between the central and noncentral moments. In the case of the Gaussian probability distribution (1.137a), the noncentral characteristic function (1.139b) is

$$D(a) = \exp\left(ia\mu - \frac{a^2\sigma^2}{2}\right). \qquad (1.144)$$

Its power series expansion is (1.145b):

$$n = k + 2m: \quad D(a) = \sum_{k=0}^{\infty}\frac{\mu^k(ia)^k}{k!}\sum_{m=0}^{\infty}\frac{(ia)^{2m}\sigma^{2m}}{m!2^m} \equiv \sum_{m=0}^{\infty}\sum_{m=0}^{\le n/2}\frac{(ia)^n\mu^{n-2m}\sigma^{2m}}{(n-2m)!m!2^m}, \qquad (1.145\text{a,b})$$

where the change of variable (1.145a) was used, so that the inner sum holds for $0 \leq m \leq n/2$. Comparison of (1.145b) with (1.141b) specifies the noncentral moments:

$$M_n = \sum_{m=0}^{\leq n/2} \frac{\mu^{n-2m}\sigma^{2m}}{(n-2m)!m!\,2^m} \frac{n!}{} = \sum_{m=0}^{\leq n/2} \frac{n!\mu^{n-2m}\sigma^{2m}}{(n-2m)!(2m)!!}; \qquad (1.146a,b)$$

The agreement of (1.146a) ≡ (1.146b) ≡ (1.127) follows from

$$\frac{1}{m!2^m} = \frac{1}{(2m)!!} = \frac{(2m-1)!!}{(2m)!}, \qquad (1.147)$$

using single and double factorials.

NOTE 1.7: Equivalent Descriptions of a General (Normal) Probability Distribution

There are *six equivalent descriptions of a general* **continuous univariate probability distribution,** *namely by the (1) probability density function; (2) the cumulative probability function that is its integral (1.148a):*

$$P(x) = \int^x p(y)\mathrm{d}y, \quad p(x) = \frac{\mathrm{d}P}{\mathrm{d}x}, \qquad (1.148a,b)$$

so that (1) is obtained inversely by differentiation (1.148b); (3) the noncentral characteristic function that is the Fourier transform (1.140a), so that (1) is specified by the inverse Fourier transform (1.140b); (4) the noncentral moments of all orders (1.123b) that specify the noncentral characteristic function (1.141b) and conversely are specified by its derivatives (1.142c); (5) the central characteristic function (1.133) that is related to (3) by (1.139b); (6) the central moments of all orders (1.113) that specify the central characteristic function (1.133) and conversely are determined (1.135c) by it.

In the case of the **normal probability distribution** *(Gauss, 1809) with mean* μ *and r.m.s. value* σ: *(1), the probability density function (1.120a) is a Gaussian, hence the name Gaussian distribution; (2) the cumulative probability function (1.120b) is an error function (1.25b); (3) the central characteristic function is another Gaussian (1.137a); (4) it depends only on the r.m.s. value* σ, *as do all the central moments (1.115a,b) ≡ (1.137b, 1.138); (5) the mean value* μ *also appears in the noncentral characteristic function (1.144); and (6) the same two parameters* (μ, σ) *appear in the noncentral moments (1.146a) ≡ (1.146b) ≡ (1.127).* The Gaussian integrals (Section 1.1.) were used in connection with (1) generalized functions, namely the unit jump (Section 1.2), unit impulse (Section 1.3), and all its derivatives (Sections 1.4 through 1.9); (2) the normal probability distribution (Notes 1.3 through 1.7). The Gaussian probability distribution may be generalized to the exponential (Johnson et al., 1995) and combined Gamma and exponential (Campos and Marques, 2004) distributions that involve generalized Gaussian integrals (Notes 1.9 and 1.12); the latter are evaluated using the Gamma function (Note 1.8).

NOTE 1.8: Gamma Function (Euler, 1729) in the Complex Plane

The Gamma function is defined by the **Euler (1729) integral of first kind** (1.149b),

$$\mathrm{Re}(\alpha) > 0: \quad \Gamma(\alpha) \equiv \int_0^\infty x^{\alpha-1}e^{-x}\mathrm{d}x, \qquad (1.149a,b)$$

and converges at the origin in the right-hand complex-α plane (1.149a). The integral (1.149b) ≡ (1.150a) exists,

$$\Gamma(\alpha) = \lim_{\substack{\varepsilon \to 0 \\ X \to \infty}} \int_\varepsilon^X x^{\alpha-1}e^{-x}\mathrm{d}x, \quad \int^\varepsilon x^{\alpha-1}\mathrm{d}x = \frac{\varepsilon^\alpha}{\alpha}, \qquad (1.150a,b)$$

at the lower limit (1.150b), bearing in mind (I.5.28) that

$$\left|x^{\alpha}\right| \equiv \left|x^{\text{Re}(\alpha)+i\,\text{Im}(\alpha)}\right| = x^{\text{Re}(\alpha)}\left|\exp\left[i\,\text{Im}(\alpha)\log x\right]\right| = x^{\text{Re}(\alpha)} \tag{1.151}$$

vanishes as $x \to 0$ if (1.149a) is met. The simplest value of the Gamma function is unity at the point unity:

$$\Gamma(1) = \int_0^{\infty} e^{-x}\mathrm{d}x = \left[-e^{-x}\right]_0^{\infty} = 1 = 0!. \tag{1.152}$$

An integration by parts,

$$\Gamma(\alpha+1) = \int_0^{\infty} x^{\alpha}e^{-x}\mathrm{d}x = -\int_0^{\infty} x^{\alpha}\mathrm{d}\left(e^{-x}\right) = -\left[x^{\alpha}e^{-x}\right]_0^{\infty} + \int_0^{\infty} e^{-x}\mathrm{d}\left(x^{\alpha}\right), \tag{1.153}$$

leads to **the recurrence property**:

$$\text{Re}(\alpha) > 0: \quad \Gamma(\alpha+1) = \alpha\int_0^{\infty} x^{\alpha-1}e^{-x}\mathrm{d}x = \alpha\Gamma(\alpha). \tag{1.154a,b}$$

This may be applied iteratively (1.155b) an integer number of times (1.155a):

$$n \in | N: \quad \Gamma(\alpha+n) = (\alpha+n-1)(\alpha+n-2)...(\alpha+1)\alpha\Gamma(\alpha). \tag{1.155a,b}$$

leading to the **descending relation** (1.155b).

From (1.155b), it follows that the Gamma function reduces to the factorial (1.156b) for positive integer values (1.156a) of the argument,

$$n \in | N: \quad \Gamma(n) = (n-2)...2.1.\Gamma(1) = (n-1)!, \tag{1.156a,b}$$

and generalizes the factorial to complex values of the argument. Inverting (1.155b) algebraically (1.157b) provides the analytic continuation to the left-hand complex $-\alpha$ half-plane (1.157a):

$$\text{Re}(\alpha) > -n: \quad \Gamma(\alpha) = \frac{\Gamma(\alpha+n)}{(\alpha+n-1)(\alpha+n-1)...(\alpha+1)\alpha}. \tag{1.157a,b}$$

It follows that *the Gamma function (1.149a,b) generalizes the factorial (1.156b) from positive integers (1.156a) to complex values (1.157a,b) of the argument. It is an analytic function in the whole complex α-plane except (Figure 1.14) for simple poles (1.158b) at the origin and negative integer values of the argument (1.158a) where it has residues (1.158c):*

$$m = 0,-1,-2,...: \quad \Gamma(\alpha) = \frac{A_{-1}}{\alpha+m} + O(1), \quad A_{-1} = \frac{(-)^m}{m!}. \tag{1.158a–c}$$

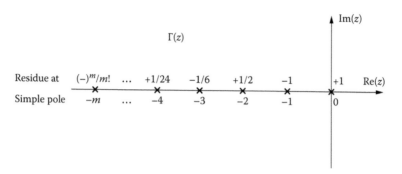

FIGURE 1.14 The Gamma function $\Gamma(z + 1) = z!$ is the generalization of the factorial from $z = n$ a positive integer to the complex z-plane. The factorial function $\Gamma(z)$ diverges only for $z = -m$ zero or a negative integer $(-m)! = (-m)$ $(-m-1)...$, corresponding to simple poles with residue $(-)^m/m!$.

The residue (1.158c) at the simple pole at (1.158a) is calculated by (I.15.24b):

$$A_{-1} \equiv \Gamma_{(1)}(-m) = \lim_{\alpha \to -m}(\alpha + m)\Gamma(\alpha) = \lim_{\alpha \to -m}\frac{\Gamma(\alpha + m + 1)}{(\alpha + m - 1)...\alpha} = \frac{\Gamma(1)}{(-m)(-m+1)...(-2)(-1)} = \frac{(-)^m}{m!}, \quad (1.159)$$

where (1.157b) was used with $n = m + 1$ and also (1.152).

NOTE 1.9: Evaluation of Generalized Gaussian Integrals

A generalization of the Gaussian probability density function (1.111b) is

$$p(x) = ax^d \exp(-bx^c), \quad (1.160)$$

which may designated the combined Gamma and generalized exponential probability distribution (Campos and Marques, 2004). It is taken with zero mean since this corresponds to a translation that would multiply by a polynomial in x; this leads to terms of the form (1.160) as do the moments of all orders either central (1.113) or noncentral relative to the origin (1.123b). The consideration of this more general probability distribution thus involves generalized Gaussian integrals (1.161b):

$$\text{Re}(\gamma) > -1: \quad I(\alpha, \gamma, \delta) \equiv \int_0^\infty x^\gamma \exp(-\alpha x^{2\delta})dx. \quad (1.161a,b)$$

The integral (1.161a,b) generalizes (1.17d) that is the particular case $\delta = 1$ and $\gamma = 2m$ an even positive integer, whereas in (1.161b), all three parameters (α, γ, δ) can be complex, provided that the integral exists. The convergence of the integral (1.161b) at the lower limit (1.150a) requires (1.150b, 1.151) the condition (1.161a) and is not affected by the exponential factor provided that (1.163a) be met. The convergence of the integral at the upper limit follows from the transformation into a Gamma function that follows.

The change of variable,

$$y = \alpha x^{2\delta}: \quad x = \left(\frac{y}{\alpha}\right)^{\frac{1}{2\delta}}, \quad dx = \frac{\alpha^{-\frac{1}{2\delta}}}{2\delta}y^{\frac{1}{2\delta}-1}dy, \quad (1.162a-c)$$

together with (1.163a), transforms the generalized Gaussian integral (1.161b) to (1.163b):

$$\mathrm{Re}(\delta) > 0: \quad I(\alpha,\gamma,\delta) = \frac{\alpha^{-\frac{\gamma+1}{2\delta}}}{2\delta} \int_0^\infty y^{\frac{\gamma+1}{2\delta}-1} e^{-y} dy. \tag{1.163a,b}$$

Use of the Gamma function (1.149a,b) in (1.163b) leads to (1.164b,c):

$$\mathrm{Re}(\gamma) > -1, \quad \mathrm{Re}(\delta) > 0: \quad I(\alpha,\gamma,\delta) = \frac{\alpha^{-\frac{\gamma+1}{2\delta}}}{2\delta} \Gamma\left(\frac{\gamma+1}{2\delta}\right). \tag{1.164a–c}$$

Thus, *the generalized Gaussian integral (1.161b) is evaluated in terms of the Gamma function (1.149b) by (1.164c), where the complex parameters (α, β, δ) are restricted by (1.164a,b)* that exclude poles of the Gamma function. The simplest Gaussian integral (1.1c) leads

$$\frac{\sqrt{\pi}}{2} = \int_0^\infty e^{-x^2} dx = I(1,0,1) = \frac{1}{2}\Gamma\left(\frac{1}{2}\right), \quad \Gamma\left(\frac{1}{2}\right) = \sqrt{\pi} \tag{1.165a,b}$$

to the *particular values (1.165b, 1.166) for the Gamma function*:

$$\Gamma\left(n+\frac{1}{2}\right) = \left(n-\frac{1}{2}\right)\dots\frac{1}{2}\Gamma\left(\frac{1}{2}\right) = 2^{-n}(2n-1)\dots3.1.\sqrt{\pi} = (2n-1)!!2^{-n}\sqrt{\pi}, \tag{1.166}$$

where the descending relation (1.155b) was used. The latter result (1.166) appears in the Gaussian integral:

$$\int_0^\infty x^{2m} e^{-\alpha x^2} dx = I(\alpha,2m,1) = \frac{1}{2}\alpha^{-m-\frac{1}{2}}\Gamma\left(m+\frac{1}{2}\right) = (2m-1)!!2^{-m-1}\alpha^{-m-\frac{1}{2}}\sqrt{\pi}; \tag{1.167}$$

this is one-half of (1.17d) because in the latter the integration is over the whole real axis and in (1.167) over the positive part.

NOTE 1.10: Oscillatory Generalized Gaussian Integral

The characteristic function (1.140a) of the combined Gamma and exponential distribution (1.160) is a particular case of the **oscillatory generalized Gaussian integral** (1.168b):

$$\mathrm{Re}(\gamma) > -1: \quad J(\alpha,\beta,\gamma,\delta) \equiv \int_0^{+\infty} x^\gamma \exp\left(i\beta x^\delta - \alpha x^{2\delta}\right) dx. \tag{1.168a,b}$$

The integral (1.168a,b) generalizes (1.14) that is the particular case $\delta = 1$ and $\gamma = n$ a positive integer, allowing all four parameters (α, β, γ, δ) to have complex values, subject to the convergence of the integral: (1) the conditions (1.161a) ≡ (1.168a) and (1.163a) ensure the convergence at the lower limit as before (1.164a,b); and (2) the convergence at the upper limit is established by transformation to a Gamma function. Instead of one step for (1.161b) in the case of (1.168b), two steps are needed. Since the second step is similar, just one step is needed to reduce the oscillatory (1.168b) to the generalized (1.161b) Gaussian integral, namely use of the power series (I.23.60) ≡ (II.1.26b) for the exponential:

$$J(\alpha,\beta,\gamma,\delta) = \sum_{k=0}^{\infty} \frac{(i\beta)^k}{k!} \int_{0}^{+\infty} x^{\gamma+k\delta} \exp(-\alpha x^{2\delta}) dx. \qquad (1.169)$$

The same change of variable (1.162a–c) applies since the integrals (1.169) are of the form (1.161b) with γ replaced by γ + kδ:

$$J(\alpha,\beta,\gamma,\delta) = \sum_{k=0}^{\infty} \frac{(i\beta)^k}{k!} I(\alpha, \gamma + k\delta, \delta). \qquad (1.170)$$

The condition Re(γ + kδ) > −1 is met for all k = 0, 1, … by (1.171a,b), and substitution of (1.164c) in (1.170) yields (1.171c):

$$\mathrm{Re}(\gamma) > -1, \quad \mathrm{Re}(\delta) > 0: \quad I(\alpha,\beta,\gamma,\delta) = \sum_{k=0}^{\infty} \frac{(i\beta)^k}{k!} \frac{\alpha^{-\frac{\gamma+1}{2\delta} - \frac{k}{2}}}{2\delta} \Gamma\left(\frac{\gamma+1}{2\delta} + \frac{k}{2}\right). \qquad (1.171a\text{-}c)$$

Thus, *the oscillatory generalized Gaussian integral (1.168b) is evaluated (1) by (1.170) in terms of the non-oscillatory generalized Gaussian integral (1.161b); and (2) by (1.171c) in terms of the Gamma function (1.149b), with the restrictions (1.171a,b) that ensure that the Gamma function is analytic.* When β = 0, then (1.171c) reduces to the first term k = 0 of the sum that coincides with (1.164c).

NOTE 1.11: Bilateral and Unilateral Generalized Gaussian Integrals

The unilateral (bilateral) generalized Gaussian integral (1.161b) [(1.172)] involves an integration over the positive (whole) real line:

$$\int_{-\infty}^{+\infty} x^\gamma \exp(-\alpha x^{2\delta}) dx = \int_{0}^{\infty} \left[x^\gamma + (-x)^\gamma\right] \exp(-\alpha x^{2\delta}) = \left(1 + e^{i\pi\gamma}\right) \int_{0}^{\infty} x^\gamma \exp(-\alpha x^{2\delta}) dx. \qquad (1.172)$$

Thus, the **bilateral generalized Gaussian integral** *is given by (1.172):*

$$\int_{-\infty}^{+\infty} x^\gamma \exp(-\alpha x^{2\delta}) dx = \left(1 + e^{i\pi\gamma}\right) I(\alpha,\gamma,\delta), \qquad (1.173)$$

in terms of the corresponding (1.164a–c) unilateral generalized Gaussian integral:

$$\mathrm{Re}(\gamma) > -1, \quad \mathrm{Re}(\delta) > 0: \quad \int_{0}^{\infty} x^\gamma \exp(-\alpha x^{2\delta}) dx = \left(1 + e^{i\pi\gamma}\right) \frac{\alpha^{-\frac{\gamma+1}{2\delta}}}{2\delta} \Gamma\left(\frac{\gamma+1}{2\delta}\right). \qquad (1.174a\text{-}c)$$

The same transformation (1.171) with γ replaced by $\gamma + k\delta$ applies in (1.169) each term (1.170) of the corresponding oscillatory generalized Gaussian integrals:

$$\int_{-\infty}^{+\infty} x^{\gamma} \exp\left(i\beta x^{\delta} - \alpha x^{2\delta}\right) dx = \sum_{k=0}^{\infty} \frac{(i\beta)^{k}}{k!}\left[1 + e^{i\pi(\gamma + k\delta)}\right] I\left(\alpha, \gamma + k\gamma, \delta\right). \tag{1.175}$$

Thus, *the bilateral oscillatory generalized Gaussian integral (1.175) is expressible in terms of (1.164a–c) unilateral nonoscillatory generalized Gaussian integrals by*

$$\mathrm{Re}(\gamma) > -1, \mathrm{Re}(\delta) > 0: \quad \int_{0}^{\infty} x^{\gamma} \exp\left(i\beta x^{\delta} - \alpha x^{2\delta}\right) dx = \sum_{k=0}^{\infty} \frac{(i\beta)^{k}}{k!}[1 + e^{i\pi(\gamma + k\delta)}] \frac{\alpha^{-\frac{\gamma+1}{2\delta} - \frac{k}{2}}}{2\delta} \Gamma\left(\frac{\gamma+1}{2\delta} + \frac{k}{2}\right).$$

$$\tag{1.176a–c}$$

If $\delta = 0$, then (1.176c) reduces to the first term of the sum $k = 0$ that coincides with (1.174c). If $\gamma = 2m + 1$ is an odd integer, then $e^{i\pi\gamma} = -1$, and (1.174c) vanishes, because the integrand is an odd function of x.

Setting (1.177a–c) [(1.178a–c)] in (1.176c) leads to (1.177d) [(1.178d)]:

$$\delta = 1, \gamma = 2m, m \in |N: \quad \int_{-\infty}^{+\infty} x^{2m} \exp\left(i\beta x - \alpha x^{2}\right) dx = \sum_{k=0}^{\infty} \frac{(i\beta)^{k}}{k!} \frac{1 + e^{i\pi k}}{2} \alpha^{-m - \frac{1}{2} - \frac{k}{2}} \Gamma\left(m + \frac{k}{2} + \frac{1}{2}\right),$$

$$\tag{1.177a–d}$$

$$\delta = 1, \gamma = 2m + 1, m \in |N: \quad \int_{-\infty}^{+\infty} x^{2m+1} \exp\left(i\beta x - \alpha x^{2}\right) dx = \sum_{k=0}^{\infty} \frac{(i\beta)^{k}}{k!} \frac{1 - e^{i\pi k}}{2} \alpha^{-m-1-\frac{k}{2}} \Gamma\left(m + \frac{k}{2} + 1\right),$$

$$\tag{1.178a–d}$$

where the following equations are used:

$$\exp\left[i\pi(\gamma + k\delta)\right] = \exp\left[i\pi(k + 2m)\right] = \exp\left(i\pi k\right), \tag{1.179}$$

$$\exp\left[i\pi(\gamma + k\delta)\right] = \exp\left[i\pi(2m + k + 1)\right] = -\exp\left(i\pi k\right). \tag{1.180}$$

These (1.179) [(1.180)] suppress the odd (even) terms in (1.177d) [(1.178d)] leading to (1.181) [(1.182)]:

$$\int_{-\infty}^{+\infty} x^{2m} \exp\left(-\alpha x^{2}\right) \cos\left(\beta x\right) dx = \alpha^{-m-1/2} \sum_{k=0}^{\infty} \frac{(-)^{k}}{(2k)!} \left(\frac{\beta^{2}}{\alpha}\right)^{k} \Gamma\left(m + k + \frac{1}{2}\right)$$

$$= \sqrt{\frac{\pi}{\alpha}} (2\alpha)^{-m} \sum_{k=0}^{\infty} \frac{(-)^{k}}{(2k)!} \left(\frac{\beta^{2}}{2\alpha}\right)^{k} (2m + 2k - 1)!!, \tag{1.181}$$

$$\int_{-\infty}^{+\infty} x^{2m+1} \exp\left(-\alpha x^2\right) \sin\left(\beta x\right) dx = \alpha^{-m-1}\beta \sum_{k=0}^{\infty} \frac{(-)^k}{(2k+1)!} \left(\frac{\beta^2}{\alpha}\right)^k \Gamma\left(m+k+\frac{3}{2}\right)$$

$$= \sqrt{\pi}\,(2\alpha)^{-m-1}\beta \sum_{k=0}^{\infty} \frac{(-)^k}{(2k+1)!} \left(\frac{\beta^2}{2\alpha}\right)^{2k+1} (2m+2k+1)!!, \qquad (1.182)$$

where only the real (imaginary) part is nonzero, and (1.166) was used. Thus, *the oscillatory Gaussian integrals with even (odd) powers are given by the series (1.181) [(1.182)] involving Gamma functions (1.149b, 1.166), as an alternative to the product of Gaussian by Hermite polynomials of even (odd) order (1.20a) [(1.20b)].* The Hermite polynomials (Note 1.13) also arise from the bilateral oscillatory generalized Gaussian integral (1.176a–c) evaluated in an alternative way (Note 1.12).

NOTE 1.12: Bilateral Oscillatory Generalized Gaussian Integral

The condition (1.168a) ≡ (1.183a) remains relevant to the oscillatory generalized Gaussian integral in the unilateral (1.168b) [bilateral (1.183b)] cases because the origin is at the lower limit of (inside) the range of integration:

$$\text{Re}(\gamma) > -1: \quad K\left(\alpha,\beta,\gamma,\delta\right) = \int_{-\infty}^{+\infty} x^{\gamma} \exp\left(i\beta x^{\delta} - \alpha x^{2\delta}\right) dx. \qquad (1.183a,b)$$

The integral (1.183b) is evaluated in five steps, of which the first uses a change of variable (1.184a–c):

$$z = x^{\delta}: \quad x = z^{\frac{1}{\delta}}, \quad dx = \frac{1}{\delta} z^{\frac{1}{\delta}-1} dz. \qquad (1.184a–c)$$

The change of variable (1.184a) is the square root of (1.162a) so that the argument of the exponential remains free from radicals in the passage from (1.183b) to (1.185):

$$K\left(\alpha,\beta,\gamma,\delta\right) = \frac{1}{\delta} \int_{-\infty}^{+\infty} z^{\frac{\gamma+1}{\delta}-1} \exp\left(i\beta z - \alpha z^2\right) dz. \qquad (1.185)$$

The factor $\sqrt{\alpha}$ was not included in the change of variable (1.184a–c) unlike the square root of (1.162a–c), and in addition the parameter δ is restricted to positive odd integer values (1.186a,b) so that the limits of integration are the same (1.186c) in (1.184c) and (1.185):

$$n \in N, \quad \delta = 2n+1: \quad \lim_{x\to\pm\infty} z = \lim_{x\to\pm\infty} x^{\delta} = \pm\infty. \qquad (1.186a–c)$$

The condition (1.163a) for convergence of the integral at the origin is met by (1.186a,b).

The second step is another change of variable (1.187a–c) similar to (1.11a–c) without the factor α:

$$u = z - \frac{i\beta}{2\alpha}: \quad -\alpha u^2 = -\alpha z^2 + i\beta z + \frac{\beta^2}{4\alpha}, \quad du = dz, \qquad (1.187a–c)$$

that transforms the exponential in (1.185) to a Gaussian in (1.188):

$$K\left(\alpha,\beta,\gamma,\delta\right)=\frac{2}{\delta}\exp\left(-\frac{\beta^2}{4\alpha}\right)\int\limits_{-\infty}^{+\infty}\left(u+\frac{i\beta}{2\alpha}\right)^{\frac{\gamma+1}{\delta}-1}\exp\left(-\alpha u^2\right)\mathrm{d}u. \tag{1.188}$$

The third step is a further change of variable (1.189c–e) for α real and positive (1.189a,b):

$$\alpha\in R,\quad \alpha>0:\quad v=\alpha u^2,\quad u=\sqrt{\frac{v}{\alpha}},\quad \mathrm{d}u=\frac{\mathrm{d}v}{2\sqrt{v\alpha}}, \tag{1.189a–e}$$

that reduces the Gaussian in (1.188) to a simple exponential in (1.190):

$$K\left(\alpha,\beta,\gamma,\delta\right)=\frac{\alpha^{-\frac{\gamma+1}{2\delta}}}{\delta}\exp\left(-\frac{\beta^2}{4\alpha}\right)\int\limits_{0}^{\infty}\left(\sqrt{v}+\frac{i\beta}{2\sqrt{\alpha}}\right)^{\frac{\gamma+1}{\delta}-1}e^{-v}\frac{\mathrm{d}v}{\sqrt{v}}. \tag{1.190}$$

The fourth step is to use the binomial series (I.25.37a–c) in the integrand of (1.190) leading to (1.191):

$$K\left(\alpha,\beta,\gamma,\delta\right)=\frac{\alpha^{-\frac{\gamma+1}{2\delta}}}{\delta}\exp\left(-\frac{\beta^2}{4\alpha}\right)\sum_{k=0}^{\infty}\binom{\frac{\gamma+1}{\delta}-1}{k}\left(\frac{i\beta}{2\sqrt{\alpha}}\right)^k\int\limits_{0}^{\infty}v^{\frac{\gamma+1}{2\delta}-\frac{k}{2}-1}e^{-v}\mathrm{d}v. \tag{1.191}$$

The Gamma function (1.149b) is used to evaluate the integrals in (1.191) leading to

$$K\left(\alpha,\beta,\gamma,\delta\right)=\frac{\alpha^{-\frac{\gamma+1}{2\delta}}}{\delta}\exp\left(-\frac{\beta^2}{4\alpha}\right)\sum_{K=0}^{\infty}\binom{\frac{\gamma+1}{\delta}-1}{k}\left(\frac{i\beta}{2\sqrt{\alpha}}\right)^k\Gamma\left(\frac{\gamma+1}{2\delta}-\frac{k}{2}\right), \tag{1.192}$$

which is the fifth and final step.

If $\beta=0$ in (1.192), then the sum reduces to the first term $k=0$ that equals twice of (1.164c), because the former (latter) is a bilateral (1.183b) [unilateral (1.161b)] improper integral over the whole (positive) real line:

$$K\left(\alpha,0,\gamma,\delta\right)=\frac{\alpha^{-\frac{\gamma+1}{2\delta}}}{\delta}\Gamma\left(\frac{\gamma+1}{2\delta}\right)=2I\left(\alpha,\gamma,\delta\right). \tag{1.193}$$

The bilateral oscillatory generalized Gaussian integral (1.183b) $\equiv$ (1.194c) with the conditions (1.183a) $\equiv$ (1.194a) and (1.186a,b) $\equiv$ (1.194b),

$$\mathrm{Re}\left(\gamma\right)>-1;\quad n\in N:\quad K_n\left(\alpha,\beta,\gamma\right)\equiv K\left(\alpha,\beta,\gamma,2n+1\right)=\int\limits_{-\infty}^{+\infty}x^\gamma\exp\left(i\beta x^{2n+1}-\alpha x^{4n+2}\right)\mathrm{d}x, \tag{1.194a–c}$$

is given by (1.192) ≡ (1.195):

$$K_n(\alpha,\beta,\gamma) = \frac{\alpha^{-\frac{\gamma+1}{4n+2}}}{2n+1} \exp\left(-\frac{\beta^2}{4\alpha}\right) \sum_{k=0}^{\infty} \binom{\frac{\gamma+1}{2n+1}-1}{k} \left(\frac{i\beta}{2\sqrt{\alpha}}\right)^k \Gamma\left(\frac{\gamma+1}{4n+2}-\frac{k}{2}\right).$$

(1.195)

With the additional restriction (1.196a), it can be evaluated alternatively by (1.196b):

$$\beta \neq 0: \quad K_n(\alpha,\beta,\gamma) = \frac{\alpha^{-\frac{\gamma+1}{4n+2}}}{2n+1} \exp\left(-\frac{\beta^2}{4\alpha}\right) \sum_{k=0}^{\infty} \binom{\frac{\gamma+1}{2n+1}-1}{k} \left(\frac{i\beta}{2\sqrt{\alpha}}\right)^{\frac{\gamma+1}{2n+1}-k-1} \Gamma\left(\frac{k+1}{2}\right),$$

(1.196a,b)

where (1.156b) [(1.166)] for odd (even) k may be used. The result (1.196b) follows like (1.195) ≡ (1.192) from (1.190), also applying the binomial series (I.25.37a–c) in a way alternative to (1.191, 1.192), namely, (1.197):

$$K_n(\alpha,\beta,\gamma) = \frac{\alpha^{-\frac{\gamma+1}{4n+2}}}{2n+1} \sum_{k=0}^{\infty} \binom{\frac{\gamma+1}{2n+1}-1}{k} \left(\frac{i\beta}{2\sqrt{\alpha}}\right)^{\frac{\gamma+1}{2n+1}-k-1} \int_0^{\infty} v^{k/2-1/2} e^{-v} dv;$$

(1.197)

using the Gamma function (1.149b) in (1.197) leads to (1.196b). The bilateral oscillatory generalized Gaussian integral (1.194a–c) evaluated in the form (1.196a,b) generalizes (1.21b,c) and may be related to the Hermite polynomials (Note 1.13).

NOTE 1.13: Hermite Polynomials of Even/Odd Degree

Choosing in (1.194a–c, 1.196a,b) the values (1.198a–c) [(1.199a–c)] for the parameters leads to (1.198d) [(1.199d)]:

$$n=0; \quad m \in |N; \quad \gamma = 2m: \quad \int_{-\infty}^{+\infty} x^{2m} \exp\left(i\beta x - \alpha x^2\right) dx = K_0(\alpha,\beta,2m)$$

$$= \alpha^{-m-1/2} \exp\left(-\frac{\beta^2}{4\alpha}\right) \sum_{k=0}^{2m} \binom{2m}{k} \left(\frac{i\beta}{2\sqrt{\alpha}}\right)^{2m-k} \Gamma\left(\frac{k+1}{2}\right),$$

(1.198a–d)

$$n=0; \quad m \in |N; \quad \gamma = 2m+1: \quad \int_{-\infty}^{+\infty} x^{2m+1} \exp\left(i\beta x - \alpha x^2\right) dx = K_0(\alpha,\beta,2m+1)$$

$$= \alpha^{-m-1} \exp\left(-\frac{\beta^2}{4\alpha}\right) \sum_{k=0}^{2m+1} \binom{2m+1}{k} \left(\frac{i\beta}{2\sqrt{\alpha}}\right)^{2m+1-k} \Gamma\left(\frac{k+1}{2}\right),$$

(1.199a–d)

where the sum $k = 0,1, \ldots, 2m(2m + 1)$ is finite because the factorial $(2m - k)![(2m + 1 - k)!]$ in the denominator diverges for $k > 2m(k > 2m + 1)$. For (α, β) real and α positive (1.200a–c), the real (imaginary) part of (1.198d) [(1.199d)] leads to (1.200d) [(1.200e)]:

$\alpha, \beta \in |R, \alpha > 0$:

$$\int_{-\infty}^{+\infty} x^{2m} \exp\left(-\alpha x^2\right)\cos(\beta x)dx = \frac{(-)^m}{\sqrt{\alpha}} \alpha^{-m} \exp\left(-\frac{\beta^2}{4\alpha}\right) \sum_{k=0}^{m} \binom{2m}{2k}(-)^k \left(\frac{\beta^2}{4\alpha}\right)^{m-k} \Gamma\left(k + \frac{1}{2}\right), \quad (1.200\text{a–d})$$

$$\int_{-\infty}^{+\infty} x^{2m+1} \exp\left(-\alpha x^2\right)\sin(\beta x)dx = (-)^m \alpha^{-m-1} \exp\left(-\frac{\beta^2}{4\alpha}\right) \sum_{k=0}^{m} \binom{2m+1}{2k}(-)^k \left(\frac{\beta^2}{4\alpha}\right)^{m-k+\frac{1}{2}} \Gamma\left(k + \frac{1}{2}\right),$$

$$(1.200\text{e})$$

where (1.166) [(1.156b)] may be used for the values of the Gamma function.

Comparing (1.200d) [(1.200e)] with (1.20a) [(1.20b)] for the variable (1.21a) $\equiv$ (1.201a) leads to (1.201b) [(1.201c)]:

$$z = \frac{\beta}{2\sqrt{\alpha}}: \quad H_{2m}(z) = \frac{2^{2m}}{\sqrt{\pi}} \sum_{k=0}^{m} (-)^k \binom{2m}{2k} z^{2m-2k} \Gamma\left(k + \frac{1}{2}\right)$$

$$= 2^{2m} \sum_{k=0}^{m} \frac{(2m)!(2k-1)!!}{(2k)!(2m-2k)!} 2^{-k} (-)^k z^{2m-2k}$$

$$= \sum_{k=0}^{m} \frac{(2m)! 2^{2m-k}}{(2m-2k)!(2k)!!}(-)^k z^{2m-2k}, \quad (1.201\text{a,b})$$

$$H_{2m+1}(z) = \frac{2^{2m+1}}{\sqrt{\pi}} \sum_{k=0}^{m} (-)^k \binom{2m+1}{2k} z^{2m-2k+1} \Gamma\left(k + \frac{1}{2}\right)$$

$$= 2^{2m+1} \sum_{k=0}^{m} \frac{(2m+1)!(2k-1)!!}{(2k)!(2m-2k+1)!}(-)^k 2^{-k} z^{2m-2k+1}$$

$$= \sum_{k=0}^{m} \frac{(2m+1)! 2^{2m-k+1}}{(2m-2k+1)!(2k)!!}(-)^k z^{2m-2k+1}. \quad (1.201\text{c})$$

Thus, *the Gaussian integrals with even (odd) powers and cosine (sine) may be evaluated equivalently by (1.200d) $\equiv$ (1.20a) [(1.200e) $\equiv$ (1.20b)] leading to the explicit formulas for the Hermite polynomials (1.18) of even (1.201b) [odd (1.201c)] degree.* Setting $m = 0,1,2(m = 0,1)$ in (1.201b) [(1.201c)] leads to (1.19a,c,e) [(1.19b,d)].

NOTE 1.14: Recurrence/Differentiation Formulas and Differential Equation

An alternative way to calculate the (1.19a–e) is to use the **recurrence formula** *for Hermite polynomials*:

$$H_{n+1}(z) = 2z\, H_n(z) - 2nH_{n-1}(z). \tag{1.202}$$

The latter can be proved from the definition (1.18) using the Leibnitz rule (I.13.31) for the *n*th derivative of a product:

$$H_{n+1}(z) = (-)^{n+1} \exp(z^2) \frac{d^{n+1}}{dz^{n+1}}\left[\exp(-z^2)\right] = (-)^n \exp(z^2) \frac{d^n}{dz^n}\left[2z \exp(-z^2)\right]$$

$$= (-)^n \exp(z^2)\left\{2z \frac{d^n}{dz^n}\left[\exp(-z^2)\right] + 2n \frac{d^{n-1}}{dz^{n-1}}\left[\exp(-z^2)\right]\right\}$$

$$= 2zH_n(z) - 2nH_{n-1}(z). \tag{1.203}$$

The Hermite polynomials satisfy a **differentiation formula**:

$$H_n'(z) = 2z\, H_n(z) - H_{n+1}(z), \tag{1.204}$$

that follows from (1.18):

$$H_n'(z) = \frac{d}{dz}\left\{\exp(z^2)\frac{d^n}{dz^n}\left[\exp(-z^2)\right]\right\}$$

$$= (-)^n \exp(z^2)\left\{2z \frac{d^n}{dz^n}\left[\exp(-z^2)\right] + \frac{d^{n+1}}{dz^{n+1}}\left[\exp(-z^2)\right]\right\}$$

$$= 2z\, H_n(z) - H_{n+1}(z). \tag{1.205}$$

The simplest way to calculate the Hermite polynomials of increasing degree is (1) to use the definition (1.18) with $n = 0, 1$ to obtain the first two (1.19a,b); (2) to use the recurrence formula (1.202) to obtain all others, for example, (1.19c–e) for $n = 2, 3, 4$.

The recurrence (1.202) and differentiation (1.204) formulas imply that the Hermite polynomials satisfy a linear second-order ordinary differentiation with variable parameters:

$$H_n''(z) - 2z\, H_n'(z) + 2nH_n(z) = 0. \tag{1.206}$$

The proof of (1.206) is made in three steps: (1) differentiation of (1.204):

$$H_n''(z) = 2zH_n' + 2H_n(z) - H_{n+1}'(z); \tag{1.207}$$

(2) use of (1.204) for $n + 1$ leading to (1.208a):

$$H_n''(z) - 2zH_n'(z) = 2H_n(z) - 2zH_{n+1}(z) + H_{n+2}(z), \tag{1.208a}$$

$$= 2H_n(z) - 2(n+1)H_n(z) = -2nH_n(z); \tag{1.208b}$$

and (3) use of (1.202) for $n + 2$ leading to (1.208b) ≡ (1.206). The Hermite differential equation (1.206) can be generalized to complex values of the parameter n, leading to the Hermite functions. Since it is a linear ordinary differential equation of the second-order, it has two linearly independent solutions, namely the Hermite functions of the first and second kinds. It is shown in the theory of differential equations that the Hermite polynomials correspond to the Hermite functions of the first kind with positive integer parameters. Further instances of Gaussian and exponential integrals (Section 1.1, Notes 1.1 through 1.14, Subsection 5.9.1) appear in Examples 10.14 and 10.17.

1.10 Conclusion

The area under the Gaussian curve in (Figure 1.1) is $\sqrt{\pi}$, as can be shown by a two-dimensional integration over a square (Figure 1.2) bounded by two quarter circles. The Gaussian functions of unit area and decreasing variance σ as σ → 0 tend (Figure 1.6) to the Dirac delta or unit impulse that is a generalized function $\delta(x)$, which is zero everywhere, except at the origin, where it is infinite, with unit integral. The integral of the Gaussian functions (Figure 1.6) specifies the error functions (Figure 1.5) that go from 0 at −∞ to 1 at +∞ through a transition with a slope at the point (0,1,2) that diverges as σ → 0, leading to the generalized function unit step or Heaviside function. The derivatives of the Gaussian functions (Figure 1.6) are odd functions (Figure 1.7), with a maximum (minimum) before (after) the origin, which diverges to +∞(−∞) as σ → 0 and the origin is approached, leading to the generalized function derivate Dirac delta or unit impulse, which is zero everywhere, except at the origin, where it jumps from 0 to +∞, to −∞ and back to 0 and has integral −1 when multiplied by x. Each additional differentiation of the generalized function Dirac delta or unit impulse does not change the zero value outside the origin, but leads to one more jump between ±∞ as the origin is crossed, for example, one (two) jump for the first (second)-order derivative of the unit impulse [Figure 1.7 (Figure 1.8)]. The Heaviside (Dirac) or unit step (impulse) can be used to specify other generalized functions, such as the sign (Figure 1.9) and modulus (Figure 1.10). The sign function (Figure 1.9) has a finite discontinuity (Figure 1.11b) with jump two. Other types of discontinuities are removable (Figure 1.11a) and infinite (Figure 1.11c). The derivatives of functions with isolated finite discontinuities (Figure 1.12a) can be obtained in terms of generalized functions, including the jumps of the function and lower-order derivatives (Figure 1.12b) at the point of discontinuity. The Gaussian functions that were used to define the unit impulse (Figure 1.6) also serve as the envelope (Figure 1.3a) for sinusoidal oscillations (Figure 1.3b) that lead to the generalization of the Gaussian integrals. A further generalization has for integrand sinusoidal oscillations (Figure 1.4b) whose envelope is the product of a power and a Gaussian (Figure 1.4a). A further generalization of the Gaussian integral involving powers with complex exponents can be evaluated using the Gamma function (Figure 1.14). The Gaussian integrals are related to the Gaussian probability distribution (Figure 1.13b) whose parameters are mean value (Figure 1.13a) and variance or r.m.s. value (Figure 1.13c).

2

Shape of a Loaded String

A generalized function can be used to represent loads on structures, the simplest example being a thin elastic string under tension (Section 2.1) with fixed ends. If the transverse forces acting along the length of the string are represented by differentiable functions, the shear stress or force per unit length is an ordinary function (Section 2.2). If there are (Section 2.3) one or more concentrated forces, for example, a finite transverse force acting at a point, then (1) the shear stress is zero outside the point and infinite at the point; and (2) the integral of the shear stress, namely the transverse force, has a finite jump as it passes through the point where the concentrated force is applied. Thus the idealization of a concentrated force applied at a point leads to a transverse force (shear stress) that involves (Section 2.3) the generalized function Heaviside unit jump (Dirac unit impulse). The string is deflected under load, and its shape when a unit impulse loading is applied specifies the linear (Section 2.3) [nonlinear (Section 2.6)] influence or Green function; the string has a triangular shape when a point load is applied both in the linear (Section 2.3) and nonlinear (Section 2.6) cases. The differential equation specifying the shape of the string is nonlinear (linear) if the slope is not (is) small everywhere, for example, for the large (small) deflection of a string under its own weight [Section 2.4 (Section 2.8)]; the large deflection changes the parabolic shape of the small deflection. In the linear case, the shape of the string under arbitrary loading can be calculated by the principle of superposition, using a convolution integral with the influence or Green function (Section 2.3); also, the principle of reciprocity holds, allowing interchange of the points of application of the concentrated load and of measurement of the deflection. In contrast, the nonlinear Green function satisfies neither the principle of superposition nor the principle of reciprocity. The types of loading considered include a transverse force (shear stress) that is (1) a linear (constant) function for the uniform loading of a homogeneous string in the linear (Section 2.4) and nonlinear (Section 2.7) cases, (2) a concentrated load represented by a generalized jump (impulse) function in the linear (Section 2.3) and nonlinear (Section 2.6) cases, and (3) multiple concentrated loads both in the linear (Section 2.5) and nonlinear (Example 10.5) cases. The deflection of an elastic string under its own weight corresponds to a uniform load in the linear case (Section 2.4) but not in the nonlinear case (Section 2.8). The nonlinear deflection of an elastic string under its own weight (Section 2.8) is larger than for the catenary, which corresponds to an inelastic string (Section 2.9); the inelastic string has constant length even for a nonlinear deflection, whereas the elastic string (1) has constant length in the linear approximation (Sections 2.3 through 2.5) and (2) increases in length under a transverse load in the nonlinear case (Sections 2.6 through 2.8).

2.1 Tangential Tension and Transverse Force

A string is an elastic body of negligible cross section that can support a tension T along its length; thus the tension is tangent to the curve that specifies the shape of the string. The transverse force (Subsection 2.1.1) per unit length specifies the shear stress (Subsection 2.1.2) and causes the deflection of the string; deflections with a small (large) slope are (Subsection 2.1.3) linear (nonlinear). The elastic string is the one-dimensional analogue of an elastic membrane (Subsection 2.1.4) in two dimensions (Sections II.6.1 and II.6.2).

2.1.1 Tangential and Normal Components of the Tension

In the absence of transverse loading, the string is straight, and the x-axis can be taken along it; in the presence of transverse loads (Figure 2.1a), the string is deflected (2.1a), and the arc length is given by (2.1b)

$$y = \zeta(x): \quad (ds)^2 = (dy)^2 + (dx)^2 = (1 + \zeta'^2)(dx)^2, \quad \zeta' \equiv \frac{d\zeta}{dx}, \qquad (2.1a\text{–}c)$$

where prime denotes the derivative with regard to x; for example, (2.1c) is the slope of the shape (2.1a) of the string. Since the tension T is tangent to the string, its horizontal (T_x) and vertical (T_y) components are given (Figure 2.1a) by

$$T_x = T\frac{dx}{ds} = T\left|1 + \zeta'^2\right|^{-1/2}, \quad T_y = T\frac{dy}{ds} = T\frac{dy}{dx}\frac{dx}{ds} = T_x\zeta' = T\zeta'\left|1 + \zeta'^2\right|^{-1/2}, \qquad (2.2a,b)$$

where (2.1a through c) is used. The exact horizontal (2.2a) and vertical (2.2b) tensions are a nonlinear function of the slope, which can be linearized for small slopes (2.3a):

$$\zeta'^2 \ll 1 \Leftrightarrow ds = dx \Rightarrow (y_{\max})^2 \ll L^2, \qquad (2.3a\text{–}c)$$

and (1) this is equivalent to the statement (2.3b) that the length of the string is equal to its horizontal projection (Figure 2.1b); (2) it implies that (2.3c) the maximum transverse deflection $y_{\max}$ is small relative to the length L of the string (Figure 2.2a); and (3) the latter condition is necessary but not sufficient, that is, the relative total deflection may be small, yet the slope could be locally large if the string has steep ripples (Figure 2.2b), and in this case the deformation is nonlinear. For a small slope (2.4a) everywhere,

$$\zeta'^2 \ll 1: \quad T_x = T, \quad T_y = T\zeta', \qquad (2.4a\text{–}c)$$

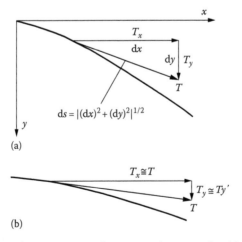

(a)

(b)

FIGURE 2.1 The tension along the tangent to an elastic string has vertical and horizontal components (a) that simplify in the linear case of small slope (b).

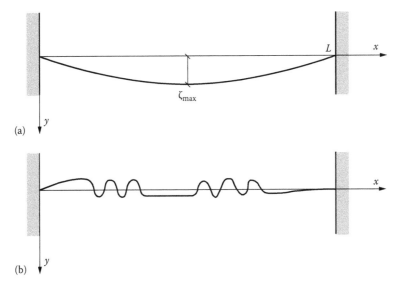

FIGURE 2.2 A linear deflection is defined by a small slope, and implies that the maximum deflection is small compared with the distance between the supports (a). The converse may not be true, for example, if the maximum deflection is small but the slope is large due to the presence of steep "ripples" (b) the deflection of the string is nonlinear.

the horizontal tension (2.4b) coincides with the tension *T* along the string, and the vertical tension (2.4c) is small but nonnegligible by comparison. Hence, *for an elastic string of shape (2.1a), under uniform or nonuniform tension T, the horizontal and vertical components are given exactly in the nonlinear case by (2.2a and b) and are approximated by (2.4b and c) in the linear case of a small slope everywhere by (2.4a)* ≡ *(2.3a through c).*

2.1.2 Transverse Force and Shear Stress

If a transverse force $F(x)$ is applied to the string, then the latter will deflect to an equilibrium shape such that, along each segment dx, the variation of the **transverse force** dF is balanced by the variation of the vertical component of the tension (2.5a):

$$dF + dT_y = 0, \quad F(x) = -T_y(x),$$ (2.5a,b)

where the constant of integration in (2.5b) is zero because the two terms must balance at the supported ends of the string, $x = 0, L$. Substituting (2.2b) into (2.5b) leads to the differential equation of first order (2.6) specifying the shape of the string:

$$-F(x) = T\zeta' \left| 1 + \zeta'^2 \right|^{-1/2},$$ (2.6)

under an arbitrary transverse force $F(x)$. The **shear stress** $f(x)$ is defined as the transverse force per unit length (2.7a):

$$f(x) \equiv \frac{dF}{dx}, \quad F(x) = F(0) + \int_0^x f(\xi) d\xi,$$ (2.7a,b)

leading by integration to (2.7b). Using the identity

$$\left\{\zeta'\left|1+\zeta'^2\right|^{-1/2}\right\}' = \zeta''\left|1+\zeta'^2\right|^{-1/2} - \zeta'^2\zeta''\left|1+\zeta'^2\right|^{-3/2} = \zeta''\left|1+\zeta'^2\right|^{-3/2}, \tag{2.8a}$$

leads from (2.6; 2.7a) to a differential equation of the second order:

$$-f(x) = \left\{T\zeta'\left|1+\zeta'^2\right|^{-1/2}\right\}' = T'\zeta'\left|1+\zeta'^2\right|^{-1/2} + T\zeta''\left|1+\zeta'^2\right|^{-3/2}, \tag{2.8b}$$

which is valid for nonuniform tension along the string; if the tension is uniform (2.9a), then (2.8b) simplifies to (2.9b)

$$T = \text{const}: \quad -\frac{f(x)}{T} = \left\{\zeta'\left|1+\zeta'^2\right|^{-1/2}\right\}' = \zeta''\left|1+\zeta'^2\right|^{-3/2} \equiv k(x), \tag{2.9a,b}$$

where appears the curvature (II.6.98a and b) of the string; thus, *the ratio of the shear stress to the uniform tension along the string equals minus the curvature both in the nonlinear and linear cases.*

2.1.3 Conditions for Linear and Nonlinear Deflection

The slope of the string is determined by (2.6) ≡ (2.10b), the transverse force:

$$\left|F(x)\right| \le T(x): \quad -\zeta'(x) = \left|\left[\frac{T(x)}{F(x)}\right]^2 - 1\right|^{-1/2}. \tag{2.10a,b}$$

The passage from (2.6) to (2.10b) involves a square root and could lead to ± signs; only the sign – in (2.10b) is consistent with (2.6) in the linear approximation (2.4a) ≡ (2.10a). A real slope (2.10b) requires that the transverse force does not exceed the tension (2.10a); an equality would imply an infinite slope because the string would have to be along the external force for the tension to balance it. In the linear case (2.4a) ≡ (2.3a through c), the slope is small, implying that the tension along the string is much larger than the transverse force:

$$\zeta'^2 \ll 1 \Leftrightarrow \left[T(x)\right]^2 \gg \left[F(x)\right]^2. \tag{2.10c,d}$$

In this case of a transverse force that is small compared to the tension along the string (2.11a), the equation specifying the shape of the spring in terms of the transverse force (2.6) [shear stress (2.8b)] simplifies to (2.11b) [(2.11c)]:

$$F^2 \ll T^2: \quad F(x) = -T\zeta', \quad f(x) = -\left(T\zeta'\right)' = -T\zeta'' - T'\zeta', \tag{2.11a-c}$$

by neglecting ζ'^2, compared to unity; (2.11b) [(2.11c)] can also be obtained by substituting (2.4c) into (2.5b) [(2.7a)]. In the linear case (2.12b) with constant tension (2.12a), (2.11b) [(2.11c)] simplifies to (2.12c) [(2.12d)]:

$$T \sim \text{const}, F^2 \ll T^2: \quad F(x) = -T\zeta', \quad f(x) = -T\zeta''. \tag{2.12a-d}$$

The equation for the transverse force (2.11b) ≡ (2.12c) [shear stress (2.11c) ≠ (2.12d)] is the same (different) for uniform and nonuniform tension. The tension will be constant along an elastic string (Chapter 2), but possibly not along a beam (Chapter 4); for example, for a chain hanging in the gravity field, the weight adds to the tension a term proportional to the length below. The present formulation encompasses both cases. Thus, *the shape of an elastic string is specified by the transverse force (2.6) [shear stress (2.8b)], for nonuniform longitudinal tension, in the nonlinear case of a large slope. This leads to two possible simplifications: (1) uniform tension (2.9a) in (2.6) [(2.9b)]; and (2) linear case of a small slope (2.3a through c) ≡ (2.11a) in (2.11b) [(2.11c)]. The two simplifications combine in the linear case (2.12c) [(2.12d)] with constant tension (2.3a through c) ≡ (2.12b) and (2.12a).*

2.1.4 One (Two)-Dimensional Elastic String (Membrane)

The elastic string (membrane) may be considered as one(two)-dimensional analogues [Chapter 2 (Sections II.6.1 and II.6.2)]. The transverse deflection of the elastic string under tension due to a shear stress or force per unit length (2.13a) [area (2.13b)] is similar exchanging the spatial derivative (2.13c) [nabla or gradient operator (2.13d)]:

$$f(x) = \frac{dF}{dx} \leftrightarrow f(x,y) = \frac{dF}{dS} = \frac{dF}{dx\,dy}, \quad \frac{d}{dx} \leftrightarrow \nabla = \vec{e}_x \frac{\partial}{\partial x} + \vec{e}_y \frac{\partial}{\partial y}. \tag{2.13a–d}$$

Substitution of (2.13b and d) in (2.8b) leads to

$$-f(x,y) = \nabla \cdot \left\{ T\nabla\zeta \left| 1 + (\nabla\zeta \cdot \nabla\zeta) \right|^{-1/2} \right\}, \tag{2.14a}$$

$$= \left| 1 + (\nabla\zeta \cdot \nabla\zeta) \right|^{-1/2} \left\{ \nabla \cdot (T\nabla\zeta) + T\nabla\zeta \cdot \left[(\nabla\zeta \cdot \nabla)\nabla\zeta \right] \left[1 + (\nabla\zeta \cdot \nabla\zeta) \right]^{-1} \right\}, \tag{2.14b}$$

$$= \left| 1 + (\nabla\zeta \cdot \nabla\zeta) \right|^{-3/2} \left\{ \left(T\nabla^2\zeta + \nabla T \cdot \nabla\zeta \right) \left[1 + (\nabla\zeta \cdot \nabla\zeta) \right] + T\nabla\zeta \cdot \left[(\nabla\zeta \cdot \nabla)\nabla\zeta \right] \right\}, \tag{2.14c}$$

which specifies *the nonlinear deflection ζ(x, y) of an elastic membrane under tension T, subject to a shear stress or shear force (2.13b)*. There are for the elastic membrane [(2.14a) ≡ (2.14b) ≡ (2.14c) ≡ (II.6.18b) ≡ (II.6.22a) ≡ (II.6.22b)] the same particular cases as for an elastic string (2.8b); for example, (1) in the linear case of a small slope (2.15a), the membrane equation (2.14a) simplifies to (2.15b):

$$(\nabla\zeta \cdot \nabla\zeta) \ll 1: \quad -f(x,y) = \nabla \cdot (T\nabla\zeta) = T\nabla^2\zeta + \nabla T \cdot \nabla\zeta, \tag{2.15a,b}$$

so that the outer (inner) derivative in (2.11c) is replaced by the divergence (gradient) operator; and (2) if, in addition (2.15a) ≡ (2.16a), the tension is uniform (2.16b), the shape of the membrane satisfies a Poisson equation (2.16c):

$$(\nabla\zeta \cdot \nabla\zeta) \ll 1, \quad T = \text{const}: \quad -\frac{f(x,y)}{T} = \nabla^2\zeta, \tag{2.16a–c}$$

so that the second-order derivative in (2.12d) is replaced by the Laplacian operator. The elastic string (2.8b) corresponds to the one-dimensional case of the elastic membrane (2.14a), suppressing the variable y and retaining only the variable x. The equilibrium equation (2.14a) ≡ (II.6.21) for the elastic membrane can be obtained by a variational energy method (Section II.6.1), which also specifies the boundary conditions.

A similar variational energy method leads to the equilibrium equation for an elastic string obtained before (Subsections 2.1.1 through 2.1.3) and also to the boundary conditions, as shown next (Section 2.2).

2.2 Shear Stress and Elastic Energy

The equations specifying the nonlinear (linear) deflection of a string can be obtained alternatively by an energy method (Subsection 2.2.1); as in the case of the nonlinear (linear) deflection of a membrane (Subsections II.6.1.1 through II.6.1.3), the energy method specifies also (Subsection 2.2.2) the boundary conditions at the ends of the string.

2.2.1 Quadratic and Higher-Order Terms in the Elastic Energy

The **elastic energy** of the string equals the tension times the extension, which is the change in length from dx in the undeflected state to ds in the deflected (2.1a) condition:

$$E_d = \int_0^L T(ds - dx) = \int_0^L T\left[\left|1 + \zeta'^2\right|^{1/2} - 1\right]dx, \tag{2.17}$$

where the integration is along the length L of the string projected on the x-axis. For a linear deflection (2.18a), that is, with a small slope (2.3), the elastic energy (2.18b) is a quadratic function of the slope:

$$\zeta'^2 \ll 1: \quad E_d = \frac{1}{2}\int_0^L T\zeta'^2 dx. \tag{2.18a,b}$$

In the nonlinear case, higher powers of the slope of even order appear in the binomial series (I.25.37a through c) in (2.17) for the elastic energy (2.19a):

$$E_d = \sum_{n=1}^{\infty} \binom{1/2}{n} \int_0^L (\zeta')^{2n} dx; \quad \binom{1/2}{n} = \left(\frac{1}{2}\right)\left(-\frac{1}{2}\right)...\left(\frac{1}{2} - n + 1\right) = \frac{(-1)(-3)...(-2n+3)}{n!2^n}, \tag{2.19a,b}$$

where the coefficients are given by (2.19b) $\equiv$ (2.20a):

$$n \geq 2: \quad \binom{1/2}{n} = (-)^{n-1}\frac{(2n-3)!!}{(2n)!!} \equiv -b_n, \tag{2.20a,b}$$

using the double factorials

$$(2n)!! \equiv 2.9,4.6...(2n-2)\cdot 2n, \quad (2n-3)!! \equiv 1.3.5...(2n-5)\cdot(2n-3) \tag{2.20c,d}$$

and leading to the values (II.6.63a and b; II.6.64a through f); since the double factorial (2.20b) applies for (2.20a), the first term $n = 1$ in (2.19a), which coincides with (2.18b), is singled out in (2.21):

$$E_d = \frac{1}{2}\int_0^L T\zeta'^2 dx - \sum_{n=2}^{\infty} \frac{(2n-3)!!}{(2n)!!} \int_0^L T\left(-\zeta'^2\right)^n dx. \tag{2.21}$$

The exact elastic energy of an elastic string of length L under tension T is (2.17) $\equiv$ (2.21), and it simplifies to a quadratic function of the slope in the linear case (2.18a and b). All expressions are valid for nonuniform tension with T under the integral sign; for uniform tension, T can be brought out of the integral.

2.2.2 Balance Equation and Boundary Conditions

The principle of virtual work states that the work performed by the transverse force in deflecting the string is balanced by the variation of the elastic energy:

$$\int_0^L f(x)\delta\zeta\,dx = \int_0^L dF\,\delta\zeta = \delta W = \delta E_d = \int_0^L T\zeta'\delta\zeta'\,dx, \tag{2.22}$$

where the case of linear deflection (2.11b), or quadratic energy (2.18a and b), is considered first. The displacement dx, which is a differential of the coordinate x, should be distinguished from $\delta\zeta$, which is a small variation of the shape of the string (Figure 2.3). The tension T is the same at the corresponding points x of the equilibrium shape ζ and its variation $\zeta + \delta\zeta$, implying (2.23a) when integrating the l.h.s. of (2.22) by parts (2.23b):

$$T\zeta'\delta\zeta'\,dx = T\zeta'\,dx\,\delta\left(\frac{d\zeta}{dx}\right) = T\zeta'\,dx\frac{d}{dx}(\delta\zeta) = T\zeta'\,d(\delta\zeta) = d(T\zeta'\delta\zeta) - \delta\zeta\,d(T\zeta'). \tag{2.23a,b}$$

Substituting (2.23b) in (2.22) leads to

$$\int_0^L f\delta\zeta\,dx = \left[T\zeta'\delta\zeta\right]_0^L - \int_0^L \delta\zeta\,d(T\zeta'). \tag{2.24}$$

This imposes two conditions: (1) the first term on the r.h.s. is the boundary condition, requiring that the variation of the deflection $\delta\zeta$ must be zero at both ends (2.25a and b), that is, the string is fixed at both ends (2.25c and d):

$$\delta\zeta(x=0) = 0 = \delta\zeta(x=L): \quad \zeta(0) = \zeta_1, \quad \zeta(L) = \zeta_2; \tag{2.25a–d}$$

(2) omitting the first term on the r.h.s., which is zero by (2.25a and b), leads to the integral

$$\int_0^L \left[f(x) + (T\zeta')'\right]\delta\zeta(x)\,dx = 0, \tag{2.26}$$

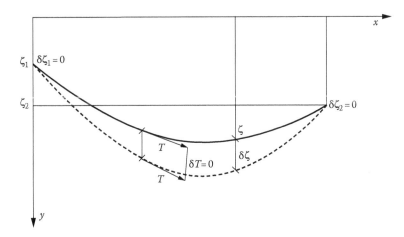

FIGURE 2.3 A variation $\delta\zeta$ of the deflection of the string: (1) changes the shape ζ to $\zeta + \delta\zeta$; (2) does not change the tension $\delta T = 0$ at the corresponding points with the same longitudinal coordinate x for the original ζ and varied $\zeta + \delta\zeta$ string; and (3) the two supports are unchanged by the variation.

that must vanish for arbitrary displacements. This implies (by the Euler or Dubois-Raymond lemmas of the calculus of variations) that the term in square brackets must vanish; this proves (2.11c), which includes (2.12d) but not (2.8b). The latter is the general nonlinear case and is treated similarly, using in (2.22) the exact elastic energy (2.17), which is

$$\int_0^L f(x)\delta\zeta dx = \delta W = \delta E_d = \int_0^L T\delta\left(\left|1+\zeta'^2\right|^{1/2}\right)dx = \int_0^L T\left|1+\zeta'^2\right|^{-1/2}\zeta'\delta\zeta' dx$$

$$= \left[T\left|1+\zeta'^2\right|^{-1/2}\zeta'\delta\zeta\right]_0^L - \int_0^L \left\{T\left|1+\zeta'^2\right|^{-1/2}\zeta'\right\}' \delta\zeta dx, \qquad (2.27)$$

where an integration by parts is used:

$$T\left|1+\zeta'^2\right|^{-1/2}\zeta'\delta\zeta' = \left\{T\left|1+\zeta'^2\right|^{-1/2}\zeta'\delta\zeta\right\}' - \left\{T\left|1+\zeta'^2\right|^{-1/2}\zeta'\right\}'\delta\zeta. \qquad (2.28)$$

The identity (2.27) leads to (1) the same boundary condition (2.25a through d) from the first term on the r.h.s. of (2.27); and (2) the remaining terms lead to (2.8b). Thus, *the exact (2.17) [quadratic (2.18a and b)] elastic energy for a nonlinear (linear) deflection leads, through the condition of balancing its variation against the work of the transverse force or shear stress (2.27) [(2.22) ≡ (2.24)], to (1) the balance of the transverse force (2.6) [(2.11b)] and shear stress (2.8b) [(2.11c)]; and (2) the boundary condition (2.25c and d) that the string must be fixed at both ends.*

2.3 Influence Function, Superposition, and Reciprocity

The deflection of an elastic string under load is considered first in the linear case (Sections 2.3 through 2.5) for subsequent comparison with nonlinear solutions (Sections 2.6 through 2.8). The small deflection under a concentrated load specifies the linear influence, or Green (1828) function, which may be considered for supports at unequal (equal) heights [Subsection 2.3.2 (Subsection 2.3.3)]. Since the influence function specifies the displacement by a point force at any position, generally at an unequal distance from the supports (Subsection 2.3.1), it includes the particular case of equal distance from the supports (Subsection 2.3.4). The principle of reciprocity (Subsection 2.3.5), allowing interchange of the positions of (1) measurement of the displacement and (2) application of the concentrated force, holds for symmetric boundary conditions, that is, supports at the same height; it is extended to supports at different heights (Subsections 2.3.7 and 2.3.8), using a reciprocal influence function (Subsection 2.3.6). In all cases of linear deflection, the principle of superposition (Subsection 2.3.5) holds, specifying the deflection under an arbitrary load by integration after multiplication by the influence function (Subsection 2.3.9).

2.3.1 Linear Influence Function of a String

The small deflection at x of a string under a concentrated load at ξ is designated the **linear inference or Green function** $G(x;\xi)$, and it satisfies the differential equation (2.12d) ≡ (2.29c):

$$T = \text{const}, \quad P^2 \ll T^2: \quad T\frac{d^2}{dx^2}\left[G(x;\xi)\right] = -P\delta(x-\xi) = -f(x), \qquad (2.29a\text{–c})$$

in the linear case of a transverse force that is small compared with the uniform longitudinal tension (2.12a and b) ≡ (2.29a and b). The concentrated force P is taken in the downward direction parallel to the y-axis in Figure 2.4a. The first integration of (2.29c):

$$T\frac{d}{dx}\big[G(x;\xi)\big]=A-PH(x-\xi)=\begin{cases}A\equiv-R_- & \text{if } 0\le x<\xi, & \text{(2.30a)}\\ A-P=R_+ & \text{if } \xi<x\le L, & \text{(2.30b)}\end{cases}$$

involves one arbitrary constant A and shows that the Green function has a discontinuous first derivative, that is, the slope of the string changes (Figure 2.4a) across the point of application of the concentrated load. The constant A equals the transverse force at the first support, that is, specifies minus the reaction force; since the reaction force is negative, $R_- < 0$, the slope $A > 0$ is positive before the point of application of the force $x < \xi$. The force is larger than the reaction $P > A$, and thus, the slope $R_+ = A - P$,

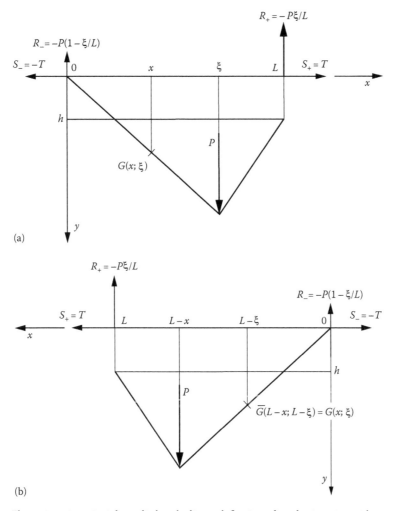

(a)

(b)

FIGURE 2.4 The reciprocity principle applied to the linear deflection of an elastic string with supports at different heights requires the consideration of two Green or influence functions, (a) the original influence function G and (b) the reciprocal $\bar{G}$; with reversed supports. This allows interchange of the point of application of concentrated force and the point of measurement of the displacement. A simpler form of the reciprocity principle applies if the supports are at the same height (Figure 2.6).

which equals the reaction at the second support, is negative, $R_+ < 0$, beyond the point of application of the force. Another two constants B, C appear in a further integration:

$$TG(x;\xi) = \begin{cases} Ax + B & \text{if } 0 \le x \le \xi, \\ (A - P)\, x + C & \text{if } \xi \le x \le L, \end{cases} \qquad (2.31a,b)$$

where the discontinuity at $\xi = x$ is "smoothed," that is, the influence function is continuous, so that the shape of the string is also continuous. In order to determine the three constants of integration A, B, C in (2.31a and b), the same three conditions (2.32a through c) apply to the nonlinear and linear cases, stating that the influence function corresponds to a string (1/2) that is fixed at the two ends (2.25a through d), possibly (Figure 2.4a) with a height difference h, so that one is taken at the origin (2.32a) and the other (2.32b) at (L, h):

$$G(0,\xi) = 0, \quad G(L,\xi) = h; \quad G(\xi - 0;\xi) = G(\xi + 0;\xi); \qquad (2.32a\text{--}c)$$

(2) the deflection is continuous (2.32c) at the point $x = \xi$, where the concentrated load is applied. The supports at different heights (Figure 2.5a) correspond to a force making an angle (2.32d) with the normal to the undeflected position of the string, corresponding to the transverse (2.32e) [longitudinal (2.32f)] force components:

$$\tan\alpha = \frac{h}{L}: \quad P_n = P\cos\alpha = \frac{PL}{\sqrt{h^2 + L^2}}, \quad P_s = P\sin\alpha = \frac{Ph}{\sqrt{h^2 + L^2}}; \qquad (2.32d\text{--}f)$$

thus, supports at the same height (Figure 2.6a), $h = 0 = \alpha = P_s$, correspond to a transverse force $P_n = P$. The influence function depends both on the differential equation (2.29c) through (P, T) and on the boundary

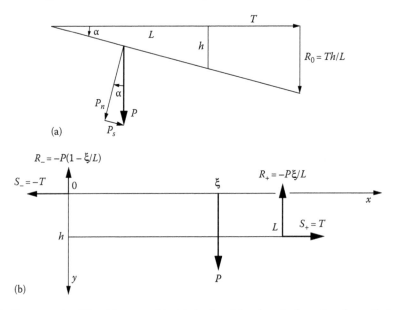

(a)

(b)

FIGURE 2.5 The supports of the string are subject to horizontal and vertical reaction forces that can be determined by a balance of forces and moments (a). For a linear deflection with small slope the horizontal reaction at the supports are due only to the tension along the string. The vertical reactions on the supports depend on the magnitude of the concentrated force and the position of the point of application relative to the supports. If the supports are not at the same height (b) there is an additional vertical reaction force with opposite signs at the supports that does not change the total vertical force.

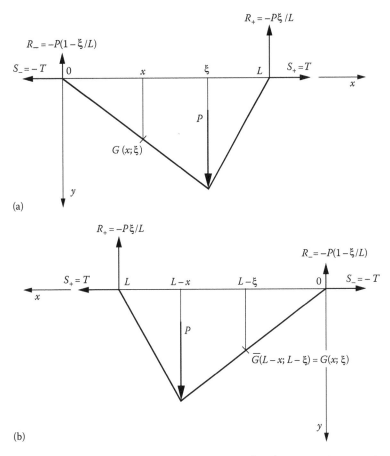

FIGURE 2.6 The reciprocity principle applied the linear deflection of an elastic string (Figure 2.4) simplifies in the case when the two supports are at the same height (Figure 2.6). It allows interchange of the point of application of the concentrated force and the point of measurement of the displacement between (a) and (b). This corresponds to a symmetric Green or influence function and to a self-adjoint differential equation with symmetric boundary conditions.

conditions (2.32a and b) through h; the continuity condition (2.32c) must be met in every case. The constants of integration are determined (1) by (2.31a; 2.32a) leading to (2.33a); (2) by (2.31a and b; 2.32c) leading to (2.33b):

$$B = 0, \quad C = P\xi, \quad A = P + \frac{Th - C}{L} = T\frac{h}{L} + P\left(1 - \frac{\xi}{L}\right); \tag{2.33a--c}$$

and (3) by (2.31b; 2.32b) leading to (2.33c).

2.3.2 String with Supports at Different Heights

Thus, *the linear (2.29b) ≡ (2.35a) influence, or Green function (2.29c), for an elastic string of length L under constant tension (2.29a), fixed at the two ends (2.32a and b) with a height difference h, is given by (2.34a and b) for a force P concentrated at an arbitrary position x = ξ:*

$$G\left(x;\xi\right) = \frac{h}{L}x + \frac{P}{TL} \times \begin{cases} \left(L - \xi\right)x & \text{if } 0 \leq x \leq \xi, \\ \left(L - x\right)\xi & \text{if } \xi \leq x \leq L. \end{cases} \tag{2.34a,b}$$

The influence function (2.34a and b) consists of (1) a straight line joining the supports (Figure 2.4a), corresponding to the first term on the r.h.s.; and (2) the second term on the r.h.s. is the influence function for supports (Figure 2.6a) at the same height (2.41a through c) for a force P concentrated in general at any position x = ξ. The shape of the string is triangular, with maximum deflection at the point where the load is applied (Figure 2.4a):

$$P^2 \ll T^2: \quad G_{max} = G(\xi,\xi) = \frac{h}{L}\xi + \frac{P}{T}\left(1 - \frac{\xi}{L}\right)\xi. \tag{2.35a,b}$$

The constant slope on either side determines the reactions at the supports:

$$0 \le x < \xi: \quad R_- = F(0) = -TG'(0;\xi) = -TG'(x < \xi;\xi) = -T\frac{h}{L} - P\left(1 - \frac{\xi}{L}\right), \tag{2.36a}$$

$$\xi < x < L: \quad R_+ = -F(L) = TG'(L;\xi) = TG'(x > \xi;\xi) = T\frac{h}{L} - P\frac{\xi}{L}. \tag{2.36b}$$

The reactions (2.37b and c) consist of (1) the reactions for the supports at the same height (2.39d and c); and (2) plus or minus a term (2.37a) involving the tension and the slope of the line joining the supports (Figure 2.5a):

$$R_0 = T\frac{h}{L}: \quad R_- = -P\left(1 - \frac{\xi}{L}\right) - R_0, \quad R_+ = -P\frac{\xi}{L} + R_0. \tag{2.37a-c}$$

The reactions at the supports can be obtained (Figure 2.5b) by (1) a balance of vertical forces (2.38a); (2) a balance of moments relative to the first support with a horizontal tension (2.38b) at the other support; (3) instead of (2), the balance of moments at the second support with horizontal tension −T at the origin (2.38c):

$$R_- + R_+ = -P, \quad -R_+L = P\xi - Th, \quad R_-L = -P(L - \xi) - Th, \quad S_\pm = \pm T; \tag{2.38a-d}$$

and (4) the horizontal reaction at the supports (2.38d) balances the tension, which is horizontal because in the linear approximation, the slope is neglected. The consistency of (2.36a and b) ≡ (2.38a through c) follows from (1/2) the coincidence of (2.38b) ≡ (2.37a and c) = (2.36b) [(2.38c) ≡ (2.37b and c) ≡ (2.36a)]; and (3) the sum of (2.36a and b) ≡ (2.37a through c) coinciding with (2.38a).

2.3.3 Reaction Forces at the Two Supports

If the supports are (Figure 2.6a) at the same height (2.39a), the boundary conditions (2.32a and b) are symmetric (2.39b and c):

$$h = 0: \quad G(0;\xi) = 0 = G(L;\xi), \quad R_- = -P\left(1 - \frac{\xi}{L}\right), \quad R_+ = -P\frac{\xi}{L}, \tag{2.39a-e}$$

and the reactions at the supports (2.36a and b) simplify to (2.39d and e). The reactions at the supports can (Figure 2.7) be determined from (1) the balance of forces (2.40a); and (2) the balance of moments about any point, for example, the origin (2.40b) or the other support (2.40c):

$$R_- + R_+ = -P, \quad -R_+L = P\xi, \quad R_-L = -P(L - \xi). \tag{2.40a-c}$$

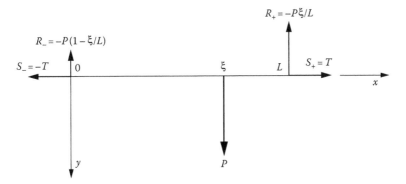

FIGURE 2.7 The force and moment balance that determines the horizontal and vertical reaction forces at the supports for the linear deflection of an elastic string (Figure 2.5) simplifies in the case when the two supports are at the same height (Figure 2.7).

In the case of supports at the same height (2.39a), there is no effect of the horizontal tension, that is, (2.38a through c) reduces to (2.40a through c), from which follow (2.39d and e). Thus, *if the supports of the string are at the same height (2.39a)* $\equiv$ *(2.41a), the reaction forces at the first (2.39d) [second (2.39e)] support divided by* $-T(T)$ *specify the slope before (after) the point of application of the concentrated force, corresponding to the influence function (2.41b and c):*

$$h = 0: \quad G(x;\xi) = \frac{P}{T} \times \begin{cases} \left(1 - \dfrac{\xi}{L}\right)x & \text{if } 0 \le x \le \xi, \\ \left(1 - \dfrac{x}{L}\right)\xi & \text{if } \xi \le x \le L, \end{cases} \tag{2.41a–c}$$

as follows from (2.34a and b) with $h = 0$.

2.3.4 Concentrated Force at Equal Distance from the Supports

Instead of supports at the same height (Subsection 2.3.3), a distinct particular case is *a concentrated force at equal distance from the supports (2.42a), leading (2.36a and b) to the reaction forces (2.42b and c):*

$$\xi = \frac{L}{2}: \quad R_- = -T\frac{h}{L} - \frac{P}{2}, \quad R_+ = T\frac{h}{L} - \frac{P}{2}, \tag{2.42a–c}$$

which add to the external force (2.38a). The corresponding (2.34a and b) influence function (2.43b and c),

$$\xi = \frac{L}{2}: \quad G\left(x;\frac{L}{2}\right) = \frac{h}{L}x + \frac{P}{2T} \times \begin{cases} x & \text{if } 0 \le x \le \dfrac{L}{2}, \\ L - x & \text{if } \dfrac{L}{2} \le x \le L, \end{cases} \tag{2.43a–c}$$

has maximum deflection at the point of application of the force:

$$\xi = \frac{L}{2}: \quad G\left(\frac{L}{2},\frac{L}{2}\right) = \frac{h}{2} + \frac{PL}{4T}. \tag{2.44a,b}$$

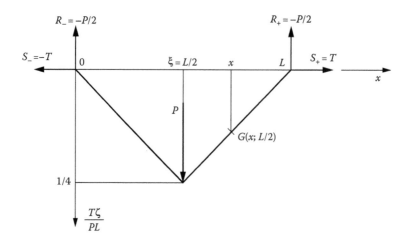

FIGURE 2.8 The linear deflection of an elastic string under constant tension T with supports at the same height and distance L, by a concentrated force P applied at equal distance from the supports, leads to an isosceles triangular shape with maximum deflection δ satisfying $T\delta = 4PL$.

Combining the two particular cases for *a concentrated force applied at equal distance (2.44a) ≡ (2.45a) from the two supports at the same height (2.45b) ≡ (2.39a), (1) the reaction forces are (Figure 2.8) one-half of the applied force (2.45c), with opposite sign*:

$$\xi = \frac{L}{2}; h = 0: \quad R_- = -\frac{P}{2} = R_+; \quad G\left(\frac{L}{2};\frac{L}{2}\right) = \frac{PL}{4T}, \tag{2.45a–e}$$

(2) the maximum deflection is (2.45c); and (3) it corresponds to the maximum of

$$\xi = \frac{L}{2}; h = 0: \quad G\left(x;\frac{L}{2}\right) = \frac{P}{2T} \times \begin{cases} x & \text{if } 0 \le x \le \dfrac{L}{2}, \\[2mm] L-x & \text{if } \dfrac{L}{2} \le x \le L, \end{cases} \tag{2.46a–d}$$

the influence function (2.46c and d).

2.3.5 Principles of Reciprocity and Superposition

The influence function with symmetric boundary conditions (2.39b and c) corresponding to suspension points (2.32a and b) at the same height (2.41a) ≡ (2.47a) is unchanged (2.41b and c), exchanging (2.47b) the point ξ where the concentrated force is applied and the point x where the deflection of the string is measured:

$$G(0;\xi) = G(L;\xi): \quad G(x;\xi) = G(\xi;x). \tag{2.47a,b}$$

Thus, (2.47a and b) states the **principle of reciprocity**: *the deflection at x due to a force concentrated at ξ, is equal to the deflection at ξ due to the same force concentrated at x for a string with suspension points at the same height (Figure 2.6a and b).* The principle of reciprocity holds with suspension points at unequal heights if the suspension points are interchanged (Figure 2.4a and b) and is made (Subsection 2.3.6) either a displacement correction (2.53b) or (Subsection 2.3.7) a change of coordinates (2.54b). The reciprocity principle is generally valid (Section 7.7) for linear, nondissipative systems, specified by differential equations with self-adjoint operators, of which (2.29c) is a simple example (7.92). *The displacement of the string for small deflection with symmetric boundary conditions under nonuniform tension satisfies self-adjoint equation (2.29c) ≡ (7.92) for the shear stress and thus satisfies the* **principle of reciprocity**, *that is, the principle of reciprocity holds both for uniform (2.12a through d) and nonuniform (2.11a through c) tension.* A property of linear systems, dissipative or not, is the **principle of superposition**: *the deflection of the string due to several loads applied simultaneously is equal to the sum of the deflections due to each load applied separately.* This principle applies, provided that the total transverse force remains small everywhere compared with the longitudinal tension, otherwise the linear equation (2.11c) would have to be replaced by the exact, nonlinear one (2.8b). *The principle of superposition holds for small deflections of a string under nonuniform (2.11a through c) [uniform (2.12a through d)] tension*, because the differential equations are linear in both cases, with variable (constant) coefficients. It will be shown (Section 2.6) that the nonlinear Green function for the exact equation (2.8b) satisfies neither the principle of reciprocity nor the principle of superposition. The symmetry property $\zeta(x) = \zeta(L - x)$ holds even in the nonlinear case, provided that the suspension points are at the same height $\zeta(0) = \zeta(L)$, and the only load is a single concentrated force at equal distance from the supports (Figure 2.8); if the suspension points are at unequal heights, if $\zeta(0) \neq \zeta(L)$, the shape of the string is nonsymmetric even in the linear case (Figure 2.4a).

2.3.6 Original and Reciprocal Influence Function

To extend the reciprocity principle (2.47b) to the case of unsymmetric boundary conditions, when (2.47a) is not satisfied, two influence functions are considered, namely the **original (reciprocal) influence function** $G(x;\xi)[\bar{G}(\xi;x)]$, that satisfies (1) the differential equation (2.29c) [with (x, ξ) interchanged in (2.48d)]; (2) the boundary conditions (2.32a and b) [interchanged (2.48a and b)]; and (3) the continuity condition (2.32c) [(2.48c)] at $\xi(x)$:

$$\bar{G}(0;x) = h, \quad \bar{G}(L;x) = 0, \quad \bar{G}(\xi-0;x) = \bar{G}(\xi+0;x): \quad T\frac{d^2}{d\xi^2}\left[\bar{G}(\xi;x)\right] = -P\delta(\xi - x). \quad (2.48\text{a–d})$$

The first integration of (2.48d) yields

$$T\frac{d}{d\xi}\left[\bar{G}(\xi;x)\right] = \bar{A} - PH(\xi - x) = \begin{cases} \bar{A} & \text{if } 0 \leq \xi < x, \\ \bar{A} - P & \text{if } x < \xi \leq L, \end{cases} \quad (2.49\text{a,b})$$

where $\bar{A}$ is an arbitrary constant; a further integration introduces two more arbitrary constants:

$$T\bar{G}(\xi;x) = \begin{cases} \bar{A}\xi + \bar{B} & \text{if } 0 \leq \xi \leq x, \\ (\bar{A} - P)\xi + \bar{C} & \text{if } x \leq \xi \leq L. \end{cases} \quad (2.50\text{a,b})$$

The continuity (2.48c) [boundary (2.48a and b)] conditions specify the three constants in the sequence (2.48a) ≡ (2.51a), (2.48c) ≡ (2.51b) and (2.48b) ≡ (2.51c):

$$\bar{B} = Th, \quad \bar{C} = \bar{B} + Px = Th + Px, \quad \bar{A} = P - \frac{\bar{C}}{L} = -T\frac{h}{L} + P\left(1 - \frac{x}{L}\right). \quad (2.51\text{a–c})$$

Substituting (2.51a through c) in (2.50a and b) yields *the reciprocal influence function (2.52a and b) satisfying (2.48a through d)*:

$$\bar{G}(\xi;x) = \frac{h}{L}(L-\xi) + \frac{P}{TL} \times \begin{cases} (L-x)\xi & \text{if } 0 \le \xi \le x, \\ (L-\xi)x & \text{if } 0 \le \xi \le L, \end{cases}$$

(2.52a,b)

to be compared with the original influence function (2.34a and b) satisfying (2.29c; 2.32a through c).

2.3.7 Extended Form of the Reciprocity Principle

The equalities (2.52a) ≡ (2.34b) and (2.52b) = (2.34a), that is, (2.53b) hold, provided that in the first term is made the substitution $x \to L - \xi$:

$$G(0;\xi) \ne G(L;\xi): \quad G(x;\xi) - \frac{h}{L}x = \bar{G}(\xi;x) - \frac{h}{L}(L-\xi),$$

(2.53a,b)

because the supports are at different heights (2.53a), and the distance of the measuring point from the support at zero height changes from x in Figure 2.4a to $L - \xi$ in Figure 2.4b. Thus the **extended reciprocity principle** has been proved: *the linear deflection of a string with ends attached at different heights (2.53a) states that (2.34a and b) the displacement at x due to a concentrated force at ξ is the same as that (2.52a and b) for a displacement at ξ due to a concentrated force at x, provided that in both cases is subtracted a term (2.53b) equal to the vertical divided by the horizontal distance between the supports, that is, the slope, multiplied by (Figure 2.4a and b) the distance of the measuring point from the support at zero height.*

2.3.8 Alternate Extended Reciprocity Principle

The equality (2.53b) may be replaced (2.34a and b; 2.52a and b) by (2.54b), with (2.53a) ≡ (2.54a) unchanged:

$$G(0;\xi) \ne G(L;\xi): \quad G(x;\xi) = \bar{G}(L-\xi;L-x).$$

(2.54a,b)

Thus, *an* **alternate extended reciprocity** *principle* states that the displacement at x due to a concentrated force at ξ is the same (Figure 2.4a) as the displacement at ξ due to a concentrated force at x even (Figure 2.4b) if the supports are at unequal heights, provided that (1) the suspension points be interchanged; and (2) the direction of the longitudinal axis be reversed. This (1,2) ensures that the positions of the displacement and force are at the same distance from the support at lower and higher heights. For example, if the point of measurement of the displacement (application of the concentrated force) is closer (farther) from the support at zero height (Figure 2.4a), the same must apply when the two are interchanged (Figure 2.4b), so that the origin is displaced by h and the direction of the x-axis reversed. *The extended reciprocity principle (2.53b) and its alternate form (2.54b) apply, regardless of whether the boundary conditions are unsymmetric (2.53a) = (2.54a) or symmetric (2.47a), that is, the supports are at the same or different heights; in contrast, the original reciprocity principle (2.47b) applies only to symmetric boundary conditions (2.47a), that is, supports at the same height.* The original (2.47a and b) [alternate (2.53a and b) or extended (2.54a and b)] reciprocity principle relates to

$$\delta(x-\xi) = \delta(\xi-x), \quad \delta(x-\xi) = \delta\big((L-\xi)-(L-x)\big),$$

(2.55a,b)

the symmetry property (1.44a) ≡ (2.55a) [the property (2.55b)] of the generalized function unit impulse.

2.3.9 Linear Deflection under an Arbitrary Load

The influence function due to a concentrated load P at ξ, in the linear case (2.11a) $\equiv$ (2.56a) with non-uniform tension (2.11c) satisfies (2.56b):

$$P^2 \ll T^2 : \quad \frac{d}{dx}\left\{T(x)\frac{d}{dx}\left[G(x;\xi)\right]\right\} = -P\delta(x-\xi), \qquad (2.56a,b)$$

which simplifies to (2.29c) for uniform tension (2.29a); in both cases is considered linear deflection (2.29b) $\equiv$ (2.56a), so that the principle of superposition holds (Subsection 2.3.5). The principle of superposition can be stated by using the integration property (1.76) $\equiv$ (2.57a) of the unit impulse:

$$-f(x) = -\int_0^L \delta(x-\xi)f(\xi)d\xi = \frac{d}{dx}\left\{T(x)\frac{d}{dx}\left[\frac{1}{P}\int_0^L G(x;\xi)f(\xi)d\xi\right]\right\}, \qquad (2.57a,b)$$

where (1) substitution of (2.56b) in (2.57a) leads to (2.57b); and (2) the terms dependent on x, including the derivatives d/dx, are taken out of the integral in $d\xi$, assuming that the latter is (I.13.40) uniformly convergent with regard to x. Comparing (2.57b) with (2.11c) $\equiv$ (2.58b), valid in the linear approximation (2.11a) $\equiv$ (2.58a)

$$P^2 \ll T^2 : \quad -f(x) = \frac{d}{dx}\left[T(x)\frac{d\zeta}{dx}\right], \qquad (2.58a,b)$$

it follows that

$$\zeta'^2 \ll 1 : \quad \zeta(x) = \frac{1}{P}\int_0^L G(x;\xi)f(\xi)d\xi. \qquad (2.59a,b)$$

Thus *in the linear approximation (2.59a), the deflection due to an arbitrary distribution of shear stress $f(\xi)$ equals its product by the influence function per unit load $G(x;\xi)/P$ integrated over the length of the string. This result holds for uniform or nonuniform tension and supports at equal or unequal heights.*

   In the case of uniform tension (2.60a) and supports at unequal heights (2.60b and c), the substitution of the influence function (2.34a and b) in the superposition principle (2.59b) leads to the displacement for arbitrary load (2.60d):

$$T = \text{const}; \quad \zeta(0) = 0, \quad \zeta(L) = h :$$

$$T\zeta(x) = \frac{Th}{PL}x\int_0^L f(\xi)d\xi + \left(1-\frac{x}{L}\right)\int_0^x \xi f(\xi)d\xi + x\int_x^L\left(1-\frac{\xi}{L}\right)f(\xi)d\xi. \qquad (2.60a\text{–}d)$$

The latter result (2.60d) can be checked, differentiating twice to regain (2.12d). A first differentiation of (2.60d) gives (2.61a):

$$T\frac{d\zeta}{dx} - \frac{Th}{PL}\int_0^L f(\xi)d\xi + \frac{1}{L}\int_0^x \xi f(\xi)d\xi - \int_x^L\left(1-\frac{\xi}{L}\right)f(\xi)d\xi = \left(1-\frac{x}{L}\right)\left[\xi f(\xi)\right]_{\xi=x} - x\left[\left(1-\frac{\xi}{L}\right)f(\xi)\right]_{\xi=x} = 0,$$

$$(2.61a)$$

where was used (I.13.46a and b) the rule of parametric differentiation of an integral by (1) the collection on the l.h.s. of the derivatives outside the integrals; and (2) leaving on the r.h.s. the derivatives at the limits of the integrals that cancel. A further differentiation

$$T\frac{d^2\zeta}{dx^2} = -\frac{1}{L}\big[\xi f(x)\big]_{\xi=x} - \Big[\Big(1-\frac{\xi}{L}\Big)f(\xi)\Big]_{\xi=x} = -f(x) \tag{2.61b}$$

confirms the coincidence of (2.61b) $\equiv$ (2.12d). The principle of superposition (2.60a through d) can be applied to the linear deflection (2.58a) of a string (2.58b), in particular when suspended from two ends at the same height (2.39a through c) for (1) continuously distributed loads like the weight (Section 2.4); (2) multiple concentrated loads (Subsections 2.5.1 through 2.5.6); and (3) a combination of both (Subsections 2.5.7 through 2.5.9). The principles of superposition and reciprocity also hold (7.92) for the linear deflection of an elastic string with nonuniform tension (2.11a through c), for instance, when the nonuniform tension is a linear function of position (Example 10.7).

2.4 Linear Deflection under Own Weight

The simplest continuous load is uniform (Subsection 2.4.1), and the deflection and slope (Subsection 2.4.2) can be obtained from either the superposition principle or the balance of forces (Subsection 2.4.3) that apply equally in the linear case (Subsection 2.4.4).

2.4.1 Deflection under a Uniform Load

The comparison of (2.34a,b) and (2.41a–c) suggests that *the difference between the linear (2.62a) deflection $\overline{\zeta}$ (ζ) of a string under a uniform tension (2.62b), between the cases of supports at different (2.62c) [the same (2.62d)] heights is a straight line (2.62e):*

$$\zeta'^2 \ll 1, T = \text{const}: \quad \overline{\zeta}(L) - \overline{\zeta}(0) = h, \quad \zeta(L) = \zeta(0) \Rightarrow \overline{\zeta}(x) - \zeta(x) = \frac{h}{L}x, \tag{2.62a–e}$$

assuming that the tension and the loading are the same. This agrees (2.62g) with the first term of the r.h.s. of (2.60a) noting that (2.62f) the total transverse load equals the integral of the shear stress along the string.

$$P = \int_0^L f(\xi)d\xi: \quad \overline{\zeta}(x) - \zeta(x) = \frac{hx}{PL}\int_{-\infty}^{+\infty} f(\xi)d\xi = \frac{h}{L}x. \tag{2.62f,g}$$

The proof of (2.62a–e) follows from: (1) the equation (2.12d) for the linear deflection with constant tension, that is, the same (2.62b) for $\overline{\zeta}$ and ζ:

$$-\frac{f(x)}{T} = \frac{d^2\zeta}{dx^2} = \frac{d^2\overline{\zeta}}{dx^2}; \quad \overline{\zeta}(L) - \overline{\zeta}(0) = \zeta(L) + h - \zeta(0) = h, \tag{2.62h,i}$$

(2) the difference in heights between supports (2.62c) follows (2.62i) from (2.62d,e). The property (2.62c–e) does not extend to nonuniform tension (nonlinear deflection) because (2.11c) [(2.9b)] is not invariant under the transformation (2.62e). The result (2.62a–e) applies for any loading; the two simplest loadings are a concentrated transverse force leading to the Green or influence function (Section 2.3) and the

uniform shear stress considered next (Section 2.4). In the case of the linear deflection (2.3a) ≡ (2.63a) under a uniform load, for example, the weight (2.63c),

$$\zeta'^2 \ll 1, \quad \rho \equiv \frac{dm}{ds} = \frac{dm}{dx}: \quad f(x) = \rho g, \tag{2.63a–c}$$

the shear stress or weight per unit length is constant (1) in a uniform gravity field with acceleration g; (2) for a homogeneous string with constant mass density (2.63b) per unit length; and (3) the latter corresponds to a constant mass density per unit horizontal projection in the linear case (2.3b). In the nonlinear case (Section 2.8) when $ds \neq dx$ in (2.1b), (3) is not true. Also in the linear approximation (2.3a), the length of the string is constant, and unlike in the nonlinear case, it does not matter whether it is elastic hence extensible (Section 2.8) or inextensible (Section 2.9).

2.4.2 Deflection, Slope, and Their Extrema

For the linear deflection (2.64a) = (2.3a) and suspension points at the same height (2.39b and c) ≡ (2.64b and c) and constant load (2.64d) under constant tension (2.64e), the deflection is given by (Figure 2.9) a parabolic shape (2.64f):

$$\zeta'^2 \ll 1, \zeta(0) = 0 = \zeta(L), \rho g = \text{const}, T = \text{const}: \quad \zeta(x) = \frac{\rho g}{2T} x(L - x). \tag{2.64a–f}$$

The maximum deflection is at the midposition (2.65a):

$$\delta \equiv \zeta_{max} = \zeta\left(\frac{L}{2}\right) = \frac{\rho g L^2}{8T}; \quad \zeta'(x) = \frac{\rho g}{T}\left(\frac{L}{2} - x\right), \tag{2.65a,b}$$

the slope (2.65b) is maximum in modulus at the suspension points (2.66a) where the reaction forces equal one-half of the weight (2.66b):

$$\theta \equiv \left|\zeta'(x)\right|_{max} = \zeta'(0) = -\zeta'(L) = \frac{\rho g L}{2T}, \quad R_- = -T\zeta'(0) = -\frac{\rho g L}{2} = T\zeta'(L) = R_+. \tag{2.66a,b}$$

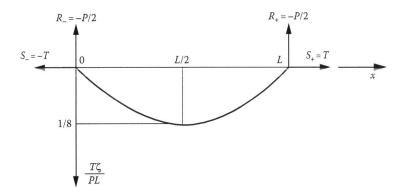

FIGURE 2.9 In the same conditions as for Figure 2.8 replacing the concentrated force by a uniformly distributed weight with the same total transverse force: (1) halves the maximum displacement; (2) changes the triangular shape to a parabola for a homogeneous string with constant cross section in a uniform gravity field; and (3) the tangent at the supports is the same in both cases (Figure 2.10).

The linearity condition (2.64a) of a small slope

$$P = mgL: \quad 1 \gg \left(|\zeta'|_{max} \right)^2 = \left(\frac{\rho g L}{2T} \right)^2 = \left(\frac{P}{2T} \right)^2, \tag{2.67a,b}$$

states that the total weight of the string (2.67a) is small compared with twice the tension (2.67b). This corresponds to the condition of a small nonlinearity parameter (Sections 2.6 through 2.9).

2.4.3 Parabolic Shape for a Homogeneous String

All the preceding results (Subsection 2.4.2) follow from the parabolic shape (2.64f) of the string that can be obtained in three ways. First, by substituting the uniform load (2.64d) and tension (2.64e) in the superposition principle (2.60d) with supports at the same height (2.68a):

$$h = 0: \quad \frac{T}{\rho g} \zeta(x) = \left(1 - \frac{x}{L} \right) \int_0^x \xi \, d\xi + x \int_x^L \left(1 - \frac{\xi}{L} \right) d\xi$$

$$= \left(1 - \frac{x}{L} \right) \frac{x^2}{2} + x(L - x) - \frac{x}{2L}(L^2 - x^2) = \frac{xL}{2} - \frac{x^2}{2}, \tag{2.68a,b}$$

leading to (2.68b) ≡ (2.64f). Second, by substituting (2.64d) in (2.12d) leading to (2.69a), whose solution is (2.69b):

$$\zeta'' = -\frac{\rho g}{T}: \quad \zeta(x) = -\frac{\rho g}{2T} x^2 + Ax + B; \quad B = 0, \quad A = \frac{\rho g L}{2T}; \tag{2.69a–d}$$

the boundary conditions (2.64b and c) determine the arbitrary constants (2.69c and d) in (2.69b) leading to (2.64f). A third way to obtain the result (2.64f) follows immediately from the equation of equilibrium (2.69a) ≡ (2.70b) and boundary conditions (2.64b and c) ≡ (2.70c and d):

$$q^2 \equiv \left(\frac{\rho g L}{2T} \right)^2 = \left(\frac{P}{2T} \right)^2 \ll 1: \quad \zeta'' = -\frac{\rho g}{T}, \quad \zeta(0) = 0 = \zeta(L); \tag{2.70a–d}$$

in the linear case (2.67b) ≡ (2.70a) when the weight of the string is small compared with twice the tension because (1) the curve (2.70b) is a parabola, with leading coefficient $-\rho g/2T$ for the quadratic term x^2; and (2) by (2.70c and d), the polynomial of second degree has as roots $x = 0, L$ corresponding to the factors $x, x - L$.

2.4.4 Small Total Weight Compared with the Tension

Retaining the assumption of small total weight compared with twice the tension, the preceding results can be extended to (1) nonhomogeneous strings, for example, a string with two sections of different materials, for which the mass density is discontinuous (Example 10.2); (2) a homogeneous string of varying cross section, for which the mass density $\rho(x)$ is a continually varying function of position (Example 10.3); (3) a uniform string with suspension points at unequal heights (Example 10.4); and (4) a uniform string under a nonuniform longitudinal tension (Example 10.7). *For a homogeneous string of uniform cross section, the mass density per unit length is a constant, and the total weight (2.67a) appears in (1) the displacement (2.64f) ≡ (2.71a) [slope (2.65b) ≡ (2.71b)]:*

$$\zeta(x) = \frac{P}{2TL}x(L-x), \quad \zeta'(x) = \frac{P}{T}\left(\frac{1}{2} - \frac{x}{L}\right); \quad (2.71\text{a,b})$$

(2) their maximum values (2.72a) [in modulus (2.72b)]:

$$\delta \equiv \left|\zeta(x)\right|_{\max} = \zeta\left(\frac{L}{2}\right) = \frac{PL}{8T} = \frac{1}{2}G\left(\frac{L}{2};\frac{L}{2}\right), \quad (2.72\text{a})$$

$$\theta = \left|\zeta'\right|_{\max} = \zeta'(0) = -\zeta'(L) = \frac{P}{2T} = G'\left(0;\frac{L}{2}\right) = -G'\left(0;\frac{L}{2}\right). \quad (2.72\text{b})$$

The factor one-half appears in the linearity condition (2.67b) because (1) the tension must be large compared with the reaction force at the supports, that is, one-half of the total weight for symmetric loading:

$$R_- = F(0) = -T\zeta'(0) = R_+ = -F(L) = T\zeta'(L) = \frac{\rho g L}{2} = -\frac{P}{2}; \quad (2.72\text{c})$$

(2) equivalently, the total weight is balanced by the vertical component of the tension at the two supports.

2.4.5 Linear Deflection under Uniform or Concentrated Loads

Comparing the linear deflection of a string under constant tension and with supports at the same height for the same total transverse force P uniformly distributed (Section 2.4) [concentrated at equal distance from the supports (Section 2.3)], (1) the reactions at the supports are the same (2.72c) [≡(2.45c and d)]; (2) the slopes (2.71b) [(2.46c and d) implies (2.73a and b)]:

$$h = 0: \quad G'\left(x;\frac{L}{2}\right) = \frac{P}{2T} \times \begin{cases} +1 & \text{if } 0 \leq x \leq \dfrac{L}{2}, \\[2mm] -1 & \text{if } \dfrac{L}{2} \leq x \leq L, \end{cases} \quad (2.73\text{a,b})$$

coincide only at the supports (2.72b); and (3) since the deflected shape is [Figure 2.9 (Figure 2.8)] a parabola (2.71a) [isosceles triangle (2.46c and d)], the maximum deflection is (2.72a) [the double (2.45e)] for the concentrated force. Thus (Figure 2.10) from the parabolic shape of the linear deflection, taking the tangent lines at the supports follows the isosceles triangular shape, due to the same total transverse force concentrated at equal distance from the supports. This simple result no longer applies to a concentrated force at unequal distances from the supports (Figure 2.11), since in that case the maximum deflection (2.35b) for supports at the same height (2.74a) has, relative to (2.44b), a factor

$$h = 0: \quad \frac{G(\xi;\xi)}{G(L/2;L/2)} = 4\frac{\xi}{L}\left(1 - \frac{\xi}{L}\right) \begin{cases} = 1 & \text{if } \xi = \dfrac{L}{2}, \\[2mm] < 1 & \text{if } \xi \neq \dfrac{L}{2}, \end{cases} \quad (2.74\text{a–c})$$

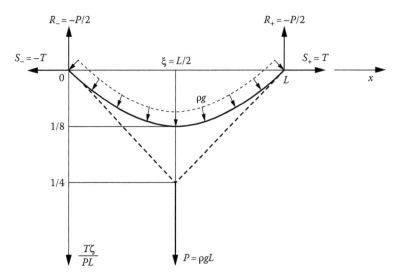

FIGURE 2.10 In the conditions of Figure 2.8, the deflection at constant weight follows the deflection by a concentrated force at equal distance from the supports provided that the total transverse force be the same $P = \rho gL$. In this case, the reaction forces at the supports and the slopes are the same; since the deflected shape of the string is a parabola (isosceles triangle) for a uniform (concentrated) load the maximum deflection is double in the latter case, as can be confirmed comparing Figures 2.8 and 2.9.

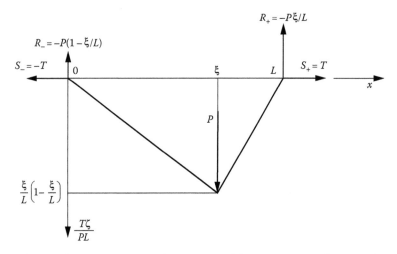

FIGURE 2.11 If the conditions of Figure 2.8, the concentrated force is not at equal distance from the supports ($\xi = L/2$ in Figure 2.10) but is at any distance $0 < \xi < L$ from the supports, the shape of the string is a nonisosceles triangle and the maximum deflection is reduced by a factor $4(\xi/L)(1 - \xi/L)$.

which (1) is unity only at the mid-position (2.74b); and (2) is smaller (2.74c) at any other position (Figure 2.11). The proof of (2.73c) follows because the derivative

$$0 \le \xi < \frac{L}{2}: \quad \frac{d}{d\xi}\left[4\frac{\xi}{L}\left(1-\frac{\xi}{L}\right)\right] = \frac{4}{L}\left(1-\frac{2\xi}{L}\right) > 0, \tag{2.75a,b}$$

is positive (2.75b) in the range (2.75a); *the maximum deflection is largest when (2.75b) vanish, that is, for $\xi = L/2$ a concentrated force at equal distance from the supports (2.74b).* For $\xi \neq L/2$, the maximum deflection is the same at ξ and $L - \xi$, so the conclusion that it is smaller in the range (2.75a) also applies in the range $L/2 < \xi \le L$. The comparison of a single concentrated (Section 2.3) and a uniform (Section 2.4) load is followed by the consideration of multiple and mixed loads (Section 2.5).

2.5 Multiple Concentrated and Distributed Loads

Two particular cases of multiple concentrated loads with equal spacing are (1) the same direction (Subsection 2.5.2), corresponding (Subsection 2.5.3) to the transverse force (shear stress) that is specified by the generalized function (Subsection 2.5.1) unit staircase (impulse haircomb); and (2) alternating direction (Subsection 2.5.5), corresponding to a transverse force (shear stress) specified by the generalized function (Subsection 2.5.4) unit (Subsection 2.5.6) parapet (alternating impulse haircomb). The cases of mixed distributed (Section 2.4) and concentrated (Section 2.3) loads include the weight balanced by opposite concentrated forces (Subsection 2.5.8) leading (Subsection 2.5.9) to a shear stress that is specified by the generalized function saw tooth (Subsection 2.5.7).

2.5.1 Unit Staircase and Impulse Haircomb

Considering N transverse forces P_n concentrated at the points $x = \xi_n$, the transverse force (shear stress) is given by (2.76a) [(2.76b)]:

$$F_1(x) = F_1(0) + \sum_{N=1}^{N} P_n H(x - \xi_n), \quad F_1'(x) = \sum_{n=1}^{N} P_n \delta(x - \xi_n). \qquad (2.76a,b)$$

The transverse force (shear stress) corresponds to the generalized function **Heaviside staircase** (2.76a) [**Dirac impulse haircomb** (2.76b)], with irregular steps of height $P_n = P$ and placed at ξ_n in Figure 2.12a (b);

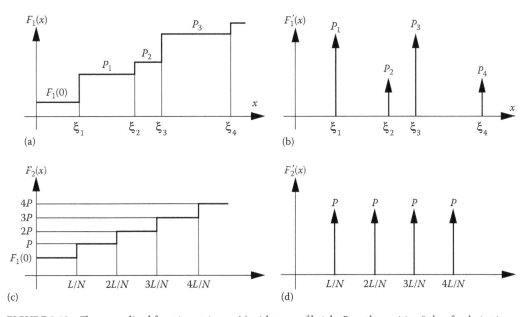

FIGURE 2.12 The generalized function staircase (a) with steps of height P_n at the position ξ_n has for derivative a Dirac impulse haircomb function (b) with impulses P_n at ξ_n. If the height P and spacing L/N of all the steps is the same, the regular stair case (c) has for derivative a uniform impulse haircomb (d).

the particular case of steps of equal height (2.77a) and spacing (2.77b) leads to Figure 2.12c (Figure 2.12d) to the generalized functions **regular staircase** (2.77c) [**haircomb** (2.77d)]:

$$P_n = P, \xi_n = n\frac{L}{N}: \quad F_2(x) = F_2(0) + P\sum_{n=1}^{N} H\left(x - n\frac{L}{N}\right), \quad F(x) = P\sum_{n=1}^{N} \delta\left(x - n\frac{L}{N}\right). \qquad (2.77\text{a-d})$$

As another interpretation of these functions, consider an electric circuit, to which are connected N condensers, each able to furnish instantaneously a charge P_n at the time ξ_n the switch connecting it to the circuit is activated; then the staircase function (2.76a) [(2.77c)] represents the charge on the circuit at all times, and its derivative, the impulse haircomb function (2.76b) [(2.77d)], represents the electric current, if the discharges occur at unequally (equally) spaced times.

2.5.2 Two Concentrated Loads with the Same Direction

By the principle of superposition (Subsection 2.3.5), the linear displacement due to multiple concentrated loads is the sum of the triangular shapes (Figure 2.11) due to each load at each position (Subsection 2.3.4). *The simplest case of multiple concentrated loads is $N = 2$ forces P concentrated at equal distances from the suspension points (2.78b)*:

$$P^2 \ll T^2: \quad f_a(x) = P\left\{\delta\left(x - \frac{L}{3}\right) + \delta\left(x - \frac{2L}{3}\right)\right\}; \qquad (2.78\text{a,b})$$

in the linear case (2.78a), the shape of the string is trapezoidal (Figure 2.13):

$$\zeta_a(x) = \frac{P}{T} \times \begin{cases} x & \text{if } 0 \le x \le \dfrac{L}{3}, \\[2mm] \dfrac{L}{3} & \text{if } \dfrac{L}{3} \le x \le \dfrac{2L}{3}, \\[2mm] L - x & \text{if } \dfrac{2L}{3} \le x \le L; \end{cases} \qquad (2.79\text{a-c})$$

from (2.79a through c) follows (2.12c) the transverse force along the string:

$$F_a(x) = -T\frac{d\zeta_a}{dx} = \begin{cases} -P = R_- & \text{if } 0 \le x < \dfrac{L}{3}, \\[2mm] 0 & \text{if } \dfrac{L}{3} < x < \dfrac{2L}{3}, \\[2mm] P = -R_+ & \text{if } \dfrac{2L}{3} < x \le L, \end{cases} \qquad (2.80\text{a-c})$$

which shows that the reactions on the supports are both equal to the forces and with opposite sign.

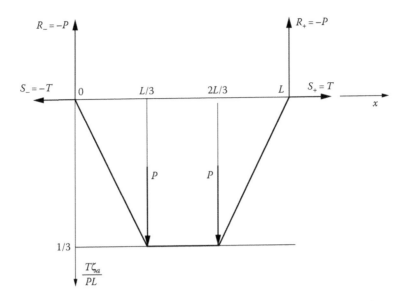

FIGURE 2.13 The loading of a string by two equal concentrated forces at equal distances between themselves and the supports leads to a transverse force (shear stress) that is a regular staircase (impulse haircomb) with [Figure 2.12c (Figure 2.12d)] two steps (singularities). The shape of the string in the conditions of Figure 2.8 is trapezoidal, with vertical reactions at the supports equal and opposite to the concentrated forces.

2.5.3 Differential Equation and Matching Conditions

The solution (2.79a through c) satisfies the differential equation

$$-\frac{T}{P}\frac{d^2\zeta_a}{dx^2} = \delta\left(x - \frac{L}{3}\right) + \delta\left(x - \frac{2L}{2}\right), \tag{2.81}$$

vanishes at the suspension points (2.64b and c) ≡ (2.82a and b), and is continuous at the points of application of the forces (2.82c and d):

$$\zeta(0) = 0 = \zeta(L), \quad \zeta\left(\frac{L}{3} - 0\right) = \zeta\left(\frac{L}{3} + 0\right), \quad \zeta\left(\frac{2L}{3} - 0\right) = \zeta\left(\frac{2L}{3} + 0\right). \tag{2.82a-d}$$

The solution can be obtained by three methods: (1) integrating (2.81) with the stated boundary and continuity conditions; and (2) substituting (2.78b) into the principle of superposition (2.60a through d), with (2.83d) supports at the same height (2.83b and c):

$$T = \text{const}; \zeta(0) = 0 = \zeta(L): \quad T\zeta(x) = \left(1 - \frac{x}{L}\right)\int_0^x \xi f(\xi)d\xi + x\int_x^L\left(1 - \frac{\xi}{L}\right)f(\xi)d\xi; \tag{2.83a-d}$$

(3) adding (2.41b and c) with ξ = L/3 and ξ = 2L/3. The results (2.79a through c; 2.80a through c) can be obtained most simply by a fourth method as follows: (1) taking moments about x = L, it is found that −R_L = PL/3 + 2PL/3 = PL, so that the reaction at x = 0 is R_ = −P; (2) the transverse force is thus −P until x = L/3, when it jumps to −P + P = 0, until x = 2L/3, when it jumps again to 0 + P = P, so the reaction at x = L is −P; (3) the slope for 0 ≤ x ≤ L/3 is P/T, so the shape is P x/T for 0 ≤ x ≤ L/3 because ζ(0) = 0;

(4) in the range $L/3 \leq x \leq 2L/3$, the string is horizontal, and the displacement equals $\zeta(L/3) = PL/3T = \zeta(2L/3)$ by continuity at $L/3$, $2L/3$; and (5) for $2L/3 < x \leq L$, the slope is $-P/T$, so the shape is $-(P/T)(x - L)$ for $2L/3 \leq x \leq L$ because $\zeta(L) = 0$.

2.5.4 Heaviside Parapet and Dirac Alternating Haircomb

Assuming that the concentrated forces are applied alternatively upward and downward, instead of always upward as in (2.74a and b), leads to the transverse force (shear stress) specified by the generalized function (2.84a) [(2.84b)] **parapet (alternating impulse haircomb)**

$$F_3(x) = F_3(0) + \sum_{n=1}^{N} (-)^n P_n H(x - \xi_n), \quad F_3'(x) = \sum_{n=1}^{N} (-)^n \delta(x - \xi_n), \tag{2.84a,b}$$

with irregular jumps P_n unequally spaced ξ_n and alternating in sign [Figure 2.14a (Figure 2.14b)]; the case of equal jumps (2.77a) equally spaced (2.77b) leads to [(Figure 2.14c (Figure 2.14d)] the generalized function (2.85a) [(2.85b)], **regular parapet (alternating impulse haircomb)**

$$F_4(x) = F_4(0) + P\sum_{n=1}^{N} (-)^n H\left(x - n\frac{L}{N}\right), \quad F_4'(x) = P\sum_{n=1}^{N} (-)^n \delta\left(x - n\frac{L}{N}\right). \tag{2.85a,b}$$

As another interpretation of (2.85a and b), consider a particle of linear momentum $p = mv$ equal to the mass times the velocity that collides perpendicularly with two moving walls, receiving a momentum P_n at times ξ_n, with alternating signs; then, the parapet function (2.84a) [(2.85a)] represents the linear momentum of the particle and the alternating haircomb (2.84b) [(2.85b)] the impulsive forces at the instants of the alternating collisions. In the theory of electrical circuits and signals, the parapet distribution shifted to have zero mean is called a **square wave**.

2.5.5 Two Concentrated Forces with Opposite Directions

In the case $N = 2$ of *two opposite transverse forces $\pm P$ at equal distances from the suspension points (2.86b)*,

$$P^2 \ll 4T^2: \quad f_b(x) = P\left\{\delta\left(x - \frac{L}{3}\right) - \delta\left(x - \frac{2L}{3}\right)\right\}, \tag{2.86a,b}$$

the small deflection of the string under tension in the linear case (2.86a) consists of two triangles (Figure 2.15) skew-symmetric relative to the mid-position:

$$\zeta_b(x) = \frac{P}{3T} \times \begin{cases} x & \text{if } 0 \leq x \leq \dfrac{L}{3}, \\[2mm] L - 2x & \text{if } \dfrac{L}{3} \leq x \leq \dfrac{2L}{3}, \\[2mm] x - L & \text{if } \dfrac{2L}{3} \leq x \leq L. \end{cases} \tag{2.87a–c}$$

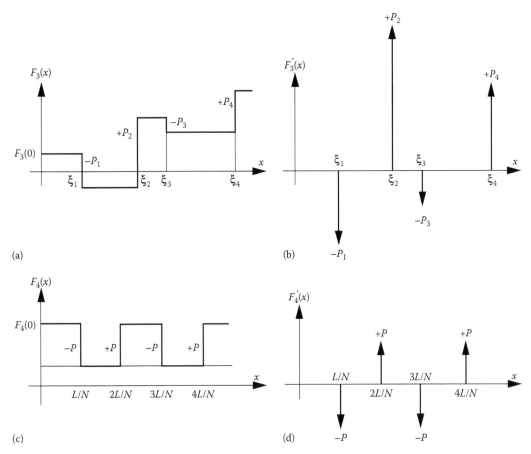

FIGURE 2.14 In the same irregular spacings ξ_n are associated with jumps with the same (alternating) sign the corresponding generalized function is [Figure 2.12a (Figure 2.14a)] the staircase (parapet) and its derivative is a monotonic (alternating) Dirac impulse haircomb [Figure 2.12b (Figure 2.14b)]. If the spacings and height of the steps is the same, the parapet function is regular (Figure 2.14c) and a shift of the x-axis to the mid position leads to a "square wave." The vertical translation does not affect the derivative that is a regular alternating Dirac impulse haircomb (Figure 2.14d) both for the parapet function and for the square wave.

The extrema, that is, maximum and minimum deflections, occur at the points where the concentrated forces are applied:

$$\zeta_{b\max} = \zeta\left(\frac{L}{3}\right) = \frac{PL}{9T} = -\zeta\left(\frac{2L}{3}\right) = -\zeta_{b\min}. \tag{2.88}$$

The transverse force is calculated from (2.11b):

$$F_b(x) = -T\zeta_b'(x) = \begin{cases} -\dfrac{P}{3} = R_- & \text{if } 0 < x \leq \dfrac{L}{3}, \\[2mm] \dfrac{2P}{3} & \text{if } \dfrac{L}{3} < x < \dfrac{2L}{3}, \\[2mm] -\dfrac{P}{3} = -R_+ & \text{if } \dfrac{2L}{3} < x \leq L, \end{cases} \tag{2.89a–c}$$

which shows that the reactions at the supports are equal in sign –P/3 and magnitude.

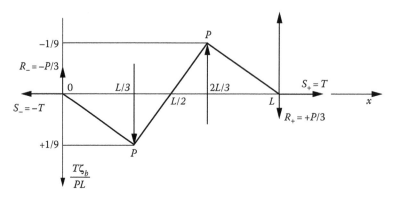

FIGURE 2.15 In the conditions of Figure 2.8, the two equidistant concentrated forces have the same (opposite) signs [Figure 2.13 (Figure 2.15)], the corresponding: (1) transverse force is [Figure 2.12c (Figure 2.14c)] a regular staircase (parapet) with two steps; and (2) shear stress is [Figure 2.12d (Figure 2.14d)] a regular monotonic (alternating) haircomb with two singularities. The two concentrated equidistant equal (opposite) forces [Figure 2.13 (Figure 2.15)] lead to a symmetric trapezoidal shape (a shape consisting of two triangles antisymmetrical relative to the mid position between the supports). The inversion of one of the concentrated forces: (1) reduces the maximum deflection by one-third; (2) reduces the modulus of the vertical reactions at the supports by the same amount; and (3) changes the direction of the vertical reaction at the support closer to the concentrated force whose direction was also changed.

2.5.6 Boundary and Continuity Conditions

The result (2.85a and b) may be obtained using three methods: (1) integrating (2.12d; 2.86b) ≡ (2.90)

$$-\frac{T}{P}\frac{d^2\zeta_b}{dx^2} = \delta\left(x - \frac{L}{3}\right) - \delta\left(x - \frac{2L}{3}\right),\tag{2.90}$$

and applying the same boundary (2.82a and b) and continuity (2.82c and d) conditions as before; (2) substituting the loads (2.86b) into (2.83d); and (3) subtracting (2.41b and c) with $\xi = 2L/3$ from the same expression with $\xi = L/3$. The results (2.87a through c; 2.89a through c) can be obtained most simply by a fourth method as follows: (1) the reaction at the support $x = 0$ satisfies the moment equation at $x = L$, that is, $R\_L = -2PL/3 + PL/3 = -PL/3$, so that $R_- = -P/3$; (2) the transverse force is $P/3$ before $x = L/3$, then $P/3 - P = -2P/3$ until $x = 2L/3$, and after $-2P/3 + P = P/3$, so that the reaction at $x = L$ is $R_+ = P/3$; (3) the slope for $0 \leq x < L/3$ is $P/3T$ and the shape $Px/3T$ for $0 \leq x \leq L/3$ since $\zeta(0) = 0$; (4) the slope for $2L/3 < x < L$ is $P/3T$ and the shape $(P/3T)(x - L)$ for $2L/3 \leq x \leq L$ since $y(L) = 0$; and (5) for $L/3 \leq x \leq 2L/3$, the shape is a straight line joining $PL/9T$ at $x = L/3$ to $-PL/9T$ at $x = 2L/3$, thus having a slope $(-2PL/9T)/(L/3) = -2P/3T$ and passing through $x = L/2$, which corresponds to the equation $-(2P/3T)(x - L/2) = (P/3T)(L - 2x)$. The method (2) is applied next to the case of two equal (opposite) forces [Subsection 2.5.2 (Subsection 2.5.5)], to confirm (2.78a and b) [(2.87a through c)].

The shape of the string (2.41b and c) for a concentrated force at (2.91a) [(2.92a)] is given by (2.91b and c) [(2.92b and c)]:

$$\xi = \frac{L}{3}: \quad G\left(x;\frac{L}{3}\right) = \frac{P}{T} \times \begin{cases} \dfrac{2x}{3} & \text{if } 0 \leq x \leq \dfrac{L}{3}, \\[2mm] \dfrac{L-x}{3} & \text{if } \dfrac{L}{3} \leq x \leq L, \end{cases}\tag{2.91a–c}$$

$$\xi = \frac{2L}{3}: \quad G\left(x;\frac{L}{3}\right) = \frac{P}{T} \times \begin{cases} \dfrac{x}{3} & \text{if } 0 \leq x \leq \dfrac{2L}{3}, \\[2mm] \dfrac{2L-2x}{3} & \text{if } \dfrac{2L}{3} \leq x \leq L. \end{cases}\tag{2.92a–c}$$

The sum of (2.91b and c) and (2.92b and c) specifies

$$
\xi_a = G\left(x; \frac{L}{3}\right) + G\left(x; \frac{2L}{3}\right) = \frac{P}{T} \times
\begin{cases}
\dfrac{2x}{3} + \dfrac{x}{3} = x & \text{if } 0 \le x \le \dfrac{L}{3}, & (2.93a) \\[2mm]
\dfrac{L-x}{3} + \dfrac{x}{3} = \dfrac{L}{3} & \text{if } \dfrac{L}{3} \le x \le \dfrac{2L}{3}, & (2.93b) \\[2mm]
\dfrac{L-x}{3} + \dfrac{2L-2x}{3} = L-x & \text{if } \dfrac{2L}{3} \le x \le L, & (2.93c)
\end{cases}
$$

the displacement (2.93a through c) ≡ (2.79a through c) due to two equal parallel concentrated forces at equal distance from the supports. Subtracting (2.92b and c) from (2.91b and c) specifies

$$
\xi_b = G\left(x; \frac{L}{3}\right) - G\left(x; \frac{2L}{3}\right) = \frac{P}{T} \times
\begin{cases}
\dfrac{2x}{3} - \dfrac{x}{3} = \dfrac{x}{3} & \text{if } 0 \le x \le \dfrac{L}{3}, & (2.94a) \\[2mm]
\dfrac{L-x}{3} - \dfrac{x}{3} = \dfrac{L-2x}{3} & \text{if } \dfrac{L}{3} \le x \le \dfrac{2L}{3}, & (2.94b) \\[2mm]
\dfrac{L-x}{3} - \dfrac{2L-2x}{3} = \dfrac{x-L}{3} & \text{if } \dfrac{2L}{3} \le x \le L, & (2.94c)
\end{cases}
$$

the displacement (2.94a through c) ≡ (2.87a through c) due to two antiparallel equal concentrated forces at equal distances from the supports.

2.5.7 Regular/Irregular Sawtooth Generalized Functions

Superimposing continuous and concentrated forces leads to a case of **mixed loading**, for example, the transverse force

$$
F_5(x) = F_5(0) + a \sum_{n=1}^{N-1} (x - \xi_n)\{H(x - \xi_n) - H(x - \xi_{n+1})\} \tag{2.95}
$$

is the generalized function **irregular sawtooth** *function with slope a (Figure 2.16a), which jumps back to zero ordinate at the points $x = \xi_n$, so that the shear stress, that is, the derivative (Figure 2.16b) is a constant a plus an impulse haircomb with jumps $a(\xi_{n+1} - \xi_n)$ at ξ_n:*

$$
F_5'(x) = a\{H(x - \xi_1) - H(x - \xi_N)\} - a \sum_{n=1}^{N} (\xi_{n+1} - \xi_n)\delta(x - \xi_{n+1}). \tag{2.96}
$$

If the jumps are equally spaced (2.77b), this leads to a transverse force, that is, the generalized function **regular sawtooth** of slope a has teeth with height L/N (Figure 2.16c):

$$
F_6(x) = F_6(0) + a \sum_{n=1}^{N-1} \left(x - L\frac{n}{N}\right)\left\{H\left(x - L\frac{n}{N}\right) - H\left(x - L\frac{n+1}{N}\right)\right\}, \tag{2.97}
$$

$$
F_6'(x) = a\{H(x) - H(x - L)\} - \frac{aL}{N} \sum_{n=1}^{N-1} \delta\left(x - L\frac{n+1}{N}\right), \tag{2.98}
$$

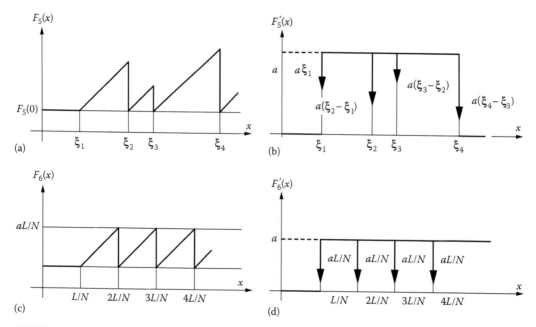

FIGURE 2.16 If in a linear function with slope a at irregular intervals ξ_n is subtracted the value $a(\xi_n - \xi_{n-1})$ so that it comes back to the same base value, the result is the generalized function saw-tooth with constant slope (Figure 2.16a); it would also be possible to have a different slope a_n for each tooth. The same a (different a_n) slopes for each tooth leads for the derivative (Figure 2.16b) to: (1) a flat floor of height a (a sequence of "plateaux" of level a_n); and (2) plus a Dirac impulse haircomb located at the jumps ξ_n, and equal to their magnitude, with negative sign because they are all downwards. For identical teeth with equal length and height, and hence slope, is obtained the regular saw tooth function (Figure 2.16c) whose derivative is (Figure 2.16d) the constant slope minus a regular Dirac impulse haircomb at the jumps equal to their height.

corresponding to the shear stress (2.98). The formula (2.98) follows from (2.95), using the rule of derivation of discontinuous functions (1.103a and b), namely

$$F_5'(x) = a \sum_{n=1}^{N-1} \left\{ H(x - \xi_n) - H(x - \xi_{n+1}) \right\} + a \sum_{n=0}^{N-1} (x - \xi_n) \left\{ \delta(x - \xi_n) - \delta(x - \xi_{n+1}) \right\}, \tag{2.99}$$

noting that (2.99) simplifies to (2.96), using the substitution property (1.78b) of the generalized function unit impulse. As another interpretation of (2.95), consider a dam that accumulates water from rain and rivers at a constant rate a per unit time and at times ξ_n, discharges instantly the accumulated water, so as to return to the initial reference level; the level of the water in the dam is a sawtooth function (2.95) [(2.97)] if the discharges occur at irregular ξ_n [regular (2.77b)] times. The volume flux of the water (2.96) [(2.98)] is a constant inflow a, with instantaneous impulsive outflows $a(\xi_{n+1} - \xi_n)$ [$a L/N$] at the times (2.77b) of the discharges.

2.5.8 Mixed, Distributed, and Concentrated Loads

As an example of mixed loading, consider *a uniform, homogeneous string of density P, whose weight P = ρgL is balanced by two opposite, equal −P/2 concentrated forces at equal distances from the supports*:

$$f(x) = \rho g - \frac{\rho g L}{2} \left\{ \delta\left(x - \frac{L}{3}\right) + \delta\left(x - \frac{2L}{3}\right) \right\}; \tag{2.100}$$

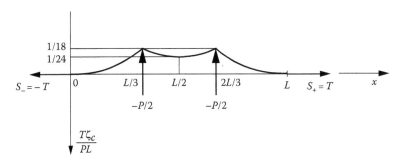

FIGURE 2.17 If in the conditions of Figure 2.8, the weight of an string as in Figure 2.9 is compensated by two equal equidistant forces as in Figure 2.13, then: (1) the transverse force (shear stress) is a sawtooth (its derivative) with [Figure 2.15c (Figure 2.15d)] the weight per unit length as slope (constant value) and half the total weight as downward jump (impulse at the two steps); (2) the shape of the string consists of three parabolas with angular points at the locations of the concentrated forces; and (3) the latter balance the total weight of the string so there are no vertical reaction forces at the supports.

$$F(x) = F(0) + \rho g x - \frac{\rho g L}{2}\left\{ H\left(x - \frac{L}{3}\right) + H\left(x - \frac{2L}{3}\right) \right\}. \tag{2.101}$$

The shape is given (Figure 2.17) by

$$\zeta(x) = -\frac{\rho g}{2T} \times \begin{cases} x^2 & \text{if } 0 \le x \le \dfrac{L}{3}, & \text{(2.102a)} \\[2ex] x^2 - xL + \dfrac{L^2}{3} & \text{if } \dfrac{L}{3} \le x \le \dfrac{2L}{3}, & \text{(2.102b)} \\[2ex] (x-L)^2 & \text{if } \dfrac{2L}{3} \le x \le L, & \text{(2.102c)} \end{cases}$$

and the deflections at the [points of application of the forces (2.103b and c)] mid-point (2.104e) by

$$P \equiv \rho g L: \quad \zeta_{\min} = \zeta\left(\frac{L}{3}\right) = \zeta\left(\frac{2L}{3}\right) = -\frac{\rho g L^2}{18T} = -\frac{PL}{18T}, \tag{2.103a–c}$$

$$P^2 \ll 4T^2: \quad \zeta_{\mathrm{mid}} = \zeta\left(\frac{L}{2}\right) = -\frac{\rho g L^2}{24T} = -\frac{PL}{24T} = \frac{3}{4}\zeta_{\min}, \tag{2.104a–c}$$

where the total weight (2.103a) is small compared with twice the tension (2.104d) for linear deflections. The transverse force is specified by

$$F(x) = -T\zeta'(x) = \rho g \times \begin{cases} x & \text{if } 0 < x < \dfrac{L}{3}, & \text{(2.105a)} \\[2ex] x - \dfrac{L}{2} & \text{if } \dfrac{L}{3} \le x < \dfrac{2L}{3}, & \text{(2.105b)} \\[2ex] x - L & \text{if } \dfrac{2L}{3} < x \le L; & \text{(2.105c)} \end{cases}$$

this shows that the reactions at the supports are zero, because they are equal and opposite for the weight P/2 = ρgL/2 and the concentrated forces −P/2. The result (2.101a through c) can be obtained by two methods: (1) substituting (2.101) in (2.12c) [(2.100) in (2.12d)] and using the boundary (2.82a and b) and continuity (2.82c and d) conditions; and (2) alternatively, (2.100) can be substituted in (2.83d) to yield (2.102a through c). A third and simplest method is (2.106b) to add (1) the deflection (2.64f) due to the weight and (2) the deflection (2.79a through c) due to two concentrated loads opposite to half the weight (2.106a) at equal distance from the supports:

$$P = -\frac{\rho g L}{2}: \quad \zeta_c(x) = \frac{\rho g}{2T} x(L-x) - \zeta_a(x). \tag{2.106a,b}$$

Substituting (2.79a through c) and (2.106a) in (2.106b) leads to

$$\zeta_c(x) = \frac{\rho g}{2T} \times \begin{cases} x(L-x) - xL = -x^2 & \text{if } 0 \le x \le \dfrac{L}{3}, & (2.107a) \\[2mm] x(L-x) - \dfrac{L^2}{3} = -x^2 + xL - \dfrac{L^2}{3} & \text{if } \dfrac{L}{3} \le x \le \dfrac{2L}{3}, & (2.107b) \\[2mm] x(L-x) - L(L-x) = -(L-x)^2 & \text{if } \dfrac{2L}{3} \le x \le \dfrac{2L}{3}, & (2.107c) \end{cases}$$

that coincides with (2.102a through c) ≡ (2.107a through c).

2.5.9 Comparison of Distributed, Concentrated, and Mixed Loads

The linear deflection of a string under constant tension with supports at the same height is compared (Table 2.1) for five loadings: (1) own weight (Section 2.4), assuming uniform weight per unit length (2.64d), for example, a homogeneous string with constant cross section, and hence constant mass density (2.63b) in a uniform gravity field (Figure 2.9); (2–4) neglect of weight compared with concentrated forces P, either (2) one (Figure 2.8) at the middle (Subsection 2.3.4) or (3) two (Figure 2.13) at equal distance from the supports (Subsection 2.5.2) or (4) the same as (2) with (Figure 2.15) the direction of the second force reversed (Subsection 2.5.5); and (5) the combination of own weight (1) with two equally spaced forces (3) compensating (Figure 2.17) the total weight (Subsection 2.5.8). The shape of the string is shown in all five cases (1–5), using the same dimensionless scales in Figures 2.8 through 2.11, 2.13, 2.15, and 2.17 to allow a visual comparison of the magnitude of the deflections. Table 2.1 indicates, for each of the five (1–5) cases, (a) the loads; (b–c) the maximum deflection (slope) in modulus and its location; and (d) the reactions at the supports. The maximum slope is proportional to the ratio P/T of the total load P to the tension T, and the maximum deflection multiplies the slope by L. Thus, the maximum deflection (slope) in modulus is PL/T (P/L) times (1) 1/4 (1/2) for a single concentrated force in the middle; (2) one-half 1/8 (the same 1/2) for a total weight equal to the concentrated force; (3) larger 1/3(1) than (1, 2) for two equally spaced forces; (4) for the same forces with opposite direction, a smaller [intermediate 2/3 between (1) ≡ (2) and (3)] value 1/9; and (5) for the mixed load (5), it is 1/24, smaller than all others [equals 1/3 of the previous smallest (3)]. Concerning the reactions at the supports, they are (1–3) half the load, that is, $P/2$ for (1, 2) and P for (3) since there are two equal applied forces; (4) one-third of the load for opposite, equally spaced forces; and (5) zero if the two forces (3) balance the weight (1). The methods of the determination of the shape of the string under various loads are (1) solution of the balance equation with boundary conditions, for example, (2.29–2.34) [(2.64; 2.69; 2.70)] for a concentrated (uniform) load; (2) substitution in the integral (2.60a through d) involving the influence function, for example, (2.68a and b); and (3) superposition of loads and deflections, either concentrated or mixed, for example, (2.91–2.94) [(2.106; 2.107)]. Of these three methods, (2) and (3) apply only in the linear case of deflection with a small slope due to a small total transverse force compared with twice the tension. The method (1) applies in all cases and is used in the nonlinear problems in the sequel (Sections 2.6 through 2.9).

TABLE 2.1 Comparison of Five Loadings for the Linear Deflection of an Elastic String

Case	I	II	III	IV	V = I + III
Subsection	2.4.4	2.3.4	2.5.1 through 2.5.3	2.5.4 through 2.5.6	2.5.7 and 2.5.8
Description	Own weight	One concentrated force	Equal concentrated forces	Opposite concentrated forces	Weight balanced by equal concentrated forces
Figure	2.9	2.8	2.13	2.15	2.17
Concentrated force	—	P	P, P	$P, -P$	$-\dfrac{\rho g L}{2}; -\dfrac{\rho g L}{2}$
Location	—	$\dfrac{L}{2}$	$\dfrac{L}{3}, \dfrac{2L}{3}$	$\dfrac{L}{3}, \dfrac{2L}{3}$	$\dfrac{L}{3}, \dfrac{2L}{3}$
Distributed load	ρg	—	—	—	ρg
Total load	$\rho g L = P$	P	$2P$	0	0
Maximum deflection $\delta = \lvert \zeta \rvert_{max} = \dfrac{PL}{T} \times \ldots$	$\dfrac{1}{8}$	$\dfrac{1}{4}$	$\dfrac{1}{3}$	$\dfrac{1}{9}$	$-\dfrac{1}{24}$
At	$\dfrac{L}{2}$	$\dfrac{L}{2}$	$\dfrac{L}{3} \le x \le \dfrac{2L}{3}$	$x = \dfrac{L}{3}, \dfrac{2L}{3}$	$x = \dfrac{L}{2}$
Maximum slope $\tan\theta \equiv \lvert \zeta' \rvert_{max} = \dfrac{P}{T} \times \ldots$	$\dfrac{1}{2}$	$\dfrac{1}{2}$	1	$\dfrac{2}{3}$	$\dfrac{1}{3}$
At	$x = 0, L$	$x = 0, L$	$x < \dfrac{L}{3}; x > \dfrac{2L}{3}$	$\dfrac{L}{3} \le x \le \dfrac{2L}{3}$	$x = \dfrac{L}{3} - 0, \dfrac{2L}{3} + 0$
Reactions at the supports: $R_{\pm} = P \times \ldots$	$-\dfrac{1}{2}$	$-\dfrac{1}{2}$	-1	$-\dfrac{1}{3}, +\dfrac{1}{3}$	0

Note: Comparison of the linear deflection of an elastic string for five different load distributions. In all cases, the tension is uniform and the supports are at the same height. The nonlinear deflections are indicated in Table 2.7.

The deflection of an elastic string with constant tension with supports at the same height due to two concentrated forces at arbitrary positions is considered both in the linear and nonlinear cases for (1) two concentrated forces at arbitrary positions (Example 10.5); and (2) one uniform and one concentrated force at arbitrary position (Example 10.6). In the linear (nonlinear) cases, the principles of reciprocity and superposition do (do not) hold. The nonlinear deflection under constant tension and supports at the same height is considered next for a concentrated force (Section 2.6) [own weight (Sections 2.7 and 2.8)] for an elastic string, followed by own weight for an inextensible string (Section 2.9).

2.6 Nonlinear Deflection by a Concentrated Force

The problem of finding the shape of a string under arbitrary loading can be reduced to quadratures, that is, integrations (Subsection 2.6.1), both in the linear and nonlinear cases [Sections 2.3 through 2.5 (Sections 2.6 through 2.9)]. The first case of loading to be considered is a concentrated force leading to the linear (nonlinear) influence or Green function [Section 2.3 (Section 2.6)]. It specifies a triangular string deflection that is different because (1) the principles of reciprocity and superposition are (are not) satisfied [Subsection 2.3.5 (Subsection 2.6.14)]; (2) the balance of forces and moments specifying the reactions at the supports does not (does) involve [Subsections 2.3.2 through 2.3.4 (Subsections 2.6.8 through 2.6.11)] the tension tangent to the string, whose slope can (cannot) be neglected [Subsection(s) 2.3.1 (Subsections 2.6.2 through 2.6.7)]; (3) the deflection and slope and their extreme values correspond to lowest-order (exact) expansions in powers of the nonlinearity parameter [Subsection 2.3.3 (Subsection 2.6.13)]; and (4) the length of the deflected string is approximately equal to (exactly always larger than) the distance between the supports [Subsection 2.3.1 (Subsection 2.6.12)], with implications for the elastic energy. Thus, the differences between the linear and nonlinear cases, for example, for the elastic energy (Subsection 2.6.13), depend on the magnitude of the nonlinearity parameter (Subsection 2.6.14), which affects the shape of the string (Panel 2.1) through the maximum deflection and slope (Table 2.2).

2.6.1 Shape of a String under Arbitrary Loading

The exact shape of the string under possible nonuniform tension $T(x)$ and arbitrary transverse loading $F(x)$ is given from (2.10b) by (2.108b):

$$|F(x)| < T(x): \quad \zeta(x) = \zeta(0) - \int_0^x \left| \left[\frac{T(\xi)}{F(\xi)} \right]^2 - 1 \right|^{-1/2} d\xi; \tag{2.108a,b}$$

This is the solution of the differential equation of the first order (2.6); for a real solution in (2.108b), the transverse force in modulus cannot exceed the tension (2.108a) at any position along the string. The solution of the differential equation of the second order (2.8b), specifying the shape of the string for an arbitrary shear stress $f(x)$, is obtained substituting (2.7b) into (2.108b) and involves two integrations. This is analogous to the problem of finding the shape of a curve ζ in (2.109c), given the curvature (2.9b) $\equiv$ (2.109a):

$$k(x) = -\frac{f(x)}{T}, T = \text{const}: \quad \zeta(x) = \zeta(0) + \int_0^x \left| \left(\int_0^\xi k(\eta) d\eta \right)^{-2} - 1 \right|^{-1/2} d\xi, \tag{2.109a–c}$$

where a uniform tension was assumed (2.9a) $\equiv$ (2.109b). Considering the particular linear case of a small slope (2.12a) $\equiv$ (2.110b) with nonuniform tension, the shape of the string is determined (2.7b) from the transverse force (2.11b) [shear stress (2.11a)] by performing one (2.110b) [two (2.110c)] integration:

$$[F(x)]^2 \ll [T(x)]^2: \quad \zeta(x) = \zeta(0) - \int_0^x \frac{F(\xi)}{T(\xi)} d\xi, \tag{2.110a,b}$$

$$\zeta(x) = \zeta(0) - F(0) \int_0^x \frac{d\xi}{T(\xi)} - \int_0^x \frac{d\xi}{T(\xi)} \int_0^\xi f(\eta) d\eta. \tag{2.110c}$$

The linear (2.110b) case follows from the nonlinear (2.108b) case using the approximation (2.110a); substitution of (2.7b) in (2.110b) leads to (2.110c). Alternatively, (1) direct integration of (2.11b) leads to (2.110b); and (2) a first integration of (2.11c) gives

$$\int_0^x f(\xi) d\xi = -[T(x)\zeta'(x)]_0^x = T(0)\zeta'(0) - T(x)\zeta'(x); \tag{2.111a}$$

(3) solving (2.111a) for $\zeta'(x)$ and integrating once more gives (2.105c):

$$F(0) = -T(0)\zeta'(0): \quad \zeta(x) = \zeta(0) + T(0)\zeta'(0) \int_0^x \frac{d\xi}{T(\xi)} - \int_0^x \frac{d\xi}{T(\xi)} \int_0^\xi f(\eta) d\eta, \tag{2.111b,c}$$

that coincides with (2.111c) ≡ (2.110c) on account of (2.111b). If in addition to the linear approximations (2.110a) ≡ (2.112a) the tension is uniform (2.112b), then (2.110b and c) simplify to (2.112c and d):

$$\left[F(x)\right]^2 \ll T^2; T = \text{const}: \quad \zeta(x) = \zeta(0) - \frac{1}{T}\int_0^x F(\xi)\,d\xi, \tag{2.112a–c}$$

$$\zeta(x) - \zeta(0) - \zeta'(0)x = \zeta(x) - \zeta(0) + \frac{F(0)}{T}x = -\frac{1}{T}\int_0^x d\xi \int_0^\eta f(\eta)\,d\xi. \tag{2.112d}$$

In the cases when the displacement is expressed in terms of the shear stress (2.109c; 2.110c; 2.112d),there are two arbitrary constants, namely (1) the deflection $\zeta(0)$ at the origin; and (2) the curvature $k(0)$ in (2.109c) or the transverse force $F(0)$ in (2.110c) or the slope $\zeta'(0)$ in (2.112d) at the origin. In the cases when the displacement is expressed in terms of the transverse force (2.108b; 2.110b; 2.112c), there is only one integration constant, namely (1) the displacement at the origin $\zeta(0)$, and (2) the transverse force at the origin $F(0)$ appears as the second arbitrary constant if the shear stress (2.7b) is used. It has been shown that *the shape of an elastic string under tension, subject to a transverse force F (shear stress f) (1) is given by (2.108b) [with (2.7b)], exactly in the nonlinear case of a large slope, allowing also for nonuniform tension; (2) simplifies to (2.110b) [(2.110c) ≡ (2.111b and c)] in the linear case of a small slope everywhere (2.110a); and (3) simplifies further to (2.112c) [(2.112d)] if the tension is uniform (2.112b) and much larger than the transverse force (2.112a).*

2.6.2 Nonlinear Influence or Green Function for a String

The nonlinear influence or Green function specifies the exact shape of the string with a force P concentrated at an arbitrary point $x = \xi$. A **concentrated force** P causes (2.30a and b) a jump P in the transverse force (2.113a) across $x = \xi$:

$$F(x) = F(0) + PH(x - \xi), \quad f(x) = P\delta(x - \xi), \tag{2.113a,b}$$

implying that it involves (1.28a through c) a Heaviside unit jump; the corresponding (2.7a) shear stress (2.113b) involves (1.34a and b) the Dirac unit impulse. The large deflection of a string under a uniform (2.114a) tension T subject to a transverse force P concentrated at ξ specifies the nonlinear influence or Green function (2.114b),

$$T = \text{const}; \zeta(x) \equiv G(x;\xi): \quad -\frac{f(x)}{T} = -\frac{P}{T}\delta(x - \xi) = \left\{\zeta'\left[1 + \zeta'^2\right]^{-1/2}\right\}', \tag{2.114a–c}$$

as the fundamental solution of the nonlinear differential equation (2.9b) ≡ (2.114c) forced by a unit impulse. The corresponding transverse force (2.6) is (2.109b and c):

$$-F(x) = T\zeta'\left|1 + \zeta'^2\right|^{-1/2} = A - PH(x - \xi) = \begin{cases} A = -R_- & \text{if } 0 \le x < \xi, \\ A - P = R_+ & \text{if } \xi < x \le L, \end{cases} \tag{2.115a–c}$$

where (2.115b) [(2.115c)] equals the vertical component of minus the reaction (the reaction) force at the first (second) support. The slope (2.10b) corresponding to (2.115b and c) is constant (2.116b and c) and discontinuous at $x = \xi$:

$$\zeta'(x) = \begin{cases} \left| \left(\dfrac{T}{A} \right)^2 - 1 \right|^{-1/2} & \text{if } 0 \leq x < \xi, & (2.116a) \\[4mm] \left| \left(\dfrac{T}{A-P} \right)^2 - 1 \right|^{-1/2} & \text{if } \xi < x \leq L. & (2.116b) \end{cases}$$

In (2.116a,b) is used the + sign instead of − sign in (2.10b) because A has the opposite sign of the reaction force and equals the transverse force $0 < x < \xi$; this ensures a positive slope for $0 < x < \xi$, since the $y = \zeta(x)$ axis is taken downward.

Integration of (2.116a and b) shows that the shape of the string is triangular:

$$G(x;\xi) = \zeta(x) = \begin{cases} \left| \left(\dfrac{T}{A} \right)^2 - 1 \right|^{-1/2} x & \text{if } 0 \leq x \leq \xi, & (2.117a) \\[4mm] \left| \left(\dfrac{T}{A-P} \right)^2 - 1 \right|^{-1/2} (L - x) & \text{if } \xi \leq x \leq L, & (2.117b) \end{cases}$$

where the constant of integration in (2.117a) [(2.117b)] was chosen to satisfy the boundary condition that the string is attached at the same height at each end (2.39b and c); the condition of continuity (2.32c) of the Green function at the point of application of the concentrated load shows that the remaining constant A in (2.117a and b) satisfies

$$(\xi - L)^2 \left[\left(\frac{T}{A} \right)^2 - 1 \right] = \xi^2 \left[\left(\frac{T}{A-P} \right)^2 - 1 \right]. \tag{2.118}$$

Thus, *the nonlinear deflection of a string under constant tension T by a concentrated force P at ξ is specified by (2.117a and b), corresponding to the slope (2.116a and b), where the constant A is a real positive root of the quartic polynomial (2.118).* The existence of a real positive root will be proved (Subsections 2.6.6 and 2.6.7) after consideration of the maximum deflection (Subsections 2.6.3 through 2.6.5) and relates to the reactions at the supports (Subsections 2.6.8 through 2.6.11).

2.6.3 Maximum Nonlinear Deflection by a Concentrated Force

The maximum deflection occurs at the point of application of the force and is given equivalently (2.119a) [(2.119b)] by (2.117a) [or (2.117b)] at $x = \xi$:

$$\delta \equiv \left[G(x; \xi) \right]_{\max} = G(\xi; \xi) = \left| \left(\frac{T}{A} \right)^2 - 1 \right|^{-1/2} \xi = \left| \left(\frac{T}{A-P} \right)^2 - 1 \right|^{-1/2} (L - \xi), \tag{2.119a,b}$$

because A is a root of the quartic polynomial (2.118) $\equiv$ (2.120):

$$0 = \xi^2 A^2 \left[T^2 - (A - P)^2 \right] - (\xi - L)^2 (A - P)^2 (T^2 - A^2) \equiv j_4(A). \tag{2.120}$$

In the nonlinear case with a concentrated force at equal distance from the supports (2.121a), the quartic (2.120) reduces to a linear form (2.121b):

$$\xi = \frac{L}{2}: \quad 0 = \left(\frac{TL}{2}\right)^2 P(P - 2A) \equiv j_1(A); \quad -\frac{P}{2} = -A = R_- = R_+, \quad \frac{2\delta}{L} = \left[\left(\frac{2T}{P}\right)^2 - 1\right]^{1/2}, \qquad (2.121a\text{-}e)$$

whose root is one-half of the concentrated force (2.121c). Thus, *in the nonlinear deflection of an elastic string by a concentrated force in the middle, the vertical reaction forces at the supports are (1) equal (2.121d) in agreement with the symmetry of the problem; and (2) since their sum is minus the concentrated force P, and are opposite to it, their value is (2.121c). From (2.121c) follows (2.119a and b) that the maximum deflection is (2.121e).*

2.6.4 Linear Limit of a Small Slope

The linear deflection (2.3a) $\equiv$ (2.4a) corresponds to a small slope and to a tension much larger than (2.122a) the concentrated force P; for any position ξ of the latter, the tension is much larger than the vertical reaction forces at the supports that are comparable to the concentrated force (2.122b). In this case, the quartic (2.120) reduces to a quadratic (2.122c):

$$T^2 \gg P^2 \sim A^2: \quad 0 = T^2\left[(\xi - L)^2(A - P)^2 - \xi^2 A^2\right]$$

$$= T^2 L^2\left[A^2\left(1 - \frac{2\xi}{L}\right) - 2AP\left(1 - \frac{\xi}{L}\right)^2 + P^2\left(1 - \frac{\xi}{L}\right)^2\right] \equiv \zeta_2(A). \qquad (2.122a\text{-}c)$$

Since this is a linear case, the roots of (2.122c) specify the reaction (2.109b) at the first support:

$$R_- = -F(0) = -A = -P\left(1 - \frac{\xi}{L}\right)\frac{1 - \xi/L \pm \xi/L}{1 - 2\xi/L} = -P\left(1 - \frac{\xi}{L}\right)\left\{1, \frac{1}{1 - 2\xi/L}\right\}. \qquad (2.123a,b)$$

The first root coincides with (2.39d), and the second root would lead to $R_- > 0$ for $L/2 < \xi < L$. The reaction at the other support (2.115c) is specified by (2.124a):

$$R_+ = A - P = -P\frac{\xi}{L}\left\{1, \frac{-1}{1 - 2\xi/L}\right\}; \quad R_+ + R_- = -P. \qquad (2.124a\text{-}c)$$

The first root coincides with (2.39e), and the second would lead to $R_+ > 0$ for $0 < \xi < L/2$. Also, both the second roots (2.123b) and (2.124b) would be infinity for $\xi = L/2$, contradicting (2.121a through c), which agrees with the first roots (2.123a; 2.124a). The reactions at both supports must be upward, that is, negative, for all positions of the concentrated force, and this is true for the first roots (2.123a) $\equiv$ (2.39d) and (2.124a) $\equiv$ (2.39e). The second set of roots is a spurious consequence of squaring when applying the continuity of the influence function in the passage from (2.117a and b) to (2.120). This can be confirmed from (2.118) in the linear case (2.125a) when it leads to (2.125b):

$$T^2 \gg P^2, A^2: \quad \pm \xi A = (\xi - L)(A - P), \quad A = P\left(1 - \frac{\xi}{L}\right)\left\{1, \frac{1}{1 - 2\xi/L}\right\}, \qquad (2.125a\text{-}c)$$

that is equivalent to (2.123a and b) $\equiv$ (2.125c).

2.6.5 Quartic Equation for the Nonlinear Unsymmetric Elastic Deflection

The first (2.123a) ≡ (2.126a) [second (2.124a) ≡ (2.127a)] root is a root of the quartic polynomial (2.120) only if the condition (2.126b) [(2.127b)] is met:

$$A = P\frac{\xi}{L}: \quad 0 = j_4\left(P\frac{\xi}{L}\right) = P^2 L (L - 2\xi)\left[T^2\left(1 - \frac{2\xi}{L} + 2\frac{\xi^2}{L^2}\right) - P^2 \frac{\xi^2}{L^2}\left(1 - \frac{\xi}{L}\right)^2 \right], \tag{2.126a,b}$$

$$A = P\left(1 - \frac{\xi}{L}\right): \quad 0 = j_4\left(P\left(1 - \frac{\xi}{L}\right)\right) = P^4 \xi^2\left(1 - \frac{\xi}{L}\right)^2\left(1 - \frac{2\xi}{L}\right). \tag{2.127a,b}$$

The polynomials (2.126b) [(2.127b)] are evaluated substituting (2.126a) [(2.127a)] in (2.120), leading to (2.128) ≡ (2.126b) [(2.129) ≡ (2.127b)]:

$$j_4\left(P\frac{\xi}{L}\right) = \left(\frac{P\xi^2}{L}\right)^2\left[T^2 - P^2\left(1 - \frac{\xi}{L}\right)^2 \right] - (\xi - L)^2 P^2\left(1 - \frac{\xi}{L}\right)^2\left(T^2 - P^2 \frac{\xi^2}{L^2}\right)$$

$$= P^2 T^2 L^2\left[\left(\frac{\xi}{L}\right)^4 - \left(1 - \frac{\xi}{L}\right)^4 \right] - P^4 \frac{\xi^2}{L^2}\left(1 - \frac{\xi}{L}\right)^2\left[\frac{\xi^2}{L^2} - \left(1 - \frac{\xi}{L}\right)^2 \right]$$

$$= P^2 L^2\left[\frac{\xi^2}{L^2} - \left(1 - \frac{\xi}{L}\right)^2 \right]\left\{ T^2\left[\frac{\xi^2}{L^2} + \left(1 - \frac{\xi}{L}\right)^2 \right] - P^2 \frac{\xi^2}{L^2}\left(1 - \frac{\xi}{L}\right)^2 \right\}$$

$$= P^2 L (2\xi - L)\left[T^2\left(1 - \frac{2\xi}{L} + 2\frac{\xi^2}{L^2}\right) - P^2 \frac{\xi^2}{L^2}\left(1 - \frac{\xi}{L}\right)^2 \right], \tag{2.128}$$

$$j_4\left(P\left(1 - \frac{\xi}{L}\right)\right) = P^2 \xi^2\left(1 - \frac{\xi}{L}\right)^2\left(T^2 - P^2 \frac{\xi^2}{L^2}\right) - (\xi - L)^2 P^2 \frac{\xi^2}{L^2}\left[T^2 - P^2\left(1 - \frac{\xi}{L}\right)^2 \right]$$

$$= P^4 \xi^2\left(1 - \frac{\xi}{L}\right)^2\left[\left(1 - \frac{\xi}{L}\right)^2 - \frac{\xi^2}{L^2} \right] = P^4 \xi^2\left(1 - \frac{\xi}{L}\right)^2\left(1 - \frac{2\xi}{L}\right). \tag{2.129}$$

The term in square brackets in (2.128) is positive:

$$T^2 > P^2, 0 < \xi < L: \quad 1 - 2\frac{\xi}{L} + 2\frac{\xi^2}{L^2} = \left(1 - \frac{\xi}{L}\right)^2 + \frac{\xi^2}{L^2} > \left(1 - \frac{\xi}{L}\right)^2 > \frac{\xi^2}{L^2}\left(1 - \frac{\xi}{L}\right)^2. \tag{2.130a,b}$$

Thus, the polynomials (2.128) [(2.129)] vanish (2.126b) ≡ (2.131a) [(2.127b) ≡ (2.131c)] only if the concentrated force is at equal distance from the supports (2.131b):

$$j_4\left(P\frac{\xi}{L}\right) = 0 \Rightarrow \xi = \frac{L}{2} \Leftarrow j_4\left(P\left(1 - \frac{\xi}{L}\right)\right); \tag{2.131a-c}$$

in this case (2.131b) ≡ (2.121a) the vertical reaction forces at the supports are equal (2.121b through d). *If the condition (2.131b) is (is not) met, the reactions at the supports are (are not) specified (2.121b and c) by* $R_- = -A$ *in (2.126a)* ≡ *(2.39d) and* $R_+ = -A$ *in (2.127a)* ≡ *(2.39e) for the linear (nonlinear) deflection of an elastic string with a concentrated force P at position* ξ.

2.6.6 Existence of Real Roots of the Quartic Polynomial

In the general case of nonlinear deflection by a concentrated force at unequal distances from the supports, the constant A is a root of the quartic polynomial (2.120) $\equiv$ (2.132a) with coefficients (2.132b and c):

$$0 = j_4(A) = \sum_{n=0}^{4} e_n A^n: \quad e_{4,3,2} = L(L - 2\xi)\{1, -2P, P^2 - T^2\} \tag{2.132a–c}$$

$$e_{1,0} = (L - \xi)^2 T^2 P\{2, -P\}.$$

The nonlinear case of a concentrated force at equal distance from the supports has already been considered (2.121a through e), so only the nonlinear case of unequal distances is considered next, for which (1) the reaction forces at the supports cannot be equal, because (2.126b; 2.127b) do not hold for $L \neq 2\xi$; and (2) the polynomial (2.120) $\equiv$ (2.132a through c) $\equiv$ (2.133) does not reduce to the second degree and remains of the fourth degree:

$$0 = j_4(A) = L(L - 2\xi)A^2\left(A^2 - 2AP + P^2 - T^2\right) + (L - \xi)^2 T^2 P(2A - P). \tag{2.133}$$

The polynomial (2.133) has opposite signs (2.134b) [(2.134d)] for $A = \pm\infty$ ($A = P/2$),

$$T > P: \quad j_4(\pm\infty) = \infty \operatorname{sgn}(L - 2\xi), \quad j_4(0) = -T^2 P^2 (L - \xi)^2 < 0, \tag{2.134a–c}$$

$$j_4\left(\frac{P}{2}\right) = (L - 2\xi)\frac{P^2 L}{4}\left(\frac{P^2}{4} - T^2\right) = -L|L - 2\xi|\frac{P^2}{4}\left(T^2 - \frac{P^2}{4}\right)\operatorname{sgn}(L - 2\xi), \tag{2.134d}$$

and is negative at the origin (2.134c). The opposite signs at $A = \pm\infty$ and $A = P/2$ imply that there are at least two real roots, that is, (1) either two real and two complex conjugate roots or (2) four real roots:

$$j_4(A) = L(L - 2\xi)(A - A_1)(A - A_2)\left(A^2 + 2aA + b\right). \tag{2.135}$$

The roots of (2.135) are considered next (Subsection 2.6.7) for arbitrary positions of the concentrated force.

2.6.7 Inequalities for the Reaction Forces at the Supports

If the concentrated force is applied closer to the first support (2.136a), then (2.134b and c) imply (2.136b and c), and one of the roots (2.136d) has a value between $+\infty$ and $P/2$:

$$L > 2\xi: \quad j_4(\infty) > 0 > j_4\left(\frac{P}{2}\right) \Rightarrow \frac{P}{2} < A_1 = -R_-. \tag{2.136a–c}$$

Thus for a concentrated force closer to the first support (2.136a) $\equiv$ (2.137a), the reaction force in modulus (2.136c) $\equiv$ (2.137b) is larger than half the concentrated force:

$$L > 2\xi: \quad |R_-| = -R_- > \frac{P}{2}; \quad |R_+| = -R_+ = P + R_- = P - |R_-| < P - \frac{P}{2} = \frac{P}{2}; \tag{2.137a–c}$$

the reaction at the second support satisfies (2.115c), implying that its modulus (2.137c) is less than half the concentrated force. The result (2.137a through c) $\equiv$ (2.138a and b),

$$\xi < \frac{L}{2} \Rightarrow |R_-| > \frac{P}{2} > |R_+|, \quad \xi > \frac{L}{2} \Rightarrow |R_+| > \frac{P}{2} > |R_-|, \tag{2.138a–d}$$

would be reversed (2.138d) if the concentrated force was applied closer to the second support (2.138c). The case intermediate between (2.138a and b) and (2.138c and d) is a concentrated force at equal distance from the supports (2.121a through e). It has been shown *that the nonlinear deflection of an elastic string under uniform tension by a concentrated force P at any position ξ leads to reaction forces at the first R_ and second R_+ supports specified by (2.115b and c) ≡ (2.144a and b), where A is a root of the quartic polynomial (2.120) ≡ (2.132a through c) = (2.133). There is at least one real root, implying that (1) if the concentrated force is closer to the first support (2.138a), the reaction force at that (the other) support is larger (smaller) in modulus than half the concentrated force (2.138b); (2) vice versa (2.138d) if the concentrated force is closer to the second support (2.138c); and (3) the inequalities in (2.138b) and (2.138d) are replaced by equalities for a concentrated force at equal distance from the two supports (2.121a through c).* The preceding analysis supplies only the vertical components of the reaction forces at the supports, because the equilibrium equation (2.8b) results from a balance of vertical forces (2.2b); the horizontal components of the reaction forces at the supports follow from the horizontal force balance (2.2a). Both the horizontal and vertical components of the reaction forces at the supports follow from the balance of forces and moments (Subsection 2.6.8), which also serves as a check on the preceding results (Subsections 2.6.2 through 2.6.7).

2.6.8 Balance of the Tension and Reaction Forces

The balance of forces (Figure 2.18) in the nonlinear case of a nonnegligible slope for a concentrated force at unequal distances from the supports consists of the external applied force plus (1/2) the vertical (horizontal) components of the reaction force at the first $R_-(S_-)$ and second $R_+(S_+)$ support; and (3) the tension along the string, which makes an angle $\alpha(\beta)$ with the horizontal before (after) the point of application of the force, satisfying (2.139a through c) [(2.140a through c)]:

$$\tan\alpha = \frac{\delta}{\xi}, \quad \{\cos\alpha, \sin\alpha\} = \frac{\{\xi, \delta\}}{\sqrt{\xi^2 + \delta^2}}, \qquad (2.139a\text{–}c)$$

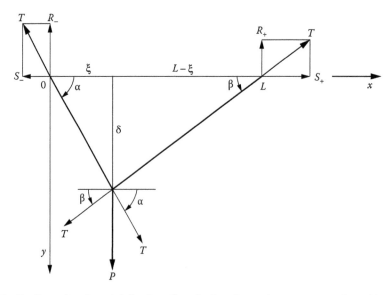

FIGURE 2.18   The linear (nonlinear) deflection of an elastic string under constant tension with supports at the same height has [Figure 2.11 (Figure 2.18)]: (1) small (nonnegligible) slope on both sides (at least one side) of the concentrated force; (2) the concentrated force is much smaller (not much smaller but no larger) than the tangential tension; and (3) maximum deflection much smaller (not much smaller) than the distance between the supports.

$$\tan\beta = \frac{\delta}{L-\xi}, \quad \{\cos\beta, \sin\beta\} = \frac{\{L-\xi, \delta\}}{\sqrt{(L-\xi)^2 + \delta^2}}, \tag{2.140a–c}$$

where δ is the maximum deflection (2.119a) $\equiv$ (2.119b). The horizontal projection of the tension along the string specifies the horizontal reaction force at the first (2.141a) [second (2.141b)] support:

$$\{S_-, S_+\} = T\{-\cos\alpha, \cos\beta\} = T\left\{-\left|\xi^2 + \delta^2\right|^{-1/2}\xi, \left|(L-\xi)^2 + \delta^2\right|^{-1/2}(L-\xi)\right\}; \tag{2.141a,b}$$

the vertical projection of the tension along the string specifies the vertical reaction force at the first (2.142a) [second (2.142b)] support:

$$\{R_-, R_+\} = -T\{\sin\alpha, \sin\beta\} = -T\delta\left\{\left|\xi^2 + \delta^2\right|^{-1/2}, \left|(L-\xi)^2 + \delta^2\right|^{-1/2}\right\}. \tag{2.142a,b}$$

The vertical force balance at the point of application of the concentrated force leads to

$$P = T(\sin\alpha + \sin\beta) = T\delta\left\{\left|\delta^2 + \xi^2\right|^{-1/2} + \left|\delta^2 + (L-\xi)^2\right|^{-1/2}\right\} = -R_- - R_+, \tag{2.143}$$

which is minus the sum of (2.142a and b).

2.6.9 Linear Deflection or Symmetric Case

Solving (2.142a) [(2.142b)] for the maximum deflection δ leads to (2.144c) [(2.144d)]:

$$R_- = -A, \; R_+ = A - P: \quad \left|\left(\frac{T}{R_-}\right)^2 - 1\right|^{-1/2}\xi = \delta = \left|\left(\frac{T}{R_+}\right)^2 - 1\right|^{-1/2}(L-\xi). \tag{2.144a–d}$$

The comparison of (2.144a) $\equiv$ (2.119a) [(2.144b) $\equiv$ (2.119b)] agrees with the vertical reaction forces at the first (2.144a) $\equiv$ (2.115b) [second (2.144b) $\equiv$ (2.115c)] support:

$$\left[\left(\frac{T}{R_+}\right)^2 - 1\right]\xi^2 = \left[\left(\frac{T}{R_-}\right)^2 - 1\right](L-\xi)^2. \tag{2.145}$$

The balance of the moments relative to the supports of the external concentrated force P and of the reactions at the supports involves only the vertical components of the latter in the case of supports at the same height (2.39b and c). Substituting (2.39d and e) in (2.145) leads to (2.146a):

$$\left(\frac{T}{P}L\right)^2 - \xi^2 = \left(\frac{T}{P}L\right)^2 - (L-\xi)^2 \quad \text{if } \xi = \frac{L}{2} \text{ or } T^2 \gg P^2, \tag{2.146a–c}$$

which is satisfied when (1) either the concentrated force is applied at equal distance from the supports (2.146b), corresponding to (2.121a through e) or (2) both sides of (2.146a) are large, that is, in the linear case (2.10d) of tension much larger than the transverse force (2.146c), corresponding to (2.39d and e). This confirms that the reaction forces for the nonlinear deflection of an elastic string by a concentrated force at unequal distances from the supports are not given by (2.39d and c) because, although the force balance (2.115c) holds, the moment balance (2.38b and c) is not valid (Subsection 2.6.10).

2.6.10 Balance of Forces and Moments

Outside the cases (1) and (2), that is, (3) for the nonlinear deflection by a concentrated force at unequal distances from the supports, (2.146a) is not satisfied, implying that the moments of the external force and reactions at the supports do not balance, as shown by (2.126a and b; 2.127a and b) not being satisfied for $\xi \neq 2L$. The reason is that the tension along the string also gives contribution to the moment of internal forces at the supports; (a) in the cases of (1) a small slope or (2) symmetric loading, the moments of internal forces balance, and the balance of moments holds only with the external forces and reactions at the supports; and (b) in the case of (3) nonlinear deflection and unsymmetric loading, the balance of the moment of the concentrated force and vertical reaction forces relative to the second (2.147a) [first (2.147b)] support specify the balance of the moments of the internal forces:

$$N_- = -P(L-\xi) - R_- L = -P(L-\xi) + TL\left|\xi^2 + \delta^2\right|^{-1/2}\delta, \tag{2.147a}$$

$$N_+ = P\xi + R_+ L = P\xi - TL\left|(L-\xi)^2 + \delta^2\right|^{-1/2}L. \tag{2.147b}$$

Thus, *the deflection of an elastic string (1) satisfies the balance of external and reaction forces at the supports in all cases, including linear or nonlinear deflections and symmetric or unsymmetric loadings, because the internal forces due to the tension always balance; and (2) may not satisfy the balance of the moments of the external forces and reactions at the supports unless the moments of the internal forces associated with the tension cancel. The balance of moments of the external and reaction forces holds only if the moments of the internal tension cancel, that is, for two cases: (a) linear deflection with a concentrated force at any position, that is, an unsymmetric loading; and (b) a concentrated force at equal distance from the supports, that is, a symmetric loading, even in the nonlinear case. The nonlinear deflection of an elastic string under a constant tension T with supports at the same height and distance L, by a concentrated transverse force P applied at a distance ξ from the first support (Figure 2.18) leads to (1) the horizontal (vertical) components of the reaction force at the first (2.141a) [(2.142a)] and second (2.141b) [(2.142b)] supports; (2) minus the vertical reaction force at the first support (2.144a) is a root of the quartic polynomial (2.120) $\equiv$ (2.133), and the vertical reaction force at the second support (2.144b) adds to the first to balance the external concentrated transverse force (2.143); and (3) in the preceding vertical reaction forces (1) and (2) appear in the maximum deflection at the point of application of the concentrated force (2.119a) $\equiv$ (2.119b) that specifies the angles of the string with the horizontal before (2.139a through c) [after (2.140a through c)] the point of application of the concentrated force.*

2.6.11 Vertical/Horizontal Reaction Forces at the Supports

If the concentrated force is applied at an equal distance from the supports (2.148a) $\equiv$ (2.121a), then (2.121c) $\equiv$ (2.148b) implies that (1) the maximum deflection (2.119a) $\equiv$ (2.119b) is (2.148c); (2) the angles of the string with the horizontal (2.139a through c; 2.140a through c) are equal (2.141d through f):

$$\xi = \frac{L}{2}: \quad A = \frac{P}{2}, \quad \delta = \left|\left(\frac{2T}{P}\right)^2 - 1\right|^{-1/2}\frac{L}{2}, \quad \tan\alpha = \frac{2\delta}{L} = \tan\beta,$$

$$\{\cos\alpha, \sin\alpha\} = \{\cos\beta, \sin\beta\} = \left|\frac{L^2}{4} + \delta^2\right|^{-1/2}\left\{\frac{L}{2}, \delta\right\}; \tag{2.148a–f}$$

(3) the horizontal (vertical) components of the reaction force at the two supports are

$$S_\pm = \pm T \cos\alpha = \pm \frac{TL}{2}\left|\delta^2 + \frac{L^2}{4}\right|^{-1/2}, \quad -\frac{P}{2} = R_+ = R_- = -T\sin\alpha = -T\delta\left|\delta^2 + \frac{L^2}{4}\right|^{-1/2}, \quad (2.149a,b)$$

opposite (2.149a) [equal (2.149b)].

In the linear case (2.150a) ≡ (2.122a) ≡ (2.150b), (1) the maximum deflection (2.119a) ≡ (2.119b) is (2.150c and d):

$$\delta^2 \ll \xi^2, (L-\xi)^2; T^2 \gg P^2, R_\pm^2: \quad \xi\frac{R_-}{T} = -\delta = \frac{R_+}{T}(L-\xi); \quad (2.150a-d)$$

$$\tan\alpha = \frac{\delta}{\xi} = \sin\alpha, \quad \tan\beta = \frac{\delta}{L-\xi} = \sin\beta, \quad \cos\alpha = 1 = \cos\beta; \quad (2.151a-f)$$

(2) the angles are small (2.151a through f); and (3) the reaction forces are given by (2.152a through d):

$$T^2 \gg P^2: \quad R_- = -T\frac{\delta}{\xi}, \quad R_+ = -T\frac{\delta}{L-\xi}, \quad S_\pm = \mp T. \quad (2.152a-d)$$

The vertical reaction forces are given by (2.53d) ≡ (2.153a) [(2.39e) ≡ (2.153b)] in the linear case:

$$R_- = -P\left(1 - \frac{\xi}{L}\right), \quad R_+ = -P\frac{\xi}{L}; \quad P(L-\xi)\xi = TL\delta, \quad (2.153a-c)$$

the comparison of (2.153a) ≡ (2.152b) and (2.153b) ≡ (2.152c) implies (2.153c).

In the linear case with a concentrated force at equal distance from the supports, the combination of (2.150a through d; 2.151a through f; 2.152a through d; 2.153a through c) and (2.148a through f; 2.149a and b) leads to

$$P^2 \ll T^2: \quad R_\pm = -\frac{P}{2}, \quad S_\pm = \mp T, \quad \frac{\delta}{L} = \frac{P}{2T}, \quad (2.154a-d)$$

$$\delta^2 \ll L^2: \quad \sin\alpha = \tan\alpha = \frac{2\delta}{L} = \frac{P}{T} = \tan\beta = \sin\beta, \quad \cos\alpha = \cos\beta. \quad (2.155a-g)$$

The relative maximum deflection (2.154d) is the nonlinearity parameter that in the nonlinear case specifies [Subsection 2.6.12 (Subsection 2.6.13)] the maximum deflection and slope (the extension of the string and the associated elastic energy).

2.6.12 Nonlinear Extension and Deflection

In the nonlinear case, the maximum deflection (2.119a and b) is not negligible compared with the distance L between the supports, and thus, the length ℓ of the stretched string is larger, corresponding to the deflection into a triangle with vertex at (ξ, δ):

$$\ell = \left|\xi^2 + \delta^2\right|^{1/2} + \left|(L-\xi)^2 + \delta^2\right|^{1/2} = \left|1 - \left(\frac{A}{T}\right)^2\right|^{-1/2}\xi + \left|1 - \left(\frac{A-P}{T}\right)^2\right|^{-1/2}(L-\xi). \quad (2.156a,b)$$

In the linear case (2.157a), the length of the string is unchanged because both terms in the square roots in (2.156b) are approximately unity, leading to (2.157b):

$$\delta^2 \ll \xi^2, (L-\xi)^2: \quad \ell \sim L; \quad \xi = \frac{L}{2}: \quad \ell = \left|1 - \left(\frac{P}{2T}\right)^2\right|^{-1/2} L. \tag{2.157a-d}$$

in the nonlinear case, with a concentrated force at equal distance from the supports (2.121a) ≡ (2.157c), and hence equal reactions (2.121c), the length (2.156b) of the deflected string is given by (2.157d). The latter may be expanded (I.25.37a through c) in a binomial series (2.158a and b):

$$\ell = L\left|1 - \left(\frac{P}{2T}\right)^2\right|^{-1/2} = L\sum_{n=0}^{\infty}\binom{-1/2}{n}\left(-\frac{P^2}{4T^2}\right)^n = L\sum_{n=0}^{\infty} a_n \left(\frac{P}{2T}\right)^{2n}, \tag{2.158a,b}$$

with coefficients (II.6.55b) ≡ (II.6.56b) ≡ (II.6.57a through g) ≡ (2.159b) valid for (II.6.56a) ≡ (2.159a):

$$n \geq 1: \quad a_n \equiv (-)^n \binom{-1/2}{n} = \frac{(-)^n}{n!}\left(-\frac{1}{2}\right)\left(-\frac{3}{2}\right)\cdots\left(-\frac{1}{2}-n+1\right)$$

$$= \frac{1}{n!}\frac{1}{2}\frac{3}{2}\cdots\left(n-\frac{1}{2}\right) = \frac{1.3\cdots(2n-1)}{n!2^n} = \frac{(2n-1)!!}{(2n)!!}. \tag{2.159a,b}$$

Substitution of (2.159b) in (2.158b) yields *the **relative extension** of the string between the deflected (undeflected) state with length ℓ(L)*:

$$\xi = \frac{L}{2}: \quad \frac{\ell}{L} - 1 = \sum_{n=1}^{\infty} 2^{-3n}\frac{(2n-1)!!}{n!}\left(\frac{P}{T}\right)^{2n} = \frac{P^2}{8T^2}\left[1 + \frac{3P^2}{16T^2} + O\left(\frac{P^4}{T^4}\right)\right], \tag{2.160a,b}$$

showing that (1) there is no extension in the linear case, that is, to O(P/T); and (2/3) the extension is a nonlinear effect that appears at O(P²/T²) and at all increasing orders. The maximum deflection (2.119a) ≡ (2.161b) in the same case (2.121a) ≡ (2.160a) ≡ (2.161a),

$$\xi = \frac{L}{2}: \quad \delta = \left|\left(\frac{2T}{P}\right)^2 - 1\right|^{-1/2}\frac{L}{2} = \frac{LP}{4T}\left|1 - \left(\frac{P}{2T}\right)^2\right|^{-1/2} = \frac{P\ell}{4T}, \tag{2.161a,b}$$

*appears at one order lower; thus, the **relative deflection**, defined by the ratio of the maximum deflection to the distance between the supports,*

$$\frac{\delta}{L} = \frac{P}{4T}\left[1 + \sum_{n=1}^{\infty} 2^{-3n}\frac{(2n-1)!!}{n!}\left(\frac{P}{T}\right)^{2n+1}\right] = \frac{P}{4T}\left[1 + \frac{P^2}{8T^2} + \frac{3P^4}{128T^4} + O\left(\frac{P^6}{T^6}\right)\right], \tag{2.162a,b}$$

consists of (1) a linear effect (2.45e) corresponding to the first term on the r.h.s. of (2.162b); and (2) the remaining terms on the r.h.s. of (2.162b) specify a nonlinear effect similar to (2.160b).

2.6.13 Elastic Energy and Slope

From (2.161b) follows a simple result: *the ratio of the maximum deflection to the length of the deflected string equals the ratio of the concentrated force to four times the tension along the string, both in the linear and nonlinear cases (2.163a):*

$$\frac{\delta}{\ell} = \frac{P}{4T}: \quad E_d = T(\ell - L) = T\left(\frac{4T\delta}{P} - L\right) = \frac{PL}{4\delta}(\ell - L). \qquad (2.163\text{a-c})$$

the elastic energy (2.163b) equals (2.17) the uniform tension times the extension of the string and involves the maximum deflection (2.161a and b); if the concentrated force is applied at equal distance from the supports (2.164a and b), the elastic energy (2.163c) simplifies (2.158a) ≡ (2.160a through 2.164c):

$$\xi = \frac{L}{2}, A = \frac{P}{2}: \quad E_d = TL\left\{\left|1 - \left(\frac{P}{2T}\right)^2\right|^{-1/2} - 1\right\} = TL\sum_{n=1}^{\infty}(-)^n\binom{-1/2}{n}\left(\frac{P}{2T}\right)^{2n}$$

$$= TL\sum_{n=1}^{\infty}\left[\frac{(2n-1)!!}{n!}\right]2^{-3n}\left(\frac{P}{T}\right)^{2n} = \frac{P^2L}{8T}\left[1 + \frac{3P^2}{16T^2} + O\left(\frac{P^4}{T^4}\right)\right]. \qquad (2.164\text{a-c})$$

The nonlinear correction factor is the same,

$$q \equiv \frac{P}{2T}: \quad 1 + \frac{E_d}{TL} = \frac{\ell}{L} = 4\frac{T\delta}{PL} = \frac{2\delta}{qL} = \frac{\tan\theta}{q}$$

$$= \left|1 - \left(\frac{P}{2T}\right)^2\right|^{-1/2} = \left|1 - q^2\right|^{-1/2} = \sum_{n=0}^{\infty}(-)^n\binom{-1/2}{n}q^{2n}$$

$$= 1 + \sum_{n=1}^{\infty}\frac{(2n-1)!!}{(2n)!!}q^{2n} = 1 + \frac{q^2}{2} + \frac{3}{8}q^4 + O(q^6), \qquad (2.165\text{a,b})$$

for (1) the elastic energy (2.164c); (2) the relative deflection (2.161a and b) ≡ (2.166c):

$$\xi = \frac{L}{2}, A = \frac{P}{2}: \quad \frac{2\delta}{L} = \left|\left(\frac{2T}{P}\right)^2 - 1\right|^{-1/2} = \tan\theta = \zeta'(0) = -\zeta'(L) = |\zeta'(x)|; \qquad (2.166\text{a-d})$$

and (3) the constant slope (2.166d). The nonlinear effects are accounted for by a nonlinearity parameter (2.165a) similar to (2.70a), replacing the weight by the concentrated load (2.67a). The elastic energy in the linear approximation, that is, the first-term of (2.164c), is proportional to the square of the concentrated force and inversely proportional to the tension. If the tension was infinite, there would be no deflection, and the elastic energy would be zero, because the relative extension scales on the inverse square of the tension (2.160b) and dominates the factor T in the elastic energy (2.163b).

2.6.14 Effect of the Nonlinearity Parameter

Table 2.2 indicates, for 11 equally spaced values of the nonlinearity parameter (2.165a) over its full range (2.167a),

$$0 \le q \equiv \frac{P}{2T} \le 1: \quad \frac{\delta}{L} = \frac{1}{2}\frac{q}{\sqrt{1-q^2}} = \frac{1}{2}\tan\theta, \quad \frac{\ell}{L} - 1 = \frac{1}{\sqrt{1-q^2}} - 1 = \frac{E_d}{TL}, \qquad (2.167\text{a-e})$$

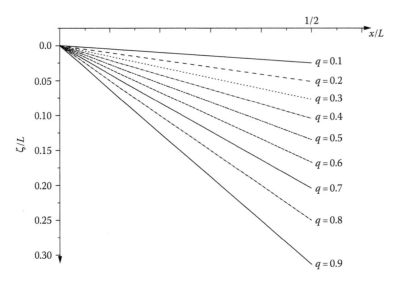

PANEL 2.1 The nonlinear deflection of an elastic string by a concentrated force applied at equal distance from the supports leads to an isosceles triangular shape for which the slope of the sides increases with the value of the nonlinearity parameter. In Panels 2.1 through 2.5, it is assumed: (1) uniform tension tangent to the string; and (2) supports at the same height.

TABLE 2.2 Nonlinear Deflection of an Elastic String by a Concentrated Force

Nonlinearity Parameter	Maximum Deflection	Slope (Angle)	Relative Extension	Normalized Elastic Energy
$q = \dfrac{P}{2T}$	$\dfrac{\delta}{L}$	$\tan\theta$	$\dfrac{\ell}{L}$	$\dfrac{E_d}{TL}$
(2.167a)	(2.167b)	(2.167c)	(2.167d)	(2.167e)
0	0	0	1.0000	0.0000
0.1	0.0525	0.1005 (5.74°)	1.0050	0.0050
0.2	0.1021	0.2041 (11.54°)	1.0206	0.0206
0.3	0.1572	0.3145 (17.46°)	1.0483	0.0483
0.4	0.2182	0.4364 (23.58°)	1.0911	0.0911
0.5	0.2887	0.5774 (30.00°)	1.1547	0.1547
0.6	0.3750	0.7500 (36.87°)	1.2500	0.2500
0.7	0.4901	0.9802 (44.43°)	1.4003	0.4003
0.8	0.6667	1.3333 (53.13°)	1.6667	0.6667
0.9	1.0324	2.0647 (64.16°)	2.2941	1.2941
1.0	∞	∞ (90.00°)	∞	∞

Notes: L, distance between the supports at the same height, equal to the undeflected length of the string; ℓ, length of the string when deflected; δ, maximum deflection at the point of application of the concentrated force; θ, maximum angle-of-inclination in modulus at the supports; *P*, concentrated force applied at equal distance from the supports; *T*, uniform tension along the string; E_d, elastic energy; *q*, nonlinearity parameter.

The nonlinear deflection of an elastic string due to a concentrated force at equal distance from the supports is considered for 11 values of the nonlinearity parameter over its full domain, namely, the unit interval. The shape of the string for the nine intermediate values is plotted in Panel 2.1.

the values of the following quantities: (1) the relative maximum deflection (2.161b) ≡ (2.167b); (2) the constant slope in modulus (2.166d) ≡ (2.167c); (3) the relative extension (2.157d) ≡ (2.167d); and (4) the normalized elastic energy (2.164b) ≡ (2.167e). The former (latter) two (2.167b and c) [(2.167d and e)] scale like the (the square of the) nonlinearity parameter (2.167a) when it is small. This is also visible from (Panel 2.1) the shape of the string as an isosceles triangle (2.268b and c):

$$\zeta(x) = \frac{2\delta}{L} \times \begin{cases} x & \text{if } 0 \le x \le \dfrac{L}{2}, \\ (L-x) & \text{if } \dfrac{L}{2} \le x \le L, \end{cases} \tag{2.168a,b}$$

so that only one half is shown for nine values of the nonlinearity parameter (2.167a). In all cases was considered a concentrated force at equal distance from the supports at the same height. The deflection for the same height by a concentrated force not much smaller than the tension specifies the nonlinear influence or Green function (2.117a and b), which does not satisfy (1) the reciprocity principle (2.47b) because in (2.117a and b), x and ξ are not interchangeable; and (2) the principle of superposition; for example, the nonlinear deflection due to two concentrated loads (Example 10.5) does not equal the sum of the deflections due to each of them. The reciprocity and superposition principles remain valid for a linear deflection with nonuniform tension (Example 10.7), but again fail in the nonlinear case. Thus, neither the principle of reciprocity nor the principle of superposition hold in the remaining cases concerning nonlinear deflection of a heavy elastic (inextensible) string [Sections 2.7 and 2.8 (Section 2.9)].

2.7 Large Deflection by a Uniform Load

The simplest continuous load is uniform and applies to a homogeneous string with constant cross section in the linear case (Section 2.4) when the length and mass are constant. In the nonlinear case, a uniform load corresponds to a constant mass per unit horizontal distance; for a homogeneous string with constant mass density per unit volume, this implies that the cross section reduces as the string is stretched (Subsection 2.7.1). The deflection and slope (Subsection 2.7.2) and their maximum values (Subsection 2.7.3) depend on the nonlinearity parameter (Subsection 2.7.5), that also appears in the extension and elastic energy (Subsection 2.7.6). The latter four appear (Subsection 2.7.7) in Table 2.3 and Panel 2.2, illustrating the shape of the string, that is, a circular arc (Subsection 2.7.8).

2.7.1 Nonlinear Deflection by Own Weight

The shear stress in a gravity field g due to the weight is proportional to the mass per unit horizontal projection of the string (2.169b):

$$\rho = \frac{dm}{ds}: \quad f(x) = g\frac{dm}{dx} = g\frac{dm}{ds}\frac{ds}{dx} = \rho g\left|1+\zeta'^2\right|^{1/2}; \tag{2.169a,b}$$

this relates to the mass per unit length of the string (2.169a) through (2.1b). Thus, several cases arise for a **homogeneous** string, that is, with constant **mass density** per unit volume (2.170a), so that the mass density per unit length (2.170b) is the product by the cross section:

$$\nu = \frac{dm}{dV}: \quad \rho = \frac{dm}{ds} = \frac{dm}{dV}\frac{dV}{ds} = \nu S(x). \tag{2.170a,b}$$

The simplest case I is the linear deflection (2.171a) with constant cross section (2.171b), leading to a constant mass per unit length (2.170b):

$$\zeta'^2 \ll 1, S(x) = \text{const} = S(0), g = \text{const}: \quad f = \rho g = vgS = \text{const}. \tag{2.171a–d}$$

In a uniform gravity field (2.171c), the shear stress, which equals (2.63c) the weight per unit length, is constant (2.171d). The uniform loading (2.171d) can be maintained (2.172b) in the case II of the nonlinear deflection of a homogeneous string (2.172a) only if the cross-sectional area varies inversely with the extension (2.172c), so that (Section 2.7) the total mass of the string remains constant (2.172d):

$$v \equiv \frac{dm}{dV} = \text{const}: \quad f = \rho g = vgS(x)\sqrt{1+\zeta'^2} = \text{const}, \tag{2.172a,b}$$

$$S(x)\sqrt{1+\zeta'^2} = \text{const}: \quad m = \int_0^L \frac{dm}{dx}dx = v\int_0^L S(x)\sqrt{1+\zeta'^2}dx = \text{const}. \tag{2.172c,d}$$

Another case III is the nonlinear deflection when the homogeneous string (2.173a) retains a constant cross section (2.173b) and hence a constant mass density per unit length (2.173c), and the shear stress (2.169b) increases with the slope, and the total mass increases with the deflected length (2.173d):

$$v \equiv \frac{dm}{dV} = \text{const}, \quad S(x) = \text{const}; \quad \rho = \text{const}, \quad m = vS\int_0^L \sqrt{1+\zeta'^2}dx = \rho\ell \neq \text{const}. \tag{2.173a–d}$$

In the last case III, the string must slide through at least one of the supports to allow the total mass to increase in proportion to the total length for nonlinear deflection (Section 2.8). The case II (III) is considered next (after) in Section 2.7 (Section 2.6).

2.7.2 Nonlinear Deflection with Constant Shear Stress

In the case of constant shear stress (2.174a), it equals the total transverse force P divided by the undeflected length L; it corresponds to the transverse force (2.174d) with the integration (2.7a) from the midposition where the slope (2.173b) and transverse force (2.174c) are zero:

$$f(x) = \text{const} = \frac{P}{L}; \quad \zeta'\left(\frac{L}{2}\right) = 0 = F\left(\frac{L}{2}\right); \quad F(x) = \int_{L/2}^x f(\xi)d\xi = P\left(\frac{x}{L}-\frac{1}{2}\right). \tag{2.174a–d}$$

Substitution of (2.174d) in (2.10b) $\equiv$ (2.175c), where the deflection is zero at the first support (2.175a), specifies the shape of the string (2.175c) under uniform tension (2.175b):

$$\zeta(0) = 0; T = \text{const}: \quad \zeta(x) = -\int_0^x \left|\left[\frac{T}{F(\xi)}\right]^2 - 1\right|^{-1/2} d\xi = -\frac{1}{T}\int_0^x F(\xi)\left|1 - \left[\frac{F(\xi)}{T}\right]^2\right|^{-1/2} d\xi$$

$$= \frac{P}{2T}\int_0^x \left(1 - \frac{2\xi}{L}\right)\left|1 - \left(\frac{P}{2T}\right)^2\left(1 - \frac{2\xi}{L}\right)^2\right|^{-1/2} d\xi. \tag{2.175a–c}$$

The deflection at an arbitrary position is given by

$$\frac{P}{TL}\zeta(x) = \left\{\left|1 - \left(\frac{P}{2T}\right)^2\left(1 - \frac{2\xi}{L}\right)^2\right|^{1/2}\right\}_0^x = \left|1 - \left(\frac{P}{2T}\right)^2\left(1 - \frac{2x}{L}\right)^2\right|^{1/2} - \left|1 - \left(\frac{P}{2T}\right)^2\right|^{1/2}. \tag{2.176}$$

The shape of the string is symmetric relative to the middle $x = L/2$, where the maximum deflection occurs:

$$\delta \equiv \left| \zeta \right|_{\max} = \zeta\left(\frac{L}{2}\right) = L\frac{T}{P}\left\{ 1 - \left| 1 - \left(\frac{P}{2T}\right)^2 \right|^{1/2} \right\}. \tag{2.177}$$

The slope is given (2.175c; 2.174c) by (2.178):

$$\operatorname{sgn}\left(\frac{L}{2} - x\right)\zeta'(x) = -\left| \left[\frac{T}{F(x)}\right]^2 - 1 \right|^{-1/2} = -\frac{F(x)}{T}\left| 1 - \left[\frac{T(x)}{T}\right]^2 \right|^{-1/2}$$

$$= \frac{P}{2T}\left(1 - \frac{2x}{L} \right)\left| 1 - \left(\frac{P}{2T}\right)^2 \left(1 - \frac{2x}{L} \right)^2 \right|^{-1/2}, \tag{2.178}$$

with opposite signs (1.89a and b) before + and after − the mid-point. The maximum slope in modulus

$$\tan\theta = \left| \zeta' \right|_{\max} = \zeta'(0) = -\zeta'(L) = \frac{P}{2T}\left| 1 - \left(\frac{P}{2T}\right)^2 \right|^{-1/2}, \tag{2.179}$$

occurs at the supports.

2.7.3 Exact and First-Order Nonlinear Corrections

The linear approximation (2.70a) corresponds to the case where the total weight of the string is much less than twice the tension. In general, the linear approximation (2.10c) requires the transverse force to be small everywhere small compared with the tension (2.10d); the maximum transverse force (2.174d) for the string under constant shear stress (2.174a) is one-half of the total transverse force $F_{\max} = P/2$. Thus, the linear results should follow from the exact ones in the limit of small $P/(2T)$; expanding to higher powers of $(P/2T)^2$, the nonlinear corrections can be obtained from the exact formulas (2.176–2.179). The latter give imaginary values if $P > 2T$, showing that equilibrium is possible only if the weight does not exceed twice the tension. For $P = 2T$, the angle of the string at the support (2.179) would be $\theta = \pi/2$, implying that the tension has to balance the reaction at the supports, which equals one-half of the total transverse force, and this is possible only if the string is vertical. Thus, $P < 2T$, implying that (1) the maximum deflection (2.177) [slope (2.179)] can be expanded in a (I.25.37a through c) binomial series leading to (2.180) [(2.181)]:

$$\frac{\delta}{L} = -\frac{T}{P}\sum_{n=1}^{\infty}\binom{1/2}{n}\left(-\frac{P^2}{4T^2} \right)^n = \frac{P}{8T} - \frac{T}{P}\sum_{n=2}^{\infty}b_n\left(\frac{P}{2T}\right)^{2n}$$

$$= \frac{P}{8T} + \frac{1}{2}\sum_{n=2}^{\infty}\frac{(2n-3)!!}{(2n)!!}\left(\frac{P}{2T}\right)^{2n-1}, \tag{2.180}$$

$$\tan\theta = \frac{P}{2T}\sum_{n=0}^{\infty}\binom{-1/2}{n}\left(-\frac{P^2}{4T^2} \right)^n = \frac{P}{2T} + \sum_{n=1}^{\infty}a_n\left(\frac{P}{2T}\right)^{2n+1}$$

$$= \frac{P}{2T} + \sum_{n=1}^{\infty}a_n\frac{(2n-1)!!}{(2n)!!}\left(\frac{P}{2T}\right)^{2n+1}, \tag{2.181}$$

where the following are used: (1) the coefficients (2.159a and b) in (2.181); (2) in (2.180) appear the coefficients (2.182b) ≡ (II.6.62b) ≡ (II.6.63b) ≡ (II.6.64a through f) valid for (II.6.63a) ≡ (2.182a):

$$n \geq 2: \quad b_n = (-)^n \binom{1/2}{n} = \frac{(-)^n}{n!} \frac{1}{2} \left(-\frac{1}{2} \right) \left(-\frac{3}{2} \right) \cdots \left(\frac{1}{2} - n + 1 \right) = -\frac{1}{n!} \frac{1}{2} \frac{1}{2} \frac{1}{2} \frac{3}{2} \left(n - \frac{3}{2} \right)$$

$$= -\frac{1}{n!} \frac{1 \cdot 3 \cdots (2n-3)}{2^n} = -\frac{(2n-3)!!}{n!2^n} = -\frac{(2n-3)!!}{(2n)!!}. \tag{2.182a,b}$$

If the total transverse force is small compared with the tension,

$$\frac{\delta}{L} = \frac{P}{8T} \left\{ 1 + \frac{P^2}{16T^2} + O\left(\frac{P^4}{T^4} \right) \right\}, \quad \tan\theta = \frac{P}{2T} \left\{ 1 + \frac{P^2}{8T^2} + O\left(\frac{P^4}{T^4} \right) \right\}, \tag{2.183a,b}$$

the first term in the deflection (2.183a) [slope (2.183b)] coincides with the linear approximation (2.72a) [(2.72b)]; and (3) the second term in (2.183a) [(2.183b)] specifies the first nonlinear correction, which increases to the maximum deflection (slope) relative to the linear prediction.

2.7.4 Comparison of Linear and Nonlinear Deflections

It has been shown that *a string under uniform loading, for example, a string of constant mass per unit length (2.170b) under uniform tension (2.175b) in a uniform gravity field (2.171c) with acceleration g, corresponds to a uniform shear stress (2.174a) equal to the total weight or transverse force P divided by the distance between the supports and (1) can support a weight less than half the tension (2.184a); (2) when suspended from two supports at equal height (2.83b and c) and distance L, the exact shape is (2.176) and can be expanded in powers (2.184b) using two binomial series (I.25.37a through c) with coefficients (2.182a and b):*

$$P < 2T: \quad \zeta(x) = L\frac{T}{P} \sum_{n=1}^{\infty} \binom{1/2}{n} \left(-\frac{P^2}{4T^2} \right)^n \left[\left(\frac{2x}{L} - 1 \right)^{2n} - 1 \right]$$

$$= -\frac{PL}{8T} \left[\left(\frac{2x}{L} - 1 \right)^2 - 1 \right] + \frac{L}{2} \sum_{n=2}^{\infty} b_n \left(\frac{P}{2T} \right)^{2n-1} \left[\left(\frac{2x}{L} - 1 \right)^{2n} - 1 \right]$$

$$= \frac{P}{2TL} x(L-x) + \frac{L}{2} \sum_{n=2}^{\infty} \frac{(2n-3)!!}{(2n)!!} \left(\frac{P}{2T} \right)^{2n-1} \left[1 - \left(\frac{2x}{L} - 1 \right)^{2n} \right]; \tag{2.184a,b}$$

(3) for a small weight compared with the tension, that is, to the third order, which is the next higher than linear, the deflection is

$$\zeta(x) = \frac{P}{2TL} x \left\{ L - x - \left(\frac{P}{2TL} \right)^2 \left[\left(x^3 - 2x^2L + \frac{3}{2}xL^2 - \frac{L^3}{2} \right) \right] + O\left(\frac{P}{2T} \right)^4 \right\}; \tag{2.185}$$

(4) the linear approximation (2.71a) is the first term of (2.185); (5) the slope (2.178) can also be expanded (2.186) in a binomial series (I.25.37a through c) with coefficients (2.159a and b).

$$\text{sgn}\left(\frac{L}{2}-x\right)\zeta'(x)=\frac{P}{2T}\left(1-\frac{2x}{L}\right)\sum_{n=0}^{\infty}\binom{-1/2}{n}\left[-\left(\frac{P}{2T}\right)^2\left(\frac{2x}{L}-1\right)^2\right]^n$$

$$=\frac{P}{2T}\left(1-\frac{2x}{L}\right)+\sum_{n=1}^{\infty}\frac{(2n-1)!!}{(2n)!!}\left[\frac{P}{2T}\left(1-\frac{2x}{L}\right)\right]^{2n+1}\;; \tag{2.186}$$

(6) the lowest-order terms are

$$\text{sgn}\left(\frac{L}{2}-x\right)\zeta'(x)=\frac{P}{T}\left(\frac{1}{2}-\frac{x}{L}\right)\left\{1+\frac{P^2}{2T^2}\left(\frac{1}{2}-\frac{x}{L}\right)^2+O\left(\left[\frac{P}{T}\left(\frac{1}{2}-\frac{x}{L}\right)\right]^4\right)\right\}; \tag{2.187}$$

and (7/8) from the deflection (2.184b; 2.185) [slope (2.186; 2.187)] at all points follow the corresponding expressions for the maximum deflection at the mid-position (2.177) ≡ (2.180) [maximum of the modulus of the slope at the supports (2.179) ≡ (2.181)].

2.7.5 Deflection, Slope, and Extrema

The **nonlinearity parameter** can be defined as the ratio of the total transverse force of the string to twice the tension and takes values (2.188a) in the unit interval corresponding to (2.70a) ≡ (2.184a). *The slope (2.178) ≡ (2.188b) ≡ (2.186) ≡ (2.187) [shape (2.176) ≡ (2.184b) ≡ (2.185) ≡ (2.188c)] of a homogeneous string with constant mass (2.172c) with uniform tension (2.175b) under a uniform load (2.174a) is specified by*

$$0\le q\equiv\frac{P}{2T}<1:\;\;\text{sgn}\left(\frac{L}{2}-x\right)\zeta'(x)=q\left(1-\frac{2x}{L}\right)\left|1-q^2\left(1-\frac{2x}{L}\right)^2\right|^{-1/2}$$

$$=q\left(1-\frac{2x}{L}\right)+\sum_{n=1}^{\infty}\frac{(2n-1)!!}{(2n)!!}q^{2n+1}\left(1-\frac{2x}{L}\right)^{2n+1}$$

$$=q\left(1-\frac{2x}{L}\right)\left[1+\frac{q^2}{2}\left(1-\frac{2x}{L}\right)^2+O\left(q^4\right)\right]. \tag{2.188a,b}$$

$$\frac{\zeta(x)}{L}=\frac{1}{2q}\left\{\left|1-q^2\left(1-\frac{2x}{L}\right)^2\right|^{1/2}-\left|1-q^2\right|^{1/2}\right\}$$

$$=q\frac{x}{L}\left(1-\frac{x}{L}\right)+\frac{1}{2}\sum_{n=2}^{\infty}\frac{(2n-3)!!}{(2n)!!}q^{2n-1}\left[1-\left(1-\frac{2x}{L}\right)^{2n}\right]$$

$$=q\frac{x}{L}\left\{1-\frac{x}{L}-q^2\left[\left(\frac{x}{L}\right)^3-2\left(\frac{x}{L}\right)^2+\frac{3x}{2L}-\frac{1}{2}\right]+O\left(q^4\right)\right\}, \tag{2.188c}$$

The nonlinearity parameter (2.188a) appears linearly in the first term of (2.188c) [(2.188b)], corresponding to (1) a linear deflection (2.71a) [slope (2.71b)] when the total transverse force is small compared with twice the tension (2.67b); (2) the remaining terms apply when the total transverse force is smaller but not much smaller than twice the tension (2.188a) and appear as a series of powers of the square of the nonlinearity parameter. Since the shape of the string is symmetric, the maximum deflection (2.189) [slope (2.190)] in modulus occurs at the mid-position (supports),

$$\frac{\delta}{L} \equiv \frac{1}{L}\left|\zeta(x)\right|_{max} = \frac{1}{L}\zeta\left(\frac{L}{2}\right) = \frac{1}{2q}\left[1-\left|1-q^2\right|^{1/2}\right] = \frac{1}{2q}\sum_{n=1}^{\infty}(-)^{n-1}\binom{1/2}{n}q^{2n}$$

$$= \frac{q}{4} + \frac{1}{2}\sum_{n=2}^{\infty}\frac{(2n-3)!!}{(2n)!!}q^{2n-1} = \frac{q}{4}\left[1+\frac{q^2}{4}+O\left(q^4\right)\right], \tag{2.189}$$

$$\tan\theta = \left|\zeta'(x)\right|_{max} = \zeta'(0) = -\zeta'(L) = q\left|1-q^2\right|^{-1/2} = \sum_{n=0}^{\infty}(-)^n\binom{-1/2}{n}q^{2n+1}$$

$$= q + \sum_{n=1}^{\infty}\frac{(2n-1)!!}{(2n)!!}q^{2n+1} = q\left[1+\frac{q^2}{2}+O\left(q^4\right)\right], \tag{2.190}$$

in agreement (2.188a) with (2.180) ≡ (2.183a) ≡ (2.189) [(2.181) ≡ (2.183b) ≡ (2.190)] in the nonlinear case and (2.72a) [(2.72b)] in the linear case.

2.7.6 Elastic Energy and Extension

The elastic energy (2.17) ≡ (2.163b) ≡ (2.191a) is the product of the tension by the extension of the string; the latter is the difference between the length of the deflected string ℓ and undeflected string L. The former is given (2.1b) in the nonlinear case by (2.191b):

$$E_d = T(\ell - L): \quad \ell = \int_0^L\left|1+\zeta'^2\right|^{1/2}\,dx = \int_0^L\left|1-\left[\frac{F(\xi)}{T}\right]^2\right|^{-1/2}\,d\xi, \tag{2.191a–c}$$

where (2.10b) is substituted, leading to (2.191c) in terms of the transverse force. In the case of uniform loading (2.174d) with a total force P in (2.174a), the extension (2.191c) is given by (2.192b), suggesting the change of variable (2.192a):

$$\sin\psi = \frac{P}{T}\left(\frac{\xi}{L}-\frac{1}{2}\right): \quad \ell = \int_0^L\left|1-\left(\frac{P}{T}\right)^2\left(\frac{\xi}{L}-\frac{1}{2}\right)^2\right|^{-1/2}\,d\xi. \tag{2.192a,b}$$

From (2.192a) follows:

$$\left|1-\left(\frac{P}{T}\right)^2\left(\frac{\xi}{L}-\frac{1}{2}\right)^2\right|^{-1/2} = \left|1-\sin^2\psi\right|^{-1/2} = \sec\psi, \quad d\xi = L\frac{T}{P}\cos\psi\,d\psi, \tag{2.193a,b}$$

which, when substituted in (2.192b), yields:

$$\frac{P\ell}{TL} = \int_{\arg\sin(-P/2T)}^{\arg\sin(P/2T)}d\psi = 2\arg\sin\left(\frac{P}{2T}\right) = 2\arg\sin q. \tag{2.194}$$

Thus, *for the nonlinear deflection of an elastic string with uniform loading (2.174a), under constant tension (2.175b), with supports (2.64b and c) at the same height at a distance L, the length under deflection in specified by (2.194) ≡ (2.195a) which involves the nonlinearity parameter (2.188a):*

$$\frac{\ell}{L} = \frac{2T}{P}\arc\sin\left(\frac{2P}{T}\right) = \frac{1}{q}\arc\sin q = 1 + \sum_{n=1}^{\infty}\frac{(2n-1)!!}{(2n)!!}\frac{q^{2n}}{2n+1} = 1 + \frac{q^2}{6} + \frac{3q^4}{40} + O\left(q^6\right); \tag{2.195a,b}$$

The power series for the arc sin (II.7.167b) was used to show that the extension appears to lowest-order $O(q^2)$; the lowest nonlinear correction is included in (2.195b) as well as in the elastic energy (2.191a).

2.7.7 Effect of the Nonlinearity Parameter

Table 2.3 indicates, for 11 equally spaced values of the nonlinearity parameter in its range (2.196a),

$$0 \le q = \frac{P}{2T} \le 1: \quad \frac{\delta}{L} = \frac{1}{2q}\left(1 - \sqrt{1-q^2}\right), \quad \tan\theta = \frac{q}{\sqrt{1-q^2}}, \quad \frac{\ell}{L} = \frac{1}{q}\text{arc}\sin q = \frac{E_d}{TL} + 1, \quad (2.196a\text{–e})$$

the following quantities: (1/2) the maximum relative deflection (2.189) $\equiv$ (2.196b) [slope in modulus (2.190) $\equiv$ (2.196c)]; and (3/4) the relative extension (2.195b) $\equiv$ (2.196d) [elastic energy (2.191a) $\equiv$ (2.196e)]. The conditions concerning Tables 2.2 (2.3) are (1) the same as regards the uniform tension (2.114a) $\equiv$ (2.175b) and supports at the same height (2.64b and c) and distance L; and (2) differ in that the total transverse force is concentrated at equal distance from (uniformly distributed between) the supports. The shapes of the deflected string for nine equally spaced values of the nonlinearity parameter in Panel 2.2 are all circular arcs, because *a uniform shear stress with constant tension leads (2.9b) to a constant curvature or radius of curvature (2.197a):*

$$-k = -\frac{1}{R} = \frac{f}{T} = \frac{P}{TL} = \frac{2q}{L}; \quad \frac{2q}{L}\zeta + \left|1-q^2\right|^{1/2} = \left|1 - q^2\left(\frac{2x}{L} - 1\right)^2\right|^{1/2}. \quad (2.197a,b)$$

This can be confirmed from the exact shape of the string (2.188c) $\equiv$ (2.197b) $\equiv$ (2.198) in the equivalent form:

$$0 = \left(\zeta + \frac{L}{2q}\sqrt{1-q^2}\right)^2 - \left(\frac{L}{2q}\right)^2\left[1 - q^2\left(\frac{2x}{L} - 1\right)^2\right]$$

$$= \left(\zeta + \frac{L}{2}\sqrt{\frac{1}{q^2} - 1}\right)^2 + \left(x - \frac{L}{2}\right)^2 - \left(\frac{L}{2q}\right)^2 = \left(\zeta - y_0\right)^2 + \left(x - x_0\right)^2 - R^2. \quad (2.198)$$

This is a circle (2.198) $\equiv$ (2.199a) with center at (2.199b and c) and radius (2.199d). In (2.198), the radius of curvature could be $R = \pm L/(2q)$. The minus sign in (2.197a) implies (2.9b) that $\zeta'' \le 0$, that is, the slope decreases with increasing x because (Figure 2.19) the circular arc is on the downward side of the circle, and the Cartesian reference frame (x, y) is left-handed. The radius of the circle $a = |R| = L/(2q)$ and is unaffected by the orientation of the Cartesian reference frame.

2.7.8 Shape of the String as a Circular Arc

It has been shown that *an elastic string under uniform tension (2.175b) with nonlinear deflection by a uniform shear stress (2.174a) has a constant curvature (2.197a), corresponding to a circular arc with radius (2.199d); in the case of supports at the same height (2.64b and c),*

$$\left(x - x_0\right)^2 + \left(\zeta - y_0\right)^2 = a^2: \quad \{x_0, y_0\} = \frac{L}{2}\left\{1, -\sqrt{\frac{1}{q^2} - 1}\right\}, \quad a \equiv |R| = \left|\left(x_0\right)^2 + \left(y_0\right)^2\right|^{1/2} = \frac{L}{2q}. \quad (2.199a\text{–d})$$

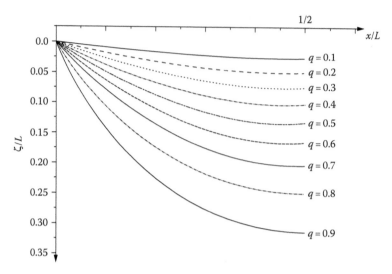

PANEL 2.2   The nonlinear deflection of a string by a uniform load, that is a constant shear stress leads to a shape that is a circular arc, with center at equal distance from the supports (Figure 2.18) and radius of curvature equal to the half-distance from the supports divided by the nonlinearity parameter. Thus the increase in curvature as the nonlinearity parameter increases. This case includes the nonlinear deflection of a string by its own weight, if the cross section reduces as the string stretches so as to keep constant the horizontal projection of the weight.

TABLE 2.3   Nonlinear Deflection of an Elastic String by a Uniform Shear Stress

Nonlinearity Parameter	Maximum Deflection	Maximum Slope (Angle)	Relative Extension	Normalized Elastic Energy
$q = \dfrac{P}{2T}$	$\dfrac{\delta}{L}$	$\tan\theta$	$\dfrac{\ell}{L}$	$\dfrac{E_d}{TL}$
(2.196a)	(2.196b)	(2.196c)	(2.196d)	(2.196e)
0	0	0	1	0
0.1	0.0251	0.1005 (5.74°)	1.0017	0.0017
0.2	0.0505	0.2041 (11.54°)	1.0068	0.0068
0.3	0.0768	0.3145 (17.46°)	1.0156	0.0156
0.4	0.1044	0.4364 (23.58°)	1.0288	0.0288
0.5	0.1340	0.5774 (30.00°)	1.0472	0.0472
0.6	0.1667	0.7500 (36.87°)	1.0725	0.0725
0.7	0.2042	0.9802 (44.43°)	1.1077	0.1077
0.8	0.2500	1.3333 (53.13°)	1.1591	0.1591
0.9	0.3134	2.0647 (64.16°)	1.2442	0.2442
1.0	∞	∞ (90.00°)	1.5708	0.5708

Notes: L, distance between the supports at the same height, equal to the undeflected length of the string; ℓ, length of the string when deflected; δ, maximum deflection at equal distance from the support; θ, maximum inclination in modulus at the supports; P, total transverse force; P/L, uniform shear stress; T, uniform tension along the string; E_d, elastic energy; q, nonlinearity parameter.

     As for Table 2.2 with the nonlinear deflection of the elastic string (Panel 2.2) due to a uniform shear stress.

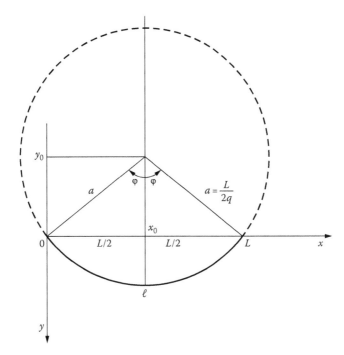

FIGURE 2.19 The nonlinear deflection of a string under uniform tension T leads to a radius of curvature $-R = T/f$ equal to minus its ratio by the transverse force. The two simplest cases are: (1) a zero transverse force $f = 0$, or infinite radius of curvature $R = \infty$ corresponding to a straight undeflected string; and (2) a constant shear stress leading to a constant curvature and thus a shape of the deflection string that is a circular arc. This applies to the nonlinear deflection of a heavy string with constant total mass provided that as it stretches the cross section reduces in such a way that the horizontal projection of the mass per unit length is a constant. For this or any other case with constant shear stress, the total transverse force is $P = fL$, and the radius of curvature of the string equals $R = TL/P = L/2q$ where q is the nonlinearity parameter. If the string has supports at the same height and positions $x = 0, L$, its deflected shape is symmetric relative to the middle, and the center of the circular arc must lie on the line $x = L/2$. This together with the radius of curvature $R = L/2q$ determines the height of the center of the circle $h = (L/2)\sqrt{q^{-2}-1}$.

the circle (2.199a) with radius (2.199d) has (Figure 2.19) a center on the line at equal distance from the supports (2.199b); this implies that the distance from the center of the circle, that is, the deflected shape of the string, to the straight line, that is, the undeflected shape of the string, satisfies (2.199c) ≡ (2.200a):

$$-y_0 = \left|a^2 - x_0^2\right|^{1/2} = \left|\left(\frac{L}{2q}\right)^2 - \left(\frac{L}{2}\right)^2\right|^{1/2} = \frac{L}{2}\left|\frac{1}{q^2}-1\right|^{1/2}. \tag{2.200a}$$

From Figure 2.19, it also follows that the length of the deflected string is

$$\ell = 2a\,\varphi = 2a \arcsin\left(\frac{L}{2a}\right) = \frac{L}{q}\arcsin q, \tag{2.200b}$$

where (2.199d) is used; this geometric result is consistent (2.200b) ≡ (2.195a) with the analytic method (Subsection 2.7.6). The problem could be solved geometrically as follows: (1) for a constant shear stress, the shape of the string is a circle of radius (2.197a) ≡ (2.201a):

$$a = |R| = \frac{1}{|k|} = \frac{T}{f} = \frac{TL}{P}, \quad x_0 = \frac{L}{2}: \quad -y_0 = \left|a^2 - (x_0)^2\right|^{1/2} = \left|\left(\frac{TL}{P}\right)^2 - \frac{L^2}{4}\right|^{1/2} = \frac{L}{2}\left|\left(\frac{2T}{P}\right)^2 - 1\right|^{1/2}; \quad (2.201\text{a–c})$$

(2) since the circle must pass through the suspension points at $x = 0$ and $x = L$, the center must be on the bisector (2.201b); (3) it follows that the distance of the center from the undeflected position of the string is (2.201c) ≡ (2.200a); and (4) substituting (2.201a through c) in the equation of the circle (2.199a) leads to the shape of the string (2.198) ≡ (2.197b) ≡ (2.188c), from which follow all other results. The principle of superposition can (cannot) be applied to the linear (nonlinear) deflection of an elastic string by a combination (Example 10.6) of a concentrated force (Section 2.6) and a uniformly distributed load (Section 2.7).

2.8 Nonlinear Deflection of a Heavy Elastic String

The nonlinear deflection of a heavy elastic string is considered in the case of constant mass density per unit volume (2.170a) and constant cross section (2.171b) so that the mass per unit length (2.170b) is constant; this implies that in a uniform gravity field, the shear stress (2.169b) increases with the slope. The equilibrium equation is written in dimensionless coordinates with the origin at the apex (Subsection 2.8.1), which simplifies the integration (Subsection 2.8.2) to determine (1) the slope and deflection (Subsection 2.8.3); (2) their maximum values (Subsection 2.8.4); and (3) the nonlinear correction to the linear approximation (Subsection 2.8.5). The change from apex coordinates to support coordinates involves (Subsection 2.8.1) a translation of the Cartesian coordinate frame from the lowest point to the first support and involves the maximum deflection (Subsection 2.8.6). Both the deflection and slope can be specified either in apex or support coordinates (Subsection 2.8.7). As before (Sections 2.6 and 2.7), the nonlinearity parameter affects (Subsection 2.8.8) all of the preceding quantities (Table 2.4) and the shape of the string (Panel 2.3).

2.8.1 Support and Dimensional/Dimensionless Apex Coordinates

The equilibrium equation (2.8b) specifying the shape or deflection of an elastic string has been solved (Sections 2.3 through 2.7) in **support coordinates**, that is, using a Cartesian coordinate system (Figure 2.20) with (1) the origin at the first or left-hand support; (2) x-coordinate along the undeflected position of the string; and (3) y-coordinate vertical downward so that the coordinate system is left-handed. This leads, for example, for the linear elastic string under its own weight, to the deflection at all points (2.64f) and its maximum value (2.65a), implying (2.202):

$$\zeta(x) - \delta = \frac{\rho g}{2T}\left(xL - x^2 - \frac{L^2}{4}\right) = -\frac{\rho g}{2T}\left(x - \frac{L}{2}\right)^2. \tag{2.202}$$

Choosing the coordinates (2.203a and b) leads to a centered parabola (2.203c) with slope (2.203d):

$$Y \equiv \frac{\rho g}{T}(\delta - y) \equiv \frac{\rho g \eta}{T}, \quad X \equiv \frac{\rho g}{T}\left(x - \frac{L}{2}\right) \equiv \frac{\rho g \xi}{T}: \quad Y = \frac{1}{2}X^2 \equiv Z(X), \quad Z' = \frac{dZ}{dX} = X. \tag{2.203\text{a–d}}$$

The first dimensionless (dimensional) apex coordinates $(X, Y)[(\xi, \eta)]$ are related (Figure 2.20) to the *support coordinates* (x, y) *by (2.203a and b), implying that (1) the origin is at the apex or lowest point; and*

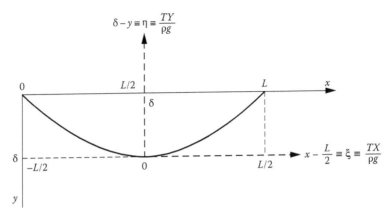

FIGURE 2.20 Considering the nonlinear deflection of a string by its own weight, if the mass density per unit length remains constant as it is deflected, the total mass increases as it stretches and the shear stress is no longer constant, that is it increases with the slope. Thus the shape of the string is no longer a circular arc for the same conditions of uniform tension and supports at the same height. The shape of the string in this case is conveniently determined using the first dimensionless apex coordinates with: (1) origin at the lowest point or apex tangent to the apex hence horizontal $\xi = x - L/2$; (2) η-axis upward so that $\eta = \delta - y$ where δ is the maximum deflection; and (3) both (ξ, η) are made dimensionless multiplying by $T/\rho g$ that is the tension divided by the weight per unit length.

(2) the $OX \equiv O\xi (OY \equiv O\eta)$ *axis is tangent, hence horizontal (vertical upward).* For symmetric loading (2.204a) and supports at the same height (2.204b and c),

$$f(\eta) = f(L - \eta); \zeta(0) = 0 = \zeta(L); Y = Z(X): \quad Z(0) = 0 = Z'(0), \qquad (2.204a\text{-}f)$$

the shape of the string (2.204d) satisfies the simple boundary conditions (2.204e and f) in the first dimensionless apex coordinates.

2.8.2 Solution of a Nonlinear Ordinary Differential Equation

Assuming that the mass per unit length of the string is constant (2.170b) ≡ (2.205a) in a uniform gravity field (2.205b), the shear stress (2.169b) leads (2.9b) to the equilibrium equation (2.205d) for uniform tension (2.205c):

$$\rho, g, T = \text{const}: \quad -\frac{\rho g}{T} = \frac{\zeta''}{\left(1 + \zeta'^2\right)^2}. \qquad (2.205a\text{-}d)$$

Changing to the dimensionless apex coordinates (2.203a and b) simplifies (2.206a and b) the balance equation (2.205d through 2.206c):

$$Z' \equiv \frac{dY}{dX} = -\frac{dy}{dx} \equiv -\zeta', Z'' \equiv \frac{d^2Z}{dX^2} = \frac{dZ'}{dX} = -\frac{dx}{dX}\frac{d\zeta'}{dx} = -\frac{T}{\rho g}\zeta'': \quad \left(1 + Z'^2\right)^2 = Z''. \qquad (2.206a\text{-}c)$$

The latter (2.206c) ≡ (2.207a) is integrated once, using the boundary conditions (2.204e and f) leading to (2.207b):

$$\left(1 + Z'^2\right)^2 = \frac{dZ'}{dX} = \frac{dZ'}{dZ}\frac{dZ}{dX} = Z'\frac{dZ'}{dZ}, \qquad (2.207a)$$

$$2Z = \int_0^{Z'} \frac{2\xi d\xi}{(1+\xi)^2} = \left[-\frac{1}{1+\xi^2}\right]_0^{Z'} = 1 - \frac{1}{1+Z'^2} = \frac{Z'^2}{1+Z'^2} = \frac{1}{1+\frac{1}{Z'^2}}. \tag{2.207b}$$

Solving (2.207b) $\equiv$ (2.208a) for Z', the sign function $\text{sgn}(X)$ is used (1.86a through c) in (2.208a) because the slope (Figure 2.20) is positive (negative) for $x > L/2$ ($x < L/2$), that is, for $X > 0$ ($X < 0$) in (2.203b). This leads to the modulus function (1.89a and b) in (2.208b):

$$Z'(X) = \frac{dZ}{dX} = \left|\frac{2Z}{1-2Z}\right|^{1/2} \text{sgn}(X), \quad |X| = X\,\text{sgn}(X) = \int_0^Z \sqrt{\frac{1-2t}{2t}}\,dt. \tag{2.208a,b}$$

The integration of (2.208a) leads to (2.208b), which suggests the change of variable:

$$\sin^2\psi = 2t: \quad \sqrt{1-2t} = \cos\psi, \quad dt = \sin\psi\cos\psi\,d\psi. \tag{2.209a–c}$$

Using (2.209a through c) transforms the integral (2.208b through 2.210b):

$$\sin u = \sqrt{2Z}: \quad |X| = \int_0^u \cos^2\psi\,d\psi = \frac{1}{2}\int_0^u \left[1 + \cos(2\psi)\right]d\psi = \frac{u}{2} + \frac{\sin(2u)}{4} = \frac{u + \cos u \sin u}{2}, \tag{2.210a,b}$$

with the upper limit (2.210a). Substitution of (2.210a) in (2.210b) leads to

$$2|X| = \arg\sin\left(\sqrt{2Z}\right) + \sqrt{2Z}\sqrt{1-2Z}, \tag{2.211}$$

as the solution (2.211) of the nonlinear second-order ordinary differential equation (2.206c) with boundary conditions (2.204e and f).

2.8.3 Nonlinear Deflection under Own Weight

The first seven of the coefficients (II.6.56a and b) $\equiv$ (2.159a and b) [(II.6.63a and b) $\equiv$ (2.182a and b)] are (II.6.12a through g) $\equiv$ (2.204a through g) [(II.6.64a through f) $\equiv$ (2.213a through g)]

$$n = 0,\ldots,6: \quad a_n = (-)^n\binom{-1/2}{n} = \left\{1, \frac{1}{2}, \frac{3}{8}, \frac{5}{16}, \frac{35}{128}, \frac{63}{256}, \frac{231}{1024}\right\}, \tag{2.212a–g}$$

$$n = 0,\ldots,6: \quad -b_n \equiv (-)^{n-1}\binom{1/2}{n} = \left\{-1, \frac{1}{2}, \frac{1}{8}, \frac{1}{16}, \frac{5}{128}, \frac{7}{256}, \frac{21}{1024}\right\}. \tag{2.213a–g}$$

The shape of the string $Z(X)$ is specified by (2.211) in inverse form, and using the binomial series (I.25.37a through c) and power series for the inverse sine (II.7.167c) leads to (2.214b):

$$|2Z| < 1: \quad 2|X| = \sqrt{2Z}\left[1 + \sum_{n=1}^{\infty}\frac{(2n-1)!!}{(2n)!!}\frac{(2Z)^n}{2n+1} + \sum_{n=1}^{\infty}\binom{1/2}{n}(-2Z)^n\right]$$

$$= \sqrt{2Z}\left[2 + \sum_{n=1}^{\infty}(b_n + c_n)(2Z)^n\right], \tag{2.214a,b}$$

which is valid for (2.214a) and involves the coefficients (2.182a and b) $\equiv$ (2.213a through g) and (2.215a through c; 2.159a and b):

$$n \geq 1: \quad c_n \equiv \frac{(2n-1)!!}{(2n)!!} \frac{1}{2n+1} = \frac{a_n}{2n+1}; \qquad (2.215\text{a--c})$$

The first seven values (2.215a through g) are (2.216a through g):

$$n = 0,\ldots,6: \quad c_n = \left\{1, \frac{1}{6}, \frac{3}{40}, \frac{5}{112}, \frac{35}{1152}, \frac{63}{2816}, \frac{231}{13312}\right\}. \qquad (2.216\text{a--g})$$

The nonlinear deflection of a homogeneous (2.172a) elastic string with uniform cross section (2.171b), under constant tension (2.205c) in a uniform gravity field (2.205b), with supports at the same height (2.64b and c), is given Z(X) in inverse form X(Z) by (2.211) $\equiv$ *(2.214a and b; 2.213a through g; 2.216a through g) in dimensionless coordinates (2.203a and b) with the origin at the apex, that is, apex coordinates.* In order to change to support coordinates, it is necessary to determine the maximum deflection.

2.8.4 Maximum Deflection and Apex/Support Coordinates

The supports (2.217a) correspond (2.203b) to (2.217b) the maximum (2.217c) deflection (2.203a):

$$x = 0, L: \quad X = \mp \frac{\rho g L}{2T} = \mp \frac{P}{2T} = \mp q, \quad Z_{\max} = \frac{\rho g \delta}{T} = 2q \frac{\delta}{L} \equiv \Delta, \qquad (2.217\text{a--c})$$

where was introduced the nonlinearity parameter (2.165a) corresponding to the weight of the undeflected string (2.67a). The maximum deflection is a root of (2.211) $\equiv$ (2.214b) for the values (2.217b and c):

$$4q^2 = \left(\arcsin\sqrt{2\Delta} + \sqrt{2\Delta}\sqrt{1-2\Delta}\right)^2 = 2\Delta \left[\sum_{n=0}^{\infty} (b_n + c_n)(2\Delta)^n\right]^2. \qquad (2.218\text{a,b})$$

Substituting (2.203a and b) in (2.211) $\equiv$ (2.214b) specifies *the shape of the string $\zeta(x)$ in support coordinates in inverse form $x(\zeta)$ by*

$$x = \frac{L}{2} + \frac{TX}{\rho g} = \frac{L}{2} + \frac{T|X|}{\rho g}\operatorname{sgn}(X)$$

$$= \frac{L}{2} + \operatorname{sgn}(X)\frac{T}{2\rho g}\left\{\arcsin\left[\sqrt{\frac{2\rho g}{T}(\delta-\zeta)}\right] + \sqrt{\frac{2\rho g}{T}(\delta-\zeta)}\sqrt{1 - \frac{2\rho g}{T}(\delta-\zeta)}\right\}$$

$$= \frac{L}{2} + \operatorname{sgn}(X)\sqrt{\frac{T}{2\rho g}(\delta-\zeta)}\left\{1 + \frac{1}{2}\sum_{n=1}^{\infty}(b_n + c_n)\left[\sqrt{\frac{2\rho g}{T}(\delta-\zeta)}\right]^n\right\} \qquad (2.219\text{a,b})$$

(2.219a) $\equiv$ (2.219b), *the latter involving the coefficients (2.182a and b; 2.213a through g) and (2.215a and b; 2.216a through g).* The maximum deflection at the apex is a root of (2.220a) $\equiv$ (2.220b):

$$\frac{\rho g L}{T} = \arcsin\left(\sqrt{\frac{2\rho g \delta}{T}}\right) + \sqrt{\frac{2\rho g \delta}{T}}\sqrt{1 - \frac{2\rho g \delta}{T}} = \frac{2\rho g \delta}{T}\sum_{n=0}^{\infty}(b_n + c_n)\left(\frac{2\rho g \delta}{T}\right)^n, \qquad (2.220\text{a,b})$$

as follows from (2.218a) $\equiv$ (2.218b) with (2.217b and c).

2.8.5 Nonlinear Correction to the Linear Shape

The exact inverse shape of the string (2.211) includes the nonlinear corrections of all orders (2.214b) with coefficients (2.182a and b; 2.213a through g) and (2.215a and b; 2.216a through g). The lowest two orders are

$$|X| = \sqrt{2Z}\left[1 + (b_1 + c_1)Z\right] = \sqrt{2Z}\left[1 + \left(\frac{1}{6} - \frac{1}{2}\right)Z\right] = \sqrt{2Z}\left(1 - \frac{Z}{3}\right). \tag{2.221}$$

To the same order of approximation as in (2.221) follows (2.222b):

$$X^2 = 2Z: \quad 2Z = \frac{X^2}{(1 - Z/3)^2} = X^2\left(1 + \frac{2Z}{3}\right) = X^2\left(1 + \frac{X^2}{3}\right), \tag{2.222a,b}$$

where in the highest-order term in (2.222b) may be substituted the lowest-order approximation (2.222a). The latter is the linear deflection (2.203c) and (2.222b) $\equiv$ (2.223a) provides the lowest-order nonlinear correction corresponding to the slope (2.223b) in apex coordinates:

$$Z = \frac{X^2}{2} + \frac{X^4}{6} + O(X^6), \quad Z' \equiv \frac{dZ}{dX} = X + \frac{2}{3}X^3 + O(X^5). \tag{2.223a,b}$$

Thus, *the shape (slope) of the elastic string under its own weight in the conditions (2.205a through c; 2.204b and c) is given by (2.223a) [(2.223b)], which consists of (1) the linear approximation (2.203c) [(2.203d)] corresponding to the first term on the r.h.s.; and (2) the second term on the r.h.s. specifies the lowest-order nonlinear correction.* To use support coordinates (x, y), the maximum deflection needs to be determined.

2.8.6 Nonlinear Shape and Maximum Deflection

In support coordinates (2.203a and b), the shape of the string is specified by

$$\zeta(x) = \delta - \frac{TZ}{\rho g} = \delta - \frac{TX^2}{2\rho g} - \frac{TX^4}{6\rho g}. \tag{2.224}$$

The maximum deflection (2.217c) is a root of (2.225a and b), leading (2.209b) to (2.225c):

$$\zeta(0) = 0 = \zeta(L): \quad \delta = \frac{T}{2\rho g}\lim_{X \to \pm \rho g L/2T}\left(X^2 + \frac{X^4}{3}\right) = \frac{\rho g L^2}{8T}\left[1 + \frac{1}{3}\left(\frac{\rho g L}{2T}\right)^2\right]. \tag{2.225a–c}$$

Substituting (2.203b) in (2.224) gives (2.226):

$$\zeta(x) - \delta = -\frac{\rho g}{2T}\left(x - \frac{L}{2}\right)^2\left[1 + \frac{1}{3}\left(\frac{\rho g}{T}\right)^2\left(x - \frac{L}{2}\right)^2\right]. \tag{2.226}$$

Adding (2.225c) to (2.226) leads to the shape of the string (2.227) in support coordinates:

$$\zeta(x) = \frac{\rho g}{8T}\left[L^2 - (2x - L)^2\right] + \frac{1}{12}\left(\frac{\rho g}{2T}\right)^3\left[L^4 - (2x - L)^4\right]. \tag{2.227}$$

Thus, *in the conditions (2.205a through c) and (2.204a and b) ≡ (2.225a and b), the shape (maximum deflection) of a string under its own weight is given by (2.227) ≡ (2.228) [(2.225c)]:*

$$\zeta(x) = \frac{\rho g}{2T} x \left[L - x - \frac{2}{3} \left(\frac{\rho g}{2T} \right)^2 \left(2x^3 - 4x^2 L + 3xL^2 - L^3 \right) \right],$$

(2.228)

where (1) the linear approximation (2.205a through c) and (2.204a and b) ≡ (2.225a and b) corresponds to the first term on the r.h.s.; (2) the second term on the r.h.s. is the lowest-order nonlinear correction.

2.8.7 Exact/Maximum Slope in Apex/Support Coordinates

The exact slope is given in dimensionless apex coordinates by (2.208a):

$$Z' \operatorname{sgn}(X) = \sqrt{2Z} \left| 1 - 2Z \right|^{-1/2} = \sqrt{2Z} \sum_{n=0}^{\infty} \binom{-1/2}{n} (-2Z)^n$$

$$= \sqrt{2Z} \sum_{n=0}^{\infty} a_n (2Z)^n = \sqrt{2Z} \left(1 + \sum_{n=1}^{\infty} \frac{(2n-1)!!}{n!} Z^n \right),$$

(2.229)

where the coefficients (2.159b) are used:

$$a_n (2Z)^n = 2^n Z^n \frac{(2n-1)!!}{(2n)!!} = \frac{(2n-1)!!}{n!} Z^n.$$

(2.230)

The maximum slope occurs at (2.217c) the supports, leading to

$$\left| Z' \right|_{\max} = \sqrt{\frac{2\Delta}{1 - 2\Delta}} = \sqrt{2\Delta} \left(1 + \sum_{n=1}^{\infty} \frac{(2n-1)!!}{n!} \Delta^n \right),$$

(2.231)

where Δ is a root of (2.218a) ≡ (2.218b). The two lowest-order terms in (2.229) [(2.231)] are (2.232a) [(2.232b)]:

$$Z' \operatorname{sgn}(X) = \sqrt{2Z} (1 + Z). \quad \left| Z' \right|_{\max} = \sqrt{2\Delta} (1 + \Delta),$$

(2.232a,b)

where the first (second) term on the r.h.s. is the linear approximation (first-order nonlinear correction).
   In support coordinates (2.203a and b), the slope (2.206a) is given (2.229) by

$$\zeta' = \sqrt{\frac{2\rho g}{T} (\delta - \zeta)} \left\{ 1 + \sum_{n=1}^{\infty} \frac{(2n-1)!!}{n!} \left[\frac{\rho g}{T} (\delta - \zeta) \right]^n \right\},$$

(2.233)

where the maximum deflection is a root of (2.220a) ≡ (2.220b). The maximum values (2.234c) occur (2.231) at the supports (2.234a and b):

$$\zeta(0) = 0 = \zeta(L): \quad \tan\theta = \left| \zeta' \right|_{\max} = \sqrt{\frac{2\rho g \delta}{T}} \left[1 + \sum_{n=1}^{\infty} \frac{(2n-1)!!}{n!} \left(\frac{2\rho g \delta}{T} \right)^n \right].$$

(2.234a–c)

The slope in apex (support) coordinates (2.229) [(2.233)] is given Z′(Z)[ζ′(ζ)] in implicit form; elimination of Z(ζ) from X(Z)[x(ζ)] in (2.211) ≡ (2.214b) [(2.219a) ≡ (2.211b)] leads to the explicit form Z′(X)[ζ′(η)]. The latter is obtained to the two lowest orders in apex (support) coordinates (2.223b) [(2.235)] as follows from (2.223a) [(2.228)]:

$$\zeta'(x) = \frac{\rho g}{2T}\left[L - 2x - \frac{2}{3}\left(\frac{\rho g}{2T}\right)^2 \left(8x^3 - 12x^2 L + 6xL^2 - L^3\right)\right], \tag{2.235}$$

where the first term on the r.h.s. is the linear approximation (2.203d) [(2.65b)] and the second term is the lowest-order nonlinear correction. The maximum slope is (2.236a) [(2.236b)] in apex (2.233b) [support (2.235)] coordinates:

$$\left| Z' \right|_{max} = \mp Z'\left(\mp \frac{\rho g L}{2T}\right) = \frac{\rho g L}{2T} + \frac{2}{3}\left(\frac{\rho g L}{2T}\right)^3, \tag{2.236a}$$

$$\tan\theta = \left|\zeta'\right|_{max} = \zeta'(0) = -\zeta'(L) = \frac{\rho g L}{2T}\left[1 + \frac{2}{3}\left(\frac{\rho g L}{2T}\right)^2\right]. \tag{2.236b}$$

Since the slope is dimensionless, it is the same (2.206a) in modulus (2.236a) ≡ (2.236b) in apex and support coordinates, with opposite signs at each support.

2.8.8 Effect of the Nonlinearity Parameter

The nonlinearity parameter (2.237a) ≡ (2.196a; 2.67a) is given for 11 equally spaced values in its range in Table 2.4, where are indicated the maximum deflections (2.225c) ≡ (2.237b) [slope in modulus (2.236a) ≡ (2.237c)] to the lowest-order nonlinear approximation:

$$q = \frac{\rho g L}{2T}: \quad \frac{\delta}{L} = \frac{q}{4}\left[1 + \frac{q^2}{3} + O\left(q^4\right)\right], \quad \tan\theta = q\left[1 + \frac{2q^2}{3} + O\left(q^4\right)\right]. \tag{2.237a–c}$$

The deflections (2.228) [slope (2.235)] are given in terms of the nonlinearity parameter (2.237a) by (2.238) [(2.239)]:

$$\frac{\zeta(x)}{L} = q\frac{x}{L}\left\{1 - \frac{x}{L} - \frac{2}{3}q^2\left[2\left(\frac{x}{L}\right)^3 - 4\left(\frac{x}{L}\right)^2 + 3\frac{x}{L} - 1\right]\right\}, \tag{2.238}$$

$$\zeta'(x) = q\left\{1 - 2\frac{x}{L} - \frac{2}{3}q^2\left[8\left(\frac{x}{L}\right)^2 - 12\left(\frac{x}{L}\right) + 6\left(\frac{x}{L}\right) - 1\right]\right\}. \tag{2.239}$$

All expressions (2.225c; 2.234b; 2.228; 2.235) consist of the linear approximation (2.65a; 2.66a; 2.64f; 2.65b) respectively, plus the lowest-order nonlinear correction in support coordinates. The shape of the string in plotted in Panel 2.3 for nine equally spaced values of the nonlinearity parameter (2.237a). The nonlinear deflection of a homogeneous string, that is, with constant mass density per unit volume (2.172a) by its own weight, with uniform tension (2.205c) in a uniform gravity field (2.205b) with supports at the same height (2.204b and c), is considered in three cases: (I) the cross section decreases as the slope increases (2.172c), so that the mass per unit length is constant, and the weight corresponds (2.172b) to a uniform shear stress (Section 2.7; Table 2.3; Panel 2.2); (II) the cross section is constant (2.173b),

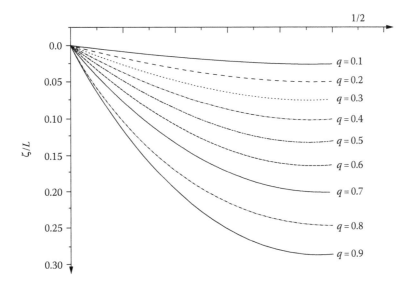

PANEL 2.3 The shape of the string due to nonlinear deflection by its own weight (Panel 2.2) is modified (Panel 2.3) if the mass density per unit length remains constant, so that the mass of the string increases at it is deflected implying that it must slide through at least one of the supports. Thus the shape of the string is different in Panels 2.2 and 2.3, for each of the nine values of the nonlinearity parameter, that are the same as in Panels 2.1 and 2.4.

TABLE 2.4 Nonlinear Deflection of an Elastic String by Its Own Weight

Nonlinearity Parameter	Maximum Deflection	Maximum Slope (Angle)
$q = \dfrac{\rho g L}{2T}$	$\dfrac{\delta}{L}$	$\zeta'(0) = \zeta'(L) \equiv \tan\theta$
(2.237a)	(2.237b)	(2.237c)
0	0	0
0.1	0.0250	0.1006 (5.75°)
0.2	0.0507	0.2053 (11.60°)
0.3	0.0773	0.3180 (17.64°)
0.4	0.1053	0.4427 (23.88°)
0.5	0.1354	0.5833 (30.26°)
0.6	0.1680	0.7440 (36.65°)
0.7	0.2036	0.9287 (42.88°)
0.8	0.2427	1.1413 (48.78°)
0.9	0.2858	1.3860 (54.19°)
1.0	∞	∞

Notes: L, length of the undeflected string; ρ, constant mass density; *T*, uniform tension along the string; *g*, acceleration of gravity; ζ, transverse deflection; *x*, longitudinal coordinate; ζ', $\equiv d\zeta/dx$ – slope; δ, maximum deflection; θ, maximum angle-of-inclination; *q*, nonlinearity parameter.

As for Table 2.1 with the nonlinear deflection of an elastic string (Panel 2.3) due to its own weight, assuming that it is homogeneous and has a constant cross section.

so that the mass per unit length is also constant (2.170b), and the shear stress increases with the slope (2.169b), implying that the total mass increases (2.173d) as the string is deflected and slides through the support (Section 2.8; Table 2.4; Panel 2.3); (III) whereas in cases I and II the string is assumed to be elastic, in case III it is assumed to be inextensible, so that the mass density per unit volume (2.172a) or per unit length (2.170b), the cross section (2.171b), and the mass (2.173d) are all constant, and the length is a fixed constant larger than the distance from the supports (Section 2.9; Table 2.5; Panel 2.4). In case

II (cases I and III), the total weight is not (is) constant and specifies the total transverse force; the latter is distributed continuously and uniformly (nonuniformly) in case I (cases II and III), in contrast to case IV of a concentrated force (Section 2.6; Table 2.2; Panel 2.1).

2.9 Comparison with an Inextensible String: The Catenary

In the case of the deflection of a heavy inextensible string, the fixed length ℓ must exceed $\ell > L$ the distance between the supports L and specifies the nonlinearity parameter (Subsection 2.9.3). The shape of the string is determined by the condition (Subsection 2.9.1) that the vertical (horizontal) component of the uniform tension tangent to the string equals the weight (is a constant). This determines (Subsection 2.9.2) the shape of the string as a hyperbolic cosine or catenary, from which follow (1) the deflection, the slope, and their maxima (Subsection 2.9.4); (2) the gravity potential energy (Subsection 2.9.5); (3) the effect of the nonlinearity parameter on the preceding (Subsection 2.9.7). The gravity potential (elastic) energy is linear (quadratic) in the nonlinearity parameter (Subsection 2.9.6). For the comparison of the non-linear deflection of strings under different loads are used the second dimensionless apex coordinates (Subsection 2.9.8); they differ from the first dimensionless apex coordinates (Subsection 2.8.1), in that the string occupies a unit interval centered at the origin. This allows the comparison of five loading cases, all with uniform tension and supports at the same height, namely (1) the linear deflection by own weight (Section 2.4) extrapolated to serve as reference (Subsection 2.9.9) with the nonlinear cases (2–5); (2) the nonlinear deflection of an inextensible string by its own weight (Subsections 2.9.1 through 2.9.7), that is, the catenary (Subsection 2.9.10); (3) the nonlinear deflection of an elastic, hence extensible, string (Section 2.8) by its own weight (Subsection 2.9.9); and (4/5) the nonlinear deflection of an elastic string (Section 2.7) by a uniform shear stress (a concentrated force at equal distance from the supports), such that [Section 2.6 (Section 2.5)] the total transverse force equals the weight (Subsection 2.9.11). The comparison (Subsections 2.9.14 and 2.9.15) of the five loading cases (1–5) is made (Subsection 2.9.12) for a nonlinearity parameter equal to one-half (Subsection 2.9.13) and includes (Table 2.6): (1) the shape of the string (Panel 2.5); (2) the slopes; (3/4) the maximum values of the deflection (1) [of the modulus of the slope (2)]; and (5) the extensions.

2.9.1 Shape of a Heavy, Inextensible String

Considering an inextensible string under uniform tension (2.240a) with constant mass density (2.240b) in a uniform gravity field (2.140c), deflected by its own weight, the horizontal (vertical) component of the tension (2.2a) [(2.2b)] is a constant equal to the horizontal reactions at the supports (2.240d) [equals the weight (2.240e) where s is the arc length (2.1b)]:

$$T, \rho, g = \text{const}: \quad T_x = T\frac{dx}{ds} \equiv b = \text{const}, \quad T_y = T\frac{dy}{ds} = \rho g s. \tag{2.240a–e}$$

From (2.240d and e), it follows that the tension is given (2.1b) by (2.141a), and substitution back into (2.240d and e) leads to the horizontal (2.241b) [vertical (2.241c)] component of the unit vector tangent to the string:

$$T = \sqrt{b^2 + (\rho g s)^2} : \quad \left\{ \frac{dx}{ds}, \frac{dy}{ds} \right\} = \frac{\{b, \rho g s\}}{\sqrt{b^2 + (\rho g s)^2}}. \tag{2.241a–c}$$

The change of variable (2.242a) leads from (2.241b and c) to (2.242b and c):

$$\psi = \frac{\rho g s}{b} : \quad \left\{ \frac{dx}{d\psi}, \frac{dy}{d\psi} \right\} = \frac{ds}{d\psi}\left\{ \frac{dx}{ds}, \frac{dy}{ds} \right\} = \frac{b}{\rho g}\frac{\{1, \psi\}}{\sqrt{1 + \psi^2}}. \tag{2.242a–c}$$

Introducing the constant (2.243a), equal to the ratio of the horizontal component of the tension (2.240d) to the weight per unit length (2.243b), the shape of the string is specified in parametric form by (2.243c and d):

$$a \equiv \frac{b}{\rho g} = \frac{T_x}{\rho g} = \text{const}: \quad x(\psi) = a \arg \sinh \psi, \quad y = a\sqrt{1 + \psi^2}, \qquad (2.243\text{a–d})$$

using (II.7.116b) in (2.242b) ≡ (2.243c). The parameter ψ may be eliminated between (2.243c and d), leading (2.244a) to the shape of the string (2.244b) as the hyperbolic cosine or **catenary**:

$$\psi = \sinh\left(\frac{x}{a}\right): \quad y = a\left|1 + \left[\sinh\left(\frac{x}{a}\right)\right]^2\right|^{1/2} = a \cosh\left(\frac{x}{a}\right), \qquad (2.244\text{a,b})$$

using (II.5.9.9a). In (2.244b) may be introduced a translation of the origin; for example, the vertical translation (2.245a and b) leads from (2.244b) to (2.245c):

$$x = \xi, \eta = y - a: \quad \eta(\xi) = a\left[\cosh\left(\frac{\xi}{a}\right) - 1\right], \quad \eta(0) = 0 = \eta'(0), \qquad (2.245\text{a–e})$$

implying (Figure 2.20) that the origin is at the apex (2.245d), and the x-axis is tangent and horizontal (2.245e). Thus, *the nonlinear deflection of an inextensible homogeneous (2.240b) string under constant tension (2.240a) in a uniform gravity field (2.240c) is specified by the catenary (2.245c) in Cartesian coordinates* (η, ξ) *with (Figure 2.20) the origin at the apex, the $O\xi$-axis horizontal and the $O\eta$-axis vertical upward. The parameter a depends on the fixed length ℓ of the string, which must exceed $\ell > L$ the distance L between the supports.* This relation is obtained next (Subsection 2.9.2).

2.9.2 Length of the String and Distance from the Supports

The length of the string is given (2.1b) by (2.246a) in general:

$$\ell = 2\int_0^{L/2}\sqrt{1 + \eta'^2}\,d\xi = 2\int_0^{L/2}\left|1 + \left[\sinh\left(\frac{\xi}{a}\right)\right]^2\right|^{1/2} d\xi = 2\int_0^{L/2}\cosh\left(\frac{\xi}{a}\right)d\xi = 2a\sinh\left(\frac{L}{2a}\right), \qquad (2.246\text{a,b})$$

and by (2.246b) in the particular case of the catenary (2.245c). *Thus, the parameter a of the catenary (2.243b) is related to the fixed length ℓ of the string and distance L from the supports at the same height (2.247a) by (2.246b) ≡ (2.247b):*

$$\eta\left(-\frac{L}{2}\right) = \eta\left(\frac{L}{2}\right): \quad \frac{\ell}{2a} = \sinh\left(\frac{L}{2a}\right) = \frac{L}{2a} + \sum_{n=1}^{\infty}\frac{1}{(2n+1)!}\left(\frac{L}{2a}\right)^{2n+1}, \qquad (2.247\text{a,b})$$

where the power series (II.7.15c) is used for the hyperbolic sine. The parameter a is proportional to the distance between the supports (2.248a) through the nonlinearity parameter q, which satisfies (2.247b) ≡ (2.248b):

$$q \equiv \frac{L}{2a}: \quad \frac{\ell}{L} = \frac{\sinh q}{q} = 1 + \sum_{n=1}^{\infty}\frac{q^{2n}}{(2n+1)!} = 1 + \frac{q^2}{6} + \frac{q^4}{120} + O(q^6). \qquad (2.248\text{a–c})$$

To confirm that the parameter q is indeed the nonlinearity parameter, the catenary is compared next (Subsection 2.9.3) with the parabolic shape of the string in the linear case (Section 2.4).

2.9.3 Exact and Approximate Nonlinearity Parameter

The linear deflection (2.3a) of an elastic string is inextensible (2.3b), and thus, a comparison with the catenary is legitimate. The equation of the catenary (2.245c) can be put into the form (2.249b), using the power series (II.7.14c) for the hyperbolic cosine:

$$\xi = x - \frac{L}{2}: \quad \frac{\eta(\xi)}{L} = \frac{a}{L}\sum_{n=1}^{\infty}\frac{1}{(2n)!}\left(\frac{\xi}{a}\right)^{2n} = \sum_{n=1}^{\infty}\frac{1}{(2n)!}\left(\frac{\xi}{L}\right)^{2n}\left(\frac{L}{a}\right)^{2n-1}$$

$$= \sum_{n=1}^{\infty}\frac{(2q)^{2n-1}}{(2n)!}\left(\frac{x}{L}-\frac{1}{2}\right)^{2n} = \frac{\delta - \zeta(x)}{L}, \tag{2.249a,b}$$

where a horizontal translation (2.249a) is used, so that the y-axis passes through the first support $x = L/2$, and δ is the maximum deflection. The two lowest-order terms in (2.249b) are

$$\zeta(x) = \delta - \frac{q}{L}\left(x - \frac{L}{2}\right)^2\left[1 + \frac{q^2}{3L^2}\left(x - \frac{L}{2}\right)^2 + O\left(\frac{q^4}{L^4}\left(x - \frac{L}{2}\right)^4\right)\right]. \tag{2.250}$$

Thus, *the shape of the catenary is given exactly by (2.250) [(2.249b)] in Cartesian coordinates with (Figure 2.20) a horizontal x-axis (ξ-axis tangent to the apex), a vertical y axis downward (η-axis upward) and the origin at the first support (apex). The first term on the r.h.s. of (2.250) is the linear approximation, which is the same for (1) an inextensible string; and (2) an elastic string that is also inextensible (2.3a). Comparison of the first term on the r.h.s. of (2.250) with (2.202) shows that (1) the nonlinearity parameter is (2.251a) in the linear approximation:*

$$q = \frac{\rho g L}{2T}, \quad a = \frac{L}{2q} = \frac{T}{\rho g}, \quad T_x = b = \rho g a = \frac{\rho g L}{2q} = T; \tag{2.251a-c}$$

(2) the parameter a is given (2.248a) by (2.251b); and (3) the constant horizontal tension (2.240d) is given (2.243a) by (2.251c), because the slope is neglected in the linear approximation (2.3a and b; 2.10c and d). The nonlinearity parameter is given by (2.251a) only in the linear case. In the nonlinear case, it is a root of (2.248b) and depends on the ratio of the fixed length of the string ℓ to the distance between the supports L. The first (first two) terms on the r.h.s. of (2.248b) lead to the lowest- (next-) order approximations (2.252a) [(2.252b)]:

$$q_0 = \sqrt{6}\left(\frac{\ell}{L}-1\right), \quad \frac{(q_1)^2}{10} = -1 + \sqrt{1 + \frac{6}{5}\left(\frac{\ell}{L}-1\right)}. \tag{2.252a,b}$$

In the linear (nonlinear) case can (must) be used (2.250) [(2.249)], with a nonlinearity parameter given approximately by (2.252a) [exactly as a positive real root of (2.248b)]. The positive root of the biquadratic (2.248c) was used in the second-order approximation (2.252b) to the nonlinearity factor, so that it equals the positive real root of (2.252a) to the lowest order.

2.9.4 Deflection, Slope, and Their Extrema

The shape of an inextensible, homogeneous (2.240b) string under uniform tension (2.240a) in a uniform gravity field (2.240c) is given (2.253) by the catenary (2.245c) in terms (2.248a) of the nonlinearity parameter, which is a root of (2.248b):

$$\frac{\eta(\xi)}{L} = \frac{1}{2q}\left[\cosh\left(2q\frac{\xi}{L}\right) - 1\right] = \sum_{n=1}^{\infty}\frac{(2q)^{2n-1}}{(2n)!}\left(\frac{\xi}{L}\right)^{2n}, \tag{2.253}$$

in Cartesian coordinates (ξ, η) with the origin at the apex (Figure 2.20). The slope is given by

$$\eta' \equiv \frac{d\eta}{d\xi} = \sinh\left(2q\frac{\xi}{L}\right) = \sum_{n=0}^{\infty} \frac{(2q)^{2n+1}}{(2n+1)!}\left(\frac{\xi}{L}\right)^{2n+1}. \tag{2.254}$$

The maximum deflection is

$$\frac{\delta}{L} = \eta\left(\pm\frac{L}{2}\right) = \frac{1}{2q}(\cosh q - 1) = \frac{1}{2}\sum_{n=1}^{\infty} \frac{q^{2n-1}}{(2n)!} = \frac{q}{4}\left[1 + \frac{q^2}{12} + O(q^4)\right] \tag{2.255}$$

and the maximum slope

$$\tan\theta = \eta'\left(\frac{L}{2}\right) = -\eta'\left(-\frac{L}{2}\right) = \sinh q = \sum_{n=1}^{\infty} \frac{q^{2n+1}}{(2n+1)!} = q\left[1 + \frac{q^2}{6} + O(q^4)\right] \tag{2.256}$$

in apex coordinates.

Choosing instead support coordinates, that is, Cartesian coordinates with the origin at the first support, the x-axis through the second support, and the y-axis downward (Figure 2.20), leads (2.249a) to (2.253; 2.255) the deflection

$$\begin{aligned}
\frac{\zeta(x)}{L} &= \frac{\delta}{L} - \frac{\eta(\xi)}{L} = \frac{\delta}{L} - \frac{1}{L}\eta\left(x - \frac{L}{2}\right) = \frac{1}{2q}\left[\cosh q - \cosh\left(2q\frac{\xi}{L}\right)\right] \\
&= \frac{1}{2}\sum_{n=1}^{\infty} \frac{q^{2n-1}}{(2n)!}\left[1 - \left(\frac{2\xi}{L}\right)^n\right] = \frac{1}{2}\sum_{n=1}^{\infty} \frac{q^{2n-1}}{(2n)!}\left[1 - \left(\frac{2x}{L} - 1\right)^{2n}\right] \\
&= q\frac{x}{L}\left\{1 - \frac{x}{L} - \frac{q^2}{6}\left[2\left(\frac{x}{L}\right)^3 - 4\left(\frac{x}{L}\right)^2 + 3\frac{x}{L} - 1\right] + O(q^4)\right\}
\end{aligned} \tag{2.257}$$

and the corresponding slope

$$\begin{aligned}
\zeta'(x) &\equiv \frac{d\zeta}{dx} = \frac{d\eta}{d\xi} = \eta'(\xi) = \sinh\left(2q\frac{\xi}{L}\right) \\
&= \sinh\left[q\left(\frac{2x}{L} - 1\right)\right] = \sum_{n=0}^{\infty} \frac{q^{2n+1}}{(2n+1)!}\left(\frac{2x}{L} - 1\right)^{2n+1} \\
&= q\left(\frac{2x}{L} - 1\right)\left[1 + \frac{q^2}{6}\left(\frac{2x}{L} - 1\right)^2 + O\left(q^4\left(\frac{2x}{L} - 1\right)^4\right)\right].
\end{aligned} \tag{2.258}$$

In (2.255; 2.256; 2.257; 2.258) the linear approximation (lowest-order nonlinear correction) has been indicated as the first (second) term on the r.h.s.

2.9.5 Potential Energy in the Gravity Field

The inextensible string does not have elastic energy. It does have **gravity potential energy** equal to the mass per unit length multiplied by the vertical coordinate and integrated over the length of the string (2.259a):

$$E_g = \int_0^\ell \rho g y \, ds = 2\rho g \int_0^{L/2} \eta \sqrt{1 + \eta'^2} \, d\xi.$$
(2.259a,b)

In (2.259b), the integration is made from the apex, implying that the deflection is given by (2.245c), which leads to

$$\sqrt{1 + \eta'^2} = \left| 1 + \sinh^2\left(\frac{\xi}{a}\right) \right|^{1/2} = \cosh\left(\frac{\xi}{a}\right) = 1 + \frac{\eta(\xi)}{a}.$$
(2.260)

Substituting (2.260; 2.245c) in (2.259b) leads to

$$E_g = \frac{2\rho g}{a} \int_0^{L/2} \eta(\eta + a) \, d\xi = 2\rho g a \int_0^{L/2} \left[\cosh^2\left(\frac{\xi}{a}\right) - \cosh\left(\frac{\xi}{a}\right) \right] d\xi$$

$$= \rho g a \int_0^{L/2} \left[1 - 2\cosh\left(\frac{\xi}{a}\right) + \cosh\left(\frac{2\xi}{a}\right) \right] d\xi$$

$$= \rho g a \left[\frac{L}{2} - 2a \sinh\left(\frac{L}{2a}\right) + \frac{a}{2} \sinh\left(\frac{L}{a}\right) \right],$$
(2.261)

that completes the integration for the gravity potential energy.

2.9.6 Comparison of the Elastic and Gravity Energies

Substitution of (2.248a) in (2.261) leads to

$$\frac{2E_g}{\rho g L^2} = \frac{1}{2q} - \frac{1}{q^2} \sinh q + \frac{1}{4q^2} \sinh(2q).$$
(2.262)

The normalized gravity potential energy (2.262) depends only on the nonlinearity parameter and is of the first order, as shown by

$$\frac{2E_g}{\rho g L^2} = \frac{1}{2q} - \sum_{n=0}^\infty \frac{q^{2n-1}}{(2n+1)!} + \sum_{n=0}^\infty \frac{(2q)^{2n-1}}{(2n+1)!}$$

$$= \sum_{n=1}^\infty \frac{2^{2n-1} - 1}{(2n+1)!} q^{2n-1} = \frac{q}{6} \left[1 + \frac{7}{20} q^2 + O(q^4) \right],$$
(2.263)

where (1) the power series (II.7.15c) is used for the hyperbolic sine; (2) the $O(1/q)$ terms in (2.263) cancel, because the gravity potential energy cannot be singular for zero nonlinearity parameter; and (3) the gravity potential energy is linear in the nonlinearity parameter as shown by the leading term in (2.263). This contrasts with the elastic energy for an extensible string that is quadratic in the nonlinearity parameter

(2.165b). Thus, *the potential gravitational (elastic) energy of a string is linear (quadratic) in the nonlinearity parameter to the lowest order, for example, for the nonlinear deflection of an inextensible (elastic) string under its own weight (2.263) [by a concentrated force at equal distance from the supports (2.165b)].*

2.9.7 Effect of the Nonlinearity Parameter

The maximum deflections (2.255) ≡ (2.264a) [slope (2.256) in modulus (2.264b)] are indicated in Table 2.5 for the same 11 equally spaced values of the nonlinearity parameter as in Tables 2.2 through 2.4:

$$\frac{\delta}{L} = \frac{1}{2q}(\cosh q - 1), \quad \tan\theta = \sinh q. \tag{2.264a,b}$$

In the case of the catenary, the nonlinearity parameter is a root of (2.248b and c). The latter specifies the "relative extension" (2.265a), understood as the ratio of the fixed length of the inextensible string to the distance between the supports:

$$\frac{\ell}{L} = \frac{\sinh q}{q}; \quad \frac{E_g}{\rho g L^2} = \frac{1}{8q^2}\left[2q - 4\sinh q + \sinh(2q)\right], \tag{2.265a,b}$$

the relative extension (2.265a) also appears in Table 2.5, together with the normalized gravity potential energy (2.262) ≡ (2.265b). The shape of the catenary (2.257) ≡ (2.266)

$$\frac{\zeta(x)}{L} = \frac{1}{2q}\left[\cosh q - \cosh q\left(\frac{2x}{L} - 1\right)\right] \tag{2.266}$$

is plotted in Panel 2.4 for the same nine values of the nonlinearity parameter as in Panels 2.1 through 2.3. The nonlinearity parameter for the heavy inextensible string is a root of (2.248b) and thus is not limited to the unit interval as in the case of the heavy elastic string (2.188a). For this reason, all values in Table 2.5 are finite for $q = 1$, in contrast to some singular values in Tables 2.2 through 2.4.

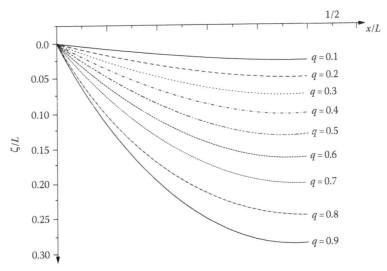

PANEL 2.4 The nonlinear deflection of an inextensible string by its own weight is specified by a catenary or hyperbolic cosine whose parameter specifies on the ratio of the length of the string ℓ to the distance L between the supports. This ratio ℓ/L depends on the nonlinearity parameter q; it is indicated in Panel 2.4 with the shapes of the deflected string for the same nine values of the nonlinearity parameter as in Panels 2.1 through 2.3.

TABLE 2.5 Nonlinear Deflection of an Inextensible String: The Catenary

Nonlinearity Parameter	Maximum Deflection	Maximum Slope (Angle)	Relative Extension	Normalized Gravity Elastic Energy
$q = \dfrac{P}{2T}$	$\dfrac{\delta}{L}$	$\tan\theta$	$\dfrac{\ell}{L}$	$\dfrac{E_g}{\rho g L^2}$
(2.248a)	(2.264a)	(2.264b)	(2.265a)	(2.265b)
0	0	0	1	0
0.1	0.0250	0.1002 (5.72°)	1.0017	0.0084
0.2	0.0502	0.2013 (11.38°)	1.0067	0.0169
0.3	0.0756	0.3045 (16.94°)	1.0151	0.0258
0.4	0.1013	0.4108 (22.33°)	1.0269	0.0352
0.5	0.1276	0.5211 (27.52°)	1.0422	0.0454
0.6	0.1546	0.6367 (32.48°)	1.0611	0.0565
0.7	0.1823	0.7586 (37.18°)	1.0837	0.0689
0.8	0.2109	0.8881 (41.61°)	1.1101	0.0826
0.9	0.2406	1.0265 (45.75°)	1.1406	0.0982
1.0	0.2715	1.1752 (49.60°)	1.1175	0.1158

Notes: L, distance between the supports at the same height; ℓ, fixed length of the inextensible string; δ, maximum deflection at the point of application of the concentrated force; θ, maximum inclination in modulus at the supports; ρ, mass density, g, acceleration of gravity; T, conform tension along the string; E_g, gravity potential energy; q, nonlinearity parameter.

As for Table 2.1 with the nonlinear deflection due to the own weight applying to an inextensible (Table 2.5) instead of an elastic (Table 2.4) string; the shape of string is a catenary (Panel 2.5).

2.9.8 Second Dimensionless Apex Coordinates

The linear deflection of an elastic string under its own weight (2.202) is given by (2.267c) using the **second dimensionless apex coordinates** (2.267a and b):

$$\bar{X} = \frac{x}{L} - \frac{1}{2} = \frac{\xi}{L} = \frac{TX}{\rho g L}, \quad \bar{Y} = \frac{T(\delta - \zeta)}{\rho g L^2} = \frac{T\eta}{\rho g L^2} = \left(\frac{T}{\rho g L}\right)^2 Y, \quad \bar{Y} = \frac{1}{2}\bar{X}^2 = \bar{Z}(\bar{X}). \tag{2.267a–c}$$

The first (2.203a and b) [second (2.267a and b)] dimensionless apex coordinates [Figure 2.20 (Figure 2.21)] both have (1) the origin at the apex; (2) distinct horizontal coordinates $X(\bar{X})$ along the tangent; (3) distinct vertical coordinates $Y(\bar{Y})$ upward. They lead to (4) the same deflection by own weight in the linear case (2.203c) [(2.267c)]; (5) a different form of the equilibrium equation (2.9b) $\equiv$ (2.268a), namely, (2.268b) [(2.268c)]:

$$-\frac{f(x)}{T} = \zeta''|1 + \zeta'|^{-3/2} \Leftrightarrow \frac{f}{\rho g} = Z''|1 + Z'^2|^{-3/2} \Leftrightarrow \frac{f}{\rho g} = \bar{Z}''\left|1 + \left(2q\bar{Z}'\right)^2\right|^{-3/2}; \tag{2.268a–c}$$

and (6) a range of the horizontal coordinates (2.269a) that depends (2.269c) on the nonlinearity parameter (2.261b) [is always the unit interval centered at the origin (2.269d)]:

$$0 \leq x \leq L: \quad q = \frac{\rho g L}{2T}: \quad -q \leq X \leq q, \quad -\frac{1}{2} \leq \bar{X} \leq \frac{1}{2}. \tag{2.269a–d}$$

Thus, the first (second) dimensionless apex coordinates lead to a simpler equilibrium equation [are more convenient to compare the shape (Panel 2.5) and deflection (Table 2.6) of the string for different loadings and cases].

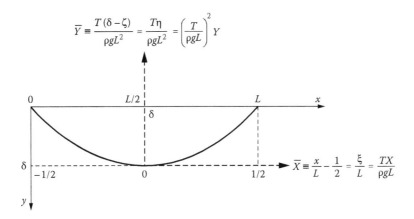

FIGURE 2.21 The problem of Figure 2.20 of the nonlinear deflection of a string by its own weight has a different solution if the string is inextensible, because there is no elastic energy, only gravity potential energy. The shape is a catenary or hyperbolic cosine and is conveniently expressed in the second dimensionless apex coordinates (Figure 2.21) that compared with the first dimensionless apex coordinates (Figure 2.20) have: (1) the same origin at the apex and the same directions for the coordinate axis: (2) different values for the coordinates, for example the horizontal coordinate is replaced by $\xi = x/L - 1/2$. This has the advantage that it maps the domain of the string $0 \leq \xi \leq L$ to the unit interval centered on the origin $-1/2 \leq \xi \leq + 1/2$. This allows the comparison of the deflections of the string for several nonlinear cases (Panel 2.5; Table 2.6).

Substituting (2.206a and b) in (2.268a) leads to (2.268b). To obtain (2.268c) from (2.268a), (2.270a and b) are used:

$$\bar{Z}' \equiv \frac{d\bar{Y}}{d\bar{X}} = -\frac{T}{\rho g L}\frac{d\zeta}{dx} = -\frac{\zeta'}{2q}, \quad \bar{Y}'' \equiv \frac{d\bar{Z}'}{d\bar{X}} = \frac{dx}{d\bar{X}}\frac{d\bar{Z}'}{dx} = -\frac{L}{2q}\zeta'' = -\frac{T}{\rho g}\zeta''. \tag{2.270a,b}$$

In the case of an elastic string with constant mass density (2.169b) ≡ (2.271a), the balance equations (2.268a through c) become (2.271b through d):

$$f(x) = \rho g \left|1 + \zeta'^2\right|^{1/2}: \quad -\frac{2q}{L} = -\frac{\rho g}{T} = \frac{\zeta''}{\left(1 + \zeta'^2\right)^2} \Leftrightarrow \frac{Z''}{\left(1 + Z'^2\right)^2} = 1 \Leftrightarrow 1 = \frac{\bar{Z}''}{\left[1 + \left(2q\bar{Z}'\right)^2\right]^2}; \tag{2.271a–d}$$

in the passage from (2.271b) to (2.371c) [(2.271d)], (2.206a and b) [(2.270a and b)] were used. The first (second) dimensionless apex coordinates (2.203a and b) [(2.267a and b)] lead to the same equation for the deflection under own weight in the linear case (2.203c) ≡ (2.267c) but not in the nonlinear (2.206c) ≡ (2.271c) ≠ (2.271d). The fixed domain of the second dimensionless apex coordinates as the unit interval centered on the origin (2.269d) allows a direct comparison of the shape of the string $Y = \bar{Z}(\bar{X})$ under different loadings and is retained in the sequel (Subsections 2.9.9 through 2.9.15).

2.9.9 Linear/Nonlinear Deflection under Own Weight

The reference case is taken to be the linear deflection under own weight (2.267c) ≡ (2.272a) [leading to the slope (2.272b)] and maximum values (2.272c) [in modulus (2.275d)]:

$$\bar{Y}_0 = \bar{Z}_0(X) = \frac{1}{2}\bar{X}^2, \quad \bar{Z}_0' = X, \quad \bar{Z}_{0\max} = \bar{Z}_0\left(\pm\frac{1}{2}\right) = \frac{1}{8}, \quad \left|Z_0'\right|_{\max} = \pm Z_0'\left(\pm\frac{1}{2}\right) = \frac{1}{2}. \tag{2.272a–d}$$

These values will be extrapolated for comparison with the nonlinear cases. In all cases, it will be assumed that the tension is uniform (2.205c) and the supports are (2.204b and c) at the same height and distance L. For the nonlinear deflection under own weight, the first-order nonlinear approximation (2.226) in the first dimensionless apex coordinates (2.267a and b) is (2.273a):

$$\bar{Y}_3 = \bar{Z}_3(\bar{X}) = \frac{\bar{X}^2}{2}\left[1+\frac{4}{3}q^2\bar{X}^2+O(q^4\bar{X}^4)\right], \quad \bar{Z}_3'(\bar{X}) = \bar{X}\left[1+\frac{8}{3}q^2\bar{X}^2+O(\bar{X}^4)\right], \qquad (2.273a,b)$$

corresponding to the slope (2.273b) and the extremum values (2.274a and b):

$$\bar{Z}_{3\max} = Z_3\left(\pm\frac{1}{2}\right) = \frac{1}{8}\left[1+\frac{q^2}{3}+O(q^2)\right], \quad \left|\bar{Z}_3'\right|_{\max} = \pm\bar{Z}_3'\left(\pm\frac{1}{2}\right) = \frac{1}{2}\left[1+\frac{2q^2}{3}+O(q^4)\right]. \qquad (2.274a,b)$$

The linear approximation or reference case (2.272a through d) is the first term on the r.h.s. of (2.273a and b; 2.274a and b) and is the same for all loadings that follow (Subsections 2.9.10 and 2.9.11); it is the remaining nonlinear terms that differ for each distinct loading that need comparison (Subsections 2.9.12 through 2.9.15).

2.9.10 Catenary in the First Dimensionless Apex Coordinates

In the case of the inextensible string, the shape of the catenary (2.253) in the first dimensionless apex coordinates (2.267a and b) is given by (2.275):

$$\bar{Y}_1 = \bar{Z}_1(\bar{X}) = \frac{1}{2q}\left[\cosh(2q\bar{X})-1\right] = \sum_{n=1}^{\infty}\frac{(2q)^{2n-1}}{(2n)!}\bar{X}^{2n} = q\bar{X}^2\left[1+\frac{q^2}{3}\bar{X}^2+O(\bar{X}^4)\right]. \qquad (2.275)$$

The corresponding slope (2.254) is

$$\bar{Z}_1' = \sinh(2q\bar{X}) = \sum_{n=0}^{\infty}\frac{(2q\bar{X})^{2n+1}}{(2n+1)!} = 2q\bar{X}\left[1+\frac{2}{3}q^2\bar{X}^2+O(\bar{X}^4)\right]. \qquad (2.276)$$

The maximum deflection is:

$$\bar{Z}_{1\max} = Z_1\left(\pm\frac{1}{2}\right) = \frac{1}{2q}(\cosh q-1), \qquad (2.277)$$

in agreement with (2.255) ≡ (2.277), and the maximum slope in modulus

$$\tan\theta = \left|Z_1'\right|_{\max} = \pm Z_1'\left(\pm\frac{1}{2}\right) = \sinh q, \qquad (2.278)$$

in agreement with (2.256) ≡ (2.278).

2.9.11 Nonlinear Deflection by a Uniform or Concentrated Load

The nonlinear deflection (2.188c; 2.189) under a constant shear stress (2.173a through d) is specified in the first dimensionless apex coordinates (2.267a and b) by (2.279):

$$
\bar{Y}_2 = \bar{Z}_2(\bar{X}) = \frac{T}{\rho g L}\frac{1}{2q}\left\{1 - \left|1 - q^2\left(1 - \frac{2x}{L}\right)^2\right|^{1/2}\right\} = \frac{1}{4q^2}\left\{1 - \left|1 - \left(2q\bar{X}\right)^2\right|^{1/2}\right\}
$$

$$
= -\frac{1}{4}\sum_{n=1}^{\infty}\binom{1/2}{n}q^{2n-2}\left(-4\bar{X}^2\right)^n = \frac{1}{2}\bar{X}^2 - \frac{1}{4}\sum_{n=2}^{\infty}b_n q^{2n-2}\left(2\bar{X}\right)^{2n}
$$

$$
= \frac{1}{2}\bar{X}^2 + \sum_{n=2}^{\infty}\frac{(2n-3)!!}{n!}2^{n-2}q^{2n-2}\bar{X}^{2n} = \frac{1}{2}\bar{X}^2\left[1 + q^2\bar{X}^2 + O\left(\bar{X}^4\right)\right], \tag{2.279}
$$

where the coefficients (2.182a and b) ≡ (2.213a through g) are used. The corresponding slope is

$$
\bar{Z}_2' = \bar{X}\left|1 - \left(2q\bar{X}\right)^2\right|^{-1/2} = \sum_{n=0}^{\infty}\binom{-1/2}{n}\left(-4q^2\right)^n \bar{X}^{2n+1} = \bar{X} + \sum_{n=1}^{\infty}a_n 2^{2n}q^{2n}\bar{X}^{2n+1}
$$

$$
= \bar{X} + \sum_{n=1}^{\infty}\frac{(2n-1)!!}{n!}2^n q^{2n}\bar{X}^{2n+1} = \bar{X}\left[1 + 2q^2\bar{X}^2 + O\left(q^4\bar{X}^4\right)\right], \tag{2.280}
$$

where the coefficients (2.158a and b; 2.212a through g) are used. The maximum deflection is

$$
\bar{Z}_{2\max} = Z_2\left(\pm\frac{1}{2}\right) = \frac{1}{4q^2}\left[1 - \left|1 - q^2\right|^{1/2}\right] = -\frac{1}{4q^2}\sum_{n=1}^{\infty}\binom{1/2}{n}\left(-q^2\right)^n
$$

$$
= \frac{1}{8} - \frac{1}{4}\sum_{n=2}^{\infty}b_n q^{2n-2} = \frac{1}{8} + \frac{1}{4}\sum_{n=2}^{\infty}\frac{(2n-3)!!}{(2n)!!}q^{2n-2} = \frac{1}{8}\left[1 + \frac{q^2}{4} + O\left(q^4\right)\right], \tag{2.281}
$$

and the maximum slope is given by

$$
\left|\bar{Z}_2'\right|_{\max} = \pm Z_3'\left(\pm\frac{1}{2}\right) = \frac{1}{2}\left|1 - q^2\right|^{-1/2} = \frac{1}{2}\sum_{n=0}^{\infty}\binom{-1/2}{n}\left(-q^2\right)^n
$$

$$
= \frac{1}{2} + \frac{1}{2}\sum_{n=1}^{\infty}a_n q^{2n} = \frac{1}{2} + \sum_{n=1}^{\infty}\frac{(2n-1)!!}{n!2^{n+1}}q^{2n} = \frac{1}{2}\left[1 + \frac{q^2}{2} + O\left(q^4\right)\right]. \tag{2.282}
$$

The last comparison concerns (Section 2.6) the nonlinear deflection by a concentrated force corresponding to the same total transverse force but applied at an equal distance from the supports (2.168a and b), leading in terms of the first dimensionless apex coordinates (2.267a and b):

$$
\mathrm{sgn}(\bar{X})\bar{Y}_4 = \mathrm{sgn}(\bar{X})\bar{Z}_4(\bar{X}) = \frac{T\delta}{PL}\left(\frac{2x}{L} - 1\right) = \frac{\bar{X}\delta}{qL} = \frac{\bar{X}}{2}\left|1 - q^2\right|^{-1/2}, \tag{2.283}
$$

using also (2.167b). The slope is constant (2.284a) and equal in modulus to the maximum value (2.284b) in the preceding case (2.282):

$$\text{sgn}\left(\bar{X}\right)\bar{Z}_4' = \frac{1}{2}\left|1-q^2\right|^{-1/2} = \left|\bar{Z}_2'\right|_{\max},\tag{2.284a,b}$$

$$\bar{Z}_{4\,\max} = \bar{Z}_4\left(\pm\frac{1}{2}\right) = \frac{1}{4}\left|1-q^2\right|^{-1/2} = \frac{1}{2}\left|\bar{Z}_2'\right|_{\max},\tag{2.285a,b}$$

the maximum deflection (2.285a) is one-half (2.285b) of the preceding value (2.284a and b).

2.9.12 Comparison of Five Loading Cases

The expansion of the nonlinear deflection (slope) in powers of the nonlinearity parameter starts with the linear term (2.272a) [(2.272b)], which is the same for all three elastic cases, with differences appearing to the lowest nonlinear order, namely (1) nonlinear deflection by own weight (2.273a) [(2.273b)], assuming constant mass density (2.173a) and cross section (2.173b) and hence shear stress increasing with the slope (2.169b); (2) nonlinear deflection (2.279) [(2.280)] by a constant shear stress (2.172b), implying a cross section varying with the slope (2.172c) for a constant mass density (2.172a); and (3) a force equal to the total weight in the case (2) of constant mass, concentrated at equal distance from the supports (2.283) [(2.284a)], corresponding to a particular case of the nonlinear influence or Green function. The comparison of the cases (1) and (2) shows that (a) the nonlinear deflection (2.286) [slope (2.287)] exceeds the extrapolated linear value:

$$\bar{Z}_3\left(\bar{X}\right) = \frac{\bar{X}^2}{2}\left(1+\frac{4q^2}{3}\bar{X}^2\right) > \frac{\bar{X}^2}{2}\left(1+q^2\bar{X}^2\right) = \bar{Z}_2\left(\bar{X}\right) > \frac{\bar{X}^2}{2} = \bar{Z}_0\left(\bar{X}\right),\tag{2.286}$$

$$\left|\bar{Z}_3'\left(\bar{X}\right)\right| = \left|\bar{X}\right|\left(1+\frac{8q^2}{3}\bar{X}^2\right) > \left|\bar{X}\right|\left(1+2q^2\bar{X}^2\right) = \left|\bar{Z}_2'\left(\bar{X}\right)\right| > \left|\bar{X}\right| = \left|\bar{Z}_0'\left(\bar{X}\right)\right|;\tag{2.287}$$

(b) the value is larger (smaller) in the case (1) [(2)] when the mass of the string increases with (is independent of) the deflection.

The case (3) of a concentrated force is not comparable at all points to the cases (1) and (2) because the shape is an exact triangle instead of a parabola to a leading order; the comparison can be made for the maximum deflection and slope (Subsection 2.9.13) in all three cases (1–3) plus the catenary, which applies to (4) an inextensible instead of (1–3) an elastic string. The case (4) of the deflection of the inelastic string by its own weight leads to the catenary (2.275), whose leading term coincides with the linear approximation (2.272a) if the nonlinearity parameter equals one-half (2.288a):

$$q = \frac{1}{2}: \quad \bar{Z}_1\left(\bar{X}\right) = \frac{\bar{X}^2}{2}\left(1+\frac{\bar{X}^2}{12}\right) < \frac{\bar{X}^2}{2}\left(1+\frac{\bar{X}^2}{4}\right) = \bar{Z}_2\left(\bar{X}\right),$$

$$\left|\bar{Z}_1'\left(X\right)\right| = \left|\bar{X}\right|\left(1+\frac{\bar{X}^2}{6}\right) < \left|\vec{X}\right|\left(1+\frac{\bar{X}^2}{2}\right) = \left|\bar{Z}_2'\left(\bar{X}\right)\right|,\tag{2.288a–c}$$

implying the deflection (2.288b) [slope (2.286c)] that is smaller than for the nonlinear deflection of the elastic strings (2.286) [(2.287)]. The nonlinearity parameter (2.288a) ≡ (2.289a) leads to the values indicated in Table 2.5 for (1/2) the maximum deflection (2.264a) [(2.264b)]; (3) the ratio of the length of the string to the distance between the supports (2.265a); and (4) the normalized elastic energy (2.265b):

$$q = 0.5: \quad \frac{\delta}{L} = 0.1276, \quad \tan\theta = 0.5211, \quad \frac{\ell}{L} = 1.0422, \quad \frac{E_g}{\rho g L^2} = 0.0454. \qquad (2.289a\text{--}c)$$

The value (2.288a) ≡ (2.289a) of the nonlinearity parameter is used next (Subsections 2.9.13 through 2.9.15) in the comparison of nonlinear elastic (1–3) and inextensible (4) deflections, with the extrapolated linear case as reference for comparison in Panel 2.5 and Table 2.6.

The deflection of the string is plotted in Panel 2.5 in the case of nonlinearity parameter $q = 0.5$ for five loadings, all with constant tension (2.205c) and for a homogeneous string (2.205a) in a uniform gravity field (2.205b) with suspension points at the same height and distance (2.204b and c). Since the loading is symmetric (2.204a), only one-half of the deflection curve is plotted in Panel 2.5 and indicated in Table 2.6 for 10 positions. Both in Panel 2.5 and Table 2.6, five loading cases are considered: (I) the reference (Section 2.4; Subsection 2.9.9) linear deflection (2.272a through d); (II) the corresponding nonlinear deflection (Section 2.8, Subsection 2.9.9) of a heavy elastic string (2.273a and b; 2.274a and b); (III) the catenary (Subsections 2.9.1 through 2.9.7 and 2.9.10) for the nonlinear deflection of an inelastic string in the case (2.288a through c) when the linear approximation coincides with the preceding cases (I and II); (IV/V) the nonlinear deflection [Section 2.7 (2.6)] due to (Subsection 2.9.11) a uniform distributed load (2.279–2.282) [a concentrated load at equal distance from the supports (2.283; 2.284a and b; 2.285a and b)]. The comparison of the magnitudes of the deflections and slopes (Subsection 2.9.13) is followed by a physical explanation of the results (Subsection 2.9.14).

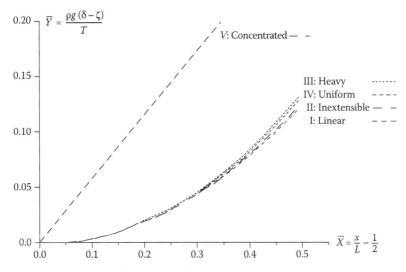

PANEL 2.5 A fixed intermediate value $q = 0.5$ of the nonlinearity parameter is used to compare the shape of the deflected string for the same total transverse force but different shear stress distributions: (1) one extreme is a concentrated force at equal distance from the supports (Panel 2.1) that leads to the largest deflection because all the force is put where it has largest moment relative to the supports, and the string deflects more easily; (2) hence the deflection is smaller for the heavy elastic string with constant mass density per unit length (Panel 2.3); (3) the case of a uniform load leads to a smaller deflection than in the case (2) when the mass increases as the string is stretched; and (4) the inextensible string or catenary has the smallest nonlinear deflection because all energy is associated with gravity, and it does not store elastic energy. The smallest deflection is the linear case (5) that neglects the effect of the slope in increasing the length of the string; thus the deflection is smaller in the linear than in all nonlinear cases (1–4). The linear deflection (5) with uniform weight is plotted as an extrapolation for comparison purposes.

TABLE 2.6 Nonlinear Deflection of a String by Different Loadings

X-Coordinate	Y-Coordinate				
Case	Linear	Nonlinear			
Order	I	II	III	IV	V
Load	Weight	Own weight	Uniform load	Own weight	Concentrated load
String	Elastic	Inextensible	Elastic	Elastic	Elastic
(2.267a)	(2.272a)	(2.275)	(2.273b)	(2.279)	(2.283)
0.05	0.0012500	0.0012503	0.0012508	0.001251	0.02887
0.1	0.005000	0.005004	0.0050125	0.005017	0.05774
0.15	0.01125	0.01127	0.01131	0.001133	0.08660
0.20	0.02000	0.02007	0.02020	0.02027	0.11547
0.25	0.03125	0.03141	0.03174	0.03190	0.14434
0.30	0.04500	0.04534	0.04601	0.04635	0.17320
0.35	0.06125	0.06188	0.06313	0.06375	0.20207
0.40	0.08000	0.08107	0.08320	0.08427	0.23094
0.45	0.10125	0.10296	0.10638	0.10808	0.25981
0.50	0.12500	0.12760	0.13281	0.13542	0.28868

Notes: P, same total transverse force in all cases; X, normalized apex coordinate: $\frac{x}{L} - \frac{1}{2}$; Y, normalized apex coordinate: $\frac{P(\delta - \zeta)}{TL^2}$; $q = \frac{P}{2T} = 0.5$, fixed value of nonlinearity parameter.

The nonlinear deflection of a string with different load distributions is compared with the linear case that is extrapolated as reference. The total transverse load is the same allowing the comparison of the shapes of the strings either numerically (Table 2.6) or graphically (Panel 2.5) for a nonlinearity parameter equal to one-half.

2.9.13 Sequence of Increasing Deflections/Slopes

Taking as reference the deflection in the linear case (2.272a), it is (1) increased in all nonlinear cases; (2) least for the inextensible string or catenary; (3) more for the elastic string; (4) less in the case of constant shear stress relative to constant mass density:

$$\bar{Z}_0(\bar{X}) < \bar{Z}_1(\bar{X}) < \bar{Z}_2(\bar{X}) < \bar{Z}_3(\bar{X}) < \bar{Z}_4(\bar{X});$$ (2.290)

and (5) the deflection in the case of the concentrated load is the largest, for example, for a nonlinearity parameter equal to one-half (2.291a) using (2.285a; 2.274a; 2.281; 2.277; 2.285a), leading to (2.291b):

$$q = \frac{1}{2}: \quad \bar{Z}_{4\max} = \frac{1}{2\sqrt{3}} = 0.2887 > \bar{Z}_{3\max} = \frac{13}{96} = 0.1354 > \bar{Z}_{2\max} = \frac{17}{128} = 0.1328$$

$$> \bar{Z}_{1\max} = \cosh(0.5) - 1 = 0.1276 > \bar{Z}_{0\max} = \frac{1}{8} = 0.1250,$$ (2.291a,b)

which is a sequence of decreasing values. A similar sequence applies to the maximum slope in modulus for the same value (2.291a) ≡ (2.292a) of the nonlinearity parameter using (2.285b; 2.274b; 2.282; 2.278; 2.285b), leading to (2.292b):

$$q = \frac{1}{2}: \quad |\bar{Z}_4'|_{\max} = \frac{1}{\sqrt{3}} = 0.5774 > |\bar{Z}_3'|_{\max} = \frac{7}{12} = 0.5833 > |\bar{Z}_2'|_{\max} = \frac{9}{16} = 0.5625$$

$$> |\bar{Z}_1'|_{\max} = \sinh(0.5) = 0.5211 \geq |\bar{Z}_0'| = \frac{1}{2} = 0.5000.$$ (2.292a,b)

The relative magnitudes of the deflection and slope are explained qualitatively next (Subsection 2.9.14).

2.9.14 Effects of the Spatial Load Distribution

The preceding results are summarized in Table 2.6 and Panel 2.5 and may be explained as follows. The largest deflection and slope is for a concentrated force at equal distance from the supports (Section 2.6; Subsection 2.9.11) since the moment relative to the supports is largest and the mid-point is where the string deflects more easily. The same total transverse force distributed uniformly as a constant shear stress (Section 2.7; Subsection 2.9.11) leads to the same slope at the supports (2.284a and b); since the shape is triangular for the concentrated load and curved for a uniform load, the slope is smaller in the latter case at all points other than the supports and also the deflection. The nonlinear deflection by own weight with uniform shear stress implies that the total mass is constant; hence as the string is stretched, the mass per unit length is reduced; if the string is homogeneous, that is, has constant mass per unit volume, the cross section must reduce. This reduction in mass per unit length and cross section leads to a smaller deflection and slope than in the case of constant cross section (Section 2.8; Subsection 2.9.10); in the latter case, the mass of the string increases as it becomes longer under nonlinear deflection, and it must slide through the support. Thus, there are two effects in comparison for strong deflection that increases the length of the string: (1) either the total mass is constant and the cross section decreases or (2) the cross section is constant and the total mass increases. The preceding analysis shows that the "thinning" effect (1) and larger mass (2) together imply that the nonlinear deflection and slope under a uniform load are smaller than by the own weight with constant mass per unit length.

The nonlinearity parameter is related to the ratio of the length of the string to the distance between the supports; in the case of the inextensible (elastic) string, the length is fixed "a priori" and independent of the load (is equal to the distance between the supports in the absence of a load and increases with the load for a nonlinear deflection). The linear approximation to the deflection of the string, which neglects the square of the slope everywhere, is dependent (independent) on the nonlinearity parameter for the inextensible (elastic) string. The linear approximation to the deflection of the inextensible string equals that of the elastic string only for the value that is one-half of the nonlinearity parameter. In this case, the nonlinear deflection is less for the inextensible than for the elastic string with the same mass density in an equal gravity field. The reason is that the inextensible string has gravity potential energy only, whereas the elastic string also has energy of deformation. The deflection of the heavy inelastic string is determined by the gravity potential energy alone; in the case of the heavy elastic string, the nonlinear deflection is associated with a stretching of the string, leading to a larger deflection.

2.9.15 Linear/Nonlinear Deflection of Elastic/Inextensible Strings

The nonlinear deflection is larger in all cases, either for the inextensible or elastic string, than the linear deflection; the latter neglects the square of the slope everywhere and implies that the length of the string is equal to the distance between the supports. The is not true for (a) the inelastic string if its fixed length, chosen "a priori" and independent of the loads, is larger than the distance between the supports; (b) the elastic string that is stretched by a nonlinear deflection, so that its length is larger than the distance between the supports for large loads. In both cases of the (a) inextensible and (b) elastic strings, the nonlinear deflection is associated with a length larger than in the linear case, and thus, nonlinearity increases the deflection. Thus, for the same total transverse force, the sequence of increasing deflections is (1) linear approximation of constant length of the string; (2) inelastic string with nonlinearity parameter 1/2; (3) elastic string with uniform shear stress; (4) heavy elastic string with constant mass density and cross section; and (5) concentrated force at equal distance from the supports of an elastic string. The comparison of the shapes of the deflected string in Panel 2.5 shows that the difference between the cases (1–4) with distributed loads is small, and the deflection is much larger in the case (5) of the concentrated load.

This is confirmed by Table 2.6, showing the deflections at 10 positions, with a maximum for the concentrated load about the double of that for distributed loads. Table 2.6 concerns the lowest-order nonlinear approximations (2.283; 2.273a; 2.279; 2.288b) compared with the linear case (2.272a) for a nonlinearity parameter equal to one-half (2.293a):

$$q = 0.5: \quad \bar{Z}_4 = \frac{\bar{X}}{\sqrt{3}} > \bar{Z}_3 = \frac{\bar{X}^2}{2}\left(1 + \frac{\bar{X}^2}{3}\right) > \bar{Z}_2 = \frac{\bar{X}^2}{2}\left(1 + \frac{\bar{X}^2}{4}\right)$$

$$> \bar{Z}_1 = \frac{\bar{X}^2}{2}\left(1 + \frac{\bar{X}^2}{12}\right) > \bar{Z}_0 = \frac{\bar{X}^2}{2}, \tag{2.293a,b}$$

using the second dimensionless apex coordinates (2.267a and b); since the $\bar{X}$-coordinate lies in the unit interval centered at the origin (2.269d), the relative error does not exceed $\bar{X}^4 \geq 0.5^4 = 0.625$ or 6%. The exact results to all orders in the nonlinearity parameter are indicated in Table 2.7 for the five cases of deflection of a string compared in Panel 2.5 and Table 2.6. These five cases are a subset of the 16 cases of the shape of a loaded string that appear in Classification 2.1.

NOTE 2.1: Variants of the Loaded String Problem

The problem of finding the shape of a string deflected by a transverse load has variants (Classification 2.1) according to at least five criteria: (1) linear (nonlinear) deflection for a small (large) slope everywhere (somewhere); (2) uniform (nonuniform) tangential tension; (3) homogeneous (inhomogeneous) string, that is, with constant (variable continuous or discontinuous) mass density per unit length; (4) suspension points at the same (different) height; and (5) one (several) concentrated load(s) or a distributed continuous load or the mixed case, that is, a combination of both. There are $2 \times 2 \times 3 \times 2 \times 3 = 72$ combinations of the five criteria (1–5) of which (a) 16 were solved explicitly (Classification 2.1); (b) to the remaining cases could be applied similar methods. The problems become more difficult as nonlinearity, nonuniform tension, nonconstant mass density, supports at unequal heights, and mixed load are combined. All these variations (1–5) concern the one-dimensional problem; two- or higher-dimensional problems like spinning or moving strings could also be considered. The simplest case treated most often is the linear deflection of a homogeneous string with constant tension under (a) a single concentrated load, specifying the Green or influence function; and (b) a uniform load, for example, the weight in a constant gravity field. The variations around these baseline cases are indicated in Classification 2.1, List 2.1, and Diagram 2.1. The linear (nonlinear) deflection of an elastic string does not (does) change the total length, because the square of the slope is (is not) neglected. Thus, nonlinear deflection of an inextensible string by its own weight is a distinct problem, leading to the catenary (Note 2.2).

    There are 17 distinct problems concerning the shape of a loaded string in List 2.1, related as indicated in Classification 2.1 and Diagram 2.1: (1) the nonlinear deflection of an inelastic string by its own weight leads to the catenary (Section 2.9); (2) the remaining problems concern the elastic string, starting with the nonlinear deflection by one concentrated force, which leads to a nonlinear influence function (Section 2.6); (3, 4) the principle of superposition does not apply to nonlinear deflections, for example, to the case of two concentrated forces (Example 10.5) or a concentrated plus a uniform load (Example 10.6); (5, 6) the nonlinear deflection of an elastic string under its own weight is considered for constant mass (mass density) [Section 2.7 (2.8)]; (7) the linear deflection of an elastic string by a concentrated load specifies the linear influence function (Section 2.3); (8–11) the superposition principle (Subsection 2.3.9) can be used to consider other loads, such as uniform (Section 2.4), two equal (opposite) concentrated forces at equal distance from the supports [Subsections 2.5.1 through 2.5.3 (Subsections 2.5.4 through 2.5.7)], or the weight balanced by two equal, opposite concentrated forces with equal spacing (Subsections 2.5.7 through 2.5.9); (12) nonuniform tension varying with the distance between the supports is considered both in the linear and nonlinear cases with a transverse concentrated force (Example 10.7); (13, 14) supports at unequal heights are considered for a concentrated force (Section 2.3) and own weight (Example 10.3); and (15, 16) the nonhomogeneous string is considered in the cases of discontinuous (continuous) nonuniform mass density [Example 10.2 (10.3)].

TABLE 2.7 Comparison of Different Loadings

Deflection	Linear		Nonlinear		
Case	I	II	III	IV	V
String	Elastic	Inextensible	Elastic	Elastic	Elastic
Load	Weight	Weight	Uniform	Weight	Concentrated
Section/Subsection	2.4; 2.9.9	2.9.1 through 2.9.8; 2.9.10	2.7; 2.9.11	2.8; 2.9.9	2.6; 2.9.11
Table	—	2.5	2.3	2.4	2.2
Panel	—	2.4	2.2	2.3	2.1
Deflection	(2.64f; 2.71a; 2.272a)	(2.245c; 2.249b; 2.250; 2.253; 2.257; 2.275)	(2.176; 2.184a; 2.185; 2.188c; 2.279)	(2.211; 2.214a and b; 2.219a and b; 2.221; 2.223a; 2.224; 2.227; 2.228; 2.238; 2.273a)	(2.116a and b; 2.168a and b; 2.283)
Slope	(2.65b; 2.71b; 2.272b)	(2.254; 2.258; 2.276)	(2.178; 2.186; 2.187; 2.188b; 2.280)	(2.223b; 2.229; 2.232a; 2.233; 2.235; 2.239; 2.273b)	(2.116a and b; 2.139a through c; 2.140a through c; 2.167c; 2.284a)
Maximum deflection	(2.65a; 2.72a; 2.272c)	(2.255; 2.264a; 2.277)	(2.177; 2.180; 2.183a; 2.189; 2.196b; 2.281)	(2.218a and b; 2.220a and b; 2.225c; 2.237b; 2.274a)	(2.119a and b; 2.144c and d; 2.148c; 2.161a and b; 2.162a and b; 2.167b; 2.285a)
Maximum slope in modulus	(2.66a; 2.72b; 2.272d)	(2.256; 2.264b; 2.278)	(2.179; 2.181; 2.183b; 2.190; 2.196c; 2.282)	(2.231; 2.232b; 2.234c; 2.236a and b; 2.237c; 2.274b)	(2.116a and b; 2.139a through c; 2.140a through c; 2.167c; 2.284a)
Extension or length deflected	—	(2.247b; 2.248a and b; 2.265a)	(2.195a and b; 2.196d)	—	(2.156a and b; 2.157c and d; 2.160a and b; 2.167d)
Energy	—	(2.261; 2.262; 2.263; 2.265b)	(2.191a; 2.196e)	—	(2.163b and c; 2.164b and c; 2.165a and b; 2.167e)
Reactions at the supports	(2.66b; 2.72c)	(2.72c)	(2.72c)	(2.72c)	(2.118; 2.120; 2.138a through d; 2.141a and b; 2.142a and b; 2.144a and b; 2.132a through c; 2.147a and b)

Note: Linear and nonlinear deflections of elastic and inelastic strings with different loadings including a concentrated force or a distributed shear stress; the latter may be either uniform (or nonuniform), such as the own weight in the linear (nonlinear) case.

LIST 2.1 Seventeen Shapes of Loaded Elastic Strings

A. 1. Nonuniform tension: Example 10.7
B. Inhomogeneous
    2. Discontinuous mass density: Example 10.2
    3. Continuous mass density: Example 10.3
C. Supports at unequal height
    4. Concentrated load: influence or Green function: Section 2.3
    5. Own weight: Example 10.4
D. Linear deflection
    6. Concentrated load: influence or Green function: Section 2.3
    7. Uniform load: own weight: Section 2.4
    8. Two parallel concentrated loads: Subsections 2.5.1 through 2.5.3
    9. Two opposite concentrated loads: Subsections 2.5.4 through 2.5.6
    10. Mixed concentrated and distributed loads: Subsections 2.5.7 through 2.5.9
    11. Arbitrary load-superposition principle: Subsection 2.3.9
E. Nonlinear deflection by
    12. One concentrated force: nonlinear influence function: Section 2.6
    13. Two concentrated forces: Example 10.5
    14. Uniform shear stress: Section 2.7
    15. Own weight: Section 2.8
    16. Concentrated force plus uniform shear stress: Example 10.6
F. 17. Nonlinear inelastic string: centenary: Section 2.9

The 16 cases in Diagram 2.1 and Table 2.8 (Classification 2.1) coincide with two differences: (1) the case of nonuniform tension (Example 10.7) is not split (is split) into linear and nonlinear cases; and (2) the linear case of uniform load is split (not split) into supports at equal and unequal heights. Since (1) and (2) compensate each other, the number of cases is the same, 16. List 2.1 has an extra case, adding to 17, namely linear deflection by an arbitrary load using the influence or Green function for supports at equal or unequal heights and uniform tension. These are a subset of the six influence or Green functions in List 2.2. The influence or Green function, that is, the deflection by a single concentrated force at an arbitrary position, is obtained in four linear cases: (1) uniform tension with supports at the same height (Subsection 2.3.3) when the reciprocity principle holds; (2, 3) for supports at unequal heights, the reciprocity principle can be extended (Subsection 2.3.7) in two alternate (Subsection 2.3.8) ways; (4) the influence function with nonuniform tension (Examples E10.7.1 and E10.7.2) also satisfies the reciprocity principle because the operator is self-adjoint (Subsection 7.7.9). The four cases (1–4) of linear influence functions extend to arbitrary loads because the principle of superposition holds. Neither the principle of superposition nor the reciprocity principle hold for the nonlinear influence or Green function corresponding to a concentrated force with (5) uniform tension (Section 2.6); and (6) nonuniform tension (Examples E10.7.3 and E10.7.4).

NOTE 2.2: Catenary, Catenoid, Variations, and Minimal Surfaces

The catenary arises in at least three contexts: (1) the shape of a homogeneous heavy string under constant tangential tension in a uniform gravity field is given by a hyperbolic cosine as a solution of the force balance (Section 2.9); (2) this shape corresponds to the lowest possible center of gravity for a given length, leading to a variational problem; and (3) the rotation of the catenary around an axis through the supports generates a surface of revolution, the catenoid (Figure II.6.7c ≡ 2.22), that is, a minimal surface (Section II.6.3). The latter is also the solution of a variational problem: of all surfaces of revolution

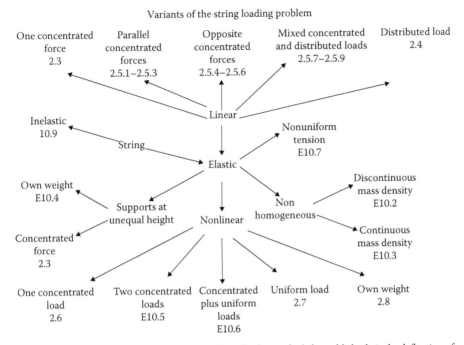

Variants of the string loading problem

DIAGRAM 2.1 Arguably the simplest problem of finding the shape of a deformable body is the deflection of a string by a transverse load. This allows the consideration of several variations of the problem, such as linear (nonlinear) deflection with small (large) slope, uniform (nonuniform) longitudinal tension, homogeneous (inhomogeneous) string with mass density constant (varying) along the length, symmetric (unsymmetric) boundary conditions for supports at equal (unequal) heights, different loads (concentrated, distributed and mixed), and elastic or inextensible strings.

generated by a curve of fixed length joining two points, it is the shape that leads to the smallest area. These three problems (1–3) all have the same solution.

NOTE 2.3: Strings, Bars, Beams, Membranes, Plates, Shells, and Bodies

The string (I) is the simplest elastic body since (1) it is one-dimensional; (2) it supports only one load, that is, a tangential tension; (3) it is straight unless a transverse force or constraint is applied; and (4) it does not resist bending. A bar (II) is also one-dimensional and has bending stiffness (Chapter 4). A bar may also support an axial traction or compression (Note 2.3), in which case it is called (Chapter 4) a beam (III). A bar may also be subject to torsion (Sections II.6.5 through II.6.9; Note 2.4) and compressions or tractions (Notes 2.5 and 2.6). The bar may be curved (IV) in the unloaded condition. A string (bar) may have nonuniform cross section and be inhomogeneous, for example, have nonuniform mass density and/or elastic modulus. The two-dimensional analogue of the string is (V) a membrane (Sections II.6.1 and II.6.2), which also supports a tension and deflects under a transverse load; the two-dimensional analogue of a bar is a plate (VI) that has bending stiffness. The plate may also be subject to compression/traction and/or torsion; a plate that is curved in the unloaded condition is called a shell (VII). The one-dimensional (I–IV) [two-dimensional (V–VII)] cases are elastic bodies in that two (one) of the three spatial dimensions are small; the three-dimensional elastic bodies (VIII) can bear all of the preceding loads. The methods used to consider the simplest elastic system, the string (I), can be extended to all other (II–VIII) elastic bodies, for example, (a) the linear (nonlinear) balance equations for small (large) rates-of-deformation or strains (Chapter II.4); (b) the use of the energy balance to obtain boundary conditions (Sections II.6.1 and 2.2); (c) the principle of superposition (Sections 2.4 and 4.4) for the linear influence or Green function; (d) the principle of reciprocity (Sections 2.4 and 4.4) for the linear influence or Green function if the differential operator is self-adjoint (Sections 7.7 through 7.9). These methods also extend

CLASSIFICATION 2.1 Sixteen Cases of Deflection of a String

Case	Section	Figure	Approximation	Tension	Mass Density	Supports	Load
1	2.3	2.4 through 2.8	Linear	Uniform	Constant	Unequal heights	Concentrated
2	2.4; 2.9.9	2.9	Linear	Uniform	Constant	Equal heights	Weight
3	E10.4	E10.4	Linear	Uniform	Constant	Unequal heights	Weight
4	2.5.1–2.5.3	2.12 and 2.13	Linear	Uniform	Constant	Equal heights	2 equal concentrated
5	2.5.4–2.5.6	2.14 and 2.15	Linear	Uniform	Constant	Equal heights	2 opposite concentrated
6	2.5.7–2.5.9	2.16 and 2.17	Linear	Uniform	Constant	Equal heights	Uniform +2 concentrated
7	E10.2	E10.2	Linear	Uniform	Discontinuous	Equal heights	Weight
8	E10.3	E10.3	Linear	Uniform	Continuous	Equal heights	Weight
9	E10.7.1, E10.7.2	E10.7	Linear	Nonuniform	Constant	Equal heights	Weight
10	2.6; 2.9.9	2.18	Nonlinear	Uniform	Constant	Equal heights	Concentrated
11	2.7; 2.9.11	2.19	Nonlinear	Uniform	Constant	Equal heights	Uniform
12	2.8; 2.9.11	2.20	Nonlinear	Uniform	Constant	Equal heights	Weight
13	2.9	2.21	Nonlinear	Uniform	Inextensible	Equal heights	Weight
14	E10.5	E10.5	Nonlinear	Uniform	Constant	Equal heights	2 concentrated
15	E10.6	E10.6	Nonlinear	Uniform	Constant	Equal heights	Uniform + concentrated
16	E10.7.3, E10.7.4	E10.7	Nonlinear	Nonuniform	Constant	Equal heights	Uniform

Note: 16 cases of deflection of a string with different assumptions such as (1) elastic (inextensible), (2) small (large) slope, (3) homogeneous (inhomogeneous), (4) supports at equal (unequal) height and (5) concentrated, distributed or mixed loads.

LIST 2.2 Six Influence or Green Functions for an Elastic String

A. 1. Baseline: linear, uniform tension, supports at the same height: Subsection 2.3.3
B. Supports at different heights: Section 2.3
    2. Extended influence function: Section 2.3.7
    3. Alternate influence function: Subsections 2.3.8
C. 4. Nonuniform tension: Example E10.7.1
D. Nonlinear
    5. Uniform tension: Section 2.6
    6. Nonuniform tension: Example E10.7.2

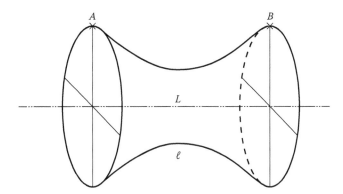

FIGURE 2.22 The catenary is connected with the solution of three problems: (1) it is the shape of a heavy inextensible uniform string of length ℓ suspended in a uniform gravity field from supports at the same height (Figure 2.21); (2) of all possible string shapes satisfying the conditions (1) it has the lowest center of gravity, and hence the least gravity potential energy; and (3) of all curves of length ℓ joining two points that which generates a surface of revolution with least area when rotated around an external parallel axis is the catenary (Figure 2.22). The surface of revolution whose generator is a catenary is called catenoid and is a minimal surface. Thus (2) and (3) are variational problems and (1) is a differential problem.

to time-dependent deformations, that is, vibrations and/or waves, for example, waves in a string or membrane or three-dimensional elastic body or the vibrations of beams and plates. The steady deflection of the elastic string is thus the simplest illustration of the methods (a–d). Although the string under tension is the simplest system, it has some analogies (Notes 2.4 through 2.7).

NOTE 2.4: Torsion of a Slender Rod

The application of an axial moment M_z to a slender rod causes (Figure II.6.10) **torsion** (Sections II.6.5 through II.6.8), that is (2.294a), the rate of rotation of the azimuthal angle φ of a piece of material along the z-axis:

$$\tau \equiv \frac{d\varphi}{dz} \equiv \varphi': \quad M_z = C\tau = C\varphi', \quad M_z' = \left(C\varphi'\right)'. \tag{2.294a–c}$$

The torsion moment (2.294b) is proportional to the torsion (2.294a) through the torsional stiffness C, which depends on the material and shape of the cross section. The torsional stiffness is constant (variable) for a homogeneous (inhomogeneous) rod with uniform (nonuniform) cross section; in both cases, the torsion moment (its rate of change along the axis) satisfies (2.294b) [(2.294c)]. This is analogous to (2.11a) [(2.11c)] with

$$\varphi(z) \leftrightarrow \zeta(x), \quad C \leftrightarrow T, \quad M'_z \leftrightarrow -f, \quad M_z \Leftrightarrow -F. \tag{2.295a–d}$$

Thus, *there is an analogy between the torsion of a slender rod (the linear deflection of an elastic string) involving (1) the variation of (2.295a) the angle of rotation φ (transverse deflection ζ) along the z axis (the length x); (2) the torsional stiffness C (tangential tension T) taken as constant in (2.295b); (3) the axial variation of the moment of torsion (minus the shear stress, that is, transverse force per unit length) in (2.295c); and (4) the moment of torsion (minus the transverse force) in (2.295d).*

NOTE 2.5: Linear Compression/Traction of a Rod

A rod may be deformed by (1) torsion (Sections II.6.5 through II.6.9; Note 2.4); (2) bending (Chapter 4); and/or (3) traction or compression (Note 2.5). If the axial displacement u depends only on the axial coordinate x, the strain tensor (Section II.4.2) has only one component (2.296a), corresponding [Figure 2.23a (Figure 2.23b)] to an extension (contraction) if positive (negative):

$$S = \frac{du}{dx} = u'; \; T = ES = Eu', \; F(x) = -T': \; -F = (ES)' = (Eu')', \tag{2.296a–d}$$

Hooke's (1678) law for a linear elastic material (Section II.4.3) states (2.296b) that the stress T is proportional to the strain S through Young's (1807) modulus E of the material (2.296b); the latter is positive, so that a traction $T > 0$ (compression $T < 0$) corresponds to an extension $S > 0$ (contraction $S < 0$). The axial rate of change of the stress equals (2.296c) minus the longitudinal force (Section II.4.4). Substituting (2.296a) into (2.296b) and this into (2.296c) leads to (2.296d). The latter can be compared with (2.11c), leading to

$$u(x) \leftrightarrow \zeta(x), \quad E \leftrightarrow T, \quad F(x) \leftrightarrow f(x). \tag{2.297a–c}$$

Thus, *there is an analogy between the longitudinal compression ≡ contraction or traction ≡ extension (transverse deflection) of an elastic rod (string) involving (1) the longitudinal displacement u (transverse*

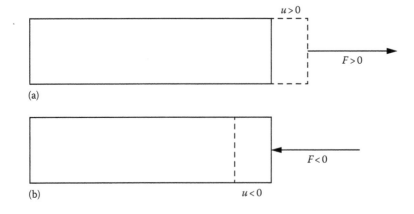

(a)

(b)

FIGURE 2.23 An analogue of the linear deflection of an elastic string is the axial or longitudinal deformation of a slender rod, that is extension (contraction) in the case [Figure 2.12a (Figure 2.12b)] of a traction (compression) force. The linear analogy holds if the constitutive relation between the stress and strain is linear, that is elastic.

deflection ζ) as a function of distance x in (2.297a): (2) Young's modulus E of the elastic material (the tension T tangential to the string), which may be uniform or not (2.297b); and (3) the longitudinal force F (the shear stress or transverse force per unit length) in (2.297c). If Young's modulus is nonuniform [uniform (2.298b)], the longitudinal deformation of a rod is specified by (2.298a) [(2.298c)]:

$$F = E'u' + Eu''; E = \text{const}: \quad F = Eu'', \qquad (2.298\text{a–c})$$

which is analogous to (2.11c) [(2.12d)]. Thus, the methods used to determine the shape of an elastic string under weak deflection (Sections 2.1 through 2.5; Examples 10.2 through 10.4) apply equally well to the torsion (compression/traction) of a rod [Note 2.4 (2.5)] in the linear case. The nonlinear case is distinct, as indicated next (Note 2.6).

NOTE 2.6: Nonlinear Hardening/Softening of a Material

The nonlinear deflection of a string is associated with a large slope in modulus $|\zeta'|$, that is, a geometric parameter related to the shape. The nonlinear longitudinal deformation of a rod is associated with a large strain in modulus $|u'|$, which generally affects the constitutive relation with the stress (2.296b). In nonlinear cases, the stress–strain relation may have, for example, a quadratic term (2.299a), involving a second constitutive parameter of the material, namely the nonlinearity modulus $\bar{E}$:

$$T = ES + \bar{E}S^2 = u'\left(E + \bar{E}u'\right); \quad -F = T' = \left[u'\left(E + \bar{E}u'\right)\right]', \qquad (2.299\text{a,b})$$

whereas Young's modulus is always positive, the nonlinearity modulus may be positive (negative) corresponding to a **hardening (softening)** material whose stress–strain relation (Figure 2.24) shows that, compared with linear elasticity, (1) the same strain requires a larger (smaller) stress; (2) the same stress corresponds to a smaller (larger) strain; (3) as the strain increases, the material resists deformation more (less).

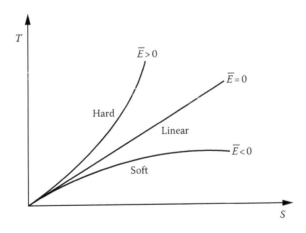

FIGURE 2.24 The linear constitutive relation between stress and strain, is modified for nonlinear materials that harden or soften as the strain increases. There is an analogy between the cases I (II) of the linear longitudinal deformation (transverse deflection) of a slender rod (Figure 2.23a and b) [elastic string (Figure 2.1b)]. This analogy is invalidated at the nonlinear level if the nonlinearity is constitutive (geometric), that is, due to the nonlinear constitutive relation (Figure 2.24) [large slope (Figure 2.1a)] for the cases I (II).

The same stress–force balance or momentum equation (2.296c) with the linear (nonlinear) constitutive relation (2.296b) [(2.299a)] leads to (2.296d) [(2.299b) ≡ (2.300)]:

$$-F(x) = u''(E + 2\bar{E}u') + E'u' + \bar{E}'u'^2. \tag{2.300}$$

For a homogeneous material with constant (2.301a) [nonlinear (2.301b)] Young's modulus, the longitudinal deformation of the rod is specified by (2.301c):

$$E, \bar{E} = \text{const}: \quad -F(x) = u''(E + 2\bar{E}u'). \tag{2.301a–c}$$

Since (2.300) [(2.301c)] is different from (2.8b) [(2.9b)], the analogy between the longitudinal deformation of a rod and the transverse deflection of a string does not extend from the linear (Note 2.5) to the nonlinear (Note 2.6) case because the nonlinearity has a different origin, that is, it arises from (1) the geometry in the first case; and (2) the constitutive properties of the material in the second case. A generalization of this problem is the nonlinear extension of an inelastic rod by its own weight in the gravity field (Notes 4.4–4.7). Besides the analogies of the linear deflection of an elastic string with the torsion (longitudinal deformation) of the slender rod [Note 2.5 (2.6)], there is a third analogy with one-dimensional accelerated motion (Note 2.7).

NOTE 2.7: Dynamic Analogy with Accelerated Motion

A particle or body with mass m, moving in one dimension x with (Figure 2.25a) velocity (2.302a), has a linear momentum (2.302b) whose derivative with regard to time is the inertial force (2.302c):

$$v \equiv \frac{dx}{dt} \equiv \dot{x}; \quad p \equiv mv = m\dot{x}: \quad F = \dot{p} = \frac{d}{dt}\left(m\frac{dx}{dt}\right) = (m\dot{x})^{\cdot} = (mv)^{\cdot}. \tag{2.302a–c}$$

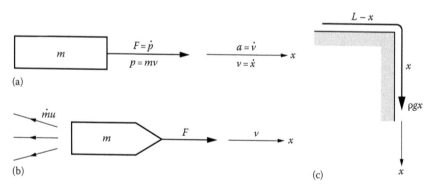

(a)

(b)

(c)

FIGURE 2.25 Besides the analogies of the linear deflection of an elastic string with the torsion (traction/compression) of a bar a third analogy is the Newton law applied to the one-dimensional motion of mass m under a force F. The analogy applies for a force $F(t)$ and mass $m(t)$ depending on time, because time in the dynamical problem (a) replaces position x in the string deflection problem. The analogy does not apply if the mass or force depends on time or position. Although most bodies have constant mass, there are important exceptions. An airplane (rocket) loses mass as it burns duel (propellant) for propulsion; this effect is significant if the fuel (propellant) mass is a significant fraction of the total mass (b). An example of a body whose mass depends on position is a chain sliding from a table (c); the motion is driven by the weight of the part of the chain that is hanging, and the mass of this part of the chain increases as the distance of the end of the chain from the edge of the table decreases.

There is an analogy between Newton's law of motion and linear deflection of an elastic string:

$$t \leftrightarrow x, \quad x(t) \leftrightarrow \zeta(x), \quad m(t) \leftrightarrow T(x), \quad F(t) \Leftrightarrow -f(x), \quad \text{(2.303a–d)}$$

involving (1) the position x of the particle or point of the body as a function of time t (the transverse deflection ζ of the string as a function of the longitudinal position x) in (2.303a and b); (2) the mass m of the particle or body (tension T along the tangent to the string) in (2.303c); and (3) inertial force F (minus the shear stress), which may depend on position (time) in (2.303d). In the particular case of a particle or body with constant mass (2.304a), the inertial force equals (2.304c) the product of the mass by the acceleration (2.304b) or the time derivative of the velocity:

$$m = \text{const}: \quad a = \dot{v} = \frac{dv}{dt} = \ddot{x} = \frac{d^2 x}{dt^2}, \quad F = ma = m\dot{v} = m\ddot{x}. \quad \text{(2.304a–c)}$$

The analogy does not apply if the force or mass depend on position, because in the elastic string analogy, the analogue (2.303a) of time is position. An example of a body whose mass depends on time (position) is [Figure 2.25b (Figure 2.25c)] an airplane or rocket that spends fuel/propellants for propulsion (a chain sliding from a table, for which only the hanging part drives the motion in the gravity field).

2.10 Conclusion

The tangential tension T along an elastic string may be decomposed into horizontal T_x and vertical T_y components for large (small) slopes [Figure 2.1a (Figure 2.1b)]; a small relative deflection from the equilibrium position $y = 0$ is linear (Figure 2.2a) if the slope is small everywhere but is (Figure 2.2b) nonlinear if the deflection is large or the slope is locally not small at some steep ripples. The shape of an elastic string deflected by transverse loads can be obtained by (1) a balance of vertical forces (Figure 2.1); (2) the principle of virtual work matching the work of the force against the variation of the elastic energy in a virtual deflection of the string (Figure 2.3). The shape of a taut string under transverse loads much smaller than the longitudinal tension is (Figure 2.9) a parabola for a uniform homogeneous string, under its own weight, with suspension points at equal height; (Figure 2.11) a triangle, that is, the response or Green function for a force P concentrated at a point ξ; (Figure 2.13) trapeze for equal concentrated forces at equal distances from the supports; (Figure 2.15) two skew-symmetric triangles for opposite forces at equal distances from the supports; (Figure 2.17) a trapeze with curved sides, that is, the difference of Figures 2.9 and 2.13 for a heavy string whose weight is balanced by equal forces with the same distance between them and the suspension points. The last three cases [(Figure 2.13/2.15/2.17)] correspond respectively to a transverse force (shear stress) that is (Figure 2.12/2.14/2.16) a staircase/parapet/sawtooth function (a monotonic/alternating/translated impulse haircomb). The deflection due to a concentrated force (Figure 2.11) specifies the Green or influence function that in the linear case satisfies the principle of reciprocity, allowing the interchange of the force and displacement both for symmetric (asymmetric) boundary conditions, that is, supports [Figure 2.6 (Figure 2.4)] at the same (unequal) height; the corresponding balances of forces and moments [Figure 2.7 (Figure 2.5)] specify the horizontal and vertical reaction forces at the supports. The simplest case is a concentrated force at equal distance from the supports (Figure 2.8); in that case, the slope at the supports is the same (Figure 2.10) as for a uniform string with total weight equal to the total force.

There are corresponding linear (nonlinear) cases [Figures 2.4 through 2.17 (Figures 2.18 through 2.21)]. The case of a single concentrated force extended from a small (to a large) slope [Figures 2.4 through 2.8 (Figure 2.18)] specifies the linear (nonlinear) influence or Green function that satisfies (does not satisfy) the principles of superposition and reciprocity. Other cases of nonlinear deflection include (1) the uniform shear stress that leads to a circular arc as string deflection (Figure 2.19); and (2/3) the deflection of a heavy elastic (inextensible) string, using the first (second) dimensionless apex coordinates [Figure 2.20 (Figure 2.21)]. In the last case, the shape of the deflected string is a catenary (Figure 2.21) that generates (Figure 2.22) a catenoid that is a minimal surface of revolution obtained by rotation around an axis parallel to the line joining the supports and below the apex. The four nonlinear cases are plotted (Panels 2.1 through 2.4) for the same set of nine values of the nonlinearity parameter (Tables 2.2 through 2.5); the deflections for the four loading cases are also compared for an intermediate value of the nonlinearity parameter (Panel 2.5; Table 2.6). The linear deflection of an elastic string (Table 2.7) has three analogues: (1/2) the torsion (longitudinal deformation) of a slender beam [Figure II.6.10 (Figure 2.23)]; and (3) Newton's law for the one-dimensional motion of a material particle (Figure 2.25a), including (not including) the case when the mass and force [Figure 2.25b (Figure 2.25c)] depend on time (position). The possible nonlinear analogy (2) fails, because the geometric (constitutive) nonlinearities for the large transverse deflection (longitudinal deformation) of a string (slender beam) are different [Figure 2.1a (Figure 2.24)]. The deflection of an elastic string by a transverse load is arguably the simplest problem of finding the shape of a deformable body; for this reason, it allows the consideration of several effects (Diagram 2.1; Classification 2.1) like nonlinearity, inhomogeneity, nonuniform tension, and unsymmetric boundary conditions (Lists 2.1 and 2.2).

3

Functionals over Test Functions

The definition of generalized function as the nonuniform limit (Chapter 1) of a family of ordinary functions is an intuitive approach that helps visualize the more important instances, such as the Heaviside unit jump, the Dirac unit impulse, and its derivatives. The proof of the properties of a generalized function using the associated sequences of ordinary functions can become tedious, suggesting that a different approach, directed at the fundamental properties, may be more practical (Chapter 3). The fundamental property of generalized functions is that they are infinitely differentiable, that is, the generalized function is a generalization of the ordinary function conceived so as to account for all singularities that arise by differentiation (Sections 1.8 through 1.9). The implication is that, starting from a generalized function and integrating it a sufficient number of times, the singularities are smoothed out and an ordinary function is obtained; this provides a second approach to the theory of generalized functions. A third approach is to multiply the generalized function by an ordinary function such that the singularity of the product becomes integrable, that is, the integral takes a definite value. This leads to the introduction of a set of reference or test functions, and the definition of a generalized function as a functional that assigns to each test function a number, namely, the integral of their product. A property, for example, differentiation, of a generalized function is defined by performing the operation on the test function that may be chosen from a class of ordinary functions where the operation is possible. In this way, the properties of generalized functions appear as functionals, calculated using the rules of integration, transferred to the test functions. To define (1) the generalized function Heaviside unit jump as a functional (Section 3.1), the test functions need only to be integrable; and (2) the derivative of the unit jump, that is, the generalized function Dirac unit impulse, is defined as a functional by passing the derivative to the test functions that should be differentiable, and can be extended to the classes of continuous functions. This approach extends to the derivatives of all orders of the unit impulse (Section 3.2). To prove more properties of generalized functions, the corresponding further restrictions are placed on the test functions; in every case, the most general class of test functions is considered for which the property of a generalized function holds. This leads to the introduction of several classes of test functions (Section 3.3); the generalized function is then defined (Section 3.4) as a linear continuous functional over a given set of test functions. The functional approach (Section 3.4) provides an alternative and simpler proof of the properties of generalized functions, including (1) series of impulses, jumps, and ramps (Section 3.5); (2) products of powers and derivative impulses (Section 3.6), including the substitution rule for the derivative of the impulse; and (3) products of moduli, powers, and logarithms (Section 3.8). Two consequences of these properties are as follows: (1) algebraic equations can have different solutions (Section 3.7) in terms of ordinary or generalized functions and (2) the evaluation of the Hadamard finite part of an integral with a power-law singularity at one end of the path of integration (Section 3.9).

3.1 Unit Jump and Unit Impulse

A generalized function may be defined multiplying by an ordinary "test" function and integrating over the real line, provided this leads to a finite unique result. For example, taking as test functions the functions integrable on the real axis, multiplication by the unit jump limits the integral to the positive real axis (Subsection 3.1.1); thus the unit jump is defined as the functional that assigns to each test function integrable on the real axis a number, namely, its integral on the positive real axis. This leads to the fundamental property, the unit jump (Subsection 3.1.2). The unit impulse is defined as the derivative of the unit jump; multiplying by the test function, and integrating by parts, requires the test function to be differentiable on the real axis. This leads to the definition of the unit impulse as the functional that assigns to each differentiable test function on the real line a number, namely, its value at the origin (Subsection 3.1.3); this is the fundamental property of the unit impulse that can be extended to continuous functions and leads to linear transformation (Subsection 3.1.4) including change of scale and translation.

3.1.1 Unit Jump and Integrable Functions

The generalized function Heaviside unit jump (Section 1.2) has the integral property

$$\int_{-\infty}^{+\infty} H(x)\Phi(x)\,dx = \int_{0}^{+\infty} \Phi(x)\,dx, \tag{3.1}$$

because (1.28a–c) it is zero for $x < 0$ and unity for $x > 0$; the only difference between (3.1) and (1.28a–c) is that the former does not depend on the value of $H(x)$ at $x = 0$, as long as it is finite. Thus, (3.1) may be used as a definition of the generalized function unit jump with $\Phi(x)$ **test functions** such that the integral exists. Thus is introduced the set of test functions, as the set of functions **integrable** over the real axis:

$$\mathcal{E}(a,b) \equiv \left\{ \Phi(x): \quad \exists \ \int_{a}^{b} \Phi(x)\,dx \right\}. \tag{3.2}$$

The Heaviside **unit jump** (3.3a) is defined as the functional,

$$H(x): \quad \Phi(x) \in \mathcal{E}(|R) \to \left[H(x), \Phi(x)\right] \equiv \int_{-\infty}^{+\infty} H(x)\Phi(x)\,dx = \int_{0}^{+\infty} \Phi(x)\,dx, \tag{3.3a–c}$$

that assigns to each function integrable on the real line (3.3b) a number (3.3c), namely, its integral from 0 to ∞. In the definition (3.3c) was used the inner product of functions (Subsections II.5.5.1 and 3.4.4). The convention is made that uppercase Latin (Greek) letters represent generalized (ordinary test) functions; for example, H for the unit jump (3.1) [Φ for the integrable functions on the real line (3.2)] appear in (3.3a–c). Ordinary functions other than test functions are designated by lowercase Latin letters. There are a few exceptions to these rules, for example, δ is retained for the generalized function unit impulse.

As an example of the use of the definition (3.1) ≡ (3.3c), consider the same integral over the whole real line (3.4) that exists for the choice of test functions integrable (3.2) on the real line (3.3a):

$$\left[1, \Phi(x)\right] \equiv \int_{0}^{+\infty} \Phi(x) dx = \int_{-\infty}^{0} \Phi(x) dx + \int_{0}^{+\infty} \Phi(x) dx$$

$$= \int_{0}^{+\infty} \left[H(-x) + H(x)\right] \Phi(x) dx \equiv \left[H(x) + H(-x), \Phi(x)\right]. \tag{3.4}$$

This proves the **symmetry property** *(1.81) ≡ (3.4) of the unit jump*, without making use of its value (1.28b) at the origin.

3.1.2 Fundamental Property of the Unit Jump

The fundamental property of the generalized function unit jump assigns to each integrable test function (3.5a) the number (3.5b):

$$\Phi \in \mathcal{E}\left(\big|R\right): \quad \int_{-\infty}^{+\infty} H(ax+b) \Phi(x) dx = \int_{-b/a}^{+\infty} \Phi(x) dx; \tag{3.5a,b}$$

The result (3.5b) follows from $H(ax + b) = 0(= 1)$ for $ax + b < 0(> 0)$. The formula (3.5b) ≡ (3.6b) can be proved by means of the change of variable (3.6a):

$$\xi \equiv ax + b: \quad \int_{-\infty}^{+\infty} H(ax+b) \Phi(x) dx = \int_{-\infty}^{+\infty} H(\xi) \Phi\left(\left(\frac{\xi - b}{a}\right)\right) \frac{d\xi}{a}$$

$$= \int_{0}^{+\infty} \Phi\left(\left(\frac{\xi - b}{a}\right)\right) \frac{d\xi}{a} = \int_{-b/a}^{+\infty} \Phi(x) dx, \tag{3.6a,b}$$

from the property (3.1). As an example of the use of the functional approach to the generalized function unit jump, consider the sign function (1.86a–c) that equals ±1 for $x > 0$ ($x < 0$), that is, the value at $x = 0$ is not needed:

$$\left[\text{sgn}(x), \Phi(x)\right] \equiv \int_{-\infty}^{+\infty} \text{sgn}(x) \Phi(x) dx = \left(\int_{0}^{\infty} - \int_{-\infty}^{0}\right) \Phi(x) dx$$

$$= \int_{-\infty}^{+\infty} \left\{H(x) - H(-x)\right\} \Phi(x) dx = \left[H(x) - H(-x), \Phi(x)\right]. \tag{3.7}$$

Since (3.7) holds for all integrable functions, it follows that *the sign function is twice the odd part (3.8a) of the unit jump (3.7)*:

$$\text{sgn}(x) = H(x) - H(-x) = 1 - 2H(-x) = 2H(x) - 1. \tag{3.8a–c}$$

The second and third expressions (3.8b,c) follow from (1.81) ≡ (3.4) because the even part (1.82) of the unit jump is 1/2.

3.1.3 Unit Impulse and Continuous Functions

The generalized function Dirac unit impulse is defined as the derivate of Heaviside's unit jump (3.9a); instead of using a family of smooth functions (Section 1.3), the approach now is to operate (Section 3.2) on integrals like (3.9c)

$$\delta(x) \equiv H'(x), \Phi(x) \in \mathcal{D}(|R): \left[\delta(x), H(x)\right]$$

$$\equiv \left[H'(x), \Phi(x)\right] \equiv \int_{-\infty}^{+\infty} H'(x)\Phi(x)dx$$

$$= \left\{H(x)\Phi(x)\right\}_{-\infty}^{+\infty} - \int_{-\infty}^{+\infty} H(x)\Phi'(x)dx$$

$$= -H(-\infty)\Phi(-\infty) + H(\infty)\Phi(\infty) - \int_0^\infty \Phi'(x)dx$$

$$= -H(-\infty)\Phi(-\infty) + \Phi(\infty) - \left[\Phi(x)\right]_0^\infty$$

$$= -H(-\infty)\Phi(-\infty) + \Phi(0) = \Phi(0), \tag{3.9a–c}$$

where an integration by parts was performed, and (1.28a,c) was used: (1) from $H(+\infty) = 1$ it follows that the term at $x = +\infty$ cancels; (2) from $H(-\infty) = 0$ it follows that the term at $x = -\infty$ vanishes if $\Phi(-\infty)$ is bounded; and (3) this leaves only the value of the function at the origin $\Phi(0)$. A differentiable function is bounded, and thus the condition $\Phi(-\infty)$ bounded in (2) is met by a test function that is differentiable (3.10):

$$\mathcal{D}(a,b) = \left\{\Phi(x): \quad \forall_{a\le c\le b} \quad \exists^! \quad f'(c) = \lim_{x\to c}\frac{f(x)-f(c)}{x-c}\right\}, \tag{3.10}$$

on the real line (3.9b). Thus, the **unit impulse** is defined as the functional (3.11a) that assigns to each continuous test function (3.11b) its value at the origin (3.11c):

$$\delta(x): \quad \Phi(x) \in C(|R) \to \left[\delta(x),\Phi(x)\right] \equiv \int_{-\infty}^{+\infty}\delta(x)\Phi(x)dx = \Phi(0). \tag{3.11a–c}$$

The test function (1) in (3.11a–c) must be **continuous**,

$$C(a,b) \equiv \left\{\Phi(x): \quad \forall_{a\le c\le b} \quad \exists^! \quad \lim_{x\to c}f(x) = f(c)\right\}, \tag{3.12}$$

on the real line (3.11b) so that the value at the origin exists (3.11c), or at any other point; and (2) must be differentiable (3.10) on the real line (3.9b) so that the ordinary rule of integration by parts holds (3.9c).

The latter requirement (2) is stricter than (1) and is adopted only if an integration by parts is involved. This fundamental property can be generalized by a linear change of variable (Subsection 3.1.4) as for the unit jump (Subsection 3.1.2).

3.1.4 Translations and Change of Scale

The property (3.11a–c) may be extended by the same change of variable (3.6a) ≡ (3.13a) as in (3.6b), leading to (3.13b)

$$d\xi = a\,dx: \quad \int_{-\infty}^{+\infty}\delta(ax+b)\Phi(x)dx = \int_{-\infty}^{+\infty}\delta(\xi)\Phi\left(\frac{\xi-b}{a}\right)\frac{d\xi}{a} = \frac{1}{a}\Phi\left(-\frac{b}{a}\right). \tag{3.13a,b}$$

This is the **fundamental property** *of the generalized function Dirac unit impulse (3.14b):*

$$\Phi \in C(|R): \quad \int_{-\infty}^{+\infty}\delta(ax+b)\Phi(x)dx = \frac{1}{a}\Phi\left(-\frac{b}{a}\right), \tag{3.14a,b}$$

with regard to continuous (3.12) test functions on the real line (3.14a): the particular cases of (3.14b) include the **translation** *(3.15a)* [**scale** *(3.15b)]* **formula** *for the unit impulse:*

$$\int_{-\infty}^{+\infty}\delta(x-c)\Phi(x)dx = \Phi(c), \quad \int_{-\infty}^{+\infty}\delta(ax)\Phi(x)dx = \frac{1}{a}\Phi(0). \tag{3.15a,b}$$

The latter shows that if the length of the interval is multiplied by a, the Dirac unit impulse is divided by a to preserve the unit integral.

3.2 Derivates of All Orders of the Unit Impulse

The consideration of the generalized function derivative of the unit impulse (Subsection 3.2.1) [derivatives of all higher orders (Subsection 3.2.2)] restricts the class of test functions to be twice (infinitely) differentiable.

3.2.1 Derivate of the Unit Impulse as a Functional

The generalized function **derivate unit impulse** is a functional that assigns to every twice-differentiable function on the real line (3.15a) a number, namely, minus its derivate at the origin:

$$\Phi \in \mathcal{D}^2(|R): \quad \left[\delta'(x),\Phi(x)\right] \equiv \int_{-\infty}^{+\infty}\delta'(x)\Phi(x)dx$$

$$= \left\{\Phi(x)\delta(x)\right\}_{-\infty}^{+\infty} - \int_{-\infty}^{+\infty}\delta(x)\Phi'(x)dx$$

$$= \Phi(+\infty)\delta(+\infty) - \Phi(-\infty)\delta(-\infty) - \left[\delta(x),\Phi'(x)\right] = -\Phi'(0), \tag{3.16a,b}$$

where (1) the terms at $x = \pm \infty$ vanish because $\delta (\pm\infty) = 0$ provided that the test function $\Phi (\pm\infty)$ is finite at infinity; and (2) this is the case since (3.9b,c) was used and hence Φ' must be differentiable, that is, Φ is twice **differentiable**:

$$\mathscr{D}''(a,b) \equiv \left\{ \Phi(x): \quad \forall_{a \leq c \leq b} \quad \exists^1 \quad \Phi^{(n)}(c) \right\}. \tag{3.17}$$

The derivative of the unit impulse (3.18a) can be defined as a functional that assigns to each test function a number, namely, minus its derivative at the origin (3.18c):

$$\delta'(x): \quad \Phi(x) \in C^1(|R) \to \left[\delta'(x), \Phi(x) \right] \equiv \int_{-\infty}^{+\infty} \delta'(x) \Phi(x) dx = -\Phi'(0). \tag{3.18a-c}$$

The test function must be (1) **continuously differentiable** (3.19),

$$C''(a,b) \equiv \left\{ \Phi(x): \quad \forall_{a \leq c \leq b} \quad \exists^1 \quad \lim_{x \to c} \Phi^{(n)}(x) = \Phi^{(n)}(c) \right\}, \tag{3.19}$$

on the real line (3.18b) so that the value of its derivative exists at the origin (3.18c) or any other point; and (2) twice differentiable (3.17) on the real line (3.16a) so that the ordinary rule of integration by parts holds in (3.16b). Performing linear change of variable (3.6a; 3.13a) leads, as for (3.13b) $\equiv$ (3.14b), from (3.16a,b) to

$$\Phi(x) \in C^1(|R): \quad \int_{-\infty}^{+\infty} \delta'(ax+b) \Phi(x) dx = -\frac{1}{a} \Phi'\left(-\frac{b}{a}\right), \tag{3.20a,b}$$

which is *the fundamental property of the derivative unit impulse (3.20b), applied to a test function (3.20a) that is continuously differentiable on the real line.*

3.2.2 *n*th Derivative of the Unit Impulse

Generalizing (3.11c) and (3.16b), it may be expected that the *n*th derivate of the unit impulse has the integral property (3.21b)

$$\Phi \in \mathscr{D}^{n+1}(|R): \quad \int_{-\infty}^{+\infty} \delta^{(n)}(x) \Phi(x) dx = (-)^n \Phi^{(n)}(0), \tag{3.21a,b}$$

that corresponds to (1.70b) and applies to test functions that are $(n + 1)$ times differentiable (3.17) on the real line (3.21a). In the present functional approach, (3.21a,b) is proved by induction: (1) it holds for $n = 1$, when it coincides with (3.16a,b); and (2) if it holds for n, it also holds for $n + 1$:

$$\int_{-\infty}^{+\infty} \delta^{(n+1)}(x) \Phi(x) dx = \left\{ \delta^{(n)}(x) \Phi(x) \right\}_{-\infty}^{+\infty} - \int_{-\infty}^{+\infty} \delta^{(n)}(x) \Phi'(x) dx = (-)^{n+1} \Phi^{(n+1)}(0), \tag{3.21c}$$

where it was assumed that $\Phi(x)$ is $(n + 2)$ times differentiable so it is finite at infinity and the term involving $\delta^{(n)}(\pm\infty) = 0$ vanishes. The ***n*th derivative unit impulse** *is the functional (3.22a) that assigns to each test function a number, namely, the value of its *n*th derivative at the origin (3.22c) with + (−) sign if n is even (odd):*

$$\delta^{(n)}(x): \quad \Phi \in C^n(|R) \to \left[\delta^{(n)}(x), \Phi(x)\right] \equiv \int_{-\infty}^{+\infty} \delta^{(n)}(x)\Phi(x)\,\mathrm{d}x = (-)^n \Phi^{(n)}(0). \tag{3.22a–c}$$

The test function must be (1) n times continuously differentiable (3.19) on the real line (3.22b) for the value (3.22c) to exist at the origin or any other point; and (2) (n + 1) times differentiable (3.20a) for the evaluation of (3.20b) using the ordinary rule of integration by parts.

3.3 Growth/Decay and Support of Test Functions

The generalized functions unit jump and impulse (other derivatives of all orders) defined as functionals [Section 3.1 (Section 3.2)] apply to different classes of test functions from which their properties are inherited. Thus before proceeding to the definition of generalized functions (Section 3.4), some possible classes of test functions are considered (Section 3.3). The two properties most important for the test functions are (1) differentiability (Subsections 3.3.3 and 3.3.4), which determines the differentiability of the generalized function, and (2) behavior at infinity (Subsections 3.3.1 and 3.3.2), which affects the convergence at infinity of the integral over the real line used as linear functional. Concerning behavior at infinity (2), there are two options: (2-1) to specify polynomial or power law or slow growth (decay) for fairly good (good) functions or decay faster than any power for very good functions (Subsection 3.3.1); and (2-2) to impose compact support, that is, vanishing outside a finite interval (Subsection 3.3.2). The functions with compact support can be (1) *n* times differentiable (Subsection 3.3.3); (2) infinitely differentiable or smooth (Subsection 3.3.4); or (3) not analytic, otherwise they vanish. The inclusion of the various classes of test functions (Subsection 3.3.5) shows that the two most important, which have more properties to be inherited by the generalized functions, are (Subsection 3.3.6) the excellent (superlative) functions that are smooth with compact support (fast decay or very good).

3.3.1 Fairly Good, Good, and Very Good Functions

Among the properties of test functions of interest are the behavior at infinity, that is, (1) the **slow growth** or **fairly good** functions of degree n that grow at infinity like (3.23a), slower than a polynomial of degree n; (2) the **slow decay** or **good** functions of degree n that decay at infinity faster than (3.23b), the inverse of a polynomial of degree n; and (3) the **fast decay** or **very good** functions that decay faster at infinity (3.23c) than any inverse power:

$$x \to \pm\infty: \quad \Phi(x) = \begin{cases} o(x^n) & \text{slow growth: } \mathcal{V}^n, \\ o(x^{-n}) & \text{slow decay: } \mathcal{V}_n, \\ o(x^{-n}) \text{ for all } n & \text{fast decay: } \mathcal{V}_\infty. \end{cases} \tag{3.23a–c}$$

The classification (3.23a,b) is equivalent for unspecified degree n to the fairly good (good) functions that grow (decay) at infinity slower (faster) than some polynomials.

$$\bar{\mathcal{V}} \equiv \left\{ \Phi(x): \quad \exists_{n\in N} \lim_{|x|\to\infty} |x|^{-n} \Phi(x) = 0 \right\}, \tag{3.24a}$$

$$\underline{\mathcal{V}} \equiv \left\{ \Phi(x): \quad \exists_{n \in |N} \quad \lim_{|x| \to \infty} |x|^{n} \, \Phi(x) = 0 \right\}, \tag{3.24b}$$

$$\mathcal{V}_{\infty} \equiv \left\{ \Phi(x): \quad \forall_{n \in |Z} \quad \lim_{|x| \to \infty} x^{n} \Phi(x) = 0 \right\}. \tag{3.24c}$$

The very good function corresponds to a good function of infinite order (3.23c) $\equiv$ (3.24c) and decays at infinity faster than any polynomial. For example, (1) a function vanishing at infinity is both of slow growth and decay of order zero; and (2) a function integrable on the real line must be $O(x^{-\alpha})$ with $\alpha > 1$ as $x \to \infty$ so it is a good function of order $n = 1$; and (3) the negative exponential $\exp(-|x|)$ and the Gaussian $\exp(-x^2)$ are very good or fast decay functions.

3.3.2 Compact Support for Ordinary or Generalized Functions

The **support** of an ordinary or generalized function is defined as the set of points where it does not vanish (3.25a):

$$S\{f(x)\} \equiv \{x: \ f(x) \neq 0\}; \quad S\{\delta^{(n)}(x)\} \equiv \{0\}, \quad S\{H(x)\} = (0, +\infty(, \tag{3.25a-c}$$

for example, (1) the support of the unit impulse and its derivates is a point (3.25b) that is the origin and (2) the support of the unit jump is (3.25c) the positive real axis, including the origin if its value there is chosen to be nonzero. An ordinary or generalized function with **compact support** vanishes outside a finite interval (3.26):

$$T^{0}(a,b) \equiv \left\{ \Phi(x): \quad x < a \text{ or } x > b \Rightarrow \Phi(x) = 0 \right\}. \tag{3.26}$$

The ordinary functions with compact support (3.26) are the particular case of order zero of the **excellent functions of order n** that have compact support and are n times differentiable:

$$T^{n}(a,b) \equiv \left\{ \Phi(x): \quad \forall_{x \in |R} \ \exists \ \Phi^{(n)}(x); \quad \exists_{o < a < b \in |R} x < a \text{ or } x > b \Rightarrow \Phi(x) = 0 \right\}. \tag{3.27}$$

The condition of compact support (3.26) possibly combined (3.27) with differentiability (3.17):

$$\mathcal{D}^{n} \supset \mathcal{D}^{n} \cap T^{0} \equiv T^{n} \subset T^{0} \subset \mathcal{V}_{\infty} \subset \mathcal{V}_{n} \subset \mathcal{V}^{n}, \tag{3.28}$$

is more restrictive than the progressively less stringent conditions of (1) fast decay of very good functions (3.23c) $\equiv$ (3.24c) and (2/3) power law or slow decay (growth) of good (fairly good) functions (3.23b) $\equiv$ (3.24b) [(3.23a) $\equiv$ (3.24a)].

3.3.3 Functions with Slow/Fast Decay/Growth

The conditions that the test functions be fairly good or slow growth (3.23a;3.24a) can be replaced by more stringent conditions (3.28) like good or slow decay (3.23b;3.24h), very good or fast decay (3.23c) $\equiv$ (3.24c), or compact support (3.26); the condition (3.27) combines differentiability with compact support, leading to the excellent functions of order n that have compact support together with their derivatives up to order n. If a function has compact support, its derivatives also vanish outside a finite interval, but they may not be continuous. An example of an n times differentiable function with compact support (3.29a,b) is (3.29c,d)

$$n+1>\alpha>n\in|N,\quad \Phi\in T^n(|R),\quad S(\Phi(x))=(-a,+a):\quad \Phi(x)=\begin{cases} 0 & \text{if } |x|\geq a, \\ (x^2-a^2)^\alpha & \text{if } |x|\leq a, \end{cases} \qquad (3.29\text{a-d})$$

because (1) it vanishes outside the interval $(-a,+a)$ and (2) it is n times differentiable in $(-\infty, +\infty)$:

$$k=0,\ldots,n:\quad \lim_{x\to \pm a}\Phi^{(k)}(x)=\lim_{x\to \pm a}O\left((x\mp a)^{\alpha-k}\right)=0, \qquad (3.30)$$

including at $x=\pm a$, where all derivates up to and including order n vanish. The derivatives of order $k=n+1,\ldots$ of (3.29c,d) are infinite as $x\to a-0$ or $x\to a+0$ and zero $\to -a-0$; the infinite discontinuities of the derivatives of order $k>n$ at $x=\pm a$ imply that the function (3.29c,d) is excellent of order n but not excellent of order $n+1$. If $\alpha=m$ was a positive integer, then the mth derivative would be finite at $x\to a-0$ and $x\to -a+0$ and zero as $x\to -a+0$ or $x\to -a-0$; the finite discontinuities of the derivative of order $\alpha=m$ at $x=\pm a$ imply that the function is excellent of order $m-1$ but not excellent of order m. Next is given an example of a function with compact support and also infinitely differentiable or smooth.

3.3.4 Smooth Functions with Compact Support

To consider the derivates of all orders of the unit impulse requires **smooth** test functions that are infinitely differentiable:

$$\mathcal{D}^\infty \equiv \left\{ \Phi(x):\quad \forall_{x\in|R}\quad \forall_{n\in|N}\quad \exists\, \Phi^{(n)}(x) \right\}. \qquad (3.31)$$

The combination of smooth functions (3.31) with compact support (3.26) leads to the excellent functions (3.27) of infinite order, or simply **excellent** functions:

$$T^\infty(a,b)\equiv \left\{\Phi(x):\quad \forall_{x\in|R\,n\in|N}\,\exists\,\Phi^{(n)}(x);\quad x>a \text{ or } x<b \Rightarrow \Phi(x)=0 \right\}. \qquad (3.32)$$

An example of a smooth function (3.33a) with compact support (3.33b) is (3.33c,d)

$$\Phi\in T^\infty(|R):S(\Phi(x))=(a,b):\quad \Phi(x)=\begin{cases} 0 & \text{if } x<a \text{ or } x>b, \\ \exp\left(\dfrac{1}{a-x}+\dfrac{1}{x-b}\right) & \text{if } a<x<b, \end{cases} \qquad (3.33\text{a-d})$$

because (1) it vanishes outside the interval (3.33b) and (2) it is infinitely differentiable in $(-\infty, +\infty)$, including at $x=a, b$:

$$\Phi^{(n)}(x)=O\left(\left((x-a)(x-b)\right)^{-n-1}\right)\exp\left(\dfrac{1}{a-x}+\dfrac{1}{b-x}\right), \qquad (3.34)$$

since (3.34) vanishes as $x \to a + 0$, $b - 0$, because $\exp(-\infty)$ dominates any inverse power; this is due to $\exp$ x having an essential singularity at infinity (Section I.27.5), hence dominating any power:

$$\alpha > 0, n \in | Z : \quad \lim_{x \to \pm\infty} |x|^n \exp\left(-\alpha |x|\right) = 0 = \lim_{x \to \pm\infty} |x|^n \exp\left(-x^2\right); \tag{3.35a–d}$$

this shows that the negative exponential of the modulus (3.35c) [Gaussian function (3.35d)] is of fast decay (2.23c) or a very good function (2.24c). Since the support is the whole real line, neither of these functions is excellent. This leads to the question of which classes of test functions are more restrictive (Subsection 3.3.5) and thus yield more properties (Subsection 3.3.6) for the generalized functions (Section 3.4).

3.3.5 Inclusion of Classes of Test Functions

The reference functions that are (a) of slow growth or fairly good (3.23a; 3.24a), (b) of slow decay or good (3.23b; 3.24b), (c) of fast decay or very good (3.23c; 3.24c), or (d) excellent of order zero or have compact support (3.26) can in addition (1) be continuous (3.12); (2) be integrable (3.2); (3) be differentiable n times (3.17); (4) have nth derivative that is continuous (3.19); (5) be sectionally continuous $\bar{C}$ or (6) differentiable $\mathcal{D}$; (7) be sectionally differentiable n times $\mathcal{D}^n$; (8) have sectionally continuous nth derivative $\bar{C}^n$; (9) be smooth $\mathcal{D}^\infty$, that is, infinitely differentiable (3.31); and (10) be **analytic**, that is, with Taylor series (I.23.32a,b) $\equiv$ (3.36):

$$A(a,b) \equiv \left\{ \Phi(x): \; \forall_{a \le x, c \le b}: \; \Phi(x) = \sum_{n=0}^{\infty} \frac{(x-c)^n}{n!} \Phi^{(n)}(c) \right\}, \tag{3.36}$$

implying that it has derivatives of all orders, that is, smooth, and besides the series converges. The imposition of more properties leads to a smaller set of test functions by both criteria, namely, (a–d) behavior at infinity (3.28); (1–10) continuity/integrability (3.37a,b):

$$\mathcal{E} \supset C \supset \mathcal{D} \supset \mathcal{D}^n \supset C^n \supset \mathcal{D}^\infty \supset A, \quad \mathcal{E} \supset \bar{C} \supset \bar{\mathcal{D}} \supset \bar{\mathcal{D}}^n \supset \bar{C}^n \supset \mathcal{D}^\infty \supset A. \tag{3.37a,b}$$

The combination of an analytic function (3.36) in an interval (α, β) with a smaller (a,b) compact support (3.26) leads to a trivial zero function:

$$b > \beta \text{ or } a < \alpha: \quad \Phi \in A(\alpha, \beta) \cap T^0(a,b) \Rightarrow \Phi(x) = 0, \tag{3.38}$$

because (1) an analytic function has a convergent Taylor series (3.36); (2) choosing a point c outside the support, that is, $a < c < \alpha$ or $\beta < c < b$, all the derivatives vanish; and (3) thus the Taylor series is zero. Thus an analytic function with compact support is zero everywhere. Nonzero smooth functions with compact support exist, for example, (3.33c,d). This function is smooth, that is, infinitely differentiable, but is not analytic: Its Taylor series about $x \ge b$ or $x \le a$ has all coefficients zero, so the function coincides with the remainder of its Taylor series, and the Taylor series does not converge. [Compare with I.27.7a–d in Subsection I.27.1.2.] The preceding discussion of possible classes of test functions (Subsections 3.3.1 through 3.3.5) leads to the two most important, namely, the excellent and superlative (Subsection 3.3.6).

3.3.6 Excellent and Superlative Test Functions

The function (3.33c,d) is an example of an excellent function of order ∞, that is, a smooth function with compact support that may be designated simply an excellent function. An excellent function is an infinitely differentiable or smooth function with compact support (3.32). The excellent functions of class n in (3.27) are n times differentiable (3.17) and compared with the slow decay or good functions of class zero (3.23b; 3.24b), vanish not just at infinity but also outside a finite interval and thus are a more restricted set (3.39a):

$$\mathcal{T}^n \subset \mathcal{D}^n \cap \mathcal{V}_0; \quad \mathcal{T}^\infty \subset \mathcal{D}^\infty \cap \mathcal{V}_\infty \equiv \mathcal{V}_\infty^\infty, \tag{3.39a,b}$$

Likewise the excellent functions are included in, but do not coincide with, the smooth functions of fast decay (3.39b) that may be designated **superlative functions**; the excellent functions are more restrictive than the superlative functions and lead to a fairly wide class of generalized functions, namely, the temperate generalized functions (Subsection 3.4.1). The superlative functions (3.39b) $\equiv$ (3.40) are smooth functions (3.31) with fast decay (3.23c) $\equiv$ (3.24c), that is, they are infinitely differentiable and decay faster than any power at infinity:

$$\mathcal{V}_\infty^\infty(|R) \equiv \left\{ \Phi(x): \quad \forall_{n\in|N} \quad \exists_{\Phi^{(n)}(x)}; \quad \lim_{|x|\to\infty} x^n \Phi(x) = 0 \right\}. \tag{3.40}$$

The Gaussian is a superlative function and remains so when (1) multiplied by a polynomial or (2) replacing in the exponential the square with any positive exponent:

$$a,b > 0: \quad \Phi_0(x) = \exp\left(-ax^{2b}\right) \sum_{n=0}^N C_n x^n; \tag{3.41a-c}$$

for example, the function (3.41a–c) is infinitely differentiable, and its product by any power is dominated by the exponential that vanishes at infinity (3.42a,b):

$$c > 0: \quad \lim_{|x|\to\infty} x^c \Phi_0(x) = 0: \quad \Phi_0(x) \in \mathcal{V}_\infty^\infty(|R), \tag{3.42a-c}$$

hence (3.41a–c) is a superlative function (3.42c). The class of excellent functions (3.32) has all the properties, except being analytic (3.43a):

$$\mathcal{V}^n \supset \mathcal{V}_n \supset \mathcal{V}_\infty \supset \mathcal{T}^0 \supset \mathcal{T}^n \supset \mathcal{T}^\infty \subset \mathcal{V}_\infty^\infty, \tag{3.43a}$$

$$\mathcal{A} \supset \mathcal{V}_\infty^\infty \subset \mathcal{D}^\infty \subset C^n \subset \overline{C}^n \subset \mathcal{D}^n \subset \overline{\mathcal{D}}^n \subset \mathcal{D} \subset \overline{\mathcal{D}} \subset C \subset \overline{C} \subset \mathcal{E}. \tag{3.43b}$$

The superlative functions (3.40) can be analytic (3.43b). The excellent functions will be used as test functions next (Section 3.4) in the general definition of temperate generalized functions, because it is sufficiently restricted to provide all the properties needed in most applications. For specific properties of some generalized functions, it is not necessary to restrict to the superlative or excellent test functions, and more general classes of test functions that provide minimum conditions of validity will be indicated.

3.4 Generalized Function as a Continuous Functional

The definition of generalized function as a linear continuous functional over the excellent test functions (Subsection 3.4.1) leads to a rich set of properties, for example, generalized functions are infinitely differentiable (Subsection 3.4.2). Many properties of specific generalized functions like the unit impulse or the sign functions hold over classes of test functions much less restrictive than the excellent functions (Subsection 3.4.6). The general method of proof of the properties of generalized functions is to pass these properties to the test functions, for example, the even (odd) generalized functions are defined (Subsection 3.4.3) using odd (even) test functions. The linear functional used to define a generalized function corresponds to the inner product of two functions (Subsection 3.4.4) and can be extended to complex functions (Subsection 3.4.5) and higher dimensions (Chapter 5). The desirable property of generalized functions of being infinitely differentiable (Subsection 3.4.2) brings the limitations of a linear operator: the square or power of a generalized function cannot be defined, otherwise serious contradictions may arise (Subsection 3.4.7).

3.4.1 Temperate Generalized Functions

A **generalized function** $F(x)$ is defined as a linear, continuous **functional** over a set of reference or test functions that assigns to each a real number:

$$F \in G: \quad \Phi(x) \in \mathcal{J}^{\infty}(|R): \quad \rightarrow \left[F(x), \Phi(x)\right] \equiv \int_{-\infty}^{+\infty} F(x)\Phi(x)\mathrm{d}x \in |R; \qquad (3.44a\text{-}c)$$

$$\alpha, \beta \in |R; \quad \Psi(x) \in \mathcal{J}^{\infty}: \quad \left[F(x), \alpha\Phi(x)+\beta\Psi(x)\right] = \alpha\left(F(x),\Phi(x)\right)+\beta\left[F(x),\Psi(x)\right], \quad (3.45a\text{-}d)$$

$$n \in |R; \quad \Phi_n(x) \in \mathcal{J}^{\infty}: \quad \lim_{n\to\infty}\Phi_n(x) = \Phi(x) \Rightarrow \lim_{n\to\infty}\left[F(x),\Phi_n(x)\right] = \left[F(x),\Phi(x)\right]. \quad (3.46a\text{-}c)$$

A **temperate** *generalized function uses as test functions the excellent functions (3.44b)* ≡ *(3.32), that is, the smooth functions with compact support.* This definition of generalized function requires the functional (3.44a–c) to be both continuous (3.46a–c) and linear (3.45a–c). The properties of generalized functions are defined by "transferring" them to the test functions. Thus a generalized function has more (less) properties, if the set of test functions over which it is defined has more (less) properties, that is more (less) restricted; for example, the generalized functions over excellent (superlative) test functions have more properties than over other test functions in (3.43a) [(3.43b)]. Also, the excellent (3.32) test functions are more restrictive than the superlative (3.40) test functions and thus lead to a wider class of generalized functions.

3.4.2 Infinitely Differentiable Generalized Functions

The approach to be followed is to consider the widest class of test functions for which the generalized function has the corresponding properties; for example, the derivate of a generalized function $F(x)$ is defined by transferring the derivate to the test function with reversed sign [e.g., in (3.18a through c) or (3.21a through c)], as in an integration by parts (3.47a):

$$\left[F'(x),\Phi(x)\right] = -\left[F(x),\Phi'(x)\right]; \quad \left[F^{(n)}(x),\Phi(x)\right] \equiv (-)^n\left[F(x),\Phi^{(n)}(x)\right], \qquad (3.47a,b)$$

it follows (3.47b) that the *n*th derivate of a generalized function $F^{(n)}(x)$ is passed over (3.22a–c) to the test function, with the same (reversed) sign if n is even (odd). The test function must be continuously differentiable [*n* times differentiable] (3.19) in (3.47a) [(3.47b)]. If the test functions are smooth (3.31), they have derivates of all orders, and thus follows the **fundamental property of generalized functions**: *the generalized functions (3.48a)* ≡ *(3.44a–c) over smooth test functions (3.48b)* ≡ *(3.31) are infinitely differentiable (3.48c)*:

$$F(x) \in \mathcal{G}; \Phi \in \mathcal{D}^{\infty}(|R): \quad \left[F^{(n)}(x), \Phi(x)\right] \equiv (-)^n \left[F(x), \Phi^{(n)}(x)\right] = (-)^n \int_{-\infty}^{+\infty} F(x)\Phi^{(n)}(x)\mathrm{d}x. \quad (3.48a\text{–}c)$$

The preceding definitions apply to multidimensional generalized functions of several variables (Chapter 5) as well as to a generalized function of one real variable (Chapter 1). Another example of borrowing properties of the generalized from the test functions is the consideration of even/odd ordinary/generalized functions (Subsection 3.4.3); also passing from ordinary to generalized functions allows differentiation at simple discontinuities (Sections 1.8 and 1.9) and at singularities such as the branch-cut of the logarithm (Subsection 3.4.5).

3.4.3 Even/Odd Test and Generalized Functions

Consider the subsets of even (odd) test functions:

$$\mathcal{V}_{e,d} \equiv \left\{\Phi(x): \quad \Phi(x) \in \mathcal{V} \wedge \Phi(-x) = \pm\Phi(x)\right\}. \tag{3.49}$$

The set of **even** $\mathcal{G}_e$ $(\mathcal{G}_d)$ **(odd) generalized functions** $E(x)$ $[D(x)]$ are those whose inner product by odd (even) integrable test functions is zero:

$$E(x) = E(-x) \in \mathcal{G}_e: \quad \Phi(x) \in \mathcal{V}_d(|R) \Rightarrow \left[E(x), \Phi(x)\right] = 0, \tag{3.50a,b}$$

$$D(x) = -D(-x) \in \mathcal{G}_d: \quad \Phi(x) \in \mathcal{V}_e(|R) \Rightarrow \left[D(x), \Phi(x)\right] = 0. \tag{3.51a,b}$$

For example, an odd test function vanishes at the origin (3.52a):

$$\Phi(x) = -\Phi(-x): \quad \left[\delta(x), \Phi(x)\right] \equiv \Phi(0) = -\Phi(0) = 0 \Rightarrow \delta(x) = \delta(-x), \tag{3.52a\text{–}c}$$

and hence its inner product with the unit impulse vanishes (3.52b); it follows that the unit impulse is an even generalized function (3.52c), and its derivatives of even (odd) order are even (odd) generalized functions (1.85a,b).

3.4.4 Relation with Inner Product and Norm of Square-Integrable Functions

The linear functional (3.44a–c) used to define a generalized function over a set of test functions corresponds (II.5.137a–d) ≡ (3.53a–d) to the **inner product** of functions:

$$f, g \in \mathcal{E}(|R^2), w \in |R: \quad \left[f(z), g(z)\right] = \int_a^b f(z)g^*(z)w(z)\mathrm{d}z, \tag{3.53a\text{–}d}$$

defined in (3.53d) for complex functions (3.53a,b) with real **weighting function** (3.53c). The simplest weighting function is unity (3.54a), and the inner product of a function by itself (3.54b) is the integral of the square of the modulus:

$$w(z) = 1: \quad [f(z), f(z)] = \int_a^b f(z) f^*(z) \, dz = \int_a^b |f(z)|^2 \, dz = \|f(z)\|_2, \qquad (3.54a,b)$$

corresponding to the L^2-norm (7.2a,b) for square integrable functions. An example of a function of complex variable (3.55a) is the logarithm represented by its principal branch (I.7.10b) ≡ (3.55b):

$$z = r e^{i\varphi}: \quad \log z = \log r + i\varphi = \log(z) + i \arg(z), \qquad (3.55a,b)$$

whose real (imaginary) part is (3.55b) ≡ (I.1.23a,b) the logarithm of the modulus (the argument of the variable). The unit impulse can be used to differentiate ordinary functions at simple discontinuities (Section 1.8 and 1.9) where the derivate does not exist in the ordinary sense; it also allows differentiation at singularities such as the branch-point and branch-cut (Chapters I.7 and I.9) of the logarithm along the negative real axis (Subsection 3.4.5).

3.4.5 Derivatives of the Logarithm on the Branch-Cut

The logarithm of a real variable is taken along the whole real axis (3.56a) as the principal branch (3.55b) of its value on the complex plane, leading to (3.56b,c)

$$x \in R: \quad \log x = \log|x| + i \arg(x) = \begin{cases} \log|x| & \text{if } x > 0, \\ \log|x| - i\pi & \text{if } x < 0. \end{cases} \qquad (3.56a\text{–}c)$$

The jump in the imaginary part across the origin can be represented by the unit jump:

$$\log x = \log|x| - i\pi H(-x). \qquad (3.57)$$

Differentiating leads to

$$\frac{d}{dx}(\log x) = \frac{d\{\log|x|\}}{d|x|} \frac{d|x|}{dx} - i\pi \frac{d[H(-x)]}{d(-x)} \frac{d(-x)}{dx}$$

$$= |x|^{-1} \operatorname{sgn} x + i\pi \delta(-x) = \frac{1}{x} + i\pi\delta(x), \qquad (3.58)$$

where the properties of the sign function (1.89a,b) are used and the unit impulse is even (3.52c). Thus

$$\frac{d^{n+1}}{dx^{n+1}}(\log x) = (-)^n x^{-1-n} + i\pi\delta^{(n)}(x), \qquad (3.59)$$

specifies *the derivates of all orders (3.59) of the logarithm (3.57) ≡ (3.56b,c) of real variable (3.56a) including on the branch-cut (– ∞, 0) along the negative real axis.*

3.4.6 Test and Other Alternative Reference Functions

The excellent test functions (3.44a) are used in the definition of temperate generalized functions (3.44b) because they yield a usefully wide range of properties. Some properties of specific generalized functions hold for classes of test functions much wider and less restrictive than the excellent functions. Numerous applications of generalized functions in physics and engineering concern test functions that are not excellent (superlative), for example, they may not vanish outside a finite interval (decay fast enough at infinity). Thus, it is of some interest to prove the properties of specific generalized functions for the widest class of test functions possible. Two examples are given next.

The first example is the derivative of the product of the unit impulse by an ordinary function:

$$f \in \mathcal{D} \cap \mathcal{V}^0(|R): \quad \{f(x)H(x-a)\}' = f'(x)H(x-a) + f(a)\delta(x-a), \qquad (3.60a,b)$$

This is analogous to the derivate of a product and was used before (1.102b; 1.106b). To prove (3.60b) from the definition (3.47a) of the derivative of a generalized function, multiply by a differentiable test function (3.10) and integrate from $-\infty$ to $+\infty$:

$$\left[\{f(x)H(x-a)\}', \Phi(x)\right] \equiv -\left[f(x)H(x-a), \Phi'(x)\right] = -\int_{-\infty}^{+\infty} f(x)H(x-a)\Phi'(x)\,dx$$

$$= -\int_a^\infty f(x)\Phi'(x)\,dx = -\left[f(x)\Phi(x)\right]_a^\infty + \int_a^\infty f'(x)\Phi(x)\,dx$$

$$= f(a)\Phi(a) + \int_{-\infty}^{+\infty} f'(x)H(x-a)\Phi(x)\,dx$$

$$= \left[\{f(a)\delta(x-a) + f'(x)H(x-a)\}, \Phi(x)\right], \qquad (3.61)$$

where the property $\Phi(\infty) = 0$ was used, assuming that $\Phi(x)$ is a slow decay or good function. Thus, *the rule of derivation of the product of the unit impulse by an ordinary function (3.60b) holds with regard to good differentiable functions of order zero (3.60a) on the real line*, that is, differentiable functions vanishing at infinity that are included in the excellent functions (3.27) of order $n = 1$.

A second example is the application of the derivative impulse in the calculation of the second derivate of the sign function:

$$\left\lfloor \text{sgn}''(x), \Phi(x)\right\rfloor \equiv \left\lfloor \text{sgn}(x), \Phi''(x)\right\rfloor dx = \int_{-\infty}^{+\infty} \text{sgn}(x)\Phi''(x)\,dx$$

$$= \left\{\text{sgn}(x)\Phi'(x)\right\}_{-\infty}^{+\infty} - \int_{-\infty}^{+\infty} \text{sgn}'(x)\Phi'(x)\,dx$$

$$= -\int_{-\infty}^{+\infty} 2\delta(x)\Phi'(x)\,dx = -2\Phi'(0) = 2\left[\delta'(x), \Phi(x)\right]; \qquad (3.62)$$

This result $(3.62) \equiv (3.63b)$

$$\Phi \in \mathcal{D}^2 \cap \mathcal{V}^1(|R): \quad \frac{d^2}{dx^2}\big[\operatorname{sgn}(x)\big] = 2\delta'(x) = 2H''(x) = 2H''(-x), \qquad (3.63a,b)$$

agrees with (1.88b) for $n = 2$ and is valid for (3.63a) fairly good twice-differentiable test functions of order one $\Phi(x) \sim o(x)$ for which $\Phi'(x) \sim o(1)$ so $\Phi'(\pm\infty) = 0$, cancelling the term in curly brackets in (3.62).

3.4.7 Advantages and Limitations of Generalized Functions

The generalized functions have, compared with ordinary functions, (1) the advantage of being infinitely differentiable (Subsection 3.4.2) and (2) the limitation of being linear functionals, thus excluding powers of a generalized function. Two examples are given next of the problems that can arise if an attempt is made to define the square (power) of a generalized function, for example, the unit impulse (jump). The square of the unit impulse applied to a continuous test function

$$\Phi(x) \in C(|R): \quad \Big[\big(\delta(x)\big)^2, \Phi(x)\Big] \equiv \int_{-\infty}^{\infty} \Phi(x)\delta(x)\delta(x)dx = \Phi(0)\delta(0), \qquad (A)$$

leads to (1) infinity $\delta(0) = \infty$ for $\Phi(0) \neq 0$; and (2) for $\Phi(0) = 0$ to an indeterminacy $0 \times \infty = ?$ The N-th power of the unit jump (1.28a–c) could be defined by

$$\big[H(x)\big]^N = \begin{cases} 0 & \text{if } x < 0 \\ 2^{-N} & \text{if } x = 0 \\ 1 & \text{if } x > 0. \end{cases} \qquad (B)$$

Bearing in mind that it has a unit jump at the origin, its derivative should be

$$\frac{d}{dx}\big\{\big[H(x)\big]\big\}^N = \delta(x). \qquad (C)$$

Applying the rule of differentiation of the power leads to

$$\frac{d}{dx}\big\{\big[H(x)\big]^N\big\} = N\big[H(x)\big]^{N-1}\delta(x) = N\big[H(0)\big]^{N-1}\delta(x) = 2^{1-N}N\delta(x). \qquad (D)$$

Comparing (C) and (D), it follows that (1) they agree in the linear case $N = 1$, as they must; (2) they also happen to agree in the quadric case $N = 2$; and (3) they disagree for all other positive integers $N = 3.4...$ The preceding examples show that it is not possible to define consistently powers or nonlinear operators of generalized functions. This has implications for multidimensional generalized functions (Chapter 5) with several independent variables.

3.5 Series of Impulses, Jumps, and Ramps

The analytic test functions lead to a Taylor series (Subsection 3.5.1) for the generalized function Dirac (Heaviside) unit impulse (jump); the successive integrals of the unit impulse or jump lead to ramp functions (Subsection 3.5.2) that can be used in a series expansion for an arbitrary function that is analytic (zero) in the positive (negative) real axis.

3.5.1 Taylor Series for the Unit Jump and Impulse

A set more restricted than the smooth functions is that of analytic functions (3.36) on the real line:

$$\mathcal{A}\left(|R\right)\equiv\left\{\Phi(x):\ \ \Phi(x)=\sum_{n=0}^{\infty}\frac{x^n}{n!}\Phi^{(n)}(0)\right\}, \tag{3.64}$$

for which the Taylor series may be expanded around the origin with an infinite radius of convergence, leading to the Stirling–MacLaurin series (I.23.34a,b). An example of an analytic fairly good (very good) function is a polynomial (the Gaussian function). The MacLaurin series of an analytic function (3.64) may be rewritten in terms of the unit impulse (3.11c) and its derivates (3.22c):

$$\left[\delta(x-a),\Phi(x)\right]=\Phi(a)=\sum_{n=0}^{\infty}\frac{a^n}{n!}\Phi^{(n)}(0)=\left[\sum_{n=0}^{\infty}\frac{a^n}{n!}(-)^n\,\delta^{(n)}(x),\Phi(x)\right]. \tag{3.65}$$

The formula (3.65) ≡ (3.66b) is a Stirling–MacLaurin series (I.23.34a,b) for the generalized function unit impulse:

$$\Phi\in\mathcal{A}\left(|R\right):\ \ \delta(x+a)=\sum_{n=0}^{\infty}\frac{a^n}{n!}\delta^{(n)}(x), \tag{3.66a,b}$$

where $-a$ was replaced by a; integrating (3.66b) leads to the unit jump (3.67b):

$$\Phi\in\mathcal{A}\left(|R\right):\ \ H(x+a)=H(x)+\sum_{n=0}^{\infty}\frac{a^n}{n!}\delta^{(n-1)}(x). \tag{3.67a,b}$$

It has been shown that the Taylor *series for the generalized function unit impulse (3.66b) [jump (3.67b)] hold over the set analytic (3.36) test functions on the real line (3.64).* The set of smooth test functions in (3.66a) ≡ (3.67a) compared with the superlative functions (3.40) (1) is more restricted in that it replaces smoothness (3.31) by analyticity (3.36) and (2) is less restricted in that it does not require fast decay (3.23c) ≡ (3.24c). The properties (3.66b; 3.67b) are not relevant for excellent test functions (3.32) because (3.38) an analytic test function with compact support is zero.

3.5.2 Expansion in a Series of Ramp Functions

The differentiations go from less to more singular generalized functions, for example, from the unit jump (Figure 3.1a) to the unit impulse (Figure 3.1b). Conversely, integration smoothes out singularities,

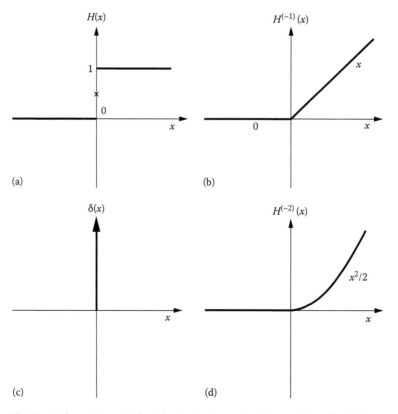

FIGURE 3.1 The Heaviside unit jump (a) leads by derivation to the Dirac unit impulse (b); by integration once (twice), the unit jump leads to the unit linear (c) [quadratic (d)] ramp function that is continuous (has continuous derivate) with unit jump at the origin for the slope (curvature).

for example, the primitive of the unit jump has a unit jump in the slope and is the **unit linear ramp** function (Figure 3.1c):

$$\delta^{(-2)}(x) \equiv H^{(-1)}(x) = \int_{-\infty}^{x} H(\xi)\mathrm{d}\xi = \begin{cases} 0 & \text{if } x < 0, \\ x & \text{if } x > 0. \end{cases} \qquad (3.68a,b)$$

Integrating once more leads to the **unit quadratic ramp** function that has a unit jump in the curvature (Figure 3.1d):

$$\delta^{(-3)}(x) \equiv H^{(-2)}(x) \equiv \int_{-\infty}^{x} H^{(-1)}(\xi)\mathrm{d}\xi = \begin{cases} 0 & \text{if } x \leq 0, \\ \dfrac{x^2}{2} & \text{if } x \geq 0. \end{cases} \qquad (3.69a,b)$$

After n integrations, **nth power ramp** is obtained:

$$\delta^{(-n-1)}(x) \equiv H^{(-n)}(x) = \begin{cases} 0 & \text{if } x \leq 0 \\ \dfrac{x^n}{n!} & \text{if } x \geq 0, \end{cases} \qquad (3.70a,b)$$

which has a unit jump in the nth derivative. Considering the following function,

$$F(x) = \begin{cases} 0 & \text{if } x < 0, \\ f(x) \equiv \sum_{n=0}^{\infty} \dfrac{x^n}{n!} f^{(n)}(0) & \text{if } x > 0. \end{cases} \tag{3.71a,b}$$

which is zero for x negative (3.71a) and analytic for x positive (3.71b), it can be represented as a **series of power ramps**:

$$f(x)H(x) = \sum_{n=0}^{\infty} f^{(n)}(0)H^{(-n)}(x) = \sum_{n=0}^{\infty} f^{(n)}(0)\delta^{(-n-1)}(x). \tag{3.72}$$

The series of impulses (3.66a,b) and jumps (3.67a,b) [ramps (3.72)] is a linear combination of derivates (primitives) of the unit jump and impulse.

3.6 Product of a Power and a Derivate of an Impulse

The power function x^α with $\alpha > 0$ vanishes only at the origin, whereas the impulse $\delta(x)$ and its derivates $\delta^{(n)}(x)$ vanish everywhere except at the origin, where they are singular; thus, the product $x^\alpha \delta^{(n)}(x)$ is zero everywhere, except possibly at the origin where a $0 \times \infty$ indeterminate product must be evaluated. The integral (Subsection 3.6.3) and nonintegral powers are considered separately, starting with the latter and distinguishing the generalized function unit impulse (its derivatives) as a factor [Subsection 3.6.1 (Subsection 3.6.2)]. The product of an integral power by the derivative of a unit impulse (Subsection 3.6.3) may be generalized, replacing (Example 10.8) the integral power by an analytic function (Subsection 3.6.4) and using the linear property of generalized functions; a particular case is the product of a derivative unit impulse by a polynomial (Subsection 3.6.5).

3.6.1 Product of a Nonintegral Power by a Unit Impulse

First is proved a generalization of (1.79d,e):

$$\alpha > 0: \quad x^\alpha \delta(x) = 0 \Leftrightarrow \delta(x) = o(x^{-\alpha}), \tag{3.73a-c}$$

which shows that *the unit impulse is a singularity weaker than any power (3.73b,c) that has a positive exponent (3.73a)*. The proof follows from the functional (3.74b):

$$\Phi \in C(|R): \quad \left[x^\alpha \delta(x), \Phi(x)\right] = \int_{-\infty}^{+\infty} x^\alpha \Phi(x)\delta(x)dx = \lim_{x \to 0} x^\alpha \Phi(x) = \Phi(0)\lim_{x \to 0} x^\alpha = 0, \tag{3.74a,b}$$

for a test function (3.74a) continuous at $x = 0$, since $x^\alpha \to 0$ as $x \to 0$ for $\alpha > 0$. Comparing with direct use of (3.11a)

$$\alpha > 0: \quad \int_{-\infty}^{+\infty} x^\alpha \delta(x)dx = \lim_{x \to 0} x^\alpha = 0, \tag{E}$$

it follows that (1) the definition of a generalized function (3.44a–c) requires the product by a test function (3.74a,b), before integration; (2) the function x^α is not differentiable at infinity for $\alpha > 1$ as required in (3.9a); and (3) the function x^α is continuous at $x = 0$ for $\alpha > 0$ as required in (3.11b).

3.6.2 Product of a Nonintegral Power by a Derivative Unit Impulse

In order to generalize the property to the nth derivate of the unit impulse, the chain rule for the derivative of the product (I.13.31) is used in (3.75b)

$$\Phi \in \mathcal{D}^n\big(|R\big): \quad \Big[x^\alpha \delta^{(n)}(x), \Phi(x)\Big]$$

$$= \int_{-\infty}^{+\infty} x^\alpha \Phi(x)\delta^{(n)}(x)\,dx = (-)^n \lim_{x\to 0}\frac{d^n}{dx^n}\Big[x^\alpha \Phi(x)\Big]$$

$$= (-)^n \sum_{k=0}^{n}\binom{n}{k}\lim_{x\to 0}\Phi^{(n-k)}(x)\frac{d^k}{dx^k}\big(x^\alpha\big)$$

$$= (-)^n \sum_{k=0}^{n}\binom{n}{k}\Phi^{(n-k)}(0)\,\alpha(\alpha-1)\cdots(\alpha-k+1)\lim_{x\to 0}x^{\alpha-k} = 0, \qquad (3.75a,b)$$

where the test function is n times differentiable (3.75a), and all terms vanish for $k = 0, 1,\ldots, n$ if (3.76a) is met. Thus

$$\alpha > n: \quad x^\alpha \delta^{(n)}(x) = 0 \Leftrightarrow \delta^{(n)}(x) = o\big(x^{-\alpha}\big), \qquad (3.76a,b)$$

the nth derivative impulse is (3.76b) a weaker singularity than $x^{-\alpha}$ if the condition (3.76a) is met. The particular case $n = 0$ of (3.76a,b) is (3.73a,b). The use of (3.22a,b)

$$\alpha > n: \quad \int_{-\infty}^{+\infty} x^\alpha \delta^{(n)}(x)\,dx = (-)^n \lim_{x\to 0}\frac{d^n}{dx^n}\big(x^\alpha\big) = (-)^n a(\alpha-1)\cdots(\alpha-n+1)\lim_{x\to 0}x^{\alpha-n} = 0, \qquad (F)$$

leads to the same comments as before that (1) the definition of a generalized function (3.44a–c) requires the inner product by a test function; (2) x^α is not $(n + 1)$ times differentiable at infinity for $\alpha > n + 1$, as required in (3.21a); and (3) x^α has continuous nth derivatives at the origin $x = 0$ for $\alpha >, n$ as required in (3.22b).

3.6.3 Product of an Integral Power by a Unit Impulse or Derivative Impulse

After the product of the nth derivative impulse by a nonintegral power (3.76a,b), it remains to consider $\alpha = m$ an integral power, in which case (3.75b) is replaced by

$$\Phi \in \mathcal{D}^\infty\big(|R\big): \quad [x^m \delta^{(n)}(x), \Phi(x)] \equiv \int_{-\infty}^{+\infty} x^m\,\Phi(x)\,\delta^{(n)}(x)\,dx$$

$$= (-)^n \lim_{x\to 0}\frac{d^n}{dx^n}[x^m\,\Phi(x)]$$

$$= (-)\sum_{k-0}^{n}\binom{n}{k}\Phi^{(n-k)}(0)\lim_{x\to 0}\frac{d^k}{dx^k}(x^m), \qquad (3.77)$$

where were used: (1) the definition of inner product (3.44c), (2) the property (3.22a–c) of the nth order derivative of the unit impulse, (3) the Leibniz rule (I.13.31) for the nth order derivative of the product of two functions. In (3.77) two cases arise: (1) if $m > n \geq k$ the kth term is $O(x^{m-k})$ and vanishes as $x \to 0$ for all $k = 0,\ldots, n$, and the result is zero; and (2) if $m \leq n$, then the terms with $k = 0, 1,\ldots, m - 1$ vanish for the same reason, those with $k = m + 1,\ldots, n$ also vanish because $d^k(x^m)/dx^k = 0$, and only the $k = m$ term remains:

$$n \geq m: \quad \left[x^m \delta^{(n)}(x), \Phi(x) \right] = (-)^n \binom{n}{m} \Phi^{(n-m)}(0) \lim_{x \to 0} \frac{d^m(x^m)}{dx^m}$$

$$= (-)^n m! \binom{n}{m} (-)^{n-m} \left[\delta^{(n-m)}(x), \Phi(x) \right]$$

$$= (-)^m \frac{n!}{(n-m)!} \left[\delta^{(n-m)}(x), \Phi(x) \right]. \tag{3.78a,b}$$

It has been proved that *the product of the mth power by the nth derivate of the unit impulse is either zero (3.79a) for m > n or a constant times an impulse (3.79b) [derivative impulse (3.79c)] for m = n (m < n):*

$$x^m \delta^{(n)}(x) = \begin{cases} 0 & \text{if } m > n, \\ n!(-)^n \delta(x) & \text{if } m = n, \\ (-)^m \dfrac{n!}{(n-m)!} \delta^{(n-m)}(x) & \text{if } m < n. \end{cases} \tag{3.79a–c}$$

The result (3.79a) is consistent with (3.76a,b) for $\alpha = m > n$, and (3.79b) agrees with (3.78b) $\equiv$ (3.79c) for $m = n$; the formula (1.80a,b) used before (Subsection 1.6.2) corresponds to (3.73a) with $m - n = 1$. Table 3.1 indicates the products of powers with exponent m and derivatives of order n of the unit impulse up to six $m = 1,\ldots, 6; n = 0,\ldots, 6$. The linear property of the generalized functions (3.45a–d), in particular applied to the derivative unit impulse, extends (3.79a–c) to any analytic function (Subsection 3.4.4), and in particular to polynomials (Subsection 3.4.5).

TABLE 3.1 Products of Powers and Derivatives of the Unit Impulse

Product	x	x^2	x^3	x^4	x^5	x^6
$\delta(x)$	0	0	0	0	0	0
$\delta'(x)$	$-\delta(x)$	0	0	0	0	0
$\delta''(x)$	$-2\delta'(x)$	$2\delta(x)$	0	0	0	0
$\delta'''(x)$	$-3\delta'(x)$	$6\delta'(x)$	$-6\delta(x)$	0	0	0
$\delta^{iv}(x)$	$-4\delta^{iii}(x)$	$12\delta''(x)$	$-24\delta'(x)$	$24\delta(x)$	0	0
$\delta^{v}(x)$	$-5\delta^{iv}(x)$	$20\delta'''(x)$	$-60\delta''(x)$	$120\delta'(x)$	$-120\delta(x)$	0
$\delta^{vi}(x)$	$-6\delta^{v}(x)$	$30\delta^{iv}(x)$	$-120\delta'''(x)$	$360\delta''(x)$	$-720\delta'(x)$	$720\delta(x)$

Note: Product of the unit impulse and its derivatives up to order six ($n = 0, 1,\ldots, 6$) by the powers of the argument with exponent up to six also ($m = 1, 2,\ldots, 6$) according to (3.79a–c).

3.6.4 Generalized Substitution Rule for the Derivative Impulse

If the function $\Phi(x)$ has continuous derivatives (3.80a) of order $n + 1$, then its Stirling–MacLaurin series (3.64) is replaced by an expansion (3.80b) consisting of the first $(n + 1)$ terms (I.23.45a,b) plus a remainder (I.23.46a–c) of higher order corresponding to an extended mean-value theorem:

$$\Phi \in \mathcal{D}^{n+1}\left(|R\right): \quad \Phi(x) = \sum_{m=0}^{n} \frac{x^m}{m!} \Phi^{(m)}(0) + O\left(x^{n+1}\right). \tag{3.80a,b}$$

Bearing in mind (3.79b,c), it follows that

$$\delta^{(n)}(x), \Phi(x) = \sum_{m=0}^{n} \frac{\Phi^{(m)}(0)}{m!} x^n \delta^{(n)}(x) + O\left(x^{n+1}\delta^{(n)}(x)\right)$$

$$= \sum_{k=0}^{n} (-)^m \, \Phi^{(m)}(0) \frac{n!}{m!(n-m)!} \delta^{(n-m)}(x). \tag{3.81}$$

A change of variable $x \to x - \xi$ leads to

$$f \in \mathcal{D}^{n+1}\left(|R\right): \quad \delta^{(n)}(x-\xi)f(x) = \sum_{m=0}^{n} \binom{n}{m} f^{(m)}(\xi)(-)^m \, \delta^{(n-m)}(x-\xi). \tag{3.82a,b}$$

The case $n = 0$ is the substitution formula (1.78b), and the following cases $n = 1$ ($n = 2$) are (3.83a and b) [(3.84a and b)]

$$f \in \mathcal{D}^{2}\left(|R\right): \quad f(x)\delta'(x-\xi) = f(\xi)\delta'(x-\xi) - f'(\xi)\delta(x-\xi). \tag{3.83a,b}$$

$$f \in \mathcal{D}^{3}\left(|R\right): \quad f(x)\delta''(x-\xi) = f(\xi)\delta''(x-\xi) - 2f'(\xi)\delta'(x-\xi) + f''(\xi)\delta(x-\xi). \tag{3.84a,b}$$

The substitution formula (1.78b) was proved for an integrable function in x uniformly continuous at ξ; both conditions are met by a differentiable function in x, which is the case when (3.82b) holds for $n = 0$. Likewise, (3.83b) [(3.84b)] hold for twice (three times)-differentiable functions, using the approximation of the function by the first three (four) terms of its Taylor series. The generalization (3.82b) holds (3.82a) for an $(n + 1)$ times differentiable function and thus also for an analytic function (3.64).

3.6.5 Product of a Derivative Impulse by a Polynomial

A particular case of an analytic function is a polynomial:

$$f(x) \equiv P_N(x) = \sum_{k=0}^{N} a_k x^k, \ P_N^{(m)}(0) = m!a_m, \tag{3.85a,b}$$

for which (3.82b) simplifies to

$$\delta^{(n)}(x)P_N(x)=\sum_{m=0}^{\leq n,N}\frac{n!}{(n-m)!}(-)^m a_m\delta^{(n-m)}(x).$$ (3.86)

This can be checked directly from

$$\Phi\in\mathcal{D}^{n+1}(|R):\quad\left[\delta^{(n)}(x)P_N(x),\Phi(x)\right]=\left[\delta^{(n)}(x),P_N(x)\Phi(x)\right]$$

$$=(-)^n\lim_{x\to0}\frac{d^n}{dx^n}\left[P_N(x)\Phi(x)\right]$$

$$=(-)^n\lim_{x\to0}\sum_{m=0}^{\leq n}\binom{n}{m}\Phi^{(n-m)}(x)\frac{d^m}{dx^m}\left[P_N(x)\right]$$

$$=(-)^n\sum_{m=0}^{\leq n}\frac{n!}{m!(n-m)!}P_N^{(m)}(0)\Phi^{(n-m)}(0)$$

$$=(-)^n\sum_{m=0}^{\leq n,N}\frac{n!}{(n-m)!}a_m(-)^{n-m}\left[\delta^{(n-m)}(x),\Phi(x)\right]$$

$$=\left[\sum_{m=0}^{\leq n,N}\frac{n!}{(n-m)!}(-)^m a_m\delta^{(n-m)}(x),\Phi(x)\right],$$ (3.87a,b)

using (1) the definition of the *n*th derivative of a generalized function (3.47b), applied to the unit impulse (3.21a,b); (2) the Leibnitz chain rule of differentiation (I.13.31); and (3) the derivatives (3.85b) of the polynomial (3.85a). It has been shown that *the product of an (n + 1) times differentiable function (3.82a) by the nth derivate of the unit impulse is given generally by (3.82b), and in particular (1) for n = 0 by the substitution rule (1.78a,b), (2) for n = 2(n = 3) by (3.83a and b) [(3.84a and b)], and (3) for polynomials (3.85a) by (3.86).*

3.7 Algebraic Equations Involving Generalized Functions

Since the product of a power by a derivative unit impulse can be zero, an algebraic equation can have nonzero solutions for generalized functions when only a trivial or zero solution would exist for ordinary functions; this is demonstrated for homogeneous and inhomogeneous algebraic equations involving [Subsections 3.7.1 (3.7.2)] the product of a generalized function by its variable (a power of its variable).

3.7.1 Product of a Generalized Function by Its Variable

If $f(x)$ is an ordinary continuous function, the only solution of $xf(x)$ is $f(x) = 0$; but this is not the case for a generalized function:

$$F(x)\in\mathcal{G};\Phi\in\mathcal{D}(|R):\quad xF(x)=0\Leftrightarrow F(x)=C\delta(x),$$ (3.88a–d)

where *C* is an arbitrary constant. The proof that (3.88d) is a solution of (3.88b) involving the generalized function (3.88a) follows from

$$F(x)=C\delta(x)\Rightarrow xF(x)=Cx\delta(x)=0,$$ (3.89a,b)

where (3.79a) was used with $m = 1, n = 0$, or (3.74b) with $\alpha = 1$. Having proved the sufficient condition $\Leftarrow$ in (3.88a–c), the necessary condition $\Rightarrow$ is considered next. To prove that (3.88d) is the general solution of (3.88c), consider the function (3.90a)

$$\Psi(x) \equiv \frac{\Phi(x) - \Phi(0)e^{-x^2}}{x}: \quad \Phi \in \mathcal{D}(|R) \Rightarrow \Psi \in C(|R); \qquad (3.90a\text{-}c)$$

If the function Φ is differentiable (3.88b) $\equiv$ (3.90b), then Ψ is also differentiable and hence continuous (3.90c) for $x \neq 0$; for $x \to 0$, the mean value theorem (I.23.40) $\equiv$ (3.91a) for the differentiable function Φ implies that (3.90a) is continuous (3.91b) at $x = 0$:

$$\Phi(x) = \Phi(0) + x\Phi'(0) + o(x): \quad \Psi(x) = \frac{\Phi(0) + x\Phi'(0) + o(x) - \Phi(0)}{x} = \Phi'(0) + o(1). \qquad (3.91a,b)$$

Using (3.90a) to calculate the functional (3.92)

$$[F(x), \Phi(x)] \equiv \int_{-\infty}^{+\infty} F(x)\Phi(x)dx = \int_{-\infty}^{+\infty} xF(x)\Psi(x)dx + \Phi(0)\int_{-\infty}^{+\infty} F(x)\exp(-x^2)dx, \qquad (3.92)$$

leads to (3.93c)

$$xF(x) = 0, C \equiv \int_{-\infty}^{+\infty} F(x)\exp(-x^2)dx: \quad [F(x), \Phi(x)] = C\Phi(0) = [C\delta(x), \Phi(x)], \qquad (3.93a\text{-}c)$$

where (1) the first term on the r.h.s. of (3.92) is zero assuming (3.93a) $\equiv$ (3.88c) and (2) the second term on the r.h.s. of (3.92) involves a constant (3.93b). This proves the necessary condition (3.88a–d) with $\Rightarrow$, that is, the reverse of the $\Leftarrow$ sufficient condition (3.89a,b). In the case of an inhomogeneous equation (3.94a–d),

$$F(x) \in \mathcal{G}; \Phi \in \mathcal{D}(|R): \quad xF(x) = J(x) \Leftrightarrow F(x) = \frac{J(x)}{x} + C\delta(x), \qquad (3.94a\text{-}d)$$

the solution of the homogeneous equation (3.88a–d) is added. Thus has been obtained *the solution of the homogeneous (3.88c,d) [inhomogeneous (3.94c,d)] algebraic equation involving the product of (3.88a) [$\equiv$(3.93a)] a generalized function F(x) by its variable x; the generalized function is defined on the set of reference functions (3.90b [(3.94b)]) that are differentiable (3.10) on the real line.*

3.7.2 Product of a Generalized Function by a Power of Its Variable

The result (3.88a–d) may be generalized to the second power:

$$x^2 F(x) = 0 \Leftrightarrow xF(x) = -C_1\delta(x) = C_1 x\delta'(x), \qquad (3.95a,b)$$

where (3.79b) was used, with $m = 1 = n$, and C_1 is an arbitrary constant. From (3.95a,b) follows:

$$x\{F(x) - C_1\delta'(x)\} = 0 \Leftrightarrow F(x) - C_1\delta'(x) = C_0\delta(x), \qquad (3.96a,b)$$

using (3.88c,d), where C_0 is another arbitrary constant. The latter result may be generalized to homogeneous equations of degree N:

$$F(x) \in G; \Phi \in \mathcal{D}^N(|R): \quad x^N F(x) = 0 \Leftrightarrow F(x) = \sum_{n=0}^{N-1} C_n \delta^{(n)}(x), \qquad (3.97\text{a–d})$$

where the $C_0, \ldots, C_{N-1}$ are arbitrary constants. The result (3.97d) is proved by induction: (1) it holds for $N = 1$ in (3.88a–d) and $N = 2$ in (3.96a,b); and (2) if it holds for N in (3.97a–d), it also holds for $N + 1$. To prove the latter (2), the starting point is (3.88c,d) that implies (3.98b,c)

$$C_N \equiv (-)^N N! C: \quad x^{N+1} F(x) = 0 \Leftrightarrow x^N F(x) = C \delta(x) = C_N x^N \delta^{(N)}(x), \qquad (3.98\text{a–c})$$

where (3.79b) with $n = N$ was used in (3.98c), and the constant C is replaced by C_N according to (3.98a). The coincidence of (3.98c) $\equiv$ (3.99a) leads (3.97d) to (3.99b)

$$x^N \left\{ F(x) - C_N \delta^{(N)}(x) \right\} = 0 \Leftrightarrow F(x) - C_N \delta^{(N)}(x) = \sum_{n=0}^{N-1} C_n \delta^{(n)}(x). \qquad (3.99\text{a,b})$$

This proves (3.97c,d) with N replaced by $N + 1$. The inhomogeneous analogue of (3.97a–d) is

$$F(x) \in G; \Phi \in \mathcal{D}^N(|R): \quad x^N F(x) = J(x) \Leftrightarrow F(x) = \frac{J(x)}{x} + \sum_{n=0}^{N-1} C_n \delta^{(N)}(x). \qquad (3.100\text{a–d})$$

Thus has been obtained *the solution of the homogeneous (3.97c,d) [inhomogeneous (3.100c,d)] algebraic equation involving the product of a power x^n by a generalized function (3.97a)[$\equiv$(3.100a)] over the set of reference functions (3.97b) [$\equiv$(3.100b)] that are N times differentiable (3.17) on the real line.*

3.8 Products of Moduli, Powers, and Logarithms

The passage from ordinary to generalized functions allows differentiation to any order of (1) a function with a finite number of finite discontinuities (Sections 1.8 and 1.9) including at the points of discontinuity and (2) the logarithm of a real variable (Subsection 3.4.6) including on the branch-cut along the negative real axis. The approach to generalized functions as nonuniform limits of families of analytic functions (Chapter 1) was used to introduce the sign (modulus) functions [Subsection 1.7.2 (1.7.3)], and they can also be introduced as linear functionals (Subsection 3.1.2). The powers of the modulus, also multiplied by the sign or unit jump, can be differentiated either with regard to the variable or exponent (Subsection 3.8.1). This leads to generalized functions consisting of products of powers, logarithms, signs, powers of the modulus, and their products that are defined and differentiated indefinitely at all points of the real axis, including the origin for (1) powers whose exponents are negative nonintegral real numbers (Subsection 3.8.2 and 3.8.3) and (2) inverse integral powers (Subsection 3.8.4).

3.8.1 Differentiation with regard to the Variable and a Parameter

The generalized functions

$$\alpha > 0: \quad F_\alpha(x) = \left\{ |x|^\alpha, |x|^\alpha \operatorname{sgn}(x), x^\alpha H(x) \right\}, \qquad (3.101\text{a–d})$$

have derivates with regard to the variable x:

$$\alpha > 1: \quad \frac{d}{dx}\left[F_\alpha(x)\right] = \left\{\alpha|x|^{\alpha-1}\operatorname{sgn}(x), \alpha|x|^{\alpha-1}, \alpha x^{\alpha-1}H(x)\right\}, \tag{3.102a–d}$$

as follows from

$$\frac{d}{dx}\left\{|x|^\alpha\right\} = \alpha|x|^{\alpha-1}\frac{d}{dx}\left\{|x|\right\} = \alpha|x|^{\alpha-1}\operatorname{sgn}(x), \tag{3.103a}$$

$$\frac{d}{dx}\left[|x|^\alpha\operatorname{sgn}(x)\right] = \alpha|x|^{\alpha-1}\left\{\left\{\operatorname{sgn}(x)\right\}^2 + 2|x|^\alpha\delta(x)\right\} = \alpha|x|^{\alpha-1}, \tag{3.103b}$$

$$\frac{d}{dx}\left[x^\alpha H(x)\right] = \alpha x^{\alpha-1}H(x) + x^\alpha\delta(x) = \alpha x^{\alpha-1}H(x), \tag{3.103c}$$

where (1.90a–c) is used in (3.103a), (1.90a–c; 1.86a–c; 3.73a–c) in (3.103b), and (3.73a–c) in (3.103c). *Instead of differentiation (3.102a–d) of the generalized functions with regard to the variable x, differentiation with regard to the parameter α leads to the new generalized functions,*

$$\frac{d}{d\alpha}\left[F_\alpha(x)\right] \equiv \left\{|x|^\alpha\log|x|, |x|^\alpha\log|x|\operatorname{sgn}(x), x^\alpha\log|x|H(x)\right\}, \tag{3.104a–c}$$

that involve logarithmic factors:

$$\frac{d}{d\alpha}\left\{|x|^\alpha\right\} = |x|^\alpha\log|x|, \tag{3.104d}$$

as follows from (II.3.79a) ≡ (3.104d).

3.8.2 Extension of Powers to Negative Exponents

Derivates can be applied to (3.101a–d) with regard to α, x, or both to higher orders beyond (3.102a–d) and (3.104a–c). Further derivates with regard to α introduce additional logarithmic factors into (3.104a–c). Derivation with regard to x, performed n times in (3.101a–d) using (3.102a–d), leads to (3.105b–d)

$$\alpha > n: \quad F_\alpha^{(n)}(x) = \alpha(\alpha-1)\cdots(\alpha-n+1)$$
$$\left\{|x|^{\alpha-n}\left[\operatorname{sgn}(x)\right]^n, |x|^{\alpha-n}\left[\operatorname{sgn}(x)\right]^{n-1}, x^{\alpha-n}H(x)\right\}, \tag{3.105a–d}$$

where (3.105a) is assumed. The change of parameter (3.106a) leads from (3.105b–d) to (3.106b–d)

$$\alpha \to \alpha + n: \quad F_{\alpha+n}^{(n)}(x) = \frac{d^n}{dx^n}\left\{|x|^{\alpha+n}, |x|^{\alpha+n}\operatorname{sgn}(x), x^{\alpha+n}H(x)\right\}$$
$$= (\alpha+1)\cdots(\alpha+n)\left\{|x|^\alpha\left[\operatorname{sgn}(x)\right]^n, |x|^{\alpha+n}\left[\operatorname{sgn}(x)\right]^{n-1}, x^\alpha H(x)\right\}; \tag{3.106a–d}$$

the conditions (3.105a; 3.106a) imply (3.107a) in (3.106d) ≡ (3.107b):

$$\alpha > -n: \quad \frac{d^n}{dx^n}\left\{x^{\alpha+n}H(x)\right\} = (\alpha+1)\cdots(\alpha+n)x^\alpha H(x). \tag{3.107a,b}$$

Thus (3.107b) extends (3.101d) from positive (3.101a) to negative (3.107a) values of α.

3.8.3 Extension of Inverse Powers to the Origin

A similar result of extending (3.101b,c) from positive (3.101a) to negative (3.108a) values of the exponent α follows from (3.106b,c) multiplying by $[\mathrm{sgn}(x)]^n$, thus leading to (3.108b,c)

$$\alpha > -n: \quad (\alpha+1)\cdots(\alpha+n)\left\{|x|^{\alpha}, |x|^{\alpha}\,\mathrm{sgn}(x)\right\} = \left[\mathrm{sgn}(x)\right]^n \frac{\mathrm{d}^n}{\mathrm{d}x^n}\left\{|x|^{\alpha+n}, |x|^{\alpha+n}\,\mathrm{sgn}(x)\right\}. \qquad (3.108\mathrm{a\text{–}c})$$

If, besides satisfying (3.107a) $\equiv$ (3.108a) $\equiv$ (3.109a), the exponent α is not a negative integer (3.109b), then (3.107b) [(3.108b,c)] leads to (3.109c) [(3.109d,e)]

$$\alpha > -n; \quad \alpha \neq -1,\ldots,-n: \quad x^{\alpha}H(x) = \frac{1}{(\alpha+1)\cdots(\alpha+n)}\frac{\mathrm{d}^n}{\mathrm{d}x^n}\left\{x^{n+\alpha}H(x)\right\}, \qquad (3.109\mathrm{a\text{–}c})$$

$$\left\{|x|^{\alpha}, |x|^{\alpha}\,\mathrm{sgn}(x)\right\} = \frac{\left[\mathrm{sgn}(x)\right]^n}{(\alpha+1)\cdots(\alpha+n)}\frac{\mathrm{d}^n}{\mathrm{d}x^n}\left\{|x|^{n+n}, |x|^{n+n}\,\mathrm{sgn}(x)\right\}. \qquad (3.109\mathrm{d\text{–}e})$$

Thus *the ordinary functions (3.101b) [(3.101c,d)] taken as generalized functions can be extended by (3.109c) [(3.109d,e)] from positive (3.101a) to negative (3.109a) noninteger (3.109b) values of the parameter α.*

3.8.4 Powers with Negative Nonintegral and Integral Exponents

The case (3.110a) in (3.101b–d) leads to (3.110b–d):

$$\alpha = -1: \quad F_{-1}(x) = \left\{|x|^{-1}, |x|^{-1}\,\mathrm{sgn}(x),\, x^{-1}H(x)\right\}, \qquad (3.110\mathrm{a\text{–}d})$$

where *x^{-1} is the odd generalized function satisfying $x^{-1}\,x = 1$. The remaining negative integral powers are defined by*

$$m \in |N: \quad x^{-m} = \frac{(-)^{m-1}}{(m-1)!}\frac{\mathrm{d}^{m-1}}{\mathrm{d}x^{m-1}}(x^{-1}). \qquad (3.111\mathrm{a,b})$$

Thus, *the ordinary integral (3.111a) inverse powers as functions x^{-m} for $x \neq 0$ can be extended as generalized functions (3.111b) to all values of x including $x = 0$.* The definition of the inverse integral power as a generalized function extends to all rational functions because the ratio of two polynomials can be represented as a sum of fractions (Sections I.31.8 and I.31.9). The generalized function (3.44c) inverse integral power (3.111b) implies considering the integral as the product by a test function:

$$\left[x^{\alpha}, \Phi(x)\right] = \int_{-\infty}^{+\infty} x^{\alpha}\Phi(x)\mathrm{d}x. \qquad (3.112)$$

For (3.110a), this integral is singular at the origin, leading to the Cauchy principal value (Sections I.17.8 and I.17.9 and I.18.6 and I.18.7). The integral is considered next (Section 3.9) as the Hadamard finite part for nonintegral negative α.

3.9 Finite Part of an Integral (Hadamard)

The evaluation of improper integrals of the second kind (Section I.17.1), that is, with singularities on the path of integration, raises questions of convergence, for example, the Cauchy principal value (Sections I.17.8 and I.17.9 and I.18.6 and I.18.7) [Hadamard finite part (Section 3.9)] of an integral with a power-type singularity in the interior (at an endpoint) of the interval of integration. The evaluation of the Hadamard finite part of an integral (Subsection 3.9.2) with a power-type singularity at one end of the path of integration (Subsection 3.9.1) uses some of the generalized functions introduced before (Subsection 3.8.3).

3.9.1 Integral with Power-Type Singularity

Consider the integral $(3.112) \equiv (3.113d)$

$$\alpha \in | R, \alpha < -1, -\alpha \notin | N: \quad J(\alpha) \equiv \int_0^a x^\alpha g(x) \, dx, \tag{3.113a–d}$$

whose integrand involves a power with a real (3.113a) negative (3.113b) nonintegral (3.113c) exponent and thus is singular at the lower boundary. The integral (3.113d) may be rewritten

$$J(\alpha) = \int_{-\infty}^{+\infty} x^\alpha \{ H(x) - H(x-a) \} g(x) \, dx \equiv [x^\alpha \{ H(x) - H(x-a) \}, g(x)], \tag{3.114}$$

using unit jumps (1.28a–c); the latter have the property (3.115b) obtained choosing (3.115a) in (3.60b)

$$f(x) = \frac{x^{\alpha+1}}{\alpha+1} : \quad \{ f(x) H(x-a) \}' = \left\{ \frac{x^{\alpha+1}}{\alpha+1} H(x-a) \right\}' = x^\alpha H(x-a) + \frac{x^{\alpha+1}}{\alpha+1} \delta(x-a). \tag{3.115a,b}$$

Setting $n = 1$ in (3.109c) yields

$$x^\alpha H(x) = \frac{1}{\alpha+1} \{ x^{\alpha+1} H(x) \}', \tag{3.116}$$

and adding (3.115b) leads to

$$x^\alpha \{ H(x) - H(x-a) \} = \frac{1}{\alpha+1} \{ x^{\alpha+1} [H(x) - H(x-a)] \}' + \frac{x^{\alpha+1}}{\alpha+1} \delta(x-a). \tag{3.117}$$

Substituting (3.117) into (3.114) gives

$$(\alpha+1) J(\alpha) = [\{ x^{\alpha+1} [H(x) - H(x-a)] \}', g(x)] + x^{\alpha+1} [\delta(x-a), g(x)]$$

$$= -[x^{\alpha+1} [H(x) - H(x-a)], g'(x)] + a^{\alpha+1} g(a), \tag{3.118}$$

where (3.15a) and the definition (3.47a) $\equiv$ (3.119c) of the first derivative of a generalized function *(3.119a)* were used over a set of differentiable test functions *(3.119b)*:

$$F \in G, g \in \mathcal{D}(|R): \quad \lfloor F'(x), g(x) \rfloor \equiv -\lfloor F(x), g'(x) \rfloor; \tag{3.119a–c}$$

the result (3.118) can be rewritten (3.120c)

$$g \in \mathcal{D}(0,a); \alpha > -2: \quad (\alpha+1)J(\alpha) = (\alpha+1)\int_a^b x^\alpha g(x)\mathrm{d}x = -\int_0^a x^{\alpha+1}g'(x)\mathrm{d}x + a^{\alpha+1}g(a), \tag{3.120a–c}$$

where it is assumed that $g(x)$ is (3.120a) differentiable (3.10) in the interval $(0, a)$. If (3.120b) holds, the integral in (3.120c) is nonsingular; the formula is similar to a formal integration by parts, retaining the term at the nonsingular endpoint $x = a$ and omitting the singular term $O(x^{\alpha+1})$ as $x \to 0 +$. It has been shown that *the integral (3.113d) involving a differentiable function (3.120a) and a power with a negative exponent satisfying (3.120b) is evaluated by (3.120c) as the sum of (1) the first integral on the r.h.s. of (3.120a) that is nonsingular; (2) the second term on the r.h.s., corresponding to an integration by parts at the nonsingular endpoint; and (3) the singular endpoint does not appear.* The restriction (3.120b) is removed next, allowing α to have any real, negative, noninteger value (Subsection 3.9.2).

3.9.2 Extension to All Negative Nonintegral Powers

The formula (3.117) may be applied recursively:

$$x^\alpha \left[H(x) - H(x-a) \right] = \frac{1}{(\alpha+1)_n} \frac{\mathrm{d}^n}{\mathrm{d}x^n} \left\{ x^{\alpha+n} \left[H(x) - H(x-a) \right] \right\} + \sum_{k=1}^n \frac{x^{k+1}}{(\alpha+1)_k} \delta^{(n-k)}(x-a), \tag{3.121}$$

where $(\alpha)_n$ is **Pochhammer's symbol** defined before $[(\mathrm{I}.29.79\mathrm{a,b}) \equiv (\mathrm{II}.1.160) \equiv (3.122)]$:

$$(\alpha+1)_n \equiv (\alpha+1)(\alpha+2)...(\alpha+n). \tag{3.122}$$

Substituting (3.121) into (3.114) leads to

$$
\begin{aligned}
J(\alpha) &= \frac{1}{(\alpha+1)_n} \left[\frac{\mathrm{d}^n}{\mathrm{d}x^n} \left\{ x^{\alpha+n} \left[H(x) - H(x-a) \right] \right\}, g(x) \right] + \sum_{k=1}^n \frac{a^{k+1}}{(\alpha+1)_k} \left[\delta^{(n-k)}(x-a), g(x) \right] \\
&= \frac{(-)^n}{(\alpha+1)_n} \left[x^{\alpha+n} \left[H(x) - H(x-a) \right], g^{(n)}(x) \right] + \sum_{k=1}^n \frac{a^{k+1}}{(\alpha+1)_k} (-)^{n-k} g^{(n-k)}(a),
\end{aligned} \tag{3.123}
$$

where the following were used (1) in the second term on the r.h.s. (3.20b); and (2) in the first term on the r.h.s. *the definition (3.47b) $\equiv$(3.124c) of nth derivative of a generalized function (3.124a), over the reference functions (3.124b) that are n times differentiable (3.17)*:

$$F \in G, g \in \mathcal{D}^n: \quad \left[F^{(n)}(x), g(x) \right] \equiv (-)^n \left[F(x), g^{(n)}(x) \right]. \tag{3.124a–c}$$

The preceding result (3.123) is rewritten (3.125e)

$$\alpha \in |R, \alpha > -n-1, -\alpha \notin |N, g \in \mathcal{D}''(0,a):$$

$$\int_a^b x^\alpha g(x)dx = \frac{(-)^n}{(\alpha+1)_n} \int_0^a x^{\alpha+n} g^{(n)}(x)dx + \sum_{k=1}^n \frac{a^{\alpha+k}}{(\alpha+1)_k}(-)^{n-k} g^{(n-k)}(a).$$ (3.125a–e)

It has been shown that *the singular integral (3.125e), where (3.125d) is n times differentiable (3.17) in the interval (0, a) and α is a negative real (3.125a) noninteger (3.125c) number satisfying (3.125b) can be evaluated as (1) the nonsingular integral involving the power $x^{\alpha+n}$ with α + n > − 1 and $g^{(n)}(x)$; (2) the* **Hadamard finite part**, *that is, a set of terms similar to an integration by parts, evaluated at the nonsingular endpoint x = a; and (3) the terms $O(x^{\alpha+k})$ with k = 0, 1,..., n − 1 as x → 0 + at the singular endpoint can be omitted.*

NOTE 3.1: Generalized Functions, Jumps, and Impulses

The generalized functions used most often are the Dirac unit impulse and its derivatives (Chapter 1) that represent singularities corresponding to multipoles (Chapters 6, 8, and 9). They allow differentiation of all orders of ordinary piecewise differentiable functions including at the points of finite discontinuity; this implies that the singularities are integrable, that is, lead back to ordinary functions after a finite number of primitives. The "limitation" of generalized functions to integrable singularities allows their alternate definition (Chapter 3 and 5) as linear functionals over a class of test functions; this also highlights the basic tradeoff that (Chapter 3) is involved: (1) generalized gain over ordinary functions the property of unlimited differentiation; and (2) however they inherit the limitations of linear operators (Chapters 3 and 5). Thus they can be (cannot be) useful to solve linear (nonlinear) ordinary or partial differential equations. The subsequent use in these and other contexts raises a question of nomenclature "generalized functions" or "distributions." The term "distribution" was coined with the formal theory (Chapter 3) of generalized functions as linear functionals (Schwartz, 1949), long after Heaviside (1876) [Dirac (1930)] introduced their functions in electrical theory (quantum mechanics) and when they were already routinely used by engineers (physicists). The rigor (Chapter 3) and generality (Chapter 5) of the formal approach justify it to be the preferred choice (Chapter 7). The designation "distribution" however was far from new even in 1949 and is commonly used in other contexts, such as (1) discrete and continuous probability distributions (Notes 1.3 through 1.7) in mathematics and statistics, including statistical mechanics and statistical physics; (2) mass distributions in mechanics and gravity field theory (Chapter I.18); and (3) electric charge, current, and other multipolar distributions (Chapters 6, 8, and 9) for potential, wave, and other problems. Some confusion could arise from a physical "distribution" of point electric charges consisting each of a mathematical impulse "distribution." To avoid confusion, the term generalized function may be preferred to distribution. The designation unit step (impulse) function is also an alternative to Heaviside unit (Dirac delta) function.

NOTE 3.2: Corresponding Generalized and Test Functions

Defining the generalized functions as functionals over a class of test or reference functions, the properties of the former arise from the properties of the latter. For example, the derivatives (derivatives of all orders) of a generalized function arise (3.119a–c) [(3.124a–c)] from the derivatives of the test function that must have derivatives of the same order; besides, the convergence of the integral over the real line may require asymptotic conditions on the reference functions, for example, (1) slow-growth or fairly good (3.23a) ≡ (3.24a), (2) slow decay or good (3.20b) ≡ (3.21b), (3) very good or fast decay (3.23c) ≡ (3.24c), or (4) compact support (3.26). The combination of differentiability any number of times or smoothness and fast decay (compact support) leads to superlative (excellent)

TABLE 3.2 Generalized Functions and Classes of Test Functions

Generalized Function	Heaviside Unit Step	Dirac Unit Impulse	nth Derivative of Unit Impulse	Taylor Series for Unit Impulse
Symbol	$H(x)$	$\delta(x)$	$\delta^{(n)}(x)$	(3.66b)
Test functions for existence	$\mathcal{E}$ (integrable)	C (continuous)	C^n (n times continuously differentiable)	$\mathcal{A}$ (analytic)
Class	(3.2)	(3.12)	(3.19)	(3.64)
Test function for inner product	$\mathcal{D}$ (differentiable)	$\mathcal{D}^2$ (twice differentiable)	$\mathcal{D}^{n+1}$ ($(n+1)$-times differentiable)	$\mathcal{A}$ (analytic)
Class	(3.10)	(3.17)	(3.17)	(3.64)

Note: As the class of test functions is more restricted by imposing more properties, the set of associated generalized functions becomes larger.

functions. *As the test functions gain more properties, more generalized functions can be defined; thus restricting the class of test functions enlarges the set of generalized functions.* An example is given in Table 3.2: (1) the integrable (3.2) test functions allow the definition (3.3a–c) of the unit jump; (2/3) the continuous (3.12) [n times continuously differentiable (3.19)] test functions allow the definition of unit impulse (3.11a–c) [its nth derivative (3.22a–c)]; (4/5) the differentiable (3.10) [$(n + 1)$ times differentiable (3.17)] test functions allow integration by parts, with the unit impulse (3.9a–c) [its nth derivative (3.21a,b)]; and (6) the analytic functions (3.64) allow the use of the Taylor series for the unit impulse (3.66a,b). It would be possible to restrict further the test functions retaining the validity of (1) to (4), for example, for analytic superlative (3.40) but not for analytic excellent (3.38) test functions. In the text, the choice was made to restrict the test functions as little as possible for the property being considered; thus the restrictions on the test functions appear gradually, as the set of generalized functions is enlarged, as shown in Table 3.2. The properties of the generalized functions are inherited from the associated class of test functions. For example, the Fourier series (integrals) of a generalized function [Note 3.3 (3.7)] arise from those of the test function, that is, of bounded oscillation and periodic (Note 3.3) [and absolutely integrable (Note 3.6)] in a finite interval (on the real line). This allows the calculation of the Fourier series (integral) of the generalized function unit impulse and its derivatives of all orders [Note 3.5 (3.8)]. The Fourier series (integral) for the unit impulse converges in the sense of generalized functions but does not converge in the sense [Note 3.4 (3.9)] of ordinary integrals (series). The forcing of a differential equation by a unit impulse specifies as the fundamental solution the influence or Green function (Note 3.10). If the differential equation is linear, the principle of superposition applied to the influence function specifies the solution for arbitrary forcing (Note 3.11). If the differential equation is linear with constant coefficients, the influence function can be obtained using the Fourier transform and contour integration (Note 3.12). This is illustrated by obtaining the response function for linear ordinary differential equations with constant coefficients of order (1) one (Note 3.13) and (2) two (Notes 3.14 through 3.18). The case (2) includes the influence function (Note 3.19) for (Chapter 2) the linear deflection of an elastic string under constant tension.

NOTE 3.3: Fourier Series of a Generalized Function

A function of **bounded oscillation**, or bounded fluctuation or bounded variation (Subsections I.27.9.5 and II.5.7.5) in an interval (3.126a), has a bounded sum of moduli of differences (3.126c) for any partition of the interval (3.126b):

$$\Phi \in \mathcal{F}(a,b): \quad a = x_0 < x_2 < \cdots < x_n < \cdots < x_N \equiv b \Rightarrow \sum_{m=0}^{M-1} \left| \Phi\left(x_{m+1}\right) - \Phi\left(x_m\right) \right| \le B(\Phi, a, b) < \infty. \quad (3.126\text{a–c})$$

The **theorem of Fourier series** *states that a periodic (3.127a) function with bounded oscillation in a period (3.127b) has Fourier series (3.127c) with coefficients (3.127d)*

$$\Phi(\theta) = \Phi(\theta + 2\pi); \quad \Phi \in \mathcal{F}) - \pi, +\pi): \quad \Phi(\theta) = \sum_{n=-\infty}^{+\infty} C_m e^{im\theta}, \; C_m = \frac{1}{2\pi} \int_{-\pi}^{+\pi} \Phi(\theta) e^{im\theta} d\theta. \qquad (3.127a\text{-}d)$$

The **Fourier series of a generalized function** *(3.128a) is obtained from (3.128b) the Fourier series of the test function assumed to be periodic (3.127a) with bounded oscillation in a period (3.127b):*

$$F(\theta) = \sum_{m=-\infty}^{+\infty} D_m e^{im\theta} \in G \Leftrightarrow \left[F(\theta), \Phi(\theta) \right] \equiv \sum_{m=-\infty}^{+\infty} D_m \left[e^{im\theta}, \Phi(\theta) \right]. \qquad (3.128a,b)$$

Considering a periodic unit impulse (3.129b), its Fourier series would have coefficients (3.129a)

$$\frac{1}{2\pi} \int_{-\pi}^{+\pi} \delta(\theta) e^{im\theta} d\theta = \frac{1}{2\pi}: \quad \sum_{\ell=-\infty}^{+\infty} \delta(\theta - 2\pi\ell) = \frac{1}{2\pi} \sum_{m=-\infty}^{+\infty} e^{im\theta}. \qquad (3.129a,b)$$

The simple derivation (3.129a) of the Fourier series for the periodic unit impulse (3.129b) is invalid because the series does not converge in an ordinary sense (Note 3.4). To justify rigorously the Fourier series for the unit impulse and its derivatives (Note 3.5), the convergence must be taken in the sense (3.128a,b) of generalized functions.

NOTE 3.4: Ordinary and Generalized Convergence of a Series

The Fourier series (3.129b), taken in the sense of ordinary functions, is absolutely divergent (Section I.21.3) because the series of moduli $|e^{in\theta}| = 1$ has constant terms $1/2\pi$. For $\theta = 0$, the Fourier series on the r.h.s. of (3.129b) has constant terms $1/2\pi$ and diverges; on the l.h.s. $\delta(0) = \infty$, implying an equivalent singularity. For (3.130a) the l.h.s. of (3.129b) is zero, but the r.h.s. is generally not zero, as follows from

$$\theta \neq 0: \quad \sum_{n=-\infty}^{+\infty} e^{in\theta} = \lim_{M,N \to \infty} \sum_{n=-M}^{N} e^{in\theta} = \lim_{N \to \infty} \sum_{m=0}^{N} e^{im\theta} + \lim_{M \to \infty} \sum_{m=0}^{M} e^{-im\theta} - 1$$

$$= \lim_{N \to \infty} \frac{1 - e^{i(N+1)\theta}}{1 - e^{i\theta}} + \lim_{M \to \infty} \frac{1 - e^{-i(M+1)\theta}}{1 - e^{-i\theta}} - 1 \equiv \lim_{N,M \to \infty} S_{N,M}(\theta), \qquad (3.130a,b)$$

where the sum of the geometric series (I.21.70a-c) were used with $N + 1$ ($M + 1$) terms, the first term unity and ratio $e^{i\theta}(e^{-i\theta})$. The convergence in the ordinary sense (Section I.21.1) of the series (3.129b) requires the **partial sums** in (3.130b) to tend to a fixed limit whatever way $N, M \to \infty$:

$$\text{ordinary convergence:} \quad \lim_{N,M \to \infty} S_{N,M}(\theta) = S(\theta). \qquad (3.130c)$$

The partial sums in (3.130b) are real as follows from

$$\left(1 - e^{i\theta}\right)\left(1 - e^{-i\theta}\right)\left[S_{N,M}(\theta) + 1\right] = \left[1 - e^{i(N+1)\theta}\right]\left(1 - e^{-i\theta}\right) + \left[1 - e^{-i(M+1)\theta}\right]\left(1 - e^{i\theta}\right), \qquad (3.131)$$

which is equivalent to

$$\left(2-e^{i\theta}-e^{-i\theta}\right)\left[S_{N,M}(\theta)+1\right] = 2-e^{i\theta}-e^{-i\theta}-e^{i(N+1)\theta}+e^{iN\theta}-e^{-i(M+1)\theta}+e^{-iM\theta}, \tag{3.132}$$

which in turn simplify to

$$2\left(1-\cos\theta\right)S_{N,M}(\theta) \equiv e^{iN\theta}+e^{-iM\theta}-e^{i(N+1)\theta}-e^{-i(M+1)\theta}. \tag{3.133}$$

The sums in (3.133) are thus oscillatory in the range

$$\left|S_{N,M}(\theta)\right| = \frac{\left|e^{iN\theta}+e^{-iM\theta}-e^{i(N+1)\theta}-e^{-i(M+1)\theta}\right|}{2\left(1-\cos\theta\right)} \leq \frac{2}{1-\cos\theta}, \tag{3.134}$$

since each term in the numerator has modulus unity. In the particular case of symmetric limits (3.135a), the partial sums (3.133) simplify to (3.135b)

$$M = N: \quad S_{N,N}(\theta) = \frac{\cos(N\theta)-\cos\left[(N+1)\theta\right]}{1-\cos\theta}, \tag{3.135a,b}$$

which is oscillatory and can be rewritten as an alternative to (3.135a,b):

$$S_{N,N}(\theta) = \frac{\cos(N\theta)-\cos(N\theta)\cos\theta+\sin(N\theta)\sin\theta}{1-\cos\theta}$$

$$= \cos(N\theta)+\frac{\sin\theta}{1-\cos\theta}\sin(N\theta). \tag{3.136}$$

It has been shown that the *Fourier series (3.129a) for the periodic unit impulse (3.129b) does not converge in the sense of ordinary series (3.130c) ≡ (I.21.3a–c) because (1) it diverges for θ = 0 when all terms are unity and (2) for θ ≠ 0, the partial sums (3.130a,b) oscillate (3.133) in the range (3.134), as confirmed by the symmetric sums (3.135a,b) ≡ (3.136). The symmetric sums (3.135a) are not sufficient to prove convergence since convergence requires the limit to be the same, unaffected by the way N, M → ∞ independently.*

NOTE 3.5: Fourier Series for the Unit Impulse and Its Derivatives

The Fourier series is valid (3.127b) in the interval (3.137a) closed (open) at the upper (lower) limit because the periodic condition (3.127a) ≡ (3.137b) imposes that the value of the function at − π + 0 must be the same as (3.137c) the value at π − 0:

$$-\pi < \theta \leq \pi: \quad \Phi(\theta) = \Phi(\theta+2\pi) \Rightarrow \Phi(-\pi+0) = \Phi(\pi-0). \tag{3.137a–c}$$

The function will be continuous (discontinuous) at the boundary point if (3.138a) [(3.138b)] holds:

$$\Phi(-\pi+0)-\Phi(-\pi-0) = \Phi(\pi-0)-\Phi(\pi+0)\begin{cases} =0 & \text{continuous} \\ \neq 0 & \text{discontinuous.} \end{cases} \tag{3.138a,b}$$

The series on the r.h.s of (3.129b) could not be convergent for $\theta = 0$ because in that case its sum would be finite and it could not equal the l.h.s. that is infinite. The Fourier series specifies a periodic function, so the l.h.s of (3.129b) must be a periodic unit impulse with period 2π. The period can be modified from 2π to any positive real number X by a change of variable (Note 3.6). Thus it is sufficient to prove that the Fourier series for the periodic unit impulse (3.129b) of period 2π is valid in the sense of generalized functions (Note 3.4), corresponding to **generalized convergence** (3.128a,b) as distinct from **ordinary convergence** (3.130c) $\equiv$ (I.21.1; I.21.2a,b; I.21.3a–c). The generalized convergence (3.128a,b) of the Fourier series for the periodic unit impulse (3.129a,b) is proved by

$$\Phi \in \mathcal{F})-\pi,+\pi): \quad \left[\delta(\theta), \Phi(\theta)\right] = \Phi(0) = \sum_{m=-\infty}^{+\infty} C_m = \frac{1}{2\pi}\sum_{m=-\infty}^{+\infty}\int_{-\pi}^{+\pi}\Phi(\theta)e^{-im\theta}\,d\theta$$

$$= \frac{1}{2\pi}\sum_{m=-\infty}^{+\infty}\left[\Phi(\theta), e^{-im\theta}\right] = \left[\frac{1}{2\pi}\sum_{m=-\infty}^{+\infty}e^{im\theta}, \Phi(\theta)\right], \quad (3.139a,b)$$

where the following were used (1) the fundamental property (3.11a–c) of the unit impulse, at a point of continuity of Φ; (2/3) the Fourier series (3.127c) for the test function with bounded fluctuation (3.127a,b) and its coefficients (3.117d); and (4) the inner product for complex functions (II.5.137a–d) $\equiv$ (3.53a–d; 3.54a) $\equiv$ (3.140a) that has the property (3.140b)

$$\left[f(\theta), \Phi(\theta)\right] \equiv \int_{-\pi}^{+\pi}f(\theta)\Phi^*(\theta)\,d\theta = \left[\int_{-\pi}^{+\pi}f^*(\theta)\Phi(\theta)\,d\theta\right]^* = \left[\Phi(\theta), f(\theta)\right]^*, \quad (3.140a,b)$$

where the asterisk denotes conjugate. Thus *the Dirac unit impulse (its derivative or order n) has the Fourier series (3.141c) $\equiv$ (3.129b) [(3.142c)] over the set of reference functions with period 2π in (3.127a) $\equiv$ (3.141a) [$\equiv$(3.142a)] and bounded oscillation (nth derivative with bounded oscillation) in a period (3.127b) $\equiv$ (3.141b) [(3.142b)]*

$$\Phi(\theta) = \Phi(\theta+2\pi); \quad \Phi \in \mathcal{F})-\pi,+\pi): \quad 2\pi\sum_{\ell=-\infty}^{+\infty}\delta(\theta-2\pi\ell) = \sum_{m=-\infty}^{+\infty}e^{im\theta}, \quad (3.141a–c)$$

$$\Phi(\theta) = \Phi(\theta+2\pi); \quad \Phi \in \mathcal{F}^n)-\pi,+\pi): \quad 2\pi\sum_{\ell=-\infty}^{+\infty}\delta^{(n)}(\theta-2\pi\ell) = \sum_{m=-\infty}^{+\infty}(im)^n e^{im\theta}. \quad (3.142a–c)$$

The series (3.142c) is obtained from (3.141c) by differentiation and could be obtained directly as (3.139b); the general term of (3.142c) has modulus m^n, and that diverges as $m \to \infty$ for $n = 1, 2, \ldots$ and thus the series does not converge in the ordinary sense (3.130c) and must be interpreted in the sense of generalized functions, that is, for generalized convergence (3.128a,b).

NOTE 3.6: Fourier Series in a Finite Interval

Substituting the coefficients (3.127d) in the Fourier series (3.127c) leads to (3.143c)

$$\Phi(\theta) = \Phi(\theta+2\pi); \quad \Phi \in \mathcal{F})-\pi,+\pi): \quad \Phi(\theta) = \frac{1}{2\pi}\sum_{m=-\infty}^{+\infty}\int_{-\pi}^{+\pi}\Phi(\phi)e^{im(\theta-\phi)}\,d\phi, \quad (3.143a–c)$$

for a test function of bounded oscillation (3.143b) in the period (3.143a). The change of variable (3.144a,b) passes from the interval (3.144c) of length 2π to the interval (3.144d) of length $2X$ both centered at the origin:

$$\{\theta, \phi\} = \frac{\pi}{X}\{x, y\}: \quad -\pi < \theta \leq +\pi \Leftrightarrow -X < x \leq +X. \tag{3.144a-d}$$

Thus *a function (3.145a) with bounded oscillation (3.145c) in a period (3.145b) has Fourier series (3.145d)*

$$\Psi(x) = \Phi(\theta), \Psi(x) = \Psi(x + 2X), \Psi \in \mathcal{F}) - X, +X):$$

$$\Psi(x) = \frac{1}{2X} \sum_{m=-\infty}^{+\infty} \int_{-X}^{+X} \Psi(y) \exp\left(im\pi\frac{x-y}{X}\right) dy. \tag{3.145a-d}$$

Stretching the interval (3.146a) to the whole real axis in the limit (3.146b) and replacing the sum by an integral (3.146c) leads to

$$\lim_{X \to \infty} \int_{-X}^{+X} = \int_{-\infty}^{+\infty}, \frac{m\pi}{X} \to k; \quad \frac{1}{2X}\sum_{m=-\infty}^{+\infty} = \frac{1}{2\pi}\sum_{m=-\infty}^{+\infty}\frac{\pi}{X} \to \frac{1}{2\pi}\int_{-\infty}^{+\infty} dk; \tag{3.146a-c}$$

substituting (3.146c) in (3.145d) yields (3.147b)

$$\Psi \in \mathcal{F} \cap L^1(|R): \quad \Psi(x) = \frac{1}{2\pi}\int_{-\infty}^{+\infty} dk \int_{-\infty}^{+\infty} dy \, e^{ik(x-y)}\Psi(y). \tag{3.147a,b}$$

The function is (3.147a) $\equiv$ (3.145a; 3.143b) of bounded oscillation on the real line, so it need not be periodic, but a condition must be imposed to ensure the existence of the integral over the real line, such as absolute integrability (Note 3.7).

NOTE 3.7: Absolute Integrability and Fourier Identity

The existence of the integral (3.147b),

$$\left|\int_{-\infty}^{+\infty} e^{ik(x-y)}\Psi(y)dy\right| \leq \int_{-\infty}^{+\infty}\left|e^{ik(x-y)}\Psi(y)\right|dy = \int_{-\infty}^{\infty+}|\Psi(y)|dy, \tag{3.148}$$

is ensured if the function is absolutely integrable:

$$L^1(a,b) \equiv \left\{\Phi(x): \; \exists \int_a^b |\Phi(x)|dx\right\}, \tag{3.149}$$

on the real line. The Fourier integral identity (3.147b) thus applies to a function absolutely integrable (3.149) and of bounded oscillation (3.126a–c) on the real line. The theorem on Fourier series (3.127a–d) [integral (3.147a,b)] will be proved rigorously in a subsequent book since the main aim here is the interpretation in terms of the generalized functions, such as the unit impulse and its derivatives. The requirements that the function be absolutely integrable (3.149) and of bounded oscillation (3.126a–c) on the real line are nonredundant in the sense that neither implies the other, as can be shown by three examples. The first example is a constant (3.150a) that has zero oscillation in any finite or infinite interval (3.150b) and is not absolutely integrable on the real line (3.150c):

$$\Psi(x) = const: \quad B\big(\Psi(x); a, b\big) = 0, \quad \int_{-\infty}^{+\infty} |\Psi(x)| dx = \infty. \tag{3.150a–c}$$

The second example of the function (3.151a) is (Subsection I.27.9.5) of unbounded oscillation (3.151d) in any finite interval including the origin (3.151b,c) and is absolutely integrable (3.151e):

$$\Psi(x) = \exp\left(\frac{i}{x}\right): \quad a < 0 < b \Rightarrow B\big(\Psi(x); a, b\big) = \infty, \quad \int_{a}^{b} \left|\exp\left(\frac{i}{x}\right)\right| dx = b - a. \tag{3.151a–e}$$

The third example extends the second example, multiplying the functions (3.151a) by a Gaussian (3.152a) that does not change the unbounded oscillation near the origin (3.152b–d) and ensures absolute integrability on the real line (3.152e) ≡ (1.1a):

$$\Psi(x) = \exp\left(-x^2 + \frac{i}{x}\right): \quad a < 0 < b \Rightarrow B\big(\Psi(x); a, b\big) = \infty,$$

$$\int_{-\infty}^{+\infty} |\Psi(x)| dx = \int_{-\infty}^{+\infty} \exp\left(-x^2\right) dx = \sqrt{\pi}. \tag{3.152a–e}$$

The first (third) example shows that a function can be of bounded oscillation and not absolutely integrable (vice versa) on the real line (3.150a–c) [(3.152a–e)], so neither restriction replaces the other.

NOTE 3.8: Fourier Transform of the Unit Impulse and Its Derivatives

The result (3.147a,b) may be stated as the **Fourier integral theorem**: *a function with bounded oscillation (3.126a–c) absolutely integrable (3.149) on the real line (3.147a)* ≡ *(3.153a) satisfies the identity (3.147b) that may be decomposed into the direct (inverse) Fourier transform (3.153b) [(3.153c)]:*

$$\Psi \in \mathcal{F} \cap L^1(|R): \quad \Psi(x) = \int_{-\infty}^{+\infty} \tilde{\Psi}(k) e^{ikx} dk, \quad \tilde{\Psi}(k) = \frac{1}{2\pi} \int_{-\infty}^{+\infty} \Psi(x) e^{-ikx} dx. \tag{3.153a–c}$$

This is equivalent to the integral property of the unit impulse (3.154b) over the test functions (3.154a) that are continuous and absolutely integrable and have bounded oscillation on the real line:

$$\Psi \in \mathcal{F} \cap C \cap L^1(|R): \quad \int_{-\infty}^{+\infty} e^{ik(x-y)} dk = 2\pi\delta(x - y); \tag{3.154a,b}$$

thus follows the property (3.155b) for the nth derivative of the unit impulse over the test functions (3.155a) that are absolutely integrable and have nth derivative of bounded oscillation on the real line:

$$\Psi \in \mathcal{F}^n \cap \mathcal{L}^1(|R): \quad \int_{-\infty}^{+\infty} k^n e^{ik(x-y)} dk = i^{-n} 2\pi \delta^{(n)}(x-y). \tag{3.155a,b}$$

The property (3.154b) arises from the comparison of (3.147b) with (3.156b) ≡ (3.15a):

$$\Psi \in C(|R): \quad \int_{-\infty}^{+\infty} \Psi(y) \delta(x-y) dy = \Psi(x), \tag{3.156a,b}$$

requiring the test function to satisfy (3.154a) both conditions (3.147a) and (3.156a) ≡ (3.14a). Differentiation of (3.154b) n times leads to (3.155b), where the test function must have nth derivative with bounded oscillation (3.155a); thus it is also continuous, so this requirement is dropped from (3.154a) to (3.155a). The integrals in (3.154b) and (3.155b) do not converge as ordinary integrals, just as the Fourier series (3.141c; 3.142c) do not converge as ordinary series. Thus the ordinary must be extended to generalized convergence when applied to Fourier series (integrals) of generalized functions [Note 3.4 (3.9)].

NOTE 3.9: Ordinary and Generalized Convergence of Fourier Integrals

As an example of the distinction between ordinary and generalized convergence for improper integrals of the first kind (Section I.17.1), consider (3.154b) that diverges in the ordinary sense for $k = 0$, corresponding to the same singularity as $2\pi\delta(0)$. For the improper integral of the second kind (I.17.2b) to converge in the ordinary sense, the limit

$$\text{ordinary convergence}: \quad \int_{-\infty}^{+\infty} e^{ik(x-y)} dk \equiv \lim_{A,B \to \infty} \int_{-A}^{+B} e^{ik(x-y)} dk \tag{3.157}$$

must exist and be unique independently of how $A, B \to \infty$. For (3.158a) the r.h.s. of (3.154b) is zero, but the integral on the r.h.s. is oscillatory (3.158b):

$$x \neq y: \quad \int_{-\infty}^{+\infty} e^{ik(x-y)} dk \equiv \lim_{A,B \to \infty} \int_{-A}^{B} e^{ik(x-y)} dk$$

$$= \frac{i}{y-x} \left\{ \lim_{B \to \infty} \exp\left[iB(x-y)\right] - \lim_{A \to \infty} \exp\left[iA(y-x)\right] \right\}; \tag{3.158a,b}$$

the oscillation simplifies to (3.159b) in the symmetric case (3.159a):

$$A = B: \quad \lim_{A \to \infty} \int_{-A}^{+A} e^{ik(k-y)} dk = \lim_{A \to \infty} \frac{e^{iA(x-y)} - e^{-iA(x-y)}}{i(x-y)} = 2 \lim_{A \to \infty} \frac{\sin\left[A(x-y)\right]}{x-y}. \tag{3.159a-c}$$

Using the identity (I.5.14b) ≡ (3.160a), the limit (3.159b) for $x = y$ becomes (3.160b)

$$\lim_{z \to 0} \frac{\sin z}{z} = 1: \quad \lim_{A \to \infty} \lim_{x \to y} \frac{\sin\left[A(x-y)\right]}{x-y} = \lim_{A \to \infty} A = \infty = \int_{-\infty}^{+\infty} dk, \tag{3.160a-c}$$

confirming divergence for $x = y$. The symmetric limit $A = B$ does not prove convergence that requires the independent limits in (3.157) at the upper and lower boundaries, but it proves divergence for $A = B$. It has been shown that *the integrals (3.154b) [(3.155b)] can be evaluated in terms of the unit impulse (its nth derivative) with generalized convergence in the sense of generalized functions as linear continuous functionals over test functions that are absolutely integrable and continuous with bounded oscillation (3.154a) [have nth derivative of bounded oscillation (3.155a)]. Taken (3.157) as an ordinary integral, (3.154b) is oscillatory for $x \neq y$ in (3.158a,b) as confirmed by the symmetric limit (3.159a–c); for $x = y$ the integral (3.154b) diverges (3.160c), as confirmed by the symmetric limit (3.160a,b).*

NOTE 3.10: Influence Function due to the Forcing by a Unit Impulse

An **ordinary differential equation** (o.d.e.) of order N is a relation between an **independent variable** x, a **dependent variable** or function $y(x)$, and its derivatives up to order N:

$$0 = F\left(x; y(x); y'(x), y''(x), \ldots, y^{(N)}(x)\right). \tag{3.161}$$

It is a linear o.d.e. (l.o.d.e.) if the function and its derivatives appear linearly, with coefficients that may depend only on the independent variable:

$$f(x) = \sum_{n=0}^{N} A_n(x) \frac{d^n y}{dx^n} \equiv \left\{ L\left(\frac{d}{dx}\right) \right\} y(x). \tag{3.162}$$

The **fundamental solution** for forcing by a unit impulse is the **influence or Green function**:

$$\delta(x - \xi) = \left\{ L\left(\frac{d}{dx}\right) \right\} G(x; \xi) = \sum_{n=0}^{N} A_n(x) \frac{d^n}{dx^n} \left[G(x; \xi) \right]. \tag{3.163}$$

The **principle of superposition** *for a l.o.d.e. (3.162) specifies the particular solution for arbitrary forcing as (Chapter 7) the integral after multiplication by the influence function:*

$$y_1(x) = \int_{-\infty}^{+\infty} f(\xi) G(x; \xi) d\xi. \tag{3.164}$$

Bearing in mind that (3.162; 3.164) imply (3.165a)

$$0 = \left\{ L\left(\frac{d}{dx}\right) \right\} \left[y(x) - y_1(x) \right] \equiv \left[L\left(\frac{d}{dx}\right) \right] y_0(x): \quad y(x) = y_1(x) + y_0(x), \tag{3.165a,b}$$

it follows that **the complete integral** *(3.165b) of the forced linear ordinary differential equation (3.162) is the sum of (i) the* **general integral** *(3.165a) without forcing and (ii) a* **particular integral** *(3.164; 3.162) with forcing.*

NOTE 3.11: Linear Differential Equation with Constant Coefficients

In the case of a *l.o.d.e.* (3.162) *with constant coefficients (l.o.d.e.c.c.),*

$$\delta(x - \xi) = \sum_{n=0}^{N} A_n \frac{d^n}{dx^n} \left[G(x; \xi) \right] = \left[P_N\left(\frac{d}{dx}\right) \right] G(x; \xi), \tag{3.166}$$

*the linear ordinary differential operator (3.162) is a **characteristic polynomial** (3.166) of derivatives with regard to the independent variable x all applied to the dependent variable y.* The influence function may be obtained using the Fourier transform (3.153b):

$$G(x;\xi) = \int_{-\infty}^{+\infty} \tilde{G}(k;\xi) e^{ikx} dk.$$ (3.167)

Substitution of (3.167) in the l.o.d.e.c.c. (3.166) leads to the algebraic relation (3.168)

$$\frac{e^{-ik\xi}}{2\pi} = \tilde{G}(k;x) \sum_{n=0}^{N} A_n (ik)^n = \tilde{G}(k;x) P_N(ik),$$ (3.168)

where the following were used (1) the differentiation property of the Fourier transform (3.167), assuming that it is uniformly convergent (Section I.13.8) with regard to k

$$\frac{d^n}{dx^n}\left[G(x;\xi)\right] = \int_{-\infty}^{+\infty} (ik)^n \tilde{G}(x;\xi) e^{ikx} dk;$$ (3.169)

and (2) the Fourier transform (3.153c) of the unit impulse

$$\tilde{\delta}(x-\xi) = \frac{1}{2\pi} \int_{-\infty}^{+\infty} \delta(x-\xi) e^{-ikx} dx = \frac{e^{-ik\xi}}{2\pi}.$$ (3.170)

which (1) is the inverse Fourier transform (3.153b,c) of (3.154b) and (2) alternatively follows from (3.11a–c) because e^{-ikx} is a differentiable function; (3.170) can also be proved from the *definition of Fourier transform of a generalized function*:

$$F \in G, \Phi \in \tilde{E} \cap (|R): \quad (\tilde{F}, \Phi) \equiv (F, \tilde{\Phi}).$$ (3.171a–c)

This definition is used to obtain in a rigorous way the Fourier transform of the unit impulse:

$$\left(\tilde{\delta}, \Phi\right) = \left(\delta, \tilde{\Phi}\right) = \int_{-\infty}^{+\infty} \delta(x-\xi) \tilde{\Phi}(x) dx = \tilde{\Phi}(\xi) = \frac{1}{2\pi} \int_{-\infty}^{+\infty} e^{-ik\xi} \Phi(k) dk = \left[\frac{e^{-ik\xi}}{2\pi}, \Phi\right],$$ (3.172)

in agreement with (3.172) ≡ (3.170).

NOTE 3.12: Evaluation of the Influence Function by Residues

Solving (3.168) for the Fourier transform of the influence function and substituting in (3.167) leads to (3.173b)

$$P_N(ik) \equiv \sum_{n=0}^{N} A_n (ik)^n : \quad 2\pi G(x;\xi) = \int_{-\infty}^{+\infty} \frac{e^{ik(x-\xi)}}{P_N(ik)} dk,$$ (3.173a,b)

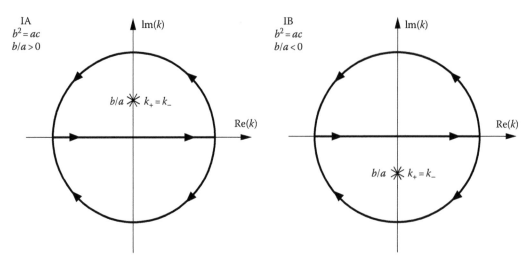

FIGURE 3.2 The influence function for a linear ordinary differential equation with constant coefficients (l.o.d.e.c.c.) of any order is the fundamental solution forced by the unit impulse $\delta(x - \xi)$. It can be obtained for $x > \xi (x < \xi)$, evaluating by residues a loop integral around a semicircle with infinite radius and centre at the origin taken in the positive (negative) or counterclockwise (clockwise) direction in the upper (lower)-half complex k-plane. The integral is evaluated by the residues at the poles in the k-plane that are the roots of the characteristic polynomial associated with the l.o.d.e.c.c. For the response function of a second-order l.o.d.e.c.c., there are five cases (nine subcases) in Table 3.2. The first case IA (IB) involves the residue at a double pole in the upper (lower)-half complex plane $a(b)$.

involving the **characteristic polynomial** (3.173a) of the l.o.d.e.c.c. (3.166). The path of integration along the real axis in (2.173b) can be closed (Figure 3.2) by a half-circle of infinite radius either in the upper or lower complex-k half plane. In both cases, the integral along the half-circle vanishes because the asymptotic condition (I.17.29) is met in all directions (3.174a):

$$\lim_{|k|\to\infty} \frac{1}{P_N(ik)} = 0; \quad \lim_{|k|\to\infty}\left|e^{ik(x-\xi)}\right| = \lim_{|k|\to\infty}\exp\left[-(x-\xi)\mathrm{Im}(k)\right]$$

$$= 0 \begin{cases} \text{if } \mathrm{Im}(k) > 0 \quad \text{for } x > \xi, \\ \text{if } \mathrm{Im}(k) < 0 \quad \text{for } \xi > x. \end{cases} \tag{3.174a-c}$$

In the case (3.174b) [(3.174c)], the path of integration must be closed in the upper (lower) complex-k half plane, and the integral (3.173b) is evaluated (Section I.17.5) by (1) $+2\pi i(-2\pi i)$ times the sum of the residues in the interior and (2) plus $+\pi i(-\pi i)$ the sum of the residues at the poles on the real axis that are indented:

$$x \gtrless \xi: \quad G(x;\xi) = \pm i \sum_{\mathrm{Im}(k)><0} \mathrm{Res}\left[\frac{e^{ik(x-\xi)}}{P_N(ik)}\right] \pm \frac{i}{2}\sum_{\mathrm{Im}(k)=0}\mathrm{Res}\left[\frac{e^{ik(x-\xi)}}{P_N(ik)}\right]. \tag{3.175}$$

Thus *the influence Green or function or fundamental solution of the l.o.d.e.c.c. of order N forced by a unit impulse (3.162) is given by (3.175) with upper (lower) signs for $x > \xi(x < \xi)$; the residues are evaluated at the poles in the upper (lower) complex-k half-plane corresponding to zeros of the characteristic polynomial (3.173a). The poles on the real axis may be indented so as to appear either for $x > \xi$ or $x < \xi$.*

NOTE 3.13: Ordinary Differential Equation of the First Order

The simplest example of the preceding method is a l.o.d.e.c.c. that is of the first order (3.176b) if (3.176a) is met:

$$a \neq 0: \quad aG'(x;\xi) + bG(x;\xi) = \delta(x;\xi). \tag{3.176a,b}$$

The characteristic polynomial (3.177a) leads (3.173b) to the integral (3.177b)

$$P_1(ik) = b + ika: \quad 2\pi G(x;\xi) = \int_{-\infty}^{+\infty} \frac{e^{ik(x-\xi)}}{ika + b} dk. \tag{3.177a,b}$$

There is one simple pole (3.178b) that for (3.178a) lies in the upper-half complex k-plane (Figure 3.2a), so that (1) for $x < \xi$ the integral (3.177b) is zero and (2) for $x > \xi$ it equals $2\pi i$ times the residue (I.15.24b) at the pole (3.178b) leading to (3.178c)

$$\frac{b}{a} > 0, \quad k_0 = i\frac{b}{a}: \quad G(x;\xi) = \frac{H(x-\xi)}{2\pi} \int_{-\infty}^{+\infty} \frac{e^{ik(x-\xi)}}{ika + b} dk$$

$$= iH(x-\xi) \lim_{k \to ib/a} \left(k - i\frac{b}{a} \right) \frac{e^{ik(x-\xi)}}{ika + b}$$

$$= \frac{H(x-\xi)}{a} \lim_{k \to ib/a} \exp\left[ik(x-\xi) \right]. \tag{3.178a–c}$$

Thus the *l.o.d.e.c.c. of first order (3.176a,b) has influence function (3.178c) $\equiv$ (3.179b,c) if (3.178a) $\equiv$ (3.179a) holds (Figure 3.2a):*

$$\frac{b}{a} > 0: \quad G(x;\xi) = \frac{H(x-\xi)}{a} \exp\left[-\frac{b}{a}(x-\xi) \right] = \begin{cases} 0 & \text{if } x < \xi, \\ \frac{1}{a} \exp\left[\frac{b}{a}(\xi-x) \right] & \text{if } x > \xi; \end{cases} \tag{3.179a–c}$$

in the (Figure 3.2b) opposite case (3.163a), the influence function is (3.163b,c)

$$\frac{b}{a} < 0: \quad G(x;\xi) = -\frac{H(x-\xi)}{a} \exp\left[-\frac{b}{a}(x-\xi) \right] = \begin{cases} -\frac{1}{a} \exp\left[\frac{b}{a}(\xi-x) \right] & \text{if } x < \xi, \\ 0 & \text{if } x > \xi. \end{cases} \tag{3.180a–c}$$

In both cases the influence function vanishes at infinity (3.181a):

$$\lim_{x \to \pm\infty} G(x;\xi) = 0; \quad G(x;\xi) \neq G(\xi;x). \tag{3.181a,b}$$

in neither case it is symmetric (3.181b), so that the principle of reciprocity does not hold, because the operator in (3.176a,b) is not self-adjoint (7.116a–c).

NOTE 3.14: Equation of the Second Order and Double Pole

In the case of the l.o.d.e.c.c. of second order,

$$a \neq 0: \quad aG''(x;\xi) + 2bG'(x;\xi) + cG(x;\xi) = \delta(x;\xi), \tag{3.182a,b}$$

the influence function (3.173b) involves (3.173a) the characteristic polynomial (3.183a) whose roots are (3.183b)

$$P_2(ik) = -k^2 a + 2ibk + c = -a(k - k_+)(k - k_-): \quad k_\pm a = ib \pm \sqrt{ac - b^2}. \tag{3.183a,b}$$

This leads to five cases and nine subcases in Table 3.3 and Figures 3.2 through 3.6. In the case I, there is (3.184a) a double root (3.184b); for (3.184c) in the subcase IA, there is a double pole on the upper-half complex k-plane (Figure 3.2a) leading (I.15.33b) to (3.184d)

$$b^2 = ac, \quad k_+ = k_- = i\frac{b}{a} = k_0, \frac{b}{a} > 0:$$

$$G(x;\xi) = -\frac{1}{2\pi a} \int_{-\infty}^{+\infty} \frac{e^{ik(x-\xi)}}{(k - ib/a)^2} = -\frac{i}{a} H(x-\xi) \lim_{k \to ib/a} \frac{d}{dk}\left[e^{ik(x-\xi)}\right], \tag{3.184a-d}$$

TABLE 3.3   Response Function for Second-Order l.o.d.e.c.c.

Case		Subcase		Influence	
Number	Parameter	Letter	Parameters	Function	Figure
I	$b^2 = ac$	IA	$\dfrac{b}{a} > 0$	(3.185a–c)	3.2a
		IB	$\dfrac{b}{a} < 0$	(3.186a–c)	3.2b
II	$b^2 < ac$	IIA	$\dfrac{b}{a} > 0$	(3.189a–c)	3.3a
		IIB	$\dfrac{b}{a} < 0$	(3.190a–c)	3.3b
III	$b^2 > ac > 0$	IIIA	$\dfrac{b}{a} > 0$	(3.193a–c)	3.4a
		IIIB	$\cdot \ \dfrac{b}{a} < 0$	(3.194a–c)	3.4b
IV	$b^2 > ac < 0$	IV	$\dfrac{b}{a} > (<)0$	(3.197a–c)	3.5a (b)
V	$b = 0 = c$	VA	Negative indentation	(3.199a–c)	3.6a
		VB	Positive indentation	(3.200a–c)	3.6b

Note: Five cases and nine subcases for the fundamental solution of a linear ordinary differential equation with constant coefficients (l.o.d.e.c.c.) of order two.

- Linear ordinary differential operator with constant coefficients:

$$L\left(\frac{d}{dx}\right) \equiv a\frac{d^2}{dx^2} + 2b\frac{d}{dx} + c$$

- Influence or Green function or fundamental solution:

$$\left\{L\left(\frac{d}{dx}\right)\right\}G(x;\xi) = \delta(x-\xi)$$

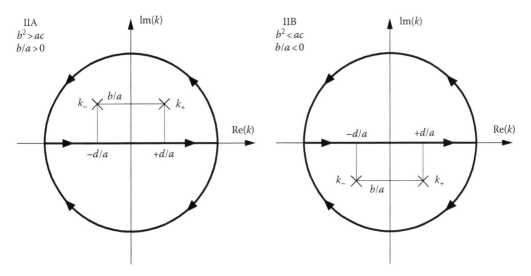

FIGURE 3.3   As for Figure 3.2, in the case IIA (IIB) with two simple poles symmetric relative to the imaginary axis in the upper (lower)-half complex-k plane $a(b)$.

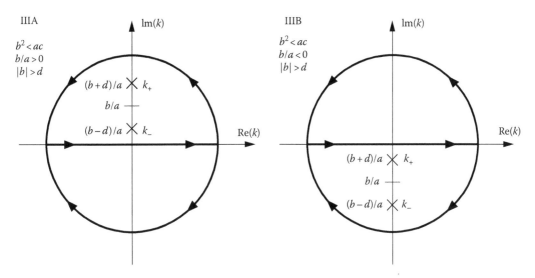

FIGURE 3.4   As for Figure 3.2, in the case IIIA (IIIB) with two simple poles on the positive (negative) imaginary axis in the complex-k plane $a(b)$.

which simplifies to

$$\text{IA:} \quad b^2 = ac, \quad \frac{b}{a} > 0: \quad G(x;\xi) = \frac{x-\xi}{a}\exp\left[\frac{b}{a}(\xi-x)\right]H(x-\xi). \tag{3.185a–c}$$

The subcase IB of double pole (3.186a) in the lower-half (Figure 3.2b) complex k-plane (3.186b) leads to (3.186c)

$$\text{IB:} \quad b^2 = ac, \quad \frac{b}{a} < 0: \quad G(x-\xi) = \frac{\xi-x}{a}\exp\left[\frac{b}{a}(\xi-x)\right]H(\xi-x). \tag{3.186a–c}$$

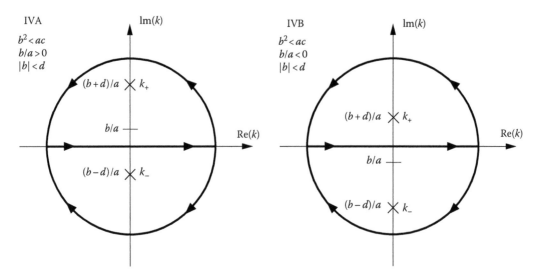

FIGURE 3.5 As for Figure 3.2, in the case IVA (IVB) with simple poles on the positive and negative imaginary axis (a), whose positions have a common translation keeping one pole in each half-plane (b).

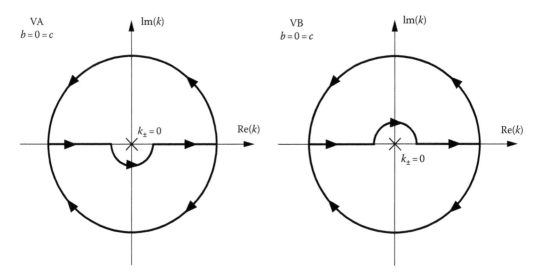

FIGURE 3.6 As for Figure 3.2, in the case VA (VB) with a double pole at the origin of the k-plane with a lower (upper) or positive (negative) or counterclockwise (clockwise) indentation $a(b)$.

Thus *the influence function for the l.o.d.e.c.c. of second order (3.182a,b) is given by (3.185c) [(3.186c)] in the case (3.185b) [(3.186b)] corresponding to a double pole (3.185a) [≡(3.186a)] in the upper (lower) complex-k half-plane [Figure 3.2a (b)].*

NOTE 3.15: Two Simple Poles Not on the Coordinate Axis

In the case II, there are (3.187a) two roots (3.187b), symmetric (Figure 3.3) relative to the imaginary axis (3.187c):

$$\text{II:} \quad ac > b^2: \quad k_\pm a = ib \pm d, \quad d = \left| ac - b^2 \right|^{1/2}. \tag{3.187a–c}$$

In the subcase IIA, the roots lie (Figure 3.3a) in the upper complex-k half-plane (3.188a) leading (3.173b) to the influence function (3.188b):

$$\text{IIA:} \quad \frac{b}{a} > 0: \quad G(x;\xi) = -\frac{H(x-\xi)}{2\pi a} \int_{-\infty}^{+\infty} \frac{e^{ik(x-\xi)}}{(k-k_+)(k-k_-)} dk$$

$$= -\frac{i}{a}\frac{H(x-\xi)}{k_+-k_-}\left\{\exp\left[ik_+(x-\xi)\right] - \exp\left[ik_-(x-\xi)\right]\right\}$$

$$= \frac{H(x-\xi)}{2id}\exp\left[\frac{b}{a}(\xi-x)\right]\left\{\exp\left[i\frac{d}{a}(x-\xi)\right] - \exp\left[-i\frac{d}{a}(x-\xi)\right]\right\}, \quad (3.188\text{a,b})$$

that simplifies to

$$\text{IIA:} \quad ac > b^2, \quad \frac{b}{a} > 0: \quad G(x;\xi) = \frac{H(x-\xi)}{d}\exp\left[\frac{b}{a}(\xi-x)\right]\sin\left[\frac{d}{a}(x-\xi)\right]. \quad (3.189\text{a–c})$$

In the opposite (Figure 3.3b) subcase IIB, the influence function is given by

$$\text{IIB:} \quad ac > b^2, \quad \frac{b}{a} < 0: \quad G(x-\xi) = -\frac{H(\xi-x)}{d}\exp\left[\frac{b}{a}(\xi-x)\right]\sin\left[\frac{d}{a}(x-\xi)\right]. \quad (3.190\text{a–c})$$

Thus the *influence function for the l.o.d.e.c.c. (3.182a,b) is given by (3.189c) [(3.190c)] in the case (3.189b) [(3.190b)] of two poles (3.189a) [≡(3.190a)], symmetric relative to the imaginary axis (3.187b,c) in the upper (lower) complex-k half-plane [Figure 3.3a (b)].*

NOTE 3.16: Two Simple Poles on the Same Side of the Imaginary Axis

In the case III of two simple poles (3.191a) on the imaginary axis (3.191c), both lie (Figure 3.4) on the imaginary axis on the same side of the real axis (3.191d) if (3.191b) is met:

$$\text{III:} \quad b^2 > ac > 0: \quad k_\pm a = ib \pm id, \quad d < |b|. \quad (3.191\text{a–d})$$

In the subcase IIIA when both lie (Figure 3.4a) on the upper half (3.192a) of the complex k-plane, the influence function (3.156b) is given by (3.175b)

$$\text{IIIA:} \quad \frac{b}{a} > 0: \quad G(x;\xi) = -\frac{H(x-\xi)}{2\pi a} \int_{-\infty}^{+\infty} \frac{e^{ik(x-\xi)}}{(k-k_+)(k-k_-)} dk$$

$$= -\frac{i}{a}\frac{H(x-\xi)}{k_+-k_-}\left\{\exp\left[ik_+(x-\xi)\right] - \exp\left[ik_-(x-\xi)\right]\right\}$$

$$= -\frac{H(x-\xi)}{2d}\exp\left[-\frac{b}{a}(x-\xi)\right]\left\{\exp\left[-\frac{d}{a}(x-\xi)\right] - \exp\left[\frac{d}{a}(x-\xi)\right]\right\}, \quad (3.192\text{a,b})$$

which simplifies to

$$\text{IIIA:} \quad b^2 > ac > 0, \quad \frac{b}{a} > 0: \quad G(x;\xi) = \frac{H(x-\xi)}{d}\exp\left[\frac{b}{a}(\xi-x)\right]\sinh\left[\frac{d}{a}(x-\xi)\right]. \quad (3.193\text{a–c})$$

In the opposite subcase IIIB when (Figure 3.4b) both poles lie on the imaginary axis (3.194a) below the real axis (3.194b), the influence function is given by (3.194c)

$$\text{IIIB:}\quad b^2 > ac > 0,\quad \frac{b}{a} < 0:\quad G(x;\xi) = -\frac{H(\xi - x)}{d}\exp\left[\frac{b}{a}(x - \xi)\right]\sinh\left[\frac{d}{a}(x - \xi)\right].\quad \text{(3.194a–c)}$$

Thus *the influence function for the l.o.d.e.c.c. of second order (3.682a,b) is given by (3.193c) [(3.194c)] in the case of two simple poles (3.187c; 3.191a–d) on the imaginary axis (3.193a) [≡(3.194a)] above (3.194b) [below (3.194b)] the real axis [Figure 3.4a (b)].*

NOTE 3.17: Two Simple Poles on Opposite Sides of the Imaginary Axis

The remaining case IV of two simple poles (Figure 3.5) on the imaginary axis (3.183b) ≡ (3.195c) corresponds to opposite sides (3.195d) in (3.187c), if (3.195a,b) holds:

$$\text{IV:}\quad b^2 > ac < 0:\quad k_\pm a = ib \pm id,\quad d > |b|.\quad\quad\quad\text{(3.195a–d)}$$

In this case, (3.195d) implies (3.195b) that there is one pole each on the positive and negative imaginary axes:

$$\text{IV:}\quad -ik_+ = \text{Im}(k_+) > 0 > \text{Im}(k_-) = -ik_-;\quad\quad\quad\text{(3.196a,b)}$$

the subcases $b/a > 0$ ($b/a < 0$) in Figure 3.5a and b correspond to a translation of the mid-position, and the k_+ (k_-) pole remains on the upper (lower) half k-plane, so the evaluation of (3.173b) the influence function is the same:

$$G(x >< \xi;\xi) = -\frac{1}{2\pi a}\int_{-\infty}^{+\infty}\frac{e^{ik(x-\xi)}}{(k - k_+)(k - k_-)}dk$$

$$= \mp\frac{i}{a}\frac{1}{(k_\pm - k_\mp)}\exp\left[ik_\pm(x - \xi)\right],\quad\quad\quad\text{(3.196c)}$$

which simplifies to

$$\text{IV:}\quad b^2 < ac < 0:\quad G(x >< \xi;\xi) - \frac{1}{2d}\exp\left[\frac{b \pm d}{a}(\xi - x)\right].\quad\quad\text{(3.197a–c)}$$

Thus *the influence function for the l.o.d.e.c.c. (3.172a,b) is given by (3.197b) [(3.197c)] for $x > \xi$ ($x < \xi$) in the case (3.196a,b) of two poles on the imaginary axis on opposite sides of the real axis (3.187c; 3.195a–d), with k_+ above (k_- below) the real axis in Figure 3.5a and b.*

NOTE 3.18: Positive/Negative Indentation of a Pole on the Real Axis

The remaining case V for the l.o.d.e.c.c. of second order (3.172a,b) is (3.198a,b) a double pole at the origin (Figure 3.6), leading (3.173b) to the influence function (3.198c):

$$\text{IV:}\quad b = 0 = c:\quad G(x;\xi) = -\frac{1}{2\pi a}\int_{-\infty}^{+\infty}\frac{e^{ik(x-\xi)}}{k^2}dk.\quad\quad\text{(3.198a–c)}$$

Indenting the path of integration along the real axis below the pole (Figure 3.6a) leads to the sub-case IVA:

$$\text{IVA:} \quad b = 0 = c: \quad G_-\left(x;\xi\right) = -H\left(x-\xi\right)\frac{i}{2a}\lim_{k\to 0}\frac{d}{dk}\left[e^{ik(x-\xi)}\right] = \frac{x-\xi}{2a}H\left(x-\xi\right); \quad (3.199\text{a–c})$$

the subcase IVB corresponds (Figure 3.6b) to the opposite indentation below the double pole:

$$\text{IVB:} \quad b = 0 = c: \quad G_-\left(x;\xi\right) = \frac{i}{2a}H\left(\xi-x\right)\lim_{k\to 0}\frac{d}{dk}\left[e^{ik(x-\xi)}\right] = \frac{\xi-x}{2a}H\left(\xi-x\right). \quad (3.200\text{a–c})$$

It has been shown that *the l.o.d.e.c.c. (3.172a,b) in the simplest case (3.198a,b) where it remains of the second order specifies the influence function (3.199c) [(3.200c)], if the double pole at the origin is positively (negatively) indented* [Figure 3.6a and b], that is, by deforming the path of integration along the real axis below (above) the pole.

NOTE 3.19: Linear Deflection of an Elastic String

The case (3.182a,b; 3.199a,b) corresponds to (2.12d) ≡ (3.201a) to the influence function (3.201b–d) for the linear deflection of an elastic string under constant tension T with a concentrated transverse force P at $x = \xi$:

$$-\frac{T}{P}G''\left(x;\xi\right) = \delta\left(x-\xi\right): \quad a = -\frac{T}{P}, \quad b = 0 = c. \quad (3.201\text{a–d})$$

The unforced equation (3.202a) has general integral (3.202b), where (A, B) are arbitrary constants:

$$\zeta_0'' = 0: \quad \zeta_0\left(x\right) = Ax + B; \quad G\left(x;\xi\right) = \frac{x-\xi}{a}H\left(x-\xi\right) = -\frac{P}{T}\left(x-\xi\right)H\left(x-\xi\right). \quad (3.202\text{a–c})$$

The particular integral of the forced equation (3.201a) is the influence function (3.199c) where the pole at the origin is taken as inside the domain, leading to (3.202c), where was substituted (3.201b). Adding (3.202b) to (3.202c) specifies the complete integral (3.203b) of (3.203a):

$$-T\zeta'' = P\delta\left(x-\xi\right): \quad \zeta\left(x\right) = \zeta_0\left(x\right) + G\left(x;\xi\right) = Ax + B - \frac{P}{T}\left(x-\xi\right)H\left(x-\xi\right). \quad (3.203\text{a,b})$$

The boundary conditions that the string has supports at the same height (3.204a,b) determine the two constants of integration (3.204c,d)

$$\zeta\left(0\right) = 0 = \zeta\left(L\right): \quad B = 0, \quad A = \frac{P}{T}\left(1-\frac{\xi}{L}\right). \quad (3.204\text{a–d})$$

Substituting (3.204c,d) in (3.203b) specifies the influence function

$$\zeta\left(x\right) = \frac{P}{T}\left[x\left(1-\frac{\xi}{L}\right) - \left(x-\xi\right)H\left(x-\xi\right)\right] = \frac{P}{T}\begin{cases} x\left(1-\dfrac{\xi}{L}\right) & \text{if } 0 \le x \le \xi, \\[2ex] \xi\left(1-\dfrac{x}{L}\right) & \text{if } \xi \le x \le L, \end{cases} \quad (3.205\text{a,b})$$

which satisfies the boundary conditions (3.204a,b). This coincides (3.205a,b) ≡ (2.41b,c) the influence function or shape of an elastic string with supports at the same height (3.204a,b) for the linear deflection under constant tension T due to a concentrated transverse force P applied at the position $x = \xi$. The influence function for the case of a simple fourth-order l.o.d.e.c.c. is considered in Example 10.9.

Conclusion 3: The most important generalized function is the unit impulse (Figure 3.1b) from which follow (1) the derivatives of all orders, namely, first (Figure 1.6), second (Figure 1.7), and third (Figure 1.8) orders; and (2) inversely, the first three primitives are the unit jump (Figure 3.1a), and unit linear (quadratic) ramp [Figure 3.1c (d)]. The fundamental solution of an o.d.e. is the influence function corresponding to forcing by a unit impulse. In the case of a l.o.d.e., the influence function specifies the solution for arbitrary forcing using the principle of superposition. A l.o.d.e.c.c. consists of a characteristic polynomial of derivatives of the independent variable, all applied to the dependent variable; in this case the influence function can be obtained, evaluating a Fourier integral by the residues at poles that are roots of the characteristic polynomial. For example, for l.o.d.e.c.c. of first order, the influence function is evaluated from the residue at a simple pole either in the upper (Figure 3.2a) or the lower (Figure 3.2b) complex half-plane. For a l.o.d.e.c.c. of the second-order, the response function can be evaluated in five (nine) cases (subcases), namely, (1) a double pole in the positive (negative) imaginary axis [Figure 3.2a (b)]; (2) two simple poles, symmetric relative to the imaginary axis in the upper (lower) half-complex plane [Figure 3.3a (b)]; (3) two simple poles in the positive (negative) imaginary half-axis [Figure 3.4a (b)]; (4) two simple poles, one in the positive and the other in the negative imaginary axis, whose position may have a translation that keeps one pole in each half-plane [Figure 3.5.a (b)]; and (5) a double pole at the origin, indented positively (negatively), that is, below (above) the real axis [Figure 3.6a (b)].

4

Bending of Bars and Beams

A bar, like a string, is an elastic body whose longitudinal dimension or length is much larger than the cross section; in the case of the string (bar), the cross section is (is not) negligible, and there is no (there is) bending stiffness. Thus, a string cannot support a bending moment; a bending moment causes a straight bar to become curved. The curvature of a string is due to a shear stress, and thus, the displacement satisfies a second-order differential equation (Chapter 2); for a bar (Chapter 4), the bending moment is proportional to the curvature (Section 4.1), leading to a second-order differential equation that becomes of third order for the transverse force and of fourth order for the shear stress. As for a string (Chapter 2), the minimum of the elastic energy specifies (Section 4.1) the balance equation and boundary conditions. Since the displacement of a string (bar) satisfies a second (fourth)-order differential equation, there are two (four) boundary conditions, that is, one (two) at each end, indicating the coordinate of the suspension point (the method of attachment). A bar may be clamped, pin-joined, or have a sliding support or it may have a free end (Section 4.2); there are four possible static combinations of these three attachment conditions, each associated with a different set of boundary conditions. Each set of boundary conditions leads to a distinct Green or influence function that specifies the shape of the elastica or neutral line for a bar subject to a concentrated transverse force (Section 4.4). The influence function can then be used with the superposition principle to specify the shape of the bar under arbitrary loads, for example, its own weight (Section 4.3) or a concentrated moment (Section 4.5). The superposition principle applies only for linear bending, that is, with a small slope; this is a particular limit of strong bending, when the general nonlinear balance equation simplifies to linear. The linear (nonlinear) bending of a bar [Sections 4.3 through 4.5 (Sections 4.7 through 4.9)] can be considered for similar loads (Section 4.1) and boundary conditions (Section 4.2), for example, (1) a concentrated transverse force [Section 4.4 (Sections 4.8 and 4.9)] specifying the linear (nonlinear) response function; (2) one or several concentrated moments [Section 4.5 (Section 4.7)]; and (3) its own weight [Section 4.3 (Section 4.8)]. A beam is a bar subject to a bending moment plus an axial traction (like a string) or a compression (Section 4.1); if a bar has a free end, the bending causes no change in length and no compression or traction. If neither end is free, a beam under strong bending cannot change the distance between the supports, and the extension is associated with axial tractions (Section 4.6). Thus, to the transverse loads may be added a longitudinal tension either (1) as an externally applied traction or compression or (2) as a consequence of the bar being pin-joined or clamped at both ends and its extension under bending giving rise to a longitudinal traction. If the deflection of the neutral line or elastica is small (large) compared with the thickness of the cross section, the one-dimensional elastic body behaves like a bar (string); if they are comparable, the two effects are superimposed, which is the case of a beam. There are two basic cases of elastic instability (Section 4.9): (1) the linear buckling of a bar under an axial load and (2) the nonlinear collapse of a bar sliding between supports due to a transverse load. Two more cases of elastic instability concern a cantilever bar that is clamped at one end, with the other supported on a linear (rotary) spring [Example 10.10 (10.11)] placed so as to increase the deflection.

4.1 Bending Moment and Curvature of the Elastica (Bernoulli, 1744; Euler, 1744)

The bending of bar is due to a bending moment (Subsection 4.1.2) to which are associated a transverse force and a shear stress (Subsection 4.1.3); these relate the curvature with the properties of the material of the bar and the shape of the cross section (Subsection 4.1.1). To the bending of a bar may be added an axial tension, leading to a beam (Subsection 4.1.6). The bar and string are the particular cases of a beam when the length scale of the cross section is not (is) small compared with the maximum deflection (Subsection 4.1.7). The bending of a uniform or nonuniform beam (Subsection 4.1.4) is weak (strong), that is, linear (nonlinear), like for a string if (Subsection 4.1.5) the slope is everywhere small (not small at some positions).

4.1.1 Radius of Curvature and Material Properties

Consider a bar that is straight (Figure 4.1a) in the absence of transverse loads; the application of the latter causes the bar to bend (Figure 4.1b), so that the fibers on the inside (outside) are shrunk (stretched), leading to a compression (traction) force. The fibers under compression and traction are separated by a fiber (dotted line in Figure 4.1b) whose length does not change and is called the neutral fiber or **elastica**. In the case of **weak bending**, the radius of curvature R of the elastica is much larger than the length L of the bar $R^2 \gg L^2$; in this case, the maximum deflection of the beam $\delta \equiv y_{max}$ is small compared with the length $\delta^2 \gg L^2$, and the square of the slope can be neglected. In the case of **strong bending**, the slope need not be small, nor the maximum deflection, leading to nonlinear balance equations. For **simple bending** (Section 4.2), the elastica remains a plane curve, that is, it gains a curvature of flexion but no torsion, implying that (1) there is a plane longitudinal section of the bar containing the elastica; and (2) the transverse cross sections of the bar remain plane, although they are rotated from the unbent position around the elastica. Each fiber is specified by its distance y from the elastica. The y-axis is taken downward, so that (1) the Cartesian reference frame is left-handed; and (2) the positive (negative) rotation direction is clockwise (counterclockwise), indicated with $+(-)$ in Figure 4.1a. Bearing in mind that for simple bending, the transverse cross sections of the bar remain plane and merely rotate through a given angle (Figure 4.1b), the length of arc on the elastica ds and on a fiber $ds \pm dL$ at a distance $|y|$ upward $-y$ (downward y) are related (Figure 4.1b) by (4.1a and b):

$$\frac{ds+dL}{R-y} = \frac{ds}{R} = \frac{ds-dL}{R+y}, \quad -\frac{y}{R} = \frac{dL}{ds} = S_{xx}, \quad T_{xx} = S_{xx}E = -E\frac{y}{R}, \quad (4.1a\text{--}d)$$

leading to (4.1c), where R is the **radius of curvature** of the elastica and S_{xx} the relative extension in the x-direction, that is, the x-component of the **strain tensor**. The product by the **Young's modulus (1807)** specifies the longitudinal component of the **stress tensor** (4.1d).

4.1.2 Linear and Nonlinear Bending Moment

The radius of curvature is the inverse of the **curvature of flexion** and is specified (II.6.98a and b) by (4.2b and c):

$$y = \zeta(x): \quad k(x) = \frac{1}{R(x)} = \zeta''\left|1+\zeta'^2\right|^{-3/2} = \left\{\zeta'\left|1+\zeta'^2\right|^{-1/2}\right\}' = \zeta''\left[1+O\left(\zeta'^2\right)\right], \quad (4.2a\text{--}c)$$

where (4.2a) is the shape of the elastica, and (4.2b) [(4.2c)] applies to strong (weak) bending. There is an extension (contraction) that is (4.1c) positive (negative) in the case of positive curvature $R > 0$,

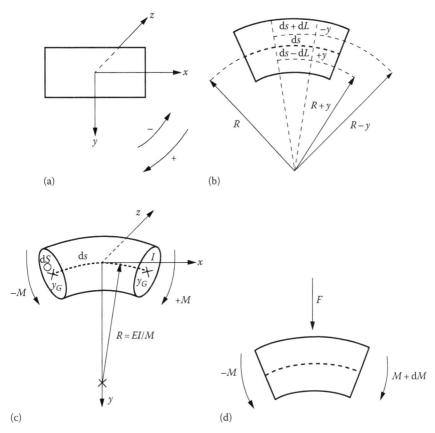

FIGURE 4.1 A bar may be bent (d) by opposite bending moments $M + dM$ and $-M$ applied at the two ends that are balanced by a transverse force F. In the case of a beam subject to a tangential tension T, its transverse component T_y is added to the transverse force F. If the bar is straight in the undeformed state (a) after bending the initially straight unstressed fibers are extended (contracted) in the outer (inner) side with (b) the exception of a separating neutral fiber called the elastica. The elastica is the line (c) joining the centers of mass of the cross section y_G with area S and area element dS. If I is the moment of inertial of the cross section the radius of curvature of the elastica is given by $R = -EI/M$, where E is the Young's modulus of the material. Since a left-handed Cartesian coordinate system with y-axis downward (a) is used, the positive (negative) direction of rotation is clockwise (counterclockwise). The simple theory of bending aims to determine the shape of the elastica.

on the outer $y < 0$ (inner $y > 0$) side of the elastica; for a bar of a material of Young's modulus $E > 0$, this corresponds to a longitudinal stress (4.1d) per unit area, that is, a traction (compression) that is positive (negative) on the outer (inner) side. The total longitudinal tension force is obtained by integration (Figure 4.1c) over the cross-sectional area dS:

$$0 = k(x)E(x)\int y\,dS = SE(x)k(x)y_G: \quad y_G \equiv \frac{\int y\,dS}{\int dS}. \tag{4.3a,b}$$

The total longitudinal tension is zero (4.3a), that is, the compressions (tractions) on the inner (outer) side of the neutral fiber balance, relative to the elastica taken as the line joining (4.3b) the centers of mass $y_G = 0$ of the sections; Young's modulus E is assumed to be constant in the cross section and varies $E(x)$ (does not vary $E = $ const) longitudinally for an **inhomogeneous (homogeneous)** bar. Unlike the tension force, the **bending moment** of an elastic bar is of the same sign on both sides of the elastica because the tractions $T_{xx} > 0$ (compression $T_{xx} < 0$) on the outer $y < 0$ (inner $y > 0$) side have in both cases a negative product (4.4a):

$$-M(x) = Ek(x) \int y^2 dA = EIk(x) = EI\zeta'' \left|1 + \zeta'^2\right|^{-3/2}, \quad I \equiv \int y^2 dS. \tag{4.4a,b}$$

The bending moment (4.4a) never vanishes for a curved bar $k(x) \neq 0$ and involves the **moment of inertia** of the cross section (4.4b) relative to the origin at the elastica. *The bending moment of a bar under simple bending is proportional (4.4a) to minus the curvature (4.2b) of the elastica (4.2a) through the Young's modulus of the material E and the moment of inertia (4.4b) of the cross section relative to the elastica (4.3a and b).* The moment of inertia (4.4b) does not (does) depend on the axial coordinate x along the bar if the cross section is constant (variable), that is, for a prismatic (tapering) bar. For a given load, the deflection of the bar may be reduced by (1) choosing a **stiffer material**, that is, one with a higher Young's modulus, for example, using a steel instead of a wooden bar; and (2) for the same cross-sectional area, choosing a larger moment of inertia I or **radius of gyration** $G = \sqrt{I/S}$, for example, taking an I-shaped cross section instead of a circular or rectangular one with the same area.

4.1.3 Transverse Force and Shear Stress

The bending moment generally varies along the length of the bar (Figure 4.1d), and its variation over a segment dx is balanced by the **transverse force** (4.5a),

$$0 = F dx + (M + dM) - M : \quad F(x) = \frac{dM}{dx} \equiv -M'(x), \tag{4.5a,b}$$

so that the equation of the elastica is given by (4.5b) where (4.4a) is used:

$$F(x) = \left\{EIk(x)\right\}' = \left\{EI\zeta''\left|1 + \zeta'^2\right|^{-3/2}\right\}' = \left\{EI\left[\zeta'\left|1 + \zeta'^2\right|^{-1/2}\right]'\right\}'. \tag{4.6a}$$

The axial rate of change of the transverse force along the bar is (2.7a) the **shear stress**:

$$f(x) = F'(x) = -M''(x) = \left\{EIk(x)\right\}'' = \left\{EI\zeta''\left|1 + \zeta'^2\right|^{-3/2}\right\}'' = \left\{EI\left[\zeta'\left|1 + \zeta'^2\right|^{-1/2}\right]'\right\}''. \tag{4.6b}$$

The Young's modulus E is a constant for a homogeneous bar made of a single material; if the cross section is constant, that is, for a cylindrical beam, the moment of inertia (4.4b) is also a constant. The **bending stiffness** (4.6c) is constant

$$B(x) = E(x)I(x) \begin{cases} = \text{constant for uniform bar,} & (4.6c) \\ \neq \text{constant for nonuniform bar} & (4.6d) \end{cases}$$

for a **uniform bar**, for example, for a homogeneous bar of constant cross section or, more exceptionally, if E, I vary along the length of the bar so as to conserve the product at each longitudinal position.

Thus, *the shape of the elastica of a bar under a bending moment M/transverse force F/shear stress f is given respectively (4.4a/4.6a/4.6b); the last two simplify for a uniform bar (4.6c) ≡ (4.7a) to (4.7b/4.7c):*

$$B \equiv EI = \text{const}: \quad F(x) = EIk'(x), \quad f(x) = EIk''(x). \tag{4.7a–c}$$

In all cases, the differential equation of the elastica is respectively of the second/third/fourth order.

4.1.4 Bending of Nonuniform and Uniform Beams

In addition to the transverse force, there may be *a longitudinal tension T such as (1) a traction T > 0 as for an elastic string (Chapter 2) or (2) a compression T < 0 that is possible for a bar but not for a string.* In the presence of a tangential tension, (4.5a and b) is replaced by (4.6a and b), stating (Figure 4.1a) that the longitudinal variation of the bending moment is balanced by the total transverse force, that is, the sum of the external transverse force and transverse component of the tension:

$$0 = (F + T_y)dx + dM, \quad F(x) = -M'(x) - T_y(x). \tag{4.8a,b}$$

In the case of a string (bar), the bending moment (tangential tension) is zero, $M = 0 (T = 0)$, and (4.8b) simplifies to (2.5b) [(4.5b)]. Substituting (4.4b) and (2.2b) in (4.8b) specifies the transverse force for a beam (4.9a):

$$F(x) = \left\{ EI\zeta'' \left|1 + \zeta'^2\right|^{-3/2} \right\}' - T\zeta' \left|1 + \zeta'^2\right|^{-1/2}; \tag{4.9a}$$

$$f(x) = \left\{ EI\zeta'' \left|1 + \zeta'^2\right|^{-3/2} \right\}'' - \left\{ T\zeta' \left|1 + \zeta'^2\right|^{-1/2} \right\}'. \tag{4.9b}$$

The derivative of the transverse force with regard to the longitudinal coordinate specifies the shear stress (2.7a) that is (4.9b) the sum of the contributions due to the bending moment (4.6b) and to the tangential tension (2.8b). Both the transverse force (4.9a) and shear stress (4.9b) involve the tangential tension T that may (1) be either traction $T > 0$ or compression $T < 0$; (2) be constant or vary along the length as in (2.6). A bar under axial tension is called a **beam**.

4.1.5 Linear and Nonlinear Bending of a Beam

It has been shown that *the equation of the elastica of a beam is specified by (4.9b) [(4.9a)] in terms of the shear stress (transverse force), where the axial tension T may be nonuniform. The Young modulus E (moment of inertia of the cross section I) may be nonuniform for an inhomogeneous material (a tapering beam with varying cross section), and the bending can be strong, that is, involve a large slope of the elastica. Thus, there are three independent approximations: (1) for a uniform beam (4.10a), that is, with constant bending stiffness (4.6c), the shear stress (4.9b) [transverse force (4.9a)] simplifies to (4.10b) [(4.10c)]:*

$$EI = \text{const}: \quad f(x) = EI \left\{ \zeta' \left|1 + \zeta'^2\right|^{-1/2} \right\}''' - \left\{ T\zeta' \left|1 + \zeta'^2\right|^{-1/2} \right\}', \tag{4.10a,b}$$

$$F(x) = EI \left\{ \zeta' \left|1 + \zeta'^2\right|^{-1/2} \right\}'' - T\zeta' \left|1 + \zeta'^2\right|^{-1/2}; \tag{4.10c}$$

(2) for uniform axial tension (4.11a), the equation for the shear stress (4.9b) ≡ (4.11b) can be written in terms of the curvature of the elastica:

$$T = \text{const}: \quad f(x) = \left\{ EIk(x) \right\}'' - Tk(x) = \left\{ EI\zeta'' \left| 1 + \zeta'^2 \right|^{-3/2} \right\}'' - T\zeta'' \left| 1 + \zeta'^2 \right|^{-3/2}, \qquad (4.11a,b)$$

and the equation (4.9a) for the transverse force is unchanged; and (3) for weak bending, when the slope of the elastica is small (4.12a), the shear stress (4.9b) ≡ (4.12b)/transverse force (4.9a) ≡ (4.12c)/bending moment (4.4a) are specified, respectively, by (4.12b/c/d):

$$\zeta'^2 \ll 1: \quad \left\{ f(x), F(x), M(x) \right\} = \left\{ \left(EI\zeta'' \right)'' - \left(T\zeta' \right)', \left(EI\zeta'' \right)' - T\zeta', -EI\zeta'' \right\}. \qquad (4.12a\text{–}d)$$

The three assumptions (1+2+3) together, in the case of weak bending (4.12a) ≡ (4.13a) of a uniform beam (4.10a) ≡ (4.13b) under uniform axial traction (4.11a) ≡ (4.13c), lead to the bending moment (4.13d)/ transverse force (4.13e)/shear stress (4.13f):

$$\zeta'^2 \ll 1, EI = \text{const}, T = \text{const}: \left\{ M(x), F(x), f(x) \right\} = \left\{ M, -M' - T_y', -M'' - T_y'' \right\}$$

$$= \left\{ -EI\zeta'', EI\zeta''' - T\zeta', EI\zeta'''' - T\zeta'' \right\}, \qquad (4.13a\text{–}f)$$

as the balance equations in their simplest form.

4.1.6 Longitudinal Tension Associated with Fixed Ends

Comparing the order of magnitude of the two terms in (4.13e) ≡ (4.14a),

$$F = EI\zeta''' - T\zeta': \quad \frac{T\zeta'}{EI\zeta'''} \sim \frac{TL^2}{EI} = \frac{T}{T_*}, \quad T_* \equiv \frac{EI}{L^2}, \qquad (4.14a\text{–}c)$$

it follows that *an elastic beam under a strong (4.15a) [weak (4.16a)] longitudinal tension acts as a string (bar) with a shape (shape of the elastica) given by (4.15b and c) [(4.16a–d)]:*

$$T \gg \frac{EI}{L^2} \equiv T_*: \quad f(x) = -T\zeta'', \quad F(x) = -T\zeta', \qquad (4.15a\text{–}c)$$

$$T \ll \frac{EI}{L^2} \equiv T_*: \quad f(x) = EI\zeta'''', \quad F(x) = EI\zeta''', \quad M(x) = -EI\zeta'', \qquad (4.16a\text{–}d)$$

in the uniform (4.10a) case. If T ~ T*, then the two effects are superimposed for a beam, leading to (4.13a–f) in the uniform case and its alternative [(4.12a–d) or (4.11a and b) or (4.10a–c)] and combined (4.9a and b) generalizations.

If a beam is fixed at both ends (4.17a and b), the increase in length of a segment dx of the beam under flexion is given by (4.17c):

$$\zeta(0) = 0 = \zeta(L): \quad ds - dx = \left| (dx)^2 + (d\zeta)^2 \right|^{1/2} - dx$$

$$= \left\{ \left| 1 + \zeta'^2 \right|^{1/2} - 1 \right\} dx = \frac{1}{2} \zeta'^2 \left[1 + O(\zeta'^2) \right] dx, \qquad (4.17a\text{–}d)$$

which simplifies to (4.17d) in the linear case of a small slope (4.12a) ≡ (4.18a):

$$\zeta'^2 \ll 1: \quad T = \frac{1}{L}\int_0^L ES(ds - dx) = \frac{1}{2L}\int_0^L ES\left[\zeta'(x)\right]^2 dx. \tag{4.18a,b}$$

The total tension is obtained multiplying (4.18b) the relative extension $(ds - dx)/L$ in (4.17b) by the Young's modulus of the material and the area S of the cross section and integrating along the length of the bar. Thus, *the weak bending (4.18a) of a beam pinned or clamped at both ends (4.17a and b) causes a longitudinal traction (4.18b), where S is the area of the cross section, E the Young's modulus of the material, L the length of the undeformed beam, and $\zeta'(x)$ the slope of the elastica. The first two (E, A) may be nonuniform in (4.18b).*

4.1.7 Bar and String as Opposite Limits of a Beam

The order of magnitude of (4.18b) can be estimated as (4.19a–c):

$$S \sim h^2, \quad \zeta' \sim \frac{\delta}{L}: \quad T \sim \frac{Eh^2\delta^2}{L^2}, \quad T\zeta' \sim \frac{Eh^2\delta^3}{L^3}, \tag{4.19a–d}$$

where the terms used are as follows: h is the length scale of the cross section and S its area; δ the maximum deflection, leading (4.18b) to (4.19c) for the tension term (4.19d) in (4.14a). The other bending term (4.20b) in (4.14a) involves the moment of inertia (4.4b) of the cross section (4.20a):

$$I \sim h^4: \quad EI\zeta''' \sim \frac{Eh^4\delta}{L^3}; \quad \frac{EI\zeta'''}{T\zeta'} \sim \frac{h^2}{\delta^2}. \tag{4.20a–c}$$

Comparing (4.20c) the two terms, it follows that *an elastic beam, clamped or pinned at both ends, under weak bending (4.12a) ≡ (4.18a), implying that the maximum deflection δ is small $\delta^2 \ll L^2$ compared with the length, acts as a* **bar (string)** *in the sense that the bending (tension) term, the first (second) on the r.h.s. of (4.12b and c) or (4.13d and e) dominates, and the equation of elastica is given by (4.16b–d) [(4.15b and c)] if the transverse dimension h of the cross section is much larger $h^2 \gg \delta^2$ (much smaller $h^2 \ll \delta^2$) than the maximum deflection δ; both terms in (4.12b–d) or (4.13b–f) are important for a beam, that is, when the maximum deflection is comparable to the transverse dimension of the cross section. In this sense, a string (bar) is a thin (thick) beam.*

4.2 Deformation and Displacement of the Cross Section (Saint-Venant, 1856) and Elastic Energy (Green, 1837)

Two assumptions are made corresponding to the Bernoulli (1744)–Euler (1744) theory of bars under simple bending (Section 4.1), that is, (1) a transverse section remains plane under bending and thus rotates, remaining orthogonal to the elastica; (2) there are no stresses at the sides and ends of the bar, so that there is only an elastic stress inside. These two assumptions specify the displacement vector (Subsection 4.2.2) that is related to (Subsection 4.2.1) the strain and stress tensors and elastic energy

(Subsection 4.2.3). The principle of virtual work stating a balance of the elastic energy against the work of the shear stresses in deflecting the elastica (Subsection 4.2.4) specifies (1) the differential equation of the elastica for weak (strong) bending [Subsection 4.2.5 (Subsection 4.2.6)] and (2) the boundary conditions (Subsection 4.2.7) that apply at each end. The combination of boundary conditions at the two ends (Subsection 4.2.9) of the beam is related to the balance of external applied and reaction forces and moments (Subsection 4.2.8).

4.2.1 Strain and Stress Tensors for Simple Bending

Consider a slender body, that is, whose length is much longer than the cross section (Figure 4.2a) and that is straight in the undeformed state, so that the x-axis may be taken in the longitudinal direction and the Cartesian (y, z) axis in the cross section (Figure 4.2b) with the y-axis downward. The bending causes a relative extension (4.1a and b) that corresponds to the longitudinal strain (4.1c) and associated stress (4.1d), where y is the distance from the elastica measured downward,

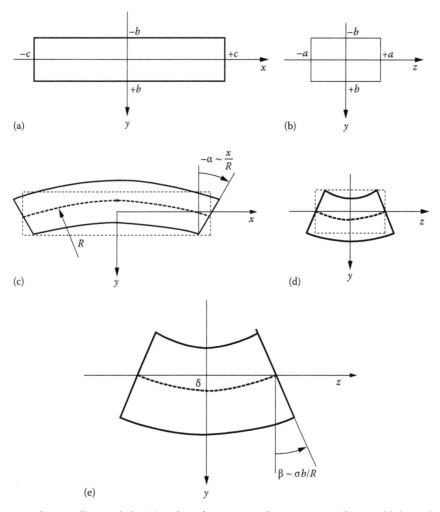

FIGURE 4.2 If an initially straight bar (a) with uniform rectangular cross section (b) is weakly bent, the elastica becomes curved (c) and the cross section is deformed with curved top and bottom and straight inclined sides (d), and its centroid is shifted downward (e).

R the radius of curvature of the elastica (4.2b and c), and *E* the Young modulus of the material. Assuming that there are no loads on the sides of the bar (4.21a),

$$T_{yy} = T_{zz} = T_{yz} = 0, \quad T_{xy} = 0 = T_{xz}, \tag{4.21a,b}$$

and no off-axis loads at the ends (4.21b), the only nonzero component of the stress tensor is a longitudinal tension along the axis (4.1d). By the inverse Hooke's law (II.4.78a–c), the uniaxial stress (4.1d) is associated with three nonzero longitudinal strains, namely (4.1c) ≡ (4.22a) and (4.22b and c), where σ is the Poisson ratio:

$$\partial_x u_x = S_{xx} = -\frac{y}{R}, \quad \partial_y u_y = S_{yy} = S_{zz} = \partial_z u_z = -\sigma S_{xx} = \sigma \frac{y}{R}, \tag{4.22a–c}$$

and the remaining three shear strains or distortions are zero:

$$2S_{xy} = \partial_x u_y + \partial_y u_x = 0, \quad 2S_{xz} = \partial_x u_z + \partial_z u_x = 0, \quad 2S_{yz} = \partial_y u_z + \partial_z u_y = 0. \tag{4.23a–c}$$

4.2.2 Longitudinal and Cross-Sectional Displacement Vectors

When integrating (4.22a–c, 4.23a–c) for the displacement vector, the constants of integration may be set to zero, by a suitable choice of the origin of the Cartesian coordinate system. The integration of (4.22a) [(4.22c)] leads to (4.24a) [(4.24b)], which satisfy (4.23b):

$$u_x = -\frac{xy}{R}, \quad u_z = \sigma \frac{yz}{R}. \tag{4.24a,b}$$

Substituting (4.24a) [(4.24b)] in (4.23a) [(4.23c)] leads to (4.25a) [(4.25b)]:

$$\partial_x u_y = -\partial_y u_x = \frac{x}{R}, \quad \partial_z u_y = -\partial_y u_z = -\sigma \frac{z}{R}; \quad \partial_y u_y = \sigma \frac{y}{R}. \tag{4.25a–c}$$

Together, (4.25a and b) and (4.22b) ≡ (4.25c) specify the *y*-component u_y of the displacement vector through all of its three partial derivatives:

$$du_y = \left(\partial_x u_y \right) dx + \left(\partial_y u_y \right) dy + \left(\partial_z u_y \right) dz = \frac{x dx + \sigma \left(y dy - z dz \right)}{R}. \tag{4.26}$$

From (4.26) follows the missing component (4.27b)

$$\vec{u} = \{u_x, u_y, u_z\} = \left\{ -\frac{xy}{R}, \frac{x^2 + \sigma \left(y^2 - z^2 \right)}{2R}, \frac{\sigma yz}{R} \right\}, \tag{4.27a–c}$$

of the displacement vector [(4.24a) ≡ (4.27a) and (4.24b) ≡ (4.27c)]. Thus, *a slender straight bar (Figure 4.2a) with ends at x = ±c with undeformed (Figure 4.2b) rectangular cross section with edges at* $(z = \pm a, y = \pm b)$ *after simple bending (Figure 4.2c through e) takes the shape*

$$x = \pm c + u_x = \pm c \left(1 - \frac{y}{R} \right), \quad z = \pm a + u_z = \pm a \left(1 + \sigma \frac{y}{R} \right), \tag{4.28a,b}$$

$$y = \pm b + u_y = \pm b - \frac{\sigma z^2}{2R} + \frac{b^2 \sigma + x_0^2}{2R}, \tag{4.28c}$$

showing that (1) as y increases, x decreases (increases) linearly on the right (left) end of the bar with $+(-)$ sign in (4.28a), implying (Figure 4.2c) that the cross section remains flat and is rotated by an angle $y/R \sim \tan \alpha \sim \alpha$ for $L^2 \ll R^2$ large radius of curvature R compared with the length L of the bar (4.1b); (2) as y increases, z increases (decreases) linearly on the right (left) vertical side of the cross section with $+(-)$ sign in (4.28b), implying (Figure 4.2d amplified in Figure 4.2e) that the initially vertical sides remain straight but are inclined at an angle $\beta \sim \sigma b/R$ to the vertical; and (3) as z increases in modulus regardless of sign, then y decreases in (4.28c) in proportion to the square, showing (Figure 4.2d ≡ 4.2e) that the initially flat horizontal upper and lower sides of the cross section become parabolas (4.28c) = (4.29a):

$$y = \pm b - \frac{1}{2}\bar{k}z^2 + \zeta, \quad \bar{k} = \frac{\sigma}{R} = \sigma k, \quad \delta = \frac{x_0^2 + b^2 \sigma}{2R}, \tag{4.29a–c}$$

with (3-1) a curvature (4.29b) that equals the curvature of the elastica times the Poisson ratio; (3-2) the elastica or "center of the cross section" is displaced downward to the position (4.29c).

4.2.3　Elastic Energy and Moment of Inertia of the Cross Section

The elastic energy per unit volume (II.4.86b) is

$$2E_d = \sum_{i,j=1}^{3} S_{ij}T_{ij} = S_{xx}T_{xx} = E(S_{xx})^2 = E\frac{y^2}{R^2} \tag{4.30}$$

for simple bending (4.1a and d, 4.21a and b); the total elastic energy is obtained (4.31b) integrating over the cross section dS and length L of the bar:

$$\zeta'^2 \ll 1: \quad \bar{E}_d = \int_0^L dx \int E_d dS = \frac{1}{2}\int_0^L \frac{E}{R^2}dx \int y^2 dS = \frac{1}{2}\int_0^L EI\zeta''^2 dx, \tag{4.31a,b}$$

using the moment of inertia of the cross section (4.4b) and curvature of the elastica (4.2c) for weak bending (4.31a). Thus, *the elastic energy of a bar under simple flexion is a quadratic function of the curvature (4.31b) for weak bending (4.31a) and is a particular case of the generalization (4.31a) ≡ (4.32a) for strong bending:*

$$2\bar{E}_d = \int_0^L EI\zeta''^2 \left|1 + \zeta'^2\right|^{-3/2} dx = \int_0^L EI\zeta'' k(x)dx; \tag{4.32a,b}$$

the latter result will be checked subsequently (4.32a and b) ≡ (4.45). The principle of virtual work, stating the balance of the elastic energy against the work of the weak (strong) bending loads, specifies not only the differential equation of the elastica (Section 4.1), but also the boundary conditions (Section 4.2). *The elastic energy of a beam is the sum of the elastic energies associated with bending (4.32a and b) and tension (2.17) in the nonlinear (4.33a) [linear (4.33b)] case:*

$$2\bar{E}_d = \int_0^L \left\{ EI\zeta''^2 \left|1 + \zeta'^2\right|^{-3/2} + 2T\left[\left|1 + \zeta'^2\right|^{-1/2} - 1\right] \right\}dx = \int_0^L \left(EI\zeta''^2 + T\zeta'^2 \right)dx. \tag{4.33a,b}$$

The second term on the r.h.s. of (4.33a and b) has already been discussed (Section 2.2) in the case of a string, and thus next (Subsections 4.1.4 and 4.2.6) is considered the first term on the r.h.s. of (4.33a and b) corresponding to a bar; in both cases of the string (bar) is used the principle of virtual work [Section 2.2 (Section 4.2)].

4.2.4 Principle of Virtual Work Applied to the Bending of a Bar

The **principle of virtual work** balances (1) the work of the shear stress, that is, its product by the deflection of elastica integrated along the length of the bar; and (2) the elastic energy (4.34b) ≡ (4.31b), considered first for weak bending (4.31a) ≡ (4.34a):

$$\zeta'^2 \ll 1: \quad \int_0^L f(x)\delta\zeta(x)dx = \delta\bar{W} = \delta\bar{E}_d = \int_0^L EI\zeta''\delta\zeta''dx. \tag{4.34a,b}$$

The r.h.s. of (4.34b) can be integrated by parts a first time:

$$EI\zeta''\delta\zeta''dx = EI\zeta''\delta\left(\frac{d\zeta'}{dx}\right)dx = EI\zeta''d(\delta\zeta') = d\left(EI\zeta''\delta\zeta'\right) - \delta\zeta'd\left(EI\zeta''\right), \tag{4.35a}$$

leading to

$$\int_0^L f\delta\zeta dx = EI\zeta''\delta\zeta'\Big|_0^L - \int_0^L \left(EI\zeta''\right)'\delta\zeta'dx. \tag{4.35b}$$

Another integration by parts is applied to the second term on the r.h.s. of (4.35b):

$$\left(EI\zeta''\right)'\delta\zeta'dx = \left(EI\zeta''\right)'\delta\left(\frac{d\zeta}{dx}\right)dx = \left(EI\zeta''\right)'d(\delta\zeta) = d\left[\left(EI\zeta''\right)'\delta\zeta\right] - \delta\zeta d\left[\left(EI\zeta''\right)'\right], \tag{4.36a}$$

leading to

$$\int_0^L \left[f - \left(EI\zeta''\right)''\right]\delta\zeta dx = EI\zeta''\delta\zeta'\Big|_0^L - \left(EI\zeta''\right)'\delta\zeta\Big|_0^L. \tag{4.36b}$$

Since $\delta\zeta$, $\delta\zeta'$ are arbitrary, all terms in (4.36b) must vanish:

$$f = \left(EI\zeta''\right)'', \quad \delta\zeta(0) = \delta\zeta(L) = 0 = \delta\zeta'(0) = \delta\zeta'(L), \tag{4.37a–e}$$

leading to three sets of conditions: (1) the equation of the elastica (4.37a) ≡ (4.12b) for weak bending (4.12a) of a nonuniform bar; and (2, 3) the boundary conditions (4.37b–e) for a bar clamped (Figure 4.3a) at both ends (Figure 4.4a). Alternative forms of the work of bending loads (Subsection 4.2.5) are considered before the extension from weak (Subsection 4.2.4) to strong (Subsection 4.2.6) bending, followed by alternative boundary conditions (Subsection 4.2.7).

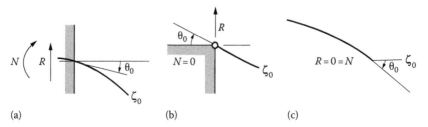

FIGURE 4.3 The three classical types of supports for a bar are: (a) clamped with fixed position and inclination implying the existence of both reaction force and moment; (b) pinned, that is free to rotate, so that the bending moment vanishes but there can be a reaction force at the fixed support; and (c) free, that is taking a position and inclination such that the bending moment and transverse force vanish.

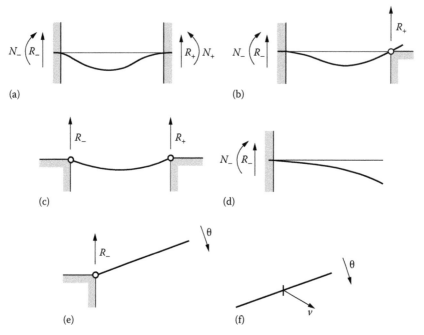

FIGURE 4.4 There are six combinations (Table 4.1) of the three types of supports (Figure 4.3) at the two ends. They may be grouped in three sets of two by considering the difference between the number n of reaction forces and moments at the supports and the number 2 of independent of force and balance equations: (I) isostatic ($n = 2$) in the (d) clamped-free and (c) pinned-pinned cases when the balance of forces and moments determines the reactions at the supports; (II) hyperstatic ($n > 2$) in the (b) clamped-pinned ($n = 3$) and (a) clamped-clamped ($n = 4$) cases when the reactions at the supports in general (apart from special cases of symmetry) can be determined only after solving the problem of elastic deformation of the bar; (III) hypostatic ($n < 2$) for the (e) pinned-free $n = 1$ [(f) free-free ($n = 0$)] bar that is generally not at rest since it can rotate around the pin (rotate and translate freely). The four classical cases of support are (a–d) and may be considered for different loads such as a uniform shear stress (Figure 4.5) or [Figure 4.6 (4.7)] a concentrated force (moment).

4.2.5 Work of the Transverse Force and Bending Moment

The work of the shear stress can also be expressed in terms of the transverse force (4.6b) [bending moment (4.5b)]:

$$\delta \bar{W} = \int_0^L f \delta \zeta \, dx = \int_0^L F' \delta \zeta \, dx = - \int_0^L M'' \delta \zeta \, dx. \qquad (4.38a\text{–}c)$$

An integration by parts of (4.38b) leads to (4.39a):

$$\delta\bar{W} = F\delta\zeta\Big|_0^L - \int_0^L F\delta\zeta'\mathrm{d}x, \quad \delta\zeta(0) = 0 = \delta\zeta(L), \tag{4.39a–c}$$

where the first term on the r.h.s. vanishes for a bar fixed at both ends (4.39b and c) ≡ (4.37b and c). A further integration by parts leads to (4.40a):

$$\delta\bar{W} = -\int_0^L F\delta\zeta'\mathrm{d}x = \int_0^L M\delta\zeta'\mathrm{d}x = M\delta\zeta'\Big|_0^L - \int_0^L M\delta\zeta''\mathrm{d}x, \quad \delta\zeta'(0) = 0 = \delta\zeta'(L), \tag{4.40a–c}$$

where the first term on the r.h.s. vanishes by (4.40b and c) ≡ (4.37d and e). The results (4.38a–c, 4.39a–c, 4.40a–c) may be combined in

$$\delta\bar{W} = \int_0^L f(x)\delta\zeta(x)\mathrm{d}x = -\int_0^L F(x)\delta\zeta'(x)\mathrm{d}x = -\int_0^L M(x)\delta\zeta''(x)\mathrm{d}x, \tag{4.41a–c}$$

three alternative expressions for the work of bending loads, in terms of the shear stress (4.41a)/transverse force (4.41b)/bending moment (4.41b), multiplying, respectively, the displacement/slope/second derivative of the elastica, with +/−/− sign, integrated along the length of the bar.

4.2.6 Principle of Virtual Work for Weak and Strong Bending

The principle of virtual work can be stated in terms of the shear stress (Subsection 4.2.4) or alternatively in terms of the transverse force or bending moment (Subsection 4.2.6) both for weak and strong bending. Retaining the elastic energy for weak bending (4.34b) balanced against the work (4.41b) of the transverse force,

$$-\int_0^L F(x)\delta\zeta'\mathrm{d}x = \delta\bar{W} = \delta\bar{E}_d = \int_0^L EI\zeta''\delta\zeta''\mathrm{d}x, \tag{4.42}$$

leads after an integration by parts (4.35a) to (4.43a):

$$-\int_0^L \left[F - (EI\zeta'')'\right]\delta\zeta\,\mathrm{d}x = EI\zeta''\delta\zeta'\Big|_0^L; \quad F = (EI\zeta'')'. \tag{4.43a,b}$$

From (4.43a) follow the boundary conditions (4.37d and e) and the equation of the elastica (4.43b) ≡ (4.12c) for weak bending. The elastic energy for weak bending (4.34b) balanced against the work (4.41c) of the bending moment (4.44a),

$$-\int_0^L M(x)\delta\zeta''\mathrm{d}x = \delta\bar{W} = \delta\bar{E}_d = \int_0^L EI\zeta''\delta\zeta''\mathrm{d}x, \quad M = -EI\zeta'', \tag{4.44a,b}$$

leads directly, without integration by parts, to the equation of the elastica (4.44) $\equiv$ (4.12d). In the case of strong bending (4.44a), it leads to

$$d\bar{E}_d = \delta\bar{W} = -\int_0^L M\delta\zeta''dx = \int_0^L EIk(x)\delta\zeta''dx = \int_0^L EI\zeta''\left|1+\zeta'^2\right|^{-3/2}\delta\zeta''dx, \qquad (4.45)$$

confirming the exact expression for the elastic energy (4.45) $\equiv$ (4.32a), where the variation applies to the second derivative.

4.2.7 Three Distinct Boundary Conditions for a Beam

One end, say $x = a$, of a beam may be **supported** in three different ways: (1) if it is **clamped** (Figure 4.3a), the coordinate and slope at the support point can be chosen (4.46a and b), and the second (third)-order derivatives determine (4.13c) [(4.13d)] the **reactions** at the support, namely, the bending moment (4.46c) [transverse force (4.46d)]:

$$\zeta(a) = \zeta_0, \zeta'(a) = \zeta_0': \quad M = -EI\zeta''(a), \quad F = EI\zeta'''(a) - T\zeta_0'; \qquad (4.46a\text{-}d)$$

(2) if it is **pin-joined** or has a **sliding support** (Figure 4.3b), the coordinate is given (4.47a), and bending moment is zero (4.47b), that is, the elastica is inclined at the support (4.47c), and there is a transverse reaction force (4.47d):

$$\zeta(a) = \zeta_0, EI\zeta''(a) = 0: \quad \tan\theta_0 = \zeta'(a), \quad F = EI\zeta'''(a) - T\zeta'(a); \qquad (4.47a\text{-}d)$$

and (3) if it is a **free end** (Figure 4.3c), it cannot support reaction forces (4.48a) or moments (4.48b), and this determines the deflection (4.48c) and slope (4.48d):

$$EI\zeta''(a) = T\zeta(a), EI\zeta'''(a) = 0: \quad \zeta_0 = \zeta(a), \quad \tan\theta_0 = \zeta'(a). \qquad (4.48a\text{-}d)$$

If at the free end there is a concentrated force F_0 (moment M_0), the boundary conditions (4.48a and b) are replaced by (4.13d and e). *The boundary conditions for a clamped (4.46a–d), pinned, or sliding (4.47a–d) or free (4.48a–d) end of a uniform beam under weak bending (4.13a–f) can be extended to nonuniform beams (4.11a and b, 4.12a–d) and strong bending (4.10b and c, 4.11a and b) or nonuniform axial tension (4.10b and c, 4.12a–d) or any combination of the three (4.4a, 4.9a and b). Other types of boundary conditions are possible, for example,*

$$p\zeta_0 = F(a) = EI\zeta'''(a) - T\zeta'(a), \quad q\zeta_0' = M(a) = -EI\zeta''(a), \qquad (4.49a,b)$$

for the straight p (rotary q) spring that exerts a transverse force (4.49a) [bending moment (4.49b)] proportional to the displacement (rotation) at the support; these cases of other types of support are considered in Example 10.10 (Example 10.11). The three basic types (Figure 4.3a through c) of boundary conditions (1–3) allow 3! = 6 distinct combinations for two supports (Figure 4.4a through f) that are considered next (Subsection 4.2.8), concerning the balance of external applied and reaction forces and moments.

4.2.8 Isostatic, Hypostatic, and Hyperstatic Beams

The boundary conditions for a beam (Subsection 4.2.7) lead to a distinct number n of reactions: (1) at a clamped end (Figure 4.3a), there is a reaction force and moment $n = 2$; (2) at a pinned or sliding end, there is only a reaction force $n = 1$ but no reaction moment; and (3) at a free end $n = 0$, there is neither a reaction moment nor a reaction force. The total number of reactions $n = n_1 + n_2$ is the sum of the values at the two supports and specifies the number of unknowns n. There are three balance equations to be satisfied: (1) external applied forces and reaction forces at the supports; and (2, 3) balance of moments at each support. Since only two of these equations are independent, it is sufficient to consider the latter (2, 3). *The balance of moments at the support supplies two equations to determine the n unknowns leading to* $2 - n$ **degrees of freedom**:

$$g = 2 - n \begin{cases} > 0 & \text{hypostatic: motion,} & (4.50a) \\ = 0 & \text{isostastic,} & (4.50b) \\ < 0 & \text{hyperstatic: elastic deformation.} & (4.50c) \end{cases}$$

This implies (Table 4.1) three cases. In the **hypostatic case** *of a bar with one end free and the other also free (pinned) since* $n = 0 (n = 1)$*, there are two (is one) degrees-of-freedom* $g = 2 (g = 1)$*, so it can have a translational and rotational (only rotational) motion [Figure 4.4f (e)]; this leads to a dynamic problem, since the bar will not generally be at rest. In the* **isostatic case**, *the bar is either clamped-free or pinned-pinned, and the two balance equations are sufficient [Figure 4.4d (c)] to determine the reactions at the supports* $n = 2$*, and the bar is static* $g = 0$. *In the* **hyperstatic case**, *the bar is clamped at one end and pinned* $n = 3$ *(clamped* $n = 4$*) at the other end* $g = -1 (g = -2)$*, and the balance equations are insufficient [Figure 4.4b (a)] to determine the reactions at the supports; the latter are specified by the elastic deformation of the bar.* Only the four isostatic and hyperstatic cases, and not the two hypostatic cases, correspond to static equilibrium; these four static combinations of two supports are considered in more detail next (Subsection 4.2.9).

4.2.9 Four Static Combinations of Three Boundary Conditions at Two Supports

Combining the three basic types (4.46a–d, 4.47a–d, 4.48a–d) of horizontal support for a bar with ends at $x = 0$ and $x = L$, there are four possible static, that is, isostatic (4.50b) or hyperstatic (4.50c) sets of boundary conditions: (1) clamped (4.51a–d) at both ends (Figure 4.4a): (2) clamped at one end and

TABLE 4.1 Types of Supports of a Bar

| Case | Figure | Support | | Reactions n | Degrees-of-Freedom g | Problem |
		First	Second			
I	4.4a	Clamped	Clamped	4	−2	Hyperstatic
II	4.4b	Clamped	Pinned[a]	3	−1	Hyperstatic
III	4.4d	Clamped	Free	2	0	Isostatic
IV	4.4c	Pinned[a]	Pinned[a]	2	0	Isostatic
V	4.4e	Free	Pinned[a]	1	1	Hypostatic
VI	4.4f	Free	Free	0	2	Hypostatic

Note: Classification of bars for six different combinations of the two supports leading to three cases: isostatic, hyperstatic and hypostatic.

[a] Or supported.

pin-joined or sliding (4.52a–d) at the other (Figure 4.4b); (3) pin-joined or sliding (4.53a–d) at both ends (Figure 4.4c); and (4) clamped at one end and free at (4.54a–d) the other (Figure 4.4d):

$$\text{Clamped-clamped}: \quad \zeta(0) = \zeta'(0) = 0 = \zeta(L) = \zeta'(L), \tag{4.51a–d}$$

$$\text{Clamped-pinned}: \quad \zeta(0) = \zeta'(0) = 0 = \zeta(L) = \zeta''(L), \tag{4.52a–d}$$

$$\text{Pinned-pinned}: \quad \zeta(0) = \zeta''(0) = 0 = \zeta(L) = \zeta''(L), \tag{4.53a–d}$$

$$\text{Clamped-free}: \quad \zeta(0) = \zeta'(0) = 0 = \zeta''(L) = \zeta'''(L) - \frac{T}{EI}\zeta'(L). \tag{4.54a–d}$$

In (4.51b,d) ≡ (4.52b) ≡ (4.54b), it is assumed that the beam is clamped horizontally, but it may instead (4.47c) be clamped at an angle θ_0; in (4.51a,c) ≡ (4.52a,c) ≡ (4.53a,c), it is assumed that both supports are at the same height (4.46a) ≡ (4.47a), taken to be zero, but they could be nonzero and different. Not included in (4.51a–d, 4.52a–d, 4.53a–d, 4.54a–d) are 3! − 4 = 6 − 4 = 2 combinations that are not statically stable, namely the hypostatic cases (4.50a), one end free and the other (5) pinned [(6) free in the Figure 4.4e (f)]. The boundary conditions like (4.53a–d) apply generally to a beam under longitudinal tension $T \neq 0$ and in particular to a bar under transverse load alone $T = 0$. The sliding (pin-joined) support, for example, in Figure 4.4c and d, lead to the same boundary conditions, but differ in one respect: as for the clamped (free) end, the pin-joint (sliding support) does not (does) allow the end point on the beam to vary, that is, the bending of the beam causes (does not cause) an extension and hence leads (does not lead) to an axial traction force. Thus, *there are two cases: (α) for a beam with at least one free or sliding end, that is, clamped-free, clamped-sliding, sliding-sliding, pinned-sliding, there is no longitudinal tension, unless it is externally applied; (β) for a beam pinned or clamped at both ends, that is, clamped-clamped, clamped-pinned, or pinned-pinned, the increase in length under bending gives rise to a longitudinal traction, in addition to any applied externally.*

4.3 Weak Bending of a Heavy Bar

The simplest transverse load on a bar is uniform, for example, corresponding to the linear bending by the weight of a uniform bar in a constant gravity field and is considered (Section 4.3) for the main combinations of support (Subsection 4.2.9), namely (1) clamped at both ends (Subsection 4.3.1); (2) clamped at one and pinned (or supported) at the other end (Subsection 4.3.2); (3) pinned at both ends (Subsection 4.3.3); and (4) clamped at one end and free at the other end (Subsection 4.3.4). The comparison of the four cases shows (Table 4.2) that they correspond to increasing maximum slopes and deflections (Subsection 4.3.5) and different reaction forces and moments at the supports (Subsection 4.3.6).

4.3.1 Heavy Bar Clamped at Both Ends

The weak bending of a uniform bar under its own weight is specified by (4.16b) ≡ (4.55b) with the shear stress (4.55a), where ρ is the mass density per unit length and g the acceleration of gravity pointing downward:

$$f = \rho g: \quad \zeta'''(x) = \frac{\rho g}{EI}, \quad \zeta(x) = \frac{\rho g x^4}{24EI} + Ax^3 + Bx^2 + Cx + D, \tag{4.55a–c}$$

leading to (4.55c), where A–D are constants of integration. The expression (4.16b) $\equiv$ (4.55b) assumes that there is no externally applied tension force as in (4.13f); if at least one end of the bar is free or sliding, no further restriction is needed, whereas if both ends are clamped or pin-joined, the maximum deflection δ must be small compared to the transverse dimension h of the cross section $\delta^2 \ll h^2$. The latter restriction applies in the case of a bar clamped at both ends (4.51a–d), for which (Figure 4.4a) the constants of integration in (4.55c) are given by

$$D = 0 = C, \quad AL + B + \frac{\rho g L^2}{24EI} = 0 = 3AL + 2B + \frac{\rho g L^2}{6EI}. \tag{4.56a–d}$$

Solving (4.56c,d) for the two remaining constants

$$A = -\frac{\rho g L}{12EI}, \quad B = \frac{\rho g L^2}{24EI} \tag{4.57a,b}$$

and substituting with (4.57a,b) in (4.55c) gives

$$\zeta_1(x) = \frac{\rho g}{24EI} x^2 (x - L)^2 = \frac{\rho g}{24EI} \left(x^4 - 2x^3 L + x^2 L^2 \right), \tag{4.58a}$$

$$\delta_1 \equiv \zeta_{1\max} = \zeta_1 \left(\frac{L}{2} \right) = \frac{\rho g L^4}{384EI} = 0.002604 \frac{\rho g L^4}{EI}. \tag{4.58b}$$

Thus, *the weak bending (4.12a) of a uniform (4.10a) clamped-clamped (4.51a–d) bar by its own weight (4.55a) leads (4.55b) to the shape of the elastica (4.58a) with maximum deflection (4.58b) in the middle (Figure 4.5d). The slope (4.59a) vanishes in the middle (4.59b) by symmetry (4.59c) $\equiv$ (4.58a):*

$$\zeta_1'(x) = \frac{\rho g x}{12EI} \left(2x^2 - 3xL + L^2 \right), \quad \zeta_1'\left(\frac{L}{2} \right) = 0, \quad \zeta_1(x) = \zeta_1(L - x). \tag{4.59a–c}$$

The bending moment (4.60a) [transverse force (4.60b)] along the bar

$$-M_1(x) = EI\zeta_1''(x) = \frac{\rho g}{12} \left(6x^2 - 6xL + L^2 \right), \quad F_1(x) = EI\zeta_1'''(x) = \rho g \left(x - \frac{L}{2} \right) \tag{4.60a,b}$$

specifies the reaction moments (4.61a,b) [forces (4.61c,d)] at the supports:

$$R_{1-} = F_1(0) = -\frac{\rho g L}{2} = -F_1(L) = R_{1+}, \quad N_{1-} \equiv -M_1(0) = \frac{\rho g L^2}{12} = -M_1(L) = -N_{1+}. \tag{4.61a–d}$$

The reaction forces (4.61a,b) [moments (4.61c,d)] are equal in modulus with the same (opposite) sign.

The maximum slope in modulus (4.62a) is shifted by (4.62b) relative to the mid-position between supports (4.62c):

$$\zeta''(bL) = 0: \quad 6b^2 - 6b + 1 = 0, \quad b_{\pm} = \frac{1}{2} \left(1 \pm \frac{1}{\sqrt{3}} \right) = 0.5 \pm 0.2887 = 0.5 \pm b_0. \tag{4.62a–c}$$

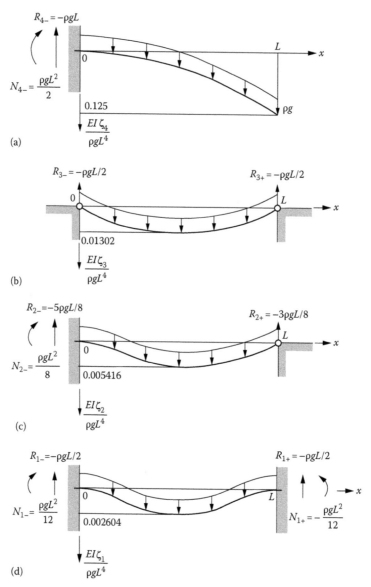

FIGURE 4.5 The four classical combinations of support, namely, (a) clamped-free, (b) pinned-pinned, (c) clamped-pinned, and (d) clamped-clamped may be considered for different loads, for example, a uniform shear stress corresponding to the weak bending of a heavy uniform bar in a constant gravity field.

The maximum slope in modulus (4.62d) corresponds to the displacement (4.62e):

$$\theta_1 \equiv \left| \zeta_1' \right|_{max} = \zeta_1'(b_-L) = -\zeta_1'(b_+L) = \frac{\rho g L^3}{12EI} b_- \left(2b_-^2 - 3b_- + 1\right)$$

$$= \frac{\rho g L^3}{12EI} \left(\frac{1}{2} - b_0\right)\left(\frac{1}{6} + b_0\right) = 0.008019 \frac{\rho g L^3}{EI}, \tag{4.62d}$$

$$\zeta(b_{\pm}L) = \frac{\rho g L^4}{24EI} \left(\frac{1}{2} \mp b_0^2\right)^2 \left(\frac{1}{2} \pm b_0\right)^2 = \frac{\rho g L^4}{24EI} \left(\frac{1}{4} - b_0^2\right)^2 = \frac{\rho g L^4}{864EI} = 0.001157 \frac{\rho g L^4}{EI}. \tag{4.62e}$$

In (4.62e) [(4.62d)], (4.62f–h) [(4.62i–k)] were used:

$$b_0 \equiv \frac{1}{2\sqrt{3}}: \quad b_0^2 = \frac{1}{12}, \quad \frac{1}{4} - b_0^2 = \frac{1}{6}, \tag{4.62f–h}$$

$$b_\pm^2 = b_\pm - \frac{1}{6}: \quad 2b_-^2 - 3b_- + 1 = \frac{2}{3} - b_- = \frac{2}{3} - \frac{1}{2} + b_0 = \frac{1}{6} + b_0, \tag{4.62i,j}$$

$$b_- \left(2b_-^2 - 3b_- + 1 \right) = \left(\frac{1}{2} - b_0 \right) \left(\frac{1}{6} + b_0 \right). \tag{4.62k}$$

The relation (4.62i) follows from (4.62b) and can be checked from (4.62c):

$$b_\pm^2 = \left(\frac{1}{2} \pm b_0 \right)^2 = \frac{1}{4} \pm b_0 + b_0^2 = \frac{1}{4} + \frac{1}{12} \pm b_0 = \frac{1}{3} \pm b_0 = \frac{1}{3} + b_\pm - \frac{1}{2} = b_\pm - \frac{1}{6}. \tag{4.62l}$$

The maximum deflection (4.58b) [slope (4.62d)] is compared in the sequel with the other three basic combinations of two supports (Subsections 4.3.2 through 4.3.4).

4.3.2 Clamped-Pinned Heavy Bar

The differential equation (4.55b) and boundary conditions at the clamped support (4.52a,b) are the same, leading to (4.55c) [(4.56a,b)]. At the pinned support, one boundary condition is the same (4.52c) [different (4.52d)], leading to (4.56c) [(4.63a)]:

$$0 = 6AL + 2B + \frac{\rho g L^2}{2EI}; \quad B = \frac{\rho g L^2}{16EI}, \quad A = -\frac{5\rho g L}{48EI}. \tag{4.63a–c}$$

Solving (4.56c, 4.63a) leads to (4.63b,c). Substitution of (4.56a,b, 4.63b,c) in (4.55c) yields

$$\zeta_2(x) = \frac{\rho g L}{24EI} x^2 \left(x^2 - \frac{5}{2} xL + \frac{3}{2} L^2 \right) = \frac{\rho g L}{24EI} x^2 (x - L) \left(x - \frac{3}{2} L \right), \tag{4.64}$$

that is, *the shape of the elastica (4.64) for the weak bending (4.12a) of a uniform (4.10a) bar under its own weight (4.55b) with clamped-pinned ends (4.52a–d). The slope is (4.65a) along the bar and (4.65b) at the pinned end:*

$$\zeta_2'(x) = \frac{\rho g x}{48EI} \left(8x^2 - 15xL + 6L^2 \right), \quad \zeta_2'(L) = -\frac{\rho g L^3}{48EI} = -0.02083 \frac{\rho g L^3}{EI}, \tag{4.65a,b}$$

and the bending moment (transverse force) is given by (4.66a) [(4.66b)]:

$$-M_2(x) = EI\zeta_2''(x) = \frac{\rho g}{8} \left(4x^2 - 5xL + L^2 \right) = \frac{\rho g}{2} (x - L) \left(x - \frac{L}{4} \right), \tag{4.66a}$$

$$F_2(x) = EI\zeta_2'''(x) = \rho g \left(x - \frac{5}{8} L \right). \tag{4.66b}$$

They specify (1) the nonzero reaction moment at the clamped support (4.67a) that is larger than in the clamped-clamped case (4.61c,d); and (2) the reaction force that is larger (4.67b) [smaller (4.67c)] than half of the weight (4.61a,b) at the clamped (pinned) support:

$$N_{2-} = -M(0) = \frac{\rho g L^2}{8}, R_{2-} = -F_2(0) = \frac{5}{8}\rho g L > \frac{3}{8}\rho g L = -F_2(L) = -R_{2+}. \qquad (4.67a\text{-}c)$$

The slope (4.65a) vanishes (4.68a,b) closer to the pinned support (4.68c):

$$\zeta_2'(aL) = 0: \quad 8a^2 - 15a + 6 = 0, \quad a = \frac{15 - \sqrt{33}}{16} = 0.5785, \qquad (4.68a\text{-}c)$$

leading to (4.64) the maximum deflection:

$$\delta_2 \equiv \zeta_{2\max} = \zeta_2(a) = \frac{\rho g L^4}{EI} \frac{a^2(2a^2 - 5a + 3)}{48} = 0.005416 \frac{\rho g L^4}{EI}, \qquad (4.68d)$$

which is larger (4.68d) than in the clamped-clamped case (4.58b). The curvature (4.66a) vanishes (4.69a) at (4.69b) the locations (4.69c,d):

$$\zeta_2''(bL) = 0: \quad 4b^2 - 5b + 1 = 0, \quad b_\pm = \frac{5}{8} \pm \frac{3}{8} = 1, \frac{1}{4}. \qquad (4.69a\text{-}d)$$

The slope is minimum (maximum) at the second support (4.65a) [at a distance from the first support equal to 1/4 of the length of the bar (4.69e)]:

$$\theta_2 \equiv \zeta_{2\max}' = \zeta_2'\left(\frac{L}{4}\right) = \frac{11\rho g L^3}{768EI} = 0.01432 \frac{\rho g L^3}{EI} < 0.02083 \frac{\rho g L^3}{EI} = -\zeta_2'(L) = -\zeta_{2\min}'. \qquad (4.69e)$$

The maximum (4.69e) is smaller than the modulus of the minimum (4.65b) in the clamped-pinned case, and both are larger than in the clamped-clamped case (4.62d).

4.3.3 Deflection by Own Weight of a Pinned-Pinned Bar

In the case of a pinned-pinned bar, the points $x = 0, L$ must be (4.53a,c) roots of the shape (4.55c) of the elastica (4.70a):

$$\zeta(x) = \frac{\rho g}{24EI}x(x-L)(x^2 + \lambda x + \mu) = \frac{\rho g x}{24EI}\left[x^3 + (\lambda - L)x^2 + (\mu - \lambda L)x - \mu L\right]. \qquad (4.70a,b)$$

The remaining two constants of integration (λ, μ) are determined by the other two boundary conditions (4.53b,d) specifying zero bending moment or curvature at the supports (4.71a,b):

$$\mu - \lambda L = 0 = \lambda + L: \quad \lambda = -L, \quad \mu = -L^2. \qquad (4.71a\text{-}d)$$

Solving (4.71a,b) for (4.71c,d) and substituting in (4.70b) gives

$$\zeta_3(x) = \frac{\rho g}{24EI}x(x-L)(x^2 - xL - L^2) = \frac{\rho g x}{24EI}(x^3 - 2x^2L + L^3). \qquad (4.72)$$

Thus, *the weak bending (4.12a) of a uniform bar (4.10a) pinned at both ends (4.53a–d) by its own weight (4.55b) leads to the shape of the elastica (4.72) with slope*

$$\zeta_3'(x) = \frac{\rho g}{24EI}\left(4x^3 - 6x^2L + L^3\right). \tag{4.73}$$

The maximum deflection (4.74a) [slope in modulus (4.74b,c)] is at the middle (supports):

$$\delta_3 \equiv \zeta_{3\max} = \zeta\left(\frac{L}{2}\right) = \frac{5\rho g L^4}{384EI} = 0.01302\frac{\rho g L^4}{EI}, \tag{4.74a}$$

$$\theta_3 \equiv |\zeta_3'|_{\max} = \zeta_3'(0) = -\zeta_3'(L) = \frac{\rho g L^3}{24EI} = 0.04167\frac{\rho g L^3}{EI}, \tag{4.74b}$$

and is larger than in (1) the clamped-pinned case (4.68d) [(4.69e)]; and (2) the clamped-clamped case by a factor of 5 (4.58b) [by a slightly larger factor (4.62d)]. The bending moment (4.75a) [transverse force (4.75b)] is

$$-M_3(x) = EI\zeta_3''(x) = \frac{\rho g}{2}x(x-L), \quad F_3(x) = EI\zeta_3'''(x) = \rho g\left(x - \frac{L}{2}\right). \tag{4.75a,b}$$

The reaction forces at the supports,

$$R_{3-} = F_3(0) = -\frac{\rho g L}{2} = -F_3(L) = R_{3+}, \tag{4.75c,d}$$

are both equal to minus one-half of the weight as in the clamped-clamped case (4.61a,b) that is also symmetric.

4.3.4 Heavy Cantilever with One Clamped and One Free End

In this case, $x = 0$ is a double root of (4.55c) on account of the first pair of boundary conditions (4.54a,b) at the clamped end leading to (4.76a):

$$\zeta(x) = \frac{\rho g}{24EI}x^2\left(x^2 + \lambda x + \mu\right) = \frac{\rho g}{24EI}\left(x^4 + \lambda x^3 + \mu x^2\right). \tag{4.76a,b}$$

The boundary conditions (4.54c,d) without axial tension (4.77a) at the free end (4.77b,c) determine (4.77d,e) the two remaining constants of integration (4.77f,g):

$$T = 0: \quad \zeta''(L) = 0 = \zeta'''(L), \quad \lambda + 4L = 0 = \mu + 3\lambda L + 6L^2, \quad \lambda = -4L, \quad \mu = 6L^2. \tag{4.77a–g}$$

Substitution of (4.77f,g) in (4.76b) gives

$$\zeta_4(x) = \frac{\rho g x^2}{24EI}\left(x^2 - 4xL + 6L^2\right), \quad \zeta_4'(x) = \frac{\rho g x}{6EI}\left(x^2 - 3xL + 3L^2\right), \tag{4.78a,b}$$

showing *that the weak bending (4.12a) of a uniform (4.10b) bar clamped at one (4.54a,b) [free at the other (4.77b,c)] end due to its own weight (4.55b) leads to the shape of the elastica (4.78a) with slope (4.78b). The maximum deflection (4.79a) [slope in modulus (4.79b)] occurs at the free ends:*

$$\delta_4 \equiv \zeta_{4\max} = \zeta_4(L) = \frac{\rho g L^4}{8EI} = 0.125\frac{\rho g L^4}{EI}, \quad \theta_4 \equiv |\zeta_4'|_{\max} = \zeta_4'(L) = \frac{\rho g L^3}{6EI} = 0.1667\frac{\rho g L^3}{EI}. \tag{4.79a,b}$$

and both are larger than for all three preceding cases, namely pinned-pinned (4.74a,b), clamped-pinned (4.68d, 4.69e), and clamped-clamped (4.58b, 4.62d). The bending moment (4.80a) [transverse force (4.80b)] along the bar

$$-M_4(x) = EI\zeta_4''(x) = \frac{\rho g}{2}(x^2 - 2xL + L^2) = \frac{\rho g}{2}(x - L)^2, \qquad (4.80a)$$

$$F_4(x) = EI\zeta_4'''(x) = \rho g(x - L), \qquad (4.80b)$$

specifies the reaction force (4.80c) [moment (4.80d)] at the clamped support:

$$M_{4-} = -M_4(0) = \frac{1}{2}\rho g L^2 = \rho g \int_0^L x\,dx, \quad F_- = F_{4-}(0) = -\rho g L = -\rho g \int_0^L dx, \qquad (4.81a,b)$$

that is due to the total weight of the bar.

4.3.5 Heavy Bar with Four Combinations of Supports

Table 4.2 lists the results concerning the bending of a bar under the following assumptions: (1) the bar is straight under no load; (2) the applied load is a uniform shear stress, for example, the bar's own weight, or a uniform load in a room above it is supporting; (3) the deflection is moderate, in the sense of much smaller than the length of the beam $\delta^2 \ll L^2$; for example, building norms usually specify $\delta/L = 0.05$; (4) there are no longitudinal applied forces, such as a horizontal bar under traction or compression; and (5) if the bar is clamped or pin-joined at each end, the deflection is much less than the transverse length scale of the section $(h/\delta)^2 \ll 1$, for example, $(h/\delta)^2 = 0.09$ or $h/\delta \equiv 0.3$ implying $h/L = (h/\delta)(\delta/L) = 0.15$ for a bar with I-shaped cross section, for a large moment of inertia, with height equal to 15% of the length. The four statically stable combinations of support using clamped, sliding, or free ends were considered, and each case was obtained: (1, 2) the shape of the elastica and its slope; (3–6) positions of maximum deflection (slope) and the value there; and (7–10) the bending moment

TABLE 4.2 Linear Bending of Elastic Bar by Its Own Weight

Case	1	2	3	4
Load	Uniform	Uniform	Uniform	Uniform
Subsection	4.3.1	4.3.2	4.3.3	4.3.4
Figure	4.4d	4.4c	4.4b	4.4a
Type of support	Clamped-clamped	Clamped-pinned	Pinned-pinned	Clamped-free
Boundary conditions	(4.51a–d)	(4.52a–d)	(4.53a–d)	(4.54a,b, 4.77b,c)
Degrees of freedom	−2	−1	0	0
Type of problem	Hyperstatic	Hyperstatic	Isostatic	Isostatic
Shape of the elastica	(4.58a)	(4.64)	(4.72)	(4.78a)
Slope of the elastica	(4.59a)	(4.65a)	(4.73)	(4.78b)
Maximum deflection	(4.58b)	(4.68c,d)	(4.74a)	(4.79a)
Extrema of slope	(4.62c,d)	(4.65b, 4.69e)	(4.74b)	(4.79b)
Bending moment	(4.60a)	(4.66a)	(4.75a)	(4.80a)
Transverse force	(4.60b)	(4.66b)	(4.75b)	(4.80b)
Shear stress	(4.55a)	(4.55a)	(4.55a)	(4.55a)
Reaction forces	(4.61a,b)	(4.67b,c)	(4.75c,d)	(4.80d)
Reaction moments	(4.61c,d)	(4.67a)	—	(4.80c)

Note: Comparison of the weak bending of a uniform bar by a uniformly distributed shear stress, namely its own weight, for the four classical cases of supports.

(transverse force) along the bar and the corresponding reaction moments [force(s)] at the supports. The maximum deflection occurs at the free end for the clamped-free case, at the middle for the symmetric cases, that is, both ends clamped or both pinned, and away from the middle toward the sliding end for the clamped-pinned case.

The maximum deflection is (1) minimum for the clamped-clamped case (4.58b) ≡ (4.82a); (2, 3) about twice (exactly five times) as large for the clamped-pinned (pinned-pinned) case (4.68d) ≡ (4.82b) [(4.74a) ≡ (4.82c)]; and (4) 48 times larger than (1) for the clamped-free case (4.79a) ≡ (4.82d):

$$\zeta_1\left(\frac{L}{2}\right) = \frac{\rho g L^4}{384 EI} = 0.002604 \frac{\rho g L^4}{EI}$$

$$< 2.080 \zeta_1\left(\frac{L}{2}\right) = \zeta_2(0.5785 L) = 0.005416 \frac{\rho g L^4}{EI}$$

$$< 0.01302 \frac{\rho g L^4}{EI} = \frac{5 \rho g L^4}{384 EI} = 5 \zeta_1\left(\frac{L}{2}\right) = \zeta_3\left(\frac{L}{2}\right)$$

$$< 0.125 \frac{\rho g L^4}{EI} = \frac{\rho g L^4}{8 EI} = \frac{48}{5} \zeta_3\left(\frac{L}{2}\right) = 48 \zeta_1\left(\frac{L}{2}\right) = \zeta_1(L). \tag{4.82a–d}$$

In the latter, a factor of $2^4 = 16$ is due to the maximum deflection being at twice the distance, that is, L instead of $L/2$, and an additional factor 3 is due to the replacement of one clamped by one free end.

The maximum slope is (1) largest for the clamped-free case (4.79b) ≡ (4.83a); (2) four times smaller for the pinned-pinned case (4.74b) = (4.83b); and (3, 4) progressively smaller replacing one (two) pinned supports by clamped supports (4.69e, 4.65b) ≡ (4.83c) [(4.62c,d) ≡ (4.83d)]:

$$\zeta_4'(L) = 0.1667 \frac{\rho g L^3}{EI} = \frac{\rho g L^3}{6 EI} = 4 \zeta_3'(0) = -4 \zeta_3'(L)$$

$$> \zeta_3'(0) = 0.04167 \frac{\rho g L^3}{EI} = \frac{\rho g L^3}{24 EI} = -\zeta_3'(L) = -1.7298 \zeta_2'(L)$$

$$> -\zeta_2'(L) = \frac{\rho g L^3}{48 EI} = 0.02083 \frac{\rho g L^3}{EI} = 1.4546 \zeta_2'\left(\frac{L}{4}\right)$$

$$> \zeta_2'\left(\frac{L}{4}\right) = 0.01432 \frac{\rho g L^3}{EI} = \frac{11 \rho g L^3}{768 EI} = 1.7858 \zeta_1'\left(\frac{L}{4}\right)$$

$$> \zeta_1'\left(\frac{L}{4}\right) = 0.008019 \frac{\rho g L^3}{EI} = \zeta_1'\left(\frac{L}{2} \pm 0.2887 L\right). \tag{4.83a–d}$$

In the case (4.83c) of the clamped-pinned bar, the maximum (4.69d) [minus the minimum (4.65b)] slopes are considered.

4.3.6 Reaction Forces and Moments at the Supports

The reactions at the supports for an elastic string (bar) consist of reaction forces (and moments) indicated in Figures 2.4 through 2.11, 2.13, 2.15, and 2.17 through 2.18 (4.5 through 4.9). The shape of the deflected elastic string (bent elastica of the bar) is shown with the same (different) vertical scales in Figures 2.8

through 2.10, 2.13, 2.15, and 2.17 (Figures 4.5a through d, 4.7a through d, and 4.9a through d) because the ratio of the largest 1/3 (1/8) to the smallest 1/18 (1/384) deflection is 6:1(48:1), that is, not so large (too large) allowing (not allowing) graphical comparison. The reaction force (1) compensates the total weight for the clamped-free case (4.81b) ≡ (4.84a, 4.85d); (2, 3) in the symmetric cases, that is, clamped-clamped (pinned-pinned), the supports share equally the weight (4.61a,b) ≡ (4.84c;4.85a) [(4.75c,d) ≡ (4.84d, 4.85b)]; and (4) in the unsymmetric case, the clamped (pinned) support carries 5/8 (3/8) of the weight (4.67b,c) ≡ (4.84b, 4.85c):

$$-R_{4-} = \rho g L > \frac{5}{8}\rho g L = -\frac{5}{8}R_{4-} = -R_{2-} > \frac{\rho g L}{2} = -\frac{1}{2}R_{4-} = R_{1-} = R_{3-}, \qquad (4.84\text{a–d})$$

$$-R_{1+} = \frac{\rho g L}{2} = -R_{3+} > \frac{3}{8}\rho g L = -\frac{3}{8}R_{1+} = -R_{2+} > 0 = R_{4+}; \qquad (4.85\text{a–d})$$

$$N_{4-} = \frac{\rho g L^2}{2} > N_{2-} = \frac{\rho g L^2}{8} = \frac{1}{4}N_{4-} > N_{1-} = \frac{\rho g L^2}{12} = \frac{1}{6}N_{4-} > 0 = N_{3-}, \qquad (4.86\text{a–d})$$

$$-N_{1+} = \frac{\rho g L^2}{12} > 0 = N_{2+} = N_{3+} = N_{4-}. \qquad (4.87\text{a–d})$$

The modulus of the bending moment at the support is (1) largest for the clamped-free case (4.81a) ≡ (4.86a) ≡ (4.87d); (2/3) one-fourth (one-sixth) with one end clamped and (4.67a) ≡ (4.86b) ≡ (4.87b) [(4.61c,d) ≡ (4.86c) ≡ (4.87c)] the other pinned (clamped); and (4) zero with both ends pinned (4.86d) ≡ (4.87d). The nonlinear (linear) or strong (weak) simple bending of a bar is considered for the three main types of loads: (1) a distributed uniform load [Section 4.9 (Section 4.3 and Table 4.2)]; (2) a concentrated force in [Section 4.8 (Section 4.4 and Tables 4.3 and 4.4)] specifying the nonlinear (linear) influence or Green function; and (3) a concentrated torque [Section 4.7 (Section 4.5 and Tables 4.5 and 4.6)], that is, a kind of load an elastic string (bar) cannot (can) support [Chapter 2 (Chapter 4)]. All the preceding cases concern a bar without axial tension; the latter case of a beam, that is, a bar under axial compression, is considered in the linear case with fixed supports (Section 4.6) and a concentrated transverse force.

4.4 Influence Functions for Pinned/Clamped/Free Ends

The weak (strong) bending [Section 4.4 (Section 4.8)] of a bar under a concentrated force specifies the nonlinear (linear) response or Green function. In the linear case, the Green function satisfies the reciprocity principle (Subsection 4.4.7), allowing the exchange of the load and measuring point; the reciprocity principle applies, regardless of whether the bar or beam is uniform or not, that is, it holds for changing cross section and varying Young's modulus and nonuniform longitudinal tension. The linear influence function specifies through the superposition principle the weak bending under any load (Subsection 4.4.7). Since the Green function depends on the boundary conditions, it is given for all the four classical pairs, namely (1) a clamped-free bar (Subsection 4.4.1); (2) a pinned-pinned bar (Subsection 4.4.2); (3) a clamped-pinned bar (Subsection 4.4.3); and (4) a clamped-clamped bar (Subsection 4.4.4). In the isostatic cases (1, 2), the reactions at the supports can be determined from the external force and moment balance and the equation of the elastica integrated directly. In the hyperstatic cases (3) [(4)], the reactions at the supports cannot be determined "a priori" and a two-stage procedure is necessary: (a) integration of the equation of the elastica with all boundary and continuity conditions; (b) once the elastica is fully determined, differentiation in reverse order of integration specifies the bending moment and transverse force along the bar and hence also at the supports. The two hyperstatic cases (3, 4) are distinct in the sense (Subsection 4.4.5) that the balance of forces and moments at the supports is (3) indeterminate for the clamped-pinned bar, that is, the balance equations are satisfied but do not determine the reactions; (4) incompatible for the clamped-clamped bar, that is, the external balance equations are not satisfied because they do not take into account internal stresses. A particular case (Subsection 4.4.6) of (1) [(2–4)] is a concentrated force at the free end (at equal distance from the supports) of the bar (Table 4.4). This particular case is not sufficient to determine the influence function that specifies the displacement for an arbitrary position of the concentrated force (Table 4.3) and relates to the reciprocity and superposition principles (Subsection 4.4.7).

4.4.1 Clamped-Free Bar with a Concentrated Force

A concentrated load P is a jump in the transverse force (4.88a) [an impulse in the shear stress (4.88b)]:

$$F(x;\xi) = PH(x-\xi) + \text{const}, \quad f(x;\xi) = P\delta(x-\xi), \tag{4.88a,b}$$

implying that the elastica of a uniform bar under weak bending (4.16b) leads to (4.89):

$$EId^4G(x;\xi)/dx^4 = P\delta(x-\xi), \tag{4.89}$$

where $G(x;\xi)$ is the influence or Green function. It has been assumed that there is no applied longitudinal tension, and none will arise from the support conditions (Figure 4.6a) for a bar (4.90a) clamped at one end (4.90b,c) and free at the other (4.90d,e):

$$T = 0: \quad \zeta(0) = 0 = \zeta'(0), \quad \zeta''(L) = 0 = \zeta'''(L). \tag{4.90a-e}$$

The only restrictions are that the bar is initially straight, and the maximum deflection δ is small compared with the length $\delta^2 \ll L^2$. Integrating (4.89) once leads to (4.91b,c):

$$G'''(L;\xi) = 0: \quad EIG'''(x;\xi) = A + PH(x-\xi) = \begin{cases} A = -P & \text{if } x \le x < \xi, \\ A + P = 0 & \text{if } \xi < x \le L, \end{cases} \tag{4.91a-c}$$

where the constant of integration A is determined from the condition (4.90e) $\equiv$ (4.91a) of zero transverse force at the free end and thus coincides with the reaction force at the clamped end.

Integrating (4.91b,c) three times yields

$$G''(L;\xi) = 0: \quad -M(x) = EIG''(x;\xi) = \begin{cases} -P(x-\xi) & \text{if } 0 \le x \le \xi, \\ 0 & \text{if } \xi \le x \le L; \end{cases} \tag{4.92a-c}$$

$$G'(0;\xi) = 0: \quad EIG'(x;\xi) = \begin{cases} -P(x^2/2 - \xi x) & \text{if } 0 \le x \le \xi, \\ P\xi^2/2 & \text{if } \xi \le x \le L; \end{cases} \tag{4.93a-c}$$

$$G(0;\xi)=0: \quad EIG(x;\xi)=\begin{cases} -P\left(x^3/6-\xi x^2/2\right) & \text{if } 0\le x\le\xi, \\ P\left(x\xi^2/2-\xi^3/6\right) & \text{if } \xi\le x\le L, \end{cases} \tag{4.94a–c}$$

where the following conditions were used (1) zero moment at the free end (4.90d) ≡ (4.92a); (2/3) clamping in (4.90c) ≡ (4.93a) and (4.90b) ≡ (4.94a); and (4) the continuity at $x=\xi$ in all cases:

$$G''(\xi-0;\xi)=G''(\xi+0;\xi), \quad G'(\xi-0;\xi)=G'(\xi+0;\xi), \quad G(\xi-0;\xi)=G(\xi+0;\xi), \tag{4.95a–c}$$

that is, for the curvature (4.95a) in (4.92b,c), slope (4.95b) in (4.93b,c), and deflection (4.95c) in (4.94b,c).

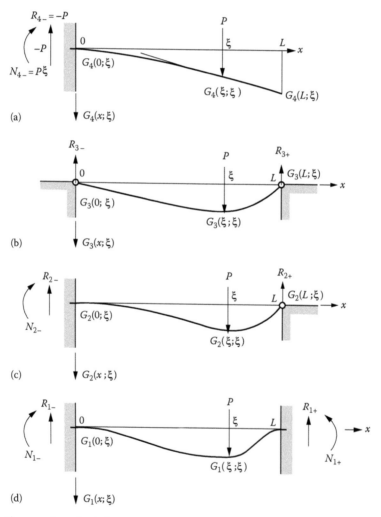

FIGURE 4.6 The same four classical support cases (a–d) as in Figure 4.5 are considered to the determine the shape of the elastica for the weak bending of a uniform bar by a concentrated force at an arbitrary position that specifies the linear influence or Green function. Its symmetry implies the reciprocity principle. Once the linear Green or influence function is known for the relevant boundary conditions the weak bending of the bar (uniform or not) can be determined for any load using the principle of superposition, for example, a uniform shear stress (concentrated moment) in Figure 4.5 (4.7).

The influence or Green function (4.94a–c) for the weak (4.12a) bending by a concentrated force (4.89) at a position ξ of a uniform (4.10a) elastic bar (4.90a) with one clamped (4.90b,c) and one free (4.90d,e) end (Figure 4.6a)

$$G_4\left(x;\xi\right)=\frac{P}{6EI}\times\begin{cases}x^2\left(3\xi-x\right) & \text{if } 0\le x\le\xi \\ \xi^2\left(3x-\xi\right) & \text{if } \xi\le x\le L\end{cases}$$

(4.96a)

(4.96b)

is a cubic parabola (4.96a) [straight line (4.96b)] before (after) the point of application of the concentrated downward force at x = ξ; the reciprocity principle (2.47b) is met by (4.96a,b). The deflection at that point (4.97a) is less than at the end (4.97b):

$$G_4\left(\xi;\xi\right)=\frac{P\xi^3}{3EI}\le G\left(L;\xi\right)=\frac{P\xi^3}{3EI}\left(\frac{3L}{2\xi}-\frac{1}{2}\right)\equiv\delta_4;\quad G_4'\left(\xi;\xi\right)=G'\left(L;\xi\right)=\frac{P\xi^2}{2EI}\equiv\theta_4.$$

(4.97a–c)

The slope (4.93c) is constant after the point of application of the force, specifying the value (4.97c) at the free end. The bending moment (4.92b,c) ≡ (4.98a) [transverse force (4.93b,c) ≡ (4.99a)]

$$-M_4\left(x;\xi\right)=EIG_4''\left(x;\xi\right)=P\left(\xi-x\right)H\left(\xi-x\right),\quad N_{4-}=-M_4\left(0;\xi\right)=P\xi,$$

(4.98a,b)

$$F_4\left(x;\xi\right)=EIG_4'''\left(x;\xi\right)=-PH\left(\xi-x\right),\quad R_{4-}=F_4\left(0;\xi\right)=-P,$$

(4.99a,b)

specifies the reaction moment (4.98b) [force (4.99b)] at the clamped end; these are determined by the moment of the (minus the) external applied force because the problem is isostatic.

In the case (Figure 4.7a) of a force applied at the free end (4.100a), the boundary condition (4.90e) is replaced by (4.100b), which specifies the transverse force and reaction force at the clamped support (4.100c) where the reaction moment (4.98b) is (4.100d):

$$\xi=L:\quad F_4\left(x;L\right)=EIG_4''\left(x;L\right)=P=-R_{4-};\quad N_{4-}=-M_4\left(0;L\right)=PL.$$

(4.100a–d)

The displacement (4.96a; 4.100a) ≡ (4.101a) is maximum at the free end (4.101c) where the concentrated force is applied (4.101a):

$$f\left(x;L\right)=P\delta\left(x-L\right):\quad G_4\left(x;L\right)=\frac{Px^2}{6EI}\left(3L-x\right),\quad G_4\left(L;L\right)=\frac{PL^3}{3EI}=\delta_4.$$

(4.101a–c)

The slope (4.93b; 4.100a) ≡ (4.102a) is also maximum at the free end (4.102b):

$$G_4'\left(x;L\right)=\frac{Px}{2EI}\left(2L-x\right),\quad G_4'\left(L;L\right)=\frac{PL^2}{2EI}\equiv\theta_4.$$

(4.102a,b)

The bending moment (4.98a; 4.100a) ≡ (4.103a) specifies the reaction moment at the clamped support (4.103b):

$$-M_4\left(x;L\right)=EIG_4''\left(x;L\right)=P\left(L-x\right),\quad N_{4-}=-M_4\left(0;L\right)=PL.$$

(4.103a,b)

The fourth boundary condition at the free end is (4.90e) [(4.100b)] if the concentrated force is not (is) at the free end. Since the present problem is isostatic, the reaction force and moment at the clamped end can be determined "a priori." The same applies to the pinned-pinned bar considered next (Subsection 4.4.2).

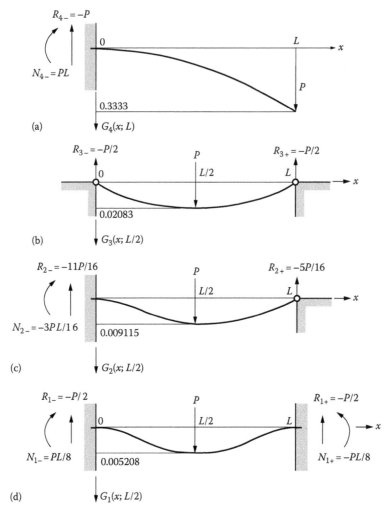

FIGURE 4.7 The weak bending of a uniform elastic beam by a concentrated force in the middle leads to a symmetric (unsymmetric) shape if the boundary conditions are symmetric (unsymmetric), for example, in the pinned-pinned (b) and (clamped-clamped (d) [clamped-pinned (c)] case, as happened for a uniform load [Figure 4.5a and c (b)]. The boundary conditions are also unsymmetric for a clamped-free bar leading to an unsymmetric shape, for example, with a uniform shear stress (Figure 4.5a) or a concentrated force at the tip (a).

4.4.2 Pinned-Pinned Bar with a Concentrated Force

In the case of a bar pinned at the two ends with concentrated force P at an arbitrary position (Figure 4.6b) holds, as for an elastic string: (1) the balance of reaction forces (2.40a); (2, 3) the balance of moments at each support (2.40b,c). Since the bar is pinned at both ends, there are no reaction moments, and the reaction forces are (2.39d,e). The influence or Green function satisfies (4.89), which is integrated in (4.91b,c):

$$F(x;\xi) = EIG'''(x;\xi) = A + PH(x-\xi) = \begin{cases} A = R_- = -P\left(1 - \dfrac{\xi}{L}\right) & \text{if } 0 \le x < \xi, \quad (4.104a) \\[2em] A + P = -R_+ = P\dfrac{\xi}{L} & \text{if } 0 < x \le L, \quad (4.104b) \end{cases}$$

with the difference that the constant of integration in (4.104a) is determined by the reaction force (4.39d) at the first support; the jump across the point of application of the concentrated force specifies (4.104b) the reaction at the second support in agreement with (4.39e). Since the bending moment vanishes at both pinned ends (4.105a,b), the integration of (4.104a) [(4.104b)] leads to (4.105c) [(4.105d)]:

$$G''(0;\xi) = 0 = G''(L;\xi): \quad -M(x;\xi) = EIG''(L;\xi) = \begin{cases} Px\left(\dfrac{\xi}{L}-1\right) & \text{if } 0 \le x \le \xi, \\[3mm] P\xi\left(\dfrac{x}{L}-1\right) & \text{if } \xi \le x \le L. \end{cases} \tag{4.105a-d}$$

The bending moment divided by the bending stiffness EI coincides (4.105c,d) $\equiv$ (2.41b,c) with the deflection of an elastic string divided by the tension T.

Integration of (4.105b) [(4.105d)] leads to (4.106a) [(4.106b)]:

$$2EIG'(x;\xi) = \begin{cases} P\left(\dfrac{\xi}{L}-1\right)x^2 + B & \text{if } 0 \le x \le \xi, \tag{4.106a} \\[4mm] P\xi x\left(\dfrac{x}{L}-2\right)+C & \text{if } 0 \le x \le L. \tag{4.106b} \end{cases}$$

The continuity of the curvature (slope) at the point of application of the force (4.95a) $\equiv$ (4.107a) [(4.95b) $\equiv$ (4.107b)] is met by (4.105c,d) in (4.107e) [relates the constants of integration B, C in (4.107b)]:

$$G''(\xi-0;\xi) = G''(\xi-0;\xi) = \frac{P\xi}{EI}\left(\frac{\xi}{L}-1\right), \tag{4.107a}$$

$$0 = 2EI\left[G'(\xi-0;\xi)-G'(\xi+0;\xi)\right] = B + P\xi^2 - C. \tag{4.107b}$$

Substituting (4.107b) in (4.106a) and integrating (4.106a) [(4.106b)] from the first (4.108a) [second (4.108c)] pinned end leads to (4.108b) [(4.108d)]:

$$G(0;\xi) = 0: \quad 6EIG(x; x \le \xi) = Px^3\left(\frac{\xi}{L}-1\right)+3\left(C-P\xi^2\right)x, \tag{4.108a,b}$$

$$G(L;\xi) = 0: \quad 6EIG(x; x \le \xi) = P\frac{\xi}{L}\left(x^3-L^3\right)-3P\xi\left(x^2-L^2\right)+3C(x-L). \tag{4.108c,d}$$

The continuity of the displacement (4.95c) $\equiv$ (4.109) at the point of application of the force,

$$0 = 6EI\left[G(\xi-0;\xi)-G(\xi+0;\xi)\right] = 3CL - P\xi^3 - 2P\xi L^2, \tag{4.109}$$

determines the remaining constant of integration.

Substituting (4.109) in (4.108b,d) leads to

$$G_3(x;\xi) = \frac{P}{6EI} \times \begin{cases} x\left(1-\dfrac{\xi}{L}\right)\left(-x^2+2L\xi-\xi^2\right) & \text{if } 0 \le x \le \xi, \tag{4.110a} \\[4mm] \xi\left(1-\dfrac{x}{L}\right)\left(-\xi^2+2Lx-x^2\right) & \text{if } \xi \le x \le L. \tag{4.110b} \end{cases}$$

Thus, *the weak (4.12a) bending by a concentrated force (4.89) at a position ξ of a uniform (4.10a) bar (4.90a) pinned at both ends (4.53a–d) is specified by the influence or Green function (4.110a,b) that satisfies the principle of reciprocity (2.47b). The corresponding slope is*

$$G_3'(x;\xi) = \frac{P}{6EI} \times \begin{cases} \left(1-\dfrac{\xi}{L}\right)\left(-3x^2+2L\xi-\xi^2\right) & \text{if } 0 \le x \le \xi, \qquad (4.111a) \\[2ex] \dfrac{\xi}{L}\left(3x^2-6xL+2L^2+\xi^2\right) & \text{if } \xi \le x \le L. \qquad (4.111b) \end{cases}$$

The deflection (slope) at the point of application of the concentrated force is (4.112a) [(4.112b)]:

$$G_3(\xi;\xi) = \frac{P\xi^2}{3EIL}(L-\xi)^2, \quad G_3'(\xi;\xi) = \frac{P\xi}{3EI}\left(1-\frac{\xi}{L}\right)(L-2\xi). \qquad (4.112a)$$

The slope at the supports is given by

$$G_3'(0;\xi) = \frac{P\xi}{6EI}\left(1-\frac{\xi}{L}\right)(2L-\xi), \quad G_3'(L;\xi) = \frac{P\xi}{6EIL}\left(\xi^2-L^2\right). \qquad (4.113b)$$

The bending moment (4.105c,d) vanishes at the pinned supports, and the transverse force (4.104a,b) leads to the reaction forces (2.39d,e) at the supports.

In the case (4.114a) of a concentrated force at equal distance from the supports (Figure 4.7b), the deflection (4.110a,b) is (4.114b,c):

$$\xi = \frac{L}{2}: \quad G_3\left(x;\frac{L}{2}\right) = \frac{P}{12EI} \times \begin{cases} x\left(\dfrac{3L^2}{4}-x^2\right) & \text{if } \xi \le x \le \dfrac{L}{2}, \\[2ex] (L-x)\left[\dfrac{3L^2}{4}-(L-x)^2\right] & \text{if } \dfrac{L}{2} \le x \le L. \end{cases} \qquad (4.114a\text{–}c)$$

The corresponding (1) slope is

$$G_3'\left(x;\frac{L}{2}\right) = \frac{P}{4EI} \times \begin{cases} \dfrac{L^2}{4}-x^2 & \text{if } \xi \le x \le \dfrac{L}{2}, \qquad (4.115a) \\[2ex] x^2-2xL+\dfrac{3L^2}{4} & \text{if } \dfrac{L}{2} \le x \le L, \qquad (4.115b) \end{cases}$$

which vanishes at the mid-point (4.116a) where the deflection is maximum (4.116b):

$$G_3'\left(\frac{L}{2};\frac{L}{2}\right) = 0, \quad G_3\left(\frac{L}{2};\frac{L}{2}\right) = \frac{PL^3}{48EI} = 0.02083\frac{PL^3}{EI} \equiv \delta_3, \qquad (4.116a,b)$$

and the concentrated force is applied (4.117a); (2) the slope takes the maximum value in modulus at the supports:

$$f\left(x;\frac{L}{2}\right) = P\delta\left(x-\frac{L}{2}\right): \quad G_3'\left(0;\frac{L}{2}\right) = -G_3'\left(0;\frac{L}{2}\right) = \frac{PL^2}{16EI} = 0.06667\frac{PL^2}{EI}; \qquad (4.117a\text{–}c)$$

(3) the bending moment (4.118a,b) vanishes at the supports:

$$-M_3\left(x;\frac{L}{2}\right) = EIG_3''\left(x;\frac{L}{2}\right) = \frac{P}{2} \times \begin{cases} -x & \text{if } 0 \leq x \leq \dfrac{L}{2}, & \text{(4.118a)} \\[2ex] x-L & \text{if } \dfrac{L}{2} \leq x \leq L; & \text{(4.118b)} \end{cases}$$

and (4) the transverse force (4.118c,d) specifies the reactions at the supports:

$$F_3\left(x;\frac{L}{2}\right) = EIG_3'''\left(x;\frac{L}{2}\right) = \begin{cases} -\dfrac{P}{2} = F_3(0) = R_{3-} & \text{if } 0 \leq x < \dfrac{1}{2}, & \text{(4.118c)} \\[2ex] \dfrac{P}{2} = F_3(L) = -R_{3+} & \text{if } \dfrac{1}{2} < x \leq L, & \text{(4.118d)} \end{cases}$$

which equal half of the concentrated load with opposite sign. The preceding (following) two cases [Subsections 4.4.1 and 4.4.2 (Subsections 4.4.3 and 4.4.4)] are isostatic (hyperstatic), implying that the reaction forces and moments at the supports (Subsection 4.4.5) can be determined "a priori" (only after integration for the shape of the elastica).

4.4.3 Clamped-Pinned Bar with a Concentrated Force

The weak bending of a uniform bar by a transverse force concentrated at $x = \xi$ specifies the influence or Green function (4.89); a first integration leads to (4.119a,b):

$$F(x;\xi) = EIG'''(x;\xi) = A + PH(x-\xi) = \begin{cases} A & \text{if } 0 \leq x < \xi, & \text{(4.119a)} \\[2ex] A+P & \text{if } \xi < x < L, & \text{(4.119b)} \end{cases}$$

where (1) for a clamped-free bar (Figure 4.6a), the transverse force is zero at the free end (4.91c), leading to the reaction force at the clamped end (4.91b); (2) for a pinned-pinned bar (Figure 4.6b), the reaction forces at the supports are not zero (4.104a,b), but can be determined by the balance of forces and moments (2.38a–c) because the system is isostatic; and (3) in the clamped-pinned case (Figure 4.6c), the reaction forces at the supports are not zero and cannot be determined "a priori" because the system is hyperstatic and follow from the equation of the elastica that is obtained next. A further integration of (4.119a) [(4.119b)] using the boundary condition at (4.120a) of zero bending moment at the pinned end leads to

$$G''(L;\zeta) = 0: \quad -M(x;\xi) = EIG''(x;\zeta) = \begin{cases} Ax + B & \text{if } 0 \leq x \leq \xi, \\[2ex] (A+P)(x-L) & \text{if } \xi \leq x \leq L; \end{cases} \quad \text{(4.120a–c)}$$

a constant of integration B appears in (4.120b), because there is generally a nonzero bending moment at the clamped end. The continuity (4.95a) $\equiv$ (4.121a) of the bending moment at the point of the concentrated force

$$G''(\xi-0;\xi) = G''(\xi+0;\xi): \quad B = -AL + P\xi - PL \qquad \text{(4.121a,b)}$$

specifies the constant of integration (4.121b). After substitution of (4.121b) in (4.120b), a further integration of (4.120b,c) leads to (4.122b,c):

$$G'(0;\xi) = 0: \quad 2EIG'(x;\xi) = \begin{cases} A(x^2 - 2xL) - 2P(L-\xi)x & \text{if } 0 \le x \le \xi, \\ (A+P)(x^2 - 2xL) + C & \text{if } \xi \le x \le L, \end{cases} \qquad (4.122a\text{–}c)$$

where (1) the clamping condition (4.122a) eliminates the constant of integration from (4.122b); and (2) at the pinned end, the slope is generally not zero, so a constant of integration C remains in (4.122c). The continuity of the slope (4.95b) $\equiv$ (4.123a),

$$G'(\xi-0;\xi) = G'(\xi+0;\xi): \quad C = P\xi^2, \qquad (4.123a,b)$$

determines the constant of integration (4.123b). Substitution of (4.123b) in (4.122c) and a final fourth integration of (4.122b,c) specifies the shape of the elastica (4.124b,c):

$$G(0;\xi) = 0: \quad 6EIG(x;\xi) = \begin{cases} A(x^3 - 3x^2L) - 3P(L-\xi)x^2 & \text{if } 0 \le x \le \xi, \\ (A+P)(x^3 - 3x^2L) + 3P\xi^2 x + D & \text{if } \xi \le x \le L, \end{cases} \qquad (4.124a\text{–}c)$$

where (1) the clamped condition (4.124a) eliminates a constant of integration from (4.124b); and (2) the continuity of the displacement (4.95c) $\equiv$ (4.125a) specifies (4.125b) the constant of integration D in (4.124c):

$$G(\xi-0;\xi) = G(\xi+0;\xi): \quad D = -P\xi^3; \quad G(L;\xi) = 0: \quad A = -P\left[1 - \frac{\xi^2}{2L^2}\left(3 - \frac{\xi}{L}\right)\right]. \qquad (4.125a\text{–}d)$$

The pinning condition (4.125c) determines the remaining constant of integration (4.125d) that specifies the reaction force (4.119a) at the clamped end.

Substitution of (4.125b,d) in (4.124b,c) leads *to the influence or Green function for the weak (4.12a) bending by a concentrated force (4.89) at a position ξ of a uniform (4.10a) bar (4.90a), clamped (4.124a, 4.122a) at one end and pinned (4.120a, 4.125c) at the other, that specifies the shape of the elastica (Figure 4.6c):*

$$G_2(x;\xi) = \frac{P}{6EI} \times \begin{cases} x^2\left[3\xi - x - \dfrac{\xi^2}{2L}\left(3 - \dfrac{x}{L}\right)\left(3 - \dfrac{\xi}{L}\right)\right] & \text{if } 0 \le x \le \xi, \qquad (4.126a) \\[2mm] \xi^2\left[3x - \xi - \dfrac{x^2}{2L}\left(3 - \dfrac{\xi}{L}\right)\left(3 - \dfrac{x}{L}\right)\right] & \text{if } \xi \le x \le L. \qquad (4.126b) \end{cases}$$

The reciprocity principle (2.47b) is satisfied by the Green function (4.126a,b). The corresponding slope,

$$G_2'(x;\xi) = \frac{P}{2EI} \times \begin{cases} x\left[2\xi - x - \dfrac{\xi^2}{2L}\left(2 - \dfrac{x}{L}\right)\left(3 - \dfrac{\xi}{L}\right)\right] & \text{if } 0 \le x \le \xi, \qquad (4.127a) \\[2mm] \xi^2\left[1 - \dfrac{x}{2L}\left(2 - \dfrac{x}{L}\right)\left(3 - \dfrac{\xi}{L}\right)\right] & \text{if } \xi \le x \le L, \qquad (4.127b) \end{cases}$$

satisfies the boundary condition (4.122a) at the clamped end and the continuity condition (4.123a). The second derivative specifies the bending moment

$$-M_2(x;\xi) = EIG''(x;\xi) = P \times \begin{cases} \xi - x - \dfrac{\xi^2}{2L}\left(1 - \dfrac{x}{L}\right)\left(3 - \dfrac{\xi}{L}\right) & \text{if } 0 \leq x \leq \xi, & (4.128a) \\[4mm] -\dfrac{\xi^2}{2L}\left(1 - \dfrac{x}{L}\right)\left(3 - \dfrac{\xi}{L}\right) & \text{if } \xi \leq x \leq L, & (4.128b) \end{cases}$$

which satisfies the continuity condition (4.121a) and takes the value (4.129a) [vanishes (4.120a)] at the clamped (pinned) support, where the slope is zero (4.122a) [nonzero (4.129b)]:

$$N_{2-} \equiv -M_2(0;\xi) = EIG_2''(0;\xi) = P\xi\left[1 - \frac{\xi}{2L}\left(3 - \frac{\xi}{L}\right)\right], \quad G_2'(L;\xi) = -\frac{P\xi^2}{4EI}\left(1 - \frac{\xi}{L}\right). \quad (4.129a,b)$$

The third derivative specifies the transverse force

$$F_2(x;\xi) = EIG_2'''(x;\xi) = P \times \begin{cases} -1 + \dfrac{\xi^2}{2L^2}\left(3 - \dfrac{\xi}{L}\right) \equiv R_{2-} = A & \text{if } 0 \leq x < \xi, & (4.130a) \\[4mm] \dfrac{\xi^2}{2L^2}\left(3 - \dfrac{\xi}{L}\right) \equiv -R_{2+} = P + R_{2-} & \text{if } \xi < x \leq L, & (4.130b) \end{cases}$$

which is constant and specifies the reactions at the supports in (4.130a) ≡ (4.125d) at the clamped end and (4.130b) at the pinned end, with jump P across the point of application of the concentrated force. The deflection (4.131a) [slope (4.131b)] at the point of application of the concentrated force,

$$G_2(\xi;\xi) = \frac{P\xi^3}{3EI}\left[1 - \frac{\xi}{4L}\left(3 - \frac{\xi}{L}\right)^2\right], \quad G_2'(\xi;\xi) = \frac{P\xi^2}{2EI}\left[1 - \frac{\xi}{2L}\left(2 - \frac{\xi}{L}\right)\left(3 - \frac{\xi}{L}\right)\right], \quad (4.131a,b)$$

follows from (4.126a) ≡ (4.126b) [(4.127a) ≡ (4.127b)] in agreement with the continuity condition (4.125a) [(4.123a)].

The preceding expressions simplify (Figure 4.7c) for a concentrated force at the middle of the bar (4.132a): (1) the transverse force (4.130a,b) ≡ (4.132c,d),

$$\xi = \frac{L}{2}: \quad N_{2-} = M_2\left(0;\frac{L}{2}\right) = \frac{3PL}{16}, \quad F_2\left(x;\frac{L}{2}\right) = \begin{cases} -\dfrac{11}{16}P = R_{2-} & \text{if } 0 \leq x < \dfrac{L}{2} \\[4mm] \dfrac{5}{16}P = -R_{2+} & \text{if } \dfrac{L}{2} < x \leq L, \end{cases} \quad (4.132a\text{-}d)$$

shows that the reaction force in modulus is larger at the clamped (4.132c) than at the pinned (4.132d) support; (2) the reaction moment at the clamped support (4.133a) is specified by the bending moment (4.128a,b) ≡ (4.133b,c):

$$\xi = \frac{L}{2}: \quad -M_2\left(x;\frac{L}{2}\right) = \frac{P}{16} \times \begin{cases} 3L - 11x & \text{if } 0 \leq x \leq \dfrac{L}{2}, \\[4mm] 5(x - L) & \text{if } \dfrac{L}{2} \leq x \leq L; \end{cases} \quad (4.133a\text{-}c)$$

(3) the slope (4.127a,b; 4.132a) ≡ (4.134a,b):

$$
G_2'\left(x;\frac{L}{2}\right) = \frac{P}{32EI} \times
\begin{cases}
x(6L - 11x) & \text{if } 0 \le x \le \dfrac{L}{2}, \qquad (4.134a) \\[2ex]
5x^2 - 10xL + 4L^2 & \text{if } \dfrac{L}{2} \le x \le L; \qquad (4.134b)
\end{cases}
$$

(4) the shape of the elastica (4.126a,b; 4.132a) ≡ (4.134a,b):

$$
G_2\left(x;\frac{L}{2}\right) = \frac{P}{96EI} \times
\begin{cases}
x^2(9L - 11x) & \text{if } 0 \le x \le \dfrac{L}{2}, \qquad (4.135a) \\[2ex]
5x^3 - 15x^2L + 12xL^2 - 2L^3 & \text{if } \dfrac{L}{2} \le x \le L; \qquad (4.135b)
\end{cases}
$$

(5) the deflection (4.135a,b) [slope (4.134a,b)] at the mid-position (4.136a) [(4.136b)]:

$$
G_2\left(\frac{L}{2},\frac{L}{2}\right) = \frac{7PL^3}{768EI} = 0.009115\frac{PL^3}{EI}, \quad G_2'\left(\frac{L}{2},\frac{L}{2}\right) = \frac{PL^2}{128EI} = 0.0078125\frac{PL^2}{EI}; \qquad (4.136a,b)
$$

and (6) the slope at the mid-position (4.136b) is 1/4 of the modulus of the slope at the pinned end (4.137):

$$
4G'\left(\frac{L}{2};\frac{L}{2}\right) = -G'\left(L;\frac{L}{2}\right) = \frac{PL^2}{32EI} = 0.03125\frac{PL^2}{EI}. \qquad (4.137)
$$

The fourth case of a bar clamped at both ends (Subsection 4.4.4) is also hyperstatic.

4.4.4 Clamped-Clamped Bar with a Concentrated Force

The weak bending of a clamped-clamped bar by a concentrated force at an arbitrary position (Figure 4.6d) is specified by the influence or Green function for the shear stress (4.89) that may be integrated as in (4.119a,b) for the transverse force. The integration for the bending moment (4.138b,c) reduces the constants to one using the continuity of the curvature (4.95a) ≡ (4.137) at the point of application of the force:

$$
G''(\xi - 0;\xi) = G''(\xi + 0;\xi): \quad M(x;\xi) = EIG''(x;\xi) =
\begin{cases}
Ax + P\xi + B & \text{if } 0 \le x \le \xi, \\[2ex]
(A + P)x + B & \text{if } \xi \le x \le L.
\end{cases} \qquad (4.138a\text{–}c)
$$

A further integration (4.139c,d) uses the clamping condition of zero slope at both ends (4.139a,b):

$$
G'(0;\xi) = 0 = G'(L;\xi): \quad 2EIG'(0;\xi) =
\begin{cases}
Ax^2 + 2(P\xi + B)x & \text{if } 0 \le x < \xi, \\[2ex]
(A + P)(x^2 - L^2) + 2B(x - L) & \text{if } \xi < x \le L.
\end{cases} \qquad (4.139a\text{–}d)
$$

The continuity of the slope at the point of application of the force (4.95b) ≡ (4.140a) determines the constant (4.140b):

$$G'(\xi-0;\xi)=G'(\xi+0;\xi):\quad 2B=-(A+P)L-P\frac{\xi^2}{L}. \tag{4.140a,b}$$

Substituting (4.140b) simplifies (4.139c,d) to (4.141a,b):

$$2EIG'(x;\xi)=\begin{cases} Ax(x-L)+P\left(2\xi-L-\dfrac{\xi^2}{L}\right)x & \text{if } 0<x\le\xi, \tag{4.141a}\\[3ex] (A+P)x^2-\left[(A+P)L+P\dfrac{\xi^2}{L}\right]x+P\xi^2 & \text{if } \xi<x\le L. \tag{4.141b} \end{cases}$$

Integrating from the supports (4.142a,b) leads to (4.142c,d):

$$G(0;\xi)=0=G(L;\xi): \tag{4.142a,b}$$

$$12EIG'(0;\xi)=\begin{cases} Ax^2(2x-3L)+3Px^2\left(2\xi-L-\dfrac{\xi^2}{L}\right) & \text{if } 0\le x<\xi, \tag{4.142c}\\[3ex] 2(A+P)(x^3-L^3)-3\left[(A+P)L+P\dfrac{\xi^2}{L}\right](x^2-L^2)+6P\xi^2(x-L) & \text{if } \xi<x\le L. \tag{4.142d} \end{cases}$$

The continuity of the displacement at the point of application of the force (4.143a) determines the remaining constant (4.143b):

$$G(\xi-0;\xi)=G(\xi-0;\xi):\quad A=-P\left[1-\frac{\xi^2}{L^2}\left(3-2\frac{\xi}{L}\right)\right]=-P\left(1+\frac{2\xi}{L}\right)\left(1-\frac{\xi}{L}\right)^2\equiv R_{1-}, \tag{4.143a--c}$$

that specifies the reaction at the first support (4.119a).

Substituting (4.143c) [(4.143b)] in (4.142c) [(4.142d)] leads to (4.144a) [(4.144b)], that is, the *influence or Green function for the weak (4.12a) bending by a concentrated force (4.83) at position ξ of a uniform (4.10a) bar (4.90a) clamped (4.51a–d) at both ends that specifies the shape of the elastica,*

$$G_1(x;\xi)=\frac{P}{6EI}\times\begin{cases} x^2\left(1-\dfrac{\xi}{L}\right)^2\left(3\xi-x-2\dfrac{x\xi}{L}\right) & \text{if } 0\le x\le\xi, \tag{4.144a}\\[3ex] \xi^2\left(1-\dfrac{x}{L}\right)^2\left(3x-\xi-2\dfrac{x\xi}{L}\right) & \text{if } \xi\le x\le L, \tag{4.144b} \end{cases}$$

and (1) satisfies the reciprocity principle (2.47b); and (2) vanishes at the supports (4.142a,b). The corresponding slope is

$$G_1'(x;\xi) = \frac{P}{2EI} \times \begin{cases} x\left(1-\dfrac{\xi}{L}\right)^2\left(2\xi - x - 2\dfrac{x\xi}{L}\right) & \text{if } 0 \le x \le \xi, & (4.145a) \\[3mm] \xi^2\left(1-\dfrac{x}{L}\right)\left(1-3\dfrac{x}{L}+2\dfrac{x\xi}{L^2}\right) & \text{if } \xi \le x \le L, & (4.145b) \end{cases}$$

which also vanishes at the supports (4.139a,b). At the point of application of the force, the displacement (slope) is (4.144a) ≡ (4.144b) ≡ (4.146a) [(4.145a) ≡ (4.145b) ≡ (4.146b)]:

$$G_1(\xi;\xi) = \frac{P\xi^3}{3EI}\left(1-\frac{\xi}{L}\right)^3, \quad G_1'(\xi;\xi) = \frac{P\xi^2}{3EI}\left(1-\frac{\xi}{L}\right)^2\left(1-\frac{2\xi}{L}\right). \qquad (4.146a,b)$$

The bending moment,

$$-M_1(x;\xi) = EIG_1''(x;\xi) = P \times \begin{cases} \left(1-\dfrac{\xi}{L}\right)^2\left(\xi - x - 2\dfrac{\xi x}{L}\right) & \text{if } 0 \le x \le \xi, & (4.147a) \\[3mm] \dfrac{\xi^2}{L^2}\left(\xi + 3x - 2L - 2\dfrac{\xi x}{L}\right) & \text{if } \xi \le x \le L, & (4.147b) \end{cases}$$

is continuous at the point of application of the force (4.138a) and specifies the reaction moments at the supports:

$$N_{1-} = -M_1(0;\xi) = P\xi\left(1-\frac{\xi}{L}\right)^2, \quad N_{1+} = -M_1(L;\xi) = -P\frac{\xi^2}{L}\left(1-\frac{\xi}{L}\right). \qquad (4.148a,b)$$

The transverse force,

$$F_1(x;\xi) = EIG_1'''(x;\xi) = P \times \begin{cases} -\left(1-\dfrac{\xi}{L}\right)^2\left(1+\dfrac{2\xi}{L}\right) = R_{1-} & \text{if } 0 \le x < \xi, & (4.149a) \\[3mm] \dfrac{\xi^2}{L^2}\left(3 - \dfrac{2\xi}{L}\right) = -R_{1+} & \text{if } \xi < x \le L, & (4.149b) \end{cases}$$

specifies the reaction forces at the first (4.143b) ≡ (4.149a) and second (4.149b) ≡ (4.150a) supports:

$$R_{1+} = -F_1(L;\xi) = -P\frac{\xi^2}{L^2}\left(3-\frac{2\xi}{L}\right), \quad R_{1-} = F_1(0;\xi) = -P\left(1-\frac{\xi}{L}\right)^2\left(1+\frac{2\xi}{L}\right) = -R_{1+}-P, \qquad (4.150a,b)$$

which add to minus the transverse force (4.150b).

In the case of a concentrated force at equal distance from the supports (4.151a), the shape of the elastica (4.144a,b) simplifies to (4.151b,c):

$$\xi = \frac{L}{2}: \quad G_1\left(x;\frac{L}{2}\right) = \frac{P}{48EI} \times \begin{cases} x^2(3L - 4x) & \text{if } 0 \le x \le \dfrac{L}{2}, \\[3mm] (L-x)^2(4x - L) & \text{if } \dfrac{L}{2} \le x \le L. \end{cases} \qquad (4.151a\text{-}c)$$

The slope is given by (4.141a,b; 4.151a) ≡ (4.152a,b):

$$G_1'\left(x;\frac{L}{2}\right) = \frac{P}{8EI} \times \begin{cases} x(L-2x) & \text{if } 0 \le x \le \dfrac{L}{2}, & \text{(4.152a)} \\[3mm] (L-x)(L-2x) & \text{if } \dfrac{L}{2} \le x \le L, & \text{(4.152b)} \end{cases}$$

and it vanishes (4.153a) at the location of the force, that is, at the mid-position where occurs the maximum deflection (4.153b):

$$G_1'\left(\frac{L}{2};\frac{L}{2}\right) = 0, \quad G_1\left(\frac{L}{2};\frac{L}{2}\right) = \frac{PL^2}{192EI} = 0.05208\frac{PL^3}{EI}. \tag{4.153a,b}$$

The bending moment (4.147a,b; 4.151a) ≡ (4.154a,b):

$$-M_1\left(x;\frac{L}{2}\right) = EIG_1''\left(x;\frac{L}{2}\right) = \frac{P}{8} \times \begin{cases} L-4x & \text{if } 0 \le x \le \dfrac{L}{2}, & \text{(4.154a)} \\[3mm] 4x-3L & \text{if } \dfrac{L}{2} \le x \le L, & \text{(4.154b)} \end{cases}$$

and transverse force (4.149a,b; 4.151a) ≡ (4.155a,b):

$$F_1\left(x;\frac{L}{2}\right) = EIG_1'''\left(x;\frac{L}{2}\right) = \begin{cases} -\dfrac{P}{2} = R_{1-} & \text{if } 0 \le x \le \dfrac{L}{2}, & \text{(4.155a)} \\[3mm] \dfrac{P}{2} = -R_{1+} & \text{if } \dfrac{L}{2} < x \le L, & \text{(4.155b)} \end{cases}$$

specify, respectively, the reaction moments (4.156a,b) [forces (4.156c,d)] at the supports:

$$N_{1-} = -M\left(0;\frac{L}{2}\right) = \frac{PL}{8} = -M(L) = -N_{1+}, \quad R_{1-} = F_1\left(0;\frac{L}{2}\right) = -\frac{P}{2} = -F_1\left(L;\frac{L}{2}\right) = R_{1+}, \tag{4.156a–d}$$

which have the same modulus and the same (opposite) sign.

4.4.5 Determinate, Indeterminate, or Incompatible Problems

The linear bending of a bar by a concentrated force in the clamped-free (pinned-pinned) case [Subsection 4.4.1 (Subsection 4.4.2)] were isostatic problems for which the balance of external forces and moments specifies uniquely the reaction forces and moments at the supports that can be determined "a priori," without obtaining first the equation of the elastica. In the linear bending of a bar by a concentrated force in the clamped-pinned case (Subsection 4.4.3), there are (Figure 4.6c) two reaction forces $R_{2\mp}$ at both supports and one reaction moment N_{2-} at the clamped end. The conditions of static equilibrium are the balance of forces and balance of moments at the two supports:

$$R_{2-} + R_{2+} + P = 0, \quad N_{2-} + R_{2-}L + P(L-\xi) = 0, \quad -N_{2-} + R_{2+}L + P\xi = 0. \tag{4.157a–c}$$

These three equations are redundant because the sum of the last two specifies the first. The transverse force is (1) equal to minus the reaction at the first support (4.130a) up to the point of application of the concentrated load; and (2) across that point, the transverse force is increased by P and afterward remains constant, specifying the reaction (4.130b) at the other support. The two independent equations cannot determine three quantities; the equations (4.157a–c) are satisfied by the reaction forces (4.130a,b) and moments (4.129a) at the supports that were obtained (Subsection 4.4.3) from the equation of the elastica after the latter was determined. Thus, *the linear bending of a clamped-pinned bar by a concentrated forces is a* **hyperstatic indeterminate problem,** *that is, the static balance is insufficient to determine the reaction forces and moments at the supports, leaving them indeterminate although they do satisfy the balance of external forces and moments.*

In the linear bending of a clamped-clamped bar by a concentrated force (Subsection 4.4.4), the moment of the concentrated force and reaction forces (4.149a,b) at the second (first) support (Figure 4.7d)] relative to the first (4.158a) [second (4.158b)] support does not coincide with (4.148a) [(4.148b)]:

$$R_{1+}L + P\xi = P\xi\left[1 - \frac{\xi^2}{L^2}\left(3 - 2\frac{\xi}{L}\right)\right] \neq N_{1-}, \qquad (4.158a)$$

$$R_{1-}L + P(L - \xi) = -P\xi\left(1 - 3\frac{\xi}{L} + 2\frac{\xi^2}{L^2}\right) \neq N_{1+}. \qquad (4.158b)$$

Thus, the system is hyperstatic incompatible (Subsection 4.4.4), that is, the internal stresses affect the balance of moments, but not the balance of forces (4.159a). Thus, the linear bending of a clamped-clamped bar by a concentrated force is a hyperstatic incompatible problem, for which the static balance equations lead to reaction forces and moments at the supports that are incompatible with the boundary and continuity/ jump conditions, because the forces and moments arising from internal stresses and deformations were not taken into account. The partial exception is the linear bending of the clamped-clamped bar by a concentrated force at equal distance from the supports, for which (1) the reaction forces (moments) are equal in modulus with the same (opposite) sign by symmetry (4.159b,c) [(4.159d)]:

$$R_{1-} + R_{1+} + P = 0: \quad R_{1-} = R_{1+} = -\frac{P}{2}, \quad N_{1+} = -N_{1-}; \qquad (4.159a\text{–}d)$$

and (2) the balance of external forces (4.159a) [moments (4.159c)] does (4.159b,c) ≡ (4.156c,d) [does not (4.156a,b)] determine the reaction forces (moments).

It is only in the case of an **isostatic** *system that the static balance equations determine all reactions at the supports, for example, the bar with one clamped and one free end, bent by a concentrated force (Subsection 4.3.1) or torque (Subsection 4.5.2). In the case of a hyperstatic system, whether indeterminate or incompatible, the reaction forces and moments at the supports cannot be predicted "a priori" from static balance equations; they are determined by the elastic stresses and deformations and arise out of the solution of the problem of the shape of the elastica.* A similar situation arises with the nonlinear deflection of an elastic string by a non-symmetric load (Section 2.6) because the vertical component of the tangential tension affects the balance of moments. When considering a nonisostatic problem, it is essential to start with the differential equation of the highest order, that is, order four for the shear stress, in order to have enough constants of integration to satisfy all boundary and continuity conditions. An invalid "a priori" assumption about the reaction forces or moments would lead to a contradiction at a later stage: there would not be enough constants of integration left to meet all boundary and continuity conditions. The general method to be followed for nonisostatic systems is the same in the case of indeterminate or incompatible (Subsections 4.4.3 and 4.4.4) hyperstatic systems. The boundary and continuity conditions should be applied at the earliest opportunity, to determine the constants of integration as early as possible, leading to expressions as simple as feasible. The weak bending of a bar by a concentrated force at the end (middle) for the clamped-free (other 3) case(s) is a particular case in which the influence function [Subsection 4.4.1 (Subsections 4.4.2 through 4.4.4)] takes a simpler form, allowing a comparison of deflections, slopes, and reaction forces and moments (Subsection 4.4.6).

4.4.6 Concentrated Force at Mid- or Extreme Positions

Table 4.3 compares the influence or Green function (Figure 4.6a through d) for the weak (4.12a) bending under a concentrated load at an arbitrary position (4.89) of a uniform (4.10a) bar (4.90a) in four cases: (1) clamped-free (Subsection 4.4.1); (2) pinned-pinned (Subsection 4.4.2); (3) clamped-pinned (Subsection 4.4.3); and (4) clamped-clamped (Subsection 4.4.4). The cases (1, 2) [(3, 4)] are isostatic (hyperstatic). Table 4.3 indicates in each of the four cases (1–4) (a) the shape of the elastica and its slope and the transverse force and bending moment before and after the point of application of the concentrated force; (b) the deflection and slope at the point of application of the concentrated force; (c) the reaction forces and moments at the supports. Table 4.4 considers the particular cases (Figure 4.7a through d) of a concentrated force at the mid-position between supports (2–4) [at the free end (1)] that allows a simpler comparison of (a–c).

The deflection (1) is maximized for a bar clamped at both ends by placing the concentrated force at equal distance from the supports (4.151a), in which case it takes twice the value (4.153b) ≡ (4.160a) for the equivalent uniform load (4.58b); (2/3) changing the one (both) end to pinning increases the deflection at the midpoint (4.136a) ≡ (4.160b) [(4.116b) ≡ (4.160c)] by a factor of 7/4 (4); (4, 5) keeping one end clamped and letting the other free increases the deflection (4.97b) ≡ (4.160d) [(4.101c) ≡ (4.160e)] at the free end by a factor of 20(64) if the concentrated force is applied in the middle (at the free end):

$$G_1\left(\frac{L}{2};\frac{L}{2}\right) = \frac{PL^3}{192EI} = 0.005208\frac{PL^3}{EI} = \frac{4}{7}G_2\left(\frac{L}{2};\frac{L}{2}\right)$$

$$< G_2\left(\frac{L}{2};\frac{L}{2}\right) = \frac{7PL^3}{768EI} = 0.009115\frac{PL^3}{EI} = \frac{7}{4}G_1\left(\frac{L}{2};\frac{L}{2}\right)$$

$$< G_3\left(\frac{L}{2};\frac{L}{2}\right) = \frac{PL^3}{48EI} = 0.02083\frac{PL^3}{EI} = 4G_1\left(\frac{L}{2};\frac{L}{2}\right)$$

$$< G_4\left(L;\frac{L}{2}\right) = \frac{5PL^3}{48EI} = 0.1042\frac{PL^3}{EI} = 20G_1\left(\frac{L}{2};\frac{L}{2}\right)$$

$$< G_4\left(L;L\right) = \frac{PL^3}{3EI} = 0.3333\frac{PL^3}{EI} = 64G_1\left(\frac{L}{2};\frac{L}{2}\right). \qquad (4.160a\text{–}e)$$

The slope is (1, 2) largest for the clamped-free bar at the point of application of the force at the free end (4.102b) ≡ (4.161a), with a reduction by 1/4 if the force is applied at the mid-position (4.93c) ≡ (4.161b); (3, 4) for the pinned-pinned (clamped-pinned) bar, the modulus of the slope at the pinned ends is (4.117b,c) ≡ (4.161c) [(4.137) ≡ (4.161d)], that is, reduces by a factor of 1/8 (1/16) relative to (1); and (5) the clamped-clamped bar has zero slope at the supports and also at the point of application of the force at the middle (4.161e):

$$G_4'\left(L;L\right) = \frac{PL^2}{2EI} > \frac{PL^2}{8EI}G_4'\left(L;\frac{L}{2}\right) = \frac{1}{4}G_4'\left(L;L\right)$$

$$> \frac{PL^2}{16EI} = G_3'\left(0;\frac{L}{2}\right) = -G_3'\left(L;\frac{L}{2}\right) = \frac{1}{8}G_4'\left(L;L\right)$$

$$> \frac{PL^2}{32EI} = -G_2'\left(L;\frac{L}{2}\right) = \frac{1}{16}G_4'\left(L;L\right) = \frac{1}{2}G_3'\left(0;\frac{L}{2}\right)$$

$$> 0 = G_1\left(\frac{L}{2};\frac{L}{2}\right) = G_1\left(0;\frac{L}{2}\right) = G_1\left(L;\frac{L}{2}\right). \qquad (4.161a\text{–}e)$$

TABLE 4.3 Linear Influence or Green Function for an Elastic Bar

Case	1	2	3	4
Load	Force P at ξ	Force P at ξ	Force P at ξ	Force P at ξ
Subsection	4.4.4	4.4.3	4.4.2	4.4.1
Figure	4.6d	4.6c	4.6b	4.6a
Type of support	Clamped-clamped	Clamped-pinned	Pinned-pinned	Clamped-free
Degrees of freedom	−2	−1	0	0
Type of problem	Hyperstatic incompatible	Hyperstatic indeterminate	Isostatic	Isostatic
Boundary conditions	(4.139a,b, 4.142a,b)	(4.120a, 4.122a, 4.124a, 4.125c)	(4.105a,b, 4.108a,c)	(4.91a, 4.92a, 4.93a, 4.94a)
Jump conditions	(4.119a,b)	(4.119a,b)	(4.104a,b)	(4.91b,c)
Continuity conditions	(4.138a, 4.140a, 4.143a)	(4.121a, 4.123a, 4.125a)	(4.107a,b, 4.109)	(4.95a–c)
Shape of the elastica	(4.144a,b)	(4.126a,b)	(4.110a,b)	(4.96a,b)
Slope of the elastica	(4.145a,b)	(4.127a,b)	(4.111a,b)	(4.93b,c)
Deflection at ξ	(4.146a)	(4.131a)	(4.112a)	(4.97a)
Slope at ξ	(4.146b)	(4.131b)	(4.112b)	(4.97c)
Bending moment	(4.147a,b)	(4.128a,b)	(4.105c,d)	(4.98a)
Transverse force	(4.149a,b)	(4.130a,b)	(4.104a,b)	(4.99a)
Shear stress	(4.88b)	(4.88b)	(4.88b)	(4.88b)
Reaction forces	(4.150a,b)	(4.130a,b)	(4.104a,b)	(4.98b)
Reaction moment	(4.148a,b)	(4.129a)	—	(4.99b)

Note: Comparison (for the same cases of supports as Table 4.2) of the weak bending of a uniform bar by a concentrated transverse force at an arbitrary position, specifying the linear influence or Green function.

TABLE 4.4 Bending of a Bar by a Concentrated Force at the Middle (or End)

Case	1	2	3	4
Load	Force P at $\xi = \dfrac{L}{2}$	Force P at $\xi = \dfrac{L}{2}$	force P at $\xi = \dfrac{L}{2}$	Force P at $\xi = L$
Subsection	4.4.4	4.4.3	4.4.2	4.4.1
Figure	4.7d	4.7c	4.7b	4.7a
Type of support	Clamped-clamped	Clamped-pinned	Pinned-pinned	Clamped-free
Degrees of freedom	−2	−1	0	0
Type of problem	Hyperstatic incompatible	Hyperstatic indeterminate	Isostatic	Isostatic
Shape of the elastica	(4.151b,c)	(4.135a,b)	(4.114b,c)	(4.101b)
Slope of the elastica	(4.152a,b)	(4.134a,b)	(4.115a,b)	(4.102a)
Deflection at ξ	(4.153a)	(4.136a)	(4.116b)	(4.101c)
Slope at ξ	(4.153b)	(4.136b)	(4.116a)	(4.102b)
Bending moment	(4.154a,b)	(4.133b,c)	(4.118a,b)	(4.103a)
Transverse force	(4.155a,b)	(4.132c,d)	(4.118c,d)	(4.100b)
Shear stress	(4.117a)	(4.117a)	(4.117a)	(4.101a)
Reaction forces	(4.156c,d)	(4.132c,d)	(4.118c,d)	(4.100c)
Reaction moments	(4.156a,b)	(4.132b)	—	(4.100d)

Note: As for Table 4.3 in the particular case when the concentrated force is applied at: (1) the tip in the clamped-free case; and (2) equal distance from the supports in all three other cases.

The reaction forces at the supports (1) equal the force with opposite sign for the clamped-free case (4.100c) ≡ (4.162a) ≡ (4.163d); (2) the force is equally divided between the supports in the symmetric cases, that is, pinned-pinned (4.118c,d) ≡ (4.162c) ≡ (4.163a) [clamped-clamped (4.155a,b) ≡ (4.162d) ≡ (4.163b)]; and (3) in the clamped-pinned case, the clamped (4.132c) ≡ (4.162b) [pinned (4.132d) ≡ (4.163c)] support carries a larger (smaller) fraction of the force:

$$-R_{4-} = P > \frac{11}{16}P = -R_{2-} > \frac{P}{2} = -R_{3-} = -R_{1-}, \tag{4.162a–d}$$

$$-R_{3-} = \frac{P}{2} = -R_{1+} > \frac{5}{16}P = -R_{2+} > 0 = R_{4+}. \tag{4.163a–d}$$

The bending moment is (1) zero at all free or pinned ends (4.164d, 4.165b–d); (2) equal at the two supports for the clamped-clamped bar (4.156a,b) ≡ (4.164c) ≡ (4.165a); and (3/4) larger (4.132b) ≡ (4.164b) [(4.103b) ≡ (4.165a)] by a factor of 3/2 (8) at the clamped end if the other end is pinned (free):

$$M_{4-} = PL > \frac{3PL}{16} = M_{2-} > \frac{PL}{8} = M_{1-} > 0 = M_{3-}, \tag{4.164a–d}$$

$$-M_{1+} = \frac{PL}{8} > 0 = M_{2+} = M_{3+} = M_{4+}. \tag{4.165a–d}$$

The influence or Green function is determined by the shape of the bar bent by a concentrated load, and in the case of linear bending, the principle of superposition specifies the shape for any distribution of loads (Subsection 4.4.7).

4.4.7 Influence Function for Four Sets of Boundary Conditions

The influence or Green function depends not only on the differential equation, but also on the boundary conditions. In the differential equations for weak deflection (4.12a) of a nonuniform bar (4.12b–d), the differential operators are self-adjoint (7.129d), so the reciprocity principle holds. Thus, the *Green influences or functions for all combinations of boundary conditions satisfy the principles of reciprocity and superposition for the weak bending of a uniform or nonuniform bar. It can be used to specify the shape of the elastica of a uniform bar under arbitrary shear stress, (4.166a) by means of the integral (4.166b) ≡ (2.59b):*

$$EI\zeta'''(x) = f(x), \quad \zeta(x) = \frac{1}{P}\int_0^L G(x;\xi) f(\xi)\,d\xi, \tag{4.166a,b}$$

which can be applied to the four classical combinations of statically stable support conditions: (1) one end clamped and one free (4.96a,b):

$$6EI\zeta_4(x) = \int_0^x (3x - \xi)\xi^2 f(\xi)\,d\xi + x^2 \int_x^L (3\xi - x) f(\xi)\,d\xi; \tag{4.167}$$

(2) both ends pinned (4.110a,b):

$$6EI\zeta_3(x) = \left(\frac{x}{L} - 1\right) \int_0^x \xi\left(\xi^2 - 2xL + x^2\right) f(\xi) d\xi$$

$$+ x \int_x^L \left(\frac{\xi}{L} - 1\right)\left(x^2 - 2\xi L + \xi^2\right) f(\xi) d\xi; \tag{4.168}$$

(3) one end clamped and one pinned (4.126a,b):

$$6EI\zeta_2(x) = \int_0^x \xi^2 \left[3x - \xi - \frac{x^2}{2L}\left(3 - \frac{\xi}{L}\right)\left(3 - \frac{x}{L}\right)\right] f(\xi) d\xi$$

$$+ x^2 \int_x^L \left[3\xi - x - \frac{\xi^2}{2L}\left(3 - \frac{x}{L}\right)\left(3 - \frac{\xi}{L}\right)\right] f(\xi) d\xi; \tag{4.169}$$

and (4) both ends clamped (4.144a,b):

$$6EI\zeta_1(x) = \left(1 - \frac{x}{L}\right)^2 \int_0^x \xi^2 \left(3x - \xi - 2\frac{x\xi}{L}\right) f(\xi) d\xi$$

$$+ x^2 \int_x^L \left(1 - \frac{\xi}{L}\right)^2 \left(3\xi - x - 2\frac{x\xi}{L}\right) f(\xi) d\xi. \tag{4.170}$$

The principle of superposition (4.167–4.170) can be used to obtain the shape of the elastica (4.166a) of a uniform (4.10a) bar (4.90a) under weak bending (4.12a) for any load, for example, (1) uniform like the weight (Section 4.3); (2) a concentrated torque (Section 4.5).

4.5 Weak Bending by a Concentrated Torque

An elastic string (bar) does not have (has) bending stiffness [Chapter 2 (Chapter 4)] and can support a concentrated force but not (and also) a concentrated moment. The shape of the elastica for the weak bending of a uniform bar can be obtained by the general method of (1) integrating the shear stress four times; and (2, 3) applying boundary (continuity) conditions at the supports (point of application of the torque). In the case of a concentrated torque (force), (1) the displacement and first (first two) derivatives, that is, slope (slope and curvature), are continuous at the point of application; and (2) across the point of application, there is a jump of the second (third)-order derivative of the displacement. The shape of the elastica of a bar bent by a concentrated torque can be obtained (Subsection 4.5.1) differentiating the influence or Green function at the point of application of the force. Using this method passes from the shape of the elastica under a concentrated torque (force) for all four classical cases of support, namely: (1) clamped-free [Subsection 4.5.2 (Subsection 4.4.1)]; (2) pinned-pinned [Subsection 4.5.3 (Subsection 4.4.2)]; (3) clamped-pinned [Subsection 4.5.4 (Subsection 4.4.3)]; and (4) clamped-clamped [Subsection 4.5.5 (Subsection 4.4.4)]. The shape of the elastica due to a concentrated torque at an arbitrary position (Table 4.5) includes as a particular case (Table 4.6) the concentrated torque at the free end (1) [equal distance from the supports (2–4)]; this leads to a comparison of deflection and slope for the four types of support (Subsection 4.5.6).

4.5.1 Derivatives of the Impulse and Influence Functions

A concentrated torque Q at position η corresponds to a jump of the bending moment (4.171a,b), where A is a constant and $H(x-\eta)$ the unit jump (1.28a–c):

$$M(x;\eta) = -EI\zeta''(x;\eta) = A + QH(x;\eta) = \begin{cases} A & \text{if } x > \eta, \quad (4.171a) \\ A+Q & \text{if } x > \eta. \quad (4.171b) \end{cases}$$

The corresponding transverse force (4.172a) [shear stress (4.172b)] involves the unit impulse (derivative of the unit impulse):

$$F(x;\eta) = EI\zeta'''(x;\eta) = -Q\delta(x-\eta), \quad f(x;\eta) = EI\zeta''''(x;\eta) = -Q\delta'(x-\eta). \quad (4.172a,b)$$

The shape of the elastica for the weak (4.12a) bending of a uniform (4.10a) bar (4.90a) by a concentrated torque is given by

$$G(x;\eta) \in \mathcal{D}^4(|R) x \mathcal{D}(|R): \quad \zeta(x;\eta) = \frac{Q}{P}\frac{d}{d\eta}\big[G(x;\eta)\big], \quad (4.173a,b)$$

using (1) the influence or Green function with the same boundary conditions at the supports; (2) differentiating with regard to the parameter at the point of application; and (3) multiplying by Q/P to change from a concentrated force P (to a torque Q). In (4.173b), it is assumed that the influence function is differentiable four times (once) in the sense of generalized (ordinary) functions with regard to the independent variable (parameter), that is, the position coordinate along the length of the bar (the location of the concentrated torque). The proof of (4.173b) follows from substitution of the shear stress (4.172b) in the superposition principle (4.166b):

$$\zeta(x;\eta) = -\frac{Q}{P}\int_0^L G(x;\xi)\delta'(\xi-\eta)d\eta = \frac{Q}{P}\lim_{\xi\to\eta}\big[G(x;\eta)\big], \quad (4.174)$$

using the fundamental property of the unit derivative impulse (3.16b) and leading to (4.174) $\equiv$ (4.173b).

4.5.2 Clamped-Free Bar with a Concentrated Torque

The shape of the elastica for the weak bending (4.12a) of a uniform (4.10a) bar (4.90a) by a concentrated torque Q at an arbitrary position (4.172b) can be obtained in the clamped-free case (Figure 4.8a) by three methods: (1) substitution of the Green function (4.96a,b) for the same boundary conditions at the supports in (4.173b):

$$0 \leq x \leq \eta: \quad \zeta_4(x;\eta) = \frac{Q}{6EI}\lim_{\xi\to\eta}\frac{d}{d\xi}\big[x^2(3\xi-x)\big] = \frac{Qx^2}{2EI}, \quad (4.175a,b)$$

$$\eta \leq x \leq L: \quad \zeta_4(x;\eta) = \frac{Q}{6EI}\lim_{\zeta\to\eta}\frac{d}{d\xi}\big[\xi^2(3x-\xi)\big] = \frac{Q\eta}{2EI}(2x-\eta); \quad (4.175c,d)$$

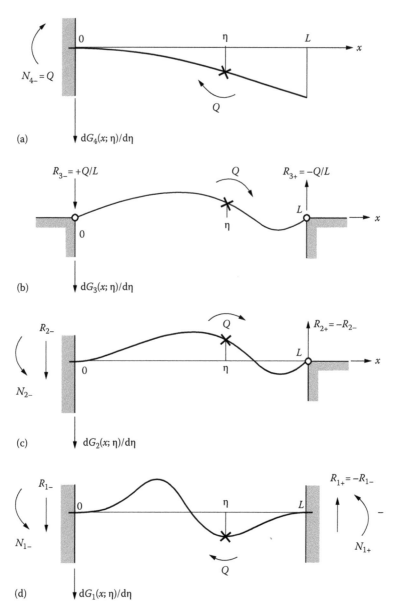

FIGURE 4.8 Weak bending of a uniform bar a concentrated moment at an arbitrary position, for the same set of four classical support conditions (a–d) as in Figure 4.5 (4.6) with a uniform shear stress (concentrated force at an arbitrary position).

(2) substitution of the shear stress (4.172b) in the superposition principle (4.167):

$$\eta \le x \le L: \quad 6EI\zeta_4(x;\eta) = -Q\int_0^x (3x-\xi)\xi^2\delta'(\xi-\eta)d\xi$$

$$= Q\frac{d}{d\eta}\left[(3x-\eta)\eta^2\right] = 6Q\eta\left(x-\frac{\eta}{2}\right), \tag{4.176a,b}$$

$$0 \le x \le \eta: \quad 6EI\zeta_4(x;\eta) = -Qx^2 \int_x^L (3\xi - x)\delta'(\xi - \eta)d\xi$$

$$= Qx^2 \frac{d}{d\eta}\big[(3\eta - x)\big] = 3Qx^2, \tag{4.176c,d}$$

where the property (3.16b) of the derivative unit impulse is used; and (3) integration of the shear stress (4.172b) four times, using the boundary and continuity conditions. The latter (3) is the general method and will be used next as a check on the methods (1) and (2); the method (1) is simplest and will be used subsequently (Subsections 4.5.3 through 4.5.5) for other types of support.

A first integration of (4.172b) leads to the transverse force (4.77a):

$$F(x;\eta) = EI\zeta'''(x;\eta) = B - Q\delta(x - \eta), \quad 0 = F(L;\eta) = B, \tag{4.177a,b}$$

where the constant of integration is zero (4.177b) because there is no transverse force at the free end. Integrating (4.177a,b) ≡ (4.172a) leads to the bending moment:

$$M(x;\eta) = -EI\zeta''(x;\eta) = A + QH(x - \eta) = \begin{cases} A = -Q = M_{4-}, & (4.178a) \\ A + Q = 0 = M_{4+}, & (4.178b) \end{cases}$$

which vanishes at the free end (4.178b), so that the constant of integration of (4.178a,b) specifies the reaction moment at the clamped support. Integrating (4.178a,b) leads to the slope (4.179c,d) that vanishes at the clamped support (4.179a) and is continuous at the point of application of the concentrated torque (4.179b):

$$\zeta'(0;\eta) = 0 = \zeta'(\eta - 0;\eta) - \zeta'(\eta + 0;\eta): \quad EI\zeta'(x;\eta) = \begin{cases} Qx & \text{if } 0 \le x \le \eta, \\ Q\eta & \text{if } \eta \le x \le L. \end{cases} \tag{4.179a-d}$$

A further integration of (4.179c,d) leads to (4.180c,d), using the zero displacement at the clamped support (4.180a) and the continuity of the displacement at the point of application of the concentrated torque (4.180b):

$$\zeta(0;\eta) = 0 = \zeta(\eta - 0;\eta) - \zeta(\eta + 0;\eta): \quad 2EI\zeta(x;\eta) = \begin{cases} Qx^2 & \text{if } 0 \le x \le \eta. \\ Q\eta(2x - \eta) & \text{if } \eta \le x \le L. \end{cases} \tag{4.180a-d}$$

The same result has been obtained for the shape of the elastica by the three methods: (1) influence function (4.175a–d); (2) superposition principle (4.176a–d); and (3) shear stress (4.180c,d).

It has been shown by three equivalent methods that *the shape of the elastica of a uniform (4.10a) bar (4.90a) due to weak bending (4.12a) by (Figure 4.8a) a moment Q concentrated (4.172b) at x = η is a parabola (4.181a) [straight line (4.181b)] before (after) the point of application of the concentrated moment:*

$$\zeta_4(x;\eta) = \frac{Q}{2EI} \times \begin{cases} x^2 & \text{if } 0 \le x \le \eta, & (4.181a) \\ \eta(2x - \eta) & \text{if } \eta \le x \le L, & (4.181b) \end{cases}$$

The slope (4.182a,b) is constant after the point of application of the concentrated torque:

$$\zeta_4'(x;\eta) = \frac{Q}{EI} \times \begin{cases} x & \text{if } 0 \le x \le \eta, \\ \eta & \text{if } \eta \le x \le L. \end{cases}$$

(4.182a)

(4.182b)

The corresponding bending moment (4.183a) [transverse force (4.184a)] involves the unit jump (impulse) and a constant (no constant) that specifies the nonzero (zero) reaction moment (force) at the clamped support (4.183b) [(4.184b)]:

$$-M_4(x;\eta) = EI\zeta_4''(x;\eta) = -Q + QH(x-\eta), \quad N_{4-} = -M(0;\eta) = Q,$$

(4.183a,b)

$$-F_4(x;\eta) = EI\zeta_4'''(x;\eta) = -Q\delta(x-\eta), \quad R_{4-} = -F(0;\eta) = 0.$$

(4.184a,b)

The deflection at the point of application of the concentrated torque (4.185a) cannot exceed the maximum (4.185b) at the free end:

$$\zeta_4(\eta;\eta) = \frac{Q\eta^2}{2EI} < \delta_{4\max} = \zeta_4(L;\eta) = \frac{Q\eta}{2EI}(2L-\eta), \quad \zeta_4'(L;\eta) = \frac{Q\eta}{EI},$$

(4.185a–c)

and the slope is constant after the point of application of the concentrated moment (4.182b) and hence is the same (4.182b) ≡ (4.185c) at the free end.

The two values coincide (4.185a) ≡ (4.185b) ≡ (4.186b) for a clamped-free or **cantilever** bar (Figure 4.9a) with a concentrated moment Q applied at the free end (4.186a):

$$\eta = L: \quad \zeta_4 \equiv \zeta_{4\max}(L;L) = \frac{QL^2}{2EI}; \quad EI\zeta_4''(x;L) = Q = -M_4(x;L), \quad \zeta_4(x;L) = \frac{Qx^2}{2EI},$$

(4.186a–d)

in the latter case of a concentrated moment at the free end (4.186c), the parabolic shape of the elastica holds throughout the bar (4.181a) ≡ (4.186d). The concentrated moment at the free end corresponds to the shear stress (4.187a) [transverse force (4.187b)]:

$$f_4(x;L) = -Q\delta'(x-L), \quad F_4(x;L) = -Q\delta(x-L),$$

(4.187a,b)

and the reaction force (moment) at the support (4.184b) [(4.183b)] is independent of the position of the concentrated moment. The shape of the elastica (4.186d) follows from the constant bending moment (4.186c) by double integration, using the two clamping boundary conditions (4.188a,b):

$$\zeta_4(0;L) = 0 = \zeta_4'(0;L); \quad \zeta_4'(x;L) = \frac{Qx}{EI}, \quad \theta_4 = \zeta_{4\max}' = \zeta_4'(L;L) = \frac{QL}{EI}.$$

(4.188a–d)

The corresponding slope (4.188c) is maximum at the free end (4.188d). For the remaining three cases of classical combinations of supports (Subsections 4.5.3 through 4.5.5), the first method (1) of the influence function (4.174) will be used to determine the shape of the elastica for the weak bending of a uniform bar by a concentrated torque, since the influence function has already been obtained for all four classical combinations of supports (Subsections 4.4.1 through 4.4.4).

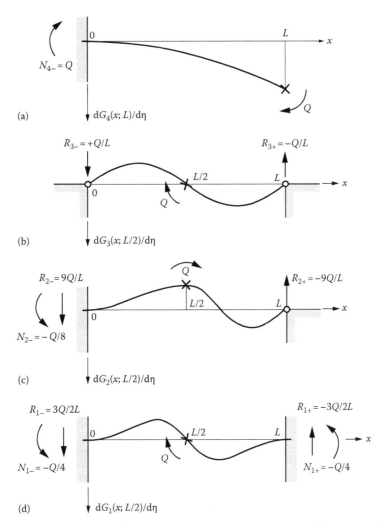

FIGURE 4.9 The weak bending of a uniform bar by a concentrated torque at the middle leads to a skew-symmetric shape if the boundary conditions are symmetric, for example, for a pinned-pinned (b) or clamped-clamped (d) bar. If the boundary conditions are unsymmetric the shape is not skew-symmetric, for example, for a concentrated torque at the middle (tip) of a clamped-pinned (c) [clamped-free (a)] bar.

4.5.3 Pinned-Pinned Bar with a Concentrated Torque

Applying (4.173b) to (4.110a,b) specifies *the shape of the elastica for the weak bending (4.12a) of a uniform (4.10a) bar (4.90a) by (4.172b) a concentrated moment Q at $x = \eta$ in the pinned-pinned (4.53a,b) case (Figure 4.8b):*

$$\zeta_3(x;\eta) = \frac{Q}{6EIL} \times \begin{cases} x(x^2 + 2L^2 - 6\eta L + 3\eta^2) & \text{if } 0 \le x \le \eta, \\ -(L-x)(x^2 - 2Lx + 3\eta^2) & \text{if } \eta \le x \le L. \end{cases} \tag{4.189a,b}$$

The corresponding slope is

$$\zeta_3'(x;\eta) = \frac{Q}{6EIL} \times \begin{cases} 3x^2 + 2L^2 - 6\eta L + 3\eta^2 & \text{if } 0 \le x \le \eta, \\ 3x^2 + 2L^2 - 6xL + 3\eta^2 & \text{if } \eta \le x \le L. \end{cases} \tag{4.190a,b}$$

The deflection (slope) at the point of application of the concentrated moment is (4.191a) [(4.191b)]:

$$\zeta_3(\eta;\eta) = \frac{Q\eta}{3EIL}(2\eta^2 - 3\eta L + L^2), \quad \zeta_3'(\eta;\eta) = \frac{Q}{3EIL}(3\eta^2 + L^2 - 3\eta L). \qquad (4.191\text{a,b})$$

The slope at the first (second) support is (4.194a) [(4.194b)]:

$$\zeta_3'(0;\eta) = \frac{Q}{6EIL}(2L^2 - 6\eta L + 3\eta^2), \quad \zeta_3'(L;\eta) = \frac{Q}{6EIL}(3\eta^2 - L^2). \qquad (4.192\text{a,b})$$

The bending moment (4.193a,b),

$$-M_3(x;\eta) = EI\zeta_3''(x;\eta) = Q \times \begin{cases} \dfrac{x}{L} & \text{if } 0 \le x < \eta, \qquad (4.193\text{a}) \\[2mm] \dfrac{x}{L} - 1 & \text{if } \eta < x \le L, \qquad (4.193\text{b}) \end{cases}$$

vanishes at the pinned supports (4.194a,b) and has a jump Q at the point of application of the torque (4.194c):

$$N_{3-} = -M_3(0;\eta) = 0 = M_3(L;\eta) = N_{3+}, \quad M_3(\eta+0;\eta) - M_3(\eta-0;\eta) = Q. \qquad (4.194\text{a–c})$$

The transverse force (4.195a)

$$F_3(x;\eta) = EI\zeta_3'''(x;\eta) = \frac{Q}{L} - Q\delta(x-\eta), \quad R_{3-} = F_3(0;\eta) = \frac{Q}{L} = F(L;\eta) = -R_{3+} \qquad (4.195\text{a–c})$$

consists of (1) a unit impulse due to the concentrated torque (4.172a); and (2) plus a constant term specifying the reactions at the supports (4.195b,c) that have the same modulus and opposite signs. The constant term disappears in the shear stress (4.172b) that is specified by a derivative of the unit impulse.

In the case of a concentrated moment at equal distance (Figure 4.9b) from the supports (4.196a), the shape of the elastica (4.189a,b) simplifies to (4.196b,c):

$$\eta = \frac{L}{2}: \quad \zeta_3\left(x;\frac{L}{2}\right) = \frac{Q}{6EIL} \times \begin{cases} x\left(x^2 - \dfrac{L^2}{4}\right) & \text{if } 0 \le x \le \dfrac{L}{2} \\[4mm] x^3 - 3x^2 L + \dfrac{11}{4}L^2 x - \dfrac{3L^3}{4} & \text{if } \dfrac{L}{2} \le x \le L. \end{cases} \qquad (4.196\text{a–c})$$

The corresponding slope is (4.197b,c):

$$f\left(x;\frac{L}{2}\right) = -Q\delta'\left(x-\frac{L}{2}\right): \quad \zeta_3'\left(x;\frac{L}{2}\right) = \frac{Q}{6EIL} \times \begin{cases} 3x^2 - \dfrac{L^2}{4} & \text{if } 0 \le x \le \dfrac{L}{2}, \\[4mm] 3x^2 - 6Lx + \dfrac{11}{4}L^2 & \text{if } \dfrac{L}{2} \le x \le L; \end{cases} \qquad (4.197\text{a–c})$$

the deflection (slope) is zero (nonzero) at the point of application (4.197a) of the concentrated moment (4.198a) [(4.198b)]:

$$\zeta_3\left(\frac{L}{2};\frac{L}{2}\right) = 0, \quad \zeta_3'\left(\frac{L}{2};\frac{L}{2}\right) = \frac{QL}{12EI}; \quad \zeta_3'\left(0;\frac{L}{2}\right) = \zeta_3'\left(L;\frac{L}{2}\right) = -\frac{QL}{24EI}. \tag{4.198a–d}$$

The slope at the supports (4.198c,d) is one-half in modulus with opposite sign. The bending moments (4.193a,b) [transverse force (4.195a)] apply for any position (4.196a) of the concentrated torque, and hence the reaction moments (4.194a,b) [forces (4.195b,c)] at the supports are the same.

4.5.4 Clamped-Pinned Bar with a Concentrated Torque

Substituting (4.126a,b) in (4.173b) specifies *the shape of the elastica for the weak (4.12a) bending of a uniform (4.10a) bar (4.90a) by (4.172b) a concentrated torque Q at x = η in (4.52a–d) the clamped-pinned case (Figure 4.8c)*:

$$\zeta_2(x;\eta) = \frac{Q}{2EI} \times \begin{cases} x^2\left[1 - \frac{\eta}{L}\left(3 - \frac{x}{L}\right)\left(1 - \frac{\eta}{2L}\right)\right] & \text{if } 0 \le x \le \eta, \tag{4.199a} \\[3mm] \eta\left[2x - \eta - \frac{x^2}{L}\left(3 - \frac{x}{L}\right)\left(1 - \frac{\eta}{2L}\right)\right] & \text{if } \eta \le x \le L. \tag{4.199b} \end{cases}$$

The corresponding slope is

$$\zeta_2'(x;\eta) = \frac{Q}{EI} \times \begin{cases} x\left[1 - 3\frac{\eta}{L}\left(1 - \frac{\eta}{2L}\right)\left(1 - \frac{x}{2L}\right)\right] & \text{if } 0 \le x \le \eta, \tag{4.200a} \\[3mm] \eta\left[1 - 3\frac{x}{L}\left(1 - \frac{x}{2L}\right)\left(1 - \frac{\eta}{2L}\right)\right] & \text{if } \eta \le x \le L, \tag{4.200b} \end{cases}$$

and satisfies the principle of reciprocity, allowing interchange of (x, η). The deflection (slope) at the point of application of the concentrated moment is (4.201a) [(4.202a)]:

$$\zeta_2(\eta;\eta) = \frac{Q\eta^2}{2EI}\left[1 - \frac{\eta}{L}\left(3 - \frac{\eta}{L}\right)\left(1 - \frac{\eta}{2L}\right)\right], \quad \zeta_2'(0;\eta) = 0, \tag{4.201a,b}$$

$$\zeta_2'(\eta;\eta) = \frac{Q\eta}{EI}\left[1 - 3\frac{\eta}{L}\left(1 - \frac{\eta}{2L}\right)^2\right], \quad \zeta_2'(L;\eta) = \frac{Q\eta}{EI}\left[1 - \frac{3}{2}\left(1 - \frac{\eta}{2L}\right)\right]; \tag{4.202a,b}$$

the slope is zero (nonzero) at the clamped (4.201b) [pinned (4.202b)] support. The bending moment is

$$-M_2(x;\eta) = EI\zeta_2''(x;\eta) = Q \times \begin{cases} 1 - 3\frac{\eta}{L}\left(1 - \frac{\eta}{2L}\right)\left(1 - \frac{x}{L}\right) & \text{if } 0 \le x < \eta, \tag{4.203a} \\[3mm] -3\frac{\eta}{L}\left(1 - \frac{\eta}{2L}\right)\left(1 - \frac{x}{L}\right) & \text{if } \eta < x \le L. \tag{4.203b} \end{cases}$$

The bending moment is nonzero (zero) at the clamped (pinned) support specifying the reaction moments (4.204a) [(4.204b)],

$$M_{2-} = -M_2(0;\eta) = Q\left[1 - 3\frac{\eta}{L}\left(1 - \frac{\eta}{2L}\right)\right], \quad M_{2+} = M_2(L;\eta) = 0, \tag{4.204a,b}$$

and has a jump Q across the point of application of the concentrated moment (4.204c):

$$M_3(\eta + 0;\eta) - M_3(\eta - 0;\eta) = Q. \tag{4.204c}$$

The transverse force (4.205)

$$F_2(x;\eta) = EI\zeta_2'''(x;\eta) = 3\frac{Q}{L}\frac{\eta}{L}\left(1 - \frac{\eta}{2L}\right) - Q\delta(x - \eta) \tag{4.205}$$

specifies the reaction forces at the supports:

$$R_{2-} = F_2(0;\eta) = 3\frac{Q}{L}\frac{\eta}{L}\left(1 - \frac{\eta}{2L}\right) = F_2(L;\eta) = -F_{2+}, \tag{4.206a,b}$$

which have the same modulus and opposite signs (4.206a,b).

   If the concentrated moment is at equal distance (Figure 4.9c) from the supports (4.207a), the shape of the elastica (4.199a,b) simplifies to (4.207b,c):

$$\eta = \frac{L}{2}: \quad \zeta_2\left(x;\frac{L}{2}\right) = \frac{Q}{16EI} \times \begin{cases} -x^2\left(1 - 3\dfrac{x}{L}\right) & \text{if } 0 \le x \le \dfrac{L}{2}, \\[4mm] L\left[8x - 2L - 3\dfrac{x^2}{L}\left(3 - \dfrac{x}{L}\right)\right] & \text{if } \dfrac{L}{2} \le x \le L. \end{cases} \tag{4.207a–c}$$

The corresponding slope is

$$\zeta_2'\left(x;\frac{L}{2}\right) = \frac{Q}{16EI} \times \begin{cases} -x\left(2 - 9\dfrac{x}{L}\right) & \text{if } 0 \le x \le \dfrac{L}{2}, & \text{(4.208a)} \\[4mm] 8L - 9x\left(2 - \dfrac{x}{L}\right) & \text{if } \dfrac{L}{2} \le x \le L. & \text{(4.208b)} \end{cases}$$

The deflection (slope) at the mid-position where the concentrated moment is applied is (4.209a) [(4.209b)]:

$$\zeta_2\left(\frac{L}{2};\frac{L}{2}\right) = \frac{QL^2}{128EI} = 0.0078125\frac{QL^2}{EI}, \quad \zeta_2'\left(\frac{L}{2};\frac{L}{2}\right) = \frac{5QL}{64EI} = 0.078125\frac{QL}{EI}. \tag{4.209a,b}$$

The slope is zero (4.210a) [nonzero (4.210b)] at the clamped (pinned) end:

$$\zeta_2'\left(0;\frac{L}{2}\right) = 0, \quad \zeta_2'\left(L;\frac{L}{2}\right) = -\frac{QL}{16EI} = -0.06667\frac{QL}{EI}. \tag{4.210a,b}$$

The bending moment (4.211a,b)

$$-M_2\left(x;\frac{L}{2}\right) = EI\zeta_2''\left(x;\frac{L}{2}\right) = -\frac{Q}{8} \times \begin{cases} 1-9\dfrac{x}{L} & \text{if } 0 \le x \le \dfrac{L}{2}, \quad (4.211a)\\[3mm] 9\left(1-\dfrac{x}{L}\right) & \text{if } \dfrac{L}{2} \le x \le L, \quad (4.211b) \end{cases}$$

determines the reaction moment at the clamped support (4.211c):

$$N_{2-} = -M_2\left(0;\frac{L}{2}\right) = -\frac{Q}{8}, \quad N_{2+} = M_2\left(L;\frac{L}{2}\right) = 0. \qquad (4.211c,d)$$

The transverse force (4.212a),

$$F_2\left(x;\frac{L}{2}\right) = \frac{9}{8}\frac{Q}{L} - Q\delta\left(x-\frac{L}{2}\right), \quad R_{2-} = F_2\left(0;\frac{L}{2}\right) = \frac{9Q}{8L} = F_2\left(L;\frac{L}{2}\right) = -R_{2+}, \qquad (4.212a\text{–}c)$$

determines the reaction force at the supports (4.212b,c) that have the same modulus and opposite signs.

4.5.5 Clamped-Clamped Bar with a Concentrated Torque

Substituting (4.144a,b) in (4.173b) specifies *the shape of the elastica (Figure 4.8d) for the weak (2.12a) bending of a uniform (2.10a) bar (2.90a) clamped at both ends (4.51a–d) by (4.172b) a concentrated torque Q at x = η:*

$$\zeta_1(x;\eta) = \frac{Q}{2EI} \times \begin{cases} x^2\left(1-\dfrac{\eta}{L}\right)\left[1-\dfrac{\eta}{L}\left(3-2\dfrac{x}{L}\right)\right] & \text{if } 0 \le x \le \eta, \quad (4.213a)\\[4mm] \eta\left(1-\dfrac{x}{L}\right)^2\left[2x\left(1-\dfrac{\eta}{L}\right)-\eta\right] & \text{if } \eta \le x \le L. \quad (4.213b) \end{cases}$$

The corresponding slope is

$$\zeta_4'(x;\eta) = \frac{Q}{EI} \times \begin{cases} x\left(1-\dfrac{\eta}{L}\right)\left[1-3\dfrac{\eta}{L}\left(1-\dfrac{x}{L}\right)\right] & \text{if } 0 \le x \le \eta, \quad (4.214a)\\[4mm] \eta\left(1-\dfrac{x}{L}\right)\left[1-3\dfrac{x}{L}\left(1-\dfrac{\eta}{L}\right)\right] & \text{if } \eta \le x \le L, \quad (4.214b) \end{cases}$$

which satisfies a reciprocity principle, allowing interchange of (η;x). The deflection (slope) at the point of application of the moment is (4.215a) [(4.215b)]:

$$\zeta^4(\eta;\eta) = \frac{Q\eta^2}{2EI}\left(1-\frac{\eta}{L}\right)^2\left(1-\frac{2\eta}{L}\right), \quad \zeta_1'(\eta;\eta) = \frac{Q\eta}{EI}\left(1-\frac{\eta}{L}\right)\left[1-3\frac{\eta}{L}\left(1-\frac{\eta}{L}\right)\right]; \qquad (4.215a,b)$$

the slope vanishes at the clamped supports. The bending moment,

$$-M_1(x;\eta) = EI\zeta_1''(x;\eta) = Q \times \begin{cases} \left(1-\dfrac{\eta}{L}\right)\left[1-3\dfrac{\eta}{L}\left(1-2\dfrac{x}{L}\right)\right] & \text{if } 0 \le x < \eta, & (4.216a) \\[3mm] -\dfrac{\eta}{L}\left[1+3\left(1-\dfrac{\eta}{L}\right)\left(1-2\dfrac{x}{L}\right)\right] & \text{if } \eta \le x \le L, & (4.216b) \end{cases}$$

specifies the reaction moments at the clamped supports (4.217a,b),

$$N_{1-} = -M_1(0;\eta) = Q\left(1-\frac{\eta}{L}\right)\left(1-3\frac{\eta}{L}\right), \quad N_{1+} = M_1(L;\eta) = -Q\frac{\eta}{L}\left(2-3\frac{\eta}{L}\right), \qquad (4.217a,b)$$

and has a jump Q at the point of application of the moment:

$$M_1(\eta+0;\eta) - M_1(\eta-0;\eta) = Q. \qquad (4.217c)$$

The transverse force (4.218),

$$F_1(x;\eta) = EI\zeta_1'''(x;\eta) = 6\frac{Q}{L}\frac{\eta}{L}\left(1-\frac{\eta}{L}\right) - Q\delta(x-\eta), \qquad (4.218)$$

specifies the reaction forces at the supports:

$$R_{1-} = F_1(0;\eta) = \frac{6Q\eta}{L^2}\left(1-\frac{\eta}{L}\right) = F_1(L;\eta) = -R_{1+}, \qquad (4.219a,b)$$

which are equal in modulus with opposite signs (4.219a,b).

   In the case (Figure 4.9d) of a concentrated moment at equal distance from the supports (4.220a), the shape of the elastica (4.213a,b) simplifies to (4.220b,c):

$$\eta = \frac{L}{2}: \quad \zeta_1\left(x;\frac{L}{2}\right) = \frac{Q}{8EI} \times \begin{cases} -x^2\left(1-2\dfrac{x}{L}\right) & \text{if } 0 \le x \le \dfrac{x}{L}, \\[3mm] \left(1-\dfrac{x}{L}\right)^2 L(2x-L) & \text{if } \dfrac{x}{L} \le x \le L. \end{cases} \qquad (4.220a\text{--}c)$$

The corresponding slope is

$$\zeta_1'\left(x;\frac{L}{2}\right) = \frac{Q}{4EI} \times \begin{cases} -x\left(1-3\dfrac{x}{L}\right) & \text{if } 0 \le x \le \dfrac{L}{2}, & (4.221a) \\[3mm] (L-x)\left(2-3\dfrac{x}{L}\right) & \text{if } \dfrac{L}{2} \le x \le L. & (4.221b) \end{cases}$$

The deflection (slope) at the point of application of the moment is zero (4.222a) [positive (4.222b)]:

$$\zeta_1\left(\frac{L}{2};\frac{L}{2}\right)=0,\quad \zeta_1'\left(\frac{L}{2};\frac{L}{2}\right)=\frac{QL}{16EI}=0.06667\frac{QL}{EI}. \tag{4.222a,b}$$

The bending moment (4.223a,b),

$$-M_1\left(x;\frac{L}{2}\right)=EI\zeta_1''\left(x;\frac{L}{2}\right)=Q\times\begin{cases}\dfrac{3x}{2L}-\dfrac{1}{4} & \text{if } 0\le x<\dfrac{L}{2}, \tag{4.223a}\\[3mm] \dfrac{3x}{2L}-\dfrac{5}{4} & \text{if } \dfrac{L}{2}<x\le L, \tag{4.223b}\end{cases}$$

has a jump Q at the point of application of the concentrated moment:

$$M_1\left(\frac{L}{2}+0;\frac{L}{2}\right)-M_1\left(\frac{L}{2}-0;0\right)=Q. \tag{4.224a}$$

The reaction moments are equal at the two clamped supports:

$$N_{1-}=-M_1\left(0;\frac{L}{2}\right)=-\frac{Q}{4}=M_1\left(L;\frac{L}{2}\right)=N_{1+}. \tag{4.224b,c}$$

The transverse force (4.225),

$$F_1\left(x;\frac{L}{2}\right)=EI\zeta_1'''\left(x;\frac{L}{2}\right)=\frac{3Q}{2L}-Q\delta\left(x-\frac{L}{2}\right), \tag{4.225}$$

determines the reaction forces at the supports:

$$R_{1-}=F_1\left(0;\frac{L}{2}\right)=\frac{3Q}{2L}=F_1\left(0;\frac{L}{2}\right)=-R_{1+}, \tag{4.226a,b}$$

which have equal modulus and opposite signs (4.226a,b).

4.5.6 Comparison of a Concentrated Torque with Different Supports

The weak bending of a uniform bar by a concentrated moment at an arbitrary position is considered in Table 4.5 for the four classical cases of supports: (1) clamped-free (Subsection 4.5.2); (2) pinned-pinned (Subsection 4.5.3); (3) clamped-pinned (Subsection 4.5.4); (4) clamped-clamped (Subsection 4.5.5). The comparison of the four cases is made in Table 4.6 with the concentrated moment at the free end (1) [at equal distance from the support (2–4)]. *In the case of symmetric boundary conditions, for example, a clamped-clamped (pinned-pinned) bar with (1) a concentrated force at the middle point (6.227a), the shape of the elastica [Figure 4.7d (b)] is symmetric (6.227b):*

$$f(x)=P\delta\left(x-\frac{L}{2}\right):\quad G\left(x;\frac{L}{2}\right)=G\left(L-x;\frac{L}{2}\right); \tag{4.227a,b}$$

TABLE 4.5   Weak Simple Bending of an Elastic Bar by a Concentrated Moment

Case	1	2	3	4
Load	Moment Q at η	Moment Q at η	Moment Q at η	Moment Q at η
Subsection	4.5.5	4.5.4	4.5.3	4.5.2
Figure	4.8d	4.8c	4.8b	4.8a
Type of support	Clamped-clamped	Clamped-pinned	Clamped-pinned	Clamped-free
Degrees of freedom	−2	−1	0	0
Type of problem	Hyperstatic incompatible	Hyperstatic indeterminate	Isostatic	Isostatic
Boundary conditions	(4.51a–d)	(4.52a–d)	(4.53a–d)	(4.90b–e)
Jump conditions	(4.171a,b)	(4.171a,b)	(4.171a,b)	(4.171a,b)
Continuity conditions	(4.179b, 4.180b)	(4.179b, 4.180b)	(4.179b, 4.180b)	(4.179b, 4.180b)
Shape of the elastica	(4.213a,b)	(4.199a,b)	(4.189a,b)	(4.181a,b)
Slope of the elastica	(4.214a,b)	(4.200a,b)	(4.190a,b)	(4.182a,b)
Deflection at η	(4.215a)	(4.201a)	(4.191a)	(4.185a)
Slope at η	(4.215b)	(4.202a)	(4.191b)	(4.185c)
Bending moment	(4.216a,b)	(4.203a,b)	(4.193a,b)	(4.183a)
Transverse force	(4.218)	(4.205)	(4.195a)	(4.184a)
Shear stress	(4.172b)	(4.172b)	(4.172b)	(4.172b)
Reaction forces	(4.219a,b)	(4.206a,b)	(4.195b,c)	(4.184b)
Reaction moments	(4.217a,b)	(4.204a,b)	(4.194a,b)	(4.183b)

Note: Comparison (for the same four cases as Tables 4.2 through 4.4) of the weak bending of a uniform bar by a concentrated moment at arbitrary position.

TABLE 4.6   Bending of a Bar by a Concentrated Moment at the Middle (or End)

Case	1	2	3	4
Load	Moment Q at $\eta = \dfrac{L}{2}$	Moment Q at $\eta = \dfrac{L}{2}$	Moment Q at $\eta = \dfrac{L}{2}$	Moment Q at $\eta = L$
Subsection	4.5.5	4.5.4	4.5.3	4.5.2
Figure	4.9d	4.9c	4.9b	4.9a
Type of support	Clamped-clamped	Clamped-pinned	Clamped-pinned	Clamped-free
Degrees of freedom	−2	−1	0	0
Type of problem	Hyperstatic incompatible	Hyperstatic indeterminate	Isostatic	Isostatic
Shape of the elastica	(4.220b,c)	(4.207b,c)	(4.196b,c)	(4.186d)
Slope of the elastica	(4.221a,b)	(4.208a,b)	(4.197b,c)	(4.188c)
Deflection at η	(4.222a)	(4.209a)	(4.198a)	(4.186b)
Slope at η	(4.222b)	(4.209b)	(4.198b)	(4.188d)
Bending moment	(4.223a,b)	(4.211a,b)	(4.193a,b)	(4.186c)
Transverse force	(4.225)	(4.212a)	(4.195a)	(4.187b)
Shear stress	(4.197a)	(4.197a)	(4.197a)	(4.187a)
Reaction forces	(4.226a,b)	(4.212b,c)	(4.194a,b)	(4.184b)
Reaction moments	(4.224b,c)	(4.211c,d)	(4.195b,c)	(4.183b)

Note: As for Table 4.5 with the concentrated moment at: (1) the tip for the clamped-free bar; and (2) at equal distance from the supports in all other three cases.

$$f(x) = -Q\delta'\left(x - \frac{L}{2}\right); \quad \zeta\left(x; \frac{L}{2}\right) = -\zeta\left(L - x; \frac{L}{2}\right); \tag{4.227c,d}$$

and (2) a concentrated torque (4.227c) at the midpoint, the shape of the elastica [Figure 4.9d (b)] is skew-symmetric (4.227d). In case (1) [(2)], the symmetry (skew-symmetry) is relative to the middle point, at equal distance from the supports, where the concentrated force (moment) is applied.

The deflection is maximum (1) at the free end of a clamped-free bar with the concentrated moment (4.187a) applied also at the free end (4.186b) ≡ (4.228a). For a concentrated moment applied at the mid position (4.197a), the maximum deflection is (2) reduced at the free end by a factor 3/4 for a clamped-free bar (4.181b) ≡ (4.228b); (3) for a clamped-pinned bar (4.209a) ≡ (4.228c), it is reduced by a factor of 1/64; and (4, 5) for symmetric supports, that is, pinned-pinned (4.198a) ≡ (4.228d) [clamped-clamped (4.222a) ≡ (4.228c)], it is zero by symmetry:

$$\zeta_4(L;L) = \frac{QL^2}{2EI} > \frac{3QL^2}{8EI} = 0.375\frac{QL^2}{EI} = \frac{3}{4}\zeta_4(L;L) = \zeta_4\left(L;\frac{L}{2}\right)$$

$$> \frac{QL^2}{128EI} = 0.0078125\frac{QL^2}{2EI} = \frac{1}{64}\zeta_4(L;L) = \zeta_2\left(\frac{L}{2};\frac{L}{2}\right)$$

$$> 0 = \zeta_3\left(\frac{L}{2};\frac{L}{2}\right) = \zeta_1\left(\frac{L}{2};\frac{L}{2}\right). \tag{4.228a-e}$$

The slope is (1, 2) largest for a clamped-free bar with a concentrated moment at the free end (4.188d) ≡ (4.229a) [reduced by a factor of 1/2 when the concentrated moment is at the mid-position (4.182b) ≡ (4.229b)]; (3/4) for a pinned-pinned (clamped-pinned) bar at the point of application of the concentrated moment (4.198b) ≡ (4.229c) [(4.209b) ≡ (4.229d)], it is reduced by a factor of 1/12 (5/64); (5/6) for a clamped-pinned (pinned-pinned) bar at the pinned end (ends), by (4.210b) ≡ (4.229e) [(4.198d) ≡ (4.229g)] a factor of 1/16 (1/24); (7) the value (5) ≡ (7) applies to the slope of a clamped-clamped bar (4.222b) ≡ (4.229f) at the point of application of the concentrated moment:

$$\zeta_4'(L;L) = \frac{QL}{EI} > \frac{QL}{2EI} = \frac{1}{2}\zeta_4'(L;L) = \zeta_4'\left(L;\frac{L}{2}\right)$$

$$> \frac{QL}{12EI} = 0.08333\frac{QL}{EI} = \frac{1}{12}\zeta_4'(L;L) = \zeta_3'\left(\frac{L}{2};\frac{L}{2}\right)$$

$$> \frac{5QL}{64EI} = 0.078125\frac{QL}{EI} = \frac{5}{64}\zeta_4'(L;L) = \zeta_2'\left(\frac{L}{2};\frac{L}{2}\right)$$

$$> \frac{QL}{16EI} = 0.0625\frac{QL}{EI} = \frac{1}{16}\zeta_4'(L;L) = -\zeta_2'\left(L;\frac{L}{2}\right) = \zeta_1'\left(\frac{L}{2};\frac{L}{2}\right)$$

$$> \frac{QL}{24EI} = 0.04167\frac{QL}{EI} = \frac{1}{24}\zeta_4'(L;L) = -\zeta_3'\left(0;\frac{L}{2}\right) = -\zeta_3'\left(L;\frac{L}{2}\right). \tag{4.229a-g}$$

The reaction moment (1) equals the applied concentrated moment (is zero) at the clamped support (free end) for a clamped-free bar (4.183b) ≡ (4.230a) [(4.178b) ≡ (4.231d)]; (2, 3) for a clamped-pinned bar (4.211c) ≡ (4.230c) [clamped-clamped bar (4.224b,c) ≡ (4.230b, 4.231a)], it is 1/8 (1/4) with opposite sign at the clamped support (supports); and (4) it is zero for all pinned supports (4.230d, 4.231b–d):

$$N_{4-} = Q > \frac{Q}{4} = -N_{1-} > \frac{Q}{8} = -N_{2-} > 0 = N_{1-}, \tag{4.230a-d}$$

$$-N_{1+} = \frac{Q}{4} > 0 = N_{2+} = N_{4-} = N_{3+} = N_{4+},$$

(4.231a–d)

$$R_{1-} = \frac{3Q}{2L} > R_{2-} = \frac{9Q}{8L} > R_{3-} = \frac{Q}{L} > 0 = R_{4-},$$

(4.232a–d)

$$-R_{1+} = \frac{3Q}{2L} > R_{2+} = \frac{9Q}{8L} > R_{3+} = \frac{Q}{L} > 0 = R_{4+}.$$

(4.233a–d)

The reaction force is (1) zero for the clamped-free bar (4.184b), except at the point of application of the moment (4.232d, 4.233d); (2) in all other cases, it has opposite signs at the two supports with the same modulus; and (3) the modulus of the reaction force increases from the pinned-pinned bar (4.195b,c) ≡ (4.232c, 4.233c) by a factor of 9/8 (3/2) for the clamped-pinned (4.212b,c) ≡ (4.232b, 4.233b) [clamped-clamped (4.226a,b) ≡ (4.232a, 4.233a) bar.

4.6 Tangential Tension along a Beam with Pinned Ends

The clamped-free bar under a concentrated moment (Subsection 4.5.2), force (Subsection 4.4.1), or uniform load (Subsection 4.3.4) is not subject to longitudinal tension arising from the supports for weak or strong bending. The bar clamped or pinned at both ends and subject to any load, for example, own weight (Section 4.3) or concentrated force (moment) [Section 4.4 (Section 4.5)], is subject to a longitudinal traction (4.18b) due to its extension; the latter can be neglected by assuming that the maximum deflection is small compared with the length scale of the cross section (4.20c). If this restriction is dropped, then the equation of the elastica (4.16b) is replaced by (4.13f), which specifies weak bending of a beam with the longitudinal tension T as a parameter (Subsection 4.6.1). The latter is determined (Subsection 4.6.2) after the solution is found, by substitution into (4.18b); the beam includes the string (bar) or opposite limits (Subsection 4.6.3) of dominant (negligible) tension to the second (fourth) order (Subsection 4.6.4).

4.6.1 Shape of a Beam under Longitudinal Tension

The method of calculation of the longitudinal tension is illustrated for the weak (4.12a) bending of a uniform (4.10a) beam (4.13f) pin-joined at both ends (4.53a–d) with a force P concentrated at the middle (Figure 4.10):

$$EI\zeta''''(x) - T\zeta''(x) = P\delta\left(x - \frac{L}{2}\right).$$

(4.234)

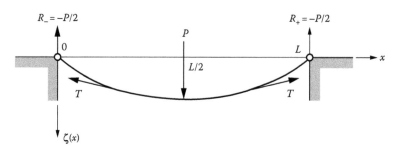

FIGURE 4.10 If a bar has at least one free (Figures 4.4d, 4.5a, 4.6a, 4.7a, 4.8a and 4.9a) or sliding end, the bending does not cause a longitudinal stress. Bending does cause a longitudinal tension if both ends of the bar are fixed, for example, in the following three out of four classical cases of support: (1) pinned-pinned (Figures 4.4c, 4.5b, 4.6b, 4.7b, 4.8b and 4.9b); (2) clamped-pinned (Figures 4.4b, 4.5c, 4.6c, 4.7c, 4.8c and 4.9c); and (3) clamped-clamped (Figures 4.4a, 4.5d, 4.6d, 4.7d, 4.8d and 4.9d).

The bending of the bar held at the supports causes an extension, and hence a traction, so that $T > 0$ in (4.234). The equation (4.234) may be integrated twice:

$$EI\zeta'''(x) - T\zeta'(x) = A + PH\left(x - \frac{L}{2}\right) = \begin{cases} A & \text{if } x < \dfrac{L}{2}, & \text{(4.235a)} \\[2ex] A + P & \text{if } x > \dfrac{L}{2}, & \text{(4.235b)} \end{cases}$$

$$EI\zeta''(x) - T\zeta(x) = \begin{cases} Ax & \text{if } 0 \le x \le \dfrac{L}{2}, & \text{(4.236a)} \\[2ex] (A+P)(x-L) & \text{if } \dfrac{L}{2} \le x \le L, & \text{(4.236b)} \end{cases}$$

where (4.236a,b) hold because the bending moment vanishes (4.53b,d) at both pin-joined ends (4.53a,c). The constant of integration A is determined (4.237a) from the continuity of the curvature at the midpoint:

$$\frac{1}{2}AL = -\frac{1}{2}(A+P)L, \quad A = -\frac{P}{2} = F(0) = R_- = -F(L) = R_+, \tag{4.237a–c}$$

and is equal (4.237b) to half the concentrated force with opposite sign and specifies the reaction force at the supports (4.237c).

Since the beam has a symmetric shape (4.238a), the boundary conditions (4.238b,c)

$$\zeta(x) = \zeta(L-x): \quad \zeta(0) = 0 = \zeta'\left(\frac{L}{2}\right), \tag{4.238a–c}$$

can be used to solve (4.236a) $\equiv$ (4.239b) in half the range (4.239a):

$$0 \le x \le \frac{L}{2}: \quad \zeta''(x) - k^2\zeta(x) = -\frac{Px}{2EI}, \quad k^2 \equiv \frac{T}{EI}, \tag{4.239a–c}$$

where the constant k in (4.239c) has the dimensions of the inverse of a length. The solution of (4.239b) with zero on the l.h.s. is the general integral specified by the l.h.s. of (4.240):

$$\zeta(x) - B\cosh(kx) - C\sinh(kx) = \frac{Px}{2EIk^2} = \frac{Px}{2T}. \tag{4.240}$$

The r.h.s. of (4.240) is a particular integral of (4.239b). The sum of the general and particular integrals form the complete integral (4.240) of the forced linear ordinary differential equation with constant coefficients (4.239b). The boundary conditions determine the constants of integration in (4.240) and lead to the relation between the load P and the tension T.

4.6.2 Relation between the Transverse Load and the Longitudinal Tension

The constants of integration are determined from the boundary conditions (4.238b,c):

$$0 = \zeta(0) = B, \quad 0 = \zeta'\left(\frac{L}{2}\right) = kC\cosh\left(\frac{1}{2}kL\right) + \frac{P}{2T}. \tag{4.241a,b}$$

Substituting (4.241a,b) in (4.240) leads to *the shape of the elastica of a uniform beam (4.10a) pinned at both ends (4.53a–d) and subject to weak (4.12a) bending (4.11a) due to a concentrated force P in the middle x = L/2 that is given by*

$$\zeta(x) = \frac{P}{2T}\left[x - \frac{\sinh(kx)}{k\cosh(kL/2)}\right],$$

(4.242)

where the longitudinal tension satisfies the transcendental relation

$$3 - \frac{6}{kL}\tanh\left(\frac{kL}{2}\right) - \tanh^2\left(\frac{kL}{2}\right) = \frac{16T^3}{P^2 ES},$$

(4.243)

involving k in (4.239c). The equation for the longitudinal tension (4.243) is obtained substituting (4.242) into (4.18b),

$$\frac{2TL}{ES} = 2\int_0^{L/2}\left\{\frac{P}{2T}\left[1 - \cosh(kx)\operatorname{sech}\left(\frac{kL}{2}\right)\right]\right\}^2 dx,$$

(4.244)

and performing the integration

$$\frac{16T^3}{P^2 ES} = \frac{4}{L}\int_0^{L/2}\left[1 - 2\cosh(kx)\operatorname{sech}\left(\frac{kL}{2}\right) + \cosh^2(kx)\operatorname{sech}^2\left(\frac{kL}{2}\right)\right]dx$$

$$= \frac{4}{L}\int_0^{L/2}\left[1 + \frac{1}{2}\operatorname{sech}^2\left(\frac{kL}{2}\right) - 2\cosh(kx)\operatorname{sech}\left(\frac{kL}{2}\right) + \frac{1}{2}\cosh(2kx)\operatorname{sech}^2\left(\frac{kL}{2}\right)\right]dx$$

$$= \frac{4}{L}\left[x + \frac{x}{2}\operatorname{sech}^2\left(\frac{kL}{2}\right) - \frac{2}{k}\sinh(kx)\operatorname{sech}\left(\frac{kL}{2}\right) + \frac{1}{4k}\sinh(2kx)\operatorname{sech}^2\left(\frac{kL}{2}\right)\right]_0^{L/2}$$

$$= 2 + \operatorname{sech}^2\left(\frac{kL}{2}\right) - \frac{8}{kL}\sinh\left(\frac{kL}{2}\right)\operatorname{sech}\left(\frac{kL}{2}\right) + \frac{2}{kL}\sinh\left(\frac{kL}{2}\right)\cosh\left(\frac{kL}{2}\right)\operatorname{sech}^2\left(\frac{kL}{2}\right)$$

$$= 3 - \frac{6}{kL}\tanh\left(\frac{kL}{2}\right) - \tanh^2\left(\frac{kL}{2}\right);$$

(4.245)

this coincides with (4.243) ≡ (4.245) specifying the relation between the transverse concentrated force and the resulting longitudinal tension.

4.6.3 Longitudinal Tension due to Bending with Fixed Supports

The case of a string (4.15a–c) corresponds (4.246a) to

$$1 \ll k^2 L^2 = \frac{TL^2}{EI} = \frac{T}{T_*} : \quad T_* = \frac{EI}{L^2},$$

(4.246a,b)

the tension T being large (4.246a) compared with the value (4.15a) $\equiv$ (4.246b); thus, the shape of the beam (4.242) for large kL is given by

$$\zeta(x) = \frac{P}{2T}\left\{x - \frac{1}{k}\exp\left[k\left(x - \frac{L}{2}\right)\right]\right\}, \tag{4.247}$$

where (1) the first term is a straight line that coincides with (2.46c) for an infinitely thin string; and (2) the second term in (4.247) indicates the order of magnitude of the first correction due to the finite thickness of the string and is small for large k on account of (4.239a). The limit opposite to the string is the bar under a small tension, that is, small k in (4.239c), to order four in (4.248a) where (4.239a,c, 4.246b) are used. This leads to the approximations (1/2) to the second order (4.248b,c) for the terms in (4.242):

$$k^4 x^4 \le \frac{k^4 L^4}{16} = \frac{T^2}{16 T_*^2} \ll 1: \quad \sinh(kx) = kx + \frac{1}{6}k^2 x^2, \tag{4.248a,b}$$

$$\cosh\left(\frac{kL}{2}\right) = 1 + \frac{k^2 L^2}{8}, \quad \tanh\left(\frac{kL}{2}\right) = \frac{kL}{2}\left(1 - \frac{k^2 L^2}{12} + \frac{k^4 L^4}{120}\right); \tag{4.248c,d}$$

and (3) to a higher order, that is, the fourth in (4.248d), instead of the second in (4.248b,c) for (4.243), since the second- and third-order terms cancel as will be shown in the sequel. Then (4.248b–d) are used as follows: (1) substitution of (4.248b,c) into (4.242) yields the elastica (4.249a) $\equiv$ (4.114b) for a bar with a concentrated force in the middle (4.114a):

$$\zeta(x) = \frac{Px}{4EI}\left(\frac{L^2}{4} - \frac{x^2}{3}\right), \quad \delta \equiv \zeta_{max} = \zeta\left(\frac{L}{2}\right) = \frac{PL^3}{48EI}, \tag{4.249a,b}$$

and the maximum deflection (4.249b), in agreement with (4.116b); (2) substitution of (4.248d) in (4.243) cancels the two terms of lowest order and leads to the next fourth order to (4.250a):

$$\frac{16T^3}{P^2 ES} = \frac{k^4 L^4}{60}; \quad T = \frac{P^2 L^4 S}{960 EI^2} = \frac{P^2}{960 ES}\left(\frac{L}{G}\right)^4, \quad I \equiv G^2 S. \tag{4.250a–c}$$

Substitution of (4.239c) in (4.250a) specifies the longitudinal traction (4.250b) that scales on the square of the transverse load, that is, a second-order effect. The **radius of gyration** of the cross section (4.250c) is the distance from the axis from which the concentrated area S or mass would produce the moment of inertia I. The longitudinal traction (4.250b) scales on the fourth power of the ratio of the length of the bar to the radius of gyration of the cross section; thus, the longitudinal traction is larger for shorter, thicker beams. For a **slender beam** *of length much larger than the radius of gyration (4.250c) of the cross section* $L^4 \gg G^4$, *the longitudinal tension is much larger (4.250b) than the transverse force* $T^2 \gg P^2$, *but its effect on the shape of the elastica can be neglected, as for a slender bar if it is sufficiently stiff, that is, has large (4.246b) bending stiffness (4.7a) and is not too long* $T_* \equiv EI/L^2 \gg T$. *The shape of the elastica of the beam (4.242) simplifies for longitudinal tension that is large (4.246a) [small (4.248a)] compared with the value (4.246b)* $\equiv$ *(4.14c) of a string (4.247) [a bar (4.249a,b)] to the order zero in $1/k$ (order three in k) in (4.239c). The latter involves the tension that is related to the load by the transcendental equation (4.243) whose simplest solution (4.250a–c) is obtained at order four. Thus, the order two (four) is needed to obtain the shape of the elastica (4.249a) [longitudinal tension (4.250a)] as shown next.*

4.6.4 Second (Third)-Order Approximation to the Elastica (Tension)

Substituting (4.248b,c) in the shape of the elastica for a beam (4.242) leads to order (4.248a) to

$$
\zeta(x) = \frac{P}{2T}\left[x - x\left(1 + \frac{k^2 x^2}{6}\right)\left(1 + \frac{k^2 L^2}{8}\right)^{-1}\right]
$$

$$
= \frac{Px}{2T}\left[1 - \left(1 + \frac{k^2 x^2}{6}\right)\left(1 - \frac{k^2 L^2}{8}\right)\right]
$$

$$
= \frac{Px}{2T}k^2\left(\frac{L^2}{8} - \frac{x^2}{6}\right) = \frac{Px}{4EI}\left(\frac{L^2}{4} - \frac{x^3}{3}\right), \tag{4.251}
$$

that is, the shape of a bar (4.251) ≡ (4.249a) ≡ (4.114b); this confirms that the bar (4.251) corresponds to a beam (4.242) with the approximation (4.248a) to second order. The same approximation fails to determine the longitudinal tension in (4.243) because the terms cancel, and thus, it necessary to use the fourth-order approximation (4.248d) as shown next: (1) the starting point is the power series for the hyperbolic sine (II.7.7b) ≡ (4.252c) [cosine (II.7.6b) ≡ (4.252d)] in the variable (4.252a) with three terms, that is, to fourth order in (4.252b) instead of second order in (4.248a):

$$
z \equiv \frac{kL}{2},\ z^5 \ll 1:\quad \sinh z = z + \frac{z^3}{6} + \frac{z^5}{120},\quad \cosh z = 1 + \frac{z^2}{2} + \frac{z^4}{24}; \tag{4.252a–d}
$$

(2) the corresponding fourth-order approximation to the hyperbolic tangent is obtained using the geometric series (I.21.62c):

$$
\tanh z = z\left(1 + \frac{z^2}{6} + \frac{z^4}{120}\right)\left(1 + \frac{z^2}{2} + \frac{z^4}{24}\right)^{-1} = z\left(1 + \frac{z^2}{6} + \frac{z^4}{120}\right)\left(1 - \frac{z^2}{2} - \frac{z^4}{24} + \frac{z^4}{4}\right)
$$

$$
= z\left[1 - z^2\left(\frac{1}{2} - \frac{1}{6}\right) + z^4\left(\frac{1}{4} - \frac{1}{24} - \frac{1}{12} + \frac{1}{120}\right)\right] = z\left(1 - \frac{z^2}{3} + \frac{2z^4}{15}\right), \tag{4.253}
$$

in agreement with (4.253) ≡ (II.7.39c); and (3) substitution of (4.253, 4.252a) in (4.243) leads to

$$
\frac{16T^3}{P^2 ES} = 3 - \frac{3}{z}\tanh z - \tanh^2 z = 3 - 3\left(1 - \frac{z^2}{3} + \frac{2z^4}{15}\right) - z^2\left(1 - \frac{z^2}{3} + \frac{2z^4}{15}\right)^2
$$

$$
= z^2 - \frac{2}{5}z^4 - z^2\left(1 - \frac{2z^2}{3}\right) = 2z^4\left(\frac{1}{3} - \frac{1}{5}\right) = \frac{4z^4}{15}, \tag{4.254}
$$

in agreement with (4.254, 4.252a) ≡ (4.250a). This confirms that *two (three) terms of the series (4.248b,c) [(4.248d)], that is, the* second (4.248a) *[fourth (4.252b)]-order approximation in (4.252a) is needed to determine the shape of the elastica [(4.251) ≡ (4.249a) ≡ (4.114b)] [the longitudinal tension (4.254)≡(4.250a) for a bar from those (4.242) [(4.243)] of a beam.*

4.7 Nonlinear Bending by Concentrated Moments

The shape of the elastica can be considered for the weak (strong) bending of a bar, that is, when [Sections 4.3 through 4.6 (Sections 4.7 through 4.9)] the slope is (is not) small. The weak (strong) bending can be considered for various loads such as uniform [Section 4.3 (Section 4.8)] or a concentrated force [Section 4.4 (Section 4.9)] or moment [Section 4.5 (Section 4.7)]. The shape of the elastica for the strong bending of a nonuniform bar can be obtained for arbitrary loading in terms of hyperelliptic integrals (Subsection 4.7.1). The simplest loading case is concentrated moments: (1) in the case of concentrated moments, between their points of application, the bending moment (4.4a) is constant; (2) for a uniform bar (4.10a), this implies a constant curvature. The shape of the elastica of a uniform bar subject to concentrated moments consists of circular arcs (Subsections 4.7.1 through 4.7.6) matched at the points of application of the moments (Subsections 4.7.7 through 4.7.9). This is analogous to the shape of an elastic string (membrane) subject to a uniform transverse shear stress (Section 2.7) [to a constant pressure difference (Subsections II.6.2.3 through II.6.2.7)], leading to a shape that is a circular arc (a spherical shell). In the case of the nonlinear bending by a single concentrated moment (Subsection 4.7.2), the maximum that a uniform bar can withstand for a vertical slope is taken as the value unity for a dimensionless nonlinearity parameter (Subsection 4.7.3). The shape of the elastica consists of a circular arc (straight segment) for constant (zero) bending moment before (after) the point of application of the concentrated moment [Subsection 4.7.4 (Subsection 4.7.5)]; the linear approximation corresponds to the lowest order in the nonlinearity parameter, and all higher orders can be obtained (Subsection 4.7.6). The case of strong bending of a uniform bar by two concentrated moments (Subsection 4.7.7) involves matching at their points of application (Subsection 4.7.8) and shows the effects of their relative magnitude (Subsection 4.7.9).

4.7.1 General Solution for Strong Bending of a Nonuniform Bar

In the absence of tension (4.255a), the shear stress (4.9b) satisfies the nonlinear equation (4.255b) for the strong bending of a nonuniform bar that may have cross section and material properties varying longitudinally:

$$T = 0: \quad f(x) = \left\{ E(x)I(x) \left[\zeta' \left| 1 + \zeta'^2 \right|^{-1/2} \right]' \right\}''. \tag{4.255a,b}$$

Two integrations yield

$$\left[\zeta' \left| 1 + \zeta'^2 \right|^{-1/2} \right]' = \frac{1}{E(x)I(x)} \left[C_1 + C_2 x + \int^x d\xi \int^\xi d\eta\, f(\eta) \right] \equiv A(x; C_1, C_2), \tag{4.256}$$

involving two constants of integration. A further integration introduces a third constant of integration in (4.257a):

$$\zeta' \left| 1 + \zeta'^2 \right|^{-1/2} = C_3 + \int^x A(\xi; C_1, C_2)\, d\xi \equiv B(x; C_1, C_2, C_3); \tag{4.257a}$$

$$\zeta'(x) = \left| \left[B(x; C_1, C_2, C_3) \right]^{-2} - 1 \right|^{-1/2}, \tag{4.257b}$$

solving (4.257a) for (4.257b) allows a fourth integration:

$$\zeta(x) = C_4 + \int\limits^{x} \left| \left[B(\xi;C_1,C_2,C_3) \right]^{-2} - 1 \right|^{-1/2} d\xi. \tag{4.258}$$

Thus, (4.258;4.257a) specifies the shape of the elastica of a nonuniform bar (4.255a) under strong bending by an arbitrary shear stress (4.256). The four constants of integration determined from the nonlinear boundary conditions are (1) the same as the linear for a clamped-clamped (4.51a–d), clamped-pinned (4.52a–d), and pinned-pinned (4.53a–d) bar; and (2) for a clamped-free bar, the boundary conditions at the free end are (4.259c) [(4.259d)] with a concentrated torque Q (force P):

$$\zeta(0) = 0 = \zeta(L): \quad Q = M(L) = -\zeta''(L) \left| 1 + \left[\zeta'(L) \right]^2 \right|^{-3/2}, \tag{4.259a–c}$$

$$P = -F(L) = M'(L) = -\zeta'''(L) \left| 1 + \left[\zeta'(L) \right]^2 \right|^{-3/2} + 3\zeta'(L) \left[\zeta''(L) \right]^2 \left| 1 + \left[\zeta'(L) \right]^2 \right|^{-5/2}$$

$$= \left| 1 + \left[\zeta'(L) \right]^2 \right|^{-5/2} \left\{ 3\zeta'(L) \left[\zeta'(L) \right]^2 - \zeta'''(L) \left\{ 1 + \left[\zeta'(L) \right]^2 \right\} \right\}, \tag{4.259d}$$

where the r.h.s. of (4.259d) ≡ (4.6a) is the derivative of the r.h.s. of (4.259c) ≡ (4.4a). In the linear (4.259e) case, (4.259a,b) ≡ (4.54a,b) are unchanged and (4.259c,d) simplify to (4.259f,g):

$$\left[\zeta'(L) \right]^2 \ll 1: \quad Q = -\zeta''(L), \quad P = -\zeta'''(L); \tag{4.259e–g}$$

$$P = 0 = Q: \quad \zeta''(L) = 0 = \zeta'''(L), \tag{4.259h–k}$$

In the nonlinear case, (4.90d,e) ≡ (4.259j,k) hold at the free end without force or moment (4.259h,i). If E, I, f are polynomials in x, the functions (4.256, 4.257a) are rational, and the shape (4.258) of the elastica is specified by a hyperelliptic integral (Section I.39.7). Several examples of nonlinear bending of a bar are given in the sequel (Sections 4.7 through 4.9).

4.7.2 Bar with One Clamped and One Free End

Consider a bar clamped at one end and free at the other end (Figure 4.11a), subject to a concentrated torque Q at position η; the bending moment Q applies at the support and must vanish at the free end, implying a jump Q across $x = \eta$; the exact nonlinear (4.2a) equation of the elastica (4.4a) is

$$EI\zeta'' \left| 1 + \zeta' \right|^{-3/2} = -M(x) = Q - QH(x - \eta) = \begin{cases} Q & \text{if } x < \eta, \tag{4.260a} \\ 0 & \text{if } x > \eta, \tag{4.260b} \end{cases}$$

since the bending moment is (4.260a) [(4.260b)] before (after) the position $x = \eta$. Considering the part of the bar between the clamped support and the concentrated torque (4.261a), the shape of the elastica is specified by the solution of the nonlinear second-order differential equation (4.261b):

$$0 \le x < \eta: \quad \frac{1}{a} \equiv \frac{Q}{EI} = \zeta'' \left| 1 + \zeta'^2 \right|^{-3/2} = \left\{ \zeta' \left| 1 + \zeta'^2 \right|^{-1/2} \right\}', \tag{4.261a,b}$$

where the constant a has the dimensions of length.

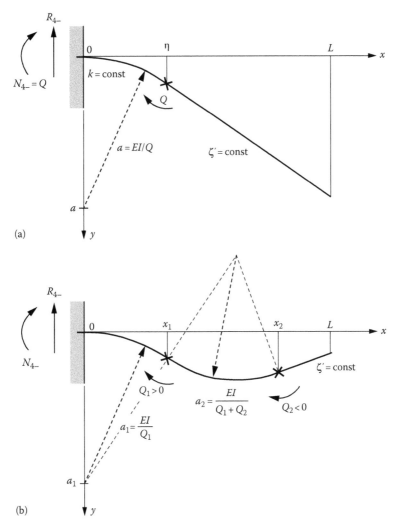

FIGURE 4.11 The strong bending of a uniform bar by concentrated moments corresponds to an elastica consisting of circular arcs, except for zero bending moment when it becomes a straight line. Thus the shape of the elastica for the strong bending of a uniform clamped-free bar by a concentrated moment at an arbitrary position (a) leads to an elastica whose shape is a circular arc before (straight line after) the point of application of the moment. In the case of two concentrated moments (b) the circular arcs may have different radii before (after) the point of application of the first moment (up to the point of application of the second moment). In both cases the tangent (curvature) is continuous (discontinuous) across the point of application of the concentrated moment. After the point of application of the last moment the elastica is straight with the same tangent.

4.7.3 Maximum Torque That the Bar Can Withstand

One method of solution is similar to the large deflection of a uniform string by a uniform load (Section 2.7), with different boundary conditions: (1) integrating (4.261b) with the first clamping boundary condition (4.262a) leads to (4.262b):

$$\zeta'(0) = 0: \quad \frac{x}{a} = \zeta' \left| 1 + \zeta'^2 \right|^{-1/2}, \quad \frac{d\zeta}{dx} \equiv \zeta' = \frac{x}{a} \left| 1 - \left(\frac{x}{a} \right)^2 \right|^{-1/2} = \frac{Qx}{EI} \left| 1 - \left(\frac{Qx}{EI} \right)^2 \right|^{-1/2}, \quad (4.262a\text{-}c)$$

which can be solved for ζ', leading to (4.262c); and (2) using the second clamping boundary condition (4.263a), the integration of (4.262b) leads to the exact shape (4.263b) of the elastica:

$$\zeta(0)=0: \quad \zeta(x)=a\left\{1-\left|1-\left(\frac{x}{a}\right)^2\right|^{1/2}\right\}=\frac{EI}{Q}\left\{1-\left|1-\left(\frac{Qx}{EI}\right)^2\right|^{1/2}\right\}. \tag{4.263a,b}$$

The solution (4.263b) is real iff the condition (4.264a) is met. Thus, *a bar of cross section with moment of inertia I made of a material with the Young's modulus E can support a concentrated torque at a distance* η *from its clamped support not exceeding (4.264a) for a finite slope in (4.262c):*

$$Q < EI\eta; \quad \zeta'^2 \ll 1 \Rightarrow \left(\zeta_{\max}\right)^2 \ll L^2 \Leftrightarrow Q^2 \ll \left(EIL\right)^2, \tag{4.264a–d}$$

The linear approximation of weak bending (4.12a) $\equiv$ (4.264b) implies that (1) the maximum deflection is small relative to the length, and (2) hence (4.263b) the torque (4.264d), much less than the valve EIL. Before the point of application of the concentrated moment (4.265a), the shape of the elastica under strong bending (4.263b) $\equiv$ (4.265a)

$$0 \le x \le \eta: \quad a^2 = x^2 + \left(\zeta - a\right)^2, \quad a \equiv \frac{EI}{Q} = \frac{B}{Q}, \quad \left(x_0, y_0\right) = \left(0, a\right), \tag{4.265a–e}$$

is a circular arc with radius (4.265c) $\equiv$ (4.261b) at a position (4.265e) on the wall at a distance a from the support (Figure 4.11a). The radius and position of the center equal (4.265d) the ratio of the bending stiffness (4.7a) to the concentrated moment. The preceding result could be obtained geometrically as follows: (1) the bending moment is $-Q$ at the support and along the bar up to the point of application of the concentrated torque Q; hence (2) the bending moment is constant and equal to $-Q$, corresponding (4.4a) to the curvature (4.261b) and radius of curvature (4.265c); and (3) thus, the shape of the elastica is an arc of a circle with radius a, tangent to the x-axis at the origin (4.265d,e). Beyond the point of application (4.260b) of the concentrated torque, the bending moment (4.260b), and hence the curvature, is zero, and the shape of the bar is a straight line (Figure 4.11a) with constant slope.

4.7.4 Two Methods of Exact Solution of the Nonlinear Equation of the Elastica

An alternative method of reaching the same result is to put the exact equation of the elastica (4.261b) in the form

$$\frac{d\zeta}{a} = \frac{Q}{EI}d\zeta = \left|1+\zeta'^2\right|^{-3/2}\frac{d\zeta'}{dx}d\zeta = \left|1+\zeta'^2\right|^{-3/2}\zeta'd\zeta'; \tag{4.266}$$

Using the two boundary conditions (4.262a;4.263a) for the clamped end, the integration of (4.266) leads to (4.267a):

$$\frac{\zeta}{a} = 1 - \left|1+\zeta'^2\right|^{-1/2}, \quad \frac{d\zeta}{dx} = \zeta' = \left|\left(1-\frac{\zeta}{a}\right)^{-2} - 1\right|^{1/2}, \tag{4.267a,b}$$

which can be solved for ζ', leading to (4.267b). The latter is integrated (4.268b) with the help of the change of variable (4.268a):

$$\xi = 1 - \frac{\zeta}{a}: \quad \frac{x}{a} = \frac{1}{a} \int \left| \left(1 - \frac{\zeta}{a}\right)^{-2} - 1 \right|^{-1/2} d\zeta = - \int \left| \xi^{-2} - 1 \right|^{-1/2} d\xi$$

$$= - \int \left| 1 - \xi^2 \right|^{-1/2} \xi d\xi = \left| 1 - \xi^2 \right|^{1/2} = \left| 1 - \left(1 - \frac{\zeta}{a}\right)^2 \right|^{1/2}, \tag{4.268a,b}$$

where the constant of integration must be zero by (4.263a). The solution obtained by the second method (4.268b) ≡ (4.269a) is equivalent to a quadratic equation

$$\left(\frac{\zeta}{a}\right)^2 - 2\frac{\zeta}{a} + \left(\frac{x}{a}\right)^2 = 0, \quad \frac{\zeta}{a} = 1 \mp \left| 1 - \left(\frac{x}{a}\right)^2 \right|^{1/2}, \tag{4.269a,b}$$

whose roots are (4.269b). On account of (4.263a), the upper sign must be chosen in (4.269b) ≡ (4.263b) proving the coincidence of the two solutions that correspond to the circular arc (4.265a–e).

4.7.5 Concentrated Moment at the Free End or at an Intermediate Position

The exact deflection (4.263b) ≡ (4.270b) [slope (4.262c) ≡ (4.270c)] of the elastica for strong bending is specified before the point of application of the concentrated torque (4.270a) by

$$0 \le x \le \eta: \quad \zeta(x) = \frac{EI}{Q} \sum_{n=1}^{\infty} (-)^{n-1} \binom{1/2}{n} \left(\frac{Qx}{EI}\right)^{2n} = -\sum_{n=1}^{\infty} b_n x^{2n} \left(\frac{Q}{EI}\right)^{2n-1}, \tag{4.270a,b}$$

$$\zeta'(x) = \frac{Qx}{EI} \left| 1 - \left(\frac{Qx}{EI}\right)^2 \right|^{-1/2} = \sum_{n=0}^{\infty} (-)^n \binom{-1/2}{n} \left(\frac{Qx}{EI}\right)^{2n+1} = \sum_{n=0}^{\infty} a_n \left(\frac{Qx}{EI}\right)^{2n+1}. \tag{4.270c}$$

The coefficients are given by (2.182a,b) ≡ (II.6.63a–b) ≡ (II.6.64a–f) [(2.159a,b) ≡ (II.6.56a,b) ≡ (II.6.57a–g)], leading to (4.271) [(4.272)]:

$$\zeta(x) = \frac{Qx^2}{EI} \left[\frac{1}{2} + \sum_{n=2}^{\infty} \frac{(2n-3)!!}{(2n)!!} \left(\frac{Qx}{EI}\right)^{2n-2} \right], \tag{4.271}$$

$$\zeta'(x) = \frac{Qx}{EI} \left[1 + \sum_{n=1}^{\infty} \frac{(2n-1)!!}{(2n)!!} \left(\frac{Qx}{EI}\right)^{2n} \right]. \tag{4.272}$$

The weak bending corresponds to the lowest-order term in

$$\zeta(x) = \frac{Qx^2}{2EI} \left[1 + \left(\frac{Qx}{2EI}\right)^2 + O\left(\left(\frac{Qx}{EI}\right)^4\right) \right], \tag{4.273}$$

$$\zeta'(x) = \frac{Qx}{EI}\left[1 + \frac{1}{2}\left(\frac{Qx}{EI}\right)^2 + O\left(\left(\frac{Qx}{EI}\right)^4\right)\right], \tag{4.274}$$

and the first nonlinear correction is included in (4.273) [(4.274)]. Thus, *deflection (slope) of the elastica is specified exactly by (4.263b) ≡ (4.270b) ≡ (4.271) [(4.262c) ≡ (4.270c) ≡ (4.272)] for strong bending, and for weak bending by the first term on the r.h.s. of (4.273) ≡ (4.181a) [(4.274) ≡ (4.182a)] with a lowest-order nonlinear correction. The exact shape (linear approximation) is a circular arc (a parabola) tangent to the x-axis at the origin (4.265a–e) [(4.181a)].* These results apply to a bar with constant cross section of moment of inertia I and material of the Young's modulus E clamped at one end and free at the other (Figure 4.11a), bent by a concentrated torque Q in two cases: (1) if the torque is at the free end, then (4.271–4.274) hold over the whole length of the bar $0 \le x \le L$; and (2) if the torque is (4.260a,b) at the position $x = \eta$, then (4.271–4.274) hold up to the point of application of the torque (4.270a), and beyond (4.275a), the bending moment is zero, and the bar is straight (4.275b):

$$\eta \le x \le L: \quad \zeta(x) = \zeta_1 + \zeta_1'(x - \eta), \quad \{\zeta_1, \zeta_1'\} \equiv \{\zeta(\eta), \zeta'(\eta)\}, \tag{4.275a–d}$$

with coefficients (4.275c,d) evaluated from (4.270b) ≡ (4.271) ≡ (4.273) [(4.270c) ≡ (4.272) ≡ (4.274)] at the point of application of the concentrated moment.

4.7.6 Nonlinearity Parameter and Exact Shape

The nonlinearity parameter (4.276b) ≡ (2.165a) for the strong deflection of a string under uniform tension by a concentrated force or a uniform distributed force like the weight (4.276a) is the ratio of the total transverse force to twice the tension and varies in the unit interval:

$$P = \rho g L: \quad 0 \le q = \frac{\rho g L}{2T} = \frac{P}{2T} < 1; \quad 0 \le q \equiv \frac{QL}{EI} = \frac{L}{a} < 1, \tag{4.276a–c}$$

In the case of bending of a bar by a concentrated *torque Q at a position x = L, the **nonlinearity parameter** (4.276c) is the ratio of their product by the bending stiffness, defined (4.7a) as the product of the Young's modulus E by the moment of inertia I of the cross section; the nonlinearity parameter takes values in the unit interval (4.264a). It appears in the dimensionless form of the displacement* (4.263b) ≡ (4.270b) ≡ (4.271) ≡ (4.277) [*slope* (4.262c) ≡ (4.270c) ≡ (4.272) ≡ (4.278)] *of the elastica:*

$$\frac{\zeta(x)}{L} = \frac{1}{q}\left\{1 - \left|1 - \left(q\frac{x}{L}\right)^2\right|^{1/2}\right\} = \frac{1}{q}\sum_{n=1}^{\infty}(-)^{n-1}\binom{1/2}{n}\left(q\frac{x}{L}\right)^{2n}$$

$$= \frac{qx^2}{2L^2} + \sum_{n=2}^{\infty}\frac{(2n-3)!!}{(2n)!!}q^{2n-1}\left(\frac{x}{L}\right)^{2n} = \frac{qx^2}{2L^2}\left[1 + \left(\frac{qx}{2L}\right)^2 + O\left(\frac{qx}{L}\right)^4\right], \tag{4.277}$$

$$\zeta'(x) = q\frac{x}{L}\left|1 - \left(q\frac{x}{L}\right)^2\right|^{-1/2} = q\frac{x}{L}\sum_{n=1}^{\infty}(-)^{n-1}\binom{-1/2}{n}\left(q\frac{x}{L}\right)^{2n}$$

$$= q\frac{x}{L} + \sum_{n=2}^{\infty}\frac{(2n-1)!!}{(2n)!!}\left(q\frac{x}{L}\right)^{2n+1} = q\frac{x}{L}\left[1 + \frac{1}{2}\left(q\frac{x}{L}\right)^2 + O\left(\left(q\frac{x}{L}\right)^4\right)\right], \tag{4.278}$$

where the first term is the linear approximation. The maximum deflection (4.279) [slope (4.280)] in modulus occurs at the farthest distance from the support:

$$\frac{\delta}{L} = \frac{1}{L}|\zeta|_{max} = \frac{\zeta(L)}{L} = \frac{1}{q}\left\{1 - \sqrt{1-q^2}\right\} = \frac{1}{q}\sum_{n=1}^{\infty}(-)^{n-1}\binom{1/2}{n}q^{2n}$$

$$= \frac{q}{2} + \sum_{n=2}^{\infty}\frac{(2n-3)!!}{(2n)!!}q^{2n-1} = \frac{q}{2}\left[1 + \frac{q^2}{4} + O(q^4)\right], \tag{4.279}$$

$$\tan\theta = |\zeta'|_{max} = \zeta'(L) = \frac{q}{\sqrt{1-q^2}} = q\sum_{n=0}^{\infty}(-)^n\binom{-1/2}{n}q^{2n}$$

$$= q + \sum_{n=1}^{\infty}\frac{(2n-1)!!}{(2n)!!}q^{2n+1} = q\left[1 + \frac{q^2}{2} + O(q^4)\right], \tag{4.280}$$

and is indicated in Table 4.7 for 10 equally spaced values of the nonlinearity parameter (4.276c). The same values are used in Panel 4.1 to plot the exact shape of the elastica of the uniform bar under nonlinear bending by a concentrated torque at the free end. The maximum slope in modulus is the same (4.280) [≡(2.167c)] in terms of the nonlinearity parameter (4.276c) [(4.276b) ≡ (2.167a)] in Table 2.2 (4.7) for nonlinear bending (deflection) of a bar (string) by a concentrated torque (force); the maximum deflection in modulus is at the tip (middle) of the bar (string) and takes distinct values [(4.279) ≠ (2.167b)].

TABLE 4.7 Nonlinear Bending of a Bar by a Torque at the Tip

Nonlinearity Parameter	Maximum Deflection	Maximum Slope (Angle)
$q = \dfrac{QL}{EI}$	$\dfrac{\zeta(L)}{L} = \dfrac{\delta}{L}$	$\zeta'(L) \equiv \tan\theta$
(4.276c)	(4.279)	(4.280)
0	0	0
0.1	0.0501	0.1005 (5.74°)
0.2	0.1010	0.2041 (11.54°)
0.3	0.1535	0.3145 (17.46°)
0.4	0.2087	0.4364 (23.58°)
0.5	0.2679	0.5774 (30.00°)
0.6	0.3333	0.7500 (36.87°)
0.7	0.4084	0.9802 (44.43°)
0.8	0.5000	1.3333 (53.13°)
0.9	0.6268	2.0647 (64.16°)
1.0	∞	∞ (90.00°)

Notes: L, length of the undeflected bar; *x* longitudinal coordinate; *E,* Young's modulus of material; *I,* moment of inertia of the cross section; ζ, transverse deflection; *Q,* concentrated torque at the tip; $\zeta' \equiv d\zeta/dx$, slope; δ, maximum deflection; θ, angle-of-inclination; *q,* nonlinearity parameter.

Strong or nonlinear bending of a uniform clamped-free bar by a concentrated moment at the free end.

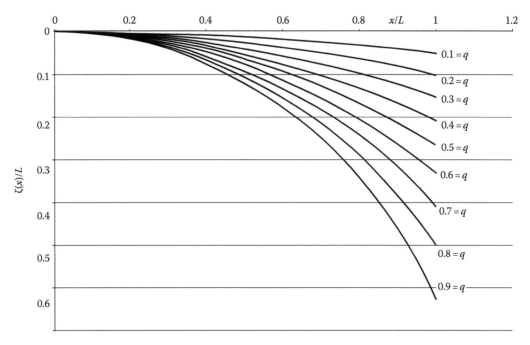

PANEL 4.1 The shape of the elastica for the strong bending of a uniform bar by a concentrated moment at the tip is a circular arc whose curvature increases with the nonlinearity parameter.

4.7.7 Nonlinear Matching of Multiple Deflections

The nonlinear solution may be applied to a bar with several concentrated torques and is illustrated next (Figure 4.11b) with two torques $Q_1(Q_2)$ at $x_1(x_2)$, for which (4.260a,b) is replaced by

$$EI\left\{\zeta'\left|1+\zeta'^2\right|^{-1/2}\right\}' = Q_1 + Q_2 - Q_1 H(x-x_1) - Q_2 H(x-x_2)$$

$$= \begin{cases} Q_1 + Q_2 & \text{if } 0 \le x < x_1, & \text{(4.281a)} \\ Q_2 & \text{if } x_1 < x < x_2, & \text{(4.281b)} \\ 0 & \text{if } x_2 < x < L. & \text{(4.281c)} \end{cases}$$

The solution (4.263b) ≡ (4.282b) holds for the clamped bar (4.262a, 4.263a) with the total torque (4.282c) up to the location (4.282a) of the first concentrated torque:

$$0 \le x \le x_1: \quad \frac{Q}{EI}\zeta(x) = 1 - \left|1 - \left(\frac{Qx}{EI}\right)^2\right|^{1/2}, \qquad\qquad (4.282\text{a,b})$$

$$Q \equiv Q_1 + Q_2: \quad \zeta'(x) = \frac{Qx}{EI}\left|1 - \left(\frac{Qx}{EI}\right)^2\right|^{1/2}; \qquad\qquad (4.282\text{c,d})$$

These specify the deflection (4.282b) [slope (4.282d)] at the point of application of the first torque (4.283a) [(4.283b)] as the basis of two integrations, to specify the slope (shape) in the next segment. A first integration of (4.281b) yields (4.283c):

$$\zeta_1 \equiv \zeta(x_1), \zeta_1' \equiv \zeta'(x_1): \quad \zeta' \left|1+\zeta'^2\right|^{-1/2} - \zeta_1' \left|1+\left(\zeta_1'\right)^2\right|^{-1/2} = \frac{Q_2}{EI}(x-x_1), \qquad (4.283\text{a-c})$$

which can be solved for the slope:

$$b \equiv \zeta_1' \left|1+\left(\zeta_1'\right)^2\right|^{-1/2}: \quad \zeta' = \left[b+\frac{Q_2}{EI}(x-x_1)\right]\left|1-\left[b+\frac{Q_2}{EI}(x-x_1)\right]^2\right|^{-1/2}. \qquad (4.284\text{a,b})$$

A second integration yields the shape (4.285b) in the next segment (4.285a) of the elastica:

$$x_1 \le x \le x_2: \quad \frac{Q_2}{EI}\left[\zeta(x)-\zeta_1\right] = \left|1-b^2\right|^{1/2} - \left|1-\left[b+\frac{Q_2}{EI}(x-x_1)\right]^2\right|^{1/2}. \qquad (4.285\text{a,b})$$

The third segment (4.286a) without torque (4.281c) is a straight (4.286b) line like (4.275b), starting with the displacement (4.286c) [slope (4.286d)] specified by (4.285b) [(4.284b)]:

$$x_2 \le x \le L: \quad \zeta(x) = \zeta_2 + \zeta_2'(x-x_2), \quad \zeta_2 \equiv \zeta(x_2), \quad \zeta'(x) = \zeta_2' \equiv \zeta'(x_2). \qquad (4.286\text{a-d})$$

Thus, *a uniform (4.10a) bar (4.90a) under strong bending by two torques $Q_1(Q_2)$ at positions $x_1 < x_2$ in Figure 4.11b has a shape (slope) specified at each of the three segments (4.281a) ≡ (4.282a)/(4.281b) ≡ (4.285a)/(4.281c) ≡ (4.286a), respectively, by (4.282b)/(4.285b)/(4.286b) [(4.282d)/(4.284b)/(4.286d)], involving the matching constants at the first (4.283a,b, 4.284a) [second (4.286c,d)] junction.* This result could be extended to any number of segments of the elastica.

4.7.8 Strong Bending of a Bar by Two Concentrated Moments

The shape (4.285b) [slope (4.284b)] would reduce to (4.263a) [(4.262c)] for zero initial displacement $\zeta_1 = 0 = x_1$ [slope $\zeta_1' = 0 = b$ in (4.284a)]. The preceding solution can be written in a dimensionless form using the variables (4.287a,b) and parameter (4.287d):

$$Y \equiv \frac{Q_2}{EI}\zeta(x), \quad X \equiv \frac{Q_2}{EI}x, \quad \lambda \equiv \frac{Q_1}{Q_2}, \quad Q = (1+\lambda)Q_2, \qquad (4.287\text{a-d})$$

that lead (4.182c) to (4.187d) and appear (1) in the first segment (4.282a,b):

$$0 \le x \le x_1: \quad (1+\lambda)Y(X) = \left\{1-\left|1-\left[(1+\lambda)X\right]^2\right|^{1/2}\right\}; \qquad (4.288\text{a,b})$$

(2) in the second segment (4.285a,b):

$$x_1 \le x \le x_2: \quad Y(X) = Y(X_1) + \left|1-b^2\right|^{1/2} - \left|1-(b+X-X_1)^2\right|^{1/2}; \qquad (4.289\text{a,b})$$

and (3) in the third segment (4.286a,b):

$$x_2 \leq x \leq L: \quad Y(X) = Y(X_2) + Y'(X_2)(X - X_2). \tag{4.290a,b}$$

These expressions involve the parameter (4.284a) where appears

$$\zeta_1' \equiv \frac{d\zeta_1}{dx_1} = \frac{dY_1}{dX_1} = X_1(1+\lambda)\left|1 - \left[X_1(1+\lambda)\right]^2\right|^{-1/2}, \tag{4.291}$$

calculated using (4.287a,b, 4.288b). The parameter (4.284a) depends on the initial slope of the second segment; for example, the angles (4.292a) correspond to the slopes (4.292b) and values (4.292c) of the parameter:

$$\theta_1 = \left\{0, \frac{\pi}{6}, \frac{\pi}{4}, \frac{\pi}{3}\right\}, \quad \zeta_1' = \tan\theta_1 = \left\{0, \frac{1}{\sqrt{3}}, 1, \sqrt{3}\right\}, \quad b = \zeta_1'\left|1 + (\zeta_1')^2\right|^{-1/2} = \left\{0, \frac{1}{2}, \frac{1}{\sqrt{2}}, \frac{\sqrt{3}}{2}\right\}. \tag{4.292a–c}$$

In the first (4.288a) [second (4.289a)] segment, the variable (4.287b) cannot exceed the value (4.293a) [(4.293b)]:

$$X_1 \equiv \frac{Q_2 x_1}{EI} \leq \frac{1}{1+\lambda} = \frac{Q_2}{Q_1 + Q_2} \equiv X_{1\max}, \quad X_2 \equiv \frac{Q_2 x_2}{EI} \leq 1 + X_1 - b = X_{2\max}. \tag{4.293a,b}$$

These inequalities arise (4.293c) [(4.293d)] from (4.288b) [(4.289b)] for the slope not to reach an infinite value:

$$-1 \leq (1+\lambda)X_1 \leq +1; \quad -1 \leq b + X_2 - X_1 \leq +1 \Leftrightarrow X_1 - b - 1 \leq X_2 \leq X_1 - b + 1. \tag{4.293c–e}$$

The condition (4.293c) implies (4.293a) and (4.293e) leads to (4.293b); since $b < 1$ in (4.284a), the second inequality (4.293d) ≡ (4.293e) is relevant for $X_2 > X_1$ and leads to (4.293b). An infinite slope is possible if the concentrated torque is large enough for the circular arc in Figure 4.11a to extend for at least a quarter of the circle; thus, the inequalities restrict the concentrated torques (Q_1, Q_2) so that the bar does not have a vertical tangent at any point in Figure 4.11b.

4.7.9 Effect of the Relative Magnitude of the Two Moments

The shape of the bar is plotted in Panel 4.2 and for five values (4.294a) of the ratio of the two moments (4.287c)

$$\lambda = \left\{2, 1, \frac{1}{2}, 0, -\frac{1}{2}\right\}, \quad X_1 \equiv \frac{1}{2(1+\lambda)} = \left\{\frac{1}{6}, \frac{1}{4}, \frac{1}{3}, \frac{1}{2}, 1\right\}, \tag{4.294a,b}$$

that include (1) first moment zero, of the same or opposite sign; and (2) first moment in modulus equal, larger, or smaller than the second. The values $\lambda < -1$ are excluded so that $X_1 > 0$ in (4.293a). The first moment Q_1 is applied at the end of the first segment (4.288a), corresponding to a maximum value (4.293a) of X_1 that would lead to infinite slope in (4.291); the value of X_1 is chosen (4.294a) to be half the maximum (4.293a), corresponding to the value (4.294b) for the dimensionless longitudinal

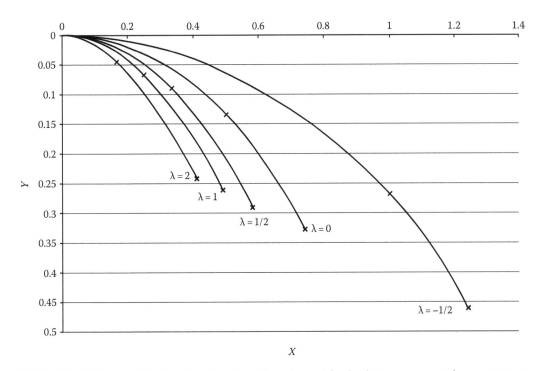

PANEL 4.2 In the case of the strong bending of a uniform clamped-free bar by two concentrated moments one at the tip and the other before, the elastica consists of circular arcs with continuous (discontinuous) tangent (curvature) at the point of application of the intermediate concentrated moment.

coordinate (4.287b). The corresponding dimensionless transverse deflection at the point of application of the first moment (4.294b, 4.288b) is

$$Y_1 \equiv Y(X_1) \equiv \frac{1 - 3/\sqrt{2}}{1 + \lambda} = \{0.04466, 0.06699, 0.08932, 0.13397, 0.26794\}. \tag{4.295}$$

The slope (4.291) at the point (4.294b) of application of the first moment is (4.296a):

$$\zeta_1' = Y_1' = Y'(X_1) = \frac{1}{\sqrt{3}} = \tan\theta_1, \quad \theta_1 = \frac{\pi}{6} = 30°, \quad b = \frac{1}{2}, \tag{4.296a–c}$$

corresponding to an angle (4.296b) and value (4.296c) of the parameter (4.284a). The point of application of the second moment Q_2 is chosen (4.297a) to be half of the maximum (4.293b) relative to (4.294b):

$$X_2 = X_1 + \frac{1-b}{2} = X_1 + \frac{1}{4}; \quad Y_2 = Y(X_2) = Y_1 + \left|1 - b^2\right|^{1/2} - \left|1 - \left(\frac{1+b}{2}\right)^2\right|^{1/2}$$

$$= Y_1 + \frac{\sqrt{3}}{2} - \frac{\sqrt{7}}{4} = Y_1 + 0.2046. \tag{4.297a,b}$$

The displacement (4.297b) at the point of application of the second moment Q_2 follows from substitution of (4.297a) in (4.289b). The corresponding (4.289b) slope (4.298b) is obtained (4.298c):

$$X_1 \leq X \leq X_2: \quad Y'(X) = \left|(b + X - X_1)^{-2} - 1\right|^{-1/2}, \tag{4.298a,b}$$

$$Y_2' = Y'(X_2) = \left| \left(\frac{1+b}{2} \right)^{-2} - 1 \right|^{-1/2} = \left| \frac{16}{9} - 1 \right|^{-1/2} = \frac{3}{\sqrt{7}} = 1.1339, \qquad (4.298c)$$

substituting (4.296c, 4.297a) in (4.298b). The shape of the elastica plotted in Panel 4.2 is specified by (4.288b) [(4.289b)] for the first (4.288a) [second (4.289a)] segment, that is, before the first torque Q_1 (between the first Q_1 and second Q_2 torques); the continuity of displacement and slope is assured at the matching point (4.294b), where is located the first torque. The location of the concentrated torques is indicated by crosses, that is, one at the tip of the bar and the other at an intermediate position; the five shapes correspond to the values (4.294a) of the ratio of torques (4.287c).

4.8 Nonlinear Influence or Green Function

The nonlinear influence or Green function specifies the shape of the elastica for strong bending of a uniform bar by a concentrated force at an arbitrary position, for example, in the cases of (1) both ends pinned (Subsection 4.8.1); and (2) one end clamped and the other free (Subsection 4.8.2). The nonlinear influence function generally satisfies neither the principle of reciprocity nor the principle of superposition. Taking as example the clamped-free or cantilever bar, the nonlinear influence or Green function (1) specifies the shape of the elastica for a concentrated force at the free end (Subsections 4.8.10 through 4.8.15) since this is a particular location of the concentrated force; (2) does not specify the shape of the elastica for a uniform weight (Subsections 4.8.4 through 4.8.9) because the principle of superposition does not hold. The determination of the shape of the elastica of a bar, either for weak or for strong bending [Sections 4.3 through 4.6 (Sections 4.7 through 4.9)], can use two coordinate systems: (1) Cartesian coordinates with the origin at one support and the ox-axis along the unbent straight bar (Sections 4.1 through 4.7 and Subsections 4.8.1 and 4.8.2); and (2) local tangential coordinates consisting of the arc length along the elastica and the angle of the tangent with the undeflected direction of the straight bar (Subsections 4.8.3 through 4.8.15 and Section 4.9). The fixed Cartesian (local tangential) coordinates (1) [(2)] are often more convenient for weak (strong) bending, although either coordinate system can be used in both cases. The equation of the elastica for strong bending of a nonuniform bar can be written in local tangential coordinates (Subsection 4.8.3) for (1) the bending moment; (2) the transverse force; and (3) the shear stress. It can be applied to various loads (Subsection 4.8.4), such as concentrated forces or moments or a uniform load. Choosing as example the nonlinear bending by a uniform weight (Subsection 4.8.5) leads to the linear approximation (Subsection 4.8.6) and first (second)-order nonlinear corrections [Subsection 4.8.7 (Subsection 4.8.8)]; the linear approximation (nonlinear corrections) are (are not) the same in Cartesian and local coordinates (Subsection 4.8.9). The exact solution including nonlinear corrections of all orders is obtained for strong bending of a uniform bar clamped at one end by a concentrated force at the free end: (1) the shape of the elastica involves elliptic integrals (Subsection 4.8.10); (2) the latter can be evaluated exactly as double series (Subsection 4.8.11) specifying, besides the leading terms that form the linear approximation, also all the following terms corresponding to nonlinear corrections of all orders (Subsection 4.8.12); (3) the lowest-order approximation (Subsection 4.8.13) coincides with weak bending (Subsection 4.3.1); (4) the next order specifies the lowest-order nonlinear effects (Subsection 4.8.14); and (5) these nonlinear effects can be analyzed as concerns, for example, the relation between the length of the bar and the slope at the tip where the concentrated force is applied (Subsection 4.8.15).

4.8.1 Nonlinear Influence Function due to a Concentrated Force

From (4.4a)/(4.6a)/(4.6b), it follows, respectively, that *the shape of the elastica of a bar (2.299a) under nonlinear bending is specified by the bending moment (2.299b)/transverse force (4.299c)/shear stress (4.299d)*:

$$T = 0: \quad \{M(x), F(x), f(x)\} = \left\{ -1, \frac{d}{dx}, \frac{d^2}{dx^2} \right\} EI\zeta'' \left| 1 + \zeta'^2 \right|^{-3/2}. \qquad (4.299a\text{–}d)$$

These simplify, respectively, to (4.300b/c/d) for a uniform bar that has constant (4.300a) bending stiffness (4.7c):

$$EI = \text{const}: \quad \{-M(x), F(x), f(x)\} = EI\left\{\frac{d}{dx}, \frac{d^2}{dx^2}, \frac{d^3}{dx^3}\right\}\zeta'\left|1+\zeta'^2\right|^{-1/2}. \tag{4.300a–d}$$

The balance equation for the bending moment (4.300b) [shear stress (4.300d)] was considered before (is considered next) for a concentrated moment (4.260a,b) [force (4.301)]:

$$EI\left\{\zeta'\left|1+\zeta'^2\right|^{-1/2}\right\}''' = f(x) = P\delta(x-\xi). \tag{4.301}$$

An integration leads to the transverse force:

$$F(x) = EI\left\{\zeta'\left|1+\zeta'^2\right|^{-1/2}\right\}'' = \begin{cases} A = R_- & \text{if } 0 \le x < \xi, & (4.302a) \\ A+P = -R_+ & \text{if } \xi < x \le L. & (4.302b) \end{cases}$$

In the case (Figure 4.6b) of a bar pinned at both ends (4.303a,b), the bending moment vanishes there, and the integration of (4.302a,b) leads to (4.303c,d):

$$\zeta''(0) = 0 = \zeta''(L): \quad -M(x) = EI\left\{\zeta'\left|1+\zeta'^2\right|^{-1/2}\right\}' = \begin{cases} Ax & \text{if } 0 \le x \le \xi, \\ (A+P)(x-L) & \text{if } \xi \le x \le L. \end{cases} \tag{4.303a–d}$$

The constant A is determined by the continuity of the bending moment at the point of application of the concentrated force (4.304a),

$$M(\xi-0) = M(\xi+0): \quad A = -P\left(1-\frac{\xi}{L}\right) = R_{3-}, \quad R_{3+} = -A-P = -P\frac{\xi}{L}, \tag{4.304a–c}$$

and specifies (4.304b,c) the reaction forces at the two supports.

Substituting (4.304b,c) in (4.303c,d) and integrating once more specifies the slopes before and after the point of application of the force:

$$EI\zeta'\left|1+\zeta'^2\right|^{-1/2} = \begin{cases} P\left(\dfrac{\xi}{L}-1\right)\dfrac{x^2}{2} + B & \text{if } 0 \le x \le \xi, & (4.305a) \\[2mm] P\xi x\left(\dfrac{x}{2L}-1\right) + C & \text{if } \xi \le x \le L, & (4.305b) \end{cases}$$

where B, C are integration constants related (4.306b) by the continuity of the slope (4.306a) at the point of application of the concentrated force:

$$\zeta'(\xi-0) = \zeta'(\xi+0): \quad C = B + \frac{1}{2}P\xi^2. \tag{4.306a,b}$$

Solving (4.305a,b) for the slopes leads to (4.307a,b):

$$\zeta'(x) = \begin{cases} \left| \left[\dfrac{P}{EI}\left(\dfrac{\xi}{L}-1\right)\dfrac{x^2}{2} + \dfrac{B}{EI}\right]^{-2} - 1 \right|^{-1/2} & \text{if } 0 \le x \le \xi \quad (4.307a) \\[4mm] \left| \left[\dfrac{P\xi x}{EI}\left(\dfrac{x}{2L}-1\right) + \dfrac{C}{EI}\right]^{-2} - 1 \right|^{-1/2} & \text{if } \xi \le x \le L. \quad (4.307b) \end{cases}$$

Integrating (4.307a) [(4.307b)] from the first (4.308a) [second (4.308c)] pinned end leads to (4.308b) [(4.308d)]:

$$G(0;\xi) = 0: \quad G_3(x < \xi;\xi) = \int_0^x \left| \left[\dfrac{P}{2EI}\left(\dfrac{\xi}{L}-1\right)z^2 + \dfrac{B}{EI}\right]^{-2} - 1 \right|^{-1/2} dz, \quad (4.308a,b)$$

$$G(L;\xi) = 0: \quad G_3(x > \xi;\xi) = \int_L^x \left| \left[\dfrac{P\xi z}{EI}\left(\dfrac{z}{2L}-1\right) + \dfrac{P\xi^2}{2EI} + \dfrac{B}{EI}\right]^{-2} - 1 \right|^{-1/2} dz, \quad (4.308c,d)$$

where (1) the constant C was substituted in (4.308d) in terms (4.306b) of B; and (2) the remaining constant of integration B is determined (4.309b) by the continuity of the displacement at the point of application of the force:

$$G_3(\xi - 0;\xi) = \int_0^\xi \left| \left[\dfrac{P}{2EI}\left(\dfrac{\xi}{L}-1\right)z^2 + \dfrac{B}{EI}\right]^{-2} - 1 \right|^{-1/2} dz$$

$$= \int_L^\xi \left| \left[\dfrac{P\xi}{EI}\left(\dfrac{z}{2L}-1\right) + \dfrac{P\xi^2}{2EI} + \dfrac{B}{EI}\right]^{-2} - 1 \right|^{-1/2} dz = G_3(\xi + 0;\xi). \quad (4.309)$$

It has been shown that *the nonlinear influence or Green function specifying the shape of the elastica of a uniform* (4.10a) *bar* (4.90a) *pinned at both ends* (4.53a–d) *under strong bending by a concentrated force at an arbitrary position* (4.301) *is given by* (4.308b,d), *where the constant B satisfies* (4.309). As a first (second) example, the nonlinear influence function for strong bending of a uniform bar is considered for a pinned-pinned (clamped-free) bar [Subsection 4.8.1 (Subsection 4.8.2)].

4.8.2 Nonlinear Bending of Pinned-Pinned and Clamped-Free Bars

Returning to the nonlinear influence function for strong bending of a uniform bar by a concentrated force at an arbitrary position (4.301), an integration for a clamped-free bar (Figure 4.6a) leads to

$$F(x) = EI \left\{ \zeta' \left| 1 + \zeta'^2 \right|^{-1/2} \right\}'' = A + PH(x-\xi) = \begin{cases} A = -P = R_{4-} & \text{if } \xi < x \le L, \quad (4.310a) \\[2mm] A + P = 0 = -R_{4+} & \text{if } \xi < x \le L, \quad (4.310b) \end{cases}$$

where (1) the transverse force at the free end is zero (4.310b); and (2) the reaction force is specified at the clamped end (4.310a). A further integration yields

$$
M_- = -M(0) = P\xi: \quad -M(x) = EI\left\{\zeta'\left|1+\zeta'^2\right|^{-1/2}\right\}' = \begin{cases} P(\xi-x) & \text{if } 0 \leq x \leq \xi, \\ 0 = M_{4+} & \text{if } \xi \leq x \leq L, \end{cases}
\tag{4.311a–c}
$$

since (1) there is no bending moment at the free end (4.311c); (2) the bending moment is continuous at the point of application of the concentrated force (4.311b); and (3) it follows that the reaction bending moment at the clamped end is (4.311a). A further integration with the clamping boundary condition (4.312a) leads to (4.312b,c):

$$
\zeta'(0) = 0: \quad 2EI\zeta'\left|1+\zeta'^2\right|^{-1/2} = \begin{cases} Px(2\xi-x) & \text{if } 0 \leq x \leq \xi, \\ P\xi^2 & \text{if } \xi \leq x \leq L, \end{cases}
\tag{4.312a–c}
$$

where a continuity condition at $x = \xi$ is used in (4.312c). Solving for the slope yields

$$
\zeta'(x) = \begin{cases} \left|\left[\dfrac{Px}{2EI}(2\xi-x)\right]^{-2} - 1\right|^{-1/2} & \text{if } 0 \leq x \leq \xi, \tag{4.313a} \\[3em] \left|\left[\left(\dfrac{P\xi^2}{2EI}\right)^2 - 1\right]^{-2} - 1\right|^{-1/2} & \text{if } \xi \leq x \leq L, \tag{4.313b} \end{cases}
$$

whose integration is immediate.

Integrating (4.313a) from the clamped end (4.314a) of the bar leads to (4.314b):

$$
G_4(0;\xi) = 0: \quad G_4(x;\xi) = \begin{cases} \displaystyle\int_0^x \left|\left[\dfrac{Pz}{2EI}(2\xi-z)\right]^{-2} - 1\right|^{-1/2} dz & \text{if } 0 \leq x \leq \xi, \tag{4.314a–c} \\[3em] \left|\left[\left(\dfrac{P\xi^2}{2EI}\right)^2 - 1\right]^{-2} - 1\right|^{-1/2} z + B & \text{if } \xi \leq x \leq L; \end{cases}
$$

The integration of (4.313b) leads to (4.314c), where the constant B is specified by the continuity of the displacement at the point of application of the concentrated force (4.315a):

$$
G_4(\xi+0;\xi) = B + \left|\left[\left(\dfrac{P\xi^2}{2EI}\right)^2 - 1\right]^{-2} - 1\right|^{-1/2} \xi
$$

$$
= \int_0^\xi \left|\left[\dfrac{Pz}{2EI}(2z-x)\right]^2 - 1\right|^{-1/2} dz = G_4(\xi-0;\xi).
\tag{4.315}
$$

It has been shown that *the nonlinear influence function specifying the shape of the elastica of a uniform (4.10a) bar (4.90a) clamped at one end and free at the other (4.259a,b,j,k), under strong bending by a concentrated force at an arbitrary position (4.301), is given by (4.314b,c), involving the constant B specified by (4.315).* Comparing the linear (nonlinear) influence function for weak (strong) bending of a uniform bar, both pinned-pinned (4.110a,b) [(4.308b,d, 4.309)] and clamped-free (4.96a,b) [(4.314b,c, 4.315)], it follows that it does (does not) satisfy the reciprocity principle (2.47) and likewise does (does not) meet the principle of superposition. The **elliptic function of degree N** (Section I.39.7) is defined as the inverse (4.316a) of the **elliptic integral** (4.316b):

$$w = E_N(z): \quad z = \int \left| P_N(w) \right|^{-1/2} R(w) dw, \quad P_N(w) = \sum_{n=0}^{N} a_n w^m, \qquad (4.316\text{a--c})$$

involving (1) in the inverse square root a polynomial of degree N; and (2) in the numerator a rational function. It follows that the nonlinear influence function for strong bending of a uniform bar, for example, in the pinned-pinned (4.308b,d, 4.309) [clamped-free (4.314b,c, 4.315)] cases, is specified by elliptical integrals (4.316a--c) of degree four, of which one example is the Jacobian elliptical functions (Section I.39.9). The elliptic integrals also appear in the strong bending of uniform bars replacing fixed Cartesian coordinates by local tangent coordinates (Subsection 4.8.3).

4.8.3 Relation between Fixed Cartesian and Local Tangent Coordinates

The weak (Sections 4.3 through 4.5) and strong (Section 4.7 and Subjections 4.8.1 and 4.8.2) bending of bars and beams (Sections 4.1 and 4.2) has so far (Figures 4.5 through 4.11) been considered using fixed Cartesian coordinates with (1) the origin at one support; (2) the x-axis along the unbent straight bar; and (3) the y-axis downward, for example, in the direction of gravity. An alternative well suited to strong bending (Subsections 4.8.3 through 4.8.15 and Section 4.9) is to use local tangent coordinates consisting (Figure 4.12a) of: (1) the arc length (2.1b) measured from the first support (4.317a):

$$s = \int_0^x \frac{ds}{dx} dx = \int_0^x \left| 1 + \zeta'^2 \right|^{1/2} dx; \quad \tan\theta = \frac{dy}{dx} = \frac{d\zeta}{dx} \equiv \zeta'; \qquad (4.317\text{a,b})$$

and (2) the angle θ of the local tangent with the direction of the unbent straight bar (4.317b) that also satisfies (4.318a,b):

$$\cos\theta = \frac{dx}{ds} = \left| 1 + \zeta'^2 \right|^{-1/2}, \quad \sin\theta = \frac{dy}{ds} = \zeta' \frac{dx}{ds} = \zeta' \left| 1 + \zeta'^2 \right|^{-1/2}. \qquad (4.318\text{a,b})$$

The curvature (4.2b) of the elastica is given by

$$k(x) = \left| 1 + \zeta'^2 \right|^{-1/2} \frac{\zeta''}{1 + \zeta'^2} = \frac{dx}{ds} \frac{d}{dx} \left(\arctan \zeta' \right) = \frac{dx}{ds} \frac{d\theta}{dx} = \frac{d\theta}{ds}, \qquad (4.319)$$

which may be substituted in (4.4a). Thus, *the curvature of a plane curve, such as the elastica of a strongly bent bar in local tangent coordinates (4.317a,b, 4.318a,b), equals (4.2b) $\equiv$ (4.320a) the derivative of the angle of the tangent with regard to the arc-length:*

$$\frac{1}{R} = k = \frac{d\theta}{ds}: \quad -M = EIk = EI \frac{d\theta}{ds}, \qquad (4.320\text{a,b})$$

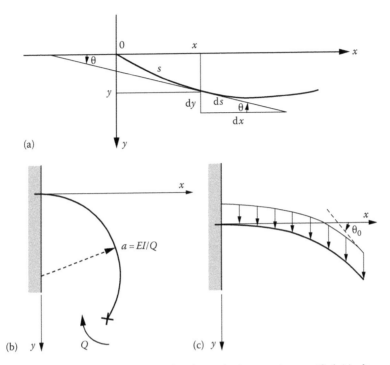

(a)

(b)

(c)

FIGURE 4.12 The shape of the elastica for the bending of a bar can be specified (a) alternatively using: (1) Cartesian coordinates, (Figures 4.5 through 4.11) with origin at the first support, x-axis along the undeformed position and y-axis downward; and (2) local tangent coordinates at each point along the elastica, using the arc length s measured from the first support and the angle of inclination θ relative to the undeformed shape. The fixed Cartesian (local tangent) coordinates are most suitable for weak (strong) bending due to their simplicity (ability to follow the curvature of the elastica). The strong bending of a bar by concentrated moment(s) [transverse force(s) or shear stress(s)] can lead to a curled elastica (b) [for own weight leads at most to a vertical slope (c) but not curling].

and specifies (1) the bending moment (4.320b); and (2/3) the transverse force (4.321a) [shear stress (4.321b)]:

$$F = -\frac{dM}{dx} = -\frac{ds}{dx}\frac{dM}{ds} = \sec\theta \frac{d}{ds}\left(EI\frac{d\theta}{ds}\right), \tag{4.321a}$$

$$f = \frac{dF}{dx} = \frac{ds}{dx}\frac{dF}{ds} = \sec\theta \frac{d}{ds}\left[\sec\theta \frac{d}{ds}\left(EI\frac{d\theta}{ds}\right)\right]. \tag{4.321b}$$

In the case of a uniform bar (4.6c) ≡ (4.322a), the transverse force (4.321a) [shear stress (4.321b)] simplifies to (4.322b) [(4.322c)]:

$$EI = \text{const}: \quad F\cos\theta = EI\frac{d^2\theta}{ds^2}, \quad f\cos\theta = EI\frac{d}{ds}\left(\sec\theta \frac{d^2\theta}{ds^2}\right). \tag{4.322a–c}$$

The bending moment (4.320b)/transverse force (4.321b)/shear stress (4.321c) in local tangent coordinates (4.317a,b–4.318a,b) are used to consider the strong bending of a uniform bar respectively by concentrated moments/own weight/a concentrated force respectively in Subsection(s) 4.8.4/4.8.5 through 4.8.9/4.8.10 through 4.8.15.

4.8.4 Strong Bending by Moments, Forces, and Weight

In the case of strong bending of a uniform bar by concentrated torques, the bending moment is constant between them (4.323a), and the angle of the tangent (4.320b) is proportional to the arc length (4.323b):

$$M = \text{const} = -Q: \quad \theta = \frac{Q}{EI} s. \tag{4.323a,b}$$

The elastica is a circular arc (Figure 4.11a) and can curl up beyond the vertical $\theta > \pi/2$ in (4.323b) if the bending moment is large, the bending stiffness (4.7a) is small, or the bar is long (Figure 4.12b). In the case of a concentrated transverse force, the transverse force is constant up to the point of application and vanishes between the last concentrated force and the tip (4.324a), implying (4.324b) by (4.322b); if in addition there is no bending moment (4.324c), then (4.320b) implies (4.324d) that the slope is constant (4.324e) and the bar is straight, as shown in Figure 4.11a and b toward the tip:

$$F = 0: \quad \frac{d^2\theta}{ds^2} = 0; \quad M = 0: \quad \frac{d\theta}{ds} = 0, \quad \theta(s) = \text{const} = \theta_0. \tag{4.324a–e}$$

The case of own weight (a concentrated force) is considered next (subsequently) in Subsections 4.8.4 through 4.8.9 (Subsections 4.8.10 through 4.8.15).

In the case of a distribution of transverse force that is an analytic function of the arc length, the slope cannot reach the vertical $\theta < \pi/2$ and the bar cannot curl-up (Figure 4.12c), that is, the elastica can have at most a vertical asymptote. Thus, *there is a significant difference between the nonlinear bending of a clamped-free or cantilever bar (Figure 4.12a) depending on the loads: (1) a concentrated (Figure 4.12b) or distributed moment can curl-up the bar beyond the vertical slope; (2) a distribution of transverse force that is an analytic function of the arc length leads to an uncurled elastica (Figure 4.12c) that has at most a vertical asymptote.* For the proof of the statement (1), it is sufficient to take the example of a concentrated moment Q for which (4.323b) has no upper bound, that is the bar will curl-up beyond the vertical $\theta > \pi/2$ if $Q > \frac{\pi EI}{2s}$. The statement (2) is rephrased in a more precise way proved next. *Consider a uniform (4.325a) bar clamped horizontally or obliquely at one end (4.325b) and with the other end free (4.325c) subject (4.322b) ≡ (4.325f) to a transverse force that is an analytic function of the arc length (4.325d):*

$$EI = \text{const}; \quad \theta(0) < \frac{\pi}{2}, M(L) = 0, F(s) \in \mathcal{A}(0, L): \quad EI \frac{d^2\theta}{ds^2} = F(s) \cos\theta. \tag{4.325a–f}$$

Then the angle of inclination of the elastica relative to the horizontal is an analytic function of the arc length (4.325g), and the corresponding Taylor series (I.23.32a) ≡ (4.325h,i) has an upper bound $\pi/2$ that it cannot reach (4.325j):

$$\theta \in \mathcal{A}(0, L); \quad 0 \le a \le L: \quad \theta(s) = \sum_{n=0}^{\infty} \frac{(s-a)^n}{n!} \theta^{(n)}(a) < \frac{\pi}{2}. \tag{4.325g–j}$$

The implication is that the elastica cannot curl-up and can at most have a vertical asymptote. The proof will be made by "reductio ad absurdum," showing that a vertical slope $\theta = \pi/2$ would lead to a contradiction, and thus cannot be reached or exceeded.

Applying the Leibnitz rule for the nth order derivative of a product (I.13.31) to (4.325f), and noting that the transverse force is an analytic function (4.225d) and hence has derivatives of all orders, it follows that the angle of inclination has derivatives of all orders with regard to the arc length:

$$\frac{d^{n+2}\theta}{ds^{n+2}} = \cos\theta \frac{d^n F}{ds^n} + \sum_{m=0}^{n-1} \binom{n}{m} \frac{d^m F}{ds^m} \frac{d^{n-m}}{ds^{n-m}} (\cos\theta), \tag{4.326a}$$

and these are (4.326c) bounded or grow at most like $O(n)$ at any point of the bar (4.326b):

$$0 \leq s \leq L: \qquad \left|\frac{d^n\theta}{ds^a}\right| \sim O(n); \qquad \lim_{n\to\infty} R_N(s,a) = 0, \qquad (4.326\text{b–d})$$

from (4.326a) also follows (4.326d) that the remainder of the Taylor series (I.23.46c) $\equiv$ (4.326e,f) vanishes as $N \to \infty$:

$$0 < \lambda < 1: \qquad R_n(s) \equiv \theta(s) - \sum_{n=0}^{N} \frac{(s-a)^n}{n!}\theta^{(n)}(a) = \frac{(s-a)^N}{N!}\theta^{(N)}(\lambda(s-a)), \qquad (4.326\text{e–f})$$

and thus the series (4.325i) converges. This proves that the angle of inclination is an analytic function of the arc length (4.325g) over the whole length of the bar (4.326a).

The condition of zero bending moment at the tip (4.325c) implies (4.320b) that the first-order derivative vanishes at the tip (4.326g); the second-order derivative (4.325f) at the tip (4.326h) would also vanish for vertical slope (4.326i):

$$\theta'(L) = 0: \qquad \theta''(L) = F(L)\cos[\theta(L)], \qquad \lim_{\theta(L)\to\pi/2} \theta''(L) = 0. \qquad (4.326\text{g–i})$$

The derivatives of all higher orders (4.326j) also vanish at the tip (4.326k) because (4.326a) consists of: (i) the first term or the r.h.s. that vanishes for $\theta(L) \to \pi/2$; (ii) all the remaining terms on the r.h.s involve $\theta'(L)$ as a factor, that vanishes too (4.326g):

$$n = 2, \cdots, \infty: \qquad \theta^{(n)}(L) = 0: \qquad \theta(s) = \theta(L) = \pi/2. \qquad (4.326\text{j–m})$$

choosing $a = L$ in the Taylor series (4.325i) it reduces to a constant (4.326l), whose value (4.326m) must be $\pi/2$. This value could only hold if the bar was hanging vertically, and contradicts the assumption that it is clamped horizontally or obliquely (4.325b). This contradiction proves that the vertical inclination cannot be reached or exceeded.

4.8.5 Nonlinear Bending or a Heavy Cantilever

In the case of a beam subject to its own weight (Figure 4.12c) the transverse force at position s is due to the weight (4.327a) up to the tip, leading (4.325f) to the equation of the elastica (4.327b):

$$F(s) = -\rho g(L-s): \qquad EI\frac{d^2\theta}{ds^2} = -\rho g(L-s)\cos\theta. \qquad (4.327\text{a,b})$$

The transverse force (4.327a) is an analytic function of the arc length (4.325d) and choosing $a = 0$ in the Taylor series (4.325f) leads to the MacLaurin series (4.328a) for the angle of inclination as a function of the arc length, where the first term is $O(s)$ in the case (4.328a) of horizontal clamping:

$$\theta(0) = 0: \qquad \theta(s) = \sum_{n=1}^{\infty} a_n s^n; \qquad \theta'(L) = 0: \qquad 0 = \sum_{n=1}^{\infty} a_n L^{n-1} n, \qquad (4.328\text{a–d})$$

since the bending moment (4.320b) vanishes at the free end (4.328c) the boundary condition (4.328d) must be met by the coefficients of the series. Thus *the shape of the elastica for the nonlinear bending of a horizontally clamped (4.328a) uniform (4.10a) cantilever bar (4.90a) in a uniform gravity field (4.55a) is given by the angle of the tangent as a power series of the arc length (4.328b) with coefficients satisfying: (1) the balance equation (4.327b); (2) the boundary condition (4.328d) at the free end. The first three (next two) terms of (4.328b) correspond [Subsection 4.8.5 (Subsections 4.8.6 and 4.8.7)] to the linear approximation for weak bending (first and second-order nonlinear corrections for strong bending).* The series (4.328b) truncated at the order $n = N$ would specify a polynomial in s of degree N that cannot be an exact solution of (4.327b) because: (i) the l.h.s. becomes a polynomial of degree $N - 2$ in s; (ii) the r.h.s. is an infinite series in s; (iii) thus (i) and (ii) cannot be exactly equal. Thus an exact solution of (4.327b) must be an infinite series (4.328b), namely the MacLaurin series (I.23.34b) obtained setting $a = 0$ in the Taylor series (4.325i). It follows (4.326m) that the sum of (4.328b) cannot reach or exceed $\pi/2$, so the series converges. Since only the first three to five terms of the series will be computed (Subsections 4.8.5 through 4.8.8.), the truncated series is valid for small arc length s such that the sum remains below $\pi/2$.

In order to obtain nonlinear corrections up to the second-order, the first five terms of the power series (4.328b) must be used:

$$\theta(s) = a_1 s + a_2 s^2 + a_3 s^3 + a_4 s^4 + a_5 s^5 + O(s^6). \tag{4.329a}$$

Substitution of (4.329a) in the r.h.s. of (4.327b) gives terms up to the third order in the arc length (4.329b).

$$\frac{d^2\theta}{ds^2} = 2a_2 + 6a_3 s + 12a_4 s^2 + 20a_5 s^3 + O(s^4); \tag{4.329b}$$

$$\cos\theta = 1 - \frac{\theta^2}{2} + O(\theta^4) = 1 - \frac{1}{2}(a_1 s + a_2 s^2)^2 + O(s^4) = 1 - \frac{1}{2}(a_1)^2 s^2 - a_1 a_2 s^3 + O(s^4). \tag{4.329c}$$

The same order of approximation is used in the cosine of the angle of inclination (4.329c) for substitution together with (4.329a) in (4.327b):

$$\frac{EI}{\rho g}\left[2a_2 + 6a_3 s + 12a_4 s^2 + 20a_5 s^3 + O(s^4)\right] = -(L-s)\left[1 - \frac{1}{2}(a_1)^2 s^2 - a_1 a_2 s^3 + O(s^4)\right]$$

$$= -L + s + \frac{L}{2}(a_1)^2 s^2 + a_1\left(a_2 L - \frac{1}{2}a_1\right)s^3 + O(s^4). \tag{4.330}$$

Equating the coefficients of powers of s on the r.h.s. of (4.330) leads to four relations

$$\frac{EI}{\rho g}\{a_2, a_3, a_4, a_5\} = \left\{-\frac{L}{2}, \frac{1}{6}, \frac{L}{24}(a_1)^2, \frac{a_1}{20}\left(a_2 L - \frac{1}{2}a_1\right)\right\}$$

$$= \left\{-\frac{L}{2}, \frac{1}{6}, \frac{L}{24}(a_1)^2, -\frac{a_1}{40}\left(\frac{\rho g L^2}{EI} + a_1\right)\right\}, \tag{4.331a--d}$$

which (1) determine the coefficients a_2 and a_3; and (2) specify the coefficients a_4 and a_5 in terms of a_1. The latter is determined from the boundary condition (4.328d) $\equiv$ (4.332a):

$$0 = \frac{EI}{\rho g}\left(a_1 + 2a_2 L + 3a_3 L^2 + 4a_4 L^3 + 5a_5 L^4\right)$$

$$= \frac{EI}{\rho g}a_1 - \frac{L^2}{2} + \frac{L^4}{6}(a_1)^2 - \frac{L^4}{8}a_1\left(\frac{\rho g L^2}{EI} + a_1\right), \tag{4.332a,b}$$

that is, quadratic equation in a_1.

4.8.6 Linear Approximation in Cartesian and Local Coordinates

The linear approximation (4.333a) corresponds to the first three terms in (4.329a) and hence (4.328d) to the first two terms in (4.332b), that is, linear in a_1, and leads to (4.333b):

$$\zeta'^2 \ll 1: \quad a_1 = \frac{\rho g L^2}{2EI}; \quad a_2 = -\frac{\rho g L}{2EI}, \quad a_3 = \frac{-\rho g L}{6EI}. \tag{4.333a–d}$$

The second (4.331a) ≡ (4.333c) and third (4.331b) ≡ (4.333d) coefficients are independent of a_1, and hence also independent of the boundary condition (4.328c). Substituting the first three coefficients (4.333b–d) in (4.329a) specifies the angle of inclination of the elastica (4.334a) as a function of the arc length:

$$\theta_4(s) = \frac{\rho g s}{6EI}\left[3L^2 - 3Ls + s^2 + O(s^3)\right]; \qquad M_4(s) = \frac{\rho g}{2E}\left[L^2 - 2Ls + s^2 + O(s^3)\right] \tag{4.334a,b}$$

the corresponding (4.320b) bending moment (4.334b) vanishes at the tip. In the linear approximation (4.333a) ≡ (4.335a):

$$\zeta'^2 \ll 1: \quad s \sim x, \quad \theta \sim \zeta', \tag{4.335a–c}$$

the arc length (angle of inclination) corresponds to the longitudinal coordinate (4.335b) [slope (4.335c)]. Thus, the shape of the elastica in the linear approximation (4.335a–c) is the same in local (4.339b) and Cartesian (4.336a,b) coordinates:

$$\zeta'^2 \ll 1: \quad \zeta'_4(x) = \frac{\rho g x}{6EI}\left[3L^2 - 3Lx + x^2 + O(x^3)\right], \tag{4.336a,b}$$

in agreement with (4.78b) ≡ (4.336b). Integrating (4.336b) from the clamped end (4.337a) specifies the shape of the elastica that in the linear case is identical in terms of Cartesian (4.78a) and local (4.337b) coordinates:

$$\zeta_4(0) = 0: \quad \zeta_4(s) = \frac{\rho g s^2}{24EI}\left[6L^2 - 4sL + s^2 + O(s^3)\right]. \tag{4.337a,b}$$

It has been shown that *the linear bending of a cantilever or clamped-free (4.90b–e) [(4.328a,c)] uniform (4.10a) bar (4.90a) by its own weight has identical displacement (4.78a) [≡ (4.337b)] and slope (4.78b) ≡ (4.334a)] in Cartesian (local) coordinates (4.335b,c) corresponding to a small slope (4.333a) ≡ (4.335a) [neglecting terms of order $O(s^4)$ in the slope (4.334b) and order $O(s^5)$ in the displacement (4.337b)].* Next are obtained the first (second) order nonlinear correction corresponding to the terms of order $O(s^4)\left[O(s^5)\right]$ in the angle of inclination θ expanded in powers (4.328b) of the arc length s.

4.8.7 First-Order Nonlinear Correction

The first-order nonlinear correction corresponds to taking the first four terms on the r.h.s. of (4.329a), leading to the first three terms in the boundary condition (4.332b), specifying a_1 as a root of the quadratic polynomial (4.338a):

$$0 = (a_1)^2 + \frac{6EI}{\rho g L^4}a_1 - \frac{3}{L^2}; \quad a_1^\pm = -\frac{3EI}{\rho g L^4}\left\{1 \pm \left|1 + \frac{\rho^2 g^2 L^6}{3E^2 I^2}\right|^{1/2}\right\}. \tag{4.338a,b}$$

The root $a_1^+ < 0$ in (4.338b) is discarded because it would lead to a displacement opposite to the weight and thus is a spurious consequence of the quadratic relation (4.338a). The remaining root (4.338b) is approximated using (I.25.37a–c) the binomial expansion

$$a_1^- = -\frac{3EI}{\rho g L^4}\left\{1 - \left[1 + \frac{\rho^2 g^2 L^6}{6E^2 I^2} - \frac{1}{8}\left(\frac{\rho^2 g^2 L^6}{3E^2 I^2}\right)^2 + O\left(\frac{\rho^6 g^6 L^{18}}{E^6 I^6}\right)\right]\right\}$$

$$= \frac{\rho g L^2}{2EI} - \frac{\rho^3 g^3 L^8}{24 E^3 I^3} + O\left(\frac{\rho^5 g^5 L^{14}}{E^5 I^5}\right), \tag{4.339}$$

consisting of (1) the linear approximation (4.333b) as the leading term; and (2) the first-order nonlinear correction as the next term. Substituting (4.339) in (4.331c,d) specifies the fourth coefficient:

$$a_4 = \frac{\rho g L}{24 EI}\left(a_1^-\right)^2 = \frac{\rho^3 g^3 L^5}{96 E^3 I^3}, \tag{4.340}$$

with the same level of approximation as (4.339); substituting (4.333c,d, 4.339, 4.340) in (4.327) specifies the angle of inclination of the elastica as a function of the arc length (4.341a) that adds the next order to (4.334a):

$$O_4(s) = \frac{\rho g s}{6EI}\left[3L^2 - 3Ls + s^2 - \frac{\rho^2 g^2 L^5}{16 E^2 I^2}(4L^3 - s^3) + O(s^4)\right]; \tag{4.341a}$$

$$M_4(s) = \frac{\rho g}{2}\left[L^2 - 2Ls + s^2 - \frac{\rho^2 g^2 L^5}{12}(L^3 - s^3) + O(s^4)\right], \tag{4.341b}$$

the corresponding (4.320b) bending moment (4.341b) vanishes at the tip.

4.8.8 Second-Order Nonlinear Correction

The second-order nonlinear correction corresponds to all five terms in (4.329a) and all four terms in the quadratic relation (4.332b) ≡ (4.342a) with roots (4.342b):

$$(a_1)^2 + 2X\,a_1 - \frac{12}{L^2} = 0, \quad a_1^\pm = -X\left\{1 \pm \left|1 + \frac{12}{L^2 X^2}\right|^{1/2}\right\}, \tag{4.342a,b}$$

where

$$X \equiv \frac{12EI}{\rho g L^4} - \frac{3\rho g L^2}{2EI} = \frac{12EI}{\rho g L^4}\left(1 - \frac{\rho^2 g^2 L^6}{8E^2 I^2}\right) \tag{4.342c}$$

to the same order of approximation as (4.339). As before, only the positive root in (4.342b) corresponding to displacement in the same direction as the weight should be considered:

$$a_1^- = X\left\{-1 + \left[1 + \frac{6}{L^4 X^4} - \frac{18}{L^4 X^4} + O\left(\frac{1}{L^4 X^4}\right)\right]\right\} = \frac{6}{L^4 X} - \frac{18}{L^4 X^3} + O\left(\frac{1}{L^6 X^5}\right), \tag{4.343a}$$

where the first three terms of the binomial series (I.25.37a–c) are used. Substitution of (4.342c) in (4.343a) leads to

$$a_1^- = \frac{\rho g L^2}{2EI}\left(1 + \frac{\rho^2 g^2 L^6}{8E^2 I^2}\right) - \frac{\rho^3 g^3 L^8}{96E^2 I^2} = \frac{\rho g L^2}{2EI} + \frac{5\rho^3 g^3 L^8}{96E^3 I^3}, \tag{4.343b}$$

which coincides with (4.339) to the lowest order but differs in the coefficient to next order, that is 5/96 in (4.343b) and $-1/24 = -4/96$ in (4.339). The remaining coefficient is (4.331d):

$$a_5 = -\frac{\rho g}{40EI}a_1^-\left(\frac{\rho g L^2}{EI} + a_1^-\right) = -\frac{\rho g}{40EI}\frac{\rho g L^2}{2EI}\frac{3}{2}\frac{\rho g L^2}{EI} = -\frac{3\rho^3 g^3 L^2}{160E^3 I^3}, \tag{4.344}$$

to the same order of approximation as (4.343b).

4.8.9 Comparison of Linear and Nonlinear Terms

Substituting (4.343b, 4.333c,d, 4.340, 4.344) in (4.329a) leads to

$$\theta_4(s) = \frac{\rho g s}{6EI}\left[3L^2 - 3Ls + s^2 + \frac{\rho^2 g^2 L^4}{16E^2 I^2}\left(5L^4 + s^3 L - \frac{9}{5}s^4\right) + O\left(s^5\right)\right]; \tag{4.345a}$$

$$M_4(s) = \frac{\rho g}{2}\left[L^2 - 2Ls + s^2 + \frac{\rho^2 g^2 L^4}{48}\left(5L^4 + 4s^3 L - 9s^4\right) + O\left(s^5\right)\right], \tag{4.345b}$$

Thus, *the angle of inclination of the elastica as a function of the arc length for nonlinear bending of a uniform (4.10a) bar (4.90a) by its own weight in a uniform gravity field (4.55a) is given by (4.341a) [(4.345a)] including the first (second) order nonlinear correction.* The corresponding (4.320b) bending moment (4.345b) vanishes at the tip. The substitution of local by Cartesian coordinates (4.335a–c) that applies (4.334a) ≡ (4.336b) in the linear approximation does not apply to the nonlinear terms of the first (second) order in (4.341a) [(4.345a)]. The comparison of the first (4.341a) and second (4.345a) nonlinear approximations shows that (1) the linear terms are the same (4.334a) and satisfy (4.346b) the boundary condition (4.228c) = (4.346a) at the free end; (2) the first-order nonlinear correction (4.341a) adds an $O(s^4)$ term and a correction to the $O(s)$ term that satisfies (4.346c) the same boundary condition (4.346a); (3) the second-order nonlinear correction (4.345a) adds an $O(s^5)$ term, retains the $O(s^4)$ term and changes the $O(s)$ term so that the boundary condition (4.346a) is again satisfied (4.346d):

$$\lim_{s \to L}\frac{d\theta}{ds} = 0: \quad \lim_{s \to L}\frac{d}{ds}\left(3L^2 s - 3Ls^2 + s^3\right) = 0, \tag{4.346a,b}$$

$$\lim_{s \to L}\frac{d}{ds}\left(4L^3 s - s^4\right) = 0, \tag{4.346c}$$

$$\lim_{s \to L}\frac{d}{ds}\left(5L^4 s + s^4 L - \frac{9}{5}s^5\right) = 0. \tag{4.346d}$$

The higher-order nonlinear corrections can be obtained similarly, adding more terms (4.328b) to (4.329a) and leading to a boundary condition (4.328d) involving cubic or higher powers of a_1, resulting in a more tedious algebra. The nonlinear corrections of all orders are obtained next from the exact solution for strong bending of a uniform cantilever bar by a concentrated force at the tip (Subsections 4.8.10 through 4.8.15).

4.8.10 Clamped-Free Bar with a Concentrated Force at the Tip

In the case of strong bending (4.322b) of a uniform (4.10a) clamped-free bar (4.90a) by a concentrated force P at the tip (Figure 4.13), the transverse force is constant and equal to the reaction force at the clamped support (4.347a) leading to (4.347b):

$$R_- = -P = \text{const}: \quad -\frac{P}{EI}\cos\theta = \frac{d^2\theta}{ds^2} = \frac{d\theta'}{ds} = \frac{d\theta'}{d\theta}\frac{d\theta}{ds} = \theta'\frac{d\theta'}{d\theta}, \quad \theta' \equiv \frac{d\theta}{ds} \equiv k, \qquad \text{(4.347a–c)}$$

where (4.347c) is the derivative of the angle of inclination of the tangent with regard to the arc length that coincides with the curvature (4.319). Integration of (4.347c) leads to (4.348c):

$$\theta(L) = \theta_0, \theta'(L) = 0: \quad \frac{EI}{2P}\theta'^2 + \sin\theta = \text{const} = \sin\theta_0, \qquad \text{(4.348a–c)}$$

where the constant of integration is determined noting that at the free end $s = L$, (1) the bending moment (4.320b) is zero (4.348b); and (2) the angle of the tangent (4.348a) is generally nonzero. For a horizontally clamped bar, the angle of inclination is zero at the clamped end (4.349a), and integration of (4.348a) leads to (4.349b):

$$s(0) = 0: \quad s = \int_0^\theta \frac{ds}{d\theta}d\theta = \int_0^\theta \frac{d\theta}{\theta'} = \sqrt{\frac{EI}{2P}}\int_0^\theta \frac{d\theta}{\sqrt{\sin\theta_0 - \sin\theta}}. \qquad \text{(4.349a,b)}$$

The Cartesian coordinates (4.318a) [(4.318b)] with the origin at the clamped end (4.350a) [(4.351a)] are given in terms of the angle of inclination by (4.350b) [(4.351b)]:

$$x(0) = 0: \quad x = \int_0^\theta \frac{dx}{ds}\frac{ds}{d\theta}d\theta = \sqrt{\frac{EI}{2P}}\int_0^\theta \frac{\cos\theta}{\sqrt{\sin\theta_0 - \sin\theta}}d\theta$$

$$= -\sqrt{\frac{2EI}{P}}\left[\sqrt{\sin\theta_0 - \sin\theta}\right]_0^\theta = \sqrt{\frac{2EI}{P}}\left[\sqrt{\sin\theta_0} - \sqrt{\sin\theta_0 - \sin\theta}\right], \qquad \text{(4.350a,b)}$$

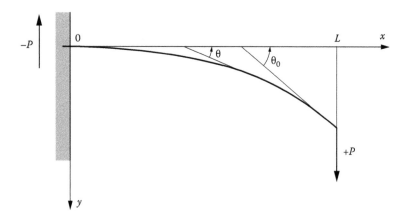

FIGURE 4.13 The weak or strong bending of a uniform clamped-free bar by a transverse force concentrated at the tip is stable in sense that the deflection is proportional to the load, and thus: (1) one cannot vanish if the other is nonzero; and (2) if one is continuous the other is too.

$$y(0) = 0: \quad y = \int_0^\theta \frac{dy}{ds}\frac{ds}{d\theta}d\theta = \sqrt{\frac{EI}{2P}} \int_0^\theta \frac{\sin\theta}{\sqrt{\sin\theta_0 - \sin\theta}}d\theta, \qquad (4.351a,b)$$

involving an elementary (elliptic) integral. It has been shown that *a clamped-free uniform (4.10a) bar (4.90a) strongly bent by a concentrated force P at the tip (Figure 4.13) has an elastica whose shape as a function of the angle of inclination is specified by(1/2) the Cartesian coordinates (4.350b) ≡ (4.352a) and (4.351b) ≡ (4.352b):*

$$x(\theta) = \left(\sqrt{\frac{2EI}{P}} \sqrt{\sin\theta_0} - \sqrt{\sin\theta_0 - \sin\theta} \right), \quad y(\theta) = \sqrt{\frac{EI}{2P}} \int_0^\theta \frac{\sin\theta}{\sqrt{\sin\theta_0 - \sin\theta}} = d\theta; \qquad (4.352a,b)$$

(3) the arc length is given by (4.349b) ≡ (4.353a):

$$s(\theta) = \sqrt{\frac{EI}{2P}} \int_0^\theta \frac{d\theta}{\sqrt{\sin\theta_0 - \sin\theta}}; \quad L = s(\theta_0) = \sqrt{\frac{EI}{2P}} \int_0^{\theta_0} \frac{d\theta}{\sqrt{\sin\theta_0 - \sin\theta}}; \qquad (4.353a,b)$$

and (4) the angle of inclination of the tangent at the tip (4.348a) is related to the length of the bar by (4.353b). The shape of the elastica in Cartesian coordinates $y = \zeta(x)$ *is obtained eliminating* θ *between (4.352a,b); this requires the evaluation of the elliptical integral (4.352b), which can be done using double series (Subsection 4.8.11); the same method applies to the evaluation of (4.353a,b).*

4.8.11 Evaluation of Elliptic Integrals by Double Series

Since the angle of inclination of the tangent is maximum at the tip (4.354a), the binomial series (I.24.37a–c) ≡ (4.354c) converges over the whole length of the bar (4.354b):

$$0 \le \theta < \theta_0; \quad \left| \frac{\sin\theta}{\sin\theta_0} \right| < 1: \quad \left| 1 - \frac{\sin\theta}{\sin\theta_0} \right|^{-1/2} = \sum_{n=0}^\infty (-)^n \binom{-1/2}{n} \left(\frac{\sin\theta}{\sin\theta_0} \right)^n = \sum_{n=0}^\infty a_n \left(\frac{\sin\theta}{\sin\theta_0} \right)^n, \qquad (4.354a-c)$$

where the coefficients are given by (2.159a,b) ≡ (II.6.56a–b) ≡ (II.6.57a–g). Substituting (4.354c) in (4.353a) specifies the arc length (4.355):

$$\sqrt{\frac{2P}{EI}} s(\theta) = \sum_{n=0}^\infty a_n \int_0^\theta \sin^n\theta \sin^{-1/2-n}\theta_0 d\theta$$

$$= \sin^{-1/2}\theta_0 \int_0^\theta \left\{ 1 + \frac{1}{2}\frac{\sin\theta}{\sin\theta_0} + \sum_{n=2}^\infty \frac{(2n-3)!!}{(2n)!!} \left(\frac{\sin\theta}{\sin\theta_0} \right)^n \right\} d\theta. \qquad (4.355)$$

The even (odd) powers of sine can be expanded in terms of cosines (sines) of multiple angles by (II.5.79a) ≡ (4.356a) [(II.5.79b) ≡ (4.356b)]:

$$\sin^{2n}\theta = 2^{-2n} \left\{ \binom{2n}{n} + (-)^n 2 \sum_{m=0}^{n-1} (-)^m \binom{2n}{m} \cos[2(n-m)\theta] \right\}, \qquad (4.356a)$$

$$\sin^{2n+1}\theta = 2^{-2n}(-)^n \sum_{m=0}^n (-)^m \binom{2n+1}{m} \sin[(2n-2m+1)\theta]. \qquad (4.356b)$$

Substituting (4.356a,b) in (4.355) and performing the integrations

$$\left|\frac{2P}{EI}\sin\theta_0\right|^{1/2} s(\theta) = \theta + \frac{1-\cos\theta}{2\sin\theta_0}$$

$$+ \sum_{n=1}^{\infty} \frac{(4n-3)!!}{(4n)!!}(2\sin\theta_0)^{-2n}\left\{\frac{(2n)!}{n!n!} + (-)^n 2\sum_{m=0}^{n-1}\frac{(-)^m(2n)!}{m!(2n-m)!}\frac{\sin[2(n-m)\theta]}{2n-2m}\right\}$$

$$+ 2\sum_{n=1}^{\infty}\frac{(4n-1)!!}{(4n+2)!!}(2\sin\theta_0)^{-2n-1}(-)^n\sum_{m=0}^{n}\frac{(-)^m(2n+1)!}{m!(2n-m+1)!}\frac{1-\cos[(2n-2m+1)\theta]}{2n-2m+1},$$

$$\text{(4.357)}$$

specifies the arc length in terms of the angle of inclination, including nonlinear terms of all orders.

4.8.12 Nonlinear Corrections to All Orders

The other elliptic integral (4.352b) is evaluated similarly:

$$\left|\frac{2P}{EI}\sin\theta_0\right|^{1/2} y(\theta) = \int_0^\theta \sin\theta\left|1-\frac{\sin\theta}{\sin\theta_0}\right|^{-1/2}d\theta = \sum_{n=0}^{\infty}a_n\sin^{-n}\theta_0\int_0^\theta\sin^{n+1}d\theta$$

$$= \int_0^\theta\left[\sin\theta + \frac{1-\cos(2\theta)}{4\sin\theta_0} + \sum_{n=2}^{\infty}\frac{(2n-3)!!}{(2n)!!}\frac{\sin^{n+1}\theta}{\sin^n\theta_0}\right]d\theta. \quad\text{(4.358)}$$

Splitting into even (4.356a) and odd (4.356b) powers leads to the integration

$$\left|\frac{2P}{EI}\sin\theta_0\right|^{1/2} y(\theta) = 1-\cos\theta + \frac{2\theta-\sin(2\theta)}{8\sin\theta_0}$$

$$+ \sum_{n=2}^{\infty}\frac{(4n-3)!!}{(4n)!!}(2\sin\theta_0)^{-2n}(-)^n\sum_{m=0}^{n}\frac{(-)^m(2n+1)!}{m!(2n-m+1)!}\frac{1-\cos[(2n-2m+1)\theta]}{2n-2m+1}$$

$$+ 2\sum_{n=2}^{\infty}\frac{(4n-5)!!}{(4n-2)!!}(2\sin\theta_0)^{-2n-1}\left\{\frac{(2n)!}{n!n!} + (-)^n\sum_{m=0}^{n-1}\frac{(-)^m(2n)!}{m!(2n-m)!}\frac{\sin[2(n-m)\theta]}{2n-2m}\right\}.$$

$$\text{(4.359)}$$

Thus, *the shape of the elastica for the strong bending of a uniform (4.10a) clamped-free bar (4.90a) by a concentrated force P at the free end (Figure 4.13) is given by (4.352a, 4.359) using the angle of inclination of the tangent as a parameter that by elimination specifies the shape in Cartesian coordinates* $y = \zeta(x)$. *The arc length is given by (4.357) and relates (4.353a) the length of the bar to the inclination at the tip:*

$$\left|\frac{2P}{EI}\sin\theta_0\right|^{1/2} L = \theta_0 + \frac{1-\cos\theta_0}{2\sin\theta_0}$$

$$+ \sum_{n=1}^{\infty}\frac{(4n-3)!!}{(4n)!!}(2\sin\theta_0)^{-2n}\left\{\frac{(2n)!}{n!n!} + (-)^n 2\sum_{m=0}^{n-1}\frac{(-)^m(2n)!}{m!(2n-m)!}\frac{\sin[2(n-m)\theta_0]}{2n-2m}\right\}$$

$$+ 2\sum_{n=1}^{\infty}\frac{(4n-1)!!}{(4n+2)!!}(2\sin\theta_0)^{-2n-1}(-)^n\sum_{m=1}^{n}\frac{(-)^m(2n+1)!}{m!(2n-m+1)!}\frac{1-\cos[2(n-m+1)\theta_0]}{2n-2n+1}.$$

$$\text{(4.360)}$$

The lowest (second) lowest-order terms in (4.352a, 4.357, 4.359, 4.360) correspond to the linear approximation (Subsections 4.4.1 and 4.8.13) [lowest-order nonlinear correction (Subsection 4.8.14)].

4.8.13 Linear Approximation to Nonlinear Bending

In the linear case of weak bending (4.12a) of a uniform (4.10a) clamped-free bar (4.90a) by a concentrated force P at the tip, the shape of the elastica is specified by the influence function (4.101b) = (4.361a):

$$G_4(x;L) = \frac{Px^2}{6EI}(3L - x), \quad \theta \sim \tan\theta = \zeta'(x) = \frac{Px}{2EI}(2L - x) \le \theta_0 = \frac{PL^2}{2EI}, \tag{4.361a–c}$$

leading to the slope (4.102a) ≡ (4.361b) and its maximum value (4.102b) ≡ (4.361c). These results for weak bending will be checked by comparison with the lowest-order term of the solution for strong bending. The length of the bar is related to the angle of inclination at the tip (4.353b) to the lowest order by (4.362):

$$L\sqrt{\frac{P}{2EI}} = \frac{1}{2}\int_0^{\theta_0} \frac{d\theta}{\sqrt{\theta_0 - \theta}} = \left[-\sqrt{\theta_0 - \theta}\right]_0^{\theta_0} = \sqrt{\theta_0}, \tag{4.362}$$

which coincides with (4.361c) ≡ (4.362). The horizontal coordinate (4.352a) ≡ (4.363a) [arc length (4.353a) ≡ (4.363b)] coincide in the linear approximation:

$$\sqrt{\frac{P}{2EI}}x(\theta) = \sqrt{\theta_0} - \sqrt{\theta_0 - \theta} = \sqrt{\theta_0}\left(1 - \sqrt{1 - \frac{\theta}{\theta_0}}\right) = \frac{\theta}{2\sqrt{\theta_0}}, \tag{4.363a}$$

$$\sqrt{\frac{P}{2EI}}s(\theta) = \frac{1}{2}\int_0^{\theta} \frac{d\theta}{\sqrt{\theta_0 - \theta}} = \left[-\sqrt{\theta_0 - \theta}\right]_0^{\theta_0} = \sqrt{\theta_0} - \sqrt{\theta_0 - \theta}. \tag{4.363b}$$

The vertical coordinate (4.352b) is given to the lowest order by

$$\sqrt{\frac{P}{2EI}}y(\theta) = \frac{1}{2}\int_0^{\theta} \frac{\theta}{\sqrt{\theta_0 - \theta}}d\theta = \frac{1}{2}\int_0^{\theta}\left(\frac{\theta_0}{\sqrt{\theta_0 - \theta}} - \sqrt{\theta_0 - \theta}\right)d\theta$$

$$= \left[-\theta_0\sqrt{\theta_0 - \theta} + \frac{1}{3}(\theta_0 - \theta)\sqrt{\theta_0 - \theta}\right]_0^{\theta}$$

$$= \frac{\theta_0^{3/2}}{3}\left(2 - 3\left|1 - \frac{\theta}{\theta_0}\right|^{1/2} + \left|1 - \frac{\theta}{\theta_0}\right|^{3/2}\right). \tag{4.364}$$

From (4.363a, 4.361c) follows

$$\sqrt{1 - \frac{\theta}{\theta_0}} = 1 - \frac{1}{2}\frac{\theta}{\theta_0} = 1 - x\sqrt{\frac{P}{2EI\theta_0}} = 1 - \frac{x}{L}, \tag{4.365}$$

which may be substituted in (4.364):

$$y = \sqrt{\frac{2EI}{P}}\,\frac{\theta_0^{3/2}}{3}\left[2 - 3\left(1 - \frac{x}{L}\right) + \left(1 - \frac{x}{L}\right)^3\right] = \frac{Px^2}{6EI}(3L - x) = G_4(x;L),$$ (4.366)

leading to (4.366), which coincides with the shape of the elastica (4.361a) ≡ (4.366) in the linear case of weak bending. The method used to check the linear approximation can be taken to the next order to specify the lowest-order nonlinear correction, for example, for the relation (4.361c) between the length of the bar and the slope at the tip.

4.8.14 Lowest-Order Nonlinear Correction

The lowest-order nonlinear correction uses the first two terms in the power series for the circular sine (II.5.7.13b) ≡ (4.367a) instead of only the first in the linear approximation (4.362):

$$\sin\theta = \theta - \frac{\theta^3}{6} + O(\theta^5); \quad \sin\theta_0 - \sin\theta = \theta_0 - \theta - \frac{1}{6}\left(\theta_0^3 - \theta^3\right)$$

$$= (\theta_0 - \theta)\left[1 - \frac{1}{6}\left(\theta_0^2 + \theta_0\theta + \theta^2\right)\right].$$ (4.367a,b)

From (4.367a) follows (4.367b), which appears in the integrals (4.353a,b, 4.352b) in the form

$$\left|\sin\theta_0 - \sin\theta\right|^{-1/2} = \left|\theta_0 - \theta\right|^{-1/2}\left[1 + \frac{1}{12}\left(\theta_0^2 + \theta_0\theta + \theta^2\right)\right].$$ (4.368)

The term in curved brackets is decomposed into a sum of powers of the difference $\theta_0 - \theta$:

$$\theta_0^2 + \theta_0\theta + \theta^2 = (\theta_0 - \theta)^2 + 3\theta_0\theta = (\theta_0 - \theta)^2 - 3\theta_0(\theta_0 - \theta) + 3\theta_0^2,$$ (4.369)

for substitution in (4.368):

$$\left|\sin\theta_0 - \sin\theta\right|^{-1/2} = \left(1 + \frac{\theta_0^2}{4}\right)\left|\theta_0 - \theta\right|^{-1/2} - \frac{\theta_0}{4}\left|\theta_0 - \theta\right|^{1/2} + \frac{1}{12}\left|\theta_0 - \theta\right|^{3/2}.$$ (4.370)

This form (4.370) leads to an immediate integration when substituted in the length of the bar (4.353b), yielding

$$L\sqrt{\frac{P}{2EI}} = \frac{1}{2}\int_0^{\theta_0}\left|\sin\theta_0 - \sin\theta\right|^{-1/2}d\theta$$

$$= \left[-\left(1 + \frac{\theta_0^2}{4}\right)\left|\theta_0 - \theta\right|^{1/2} + \frac{\theta_0}{12}\left|\theta_0 - \theta\right|^{3/2} - \frac{1}{60}\left|\theta_0 - \theta\right|^{5/2}\right]_0^{\theta_0}$$

$$= \sqrt{\theta_0}\left[1 + \theta_0^2\left(\frac{1}{4} - \frac{1}{12} + \frac{1}{60}\right)\right] = \sqrt{\theta_0}\left(1 + \frac{11}{60}\theta_0^2\right).$$ (4.371)

The factor before the bracket on the l.h.s. of (4.371) corresponds to the linear approximation (4.361c) for weak bending, and the terms in the bracket specify the lowest-order nonlinear correction for strong bending.

4.8.15 Comparison of Weak and Strong Bending of a Bar

The method used to obtain from the exact relation between the length of the bar and inclination of the tangent at the tip (4.353b) in the zero (first)-order approximation corresponding to weak bending (4.361c) [the lowest-order nonlinear correction for strong bending (4.371)] can also be applied to the arc length (4.353a) [Cartesian coordinates (4.352a,b)] to obtain the lowest-order nonlinear correction to (4.363b) [(4.363a, 4.364)]. The comparison between linear and nonlinear bending can be illustrated by the length of the bar (4.371) ≡ (4.372a) as a function of the inclination of the tangent at the tip:

$$L = \sqrt{\frac{2EI\theta_0}{P}}\left(1+\frac{11}{60}\theta_0^2\right), \quad \theta_0 = \frac{PL^2}{2EI}\left(1-\frac{11P^2L^4}{120E^2I^2}\right), \tag{4.372a,b}$$

whose inverse is (4.372b). The latter (4.372b) is obtained from (4.372a) by (1) passing the curved bracket from the r.h.s. to the l.h.s. (4.373a):

$$\sqrt{\theta_0} = L\sqrt{\frac{P}{2EI}}\left(1-\frac{11}{60}\theta_0^2\right) = L\sqrt{\frac{P}{2EI}}\left(1-\frac{11P^2L^4}{240E^2I^2}\right); \tag{4.373a,b}$$

(2) substituting the linear approximation (4.361c) in the nonlinear term in the curved brackets in (4.373a) leading to (4.373b) with the same order of approximation; and (3) the square of (4.373b) is (4.372b). Thus, *the lowest-order nonlinear bending compared with the linear bending (4.12a) for a uniform (4.10a) clamped-pinned bar (4.90a) with a concentrated force at the tip (Figure 4.13) shows that the non-linear effect (1) increases the length for the same slope at the tip (4.372a); and (2) decreases the slope at the tip for the same length of the bar (4.372b).*

4.9 Elastic Instability: Buckling and Collapse

The nonlinear or linear bending of a uniform clamped-free bar can be considered either for a transversal or axial concentrated force [Subsections 4.8.10 through 4.8.15 (Subsections 4.9.1 through 4.9.3)]. The case of an axial force (Subsection 4.9.1) leads to elliptic integrals in the nonlinear case as before for a transverse load (Subsection 4.8.10). The linear approximation is sufficient to demonstrate one type of elastic instability for an axial load; the bar does not deflect until a critical buckling load is reached; this is confirmed by the lowest-order nonlinear term (Subsection 4.9.2). The linear approximation is sufficient to specify the shape of the buckled beam (Subsection 4.9.3). Another case is nonlinear bending of a uniform bar with sliding supports by a transverse concentrated force applied at the middle (Subsections 4.9.4 through 4.9.8). The sliding support implies for strong bending a reaction force orthogonal to the bar, because a tangential reaction cannot be compensated (Subsection 4.9.4); in the linear case, the reaction forces are transverse to the undeflected position, corresponding to weak bending with pinned ends (Subsection 4.9.5). The first-order nonlinear correction (Subsection 4.9.6) shows the effect of strong bending on the shape and slope of the elastica (Subsection 4.9.7). In the case of linear bending, the maximum deflection is proportional to the force; the lowest-order nonlinear correction shows that for larger deflections the force decreases, implying instability: the bar slides between the supports and collapses (Subsection 4.9.7). The critical transverse force load for elastic collapse (Subsection 4.9.7) is smaller than the critical axial force for elastic buckling (Subsection 4.9.3) for a bar with the same length and cross section, made of the same material. Besides the buckling (collapse) elastic instabilities [Subsections 4.9.1

through 4.9.3 (Subsections 4.9.4 through 4.9.7)], two other types of instabilities concern a cantilever bar, that is, a clamped-free bar bent under its own weight; placing a linear (rotary) spring at the tip so as to increase the deflection (twist) can lead to elastic instability [Example 10.10 (Example 10.11)]. The displacement and twist instabilities (Subsection 4.9.9) can be compared dimensionally (Subsection 4.9.8) with the buckling and collapse instabilities, completing the four sets of elastic instabilities.

4.9.1 Nonlinear Bending by an Axial Force

Consider an elastic bar (or beam) subject to an axial force at the tip (Figure 4.14). Keeping the x-axis along the undeflected position of the bar and the y-axis orthogonal to it, the balance of the bending moment and the transverse force (4.5a) is replaced by (4.374a) leading to (4.374b)

$$0 = (M + dM) - M + F\,dy: \quad F = -\frac{dM}{dy}; \quad EI\frac{d^2\theta}{ds^2} = -\frac{dM}{ds} = -\frac{dM}{dy}\frac{dy}{ds} = F\sin\theta. \quad (4.374a\text{--}c)$$

Substitution of (4.320b) in (4.374b) leads to (4.374c). Rewriting the latter in the form (4.374c) $\equiv$ (4.375b) leads by integration to (4.375c):

$$F = -P: \quad -\frac{P}{EI}\sin\theta = \frac{d^2\theta}{ds^2} = \theta'\frac{d\theta'}{d\theta}, \quad \theta'^2 - \frac{2P}{EI}\cos\theta = \text{const}, \quad (4.375a\text{--}c)$$

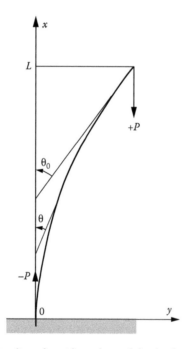

FIGURE 4.14 The weak or strong bending of a uniform clamped-free bar by an axial force concentrated at the tip is an example of buckling instability: (1) if the force is less than the critical bucking value there is no deflection at all; and (2) as the critical bucking load is exceeded the bar deflects with a slope at the tip determined by the difference between the axial force and critical load for buckling. The linear (nonlinear) elastica instabilities associated with the weak (strong) bending of a bar include buckling (collapse) by an axial (transverse) force [Figure 4.12 (4.13)]. The critical load for collapse with sliding supports is less than the buckling load for a clamped-free bar.

where (1) the transverse force is due to the concentrated force (4.375a); and (2) in (4.375b), (4.347b) is used. At the free end of the bar, the angle of inclination of the tangent is nonzero (4.376a), and the bending moment (4.320b) is zero (4.376b), determining the constant of integration in (4.375c) ≡ (4.376c):

$$\theta(L) = \theta_0, \theta'(L) = 0: \quad \frac{EI}{2P}\theta'^2 - \cos\theta = -\cos\theta_0; \quad \frac{d\theta}{ds} = \theta' = \sqrt{\frac{2P}{EI}}\sqrt{\cos\theta - \cos\theta_0}, \quad (4.376a\text{--}d)$$

Solving (4.376c) for θ' leads to (4.376d), whose integration specifies the arc length as a function of the angle of inclination of the tangent (4.377a):

$$s = \int_0^\theta d\theta = \sqrt{\frac{EI}{2P}} \int_0^\theta \frac{d\theta}{\sqrt{\cos\theta - \cos\theta_0}}; \quad L = \sqrt{\frac{EI}{2P}} \int_0^{\theta_0} \frac{d\theta}{\sqrt{\cos\theta - \cos\theta_0}}, \quad (4.377a,b)$$

In particular, the length of the bar (4.377b) is specified by the arc length (4.377a) at the tip (4.376a). The longitudinal (transversal) Cartesian coordinate (4.318a) [(4.318b)] is a function of the inclination specified by (4.378a) [(4.378b)]:

$$x = \int_0^\theta \frac{dx}{ds}\frac{ds}{d\theta}d\theta = \sqrt{\frac{EI}{2P}} \int_0^\theta \frac{\cos\theta}{\sqrt{\cos\theta - \cos\theta_0}}d\theta, \quad (4.378a)$$

$$y = \int_0^\theta \frac{dy}{ds}\frac{ds}{d\theta}d\theta = \sqrt{\frac{EI}{2P}} \int_0^\theta \frac{\sin\theta}{\sqrt{\cos\theta - \cos\theta_0}}d\theta$$

$$= \sqrt{\frac{EI}{2P}}\left[-2\sqrt{\cos\theta - \cos\theta_0}\right]_0^\theta = \sqrt{\frac{2EI}{P}}\left[\sqrt{1 - \cos\theta_0} - \sqrt{\cos\theta - \cos\theta_0}\right], \quad (4.378b)$$

which is an elliptic (4.378a) [elementary (4.378b)] integral. Thus, *the shape of the elastica for nonlinear bending of a uniform (4.10a) clamped-free bar (4.90a) by an axial concentrated force P at the tip (Figure 4.14) is specified by the Cartesian coordinates (4.378a,b) ≡ (4.379a,b) as a function of inclination of the tangent:*

$$x(\theta) = \sqrt{\frac{EI}{2P}} \int_0^\theta \frac{\cos\theta}{\sqrt{\cos\theta - \cos\theta_0}}d\theta, \quad y(\theta) = \sqrt{\frac{2EI}{P}}\left[\sqrt{1 - \cos\theta_0} - \sqrt{\cos\theta - \cos\theta_0}\right], \quad (4.379a,b)$$

where (1) elimination of θ leads to the explicit $y = \zeta(x)$ instead of the parametric form (4.379a,b); and (2) the angle of inclination θ [at the tip θ_0 of the bar (4.376a)] appears in the arc length (4.377a) [length of the bar (4.377b)].

4.9.2 Linear and Nonlinear Buckling Load

The relation (4.377b) between the length of the bar and the angle of inclination of the tangent at the tip has a zero-order approximation (4.380b) corresponding to weak bending:

$$\psi \equiv \frac{\theta}{\theta_0}: \quad L\sqrt{\frac{2P}{EI}} = \int_0^{\theta_0}\left|\frac{\theta_0^2 - \theta^2}{2}\right|^{-1/2}d\theta = \sqrt{2}\int_0^1\left|1 - \psi^2\right|^{-1/2}d\psi$$

$$= -\sqrt{2}\left[\arccos\psi\right]_0^1 = \sqrt{2}\frac{\pi}{2} = \frac{\pi}{\sqrt{2}}. \quad (4.380a,b)$$

In (4.380b) the change of variable (4.380a) was used, showing that the angle of inclination at the tip drops out of the result. The implication is that there is no deflection of the bar (4.381b) unless the concentrated force equals or exceeds the critical buckling load (4.381a):

$$P < P_b = \frac{\pi^2 EI}{4L^2}: \quad y = \zeta(x) = 0. \tag{4.381a,b}$$

This result can be checked by considering the lowest-order nonlinear correction corresponding to taking in (4.377b) the first two terms of the power series for the circular cosine (II.7.12b) $\equiv$ (4.382a) that implies (4.382b):

$$\cos\theta = 1 - \frac{\theta^2}{2} + \frac{\theta^4}{24}: \quad \cos\theta - \cos\theta_0 = \frac{\theta_0^2 - \theta^2}{2} - \frac{\theta_0^4 - \theta^4}{24} = \frac{\theta_0^2 - \theta^2}{2}\left(1 - \frac{\theta_0^2 + \theta^2}{12}\right). \tag{4.382a,b}$$

The expression (4.382b) appears in (4.377b) in the form

$$\left|\cos\theta - \cos\theta_0\right|^{-1/2} = \sqrt{2}\left|\theta_0^2 - \theta^2\right|^{-1/2}\left(1 + \frac{\theta_0^2 + \theta^2}{24}\right). \tag{4.383}$$

For the purpose of integration after substitution of (4.383) in (4.377b), it is convenient to express the term in curved brackets as a sum of powers of $\theta_0 - \theta$:

$$\theta_0^2 + \theta^2 = \left(\theta_0 - \theta\right)^2 + 2\theta_0\theta = \left(\theta_0 - \theta\right)^2 - 2\theta_0\left(\theta_0 - \theta\right) + 2\theta_0^2. \tag{4.384}$$

Substitution of (4.384) in (4.383) yields

$$\frac{1}{\sqrt{2}}\left|\cos\theta - \cos\theta_0\right|^{-1/2} = \left(1 + \frac{\theta_0^2}{12}\right)\left|\theta_0^2 - \theta^2\right|^{-1/2} - \frac{\theta_0}{12}\left|\theta_0^2 - \theta^2\right|^{1/2} + \frac{1}{24}\left|\theta_0^2 - \theta^2\right|^{3/2}, \tag{4.385}$$

to be substituted in (4.377b).

When substituting (4.385) in (4.377b) the following change of variable is made:

$$\theta = \theta_0 \sin\alpha: \quad d\theta = \theta_0\cos\alpha\, d\alpha, \quad \left|\theta_0^2 - \theta^2\right|^{1/2} = \theta_0\cos\alpha, \tag{4.386a-c}$$

leading to

$$L\sqrt{\frac{P}{EI}} = \int_0^{\pi/2}\left[1 + \frac{\theta_0^2}{12} - \frac{\theta_0^3}{12}\cos^2\alpha + \frac{\theta_0^4}{24}\cos^4\alpha\right]d\alpha$$

$$= \int_0^{\pi/2}\left\{1 + \frac{\theta_0^2}{12} - \frac{\theta_0^3}{24}\left[1 + \cos(2\alpha)\right] + \frac{\theta_0^4}{96}\left[1 + 2\cos(2\alpha) + \frac{1 + \cos(4\alpha)}{2}\right]\right\}d\alpha$$

$$= \int_0^{\pi/2}\left[1 + \frac{\theta_0^2}{12} - \frac{\theta_0^3}{24} + \frac{\theta_0^4}{64} - \frac{\theta_0^3}{48}\cos(2\alpha) + \frac{\theta_0^4}{192}\cos(4\alpha)\right]d\alpha. \tag{4.387}$$

Only the constant term in the square brackets makes a contribution to the integral, leading to

$$1 + \frac{\theta_0^2}{12} - \frac{\theta_0^3}{24} + \frac{\theta_0^4}{64} + O(\theta_0^5) = \frac{2L}{\pi}\sqrt{\frac{P}{EI}} = \sqrt{\frac{P}{P_b}}, \tag{4.388}$$

which specifies an axial force in excess of the critical buckling load (4.381a) for an inclination θ_0 at the tip of the bar. The relation is equivalent to $(4.388) \equiv (4.389)$ to the same order of approximation:

$$\frac{P}{P_b} = \frac{4L^2 P}{\pi^2 EI} = \left[1 + \frac{\theta_0^2}{12} - \frac{\theta_0^3}{24} + \frac{\theta_0^4}{64} + O(\theta_0^5)\right]^2$$

$$= 1 + \frac{\theta_0^2}{6} - \frac{\theta_0^3}{12} + \frac{\theta_0^4}{32} + \frac{\theta_0^4}{144} - \frac{\theta_0^5}{144} + O(\theta_0^6)$$

$$= 1 + \frac{\theta_0^2}{6} - \frac{\theta_0^3}{12} + \frac{11\theta_0^4}{288} - \frac{\theta_0^5}{144} + O(\theta_0^6). \tag{4.389}$$

The lowest-order nonlinear approximation to (4.389) is (4.390a), which can be inverted (4.390b):

$$P = P_b\left(1 + \frac{\theta_0^2}{6}\right) = \frac{\pi^2 EI}{4L^2}\left(1 + \frac{\theta_0^2}{6}\right), \quad \theta_0 = \left|6\left(\frac{P}{P_b} - 1\right)\right|^{1/2} = \left|6\left(\frac{4L^2 P}{\pi^2 EI} - 1\right)\right|^{1/2}. \tag{4.390a,b}$$

It has been shown that *the elastica of a uniform (4.10a) clamped-free bar (4.90a) with a concentrated axial force P at the tip (Figure 4.14) does not deform (4.381b) if the force is less than the critical buckling load (4.381a). If the load exceeds the critical value (4.381a), the bar buckles, and the angle of inclination at the tip is given by (4.390b), that is, the inverse of (4.390a). The latter is an approximation of (4.388), which specifies the load that must be applied so that the bar buckles to a given angle θ_0 of inclination at the tip. The shape of the buckled bar is considered next (Subsection 4.9.3).*

4.9.3 Linear Shape of a Buckled Bar

The shape of the buckled bar is calculated next in the linear approximation corresponding to weak bending of the lowest order. Higher orders could be calculated as in (4.382a,b–4.390a,b). The arc length (4.377a) [Cartesian coordinate along the bar (4.378a)] coincide (4.391a) [(4.392b)] in the linear approximation

$$x(\theta) = s(\theta) = \sqrt{\frac{EI}{2P}} \int_0^\theta \left|\frac{\theta_0^2 - \theta^2}{2}\right|^{-1/2} d\theta = \sqrt{\frac{EI}{P}}\left[-\arccos\left(\frac{\theta}{\theta_0}\right)\right]_0^\theta = \sqrt{\frac{EI}{P}}\left[\frac{\pi}{2} - \arccos\left(\frac{\theta}{\theta_0}\right)\right], \tag{4.391a,b}$$

which can be inverted (4.392a):

$$\theta(x) = \theta_0 \cos\left(\frac{\pi}{2} - x\sqrt{\frac{P}{EI}}\right) = \theta_0 \sin\left(x\sqrt{\frac{P}{EI}}\right); \quad \frac{\pi}{2} = L\sqrt{\frac{P_b}{EI}}. \tag{4.392a,b}$$

Thus, the angle of inclination is a sinusoidal function of position with a maximum value at the tip (4.392b), corresponding to the critical buckling load (4.381a). The transverse Cartesian coordinate (4.379b) is given by

$$y(\theta) = \sqrt{\frac{2EI}{P}} \left(\sqrt{\frac{\theta_0^2}{2}} - \sqrt{\frac{\theta_0^2 - \theta^2}{2}} \right) = \sqrt{\frac{EI}{P}}\, \theta_0 \left[1 - \left| 1 - \frac{\theta^2}{\theta_0^2} \right|^{1/2} \right]$$

$$= \sqrt{\frac{EI}{P}}\, \frac{\theta^2}{2\theta_0} = \frac{\theta_0}{2}\sqrt{\frac{EI}{P}}\, \sin^2\left(x\sqrt{\frac{P}{EI}} \right) \equiv \zeta(x). \tag{4.393}$$

The corresponding slope (4.394a) takes the value (4.394b) at the tip:

$$\zeta'(x) = \frac{\theta_0}{2}\sin\left(2x\sqrt{\frac{P}{EI}} \right), \quad \zeta'(L) = \frac{\theta_0}{2}\sin\left(2L\sqrt{\frac{P}{EI}} \right). \tag{4.394a,b}$$

The slope (4.394b) vanishes at the tip for the critical buckling load (4.381b) $\equiv$ (4.392b).

The linear approximation to the shape of the elastica (4.393) may be combined with the lowest-order nonlinear correction for the inclination at the tip (4.390b), using the critical buckling load (4.381b). Thus, *the shape of the elastica for the weak (4.12a) bending of a clamped-free uniform (4.10a) bar (4.90a) with an axial force at the tip (Figure 4.14) exceeding the critical buckling load (4.395a) is given by (4.393)* $\equiv$ *(4.395b)*:

$$P \geq P_b \equiv \frac{\pi^2 EI}{4L^2}: \quad \frac{\zeta(x)}{L} = \frac{\sqrt{6}}{\pi}\sqrt{1 - \frac{P_b}{P}}\, \sin^2\left(\frac{\pi x}{2L}\sqrt{\frac{P}{P_b}} \right), \tag{4.395a,b}$$

using the angle of inclination at the tip (4.390b). The corresponding slope is (4.394a) $\equiv$ *(4.396)*:

$$\zeta'(x) = \frac{\sqrt{6}}{2}\sqrt{\frac{P}{P_b} - 1}\, \sin\left(\frac{\pi x}{L}\sqrt{\frac{P}{P_b}} \right). \tag{4.396}$$

The deflection at the tip is (4.397):

$$\frac{\delta}{L} = \frac{\zeta(L)}{L} = \frac{\sqrt{6}}{\pi}\sqrt{1 - \frac{P_b}{P}}\, \sin^2\left(\frac{\pi}{2}\sqrt{\frac{P}{P_b}} \right). \tag{4.397}$$

The inclination at the tip is given by (4.390b), or with greater accuracy by (4.389).

4.9.4 Strong Bending with Sliding Supports

Another case of elastic instability relates to strong bending of a uniform (4.10a) elastic bar (4.90a) with a concentrated force at equal distance from the sliding supports (Figure 4.15). The **sliding support** cannot have a tangent reaction force, implying that the boundary conditions are (1) fixed position; and (2) reaction force orthogonal to the bar. In the linear case of small inclination, the boundary condition

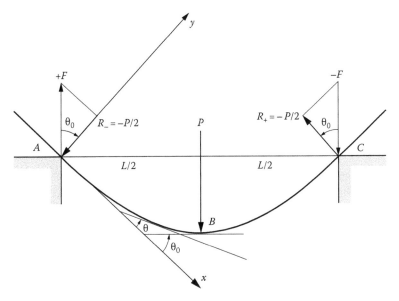

FIGURE 4.15 Besides buckling (Figure 4.14) another type of elastic instability is collapse that differs in being due to transverse rather than axial forces and is nonlinear in the sense it occurs only for strong bending. An example is the collapse of a bar with sliding supports under strong bending by a transverse concentrated force (Figure 4.15) applied at the middle: (1) the deflection is proportional to the force if it does not reach the critical collapse load; (2) when the critical collapse load is reached the bar slides between the supports and collapses; and (3) the collapse corresponds to the deflection increasing without increasing load. The collapse is possible because at a sliding support the reaction force is orthogonal to the bar. If the tangent to the bar becomes vertical at the sliding support, the reaction force is horizontal, and does not oppose the collapse of the bar between the supports. The 23 cases of bending of bars (List 4.1) include four elastic instabilities (List 4.2), namely the buckling (collapse) instabilities in Figure 4.15 (4.14) and Example 10.10 (Example 10.11) displacement (twist) instabilities in Figure 10.11d (10.12b).

(2) corresponds to a reaction force orthogonal to the undeformed position of the bar and corresponds to a pinned end. For nonlinear bending, the transverse force decreases by $-P$ when it crosses the lowest point where the concentrated force is applied. Since the transverse force is constant in the first half of the bar AB, it must equal the reaction at the support, leading to (4.398a):

$$F\cos\theta_0 = \frac{P}{2}; \quad \frac{P}{2}\frac{\cos\theta}{\cos\theta_0} = F\cos\theta = EI\frac{\mathrm{d}^2\theta}{\mathrm{d}s^2} = EI\theta'\frac{\mathrm{d}\theta'}{\mathrm{d}\theta}. \tag{4.398a,b}$$

Substitution of (4.398a) in (4.322b) leads to (4.398b), where (4.347b) is used. Integration of (4.398b) leads to (4.399c), where the constant of integration is zero because there is no bending moment (4.320b) at the first support (4.399a), and the x-axis is chosen tangent to the bar at that point (4.399b):

$$\theta'(0) = 0 = \theta(0); \quad \frac{P}{EI}\frac{\sin\theta}{\cos\theta_0} - \theta'^2 = \text{const} = 0; \quad \theta' = \frac{\mathrm{d}\theta}{\mathrm{d}s} = \sqrt{\frac{P\sin\theta}{EI\cos\theta_0}}. \tag{4.399a–d}$$

From (4.399c) $\equiv$ (4.399d) follows the arc length (4.400):

$$s(\theta) = \int_0^\theta \frac{\mathrm{d}s}{\mathrm{d}\theta}\mathrm{d}\theta = \int_0^\theta \frac{\mathrm{d}s}{\theta'} = \left|\frac{EI}{P}\cos\theta_0\right|^{1/2}\int_0^\theta \frac{\mathrm{d}\theta}{\sqrt{\sin\theta}}. \tag{4.400}$$

The longitudinal (4.318a) [transverse (4.318b)] Cartesian coordinate is given by (4.401a) [(4.401b)]:

$$x(\theta) = \int_0^\theta \frac{dx}{ds}\frac{ds}{d\theta}d\theta = \int_0^\theta \frac{\cos\theta}{\sqrt{\theta'}}d\theta = \left|\frac{EI}{P}\cos\theta_0\right|^{1/2} \int_0^\theta \frac{\cos\theta}{\sqrt{\sin\theta}}d\theta, \tag{4.401a}$$

$$y(\theta) = \int_0^\theta \frac{dy}{ds}\frac{ds}{d\theta}d\theta = \int_0^\theta \frac{\sin\theta}{\theta'}d\theta = \left|\frac{EI}{P}\cos\theta_0\right|^{1/2} \int_0^\theta \sqrt{\sin\theta}\,d\theta. \tag{4.401b}$$

The distance L between the sliding supports is related to the angle θ_0 with the x-axis by the condition that it equals the projection of the bar:

$$\frac{L}{2} = \int_0^\theta \cos(\theta - \theta_0)\frac{ds}{d\theta}d\theta = \left|\frac{EI}{P}\cos\theta_0\right|^{1/2} \int_0^{\theta_0} \frac{\cos(\theta - \theta_0)}{\sqrt{\sin\theta}}d\theta. \tag{4.402}$$

Thus, *the shape of the elastica for strong bending of a uniform (4.10a) bar (4.90a) with a transverse concentrated force P at equal distance from the sliding supports (Figure 4.15) is specified by the longitudinal (4.401a) and transversal (4.401b) Cartesian coordinate as a function of the inclination of the tangent; the arc length is given by (4.400), and the distance between the supports L and angle of the bar θ_0 with the line joining the supports at the first support are related by (4.402).* The shape of the elastica is considered next in the linear case (Subsection 4.9.5).

4.9.5 Shape of the Elastica for Linear Bending

The lowest-order approximation corresponds to weak bending, for which the arc length (4.400) and longitudinal coordinate (4.401a) coincide:

$$s(\theta) = x(\theta) = \sqrt{\frac{EI}{P}} \int_0^\theta \frac{d\theta}{\sqrt{\theta}} = 2\sqrt{\frac{EI\theta}{P}}. \tag{4.403a,b}$$

The transverse Cartesian coordinate (4.401b) is given by (4.404a):

$$y(\theta) = \sqrt{\frac{EI}{P}} \int_0^\theta \sqrt{\theta}\,d\theta = \frac{2}{3}\sqrt{\frac{EI}{P}}\theta^{3/2} = \frac{2}{3}\sqrt{\frac{EI}{P}}\left(\frac{x}{2}\sqrt{\frac{P}{EI}}\right)^3 = \frac{Px^3}{12EI} = \bar{\zeta}_3(x). \tag{4.404a,b}$$

Substitution of (4.403b) in (4.404a) specifies the shape of the elastica (4.404b) $\equiv$ (4.405a) and the maximum deflection (4.405b):

$$\bar{\zeta}_3(x) = \frac{Px^3}{12EI}, \quad \delta = \bar{\zeta}_{3\max} = \bar{\zeta}_3\left(\frac{L}{2}\right) = \frac{PL^3}{96EI}. \tag{4.405a,b}$$

The slope (4.406a) has a maximum (4.406b):

$$\bar{\zeta}_3'(x) = \frac{Px^2}{4EI}, \quad |\bar{\zeta}'|_{3\max} = \bar{\zeta}_3'\left(\frac{L}{2}\right) = \frac{PL^2}{16EI}. \tag{4.406a,b}$$

The corresponding bending moment is (4.407a):

$$-\bar{M}_3(x) = EI\bar{\zeta}_3''(x) = \frac{Px}{2}; \quad F_3(x) = EI\bar{\zeta}_3'''(x) = \frac{P}{2}, \qquad (4.407a,b)$$

The transverse force (4.407b) agrees with (4.398a) for a small slope $\theta_0 \sim 0$ and $\cos\theta_0 \sim 1$. The linear (4.12a) bending of a uniform (4.10a) bar (4.90a) with a concentrated force at the middle should coincide for pinned (4.114b) [sliding (4.405a)] supports, taking into account the different choices of coordinate axis: (1) for the pinned-pinned case (Figure 4.7b), the x-axis passes through both supports, and the y-axis is in the direction of the concentrated force; and (2) for the sliding-sliding case (Figure 4.15), the y-axis is opposite to the concentrated force, and the x-axis is tangent to the bar at the first support. Thus, the transformation from Figure 4.15 to Figure 4.7b consists of a rotation by the angle (4.406b) $\equiv$ (4.408a) and a change of sign (4.408b):

$$\theta_1 = \frac{PL^2}{16EI}: \quad \zeta_3\left(x; \frac{L}{2}\right) = -\left[\bar{\zeta}_3(x) - x\sin\theta_1\right] = \theta_1 x - \bar{\zeta}_3(x)$$

$$= -\frac{Px^3}{12EI} + x\frac{PL^2}{16EI} = \frac{Px}{12EI}\left(\frac{3L^2}{4} - x^3\right). \qquad (4.408a,b)$$

The coincidence of (4.408b) $\equiv$ (4.114b) confirms that for weak bending, the pinned and sliding supports are equivalent. This is no longer the case for strong bending because then the reaction force has a different direction, namely orthogonal to the bent (unbent) position of the bar for sliding (pinned) supports. Another example of the difference between weak and strong bending is that the arc length and the longitudinal coordinate do (do not) coincide as shown next (Subsection 4.9.6).

4.9.6 Nonlinear Effect on the Arc Length

The difference between the arc length (4.400) and the longitudinal Cartesian coordinate (4.401a) is given by (4.409), which shows that it is a nonlinear effect:

$$\left|\frac{EI}{P}\cos\theta_0\right|^{-1/2}\left[s(\theta) - x(\theta)\right] = \int_0^\theta \frac{1-\cos\theta}{\sqrt{\sin\theta}}\,d\theta, \qquad (4.409)$$

that does not vanish if the circular cosine (4.382a) is taken to the second order. The same second order must be used for consistency in all terms of (4.409), leading to

$$\left|\frac{EI}{P}\left(1 - \frac{\theta_0^2}{2}\right)\right|^{-1/2}\left[s(\theta) - x(\theta)\right] = \int_0^\theta \frac{\theta^2}{2}\left[\theta\left(1 - \frac{\theta^2}{6}\right)\right]^{-1/2}d\theta$$

$$= \frac{1}{2}\int_0^\theta \theta^{3/2}\left(1 + \frac{\theta^2}{12}\right)d\theta = \frac{1}{5}\theta^{5/2} + \frac{1}{108}\theta^{9/2}. \qquad (4.410)$$

Thus, the difference between the arc length and longitudinal coordinate is given by

$$s(\theta) - x(\theta) = \sqrt{\frac{EI}{P}}\left(1 - \frac{\theta_0^2}{4}\right)\frac{\theta^{5/2}}{5}\left(1 + \frac{5}{108}\theta^2\right), \qquad (4.411)$$

which involves the maximum inclination θ_0. The latter is related to the distance L between the supports by (4.402), which for substitution in (4.411) need be evaluated only to the lowest order (4.412a), leading to (4.412b):

$$\frac{L}{2}\sqrt{\frac{P}{EI}} = \int_0^{\theta_0} \frac{d\theta}{\sqrt{\theta}} = 2\sqrt{\theta_0}, \quad \theta_0 = \frac{PL^2}{16EI}, \tag{4.412a,b}$$

which coincides with the maximum slope (4.406b) $\equiv$ (4.412b).

Substituting (4.412b) in (4.411) specifies the difference between the arc length and the longitudinal tangential coordinate at the midpoint:

$$s(\theta_0) - x(\theta_0) = \frac{L}{20}\theta_0^2\left(1 - \frac{\theta_0^2}{4}\right)\left(1 + \frac{5}{108}\theta_0^2\right) = \frac{L}{20}\theta_0^2\left[1 - \frac{11}{54}\theta_0^2 + O\left(\theta_0^4\right)\right], \tag{4.413}$$

to the second order in the angle θ_0. Thus, *for the nonlinear bending to the first order of a uniform (4.10a) bar (4.90a) by a concentrated transverse force P at equal distance from the sliding supports (Figure 4.15), the difference between the arc length and the Cartesian coordinate tangent to the bar at the first support is given by (4.411, 4.412b)* $\equiv$ *(4.414):*

$$s(\theta) - x(\theta) = \sqrt{\frac{EI}{P}}\left(1 - \frac{P^2L^4}{1024E^2I^2}\right)\frac{\theta^{5/2}}{5}\left(1 + \frac{5}{108}\theta^2\right). \tag{4.414}$$

The value at the midpoint is (4.413, 4.412b) $\equiv$ *(4.415):*

$$s(\theta_0) - x(\theta_0) = \frac{L}{20}\theta_0^2\left(1 - \frac{11}{54}\theta_0^2\right) = \frac{P^2L^5}{5120E^2I^2}\left(1 - \frac{11P^2L^4}{13824E^2I^2}\right). \tag{4.415}$$

The difference between the arc length and Cartesian coordinate (4.415) (1) increases with the load in the linear approximation, that is, the first term on the r.h.s. of (4.415); and (2) the second term on the r.h.s. of (4.415) has opposite sign, showing that for larger loads, the nonlinear effect is the opposite, implying an instability that is considered next (Subsection 4.9.7).

4.9.7 Collapse of a Bar between Sliding Supports

The lowest-order nonlinear approximation to the relation (4.402) between the distance between the sliding supports and the angle of inclination of the bent relative to the unbent bar is

$$\frac{L}{2}\sqrt{\frac{P}{EI}} = \frac{\sqrt{\cos\theta_0}}{2}\int_0^{\theta}\left(\theta_0^2\theta^{-1/2} - \theta^{3/2}\right)d\theta$$

$$= \frac{1}{2}\left(1 - \frac{\theta_0^2}{2}\right)^{1/2}\left[2\theta_0^2\theta^{1/2} - \frac{2}{5}\theta^{5/2}\right]_0^{\theta_0}$$

$$= \frac{1}{2}\left(1 - \frac{\theta_0^2}{4}\right)\theta_0^{5/2}\left(2 - \frac{2}{5}\right) = \frac{4}{5}\theta_0^{5/2}\left(1 - \frac{\theta_0^2}{4}\right). \tag{4.416}$$

LIST 4.1 Instabilities of an Elastic Bar

A. *Buckling* by an axial load: Subsections 4.9.1 through 4.9.3
B. *Collapse* under a transverse load: Subsections 4.9.4 through 4.9.7
C. *Deflection* at a support with a displacement spring: Example 10.10 and Subsection 4.9.9
D. *Twist* at a support with a rotary spring: Example 10.11 and Subsection 4.9.9

The concentrated transverse force is related to the inclination of the bar at the first support by squaring (4.416) with the same order of accuracy (4.417a) also for the derivative (4.417b):

$$P = \frac{64EI}{25L^2}\theta_0^5\left(1 - \frac{\theta_0^2}{2}\right); \quad \frac{dP}{d\theta_0} = \frac{64EI}{5L^2}\theta_0^4\left(1 - \frac{7}{10}\theta_0^2\right), \quad (4.417\text{a,b})$$

The concentrated transverse force increases (decreases) with the angle of inclination of the bar at the support θ_0 for small (larger) values, corresponding to the first (second) term on the r.h.s. of (4.417a), that is, for weak (strong) bending, implying stability (instability). When the load P starts to decrease with increasing angle θ_0, the bar slides through the supports. The **collapse instability** first occurs (4.418c) when (4.418b) the derivative (4.417b) vanishes (4.418a), and the bar falls through the gap between the sliding supports:

$$\frac{dP}{d\theta_c} = 0: \quad \theta_c = \sqrt{\frac{10}{7}}; \quad P_c = P(\theta_c) = \frac{128}{175}\left(\frac{10}{7}\right)^{5/2}\frac{EI}{L^2}. \quad (4.418\text{a–c})$$

The critical axial (transverse) concentrated force applied at the tip (middle) of a uniform (4.10a) bar (4.90a) with clamped-free (sliding) supports [Figure 4.14 (4.15)] is (4.395a) ≡ (4.410) [(4.418c) ≡ (4.419b)], leading to the ratio (4.419c):

$$P_b = 2.46740\frac{EI}{L^2}, \quad P_c = 1.78413\frac{EI}{L^2}, \quad \frac{P_c}{P_b} = 0.72308 < 1. \quad (4.419\text{a–c})$$

For bars made of the same material, with the same shape of cross section, and the same length (as the distance between the sliding supports), the collapse due to strong bending by a transverse force requires a 27.7% smaller load than the buckling associated with weak bending. Thus, it is more difficult to buckle a bar subjecting it to an axial force at the tip than to cause it to collapse between sliding supports using a concentrated transverse force in the middle; buckling requires a larger load than collapse by a factor of 1.38297 than is inverse to (4.419c). The elastic instabilities of a bar (List 4.1) include, besides buckling (collapse) by an axial (transverse) load [Subsections 4.9.1 through 4.9.3 (Subsections 4.9.4 through 4.9.7)], also (Table 10.7) the displacement (twist) instability at a support with a linear (rotary) spring [Example 10.10 (Example 10.11)], leading (Subsection 4.9.9) to a comparison of dimensions (Subsection 4.9.8).

4.9.8 Dimensional Scalings of the Displacement, Slope, and Curvature

The curvature of a bent elastic bar (4.13d) scales (4.420a) like the bending moment divided by the bending stiffness (4.7a), where the bending moment scales like (1) the transverse force (4.5b) multiplied by a length

(4.420b); and (2) the shear stress (4.6b), for example, the weight per unit length (4.55a) multiplied by the square of the length (4.420c):

$$\zeta'' = \frac{Q}{EI} \sim \frac{PL}{EI} \sim \frac{\rho g L^2}{EI};$$
(4.420a–c)

$$\zeta' \sim \frac{QL}{EI} \sim \frac{PL^2}{EI} \sim \frac{\rho g L^3}{EI},$$
(4.421a–c)

$$\zeta \sim \frac{QL^2}{EI} \sim \frac{PL^3}{EI} \sim \frac{\rho g L^4}{EI},$$
(4.422a–c)

The corresponding dimensional scalings are (4.420a–c) [(4.421a–c)] for the slope (displacement) that is dimensionless (has the dimensions of length). These dimensional scalings can be confirmed for the curvature/slope/displacement due to (1) a concentrated torque (4.420a)/(4.421a)/(4.422a), for example, respectively (4.178a)/(4.179c,d)/(4.176a–d) for a clamped-free bar, (4.193a,b)/(4.190a,b)/(4.189a,b) for a pinned-pinned bar, (4.203a,b)/(4.200a,b)/(4.199a,b) for a clamped-pinned bar, and (4.216a,b)/(4.214a,b)/(4.213a,b) for a clamped-clamped bar; (2) a concentrated force (4.420b)/(4.421b)/(4.422b), for example, respectively (4.98a)/(4.93b,c)/(4.96a,b) for a clamped-free bar, (4.105c,d)/(4.111a,b)/(4.110a,b) for a pinned-pinned bar, (4.128a,b)/(4.127a,b)/(4.126a,b) for a clamped-pinned bar, and (4.138b,c)/(4.145a,b)/(4.144a,b) for a clamped-clamped bar; and (3) uniform shear stress, such as the linear bending of a uniform bar in a constant gravity field (4.420c)/(4.421c)/(4.422c), for example, respectively (4.80a)/(4.78b)/(4.78a) for a clamped-free bar, (4.75a)/(4.73)/(4.72) for a pinned-pinned bar, (4.66a)/(4.65a)/(4.64) for a clamped-pinned bar, and (4.60a)/(4.59a)/(4.58a) for a clamped-clamped bar.

4.9.9 Buckling/Collapse/Displacement/Twist Instabilities

Since the slope is dimensionless (4.421a–c), it implies that the bending moment (4.423a), transverse force (4.423b), and shear stress (4.423c) scale respectively as

$$M \sim \frac{EI}{L}, \quad F \sim \frac{EI}{L^2}, \quad f \sim \frac{EI}{L^3}.$$
(4.423a–c)

For example, the critical load for the buckling (4.419a) [collapse (4.419b)] elastic instability scales like (4.423b) and corresponds [Figure 4.14 (4.15)] to an axial (transverse) load at the tip (middle) of a clamped-free (sliding-sliding) bar. A third (fourth) type of elastic instability corresponds to a **cantilever bar**, that is, a clamped-free bar [Example 10.10 (Example 10.11)] with a linear (rotary) spring at the free end that (1) always ensures stability if [Figure 10.11a through c (10.12a)] it opposes the deflection (twist); (2) if [Figure 10.11d (10.12b)] it acts to increase the displacement (twist) at the tip, a **displacement (twist) instability** occurs for a resilience of the straight (rotary) spring given by (10.145c) ≡ (4.424a) [(10.154e) ≡ (4.424b)]:

$$-p = \frac{EI}{3L^3}, \quad -q = \frac{EI}{L}; \quad P_d = -pL = \frac{EI}{3L^2}, \quad P_e = \frac{M_e}{L} = -\frac{q}{L} = \frac{EI}{L^2};$$
(4.424a–d)

the straight (rotary) spring with resilience $p(q)$ exerts a force (4.424c) [moment (4.424d)] proportional to the longitudinal displacement (angle of rotation) that has the dimensions of length (is dimensionless). The **reference loads** *for the displacement (4.424c) ≡ (4.425d) [twist (4.424d) ≡*

(4.425c)] elastic instabilities can be compared with the critical load for the buckling (4.419a) ≡ (4.425a) [collapse (4.419b) ≡ (4.425b)] elastic instabilities:

$$P_b = 2.46740\frac{EI}{L^2} > P_c = 1.78413\frac{EI}{L^2} > P_e = \frac{EI}{L^2} > P_d = \frac{EI}{3L^2}, \qquad (4.425\text{a–d})$$

leading to the following ordering: (1) the critical buckling load (4.425a) for an axial force at the tip of a clamped-free bar (Figure 4.14) is the largest; (2) the critical collapse load (4.425b) for an axial load on a sliding-sliding bar (Figure 4.15) is smaller; and (3, 4) the comparison is less direct with the reference load for the displacement (4.425d) [twist (4.425c)] elastic instability that is the smallest (second smallest) and applies to a cantilever or heavy clamped-free bar with [Figure 10.11d (10.12b)] a linear (rotary) spring placed at the free end so as to increase the displacement (rotation). A linear (rotary) spring placed at the free end to oppose [Figure 10.11a through c (10.12a)] the displacement (rotation) ensures stability and could be used to prevent the buckling instability as a form of active suppression of deformation.

NOTE 4.1: Linear and Nonlinear Deflection of a String, Bar, or Beam

The linear deflection, that is, with a small slope, of a string (bar) is a simple integration problem (2.11c) [(4.12b)] that can be solved readily for most (1) loads and (2) boundary conditions. Concerning (1) the loads, the string (bar) supports concentrated forces (also moments); concerning (2) boundary conditions, they are a simple attachment for strings, whereas for a bar, several possibilities exist, such as (a) the classical clamped, sliding, pinned, and free ends; (b) others, like translational and rotational strings. The nonlinear deflection of a string is a problem of quadratures, solvable analytically for some simple loads, for example, its own weight. The nonlinear bending of a bar also leads to quadratures solvable for several loads and boundary conditions, for example, a concentrated torque on a bar with one clamped and one free end. A beam is a combination of a string and a bar and inherits mostly the properties of the latter, sometimes with additional complications: (1) the boundary conditions are similar for a bar and a beam except at a free end; (2) they may give rise to longitudinal tension if the beam has both ends fixed; (3) the elastica of a beam for weak bending is the solution of a linear differential equation that may lead to elastic instability; and (4) the strong bending of a beam is not the superposition of the nonlinear deflection of a string and a bar.

NOTE 4.2: Elastic Bodies without or with Bending Stiffness

The concept of bending applies most clearly to elastic bodies like rods (shells) for which [Figure 4.16a (b)] two (one) transversal dimensions (dimension) are (is) much smaller than the one (two) longitudinal dimension(s). The rod (shell) as a one (two)-dimensional thin elastic body (Diagram 4.1) is generally curved in the undeformed state; if it is plane (straight), it is called a beam (plate). A beam (stressed plate) may combine (1) a tangential tension like a string (membrane); and (2) a bending stiffness like a bar (unstressed plate). Besides the bending and longitudinal stresses, there may be torsion. The torsion is decoupled from bending and traction–compression in the linear case for a straight rod with a circular cross section, but more generally, coupling occurs for nonlinear cases, curved geometries, and noncircular cross sections. A case in some sense intermediate between a rod and a plate is a Vlasov (1961) rod, for which the longitudinal dimension is much larger than the transversal dimensions, and the latter are not comparable, for example, stringers (Figure 4.16c) reinforcing a shell.

NOTE 4.3: Elasticity of General and Thin Bodies

The theory of simple bending of elastic bars presented is based on the Bernoulli (1744)–Euler (1744) assumptions: (1) the transverse cross sections remain plane under bending; and (2) there are no shear stresses. In general, the bending of bars does not satisfy these assumptions, and there are multiple generalizations, like the Timoshenko (1921) bars that allow for shear stresses. The common feature of these theories is that they make "a priori" assumptions, raising the question of whether they are compatible with general three-dimensional elasticity. One way to address this question is the asymptotic theory (Campos and Viaño 1992) that represents a bar like a long thin elastic body with two scales, namely,

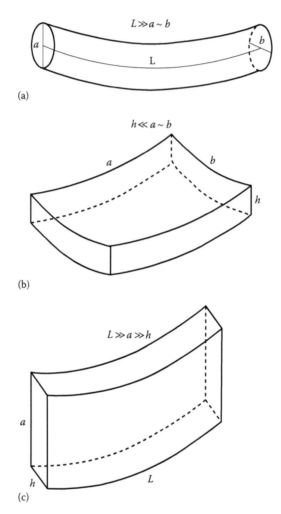

FIGURE 4.16 The thin elastic bodies are the rod (a) [plate (b)] that has one (two) large and two (one) small dimension(s), that is the rod (plate) is long (large in surface dimensions) relative to the dimensions of the cross section (thickness). An intermediate case is a Vlasov rod (c) like a stringer or frame reinforcing a shell or monocoque structure: (1) the length is much larger than the dimensions of the cross section as for a rod; and (2) the dimensions of the cross section are not comparable in orthogonal directions so that there are three scales instead of two for the plate or rod.

a large longitudinal scale and a small transversal scale. The assumptions (1–2) are then proved in the limit of a small thickness-to-length ratio and to the lowest order, leading to the theory of the elastica developed in this chapter. The asymptotic theory can be used to obtain higher-order approximations as extensions of the Bernoulli–Euler bars that are consistent with three-dimensional elasticity. The asymptotic theory leads also to zero-order longitudinal deformations decoupled from bending. The cases that can be considered include bars that are (1) homogeneous (nonhomogeneous) in composition, that is, have uniform (nonuniform) Young's modulus; (2) uniform (nonuniform) cross section, that is, are parallel-sided (tapered); and (3) linear (nonlinear) bending if the strains are all (are not all) small. The asymptotic theory can be applied to bars and plates (Vlasov rods) using two (three) scales. The chapter has been devoted to the theory of the elastica or Bernoulli–Euler bars, including weak (strong) bending for a small (larger) slope, when the shape of the elastica is specified by linear (nonlinear) differential equations [Sections 4.3 through 4.6 (Sections 4.7 through 4.9)]. These nonlinearities are geometric since they are due to the curvature of the elastica. Another type of nonlinearity arises from the constitutive

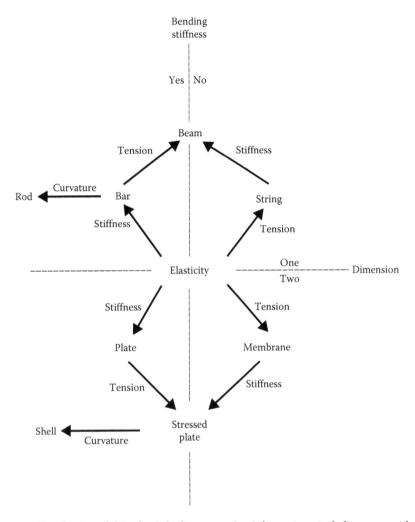

Tension and stiffness of elastic bodies

DIAGRAM 4.1 Classification of thin elastic bodies in one (two) dimensions, including cases without or with curvature in the undeformed state, and without or with bending stiffness.

equation, for example, for soft (hard) inelastic bodies (Note 2.6). The nonlinear constitutive equation is considered for the longitudinal deformation of a soft or hard inelastic bar by its own weight (Notes 4.4 through 4.7). The differential equation for the displacement (Note 4.4) has one (two) solution in the linear (nonlinear) case [Note 4.5 (4.6)]. Consideration of the elastic energy (Note 4.7) shows that the nonlinear solution that matches the linear solution is stable, and the other nonlinear solution is unstable.

NOTE 4.4: Longitudinal Deformation of a Bar in the Gravity Field

The longitudinal deformation of a bar is specified by the balance of the force per unit length and gradient of the stress (4.426a):

$$-f(x) = T' \equiv \frac{dT}{dx}; \quad S = u', \quad T = ES + \bar{E}S^2 = Eu'(1 + \nu u'). \tag{4.426a–d}$$

The stress is specified in terms of the strain (4.426b) by a constitutive relation (4.426c) ≡ (2.299a) consisting of (1) a linear term (**Hooke's Law**, 1678) for an elastic material with the Young's modulus E; and

(2) a nonlinear term for a soft $\bar{E} < 0$ or hard $\bar{E} > 0$ inelastic material (Figure 2.24) with inelastic modulus $\bar{E}$ or inelastic coefficient (4.427a), leading to (4.427b):

$$v = \frac{\bar{E}}{E} : \quad -f(x) = T' = \left(ES + \bar{E}S^2 \right)' = \left[ES(1 + vS) \right]' = \left[Eu'(1 + vu') \right]'. \qquad (4.427a,b)$$

In the case of a material with uniform elastic (4.428a) and inelastic (4.428b) moduli, the balance equation (4.427b) simplifies to (4.428c):

$$E = \text{const}; \quad \bar{E} = vE = \text{const} : \quad -f(x) = Eu' + vE\left(u'^2 \right)'. \qquad (4.428a\text{--}c)$$

For a homogeneous bar with a constant cross section hanging vertically (4.429a) in a uniform gravity field (Figure 4.17), the force per unit length is the constant weight (4.429b) ≡ (4.55a):

$$0 \le x \le L; \quad f(x) = \rho g; \quad u(0) = 0, \quad 0 = S(L) = u'(L). \qquad (4.429a\text{--}d)$$

At the fixed (free) end, the displacement (strain) is zero (4.429c) [(4.429d)]. Thus, *the longitudinal deformation of a homogeneous rod with a constant cross section in a uniform gravity field (4.429b) is specified*

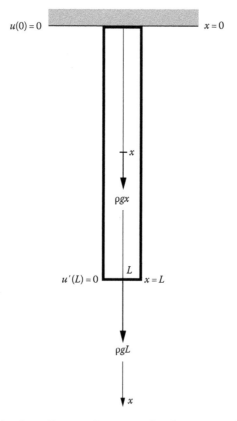

FIGURE 4.17 The nonlinear bending of bars can be associated with two types of nonlinearities: (1) geometric, due to the large slope of the elastica (Figures 4.11 through 4.15); and (2) constitutive for inelastic materials with stresses not proportional to the strain. The simplest illustration of constitutive nonlinearity is an inelastic hard or soft material for which the stress is a quadratic function of the strain (Figure 2.24). A simple configuration of nonlinear deformation for this type of material is a heavy rod hanging vertically in a uniform gravity field (Figure 4.17).

by the second-order nonlinear ordinary differential equation for the displacement (4.430) in the interval (4.429a) with the boundary conditions (4.429c,d):

$$-\rho g = Eu' + Ev\left(u'^2\right)',$$ (4.430)

where the Young's modulus of elasticity (4.428a) and inelastic coefficient (4.428b) are assumed to be constants.

NOTE 4.5: Single (Double) Solution in the Linear (Nonlinear) Cases

Integration of (4.430) from L to x with the boundary condition (4.429d) leads to

$$\rho g(L-x) = -\rho g \int_L^x dx = \left[E\left(u'+vu'^2\right)\right]_L^x = Eu'(x)\left[1+vu'(x)\right].$$ (4.431)

In the elastic case (4.432a), the ordinary differential equation (4.431) is linear and has a unique solution (4.432b) subject to the boundary condition (4.429c):

$$v=0: \quad u_0(x) = \frac{\rho g x}{2E}(2L-x); \quad u_0'(0) = \frac{\rho g L}{E}, \quad u_0(L) = \frac{\rho g L^2}{2E}.$$ (4.432a-d)

This specifies (4.432c) [(4.432d)] the strain (displacement) at the clamped (free) end. In the inelastic case (4.433a), the ordinary differential equation (4.431) $\equiv$ (4.433b) is a quadratic in the strain (4.433c):

$$v \neq 0: \quad vu'^2 + u' - \frac{\rho g}{E}(L-x) = 0, \quad 2vu'_\pm(x) = -1 \pm \left|1+\frac{4v\rho g}{E}(L-x)\right|^{1/2},$$ (4.433a-c)

leading to two distinct solutions (4.433c) $\equiv$ (4.434) meeting the same boundary condition (4.429c):

$$2vu_\pm(x) = -x \pm \frac{E}{6v\rho g}\left\{\left|1+\frac{4v\rho g L}{E}\right|^{3/2} - \left|1+\frac{4v\rho g}{E}(L-x)\right|^{3/2}\right\}.$$ (4.434)

Thus, *the longitudinal deformation of a uniform rod due to its own weight (4.430) leads to one (two) solution(s) for the displacement (4.432b) [(4.434)] in the elastic (4.432a) [inelastic (4.433a)] case when the ordinary differential equation is linear (nonlinear), implying (1) the strain (4.432c) [(4.435a)] at the fixed end:*

$$2vu'_\pm(0) = -1 \pm \left|1+\frac{4v\rho g L}{E}\right|^{1/2}; \quad 2vu_\pm(L) = -L \pm \frac{E}{6v\rho g}\left[\left|1+\frac{4v\rho g L}{E}\right|^{3/2}-1\right],$$ (4.435a,b)

and (2) the displacement (4.432d) [(4.435b)] at the free end. The possibility of matching the linear and nonlinear solutions is discussed next.

NOTE 4.6: Matching/Mismatch of the Linear and Nonlinear Solutions

The restriction (4.436b) on the total weight (4.436a) relative to the Young's modulus,

$$P = \rho g L, 1 > \frac{4v\rho g L}{E} = \frac{4vP}{E}: \quad \left|1+\frac{4v\rho g L}{E}\left(1-\frac{x}{L}\right)\right|^{3/2} = \sum_{n=0}^{\infty}\binom{3/2}{n}\left[\frac{4v\rho g L}{E}\left(1-\frac{x}{L}\right)\right]^n$$

$$= \sum_{n=0}^{\infty} d_n \left(\frac{4v\rho g L}{E}\right)^n\left(1-\frac{x}{L}\right)^n,$$ (4.436a-c)

ensures the convergence (I.25.37a–c) of the binomial series (4.436c) with coefficients

$$n \geq 3: \quad d_n \equiv \binom{3/2}{n} = \frac{1}{n!}\frac{3}{2}\frac{1}{2}\cdots\left(\frac{3}{2}-n+1\right) = \frac{(-)^{n-2}}{n!2^n}3.1.1.3\ldots(2n-5) = 3(-)^n\frac{(2n-5)!!}{(2n)!!}, \qquad (4.437a,b)$$

namely,

$$d_{0-8} = \left\{1,\frac{3}{2},\frac{3}{8},-\frac{1}{16},\frac{3}{128},-\frac{3}{256},\frac{7}{1024},-\frac{9}{2048},\frac{99}{32768}\right\}. \qquad (4.438a\text{–}i)$$

Substituting (4.436c) in the nonlinear displacement (4.434) and using the coefficients (4.419a–c) leads to

$$2vu_{\pm}(x) = -x \pm \frac{E}{6v\rho g}\sum_{n=1}^{\infty}\left(\frac{4v\rho g L}{E}\right)^n d_n\left[1-\left(1-\frac{x}{L}\right)^n\right]$$

$$= -x \pm x \pm \frac{v\rho g L}{E}x\left(2-\frac{x}{L}\right)\pm\frac{2}{3}L\sum_{n=3}^{\infty}d_n\left(\frac{4v\rho g L}{E}\right)^{n-1}\left[1-\left(1-\frac{x}{L}\right)^n\right]. \qquad (4.439)$$

Thus, *the nonlinear displacements of a uniform inelastic heavy rod hanging in a uniform gravity field (Figure 4.17) allow (Figure 4.18) two possibilities: (1) a nonlinear solution (4.440a, 4.441) that matches the linear solution (4.432b) and must be stable; (2) another nonlinear solution (4.440b) that is singular for $x = 0 \neq v$ and thus annot match the linear solution and must be unstable:*

$$u_{+}(x) = \frac{\rho g x}{2E}(2L-x) - u_{\ast}(x) = u_0(x) - u_{\ast}(x), \qquad (4.440a)$$

$$u_{-}(x) = -\frac{x}{v} - \frac{\rho g x}{2E}(2L-x) + u_{\ast}(x) = -\frac{x}{v} - u_0(x) + u_{\ast}(x). \qquad (4.440b)$$

Both nonlinear solutions (4.440a,b) ≡ (4.439) involve opposite signs: (1) the linear displacement (4.432b); and (2) the complementary displacement (4.441) associated with nonlinearity:

$$u_{\ast}(x) = \frac{x}{v}\sum_{n=3}^{\infty}\frac{(2n-5)!!}{(2n)!!}\left(-\frac{4v\rho g L}{E}\right)^n\sum_{m=0}^{n-1}\frac{n!(-)^m}{(m+1)!(n-m-1)!}\left(\frac{x}{L}\right)^m, \qquad (4.441)$$

that is, $O(v^2)$, and hence vanishes for $v = 0$. The complementary displacement (4.441) is the third or last term on the r.h.s. of (4.439) and is given by

$$u_{\ast}(x) = \frac{L}{v}\sum_{n=3}^{\infty}\frac{(2n-5)!!}{(2n)!!}\left(-\frac{4v\rho g L}{E}\right)^{n-1}\left[1-\left(1-\frac{x}{L}\right)^n\right], \qquad (4.442)$$

where the coefficients (4.437a,b) are used. In the last factor in (4.442) can be used the binomial series (4.443b) that converges for (4.443a):

$$0 \leq \frac{x}{L} < 1: \quad 1-\left(1-\frac{x}{L}\right)^n = 1-\sum_{k=0}^{n}\binom{n}{k}\left(-\frac{x}{L}\right)^k = -\sum_{k=1}^{n}\binom{n}{k}\left(\frac{x}{L}\right)^k(-)^k, \qquad (4.443a,b)$$

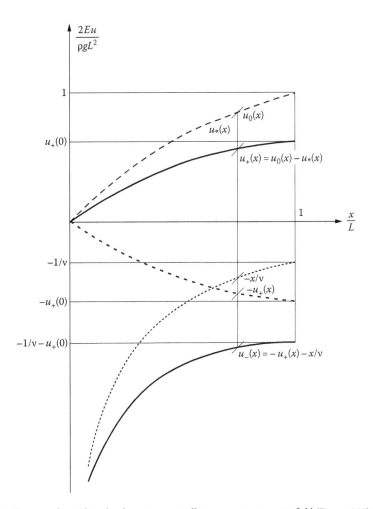

FIGURE 4.18 In the case of a uniform bar hanging vertically in a constant gravity field (Figure 4.17) the longitudinal deformation u_0 is a quadratic function of position, for an elastic constitutive relation that is stress and strain related linearly. Adding to (subtracting from) the stress a term proportional to the square of the strain leads to the constitutive relation for a hard (soft) inelastic material (Figure 2.24). In the nonlinear case there are two solutions for the displacement. The stable solution u_+ subtracts from the linear solution u_0 a set of nonlinear terms u_*; it is clearly continuous since $u_+ = u_0$ in the linear approximation $u_* = 0$. The second solution consists of: (1) an opposite displacement $-u_*$; and (2) plus a term $-\nu/x$ that is singular. Thus the first u_- (second u_+) nonlinear solution can (cannot) be matched to the linear solution, and it is stable (unstable) because it corresponds (does not correspond) to a minimum of the elastic energy.

and simplifies to (4.444b) using the change of summation variable (4.444a):

$$m = k-1: \quad 1-\left(1-\frac{x}{L}\right)^n = -\sum_{m=0}^{n-1}\binom{n}{m+1}\left(\frac{x}{L}\right)^{m+1}(-)^{m+1}$$

$$= \frac{x}{L}\sum_{m=0}^{n-1}\frac{n!(-)^m}{(m+1)!(n-m-1)!}\left(\frac{x}{L}\right)^m. \qquad (4.444a,b)$$

Substitution of (4.444b) in (4.442) leads to the complementary displacement (4.441). The stability (instability) of the nonlinear solution (4.440a) [(4.440b)] that converges (does not converge) to the linear solution (4.432b) for $\nu \to 0$ is proved next (Note 4.7).

NOTE 4.7: Linear and Nonlinear Stability or Instability

The equilibrium solutions correspond to a stationary elastic energy and are stable (unstable) if it is minimum (maximum or has an inflexion). Thus, the starting point to consider stability is the variation of the total elastic energy (4.445b), that is, the elastic energy density (II.4.86b) ≡ (4.445a) integrated over the length of the rod:

$$\delta E_d = T\delta S = T\delta u': \quad \delta\bar{E}_d = \int_0^L \delta E_d dz = \int_0^L T\delta u' dz. \tag{4.445a,b}$$

Using the nonlinear constitutive relation (4.426c, 4.427a) relating stress to strain leads to (4.446a):

$$\delta\bar{E}_d = \int_0^L Eu'(1+vu')\delta u' dz, \quad \delta W = \int_0^L f\delta u dz, \tag{4.446a,b}$$

which must balance (4.447) the work of the force per unit length (4.446b):

$$0 = \delta\bar{E}_d - \delta W = \int_0^L \Big[Eu'(1+vu')\delta u' - f\delta u\Big]dz$$

$$= \Big[Eu'(1+vu')\delta u\Big]_0^L - \int_0^L \Big\{\Big[Eu'(1+vu')\Big]' + f\Big\}\delta u dz. \tag{4.447}$$

In (4.447) was made an integration by parts, leading to two distinct terms that must vanish separately: (1) the integral along the rod [second term on the r.h.s. of (4.447)] must vanish for arbitrary variations of the displacement, leading (4.448a) to the equilibrium equation (4.448b) ≡ (4.427b) valid for a nonuniform rod:

$$0 \le x \le L: \quad 0 = f + \Big[Eu'(1+vu')\Big]'; \quad x = 0, L: \quad \delta u = 0 \text{ or } u' = 0; \tag{4.448a–e}$$

and (2) the terms at the ends of the rod [first on the r.h.s. of (4.447)] lead to the boundary conditions (4.448c) applying to fixed (4.448d) ≡ (4.429c) [for free (4.448e) ≡ (4.429d)] ends. Thus, *the first variation of the total elastic energy (4.446a) for the inelastic stress–strain relation (4.426d) balances (4.447) the work of the force per unit length (4.446b), leading to the force-stress balance (4.448b) [boundary conditions (4.448d,e)] in the interior (4.448a) [at the ends (4.448c)] of the rod. The second variation*

$$\delta^2\bar{E}_d = \int_0^L E(1+2vu')(\delta u')^2 dz \ge 0: \begin{cases} E > 0 & \text{linear}(v=0) \tag{4.449a} \\ u'v > 0 & \text{nonlinear}(v \ne 0) \tag{4.449b} \end{cases}$$

leads to a minimum of the energy, ensuring stability: (1) in the linear case (4.449a), for a positive Young's modulus, ensuring that the stress increases with the strain; and (2) in the nonlinear case (4.449b), for the extension u' > 0 (contraction u' < 0) of a hard v > 0 (soft v' < 0) material, so that vu' > 0 there is an increasing contribution to the stress. Applying the criterion (4.449b) to the two nonlinear solutions (4.433c), it follows (4.450a,b) that in (4.434), the stable (unstable) solution is (4.440a) [(4.440b)]:

$$vu'_+(x) > 0 > vu'_-(x): \begin{cases} u_+(x) & \text{stable} \tag{4.450a} \\ u_-(x) & \text{unstable} \tag{4.450b} \end{cases}$$

as was predicted from the continuity with (singularity relative to) the linear solution.

4.10 Conclusion

An elastic bar (Figure 4.1) that is straight in the absence of transverse loads (Figure 4.1a) gains a curvature of flexion (Figure 4.1b) when acted on by a bending moment (Figure 4.1c) associated with a transverse force (Figure 4.1d). In the Bernoulli–Euler theory (Figure 4.2) of simple flexion of a straight bar (Figure 4.2a), the cross sections are rotated, remaining plane and orthogonal to the elastica (Figure 4.2c); a rectangular cross section (Figure 4.2b) is deformed by bending with (Figure 4.2d) inclined straight (parabolic) sides (top and bottom) and centroid shifted down (Figure 4.2e). A bar may have (Figure 4.3) three types of support: (1) clamping (Figure 4.3a) with a reaction force and moment; and (2/3) pinning (free) with a reaction force only (Figure 4.3b) [neither reaction force nor moment (Figure 4.3c)]. The six combinations of two supports lead (Figure 4.4) to three sets of two cases: (1/2) clamped-clamped (clamped-pinned) that are hyperstatic [Figure 4.4a (b)]; (3/4) pinned-pinned (clamped-free) that are isostatic [Figure 4.4c (d)]; and (5/6) pinned-free (free-free) that are [Figure 4.4e (f)] hypostatic. The hypostatic cases allow motion and are dynamic problems. The other four classical combinations are static, and the balance of forces and moments is sufficient (not sufficient) to determine all reactions at the supports in the isostatic (hyperstatic) cases (Table 4.1). The four classical cases of support (Figure 4.4a through d) are considered for the three main types of load: (1) a uniform shear stress (Figure 4.5a through d), such as the weight of a uniform bar (Table 4.2); (2) a concentrated force at any position (Figure 4.6a through d), specifying the linear influence or Green function (Table 4.3); and (3) a concentrated moment at any position (Figure 4.8a through d), in which case the elastica is specified by the derivative of the influence function (Table 4.5). The latter (3) is an instance of the principle of superposition that specifies the shape of the elastica for any load once the influence function is known with the same boundary conditions. The symmetry of the influence function expresses the reciprocity principle allowing interchange of the source point of application of the concentrated force and the observation point of measurement of the displacement. The cases (2) [(3)] of a concentrated force (moment) at an arbitrary position simplify [Figure 4.7 (4.9)] if the point of application is a free end or at equal distance from the supports [Table 4.4 (4.6)]. A bar may have reaction forces and/or moments at the supports, and a beam has, in addition, a longitudinal tension, as for a string; the longitudinal tension may be applied externally or result from stretching, or both; for example, a bar fixed at both ends is stretched under bending, causing the appearance of a longitudinal traction (Figure 4.10).

The distinction between weak and strong bending is that [Figures 4.1 through 4.10 (4.11 through 4.15)] the slope of the elastica is (is not) everywhere small. The simplest case of strong bending (Table 4.7) is by concentrated moments (Figure 4.11), for example, one (two) leading [Figure 4.11a (b)] to an elastica consisting of circular arcs with continuous tangents and a straight line after the last concentrated moment [Panel 4.1 (4.2)]. The weak (strong) bending is conveniently described in terms of fixed Cartesian (local tangent) coordinates (Figure 4.5 (4.12a). The strong bending by moments (own weight) can lead to curling (leads to less than a vertical slope), as shown in Figure 4.12b (c). The bending of a clamped-free bar by an transverse (axial) force applied at the tip [Figure 4.13 (4.14)] is an example of elastic stability (instability) since the deflection is proportional to the load (is zero until a critical buckling load is reached). Two types of elastic instability are buckling (collapse) of a bar subject to an axial (transverse) load, for example, clamped-free (with sliding supports) in Figure 4.14 (4.15); the buckling (collapse) instabilities apply to weak (strong) bending and lead to a larger critical load in the former case. The bars (plate) are one (two)-dimensional thin elastic bodies [Figure 4.16a (b)] having as intermediate case the Vlasov bar with distinct cross-sectional dimensions (Figure 4.16c). The bar (plate) is distinguished from the string (membrane) by the bending stiffness and may be combined in a beam (stressed plate) that could be (Diagram 4.1) curved in the undeformed state leading to a rod (shell). Besides the geometric nonlinearity associated with the curvature of the elastica (Figures 4.1 through 4.15) for strong bending (Figures 4.11 through 4.15), another possibility is a constitutive nonlinearity for an inelastic material for which the stresses are not proportional to the strains. The material or constitutive nonlinearity is illustrated by the longitudinal deformation of a heavy rod hanging in the gravity field (Figure 4.17); the linear

LIST 4.2 Twenty-Three Cases of Bending of Bars and Beams

A. Linear
  a. Bar with uniform load: Section 4.3, Table 4.2, Figure 4.5
     1. Clamped-clamped: Subsection 4.3.1, Figure 4.5a
     2. Clamped-pinned: Subsection 4.3.2, Figure 4.5b
     3. Pinned-pinned: Subsection 4.3.3, Figure 4.5c
     4. Clamped-free: Subsection 4.3.4, Figure 4.5d
  b. Bar with concentrated force: influence or Green function: Section 4.4, Tables 4.3 and 4.4, Figures 4.5 and 4.6
     5. Clamped-clamped: Subsection 4.4.4, Figures 4.6d and 4.7d
     6. Clamped-pinned: Subsection 4.4.3, Figures 4.6c and 4.7c
     7. Pinned-pinned: Subsection 4.4.2, Figures 4.6b and 4.7b
     8. Clamped-free: Subsection 4.4.1, Figures 4.6a and 4.7a
  c. Bar with concentrated torque: Section 4.5; Tables 4.5 and 4.6, Figures 4.8 and 4.9
     9. Clamped-free: Subsection 4.5.2, Figures 4.8a and 4.9a
    10. Pinned-pinned: Subsection 4.5.3, Figures 4.8b and 4.9b
    11. Clamped-pinned: Subsection 4.5.4, Figures 4.8c and 4.9c
    12. Clamped-clamped: Subsection 4.5.5, Figures 4.8d and 4.9d
  d. Clamped-free heavy bar with free-end supported on a spring: Table 10.7
    13. Linear spring: Example 10.10, Figure 10.11
    14. Rotary spring: Example 10.11, Figure 10.12
  e. Beam (bar under longitudinal compression)
    15. Pinned-pinned: Section 4.6, Figure 4.10
B. Nonlinear
  f. Clamped-free bar with concentrated torques: Section 4.7, Figure 4.11
    16. One concentrated torque: Subsections 4.7.1 through 4.7.6, Table 4.7, Figure 4.11a, Panel 4.1
    17. Two concentrated torques: Subsections 4.7.7 through 4.7.9, Figure 4.11b, Panel 4.2
  g. One concentrated force nonlinear influence or Green function: Sectiion 4.8
    18. Pinned-pinned:Subsection 4.8.1
    19. Clamped-free: Subsection 4.8.2
  h. In local tangential coordinates: Subsections 4.8.3 through 4.8.9, Section 4.9, Figure 4.12a
    20. Under own weight: Subsections 4.8.3 through 4.8.9, Figure 4.12c
    21. Clamped-free: Subsections 4.8.10 through 4.8.15
    22. Buckling of clamped-free beam: Subsections 4.9.1 through 4.9.3, Figure 4.14
    23. Collapse of sliding-sliding bar: Subsections 4.9.4 through 4.9.7, Figure 4.15

solution is unique (Figure 4.18) and matches (does not match) the stable (unstable) nonlinear solution. The 23 cases of bending of rods (List 4.2) include four elastic instabilities (List 4.1) besides the buckling (collapse) instability [Figure 4.14 (4.15)], also the displacement (twist) instability [Figure 10.11d (10.12b)] of a heavy clamped-free bar with a linear (rotary) spring at the free end (Example 10.10 (Example 10.11)] placed so as to increase the deflection. Together with torsion (Sections II.6.4 through II.6.8), longitudinal deformation (Notes 4.4 through 4.7) and bending (Sections 4.1 through 4.9) are the three types of displacements of a rod that are decoupled in simple cases, like small deformations of a rod that is straight in the undeformed state.

5

Differential Operators and Geometry

The definition of generalized function as the nonuniform limit of a sequence of analytic functions (Chapter 1) [a linear continuous functional over a set of reference functions (Chapter 3)] was presented first in the simplest case of one generalized function of one variable that extends the ordinary real function of one real variable. The functional approach (Section 3.1) is chosen for the extension to vector functions of vector variables, that is, to multidimensional generalized functions of several variables. Moreover, ordinary functions may appear as arguments of generalized functions: (1) a unit jump of a real function indicates the intervals on the real axis where it is positive, negative, or zero (Section 5.1); (2) the unit impulse, for example, of the same function (Section 5.1), is zero everywhere, except for singularities where the function vanishes; (3) the unit jump of a function of two variables indicates (Section 5.2) the regions of the plane where the function is positive or negative, and the line separating them, where the function is equal to 1/2; and (4) the latter is a line of singularities for the unit impulse of the function of two variables that vanishes everywhere else on the plane (Section 5.2). A generalized function is a linear operator, thus excluding powers, but not (1) the product of two generalized functions of distinct variables (Section 5.3); and (2) the product of two generalized functions whose arguments are two linearly independent functions of two variables (Section 5.4). The preceding can be extended to the product of M generalized functions whose arguments are M independent variables or M linearly independent functions of $N \geq M$ variables (Section 5.5). A generalized function of dimension M of $N \geq M$ can be used (Section 5.6) for integration over a subspace of dimension $N - M$, including a hypersurface (hypercurve) for $M = 1$ ($M = N - 1$), for example, a curve $M = 1$ (surface $M = 2$) in a three-dimensional space $N = 3$. The generalized functions may be used to prove the Gauss (Stokes) integral theorems for the divergence, curl, and gradient (curl) in three (two) dimensions (Section 5.7); these lead to the existence of a scalar (vector) potential for an irrotational (solenoidal) vector field (Section 5.8). The invariant differential operators (divergence and curl) can be generalized to higher dimensions and multivector fields and potentials (Notes 9.3 through 9.52). Another N-dimensional problem is the replacement of the product of N generalized functions by a single generalized function; for example, a product of N unit impulses in N variables is centrally symmetric, that is, it depends only on the distance R from the origin and thus reduces (Section 5.9) to a single unit impulse of the variable R; the factor multiplying is the area of the unit hypersphere, that is, the sphere of radius unity in N-dimensional space (Section 5.9). All these properties of multidimensional generalized functions of several variables (Sections 5.3 through 5.9) are extensions of the case of a one-dimensional generalized function of one variable (Chapters 1 and 3) replacing the functional over the real line by a product of functionals over a multidimensional domain (Chapter 5).

5.1 Generalized Function with an Ordinary Function as the Argument

An extension of a real generalized function with one real variable (Chapters 1 and 3) is to take an ordinary function as the variable or argument (Section 5.1), for example, the unit jump (Subsection 5.1.1) [impulse (Subsection 5.1.2)] and their derivatives (Subsection 5.1.3). The derivative of the unit jump of an ordinary function is related to the unit impulse of the same function (Subsection 5.1.4).

5.1.1 Unit Jump of an Ordinary Function

The generalized function unit jump in earlier expressions, such as (1.28a and b), has the variable x as argument. An extension is to consider *the* generalized function *unit jump whose argument is a continuous real function*:

$$f \in C(|R): \quad H(f(x)) = \begin{cases} 1 & \text{if } f(x) > 0, & (5.1a) \\ 1/2 & \text{if } f(x) = 0, & (5.1b) \\ 0 & \text{if } f(x) < 0. & (5.1c) \end{cases}$$

The unit jump (5.1a through c) distinguishes the points where the function is positive, negative, and zero and thus has the integral property (5.2b) relative to integrable test function (5.2a):

$$\Phi \in \mathcal{E}(|\mathcal{R}): \quad \left[H(f(x)), \Phi(x) \right] \equiv \int_{-\infty}^{+\infty} H(f(x)) \Phi(x) dx$$

$$= \int_{f(x)>0} \Phi(x) dx + \frac{1}{2} \int_{f(x)=0} \Phi(x) dx. \quad (5.2a,b)$$

For example, $H(a^2 - x^2)$ is unity (Figure 5.1a) in the interval$)-a, +a($, one-half at the points $x = \pm a$. and zero for $|x| > a$, and thus

$$\left[H(a^2 - x^2), \Phi(x) \right] \equiv \int_{-\infty}^{+\infty} H(a^2 - x^2) \Phi(x) dx = \int_{-a}^{+a} \Phi(x) dx, \quad (5.3)$$

it has the integral property (5.3) with regard to the test functions (5.2a) integrable on the real line.

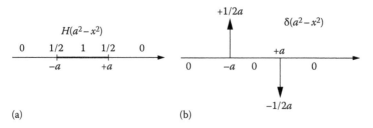

(a) (b)

FIGURE 5.1 The generalized function unit jump (a) of an ordinary function of one real variable, is unity (zero) in the intervals of the real axis where the function is positive (negative) and equals one-half at its zeros. The generalized function unit impulse of an ordinary function of one real variable (b) is singular at the zeros of the function and zero elsewhere.

5.1.2 Ordinary Function as the Argument of a Unit Impulse

Having considered the case of a continuous function $f(x)$ as argument for the unit jump (5.1a through c), next is considered the unit impulse whose argument is also a continuous function:

$$f \in C(|R): \quad \delta(f(x)) = \begin{cases} 0 & \text{if } f(x) \neq 0, \\ \\ \infty & \text{if } f(x) = 0; \end{cases}$$

(5.4a)

(5.4b)

it vanishes everywhere (5.4a) except at the zeros of $f(x)$, where it is singular (5.4b). The nature of the singularities is specified by evaluating the integral,

$$I \equiv \left[\delta(f(x)), \Phi(x) \right] \equiv \int_{-\infty}^{+\infty} \delta(f(x)) \Phi(x) dx,$$

(5.5)

by means of the change of variable (5.6b and c)

$$f \in C^1(|R): \quad \xi = f(x), \quad d\xi = f'(x) dx,$$

(5.6a–c)

where the function is assumed to be continuously differentiable (5.6a); this leads to

$$f(x_n) = 0 \neq f'(x_n): \quad I = \int_{-\infty}^{+\infty} \left[\frac{\Phi(x)}{f'(x)} \right]_{x = f^{-1}(\xi)} \delta(\xi) d\xi = \sum_{f(x)=0} \left\{ \frac{\Phi(x)}{f'(x)} \right\},$$

(5.7a–c)

where the sum (5.7c) extends over the N zeros x_n of the function (5.7a) and assumes that they are simple zeros (5.7b). The integral property (5.7c) is equivalent to

$$I = \sum_{n=1}^{N} \frac{\Phi(x_n)}{f'(x_n)} = \sum_{n=1}^{N} \left[\delta(x - x_n), \frac{\Phi(x)}{f'(x)} \right] = \sum_{n=1}^{N} \left[\frac{\delta(x - x_n)}{f'(x_n)}, \Phi(x) \right],$$

(5.8a,b)

a decomposition of the impulse of an ordinary function (5.8) into simple impulses concentrated at each zero of the function.

The preceding result may be stated as follows: *consider a continuously differentiable function (5.6a) ≡ (5.11a) with N zeros (5.9a and b), all simple (5.9c):*

$$n = 1, \dots, N: \quad f(x_n) = 0 \neq f'(x_n).$$

(5.9a–c)

*The generalized function unit **impulse with argument** f(x) has the integral property (5.10b) relative to a continuous test function (5.10a):*

$$\Phi \in C(|R): \quad \left[\delta(f(x)), \Phi(x) \right] \equiv \int_{-\infty}^{+\infty} \Phi(x) \delta(f(x)) dx = \sum_{n=1}^{N} \frac{\Phi(x_n)}{f'(x_n)}.$$

(5.10a,b)

This is equivalent (Figure 5.2) to the decomposition (5.11b) into N simple unit impulses with amplitudes (5.11c):

$$f \in C^1(|R): \quad \delta(f(x)) = \sum_{n=1}^{N} A_n \delta(x - x_n), \quad \frac{1}{A_n} = f'(x_n).$$

(5.11a–c)

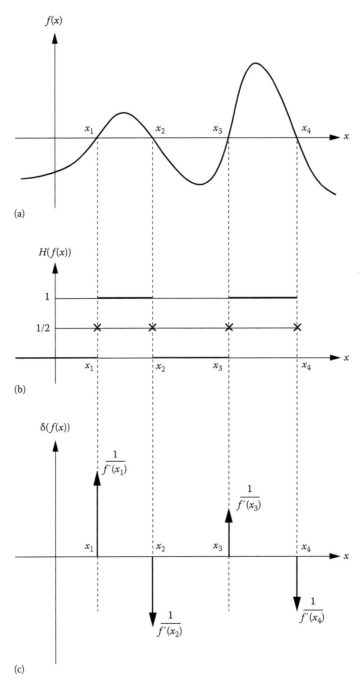

FIGURE 5.2 Consider an ordinary function (a) with: (1) four simple zeros (x_1, x_2, x_3, x_4); (2) alternating positive (negative) slopes at x_1, x_3 (x_2, x_4); and (3) larger slope in modulus at (x_3, x_4) than at (x_1, x_2). It is used as argument of a unit jump (b) that is unity (zero) in the intervals where the function is positive (negative) and one-half at the end-points where the function vanishes. It is also used as argument in the unit impulse (c) that: (1) is zero everywhere except at the zeros of the function; (2) it is infinite at the zeros of the function, corresponding to a superposition of simple unit impulses at (x_1, x_2, x_3, x_4); (3) the amplitudes of the unit impulses are positive (negative) at (x_1, x_3) $[(x_2, x_4)]$ where the slope is positive (negative); and (4) the modulus of the amplitudes of the unit jumps is larger at (x_1, x_2) [smaller at (x_3, x_4)] where the modulus of the slope is smaller (larger).

The unit impulses are located (5.11b) at the simple zeros (5.7a) of the ordinary differentiable function that appears as an argument (5.6b), and the inverse of the nonzero slope (5.7b) specifies the amplitudes (5.11c). The linear change of variable (3.13b) uses the function (5.12a) as the argument of the unit impulse

$$f(x) \equiv ax + b, \quad f'(x) = a, \quad x_1 = -\frac{b}{a}: \quad \delta(ax + b) = \frac{1}{a}\delta\left(x + \frac{b}{a}\right). \tag{5.12a–d}$$

The function (5.12a) has a constant slope (5.12b) and zero (5.12c), which, substituted in (5.11a through c), lead to (5.12d) ≡ (3.14b). Another example is the ordinary continuous quadratic function (5.13a) that has a derivative (5.13b) and slope (5.13d) at the zeros (5.13c):

$$f(x) = a^2 - x^2: \quad f'(x) = -2x, \quad x_{1,2} = \pm a, \quad f'(x_{1,2}) = \mp 2a, \tag{5.13a–d}$$

$$\delta(a^2 - x^2) = \frac{\delta(x + a) - \delta(x - a)}{2a}. \tag{5.13e}$$

Substitution of (5.13c and d) in (5.11a through c) gives (5.13e). This is equivalent to (5.13g) for a continuous test function (5.13f):

$$\Phi \in C(|R): \quad \int_{-\infty}^{+\infty} \delta(a^2 - x^2)\Phi(x)\,dx = \frac{\Phi(-a) - \Phi(a)}{2a}. \tag{5.13f,g}$$

The unit impulse (5.13e) with the ordinary function (5.13a) as the argument vanishes everywhere on the real axis except (Figure 5.1b) at the points $x = \pm a$, where it has the amplitudes $\mp 1/2\,a$, corresponding to the jumps in Figure 5.1a.

5.1.3 Derivative Impulse of an Ordinary Function

The property (5.11a through c) may be generalized (5.7c) to the nth derivative of the unit impulse with an ordinary function as the argument (5.14c):

$$f \in C^{n+1}(|R), \Phi \in C^n(|R): \quad \left[\delta^{(n)}(f(x)), \Phi(x)\right] \equiv \int_{-\infty}^{+\infty} \Phi(x)\delta^{(n)}(f(x))\,dx$$

$$= \int_{-\infty}^{+\infty}\left[\frac{\Phi(x)}{f'(x)}\right]_{x=f^{-1}(\xi)}\delta^{(n)}(\xi)\,d\xi = (-)^n \sum_{f(x)=0}\frac{d^n}{dx^n}\left[\frac{\Phi(x)}{f'(x)}\right], \tag{5.14a–c}$$

where (1) the ordinary function (5.6b) used as argument has a continuous derivative (5.14a) of order $n + 1$; and (2) the test function is n times continuously differentiable (5.14b). For example, for the first derivative of the unit impulse of an ordinary function with a continuous second-order derivative,

$$f \in C^2(|R), \Phi \in C^1(|R):$$

$$\left[\delta'(f(x)), \Phi(x)\right] \equiv \int_{-\infty}^{+\infty} \Phi(x)\delta'(f(x))\,dx = \int_{-\infty}^{+\infty}\left[\frac{\Phi(x)}{f'(x)}\right]_{x=f^{-1}(\xi)}\delta'(\xi)\,d\xi$$

$$= -\sum_{f(x)=0}\lim_{x \to x_n}\frac{d}{dx}\left[\frac{\Phi(x)}{f'(x)}\right] = \sum_n\left\{-\frac{\Phi'(x_n)}{f'(x_n)} + \frac{\Phi(x_n)f''(x_n)}{\left[f'(x_n)\right]^2}\right\}. \tag{5.15a–c}$$

Thus, **the *n*th derivative** of the unit **impulse** with argument a real function (5.6b) with a continuous derivative of order (n + 1) in (5.14a) with simple zeros (5.9a through c) has the integral property (5.14c) with regard to n times continuously differentiable test function (5.14b); the case n = 1 corresponds to (5.15a through c), which is equivalent to

$$f \in C^2(|R): \quad \delta'\big(f(x)\big) = \sum_n \Big[A_n \delta'(x - x_n) + B_n(x - x_n) \Big], \tag{5.16a,b}$$

$$\frac{1}{A_n} = f'(x_n), \quad \frac{1}{B_n} = \frac{\big[f'(x_n)\big]^2}{f''(x_n)}, \tag{5.16c,d}$$

as a decomposition of the first derivative of the unit impulse with argument a function with a continuous second-order derivative (5.16a) into a sum (5.16b) of impulses (and derivative impulses) at the simple zeros (5.9a through c) of the function with coefficients (5.16c) ≡ (5.11c) [(5.16d)]. An example corresponding to (5.13a through e) is

$$f(x) = a^2 - x^2, \, f''(x) = -2:$$

$$\delta'\big(a^2 - x^2\big) = \frac{1}{2a}\Big[\delta'(x+a) - \delta'(x-a)\Big] - \frac{1}{2a^2}\Big[\delta(x+a) + \delta(x-a)\Big], \tag{5.17a–c}$$

where substitution of (5.13c and d; 5.17b) in (5.15c) leads to (5.17c). This is equivalent to (5.17e) for a continuously differentiable test function (5.17d):

$$\Phi \in C^1(|R): \quad \int_{-\infty}^{+\infty} \delta'\big(a^2 + x^2\big)\Phi(x)\mathrm{d}x = \frac{\Phi'(a) - \Phi'(-a)}{2a} - \frac{\Phi(a) + \Phi(-a)}{2a^2}. \tag{5.17d,e}$$

The results (5.13e through g) and (5.17c through e) should be compared and could be extended to (1) higher-order derivatives n = 2, 3, ...; and (2) multidimensional generalized functions of several variables. Since the method is the same, only the case n = 0 will be treated in the sequel. Five (three) instances of the unit impulse (derivative unit impulse) with ordinary functions as arguments are indicated in Example 10.13.

5.1.4 Unit Jump of Intervals of the Real Line

The unit jump (5.1a through c) whose argument is a function with a continuous derivative (5.18a) has the property (5.18b):

$$f \in C^1(|R): \quad \frac{\mathrm{d}}{\mathrm{d}x}\Big[H\big(f(x)\big)\Big] = \frac{\mathrm{d}H}{\mathrm{d}f}\frac{\mathrm{d}f}{\mathrm{d}x} = f'(x)\delta\big(f(x)\big). \tag{5.18a,b}$$

If the zeros are all simple (5.9a through c), the property (5.11b and c) may be used, leading to

$$f'(x)\delta\big(f(x)\big) = \sum_{n=1}^{N} \frac{f'(x)}{f'(x_n)}\delta(x - x_n) = \sum_{n=1}^{N} \delta(x - x_n), \tag{5.19a,b}$$

where the substitution property (1.78a and b) was used. Thus is proved that *the derivative of the generalized function unit jump whose argument is a continuously differentiable function (5.20a) whose zeros are all simple (5.9a through c) is a sum (5.20b) of unit impulses at the zeros:*

$$f \in C^1(|R): \quad \frac{d}{dx}\left[H\left(f(x)\right)\right] = \sum_{n=1}^{N} \delta(x - x_n). \qquad (5.20a,b)$$

An example of this property is

$$\frac{d}{dx}\left[H\left(a^2 - x^2\right)\right] = \delta(x - a) + \delta(x + a), \qquad (5.21)$$

which can be proved from (5.13e):

$$\frac{d}{dx}\left[H\left(a^2 - x^2\right)\right] = -2x\delta\left(a^2 - x^2\right) = -\frac{x}{a}\left[\delta(x + a) - \delta(x - a)\right] = \delta(x + a) + \delta(x - a), \qquad (5.22)$$

using the substitution property (1.78a and b). The derivatives of the unit jump of an ordinary function may be considered for orders higher than one (5.18a and b), for example, two (5.23b) [three (5.24b)],

$$f \in C^2(|R): \quad \frac{d^2}{dx^2}\left[H\left(f(x)\right)\right] = f''(x)\delta\left(f(x)\right) + \left[f'(x)\right]^2 \delta'\left(f(x)\right), \qquad (5.23a,b)$$

$$f \in C^3(|R): \quad \frac{d^3}{dx^3}\left[H\left(f(x)\right)\right] = f'''(x)\delta\left(f(x)\right) + 3f'(x)f''(x)\delta\left(f(x)\right) + \left[f'(x)\right]^3 \delta''\left(f(x)\right), \qquad (5.24a,b)$$

if the function has a continuous second- (5.23a) [third- (5.24a)] order derivative. Following the consideration of generalized functions whose argument is an ordinary real function of one real variable, next is considered the case where the argument is an ordinary real function of two (Section 5.2) [several (Section 5.5)] variables.

5.2 Argument a Function of Two Variables

The generalized functions unit jump (Subsection 5.2.1) [impulse (Subsection 5.2.2)] with argument an ordinary continuous (differentiable) function are related to each other (Subsection 5.2.5) for the case of two independent variables (Subsection 5.2), using methods similar to one variable (Section 5.1). In the case of the unit impulse with argument an ordinary function, there are two alternative equivalent decompositions into simple impulses (Subsections 5.2.3 and 5.2.4).

5.2.1 Unit Jump of a Two-Dimensional Domain

Next is considered, instead of a function of one variable $f(x)$ that specifies intervals on the real axis, a function $f(x, y)$ of two variables that specifies (Table 5.1) a closed region D:

$$f(x,y)\begin{cases} > 0 & \text{if } (x, y) \in D - \partial D : \quad \text{interior of } D, \\ = 0 & \text{if } (x, y) \in \partial D : \qquad \text{boundary of } D, \\ < 0 & \text{if } (x, y) \in | R^2 - D : \quad \text{exterior of } D, \end{cases}$$

$\qquad\qquad\qquad (5.25a)$
$\qquad\qquad\qquad (5.25b)$
$\qquad\qquad\qquad (5.25c)$

TABLE 5.1 Region Specified by a Unit Jump of a Function of Two Variables

$f(x, y)$	>0	≥0	=0	≤0	<0
Set of points	$D - \partial D$	D	∂D	$\|R^2 - D + \partial D$	$\|R^2 - D$
Region	Open interior	Closed interior	Boundary	Closed exterior	Open exterior
$H(f(x, y))$	1	1, 1/2	1/2	1/2, 0	0
$\delta(f(x, y))$	0	0, ∞	∞	∞, 0	0

Note: The unit jump (impulse) of an ordinary function of two variables specifies regions in the plane.

that is, its boundary ∂D and its closed D (open $D - \partial D$) interior and open $\|R^2 - D$ (closed $\|R^2 - D + \partial D$) exterior. The generalized function unit jump of a continuous function of two variables $f(x, y)$ is unity (5.26a) [zero (5.26c)] in the open interior (exterior) and one-half (5.26b) on the boundary:

$$f \in C(\|R^2): \quad H(f(x,y)) = \begin{cases} 1 & \text{if } f(x,y) > 0, \quad \text{i.e., in } D - \partial D, & (5.26a) \\ 1/2 & \text{if } f(x, y) = 0, \quad \text{i.e., on } \partial D, & (5.26b) \\ 0 & \text{if } f(x, y) < 0, \quad \text{i.e., in } \|R^2 - D. & (5.26c) \end{cases}$$

The inner product with an integrable test function (5.27a) reduces to the sum (5.27b) of the integral over the interior plus one-half of the integral over the boundary:

$$\Phi \in \mathcal{E}(\|R^2): \quad \left[H(f(x,y)), \Phi(x,y) \right] \equiv \int\limits_{-\infty}^{+\infty} \int H(f(x,y)) \Phi(x,y) dxdy$$

$$= \int\limits_{D:f(x,y)>0} \Phi(x,y) dxdy + \frac{1}{2} \int\limits_{\partial D:f(x,y)=0} \Phi(x,y) ds, \quad (5.27a,b)$$

where $dxdy$ is the area element and ds the arc length along the boundary. For example,

$$H(a^2 - x^2 - y^2) = H(a^2 - r^2) = \begin{cases} 1 & \text{if } r < a: \quad \text{interior of circle}, & (5.28a) \\ 1/2 & \text{if } r = a: \quad \text{on circle}, & (5.28b) \\ 0 & \text{if } r > a: \quad \text{outside circle}, & (5.28c) \end{cases}$$

where r, the polar coordinate distance to the origin, is unity (5.28a) [one-half (5.28b)] inside (on) the circle of radius $r = a$ and center at the origin, zero (5.28c) outside (Figure 5.3a). The integral property with regard to integrable test functions (5.29a)

$$\Phi(x,y) \equiv \Psi(r,\varphi) \in \mathcal{E}(\|R^2): \quad \left[H(a^2 - x^2 - y^2), \Phi(x,y) \right] \equiv \int\limits_{-\infty}^{+\infty} \int H(a^2 - x^2 - y^2) \Phi(x,y) dxdy$$

$$= \int\limits_0^{2\pi} d\varphi \int\limits_0^a r\Psi(r,\varphi) dr + \frac{a}{2} \int\limits_0^{2\pi} \Psi(a,\varphi) d\varphi, \quad (5.29a\text{--}c)$$

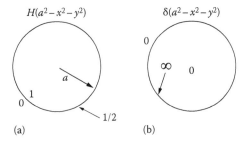

FIGURE 5.3 The generalized function unit jump of an ordinary function of two real variables (a) is unity (zero) in the regions of the plane where the function is positive (negative) and takes the value one-half on the boundary. On the same boundary curves where the ordinary function of two real variables vanishes, the generalized function unit impulse (b) is singular, taking zero value elsewhere.

is most conveniently expressed (5.29a) ≡ (5.29c) in polar coordinates (r, φ) as the area integral over the circle plus one-half the line integral over the boundary. The passage from (5.29a) to (5.29b) uses the relation between Cartesian (x, y) and polar (r, φ) coordinates (5.30a and b):

$$\{x, y\} = r\{\cos\varphi, \sin\varphi\}, \quad dxdy = r\,drd\varphi, \quad a^2 = x^2 + y^2 = r^2, \quad ds = a\,d\varphi, \tag{5.30a–e}$$

including the area element (5.30c), the circle (5.30d) of radius a, and the arc length (5.30e) along it.

5.2.2 Unit Impulse of a Function of Two Variables

Considering a continuous function of two variables (5.31a) as argument of a unit impulse, the latter vanishes everywhere (5.31b) except at the boundary where it is singular (5.31c):

$$f \in C(|R^2): \quad \delta(f(x, y)) = \begin{cases} 0 & \text{if } f(x, y) \neq 0, \quad \text{i.e., in } |R^2 - \partial D, \\ \infty & \text{if } f(x, y) = 0, \quad \text{i.e., in } \partial D, \end{cases} \tag{5.31a–c}$$

as indicated in Table 5.1; the nature of the singularity is specified by the integral (5.32b) corresponding to the inner product by a continuous test function (5.32a):

$$\Phi \in C(|R^2): \quad I \equiv \left[\delta(f(x, y)), \Phi(x, y)\right] \equiv \int\limits_{-\infty}^{+\infty}\int \delta(f(x, y))\Phi(x, y)dxdy. \tag{5.32a,b}$$

The change of variables from (x, y) to (x, f) specifies the area element through (5.33a) the Jacobian (5.33b):

$$J\,dxdy = dfdx, \quad J \equiv \frac{\partial(x, f)}{\partial(x, y)} = \begin{vmatrix} 1 & 0 \\ \partial f/\partial x & \partial f/\partial y \end{vmatrix} = \frac{\partial f}{\partial y}. \tag{5.33a,b}$$

The latter is simply $\partial f/\partial y$, so that the integral (5.32b)

$$I = \int\limits_{-\infty}^{+\infty}\int \delta(f)\frac{\Phi(x, y)}{J(x, f)}dfdx = \int\limits_{\partial D: f(x,y)=0} \frac{\Phi(x, y)}{\partial f/\partial y}dx, \tag{5.34}$$

is evaluated along the boundary ∂D. The latter is the curve specified by the implicit equation (5.35b)

$$n = 1,...,N: \quad f(x,y) = 0 \Rightarrow y = h_n(x), \tag{5.35a-c}$$

whose N roots (5.35a) are (5.35c) the functions $h_n(x)$ for $n = 1, ..., N$ that specify explicitly the boundary consisting of N arcs. Thus, (5.34) can be rewritten (5.36b)

$$\frac{\partial f}{\partial y}(x, h_n(x)) \neq 0: \quad \left[\delta(f(x,y)), \Phi(x,y) \right] \equiv \sum_{n=1}^{N} \frac{\Phi(x, h_n(x))}{\partial f / \partial y(x, h_n(x))} dx$$

$$= \sum_{n=1}^{N} \iint \frac{\Phi(x,y)}{\partial f / \partial y(x,y)} \delta(y - h_n(x)) dx dy, \tag{5.36a,b}$$

where it is assumed that $\partial f / \partial y$ does not vanish for (5.35c), that is, the roots of (5.35b) are simple (5.36a).

5.2.3 Decomposition of the Impulse of a Function into Simple Impulses

The preceding deduction (5.32a and b) through (5.36a and b) is a generalization of (5.5 through 5.8a,b) from one to two variables, and the result may be stated in a manner similar to (5.9a through c through 5.11a through c), that is, *if $f(x, y)$ is a function of two variables with continuous derivatives (5.39a) that do not vanish along the curves (5.35a through c) where the function is zero:*

$$n = 1,...,N: \quad f(x, h_n(x)) = 0 \neq \frac{\partial f}{\partial y}(x, h_n(x)); \tag{5.37a-c}$$

the generalized function unit impulse with argument $f(x, y)$ has the integral property (5.38b):

$$\Phi \in C(|R^2): \quad \left[\delta(f(x,g)), \Phi(x,y) \right] \equiv \int_{-\infty}^{+\infty}\!\!\int \delta(f(x,y)) \Phi(x,y) dx dy$$

$$= \sum_{n=1}^{N} \int_{-\infty}^{+\infty} \frac{\Phi(x, h_n(x))}{\partial f / \partial y(x, h_n(x))} dx, \tag{5.38a,b}$$

relative to continuous test functions (5.38a); this is equivalent to a decomposition into simple impulses (5.39b):

$$f \in C^1(|R^2): \quad \delta(f(x,y)) = \sum_{n=1}^{N} A_n(x) \delta(y - h_n(x)), \quad \frac{1}{A_n(x)} \equiv \frac{\partial f}{\partial y}(x, h_n(x)), \tag{5.39a-c}$$

with amplitudes (5.39c).

For example, the unit impulse

$$\delta(a^2 - x^2 - y^2) = \delta(a^2 - r^2) = \begin{cases} 0 & \text{if } r \neq a: \quad \text{not on the circle,} & \text{(5.40a)} \\ \infty & \text{if } r = a: \quad \text{on the circle,} & \text{(5.40b)} \end{cases}$$

is zero (5.40a) outside and inside the circle of radius a and is singular (5.4b) on the circle of radius $r = a$ with center at the origin (Figure 5.3b). Using polar coordinates and bearing in mind that $\delta(r + a) = 0$ on account of (5.41a) simplifies (5.13e) to (5.41b),

$$r + a > 0: \quad \delta\left(a^2 - x^2 - y^2\right) \equiv \delta\left(a^2 - r^2\right) = -\frac{\delta(a - r)}{2a} = -\frac{\delta(r - a)}{2a}; \qquad (5.41a,b)$$

considering a continuous test function in Cartesian or polar coordinates (5.42a) leads to (5.42b):

$$\Phi(x, y) = \Psi(r, \varphi) \in C\left(| R^2\right): \quad \left[\delta\left(a^2 - x^2 - y^2\right), \Phi(x, y)\right] \equiv \int_{-\infty}^{+\infty}\int \Phi(x, y)\delta\left(a^2 - x^2 - y^2\right) dx dy$$

$$= \int_0^{2\pi} d\varphi \int_0^{\infty} \Psi(r, \varphi)\delta\left(a^2 - r^2\right)r\, dr = -\frac{1}{2}\int_0^{2\pi} d\varphi\, \Psi(r, \varphi)\frac{r}{a}\delta(r - a) dr$$

$$= -\frac{1}{2}\int_0^{2\pi} \Psi(a, \varphi) d\varphi, \qquad (5.42a,b)$$

as the integral property of the unit impulse (5.41a and b).

5.2.4 Two Equivalent Decompositions into Simple Impulses

The roots of the ordinary continuously differentiable function of two variables (5.39a) could be determined for y in (5.43a through c) instead of x in (5.37a through c):

$$m = 1, \dots, M: \quad f\left(g_m(y), y\right) = 0 \neq \frac{\partial f}{\partial x}\left(g_m(y), y\right), \qquad (5.43a\text{-}c)$$

where the number of roots need not be the same M(N) in y(x). For the same continuous test function $(5.38a) \equiv (5.44a)$*, this would lead to the integral property (5.44b) instead of (5.38a and b):*

$$\Phi \in C\left(| R^2\right): \quad \left[\delta\left(f(x, y)\right), \Phi(x, y)\right] \equiv \int_{-\infty}^{+\infty}\int \delta\left(f(x, y)\right)\Phi(x, y) dx dy$$

$$= \sum_{m=1}^{M} \int_{-\infty}^{+\infty} \frac{\Phi\left(g_m(y), y\right)}{\partial f / \partial x\left(g_m(y), y\right)} dy. \qquad (5.44a,b)$$

This corresponds to (5.45a through c):

$$f \in C^1\left(| R^2\right): \quad \delta\left(f(x, y)\right) = \sum_{m=1}^{M} B_m(y)\delta\left(x - g_m(y)\right), \quad \frac{1}{B_m(y)} \equiv \frac{\partial f}{\partial x}\left(g_m(y), y\right), \qquad (5.45a\text{-}c)$$

as an alternative to (5.39a through c). For example, (5.46a) has two roots (5.46b) [(5.46c)] with regard to y(x):

$$0 = a^2 - x^2 - y^2 \equiv f(x, y): \quad y = \pm\sqrt{a^2 - x^2} = \pm h(x), \quad x = \pm\sqrt{a^2 - y^2} = \pm g(y). \qquad (5.46a\text{-}c)$$

Using (5.46d) [(5.46f)] in (5.39a through c) [(5.45a through c)] leads to the decomposition into simple impulse (5.46e) [(5.46g)]:

$$\frac{\partial f}{\partial y} = -2y: \quad \delta\left(a^2 - x^2 - y^2\right) = \frac{\delta\left(y + \sqrt{a^2 - x^2}\right) - \delta\left(y - \sqrt{a^2 - x^2}\right)}{2\sqrt{a^2 - x^2}}, \qquad (5.46d,e)$$

$$\frac{\partial f}{\partial x} = -2x: \quad \delta\left(a^2 - x^2 - y^2\right) = \frac{\delta\left(x + \sqrt{a^2 - y^2}\right) - \delta\left(x - \sqrt{a^2 - y^2}\right)}{2\sqrt{a^2 - y^2}}, \qquad (5.46f,g)$$

for the unit impulse of argument (5.46a).

5.2.5 Gradient of the Unit Jump of an Ordinary Function

The partial derivatives may be applied to (5.26a through c) the unit jump of a function of two variables, for example, to the first (5.47b) [second (5.48b)] order if the function has continuous first (5.47a) [second (5.48a)] derivatives:

$$f \in C^1\left(|R^2\right): \quad \left\{\frac{\partial}{\partial x}, \frac{\partial}{\partial y}\right\}\left[H\left(f(x,y)\right)\right] = \left\{\frac{\partial f}{\partial x}, \frac{\partial f}{\partial y}\right\}\delta\left(f(x,y)\right), \qquad (5.47a,b)$$

$$f \in C^2\left(|R^2\right): \quad \left\{\frac{\partial^2 f}{\partial x^2}, \frac{\partial^2 f}{\partial y^2}, \frac{\partial^2 f}{\partial x \partial y}\right\}\left[H\left(f(x,y)\right)\right] = \left\{\frac{\partial^2 f}{\partial x^2}, \frac{\partial^2 f}{\partial y^2}, \frac{\partial^2 f}{\partial x \partial y}\right\}\delta\left(f(x,y)\right)$$

$$+ \left\{\left(\frac{\partial f}{\partial x}\right)^2, \left(\frac{\partial f}{\partial y}\right)^2, \frac{\partial f}{\partial x}\frac{\partial f}{\partial y}\right\}\delta'\left(f(x,y)\right). \qquad (5.48a,b)$$

The first-order derivatives (5.47b) may be combined in the gradient operator (5.49b) for a function with continuous first-order derivatives (5.49a) and its unit jump (5.49c):

$$f \in C^1\left(|R^2\right): \quad \nabla \equiv \vec{e}_x \frac{\partial}{\partial x} + \vec{e}_y \frac{\partial}{\partial y}, \quad \nabla\left[H\left(f(x,y)\right)\right] = \delta\left(f(x,y)\right)\nabla f. \qquad (5.49a\text{–}c)$$

Using the property (5.39a through c) for a function with N simple zeros (5.37a through c) leads to

$$\nabla\left[H\left(f(x,y)\right)\right] = \sum_{n=1}^{N} \frac{\delta\left(y - h_n(x)\right)}{\partial f/\partial y}\left(\vec{e}_x \frac{\partial f}{\partial x} + \vec{e}_y \frac{\partial f}{\partial y}\right), \qquad (5.50a)$$

which simplifies to (5.50c) using the theorem (5.50b) of the implicit function:

$$\frac{\partial f/\partial x}{\partial f/\partial y} = -\left(\frac{dy}{dx}\right)_f = -h_n'(x): \quad \nabla\left[H\left(f(x,y)\right)\right] = \sum_{n=1}^{N} \delta\left(y - h_n(x)\right)\left[\vec{e}_y - \vec{e}_x h_n'(x)\right]. \qquad (5.50b,c)$$

Thus, *the unit jump of a function of two variables with continuous first-order derivatives (5.49a) and simple roots (5.35a through c) [(5.43a through c)] has a gradient given equivalently (5.50c) or (5.50d)*:

$$\nabla\left[H\left(f(x,y)\right)\right] = \sum_{m=1}^{M} \delta\left(x - g_m(y)\right)\left[\vec{e}_x - \vec{e}_y g_n'(y)\right], \qquad (5.50d)$$

corresponding to (5.39a through c) [(5.45a through c)].

In the particular case of the gradient of the unit jump with the function (5.46a) as argument, use of (5.46b and e) [(5.46c and f)] in (5.50c) [(5.50d)] leads to (5.51a) [(5.51b)]:

$$\nabla\left[H\left(a^2 - x^2 - y^2\right)\right] = \delta\left(y - \sqrt{x^2 - a^2}\right)\left[\vec{e}_y + \frac{x}{\sqrt{a^2 - x^2}}\vec{e}_x\right]$$

$$+ \delta\left(x + \sqrt{x^2 - a^2}\right)\left[\vec{e}_y - \frac{x}{\sqrt{a^2 - x^2}}\vec{e}_x\right] \tag{5.51a}$$

$$\nabla\left[H\left(a^2 - x^2 - y^2\right)\right] = \delta\left(x - \sqrt{y^2 - a^2}\right)\left[\vec{e}_x + \frac{y}{\sqrt{a^2 - y^2}}\vec{e}_y\right]$$

$$+ \delta\left(x + \sqrt{y^2 - a^2}\right)\left[\vec{e}_x - \frac{y}{\sqrt{a^2 - y^2}}\vec{e}_y\right]. \tag{5.51b}$$

The result (5.51a) can be checked from

$$\nabla\left[H\left(a^2 - x^2 - y^2\right)\right] = \delta\left(a^2 - x^2 - y^2\right)\left(-\vec{e}_x 2x - \vec{e}_y 2y\right)$$

$$= \frac{\delta\left(y - \sqrt{a^2 - x^2}\right) - \delta\left(y + \sqrt{a^2 - x^2}\right)}{\sqrt{a^2 - x^2}}\left(\vec{e}_x x + \vec{e}_y y\right)$$

$$= \delta\left(y - \sqrt{a^2 - x^2}\right)\left[\vec{e}_y + \frac{x}{\sqrt{a^2 - x^2}}\vec{e}_x\right]$$

$$+ \delta\left(y + \sqrt{a^2 - x^2}\right)\left[\vec{e}_y - \frac{x}{\sqrt{a^2 - x^2}}\vec{e}_x\right], \tag{5.51c}$$

using (5.46f) to obtain (5.51c) ≡ (5.51a); likewise, (5.51b) can be confirmed from (5.46g). The consideration of one-dimensional generalized functions whose argument is an ordinary function of one (two) variable [Section 5.1 (Section 5.2)] is followed by the product of two generalized functions that must have as arguments either (1) two distinct variables (Section 5.3) or and (2) two linearly independent ordinary functions (Section 5.4), so that the generalized function (Section 3.4) remains a linear functional (Subsection 3.4.7).

5.3 Products of Generalized Functions of Different Variables

The definition of generalized function (Section 3.4) as a linear functional implies that it is generally not possible to define squares or higher powers; for example, $\{\delta(x)\}^2$ does not exist (Subsection 3.4.7); this is the nonlinear property lost in exchange for the ability to have differentiation of all orders. Although a generalized function (1) cannot be multiplied by itself nor by another generalized function of the same variable or a constant multiple of the same variable, (2) it can be multiplied by a generalized function of another variable; for example (Section 5.3), it is permissible to multiply unit impulses of different variables $\delta(x)\delta(y)$; (3) it is also permissible to multiply two generalized functions whose arguments are two ordinary functions of two or more variables, provided they are not constant multiples (Section 5.4); and (4) the product can also be applied to any number of generalized functions, provided that their arguments are linearly independent ordinary functions (Section 5.5). Next are considered two-dimensional generalized functions unit jump (impulse) that are the product of two unit jumps (impulses) whose arguments are [Subsection 5.3.1 (Subsection 5.3.2)] distinct variables. The product of a unit jump by a unit impulse can also be considered (Subsection 5.3.3).

5.3.1 Two-Dimensional Unit Jump

Starting with the unit jump (3.1) in one variable applied to a test function integrable in the plane (5.52a), it leads to (5.52b) a two-dimensional integral,

$$\Phi \in \mathcal{E}(|R^2): \quad \int\limits_{-\infty}^{+\infty}\int H(x)\Phi(x,y)\,dxdy = \int\limits_{0}^{\infty}dx\int\limits_{-\infty}^{+\infty}dy\,\Phi(x,y)+\frac{1}{2}\int\limits_{-\infty}^{+\infty}\Phi(0,y)\,dy, \qquad (5.52a,b)$$

that equals an area integral over the right-hand-half plane plus half the line integral along the *OY*-axis. The **two-dimensional unit jump**

$$H(x)H(y)=\begin{cases} 1 & \text{if } x>0 \text{ and } y>0, & (5.53a)\\[4pt] 1/2 & \text{if } x>0=y \text{ or } y>0=x, & (5.53b)\\[4pt] 1/4 & \text{if } x=0=y, & (5.53c)\\[4pt] 0 & \text{if } x<0 \text{ or } y<0, & (5.53d) \end{cases}$$

is unity in the first quadrant (5.53a), one-half (5.53b) on the positive coordinate half-axis, one-quarter (5.53c) at the origin, and zero (5.53d) elsewhere (Figure 5.4a). When applied to a test function integrable in the plane (5.54a), it has the property (5.54b):

$$\Phi \in \mathcal{E}(|R^2): \quad \big[H(x)H(y),\Phi(x,y)\big] \equiv \int\limits_{-\infty}^{+\infty}\int H(x)H(y)\Phi(x,y)\,dxdy$$

$$= \int\limits_{0}^{+\infty}\int\limits_{0}^{+\infty}\Phi(x,y)\,dxdy + \frac{1}{2}\int\limits_{0}^{+\infty}\big[\Phi(x,0)+\Phi(0,x)\big]\,dx. \qquad (5.54a,b)$$

This can be compared with

$$\big[H(x),\Phi(x,y)\big] \equiv \int\limits_{-\infty}^{+\infty}H(y)\Phi(x,y)\,dx = \int\limits_{0}^{+\infty}\Phi(x,y)\,dx, \qquad (5.54c)$$

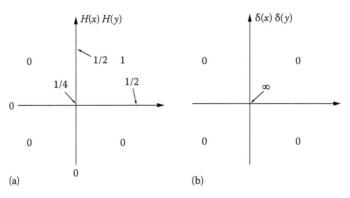

                        (a)   (b)

FIGURE 5.4 A two-dimensional generalized function is the product of two generalized functions with distinct or linearly independent variables, for example the product of two unit jumps (a) or impulses (b). The two-dimensional unit jump (a) is unity on the first quadrant, one-half on the positive coordinate axis, one-quarter at the origin and zero elsewhere. The two-dimensional unit impulse (b) is zero everywhere except at the origin where it has a singularity related to the unit impulse with cylindrical symmetry (Figure 5.5).

that is, one (5.54c) [two (5.54b)] inner product(s) of one (1.28a through c) [two (5.53a through d)] unit jump(s) with a test function (5.54a) integrable on the real plane.

5.3.2 Two-Dimensional Cartesian and Polar Impulse

The one-dimensional unit impulse (3.10) may also be used in two-dimensional integrals (5.55b):

$$\Phi(x,y) \in \mathcal{E}(|R) \times C(|R): \quad \left[\delta(y), \Phi(x,y)\right] \equiv \int_{-\infty}^{+\infty}\int \delta(y)\Phi(x,y)dy = \int_{-\infty}^{+\infty}\Phi(x,0)dx, \quad (5.55a,b)$$

where the test function (5.55a) is integrable in x and continuous in y. The **two-dimensional unit impulse**

$$\delta(x)\delta(y) = \begin{cases} 0 & \text{if } x \neq 0 \text{ or } y \neq 0, & (5.56a) \\ \infty & \text{if } x = 0 = y, & (5.56b) \end{cases}$$

is zero in the whole plane (5.56a) except (5.56b) at the origin (Figure 5.4b), where it is singular; the nature of the singularity is specified by the integral property (5.57b) relative to a test function (5.57a) continuous on the plane:

$$\Phi \in C(|R^2): \quad \left[\delta(x)\delta(y), \Phi(x,y)\right] \equiv \int_{-\infty}^{+\infty}\int \delta(x)\delta(y)\Phi(x,y)dxdy = \Phi(0,0), \quad (5.57a,b)$$

that is, the particular case $n = 0 = m$ of the integral property (5.58b) relative to a test function $\Phi(x,y)$ with (5.58a) continuous derivative of order $n(m)$ with regard to $x(y)$:

$$\Phi(x,y) \in C^n(|R) \times C^m(|R): \quad \left[\delta^{(n)}(x)\delta^{(m)}(y), \Phi(x,y)\right] \equiv \int_{-\infty}^{+\infty}\int \delta^{(n)}(y)\delta^{(m)}(y)\Phi(x,y)dxdy$$

$$= (-)^{n+m} \lim_{x,y \to 0} \frac{\partial^{n+m}\Phi}{\partial x^n \partial y^n}. \quad (5.58a,b)$$

The property (5.57b) may be rewritten,

$$\Phi(x,y) = \Psi(r,\varphi): \quad \int_0^{2\pi}d\varphi \int_0^{\infty} r\delta(x)\delta(y)\Phi(x,y)dr = \Phi(0,0)$$

$$= \Psi(0,\varphi) = \frac{1}{2\pi}\int_0^{2\pi}d\varphi \int_0^{\infty} \delta(r)\Psi(r,\varphi)dr, \quad (5.59a,b)$$

in polar coordinates

$$r \equiv |x^2 + y^2|^{1/2}: \quad \delta(r) = 2\pi r\delta(x)\delta(y). \quad (5.60a,b)$$

The relation (5.60a and b) between (Figure 5.5) the one-dimensional polar unit impulse $\delta(r)$ and (Figure 5.4b) the two-dimensional Cartesian unit impulse $\delta(x)\delta(y)$ involves the perimeter $2\pi r$ of the circle of radius r and will be generalized subsequently to arbitrary dimensions (Section 5.9). Both sides of (5.60b) vanish everywhere except at the origin, where they have a comparable singularity; this is shown in Example 2.12 using the alternate approach to generalized functions, that is, the nonuniform limit of a family of ordinary functions instead of a linear continuous functional.

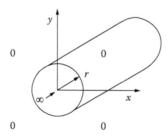

FIGURE 5.5   The generalized function unit jump with cylindrical symmetry is zero everywhere except on the axis where it is singular; thus, it relates to the product of two Cartesian unit impulses multiplied by the area of the cylinder per unity length that corresponds to the perimeter of a circle. These can be generalized to the unit jump and impulse in three (higher) dimensional spaces. The unit jump with spherical (hyperspherical) symmetry involves the area of the sphere (hypersphere).

5.3.3 Product of a Unit Jump by a Unit Impulse

The product of a unit jump (3.1) and nth derivative of the unit impulse (3.10) has the integration property (5.61b) relative to a test function that is integrable (n times continuously continuous) in the first (second) variable (5.61a):

$$\Phi(x,y) \in E(|R) \times C^n(|R): \quad \left[H(x)\delta^{(n)}(y), \Phi(x,y)\right] \equiv \int\limits_{-\infty}^{+\infty} \int H(x)\delta^{(n)}(y)\Phi(x,y)dxdy$$

$$= (-)^n \int\limits_{0}^{\infty} \partial^n \Phi/\partial y^n(x,0)dx. \qquad (5.61a,b)$$

The proof is immediate using the properties (3.3c) [(3.22c)] of the unit jump (nth derivative of the unit impulse):

$$\int\limits_{-\infty}^{+\infty} dx\, H(x) \int\limits_{-\infty}^{+\infty} \delta^{(n)}(y)\Phi(x,y)dy = (-)^n \int\limits_{-\infty}^{+\infty} H(x)\frac{\delta^{(n)}\Phi}{\partial y^n}(x,0)dx = (-)^n \int\limits_{0}^{+\infty} \frac{\partial^n \Phi}{\partial y^n}(x,0)dx. \qquad (5.62)$$

The differentiation can be applied to the product of two unit jumps, for example, to the first (5.63) [second (5.64)] order:

$$\left\{\frac{\partial}{\partial x}, \frac{\partial}{\partial y}\right\}\left[H(x)H(y)\right] = \left\{\delta(x)H(y), \delta(y)H(x)\right\}, \qquad (5.63)$$

$$\left\{\frac{\partial^2}{\partial x^2}, \frac{\partial}{\partial y^2}, \frac{\partial}{\partial x\partial y}\right\}\left[H(x)H(y)\right] = \left\{\delta'(x)H(y), \delta'(y)H(x), \delta(x)\delta(y)\right\}, \qquad (5.64)$$

which are particular cases of

$$\frac{\partial^{n+m}}{\partial x^n \partial x^m}\left[H(x)H(y)\right] = \delta^{(n)}(x)\delta^{(m)}(y). \qquad (5.65)$$

The product of a unit jump and a unit impulse inherits (1.78a and b) the substitution property (5.66b) relative to a uniformly continuous function (5.66a):

$$f \in \tilde{C}\left(\left| R^2 \right.\right): \quad f(x,y)H(x)\delta(y-\xi) = H(x)\delta(y-\xi)f(x,\xi). \tag{5.66a,b}$$

5.4 Two-Dimensional Generalized Functions with Two Ordinary Functions as Arguments

Following the one-dimensional generalized functions of one variable (Section 3.4) and one ordinary function of one (two) variable [Section 5.1 (Section 5.2)] is the two-dimensional generalized functions, that is, the product of two generalized functions with arguments two distinct variables (Section 5.3) [two linearly independent functions of two variables (Section 5.4)]. One example is the two-dimensional unit jump (impulse) of [Subsection 5.4.2 (Subsection 5.4.3)] of two linearly independent functions of two variables (Subsection 5.4.1); the latter two-dimensional unit impulse can be decomposed (Subsection 5.4.4) into a sum of products of one-dimensional impulses.

5.4.1 Two Linearly Independent Functions of Two Variables

The square of the unit impulse is not defined (Subsection 3.4.7), and likewise for the product of unit impulses whose arguments are functions of the same variable. For example, if the function $f(g)$ has $N(M)$ simple zeros and is used as argument (5.11a through c) of a unit impulse,

$$f(a_n) = 0 \neq f'(a_m): \quad \delta(f(x)) = \sum_{n=1}^{N} \frac{\delta(x-a_n)}{f'(a_n)}, \tag{A}$$

$$g(b_m) = 0 \neq g'(b_m): \quad \delta(g(x)) = \sum_{m=1}^{M} \frac{\delta(x-b_n)}{g'(b_m)}, \tag{B}$$

the product would be

$$\delta(f(x))\delta(g(x)) = \sum_{n=1}^{N}\sum_{m=1}^{M} \frac{\delta(x-a_n)\delta(x-b_m)}{f'(a_n)g'(b_m)}. \tag{C}$$

The inner product by continuous test function would be

$$\Phi \in C\left(\left| R \right.\right): \quad \left[\delta(f(x))\delta(g(x)),\Phi(x)\right]$$

$$= \sum_{n=1}^{N}\sum_{m=1}^{M} \int_{-\infty}^{+\infty} \frac{\delta(x-a_n)\delta(x-b_m)}{f'(a_n)g'(b_m)}\Phi(x)\,dx$$

$$= \sum_{n=1}^{N}\sum_{m=1}^{M} \frac{\Phi(a_n)\delta(x-b_m)}{f'(a_n)g'(b_m)}$$

$$= \sum_{n=1}^{N}\sum_{m=1}^{M} \frac{\Phi(b_m)\delta(x-a_m)}{f'(a_n)g'(b_m)}. \tag{D}$$

If $a_m \neq b_n$, the last two expressions (D) are inconsistent because one is singular at $a_n(b_n)$ whereas the other is zero at the same point; if $a_n = b_m$, then (D) involves the square of a unit impulse that is not permissible (Subsection 3.4.7). In the case of two functions of two variables that are constant multiples

$$\delta(x,y) = \lambda f(x,y): \quad \delta(f(x,y))\delta(g(x,y)) = \lambda^{-1}\left[\delta(f(x,y))\right]^2, \tag{E}$$

if they are used as argument of unit impulses, their product leads (5.12d) to the square of a unit impulse, again not permissible. Thus, *the product of two unit impulses exists if the arguments are two* **linearly independent** *functions of two variables*:

$$\alpha f(x,y) + \beta g(x,y) = 0 \Leftrightarrow \alpha = 0 = \beta. \tag{5.67a,b}$$

The functions (f,g) are linearly independent if their linear combination (5.67a) vanishes only if both coefficients (5.67b) are zero. If functions are continuously differentiable (5.68a), they are linearly independent if the **Jacobian** *is not zero (5.68b):*

$$f,g \in C^1\left(|R^2\right): \quad J \equiv \frac{\partial(f,g)}{\partial(x,y)} \equiv \begin{vmatrix} \partial f/\partial x & \partial f/\partial y \\ \partial g/\partial x & \partial g/\partial y \end{vmatrix} \equiv \frac{\partial f}{\partial x}\frac{\partial g}{\partial y} - \frac{\partial f}{\partial y}\frac{\partial g}{\partial x} \neq 0. \tag{5.68a,b}$$

The proof (5.68b) follows differentiating (5.67a) with regard to x and y, to form a linear homogeneous system of equations:

$$\begin{bmatrix} \partial f/\partial x & \partial f/\partial y \\ \partial g/\partial x & \partial g/\partial y \end{bmatrix}\begin{bmatrix} \alpha \\ \beta \end{bmatrix} = 0. \tag{5.69}$$

In order that the only solution of (5.69) be (5.67b), the determinant of the matrix of coefficients, that is, the Jacobian (5.68b), must be nonzero.

5.4.2 Product of Unit Jumps of Functions of Two Variables

The two-dimensional generalized function specified by the product of two unit jumps of two continuous functions (5.70a and b) that are not constant multiples (5.67a and b),

$$f,g \in C\left(|R^2\right): \quad H(F(x,y))H(g(x,y)) = \begin{cases} 1 & \text{if } f(x,y) > 0 < g(x,y): (x,y) \in D_1 \cap D_2, \\ 1/2 & \text{if } f(x,y) = 0 < g(x,y): (x,y) \in D_2 \cap \partial D_1, \\ & \text{or } f(x,y) > 0 = g(x,y): (x,y) \in D_1 \cap \partial D_2, \\ 1/4 & \text{if } f(x,y) = 0 = g(x,y): (x,y) \in D_1 \cap \partial D_2, \\ 0 & \text{if } f(x,y) < 0 \text{ or } g(x,y) < 0, \end{cases} \tag{5.70a–e}$$

is unity if both are positive (5.70b), zero if either is negative (5.70e), 1/4 if both are zero (5.70d), and 1/2 in the remaining case of one zero and the other positive (5.70c). If $f > 0 (g > 0)$ in the interior of a region

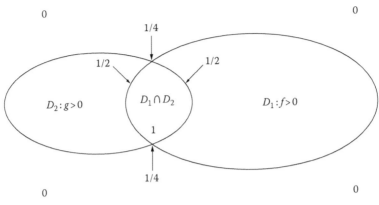

FIGURE 5.6 A two-dimensional generalized function equal to the product of two unit jumps of two linearly independent ordinary functions of two real variables takes the value: (1) unity in the intersection of the regions where each function is positive; (2) one-half on the boundary of one region contained in the other; (3) one-quarter at the points of intersection of two boundaries; and (4) zero elsewhere. An example would be one function positive in a horizontal (vertical) strip $|x| < a$ ($|y| < b$). The points of intersection of the boundaries $(x, y) = (\pm a, \pm b)$ are singularities of the product of unit impulses of the two functions (Figure 5.7).

$D_1(D_2)$, $f = 0(g = 0)$ on the boundary $\partial D_1(\partial D_2)$, and $f < 0(g < 0)$ on the exterior, then (Figure 5.6) the two-dimensional unit jump is (1) unity in the interior of the intersection of the two regions (5.70b); (2) one-half on the part of the boundary of one region contained in the other (5.70c); (3) one-quarter at the points where the boundaries of the two regions intersect (5.70c); and (4) zero elsewhere (5.70e). From this follows the corresponding integration property. The differentiation property (5.71b),

$$f, g \in C^1\left(|R^2\right): \quad \left\{\frac{\partial}{\partial x}, \frac{\partial}{\partial y}\right\}\left[H\big(f(x,y)\big)H\big(g(x,y)\big)\right]$$

$$= \left\{\delta(f)H(g)\frac{\partial f}{\partial x} + H(f)\delta(g)\frac{\partial g}{\partial x}, \delta(f)H(g)\frac{\partial f}{\partial y} + H(f)\delta(g)\frac{\partial g}{\partial y}\right\}, \qquad (5.71\text{a,b})$$

applies to continuously differentiable functions (5.71a) and can be restated in terms of the gradient operator (5.48b):

$$\nabla\left[H\big(f(x,y)\big)H\big(g(x,y)\big)\right] = \delta(f)H(g)\nabla f + H(f)\delta(g)\nabla g. \qquad (5.72)$$

5.4.3 Two-Dimensional Impulse of Two Functions of Two Variables

After considering generalized functions whose argument is an ordinary function of two variables (Section 5.2) and also two-dimensional generalized functions whose arguments are distinct variables (Section 5.3), the two aspects can be taken together (Section 5.4), proceeding to (Section 5.5) multidimensional generalized functions whose arguments are distinct functions of several variables (Diagram 5.1). Before proceeding to the general case of M-dimensional generalized functions of M functions of N variables (Section 5.5), one more two-dimensional case is given (Section 5.4), namely a two-dimensional generalized function of two ordinary functions. This follows from (Section 5.2) the one-dimensional generalized function of one function $M = 1$ of two variables $N = 2$ and (Section 5.4) two-dimensional generalized function $M = 2$ of distinct variables $N = 2$. The two-dimensional generalized function of two functions of two variables (Subsection 5.4.1) as before [Subsection 5.2.1 (Subsection 5.3.1)] leads [Subsection 5.2.2 (Subsection 5.4.3)] to a decomposition (Subsection 5.4.3) into simple impulses with suitable amplitudes (Subsection 5.4.4). An example is the product of two impulses whose arguments are linearly independent (5.67a and b) distinct functions of two variables:

$$\delta\big(f(x,y)\big)\delta\big(g(x,y)\big) = \begin{cases} 0 & \text{if } x \neq x_n \text{ or } y \neq y_n, \\ \infty & \text{if } x = x_n \text{ and } y = y_n; \end{cases}$$

(5.73a)

(5.73b)

it is zero everywhere (5.73a) except for singularities (5.73b) at the common zeros (x_n, y_n) of the continuously differentiable functions:

$$f,g \in C^1\big(|R^2\big): \quad f(x_n,y_n) = g(x_n,y_n) = 0 \neq J(x_n,y_n).$$

(5.74a–d)

In the case of Figure 5.6, the zeros lie at the intersection of the two boundaries corresponding to $f = 0 = g$. The zeros (5.74b and c) are simple (5.74d) because the Jacobian cannot vanish (5.68b) in order that the continuously differentiable functions (5.67a) be linearly independent (5.67b); then follows the integral property (5.75b) relative to test functions that are continuous in the plane (5.75a):

$$\Phi \in C\big(|R^2\big): \quad \Big[\delta\big(f(x,y)\big)\delta\big(g(x,y)\big), \Phi(x,y)\Big] \equiv \int_{-\infty}^{+\infty}\int \delta\big(f(x,y)\big)\delta\big(g(x,y)\big)\Phi(x,y)\,dxdy$$

$$= \sum_{n=1}^{N}\left\{\frac{g(x_n,y_n)}{J(x_n,y_n)}\right\}.$$

(5.75a,b)

The same result is obtained by

$$\sum_{n=1}^{N}\left\{\frac{g(x_n,y_n)}{J(x_n,y_n)}\right\} = \left[\sum_{n=1}^{N}\frac{\delta(x-x_n)\delta(y-y_n)}{J(x_n,y_n)}, \Phi(x,y)\right]$$

$$= \sum_{n=1}^{N}\int_{-\infty}^{+\infty}\int \delta(x-x_n)\delta(y-y_n)\left\{\frac{\Phi(x,y)}{J(x,y)}\right\}dxdy,$$

(5.76)

which yields a decomposition into simple impulses, generalizing (5.10a and b) and (5.39a through c).

5.4.4 Amplitudes of the Decomposition into Two-Dimensional Unit Impulses

The result may be stated as follows: *the two-dimensional* **unit impulse with arguments two functions** *f,g(x, y), continuously differentiable with simple zeros (5.74a through d), has the integral property (5.75a and b), which is equivalent (5.76) to the decomposition (5.77a) into simple two-dimensional impulses*

$$\delta\big(f(x,y)\big)\delta\big(h(x,y)\big) = \sum_{n=1}^{N} A_n \delta(x-x_n)\delta(y-y_n),$$

(5.77a)

$$\frac{1}{A_n} \equiv J(x_n,y_x) \equiv \left[\frac{\partial f}{\partial x}\frac{\partial g}{\partial y} - \frac{\partial f}{\partial y}\frac{\partial g}{\partial x}\right]_{x=x_n,y=y_n},$$

(5.77b)

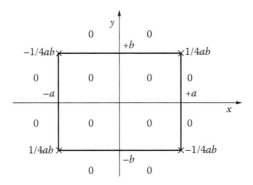

FIGURE 5.7 The generalized function product of unit impulses whose arguments are linearly independent functions of two variables is zero everywhere except at the common zeros of the two functions where it is singular.

with amplitudes (5.77b) specified by the inverse of the Jacobian (5.68b) evaluated at the zeros (5.74b and c) where the Jacobian does not vanish (5.74d) because the functions must be linearly independent (5.67a and b), implying that the zeros are simple (5.74d). As an example, consider the intersection of two circular cylinders (5.78a) [(5.78b)] with radii $a(b)$ and $x(y)$-axis, corresponding (5.68b) to the Jacobian (5.78c):

$$f(x,y) = a^2 - x^2, \quad g(x,y) = b^2 - y^2, \quad J(x,y) = 4xy, \tag{5.78a-c}$$

so that (5.77a) leads to the decomposition

$$\delta(a^2 - x^2)\delta(b^2 - y^2) = \frac{\delta(x-a)\delta(y-b) + \delta(x+a)\delta(y+b) - \delta(x+a)\delta(y-b) - \delta(x-a)\delta(y+b)}{4ab}. \tag{5.79}$$

This coincides with (5.79) ≡ (5.80) with

$$\delta(a^2 - x^2)\delta(b^2 - y^2) = \frac{\delta(x+a) - \delta(x-a)}{2a} \frac{\delta(y+b) - \delta(y-b)}{2b}, \tag{5.80}$$

that is, (Figure 5.7) is the product of two expressions like (5.13e). Four more cases of the product of two unit impulses with two ordinary functions of two variables as arguments are given in Example 10.13.

5.5 Multidimensional Generalized Functions with Several Ordinary Functions as Arguments

The preceding cases of one- or two-dimensional generalized functions whose arguments are ordinary functions of one or two variables $M, N = 1, 2$ serve as an introduction (Sections 5.1 through 5.4) to the extension to M-dimensional generalized functions whose arguments are $M \leq N$ ordinary functions of N variables (Sections 5.5 and 5.6) that must be linearly independent (Subsection 5.5.1). Using the notation in Diagram 5.1, it is possible to generalize the preceding integral properties from (1) two (x, y) to N-dimensions $(x_1,..., x_N)$; and (2) one or two functions to M functions $(f_1,..., f_M)$ as arguments. In order that the generalized function, for example, the unit impulse, can be defined, the

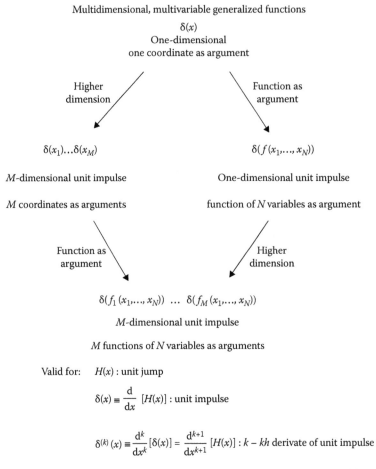

Multidimensional, multivariable generalized functions

$\delta(x)$
One-dimensional
one coordinate as argument

Higher dimension

Function as argument

$\delta(x_1)...\delta(x_M)$

M-dimensional unit impulse

M coordinates as arguments

$\delta(f(x_1,..., x_N))$

One-dimensional unit impulse

function of N variables as argument

Function as argument

Higher dimension

$\delta(f_1(x_1,..., x_N))\ ...\ \delta(f_M(x_1,..., x_N))$

M-dimensional unit impulse

M functions of N variables as arguments

Valid for: $H(x)$: unit jump

$$\delta(x) \equiv \frac{d}{dx}[H(x)] : \text{unit impulse}$$

$$\delta^{(k)}(x) \equiv \frac{d^k}{dx^k}[\delta(x)] = \frac{d^{k+1}}{dx^{k+1}}[H(x)] : k - kh \text{ derivate of unit impulse}$$

DIAGRAM 5.1   A generalized function may have as argument one or N independent variables x_n or an ordinary function f of N variables; a generalized function of dimension M is the product of M generalized functions where $M \leq N$ cannot exceed the number of independent variables. The generalized functions of dimension M in $N \geq M$ variables relates to subspaces of dimension $P = M - N$ of an N-dimensional space, for example hypercurves $P = 1$ [hypersurfaces $P = N - 1$].

M functions of N variables must be linearly independent (Subsection 5.5.1), implying that $M \leq N$, and the matrix of partial derivatives must have the maximum possible rank M. There are three cases, taking as examples the unit impulse: (a) the argument is a function of N variables (Subsection 5.5.2) instead of one (two) [Section 5.1 (5.2)]; (b) the M-dimensional generalized function is the product of M one-dimensional generalized functions (Subsection 5.6.2) instead of two (Section 5.3); (c) the M-dimensional generalized function has arguments that are M functions of N variables (Subsection 5.5.5) instead of $N = 2 = M$ in Section 5.4. Four progressively more general cases are considered: (1) a one-dimensional generalized function whose argument is an ordinary function of N variables (Subsection 5.5.2); (2) an M-dimensional generalized function whose arguments are M coordinates (Subsection 5.5.3); (3) an M-dimensional generalized function whose arguments are M ordinary functions of N variables (Subsection 5.5.4); and (4) an M-dimensional generalized function whose arguments are M functions of $N \geq M$ variables (Subsection 5.5.5). The cases (2) [(3)] lead to a point in an M-dimensional (N-dimensional) space; the cases (1) [(4)] lead to a hypersurface $M = 1$ (subspace of dimension M) of an N-dimensional space.

5.5.1 Multidimensional Linearly Independent Functions

An example of multidimensional multivariable generalized function is the M-dimensional unit impulse consisting (5.81c) of the product of M unit impulses whose arguments are M ordinary functions of N variables (5.81a and b):

$$m = 1,\dots,M; n = 1,\dots,N: \quad \prod_{n=1}^{M} \delta\big(f_m(x_n)\big) \equiv \delta\big(f_1(x_1,\dots,x_N)\big)\cdots\delta\big(f_M(x_1,\dots,x_N)\big). \quad (5.81\text{a--c})$$

Since the unit impulse, like any generalized function, is a linear functional, there can be no powers or products (Subsections 3.4.7 and 5.4.1), and thus the M functions must be **linearly independent**:

$$\sum_{m=1}^{M} \lambda_m f_m(x_n) = 0 \Leftrightarrow (\lambda_1,\dots,\lambda_M) = (0,\dots,0), \quad (5.82\text{a,b})$$

that is, their linear combination (5.82a) is zero if and only if all the coefficients are zero (5.82b). If the M functions are continuously differentiable with regard to the N variables (5.83a), from (5.82a) follows the linear system of M equations in N unknowns (5.83b)

$$f_M \in C^1\big(|R^N\big): \quad \sum_{n=1}^{N} \lambda_m \frac{\partial f_m}{\partial x_n} = 0, \quad (5.83\text{a,b})$$

whose coefficients form the $M \times N$ matrix of partial derivatives (5.84b):

$$N \geq M: \quad Ra\left(\frac{\partial f_m}{\partial x_n}\right) = M. \quad (5.84\text{a,b})$$

The **rank** of a matrix is the size of the largest nonzero determinant it contains, that is, formed by elements from its lines and columns. *In order that M functions (5.81c) of N variables (5.81a and b) be linearly independent (5.82a and b), it is necessary that the number of variables is not less than the number of unknowns (5.84a). If the functions are continuously differentiable (5.83a), a sufficient condition for linear independence (5.82a and b) is that the matrix of partial derivatives (5.84b) has the maximum possible rank M. The generalized function unit impulse (5.81c) of M ordinary functions (5.81a and b) of N variables exists if the conditions (5.84a and b) are met.*

5.5.2 One-Dimensional Generalized Function of an Ordinary Function of Several Variables

Consider the continuously differentiable function (5.85b) of N real variables (5.85a) that is not a constant, so that its partial derivatives are not all zero; choosing a nonzero partial derivative and denoting the corresponding variable by x_N leads to (5.85c):

$$n = 1,\dots,N: \quad f(x_n) \in C^1\big(|R^N\big), \quad \frac{\partial f}{\partial x_N} \neq 0; \quad (5.85\text{a--c})$$

The function is positive (negative) inside (5.86a) [outside (5.86c)] a domain and zero (5.86b) on the boundary that is a regular surface (5.86a through c):

$$f\left(x_1,\ldots,x_N\right)\begin{cases} >0 \text{ in } D: & \text{interior of domain,} & (5.86a) \\ =0 \text{ on } \partial D: & \text{boundary of domain,} & (5.86b) \\ <0 \text{ on } |R^N - \partial D: & \text{exterior of domain.} & (5.86c) \end{cases}$$

*The **unit jump of the function** is unity (zero) in the interior (5.87a) [exterior (5.87c)] and zero on the boundary (5.87b):*

$$H\left(f\left(x_1,\ldots,x_N\right)\right) = \begin{cases} 1 & \text{in interior of domain}: & \vec{x} \in D, & (5.87a) \\ \dfrac{1}{2} & \text{on the boundary}: & \vec{x} \in \partial D, & (5.87b) \\ 0 & \text{in the exterior}: & \vec{x} \in| R^2 - D - \partial D. & (5.87c) \end{cases}$$

*The **unit impulse of the function** is zero everywhere (5.88b) except on the boundary, where it is singular (5.88a):*

$$\delta\left(f\left(x_1,\ldots,x_N\right)\right) = \begin{cases} \infty & \text{on the boundary}: & \vec{x} \in \partial D, & (5.88a) \\ 0 & \text{elsewhere}: & \vec{x} \in| R^N - D. & (5.88b) \end{cases}$$

The boundary of the domain is a regular hypersurface specified by (5.89a) that can be solved for x_N leading to (5.89c), which may have several roots (5.89b) if the boundary consists of several sheets:

$$f\left(x_1,\ldots,x_N\right) = 0 \Rightarrow p = 1,\ldots,P: \quad x_N = h_p\left(x_1,\ldots,x_{N-1}\right). \qquad (5.89a\text{--}c)$$

*The **unit impulse of the function** has the property (5.90b) with regard to N-dimensional continuous test function (5.90a),*

$$\Phi\left(x_n\right) \in C\left(| R^N\right): \quad \left[\delta\left(f\left(x_1,\ldots,x_N\right)\right), \Phi\left(x_1,\ldots,x_N\right)\right] \equiv \int_{-\infty}^{+\infty} dx_1 \ldots \int_{-\infty}^{+\infty} dx_N \Phi\left(x_1,\ldots,x_N\right)\delta\left(f\left(x_1,\ldots,x_N\right)\right)$$

$$= \sum_{p=1}^{P} \int_{\partial D} \frac{\Phi\left(x_1,\ldots,x_{N-1},h_p\right)}{\partial f/\partial x_N} dx_1 \ldots dx_{N-1}, \qquad (5.90a\text{--}c)$$

of transforming the volume integral over all space into a surface integral over the boundary of the domain, the latter involving the division by the factor (5.85c) assumed not vanishing on the regular boundary. If (5.89a) has several roots (5.89c), then the r.h.s. of (5.90c) is summed (5.89b) over all roots corresponding to all sheets of the hypersurface. This is equivalent to

$$\delta\left(f\left(x_1,\ldots,x_N\right)\right) = \sum_{p=1}^{P} A_P\left(x_1,\ldots,x_{N-1}\right)\delta\left(x_N - h_p\left(x_1,\ldots,x_{N-1}\right)\right), \qquad (5.91a)$$

$$\frac{1}{A_p\left(x_1,\ldots,x_{N-1}\right)} = \frac{\partial f}{\partial x_N}\left(x_1,\ldots,x_{N-1},h_p\left(x_1,\ldots,x_{N-1}\right)\right), \tag{5.91b}$$

the decomposition of the unit impulse of a function of N variables into a linear combination (5.91a) of simple unit impulses with amplitudes (5.91b).

The proof of (5.90b) follows from the transformation of volume elements (5.92a):

$$dx_1\ldots dxdf = J\,dx_1\ldots dx_N, \quad J \equiv \frac{\partial f}{\partial x_N} \neq 0, \tag{5.92a,b}$$

where the Jacobian reduces to (5.92b) as follows from (5.91b) ≡ (5.92c):

$$J\left(x_1,\ldots,x_N\right) \equiv \frac{\partial\left(x_1,\ldots,x_{N-1},f\right)}{\partial\left(x_1,\ldots,x_{N-1},x_N\right)} = \begin{vmatrix} 1 & 0 & \ldots & 0 \\ 0 & 1 & \ldots & 0 \\ \vdots & \vdots & \ddots & 1 \\ \dfrac{\partial f}{\partial x_1} & \dfrac{\partial f}{\partial x_2} & \ldots & \dfrac{\partial f}{\partial x_N} \end{vmatrix} = \frac{\partial f}{\partial x_N}. \tag{5.92c}$$

The use of (5.92b) in (5.90b) leads to the inner product

$$\int_{-\infty}^{+\infty}dx_1\ldots\int_{-\infty}^{+\infty}dx_N\Phi\left(x_n,\ldots,x_N\right)\delta\left(f\left(x_n\right)\right)$$

$$= \int_{-\infty}^{+\infty}dx_1\ldots\int_{-\infty}^{+\infty}dx_{N-1}\int_{-\infty}^{+\infty}df\,\frac{\Phi\left(x_n,\ldots,x_N\right)}{J\left(x_1,\ldots,x_N\right)}\delta\left(f\left(x_1,\ldots,x_N\right)\right)$$

$$= \sum_{p=1}^{P}\int_{-\infty}^{+\infty}dx_1\ldots\int_{-\infty}^{+\infty}dx_{N-1}\,\frac{\Phi\left(x_1,\ldots x_{N-1},h_p\right)}{\partial f/\partial x_N\left(x_1,\ldots,x_{N-1},h_p\right)}, \tag{5.93}$$

in agreement with (5.93) ≡ (5.90c). The particular cases of (5.90a through c) [(5.91a and b)] include (1) for $N = 1$, one variable (5.10a and b) [(5.11a through c)] in Subsection 5.1.2; and (2) for $N = 2$, two variables (5.36a and b) [(5.39a through c)] in Subsection 5.1.2.

As a three-dimensional example, the function (5.94a) vanishes on the sphere of radius a and center at the origin:

$$0 = f\left(x,y,z\right) = a^2 - x^2 - y^2 - z^2: \quad z = \pm\left|a^2 - x^2 - y^2\right|^{1/2} \equiv h_\pm\left(x,y\right), \tag{5.94a,b}$$

$$\frac{\partial f}{\partial z} = -2z = \mp 2\left|a^2 - x^2 - y^2\right|^{1/2} \equiv -2h_\pm\left(x,y\right) = \frac{1}{A_\pm\left(x,y\right)}; \tag{5.94c}$$

The roots (5.94b) correspond to two sheets that are half-spheres and lead to the factor (5.94c). Used as argument of a unit impulse, it transforms an integral over all space into an integral over the sphere:

$$\left[\delta\left(a^2-x^2-y^2-z^2\right),\Phi(x,y,z)\right] \equiv \int\limits_{-\infty}^{+\infty}dx \int\limits_{-\infty}^{+\infty}dy \int\limits_{-\infty}^{+\infty}dz\ \Phi(x,y,z)\delta\left(a^2-x^2-y^2-z^2\right)$$

$$=\frac{1}{2}\int\limits_{x^2+y^2+z^2=a^2}\left|a^2-x^2-y^2\right|^{1/2}\times\left\{\Phi\left(x,y,-\left|a^2-x^2-y^2\right|^{1/2}\right)-\Phi\left(x,y,\left|a^2-x^2-y^2\right|^{1/2}\right)\right\}dxdy. \qquad (5.95)$$

The integrand on the r.h.s. of (5.95) involves the sum over the two roots (5.94b) of (5.94a) that correspond to the upper and lower half-spheres. The integral property (5.95) corresponds to

$$\delta\left(a^2-x^2-y^2-z^2\right)=\frac{\delta\left(z+\left|a^2-x^2-y^2\right|^{1/2}\right)-\delta\left(z-\left|a^2-x^2-y^2\right|^{1/2}\right)}{2\left|a^2-x^2-y^2\right|^{1/2}}, \qquad (5.96)$$

as a decomposition of unit impulses and can be compared with (5.46e).

5.5.3 Multidimensional Generalized Function with the Coordinates as the Arguments

For the test functions (5.97c) of M variables (5.97a) that have continuous (5.97d) derivatives of orders (5.97b),

$$m=1,...,M; k \equiv \sum_{m=1}^{M}k_m: \quad \mathcal{D}^k \equiv \left\{\Phi(x_m): \quad \Phi \in| R^M \frac{\partial^k\Phi}{\partial x_1^{k_1}...\partial x_M^{k_M}} \in C\left(|R^M\right)\right\}, \qquad (5.97\text{a–d})$$

the **multidimensional derivative unit impulse** *has the integral property*

$$\Phi \in \mathcal{D}^k\left(|R\right): \quad \left[\delta^{(k_1)}(x_1)...\delta^{(k_M)}(x_M),\Phi(x_1,...,x_M)\right]$$

$$\equiv \int\limits_{-\infty}^{+\infty}dx_1...\int\limits_{-\infty}^{+\infty}dx_M\delta^{(k_1)}(x_1)...\delta^{(k_M)}(x_M)\Phi(x_1,...,x_M)$$

$$=(-)^k\left[\frac{\partial^k\Phi}{\partial x_1^{k_1}...\partial x_M^{k_M}}\right]_{x_1=...=x_M=0}, \qquad (5.98\text{a,b})$$

which specifies (1) the value at the origin of the test function (5.97c); (2) after differentiation k_m times with regard to the variable x_m; and (3) with +(−) sign if the sum of orders of differentiation (5.97b) is even (odd).

The particular cases of (5.98) include (5.58a and b) for $N = 2$. The following is an example for $N = 3$ that is in three dimensions:

$$\int\limits_{-\infty}^{+\infty} dx \int\limits_{-\infty}^{+\infty} dy \int\limits_{-\infty}^{+\infty} dz\, \delta(x)\delta'(y)\delta''(z)\, y\, \exp\left(-x^3 - y^3 - z^2\right) = (-)^3 \lim_{x,y,z \to 0} \frac{\partial^3}{\partial y \partial z^2}\, y\, \exp\left(-y^3 - z^2\right)$$

$$= -\lim_{y \to 0}\left(1 - 3y^3\right)\exp\left(-y^2\right)\lim_{z \to 0}\left(-2 + 4z^2\right)\exp\left(-z^2\right) = 2. \tag{5.99}$$

In (5.98b) each of the M variables of the test function (5.98a) was involved in an inner product with a derivative of the unit impulse. In the case

$$\Phi \in C\left(|R^N\right); M \leq N: \quad \left[\delta(x_1)\ldots\delta(x_M), \Phi(x_1,\ldots,x_N)\right]$$

$$= \int\limits_{-\infty}^{+\infty} dx_1 \ldots \int\limits_{-\infty}^{+\infty} dx_M\, \delta(x_1)\ldots\delta(x_M)\Phi(x_1,\ldots,x_N) = \Phi(0,\ldots,0,x_{M+1},\ldots,x_N),$$

$$\tag{5.100a,b}$$

the first $M < N$ variables of the continuous test function in N-dimensions (5.100a) are involved in the inner product (5.100b) by unit impulse. More cases with the inner product over the whole space (only a subspace) like (5.98a and b) [(5.100a and b)] are considered next [Subsection 5.5.4 (Subsection 5.5.5)].

5.5.4 Multidimensional Generalized Function with Ordinary Functions as Arguments

The N continuously differentiable functions of N variables (5.101a through c) have gradients appearing in the Jacobian (5.101d):

$$m,n = 1,\ldots,N: \quad f_m(x_n) \in \mathcal{D}\left(|R^N\right), \quad J \equiv Det\left(\frac{\partial f_m}{\partial x_n}\right). \tag{5.101a–d}$$

The Jacobian is nonzero (5.101d) if (5.84b) the N functions (5.101c) are linearly independent (5.82a and b), that is, the corresponding N hypersurfaces intersect at a set of points, not on curves, surfaces, or subspaces. If the functions (5.101b) are continuously differentiable (5.102a), then (5.102b) holds:

$$f_m(x_n) \in C^1\left(|R^N\right): \quad \prod_{n=1}^{N} df_n = J(x_n)\prod_{n=1}^{N} dx_n, \quad J(x_n) \equiv Det\left(\frac{\partial f_m}{\partial x_n}\right) \in C\left(|R^N\right), \tag{5.102a–c}$$

relating the volume elements in f-space and x-space through the continuous Jacobian (5.102c). If the system of equations (5.103a) has P simple (5.103c) roots (5.103b),

$$f_m(x_n) = 0 \Leftrightarrow p = 1,\ldots,P: \quad x_n = x_{n,p}, \quad J_p \equiv J(x_{n,p}) \neq 0, \tag{5.103a–d}$$

the N hypersurfaces intersect at P points. It then follows for any N-dimensional continuous test function (5.104a) the integral property (5.104b and c):

$$\Phi(x_n) \in C^1(|R^N): \quad \left[\prod_{m=1}^{N} \delta(f_m(x_n)), \Phi(x_m) \right] \equiv \left[\delta(f_1(x_n))...\delta(f_N(x_n)), \Phi(x_1,...,x_N) \right]$$

$$\equiv \int_{-\infty}^{+\infty} dx_1 ... \int_{-\infty}^{+\infty} dx_N \Phi(x_n) \prod_{m=1}^{N} \delta(f_m(x_n)) = \int_{-\infty}^{+\infty} df_1 ... \int_{-\infty}^{+\infty} df_N \frac{\Phi(x_n)}{J(x_n)} \delta(f_1)...\delta(f_N)$$

$$= \sum_{p=1}^{P} \frac{\Phi(x_{n,p})}{J(x_{n,p})}. \qquad (5.104\text{a--d})$$

The integral property (5.104d) is equivalent to,

$$\prod_{m=1}^{N} \delta(f_m(x_n)) = \sum_{p=1}^{P} A_p \prod_{n=1}^{N} \delta(x_n - x_{n,p}), \quad \frac{1}{A_p} = J(x_{n,p}), \qquad (5.105\text{a,b})$$

the decomposition (5.105a) of the product of N unit impulses of N continuous functions (5.102a) of N variables (5.101a and b) into a product of simple impulses, one set for each simple root (5.103b and c) of (5.103a), with the inverse of the Jacobian (5.103d) as coefficient or amplitude (5.105b). The particular cases of (5.104a through d) [(5.105a and b)] include (5.76) [(5.77a and b)] for $N = 2$. The preceding result assumes that the number M of functions equals the number N of dimensions; in general, $M \leq N$, otherwise the system (5.103a) is either redundant or impossible. Thus besides the case $N = M$ already considered (Subsection 5.5.4), there is the case $M < N$, for example, $M = 1 < N$ in Subsection 5.5.2; the remaining cases $1 < M < N$ are considered (Subsection 5.5.5) after the example of (5.106a and b) that follows.

As an example, consider (1) the sphere (5.94a) of radius a and center at the origin; (2) the circular cylinder (5.106a) of radius b with generators parallel to the OZ-axis; and (3) the plane (5.106b) passing through the OY-axis and making equal angles with the x- and z-planes:

$$0 = g(x,y) = b^2 - x^2 - y^2, \quad 0 = h(x,z) = x - z. \qquad (5.106\text{a,b})$$

The system of equations (5.94a;5.106a and b) (1) has the Jacobian:

$$J(x,y,z) = \frac{\partial(f,g,h)}{\partial(x,y,z)} = \begin{vmatrix} -2x & -2y & -2z \\ -2x & -2y & 0 \\ 1 & 0 & -1 \end{vmatrix} = -4yz; \qquad (5.107)$$

(2) has roots:

$$a^2 = x^2 + y^2 + z^2 = b^2 + z^2, \quad z = \pm\sqrt{a^2 - b^2} = x, \qquad (5.108\text{a,b})$$

$$y = \pm\sqrt{a^2 - x^2 - z^2} = \sqrt{a^2 - 2z^2} = \pm\sqrt{2b^2 - a^2}, \qquad (5.108\text{c})$$

which are real if the three surfaces intersect; (3) the condition (5.109a) for real roots (5.108b) is that the cylinder (5.106a) has a smaller radius than the sphere (5.94a), and thus they intersect (Figure 5.8a) along two circles with radius b and center on the z-axis (5.109c) at the positions (5.109b), implying (5.109d) by (5.106b):

$$a > b: \quad c \equiv \left| a^2 - b^2 \right|^{1/2}, \quad z_\pm = \pm c = x_\pm; \qquad (5.109\text{a--d})$$

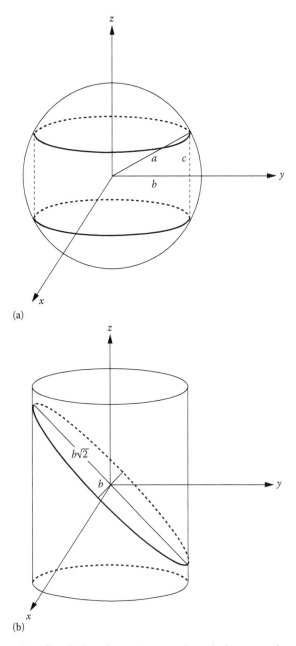

FIGURE 5.8 The intersection of a cylinder whose axis passes through the center of a sphere with larger radius specifies two parallel circles (a). A plane oblique to the axis of the cylinder intersects it along an ellipse (b). If the three surfaces that is the plane, cylinder, and sphere intersect it is at the four points common to the ellipse and two circles (Figure 5.8).

and (4) the plane (5.106b) intersects the cylinder (5.106a) along an ellipse (Figure 5.8b) with one half axis b and the other $b\sqrt{2}$ because the inclination is $\pi/4$; (5) the circle (Figure 5.8a) and ellipse (Figure 5.8b) intersect (5.110b and c) at four points (Figure 5.9) if the condition (5.110a) is met:

$$a < b\sqrt{2}: \quad d \equiv \left|2b^2 - a\right|^{1/2}, \quad y_\pm = \pm d. \tag{5.110a–c}$$

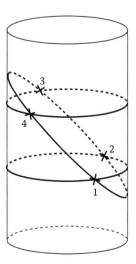

FIGURE 5.9 The intersection of a circular cylinder by [Figure 5.8b (Figure 5.8a)] a one (two) plane(s) oblique (orthogonal) to the axis specifies an ellipse (two circles). If the planes intersect they specify the four points (Figure 5.9) common to the ellipse and one of the two circles.

In conclusion, the sphere (5.94a), cylinder (5.106a), and plane (5.106b) intersect at four points (5.111c and d) ≡ (5.109c and d; 5.110c) if two conditions (5.111a and b) ≡ (5.109a; 5.110a) are met:

$$b\sqrt{2} > a > b: \quad \{x,y,z,J\} = \{\pm c,d,\pm c,\mp 4cd\},\{\pm c,-d,\pm c,\pm 4cd\}, \qquad (5.111a\text{–}d)$$

and the Jacobian (5.107) is not zero, showing that there is no intersection at neighboring points. The integral of a continuous test function (5.112a) over the intersection of the three surfaces is given by (5.112b):

$$\Phi(x,y,z) \in C(|R^2|): \quad \int\limits_{-\infty}^{+\infty} dx \int\limits_{-\infty}^{+\infty} dy \int\limits_{-\infty}^{+\infty} dz \, \Phi(x,y,z)\delta\left(a^2 - x^2 - y^2 - z^2\right)\delta\left(b^2 - x^2 - y^2\right)\delta(x - z)$$

$$= \frac{\Phi(-c,d,-c) + \Phi(c,-d,c) - \Phi(c,d,c) - \Phi(-c,-d,-c)}{4cd}. \qquad (5.112a,b)$$

This is equivalent to

$$4cd\delta\left(a^2 - x^2 - y^2\right)\delta\left(a^2 - x^2 - y^2\right)\delta(x - z)$$

$$= \delta(x + c)\delta(y - d)\delta(z + c) + \delta(x - c)\delta(y + d)\delta(z - c)$$

$$- \delta(x - c)\delta(y - d)\delta(z - c) - \delta(x + c)\delta(y + d)\delta(z + c), \qquad (5.113)$$

as a decomposition of the product of three impulse distributions with arguments (5.94a; 5.106a and b) into a linear combination of products of simple impulses involving (5.109b; 5.110b).

5.5.5 Generalized Function over a Subspace of Dimension $N - M$ of an N-Dimensional Space

The case of one (N) function(s) in Subsection 5.5.2 (Subsection 5.5.4) specifies hypersurfaces (points), that is, subspaces with $N - 1$ ($N - N = 0$) dimensions $N - M$ that are the intersection of one $M = 1$ ($M = N$) hypersurfaces. The remaining cases $1 < M < N - 1$ lead to subspaces of dimension $N - M$ that are the intersection of M hypersurfaces. *In the general case (5.81a) $\equiv$ (5.114a); the M continuously differentiable functions (5.114c) of N variables (5.114c) have a matrix of partial derivatives (5.114d) of maximum rank M where $1 \leq M \leq N$ in (5.114b) $\equiv$ (5.81c; 5.84a):*

$$m = 1,\ldots,M; n = 1,\ldots,N \geq M: \quad f_m(x_n) \in C^1(|R^N), \quad Ra\left(\frac{\partial f_m}{\partial x_n}\right) = M \leq N, \qquad (5.114a\text{–}d)$$

*that is, (5.114d) contains at least one determinant of maximum possible size $M \times M$ that is nonzero; the first M coordinates may be chosen as the M **external coordinates** (5.114a) that are eliminated between (5.115b), leading to (5.115d), where (5.115a) are $N - M$ the **internal coordinates** of the subspace of dimension $N - M$ with P elements or **sheets** (5.115c):*

$$\ell = M + 1, M + 2,\ldots,N: \quad f_m(x_n) = 0 \Leftrightarrow p = 1,\ldots,P: \quad x_m = h_{m,p}(x_\ell). \qquad (5.115a\text{–}d)$$

*The systems of M equations (5.115b) [(5.115d)] are equivalent and specify $p = 1, \ldots, P$ sheets (5.115c) of the **subspace** of dimension $N - M$ with $N - M$ internal (5.115a) [M external (5.114a)] coordinates. The Jacobian relative to the external coordinates (5.116a) must be nonzero (5.116b),*

$$m, j = 1,\ldots,M: \quad J_{M,p}(x_{M+1},\ldots,x_N) \equiv \left[Det\left(\frac{\partial f_m}{\partial x_j}\right)\right]_{x_m = h_{m,p}(x_\ell)} \neq 0, \qquad (5.116a,b)$$

in the definition of M-dimensional unit impulse (5.117b) whose arguments are M continuously differentiable ordinary functions (5.114c) of N variables. It is defined by the inner product with an N-dimensional continuous test function (5.177a) leading to the integral (5.117b):

$$\Phi(x_n) \in C^1(|R^N): \quad \left[\delta(f_1(x_n))\ldots\delta(f_M(x_n)),\Phi(x_1,\ldots,x_N)\right] \equiv \int_{-\infty}^{+\infty}dx_1\ldots\int_{-\infty}^{+\infty}dx_N\Phi(x_n)\prod_{m=1}^{M}\delta(f_m(x_n))$$

$$= \sum_{p=1}^{P}\int_{-\infty}^{+\infty}dx_{M+1}\ldots dx_N\frac{\Phi(h_{1,p},\ldots,h_{M,p},x_{M+1},\ldots,x_N)}{J_{M,p}(x_{M+1},\ldots,x_N)}$$

$$= \sum_{p=1}^{P}\int_{-\infty}^{+\infty}dx_1\ldots\int_{-\infty}^{+\infty}dx_1\ldots\int_{-\infty}^{+\infty}dx_N\frac{\Phi(x_1,\ldots,x_N)}{J_{M,p}(x_{M+1},\ldots,x_N)}\delta(x_1 - h_{1,p})\ldots\delta(x_M - h_{M,p}),$$

$$\qquad (5.117a\text{–}d)$$

which reduces to (5.117c) an integral over the subspace of dimension $N - M$. This is equivalent (5.117d) $\equiv$ (5.118a and b) to

$$\prod_{m=1}^{M}\delta(f_m(x_n)) = \sum_{p=1}^{P}A_p(x_{M+1},\ldots,x_N)\prod_{m=1}^{M}\delta(x_m - h_{m,p}(x_{M+1},\ldots,x_N)), \qquad (5.118a)$$

$$\frac{1}{A_p\left(x_{M+1},\ldots,x_N\right)}=\left[Det\left(\frac{\partial f_m}{\partial x_j}\right)\right]_{x_m=h_{m,p}\left(x_\ell\right)}=J_{M,p}\left(x_{M+1},\ldots,x_N\right),\tag{5.118b}$$

a decomposition of the product of M impulses of M linearly independent functions of N variables (5.114a through d) into a simple product of M impulses (5.118a) of the simple roots (5.115c and d) of (5.115b) with amplitudes (5.118b) specified by the Jacobian (5.116b) of the external coordinates (5.116a) of (5.115a and b). The particular cases of (5.117a through d) [(5.118a and b)] include (5.38a and b) [(5.39a through c)] and (5.44a and b) [(5.45a through c)] for $N = 2$, $M = 1$.

The proof of (5.117c) for each root (5.115c) follows from

$$m,\,j=1,\ldots,M:\quad \prod_{m=1}^{M}dx_m = J_M\prod_{m=1}^{m}dh_m,\quad J_M\equiv Det\left(\frac{\partial f_m}{\partial x_j}\right),\tag{5.119a–c}$$

where the Jacobian (5.119c) appears in

$$\left[\delta\left(f_1\left(x_n\right)\right)\ldots\delta\left(f_M\left(x_n\right)\right),\,\Phi\left(x_1,\ldots,x_N\right)\right]\equiv\int_{-\infty}^{+\infty}dx_1\ldots\int_{-\infty}^{+\infty}dx_M\Phi\left(x_1,\ldots,x_N\right)\delta\left(f_1\right)\ldots\delta\left(f_M\right)$$

$$=\int_{-\infty}^{+\infty}df_1\ldots\int_{-\infty}^{+\infty}df_M\frac{\Phi\left(x_1,\ldots,x_N\right)}{J\left(x_1,\ldots,x_N\right)}\delta\left(f_1\right)\ldots\delta\left(f_M\right)=\frac{\Phi\left(h_1,\ldots,h_M,x_{M+1},\ldots,x_N\right)}{J\left(h_1,\ldots,h_M,x_{M+1},\ldots,x_N\right)}.\tag{5.120}$$

Integrating (5.120) over $dx_{M+1}\ldots dx_N$ leads to (5.117c), followed by the sum over all roots (5.115c and d). The expressions (5.117a through d; 5.118a and b) [(5.104a through d; 5.105a and b)] are inner products involving $M(N)$ integrations over the dimension of the generalized (test) function. The two coincide (5.101a through d) if the generalized and test functions have the same dimension $M = N$ in (5.104a through d) $\equiv$ (5.117a through d) and (5.105a and b) $\equiv$ (5.118a and b).

As an example of a surface $M = 2$ in a three-dimensional space $N = 3$, consider a circular cylinder (5.106a) with radius b and OZ-axis and the plane (5.106b) bisecting the x- and z-planes; they intersect (Figure 5.8b) along the ellipse (5.121a and b):

$$y\left(x\right)=\pm\left|b^2-x^2\right|^{1/2},\quad z\left(x\right)=x:\quad J=\frac{\partial\left(g,h\right)}{\partial\left(y,z\right)}=\begin{vmatrix}-2y & 0\\ 0 & -1\end{vmatrix}=2y,\tag{5.121a–c}$$

where x is [(y, z) are] taken as internal (external) coordinate(s), and the Jacobian (5.121c) is given by

$$k\left(x\right)\equiv\left|b^2-x^2\right|^{1/2},\quad y_\pm=\pm k\left(x\right),\quad J_\pm=\pm 2k\left(x\right).\tag{5.122a–c}$$

The integral over all space (5.123b) of a continuous function (5.123a) is given by:

$$\Phi\in C\left(R^3\right):\quad\left[\delta\left(b^2-x^2-y^2\right)\delta\left(x-z\right),\,\Phi\left(x,y,z\right)\right]\equiv\int_{-\infty}^{+\infty}dx\int_{-\infty}^{+\infty}dy\int_{-\infty}^{+\infty}\Phi\left(x,y,z\right)\delta\left(b^2-x^2-y^2\right)\delta\left(x-z\right)$$

$$=\frac{1}{2}\int_{-a}^{+a}\frac{\Phi\left(x,k\left(z\right),x\right)-\Phi\left(x,-k\left(x\right),x\right)}{k\left(x\right)}dx.\tag{5.123a,b}$$

This is equivalent to the decomposition into simple impulses,

$$2k(x)\delta(b^2 - x^2 - y^2)\delta(x - z) = \left[\delta(y + k(x)) - \delta(y - k(x))\right]\delta(x - z), \qquad (5.124)$$

of the product of impulses with arguments (5.106a and b) involving the function (5.122a). Seven instances of double integrals of ordinary and generalized functions are considered in Example 10.14.

5.6 Generalized Functions for Hypersurface and Line Integration

The result (5.117a through d) [(5.118a and b)] in Subsection 5.5.5 is the most general and includes the preceding (1) for $M = 1$, hypersurfaces (5.90a through c) [(5.91a and b)] in Subsection 5.5.2; (2) for $M = N$, the set of points that is the intersection of N hypersurfaces (5.104a through d) [(5.105a and b)] in Subsection 5.5.4; and (3) in Subsection 5.5.3 is the case of $M = N$ coordinates $x_m = f_m$ in (5.98) that coincides with (5.104d) in the case without derivatives $k_m = 0$. The N-dimensional integrals involving M-dimensional impulses are integrals over one hypersurface (along hypercurves) for $M = 1(M = N - 1)$, that is, subspaces of dimension $N - M = N - 1[N - (N - 1) = 1]$, and are related to generalized function hypercurve (hypersurface) integration [Subsection 5.6.1 (Subsection 5.6.2)]. The unit impulse over a subspace such as a hypercurve (hypersurface) corresponds to a monopole and leads by differentiation to multipoles (Subsection 5.6.3).

5.6.1 Generalized Function for Integration along a Hypercurve

The case $M = N - 1$ in (5.115a through d) is a system *of $N - 1$ continuously differentiable functions of N variables (5.125a through c) with a matrix of partial derivatives (5.125d) of maximum rank $N - 1$:*

$$n = 1, \ldots, N; m = 1, \ldots, N - 1: \quad f_m(x_n) \in C^1(|R), \quad Ra\left(\frac{\partial f_m}{\partial x_n}\right) = N - 1, \qquad (5.125a–d)$$

which specifies the intersection of $N - 1$ hypersurfaces (5.126a) into a set P of hypercurves (5.126b) with parameter x_N:

$$f_m(x_n) = 0 \Leftrightarrow p = 1, \ldots, P: \quad x_m = h_{m,p}(x_N). \qquad (5.126a,b)$$

The arc length,

$$\mathrm{d}s_p = \left|\sum_{n=1}^{N}(\mathrm{d}x_n)^2\right|^{1/2} = \left|1 + \sum_{m=1}^{N-1}\left(\frac{\mathrm{d}h_{m,p}}{\mathrm{d}x_N}\right)^2\right|^{1/2}\mathrm{d}x_N, \qquad (5.127)$$

*appears in the **line integral** (5.128b) for an integrable test function (5.128a):*

$$\Phi(x_n) \in \mathcal{E}(|R^N): \quad \left[C(x_n), \Phi(x_n)\right] \equiv \int\Phi(x_n)\mathrm{d}s = \sum_{p=1}^{P}\int\Phi(h_{1,p}, \ldots, h_{N-1,p}, x_N)\mathrm{d}s_p. \qquad (5.128a,b)$$

Replacing the requirement of integrability (5.128a) by the stronger requirement of continuity (5.129a) of the test function introduces the unit impulse in (5.128b), leading to (5.128d):

$$[C(x_n),\Phi(x_n)] = \sum_{p=1}^{P} \int_{-\infty}^{+\infty} dx_1 \dots \int_{-\infty}^{+\infty} dx_{N-1} \Phi(x_n) \prod_{m=1}^{N-1} \delta(x_m - h_{m,p}(x_N)) \left| 1 + \sum_{m=1}^{N-1} \left(\frac{dh_{m,p}}{dx_N}\right)^2 \right|^{1/2} dx_N$$

$$\equiv \int_{-\infty}^{+\infty} dx_1 \dots \int_{-\infty}^{+\infty} dx_N C(x_1,\dots,x_N) \Phi(x_1,\dots,x_N). \qquad (5.128c,d)$$

Then, (5.128d) is equivalent to the inner product of the continuous test function (5.129a):

$$\Phi(x_N) \in C(|R^N): \quad C(x_n) \equiv \sum_{p=1}^{P} \left| 1 + \sum_{m=1}^{N-1} \left(\frac{dh_{m,p}}{dx_N}\right)^2 \right|^{1/2} \prod_{m=1}^{N-1} \delta(x_m - h_{m,p}(x_N)), \qquad (5.129a,b)$$

by the **generalized function line integral** *(5.129b).*

The generalized function line integral (5.129a and b)≡(5.130a and b),

$$C(x_1,\dots,x_N) = \sum_{p=1}^{P} D_p(x_N) \prod_{m=1}^{N-1} \delta(x_m - h_{m,p}(x_1,\dots,x_{N-1})), \qquad (5.130a)$$

$$D_p(x_N) \equiv \left| 1 + \sum_{m=1}^{N-1} \left(\frac{dh_{m,p}}{dx_N}\right)^2 \right|^{1/2}, \qquad (5.130b)$$

is related to, but distinct from, the $(N-1)$-dimensional unit impulse (5.118a and b) for the $N-1$ linearly independent hypersurfaces (5.124a through d) that intersect along the hypercurve because (1) the basic unit impulses are the same in (5.118a) and (5.130a); and (2) the amplitudes (5.118b) and (5.130b) are related by

$$A_p(x_N) \left[Det\left(\frac{\partial f_n}{\partial x_m}\right) \right]_{x_m = h_m(x_N)} = 1 = D_p(x_N) \left| 1 + \sum_{m=1}^{N-1} \left(\frac{dh_{m,p}}{dx_N}\right)^2 \right|^{-1/2} \qquad (5.131)$$

and are generally distinct. An example is the line integral along the ellipse (Figure 5.8b) that is the intersection of the cylinder (5.106a) with radius b and OZ-axis and the plane (5.106b) bisecting the x- and z-planes; its arc length is

$$y(x) = \pm|b^2 - x^2|^{1/2} \equiv \pm k(x): \quad \frac{ds}{dx} = \left| 1 + \left(\frac{dy}{dx}\right)^2 + \left(\frac{dz}{dx}\right)^2 \right|^{1/2} = \left| 2 + \frac{x^2}{y^2} \right|^{1/2} = \left| \frac{2b^2 - x^2}{b^2 - x^2} \right|^{1/2}. \qquad (5.132a,b)$$

It appears in the line integral summing the two arcs of the ellipse:

$$\int_L \Phi(x,y,z) ds \equiv [C(x,y,z), \Phi(x,y,z)]$$

$$= \int_{-b}^{+b} \{\Phi(x,k(x),x) + \Phi(x,-k(x),x)\} \left| \frac{2b^2 - x^2}{b^2 - x^2} \right|^{1/2} dx. \qquad (5.133)$$

The latter is equivalent to the inner product by the generalized function line integral

$$C(x, y, z) = \left| 1 + \left[\frac{b}{k(x)} \right]^2 \right|^{1/2} \delta(x - z) \left[\delta(y - k(x)) + \delta(y + k(x)) \right],$$ (5.134)

which is distinct from (5.124) in the amplitude coefficients.

5.6.2 Generalized Function for Integration over a Hypersurface

The case $M = 1$ in Subsection 5.5.5 corresponds to a single hypersurface (5.85a through c) in Subsection 5.5.2. Considering the function (5.86a through c), each of its roots (5.89a) corresponds to one sheet (5.89b and c). Any displacement $d\vec{x}_p$ along one sheet satisfies (5.135):

$$0 = \sum_{n=1}^{N} dx_n \frac{\partial}{\partial x_n} \left[x_N - h_p(x_1, \ldots, x_{N-1}) \right] = dx_N - \sum_{m=1}^{N-1} \frac{\partial h_p}{\partial x_m} dx_m = d\vec{x}_p \cdot \vec{M}_P,$$ (5.135)

showing that it is orthogonal to the vector (5.136a) that must lie in the normal direction:

$$\vec{M}_p = \left\{ -\frac{\partial h_p}{\partial x_1}, \ldots, -\frac{\partial h_p}{\partial x_{M-1}}, 1 \right\}; \quad \vec{N}_p = \frac{\vec{M}_p}{\left| \vec{M}_p \right|}.$$ (5.136a,b)

The vector (5.136a) divided by its modulus specifies the unit normal (5.136b). For example (Figure 5.10), the component of the unit normal perpendicular to the x_N-plane specifies the **direction cosine**:

$$\cos \gamma_N = N_N = \left| \vec{M}_p \right|^{-1} = \left| 1 + \sum_{m=1}^{N-1} \left(\frac{\partial h_p}{\partial x_m} \right)^2 \right|^{-1/2}$$

$$= \left| 1 + \sum_{m=1}^{N-1} \left(\frac{\partial f / \partial x_m}{\partial f / \partial h_p} \right)^2 \right|^{-1/2} = \left| 1 + \sum_{m=1}^{N-1} \left(\frac{\partial f / \partial x_m}{\partial f / \partial x_N} \right)^2 \right|^{-1/2},$$ (5.137)

leading to (5.137) ≡ (5.138b), where (5.138c) is the modulus of the gradient:

$$\frac{\partial f}{\partial x_N} \neq 0: \quad \cos \gamma_N = \frac{\partial f / \partial x_N}{|\nabla f|}, \quad |\nabla f| = \left| \sum_{m=1}^{N-1} \left(\frac{\partial f}{\partial x_n} \right)^2 \right|^{1/2}.$$ (5.138a–c)

It is assumed that the normal is not parallel (5.138a) to the x_N-plane. If it is, then another coordinate x_n is chosen such that $\partial f / \partial x_n \neq 0$; this must be possible for at least one value of $n = 1, \ldots, N$, since the function (5.85b) is not a constant and hence its gradient is not zero, so at least one coordinate exists for which the partial derivative is nonzero; that coordinate is designated x_N in (5.85c). The **area element** (*Figure 5.10*) projected on the x_N-plane is

$$dS = \frac{dx_1 \ldots dx_{N-1}}{\cos \gamma_N} = \frac{|\nabla f|}{\partial f / \partial x_N} dx_1 \ldots dx_{N-1}.$$ (5.139)

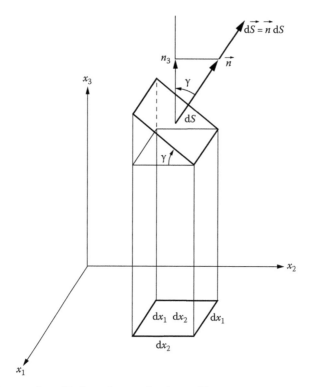

FIGURE 5.10 The element of area dS of a surface may be obtained by projection on one coordinate plane correcting for the director cosine; the normal to the surface should not be parallel to the plane of projection, and there is always at least one plane of projection for which this condition is met.

The integral of an integrable test function (5.140a) over the hypersurface (5.89a through c) is given by (5.140b):

$$\Phi(x_n) \in \mathcal{E}(| R^N): \quad \left[S(x_n), \Phi(x_n)\right] = \int \Phi(x_1,\ldots,x_N) \, dS$$

$$\equiv \sum_{p=1}^{P} \int_{-\infty}^{+\infty} dx_1 \ldots \int_{-\infty}^{+\infty} dx_{N-1} \Phi\left(x_1,\ldots,x_{N-1}, h_p(x_1,\ldots,x_N)\right) \left[\frac{|\nabla f|}{\partial f / \partial x_N}\right]_{x_N = h_p}. \qquad (5.140a,b)$$

Replacing the requirement on the test function from integrability (5.140a) to the stronger requirement of continuity (5.140c), the unit impulse may be introduced in the integral (5.140b) leading to (5.140d):

$$\Phi \in C(| R^N): \quad \left[S(x_n), \Phi(x_n)\right] = \sum_{p=1}^{N} \int_{-\infty}^{+\infty} dx_1 \ldots \int_{-\infty}^{+\infty} dx_N \Phi(x_1,\ldots,x_N)$$

$$\times \frac{|\nabla f|}{\partial f / \partial x_N} \delta\left(x_N - h_p(x_1,\ldots,x_N)\right). \qquad (5.140c,d)$$

The integral (5.140d) corresponds to the inner product of the continuous test function (5.140c) by the **generalized function hypersurface integration:**

$$S(x_1,\ldots,x_N) = \sum_{p=1}^{P} E_p(x_1,\ldots,x_{N-1}) \delta\left(x_N - h_p(x_1,\ldots,x_{N-1})\right), \qquad (5.141a)$$

$$\frac{1}{E_p\left(x_1,\ldots,x_{N-1}\right)}=\left[\frac{\partial f/\partial x_N}{|\nabla f|}\right]_{h_p\left(x_1,\ldots,x_{N-1}\right)},\qquad(5.141\mathrm{b})$$

involving in the amplitudes (5.141b) of the unit impulses (5.141a) the Nth component of the gradient (5.138a) divided by the modulus of the gradient (5.138c), that is, the inverse of the direction cosine (5.138b) in the x_N-direction. The generalized function hypersurface integration (5.141a and b) and unit impulse of a hypersurface (5.91a and b) do not coincide because (1) the basic unit impulses in (5.141a) and (5.91a) are the same; and (2) their amplitudes (5.141b) and (5.90b) are related by

$$\frac{1}{\partial f/\partial x_N\left(x_1,\ldots,x_{N-1}\right)}=A_p\left(x_1,\ldots,x_{N-1}\right)=\frac{E_p\left(x_1,\ldots,x_{N-1}\right)}{|\nabla f|},\qquad(5.142)$$

and hence generally distinct.

In the case (5.94a) of the sphere of radius a and center at the origin, the gradient is (5.143a) and has modulus (5.143b):

$$\nabla f=-2\left(x,y,z\right),\quad |\nabla f|=2\left|x^2+y^2+z^2\right|^{1/2}=2a,\qquad(5.143\mathrm{a\text{--}c})$$

leading (5.141a and b) to the generalized function surface integral (5.144a):

$$S\left(x,y,z\right)=\frac{a}{z}\left[\delta\left(z+\left|a^2-x^2-y^2\right|^{1/2}\right)-\delta\left(z-\left|a^2-x^2-y^2\right|^{1/2}\right)\right]$$

$$=2a\ \delta\left(a^2-x^2-y^2\right),\qquad(5.144\mathrm{a,b})$$

that is distinct (5.144b) from the unit impulse (5.96) of argument (5.94a). This specifies the surface integral over the upper and lower hemispheres (5.145b):

$$\iiint\limits_{x^2+y^2+z^2=a^2}\Phi\left(x,y,z\right)\mathrm{d}x\mathrm{d}y\mathrm{d}z$$

$$=a\int\limits_{-\infty}^{+\infty}\mathrm{d}x\int\limits_{-\infty}^{+\infty}\mathrm{d}y\int\limits_{-\infty}^{+\infty}\mathrm{d}z\,\frac{\Phi\left(x,y,z\right)}{z}\left[\delta\left(z+\left|a^2-x^2-y^2\right|^{1/2}\right)-\delta\left(z-\left|a^2-x^2-y^2\right|^{1/2}\right)\right].\qquad(5.145)$$

The relation (5.96) is equivalent, using the distance from the z-axis (5.146a):

$$r\equiv\left|a^2-x^2-y^2\right|^{1/2}:\quad \delta\left(r^2-z^2\right)=\frac{\delta\left(z+r\right)-\delta\left(z-r\right)}{2r},\qquad(5.146\mathrm{a,b})$$

in (5.13e) with $a = r$ and $x = z$. The surface integral (5.145) is given by

$$
\iiint\limits_{x^2+y^2+z^2=a^2} \Phi(x,y,z)\,dxdydz
$$

$$
= \int\limits_{-\infty}^{+\infty} dx \int\limits_{-\sqrt{a^2-x^2}}^{\sqrt{a^2-x^2}} dy \left|a^2 - x^2 - y^2\right|^{-1/2} \left[\Phi\left(x, y, -\left|a^2 - x^2 - y^2\right|^{1/2}\right) - \Phi\left(x, y, \left|a^2 - x^2 - y^2\right|^{1/2}\right) \right], \quad (5.147)
$$

where (5.146a) appears.

5.6.3 Generalized Function for Multipoles in Subspaces

The integration in an N-dimensional space over a hypercurve $M = N - 1$ (hypersurface $M = 1$) is speci-fied by the generalized function line (hypersurface) integral (5.130a and b) [(5.141a and b)]; the case of M-dimensional subspaces with $2 \le M \le N - 2$ requires the use of multivectors (Notes 9.3 through 9.52) or differential forms. The unit impulse and its derivatives can be used to specify multipole distribu-tions along curves or on surfaces or domains of the three-dimensional space (Notes 5.4 through 5.30; Chapters 6 and 8); the general case of multipoles in a subspace of dimension M of an N-dimensional space is illustrated next for a hypersurface $M = N - 1$. The two-dimensional identity (5.47a and b) is generalized to N-dimensions for a regular hypersurface (5.148a) by (5.148b):

$$
f \in C^1\left(|R^N\right): \quad \frac{\partial}{\partial x_m}\left[H(f(x_n))\right] = \frac{dH}{df}\frac{\partial f}{\partial x_m} = -\delta(f)|\nabla f|N_m, \quad (5.148a,b)
$$

where N_m is the unit normal (5.136a and b), as follows from (5.149b and c) for a displacement on the hypersurface (5.149a):

$$
0 = df = \sum_{n=1}^{N} \frac{\partial f}{\partial x_n} dx_n = \vec{M} \cdot d\vec{x}, \quad \vec{M} = \nabla f, \quad -\vec{N} = \frac{\vec{M}}{|\vec{M}|} = \frac{\nabla f}{|\nabla f|}. \quad (5.149a\text{–}c)
$$

For a closed region (5.86b), if f is positive (negative) in the interior (5.86b) [exterior (5.86b)], the gradient ∇f points in the direction of increasing f and corresponds to the inward normal; in order to have the **outward normal,** the minus sign must be introduced in (5.149c) and (5.148b).

The second (5.158a and b) [third (5.159a and b)]-order derivative involves dipoles (quadrupoles), that is, first(second)-order derivative impulses:

$$
f \in C^2\left(|R^N\right): \quad \frac{\partial^2}{\partial x_n \partial x_m}\left[H(f)\right] = \delta(f)\frac{\partial^2 f}{\partial x_n \partial x_m} + \delta'(f)\frac{\partial f}{\partial x_m}\frac{\partial f}{\partial x_n}, \quad (5.150a,b)
$$

$$
f \in C^3\left(|R^N\right): \quad \frac{\partial^3}{\partial x_n \partial x_m \partial x_\ell}\left[H(f)\right] = \delta(f)\frac{\partial^3 f}{\partial x_n \partial x_m \partial x_\ell}
$$

$$
+ \delta'(f)\sum_{cyl}^{m,n,\ell} \frac{\partial f}{\partial x_m}\frac{\partial^2 f}{\partial x_n \partial x_\ell} + \delta''(f)\frac{\partial f}{\partial x_n}\frac{\partial f}{\partial x_m}\frac{\partial f}{\partial x_\ell}, \quad (5.151a,b)
$$

where in (5.151b) appear cyclic permutations of (n, m, ℓ). The multipole distributions on a hypersurface (5.150a and b; 5.151a and b) can be generalized to subspaces of dimension M, for example, for a dipole:

$$\ell, m = 1, \ldots, M; \, n, k = 1, \ldots, N; \, f_m(x_n) \in C^1\left(\left| R^N \right|\right): \quad \frac{\partial}{\partial x_n}\left[\prod_{m=1}^{M} H\left(f_m(x_k)\right)\right]$$

$$= \sum_{m=1}^{M} \delta\left(f_m(x_k)\right) \frac{\partial f_m}{\partial x_n} \prod_{\substack{\ell=1 \\ \ell \neq m}}^{M}\left[H\left(f_\ell(x_k)\right)\right]. \tag{5.152a,b}$$

The kth-order derivatives specify multipoles (Chapters 6, 8, and 9); a particular case is the monopoles that have central symmetry. The unit impulse with central symmetry in N-dimensions is considered in Section 5.9. Before are considered the invariant differential operators gradient, divergence, and curl in a three-dimensional space (Section 5.7) that lead to scalar and vector potentials (Section 5.8). The extension to M-dimensional subspaces of N-dimensional spaces involves multivectors or differential forms (Notes 9.3 through 9.52).

5.7 Divergence, Gradient, and Curl Theorems

The generalized functions may be used to prove the divergence theorem (Subsection 5.7.1) equating the integral of the divergence of a continuously differentiable vector over a domain to the flux of the same vector across the closed regular boundary. There are analogue theorems [Subsection 5.7.2 (Subsection 5.7.3)] for the gradient (curl) of a continuously differentiable scalar (vector). The Stokes' theorem (Subsection 5.7.7) is broadly similar, equating the circulation of a continuously differentiable vector around a closed regular loop to the flux of its curl across any regular surface supported on the loop as boundary (Subsection 5.7.6). The gradient (curl) of a scalar (vector) is a polar (axial) vector (Subsection 5.7.5); a polar vector is a three-dimensional representation of a bivector or skew-symmetric matrix using the three-index permutation symbol (Subsection 5.7.4). The divergence (Stokes) theorem [Subsection 5.7.1 (Subsection 5.7.7)] can be applied to a plane domain with a closed regular boundary (Subsection 5.7.8). The divergence theorem (Subsection 5.7.1) leads to the first and second Green identities (Subsection 5.7.9) that can be used to prove several properties of harmonic fields (Sections 9.1 through 9.3).

5.7.1 Divergence Integral Theorem (Gauss, 1809; Ostrogradski, 1828; Hankel, 1861; Thomson, 1869)

The **divergence** of a vector whose components are differentiable functions of the coordinates in an N-dimensional space (5.153a) is defined as the sum of partial derivatives (5.153c):

$$A_\ell(x_n) \in D\left(\left| R^N \right|\right): \quad \nabla \cdot \vec{A} \equiv \partial_\ell A_\ell \equiv \sum_{\ell=1}^{N} \frac{\partial A_\ell}{\partial x_\ell}, \tag{5.153a–c}$$

where is used the **summation convention** (Einstein, 1916) that a repeated index implies a summation over its range of variation (5.153b). The divergence theorem can be proved from the following **divergence lemma**: *if (5.154a) is a continuously differentiable function that vanishes (5.86b) on a closed surface (Figure 5.11a) and is positive (negative) inside (5.86a) [outside (5.86c)], then the integral over all space of the divergence of the product of its unit jump (5.87a through c) by a continuously differentiable vector (5.154b) vanishes (5.154c):*

$$f(x_n), A_\ell(x_n) \in C^1\left(\left| R^N \right|\right): \quad 0 = \int_{-\infty}^{+\infty} dx_1 \ldots \int_{-\infty}^{+\infty} dx_N \frac{\partial}{\partial x_\ell}\left[A_\ell H\left(f(x_n)\right)\right]. \tag{5.154a–c}$$

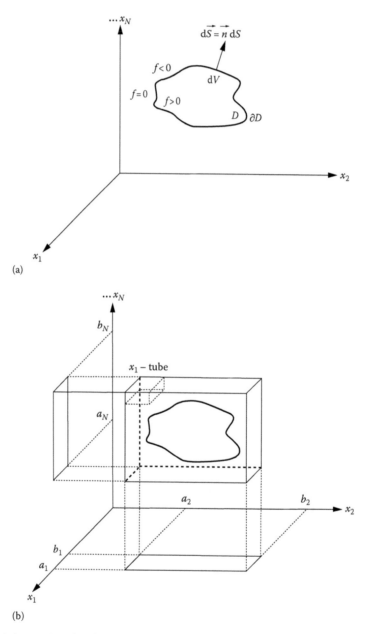

FIGURE 5.11 A domain D with a closed regular boundary ∂D may be specified by a continuously differentiable function that vanishes on the boundary, is positive in the interior and negative in the exterior (a). The regular boundary has a continuous normal that is unique at each point, thus excluding edges or corners. The N-dimensional domain with closed regular $(N-1)$-dimensional boundary appears in the divergence, gradient, and curl integral theorems; each of these three theorems is proved from the corresponding lemma, that considers a hyperparallel-epiped containing the domain (b).

To prove the lemma, (1) the domain is enclosed (Figure 5.11b) in a large hyperparallelepiped (5.155):

$$P \equiv \left\{ \vec{x} : \forall_{1 \leq n \leq N} : a_n \leq x_n \leq b_n \right\} \supset D; \tag{5.155}$$

(2) the hyperparallelepided is divided into tubes parallel to the x_1-axis, and the integral over one tube

$$\int_{a_1}^{b_1} dx_1 \ldots \int_{a_N}^{b_N} dx_N \frac{\partial}{\partial x_1} \left[A_1 H(f) \right] = \int_{a_2}^{b_2} dx_2 \ldots \int_{a_N}^{b_N} dx_N \left[A_j H(f) \right]_{a_1}^{b_1} = 0, \tag{5.156}$$

vanishes because the primitive vanishes (5.87c) outside the domain; (3) summing (5.156) for all tubes proves that the integral is zero for the x_1 term, and likewise for all x_n:

$$\sum_{\ell=1}^{N} \int_{a_1}^{b_1} dx_1 \ldots \int_{a_N}^{b_N} dx_N \frac{\partial}{\partial x_\ell} \left[A_\ell H(f(x_1,\ldots,x_N)) \right] = 0; \tag{5.157}$$

and (4) taking the limit $a_n \to -\infty$ and $b_n \to +\infty$ proves (5.154c).

The identity (5.154c) implies

$$\int_{-\infty}^{+\infty} dx_1 \ldots \int_{-\infty}^{+\infty} dx_N \frac{\partial A_\ell}{\partial x_\ell} H(f(x_n)) = - \int_{-\infty}^{+\infty} dx_1 \ldots \int_{-\infty}^{+\infty} dx_N A_\ell \frac{\partial}{\partial x_\ell} \left[H(f) \right]$$

$$= \int_{-\infty}^{+\infty} dx_1 \ldots \int_{-\infty}^{+\infty} dx_N A_\ell N_\ell \left| \nabla f \right| \delta(f). \tag{5.158}$$

where (5.148b) was used, involving the unit outer normal vector. From (5.91a through c) follows, as in (5.11a through c), that

$$\delta\big(f(x_1,\ldots,x_N) \big) = \sum_{p=1}^{P} \big(\partial f / \partial x_N \big)^{-1} \delta\big(x_N - h_p(x_1,\ldots,x_{N-1}) \big). \tag{5.159}$$

Substituting (5.159) in (5.140d), it follows that *the hypersurface integration is given by (5.160b)*:

$$\Phi(x_n) \in C\big(|R^N\big) : \quad \big[S(x_n), \Phi(x_n) \big] = \int_{-\infty}^{+\infty} dx_1 \ldots \int_{-\infty}^{+\infty} dx_N \Phi(x_1,\ldots,x_N) \big| \nabla f(x_n) \big| \delta\big(f(x_n) \big), \tag{5.160a,b}$$

corresponding to the inner product by an integrable test function (5.160a) of the generalized function:

$$S(x_n) = \big| \nabla f(x_n) \big| \delta\big(f(x_n) \big) = \Big| \nabla \big[H(f(x_n)) \big] \Big|. \tag{5.161}$$

Using (5.160b), the r.h.s. of (5.158) becomes a hypersurface integral over the boundary (5.162a) specified by the unit impulse:

$$\int_{-\infty}^{+\infty} dx_1 \ldots \int_{-\infty}^{+\infty} dx_N A_\ell N_\ell \big| \nabla f \big| \delta(f) = \int_{\partial D_N \equiv D_{N-1}} \big(\vec{A} \cdot \vec{N} \big) dS. \tag{5.162a}$$

$$\int_{-\infty}^{+\infty} dx_1 \ldots \int_{-\infty}^{+\infty} dx_N \left(A_\ell N_\ell \right) H(f) = \int_{D_N} \left(\vec{A} \cdot \vec{N} \right) dV. \tag{5.162b}$$

The unit jump on the l.h.s. of (5.158) limits the integration to the domain (5.162b) [the boundary term like in (5.27b) is not considered].

Equating (5.162a) ≡ (5.162b) in agreement with (5.158) leads to

$$f(x_n), A_j(x_n) \in C^1\left(|R^N \right): \quad \int_{D_N} \left(\nabla \cdot \vec{A} \right) dV = \int_{\partial D_N \equiv D_{N-1}} \left(\vec{A} \cdot \vec{N} \right) dS, \tag{5.163a–c}$$

the **divergence theorem** *stating the equality (5.163c) of (1) the integral of the divergence of a continuously differentiable vector (5.163b) in a domain (5.86a through c); and (2) the flux of the vector across the regular (5.163a) closed boundary (5.88a and b).* As a simple example of the divergence integral theorem is considered the position vector relative to the origin (5.164a), whose divergence is a constant equal to the dimension of the space (5.164b):

$$\vec{x} = \vec{e}_R R = \vec{e}_x x + \vec{e}_y y + \vec{e}_z z, \quad \nabla \cdot \vec{x} = 3. \tag{5.164a,b}$$

Thus, (1) the integral (5.165b) of the divergence (5.164b) over the sphere of radius a is the constant times the volume V in (5.165a):

$$V = \frac{4\pi}{3} a^3: \quad \int_{R=a} \left(\nabla \cdot \vec{x} \right) dV = 3 \int_{R=a} dV = 3V = 4\pi a^3; \tag{5.165a,b}$$

(2) the radial projection equals the radius of the sphere on its surface (5.163b) of area S in (5.163a):

$$S = 4\pi a^3: \quad \int_{R=a} \left(\vec{x} \cdot \vec{N} \right) dS = a \int_{R=a} dS = aS = 4\pi a^3; \tag{5.166a,b}$$

and (3) the equality (5.162b) ≡ (5.163b) agrees with the divergence integral theorem (5.163c).

5.7.2 Gradient Integral Theorem

The **gradient** of a scalar function differentiable with regard to its N variables (5.167a) is defined as the vector whose components are its partial derivatives (5.167b):

$$\Phi \in D\left(|R^N \right): \quad \nabla \Phi = \sum_{\ell=1}^{N} \vec{e}_\ell \frac{\partial \Phi}{\partial x_\ell} \equiv \vec{e}_\ell \partial_\ell \Phi, \tag{5.167a–c}$$

using the summation convention (5.153b and c) in (5.167b and c). Replacing in (5.154a through c) the continuously differentiable vector by a continuously differentiable scalar (5.168b), the lemma still holds:

$$f(x_n), \Phi(x_n) \in C^1\left(|R^N \right): \quad 0 = \int_{-\infty}^{+\infty} dx_1 \ldots \int_{-\infty}^{+\infty} dx_N \frac{\partial}{\partial x_j} \left[\Phi H\left(f(x_n) \right) \right]. \tag{5.168a–c}$$

The **gradient lemma** *(5.168c) applies to two continuously differentiable functions (5.164a and b), and* from (5.168c) follows

$$
\int_{-\infty}^{+\infty} dx_1 \dots \int_{-\infty}^{+\infty} dx_N H\big(f(x_n)\big) \frac{\partial \Phi}{\partial x_j} = -\int_{-\infty}^{+\infty} dx_1 \dots \int_{-\infty}^{+\infty} dx_N \Phi \frac{\partial H}{\partial x_j}
$$

$$
= \int_{-\infty}^{+\infty} dx_1 \dots \int_{-\infty}^{+\infty} dx_N \Phi |\nabla f| N_j \delta(f), \tag{5.169}
$$

where (5.148b) was used. The unit jump (impulse) on the (l.h.s.) of (5.166) limits the integral to the domain (boundary) as in (5.162a) [(5.162b)] leading to

$$
f(x_n), \Phi(x_n) \in C^1\big(|R^N\big): \quad \int_{D_N} \nabla\Phi \, dV = \int_{\partial D_N \equiv D_{N-1}} \Phi \vec{N} \, dS, \tag{5.170a–c}
$$

the **gradient theorem** *stating the equality (5.170c) of (1) the gradient of a continuously differentiable scalar (5.170b) integrated over the domain (5.86a through c); and (2) the integral of the scalar over the regular (5.170a) closed boundary (5.88a and b).*

As a simple example of the gradient integral theorem, consider the scalar (5.171a) whose gradient is unity and in the *x*-direction (5.171b), so that the integral over the sphere of radius *a* equals its volume (5.165a) ≡ (5.171c) in the same direction:

$$
\Phi = x, \quad \nabla\Phi = \vec{e}_x, \quad \int_{R=a} \nabla\Phi \, dV = \vec{e}_x \int_{R=a} dV = \vec{e}_x V = \vec{e}_x \frac{4}{3}\pi a^3. \tag{5.171a–c}
$$

The unit normal to the sphere is the unit radial vector (5.164a), and thus the surface integral of (5.171a) is (5.172b):

$$
\vec{e}_R R = \vec{e}_x x + \vec{e}_y y + \vec{e}_z z: \quad \int_{R=a} \Phi \vec{N} \, dS = \int_{R=a} \vec{e}_R x \, dS = \frac{1}{a} \int_{R=a} \big(\vec{e}_x x^2 + \vec{e}_y y + \vec{e}_z zx\big) dS. \tag{5.172a,b}
$$

Using the symmetry of the sphere with regard to all coordinate axes it follows that *either for surface (5.173a) or volume (5.173b) integrals,*

$$
\int_{R=a} \dots dS \equiv \iint_{x^2+y^2+z^2=a^2} \dots dx dy, \tag{5.173a}
$$

$$
\int_{R<a} \dots dV \equiv \iint_{x^2+y^2+z^2 \le a^2} \dots dx dy dz, \tag{5.173b}
$$

if (1) the integrand involves odd powers of any coordinate, the integral is zero, for example,

$$
0 = \int x = \int y = \int z = \int xy = \int xy = \int yz = \int x^3 = \dots; \tag{5.174}
$$

$$\int x^2 = \int y^2 = \int z^2 = \frac{1}{3}\int \left(x^2 + y^2 + z^2\right) = \frac{1}{3}\int R^2; \qquad (5.175)$$

and (2) *if even powers of one coordinate appear, they can be substituted by the same even power of any other coordinate, for example, (5.175).* From (5.174), it follows that the last two terms in curved brackets in (5.172b) have zero integral, and the first term is evaluated using (5.175):

$$\int_{R=a} \Phi\vec{N}\,dS = \frac{\vec{e}_x}{a}\int_{R=a} x^2\,dS = \frac{\vec{e}_x}{3a}\int_{R=a}\left(x^2 + y^2 + z^2\right)dS$$

$$= \vec{e}_x\frac{a}{3}\int_{R=a}dS = \vec{e}_x\frac{a}{3}S = \vec{e}_x\frac{4\pi a^3}{3}. \qquad (5.176)$$

The equality (5.171c) [(5.176)] uses the volume (5.165a) [area (5.164a)] of the sphere of radius a and agrees with the gradient integral theorem (5.170c).

5.7.3 Curl Volume Theorem for a Domain

The **curl** of a differentiable vector (5.177b) is defined in an N-dimensional space (5.177a) as the skew-symmetric matrix or **bivector** that is the **commutator** of partial derivatives (5.177c):

$$j,k,\ell,n = 1,\dots,N;\ A_\ell(x_n) \in \mathcal{D}\left(\left|R^N\right|\right):\quad \left(\nabla \wedge \vec{A}\right)_{k\ell} = \frac{\partial A_\ell}{\partial x_k} - \frac{\partial A_k}{\partial x_\ell} \equiv \partial_k A_\ell - \partial_\ell A_k. \qquad (5.177a\text{-}c)$$

In this case (5.177a through c), the analogue of the divergence (5.154a through c) [gradient (5.168a through c)] lemmas is the **curl lemma** (5.178c) *that applies to a scalar (5.178a) and a vector (5.178b) with continuous first-order derivatives*:

$$f(x_n),\,A_\ell(x_n) \in C^1\left(\left|R^N\right|\right):\quad 0 = \int_{-\infty} dx_1 \dots \int_{-\infty} dx_N \left\{\partial_k\left[A_\ell H(f)\right] - \partial_\ell\left[A_k H(f)\right]\right\}. \qquad (5.178a\text{-}c)$$

From (5.178c) follows

$$\int_{-\infty}^{-\infty} dx_1 \dots \int_{-\infty}^{-\infty} dx_N H(f)\left(\partial_k A_\ell - \partial_k A_\ell\right) = -\int_{-\infty}^{-\infty} dx_1 \dots \int_{-\infty}^{-\infty} dx_N \left\{A_\ell\partial_k\left[H(f)\right] - A_k\partial_\ell\left[H(f)\right]\right\}$$

$$= \int_{-\infty}^{-\infty} dx_1 \dots \int_{-\infty}^{-\infty} dx_N \left(A_\ell N_k - A_k N_\ell\right)\left|\nabla f\right|\delta(f), \qquad (5.179)$$

where (5.148b) is used. Again the unit jump (impulse) restricts the integration on the l.h.s. (r.h.s.) of (5.179) to the domain (boundary) as in (5.162a) [(5.162b)] leading to

$$f(x_n),\,A_\ell(x_n) \in C^1\left(\left|R^N\right|\right):\quad \int_{D_N}\left(\partial_x A_\ell - \partial_x A_k\right)dV = \int_{\partial D_N\,=\,D_{N-1}}\left(N_k A_\ell - N_\ell A_k\right)dS, \qquad (5.180a\text{-}c)$$

the **curl theorem** stating *the equality (5.180c) of (1) the integral of the curl of a continuously differentiable vector (5.180b) over a domain (5.86a through c); (2) the integral over the regular (5.180a) boundary (5.88a and b) of the outer product by the unit outer normal. In all three cases of the divergence (5.163a through c), gradient (5.170a through c), and curl (5.180a through c), theorems hold*

$$\int_{D_N} \ldots dV = \int_{\partial D_N \equiv D_{N-1}} \ldots dS: \quad \nabla \equiv \partial_i \leftrightarrow \vec{N} = N_i, \tag{5.181a–d}$$

the substitution of (1, 2) the volume (surface) element (5.181a) over (5.181b) the domain D (regular closed boundary ∂D) of dimension N(N − 1) in (5.86a through c) [(5.88a and b)]; and (3) the nabla or gradient operator (5.181c) by the unit outward normal (5.181d). There is a difference: (1) the divergence theorem (5.163a through c) is a scalar relation; (2) the gradient theorem (5.170a through c) is a vector relation; and (3) the curl theorem (5.180a through c) is a bivector relation. All three theorems (1, 2, 3) hold for a domain in an N-dimensional space D_N with a closed regular boundary ∂D_N that is a subspace D_{N-1} of dimension N − 1. The curl theorem becomes a vector (scalar) relation in a three(two)-dimensional space [Subsection 5.7.4 (Subsection 5.7.6)]. All three theorems (1, 2, 3) are particular cases of two multivector theorems for subspaces of dimension $M < N$ of an N-dimensional space (Notes 9.3 through 9.52).

5.7.4 Permutation Symbol and Polar/Axial Vectors

In a three-dimensional space can be defined the **permutation symbol** that equals +1(−1) for even (odd) permutations (5.182a) [(5.182b)] of (1, 2, 3) and is zero otherwise (5.182c), that is, for repeated indices:

$$j, m, n = 1, 2, 3 \equiv N: \quad e_{jk\ell} = \begin{cases} +1 & \text{if } (j, k, \ell) \text{ is even permutation of } (1, 2, 3) & (5.182a) \\ -1 & \text{if } (j, k, \ell) \text{ is odd permutation of } (1, 2, 3) & (5.182b) \\ 0 & \text{otherwise}: j = m \text{ or } j = n \text{ or } m = n. & (5.182c) \end{cases}$$

The permutation symbol (5.182a through c) may be used to assign to a bivector or skew-symmetric matrix (5.183a) in a three-dimensional space an **axial vector** *(5.183b):*

$$C_{k\ell} = -C_{\ell k}: \quad B_j = e_{jk\ell} C_{j\ell}. \tag{5.183a,b}$$

The **curl** *(5.177c) of a differentiable vector (5.177b) is a skew-symmetric matrix (5.184b) in a space of any dimension (5.184a):*

$$A_\ell(x_n) \in \mathcal{D}\left(|R^N\right): \quad \left(\nabla \wedge \vec{A}\right)_{k\ell} = \partial_k A_\ell - \partial_\ell A_k = -\left(\nabla \wedge \vec{A}\right)_{\ell k}. \tag{5.184a,b}$$

In a three-dimensional space, it thus specifies an axial vector (5.182b) if (5.182a) is a polar vector:

$$A_n(x_m) \in \mathcal{D}\left(|R^3\right): \quad \left(\nabla \wedge \vec{A}\right)_j = e_{jk\ell} \partial_k A_\ell. \tag{5.185a,b}$$

This should be distinguished from a **polar vector** *(5.152a), that is, a vector for any space dimension N. Thus, the curl of a polar vector (5.185a) is an axial vector (5.185b) only in a three-dimensional space.*

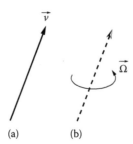

(a)   (b)

FIGURE 5.12   In a three-dimensional space exist two types of vectors the polar (axial) vector (a) [(b)] that has an inner (outer) one-direction like the linear (angular) velocity of translation (rotation).

An example of a polar vector (Figure 5.12a) is the velocity (5.188a) that specifies an **inner one-direction** *by the unit vector in its direction (5.186d):*

$$\vec{v} = \frac{d\vec{x}}{dt} = v\vec{s}: \quad ds = \left|d\vec{x}\right|, \quad \left|\vec{v}\right| = \frac{ds}{dt}, \quad \vec{s} = \frac{\vec{v}}{v} = \frac{d\vec{x}}{\left|d\vec{x}\right|} = \frac{d\vec{x}}{ds}, \tag{5.186a–d}$$

where (5.186b) is the arc length; thus, the velocity equals in modulus the arc length per unit time (5.186c) and is tangent to the trajectory, that is, a tangent curve always exists. In the case of a **rigid body rotation** *(5.187b)* ≡ *(I.4.7), the constant angular velocity (5.187a) is one-half of its curl (5.187c) and hence a polar vector:*

$$\vec{\Omega} = \text{const}: \quad \vec{V} = \vec{\Omega} \wedge \vec{x} \Rightarrow \nabla \wedge \vec{V} = 2\vec{\Omega}. \tag{5.187a–c}$$

An axial vector specifies an **outer one-direction** *that is a direction of rotation around an axis (Figure 5.11b), and a curve tangent to it may not exist.* The proof of (5.187c) from (5.187b) can be made in two steps:

$$\vec{\Omega} = \vec{e}_z \Omega: \quad \vec{\Omega} \wedge \vec{x} = \begin{vmatrix} \vec{e}_x & \vec{e}_y & \vec{e}_z \\ 0 & 0 & \Omega \\ x & y & z \end{vmatrix} = \Omega \left(\vec{e}_y x - \vec{e}_x y \right), \tag{5.188a,b}$$

$$\vec{\Omega} = \text{const}: \quad \nabla \wedge \left(\vec{\Omega} \wedge \vec{x} \right) = \begin{vmatrix} \vec{e}_x & \vec{e}_y & \vec{e}_z \\ \dfrac{\partial}{\partial x} & \dfrac{\partial}{\partial y} & \dfrac{\partial}{\partial z} \\ -\Omega y & \Omega x & 0 \end{vmatrix} = 2\vec{e}_z \Omega = 2\vec{\Omega}, \tag{5.189a,b}$$

using (1) the outer product (5.190a) of vectors (5.188b) with z-axis along $\vec{\Omega}$ in (5.188a):

$$\vec{N} \wedge \vec{A} = e_{jk\ell} \vec{e}_j N_k A_\ell; \nabla \wedge \vec{A} = e_{ijk} \vec{e}_j \partial_k A_\ell, \tag{5.190a,b}$$

and (2) replacing the first vector by the nabla or gradient operator (5.181c and d), the definition of the three-dimensional curl vector (5.190b) with constant angular velocity (5.189a) leads to (5.189b). For an arbitrary constant angular velocity along the z-axis (5.189a), the corresponding linear velocity (5.187b) lies in the transverse directions with components (5.188b).

5.7.5 Invariant Integral Theorems in Three Dimensions

The divergence (5.163a through c) [gradient (5.170a through c)] theorems take the same form in a space of any dimension. The curl theorem (5.180a through c) is a bivector relation in a space of N-dimensions and using (5.190a and b) in three dimensions,

$$A_\ell(x_n) \in C^1|R^2: \quad \int\limits_{D_3} e_{jk\ell}\vec{e}_j\partial_k A_\ell \,\mathrm{d}V = \int\limits_{D_3 \equiv D_2} e_{jk\ell}\vec{e}_j N_k A_\ell \,\mathrm{d}S, \tag{5.191a,b}$$

it becomes a vector relation:

$$A_\ell(\vec{x}) \in C^1\left(|R^3\right): \quad \int\limits_{D_3}\left(\nabla \wedge \vec{A}\right)\mathrm{d}V = \int\limits_{D_3 \equiv D_2}\left(\vec{N} \wedge \vec{A}\right)\mathrm{d}S. \tag{5.192a–c}$$

This corresponds to the **three-dimensional curl theorem** *stating the equality (5.192c) of (1) the curl of a three-dimensional continuously differentiable vector (5.192b) integrated over a domain (5.86a through c); and (2) the outer product by the area element integrated over the regular (5.191a) closed boundary (5.88a and b) that is a regular surface.* As a simple example of the three-dimensional curl vector theorem (5.192a and b), consider the vector (5.193a) corresponding to the velocity (5.188b) in the (x, y)-plane due to a rotation with unit angular velocity (5.188a; 5.189a) around the z-axis leading (5.189b) to the curl (5.192b):

$$\vec{A} = \vec{e}_y x - \vec{e}_x y, \quad \nabla \wedge \vec{A} = \vec{e}_z\left(\partial_x A_y - \partial_y A_x\right) = 2\vec{e}_z. \tag{5.193a,b}$$

The integral over the sphere of radius a equals (5.193c) twice the volume (5.165a) in the same direction:

$$\int\limits_{R=a}\left(\nabla \wedge \vec{A}\right)\mathrm{d}V = 2\vec{e}_z \int\limits_{R=a}\mathrm{d}V = 2\vec{e}_z V = \vec{e}_z\frac{8\pi a^3}{3}. \tag{5.193c}$$

The unit outward normal to the sphere is the unit vector (5.172a) leading to (5.194) for the outer product with the vector (5.193a):

$$R\left(\vec{N} \wedge \vec{A}\right) = \begin{vmatrix} \vec{e}_x & \vec{e}_y & \vec{e}_z \\ x & y & z \\ -y & x & 0 \end{vmatrix} = -\vec{e}_x xz - \vec{e}_y yz + \vec{e}_z\left(x^2 + y^2\right); \tag{5.194}$$

Using again the symmetry relative to the coordinate axis (5.174), the integral of (5.194) over the surface (5.173a) sphere of radius a is given (5.175) by

$$\int\limits_{R=a}\left(\vec{N} \wedge \vec{A}\right)\mathrm{d}S = \frac{1}{a}\int\limits_{R=a}\left[-\vec{e}_x xz - \vec{e}_y yz + \vec{e}_z\left(x^2 + y^2\right)\right]\mathrm{d}S$$

$$= \frac{\vec{e}_z}{a}\int\limits_{R=a}\left(x^2 + y^2\right)\mathrm{d}S = \frac{2}{3a}\vec{e}_z\int\limits_{R=a}R^2\,\mathrm{d}S$$

$$= \frac{2a}{3}\vec{e}_z\int\limits_{R=a}\mathrm{d}S = \frac{2a}{3}\vec{e}_z S = \vec{e}_z\frac{8\pi a^3}{3}, \tag{5.195}$$

using the area (5.166a) of the sphere. The equality (5.193) $\equiv$ (5.194) agrees with the curl integral theorem (5.193c). The divergence/gradient/curl theorems (respectively Subsection 5.7.1/5.7.2/5.7.3) apply to an N-dimensional domain with an $(N-1)$-dimensional closed regular boundary that is a hypersurface; in particular, the curl theorem in three dimensions (Subsections 5.7.4 and 5.7.5) applies to a domain with a closed regular surface as the boundary; another important result is the Stokes' theorem (Subsection 5.7.7) that applies to a regular surface supported on a closed curve or loop (Subsection 5.7.6).

5.7.6 Regular Surface Supported on a Loop

Consider a regular (5.196a) surface $f = 0$ supported (Figure 5.13) on a loop L obtained by intersection with another regular (5.196b) surface $g = 0$ such that the two surfaces are not tangent (5.196c):

$$f(x,y,z), g(x,y,z) \in C^1(|R^3|): \quad Ra\left[\frac{\partial(f,g)}{\partial(x,y,z)}\right] = 2. \tag{5.196a–c}$$

The condition (5.196c) $\equiv$ (5.197a) implies that the gradients of the scalar functions (f, g) are not parallel, and hence the outward normals (5.197b) [(5.197c)] to the surfaces $f = $ const ($g = $ const) are not parallel (5.197d):

$$\nabla f \wedge \nabla g \neq 0: \quad \vec{N}_f = \frac{\nabla f}{|\nabla f|}, \quad \vec{N}_g = -\frac{\nabla g}{|\nabla g|}, \quad \vec{N}_f \wedge \vec{N}_g \neq 0. \tag{5.197a–d}$$

Thus, the surfaces are not tangent and intersect along a curve orthogonal to both normals:

$$\vec{S} = \vec{N}_f \wedge \vec{N}_0 = \frac{\nabla f \wedge \nabla g}{|\nabla f||\nabla g|}: \quad \vec{S} \cdot \vec{N}_f = 0 = \vec{S} \cdot \vec{N}_g, \quad \vec{s} = \frac{\vec{S}}{|\vec{S}|} = \frac{\nabla f \wedge \nabla g}{|\nabla f \wedge \nabla g|}, \tag{5.198a–d}$$

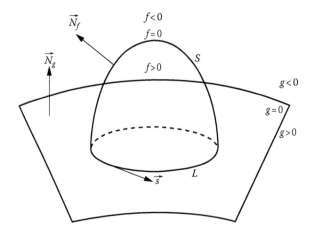

FIGURE 5.13 The Stokes theorem equates the circulation of a continuously differentiable vector along a closed regular loop to the flux of its curl across a regular surface "supported" on the loop, meaning that: (1) the regular boundary loop is the intersection of two regular surfaces; (2) the regular surface "supported" on the loop is the part of one surface lying on one side of the other surface; and (3) if both surfaces are regular, that is have continuous normals, the curve that is their intersection is also regular, that is has continuous tangent, excluding corners.

implying that (1) the vector (5.198a) is orthogonal to the normals to the two surfaces (5.198b and c) and hence tangent to their intersection; and (2) the corresponding unit tangent vector is (5.198d).

The loop or closed curve is specified by the intersection of the two surfaces (5.199a):

$$L \equiv \left\{ \vec{x} : \; f(\vec{x}) = 0 = g(\vec{x}) \right\} \equiv \partial D_2, \quad S \equiv \left\{ \vec{x} : \; f(\vec{x}) = 0 < g(\vec{x}) \right\} \equiv D_2. \qquad (5.199\text{a,b})$$

The part of the first surface $f = 0$ supported on the loop lies outside $g > 0$ the second surface (5.199b). If the surface (5.199b) is regular (5.196a), it has a continuous normal at all points, including on the boundary loop (5.199a); since the normal to the loop is continuous, the tangent is also continuous, and the loop is regular. Thus, the intersection of two regular surfaces specifies a regular boundary, that is, excluding cusps or edges on the surfaces excludes angular points on the boundary. The converse is not true: a loop may be regular, that is, devoid of angular points, and the surface supported on it may have cusps or edges outside the boundary and not be regular. It has been shown *that the intersection of two regular surfaces (5.196a and b) that are not tangent to each other (5.196c) is a closed regular loop (Figure 5.13). To obtain a regular surface supported on the loop (5.199a), it is sufficient to choose the part of one surface outside the other (5.199b). The normals to the two surfaces (5.197b and c) are orthogonal to the vector tangent to the intersection (5.198a), and thus the unit tangent is (5.198d).*

The **circulation** *of an integrable vector (5.200a) along the loop* $L \equiv \partial D_1$ *is the integral (5.200b) of its projection along the unit tangent vector* $\vec{s}$:

$$A_j(x, y, z) \in \mathcal{E}\big(|R^3|\big): \quad \int_L \vec{A} \cdot d\vec{x} = \int_L \big(\vec{A} \cdot \vec{s}\big) ds, \qquad (5.200\text{a,b})$$

where ds is the arc length (5.186b). Substituting (5.198a) in (5.200b) leads to the integral along the loop:

$$\int_L \vec{A} \cdot d\vec{x} = \int_L e_{jk\ell} A_j \frac{\big(\partial f/\partial x_k\big)\big(\partial g/\partial x_\ell\big)}{\nabla f \wedge \nabla g} ds. \qquad (5.201)$$

The integral of the function

$$\big|\nabla f \wedge \nabla g\big| \Psi(x, y, z) \equiv e_{yk\ell} A_j \big(\partial_k f\big)\big(\partial_\ell g\big) \qquad (5.202)$$

along the loop is the integral (5.162a) along the intersection of the two surfaces $f = 0 = g$ *leading to (5.203):*

$$\int_L \vec{A} \cdot d\vec{x} = \int_{-\infty}^{+\infty} dx \int_{-\infty}^{+\infty} dy \int_{-\infty}^{+\infty} dz \, \vec{s} \, \Psi \big|\nabla f\big\|\nabla g\big| \delta(f)\delta(g)$$

$$= \int_{-\infty}^{+\infty} dx \int_{-\infty}^{+\infty} dy \int_{-\infty}^{+\infty} dz \, \Psi \big|\nabla f \wedge \nabla g\big| \delta(f)\delta(g), \qquad (5.203)$$

where (5.198d) was used. Substitution of (5.202) in (5.203) leads to

$$\int_L \vec{A} \cdot d\vec{x} = \int_{-\infty}^{+\infty} dx \int_{-\infty}^{+\infty} dy \int_{-\infty}^{+\infty} dz \, e_{yk\ell} A_j \big(\partial_k f\big)\big(\partial_\ell f\big) \big|\nabla f\big\|\nabla g\big| \delta(f)\delta(g)$$

$$= \int_{-\infty}^{+\infty} dx \int_{-\infty}^{+\infty} dy \int e_{yk\ell} A_j \partial_k \big[H(f)\big] \partial_\ell \big[H(g)\big], \qquad (5.204)$$

where (5.148b) was used. The evaluation of the integral (5.204) leads to the Stokes' theorem for a vector in three dimensions. This is (Note 9.29) the particular case ($M = 1$, $N = 3$) of a more general integral theorem (Note 9.27) on M-vectors in N-dimensional spaces. The latter theorem is proved independently and thus confirms the preceding particular results (5.201; 5.202; 5.203). The evaluation of the integral (5.204) leads next to the Stokes' theorem for a vector in a three-dimensional space without referring to higher dimensions.

5.7.7 Curl Theorem for a Loop

The integrand in (5.204) is

$$e_{jk\ell}A_j\partial_k\big[H(f)\big]\partial_\ell\big[H(g)\big] = \partial_\ell\big\{H(g)e_{jk\ell}A_j\partial_k\big[H(f)\big]\big\} - H(g)\partial_\ell\big\{e_{jk\ell}A_j\partial_k\big[H(f)\big]\big\}. \quad (5.205)$$

The first term on the r.h.s. of (5.205) vanishes on integration over all space by the divergence lemma (5.154a through c), and thus only the second term on the r.h.s. of (5.205) remains in (5.206),

$$\int_L \vec{A}\cdot d\vec{x} = -\int_{-\infty}^{+\infty}dx\int_{-\infty}^{+\infty}dy\int_{-\infty}^{+\infty}dz\, e_{jk\ell}H(g)\partial_\ell\big\{A_j\partial_k\big[H(f)\big]\big\}, \quad (5.206a)$$

after integration. The integrand in (5.206a) involves

$$e_{jk\ell}H(g)\partial_\ell\big\{A_j\partial_k\big[H(f)\big]\big\} = e_{jk\ell}\big(\partial_\ell A_j\big)\partial_k\big[H(f)\big] + e_{jk\ell}A_j\frac{\partial^2}{\partial x_k\partial x_\ell}\big[H(f)\big], \quad (5.206b)$$

since $e_{jk\ell}(\partial^2/\partial x_k\partial x_\ell)$ is skew-symmetric (symmetric) in $k\ell$ (5.207c) [(5.207b)] for f with continuous second-order derivatives (5.207a)]:

$$f(x,y,z)\in C^2\big(|R^3\big): \quad \partial_m\partial_n f = \partial_m\partial_n f, \quad e_{\ell mn} = -e_{\ell mn}, \quad e_{\ell mn}\partial_m\partial_n\big[H(f)\big] = 0, \quad (5.207a\text{–}d)$$

the product with contraction, that is, summation in k, $\ell = 1, 2, 3$ vanishes (5.207d).

The proof that the double inner product of a symmetric (5.208a) and skew-symmetric (5.208b) matrix is zero (5.208c),

$$E_{k\ell} = E_{k\ell}, D_{k\ell} = -D_{k\ell}: \quad E_{k\ell}D_{k\ell} = 0, \quad (5.208a\text{–}c)$$

follows from

$$E_{k\ell}D_{k\ell} = E_{\ell k}D_{\ell k} = -D_{k\ell}E_{k\ell} = 0. \quad (5.208d)$$

In (5.207d), the symmetric (skew-symmetric) matrix is the second-order derivative (5.208a) $\equiv$ (5.207b) [two indices of the permutation symbol (5.208b) $\equiv$ (5.207c)]. Thus, (5.206b) simplifies to the first term on the r.h.s. when substituted in (5.206a) leading to

$$\int_L \vec{A}\cdot d\vec{x} = \int_{-\infty}^{+\infty}dx\int_{-\infty}^{+\infty}dy\int_{-\infty}^{+\infty}dz\, H(g)e_{jk\ell}\big(\partial_\ell A_j\big)\delta(f)N_k, \quad (5.209a)$$

where (5.148b) was used. On the r.h.s. of (5.209a), (1) the unit impulse limits the integration to the surface (5.196a) with normal (5.148b); (2) the unit jump limits the integration to the *f*-surface above the boundary loop (Figure 5.13); and (3) the curl as an axial vector (5.185b) appears in an inner product with the unit normal:

$$\left(\nabla \wedge \vec{A}\right) \cdot \vec{N} = \left(\nabla \wedge \vec{A}\right)_k N_k = e_{k\ell j} N_k \partial_\ell A_j = e_{jk\ell} N_k \partial_\ell A_j. \tag{5.209b}$$

Substituting (5.209b) in (5.209a) leads to

$$\vec{A}_n\left(x, y, z\right) \in C^1\left(| R^3 \right): \quad \int_{\partial D_2 = D_1} \vec{A} \cdot d\vec{x} = \int_{D_2} \left(\nabla \wedge \vec{A}\right) \cdot \vec{N} \, dS, \tag{5.210a,b}$$

the **curl loop theorem (Stokes, 1854; Hankel, 1861)** *that states the equality (5.210b) of (1) the circulation of a continuously differentiable vector (5.210a) along a closed regular loop (Figure 5.13); and (2) the flux of its curl across a regular surface supported on the loop.*

As a simple example of the Stokes' theorem, consider again the vector (5.193a) in the (x, y)-plane, whose curl (5.193b) projected on the normal $\vec{e}_R$ to (5.172a) the half-sphere of radius a and center at the origin equals

$$I \equiv \int_{\substack{R=a \\ z\geq 0}} \left[\left(\nabla \wedge \vec{A}\right) \cdot \vec{N}\right] dS = 2 \int_{\substack{R=a \\ z\geq 0}} \left(\vec{e}_z \cdot \vec{e}_R\right) \frac{dS}{R} = \frac{2}{a} \int_{\substack{R=a \\ z\geq 0}} z \, dS, \tag{5.211}$$

using spherical coordinates (5.212a through c) $\equiv$ (6.13a and c,d):

$$x = R\sin\theta\cos\varphi, \quad y = R\sin\theta\sin\varphi, \quad z = R\cos\theta, \tag{5.212a–c}$$

with scale factors (5.204a through c) $\equiv$ (6.14a through c) and area element (5.213d):

$$h_R = 1, \quad h_\theta = R, \quad h_\varphi = R\sin\theta, \quad dS = h_\theta h_\varphi \, d\theta d\varphi = R^2 \sin\theta \, d\theta d\varphi, \tag{5.213a–d}$$

the integral (5.211) over the upper half of the sphere (5.214a) of radius a equals (5.214b):

$$R = a: \quad I = 2a \int_0^{2\pi} d\varphi \int_0^{\pi/2} d\theta \, z\sin\theta = 4\pi a^2 \int_0^{2\pi} \cos\theta\sin\theta \, d\theta = 2\pi a^2 \left[\sin^2\theta\right]_0^{\pi/2} = 2\pi a^2. \tag{5.214a,b}$$

The circulation of the vector (5.193a) $\equiv$ (5.215b) around the circle (5.214a) of radius a in the (x, y)-plane (5.215a),

$$\theta = \frac{\pi}{2}: \quad \vec{A} = \vec{e}_y x - \vec{e}_x y = a\left(\vec{e}_y \cos\varphi - \vec{e}_x \sin\varphi\right), \tag{5.215a,b}$$

is given by (5.216b):

$$S_2 = \pi a^2: \quad \int_{r=a} \vec{A} \cdot d\vec{x} = a \int_{r=a} \left(-\sin\varphi \, dx + \cos\varphi \, dy \right)$$

$$= a \int_{r=a} \left[-\sin\varphi \, d\left(a\cos\varphi \right) + \cos\varphi \, d\left(a\sin\varphi \right) \right]$$

$$= a^2 \int_0^{2\pi} d\varphi = 2\pi a^2 = \int_{x^2+y^2=a^2} \left(x\,dy - y\,dx \right) = 2S_2, \qquad (5.216\text{a,b})$$

which equals twice the area of the circle (5.216a). The equality (5.216b) ≡ (5.214b) agrees with the curl loop theorem (5.210b).

5.7.8 Divergence and Curl Integral Theorems in the Plane

Consider (Figure 5.14) a domain D in the plane (5.25a through c; 5.26a through c) whose boundary (5.31a and b) is a closed regular (5.217a) curve $f = 0$, with unit outward normal (5.217b) and tangent vector (5.217c):

$$f(x,y) \in C^1\left(|R^2| \right): \quad \{n_x, n_y\} = -|\nabla f|^{-1} \{\partial_x f, \partial_y f\}, \quad \{s_x, s_y\} = |\nabla f|^{-1} \{-\partial_y f, \partial_x f\}. \qquad (5.217\text{a–c})$$

Using the arc length (5.218a), the unit tangent (outward normal) are given by (5.218b) [(5.218c)]:

$$ds = \left| (dx)^2 + (dy)^2 \right|^{1/2}: \quad \{s_x, s_y\} = \left\{ \frac{dx}{ds}, \frac{dy}{ds} \right\}, \quad \{n_x, n_y\} = \left\{ \frac{dy}{ds}, -\frac{dx}{ds} \right\}. \qquad (5.218\text{a–c})$$

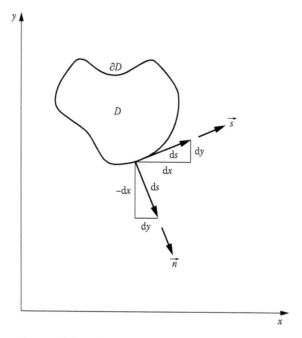

FIGURE 5.14 The Gauss, Ostrogradski, or divergence (Stokes or loop curl) theorem [Figure 5.11 (Figure 5.13)] can be applied to a domain in the plane (Figure 5.14) with a regular loop as boundary, that is a closed curve with continuous tangent, and hence also continuous normal, taken outward.

The vectors (5.217c) ≡ (5.218b) and (5.217b) ≡ (5.218c) have unit modulus (5.219a and b) and are orthogonal (5.219c):

$$\left|\vec{n}\right|^2 = \left(n_x\right)^2 + \left(n_y\right)^2 = 1 = \left(s_x\right)^2 + \left(s_y\right)^2 = \left|\vec{s}\right|^2, \quad \vec{n}\cdot\vec{s} = n_x s_x + n_y s_y = 0. \tag{5.219a–c}$$

Considering a *continuously differentiable (5.220b) vector in a plane (5.220a) domain in (5.25a through c) with closed regular boundary, the* **plane divergence (curl) theorems** *(5.220c) [(5.220d)] hold*:

$$\vec{A} = \vec{e}_x U + \vec{e}_y V; \quad U, V\left(x,y\right) \in C^1\left(\left|R^2\right|\right): \tag{5.220a,b}$$

$$\int\int_{D_2}\left(\partial_x U + \partial_y V\right)\mathrm{d}x\mathrm{d}y = \int_{\partial D_2 = D_1}\left(U\mathrm{d}y - V\mathrm{d}x\right), \tag{5.220c}$$

$$\int\int_{D_2}\left(\partial_x V - \partial_y U\right)\mathrm{d}x\mathrm{d}y = \int_{\partial D_2 \equiv D_1}\left(U\mathrm{d}x + V\mathrm{d}y\right). \tag{5.220d}$$

The proof of (5.220c) [(5.220d)] follows from the divergence (curl) theorem (5.163c) [(5.210b)] applied to the plane where (1) in the integrand of the l.h.s. of (5.220c) [(5.220d)] appears the two-dimensional divergence (5.153c) [curl (5.177c)] of the continuously differentiable (5.220b) vector (5.220a); and (2) in the integrand on the r.h.s. of (5.220c) [(5.220d)] appears the inner product of the vector (5.220a) by the unit normal (5.218d) [tangent (5.218c)] vector integrated along the loop leading to (5.221a) [(5.221b)]:

$$\left(\vec{A}\cdot\vec{n}\right)\mathrm{d}s = \left(n_x U + n_y V\right)\mathrm{d}s = U\mathrm{d}y - V\mathrm{d}x, \tag{5.221a}$$

$$\left(\vec{A}\cdot\vec{s}\right)\mathrm{d}s = \vec{A}\cdot\mathrm{d}\vec{x} = U\mathrm{d}x + V\mathrm{d}y. \tag{5.221b}$$

One example each of (5.220c and d) is given next, using polar coordinates (5.222a and b) ≡ (I.1.13a and b):

$$\left\{x,y\right\} = r\left\{\cos\varphi, \sin\varphi\right\}, \quad \mathrm{d}S = r\mathrm{d}r\mathrm{d}\varphi, \quad \mathrm{d}s = \left|\left(\mathrm{d}r\right)^2 + r^2\left(\mathrm{d}\varphi\right)^2\right|^{1/2}, \tag{5.222a–d}$$

with area element (5.222c) and arc length (5.222d).

The plane vector (5.220a) with components (5.223a and b) has divergence (5.223c):

$$U = y^2 x, V = x^2 y: \quad \nabla\cdot\vec{A} = \partial_x U + \partial_y V = x^2 + y^2 = r^2, \tag{5.223a–c}$$

and leads to the integrals (5.224a) [(5.224b)] on the l.h.s. (r.h.s.) of (5.220c) for a circle of radius *a* and center at the origin:

$$\int\int_{x^2+y^2\le a^2}\left(\partial_x U + \partial_y V\right)\mathrm{d}x\mathrm{d}y = \int_0^{2\pi}\mathrm{d}\varphi\int_0^a r^2 r\mathrm{d}r = \frac{\pi a^4}{2}, \tag{5.224a}$$

$$\int\limits_{x^2+y^2=a^2}(Udy-Vdx)=\int\limits_{r=a}r^4\Big[\cos\varphi\sin^2\varphi\,d(\sin\varphi)-\sin\varphi\cos^2\varphi\,d(\cos\varphi)\Big]$$

$$=2a^4\int\limits_0^{2\pi}\cos^2\varphi\sin^2\varphi\,d\varphi=\frac{a^4}{2}\int\limits_0^a\sin^2(2\varphi)d\varphi=\frac{a^4}{4}\int\limits_0^{2\pi}\Big[1-\cos(4\varphi)\Big]d\varphi=\frac{\pi a^4}{2}.\quad(5.224b)$$

The equality (5.224a) ≡ (5.224b) confirms the plane divergence theorem (5.220c). Concerning the plane curl theorem (5.220d), it is applied to the plane vector (5.225a and b) whose curl has components (5.225c) normal to the plane:

$$U=-x^2y,\,V=y^2x:\ \big(\nabla\wedge\vec{A}\big)\cdot\vec{e}_z=\partial_xV-\partial_yU=x^2+y^2=r^2.\qquad(5.225a\text{–}c)$$

Considering again the circle of radius a and center at the origin, the integrals on the l.h.s. (r.h.s.) of (5.220d) coincide with (5.224a) ≡ (5.226a) [(5.224b) ≡ (5.226b)]:

$$\iint\limits_{x^2+y^2\le a^2}\big(\partial_xV-\partial_yU\big)dxdy=\int\limits_0^{2\pi}d\varphi\int\limits_0^a r^2r\,dr=\frac{\pi a^4}{2},\qquad(5.226a)$$

$$\int\limits_{x^2+y^2=a^2}(Udy-Vdx)=\int\limits_{r=a}r^4\Big[-\cos^2\varphi\sin\varphi\,d(\cos\varphi)+\cos\varphi\sin^2\varphi\,d(\sin\varphi)\Big]$$

$$=2a^4\int\limits_0^{a\pi}\cos^2\varphi\sin^2\varphi\,d\varphi=\frac{\pi a^4}{2},\qquad(5.226b)$$

and confirm (5.226a) ≡ (5.226b) the plane curl theorem (5.220d). The integrals (5.224a and b) ≡ (5.226a and b) coincide because the divergence (curl) of the vector (5.223a and b) [(5.225a and b)] coincide (5.223c) ≡ (5.225c), as a consequence of the interchange of components $(U,V)\leftrightarrow(-V,U)$.

5.7.9 First and Second Green Identities

The vectors (5.227c) [(5.228c)] are continuously differentiable if the functions Ψ/Φ have continuous derivatives of first/second order respectively (5.227a and b) [both of second order (5.228a and b)]:

$$\Psi\in C^1\big(\|R^N\big),\,\Phi\in C^2\big(\|R^N\big):\ \ \vec{A}=\nabla\cdot\big(\Psi\nabla\Phi\big)=\Psi\nabla^2\Phi+\nabla\Psi\cdot\nabla\Phi,\qquad(5.227a\text{–}c)$$

$$\Psi,\Phi\in C^2\big(\|R^N\big):\ \ \vec{A}=\nabla\cdot\big(\Psi\nabla\Phi-\Phi\nabla\Psi\big)=\Psi\nabla^2\Phi-\Phi\nabla^2\Psi.\qquad(5.228a\text{–}c)$$

The identity (5.228c) follows from (5.227c) interchanging (Φ,Ψ) and subtracting, and both involve the Laplacian

$$\nabla\cdot(\nabla\Phi)=\partial_j\big(\partial_j\Phi\big)=\frac{\partial^2\Phi}{\partial x_j\partial x_j}\equiv\sum_{J=1}^N\frac{\partial^2\Phi}{\partial x_j^2}=\nabla^2\Phi.\qquad(5.229)$$

Substituting (5.227a through c) [(5.228a through c)] in the divergence theorem (5.163a through c) leads to (5.230a through c) [(5.231a through c)]:

$$\Psi \in C^1\left(\left|R^N\right|\right), \Phi \in C^2\left(\left|R^N\right|\right): \quad \int_{D_N}\left(\Psi\nabla^2\Phi + \nabla\Psi\cdot\nabla\Phi\right)dV = \int_{\partial D_N \equiv D_{N-1}} \Psi\frac{\partial\Phi}{\partial N}dS, \quad (5.230a\text{–}c)$$

$$\Psi, \Phi \in C^2\left(\left|R^N\right|\right): \quad \int_{D_N}\left(\Psi\nabla^2\Phi - \Phi\nabla^2\Psi\right)dV = \int_{\partial D_N \equiv D_{N-1}}\left(\Psi\frac{\partial\Phi}{\partial N} - \Phi\frac{\partial\Psi}{\partial N}\right)dS, \quad (5.231a\text{–}c)$$

the **first (second) Green identity** *(5.230c) [(5.231c)] in a domain with a closed regular boundary, where the function Φ has continuous second-order derivatives (5.230b) [(5.231b)] and the function Ψ has a continuous first (5.230a) [second (5.231a)] derivative.*

As an example are considered the functions (5.232a) [(5.233a)] with gradients (5.232b) [(5.233b)] and Laplacian (5.232c) [(5.233c)]:

$$\Phi = x^2 + y^2 + z^2: \quad \nabla\Phi = 2\{x, y, z\}, \quad \nabla^2\Phi = 6, \quad (5.232a\text{–}c)$$

$$\Psi = z^2: \quad \nabla\Psi = \{0, 0, 2z\}, \quad \nabla^2\Psi = 2. \quad (5.233a\text{–}c)$$

The integrals on the l.h.s./r.h.s. of the first (5.230c) [second (5.231c)] Green identity are evaluated for the sphere of radius a and center at the origin (5.214a) $\equiv$ (5.234a), using spherical coordinates (5.212a through c) with area (volume) elements (5.213d) [(5.234b)]:

$$x^2 + y^2 + z^2 = a^2, \quad dV = dR\,dS = R^2\sin\theta\,dR\,d\theta d\varphi. \quad (5.234a,b)$$

The symmetry properties (5.175) for the surface (5.173a) [volume (5.173b)] integrals leads to (5.235a) [(5.235b)]:

$$\iiint_{x^2+y^2+z^2 \leq a^2} z^2 dx\,dy\,dz = \frac{1}{3}\int_{R\leq a} dV = \frac{1}{3}\int_0^{2\pi} d\varphi \int_0^\theta \sin\theta\,d\theta \int_0^\theta R^4 dR$$

$$= \frac{2\pi}{3}\left[-\cos\theta\right]_0^\pi \frac{a^5}{5} = \frac{4\pi a^5}{15}, \quad (5.235a)$$

$$\iiint_{x^2+y^2+z^2 = a^2} z^2 dx\,dy = \frac{1}{3}\int_{R=a} R^2 dS = \frac{a^2}{3}S = \frac{4\pi a^4}{3}, \quad (5.235b)$$

where the integration (5.235b) reassembles (5.176).

Substituting (5.232b and c; 5.233a and b) on the l.h.s. (r.h.s.) of (5.230c) leads to (5.236a) [(5.236b)]:

$$\iiint_{x^2+y^2+z^2 \leq a^2}\left(\Psi\nabla^2\Phi + \nabla\Psi\cdot\nabla\Phi\right)dx\,dy\,dz = 10\int_{R\leq a} z^2 dV = \frac{8\pi a^5}{3}, \quad (5.236a)$$

$$\iint_{x^2+y^2+z^2=a^2} \Psi\frac{\partial\Phi}{\partial N}dxdy = \int_{R=a} \Psi(\vec{e}_R\cdot\nabla\Phi)dS$$

$$= 2\int_{R=a} z^2\frac{x^2+y^2+z^2}{R}dS = 2a\int_{R=a} z^2dS = \frac{8\pi a^5}{3}, \qquad (5.236b)$$

where (1) the unit outward normal to the sphere is the unit radial vector (5.164a); and (2) (5.235a) [(5.235b)] was used in (5.236a) [(5.236b)]. The coincidence of (5.236a) ≡ (5.236b) confirms the first Green identity (5.230c) for the functions (5.232a; 5.233a).

Using the same (5.233a through c) [a different (5.237a through c)] function,

$$\Phi = z^4: \quad \nabla\Phi = \{0,0,4z^3\}, \quad \nabla^2\Phi = 12z^2, \qquad (5.237a\text{--}c)$$

the l.h.s. (r.h.s.) of (5.231c) is evaluated by (5.238a) [(5.238b)] over the volume (surface) of the sphere with radius a and center at the origin:

$$\iiint_{x^2+y^2+z^2\le a^2} (\Psi\nabla^2\Phi - \Phi\nabla^2\Psi)dxdydz = 10\int_{R\le a} z^4 dV = 10\int_0^{2\pi}d\varphi\int_0^\pi\cos^4\theta\sin\theta\,d\theta\int_0^a R^6 dR$$

$$= 20\pi\left[-\frac{\cos^5\theta}{5}\right]_0^\pi\frac{a^7}{7} = \frac{8\pi a^7}{7}, \qquad (5.238a)$$

$$\iint_{x^2+y^2+z^2=a^2}\left(\Psi\frac{\partial\Phi}{\partial N} - \Phi\frac{\partial\Psi}{\partial N}\right)dxdy = \int_{R\le a}\left[\Psi(\vec{e}_R\cdot\nabla\Phi) - \Phi(\vec{e}_R\cdot\nabla\Psi)\right]dS$$

$$= 2\int_{R=a}\frac{z^6}{R}dS = 2a^5\int_0^{2\pi}d\varphi\int_0^\pi\cos^6\theta\sin\theta\,d\theta = 4\pi a^5\left[-\frac{\cos^7\theta}{7}\right]_0^\pi = \frac{8\pi a^5}{7}. \qquad (5.238b)$$

The coincidence of (5.238a) ≡ (5.238b) confirms the second Green identity (5.231c) for the functions (5.237a; 5.233a). The relation between the nine integral theorems is indicated in Diagram 5.2 and Table 5.2: (1) the volume divergence (or Gauss or Ostrogradski) theorem (Subsection 5.7.1) has the volume gradient (curl) theorems as analogues [Subsection 5.7.2 (Subsection 5.7.3)]; (2) the bivector curl theorem has a vector form (Subsection 5.7.4), namely space curl theorem in three-dimensional space (Subsection 5.7.5); (3) the curl loop (or Stokes) theorem is distinct (Subsection 5.7.7) in that it involves a regular surface supported on a loop (Subsection 5.7.6); (4) both the volume divergence (surface curl) or Gauss (Stokes) theorems have a two-dimensional particular case, namely the plane divergence (curl) theorem (Subsection 5.7.8); and (5) the volume divergence theorem leads to two Green theorems in an N-dimensional space (Subsection 5.7.9). Of the nine integral theorems, two apply to the plane (three-dimensional space) and five apply in any dimension. These nine theorems are particular cases (Note 9.29) of two theorems for multivectors in multidimensional spaces (Notes 9.25 through 9.28).

5.8 Scalar/Vector Potentials for Irrotational/Solenoidal Fields

A vector field with continuous first-order derivatives is irrotational (solenoidal) if its curl (divergence) is zero; in this case, it equals the gradient (curl) of a scalar (vector) potential that has continuous second-order derivatives [Subsection 5.8.1 (Subsection 5.8.2)]. This theorem has been applied to several vector fields,

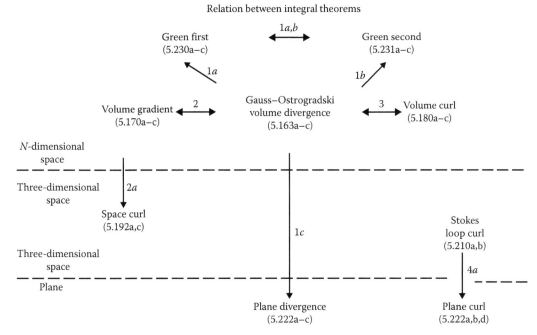

Relation between integral theorems

DIAGRAM 5.2   The two main integral theorems are the (1–3) Gauss–Ostrogradski [(4) Stokes] theorem for the divergence/gradient/curl (curl) in an *N*-dimensional domain with a closed regular boundary (a regular surface supported on a closed curve or loop in a three-dimensional space). From the divergence theorem follow (5, 6) the two Green theorems in a *N*-dimensional space; the curl theorem takes (7) a vector form in a three-dimensional space. The divergence (Stokes) theorems lead in the plane to relations (8, 9) for the integral over a domain with a closed regular boundary. The ensemble is 3 + 1 = 4 four basic theorems (1–4) with five corollaries (5–9).

TABLE 5.2   Nine Integral Theorems

Name	Domain	Closed Regular Boundary	Equation
Volume divergence (Gauss, Ostrogradski)	N-dimensions	$(N-1)$-dimensions	(5.163a through c)
Volume gradient	N-dimensions	$(N-1)$-dimensions	(5.170a through c)
Volume curl	N-dimensions	$(N-1)$-dimensions	(5.180a through c)
Loop curl (Stokes)	$N=2$: surface	$N-1=1$: loop	(5.210a through c)
Vector curl	$N=3$: domain	$N-1=2$: surface	(5.192a through c)
First Green	N-dimensions	$(N-1)$-dimensions	(5.230a through c)
Second Green	N-dimensions	$(N-1)$-dimensions	(5.231a through c)
Plane divergence	$N=2$: plane	$N-1=1$: loop	(5.220a through c)
Plane curl	$N=2$: plane	$N-1=1$: loop	(5.220a and b,d)

Note: The four basic integral theorems of Gauss–Ostrogradski (Stokes) or volume divergence/gradient/curl (loop curl) theorems lead to five corollaries as indicated in Diagram 5.2.

such as the velocity in an irrotational (incompressible) flow (Chapters I.12, 14, 16, 28, 34, 36, 38; II.2, 8; III.2) and the electromagnetostatic field (Chapters I.24, 26, 34, 36; III.8). A vector field that has continuous first-order derivatives and is neither irrotational nor solenoidal has both scalar and vector potentials (Subsections 5.8.4 and 5.8.5). These theorems have extensions to multivectors in higher-dimensional spaces (Notes 9.3 through 9.52). The vector potential reduces to a scalar field function in the case of a plane (Subsection 5.8.3) [axisymmetric (Section 6.2)] field. The proof of the existence of a scalar (vector) potential of an irrotational (solenoidal) vector field [Subsection 5.8.1 (Notes 9.31–9.33)] is fairly simple (less elementary). The continuously differentiable vector fields may be classified into four combinations (Subsection 5.8.5) of irrotational (or not) and solenoidal or not. An arbitrary continuously differentiable vector field can be decomposed into the sum of an irrotational (and a solenoidal) part (Subsection 5.8.8), specified by the scalar (vector) potential (Subsection 5.8.7) as the solution of a scalar Poisson (vector double curl) equation (Subsection 5.8.6) forced by the divergence (minus the curl) of the vector field.

5.8.1 Scalar Potential for an Irrotational Field

A differentiable vector field (5.239a) is **irrotational (solenoidal)** if its curl (divergence) is zero (5.239b) [(5.239c)]:

$$U_n(x_m) \in \mathcal{D}(|R^n): \quad \partial_n U_m = \partial_m U_n, \quad \partial_n U_n = 0. \tag{5.239a–c}$$

This leads (5.239a and b) to the **theorem of irrotational fields**: *a continuously differentiable vector (5.240a) in N-dimensional space is irrotational (5.240b) $\equiv$ (5.239b) if and only if it equals the gradient (5.240d) of a continuously twice differentiable **scalar potential** (5.240c)*

$$U_n(x_m) \in C^1(|R^N): \quad \partial_n U_m = \partial_m U_n \Leftrightarrow \Phi(x_n) \in C^2(|R^N): \quad U_m = \partial_m \Phi, \tag{5.240a–d}$$

*which is unique to within an arbitrary added constant in the **exact differential** (5.241)*:

$$d(\Phi + \text{const}) = d\Phi = dx_m(\partial_m \Phi) = \nabla\Phi \cdot d\vec{x} = \vec{U} \cdot d\vec{x}. \tag{5.241}$$

*An arbitrary displacement on an **equipotential surface** (5.242a) is orthogonal to the vector field (5.242b)*:

$$\Phi = \text{const}: \quad 0 = d\Phi = \nabla\Phi \cdot d\vec{x} = \vec{U} \cdot d\vec{x}_\Phi. \tag{5.242a,b}$$

Thus, a necessary and sufficient condition for the existence of a family of hypersurfaces orthogonal to a vector field (Figure 5.15a) is that the vector field be irrotational.

The sufficient condition assumes (5.240c) $\equiv$ (5.243a) and (5.240d), leading to (5.243b):

$$\Phi(x_n) \in C^2(|R^N): \quad \partial_n U_m - \partial_m U_n = \partial_n \partial_m \Phi - \partial_m \partial_n \Phi = 0, \tag{5.243a,b}$$

which proves (5.240b). The condition (5.243a and b) states that *the curl of the gradient of a scalar with continuous second-order derivatives (5.243a) is zero (5.243b)*. The necessary condition assumes (5.240a) and (5.240b) $\equiv$ (5.244a), leading by the Stokes' theorem (5.210b) to (5.244b):

$$\nabla \wedge \vec{U} = 0: \quad 0 = \int_S (\nabla \wedge \vec{U}) \cdot \vec{N} \, dS = \int_L \vec{U} \cdot d\vec{x} = \int_L d\Phi, \tag{5.244a,b}$$

showing that *the circulation of a continuously differentiable irrotational vector (5.244a) around a closed loop is always zero (5.243b)*; this implies that the integrand in (5.243b) is the exact differential of a scalar potential (5.241) proving that (1) the irrotational vector field (5.244a) is the gradient of a scalar potential (5.240d); and (2) the potential is determined to within an added constant. The sufficient condition was proved (5.243a and b) in an N-dimensional space; the necessary condition (5.244a and b) used the Stokes' theorem (5.210a and b) in a three-dimensional space, and the proof

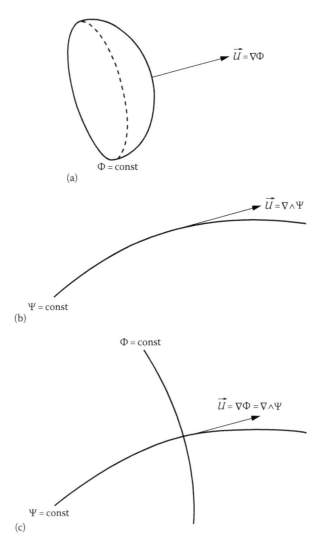

FIGURE 5.15 A continuously differentiable irrotational vector field in N dimensions is the gradient of a scalar potential, that has continuous second-order derivatives; the constancy of the potential specifies a family of regular hypersurfaces of dimension $N - 1$ orthogonal to the vector field (a). In the case $N = 2$ of a plane (c) a continuously differentiable irrotational vector field is orthogonal to the family equipotential curves $N - 1 = 1$. A plane continuously differentiable solenoidal vector field is the curl of a field function that has continuous second-order derivatives; the constancy of the field function specifies a family of field lines tangent to the vector field (b). If the plane continuously differentiable vector field is both irrotational (and solenoidal) it is orthogonal (tangent) to the equipotentials (field lines), so that the two families of curves are mutually orthogonal (c).

can be extended (Note 9.30) to any dimension N. Given a scalar potential, the irrotational vector field is obtained by differentiation (5.240d); given an irrotational vector field, the scalar potential is obtained integrating the exact differential (5.241). The simplest example is *a constant vector field (5.245a) that is irrotational (5.245b) and is the gradient (5.245c) of a potential obtained by its inner product with the position vector (5.245d)*:

$$\vec{U} = \text{const}: \quad \nabla \wedge \vec{U} = 0, \quad d\Phi = \vec{U} \cdot d\vec{x} = d(\vec{U} \cdot \vec{x}), \quad \Phi = \vec{U} \cdot \vec{x}. \qquad (5.245a\text{–}d)$$

An example of a nonconstant vector field is (5.246a) that is irrotational (5.246b) and has potential (5.246c):

$$\vec{U} = \vec{e}_x y + \vec{e}_y x, \partial_y V_x = 1 = \partial_x U_y: \quad d\Phi = y dx + x dy = d(xy), \quad \Phi = xy + \text{const.} \quad (5.246a\text{--}d)$$

A constant can be added (5.246d), and the gradient leads back to (5.246a).

5.8.2 Vector Potential of a Solenoidal Field

There is an analogue **theorem on solenoidal fields:** *a continuously differentiable vector field in three dimensions (5.247a) is solenoidal (5.239c) ≡ (5.247b) if and only if it is the curl (5.247d) of a twice continuously differentiable (5.247c)* **vector potential**:

$$U_n(x_m) \in C^1(|R^3): \quad \nabla \cdot \vec{U} = 0 \Leftrightarrow A_\ell(x_n) \in C^2(|R^3): \quad \vec{U} = \nabla \wedge \vec{A}. \quad (5.247a\text{--}d)$$

The vector potential is not uniquely defined by (5.247d), and a **gauge condition** *specifying its divergence (5.248a) can be imposed:*

$$\nabla \cdot \vec{A} = 0; \quad \vec{A}' = \vec{A} + \nabla\chi, \quad \nabla^2\chi = 0. \quad (5.248a\text{--}c)$$

Even with the gauge condition, the vector potential is defined to within the gradient (5.248b) of a harmonic function (5.148c). The theorem is limited to a three-dimensional space in this form because only in that case the curl is an axial vector (5.185a and b). It also holds in two dimensions when the vector potential reduces to a scalar field function (Subsection 5.8.3); the stream function also exists for three-dimensional axisymmetric fields (Section 6.2). There is an extension to multivectors in a multidimensional space (Notes 9.30 through 9.34). The following is the proof of the statements (5.247a through d) and (5.248a through c) in a three-dimensional space.

The sufficient condition assumes (5.247c) ≡ (5.249a) and (5.247d), implying (5.247a) and (5.249b):

$$A_\ell(x_n) \in C^2(|R^3): \quad \nabla \cdot \vec{U} = \partial_\ell A_\ell = \partial_\ell(e_{\ell mn}\partial_m A_n) = e_{\ell mn}\partial_\ell\partial_n A_n = 0, \quad (5.249a,b)$$

using (5.208a through d) and proving (5.247c). The condition (5.249) states that *the divergence of the curl of a vector with continuous second-order derivatives (5.249a) is zero (5.249b).* The sufficient condition, that is, the inverse implication from (5.247a and b) to (5.247c and d), is proved subsequently (Note 9.34) as a particular case of a theorem on multivectors in a multidimensional space (Notes 9.31–9.33). The vector field (5.247d) is unchanged adding to the vector potential the gradient of a scalar **gauge potential** (5.250b):

$$\nabla \wedge (\nabla\chi) = 0: \quad \nabla \wedge \vec{A}' = \nabla \wedge (\vec{A} + \nabla\chi) = \nabla \wedge \vec{A} + \nabla \wedge (\nabla\chi) = \nabla \wedge \vec{A}, \quad (5.250a,b)$$

where the property (5.243a and b) ≡ (5.250a) that the curl of the gradient is zero was used; the gauge condition (5.248a) is met by the vector potential (5.248b) if

$$0 = \nabla \cdot \vec{A}' = \nabla \cdot (\vec{A} + \nabla\chi) = \nabla \cdot \vec{A} + \nabla \cdot (\nabla\chi) = \nabla^2\chi, \quad (5.251)$$

the Laplacian of the potential χ is zero (5.218), that is, the potential χ is a harmonic function (5.251) $\equiv$ (5.248c). This completes the proof of the theorem on solenoidal fields (5.247a through d; 5.248a through c) apart from the sufficient condition, whose proof is deferred (to Notes 9.31–9.34).

The case of rigid body rotation (5.187a through c; 5.189a and b) shows that *a constant vector (5.189a)* $\equiv$ *(5.244a)* $\equiv$ *(5.252a) that is irrotational (5.252b) is (5.189b)* $\equiv$ *(5.252c and d) the curl of a vector potential (5.252c),*

$$\vec{U} = \text{const}: \quad \nabla \wedge \vec{U} = 0, \quad \vec{U} = \nabla \wedge \vec{A}, \quad \vec{A} = \frac{1}{2}\vec{U} \wedge \vec{x}, \qquad (5.252a\text{–}d)$$

specified by one-half of its outer product by the position vector (5.252d). In general, the determination of the vector potential of a solenoidal vector field is a nonelementary **Pfaff problem** in the theory of partial differential equations concerning the solution of the simultaneous system

$$U_x = \partial_y A_z - \partial_z A_y, \quad U_y = \partial_z A_x - \partial_x A_z, \quad U_z = \partial_x A_y - \partial_y A_x, \qquad (5.253a\text{–}c)$$

with $\vec{U}$ given and $\vec{A}$ to be determined in the three-dimensional case. In the two-dimensional case (Subsection 5.8.3), the problem simplifies to an exact differential, as was the case for the scalar potential of an irrotational vector field in any dimension (Subsection 5.8.1). The exact differential also exists for a three-dimensional axisymmetric solenoidal vector field because, as in the plane case, the vector potential reduces to a scalar field function (Section 6.2).

5.8.3 Two-Dimensional Curl and Field Function

In analogy with the three-dimensional permutation symbol with three indices (5.182a through c), in the plane, the permutation symbol has two indices (5.254a), and its components are +1(−1) for even (odd) permutations of (1, 2) and zero otherwise:

$$\{e_{11}, e_{12}, e_{21}, e_{22}\} = \{0, +1, -1, 0\}, \quad e_{ij} = \begin{bmatrix} 0 & +1 \\ -1 & 0 \end{bmatrix}, \qquad (5.254a,b)$$

corresponding to the unit two-dimensional skew-symmetric matrix (5.254b). The corresponding (5.254a) [(5.254b)] two-dimensional curl is given alternatively by (5.255a through c) [$\equiv$(5.256a and b)]:

$$U_\ell = \left(\nabla \wedge \Psi\right)_\ell = e_{\ell m} \partial_m \Psi: \quad U_x = \partial_y \Psi, \quad U_y = -\partial_x \Psi, \qquad (5.255a\text{–}c)$$

$$\vec{U} = \begin{vmatrix} \vec{e}_x & \vec{e}_y \\ \partial_x & \partial_y \end{vmatrix} \Psi = \vec{e}_y \partial_y \Psi - \vec{e}_y \partial_x \Psi. \qquad (5.256a,b)$$

Thus, *a continuously differentiable two-dimensional vector field (5.257a) is solenoidal (5.257b) iff it is the curl (5.257d) of a twice continuously differentiable field function (5.257c):*

$$\vec{U} \in C^1\left(|R^2\right): \quad 0 = \nabla \cdot \vec{U} = \partial_x U_x + \partial_y U_y \Leftrightarrow \Psi \in C^2\left(|R^2\right): \quad \{U_x, U_y\} = \{\partial_y \Psi, -\partial_x \Psi\}. \qquad (5.257a\text{–}d)$$

If the solenoidal vector field lies in the (x, y)-plane (5.258b), the vector potential is the field function orthogonal to the plane (5.258a):

$$\vec{A} = \vec{e}_z \Psi : \quad \nabla \wedge \vec{A} = \begin{vmatrix} \vec{e}_x & \vec{e}_y & \vec{e}_z \\ \dfrac{\partial}{\partial x} & \dfrac{\partial}{\partial y} & \dfrac{\partial}{\partial z} \\ 0 & 0 & \Psi \end{vmatrix} = \vec{e}_x \partial_y \Psi - \vec{e}_y \partial_x \Psi = \vec{U}, \qquad (5.258a,b)$$

as follows from the curl (5.258b) ≡ (5.257d). The field function is specified by the exact differential

$$d(\Psi + \text{const}) = d\Psi = (\partial_x \Psi) dx + (\partial_y \Psi) dy = U_x dy - U_y dx, \qquad (5.259)$$

*to within an added arbitrary constant. The vector field is tangent (5.260c) to the **field lines**, that is, the curves specified by a constant field function (5.260a and b):*

$$\Psi = \text{const}: \quad 0 = d\Psi = U_x dy - U_y dx \Leftrightarrow \left(\frac{dy}{dx} \right)_\Psi = \frac{U_y}{U_x}. \qquad (5.260a\text{--}c)$$

*Thus, a necessary and sufficient condition that a plane vector field has a tangent family of curves or **paths** (Figure 5.15b) is that it has zero divergence.*

The proof of the sufficient condition from (5.257c and d) to (5.257a and b) is immediate:

$$\{U_x, U_y\} = \{\partial_y \Psi, -\partial_z \Psi\} \Rightarrow \nabla \cdot \vec{U} = \partial_x U_x + \partial_y U_y = \partial_x(\partial_y \Psi) - \partial_x(\partial_y \Psi) = 0. \qquad (5.261a,b)$$

The proof of the necessary condition assumes (5.257a) and (5.257b) ≡ (5.261a), implying (5.261b) by the plane curl theorem (5.220c):

$$\partial_x U + \partial_y V = 0 \Rightarrow 0 = \int_L (U_x dy - U_y dx) = \int_L d\Psi. \qquad (5.262a,b)$$

It follows that the integrand in (5.262b) is an exact differential (5.259) leading to (5.257c and d). The vector potential of a solenoidal vector field also reduces to a scalar field function in the axisymmetric case (Section 6.2). The most general case of a vector field that is neither irrotational nor solenoidal is considered next (Subsection 5.8.4).

5.8.4 Scalar and Vector Potentials of a General Field

The theorem on the existence of a scalar (vector) potential of an irrotational (solenoidal) vector field [Subsection 5.8.1 (Subsection 5.8.2)] can be generalized to a vector field that is neither irrotational nor solenoidal and has both scalar and vector potentials, as the **theorem of scalar and vector potentials**: *a vector field with continuous first-order derivatives (5.263a) can be decomposed into the sum (5.263b) of an irrotational (5.263c) [and a solenoidal (5.263d)] vector part:*

$$\vec{U} \in C^1(|R^2): \quad \vec{U} = \vec{U}_i + \vec{U}_s, \quad \nabla \wedge \vec{U}_i = 0 = \nabla \cdot \vec{U}_s, \qquad (5.263a\text{--}d)$$

and the irrotational (solenoidal) that has a scalar (5.264a) [vector (5.265a)] potential:

$$\vec{U}_i = \nabla\Phi: \quad \nabla\cdot\vec{U} = \nabla\cdot\vec{U}_i + \nabla\cdot\vec{U}_s = \nabla\cdot\left(\nabla\Phi\right) + \nabla\cdot\left(\nabla\wedge\vec{A}\right) = \nabla^2\Phi, \qquad (5.264\text{a,b})$$

$$\vec{U}_s = \nabla\wedge\vec{A}: \quad \nabla\wedge\vec{U} = \nabla\wedge\vec{U}_i + \nabla\wedge\vec{U}_s = \nabla\wedge\left(\nabla\Phi\right) + \nabla\wedge\left(\nabla\wedge\vec{A}\right) = \nabla\wedge\left(\nabla\wedge\vec{A}\right). \quad (5.265\text{a,b})$$

The scalar (vector) potential satisfies a scalar Poisson (5.264a) [vector double curl (5.265b)] equation forced by the divergence (the curl) of the total vector field. Thus, the total vector field (5.263b) is the sum (5.266a) of the gradients (curl) of a scalar (vector) potential (5.264a) [(5.265a)],

$$\Phi, \vec{A} \in C^2\left(\left|R^3\right|\right): \quad \vec{U} = \nabla\Phi + \nabla\wedge\vec{A}, \qquad (5.266\text{a–c})$$

which have continuous second-order derivatives (5.266a and b). Imposing a gauge condition (5.248a) ≡ *(5.267a) on the vector potential, it satisfies (5.265b) a vector Poisson equation forced by minus the curl of the vector field (5.267b):*

$$\nabla\cdot\vec{A} = 0: \quad -\nabla\wedge\vec{U} = \nabla^2\vec{A} - \nabla\left(\nabla\cdot\vec{A}\right) = \nabla^2\vec{A}. \qquad (5.267\text{a,b})$$

The scalar (vector) potential is unique to within an added constant (5.241) [added gradient (5.248b) of a harmonic function (5.248c)]. The proof the theorem follows from preceding results; for example, in (5.264b) [(5.265b)] the property that the divergence of the curl (5.249a and b) [curl of the gradient (5.243a and b)] is zero was used. The determination of the scalar and vector potentials corresponding to a general continuously differentiable vector field involves the solution of the simultaneous system of partial differential equations

$$U_x = \partial_x\Phi + \partial_y A_z - \partial_z A_y, \quad U_y = \partial_y\Phi + \partial_z A_x - \partial_x A_z, \quad U_z = \partial_z\Phi + \partial_x A_y - \partial_y A_x. \qquad (5.267\text{c–e})$$

The preceding theorem provides the solution of this problem because (1) a continuously differentiable vector field specifies its divergence and curl (Subsection 5.8.6); (2) the latter determine the scalar (vector) potential by forcing a Laplace (double curl) differential equation (Subsection 5.8.7); and (3) the gradients (curl) of the scalar (vector) potential specify the irrotational (solenoidal) part of the vector field (Subsection 5.8.8). These properties (1–3) apply to an arbitrary continuously differentiable vector field that is neither irrotational nor solenoidal and hence also to the particular irrotational or/and solenoidal cases (Subsection 5.8.5).

5.8.5 Classification of Continuously Differentiable Vector Fields

The continuously differentiable vector fields can be classified into four cases. The first case (I) is the general vector field that is neither irrotational nor solenoidal such as a flow (Notes II.4.4 and II.4.5) that is rotational (compressible) with vorticity (5.268a) [dilatation (5.268b)]:

$$\nabla\wedge\vec{v} = \vec{\varpi}, \quad \nabla\cdot\vec{v} = \dot{D}: \quad \vec{v} = \nabla\Phi + \nabla\wedge\vec{A}, \qquad (5.268\text{a–c})$$

so that the vector field, in this case the velocity, is the sum of the gradient of a scalar potential and the curl of a vector potential (5.268c). The second case (II) is a solenoidal vector field (5.269b) that can be represented in terms of a vector potential alone; for example, for an incompressible (5.269b) rotational (5.269a) flow (Sections II.2.2 through II.2.6, II.8.9; III.6.2, III.6.6 through III.6.8), the velocity is the curl of a vector potential (5.269c):

$$\nabla \wedge \vec{v} = \vec{\omega}, \nabla \cdot \vec{v} = 0 : \quad \vec{v} = \nabla \wedge \vec{A}. \tag{5.269a–c}$$

This applies also to the magnetostatic field (Chapter I.26; Sections 6.4 through 6.6). The third case (III) is an irrotational vector field that can be represented in terms of a scalar potential alone, for example, an irrotational (5.270a) compressible (5.270b) flow (Sections I.14.6, I.14.7; II.2.1) for which the velocity is the gradient of a potential (5.270c):

$$\nabla \wedge \vec{v} = 0, \nabla \cdot \vec{v} = D : \quad \vec{v} = \nabla \Phi. \tag{5.270a–c}$$

This applies also to the electrostatic field (Chapter I.24; Sections 6.1 through 6.3), to the gravity field (Chapter I.18; Note 8.3), to steady heat condition (Chapter I.32; Note 8.5), and to the weak transverse deflection of an elastic membrane (Sections II.6.1 through II.6.3; Note 8.6). The simplest case (IV) is a **potential field** that is both irrotational (5.271a) and solenoidal (5.271b) and can be represented either by a scalar or a vector potential, for example, an irrotational (5.271a) incompressible (5.271b) flow for which the velocity is both the gradient of a scalar potential (5.271c) and the curl of a vector potential (5.271d):

$$\nabla \wedge \vec{v} = 0 = \nabla \cdot \vec{v} : \quad \vec{v} = \nabla \Phi = \nabla \wedge \vec{A}. \tag{5.271a–d}$$

This applies to a potential flow (Chapters I.12, 14, 16, 28, 34, 36, 38; II.2, 8, 6), to the electro(magneto)static field outside electric charges (currents), to the gravity field outside masses, and to steady heat conduction outside heat sources or sinks. The case III of the scalar potential applies in any dimension, and the case II of the vector potential applies in three dimensions; hence, cases I and IV also apply in three dimensions. In two dimensions, the vector potential reduces to a stream or field function, for example, for the velocity of a potential flow:

$$\{\partial_x \Phi, \partial_y \Phi\} = \nabla \Phi = \vec{v} = \nabla \wedge \Psi = \{\partial_y \Psi, -\partial_x \Psi\}. \tag{5.272a}$$

The field is orthogonal (tangent) to the equipotentials (field lines) that are (Figure 5.15c) orthogonal families of curves:

$$\nabla \Phi \cdot \nabla \Psi = (\partial_x \Phi)(\partial_x \Psi) + (\partial_y \Phi)(\partial_y \Psi) = 0. \tag{5.272b}$$

The Cauchy–Riemann conditions (5.272a) not only prove the orthogonality relation (5.272b), but also show with a continuously differentiable potential (5.273a) [and stream function (5.273b)],

$$\Phi, \Psi \in C^1(|R^2|): \quad \Phi(x,y) + i\Psi(x,y) = f(x+iy) \in \mathcal{D}(|C), \tag{5.273a–c}$$

that they are the real (imaginary) part of a complex differentiable function (5.273c).

5.8.6 Specification of a Vector Field by Its Curl and Divergence

A continuously differentiable vector field (5.263a) can be determined from the scalar (and vector) potentials (5.266c) obtained as solutions of the scalar Poisson (vector double curl) equation forced by the divergence (5.264b) [the curl (5.265b)]. This is illustrated next is the simplest case of a vector field with constant divergence (5.274a) [and curl (5.274b)] such as a flow with constant dilatation [and vorticity]:

$$\nabla \cdot \vec{U} = D = \text{const}, \nabla \wedge \vec{U} = \vec{E} = \text{const}: \quad \nabla^2 \Phi = D, \quad \nabla \wedge \left(\nabla \wedge \vec{A} \right) = \vec{E}, \qquad \text{(5.274a–d)}$$

leading to the scalar Poisson (5.230c) [vector double curl (5.230d)] equation for the scalar (vector) potential. The scalar Poisson equation (5.274c) ≡ (5.275a) in spherical coordinates,

$$\frac{1}{R^2} \frac{d}{dR} \left(R^2 \frac{d\Phi}{dR} \right) = D, \quad R^2 \frac{d\Phi}{dR} = \frac{1}{3} DR^3 + C_1, \qquad \text{(5.275a,b)}$$

leads to (5.275b), where C_1 is an arbitrary constant of integration. A further integration leads to (5.276a), where C_2 is another arbitrary constant of integration:

$$\Phi(R) = \frac{1}{6} DR^2 - \frac{C_1}{R} + C_2; \quad \nabla^2 \left(\frac{1}{R} \right) = 0. \qquad \text{(5.276a,b)}$$

The constant $C_2 = 0$ may be set to zero since the potential is defined to within an added constant, and the next term is a solution of the Laplace equation (5.276b) corresponding to a point flow source or sink (6.130a). The simplest solution omitting arbitrary constants is *the simplest scalar (vector) potential for an irrotational (solenoidal) vector field with constant divergence (5.274a) [curl (5.274b)] is (5.277a) [(5.277b)]*:

$$\Phi = \frac{D}{6} R^2 = \frac{D}{6} x_\ell x_\ell, \quad \vec{A} = -\frac{\vec{E}}{4} R^2 = -\frac{\vec{E}}{4} x_\ell x_\ell. \qquad \text{(5.277a,b)}$$

The first result (5.277a) has already been proved, and the second (5.277b) is proved next.

The proof uses the **identity matrix** whose diagonal (nondiagonal) elements are unity (zero):

$$i,j = 1,\ldots,N: \quad \delta_{ij} = \frac{\partial x_i}{\partial x_j} = \begin{cases} 1 & \text{if } i = j, \\ 0 & \text{if } i \neq j. \end{cases} \qquad \text{(5.278a–c)}$$

This implies the property (5.279a):

$$\delta_{ij} A_j = A_i: \quad x_\ell \partial_k x_\ell = x_\ell \delta_{k\ell} = x_k, \qquad \text{(5.279a,b)}$$

for example, in (5.279b). It is assumed that the vector potential scales on R^2 like (5.280a) the scalar potential (5.277a) with a vector coefficient $\vec{F}$ to be determined. It can be checked that (5.280a) is a solution, and the constant vector $\vec{F}$ is determined by substitution in (5.274d) ≡ (5.280b):

$$\vec{A} = \vec{F} R^2: \quad \vec{E} = \nabla \wedge \left(\nabla \wedge \vec{A} \right) = \nabla \left(\nabla \cdot \vec{A} \right) - \nabla^2 \vec{A}. \qquad \text{(5.280a,b)}$$

The substitution of (5.280a) on the r.h.s. of (5.280b) leads to

$$\partial_\ell A_\ell = \partial_\ell \left(F_\ell x_k x_k \right) = 2 F_\ell x_k \partial_\ell x_k = 2 F_\ell x_k \delta_{k\ell} = 2 F_\ell x_\ell, \qquad \text{(5.281a)}$$

$$\partial_j\left(\partial_\ell A_\ell\right) = \partial_j\left(2F_\ell x_\ell\right) = 2F_\ell \delta_{j\ell} = 2F_j = 2\vec{F}, \tag{5.281b}$$

$$\nabla^2 A_j = \partial_k \partial_k \left(F_j x_\ell x_\ell\right) = 2F_j \partial_k \left(x_\ell \partial_k x_\ell\right)$$

$$= 2F_j \partial_k \left(x_\ell \delta_{k\ell}\right) = 2F_j \partial_k x_k = 6F_j. \tag{5.281c}$$

Substituting (5.281b and c) ≡ (5.282a and b) in (5.280b) gives (5.282c):

$$\nabla\left(\nabla\cdot\vec{A}\right) = 2\vec{F}, \quad \nabla^2\vec{A} = 6\vec{F}: \quad \vec{E} = \nabla\left(\nabla\cdot\vec{A}\right) - \nabla^2\vec{A} = -4\vec{F}. \tag{5.282a–c}$$

Substituting (5.282c) in (5.280a) gives (5.277b) as the simplest solution of (5.274d). The vector potential (5.277b) for a solenoidal field with constant curl (5.274b) agrees with the stream function (II.2.160a and b) for an incompressible plane flow constant vorticity.

5.8.7 Vector Field with Constant Curl and Divergence

The decomposition of a continuously differentiable vector field into the sum of irrotational (solenoidal) parts (5.263b) is made determining the scalar (vector) potential as a solution of the scalar Poisson (vector double curl) equation forced by the divergence (5.264b) [minus the curl (5.265b)]. If the divergence (curl) is constant (5.274a) [(5.274b)], the scalar (vector) potential is given by (5.277a) [(5.277b)] as can be confirmed (5.283a) [(5.283b)] by substitution back in the scalar Poisson (5.264b) [vector forced double curl (5.265b)] equation:

$$\nabla^2\Phi = \partial_j\partial_j\left(\frac{D}{6}x_\ell x_\ell\right) = \frac{D}{3}\partial_j\left(x_\ell \partial_j x_\ell\right) = \frac{D}{3}\partial_j\left(x_\ell \delta_{j\ell}\right) = \frac{D}{3}\partial_j x_j = D, \tag{5.283a}$$

$$\left[\nabla\wedge\left(\nabla\wedge\vec{A}\right)\right]_k = -\left(\nabla^2\vec{A}\right)_k + \left[\nabla\left(\nabla\cdot\vec{A}\right)\right]_k = -\partial_j\partial_j A_k + \partial_k\partial_j A_j$$

$$= \frac{1}{4}\partial_j\partial_j\left(E_k x_\ell x_\ell\right) - \frac{1}{4}\partial_k\partial_j\left(E_j x_\ell x_\ell\right)$$

$$= \frac{E_k}{2}\partial_j\left(x_\ell \partial_j x_\ell\right) - \frac{E_j}{2}\partial_k\left(x_\ell \partial_j x_\ell\right)$$

$$= \frac{E_k}{2}\partial_j\left(x_\ell \delta_{j\ell}\right) - \frac{E_j}{2}\partial_k x_j$$

$$= \frac{E_k}{2}\partial_j x_j - \frac{E_j}{2}\delta_{kj} = \frac{E_k}{2}(3-1) = E_k. \tag{5.283b}$$

The gradient (5.284a) [curl (5.284b)] of the scalar (5.283a) [vector (5.283b)] potential specifies the irrotational (solenoidal) part of the vector field:

$$\partial_j\Phi = \frac{D}{6}\partial_j\left(x_\ell x_\ell\right) = \frac{D}{3}x_\ell \partial_j x_\ell = \frac{D}{3}x_\ell \delta_{j\ell} = \frac{D}{3}x_j, \tag{5.284a}$$

$$e_{jk\ell}\partial_k A_\ell = -\frac{1}{4}E_\ell e_{jk\ell}\partial_k\left(x_n x_n\right) = -\frac{1}{2}E_\ell e_{jk\ell}x_n \partial_k x_n$$

$$= -\frac{1}{2}E_\ell e_{jk\ell}x_n \delta_{kn} = -\frac{1}{2}e_{jk\ell}x_k E_\ell = \frac{1}{2}e_{jk\ell}E_k x_\ell. \tag{5.284b}$$

The preceding calculations have been made in index notation (5.281a through c; 2.283a and b; 5.284a and b) that is always possible and is most clear; the vector notation is possible in some cases as shown next (Subsection 5.8.8).

5.8.8 Decomposition of a Vector into Irrotational and Solenoidal Parts

The vector expressions corresponding to (5.284a) ≡ (5.285a) [(5.284b) ≡ (5.285b)] are

$$\vec{U}_i = \nabla\Phi = \frac{D}{3}, \quad \vec{U}_s = \nabla \wedge \vec{A} = \frac{1}{2}\vec{E} \wedge \vec{x}, \tag{5.285a,b}$$

which specify the irrotational (solenoidal) part of a vector field with constant divergence (5.274a) [curl (5.274b)]:

$$\vec{U} = \vec{U}_i + \vec{U}_s = \nabla\Phi + \nabla \wedge \vec{A} = \frac{D}{3}\vec{x} + \frac{1}{2}\vec{E} \wedge \vec{x}. \tag{5.286}$$

It has been shown that *a vector field with constant divergence (5.274a) [and curl (5.274b)] is given by the sum (5.286) of an irrotational (5.285a) and solenoidal (5.285b) part that is the gradient (curl) of the scalar (5.277a) [vector (5.277b)] potential*. It can be checked that the divergence (5.287) [curl (5.288)] of the total vector field (5.286)

$$\nabla \cdot \vec{U} = \frac{D}{3}\partial_\ell x_\ell + \frac{1}{2}\partial_j\left(e_{jk\ell}E_k x_\ell\right) = D + \frac{1}{2}e_{jk\ell}E_j\delta_{j\ell} = D, \tag{5.287}$$

$$\nabla \wedge \vec{U} = \frac{D}{3}\nabla \wedge \vec{x} + \frac{1}{2}\nabla \wedge \left(\vec{E} \wedge \vec{x}\right) = \vec{E}, \tag{5.288}$$

coincides with (5.287) ≡ (5.274a) [(5.288) ≡ (5.274b)]. In (5.287) and (5.288) was used (5.289a and b):

$$\delta_{\ell\ell} = 3, \quad \left(\nabla \wedge \vec{x}\right)_j = e_{jk\ell}\partial_k x_\ell = e_{jk\ell}\delta_{k\ell} = 0; \quad \nabla \wedge \left(\vec{E} \wedge \vec{x}\right) = 2\vec{E}, \tag{5.289a–c}$$

and in (5.288) was used (5.189a and b) ≡ (5.289c).

5.9 Cylindrical, Spherical, and Hyperspherical Symmetry

Some examples related to the gradient, divergence, and curl (Sections 5.7 and 5.8) involved the area (volume) of the circle (sphere). These can be extended to higher dimensions by considering the area of the unit circle/sphere/hypersphere that can be evaluated (Subsection 5.9.1) using Gaussian integrals (Section 2.1). The solution of a linear ordinary or partial differential equation with forcing can be obtained as the integral of the product of the forcing function by the influence or Green function, for example, for the weak deflection (bending) (2.59a and b) [(4.166a and b)] of an elastic string (bar). The influence function is the solution forced by a unit impulse and in the case of the Laplace operator can be considered in 1, 2, 3, or N-dimensions. This requires the unit impulse with cylindrical/spherical/hyperspherical symmetry that reduces 2/3/N variables to one variable (Subsection 5.9.2); the case of the unit impulse with hyperspherical symmetry involves the area of the unit hypersphere (Subsection 5.9.1).

5.9.1 Area and Volume of the *N*-Dimensional Hypersphere

Having considered generalized functions with two variables and two dimensions (Sections 5.3 and 5.4) and the extensions to a larger number of variables and/or dimensions (Sections 5.5 and 5.6), a particularly important case in *N*-dimensions is central symmetry; it is convenient to establish some properties of hyperspheres (Subsection 5.9.1) before considering the associated generalized functions (Subsection 5.9.2). The **hypersphere** of radius *R* in an *N*-dimensional space of coordinates $(x_1,\ldots,x_N)$ is the surface of equation (5.290a):

$$R = \left|\sum_{n=1}^{N}(x_n)^2\right|^{1/2} \quad : \quad \Phi(x_1,\ldots,x_N) = \Psi(R), \tag{5.290a,b}$$

and a function $\Phi(x_1,\ldots,x_n)$ is **centrally symmetric** if (5.290b) is constant on hyperspheres $\Psi(R)$. Knowing the **solid angle** σ_N of the unit hypersphere in *N*-dimensional space, the area S_N (volume V_N) of the hypersphere of radius *R* is given by (5.291a) [(5.291b)]:

$$S_N = \sigma_N R^{N-1}, \quad V_N = \int_0^R \sigma_N(\xi)\,d\xi = \sigma_N \int_0^R \xi^{N-1}\,d\xi = N^{-1}\sigma_N R^N. \tag{5.291a,b}$$

A spherically symmetric integrable function (5.290b) ≡ (5.292a) has the integral property (5.292b):

$$\Phi(x_n) \in \mathcal{E}(|R^N): \quad \int_{-\infty}^{+\infty}dx_1\ldots\int_{-\infty}^{+\infty}dx_1\Phi(x_1,\ldots,x_N) = \sigma_N\int_0^{\infty} R^{N-1}\Psi(R)\,dR. \tag{5.292a,b}$$

Substituting in (5.292b) the spherically symmetric Gaussian function

$$\Phi(x_1,\ldots,x_N) \equiv \prod_{n=1}^{N}\exp\left[-(x_n)^2\right] = \exp\left\{-\sum_{n=1}^{N}(x_n)^2\right\} = \exp(-R^2) \equiv \Psi(R), \tag{5.293}$$

the l.h.s. of (5.292b) is a product of *N* identical Gaussian (1.1a) integrals:

$$\left\{\int_{-\infty}^{+\infty}\exp(-x^2)\,dx\right\}^N = \sigma_N\int_0^{\infty} R^{N-1}\exp(-R^2)\,dR. \tag{5.294}$$

Using the basic Gaussian integral (1.1a), the area of the unit hypersphere is given by

$$\sigma_N = \pi^{N/2}\left\{\int_0^{+\infty} x^{N-1}\exp(-x^2)\,dx\right\}^{-1}, \tag{5.295}$$

in terms of a generalized Gaussian integral, involving a power in the integrand.

The integral in (5.295) is a particular case of the Gamma function (Note 1.8) and can also be evaluated by elementary methods similar to Subsection 1.1.5 as shown next. In order to evaluate the generalized Gaussian integral (5.295), a change of variable (5.296a) is performed in (1.1b) leading to (5.296b):

$$x = y\sqrt{a}: \quad I \equiv \int_0^{\infty}\exp(-ay^2)\,dy = \frac{1}{2}\sqrt{\frac{\pi}{a}}. \tag{5.296a,b}$$

The integral is uniformly convergent with regard to the parameter a, and thus (I.13.40) it can be differentiated p times with regard to a:

$$\frac{\partial^p I}{\partial a^p} = \int_0^\infty \left(-y^2\right)^p \exp\left(-ay^2\right) dy = \frac{\sqrt{\pi}}{2}\left(-\frac{1}{2}\right)\left(-\frac{3}{2}\right)\cdots\left(-\frac{1}{2}-p+1\right)a^{-1/2-p}. \tag{5.297}$$

This specifies the Gaussian integral (5.295) in the case (1) of an even power in the integrand:

$$\int_0^\infty y^{2p} \exp\left(-ay^2\right) dy = (2p-1)!!\,2^{-p-1}\sqrt{\pi}\,a^{-p-1/2}, \tag{5.298}$$

using the double factorial. In the case and (2) of an odd power in the integrand, the starting point is the elementary integral:

$$J \equiv \int_0^\infty y\exp\left(-ay^2\right) dy = \left[\frac{\exp\left(-ay^2\right)}{-2a}\right]_0^\infty = \frac{1}{2a}. \tag{5.299}$$

Again, differentiation with regard to the parameter a is permissible:

$$(-)^p \frac{\partial^p J}{\partial a^p} = \int_0^\infty y^{2p+1} \exp\left(-ay^2\right) dy = (-)^p \frac{\partial^p}{\partial a^p}\left(\frac{1}{2a}\right) = \frac{p!}{2}a^{-p-1}. \tag{5.300}$$

Thus has been *generalized the elementary Gaussian integral (5.296b) in the case of a Gaussian integral where the integrand contains a power with even (5.298) [odd (5.300)] exponents.* Substituting (5.300) [(5.298)] with $a = 1$ in (5.295) specifies the area of the unit hypersphere in a space of even (5.301a) [odd (5.301b)] dimension:

$$\sigma_N = \begin{cases} \dfrac{\pi^{p+1}2}{p!} & \text{if } N = 2p+2, \tag{5.301a} \\[3mm] \dfrac{\pi^p 2^{p+1}}{(2p-1)!!} & \text{if } N = 2p+1. \tag{5.301b} \end{cases}$$

The area (5.291a) [volume (5.291b)] of the N-dimensional hypersphere is indicated in Table 5.3 for the lowest four dimensions $2 \leq N \leq 5$. In the plane (space) of $N = 2(N = 3)$ dimensions, (1) the circle (sphere)

TABLE 5.3 Area and Volume of N-Dimensional Sphere

Quantity	Dimension	N	$N = 2$	$N = 3$	$N = 4$	$N = 5$
Meaning	Case	Equation	Circle	Sphere	Hypersphere	
Area with unit radius	σ_N	(5.301a and b)	2π	4π	$2\pi^2$	$\dfrac{8}{3}\pi^2$
Area with radius R	S_N	(5.291a)	$2\pi R$	$4\pi R^2$	$2\pi^2 R^3$	$\dfrac{8}{3}\pi^2 R^4$
Volume with radius R	V_N	(5.291b)	πR^2	$\dfrac{4}{3}\pi R^3$	$\dfrac{1}{2}\pi^2 R^4$	$\dfrac{8}{15}\pi^2 R^5$

Note: The area of the hypersphere of radius unity in an N-dimensional space specifies the area (volume) of the hypersphere of radius R, including for $N = 3$ the sphere; for $N = 2$ the circle the corresponding quantities are the perimeter (area).

has perimeter (area) $2\pi(4\pi)$ for unit radius and $2\pi r(4\pi R^2)$ for radius $r(R)$ and encloses an area πr^2 (volume $4\pi R^3/3$); and (2) in a space of dimension $N = 4$ ($N = 5$), the unit hypersphere has area $2\pi^2(8\pi^2/3)$, and the hypersphere of radius R has area $2\pi^2R^3(8\pi^2R^4/3)$ and volume $\pi^2R^4/2(8\pi^2R^5/15)$.

5.9.2 Impulse with Cylindrical, Spherical, and Hyperspherical Symmetry

The centrally symmetric N-dimensional impulse is zero everywhere (5.302a) except at the origin (5.302b):

$$\delta(R) = \begin{cases} 0 & \text{if } (x_1)^2 + \ldots + (x_N)^2 \equiv R^2 \neq 0, & (5.302a) \\[2mm] \infty & \text{if } (x_1)^2 + \ldots + (x_N)^2 \equiv R^2 = 0, & (5.302b) \end{cases}$$

where it is singular, and it has the integral property (5.303b):

$$\Psi(R) \in C^1(|R): \quad \int_0^\infty \delta(R)\Psi(R)dR = \Psi(0) = \Phi(0,\ldots,0), \qquad (5.303a,b)$$

with regard to centrally symmetric (5.290a and b) continuous functions. Like (5.302a and b), the N-dimensional impulse (5.304a),

$$\delta(\vec{x}) \equiv \delta(x_1)\ldots\delta(x_n) = \begin{cases} 0 & \text{if } (x_1,\ldots,x_N) \neq (0,\ldots,0), & (5.304a) \\[2mm] \infty & \text{if } x_1 = \ldots = x_N = 0, & (5.304b) \end{cases}$$

vanishes everywhere (5.302a) $\equiv$ (5.304a) except at the origin (5.302b) $\equiv$ (5.304b), where it is singular; it has the integral property (5.305b):

$$\Phi(x_1,\ldots,x_N) \in C^1(|R^N): \quad \Phi(0,\ldots,0) = \int_{-\infty}^{+\infty}\ldots\int\delta(x_1)\ldots\delta(x_N)\ \Phi(x_1,\ldots,x_N)dx_1\ldots dx_N$$

$$= \int_0^\infty \delta(x_1)\ldots\delta(x_N)\sigma_N R^{N-1}\Psi(R)dR, \qquad (5.305a,b)$$

with regard to continuous functions (5.305a).

Comparing (5.305b) with (5.303b) leads to *the relation between the centrally symmetric (5.302a and b) and N-dimensional (5.304a and b) impulses*:

$$\delta(R) = \sigma_N R^{N-1}\delta(x_1)\ldots\delta(x_N). \qquad (5.306)$$

Both sides of (5.306) vanish for $R \neq 0$, and for $R = 0$, the l.h.s. is ∞ and the r.h.s. is $0^{N-1} \times \infty^N$. *In the plane (spatial) case, the unit impulse with cylindrical (spherical) symmetry is related to the two- (three-) dimensional impulse by (5.60a and b) [(5.307a and b)]:*

$$R \equiv |x^2 + y^2 + z^2|^{1/2}: \quad \delta(R) = 4\pi R^2\delta(x)\delta(y)\delta(z). \qquad (5.307a,b)$$

The next impulses with hyperspherical symmetry apply in four (5.308a and b) [five (5.309a and b)] dimensions:

$$R = \left| x^2 + y^2 + z^2 + u^2 \right|^{1/2} : \quad \delta(R) = 2\pi^2 R^3 \delta(x)\delta(y)\delta(z)\delta(u), \qquad (5.308\text{a,b})$$

$$R \equiv \left| x^2 + y^2 + z^2 + u^2 + v^2 \right|^{1/2} : \quad \delta(R) = \frac{8}{3}\pi^2 R^4 \delta(x)\delta(y)\delta(z)\delta(u)\delta(v). \qquad (5.309\text{a,b})$$

In an N-dimensional space, the area of the unit hypersphere is given by (5.301a and b), the surface (volume) by (5.291a) [(5.291b)] for a radius R, and the centrally symmetric impulse (5.302a and b) is related to the N-dimensional impulse (5.304a and b) by (5.306).

NOTE 5.1: Geometry of Generalized Functions with Several Variables and Dimensions

The relation between generalized functions and geometry is more significant for larger dimensions and more variables; the unit impulse among other generalized functions (List 5.1; Table 5.4) may serve as an example of the associated geometric interpretations. The generalized function one-dimensional impulse with argument one real variable (a real function of one variable) is singular at the origin (the set of points where the function vanishes); its integral multiplied by a test function evaluates the test function at the origin (sums the values of the test function at the simple zeros of the argument function divided by the slope of the latter). In an N-dimensional space, the generalized function one-dimensional impulse with argument one real variable (one real function of N variables) (1) is singular at the hyperplane [hypersurface(s)] where the variable (function) vanishes; and (2) it reduces the integral over all space of a test function to an integral over the hyperplane [hypersurface(s)], involving the gradient of the argument function that is normal to the hyperplane [hypersurface(s)]. In an N-dimensional space, the generalized function unit impulse with M arguments, with $1 \leq M \leq N$, is singular where all its arguments vanish, that is, at the intersection of M hypersurfaces, corresponding to (1) one hypersurface if $M = 1$; (2) one point if $M = N$; and (3) a subspace of dimension $P = N - M$ for any M, for example, $P = 0$ for a point $M = N$ and $P = N - 1$ for a hypersurface $M = 1$. The subspace may be multiply connected, that is, consist of several elements. The integral over all space when multiplied by a test function then reduces to an integral over the subspace; the latter involves the gradients, representing the volume element of the subspace. If the unit impulse was replaced by the derivative impulse, there would be a derivative of the test function divided by the gradients; the gradients specify the normals, and their derivatives relate to the curvatures. Thus, the generalized functions with the coordinates (functions of the coordinates) as variables relate to Cartesian (differential) geometry. An example is the use of the generalized function line (surface) integral to evaluate line (surface) integrals. The latter appear, for example, in potential theory when calculating the fields due to source distributions (Notes 5.2 through 5.30).

LIST 5.1 Thirty Generalized Functions

    A. Unit jump
        a. One-dimensional
            1. Of a single variable: Section 1.2, Subsections 3.1.1 and 3.1.2
            2. Of an ordinary function of one variable: Subsection 5.1.1
            3. Of an ordinary function of two variables: Subsection 5.2.1
            4. Of an ordinary function of N variables: Subsection 5.5.2

(continued)

LIST 5.1 (continued) Thirty Generalized Functions

    b. Two-dimensional
       5. With distinct variables: Subsection 5.3.1
       6. With arguments two linearly independent functions of two variables: Subsection 5.4.2

B. Unit impulse
    c. One-dimensional
       7. Of a single variable: Section 1.3, Subsections 3.1.3 and 3.1.4
       8. Of a ordinary function of one variable: Subsection 5.1.2
       9. Of an ordinary function of two variables: Subsections 5.2.2 through 5.2.4
      10. Of a ordinary function of N variables: Subsection 5.5.2

    d. Multidimensional
      11. Two-dimensional of distinct variables: Subsection 5.3.2
      12. Two-dimensional with arguments two linearly independent functions of two variables: Subsection 5.4.3
      13. N-dimensional of N distinct variables: Subsection 5.5.3
      14. $(N-1)$-dimensional of $N-1$ linearly independent functions of N variables: Subsection 5.5.4
      15. M-dimensional of M linearly independent functions of $N \geq M$ variables: Subsection 5.5.5

    e. Centrally symmetric
      16. With cylindrical symmetry: Subsection 5.3.2
      17. With spherical symmetry: Subsection 5.9.2
      18. With hyperspherical symmetry: Subsection 5.9.2

C. Derivatives of the unit impulse
    f. One-dimensional
      19. Of a single variable: Sections 1.4, 1.5, and 3.2
      20. Of an ordinary function of one variable: Subsection 5.1.3

    g. Multidimensional
      21. N-dimensional of N distinct variables: Subsection 5.5.3

D. Other generalized functions
    h. One-dimensional
      22. Sign function: Subsection 1.7.2
      23. Modulus function: Subsection 1.7.3
      24. Logarithm: Subsection 3.4.5
      25. Products of powers, signs, and moduli: Section 3.8

    i. Products by derivatives of the unit impulse by
      26. A power: Subsection 3.6.3
      27. By a polynomial: Subsection 3.6.5
      28. By an analytic function: Subsection 3.6.4

    j. Integration
      29. Along hypercurve: Subsection 5.6.1
      30. Over hypersurface: Subsection 5.6.2

TABLE 5.4 Multidimensional Multivariable Generalized Functions

Generalized Function		Unit Jump	Unit Impulse	Derivative of Unit Impulse	kth Derivative of Unit Impulse
Dimension	Variable				
1	One	$H(x)$	$\delta(x)$	$\delta'(x)$	$\delta^{(k)}(x)$
		1.2, 3.1.1 and 3.1.2	1.3, 3.1.3 and 3.1.4	1.4, 3.2	1.5
1	One function of one variable	$H(f(x))$	$\delta(f(x))$	$\delta'(f(x))$	$\delta^{(k)}(f(x))$
		5.1.1	5.1.2	5.1.3	5.1.3
1	One function of two variables	$H(f(x,y))$	$\delta(f(x,y))$	$\delta'(f(x,y))$	$\delta^{(k)}(f(x,y))$
		5.2.1	5.2.2		
1	One function of M variables	$H(f(x_1,\ldots,x_M))$	$\delta(f(x_1,\ldots,x_M))$	$\delta'(f(x_1,\ldots,x_M))$	$\delta^{(k)}(f(x_1,\ldots,x_M))$
		5.5.2	5.5.2		
2	Two	$H(x)H(y)$	$\delta(x)\delta(y)$	$\delta'(x)\delta'(y)$	$\delta^{(k)}(x)\delta^{(\ell)}(y)$
		5.3.1	5.3.2		
2	Two functions of two variables	$H(f(x,y))H(g(x,y))$	$\delta(f(x,y))\delta(g(x,y))$	$\delta'(f(x,y))\delta'(g(x,y))$	$\delta^{(k)}(f(x,y))\delta^{(\ell)}(g(x,y))$
		5.4.1	5.4.2		
N	N variables	$\displaystyle\prod_{n=1}^{N} H(x_n)$	$\displaystyle\prod_{n=1}^{N} \delta(x_n)$	$\displaystyle\prod_{n=1}^{N} \delta'(x_n)$	$\displaystyle\prod_{n=1}^{N} \delta^{(k_n)}(x_n)$
		5.5.3	5.5.3	5.5.3	5.5.3
N	N functions of N variables	$\displaystyle\prod_{n=1}^{N} H(f_n(x_1,\ldots,x_N))$	$\displaystyle\prod_{n=1}^{N} \delta(f_n(x_1,\ldots,x_N))$	$\displaystyle\prod_{n=1}^{N} \delta'(f_n(x_1,\ldots,x_N))$	$\displaystyle\prod_{n=1}^{N} \delta^{(k_n)}(f_n(x_1,\ldots,x_N))$
		5.5.4	5.5.4		
N	M functions of $N \geq M$ variables	$\displaystyle\prod_{m=1}^{M} H(f_m(x_1,\ldots,x_N))$	$\displaystyle\prod_{m=1}^{M} \delta(f_m(x_1,\ldots,x_N))$	$\displaystyle\prod_{m=1}^{M} \delta'(f_m(x_1,\ldots,x_N))$	$\displaystyle\prod_{m=1}^{M} \delta^{(k_m)}(f_m(x_1,\ldots,x_N))$
		5.5.5	5.5.5		
N	One spherically symmetric	$H(R)$	$\delta(R)$	$\delta'(R)$	$\delta^{(k)}(R)$
		5.9.1	5.9.2		

Note: The generalized functions used most often are the unit jump, unit impulse and its derivatives, that may be considered (Diagram 5.1) for: (1) one or several independent variables as argument; and (2) argument one or several functions of several variables; (3) one or several dimensions. The number of dimensions M cannot exceed $M \leq N$ the number of variables. It is indicated the section or subsection where each generalized function is considered explicitly.

NOTE 5.2: Potential Fields due to Source Distributions

The theory of the potential in the classical sense concerns the (1) harmonic functions that are solutions of the Laplace equation; and (2) also the solutions of the Poisson equation representing fields due to source distributions. Both (1) and (2) concern second-order partial differential equations (Laplace and Poisson), and an extension would be fourth-order partial differential equations such as (1) the biharmonic equation applying to the stress function in plane elasticity (Chapter II.4) with constant external forces; (2/3) the scalar (vector) biharmonic equation applying to the stream function (vector potential) in an incompressible, steady, creeping, viscous flow in the plane (Note II.4.11) [in space (Note 6.7)]; and (4/5) in the spatial axisymmetric case, the stream function satisfies a modified Laplace (modified biharmonic) equation for an incompressible, inviscid (viscous creeping) flow [Section 6.2 (Note 6.9)]. The classical second-order potential fields are classified into irrotational (solenoidal) if the curl (divergence) is zero. The irrotational fields include (1) the flow due to sources and sinks (Chapter I.12); (2) the gravity field due to masses (Chapter I.18); (3) the electrostatic field due to electric charges (Chapter I.24); (4) steady heat conduction due to heat sources (Chapter I.32); (5) the linear deflection of an elastic membrane due to transverse loads (Sections II.6.1 through II.6.3). The solenoidal fields include (6) the flow due to vorticity (Chapter II.2); and (7) the magnetostatic field due to electric currents (Chapter I.26). The comparison of irrotational (solenoidal) fields could be made (1, 6) for flows (Chapters I.12, 14, 16, 28, 34, 36, 38; II.2, 8; 6) or for (3, 7) the electro(magneto)static field (Chapter 8); the latter possibility is chosen next (Notes 5.3 through 5.30), considering the potential fields due to source distributions.

NOTE 5.3: Three Methods in the Potential Theory

The potential theory can be approached by three methods: (1) complex functions for plane problems (Chapter I.11) in particular singularities like line multipoles (Chapter I.12); (2) the invariant differential operators gradient, divergence, curl, and Laplacian apply in the plane and space (Chapters 6 and 8) and extend to higher dimensions (Chapter 9); and (3) the generalized functions also apply in any dimension in particular to represent singularities like multipole distributions at isolated points, curves, surfaces, or subspaces (Chapters 1, 3, and 5). The theory of generalized functions was developed in the twentieth century, much later than the potential theory initiated in the seventeenth century, and fully developed by then; nevertheless, the properties of generalized functions are closely related to potential theory so that both alternatives can be used to develop potential theory in three or higher dimensions. The two approaches are demonstrated for (1) incompressible irrotational and rotational flows (Chapter 6) using the differential operators divergence, gradient, curl, Laplacian, and their modifications; and (2) the electro- and magnetostatic fields (Chapter 8) using multidimensional generalized functions. It would be possible to use either method for both applications. The plane potential fields are treated most simply using functions of a complex variable (Books 1 and 2); the potential fields of dimensions higher than three (Chapter 9) lead to multivectors. The combination of the two methods of (1) differential operators and (2) generalized functions is used next (Notes 5.4 through 5.30) to determine the electro (magneto) static field due to simple distributions of electric charges (currents) such as (1) a point electric charge (Notes 5.4 and 5.5) [current (Note 5.6)]; (2) a uniform electric charge (current) distribution on a (flowing along a) straight segment [Notes 5.7 through 5.9 (Notes 5.10 and 5.11)]; (3) the straight segment may be closed into a circular loop [Notes 5.12 and 5.13 (Notes 5.14 through 5.16)]; and (4) the latter may be stretched axially into a circular helix [Notes 5.17 through 5.21 (Notes 5.22 through 5.30)].

NOTE 5.4: Electrostatic Field due to Electric Charges

The electrostatic field $\tilde{E}$ in an isotropic homogeneous medium with dielectric permittivity ε satisfies the Maxwell equations (I.24.1b) $\equiv$ (5.310a) and (I.24.5b) $\equiv$ (5.310b), where q is the density of electric charge:

$$\nabla \wedge \vec{E} = 0, \quad \nabla \cdot \vec{E} = \frac{q}{\varepsilon}; \quad \vec{E} = -\nabla\Phi, \quad -\frac{q}{\varepsilon} = \nabla \cdot (\nabla\Phi) = \nabla^2\Phi. \qquad (5.310a\text{--}d)$$

These lead to the existence of an electrostatic potential (5.310c) satisfying a Poisson equation (5.310d). The second Maxwell equation (5.310b) through the divergence theorem (5.163a through c) implies

$$\varepsilon = \text{const}: \quad e = \int_D q\, dV = \varepsilon \int_D \left(\nabla \cdot \vec{E}\right) dV = \varepsilon \int_D \left(\vec{E} \cdot \vec{N}\right) dS, \qquad (5.311a,b)$$

stating that *the total electric charge e in a domain D with volume element dV equals (5.311b) the product of the constant dielectric permittivity (5.311a) by the flux of the electric field* $\vec{E}$ *across its closed regular* ∂D *boundary with area element dS and unit outward normal* $\vec{N}$. In the case (Figure 5.16) of a point electric charge at the origin, since the electric field by symmetry is radial and uniform over a sphere of radius R and area $4\pi R^2$, from (5.311a and b) follows (5.312a):

$$e = \varepsilon 4\pi R^2 E_R; \quad \vec{E}(R) = \vec{e}_R \frac{e}{4\pi\varepsilon R^2} = -\vec{e}_R \frac{d\Phi}{dR}, \quad \Phi(R) = \frac{e}{4\pi\varepsilon R}. \qquad (5.312a\text{-}c)$$

The radial electric field (5.312b) due to a point electric charge e in a medium with constant dielectric permittivity (5.311a) corresponds to the electrostatic potential (5.312c). In the case of a distribution of electric charges with density (5.313a) in a domain D with volume element $d^3\vec{y}$, *the electrostatic potential (electric field) is given by (5.313c) [(5.313d)], where (5.313b) is the position vector from the source to the observer:*

$$q \equiv \frac{de}{dV}, \ \vec{\ell} \equiv \vec{x} - \vec{y}: \quad \Phi(\vec{x}) = \frac{1}{4\pi\varepsilon}\int_D \frac{q(\vec{y})}{|\vec{x} - \vec{y}|}d^3\vec{y}, \quad \vec{E}(\vec{x}) = \frac{1}{4\pi\varepsilon}\int \frac{\vec{x} - \vec{y}}{|\vec{x} - \vec{y}|^3}q(\vec{y})d^3\vec{y}. \quad (5.313a\text{-}d)$$

The result [5.313c (d)] follows from [5.312c (b)], using the principle of superposition that is valid because the Poisson equation (5.310d) is linear, proving that (5.313c) is its forced solution. The preceding result specifies the influence or Green function for the Laplace equation in three dimensions (Note 5.5).

NOTE 5.5: Influence Function for the Three-Dimensional Laplacian

Setting (5.314a) in (5.310d) [(5.312c)] leads to (5.314b) [(5.314c)]:

$$-\frac{q}{\varepsilon} = 1: \quad \sum_{j=1}^{3} \frac{\partial^2}{\partial x_j^2}\left[G(x_i; y_i)\right] = \delta(x_i - y_i), \qquad (5.314a,b)$$

$$G(x_i; y_i) = -\frac{1}{4\pi|x_i - y_i|} = -\frac{1}{\sigma_3|\vec{x} - \vec{y}|} = G(y_i; x_i) = G(\vec{x} - \vec{y}), \qquad (5.314c\text{-}e)$$

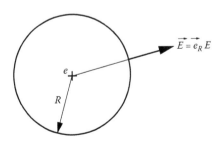

FIGURE 5.16 A point electric charge creates a radial electric field whose flux across a sphere is independent of the radius, that is, the electric charge equals the product of the radial electric field by the area of the sphere (multiplied by dielectric permittivity of the medium).

showing that *the influence or Green function for the Laplace operator in three dimensions (5.314b) equals (5.314c) minus the area of a unit sphere times the distance from observer to source. Since the influence function (5.314c) is symmetric (5.314d), the principle of reciprocity holds, allowing interchange of observer and source position, because the Laplacian is a self-adjoint operator (7.213a through d). The influence function depends only on the relative position of observer and source (5.313b), and hence the solution of the Poisson equation with arbitrary forcing (5.315a),*

$$\frac{\partial^2 \Phi}{\partial x_j^2} = f(\vec{x}), \quad \Phi(x) \equiv \int_{-\infty}^{+\infty} G(\vec{x} - \vec{y}) f(\vec{y}) d^3\vec{y} = f * G(\vec{x}), \tag{5.315a,b}$$

is given by the principle of superposition (5.315b) because the Laplace operator is linear; the integral (5.315b) is a three-dimensional convolution (7.36). The influence function for the Laplacian can be generalized to any dimension (Subsection 9.2.2).

The influence function (5.314c) for the three-dimensional Laplace operator can be obtained as follows: (1) in three-dimensional Cartesian coordinates, the Poisson equation for the influence function forced by a unit impulse is (5.316a):

$$\frac{\partial^2 G}{\partial x^2} + \frac{\partial^2 G}{\partial y^2} + \frac{\partial^2 G}{\partial z^2} = \delta(x)\delta(y)\delta(z); \tag{5.316a}$$

(2) in spherical coordinates with central symmetry (5.316b) $\equiv$ (5.307a), the unit impulse is (5.307b), and the Laplacian (6.46a) simplifies to the first term leading to (5.316c):

$$R = \left| x^2 + y^2 + z^2 \right|^{1/2} : \quad \frac{1}{R^2} \frac{d}{dR} \left(R^2 \frac{dG}{dR} \right) = \frac{\delta(R)}{4\pi R^2}; \tag{5.316b,c}$$

(3) a first integration of (5.316c) introduces the unit jump (5.317b) that is unity outside the origin (5.317a):

$$R > 0: \quad \frac{dG}{dR} = \frac{H(R)}{4\pi R^2} = \frac{1}{4\pi R^2}; \quad G(R) = -\frac{1}{4\pi R}; \tag{5.317a-c}$$

and (4) the primitive of (5.317b) is (5.317c) $\equiv$ (5.314c), bearing in mind that the influence function is defined to within an added constant and choosing the latter so it vanishes at infinity.

NOTE 5.6: Magnetostatic Field due to Electric Currents

The magnetostatic induction $\vec{B}$ in an isotropic homogeneous medium with magnetic permeability μ satisfies the Maxwell equations (I.26.1a through c, I.26.2a) $\equiv$ (5.318a and b), where $\vec{j}$ is the density of electric current and c the speed of light in vacuo:

$$\nabla \cdot \vec{B} = 0, \quad \nabla \wedge \vec{B} = \frac{\mu}{c} \vec{j}; \quad \vec{B} = \nabla \wedge \vec{A}, \quad \nabla \cdot \vec{A} = 0, \quad -\frac{\mu}{c}\vec{j} = -\nabla \wedge \left(\nabla \wedge \vec{A} \right) = \nabla^2 \vec{A}. \tag{5.318a-e}$$

These lead to the existence of a vector potential (5.318c) satisfying a vector Poisson equation (5.318e) if the gauge condition (5.318d) is imposed. The analogy between the scalar (vector) Poisson equation (5.310d) [(5.318e)] extends to their solutions (5.313c) [(5.319b)]:

$$\mu = \text{const}: \quad \vec{A}(\vec{x}) = \frac{\mu}{4\pi c} \int_D \frac{\vec{j}(\vec{y})}{|\vec{x} - \vec{y}|} d^3\vec{y}, \quad \vec{B}(\vec{x}) = \frac{\mu}{4\pi c} \int_D \frac{\vec{j}(\vec{y}) \wedge (\vec{x} - \vec{y})}{|\vec{x} - \vec{y}|^3} d^3\vec{y}, \tag{5.319a-c}$$

and thus *a distribution of electric currents with density $\vec{j}$ in a domain D with volume element $\mathrm{d}^3\vec{y}$ in a medium with constant magnetic permeability (5.319a) creates at an observation point $\vec{x}$ a vector potential (5.319b) and magnetic induction (5.319c).*

The electric field (5.313d) [magnetic induction (5.319c)] follows from the scalar electrostatic (5.313a) [vector magnetostatic (5.319b)] potential (5.320) [(5.321)]:

$$\vec{E}(\vec{x}) = -\frac{1}{4\pi\varepsilon} \int_D q(\vec{y}) \nabla\left(\frac{1}{|\vec{x}-\vec{y}|}\right) \mathrm{d}^3\vec{y} = \frac{\mu}{4\pi\varepsilon} \int_D q(\vec{y}) \frac{\vec{x}-\vec{y}}{|\vec{x}-\vec{y}|^3} \mathrm{d}^3\vec{y}, \tag{5.320}$$

$$\vec{B}(\vec{x}) = -\frac{\mu}{4\pi c} \int_D \vec{j}(\vec{y}) \wedge \nabla\left(\frac{1}{|\vec{x}-\vec{y}|}\right) \mathrm{d}^3\vec{y} = \frac{\mu}{4\pi c} \int_D \frac{\vec{j}(\vec{y}) \wedge (\vec{x}-\vec{y})}{|\vec{x}-\vec{y}|^3} \mathrm{d}^3\vec{y}, \tag{5.321}$$

bearing in mind that

$$|\vec{x}-\vec{y}| \frac{\partial}{\partial x_i}\left(|\vec{x}-\vec{y}|\right) = \frac{1}{2}\frac{\partial}{\partial x_i}\left(|\vec{x}-\vec{y}|^2\right) = \frac{1}{2}\frac{\partial}{\partial x_i}\left[(x_j-y_j)(x_j-y_j)\right]$$

$$= (x_j-y_j)\frac{\partial}{\partial x_i}(x_j-y_j) = (x_j-y_j)\frac{\partial x_j}{\partial x_i}$$

$$= (x_j-y_j)\delta_{ij} = x_i - y_i. \tag{5.322a}$$

Thus, *the derivative of the distance from the source to the observer with regard to the observer's (source) position is the unit vector from the source to the observer (5.322b) [vice-versa (5.322c)]:*

$$\frac{\partial}{\partial x_i}\{|\vec{x}-\vec{y}|\} = \frac{x_i-y_i}{|\vec{x}-\vec{y}|} = -\frac{y_i-x_i}{|\vec{x}-\vec{y}|} = -\frac{\partial}{\partial y_i}\{|\vec{x}-\vec{y}|\}. \tag{5.322b,c}$$

Using the nabla or gradient in (5.322b) $\equiv$ (5.323a) leads to (5.323b):

$$\nabla\{|\vec{x}-\vec{y}|\} = \frac{\vec{x}-\vec{y}}{|\vec{x}-\vec{y}|} : \quad \nabla\left(\frac{1}{|\vec{x}-\vec{y}|}\right) = -\frac{\nabla\{|\vec{x}-\vec{y}|\}}{|\vec{x}-\vec{y}|^2} = -\frac{\vec{x}-\vec{y}}{|\vec{x}-\vec{y}|^3}, \tag{5.323a,b}$$

implying (5.323b). Substituting (5.323b) in proves (5.320) $\equiv$ (5.313d) [(5.321) $\equiv$ (5.319c)]. In (5.321) the property of the curl (5.324a and b) was used:

$$\vec{C} = \text{const}: \quad \nabla \wedge \left(\vec{C}\Phi\right) = -\vec{C} \wedge \nabla\Phi, \tag{5.324a,b}$$

$$e_{ijk}\partial_j\left(C_k\Phi\right) = e_{ijk}\left(\partial_j\Phi\right)C_k = -e_{ijk}C_j\left(\partial_k\Phi\right), \tag{5.324c}$$

which is proved by (5.324c) in index notation.

In the case of a point electric current J along the z-axis at the origin (Figure 5.17), the electric current density is (5.325a) leading to the vector potential (5.325b):

$$\vec{j}(R) = \vec{e}_z J \frac{\delta(R)}{4\pi R^2}, \quad \vec{A}(R) = \vec{e}_z \frac{\mu J}{4\pi cR}. \tag{5.325a,b}$$

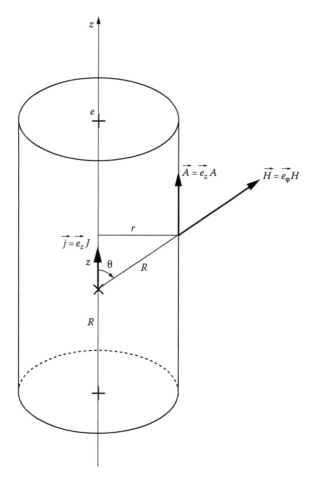

FIGURE 5.17 A point electric current creates a parallel vector potential and an azimuthal magnetic field such that the electric current equals the azimuthal magnetic induction times the perimeter of the circle that is the section of the cylinder orthogonal to the axis (multiplied by the magnetic permeability of the medium and divided by the speed of light in vacuo).

The vector potential (5.315b) is given in cylindrical coordinates (5.326a and b) by (5.326c):

$$R^2 = r^2 + z^2, r = R\sin\theta: \quad \vec{A}(r,z) = \vec{e}_z \frac{\mu J}{4\pi c}\left|r^2 + z^2\right|^{-1/2}. \qquad (5.326\text{a–c})$$

The curl in cylindrical coordinates (6.41b) specifies (5.326c) the magnetic induction:

$$\vec{B}(r,z) = -\vec{e}_\varphi \partial_r A_z = \vec{e}_\varphi \frac{\mu r J}{4\pi c}\left|r^2 + z^2\right|^{-3/2} = \vec{e}_\varphi \frac{\mu J}{4\pi c}\frac{\sin\theta}{R^2} = \vec{B}(R,\theta). \qquad (5.327\text{a,b})$$

This can also be obtained (5.328b) ≡ (5.327b) from (5.319c), using the position vector (5.327a) in cylindrical coordinates:

$$\vec{x} = \vec{e}_r r + \vec{e}_z z: \quad \vec{B} = \frac{\mu}{4\pi c}\frac{\vec{J} \wedge \vec{x}}{\left|\vec{x}\right|^3} = \frac{\mu}{4\pi c R^3}\left[\vec{e}_z J \wedge \left(\vec{e}_r r + \vec{e}_z z\right)\right] = \frac{\mu J r}{4\pi c R^3}\vec{e}_\varphi. \qquad (5.328\text{a,b})$$

Thus, *a point electric current J along the z-axis at the origin (5.325a) causes a parallel vector potential (5.325b) [(5.326c)] in spherical (cylindrical) coordinates; the corresponding magnetic induction (5.327b) [(5.328b)] is azimuthal, vanishes along the axis, and is maximum in the equatorial plane.*

NOTE 5.7: Electric Charge Distribution along a Straight Segment

Next is considered an electric distribution (5.329a) with generally nonuniform density per unit length $\vec{e}(z)$ along a segment $(-a, +a)$ in Figure 5.18:

$$q(r,z) = \vec{e}(z)\delta(r)H(a^2 - r^2), \quad \Phi(r,z) = \frac{1}{4\pi\varepsilon}\int_{-a}^{+a}\frac{\vec{e}(\xi)}{D(r,z;\xi)}d\xi, \tag{5.329a,b}$$

corresponding (5.313c) to the potential (5.329b), where D is the distance (5.330d) of the observer at (5.330c) from the source at (5.330a and b):

$$-a \leq \xi \leq a: \quad \vec{y} = \vec{e}_z\xi, \quad \vec{x} = \vec{e}_z z + \vec{e}_r r: \quad D \equiv |\vec{x} - \vec{y}| = \left|(z-\xi)^2 + r^2\right|^{1/2}. \tag{5.330a–d}$$

The two-dimensional analogue is a source distribution along a segment (5.330a) that corresponds (Sections I.18.6 and I.18.7) to infinite slab (Figure I.18.5), for example, for the gravity field. The present three-dimensional problem (Figure 5.18) is axisymmetric and (5.329b) for a uniform electric charge per unit length (5.331a) leads to the potential (5.331b):

$$\bar{e} \equiv \frac{de}{dz} = \text{const}: \quad \Phi(r,z) = \frac{\bar{e}}{4\pi\varepsilon}\int_{-a}^{+a}\left|(z-\xi)^2 + r^2\right|^{-1/2}d\xi. \tag{5.331a,b}$$

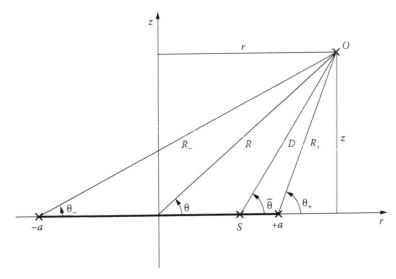

FIGURE 5.18 A uniform distribution of electric charges (currents) along a straight segment leads to a scalar (vector) potential and electric field (magnetic induction) that depend only on the distances $R_\pm$ and polar angles $\theta_\pm$ of the observer from the end-points.

The change of variable (5.332a and b) is used to perform (II.7.122b) the integration (5.295c):

$$\eta \equiv \frac{z - \xi}{r}, d\eta = -\frac{d\xi}{r}: \quad \Phi(r,z) = \frac{\bar{e}}{4\pi\varepsilon} \int\limits_{(z-a)/r}^{(z+a)/r} \left|1 + \eta^2\right|^{-1/2} d\eta = \frac{\bar{e}}{4\pi\varepsilon}[\arg\sinh\eta]_{(z-a)/r}^{(z+a)/r}. \quad (5.332\text{a–c})$$

The derivatives

$$\left\{\frac{\partial}{\partial z}, \frac{\partial}{\partial r}\right\}\arg\sinh\left(\frac{z \pm a}{r}\right) = \left|1 + \left(\frac{z \pm a}{r}\right)^2\right|^{-1/2}\left\{\frac{1}{r}, -\frac{z \pm a}{r^2}\right\} = \left|r^2 + (z \pm a)^2\right|^{-1/2}\left\{1, -\frac{z \pm a}{r}\right\} \quad (5.333\text{a,b})$$

will appear in the calculation of the electric field.

Thus, *an electric charge distribution (5.329a) with density per unit length $\bar{e}(z)$ along a straight segment (5.330a) in a medium with constant dielectric permittivity ε creates (Figure 5.18) an electrostatic potential (5.329b); the distance of the observer (5.330c) from the end points is (5.334a), and the angles with the axis satisfy (5.334b):*

$$R_\pm \equiv D(r,z;\pm a) = \left|(z \mp a)^2 - r^2\right|^{1/2}, \quad \cot\theta_\pm = \frac{z \mp a}{r}. \quad (5.334\text{a,b})$$

If the electric charge distribution is uniform (5.331a), the electrostatic potential (5.332c) depends only on the angles:

$$\Phi(r,z) = \frac{\bar{e}}{4\pi\varepsilon}\left[\arg\sinh\left(\frac{z+a}{r}\right) - \arg\sinh\left(\frac{z-a}{r}\right)\right]$$

$$= \frac{\bar{e}}{4\pi\varepsilon}\left[\arg\sinh(\cot\theta_-) - \arg\sinh(\cot\theta_+)\right]. \quad (5.335)$$

The corresponding (5.333a and b) electric field,

$$\left\{E_z(r,z), E_r(r,z)\right\} = -\left\{\frac{\partial\Phi}{\partial z}, \frac{\partial\Phi}{\partial r}\right\}$$

$$= \frac{\bar{e}}{4\pi\varepsilon r}\left[\left|(z+a) + r^2\right|^{-1/2}\{-r, z+a\} - \left|(z-a) + r^2\right|^{-1/2}\{-r, z-a\}\right], \quad (5.336\text{a,b})$$

has (1) an axial component (5.337a) that depends only on the distances; and (2) a transverse component (5.337b) that depends also on the angles:

$$E_z(r,z) = \frac{\bar{e}}{4\pi\varepsilon}\left(\frac{1}{R_+} - \frac{1}{R_-}\right), \quad (5.337\text{a})$$

$$E_r(r,z) = \frac{\bar{e}}{4\pi\varepsilon r}\left(\frac{z+a}{R_-} - \frac{z-a}{R_+}\right) = \frac{\bar{e}}{4\pi\varepsilon}\left(\frac{\cot\theta_-}{R_-} - \frac{\cot\theta_+}{R_+}\right). \quad (5.337\text{b})$$

There are two opposite limits: (1) for an infinitely long wire, the electric field (5.336a and b) is normal to the line electric charge with uniform density $\bar{e}$ in (5.338a and b):

$$\lim_{a\to\infty} E_z(r,z) = 0, \quad \lim_{a\to\infty} E_r(r,z) = \frac{\bar{e}}{2\pi\varepsilon r} = -\frac{d\Phi}{dr}, \quad \Phi(r) = -\frac{\bar{e}}{2\pi\varepsilon}\log r, \qquad (5.338a\text{--}c)$$

corresponding to the logarithmic potential (5.338c) in two dimensions (I.24.13a); and (2) in the limit of a point electric charge e in (5.339a), the electric field is radial (5.339b):

$$e \equiv 2a\bar{e}: \quad \lim_{a\to 0}\vec{E} = \vec{e}_R \lim_{a\to 0}\left|(E_z)^2 + (E_z)^2\right|^{1/2} = \frac{e}{4\pi\varepsilon R^2}\vec{e}_R, \qquad (5.339a,b)$$

and coincides with (5.312b). The limits of the point (line) electric charge (5.339a through c) [(5.338a through c)] are proved in detail next [Note 5.8 (Note 5.9)].

NOTE 5.8: Limit of a Point Electric Charge

The point electric charge e (Figure 5.16) may be considered as the limit of a uniform electric charge distribution with density $\bar{e}$ along a segment (5.330a) of length $2a$ (Figure 5.18) as $a \to 0$ and $\bar{e} \to \infty$ to preserve the total charge (5.340a):

$$\lim_{\substack{a\to 0 \\ \bar{e}\to\infty}} 2a\bar{e} = e = \text{const}: \quad \{E_z, E_r\} = \frac{e}{8\pi\varepsilon a}\left\{\frac{1}{R_+} - \frac{1}{R_-}, \frac{1}{r}\left(\frac{z+a}{R_-} - \frac{z-a}{R_+}\right)\right\}, \qquad (5.340a\text{--}c)$$

which appears in the electric field (5.337a and b) $\equiv$ (5.340b and c). The limit (5.340a) is applied to (5.340b and c) in stages: (1) the distance (5.334a) of the observer from the end points is evaluated (5.341b) to order (5.341a):

$$a^2 \ll r^2 + z^2 = R^2: \quad R_\pm = \left|a^2 + z^2 + r^2 \mp 2az\right|^{1/2} = \left|a^2 + R^2 \mp 2az\right|^{1/2}, \qquad (5.341a,b)$$

leading to

$$R_\pm = R\left|1 \mp \frac{2az}{R^2} + \frac{a^2}{R^2}\right|^{1/2} = R\left[1 \mp \frac{az}{R} + O\left(\frac{a^2}{R^2}\right)\right]; \qquad (5.341c)$$

(2) the inverses of (5.341c) are (5.342a):

$$\frac{1}{R_\pm} = \frac{1}{R}\left[1 \mp \frac{az}{R^2} + O\left(\frac{a^2}{R^2}\right)\right]^{-1} = \frac{1}{R} \pm \frac{az}{R^3} + O\left(\frac{a^2}{R^3}\right), \qquad (5.342a)$$

and their sum (difference) is (5.342b) [(5.342c)]:

$$\frac{1}{R_+} + \frac{1}{R_-} = \frac{2}{R} + O\left(\frac{a^2}{R^2}\right), \quad \frac{1}{R_+} - \frac{1}{R_-} = \frac{2az}{R^3} + O\left(\frac{a^2}{R^3}\right); \qquad (5.342b,c)$$

(3) substitution of (5.342b and c) in (5.337a) [(5.337b)] specifies the axial (5.343a) [transverse (5.343c)] electric field:

$$E_z = \lim_{a \to 0} \frac{e}{8\pi\varepsilon a}\left[\frac{2az}{R^3} + O\left(\frac{a^2}{R^3}\right)\right] = \lim_{a \to 0}\frac{ez}{4\pi\varepsilon R^3 r} + O\left(\frac{a}{R^3}\right) = \frac{ez}{4\pi\varepsilon R^3 r}, \tag{5.343a}$$

$$E_r = \lim_{a \to 0}\frac{e}{8\pi\varepsilon a r}\left[a\left(\frac{1}{R_+} + \frac{1}{R_-}\right) + z\left(\frac{1}{R_-} - \frac{1}{R_+}\right)\right] = \lim_{a \to 0}\frac{e}{8\pi\varepsilon a r}\left[\frac{2a}{R} - \frac{2az^2}{R^3} + O\left(\frac{a^2}{R^3}\right)\right]$$

$$= \lim_{a \to 0}\frac{e}{4\pi\varepsilon R r}\left[1 - \frac{z^2}{R^2} + O\left(\frac{a}{R^3}\right)\right] = \frac{e}{4\pi\varepsilon R r}\frac{R^2 - z^2}{R^2} = \frac{e}{4\pi\varepsilon R r}\frac{r^2}{R^2} = \frac{er}{4\pi\varepsilon R^3}, \tag{5.343b,c}$$

where the higher-order terms from (5.341b and c; 5.342a through c) vanish; (4) from (5.343b and c), it follows that the electric field is radial:

$$E_r \vec{e}_r + E_z \vec{e}_z = \frac{e}{4\pi\varepsilon R^3}\left(\vec{e}_r r + \vec{e}_z z\right) = \vec{e}_R \frac{e}{4\pi\varepsilon R^2} = \vec{E}(R), \tag{5.344}$$

in agreement with (5.312b) ≡ (5.344).

NOTE 5.9: Limit of a Line Electric Charge

In the limit of a line electric charge (Figure 5.19) with uniform density $\bar{e}$ per unit length (5.331a), the electric field is taken in the form (5.337a and b), and the infinitely long segment (5.330a) with $a \to \infty$ is again applied in stages: (1) the approximation of distance from the origin, large compared with the length of the segment for a point electric charge (5.341a and b), is replaced for a line charge by length much larger than distance from the line (5.345a and b):

$$a^2 \gg z^2: \quad \lim_{a \to \infty} E_z = \frac{\bar{e}}{4\pi\varepsilon}\lim_{a \to \infty}\left\{\left|(z+a)^2 + r^2\right|^{-1/2} - \left|(z-a)^2 + r^2\right|^{-1/2}\right\} = 0, \tag{5.345a,b}$$

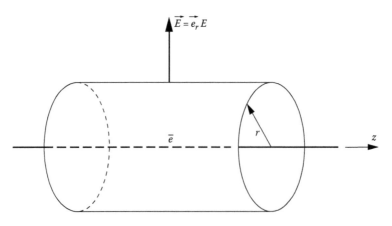

FIGURE 5.19   A uniform electric charge along an infinitely long straight wire creates a radial electric field corresponding to the two-dimensional logarithmic potential.

leading (5.336a) to zero axial field; (2) the first term of the transverse electric field (5.336b) also vanishes for a line electric charge:

$$\lim_{a \to \infty} \frac{\bar{e}z}{4\pi\varepsilon r} \left\{ \left| (z+a)^2 + r^2 \right|^{-1/2} - \left| (z-a)^2 + r^2 \right|^{-1/2} \right\} = 0; \tag{5.346}$$

and (3) the transverse electric field for a line electric charge is given by the second term in (5.336b):

$$E_r = \lim_{a \to \infty} \frac{\bar{e}a}{4\pi\varepsilon r} \left\{ \left| z^2 + r^2 + a^2 - 2az \right|^{-1/2} + \left| z^2 + r^2 + a^2 - 2az \right|^{-1/2} \right\}$$

$$= \lim_{a \to \infty} \frac{\bar{e}a}{4\pi\varepsilon r} \left[\frac{2}{a} + O\left(\frac{R}{a^2} \right) \right] = \lim_{a \to \infty} \left[\frac{\bar{e}}{2\pi\varepsilon r} + O\left(\frac{R}{a} \right) \right] = \frac{\bar{e}}{2\pi\varepsilon r}, \tag{5.347}$$

showing that the electric field is transverse, as could be expected by symmetry. The simplest way to obtain this result, that is, in the case of an infinitely long straight electric charge distribution with density $\bar{e}$ per unit length, the electric field is radial (Figure 5.19), and the divergence theorem (5.163a through c) applied per unit length to a cylinder of radius r gives

$$\bar{e} = \int_{D_3} q \, dV = \varepsilon \int_{D_3} (\nabla \cdot \vec{E}) dV = \varepsilon \int_{\partial D_3} \vec{E} \cdot d\vec{S} = \varepsilon E_r 2\pi r. \tag{5.348}$$

This specifies (5.348) ≡ (5.347) the radial electric field, leading to the logarithmic potential (5.338c) ≡ (I.24.13a) in two dimensions.

NOTE 5.10: Electric Current along a Straight Segment

In the case (Figure 5.18) of a generally nonuniform electric current (5.349a) with density $\bar{J}(z)$ per unit length along a straight segment (5.330a), the magnetic potential (5.319b) is given by (5.349b) using the same distance (5.330d) from observer to source:

$$\vec{j}(r,z) = \vec{e}_z \bar{J}(z) \delta(r) H(a^2 - z^2), \quad A(r,z) = \vec{e}_z \frac{\mu}{4\pi c} \int_{-a}^{a} \frac{\bar{J}(\xi)}{D(r,z;\xi)} d\xi. \tag{5.349a,b}$$

For a uniform current density per unit length (5.350a), the integral (5.349b) is evaluated as in (5.331b; 5.332a through c), leading to the vector potential (5.350b) similar to (5.335):

$$\bar{J} = \frac{dj}{dz} = \text{const} = \frac{J}{2a}: \quad \vec{A}(r,z) = \frac{\mu\bar{J}}{4\pi c} \vec{e}_z \left[\arg\sinh\left(\frac{z+a}{r} \right) - \arg\sinh\left(\frac{z-a}{r} \right) \right]. \tag{5.350a,b}$$

Using the curl in cylindrical coordinates (6.41b) and (5.333b), it follows that the magnetic induction is azimuthal (5.351):

$$\vec{B}(r,z) = -\vec{e}_\varphi \partial_r A_z = \vec{e}_\varphi \frac{\mu\bar{J}}{4\pi cr} \left(\frac{z+a}{R_-} - \frac{z-a}{R_+} \right). \tag{5.351}$$

Thus, *a distribution of electric currents (5.349a) along a segment (5.330a) with density per unity length $\bar{J}(z)$ leads to the vector potential (5.349b; 5.330d). In the case of a uniform distribution (5.350a), the vector potential is given by (5.350b) and leads to magnetic induction (5.351) that is azimuthal. The latter has two limits: (1) for an infinitely long wire, the magnetic induction (5.352a) corresponds to a logarithmic potential (5.352b)* ≡ *(I.26.13a):*

$$\lim_{a\to\infty}\vec{B}(r,z)=\vec{e}_\varphi\frac{\mu\bar{J}}{2\pi cr}=\vec{e}_\varphi\bar{B}(r)=-\vec{e}_\varphi\partial_r A_z,\quad \vec{A}=-\vec{e}_z\frac{\mu\bar{J}}{2\pi c}\log r; \tag{5.352a,b}$$

and (2) in the limit of a point current (5.353a), the magnetic induction simplifies to (5.352b):

$$J\equiv 2a\bar{J}:\quad \lim_{a\to 0}\vec{B}(r,z)=\vec{e}_\varphi\frac{\mu J}{4\pi c}\frac{r}{R^3}=\frac{\mu J}{4\pi c}\frac{\sin\theta}{R^2}, \tag{5.353a,b}$$

in agreement with (5.327b). The limits of a point (line) electric current (5.353a and b) [(5.352a and b)] are proved next (Note 5.11).

NOTE 5.11: Comparison of Point and Line Electric Currents

In the case of a point electric current (5.354a), the total electric current (5.350a) is used in the magnetic induction (5.351) leading to (5.354b):

$$\lim_{\substack{a\to 0\\ \bar{J}\to\infty}}2a\bar{J}=J=\text{const}:\quad \vec{B}=\vec{e}_\varphi\frac{\mu J}{8\pi acr}\left(\frac{z+a}{R_+}-\frac{z-a}{R_+}\right). \tag{5.354a,b}$$

Using (5.342a through c) the limit (5.354a) leads from (5.354b) to

$$B_\varphi=\lim_{a\to 0}\vec{e}_\varphi\frac{\mu\bar{J}}{8\pi acr}\left[a\left(\frac{1}{R_+}+\frac{1}{R_-}\right)-z\left(\frac{1}{R_+}-\frac{1}{R_-}\right)\right]$$

$$=\lim_{a\to 0}\frac{\mu\bar{J}}{8\pi acr}\left[\frac{2a}{R}-\frac{2az^2}{R^3}+O\left(\frac{a^2}{R^3}\right)\right]=\frac{\mu\bar{J}}{4\pi cr}\frac{R^2-z^2}{R^3}=\frac{\mu\bar{J}}{4\pi c}\frac{r}{R^3}, \tag{5.355}$$

in agreement with the magnetic induction of a point current (5.355) ≡ (5.353b) ≡ (5.327b) corresponding to the vector potential (5.325b) ≡ (5.326c). In the case of an infinitely long straight wire with (5.350a) with uniform current density $\bar{J}$, the magnetic induction (5.351) with (5.334a) simplifies in the limit $a\to\infty$ to

$$\lim_{a\to\infty}\vec{B}=\vec{e}_\varphi\frac{\mu\bar{J}}{4\pi cr}\lim_{a\to\infty}\left\{(z+a)\left|(z+a)^2+r^2\right|^{-1/2}-(z-a)\left|(z-a)^2+r^2\right|^{-1/2}\right\}=\vec{e}_\varphi\frac{\mu\bar{J}}{2\pi cr}, \tag{5.356}$$

using the same limits as in (5.346; 5.347). The same result can be obtained most simply noting that for a uniform line electric current (5.357a), the vector potential (magnetic induction) is axial (5.357b) [azimuthal (5.357c)] by symmetry:

$$\vec{j}(r)=\bar{J}\delta(r)\vec{e}_z,\quad \vec{A}=\vec{e}_z A(r),\quad \vec{B}=\vec{e}_\varphi B(r). \tag{5.357a–c}$$

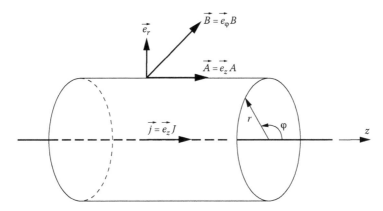

FIGURE 5.20 A uniform electric current along an infinitely long straight wire creates an azimuthal magnetic induction corresponding to a two-dimensional logarithmic vector potential specified by a scalar field function and parallel to the wire.

The curl theorem (5.192a through c) applied (Figure 5.20) per unit length of the cylinder of radius r leads to

$$\vec{e}_z \bar{J} = \int_{D_3} \vec{j}\, dV = \frac{c}{\mu} \int_{D_3} \left(\nabla \wedge \vec{B} \right) dV = \frac{c}{\mu} \int_{\partial D_3} \left(\vec{N} \wedge \vec{B} \right) dS$$

$$= \frac{c}{\mu} \int_{\partial D} \left(\vec{e}_r\, dS \wedge \vec{e}_\varphi B \right) = \frac{c}{\mu} \vec{e}_z B \int_{\partial D} dS = \frac{c}{\mu} 2\pi r B \vec{e}_z. \tag{5.358}$$

This specifies $(5.358) \equiv (5.356) \equiv (5.352a)$ the magnetic induction that is azimuthal (5.357c) and leads to the vector potential (5.352b).

NOTE 5.12: Electric Charge along a Circular Loop

In the case of an *electric charge distribution (Figure 5.21) with density $\vec{e}(\varphi)$ per unit radian on a circular loop (5.359a) with radius a,*

$$q(r,\varphi,z) = \vec{e}(\varphi)\delta(r-a)\delta(z), \quad \Phi(r,z) = \frac{a}{4\pi\varepsilon} \int_0^{2\pi} \frac{\vec{e}(\varphi)}{D(r,z,\varphi)}\, d\varphi, \tag{5.359a,b}$$

the electrostatic potential is (5.313c) given in a medium with constant dielectric permittivity ε by (5.359b), where (5.360c) is the distance of the source (5.360b) from the observer (5.360a), choosing the (x, z)-plane to pass through the observer (Figure 5.22b):

$$\vec{x} = \vec{e}_x r + \vec{e}_z z,\ \vec{y} = a\left(\vec{e}_x \cos\varphi + \vec{e}_y \sin\varphi \right): D(r,z;\varphi) = |\vec{x} - \vec{y}| = \left| \left(r - a\cos\varphi \right)^2 + a^2 \sin^2\varphi + z^2 \right|^{1/2}$$

$$= \left| z^2 + a^2 + r^2 - 2ar\cos\varphi \right|^{1/2}. \tag{5.360a–c}$$

In the case of a uniform electric charge distribution (5.361a) with total electric charge e, the potential is given by (5.361b):

$$\text{const} = \bar{e} \equiv \frac{de}{d\varphi} = \frac{e}{2\pi a}: \quad \Phi(r,z) = \frac{e}{8\pi^2\varepsilon} \int_0^{2\pi} \frac{d\varphi}{\sqrt{r^2 + z^2 + a^2 - 2ar\cos\varphi}}. \tag{5.361a,b}$$

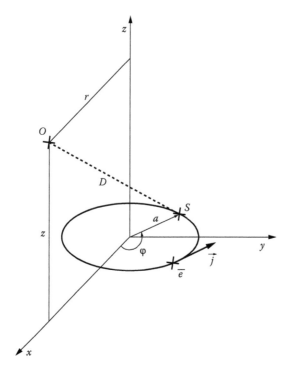

FIGURE 5.21 Another example of uniform or nonuniform electric charge or current distribution with finite extent, besides the straight segment (Figure 5.18) is a circular loop. The electric charge (current) density per unit length is a scalar (vector), and has no direction (is directed along the tangent).

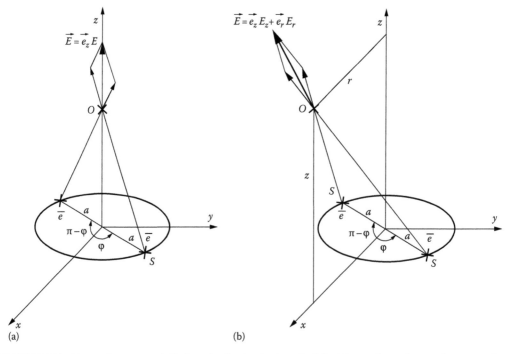

FIGURE 5.22 For an observer on (off) the axis of a circular loop [a (b)] with a uniform electric charge, two diametrically opposite charge elements create electric fields whose sum lies along the axis (not along the axis but on the plane of the observer and the axis).

An observer on the axis (5.362a) is at equal distance (5.362b) from all points on the loop (Figure 5.22a), leading to the potential (5.362c):

$$r = 0: \quad D(0, z; \varphi) = \sqrt{a^2 + z^2} \equiv D_0(z): \quad \Phi(0, z) = \frac{e}{4\pi\varepsilon} \frac{1}{\sqrt{a^2 + z^2}} = \frac{\overline{e}a}{2\varepsilon} \frac{1}{\sqrt{a^2 + z^2}}. \quad (5.362a\text{-}c)$$

The corresponding electric field on axis by symmetry has only an axial component:

$$\vec{E}(0, z) = -\vec{e}_z \frac{d\Phi(0, z)}{dz} = \frac{e}{4\pi\varepsilon} z \left| a^2 + z^2 \right|^{-3/2} \vec{e}_z, \quad (5.363)$$

because diametrically opposite charge elements (Figure 5.22a) lead to an electric field with (1) opposite radial components that cancel; and (2) equal axial components that add. If the observer is off axis (5.364a), it is only the components of the electric field normal to the plane through the observer and axis that cancel (5.364b):

$$r \neq 0: \quad E_\varphi(r, z) = 0; \quad \vec{E}(r, z) = \vec{e}_r E_r(r, z) + \vec{e}_z E_z(r, z). \quad (5.364a\text{-}c)$$

Thus, the electric field lies (Figure 5.22b) in the plane of the observer and axis (5.364c) as confirmed by the electrostatic potential (8.361b) that does not depend on the azimuthal angle. These results will be confirmed next by calculating the electrostatic potential and electric field at an arbitrary observer position including off axis, which is not on the straight line perpendicular to the plane of the circle and passing through its center.

NOTE 5.13: Electric Field on and off Axis

The calculation of the potential at all observer positions involves the evaluation of the integral in (5.361b) that can be performed by three methods: (a) using elliptic integrals (Section I.39.9) via the change of variable $t = \cos \varphi$ that leads to (I.39.115b); (b) expanding the inverse square root in a power series with Legendre polynomials as coefficients (8.74a,b) prior to the $d\varphi$– integration; (c) an equivalent method of expansion in a binomial series followed by direct integration that is similar to (4.354–4.360) is used next. The method (c) consists of three steps. The first step is based on the inequalities

$$(r - a)^2 \geq 0, \, a^2 + r^2 \geq 2ar > 2ar|\cos\varphi|: \quad b^2 \equiv a^2 + z^2 + r^2 > 2ar|\cos\varphi|, \quad (5.365a\text{-}c)$$

where (5.365c) allows the expansion of the integrand in (5.361b):

$$\Phi(r, z) = \frac{e}{8\pi^2\varepsilon} \int_0^{2\pi} \left| b^2 - 2ar\cos\varphi \right|^{-1/2} d\varphi$$

$$= \frac{e}{8\pi^2\varepsilon b} \int_0^{2\pi} \left| 1 - \frac{2ar}{b^2}\cos\varphi \right|^{-1/2} d\varphi$$

$$= \frac{e}{8\pi^2\varepsilon b} \sum_{m=0}^{\infty} a_m \left(\frac{2ar}{b^2} \right)^m \int_0^{2\pi} \cos^m \varphi \, d\varphi, \quad (5.366)$$

in a binomial series (I.25.37a through c) with coefficients (2.159a and b). Second, the integrals in (5.366) vanish for m odd by symmetry; for m even, they are given by

$$I_{2n} \equiv \int_0^{2\pi} \cos^{2n}\varphi \, d\varphi = \int_0^{2\pi} \left(\frac{e^{i\varphi} + e^{-i\varphi}}{2} \right)^{2n} d\varphi$$

$$= 2^{-2n} \sum_{k=0}^{2n} \int_0^{2\pi} \binom{2n}{k} e^{-ik\varphi} e^{i(2n-k)\varphi} d\varphi = \binom{2n}{n} \frac{2\pi}{2^{2n}}, \tag{5.367}$$

where (1) the finite binomial expansion (I.25.38) was used; (2) for $k \neq n$, the integrand has period 2π, and the corresponding integrals vanish (5.368a):

$$\int_0^{2\pi} e^{i2(n-k)\varphi} d\varphi \begin{cases} 0 & \text{if } k \neq n, \tag{5.368a} \\ 2\pi & \text{if } k = n, \tag{5.368b} \end{cases}$$

(3) the only nonvanishing term (5.368b) is $k = n$, in agreement with (I.28.24a through c); and (4) thus, the sum in (5.367) reduces to the term $k = n$, in agreement with (5.367) $\equiv$ (I.20.69).

The third step uses (2.159a) $\equiv$ (5.369a) and (5.367) $\equiv$ (5.369b):

$$a_{2n} \equiv \binom{-1/2}{2n} = \frac{(4n-1)!!}{(2n)! \, 2^{2n}}, \quad I_{2n} \equiv \int_0^{2\pi} \cos^{2n}\varphi \, d\varphi = \frac{\pi}{2^{2n-1}} \frac{(2n)!}{(n!)^2}, \tag{5.369a,b}$$

in (5.366) = (5.370c):

$$a_0 = 1, \, I_0 = 2\pi : \quad \Phi(r,z) = \frac{e}{8\pi^2 \varepsilon b} \sum_{n=0}^{\infty} \left(\frac{2ar}{b^2} \right)^{2n} a_{2n} I_{2n}, \tag{5.370a-c}$$

with leading term (5.370a and b). This specifies (5.365c) the electric potential:

$$\Phi(r,z) = \frac{e}{4\pi\varepsilon} \left\{ \left| a^2 + z^2 + r^2 \right|^{-1/2} + \sum_{n=1}^{\infty} \frac{2^{-2n}}{n!} \frac{(4n-1)!!}{n!} (ar)^{2n} \left| a^2 + z^2 + r^2 \right|^{-2n-1/2} \right\}. \tag{5.371}$$

The corresponding electric field has axial (5.372a) and radial (5.372b) components:

$$E_z(r,z) = -\frac{\partial\Phi}{\partial z} = \frac{ez}{4\pi\varepsilon} \left\{ \left| a^2 + z^2 + r^2 \right|^{-3/2} + \sum_{n=1}^{\infty} \frac{2^{-1-2n}}{n!} \frac{(4n+1)!!}{n!} (ar)^{2n} \left| a^2 + z^2 + r^2 \right|^{-2n-3/2} \right\}, \tag{5.372a}$$

$$E_r(r,z) = -\frac{\partial\Phi}{\partial r} = \frac{ea}{4\pi\varepsilon} \sum_{n=1}^{\infty} \frac{2^{1-2n}}{(n-1)!} \frac{(4n-1)!!}{n!} (ar)^{2n-1} \left| a^2 + z^2 + r^2 \right|^{-2n-1/2}$$

$$+ \frac{er}{4\pi\varepsilon} \left\{ \left| a^2 + z^2 + r^2 \right|^{-3/2} + \sum_{n=1}^{\infty} \frac{2^{1-2n}}{n!} \frac{(4n+1)!!}{n!} (ar)^{2n} \left| a^2 + z^2 + r^2 \right|^{-2n-3/2} \right\}. \tag{5.372b}$$

Thus, *a total electric charge e distributed uniformly along (Figure 5.21) a circle of radius a in a medium with constant dielectric permittivity ε creates an electrostatic potential (5.371) [electric field (5.372a and b)] that simplifies to (5.362c) [(5.363)] on the axis. If the loop shrinks to zero a → 0, the electrostatic potential (5.362c) [electric field (5.363)] with z = R corresponds to a point charge (5.312c) [(5.312b)]. The leading term (5.372c) of the electric field (5.372a and b) is radial (5.372d),*

$$b^2 \gg a^2 + r^2: \quad \vec{E}(r,z) \sim \frac{e}{4\pi\varepsilon b^3}\left(\vec{e}_z z + \vec{e}_r r\right) = \frac{eR}{4\pi\varepsilon b^3}\vec{e}_R = \vec{e}_R \frac{e}{4\pi\varepsilon d^2}$$

$$d \equiv \sqrt{\frac{b^3}{R}} = \frac{\left|a^2 + r^2 + z^2\right|^{3/2}}{\left|r^2 + z\right|^{1/2}}, \tag{5.372c–e}$$

corresponding to a monopole, that is, the total charge at a distance (5.372e).

NOTE 5.14: Electric Current along a Circular Loop

Replacing electric charge distribution by an electric current distribution (5.373b) with density per radian (5.373a),

$$\bar{J}(\varphi) = \frac{dJ}{d\varphi}: \quad \vec{j}(r,\varphi,z) = \vec{e}_\varphi \bar{J}(\varphi)\delta(z)\delta(r-a), \tag{5.373a,b}$$

along a circular loop (Figure 5.21) with radius a, in a medium with constant magnetic permeability μ, leads (5.319b) to the vector potential

$$\vec{A}(r,z) = \frac{\mu}{4\pi c}\int \frac{\bar{J}(\varphi)}{D(r,z,\varphi)}\vec{e}_\varphi \, ds = \frac{\mu a}{4\pi c}\int_0^{2\pi} \frac{-\vec{e}_x \sin\varphi + \vec{e}_y \cos\varphi}{\sqrt{a^2 + z^2 + r^2 - 2ar\cos\varphi}}\bar{J}(\varphi)d\varphi. \tag{5.374}$$

where (1) the distance from the source to the observer is given by (5.360c); and (2) the source (5.360b) lies on the circle with arc length (5.375a) and unit tangent vector (5.329b):

$$ds = |d\vec{y}| = a \, d\varphi, \quad \vec{e}_\varphi = \frac{d\vec{y}}{ds} = \frac{d\varphi}{ds}\frac{d\vec{y}}{d\varphi} = -\vec{e}_x \sin\varphi + \vec{e}_y \cos\varphi, \tag{5.375a,b}$$

appearing in (5.374). In the case of a uniform electric current distribution (5.376a) with total current J, the vector potential (5.376b),

$$const = \frac{dJ}{d\varphi} = \bar{J} = \frac{J}{2\pi a}: \quad \vec{A}(r,z) = \frac{\mu J}{8\pi^2 c}\int_0^{2\pi} \frac{\vec{e}_y \cos\varphi - \vec{e}_x \sin\varphi}{\sqrt{a^2 + z^2 + r^2 - 2ar\cos\varphi}}d\varphi, \tag{5.376a,b}$$

lies in the direction orthogonal to the position vector of the observer relative to the origin:

$$\vec{A}(r,z) = \vec{e}_\varphi \frac{\mu J}{8\pi c}\sum_{n=0}^{\infty} \frac{2^{-2n}}{n!}\frac{(4n+1)!!}{(n+1)!}(ar)^{2n+1}\left|a^2 + z^2 + r^2\right|^{-2n-3/2}. \tag{5.377}$$

This implies that the magnetic field lies in the plane through the observer and the axis, as shown in the sequel. The integration for the vector potential is (1) immediate for a linear current distribution (Figure 5.17) because the direction in (5.349b) is independent of position; and (2) in the case of the circular current loop (Figure 5.20), the direction of the current (5.373b) varies with position (5.374), and a change from polar to Cartesian unit vectors is made (5.375b). The Cartesian unit vectors are constant, showing that the vector potential (5.376b) (1) has no axial $\vec{e}_z$ component; (2) the $\vec{e}_x \equiv \vec{e}_r$ component vanishes:

$$\int_0^{2\pi} \frac{\sin\varphi}{\sqrt{a^2 + z^2 + r^2 - 2ar\cos\varphi}} \, d\varphi = \frac{1}{ar}\left[\sqrt{r^2 + z^2 + a^2 - 2ar\cos\varphi} \right]_0^{2\pi} = 0, \tag{5.378}$$

and thus is absent from (5.377); and (3) the $\vec{e}_y \equiv \vec{e}_\varphi$ component does not vanish, and its value (5.377) is calculated next (Note 5.15).

NOTE 5.15: Magnetic Potential on and off Axis

From (5.378), the nonzero $\vec{e}_\varphi$ component of the vector potential (5.376b) is

$$\vec{A}(r,z) = \frac{\vec{e}_\varphi \mu J}{8\pi^2 c} \int_0^{2\pi} \frac{\cos\varphi}{\sqrt{a^2 + z^2 + r^2 - 2ar\cos\varphi}} \, d\varphi. \tag{5.379}$$

This integral is evaluated (1) as (5.361b) before in (5.366) through (5.371) with an extra factor $\cos\varphi$:

$$\vec{A}(r,z) = \vec{e}_\varphi \frac{\mu J}{8\pi^2 c} \int_0^{2\pi} \cos\varphi \left| b^2 - 2ar\cos\varphi \right|^{-1/2} d\varphi$$

$$= \vec{e}_\varphi \frac{\mu J}{8\pi^2 cb} \int_0^{2\pi} \left| 1 - \frac{2ar}{b^2}\cos\varphi \right|^{-1/2} \cos\varphi \, d\varphi$$

$$= \vec{e}_\varphi \frac{\mu J}{8\pi^2 cb} \sum_{m=0}^{\infty} (-)^m \binom{-1/2}{m} \left(\frac{2ar}{b^2} \right)^m \int_0^{2\pi} \cos^{m+1}\varphi \, d\varphi; \tag{5.380}$$

(2) the integrals vanish for even m, and for (5.381a) odd m in (5.381b)

$$m = 2n+1: \quad \vec{A}(r,z) = \vec{e}_\varphi \frac{\mu J}{8\pi^2 cb} \sum_{n=0}^{\infty} a_{2n+1} I_{2n+2} \left(\frac{2ar}{b^2} \right)^{2n+1}, \tag{5.381a,b}$$

$(5.369a) \equiv (5.382a)$ and $(5.369b) \equiv (5.382b)$ are used:

$$a_{2n+1} = \frac{(4n+1)!!}{(2n+1)! \, 2^{2n+1}}, \quad I_{2n+2} = \frac{\pi}{2^{2n+1}} \frac{(2n+2)!}{\left[(n+1)! \right]^2}; \tag{5.382a,b}$$

and (3) substitution of (5.382a and b) in (5.381b) specifies the electrostatic potential

$$\vec{A}(r,z) = \vec{e}_\varphi \frac{\mu J}{8\pi c} \sum_{n=0}^{\infty} \frac{2^{-2n}}{n!} \frac{(4n+1)!!}{(n+1)!} (ar)^{2n+1} \left| a^2 + z^2 + r^2 \right|^{-2n-3/2}, \tag{5.383}$$

which is orthogonal to the plane of the observer and the axis and thus corresponds to the φ-component in cylindrical coordinates. The curl in cylindrical coordinates (6.41b) specifies the nonzero components of the magnetic induction that are radial (5.384a) and axial (5.384b):

$$B_r(r,z) = -\partial_z A_\varphi = \frac{\mu Jz}{8\pi c} \sum_{n=0}^{\infty} \frac{2^{-2n}}{n!} \frac{(4n+3)!!}{(n+1)!} (ar)^{2n+1} \left| a^2 + z^2 + r^2 \right|^{-2n-5/2}, \tag{5.384a}$$

$$B_z(r,z) = r^{-1}\partial_r(rA_\varphi) = \frac{\mu Ja}{4\pi c} \sum_{n=0}^{\infty} \frac{2^{-2n}}{n!} \frac{(4n+1)!!}{n!} (ar)^{2n} \left| a^2 + z^2 + r^2 \right|^{-2n-3/2}$$

$$- \frac{\mu Jz}{8\pi c} \sum_{n=0}^{\infty} \frac{2^{-2n}}{n!} \frac{(4n+3)!!}{(n+1)!} (ar)^{2n+1} \left| a^2 + z^2 + r^2 \right|^{-2n-3/2}. \tag{5.384b}$$

Thus, *a uniform electric current (5.376a) with total current J flowing (5.373b) along (5.375b) a circular loop (5.375a) with radius a, in a medium (Figure 5.21) with constant magnetic permeability μ, creates a vector potential (5.377), corresponding to the magnetic induction (5.384a and b). On the axis, the vector potential (5.377) vanishes (5.385a):*

$$\vec{A}(0,z) = 0, \quad \vec{B}(0,z) = \vec{e}_z \frac{\mu Ja}{4\pi c} \left| a^2 + z^2 \right|^{-3/2}, \tag{5.385a,b}$$

and the magnetic induction is axial (5.385b). The values of the magnetic potential (5.385a) [induction (5.385b)] on axis can be obtained more simply (Note 5.15) without calculating their values off axis.

NOTE 5.16: Magnetic Induction on and off Axis

For a circular current loop (Figure 5.21), the diametrically opposite elements φ and φ + π have (5.373b) opposite electric currents (5.375b), and if the observer is on the axis, the distance (5.362b) is the same, and the vector potential (5.319b) cancels (5.385a). This is not the case for the magnetic induction (5.385b) because the position vector (5.386) of the observer (5.360a) relative to the source (5.360b) also changes:

$$\vec{x} - \vec{y} = \vec{e}_x(r - a\cos\varphi) - \vec{e}_x a\sin\varphi + \vec{e}_x z. \tag{5.386}$$

The outer product by the electric current (5.373b; 5.375b) is

$$\vec{e}_\varphi \wedge (\vec{x} - \vec{y}) = \begin{vmatrix} \vec{e}_x & \vec{e}_y & \vec{e}_z \\ -\sin\varphi & \cos\varphi & 0 \\ r - a\cos\varphi & -a\sin\varphi & z \end{vmatrix}$$

$$= z(\vec{e}_x\cos\varphi + \vec{e}_y\sin\varphi) + \vec{e}_z(a - r\cos\varphi) = \vec{e}_r z + \vec{e}_z(a - r\cos\varphi). \tag{5.387a}$$

Substituting (5.360a; 5.373b; 5.387a) in (5.319c) specifies the magnetic induction:

$$\vec{B}(r,z) \equiv \frac{\mu a}{4\pi c} \int_0^{2\pi} \vec{J}(\varphi) \frac{\vec{e}_\varphi \wedge (\vec{x} - \vec{y})}{\left| \vec{x} - \vec{y} \right|^3} d\varphi = \frac{\mu a}{4\pi c} \int_0^{2\pi} \vec{J}(\varphi) \frac{z(\vec{e}_x\cos\varphi + \vec{e}_y\sin\varphi) + \vec{e}_z(a - r\cos\varphi)}{\left| a^2 + z^2 + r^2 - 2ar\cos\varphi \right|^{3/2}} d\varphi. \tag{5.387b}$$

In the case of uniform electric current (5.376a) ≡ (5.388a), the magnetic induction lies in the plane of the observer and axis (5.388b):

$$\vec{J}(\varphi) = \frac{J}{2\pi a}: \quad \vec{B}(r,z) \equiv \frac{\mu J}{8\pi^2 c} \int_0^{2\pi} \frac{\vec{e}_x z \cos\varphi + \vec{e}_z(a - r\cos\varphi)}{\left|a^2 + z^2 + r^2 - 2ar\cos\varphi\right|^{3/2}} d\varphi, \qquad (5.388\text{a,b})$$

because the y-component of the integrand is an exact differential:

$$\int_0^{2\pi} \frac{\sin\varphi}{\left|a^2 + z^2 + r^2 - 2ar\cos\varphi\right|^{3/2}} d\varphi = -\left[\frac{\left|a^2 + z^2 + r^2 - 2ar\cos\varphi\right|^{-1/2}}{ar}\right]_0^{2\pi} = 0. \qquad (5.388\text{c})$$

For an observer on axis, the magnetic induction (5.388b) is axial:

$$\vec{B}(0,z) \equiv \frac{\mu J}{8\pi^2 c}\left|a^2 + z^2\right|^{-3/2} \int_0^{2\pi}\left[\vec{e}_x z \cos\varphi + \vec{e}_z(a - r\cos\varphi)\right] d\varphi = \frac{\mu J}{8\pi^2 c}\left|a^2 + z^2\right|^{-3/2} 2\pi a = \frac{\mu J a}{4\pi c}\left|a^2 + z^2\right|^{-3/2},$$

$$\qquad (5.389)$$

in agreement with (5.385b) ≡ (5.389). Thus, *an electric current (5.373b) along a circular loop of radius a creates a magnetic induction (5.387b). If the current is uniform (5.388a), the magnetic induction (5.388b) lies in the plane through the observer and axis. If the observer lies on the axis, the magnetic induction is axial (5.389).*

NOTE 5.17: Electric Charges along a Cylindrical Helix

Consider next a distribution of electric currents (5.390b) with density $\vec{e}(\varphi)$ per radian along (Figure 5.23) a cylindrical circular helix with radius a and inclination γ to the horizontal (5.343a):

$$k = a\tan\gamma; \quad q(r,\varphi,z) = \vec{e}(\varphi)\delta(r-a)\delta(z - k\varphi). \qquad (5.390\text{a,b})$$

The **step** k of the helix (5.390a) is the distance along the axis per radian. A helix may (1) lie on a cylinder, cone, paraboloid, or other surfaces of revolution or not; and (2) in the case of a cylindrical helix, depending on the directrix of the cylinder, it could be a circular, elliptical, or other shape. The present case is the simplest of a circular cylindrical helix. The position of the observer remains (5.360a) on the (x, z)-plane, but the position of the source (5.360b) is changed to (5.391a):

$$\vec{y} = a(\vec{e}_x \cos\varphi + \vec{e}_y \sin\varphi) + \vec{e}_z k\varphi, \quad ds = |d\vec{y}| = \sqrt{k^2 + a^2}\,d\varphi \equiv \bar{a}\,d\varphi, \qquad (5.391\text{a,b})$$

and hence, (1) the arc length along the cylindrical helix is (5.391b), corresponding to an equivalent radius (5.392a); and (2) the unit tangent vector is (5.392b):

$$\frac{ds}{d\varphi} = \bar{a} = \sqrt{a^2 + k^2} = a\sec\gamma, \quad \vec{s} = \frac{d\vec{y}}{ds} = \frac{d\vec{y}/d\varphi}{ds/d\varphi} = \frac{1}{\bar{a}}\{-a\sin\varphi, a\cos\varphi, k\}. \qquad (5.392\text{a,b})$$

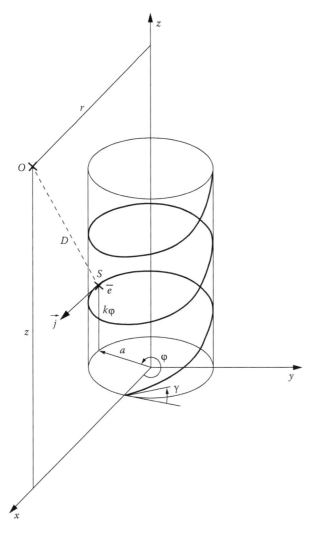

FIGURE 5.23 The electric charge (current) distribution on (along) a circular loop (Figure 5.21) is a particular case of the circular cylindrical helix (Figure 5.23) for which the axial displacement is zero instead of proportional to the azimuthal angle. For an observer at a large axial distance compared with the height of the helix with an integer number of turns N, the electric (magnetic) field is the same as for electric charges (currents) on N parallel circular loops.

The relative position vector of observer (5.360a) and source (5.391b),

$$\vec{x} - \vec{y} = \vec{e}_x \left(r - a\cos\varphi \right) - \vec{e}_y a \sin\varphi + \vec{e}_z \left(z - k\varphi \right),$$ (5.393)

specifies their mutual distance:

$$\bar{D}(r,z;\varphi) = \left| \vec{x} - \vec{y} \right| = \left| \left(r - a\cos\varphi \right)^2 + a^2 \sin^2\varphi + \left(z - k\varphi \right)^2 \right|^{1/2}$$

$$= \left| r^2 + a^2 - 2ar\cos\varphi + \left(z - k\varphi \right)^2 \right|^{1/2},$$ (5.394)

which appears in (5.313b) the electrostatic potential (5.395b):

$$N = \frac{\beta - \alpha}{2\pi}: \quad \Phi(r,z) = \frac{1}{4\pi\varepsilon} \int\limits_{s(\alpha)}^{s(\beta)} \frac{\bar{e}(\varphi)}{\bar{D}(r,z;\varphi)} ds = \frac{\bar{a}}{4\pi\varepsilon} \int\limits_{\alpha}^{\beta} \frac{\bar{e}(\varphi)}{\bar{D}(r,z;\varphi)} d\varphi, \quad (5.395a,b)$$

where $\alpha(\beta)$ is the initial (final) angle of the helix; the number of turns of the helix (5.395a) can be an integer or real number. From (5.394) follows (5.396a through d):

$$\left\{ \frac{\partial \bar{D}}{\partial r}, \frac{\partial \bar{D}}{\partial z} \right\} = \frac{1}{\bar{D}} \{r - a\cos\varphi, z - k\varphi\}, \quad (5.396a,b)$$

$$\left\{ \frac{\partial}{\partial r}, \frac{\partial}{\partial z} \right\} \frac{1}{\bar{D}} = -\frac{1}{\bar{D}^2} \left\{ \frac{\partial \bar{D}}{\partial r}, \frac{\partial \bar{D}}{\partial z} \right\} = -\frac{1}{\bar{D}^3} \{r - a\cos\varphi, z - k\varphi\}, \quad (5.396c,d)$$

leading to

$$\{E_r(r,z), E_z(r,z)\} = -\left\{ \frac{\partial\Phi}{\partial r}, \frac{\partial\Phi}{\partial z} \right\} = \frac{\bar{a}}{4\pi\varepsilon} \int\limits_{\alpha}^{\beta} \frac{\bar{e}(\varphi)}{\bar{D}^3} \{r - a\cos\varphi, z - k\varphi\} d\varphi, \quad (5.397a,b)$$

the electric field (5.397a and b) corresponding to the potential (5.395b).

NOTE 5.18: Infinite and Semi-Infinite Circular Cylindrical Helix

In the case of a uniform electric charge distribution (5.398a) with total charge e in one turn of the helix and observer on axis (5.398b),

$$\text{const} = \bar{e} = \frac{de}{d\varphi} = \frac{e}{2\pi\bar{a}}, \quad D(0,z;\varphi) = \left| a^2 + (z - k\varphi)^2 \right|^{1/2}, \quad (5.398a,b)$$

the potential (5.395b) [axial electric field (5.397b)] simplifies to (5.399a) [(5.399b)]:

$$\Phi(0,z) = \frac{e}{8\pi^2\varepsilon} \int\limits_{\alpha}^{\beta} \left| a^2 + (z - k\varphi)^2 \right|^{-1/2} d\varphi, \quad (5.399a)$$

$$\{E_r(0,z), E_z(0,z)\} = \frac{e}{8\pi^2\varepsilon} \int\limits_{\alpha}^{\beta} \left| a^2 + (z - k\varphi)^2 \right|^{-3/2} \{-a\cos\varphi, z - k\varphi\} d\varphi. \quad (5.399b,c)$$

The change of variable,

$$\zeta = \frac{z - k\varphi}{a}, \quad d\varphi = -\frac{a}{k} d\zeta, \quad \zeta_{1,2} = \frac{\{z - k\alpha, z - k\beta\}}{a}, \quad (5.400a\text{-}c)$$

can be used to perform the integrations (5.399a) ≡ (5.401a) [(5.399b) ≡ (5.401b)]:

$$\Phi(0,z) = -\frac{e}{8\pi^2\varepsilon k}\int_{\zeta_1}^{\zeta_2}\left|1+\zeta^2\right|^{-1/2}\,d\zeta = \frac{e}{8\pi^2\varepsilon k}\left[\arg\sinh\zeta\right]_{\zeta_2}^{\zeta_1}, \tag{5.401a}$$

$$E_z(0,z) = -\frac{e}{8\pi^2\varepsilon ka}\int_{\zeta_1}^{\zeta_2}\left|1+\zeta^2\right|^{-3/2}\zeta\,d\zeta = \frac{e}{8\pi^2\varepsilon ka}\left[\left|1+\zeta^2\right|^{-1/2}\right]_{\zeta_1}^{\zeta_2}. \tag{5.401b}$$

Thus, *an electric charge distribution (5.390b) with density $\bar{e}(\varphi)$ per radian along (5.391a) a circular cylindrical helix with radius a and inclination γ or step (5.390a) in a medium with constant dielectrix permittivity ε has the electrostatic potential (5.395b) [electric field (5.397a and b)], where (5.394) is the distance from the observer to the source and (5.392a) the equivalent radius. In the case (5.398a) of a uniform electric charge distribution with total charge e per turn of the helix (5.391b), the potential (electric field) simplifies to (5.399a) [(5.399b and c)] for an observer on axis. The integrals for the potential (5.401a) [axial component of the electric field (5.401b)] on axis is given by (5.402a) [(5.402b)].*

$$\Phi(0,z) = \frac{e}{8\pi^2\varepsilon k}\left[\arg\sinh\left(\frac{z-k\alpha}{a}\right) - \arg\sinh\left(\frac{z-k\beta}{a}\right)\right], \tag{5.402a}$$

$$E_z(0,z) = \frac{e}{8\pi^2\varepsilon k}\left\{\left|a^2+(z-k\beta)^2\right|^{-1/2} - \left|a^2+(z-k\alpha)^2\right|^{-1/2}\right\}. \tag{5.402b}$$

If the helix is semi-infinite along the negative real axis (5.403a), the axial electric field (5.402b) simplifies to (5.403b):

$$-\infty = \alpha < \varphi \le \beta = 0: \quad E_z(0,z) = \frac{e}{8\pi^2\varepsilon k}\left|a^2+z^2\right|^{-1/2}. \tag{5.403a,b}$$

The axial electric decays from its value at the origin (5.404a) like (5.404b):

$$E_z(0,0) = \frac{e}{8\pi^2\varepsilon ka}, \quad E_z(z,0) = E_z(0,0)\left|1+\frac{z^2}{a^2}\right|^{-1/2}. \tag{5.404a,b}$$

The radial electric field (5.397a) does not vanish on axis, but should be weak for small inclination γ, since it vanishes for zero inclination (5.363), that is, when the helix (Figure 5.23) collapses to the circle (Figure 5.21). Both components of the asymptotic electric field are calculated next (Note 5.19) asymptotically at large distance without restriction on the angle of inclination of the helix.

NOTE 5.19: Helix with Small or Large Inclination

The exact potential (5.395a) for a uniform electric charge along the helix (5.398a) and observer at an arbitrary position (5.394) is given by

$$\Phi(r,z) = \frac{e}{8\pi^2\varepsilon}\int_{\alpha}^{\beta}\frac{d\varphi}{\sqrt{a^2+z^2+r^2-2ar\cos\varphi+(z-k\varphi)^2}}. \tag{5.405a}$$

The electrostatic potential and both components of the electric field are considered next for an observer at large axial distance compared with the length of the helix (5.405b) when the scalar potential (5.405a) for a uniform charge distribution (5.398a) simplifies to (5.405c):

$$|z| \gg k|\beta|, k|\alpha|: \quad \Phi(r,z) = \frac{e}{8\pi^2\varepsilon} \int_{\alpha}^{\beta} \left| r^2 + a^2 + z^2 - 2ar\cos\varphi \right|^{-1/2} d\varphi. \tag{5.405b,c}$$

Using the preceding expansion (5.365c; 5.366) with coefficients (2.159a) leads to

$$\Phi(r,z) = \frac{e}{8\pi^2\varepsilon b} \left\{ \beta - \alpha + \sum_{m=1}^{\infty} \frac{(2m-1)!!}{m!} \left(\frac{ar}{b^2} \right)^m \int_{\alpha}^{\beta} \cos^m \varphi \, d\varphi \right\}. \tag{5.406}$$

If the helix has an integer number of turns (5.407a), an observer at large axial distance (5.407b) experiences a potential (5.407d):

$$\beta - \alpha = 2N\pi, |z| \gg 2N\pi k, m = 2n:$$

$$\Phi(r,z) = \frac{e}{8\pi^2\varepsilon b} \left\{ 2N\pi + \sum_{m=1}^{\infty} \frac{(2m-1)!!}{m!} \left(\frac{ar}{b^2} \right)^m \int_{0}^{2N\pi} \cos^m \varphi \, d\varphi \right\}$$

$$= \frac{eN}{8\pi^2\varepsilon b} \left\{ 2\pi + \sum_{n=1}^{\infty} \frac{(4n-1)!!}{(2n)!} \left(\frac{ar}{b^2} \right)^{2n} \int_{0}^{2\pi} \cos^{2n} \varphi \, d\varphi \right\}, \tag{5.407a–d}$$

where only the even terms (5.407c) of the series do not vanish. Using (5.369b) in (5.407d) it follows that,

$$\Phi(r,z) = \frac{eN}{4\pi\varepsilon} \left| a^2 + z^2 + r^2 \right|^{-1/2} \left\{ 1 + \sum_{n=1}^{\infty} \frac{2^{-2n}}{n!} \frac{(4n-1)!!}{n!} (ar)^{2n} \left| a^2 + r^2 + z^2 \right|^{-2n} \right\} \tag{5.408}$$

the potential (5.407) is N times the potential of a circular charge (5.371), that is, an axially distant observer experiences the same potential from N turns of a helix as for N circular electric charges, both uniform and with the same radius. The corresponding electric field is N times (5.372a and b) and is axial on the axis. Next are considered (Note 5.20) both components of the electric field to leading order without restricting the angle of inclination of the helix and allowing for a noninteger number of turns.

NOTE 5.20: Integer or Noninteger Number of Turns of the Helix

In the case when the number of turns of the helix is not an integer, the integrals (5.406) are evaluated by (5.431a and b) in Note 5.25. The two leading terms (5.409a) of the potential (5.405c) are (5.409b):

$$(ar)^4 \ll b^4 = \left(a^2 + z^2 + r^2 \right)^2: \quad \Phi(r,z) = \frac{e}{8\pi^2\varepsilon b} \int_{\alpha}^{\beta} \left(1 + \frac{ar}{b^2}\cos\varphi \right) d\varphi. \tag{5.409a,b}$$

The electrostatic potential (5.409b) ≡ (5.410) has the following leading terms:

$$\Phi(r,z) = \frac{e}{8\pi^2\varepsilon} \left[\left| a^2 + z^2 + r^2 \right|^{-1/2} (\beta - \alpha) + ar \left| a^2 + z^2 + r^2 \right|^{-3/2} (\sin\beta - \sin\alpha) \right], \tag{5.410}$$

where (1) the slowest decaying term $O(1/b)$ in (5.365c) corresponds to the leading term of (5.407d) in the case (5.407a) of integer number of turns of the helix; (2) the second term on the r.h.s. of (5.410) vanishes for an integer number of turns of the spiral and does not appear in (5.407d); and (3) the next term would be of order b^{-5} in agreement with (5.407d). The electric field corresponding to (5.410) is given by

$$E_r(r,z) = -\frac{\partial \Phi}{\partial r} = \frac{e}{8\pi^2 \varepsilon} \left|a^2 + z^2 + r^2\right|^{-3/2} \left[r(\beta - \alpha) + a(\sin\beta - \sin\alpha)\frac{2r^2 - a^2 - z^2}{a^2 + z^2 + r^2} \right], \qquad (5.411a)$$

$$E_z(r,z) = -\frac{\partial \Phi}{\partial z} = \frac{ez}{8\pi^2 \varepsilon} \left|a^2 + z^2 + r^2\right|^{-3/2} \left[\beta - \alpha + \frac{3ar}{a^2 + z^2 + r^2}(\sin\beta - \sin\alpha) \right]. \qquad (5.411b)$$

The electric field decays like r/b^3 (z/b^3) for the axial (5.411a) [radial (5.411b)] components. Thus, *a uniform (5.398a) electric charge distribution along a circular cylindrical helix (5.390b) with radius a and (5.390a) inclination γ or step k and total electric charge e per turn in a medium of constant dielectric permittivity ε creates at an observer at a large radial distance compared with the length of the helix (5.407a and b) (1) the potential (5.407d) ≡ (5.408) involving the integrals (5.431a and b), and the leading terms are (5.410); and (2) the latter corresponds to the electric field (5.411a and b). The preceding results hold (Figure 5.23) for any starting α or ending β angle of the helix.* Several of the preceding simplifications may be considered together, for example, the asymptotic electric field on axis for any number of turns of the helix (Note 5.21).

NOTE 5.21: Near and Far Fields of an Electrically Charged Helix

The axial electric field on axis (5.402b) simplifies at large axial distance (5.412a and b) to (5.412c):

$$(z - k\alpha)^2 \gg a^2 \ll (z - k\beta)^2: \quad E_z(0,z) = \frac{e}{8\pi^2 \varepsilon k}\left(\frac{1}{z - k\beta} - \frac{1}{z - k\alpha} \right), \qquad (5.412a\text{--}c)$$

which can be approximated:

$$E_z(0,z) = \frac{e}{8\pi^2 \varepsilon k z}\left[\left(1 - \frac{k\beta}{z}\right)^{-1} - \left(1 - \frac{k\alpha}{z}\right)^{-1} \right] = \frac{e}{8\pi^2 \varepsilon}\frac{\beta - \alpha}{z^2}; \qquad (5.413a)$$

this result (5.413a) agrees (5.414b) with (5.411b) on axis for (5.414a) an observer at an axial distance large compared with the radius:

$$z^2 \gg a^2: \quad E_z(0,z) = \frac{e}{8\pi^2 \varepsilon}\frac{\beta - \alpha}{z^2}; \quad E_r(0,z) = \frac{e}{8\pi^2 \varepsilon}\frac{a}{z^3}(\sin\alpha - \sin\beta). \qquad (5.414a\text{--}c)$$

In the same conditions, the radial component of the electric field (5.411a) on axis decays faster like z^{-3}, with a coefficient that vanishes (5.414c) for an integer number of turns of the helix (5.407a). Thus, *the asymptotic electric field at large distance (5.412a) along the axis of a uniformly charged (5.398a) cylindrical helix (5.391a) has to leading order z^{-2} only an axial component (5.413a) ≡ (5.414b) that does not vanish for any number of turns of the helix. The asymptotic radial component on axis (5.414c) is of lower order z^{-3} for a helix with a noninteger number of turns; the term of order z^{-3} vanishes for an integer number of turns of the helix, implying that the asymptotic radial electric field on axis decays like z^{-5}.* This is an analogy between the electrostatic field (velocity of an irrotational flow) due to positive/negative electric charges (flow source/sinks) that applies to any distributions, such as along (1) a straight segment (Notes 5.7 through 5.9); (2) a circular loop (Notes 5.12 through 5.14); (3) a circular cylindrical helix (Notes 5.17 through 5.21). There is another analogy between the magnetic induction (velocity) of a magnetostatic field (incompressible flow) due to electric currents (vorticity) that applies to any distribution, for example, along (4) a straight segment (Notes 5.10 and 5.11); (5) a circular loop (Notes 5.15 and 5.16); and (6) a circular cylindrical helix (Notes 5.22 through 5.30). The case (6) is considered next and applies both to (1) the magnetic field of a helical coil (Notes 5.22 through 5.28); and (2) the velocity (Notes 5.29 and 5.30) field of helicoidal vorticity, such as the wake of the propeller (rotor) of an aircraft (helicopter).

NOTE 5.22: Electric Currents along a Cylindrical Helix

Replacing the electric charge *by electric currents along the tangent (5.392b) to the circular cylindrical helix (Figure 5.23) leads to the electric current distribution:*

$$\vec{j}(r,\varphi,z) = \overline{J}(\varphi)\vec{s}\,\delta(r-a)\,\delta(z-k\varphi), \tag{5.415}$$

where $\overline{J}(\varphi)$ is the density per radian. The corresponding vector potential (5.319b) is

$$\vec{A}(r,z) = \frac{\mu\overline{a}}{4\pi c} \int_{\alpha}^{\beta} \vec{s}\, \frac{\overline{J}(\varphi)}{\overline{D}(r,z;\varphi)}\, d\varphi$$

$$= \frac{\mu}{4\pi c} \int_{\alpha}^{\beta} \overline{J}(\varphi)\, \frac{\vec{e}_z k + a(\vec{e}_y \cos\varphi - \vec{e}_x \sin\varphi)}{\sqrt{a^2 + r^2 - 2ar\cos\varphi + (z-k\varphi)^2}}\, d\varphi, \tag{5.416}$$

and the magnetic induction is

$$\vec{B}(r,z) = \frac{\mu\overline{a}}{4\pi c} \int_{\alpha}^{\beta} \overline{J}(\varphi)\, \frac{\vec{s} \wedge (\vec{x}-\vec{y})}{\left[\overline{D}(r,z;\varphi)\right]^3}\, d\varphi = \frac{\mu a}{4\pi c} \int_{\alpha}^{\beta} \overline{J}(\varphi)\left| a^2 + r^2 - 2ar\cos\varphi + (z-k\varphi)^2 \right|^{-3/2}$$

$$\times \left\{ \vec{e}_x \left[(z-k\varphi)\cos\varphi + k\sin\varphi \right] + \vec{e}_y \left[(z-k\varphi)\sin\varphi + k\left(\frac{r}{a} - \cos\varphi\right) \right] + \vec{e}_z (a - r\cos\varphi) \right\} d\varphi. \tag{5.417}$$

The latter is obtained from the position vector (5.393) of the observer (5.360a) relative to the source (5.391a) whose modulus is the distance (5.394). The outer product of (5.393) with the direction (5.392b) of the electric current is

$$\vec{s} \wedge (\vec{x}-\vec{y}) = \frac{a}{\overline{a}} \begin{vmatrix} \vec{e}_x & \vec{e}_y & n\vec{e}_z \\ -\sin\varphi & \cos\varphi & \dfrac{k}{a} \\ r - a\cos\varphi & -a\sin\varphi & z-k\varphi \end{vmatrix}$$

$$= \frac{a}{\overline{a}} \left\{ \vec{e}_x \left[(z-k\varphi)\cos\varphi + k\sin\varphi \right] + \vec{e}_y \left[(z-k\varphi)\sin\varphi + k\left(\frac{r}{a} - \cos\varphi\right) \right] + \vec{e}_z (a - r\cos\varphi) \right\}, \tag{5.418}$$

which simplifies to (5.387a) if $k = 0$, that is, if the helix collapses to a circular loop $\overline{a} = a$. Substituting (5.415) and (5.418) in (5.319c) specifies the magnetic induction (5.417).

NOTE 5.23: Vector Potential and Magnetic Induction on Axis

In the case of a uniform electric current (5.419a) corresponding (5.392a) to a total electric current J per turn, the axial component of the vector potential (5.416) is given on axis by (5.419b):

$$\text{const} = \frac{dJ}{d\varphi} = \overline{J} = \frac{J}{2\pi\overline{a}}: \quad A_z(0,z) = \frac{\mu J}{8\pi^2 c\overline{a}} \int_\alpha^\beta \frac{d\varphi}{\sqrt{a^2+(z-k\varphi)^2}}$$

$$= \vec{e}_z \frac{\mu J}{8\pi^2 c\overline{a}}\left[\text{arg}\sinh\left(\frac{z-k\alpha}{a}\right) - \text{arg}\sinh\left(\frac{z-k\beta}{a}\right)\right], \qquad (5.419\text{a,b})$$

where the integral is evaluated as in (5.399a through c; 5.400a through c; 5.401a; 5.402a). The axial component of the vector potential is zero (nonzero) for a circular (helical) current [Figure 5.21 (Figure 5.23)]. Also for a uniform electric current (5.419a), the axial component of the magnetic induction (5.417) is given on axis by (5.420a):

$$B_z(0,z) = \frac{\mu J a}{8\pi^2 c\overline{a}} \int_\alpha^\beta \left|a^2+(z-k\varphi)^2\right|^{-3/2} d\varphi = -\frac{\mu J}{8\pi^2 c\overline{a}k} \int_{\zeta_1}^{\zeta_2} \left|1+\zeta^2\right|^{-3/2} d\zeta, \qquad (5.420\text{a,b})$$

leading to (5.420b) by the change of variable (5.400a through c). The integral is evaluated using the change of variable (5.421a and b) that leads to (5.421c):

$$\zeta = \sinh\eta, \; d\zeta = \cosh\eta \, d\eta: \quad \int_{\zeta_1}^{\zeta_2} \left|1+\zeta^2\right|^{-3/2} d\zeta = \int_{\eta_1}^{\eta_2} \text{sech}^2\eta \, d\eta = \left[\tanh\eta\right]_{\eta_1}^{\eta_2} = \left[\frac{\sinh\eta}{\sqrt{1+\sinh^2\eta}}\right]_{\eta_1}^{\eta_2}$$

$$= \left[\frac{\zeta}{\sqrt{1+\zeta^2}}\right]_{\zeta_1}^{\zeta_2} = \left[\frac{z-k\varphi}{\sqrt{a^2+(z-k\varphi)^2}}\right]_\alpha^\beta, \qquad (5.421\text{a–c})$$

where the primitive (II.7.107) was used followed by the changes of variable (5.421a) and (5.400a).

Substituting (5.421c) in (5.420b), it follows that *a uniform (5.419a) electric current (5.415) along a circular cylindrical helix of radius a and (5.390a) inclination γ or step k creates a magnetostatic potential (magnetic induction) with axial component (5.419b) [(5.422)]:*

$$B_z(0,z) = \frac{\mu J}{8\pi^2 c\overline{a}k}\left[\frac{z-k\alpha}{\sqrt{a^2+(z-k\alpha)^2}} - \frac{z-k\beta}{\sqrt{a^2+(z-k\beta)^2}}\right]. \qquad (5.422)$$

In the case (5.403a) ≡ (5.423a) of semi-infinite helix along the negative z-axis, the axial component of the magnetic induction (5.422) simplifies to (5.423b):

$$-\infty \equiv \alpha < \varphi \leq \beta = 0: \quad B_z(0,z) = -\frac{\mu J}{8\pi^2 c\overline{a}k}\left(1 + \frac{z}{\sqrt{a^2+z^2}}\right), \qquad (5.423\text{a,b})$$

which is related by (5.424b) to the value at the origin (5.424a):

$$B_z(0,0) = -\frac{\mu J}{8\pi^2 c \bar{a} k} = -\frac{\mu J}{8\pi^2 c k \sqrt{a^2 + k^2}}, \quad B_z(0,z) = B_z(0,0)\left(1 + \frac{z}{\sqrt{a^2 + z^2}}\right). \tag{5.424a,b}$$

The magnetic induction far into (away from) the helix $z \to -\infty$ ($z \to +\infty$) tends to zero (5.425a) [twice the value at the origin (5.425b)]:

$$\lim_{z \to -\infty} B_z(0,z) = 0, \quad \lim_{z \to +\infty} B_z(0,z) = 2B_z(0,0), \tag{5.425a,b}$$

Thus, (1) a semi-infinite circular helical coil with uniform electric current per unit length causes on axis a magnetic induction whose axial component is zero far into the coil (5.425a) and in the opposite direction far away from the coil, doubles the value at the origin (5.425b); and (2) a vortical incompressible flow due to the wake of an aircraft propeller or helicopter rotor, represented by a circular helical vortex with constant vorticity per unit length, leads to a velocity field on axis, whose axial component vanishes far downstream into the vortex, and in the opposite direction far upstream, the velocity is the double of the value at the origin. In both cases (1) and (2), this corresponds to a magnetic flux tube (flow stream tube) contracting far above to half the area at the start of the helix and expanding to an infinite radius far below. The velocity at the origin is the arithmetic mean of the velocities far upstream and downstream:

$$v_z(0,z) = \frac{1}{2}\left[v_z(0,+\infty) + v_z(0,-\infty)\right]. \tag{5.425c}$$

The latter result can be obtained from momentum theory applied to an actuator disk (Notes 5.28 and 5.29) that specifies the thrust of an aircraft propeller or helicopter rotor (Note 5.30).

NOTE 5.24: Off Axis Asymptotic Magnetic Potential of a Helical Current

For an arbitrary observer position at an axial distance large compared with the length of the helix (5.426a), the vector potential (5.416) simplifies for a uniform current (5.419a) to (5.426b):

$$|z| \gg k|\beta|, k|\alpha|: \quad \vec{A}(r,z) = \frac{\mu J}{8\pi^2 c \bar{a}} \int_\alpha^\beta \frac{a(\vec{e}_y \cos\varphi - \vec{e}_x \sin\varphi) + \vec{e}_z k}{\sqrt{a^2 + r^2 + z^2 - 2ar\cos\varphi}} d\varphi, \tag{5.426a,b}$$

with no restriction on the starting α and ending β angle of the helix. The vector potential has three components. The component in the direction of the observer corresponds in cylindrical coordinates to the radial component in (5.426b) and is given by

$$A_r(r,z) = -\frac{\mu J a}{8\pi^2 c \bar{a}} \int_\alpha^\beta \frac{\sin\varphi}{\sqrt{a^2 + z^2 + r^2 - 2ar\cos\varphi}} d\varphi$$

$$= \frac{\mu J}{8\pi^2 c \bar{a} r}\left[\sqrt{a^2 + z^2 + r^2 - 2ar\cos\alpha} - \sqrt{a^2 + z^2 + r^2 - 2ar\cos\beta}\right], \tag{5.427a,b}$$

leading to (5.427b) as in (5.378); this component vanishes for an integer number of turns of the helix (5.407a). The axial component in (5.426b) is given by (5.428a):

$$A_z(r,z) = \frac{\mu Jk}{8\pi^2 c\bar{a}} \int_\alpha^\beta \frac{d\varphi}{\sqrt{a^2 + z^2 + r^2 - 2ar\cos\varphi}}$$

$$= \frac{\mu Jk}{8\pi^2 c\bar{a}} \left|a^2 + z^2 + r^2\right|^{-1/2} \left[\beta - \alpha + \sum_{n=1}^{\infty} \frac{(2n-1)!!}{n!} (ar)^n \left|a^2 + z^2 + r^2\right|^{-2n} \int_\alpha^\beta \cos^n\varphi \right], \qquad (5.428\text{a,b})$$

leading to (5.428b) by (5.365c; 5.366; 2.159b) as in (5.406). The component normal to the plane through the observer and the axis corresponds to the azimuthal component in cylindrical coordinates (5.426b) and is given

$$A_\varphi(r,z) = \frac{\mu Ja}{8\pi^2 c\bar{a}} \int_\alpha^\beta \frac{\cos\varphi}{\sqrt{a^2 + z^2 + r^2 - 2ar\cos\varphi}} d\varphi = \frac{\mu Ja}{8\pi^2 c\bar{a}} \left|a^2 + z^2 + r^2\right|^{-1/2}$$

$$\times \left\{ \sin\beta - \sin\alpha + \sum_{n=0}^{\infty} \frac{(2n-1)!!}{(2n)!} (ar)^n \left|a^2 + z^2 + r^2\right|^{-n} \int_\alpha^\beta \cos^{n+1}\varphi \, d\varphi \right\}, \qquad (5.429\text{a,b})$$

where (5.429a and b) differs from (5.428a and b) in having an extra factor $\cos\varphi$ in (1) the terms $n = 1$, 2, …; and (2) for the term $n = 0$, the integrand changes from 1 to $\cos\varphi$; hence, the leading term $\beta - \alpha$ changes to $\sin\beta - \sin\alpha$.

NOTE 5.25: Integer and Noninteger Number of Turns of the Helix

The azimuthal (5.429a and b) [axial (5.428a and b)] magnetic induction involves integrals of powers of the circular cosine. Using the expansion of even (odd) powers of the circular cosine as a sum of cosines of multiple angles (II.5.78a) ≡ (5.430a) [(II.5.78b) ≡ (5.430b)],

$$\cos^{2n}\varphi = 2^{-2n}\left\{ \binom{2n}{n} + 2\sum_{m=0}^{n-1} \binom{2n}{m} \cos[2(n-m)\varphi] \right\}, \qquad (5.430\text{a})$$

$$\cos^{2n+1}\varphi = 2^{-2n} \sum_{m=0}^{n} \binom{2n+1}{m} \cos[(2n-2m+1)\varphi], \qquad (5.430\text{b})$$

the corresponding integrals are evaluated by (5.431a) [(5.431b)]:

$$\int_\alpha^\beta \cos^{2n}\varphi \, d\varphi = \frac{(2n)!}{n!} \frac{2^{-2n}}{n!} (\beta - \alpha)$$

$$+ 2^{-2n} \sum_{m=0}^{n-1} \frac{(2n)!}{m!} \frac{\sin[2(n-m)\beta] - \sin[2(n-m)\alpha]}{(n-2m)!(n-m)}, \qquad (5.431\text{a})$$

$$\int_\alpha^\beta \cos^{2n+1}\varphi \, d\varphi = 2^{-2n} \sum_{m=0}^{n} \frac{(2n+1)!}{m!} \frac{\sin\left[(2n-2m+1)\beta\right] - \sin\left[(2n-2m+1)\alpha\right]}{(2n-m+1)!(2n-2m+1)}. \tag{5.431b}$$

Thus, *the vector potential of a uniform (5.419a) electric current along a circular cylindrical spiral (5.415) is given asymptotically at large distance (5.426a) by (5.426b) implying that (1) the radial component (5.427a and b) vanishes (does not vanish) for an integer (noninteger) number of turns; and (2/3) the axial (5.428a and b) [azimuthal (5.429a and b)] components involve the integrals (5.431a and b), where for an integer number of turns, only the first term on the r.h.s of (5.431a) is nonzero. The vector potential (5.427a and b; 5.428a and b; 5.429a and b) for a helix differs from that of a circle (5.380) ≡ (5.383) with a uniform electric current density per radian (5.419a; 5.376a) because (1) instead of one turn for a complete circle (5.432a), there is any real number of turns (5.432c) for a spiral with (5.432b) arbitrary starting and ending angle:*

$$0 \le \varphi \le 2\pi; \quad \alpha \le \varphi \le \beta, \quad N = \frac{\beta-\alpha}{2\pi}; \tag{5.432a–c}$$

(2) for one (an integer number N) of turns of the helix, the vector potential has the same component (5.380) ≡ (5.429b) [multiplied by N] normal to the plane through the observer and the axis; (3) the helix case adds an axial component to the vector potential (5.428b) that is nonzero in all cases including the case of an integer number of turns of the helix; and (4) in the case of a noninteger number of turns of the helix, there is a third component of the vector potential (5.427b) in the direction of the observer. The asymptotic vector potential of a uniform helical current (Note 5.25) simplifies in two cases: (1) to the two lowest orders for arbitrary starting and ending angles of the helix (Note 5.26); and (2) to all orders for an integer number of turns of the helix (Note 5.27).

NOTE 5.26: Helical Coil with a Noninteger Number of Turns

Using (5.365c) the denominator of the vector potential (5.426b)

$$\left|a^2 + r^2 + z^2 - 2ar\cos\varphi\right|^{-1/2} = \left|b^2 - 2ar\cos\varphi\right|^{-1/2} = \frac{1}{b} + \frac{ar}{b^2}\cos\varphi + O\left(\frac{a^2r^2}{b^5}\right), \tag{5.433}$$

leads to the lowest orders in b to

$$\vec{A}(r,z) = \frac{\mu J}{8\pi^2 c \bar{a} b} \int_\alpha^\beta \left[a\left(\vec{e}_y \cos\varphi - \vec{e}_x \sin\varphi\right) + \vec{e}_z k\right]\left(1 + \frac{ar}{b^2}\cos\varphi\right) d\varphi. \tag{5.434}$$

Bearing in mind that

$$\int_\alpha^\beta \{\sin\varphi, \cos\varphi\}\cos\varphi \, d\varphi = \frac{1}{2}\int_\alpha^\beta \{\sin(2\varphi), 1+\cos(2\varphi)\} d\varphi = \left[-\frac{\cos(2\varphi)}{4}, \frac{\varphi}{2} + \frac{\sin(2\varphi)}{4}\right]_\alpha^\beta, \tag{5.435}$$

the vector potential has the cylindrical components $(x, y, z) \to (r, \varphi, z)$:

$$A_r(r,z) = \frac{\mu J a}{8\pi^2 c \bar{a} b}\left\{\cos\beta - \cos\alpha + \frac{ar}{4b^2}\left[\cos(2\beta) - \cos(2\alpha)\right]\right\}, \tag{5.436a}$$

$$A_\varphi\left(r,z\right)=\frac{\mu Ja}{8\pi^2c\bar a b}\left\{\sin\beta-\sin\alpha+\frac{ar}{4b^2}\left[2\left(\beta-\alpha\right)+\sin\left(2\beta\right)-\sin\left(2\alpha\right)\right]\right\}, \tag{5.436b}$$

$$A_z\left(r,z\right)=\frac{\mu Jk}{8\pi^2c\bar a b}\left[\beta-\alpha+\frac{ar}{b^2}\left(\sin\beta-\sin\alpha\right)\right]. \tag{5.436c}$$

These results apply to any starting α and ending angle β of the helix and simplify for an integer number of turns.

Thus, *for an observer (5.360a) at a large distance (5.426a) from a circular cylindrical helix (5.391a) with a uniform (5.419a) electric current (5.415), the asymptotic vector potential (5.436a through c) has a leading term of order b^{-1} for all components (5.365c) if the number of turns is not an integer. For an integer number of turns (5.437a) of the helix, (1) the slow decaying term $O(b^{-1})$ given by (5.436c) appears only in the axial component of the vector potential (5.346c), and the next term is two orders higher (5.437b); (2) the next order $O(b^3)$ appears in the azimuthal component (5.436b) transverse to the plane through the observer and the axis (5.437c); and (3) the radial component in the direction of the observer is zero (5.437d) to all orders (5.436a) by (5.427b):*

$$\beta-\alpha=2N\pi:\quad A_z\left(r,z\right)=\frac{\mu JkN}{4\pi c\bar a b}\left[1+O\left(\frac{a^2r^2}{b^4}\right)\right],\quad A_\varphi\left(r,z\right)=\frac{\mu JNa^2r}{8\pi c\bar a b^3},\quad A_r\left(r,z\right)=0. \tag{5.437a–d}$$

In the case of zero step (5.438a), the vector potential (5.438c),

$$k=0,\bar a\equiv\sqrt{a^2+k^2}=a:\quad \vec A\left(r,z\right)=\vec e_\varphi\frac{\mu JNar}{8\pi c}\left|a^2+z^2+r^2\right|^{-3/2}, \tag{5.438a–c}$$

coincides with the leading term of the circle (5.383) for one turn $N=1$ of the helix. The vector potential of a helical electric current (5.436a through c) could also be obtained from the lowest-order terms of (5.427a and b; 5.428a and b; 5.429a and b) using (5.431a and b). In the latter (5.431a and b), only the first term on the r.h.s. of (5.431a) does not vanish for an integer number of turns of the helix; this allows evaluation of the asymptotic vector potential and magnetic induction to all orders (Note 5.27).

NOTE 5.27: Electric Current with an Integer Number of Turns

The vector potential for an observer at large axial distance (5.426a) ≡ (5.439b) for a helix with N turns (5.439a) is given (5.426b; 5.427b) by (5.439c):

$$\beta-\alpha=2N\pi,\left|z\right|\gg 2\pi Nk:$$

$$\vec A\left(r,z\right)=\frac{\mu JN}{8\pi^2c\bar a b}\left[\vec e_z k2\pi+\sum_{m=1}^\infty\frac{\left(2m-1\right)!!}{m!}\left(\frac{ar}{b^2}\right)^m\int_0^{2\pi}\left\{\vec e_z k\cos^m\varphi+\vec e_y a\cos^{m+1}\varphi\right\}d\varphi\right], \tag{5.439a–c}$$

using the same series expansion as in (5.428b; 5.365b). The vanishing of (5.427b) for an integer number of turns of the helix (5.437a) implies that the vector potential (5.437c) has two components: (1) the axial component (5.440b) involves even powers (5.440a):

$$m=2n:\quad A_z\left(r,z\right)=\frac{\mu JNk}{8\pi^2c\bar a b}\left[2\pi+\sum_{n=1}^\infty\frac{\left(4n-1\right)!!}{\left(2n\right)!}I_{2n}\left(\frac{ar}{b^2}\right)^{2n}\right], \tag{5.440a,b}$$

like the electric potential (5.366) ≡ (5.369a and b) of uniform electric charges, but due to the helicity $k \neq 0$; and (2) the azimuthal component orthogonal to the plane through the axis and the observer (5.441b) that involves only odd powers (5.441a):

$$m = 2n+1: \quad A_\varphi(r,z) = \frac{\mu J N a}{8\pi^2 c \bar{a} b} \sum_{n=0}^{\infty} \frac{(4n+1)!!}{(2n+1)!} I_{2n+2} \left(\frac{ar}{b^2}\right)^{2n+1}, \quad (5.441a,b)$$

like the vector potential of uniform electric currents along a circle (5.381b) ≡ (5.383). Using (5.369b) [(5.382b)] in (5.440b) [(5.441b)] leads to

$$\vec{A}(r,z) = \vec{e}_z \frac{\mu J N k}{4\pi c \bar{a}} \left|a^2 + z^2 + r^2\right|^{-1/2} \left\{ 1 + \sum_{n=1}^{\infty} \frac{2^{-2n}}{n!} \frac{(4n-1)!!}{n!} (ar)^{2n} \left|a^2 + z^2 + r^2\right|^{-2n} \right\}$$

$$+ \vec{e}_\varphi \frac{\mu J N a}{8\pi c \bar{a}} \sum_{n=1}^{\infty} \frac{2^{-2n}}{n!} \frac{(4n+1)!!}{(n+1)!} (ar)^{2n+1} \left|a^2 + z^2 + r^2\right|^{-2n-3/2}, \quad (5.442)$$

as the total vector potential.

The curl (6.41b) in cylindrical coordinates applied to (5.442)

$$\nabla \wedge \vec{B} = -\vec{e}_r \partial_z A_\varphi - \vec{e}_\varphi \partial_r A_z + \vec{e}_z r^{-1} \partial_r (A_\varphi r) \quad (5.443)$$

specifies the magnetic induction. Using also

$$m = n, n+1: \quad \left\{ \frac{\partial}{\partial r}, \frac{\partial}{\partial z} \right\} \left[\left|a^2 + z^2 + r^2\right|^{-m-1/2} \right] = -\{r,z\} \left|a^2 + z^2 + r^2\right|^{-m-3/2} (2m+1) \quad (5.444a-c)$$

leads to the three cylindrical components of the magnetic induction:

$$B_r(r,z) = \frac{\mu J N a z}{8\pi c \bar{a}} \sum_{n=0}^{\infty} \frac{2^{-2n}}{n!} \frac{(4n+3)!!}{(n+1)!} (ar)^{2n+1} \left|a^2 + z^2 + r^2\right|^{-2n-5/2}, \quad (5.445a)$$

$$B_\varphi(r,z) = -\frac{\mu J N k a}{2\pi c \bar{a}} \sum_{n=1}^{\infty} \frac{2^{-2n}}{n!} \frac{(4n+1)!!}{(n-1)!} (ar)^{2n-1} \left|a^2 + z^2 + r^2\right|^{-2n-1/2}$$

$$+ \frac{\mu J N k r}{2\pi c \bar{a}} \left|a^2 + z^2 + r^2\right|^{-3/2} \left\{ 1 + \sum_{n=1}^{\infty} \frac{2^{-2n}}{n!} \frac{(4n+1)!!}{(n)!} (ar)^{2n} \left|a^2 + z^2 + r^2\right|^{-2n} \right\}, \quad (5.445b)$$

$$B_z(r,z) = \frac{\mu J N a^2}{4\pi c \bar{a}} \sum_{n=0}^{\infty} \frac{2n+1}{2^{2n}} \frac{(4n+1)!!}{(n!)^2} (ar)^{2n} \left|a^2 + z^2 + r^2\right|^{-2n-3/2}$$

$$- \frac{\mu J N a r}{8\pi c \bar{a}} \sum_{n=0}^{\infty} \frac{2^{-2n}}{n!} \frac{(4n+3)!!}{(n+1)!} (ar)^{2n+1} \left|a^2 + r^2 + z^2\right|^{-2n-5/2}. \quad (5.445c)$$

On axis (5.446a), the vector potential (5.442) [magnetic induction (5.445a through c)] is axial (5.446b) [(5.445c)] and due to the helicity, hence proportional to the number of turns and thus large for a long helix:

$$r = 0: \quad \vec{A}(0,z) = \vec{e}_z \frac{\mu J N k}{4\pi c \bar{a}} \left| a^2 + z^2 \right|^{-1/2}, \quad \vec{B}(0,z) = \vec{e}_z \frac{\mu J N a^2}{4\pi c \bar{a}} \left| a^2 + z^2 \right|^{-3/2}. \tag{5.446a–c}$$

Thus, *a uniform (5.419a) current distribution (5.415) with total current J per turn along a cylindrical helix (Figure 5.23) with radius a and inclination γ, hence step (5.390a) in a medium with constant magnetic permeability μ, creates at an observer at a large axial distance compared with the length of the helix (5.426a) a vector potential (5.426b; 5.427a and b; 5.428a and b; 5.429a and b) whose leading terms are (5.436a through c), for arbitrary starting α and ending β angle of the helix. If the helix has an integer N number of turns (5.437a), then (5.437b through d) are the leading terms of the vector potential. If the helix has an integer number of turns (5.439a), the asymptotic vector potential (magnetic induction) is given to all orders by (5.442) [(5.445a through c)]; these simplify on axis (5.446a) to (5.446b) [(5.446c)] that are proportional to the number of turns of the helix (Figure 5.23) seen as N parallel circular electric currents (Figure 5.21). In the case k = 0 of a circular current, the vector potential (5.346b) vanishes on the axis (5.385a), and the magnetic induction (5.346c) corresponds to (5.385b) with one turn N = 1.*

NOTE 5.28: Pressure Jump across an Actuator Disk

The result (5.425c) that the axial velocity on the propeller or rotor plane is the arithmetic mean of the axial velocities far upstream and downstream (Figure 5.24) can be obtained more simply by applying momentum theory to an actuator disk (Figure 5.25a and b). The propeller is replaced by an **actuator disk** (Figure 5.25a) with the same radius *a*, where the axial velocity v_0 is uniform, corresponding to an infinite number of blades. Far upstream (downstream), the jet passing through the actuator disk has lower (higher) velocity $v_\infty(v)$ and larger (smaller) radius $r_-(r_+)$ to conserve the mass flux (5.447b through d) that for an incompressible flow (5.447a) corresponds to the conservation of the volume flux:

$$\rho = \text{const}: \quad \dot{m} = \rho \pi a^2 v_0 = \rho \pi r_-^2 v_\infty = \rho \pi r_+^2 v. \tag{5.447a–d}$$

On either side of the actuator disk, the flow is potential, and thus the stagnation pressure (I.14.27a through c) ≡ (II.2.12a and b) ≡ (6.101a through e) is constant and thus (1/2) the same (5.448a) [(5.448b)] far upstream (downstream) and above (below) the actuator disk:

$$p_\infty + \frac{1}{2}\rho v_\infty^2 = p + \frac{1}{2}\rho v_0^2, \quad p_\infty + \frac{1}{2}\rho v^2 = p + \Delta p + \frac{1}{2}\rho v_0^2, \tag{5.448a,b}$$

where the pressure at infinity p_∞ is the same up and downstream; and (3) the pressure has a jump from *p* above to $p + \Delta p$ below the actuator disk, that is, the jump of stagnation pressure Δp across the actuator disk corresponds to all vorticity being concentrated there. This is consistent with the asymptotic approximation (5.439a and b) of observer at large axial distance compared with the extent of the helicoidal vortex. The **thrust** of the propeller or rotor equals the area (5.449a) multiplied by the pressure jump (5.448a and b), leading to (5.449b):

$$A = \pi a^2: \quad T = A\Delta p = \pi a^2 \frac{1}{2}\rho\left(v^2 - v_\infty^2\right) = \frac{1}{2}A\rho\left(v - v_\infty\right)\left(v + v_\infty\right). \tag{5.449a–d}$$

The thrust is calculated next in an alternative way.

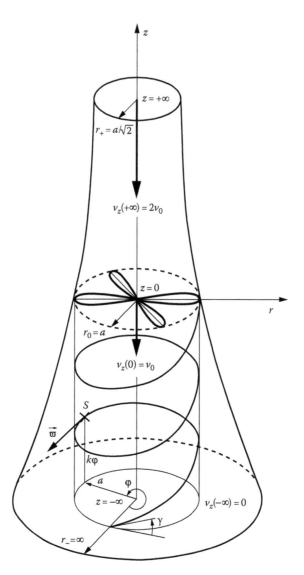

FIGURE 5.24   The circular cylindrical helical vortex wake of the propeller (rotor) of an aircraft (helicopter) taken as semi-infinite induces on the axis an axial velocity component that compared with the value on the propeller (rotor) plane: (1) decays into the vortex wake, vanishing far downstream; and (2) is increased above the vortex wake to the double of the value far upstream. This confirms that the axial velocity on the propeller (rotor) plane is the arithmetic mean of the values far up and downstream.

NOTE 5.29: Balance of Volume and Momentum Fluxes

Consider (Figure 5.25b) a control cylinder of large radius r around the stream tube of the axial velocity. The mass flux (5.447b) times the velocity specifies the momentum flux out of (5.450a) [into the (5.450b)] **control volume**:

$$T_{out} = \pi r_+^2 \rho v^2 + \pi\left(r^2 - r_+^2\right)\rho v_\infty^2, \tag{5.450a}$$

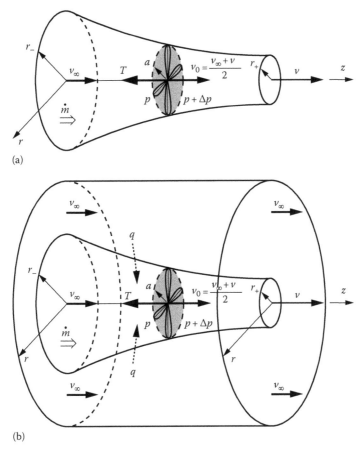

FIGURE 5.25 The preceding result (Figure 5.24) can be proved most simply replacing the aircraft propeller or helicopter rotor by an actuator disk across which there is a pressure jump (a). Comparing the thrust with that calculated using a large cylindrical control volume (b) relates the three velocities: upstream, downstream and on the propeller/rotor plane. This leads to several alternative formulas for the thrust of an aircraft propeller or helicopter rotor.

$$T_{in} = \pi r^2 \rho v_\infty^2 + \pi r_+^2 \rho (v - v_\infty) v_\infty, \tag{5.450b}$$

where the free stream velocity v_∞ is the same upstream and downstream outside the jet (inside the jet, the velocity is v). The thrust of the propeller or rotor is the difference of the momentum flux out (5.450a) and in (5.450b), leading to (5.451a):

$$T = T_{out} - T_{in} = \pi r_+^2 \rho v (v - v_\infty) = \pi a^2 \rho v_0 (v - v_\infty) = \rho A v_0 (v - v_\infty). \tag{5.451a-c}$$

Using mass conservation (5.447b) leads to (5.451b) or (5.451c) introducing the area (5.449a) of the actuator disk. Comparing (5.451c) $\equiv$ (5.449d) follows

$$2v_0 = v_\infty + v = v(-\infty) + v(+\infty) = 2v(0), \tag{5.452}$$

confirming that the axial velocity in the propeller disk is the arithmetic mean of the velocities far up and downstream, in agreement with (5.425c) $\equiv$ (5.452).

NOTE 5.30: Thrust of an Aircraft Propeller or Helicopter Rotor

The jet passing through the actuator disk is not a stream tube, because conservation of the mass flux implies that there must be a volume flux through the sides of the jet:

$$q = \pi r_+^2 v + \pi\left(r^2 - r_+^2\right)v_\infty - \pi r^2 v_\infty = \pi r_+^2\left(v - v_\infty\right). \tag{5.453}$$

Thus, *an aircraft propeller or helicopter rotor represented as an actuator disk (Figure 5.25a) across which there is a pressure jump has a thrust*

$$T = \frac{\rho}{2}A\left(v^2 - v_\infty^2\right) = \dot{m}\left(v - v_\infty\right) = \rho q v, \tag{5.454a–c}$$

which equals (1) the area of the actuator disk (5.449a) times the difference between the dynamic pressures of the jet and the free stream (5.449c) ≡ (5.454a); (2) the mass flow rate (5.447a; 5.449a) times the difference between the jet and free stream velocities (5.451b) = (5.454b); and (3) the volume influx due to mean flow entrainment through the sides of the jet (5.453) times the mass density and jet velocity (5.451a) ≡ (5.454c). The velocity in the propeller plane is the arithmetic mean of the jet and free stream velocities (5.452), and thus the thrust (5.451a) ≡ (5.455a) can be expressed by (5.455b) ≡ (5.455c):

$$T = \rho A\left(v - v_\infty\right)\frac{v + v_\infty}{2} = \rho A v_0\left(v - v_\infty\right) = 2\rho A v_0\left(v_0 - v_\infty\right) = 2\rho A v_0\left(v - v_0\right), \tag{5.455a–c}$$

in terms of any two (5.454a) ≡ (4.455b) ≡ (4.455c) of the three velocities (v_0, v_∞, v). From (5.455a) [(5.451a)], using (5.447a) [(5.453)] follows (5.456) [(5.457)]:

$$T = \rho A v_0\left(v - v_\infty\right) = \rho \pi a^2 v_0\left(v - v_\infty\right) = \dot{m}\left(v - v_\infty\right), \tag{5.456}$$

$$T = \pi r_+^2 \rho v\left(v - v_\infty\right) = \rho q v, \tag{5.457}$$

proving (4.454b) ≡ (4.456) [(4.454c) ≡ (4.457)]. The simple actuator disk model provides only one result (5.542) = (5.425c) of vortex wave theory (Notes 5.22 through 5.27). A generalization of the actuator disk distinct from the helical vortex is the general axisymmetric flow (Chapter 6).

5.10 Conclusion

The generalized function unit jump (Figure 5.1a) of argument $a^2 - x^2$ is unity in the open interval $|x| < a$, one-half at the end points $x = \pm a$, and zero outside $|x| > a$; the generalized function unit impulse of the same argument (Figure 5.1b) vanishes everywhere except at the end points $x = \pm a$, where it coincides with a simple impulse of magnitude $\mp 1/2a$. Proceeding from argument a function of one variable to a function of two variables, the generalized function unit jump (Figure 5.3a) of argument $a^2 - x^2 - y^2$ is unity inside the circle of radius a and center at the origin, one-half on the boundary, and zero outside; the corresponding generalized function impulse (Figure 5.3b) is zero everywhere except on the circle, where it coincides with a single impulse of amplitude $-1/2a$. Continuing from one- to two-dimensional generalized functions, the product (Figure 5.4a) of two unit jumps is unity on the first quadrant, one-half on the positive coordinate half-axis, one-quarter at the origin, and zero elsewhere; the corresponding product of two unit impulses (Figure 5.4b) is zero everywhere except at the origin, where it is singular like a cylindrically symmetric unit impulse (Figure 5.5), with amplitude $1/2\pi r$ as $r \to 0$. The product of unit jumps of two functions (Figure 5.6) is unity in the region where both are positive, one-half where one is positive and the other zero, one-quarter where both vanish, and zero elsewhere. The two-dimensional impulse with arguments $a^2 - x^2$, $b^2 - y^2$, is zero everywhere, except at the points $(x, y) = (\pm a, \pm b)$, where (Figure 5.7) it coincides with a simple impulse of amplitude $\{4(\pm a)(\pm b)\}^{-1}$. Considering an ordinary function with several simple zeros (Figure 5.2a), it may be used as argument in (1) the unit jump (Figure 5.2b) that is unity (zero) where the function is positive (negative) and one-half where the function is zero; and (2) the unit impulse (Figure 5.2c) that is zero (infinite) where the function is nonzero (zero). The unit jump and impulse (Table 5.4) can be used in the construction of other generalized functions (List 5.1).

The use of the generalized functions unit jump and unit impulse (Diagram 5.1) to identity domains, boundaries, interiors, and exteriors in an N-dimensional space (Figures 5.1, 5.3, 5.4, 5.6, and 5.7) is summarized in Table 5.1. The generalized function unit impulse with cylindrical symmetry (Figure 5.5) is the particular two-dimensional case of radial symmetry that can be considered in any dimension (Table 5.3). The radially symmetric unit impulses are an example of a one-dimensional generalized function of one variable equivalent to a particular combination of N-dimensional generalized functions of N variables (Diagram 5.1). The generalized functions relate to line, surface (Figure 5.10), and volume integrals and can be used to prove (1) the divergence, gradient, and curl integral theorems (Figure 5.11a) for a domain with a closed regular boundary using the corresponding lemmas (Figure 5.11b) for a surrounding hyperparallelepiped; and (2) the Stokes' theorem concerning (Figure 5.13) a regular loop that is the intersection of two regular surfaces and the part of one surface that is supported on the loop, that is, lies to one side of the other surface. The use of generalized functions to evaluate integrals on surfaces and along curves (Figure 5.8a and b) finds application in the proofs of the curl (divergence, gradient, and curl) theorem on a surface (Figure 5.13) [in a N-dimensional domain (Figure 5.11a and b)] the divergence and curl theorems in the plane (Figure 5.14) and other related theorems (Table 5.2; Diagram 5.2), such as the theorems of existence of a scalar (vector) potential of an irrotational (solenoidal) vector field.

An irrotational vector field is orthogonal to the equipotential surfaces (Figure 5.15a); a plane solenoidal field is tangent to field lines (Figure 5.15b). Thus, a plane potential field that is both irrotational and solenoidal has equipotentials and field lines that are orthogonal (Figure 5.15c). The gradient (curl) is an example of a polar (axial) vector [Figure 5.12a (Figure 5.12b)] like the linear (angular) velocity. The electro(magneto) static field is a polar (axial) vector because it is an irrotational (solenoidal) vector field with a scalar electric (vector magnetic) potential. The three-dimensional generalized function unit impulse specifies the density of electric charge (current) corresponding to a point charge (current) in Figure 5.16 (Figure 5.17) and the associated scalar electrostatic (vector magnetostatic) potential and electric field (magnetic induction). The electric charge (current) distributions along curves are specified by two-dimensional unit impulse for (1) a straight segment (Figure 5.18) that may be extended to an infinite wire [Figure 5.19 (Figure 5.20)]; (2) the latter represents a line [Figure 5.19 (Figure 5.20)] electric charge (current) in contrast [Figure 5.16 (Figure 5.17)] with a point electric charge (current); (3) a circular loop (Figure 5.21) with the observer on (off)-axis [Figure 5.22a (Figure 5.22b)]; and (4) a circular cylindrical helix (Figure 5.23) with any number of turns. A semi-infinite helicoidal vortex induces a zero velocity far into its interior and an axial velocity far outside equal to the double of the velocity at the surface (Figure 5.24). This result can be obtained more simply, replacing the helical vortex wake (Figure 5.24) of an aircraft propeller or helicopter rotor by an actuator disk (Figure 5.25a) that also specifies the thrust (Figure 5.25b).

6

Axisymmetric Flows and Four Sphere Theorems

The velocity of an irrotational flow is the gradient of a scalar potential, which can be used to study the field in any number of dimensions, for example, in the plane (Chapters I.12, 14, 16, 18, 24, 26, 28, 32, 34, 36, 38; II.2, 8) or in space (Chapters 6 and 8) or in higher dimensions (Chapter 9); the velocity of an incompressible flow is the curl of a vector potential, which reduces to a scalar, namely, the stream function in the plane case (Chapter I.12). In a three-dimensional space, the vector potential does not generally reduce to a scalar, except in the case of incompressible axisymmetric flow, as can be shown from the curl in cylindrical or spherical coordinates (Section 6.1). The incompressible axisymmetric flow has a stream function (Section 6.2) because the volume flux through a surface of revolution about the axis does not depend on the shape of the surface, but only on its boundary, which is the same for different surfaces of revolution supported on the same two circular boundary arcs on planes perpendicular to the axis. Thus a three-dimensional axisymmetric irrotational incompressible flow has potential and stream functions similar to plane potential flow; the axisymmetric potential and stream functions are related by formulas that are distinct from the Cauchy–Riemann relations for plane potential flow. Also, (1) the potential function satisfies the Laplace equation in all cases, that is, for plane, axisymmetric, and general three- (higher) dimensional flows; (2) the stream function satisfies the Laplace equation in the plane and a modified Laplace equation in the axisymmetric case; and (3) the stream function does not exist in the general three-dimensional or higher-dimensional cases when it is replaced by the vector (multivector) potential [Chapter 8 (Notes 9.3 through 9.52)]. Case (2) leads to a spherical vortex without singularity in the velocity, for which the flow is rotational and incompressible (Section 6.6). The spherical vortex can exist in a uniform stream—this flow remains axisymmetric, that is, the velocity is along planes through the axis (in analogy with plane flow), as for the flow past (1) a source or sink representing an infinite fairing [Section 6.3 (Section I.28.4)]; (2) a source–sink pair representing a finite body [Section 6.4 (Section I.28.5)]; and (3) with the limit of dipolar coincidence in (2) representing a sphere (Section 6.5) [cylinder (Sections I.28.6 through 9)]. The potential flow past a sphere is also specified by an axisymmetric sphere theorem (Section 6.7) analogous to the circle theorem for the plane (Section I.24.7); an extension is the nonaxisymmetric sphere theorem (Section 6.7), which can be used to specify the electric current distribution on a sphere between two poles (Section 6.8). There are two circle (Sections I.24.7 and II.2.5) [four sphere (Section 6.7)] theorems for the plane (space). Many plane problems have axisymmetric spatial analogues (Section 2.9), for example, (1) a sphere (cylinder) moving in a spherical (Section 6.5) [cylindrical (Section II.8.1)] cavity; and (2) two spheres (cylinders) moving along [Example 10.15 (Section II.8.2)] the line of centers. The flow is not axisymmetric (rather, it is plane) for two spheres (Section 6.9) [cylinders Example II.10.10)] moving orthogonally to the line of centers. The axisymmetric flows, for which the stream function exists, include mono-, di-, quadru-, and multipoles (Section 6.4), their superposition with uniform flow (Section 6.5), images on planes and spheres (Sections 6.8 and 8.9), and infinite images between parallel planes (Section 8.8). The multipoles have analogues in hydrodynamic (Section 6.4), electro(magneto)static [Sections 8.1 through 8.3 (Sections 8.4 through 8.6)], and other fields (Notes 8.1 through 8.16). The general triaxial multipoles (Section 6.4) are not axisymmetric, and can be considered using the invariant differential operators' gradient, divergence, and curl (Sections 5.7 and 5.8) in orthogonal curvilinear coordinates (Section 6.1).

6.1 Invariant Differential Operators in Space

The invariant differential operators (Section 5.7) are (1) the gradient (Subsection 6.1.3) of a scalar; (2/3) the curl (Subsection 6.1.4) and the divergence (Subsection 6.1.5) of a vector; and (4/5) the Laplacian of a scalar (Subsection 6.1.6) or of a vector (Subsection 6.1.7). (a) They can be introduced using the invariance properties of the tensor calculus in any space and subspace dimension (Notes 9.1 through 9.52); (b) in the simplest plane case the theory of complex holomorphic functions can be used as an elementary approach (Sections I.11.6 and 7; Subsection II.4.4.5) to the invariant differential operators in orthogonal curvilinear coordinates; and (c) the extension to orthogonal curvilinear coordinates in space (Section 6.1) uses the scale factors that appear in the arc length (Subsection 6.1.1), together with the divergence, curl, and gradient integral theorems (Section 5.8); the simplest examples of orthogonal curvilinear coordinates (Subsection 6.1.2) are polar (cylindrical and spherical) in the plane (in space).

6.1.1 Arc Length and Scale Factor

In a three-dimensional (6.1a) orthogonal curvilinear **coordinate system** (6.1b) may be considered at any point an infinitesimal parallelepiped (Figure 6.1a) with curved sides with scale factors (6.1c) along the **unit vectors** (6.1d):

$$i = 1,2,3: \quad x_i \equiv (x_1, x_2, x_3), \quad h_i \equiv (h_1, h_2, h_3), \quad \vec{e}_i = (\vec{e}_1, \vec{e}_2, \vec{e}_3). \tag{6.1a–d}$$

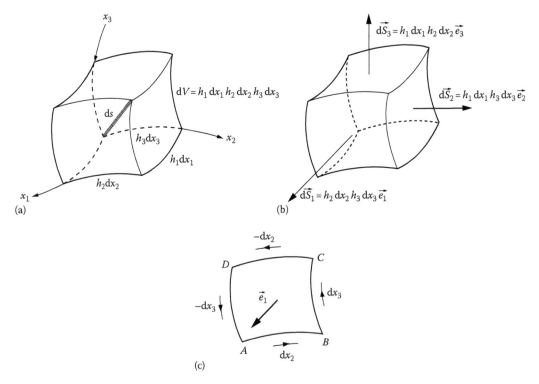

FIGURE 6.1   The orthogonal curvilinear coordinates in space are specified by the scale factors along the three axes that lead to an infinitesimal curvilinear parallelepiped, and determine: (a) its volume; (b) the area of each face (c).

The unit vectors have constant (varying) direction in a **rectilinear (curvilinear) coordinate system**. The position of the farthest vertex relative to the closest specifies the **infinitesimal relative position vector**:

$$d\vec{x} = \sum_{i=1}^{3} \vec{e}_i h_i \, dx_i = \vec{e}_1 h_1 \, dx_1 + \vec{e}_1 h_2 \, dx_2 + \vec{e}_3 h_3 \, dx_3.$$ (6.2)

In the case of an **orthogonal coordinate system** the inner product of two unit vectors is zero (unity) if they are (are not) distinct, and thus coincides with the identity matrix (5.278a–c) $\equiv$ (6.3a,b):

$$\left(\vec{e}_i \cdot \vec{e}_j\right) = \delta_{ij} \equiv \begin{cases} 1 & \text{if } i = j, \\ 0 & \text{if } i \neq j. \end{cases}$$ (6.3a,b)

The modulus of (6.2) that specifies the distance between the opposite vertices or **arc length**:

$$\left(ds\right)^2 = \left|d\vec{x}\right|^2 = \sum_{i=1}^{3} \left(h_i \, dx_i\right)^2 = \left(h_1 \, dx_1\right)^2 + \left(h_2 \, dx_2\right)^2 + \left(h_3 \, dx_3\right)^2,$$ (6.4)

does not involve cross-products:

$$\left|d\vec{x}\right|^2 = d\vec{x} \cdot d\vec{x} = \sum_{i,j=1}^{3} h_i \, dx_i \, h_j \, dx_j \left(\vec{e}_i \cdot \vec{e}_j\right) = \sum_{i,j=1}^{3} h_i h_j \delta_{ij} \, dx_i dx_j = \sum_{i=1}^{3} \left(h_i \, dx_i\right)^2.$$ (6.5)

This is because the identity matrix (6.3a and b) leads to a sum of squares in the case of an orthogonal curvilinear coordinate system.

The orthogonal curvilinear coordinate system leads to (1) the **volume element** (6.6), which involves the product of scale factors:

$$dV = h_1 \, dx_1 \, h_2 \, dx_1 \, h_3 \, dx_3 = \prod_{i=1}^{3} h_i \, dx_i,$$ (6.6)

(2) the **surface element** with facet with normal in the $\vec{e}_i$ direction, which involve the scale factors in the transverse directions (Figure 6.1b):

$$d\vec{S}_1 = \vec{e}_1 h_2 \, dx_2 \, h_3 \, dx_3, \quad d\vec{S}_2 = \vec{e}_2 h_1 \, dx_1 \, h_3 \, dx_3, \quad d\vec{s}_3 = \vec{e}_3 h_1 \, dx_1 \, h_2 \, dx_2,$$ (6.7a–c)

or alternatively:

$$i \neq j \neq k \neq i: \quad d\vec{S}_i = \vec{e}_i h_j h_k \, dx_j dx_k \, (\text{cyclic}),$$ (6.8)

where (i, j, k) can be changed cyclically; and (3) a loop integral around one face (Figure 6.1c):

$$\int_{ABCD} \Phi\left(x_1, x_2, x_3\right) ds = \int_{x_2}^{x_2 + dx_2} \left[\Phi\left(x_1, x_2, x_3\right) - \Phi\left(x_1, x_2, x_3 + dx_3\right)\right] dx_2$$

$$- \int_{x_3}^{x_3 + dx_3} \left[\Phi\left(x_1, x_2, x_3\right) - \Phi\left(x_1, x_2 + dx_2, x_3\right)\right] dx_3,$$ (6.9)

taken in the positive or counterclockwise direction that involves two opposite displacements.

6.1.2　Cartesian, Cylindrical, and Spherical Coordinates

The simplest rectilinear coordinate system is Cartesian coordinates (Figure 6.2) with the same scale factors (6.10a) equal to unity along all three straight orthogonal coordinate axes leading (6.6) [(6.4)] to (Figure 6.3a) the volume element (6.10b) [arc length (6.10c)]:

$$h_x = h_y = h_z = 1: \quad dV = dx\, dy\, dz, \quad (ds)^2 = (dx)^2 + (dy)^2 + (dz)^2. \tag{6.10a–c}$$

The plane Cartesian coordinates correspond to suppressing the third coordinate in the arc length (6.10c) and (6.10b) becomes the area element $dS = dx\, dy$ normal to the (x, y)-plane. Two examples of orthogonal curvilinear coordinate systems are cylindrical and spherical coordinates in three dimensions (Subsection 6.1.2) that can be extended to higher dimensions (Subsections 9.7.2 and 9.7.3). The **cylindrical coordinates** (Figure 6.2) replace the plane Cartesian by **polar coordinates** (6.11a through c):

$$x = r\cos\varphi, \quad y = r\sin\varphi, \quad z = z: \quad h_r = 1 = h_z, \quad h_\varphi = r, \tag{6.11a–f}$$

so that (Figure 6.3b) the scale factor (6.11f) corresponding to the polar angle is the distance from the axis, while the other two remain unity (6.11d and e). The volume element (6.12a) and arc length (6.12b) take into account that an azimuthal displacement $d\varphi$ corresponds to a circle of radius r:

$$dV = r\, dr\, d\varphi\, dz, \quad (ds)^2 = (dr)^2 + r^2 (d\varphi)^2 + (dz)^2. \tag{6.12a,b}$$

Suppressing the z-coordinate reduces the cylindrical coordinates to polar coordinates; a translation of the polar coordinates along the z-axis leads to cylindrical coordinates again.

The **spherical coordinates** (Figure 6.2) project the **radial distance** from the origin on the z-axis (6.13a) and the projection on the (x, y)-plane (6.13b):

$$z = R\cos\theta, \quad r = R\sin\theta; \quad x = R\sin\theta\cos\varphi, \quad y = R\sin\theta\sin\varphi, \tag{6.13a–d}$$

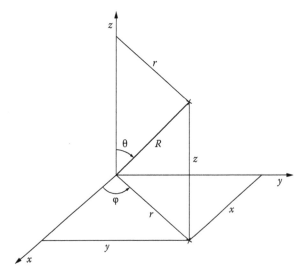

FIGURE 6.2　Comparison of Cartesian (x, y, z) with cylindrical (r, φ, z) [spherical (R, θ, φ)] coordinates with a common axis specified equivalently by $x = 0 = y$ or $r = 0$ or $\theta = 0$.

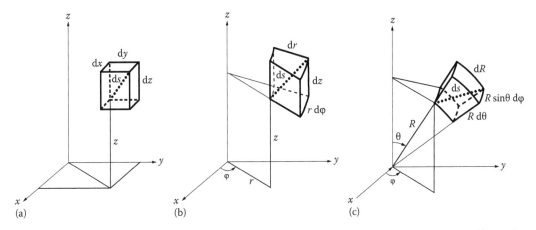

FIGURE 6.3 The infinitesimal volume element in orthogonal curvilinear coordinates (Figure 6.1a and b) may be illustrated (Figure 6.2), respectively, in Cartesian (a), cylindrical (b), and spherical (c) coordinates, specifying the scale factors along each of the three axes.

is treated (6.13c and d) as polar coordinates (6.11a and b). (1) The scale factors (Figure 6.3c) are unity (6.14b) for a radial displacement; (2) for a displacement $d\theta$ along the meridian R, φ = const the radius is (6.14b); and (3) for a displacement $d\varphi$ along a parallel the radius is (6.14c):

$$h_R = 1, \quad h_\theta = R, \quad h_\varphi = R\sin\theta; \quad dV = R^2 \sin\theta \, dR \, d\theta \, d\varphi. \tag{6.14a–d}$$

This corresponds (6.6) [(6.4)] to the volume element (6.14d) [arc length (6.15)] in spherical coordinates:

$$\left(ds\right)^2 = \left(dR\right)^2 + R^2\left(d\theta\right)^2 + R^2 \sin^2\theta\left(d\varphi\right)^2. \tag{6.15}$$

The spherical coordinates (6.14a, c, and d) reduce in the equatorial plane (6.16a) to (6.16b and c):

$$\theta = \frac{\pi}{2}: \quad z = 0, \quad r = R; \quad x = r\cos\varphi, \quad y = r\sin\varphi, \tag{6.16a–e}$$

which coincide with the polar coordinates (6.16d and e) ≡ (6.11a and b).

The volume element in cylindrical (6.12a) [spherical (6.14d)] coordinates is specified by the Jacobian of the transformation (6.11a through c) [(6.13a, c, and d)] from Cartesian coordinates (6.17) [(6.18)]:

$$\frac{\partial(x,y,z)}{\partial(R,\varphi,z)} \equiv \begin{vmatrix} \dfrac{\partial x}{\partial r} & \dfrac{\partial x}{\partial \varphi} & \dfrac{\partial x}{\partial z} \\[2mm] \dfrac{\partial y}{\partial r} & \dfrac{\partial y}{\partial \varphi} & \dfrac{\partial y}{\partial z} \\[2mm] \dfrac{\partial z}{\partial r} & \dfrac{\partial z}{\partial \varphi} & \dfrac{\partial z}{\partial z} \end{vmatrix} = \begin{vmatrix} \cos\varphi & -r\sin\varphi & 0 \\ \sin\varphi & r\cos\varphi & 0 \\ 0 & 0 & 0 \end{vmatrix} = r\left(\cos^2\varphi + \sin^2\varphi\right) = r, \tag{6.17}$$

$$\frac{\partial(x,y,z)}{\partial(R,\theta,\varphi)} \equiv \begin{vmatrix} \dfrac{\partial x}{\partial R} & \dfrac{\partial x}{\partial \theta} & \dfrac{\partial x}{\partial \varphi} \\ \dfrac{\partial y}{\partial R} & \dfrac{\partial y}{\partial \theta} & \dfrac{\partial y}{\partial \varphi} \\ \dfrac{\partial z}{\partial R} & \dfrac{\partial z}{\partial \theta} & \dfrac{\partial z}{\partial \varphi} \end{vmatrix} = \begin{vmatrix} \sin\theta\cos\varphi & R\cos\theta\cos\varphi & -R\sin\theta\sin\varphi \\ \sin\theta\sin\varphi & R\cos\theta\sin\varphi & R\sin\theta\cos\varphi \\ \cos\theta & -R\sin\theta & 0 \end{vmatrix}$$

$$= R^2\sin\theta\left(\cos^2\theta\cos^2\varphi + \sin^2\theta\sin^2\varphi + \cos^2\theta\sin^2\varphi + \sin^2\theta\cos^2\varphi\right)$$

$$= R^2\sin\theta\left(\cos^2\varphi + \sin^2\varphi\right)\left(\cos^2\theta + \sin^2\theta\right) = R^2\sin\theta. \tag{6.18}$$

The arc length in cylindrical (6.12b) [spherical (6.15)] coordinates can be obtained substituting the transformation (6.11a through c) [(6.13a, c, and d) from the Cartesian coordinates in the Cartesian arc length (6.10c); for example:

$$\begin{aligned} \left(ds\right)^2 - \left(dz\right)^2 &= \left[d(r\cos\varphi)\right]^2 + \left[d(r\cos\varphi)\right]^2 \\ &= \left(\cos\varphi\,dr - r\sin\varphi\,d\varphi\right)^2 + \left(\sin\varphi\,dr + r\cos\varphi\,d\varphi\right)^2 \\ &= \left[\left(dr\right)^2 + \left(r\,d\varphi\right)^2\right]\left(\cos^2\varphi + \sin^2\varphi\right) = \left(dr\right)^2 + r^2\left(d\varphi\right)^2, \end{aligned} \tag{6.19}$$

for cylindrical coordinates (6.19) ≡ (6.12b). Thus *the volume element (6.6) [arc length (6.4)] in orthogonal curvilinear coordinates (6.3a and b) with scale factors (6.1c), including in particular (6.12a) [(6.12b)] for cylindrical coordinates (6.11a through f) and (6.14d) [(6.15)] for spherical coordinates (6.13a, c, and d; 6.14a through c), has been obtained. The area element of the sphere (6.20a) omits the factor dR from the volume element (6.14d):*

$$dS = \frac{dV}{dR} = R^2\sin\theta\,d\theta\,d\varphi = R^2\,d\Omega, \quad d\Omega \equiv \sin\theta\,d\theta\,d\varphi, \tag{6.20a,b}$$

and for the sphere of unit radius the **solid angle** *is specified by (6.20b).* The scale factors are used next to obtain the invariant differential operators in orthogonal curvilinear coordinates.

6.1.3 Gradient of a Scalar in Orthogonal Curvilinear Coordinates

The projection of the gradient of a scalar on the infinitesimal relative position vector (6.2) is the differential of the scalar (5.241) ≡ (6.21):

$$d\Phi = \nabla\Phi \cdot d\vec{x} = \sum_{i=1}^{3}\left(\nabla\Phi\right)_i h_i\,dx_i = \left(\nabla\Phi\right)_1 h_1\,dx_1 + \left(\nabla\Phi\right)_2 h_2\,dx_2 + \left(\nabla\Phi\right)_3 h_3\,dx_3; \tag{6.21}$$

comparing with the exact differential:

$$d\Phi = \sum_{i=1}^{3}\frac{\partial\Phi}{\partial x_i}\,dx_i = \frac{\partial\Phi}{\partial x_1}\,dx_1 + \frac{\partial\Phi}{\partial x_2}\,dx_2 + \frac{\partial\Phi}{\partial x_3}\,dx_3, \tag{6.22}$$

it follows that the **gradient** *of a scalar is the vector specified by its partial derivatives (6.23a) divided by the corresponding scale factors to ensure dimensional consistency*:

$$\partial_i \equiv \frac{\partial}{\partial x_i}: \quad \nabla\Phi \equiv \sum_{i=1}^{3} \vec{e}_i h_i^{-1} \partial_i \Phi = \vec{e}_x \partial_x \Phi + \vec{e}_y \partial_y \Phi + \vec{e}_z \partial_z \Phi$$

$$= \vec{e}_r \partial_r \Phi + \vec{e}_\varphi r^{-1} \partial_\varphi \Phi + \vec{e}_z \partial_z \Phi$$

$$= \vec{e}_R \partial_R \Phi + \vec{e}_\theta R^{-1} \partial_\theta \Phi + \vec{e}_\varphi R^{-1} \csc\theta \partial_\varphi \Phi, \qquad (6.23a\text{–}e)$$

that is, the gradient of a scalar is specified by (6.23b) in orthogonal curvilinear coordinates in space (6.1a through c), and in particular by (6.23c/d/e) in Cartesian (6.10a), cylindrical (6.11d through f)/spherical (6.14a through c) coordinates. In the two-dimensional case only the first two terms appear in (6.23b) ≡ (I.11.50), for example, in polar coordinates (6.23d) ≡ (I.11.31b).

6.1.4 Divergence of a Vector in Two and Three Dimensions

The divergence of a vector in orthogonal curvilinear coordinates can be obtained by applying the divergence theorem (5.163a through c) to an infinitesimal curvilinear parallelepiped (Figure 6.1a) involving: (1) the integral of the divergence of the vector over (Figure 6.1a) the volume element (6.6):

$$\int_{D_3} \left(\nabla \cdot \vec{A}\right) dV = \left(\nabla \cdot \vec{A}\right) h_1 h_2 h_3 \, dx_1 \, dx_2 \, dx_3; \qquad (6.24)$$

(2) the flux of the vector across (Figure 6.1b) the left-hand face orthogonal to the *x*-axis with area element (6.7a) and normal opposite to the unit vector $\vec{e}_1$:

$$\int_{x_1=\text{const}} \vec{A} \cdot d\vec{S} = -\left[\left(\vec{e}_1 \cdot \vec{A}\right) h_2 h_3\right]_{x_1 x_2 x_3} dx_2 \, dx_3; \qquad (6.25)$$

(3) the flux of the vector across the other parallel face at a distance dx_1 with the same area element and opposite normal:

$$\int_{x_1+dx_1=\text{const}} \vec{A} \cdot d\vec{S} = \left[\left(\vec{e}_1 \cdot \vec{A}\right) h_2 h_3\right]_{x_1+dx_1, x_2 x_3} dx_2 \, dx_3; \qquad (6.26)$$

(4) the flux of the vector across the two facets orthogonal to the x_1-axis:

$$\left(\int_{x_1+dx_1=\text{const}} + \int_{x_1=\text{const}}\right) \vec{A} \cdot d\vec{S} = \left[A_1 h_2 h_3\right]_{x_1, x_2 x_3}^{x_1+dx_1, x_2 x_3} dx_2 \, dx_3$$

$$= \left[\frac{\partial}{\partial x_1}\left(A_1 h_2 h_3\right)\right] dx_1 \, dx_2 \, dx_3, \qquad (6.27)$$

where higher-order infinitesimals are not taken into account; (5) the flux of the vector across the boundary of the curvilinear parallelepiped is the sum of the three pairs of faces orthogonal to each axis:

$$\int_{\partial D_3 \equiv D_2} \vec{A} \cdot d\vec{S} = \left[\frac{\partial}{\partial x_1}\left(A_1 h_2 h_3\right) + \frac{\partial}{\partial x_2}\left(A_2 h_1 h_3\right) + \frac{\partial}{\partial x_3}\left(A_3 h_1 h_3\right)\right] dx_1, dx_2 \, dx_3. \qquad (6.28)$$

Equating (6.28) ≡ (6.24) by using the divergence theorem leads to

$$\nabla \cdot \vec{A} = \left(h_1 h_2 h_3\right)^{-1}\left[\partial_1\left(h_2 h_3 A_1\right) + \partial_2\left(h_1 h_3 A_2\right) + \partial_3\left(h_1 h_2 A_3\right)\right]. \tag{6.29}$$

Thus the **divergence** *of a vector is specified by (6.29) in orthogonal curvilinear coordinates (6.1a through c) and in particular in Cartesian (6.10a through c):*

$$\nabla \cdot \vec{A} = \partial_x A_x + \partial_y A_y + \partial_z A_z, \tag{6.30}$$

cylindrical (6.11d through f):

$$\nabla \cdot \vec{A} = r^{-1}\partial_r\left(rA_r\right) + r^{-1}\partial_\varphi A_\varphi + \partial_z A_z = \partial_r A_r + r^{-1}A_r + r^{-2}\partial_\varphi A_\varphi + \partial_z A_z, \tag{6.31a,b}$$

and spherical (6.14a through c):

$$\nabla \cdot \vec{A} = R^{-2}\partial_R\left(R^2 A_R\right) + R^{-1}\csc\theta\,\partial_\theta\left(\sin\theta A_\theta\right) + R^{-1}\csc\theta\,\partial_\varphi A_\varphi,$$

$$= \partial_R A_R + 2R^{-1}A_R + R^{-1}\partial_\theta A_\theta + R^{-1}\cot\theta A_\theta + R^{-1}\csc\theta\,\partial_\varphi A_\varphi, \tag{6.32a,b}$$

coordinates. In two dimensions, the divergence simplifies to (6.33) ≡ (I.11.53):

$$\nabla \cdot \vec{A} = \left(h_1 h_2\right)^{-1}\left[\partial_1\left(h_2 A_1\right) + \partial_2\left(h_1 A_2\right)\right], \tag{6.33}$$

for example, corresponding to the first two terms of (6.31a) ≡ (I.11.33b) in polar coordinates.

6.1.5 Two- and Three-Dimensional Curl of a Vector

The curl of a vector in three-dimensional orthogonal curvilinear coordinates is obtained by applying the Stokes theorem (5.210a and b) to one face (Figure 6.1c) of the infinitesimal curvilinear parallelepiped (Figure 6.1a), for example, the face orthogonal to the x_1-axis, involving: (1) the flux of the curl of the vector across the face with area element (6.7a):

$$\int_{D_2}\left(\nabla \wedge \vec{A}\right)\cdot d\vec{S} = \left(\nabla \wedge \vec{A}\right)_1 h_2 h_3 \, dx_2 \, dx_3; \tag{6.34}$$

(2) the integral of the vector projected along the side *AB*:

$$\int_{AB}\vec{A}\cdot d\vec{x} = \left[A_2 h_2\right]_{x_1,x_2,x_3} dx_2; \tag{6.35}$$

(3) the integral along the opposite face at a distance dx_3, which is taken in the opposite direction:

$$\int_{CD}\vec{A}\cdot d\vec{x} = -\left[A_2 h_2\right]_{x_1,x_2,x_3+dx_2} dx_2; \tag{6.36}$$

(4) the sum of the integrals (6.35) and (6.36) along the boundary segments parallel to the x_2-axis:

$$\left(\int\limits_{AB}+\int\limits_{CD}\right)\vec{A}\cdot d\vec{x} = \left[A_2 h_2\right]_{x_1,x_2 x_3+dx_3}^{x_1,x_2 x_3}\, dx_2 = -\left[\frac{\partial}{\partial x_1}\left(A_2 h_2\right)\right]dx_2\, dx_3; \tag{6.37}$$

(5) the sum of the integrals along the sides parallel to the x_3-axis is similar, with reversed sign because AD (CB) corresponds to a negative (positive) displacement:

$$\left(\int\limits_{DA}+\int\limits_{BC}\right)\vec{A}\cdot d\vec{x} = \left[\frac{\partial}{\partial x_2}\left(h_3 A_3\right)\right]dx_2\, dx_3; \tag{6.38}$$

(6) the loop integral around the boundary that is the sum of (6.37) and (6.38):

$$\int\limits_{\partial D_2\equiv D_1}\vec{A}\cdot d\vec{x} = \int\limits_{ABCD}\vec{A}\cdot d\vec{x} = \left[\frac{\partial}{\partial x_2}\left(h_3 A_3\right)-\frac{\partial}{\partial x_3}\left(h_2 A_2\right)\right]dx_2\, dx_3. \tag{6.39}$$

The Stokes theorem (5.210a and b) equates (6.34) $\equiv$ (6.39) leading to the first component of the curl:

$$\nabla\wedge\vec{A} = \sum_{1,2,3}^{\text{cycl}}\left(h_2 h_3\right)^{-1}\left[\partial_2\left(h_3 A_3\right)-\partial_3\left(h_2 A_2\right)\right]\vec{e}_1 = \begin{vmatrix} \dfrac{\vec{e}_1}{h_2 h_3} & \dfrac{\vec{e}_2}{h_1 h_3} & \dfrac{\vec{e}_3}{h_1 h_2} \\[2mm] \dfrac{\partial}{\partial x_1} & \dfrac{\partial}{\partial x_2} & \dfrac{\partial}{\partial x_3} \\[2mm] h_1 A_1 & h_2 A_2 & h_3 A_3 \end{vmatrix}; \tag{6.40a,b}$$

the remaining components can be obtained considering other faces of the curvilinear parallelepiped, that is, equivalent to cyclic permutation of (1, 2, 3) in (6.40). Thus *the curl of a vector is given in orthogonal curvilinear coordinates by (6.40), and in particular by (6.41a/b/c), respectively, for Cartesian (6.10a through c)/cylindrical (6.11d through f)/spherical (6.14a through c) coordinates:*

$$\begin{aligned} \nabla\wedge\vec{A} &= \vec{e}_x\left(\partial_y A_z-\partial_z A_y\right)+\vec{e}_y\left(\partial_z A_x-\partial_x A_z\right)+\vec{e}_z\left(\partial_x A_y-\partial_y A_x\right),\\ &= \vec{e}_r\left(r^{-1}\partial_\varphi A_z-\partial_z A_\varphi\right)+\vec{e}_\varphi\left(\partial_z A_r-\partial_r A_z\right)+\vec{e}_z r^{-1}\left[\partial_r\left(rA_\varphi\right)-\partial_\varphi A_r\right],\\ &= \vec{e}_R R^{-1}\csc\theta\left[\partial_\theta\left(\sin\theta A_\varphi\right)-\partial_\varphi A_\theta\right]+\vec{e}_\theta R^{-1}\left[\csc\theta\partial_\varphi A_R-\partial_R\left(RA_\varphi\right)\right]\\ &\quad +\vec{e}_\varphi R^{-1}\left[\partial_R\left(RA_\theta\right)-\partial_\theta A_R\right]. \end{aligned} \tag{6.41a–c}$$

In two dimensions, the curl is orthogonal to the plane (6.42) $\equiv$ (I.11.55):

$$\nabla\wedge\vec{A} = \frac{\vec{e}_3}{h_1 h_2}\left[\frac{\partial}{\partial x_1}\left(h_2 A_2\right)-\frac{\partial}{\partial x_2}\left(h_1 A_1\right)\right], \tag{6.42}$$

for example, in polar coordinates it corresponds to the last term in (6.41b) $\equiv$ (I.11.35b).

6.1.6 Cartesian, Cylindrical, and Spherical Laplacian

The **scalar Laplacian** *is the divergence (6.29) of the gradient (6.23b), that is, in orthogonal curvilinear coordinates:*

$$\nabla^2 \Phi = \nabla \cdot (\nabla \Phi) = (h_1 h_2 h_3)^{-1} \sum_{1,2,3}^{cycl} \partial_1 \left(\frac{h_2 h_3}{h_1} \partial_1 \Phi \right), \tag{6.43}$$

and in particular in Cartesian (6.44):

$$\nabla^2 \Phi = \partial_{xx} \Phi + \partial_{yy} \Phi + \partial_{zz} \Phi, \tag{6.44}$$

cylindrical (6.45a and b):

$$\begin{aligned} \nabla^2 \Phi &= r^{-1} \partial_r (r \partial_r \Phi) + r^{-2} \partial_{\varphi\varphi} \Phi + \partial_{zz} \Phi \\ &= \partial_{rr} \Phi + r^{-1} \partial_r \Phi + r^{-2} \partial_{\varphi\varphi} \Phi + \partial_{zz} \Phi, \end{aligned} \tag{6.45a,b}$$

and spherical (6.46a and b):

$$\begin{aligned} \nabla^2 \Phi &= R^{-2} \partial_R (R^2 \partial_R \Phi) + R^{-2} \csc \theta \partial_\theta (\sin \theta \partial_\theta \Phi) + R^{-2} \csc^2 \theta \partial_{\varphi\varphi} \Phi \\ &= \partial_{RR} \Phi + 2 R^{-1} \partial_R \Phi + R^{-2} \partial_{\theta\theta} \Phi + R^{-2} \cot \theta \partial_\theta \Phi + R^{-2} \csc^2 \theta \partial_{\varphi\varphi} \Phi, \end{aligned} \tag{6.46a,b}$$

coordinates that can be obtained by (1) substituting in (6.43) the scale factors (6.10a and b)/(6.11d through f)/(6.14a through c); and (2) applying the divergence (6.30)/(6.31a and b)/(6.32a and b) to the gradient (6.23c/d/e). *The two-dimensional Laplacian of a scalar (6.47) ≡ (I.11.52):*

$$\nabla^2 \Phi = \frac{1}{h_1 h_2} \left[\frac{\partial}{\partial x_1} \left(\frac{h_2}{h_1} \frac{\partial \Phi}{\partial x_1} \right) + \frac{\partial}{\partial x_2} \left(\frac{h_1}{h_2} \frac{\partial \Phi}{\partial x_2} \right) \right], \tag{6.47}$$

includes in particular the first two terms of (6.45a and b) ≡ (I.11.28b and c) ≡ (II.4.111) for polar coordinates.

6.1.7 Relation between the Vector and Scalar Laplacians

The vector Laplacian is related to the curl and divergence of a vector by (6.48a) ≡ (II.4.112), which involves the gradient as well:

$$\nabla^2 \vec{A} = \nabla (\nabla \cdot \vec{A}) - \nabla \wedge (\nabla \wedge \vec{A}) = \vec{e}_x \nabla^2 A_x + \vec{e}_y \nabla^2 A_y + \vec{e}_z \nabla^2 A_z; \tag{6.48a,b}$$

in Cartesian coordinates the vector Laplacian (6.48b) is the scalar Laplacian applied to each component of a vector. The relation (6.48a) can be proven in Cartesian coordinates (II.4.113), for example, for the first component in (6.48b):

$$\left[\nabla (\nabla \cdot \vec{A}) \right]_1 = \partial_1 (\nabla \cdot \vec{A}) = \partial_1 (\partial_1 A_1 + \partial_2 A_2 + \partial_3 A_3) = \partial_{11} A_1 + \partial_{12} A_2 + \partial_{13} A_3, \tag{6.48c}$$

$$\left[\nabla \wedge \left(\nabla \wedge \vec{A}\right)\right]_1 = \partial_2\left(\nabla \wedge \vec{A}\right)_3 - \partial_3\left(\nabla \wedge \vec{A}\right)_2 = \partial_2\left(\partial_1 A_2 - \partial_2 A_1\right) - \partial_3\left(\partial_3 A_1 - \partial_1 A_3\right)$$

$$= \partial_{21}A_2 - \partial_{22}A_1 - \partial_{33}A_1 + \partial_{31}A_3, \tag{6.48d}$$

$$\left[\nabla\left(\nabla \cdot \vec{A}\right) - \nabla \wedge \left(\nabla \wedge \vec{A}\right)\right]_1 = \partial_{11}A_1 + \partial_{22}A_2 + \partial_{33}A_3 + \left(\partial_{12} - \partial_{21}\right)A_2 + \left(\partial_{13} - \partial_{31}\right)A_3$$

$$= \left(\partial_{11} + \partial_{22} + \partial_{33}\right)A_1 = \left(\nabla^2 \vec{A}\right)_1, \tag{6.48e}$$

using the symmetry of continuous second-order derivatives; the proof for the remaining components of the vector Laplacian follows by cyclic permutation of (1, 2, 3). In the case of orthogonal curvilinear coordinates, the vector Laplacian does not generally reduce to a sum of scalar Laplacians. For example, the vector Laplacian (6.48a) in cylindrical (spherical) coordinates involves the gradient (6.23d) [(6.23e)], divergence (6.31a and b) [(6.32a and b)], and curl (6.41b) [(6.41c)]:

$$\nabla^2 \vec{A} = \vec{e}_r\left(\nabla^2 A_r - r^{-2}A_r - 2r^{-2}\partial_\varphi A_\varphi\right) + \vec{e}_\varphi\left(\nabla^2 A_\varphi - r^{-2}A_\varphi + 2r^{-2}\partial_\varphi A_r\right) + \vec{e}_z \nabla^2 A_z, \tag{6.49}$$

$$\nabla^2 \vec{A} = \vec{e}_R\left[\nabla^2 A_R - 2R^{-2}A_R - 2R^{-2}\cot\theta A_\theta - 2R^{-2}\csc\theta\partial_\varphi A_\varphi\right]$$

$$+ \vec{e}_\theta\left[\nabla^2 A_\theta + 2R^{-2}\partial_\theta A_R - R^{-2}\csc^2\theta A_\theta - 2R^{-2}\csc\theta\cot\theta\partial_\varphi A_\varphi\right]$$

$$+ \vec{e}_\varphi\left[\nabla^2 A_\varphi + 2R^{-2}\csc\theta\partial_\varphi A_R + 2R^{-2}\csc\theta\cot\theta\partial_\varphi A_\theta - R^{-2}\csc^2\theta A_\varphi\right]. \tag{6.50}$$

Thus the **vector Laplacian** *(6.48a) is specified by (6.23b; 6.29; 6.40) in orthogonal curvilinear coordinates, and in particular relates to the scalar Laplacian in Cartesian (6.44; 6.48b), cylindrical (6.49; 6.45a and b), and spherical (6.50; 6.46a and b) coordinates.* In the case of polar coordinates, the Laplacian of a vector reduces to the first two terms of (6.45a and b; 6.49) ≡ (II.4.110a and b).

6.2 Stream Function for an Axisymmetric Flow

An axisymmetric flow is conveniently described in cylindrical (spherical) coordinates because (Subsection 6.2.1): (1) the azimuthal velocity is zero in both cases; and (2) the two remaining cylindrical and spherical components of the velocity are related by a local rotation. (1) The components of the velocity can be expressed in terms of a stream function for an incompressible axisymmetric flow (Subsection 6.2.2); (2) the latter is a particular case of the vector potential for a general three-dimensional incompressible flow (Subsection 6.2.3); and (3) the scalar potential exists for an irrotational flow, regardless of the dimension or symmetry (Subsection 6.2.4). For a potential flow that is incompressible and irrotational (Subsection 6.2.5) (1) in the plane case both the potential and stream function satisfy the Laplace equation; (2) in the axisymmetric case the potential (stream function) satisfies the Laplace (modified Laplace) equation (Subsection 6.2.6), which are both particular cases of the vector Laplace equation (Subsection 6.2.8); and (3) these are generalized in the presence of dilatation (vorticity) to the Poisson (the modified Poisson) equation (Subsection 6.2.7), which are particular cases of the vector Poisson equation. Considering an axisymmetric incompressible rotational flow, which has a stream function but no potential: (1) the streamlines and vortex lines, to which the velocity and vorticity are tangent, respectively, form a stream surface (Subsection 6.2.9), which is also a surface of constant stagnation enthalpy and stagnation pressure for a steady homentropic flow (Subsection 6.2.10); (2) in this case, the vorticity is a function of the stream function only, leading to a generalization of the Bernoulli theorem (Subsection 6.2.11) involving the vorticity and stream function; and (3) the latter are related by a modified Poisson equation (Subsection 6.2.12), which is distinct from the scalar Poisson equation, though both are particular cases of the vector Poisson equation.

6.2.1 Coordinates and Velocity for Axisymmetric Flow

An axisymmetric flow is defined as one in which the velocity is identical in all planes through the axis, and lies on that plane. Taking the axis of symmetry as the z-axis, $r = 0$ ($\theta = 0$ axis) in cylindrical (spherical) coordinates, (1) an axisymmetric field has no azimuthal component, perpendicular to a plane through the axis (6.51a); and (2) the in-plane components of the velocity to not depend on the azimuthal angle (6.51b). The possibly nonvanishing components (Figure 6.4) are radial v_R and meridional v_θ in spherical coordinates and axial v_z and transverse v_r in cylindrical coordinates, and these are related (Figure 6.5a and b) by a rotation (I.16.55a and b) through angles $\pm\theta$:

$$v_\varphi = 0 = \partial_\varphi : \quad \begin{bmatrix} v_z \\ v_r \end{bmatrix} = \begin{bmatrix} \cos\theta & -\sin\theta \\ \sin\theta & \cos\theta \end{bmatrix} \begin{bmatrix} v_R \\ v_\theta \end{bmatrix}, \quad \begin{bmatrix} v_R \\ v_\theta \end{bmatrix} = \begin{bmatrix} \cos\theta & \sin\theta \\ -\sin\theta & \cos\theta \end{bmatrix} \begin{bmatrix} v_z \\ v_r \end{bmatrix}, \quad (6.51a–d)$$

namely, $+\theta$ from spherical to cylindrical (6.51c) and $-\theta$ from cylindrical to spherical (6.51d). The distance from the z-axis (6.11a and b) [radial distance from the origin (6.13a and b)] is denoted by $r(R)$ in (6.52a) [(6.52b)]:

$$r = \left| x^2 + y^2 \right|^{1/2}, \quad R = \left| r^2 + z^2 \right|^{1/2} = \left| x^2 + y^2 + z^2 \right|^{1/2}, \quad (6.52a,b)$$

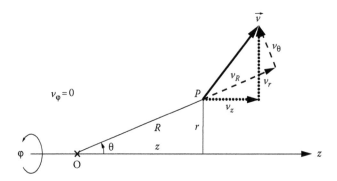

FIGURE 6.4 An axisymmetric velocity field lying on a plane through the axis of symmetry has no azimuthal component, leading to at most two nonzero components: (1) in cylindrical coordinates, namely, $v_z(v_r)$ parallel (orthogonal) to the axis along (away from) it; and (2) in spherical coordinates, namely, $v_R(v_\theta)$ radially outward from the origin (in the positive orthogonal direction, that is, counterclockwise).

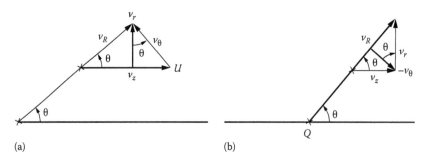

(a)   (b)

FIGURE 6.5 A uniform flow (a) with velocity U and a radial flow from a source (b) or sink are examples of the use (Figure 6.4) of cylindrical (spherical) components of the velocity.

where (x, y) are Cartesian coordinates in a plane perpendicular to the z-axis. The angle θ between the position vector r and the z-axis (Figure 6.2) satisfies:

$$\cos\theta = \frac{z}{R}, \quad \sin\theta = \frac{r}{R}, \quad \tan\theta = \frac{r}{z}. \tag{6.53a--c}$$

The relations (6.52a and b; 6.53a through c) are consistent with (6.11a and b; 6.13a through d) for Cartesian (x, y, z), cylindrical (r, φ, z), and spherical (R, θ, φ) coordinates.

6.2.2 Stream Function in Cylindrical/Spherical Coordinates

A curve passing through a point A on the axis and a point B outside the axis generates by rotation a surface of revolution (Figure 6.6). The mass flux through such a surface of revolution does not depend on its shape; as long as it goes through point B and the flow is steady (6.54a), it does not depend on time because mass cannot change over time between two surfaces passing through the same points (A, B). Also, the conservation of volume flux equates to conservation of mass flux if the flow is incompressible (6.54b); that is, the mass density is constant. The volume flux across a surface of revolution is specified in cylindrical coordinates (6.54c) by the sum of (1) the inflow due to the axial velocity v_z times the area of circular ring $d(\pi r^2) = 2\pi r\,dr$ of radius r in a plane orthogonal to the axis; and (2) the outflow due to the transverse velocity v_r across an axial slice of cylinder of length dz with radius r and area $2\pi r\,dz$. The volume flux equals (6.54d) in spherical coordinates; $r = R \sin\theta$ in (6.53b), the sum of (1) the inflow due to the radial velocity v_R across an arc of parallel of length $R\,d\theta$ and perimeter $2\pi r = 2\pi R \sin\theta$; and (2) the outflow due to the meridional velocity v_θ across an axial slice of a cone of radial length dR, radius $r = R \sin\theta$ and area $2\pi r\,dR = 2\pi \sin\theta\,R\,dR$:

$$\frac{\partial}{\partial t} = 0, \ \rho = \text{const}: \quad 2\pi d\Psi = 2\pi r\left(-v_r dz + v_z dr\right) = 2\pi R \sin\theta\left(v_R R\,d\theta - v_\theta\,dR\right). \tag{6.54a--d}$$

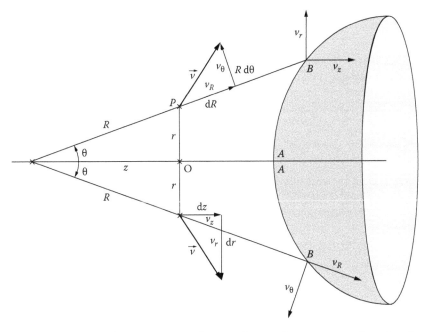

FIGURE 6.6 A line joining an arbitrary point B to the axis A becomes by rotation a surface of revolution around the latter. For a steady incompressible axisymmetric flow the volume flux is the same across all surfaces of revolution generated by different curves joining the same point to the axis. This proves the existence of a stream function.

The two expressions for the stream function (6.54c and d) are consistent with the relations (6.53a and b) between the cylindrical and spherical components of the velocity (6.51c):

$$-v_r dz + v_z dr = -\left(v_R \sin\theta + v_\theta \cos\theta\right)\left(\cos\theta \, dR - R\sin\theta \, d\theta\right)$$
$$+ \left(v_R \cos\theta - v_\theta \sin\theta\right)\left(\sin\theta \, dR + R\cos\theta \, d\theta\right) = v_R R \, d\theta - v_\theta \, dR. \qquad (6.55)$$

Comparing (6.54c) [(6.54d)] with (6.56a) [(6.56b)]:

$$d\Psi = \left(\partial_r \Psi\right) dr + \left(\partial_z \Psi\right) dz = \left(\partial_R \Psi\right) dR + \left(\partial_\theta \Psi\right) d\theta, \qquad (6.56a,b)$$

it follows that the cylindrical (spherical) components of the velocity are given by (6.57a through c) [(6.58a through c)] for an axisymmetric flow (6.51a and b):

$$\left\{v_r, v_\varphi, v_z\right\} = r^{-1}\left\{-\partial_z \Psi, 0, \partial_r \Psi\right\}, \qquad (6.57a\text{–}c)$$

$$\left\{v_R, v_\theta, v_\varphi\right\} = R^{-1} \csc\theta \left\{R^{-1}\partial_\theta \Psi, -\partial_R \Psi, 0\right\}. \qquad (6.58a\text{–}c)$$

Thus, *a steady (6.54a) incompressible (6.54b), axisymmetric (6.51a and b) flow has a* **stream function**, *that is, a function of position, and is equal to* $1/(2\pi)$ *times the volume flux across a surface of revolution generated by a curve joining the point to the axis (Figure 6.6). The velocity is specified by the stream function* $\Psi(r, z)[\Psi(R, \theta)]$ *in cylindrical (6.57a through c) [spherical (6.58a through c)] coordinates.*

6.2.3 Vector Potential for an Incompressible Flow

Using the theorem on solenoidal vector fields (5.247a through d) it follows that *the velocity of an incompressible flow (6.54b)* $\equiv$ *(6.59a) derives from a* **vector potential** *(6.59b):*

$$\nabla \cdot \vec{v} = 0 \Leftrightarrow \vec{v} = \nabla \wedge \vec{A}; \qquad (6.59a,b)$$

in the axisymmetric case (6.51a and b) $\equiv$ *(6.60a and b) the vector potential reduces to a pseudoscalar, that is, it has only one nonzero component, which is azimuthal in cylindrical (6.60c) [spherical (6.60d)] coordinates, and is specified by the stream function:*

$$v_\varphi = 0 = \partial_\varphi : \quad \vec{e}_\varphi r^{-1}\Psi(r,z) = \vec{A} \equiv \vec{e}_\varphi R^{-1}\csc\theta\,\Psi(R,\theta). \qquad (6.60a\text{–}d)$$

It can be checked from the curl in cylindrical (6.41b) [spherical (6.41c)] coordinates that the purely azimuthal vector potential (6.60c) [(6.60d)] leads to the cylindrical (6.61) $\equiv$ (6.57a through c) [spherical (6.62) $\equiv$ (6.58a through c)] components of the velocity:

$$\vec{v} = \nabla \wedge \left[\vec{e}_\varphi r^{-1}\Psi(r,z)\right] = -\vec{e}_r r^{-1}\partial_z \Psi + \vec{e}_z r^{-1}\partial_r \Psi, \qquad (6.61)$$

$$\vec{v} = \nabla \wedge \left[\vec{e}_\varphi R^{-1}\csc\theta\,\Psi(R,\theta)\right] = \vec{e}_R R^{-2}\csc\theta\,\partial_\theta \Psi - \vec{e}_\theta R^{-1}\csc\theta\,\partial_R \Psi. \qquad (6.62)$$

In a plane (6.63a and b) incompressible flow (6.63c) there is a stream function (6.63d):

$$v_z = 0 = \partial_z : \quad 0 = \nabla \cdot \vec{v} = \partial_x v_x + \partial_y v_y, \quad \left\{v_x, v_y\right\} = \left\{\partial_y \Psi, -\partial_x \Psi\right\}; \qquad (6.63a\text{–}d)$$

the vector potential corresponds to the stream function normal to the plane (6.64a) as follows from the curl (6.41a) in Cartesian coordinates (6.64b) ≡ (6.63b):

$$\vec{A} = \vec{e}_z \Psi(x,y): \quad \vec{v} = \nabla \wedge \left[\vec{e}_z \Psi(x,y) \right] = \vec{e}_x \partial_y \Psi - \vec{e}_y \partial_x \Psi. \tag{6.64a,b}$$

The vorticity (6.65a) for a plane (axisymmetric) flow:

$$\vec{\omega} = \nabla \wedge \vec{v}: \quad \vec{\omega} = \vec{e}_z \left(\partial_x v_y - \partial_y v_x \right) = -\vec{e}_z \left(\partial_{xx} \Psi + \partial_{yy} \Psi \right) = -\vec{e}_z \nabla^2 \Psi, \tag{6.65a–d}$$

$$\vec{\omega} = \vec{e}_\varphi \left(\partial_z v_r - \partial_r v_z \right) = \vec{e}_\varphi \left[-r^{-1} \partial_{zz} \Psi - \partial_r \left(r^{-1} \partial_r \Psi \right) \right] = -\vec{e}_\varphi r^{-1} \bar{\nabla}^2 \Psi, \tag{6.66a–c}$$

$$\vec{\omega} = \vec{e}_\varphi R^{-1} \left[\partial_R (R v_\theta) - \partial_\theta v_R \right] = -\vec{e}_\varphi R^{-1} \csc\theta \partial_{RR} \Psi - R^{-3} \partial_\theta \left(\csc\theta \partial_\theta \Psi \right)$$
$$\equiv -\vec{e}_\varphi R^{-1} \csc\theta \bar{\nabla}^2 \Psi, \tag{6.67a–c}$$

*is minus the Laplacian (6.65b through d) [**modified Laplacian** (6.66a through 6.66c; 6.67a through 6.67c)] applied to the stream function; the modified Laplacian is given by (6.66b and c) [(6.67b and c)] in cylindrical (spherical) coordinates by (6.68a) [(6.68b)]:*

$$\bar{\nabla}^2 \Psi = r \partial_r \left(r^{-1} \partial_r \Psi \right) + \partial_{zz} \Psi = \partial_{RR} \Psi + R^{-2} \sin\theta \partial_\theta \left(\csc\theta \partial_\theta \Psi \right). \tag{6.68a,b}$$

These results follow from the substitution of the cylindrical (6.57a through c) [spherical (6.58a through c)] components of the velocity in the curl (6.65a) using (6.41b) [(6.41c)], just as (6.65b) follows from (6.41a; 6.63b).

6.2.4 Scalar Potential for an Irrotational Flow

From the theorem on irrotational vector fields (5.240a through d) it follows that *for an irrotational flow (6.69a) the velocity is the gradient of a scalar potential (6.69b):*

$$\nabla \wedge \vec{v} = 0 \Leftrightarrow \vec{v} = \nabla \Phi, \quad \dot{D} = \nabla \cdot \vec{v} = \nabla \cdot (D\Phi) = \nabla^2 \Phi, \tag{6.69a–c}$$

which satisfies the Poisson equation (6.69c) forced by the dilatation or divergence of the velocity.

   In the case of a plane (6.63a and b) ≡ (6.70a and b) irrotational (6.70c) flow, the Cartesian components of the velocity are given by (6.70d):

$$v_z = 0 = \partial_z: \quad \partial_x v_y = \partial_y v_x \Leftrightarrow \{v_x, v_y\} = \{\partial_x \Phi, \partial_y \Phi\}; \tag{6.70a–d}$$

in the case of an axisymmetric (6.60a and b) irrotational (6.69b) flow, the cylindrical (spherical) components of the velocity are given (6.23d) [(6.23e)] by (6.71a through c) [(6.72a through c)]:

$$\{v_r, v_\varphi, v_z\} = \{\partial_r \Phi, 0, \partial_z \Phi\}, \tag{6.71a–c}$$

$$\{v_R, v_\theta, v_\varphi\} = \{\partial_R \Phi, R^{-1} \partial_\theta \Phi, 0\}. \tag{6.72a–c}$$

For an irrotational flow the potential is always a scalar regardless of (1) dimension, namely, plane (6.70a through d), space (6.73a through c), or higher-dimensional flow:

$$\{v_x, v_y, v_z\} = \{\partial_x \Phi, \partial_y \Phi, \partial_z \Phi\}; \tag{6.73a–c}$$

$$\{v_r, v_z, v_\varphi\} = \{\partial_r \Phi, r^{-1}\partial_\varphi \Phi, \partial_z \Phi\}, \tag{6.74a–c}$$

$$\{v_R, v_\theta, v_\varphi\} = \{\partial_R \Phi, R^{-1}\partial_\theta \Phi, R^{-1} \csc\theta \partial_\varphi \Phi\}; \tag{6.75a–c}$$

(2) for nonaxisymmetric [axisymmetric (6.60a and b)], for example, for the cylindrical (6.74a through c) [(6.71a through c)] or spherical (6.75a through c) [(6.72a through c)], components of the velocity.

6.2.5 Potential Flow: Irrotational and Incompressible

A **potential flow** is both irrotational (6.69a) and incompressible (6.59a). In the plane case (6.63a and b) ≡ (6.70a and b), the potential (6.70d) and stream function (6.63d) satisfy the **Cauchy–Riemann conditions** (6.76a and b) ≡ (I.11.10a and b):

$$\partial_x \Phi = v_x = \partial_y \Psi, \quad \partial_y \Phi = v_y = -\partial_x \Psi; \tag{6.76a,b}$$

these imply that if the potential and stream function have continuous second-order derivatives (6.77a) they are the real and imaginary parts, respectively, of a complex holomorphic or differentiable function, namely, the **complex potential** (6.77b):

$$\Phi, \Psi \in C^2(|R^2): \quad \Phi(x,y) + i\Psi(x,y) = f(x+iy) \in \mathcal{D}(|C). \tag{6.77a,b}$$

For an axisymmetric potential flow the Cauchy–Riemann conditions (6.76a and b) are replaced by the relation between the potential and stream function in cylindrical (6.57a through c) ≡ (6.71a through c) ≡ (6.78a and b) [spherical (6.58a through c) ≡ (6.72a through c) ≡ (6.79a and b)] coordinates:

$$\partial_r \Phi = v_r = -r^{-1}\partial_z \Psi, \quad \partial_z \Phi = v_z = r^{-1}\partial_r \Psi, \tag{6.78a,b}$$

$$\partial_R \Phi = v_R = R^{-2} \csc\theta \partial_\theta \Psi, \quad R^{-1}\partial_\theta \Phi = v_\theta = -R^{-1} \csc\theta \partial_R \Psi. \tag{6.79a,b}$$

The relations (6.76a and b) could not possibly hold in an axisymmetric flow, because the potential and stream functions have the same dimensions $[\Phi] = L^2 T^{-1} = [\Psi]$ (distinct dimensions $[\Phi] = L^2 T^{-1} \neq L^3 T^{-1} = [\Psi]$) in plane (axisymmetric) potential flow. The reason is that the plane (axisymmetric) stream function represents the flux of the velocity $[v] = LT^{-1}$ across a curve $\sim L$ (surface $\sim L^2$), so that $[\Psi] = L^2 T^{-1}([\Psi] = L^3 T^{-1})$, where L and T are the units of length and time, respectively; the potential has the dimensions $[\Phi] = [v]L = L^2 T^{-1}$ either in the plane or in space.

6.2.6 Laplace and Modified Laplace Operators

For a potential flow that is both irrotational (6.69a) ≡ (6.80a) and incompressible (6.59a) ≡ (6.80c), the scalar potential (6.69b) ≡ (6.80b) satisfies the Laplace equation (6.80d):

$$\nabla \wedge \vec{v} = 0: \quad \vec{v} = \nabla\Phi, \quad 0 = \nabla \cdot \vec{v} = \nabla \cdot (\nabla\Phi) = \nabla^2 \Phi. \tag{6.80a–d}$$

For example, in the plane (6.81a) or for an axisymmetric flow in cylindrical (6.81b) or spherical (6.81c) coordinates:

$$0 = \partial_{xx}\Phi + \partial_{yy}\Phi = r^{-1}\partial_r\left(r\partial_r\Phi\right) + r^{-2}\partial_{\varphi\varphi}\Phi$$

$$= R^{-2}\partial_R\left(R^2\partial_R\Phi\right) + R^{-2}\csc\theta\partial_\theta\left(\sin\theta\partial_\theta\Phi\right). \tag{6.81a–c}$$

For the plane potential flow the Laplace equation for the scalar potential in Cartesian coordinates (6.82b and c) ≡ (6.44) follows from the Cauchy–Riemann conditions (6.76a and b):

$$\Psi \in C^2\left(|R^2\right): \quad 0 = \partial_x\left(\partial_y\Psi\right) - \partial_y\left(\partial_x\Psi\right) = \partial_{xx}\Phi + \partial_{yy}\Phi = \nabla^2\Phi; \tag{6.82a–c}$$

$$0 = -r^{-1}\left[\partial_r\left(\partial_z\Psi\right) - \partial_z\left(\partial_r\Psi\right)\right] = r^{-1}\partial_r\left(r\partial_r\Phi\right) + \partial_{zz}\Phi = \nabla^2\Phi, \tag{6.83a,b}$$

$$0 = R^{-2}\csc\theta\left[\partial_R\left(\partial_\theta\Psi\right) - \partial_\theta\left(\partial_R\Psi\right)\right]$$

$$= R^{-2}\partial_R\left(R^2\partial_R\Phi\right) + R^{-2}\csc\theta\partial_\theta\left(\sin\theta\partial_\theta\Phi\right) \equiv \nabla^2\Phi. \tag{6.84a,b}$$

For the axisymmetric potential flow the Laplace equation for the potential in cylindrical (6.83a and b) ≡ (6.45a and b) [spherical (6.84a and b) ≡ (6.46a and b)] coordinates follows from the relations (6.78a and b) [(6.79a and b)] between the scalar potential and the stream function.

For a plane potential flow if the stream function (6.82a) [scalar potential (6.85a)] has continuous second-order derivatives then the Cauchy–Riemann conditions (6.76a and b) imply that both satisfy the Laplace equation (6.82b) [(6.85b)]:

$$\Phi \in C^2\left(|R^2\right): \quad 0 = -\partial_x\left(\partial_y\Phi\right) + \partial_y\left(\partial_x\Phi\right) = \partial_{xx}\Psi + \partial_{yy}\Psi = \nabla^2\Psi. \tag{6.85a,b}$$

For a potential axisymmetric flow, using cylindrical (6.78a and b) [spherical (6.79a and b)] coordinates it follows that the stream function satisfies a modified Laplace equation (6.86a and b) [(6.87a and b)] instead of the Laplace equation (6.83a and b) [(6.84a and b)] for the scalar potential:

$$0 = r\left[\partial_r\left(\partial_z\Phi\right) - \partial_z\left(\partial_r\Phi\right)\right] = r\partial_r\left(r^{-1}\partial_r\Psi\right) + \partial_{zz}\Psi = \bar{\nabla}^2\Psi, \tag{6.86a,b}$$

$$0 = -\sin\theta\left[\partial_R\left(\partial_\theta\Phi\right) - \partial_\theta\left(\partial_R\Phi\right)\right] = \partial_{RR}\Psi + R^{-2}\sin\theta\partial_\theta\left(\csc\theta\partial_\theta\Psi\right) = \bar{\nabla}^2\Psi. \tag{6.87a,b}$$

The Laplace equation for the potential of an axisymmetric flow in cylindrical (6.83a) ≡ (6.83b) [spherical (6.84a) ≡ (6.84b)] coordinates agrees with the Laplace operator (6.45a) [(6.46a)] if the azimuthal term is omitted (6.60b). The modified Laplace equation for the stream function in an axisymmetric potential flow in cylindrical (6.86a) ≡ (6.86b) [spherical (6.87a) ≡ (6.87b)] coordinates corresponds to the case of zero vorticity ϖ in (6.66a through c) [(6.67a through c)].

6.2.7 Scalar and Vector Poisson Equations

It has been shown that *if the stream function (6.82a) [scalar potential (6.84a)] of an incompressible (irrotational) flow has continuous second-order derivates, then (1) the potential always satisfies the scalar Poisson equation (6.69c) involving the scalar Laplace operator, for example, for a plane flow (6.82b) ≡ (6.88a) ≡ (6.44) and in an axisymmetric flow in cylindrical (6.83b) ≡ (6.88b) ≡ (6.45a) [spherical (6.84b) ≡ (6.88c) ≡ (6.46b)] coordinates:*

$$\dot{D} = \nabla^2\Phi \equiv \frac{\partial^2\Phi}{\partial x^2} + \frac{\partial^2\Phi}{\partial y^2} = \frac{1}{r}\frac{\partial}{\partial r}\left(r\frac{\partial\Phi}{\partial r}\right) + \frac{\partial^2\Phi}{\partial z^2}$$

$$= \frac{1}{R^2}\frac{\partial}{\partial R}\left(R^2\frac{\partial\Phi}{\partial R}\right) + \frac{1}{R^2\sin\theta}\frac{\partial}{\partial\theta}\left(\sin\theta\frac{\partial\Phi}{\partial\theta}\right); \tag{6.88a–c}$$

(2) the stream function also satisfies the scalar Poisson equation (6.88d) involving the Laplace operator (6.85b) ≡ (6.88d) ≡ (6.44) in the plane case (6.70a and b):

$$-\varpi = \nabla^2\Psi = \frac{\partial^2\Psi}{\partial x^2} + \frac{\partial^2\Psi}{\partial y^2}; \tag{6.88d}$$

and (3) for an axisymmetric flow the stream function does not satisfy the scalar Poisson equation involving the Laplace operator, but rather the modified Poisson equation involving the modified Laplace operator, namely, (6.86b) ≡ (6.89a) [(6.87b) ≡ (6.89b)] in cylindrical (spherical) coordinates:

$$-r\varpi = \bar{\nabla}^2\Psi = r\frac{\partial}{\partial r}\left(\frac{1}{r}\frac{\partial\Psi}{\partial r}\right) + \frac{\partial^2\Psi}{\partial z^2} = \frac{\partial^2\Psi}{\partial R^2} + \frac{\sin\theta}{R^2}\frac{\partial}{\partial\theta}\left(\frac{1}{\sin\theta}\frac{\partial\Psi}{\partial\theta}\right), \tag{6.89a,b}$$

which coincides with (6.66b) [(6.67b)]. The reason for the distinction is that: (1) in an irrotational flow the scalar potential satisfies the scalar Poisson equation forced by the dilatation (6.69a through c) regardless of dimensions, for example, for plane (6.70a through d) or three-dimensional (6.73a through c) flow; (2) the vector potential of an incompressible flow (6.90a) satisfies (6.90b) forced by the vorticity:

$$\vec{v} = \nabla\wedge\vec{A}: \quad \vec{\varpi} = \nabla\wedge\vec{v} = \nabla\wedge\left(\nabla\wedge\vec{A}\right) = \nabla\left(\nabla\cdot\vec{A}\right) - \nabla^2\vec{A}, \tag{6.90a,b}$$

*where (6.48a) is used; (3) the vector potential is not uniquely defined by its curl, and a **gauge condition** (6.91a) whose divergence is zero can be imposed:*

$$\nabla\cdot\vec{A} = 0: \quad \nabla^2\vec{A} = -\vec{\varpi}; \tag{6.91a,b}$$

(4) it follows that the vector potential satisfies the vector Poisson equation (6.91b) forced by minus the vorticity; and (5, 6) the vector Poisson equation (6.91b) for a plane (6.70a and b) [axisymmetric (6.60a and b)] flow involves the Laplace (6.88d) [modified Laplace (6.89a and b)] operator for the stream function, using Cartesian (6.64a) [cylindrical (6.60c) or spherical (6.60d)] coordinates.

6.2.8 Scalar and Vector Laplace Operators

Statements (5, 6) are proven as follows: (a) for a plane incompressible flow (6.63a through c) the vector potential is related to the stream function by (6.64a), and the vector Laplacian (6.91b) reduces (6.48b) to the scalar Laplacian (6.92):

$$-\vec{e}_z\varpi = \nabla^2\vec{A} = \nabla^2\left(\vec{e}_z\Psi\right) = \vec{e}_z\nabla^2\Psi; \tag{6.92}$$

(b) for the axisymmetric (6.60a and b) incompressible (6.59a) flow, using cylindrical (spherical) coordinates, the vector potential is related to the stream function by (6.60c) [(6.60d)], and the vector Laplacian (6.49) [(6.50)] in (6.91b) leads to (6.93a) [(6.93b)]:

$$-\vec{e}_\varphi \varpi = \nabla^2 \left[\frac{\vec{e}_\varphi}{r} \Psi(r,z) \right] = \vec{e}_\varphi \left\{ \nabla^2 \left[\frac{\Psi(r,z)}{r} \right] - \frac{\Psi}{r^3} \right\}$$

$$= \vec{e}_\varphi \left\{ \frac{1}{r} \frac{\partial}{\partial r} \left[r \frac{\partial}{\partial r} \left(\frac{\Psi}{r} \right) \right] + \frac{\partial^2}{\partial z^2} \left(\frac{\Psi}{r} \right) - \frac{\Psi}{r^3} \right\}$$

$$= \frac{\vec{e}_\varphi}{r} \left[\frac{\partial}{\partial r} \left(\frac{\partial \Psi}{\partial r} - \frac{\Psi}{r} \right) - \frac{\Psi}{r^2} + \frac{\partial^2 \Psi}{\partial z^2} \right]$$

$$= \frac{\vec{e}_\varphi}{r} \left(\frac{\partial^2 \Psi}{\partial r^2} - \frac{1}{r} \frac{\partial \Psi}{\partial r} + \frac{\partial^2 \Psi}{\partial z^2} \right) = \frac{\vec{e}_\varphi}{r} \left[r \frac{\partial}{\partial r} \left(\frac{1}{r} \frac{\partial \Psi}{\partial r} \right) + \frac{\partial^2 \Psi}{\partial z^2} \right] = \frac{\vec{e}_\varphi}{r} \bar{\nabla}^2 \Psi, \tag{6.93a}$$

$$-\vec{e}_\varphi \varpi = \nabla^2 \left[\frac{\vec{e}_\varphi}{R \sin\theta} \Psi(R,\theta) \right] = \vec{e}_\varphi \left\{ \nabla^2 \left[\frac{\Psi(R,\theta)}{R \sin\theta} \right] - \frac{\Psi(R,\theta)}{R^3 \sin^3\theta} \right\}$$

$$= \frac{\vec{e}_\varphi}{R^2} \frac{\partial}{\partial R} \left[R^2 \frac{\partial}{\partial R} \left(\frac{\Psi}{R \sin\theta} \right) \right] + \frac{\vec{e}_\varphi}{R^2 \sin\theta} \frac{\partial}{\partial\theta} \left[\sin\theta \frac{\partial}{\partial\theta} \left(\frac{\Psi}{R \sin\theta} \right) \right] - \frac{\Psi}{R^3 \sin^3\theta}$$

$$= \frac{\vec{e}_\varphi}{R^2 \sin\theta} \left[\frac{\partial}{\partial R} \left(R \frac{\partial \Psi}{\partial R} - \Psi \right) + \frac{1}{R} \frac{\partial}{\partial\theta} \left(\frac{\partial \Psi}{\partial\theta} - \cot\theta \Psi \right) - \frac{\Psi}{R \sin^2\theta} \right]$$

$$= \frac{\vec{e}_\varphi}{R \sin\theta} \left[\frac{\partial^2 \Psi}{\partial R^2} + \frac{1}{R^2} \left(\frac{\partial^2 \Psi}{\partial\theta^2} - \frac{\cos\theta}{\sin\theta} \frac{\partial \Psi}{\partial\theta} \right) \right]$$

$$= \frac{\vec{e}_\varphi}{R \sin\theta} \left[\frac{\partial^2 \Psi}{\partial R^2} + \frac{\sin\theta}{R^2} \frac{\partial}{\partial\theta} \left(\frac{1}{\sin\theta} \frac{\partial \Psi}{\partial\theta} \right) \right] = \frac{\vec{e}_\varphi}{R \sin\theta} \bar{\nabla}^2 \Psi; \tag{6.93b}$$

(c) in (6.93a) [(6.93b)] , the scalar Laplacian (6.45a) [6.46a)] was used, showing that the stream function of an incompressible axisymmetric flow satisfies the modified Poisson equation (6.89a) [(6.89b)] forced by minus the vorticity and involving the modified Laplace operator. It has been shown that *the vector Laplace operator leads: (1) in the plane case (6.94a) to the original (6.94b) scalar Laplace operator*:

$$\vec{A} = \vec{e}_z \Psi : \quad \nabla^2 \vec{A} = \vec{e}_z \nabla^2 \Psi, \tag{6.94a,b}$$

$$\vec{A} = \vec{e}_\varphi \frac{\Psi}{r} : \quad \nabla^2 \vec{A} = \frac{\vec{e}_\varphi}{r} \nabla^2 \Psi = \frac{\vec{e}_\varphi}{R \sin\theta} \bar{\nabla}^2 \Psi; \tag{6.95a-c}$$

and (2) in the axisymmetric case (6.95a) to the modified scalar Laplace operator (6.95b and c). The original and modified scalar Laplace operators are compared in Subsection 6.6.1 and some solutions are given in Section 6.6.

6.2.9 Streamlines, Vortex Lines, and Stream Surfaces

In the case of the rotational flow of an inviscid fluid (the viscous case is considered in Notes 6.6 through 6.10) the (Euler, 1752, 1759) momentum equation (6.96a) ≡ (I.14.9) ≡ (II.2.7) balances minus the pressure gradient against the inertial force per unit volume, which equals the mass density times the acceleration:

$$-\frac{1}{\rho} \nabla p = \vec{a} = \frac{d\vec{v}}{dt} = \frac{\partial \vec{v}}{\partial t} + (\vec{v} \cdot \nabla)\vec{v} = \frac{\partial \vec{v}}{\partial t} + \nabla \left(\frac{v^2}{2} \right) + (\nabla \wedge \vec{v}) \wedge \vec{v}; \tag{6.96a-c}$$

the **acceleration** (6.96b) ≡ (I.14.16) consists of two terms: (i) the local acceleration specifying the rate of change of the velocity with time at a fixed position; and (ii) the convective acceleration due to the variation of velocity with position at a fixed time, which involves (6.96c) two terms: (ii-1) the gradient of one-half of the modulus of the velocity, that is, the kinetic energy per unit mass; (ii-2) the outer vector product of the vorticity by the velocity. The r.h.s. of (6.96a) ≡ (6.96b) ≡ (6.96c) is the acceleration or **material derivative** of the velocity and the l.h.s. involves the enthalpy (6.97a), where T is the temperature and S the entropy, specifying dQ, the heat exchanged:

$$\mathrm{d}H - \frac{1}{\rho}\mathrm{d}p = T\,\mathrm{d}S = \mathrm{d}Q; \quad S = \mathrm{const}: \quad \nabla H = \frac{1}{\rho}\nabla p. \tag{6.97a–c}$$

In the absence of heat exchange, that is, for a **homentropic flow** with uniform entropy (6.97b), the gradient of the enthalpy (6.97c) appears in the momentum equation (6.96a) ≡ (6.96b) ≡ (6.96c) ≡ (6.98a):

$$\frac{\partial \vec{v}}{\partial t} + \left(\nabla \wedge \vec{v}\right) \wedge \vec{v} = -\nabla\left(H + \frac{v^2}{2}\right) = -\nabla H_0, \quad H_0 \equiv H + \frac{v^2}{2}, \tag{6.98a,b}$$

where the following may be introduced: (1) the **stagnation enthalpy** (6.98b), that is, the enthalpy plus the kinetic energy per unit mass; and (2) the vorticity (6.65a) ≡ (6.99a):

$$\vec{\varpi} \equiv \nabla \wedge \vec{v}: \quad \partial\vec{v}/\partial t + \vec{\varpi} \wedge \vec{v} = -\nabla H_0, \tag{6.99a,b}$$

leading to (6.99b).

Thus *in a homentropic flow (6.97b), that is, without heat exchanges, the gradient of the stagnation enthalpy (6.98b) is related to the velocity and vorticity (6.99a) by (6.99b). If in addition the flow is steady (6.100a) then the stagnation enthalpy is constant (6.100b) on the* **stream surfaces** *spanned by the streamlines (6.100c) [vortex lines (6.100d)]:*

$$\frac{\partial \vec{v}}{\partial t} = 0: \quad \nabla H_0 = \vec{v} \wedge \vec{\varpi}, \quad \vec{v} \cdot \nabla H_0 = 0 = \vec{\varpi} \cdot \nabla H_0, \tag{6.100a–d}$$

because (Figure 6.7): (1) the normal $\vec{N}$ to the surfaces of constant stagnation enthalpy, that is, $H_0 = $ const, or $0 = \mathrm{d}H_0 = \mathrm{d}\vec{x} \cdot \nabla H_0$ for any displacement, is parallel to ∇H_0; and (2) the projection of the velocity (6.100c) [vorticity (6.100d)] on this direction is zero. *The surfaces of constant stagnation enthalpy coincide with the* **stream surfaces** *of the axisymmetric flow, which have a "bell" shape (Figure 6.7) because (1) the velocity lies in planes through the axis since $v_\varphi = 0$ in (6.60a and b) ≡ (6.58c); (2) the vorticity (6.66a through c; 6.67a through c) lies along parallels and cuts the velocity at right angles; and (3) the stream surfaces are thus obtained by rotating the stream lines around the axis.*

6.2.10 Relation of the Stagnation Enthalpy or Pressure to the Stream Function

The result (6.100a through d) is the axisymmetric analogue of the statement (Subsection II.2.3.3) that the stagnation pressure is constant along the streamlines of a plane steady incompressible homentropic rotational flow (Figure 6.8) because (1) for an incompressible (6.101a), homentropic (6.101b) flow the surfaces of constant stagnation enthalpy (6.101c) coincide with the surfaces of constant stagnation pressure (6.101d):

$$\rho = \mathrm{const}, S = \mathrm{const}: \quad p_0 \equiv \rho H_0 = p + \frac{1}{2}\rho v^2, \quad \vec{v} \cdot \nabla p_0 = 0 = \vec{\varpi} \cdot \nabla p_0; \tag{6.101a–e}$$

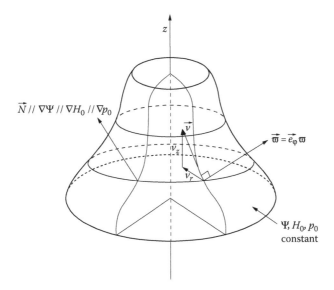

FIGURE 6.7 For an axisymmetric flow the velocity lies on planes through the axis and does not depend on the azimuthal angle; in this case the vorticity is orthogonal to the planes through the axis, and together with the velocity forms a stream surface of revolution. In the case of a steady homentropic flow the vorticity, stagnation pressure, and stagnation enthalpy are functions only of the stream function implying that: (1) they are constant on the stream surface; and (2) they can vary only orthogonally to the stream surface. In the particular case when the flow is also irrotational, the vorticity is zero and the stagnation pressure and stagnation enthalpy are uniform in all space.

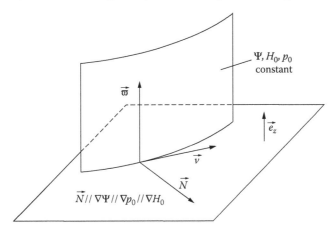

FIGURE 6.8 A plane rotational flow has a vorticity orthogonal to the plane. In this case the stream surfaces to which the velocity and vorticity are tangent are cylindrical surfaces with generators orthogonal to the plane of the streamlines that act as directrix of the cylindrical surface. In this case the stagnation pressure is conserved along streamlines and can vary only orthogonally to them for a plane, steady, homentropic, incompressible flow. If the flow is also irrotational, the vorticity is zero, and the stagnation pressure is constant in the whole plane.

*(2) for a plane flow the vorticity is orthogonal to the plane (6.101e); and (3) thus the surfaces of constant stagnation pressure or constant stagnation enthalpy are cylinders, with the streamline as directrix, and generators orthogonal to the plane. In the particular case of an irrotational flow (6.102a) the stagnation enthalpy (6.102b) and hence the **stagnation enthalpy** (6.102b) and **stagnation pressure** (6.102c) are constant everywhere:*

$$0 = \vec{\varpi} = \nabla \wedge \vec{v}: \quad H_0 = \frac{p_0}{\rho} = \text{const}, \quad p_0 = p + \frac{1}{2}\rho v^2 = \text{const}. \qquad (6.102a\text{–}c)$$

This is the simplest form of the **Bernoulli equation** *(6.102c) ≡ (I.14.27c) ≡ (II.2.12b): in an incompressible (6.101a), irrotational (6.102a), homentropic (6.101b), steady (6.100a) flow the stagnation pressure (6.102c) is constant, which is the sum of (1) the pressure; and (2) the dynamic pressure or kinetic energy per unit volume. The difference between a rotational (irrotational) steady, incompressible, homentropic flow is that the stagnation enthalpy and stagnation pressure are constant on stream surfaces (6.100c and d; 6.101d and e) [are constant everywhere (6.102b and c)].*

The preceding conclusions (6.100a through d; 6.101a through e; 6.102a through c) based on geometric reasonings are amenable analytic proof, for example, in cylindrical coordinates, bearing in mind that the velocity lies in planes through the axis (6.103a) ≡ (6.58a through c), and the vorticity along parallels (6.103b) ≡ (6.66a), so that the stagnation enthalpy satisfies (6.100b) ≡ (6.103c):

$$\vec{v} = \vec{e}_z v_z + \vec{e}_r v_r; \quad \vec{\omega} = \vec{e}_\varphi \omega: \quad \nabla H_0 = \left(\vec{e}_z v_z + \vec{e}_r v_r\right) \wedge \vec{e}_\varphi \omega; \tag{6.103a–c}$$

which simplifies to (6.103c) ≡ (6.104a) ≡ (6.104b):

$$\nabla H_0 = \omega\left(-\vec{e}_r v_z + \vec{e}_z v_r\right) \Leftrightarrow \left\{\partial_r H_0, \partial_z H_0\right\} = \omega\left\{-v_r, v_z\right\}. \tag{6.104a,b}$$

From (6.104b) it follows that on a stream surface (6.105a) holds (6.105b):

$$0 = dH_0 = \left(\partial_r H_0\right) dr + \left(\partial_z H_0\right) dz: \quad \left(\frac{dz}{dr}\right)_{H_0} = -\frac{\partial_r H_0}{\partial_z H_0} = \frac{v_z}{v_r} = -\frac{\partial_r \Psi}{\partial_z \Psi} = \left(\frac{dz}{dr}\right)_\Psi, \tag{6.105a,b}$$

where (6.57a and c) was used. Thus *the surfaces of constant stagnation enthalpy coincide with the stream surfaces.* The stagnation enthalpy $H_0(z, r)$ and stream function $\Psi(z, r)$ are generally functions of two variables; in the case of steady homentropic flow they are a function of each other, that is, $H_0 = H_0(\Psi)$, whose derivative is specified by

$$dH_0 = d\vec{x} \cdot \nabla H_0 = \omega\left(v_r\, dz - v_z\, dr\right) = -\frac{\omega}{r}\left[\left(\partial_z \Psi\right) dz + \left(\partial_r \Psi\right) dr\right] = -\frac{\omega}{r} d\Psi, \tag{6.105c}$$

where (6.104b) and (6.57a and c) were used.

6.2.11 Bernoulli Equation and Extension to Vortical Flows

Substituting (6.102b and c) in (6.105c) leads to (6.107a through h):

$$\frac{\partial}{\partial t} = 0, \frac{\partial}{\partial z} = 0 = v_z, S = \text{const}, \rho = \text{const}:$$

$$\text{const} = H_0 + \int \omega\, d\Psi = \frac{p_0}{\rho} + \int \omega\, d\Psi = \frac{p}{\rho} + \frac{v^2}{2} + \int \omega\, d\Psi, \tag{6.106a–h}$$

$$\frac{\partial}{\partial t} = 0, \frac{\partial}{\partial \varphi} = 0 = v_\varphi, S = \text{const}, \rho = \text{const}:$$

$$\text{const} = H_0 + \int \frac{\omega}{r} d\Psi = \frac{p_0}{\rho} + \int \frac{\omega}{r} d\Psi = \frac{p}{\rho} + \frac{v^2}{2} + \int \frac{\omega}{r} d\Psi, \tag{6.107a–h}$$

which is the **generalized Bernoulli (1738) equation** *relating the stagnation enthalpy (6.107f), stagnation pressure, and mass density (6.107g), pressure, velocity, and mass density (6.107h), to the vorticity and* *stream function, for a steady (6.107a), axisymmetric (6.107b and c), homentropic (6.107d), incompressible (6.107e) flow. In the case of a plane flow (6.106b and c) the same relation holds (6.106a through h) without the factor 1/r in the integral in the last term on the r.h.s. of (6.107f through h). From (6.66b and c) [(6.67b and c)] it follows that in a steady (6.108a), axisymmetric (6.108b and c), homentropic (6.108d), incompressible (6.108e) rotational flow the modified Laplace operator applied to the stream function in cylindrical (6.68a) ≡ (6.108f) spherical [(6.68b) ≡ (6.108g)] coordinates:*

$$\frac{\partial}{\partial t} = 0, \quad \frac{\partial}{\partial \varphi} = 0 = v_\varphi, \quad S = \text{const}, \quad \rho = \text{const}:$$

$$\frac{r^2}{\rho}\frac{dp_0}{d\Psi} = r^2\frac{dH_0}{d\Psi} \equiv r^2 f(\Psi) = -\varpi r = \bar{\nabla}^2\Psi = \frac{\partial^2\Psi}{\partial r^2} - \frac{1}{r}\frac{\partial\Psi}{\partial r} + \frac{\partial^2\Psi}{\partial z^2}, \tag{6.108a--f}$$

$$-\frac{R^2}{\rho}\sin^2\theta\frac{dp_0}{d\Psi} = -R^2\sin^2\theta\frac{dH_0}{d\Psi} = R^2\sin^2\theta f(\Psi)$$

$$= -\varpi R\sin\theta = \bar{\nabla}^2\Psi = \frac{\partial^2\Psi}{\partial R^2} + \frac{\sin\theta}{R^2}\frac{\partial}{\partial\theta}\left(\frac{1}{\sin\theta}\frac{\partial\Psi}{\partial\theta}\right), \tag{6.108g}$$

is forced by an arbitrary function f(Ψ) of the stream function specified equivalently by: (1) minus the vorticity multiplied by the distance from the axis; and (2/3) the derivative of the stagnation enthalpy (6.98b) [stagnation pressure (6.101c) divided by the mass density] with regard to the stream function times minus the square of the distance from the axis, that is (6.108f) [(6.108g)], in cylindrical (spherical) coordinates. For an irrotational flow (6.109a) the vorticity is zero and (6.108f and g) reduces to the scalar modified Laplace equation (6.109b):

$$\varpi = 0: \quad \bar{\nabla}^2\Psi = 0, \tag{6.109a,b}$$

involving the modified Laplace operator, specified by (6.68a) [(6.68b)] in cylindrical (spherical) coordinates. Both (6.108f and g) and (6.68a and b) are particular cases of the general equation for the stream function of an axisymmetric incompressible rotational flow, valid for the unsteady flow of a viscous fluid (Note 6.7).

6.2.12 Scalar, Modified, and Vector Poisson Equations

*A source Q > 0 or sink Q < 0 with flow rate Q equal to the dilatation (6.110c) in an irrotational flow (6.110a) leads to a **Poisson equation** (6.110d) for the scalar potential (6.110b):*

$$\nabla \wedge \vec{v} = 0: \quad \vec{v} = \nabla\Phi, \quad Q = \nabla \cdot \vec{v} = \nabla \cdot (\nabla\Phi) = \nabla^2\Phi, \tag{6.110a--d}$$

*regardless of space dimension or flow symmetry. A distribution of vorticity (6.111d) in an incompressible flow (6.111a) leads to a **vector Poisson equation** (6.111e) for a vector potential (6.111b) satisfying the gauge condition (6.91a) ≡ (6.111c):*

$$\nabla \wedge \vec{v} = 0: \quad \vec{v} = \nabla \wedge \vec{A}, \quad \nabla \cdot \vec{A} = 0, \quad -\vec{\varpi} = -\nabla \wedge \vec{v} = -\nabla \wedge (\nabla \wedge \vec{A}) = \nabla^2\vec{A}. \tag{6.111a--e}$$

In the case of a plane incompressible flow (6.70a and b) ≡ (6.112a and b) the vorticity is orthogonal to the plane (6.65b) ≡ (6.112c) like the vector potential (6.64a) ≡ (6.112d) that reduces to a stream function satisfying a scalar Poisson equation (6.112e):

$$v_z = 0 = \partial_z: \quad \vec{\varpi} = \vec{e}_z\varpi, \quad \vec{A} = \vec{e}_z\Psi: \quad \nabla^2\Psi = -\varpi, \tag{6.112a--e}$$

similar to the potential (6.110d) replacing (6.113a), the flow rate by minus the vorticity (6.113b).

$$\Phi \leftrightarrow \Psi, \quad Q \leftrightarrow -\varpi, \quad \nabla^2 \leftrightarrow \bar{\nabla}^2; \tag{6.113a–c}$$

the Laplacian is replaced by the modified Laplacian (6.113c) in the case of axisymmetric flow, which is considered next. In the case of axisymmetric incompressible flow (6.60a and b) ≡ (6.114a and b) using cylindrical (spherical) coordinates the vorticity is an azimuthal vector (6.66a) ≡ [(6.67b) ≡ (6.115a)], like the vector potential (6.60c) [(6.60d)], that reduces to a stream function (6.114c) [(6.115b)] satisfying the modified Poisson equation (6.114d) [(6.115c)]:

$$v_\varphi = 0 = \partial_\varphi: \quad \vec{A} = \vec{e}_\varphi r^{-1}\Psi, \quad -\varpi r = \bar{\nabla}^2\Psi, \tag{6.114a–d}$$

$$\vec{\varpi} = \vec{e}_\varphi \varpi: \quad \vec{A} = \vec{e}_\varphi R^{-1} \csc\theta\,\Psi, \quad -\varpi R\sin\theta = \bar{\nabla}^2\Psi; \tag{6.115a–c}$$

the modified equation differs from the original Poisson equation (6.112e) in using (6.113c) the modified Laplace (6.68a) [(6.68b)] instead of the Laplace (6.45a and b) [(6.46a and b)] operator. In the case of an incompressible (6.116a), steady (6.116b), homentropic (6.116c), axisymmetric (6.114a and b) flow, the vorticity is a function of the stream function alone leading to the nonlinear Poisson equation:

$$\rho = const, \quad \partial_t = 0, \quad S = const: \quad \varpi = f(\Psi) = -r^{-1}\bar{\nabla}^2\Psi = -R^{-1}\csc\theta\,\bar{\nabla}^2\Psi, \tag{6.116a–e}$$

in cylindrical (6.116d; 6.70a) [spherical (6.116e; 6.70b)] coordinates. The stream function can also be used in the more general case of an unsteady viscous incompressible flow, either plane (Note II.4.11) or axisymmetric (Notes 6.8 and 6.9).

Diagram 6.1 summarizes the incompressible flows (6.59a) for which the velocity has a vector potential (6.59b) that satisfies a double curl vector equation (6.90b) forced by the vorticity; imposing a gauge condition (6.91a) on the vector potential leads to a vector Laplace equation (6.91b) forced by minus the vorticity. In the plane (6.70a and b) [axisymmetric (6.60a and b)] case the vector potential reduces to a stream function (6.64a) [(6.60c and d)] that satisfies a nonlinear Poisson (6.117) [modified Poisson (6.116e)] equation:

$$\varpi = f(\Psi) = \nabla^2\Psi = \partial_{xx}\Psi + \partial_{yy}\Psi, \tag{6.117}$$

assuming a steady (6.116b) homentropic (6.116c) flow. In the absence of vorticity the flow is both incompressible and irrotational, and the stream function of a potential flow satisfies the Laplace (6.88d) [modified Laplace (6.89a and b)] equation for the plane (axisymmetric) case. The potential (6.80b) of an irrotational (6.80a) flow satisfies the scalar Laplace equation (6.80d) in all cases, including plane flow (6.88a) and three-dimensional axisymmetric (6.88b and c) [nonaxisymmetric (6.44; 6.45a and b; 6.46a and b)] flow.

6.3 Point Source/Sink in a Uniform Stream

The simplest potential flows in the plane (axisymmetric) flow are: (1) the uniform flow [Subsection I.14.8.1 (Subsection 6.3.2)]; (2) the line (point) source or sink [Section I.12.4 (Subsection 6.3.3)]; (3) their superposition leading to a Rankine semi-infinite cylindrical (Section I.28.4) [axisymmetric (Subsection 6.3.4)] fairing. In all three cases (1–3) are used the relations (Subsection 6.3.1) among the scalar potential, stream function, and cylindrical and spherical components of the velocity; the Bernoulli theorem leads to the pressure distribution on the fairing (Subsection 6.3.5), whose resultant is a drag force.

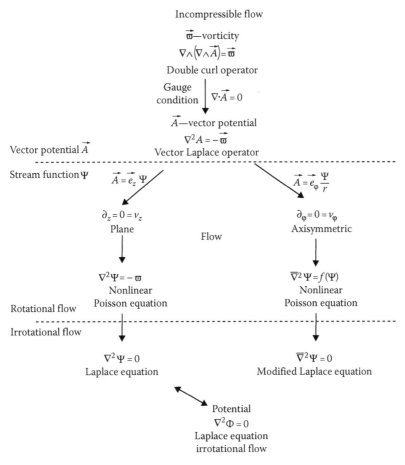

DIAGRAM 6.1 Incompressible (irrotational) flows that have a scalar (vector) potential in a space with any (three) dimension(s). Also, the cases of incompressible plane (axisymmetric) flow when the vector potential reduces to a stream function.

6.3.1 Relation between the Potential and the Stream Function

In a potential axisymmetric flow there exists a scalar potential (stream function) whose exact differential is specified by the components of the velocity, for example, in: (1) cylindrical coordinates (6.71a through c) [(6.57a through c)] by (6.118a and b) [(6.119a and b)]:

$$d\Phi = (\partial_r \Phi)dr + (\partial_z \Phi)dz = v_r \, dr + v_z \, dz,$$
(6.118a,b)

$$d\Psi = (\partial_r \Psi)dr + (\partial_z \Psi)dz = r(v_z \, dr - v_r \, dz),$$
(6.119a,b)

and (2) in spherical coordinates (6.72a through c) [(6.58a through c)] by (6.120a and b) [(6.121a and b)]:

$$d\Phi = (\partial_R \Phi)dR + (\partial_\theta \Phi)d\theta = v_R \, dR + v_\theta R \, d\theta,$$
(6.120a,b)

$$d\Psi = (\partial_R \Psi)dR + (\partial_\theta \Psi)d\theta = R\sin\theta(v_R R \, d\theta - v_\theta \, dR).$$
(6.121a,b)

Thus the scalar potential and stream function are alternative descriptions of an axisymmetric potential flow and can be determined from each other using the exact differentials:

$$\Psi = \int d\Psi = \int \{(\partial_r \Psi) dr + (\partial_z \Psi) dz\} = \int r \{(\partial_z \Phi) dr - (\partial_r \Phi) dz\}$$

$$= \int \{(\partial_R \Psi) dR + (\partial_\theta \Psi) d\theta\} = \int \sin\theta \{R^2 (\partial_R \Phi) d\theta - (\partial_\theta \Phi) dR\} \qquad (6.122\text{a–d})$$

$$\Phi = \int d\Phi = \int \{(\partial_r \Phi) dr + (\partial_z \Phi) dz\} = \int r^{-1} \{(\partial_r \Psi) dz - (\partial_z \Psi) dr\}$$

$$= \int \{(\partial_R \Phi) dR + (\partial_\theta \Phi) d\theta\} = \int \csc\theta \{R^{-2} (\partial_\theta \Psi) dR - (\partial_R \Psi) d\theta\} \qquad (6.123\text{a–d})$$

for example, (1) in (6.78a and b) cylindrical coordinates (6.122a and b) [(6.123a and b)]; and (2) in (6.79a and b) spherical coordinates (6.122c and d) [(6.123c and d)]. These relations are applied next to axisymmetric potential flows, starting with the simplest, namely, the uniform flow (Subsection 6.3.2), the line source/sink (Subsection 6.3.3), and their superposition (Subsections 6.3.4 and 6.3.5).

6.3.2 Uniform Flow Parallel to the Axis

The simplest axisymmetric flow is the *uniform stream parallel to the axis (Figure 6.5a) that has: (1) the cylindrical components of the velocity (6.124a); and (2) the corresponding (6.51d) spherical components of the velocity (6.124b):*

$$\{v_z, v_r\} = \{U, 0\}, \{v_R, v_\theta\} = U\{\cos\theta, -\sin\theta\}; \qquad (6.124\text{a,b})$$

(3, 4) the corresponding scalar potential (6.125a) [stream function (6.125b)]:

$$\Phi(z) = Uz = UR\cos\theta \equiv \Phi(R,\theta), \quad 2\Psi(r) = Ur^2 = UR^2\sin^2\theta \equiv 2\Psi(R,\theta), \qquad (6.125\text{a,b})$$

can be obtained:

$$\Phi = \int U \, dz = Uz, \quad \Psi = \int Ur \, dr = \frac{1}{2}Ur^2, \qquad (6.126\text{a,b})$$

from (6.118b) [(6.119b)].

6.3.3 Radial Flow from a Point Source or Sink

The only radially symmetric potential flow, that is, solution of the Laplace equation (6.127b) for a scalar potential depending only on the radius (6.127a), decays inversely with the radius (6.127c and d):

$$\Phi = \Phi(R): \quad 0 = \frac{d}{dR}\left(R^2 \frac{d\Phi}{dR}\right), \quad \frac{d\Phi}{dR} = \frac{C_1}{R^2}, \quad \Phi(R) = -\frac{C_1}{R} + C_2; \qquad (6.127\text{a–d})$$

the constant C_2 can be omitted (6.128a) in (6.127d) because the scalar potential is defined to within an added constant, which does not affect the velocity (6.128b and c):

$$C_2 = 0: \quad \{v_R, v_\theta\} = \left\{\frac{C_1}{R^2}, 0\right\}, \quad 4\pi C_1 = 4\pi R^2 v_R = Q; \qquad (6.128\text{a–d})$$

the velocity is radial (6.128b and c), and the remaining constant is specified by the mass flux (6.128d) across a sphere of radius R. Thus *a* **point source (sink)** *with isotropic positive (negative)* ***flow rate*** *Q has a radial outward (inward) velocity (6.129a and b) corresponding (6.51c) to the cylindrical components (6.129c and d)*:

$$v_R = \frac{Q}{4\pi R^2}, v_\theta = 0: \quad v_z = \frac{Q\cos\theta}{4\pi R^2} = \frac{Qz}{4\pi R^3}, \quad v_r = \frac{Q\sin\theta}{4\pi R^2} = \frac{Qr}{4\pi R^3}. \tag{6.129a–d}$$

The corresponding potential (stream function) is (6.130b) [(6.131b)] in cylindrical and (6.130a) [(6.131a)] in spherical coordinates:

$$\Phi(R) = -\frac{Q}{4\pi R} = -\frac{Q}{4\pi}\left|r^2 + z^2\right|^{-1/2} = \Phi(r,z), \tag{6.130a,b}$$

$$\Psi(\theta) = -\frac{Q}{4\pi}\cos\theta = -\frac{Qz}{4\pi}\left|r^2 + z^2\right|^{-1/2} = \Psi(r,z), \tag{6.131a,b}$$

which follows by substituting (6.129a and b) in (6.120b) [(6.121b)], leading to (6.132a) $\equiv$ (6.130a) [(6.132b) $\equiv$ (6.131a)]:

$$d\Phi = \frac{Q}{4\pi R^2}dR = -d\left(\frac{Q}{4\pi R}\right), \quad d\Psi = \frac{Q}{4\pi}\sin\theta\, d\theta = -d\left(\frac{Q}{4\pi}\cos\theta\right). \tag{6.132a,b}$$

Since the potential (stream function) of an axisymmetric potential flow satisfies a Laplace (6.102e) [modified Laplace (6.102f)] equation that is linear the principle of superposition holds, for example, a point (line) source/sink [Subsection 6.3.3 (Section I.12.3)] in a uniform stream [Subsection 6.3.2 (Section I.14.7)] leads to the Rankine axisymmetric (cylindrical) semi-infinite fairing [Subsections 6.3.4 and 6.3.5 (Section I.28.4)].

6.3.4 Rankine Semi-Infinite Axisymmetric Fairing

Since the potential Φ (stream Ψ) function satisfies a linear partial differential equation (6.88a through c) [(6.88d; 6.89a and b)], the principle of superposition can be used to add potential axisymmetric flows, for example, for a point source (sink) of volume flux $Q > 0 (Q < 0)$, in a uniform incident stream of velocity U at infinity, the sum of (6.125a) [(6.125b)] and (6.130a) [(6.131a)] specifies the potential (6.133a) [stream function (6.133b)]:

$$\Phi(R,\theta) = UR\cos\theta - \frac{Q}{4\pi R}, \quad \Psi(R,\theta) = \frac{1}{2}UR^2\sin^2\theta - \frac{Q}{4\pi}\cos\theta; \tag{6.133a,b}$$

the spherical components of velocity:

$$v_R(R,\theta) = U\cos\theta + \frac{Q}{4\pi R^2}, \quad v_\theta(\theta) = -U\sin\theta, \tag{6.134a,b}$$

can be calculated by: (1) substituting (6.133a) in (6.72a through c); (2) substituting (6.133b) in (6.58a through c); and (3) adding (6.124b) to (6.129a and b). The flow (6.134a and b) with source (6.135a) at the origin has a stagnation point at (6.135b and c) on the axis at a distance (6.135d) from the origin:

$$Q > 0; \quad \theta = \pi, R = a: \quad a^2 = \frac{Q}{4\pi U}; \quad -Q\cos\theta + 2\pi UR^2\sin^2\theta = Q, \tag{6.135a–e}$$

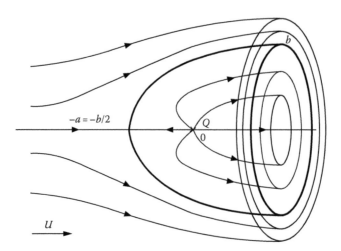

FIGURE 6.9 A point (line) source in a uniform stream corresponds to the potential flow past an axisymmetric (cylindrical) Rankine semi-infinite fairing [Figure 6.7 (Figure I.28.5a)] in the downstream direction like the rounded nose of an aircraft or airship (rounded leading-edge of an airfoil or wing). Replacing the source by a sink would lead to a similar Rankine semi-infinite fairing in the upstream direction (Figure I.28.5b), that is, the fairing (Figure 6.9) in the reverse directions represents the rounded tail of an aircraft or airship.

the stream surface (6.133b) through the stagnation point (6.135b) has equation (6.135e) ≡ (6.136):

$$\frac{Q}{2\pi U}(1+\cos\theta) = R^2 \sin^2\theta = r^2; \tag{6.136}$$

at infinity $R \to \infty$, $\theta \to 0$ in the downstream direction, the distance of the streamline from the axis is:

$$b = \lim_{R\to\infty} r = \lim_{R\to\infty} R\sin\theta = \lim_{\theta\to 0}\left|\frac{Q}{2\pi U}(1+\cos\theta)\right|^{1/2} = \sqrt{\frac{Q}{\pi U}}, \quad Q = \pi b^2 U, \tag{6.137a,b}$$

implying that (1) the stream surface corresponds to a semi-infinite axisymmetric fairing (Figure 6.9) with asymptotic radius (6.137a); (2) far downstream the flow rate (6.137b) is the uniform free stream velocity multiplied by the cross-sectional area; and (3) the flow rate from the source remains inside the fairing, separated from the external flow by the stream surface, to which the velocity is tangent. In the case of a source (6.135a) the stagnation point is upstream (6.135b and c) at the nose of a rounded axisymmetric fairing (Figure 6.9); in the plane case of a line source the same directrix applies to a cylinder (Figure I.28.5a) like the rounded leading edge of a straight wing. In the case of a sink $Q < 0$, the stagnation point is downstream at $\theta = 0$ at the same distance (6.135d) from the origin, and the axisymmetric fairing is a rounded aft-end, as in Figure I.28.5b. In both cases *the shape of the Rankine axisymmetric fairing (6.136) is given (6.137b) by (6.138a):*

$$r(\theta) = \left|\frac{Q}{2\pi U}(1+\cos\theta)\right|^{1/2} = a\sqrt{2(1+\cos\theta)} = b\sqrt{\frac{1+\cos\theta}{2}}, \quad b = 2a, \tag{6.138a,b}$$

with the asymptotic radius (6.137a) ≡ (6.137b) equal (6.138b) to twice the distance of the source from the stagnation point (6.135d). The shape of the Rankine semi-infinite fairing is also given by (6.139):

$$R(\theta) = \frac{r(\theta)}{\sin\theta} = \left|\frac{Q}{2\pi U}\frac{1+\cos\theta}{\sin^2\theta}\right|^{1/2} = \left|\frac{Q}{2\pi U}\frac{1}{1-\cos\theta}\right|^{1/2} = a\sqrt{\frac{2}{1-\cos\theta}} = \frac{b}{\sqrt{2(1-\cos\theta)}}, \tag{6.139}$$

as an alternative to (6.138a).

6.3.5 Pressure Distribution and Drag Force

The volume flux of the source is specified by the scale of the body (6.137a) ≡ (6.140a) or location of the stagnation point (6.135d) ≡ (6.140b) and leads (6.134a and b) to the velocity (6.140c and d):

$$Q = \pi U b^2 = 4\pi U a^2 : \quad v_R = U\left(\cos\theta + \frac{a^2}{R^2}\right), \quad v_\theta = -U\sin\theta; \tag{6.140a–d}$$

in a reference frame for which the fairing is at rest (Figure 6.9), the flow is steady, and the pressure is given (6.140c and d), by the Bernoulli law (6.102c):

$$p(R,\theta) = p_0 - \frac{\rho}{2}\left(v_R^2 + v_\theta^2\right) = p_0 - \frac{\rho}{2}U^2\left\{\left(\cos\theta + \frac{a^2}{R^2}\right)^2 + \sin^2\theta\right\}, \tag{6.141}$$

where ρ is the constant mass density of the incompressible fluid and p_0 the stagnation pressure that is the pressure at the position (6.135b and c). *The dynamic pressure of the free stream (6.142a) and the flow pressure at infinity (6.142b)*:

$$p_\star \equiv \frac{1}{2}\rho U^2, \quad p_\infty = p_0 - \frac{\rho}{2}U^2 = p_0 - p_\star = p(\infty,\theta), \tag{6.142a,b}$$

together with the stagnation pressure (6.102c) lead to several alternate expressions for the dimensionless **pressure coefficient**:

$$C_p(R,\theta) \equiv \frac{p_0 - p(R,\theta)}{p_\star} = \frac{p_0 - p(R,\theta)}{p_0 - p_\infty} = 2\frac{p_0 - p(R,\theta)}{\rho U^2} = \frac{|\vec{v}|^2}{U^2}; \tag{6.143a–d}$$

the latter provides a dimensionless representation of the pressure distribution:

$$p(R,\theta) = p_0 - p_\star C_p(R,\theta) = p_0 - \frac{1}{2}\rho U^2 C_p(R,\theta). \tag{6.144a,b}$$

In the case of the Rankine semi-infinite fairing (6.141) the pressure coefficient is the term in curly brackets (6.145a):

$$C_p(R,\theta) = \frac{(v_R)^2 + (v_\theta)^2}{U^2} = 1 + 2\frac{a^2}{R^2}\cos\theta + \frac{a^4}{R^4}; \quad C_p(a,\theta) = \frac{5}{9} + \frac{1}{2}\cos^2\theta - \frac{3}{4}\cos^2\theta; \tag{6.145a,b}$$

the pressure coefficient (6.145a) applies at all points of the flow past a **Rankine axisymmetric fairing** (Figure 6.9); it simplifies using (6.139) to:

$$C_p(\theta) = 1 + \cos\theta(1 - \cos\theta) + \left(\frac{1 - \cos\theta}{2}\right)^2 = \frac{5}{4} + \frac{1}{2}\cos\theta - \frac{3}{4}\cos^2\theta, \tag{6.146}$$

the pressure coefficient (6.146) ≡ (6.145b) on the fairing. Thus, *(1) the scalar potential (stream function) is given by (6.133a) [(6.133b)] and the spherical components of the velocity by (6.134a and b) ≡ (6.140c and d); (2) the shape of the fairing is given by (6.136) ≡ (6.138a) ≡ (6.139), which specifies in particular the asymptotic radius (6.137a) ≡ (6.137b) ≡ (6.140a) and the position of the stagnation point (6.135d) ≡ (6.138b) ≡ (6.140b);*

*(3) in a reference frame in which the fairing is at rest, the pressure coefficient (6.143a through c) is given by (6.145a) in the flow; and (4) on the fairing it simplifies to (6.145b), resulting (I.28.29a) in a **drag force** (6.147a):*

$$D = \rho U Q = \rho U^2 \pi b^2 = \rho U^2 A = \frac{\rho}{2} U^2 C_D A, \quad A = \pi b^2, \quad C_D = \frac{2D}{\rho U^2 A} = 2, \qquad (6.147a\text{–}c)$$

*corresponding to a **drag coefficient** (6.147c) for the asymptotic frontal area (6.147b).*

6.4 Fairings, Bodies, and Multipoles

A line (point) source in a uniform plane (axisymmetric) incident potential flow leads to the Rankine two- (three-) dimensional fairing with the same directrix (Figure 6.9) for a cylinder (Section I.28.4) [generator of a body of revolution (Subsections 6.3.4 and 6.3.5)]. Placing an opposite line (point) sink in the direction of the free stream, a Rankine finite body or oval is obtained (Figure 6.10) as the directrix of a cylinder (Subsections I.28.5.1and 2) [generator of a surface of revolution (Subsections 6.4.1 and 6.4.2)]. If the positions of the source and sink were reversed, with the sink facing the flow and the source in line behind (Subsection I.28.5.3), the result would be: (1) a gap between aligned fairings (Figure I.28.10a) or (2) a throated nozzle (valley between mountains) from (Figure I.28.10b) the inside (outside). The limit of opposite line (point) monopoles, that is, a source and a sink as their distance decreases inversely with the flow rate is a line (point) dipole [Section I.12.7 (Subsection 6.4.3)]. A similar limit for opposite line (point) dipoles leads to line (point) quadrupoles [Section I.12.8 (Subsection 6.4.4)]. Higher-order line (point) multipoles can be considered the same way, for example, (1) axial multipoles [Section I.12.9 (Subsection 6.4.5)]; (2) axial, transverse, and mixed multipoles (Subsection 6.4.6); (3) triaxial multipoles (Subsection 6.4.7). The potential of a dipole or quadrupole can be written in Cartesian, cylindrical, or spherical coordinates (Subsection 6.4.8); the number of moments increases with the order of the multipole.

6.4.1 Source and Sink Pair in a Uniform Stream

To obtain a Rankine finite body, to the uniform flow in (6.125a) [(6.125b)] are added opposite point sources/sinks (6.130a) [(6.131a)] of volume flux $\pm Q$ with $Q > 0$ on the axis at $z = \mp c$ for the scalar potential (6.148a) [stream function (6.148b)]:

$$\Phi(r,z) = Uz + \frac{Q}{4\pi}\left(\frac{1}{R_+} - \frac{1}{R_-}\right), \quad \Psi(r,z) = \frac{1}{2}Ur^2 + \frac{Q}{4\pi}\left(\frac{z-c}{R_+} - \frac{z+c}{R_-}\right), \qquad (6.148a,b)$$

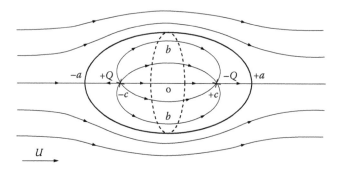

FIGURE 6.10 A source and a sink with equal flow rates aligned to a uniform stream lead to the potential flow past an axisymmetric (cylindrical) Rankine finite oval body [Figure 6.10 (Figure I.28.7)]. If the position of the source and sink is reversed the result is either a gap between aligned semi-infinite fairing (Figure I.28.10a) or a throated nozzle on the inside (Figure I.28.10b) with a valley between mountains outside.

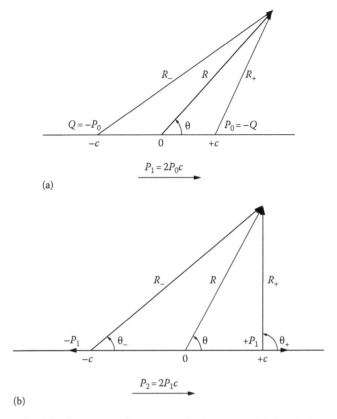

(a)

(b)

FIGURE 6.11 A point (line) dipole corresponds to a point (line) source and sink with the same flow rate increasing in inverse proportion to their distance, to keep the dipole moment constant in the limit of coincidence [Figure 6.11a (Figure I.12.3)]. The passage from a source–sink pair (Figure 6.10) to a dipole (a) can be extended to a quadrupole (b) or multipole of any order. In the case of a point (line) quadrupole it is the limit of opposite dipoles whose moment increases inversely with the mutual distance, as they coincide [Figure 6.11b (Figure I.12.6)].

where cylindrical coordinates (r, z) are used and $R_{\pm}$ in (6.149a) denotes the distance [$\theta_{\pm}$ in (6.149b), the angles with the axis measured from the position of the source (sink) at $(0, \mp c)$] in Figure 6.11a:

$$\left(R_{\pm}\right)^2 = r^2 + \left(z \mp c\right)^2 = r^2 + z^2 + c^2 \mp 2zc, \quad \cos\theta_{\pm} = \frac{z \mp c}{R_{\pm}}. \qquad (6.149a,b)$$

Bearing in mind that:

$$\left\{\partial_z R_{\pm}, \partial_r R_{\pm}\right\} = \left\{\frac{\partial}{\partial r}, \frac{\partial}{\partial z}\right\}\left|r^2 + \left(z \mp c\right)^2\right|^{1/2} = \left\{\frac{r}{R_{\pm}}, \frac{z \mp c}{R_{\pm}}\right\}, \qquad (6.150a,b)$$

$$\left\{\partial_z, \partial_r\right\}\frac{1}{R_{\pm}} = -\frac{1}{R_{\pm}^2}\left\{\partial_z, \partial_r\right\}R_{\pm} = -\left\{\frac{r}{R_{\pm}^2}, \frac{z \mp c}{R_{\pm}^2}\right\}, \qquad (6.151a,b)$$

the cylindrical components of velocity are:

$$v_z = \partial_z \Phi = U + \frac{Q}{4\pi}\left(\frac{z+c}{R_-^3} - \frac{z-c}{R_+^3}\right), \quad v_r = \partial_r \Phi = \frac{Qr}{4\pi}\left(\frac{1}{R_-^3} - \frac{1}{R_+^3}\right). \qquad (6.152a,b)$$

The stagnation points can only occur on the axis, that is, $v_r = 0$ with $R_+ \neq R_-$ implies (6.153a) in (6.152b); this leads (6.149a) to two symmetrically placed points (6.153b and c) where the axial velocity due to the pair of opposite source and sink balances the incident stream velocity (6.153d):

$$r = 0: \quad z = \pm a, \quad R_\pm = \pm a \mp c: \quad 0 = \lim_{\substack{z \to \pm a \\ r \to 0}} v_z = U + \frac{Q}{4\pi} \left[\frac{1}{(a+c)^2} - \frac{1}{(a-c)^2} \right]; \qquad (6.153\text{a–d})$$

the condition (6.153d) $\equiv$ (6.154a):

$$\frac{4\pi U}{Q} = \frac{1}{(a-c)^2} - \frac{1}{(a+c)^2} = \frac{4ac}{(a^2 - c^2)^2}: \quad a^4 - 2a^2 c^2 + c^4 - \frac{Qac}{\pi U} = 0, \qquad (6.154\text{a,b})$$

specifies the positions $(r, z) = (0, \pm a)$ of the stagnation points as roots of the quartic polynomial (6.154b).

6.4.2 Rankine Oval Finite Body

The streamline $\Psi = \text{const}$ passing (6.148b) through the stagnation points $(r, z) = (0, \pm a)$ is (6.155a) and from (6.148b) this specifies the shape (6.155b) of the Rankine finite body:

$$\Psi(0, \pm a) = 0: \quad \frac{2\pi U}{Q} r^2 = \frac{z+c}{R_-} - \frac{z-c}{R_+}, \qquad (6.155\text{a,b})$$

with $R_\pm$ given by (6.149a). The body is finite (Figure 6.10) with: (1) length equal to the distance $2a$ between the stagnation points with a root of (6.154b); and (2) width equal to $2b$, where $(r, z) = (\pm b, 0)$ is a point satisfying (6.156a) and (6.155b), leading to (6.156b):

$$R_\pm^2 = b^2 + c^2, \quad \frac{\pi b^2 U}{Q} = \frac{c}{R_\pm} = \frac{c}{\sqrt{c^2 + b^2}}. \qquad (6.156\text{a,b})$$

*For a given incident stream velocity U, the volume flux Q and locations ±c of the source and sink can be chosen, to have a finite Rankine body (Figure 6.10) of a given length 2a in (6.154b) and width 2b in (6.156b) and hence **fineness ratio** b/a. The body has two orthogonal axes of symmetry, and the same applies to the pressure distribution, on its surface, and in the flow:*

$$p(r, z) = p_0 - \frac{\rho}{2} \left(v_z^2 + v_r^2 \right) = p_0 - \frac{\rho}{2} U^2 C_p(r, z). \qquad (6.157)$$

The pressure coefficient is given by

$$C_p(r, z) = 1 + \frac{Q}{2\pi U} \left(\frac{z+c}{R_+^3} - \frac{z-c}{R_-^3} \right) + \left(\frac{Q}{4\pi U} \right)^2 \left(\frac{1}{R_+^4} + \frac{1}{R_-^4} + 2 \frac{c^2 - z^2 + r^2}{R_+^3 R_-^3} \right), \qquad (6.158)$$

using the cylindrical components of the velocity (6.152a and b), derived from the scalar potential (6.148a) or stream function (6.148b). The result (6.158) follows from (6.143d; 6.152a and b):

$$C_p(r,z) = 2\frac{p_0 - p(r,z)}{\rho U^2} = \frac{(v_z)^2 + (v_r)^2}{U^2}$$

$$= \left[1 + \frac{Q}{4\pi U}\left(\frac{z+c}{R_-^3} - \frac{z-c}{R_+^3}\right)\right]^2 + \left[\frac{Qr}{4\pi U}\left(\frac{1}{R_+^3} - \frac{1}{R_-^3}\right)\right]^2$$

$$= 1 + \frac{Q}{2\pi U}\left(\frac{z+c}{R_-^3} - \frac{z-c}{R_+^3}\right) + \left(\frac{Q}{4\pi U}\right)^2\left[\frac{r^2 + (z+c)^2}{R_-^6} + \frac{r^2 + (z-c)^2}{R_+^6} - 2\frac{z^2 - c^2 - r^2}{R_-^3 R_+^3}\right], \quad (6.159)$$

which simplifies to (6.158) using (6.149a). In an axisymmetric flow past a body of revolution placed with zero incidence in a stream the transverse pressures balance, and there is no lift force. In the case of the Rankine body (Figure 6.10) the longitudinal pressure components also balance, and there is no drag in a potential flow; that is, the drag can only arise from viscosity effects (Notes 6.5 and 6.6) and boundary layer separation. In the case of the Rankine axisymmetric fairing (Figure 6.9), the longitudinal components of pressure all have the same direction, and add up to a drag force (6.147a through c).

6.4.3 Dipole as a Limit of Opposite Monopoles

The Rankine finite body is represented (Figure 6.10) by the introduction in a uniform stream of opposite point source and sink of volume flux $\pm Q$, at the points $\mp c$, corresponding (6.130a) to the potentials (6.160a) in spherical coordinates:

$$\Phi_{\pm}(R,\theta) = \pm\frac{Q}{4\pi R_{\pm}}, \quad R_{\pm}^2 = R^2\sin^2\theta + (R\cos\theta \mp c)^2 = R^2 + c^2 \mp 2cR\cos\theta, \quad (6.160a,b)$$

in terms of the distance from (Figure 6.11a) the arbitrary point (R, θ) to the singularities $(6.160b) \equiv (6.149a)$. The inverse of the distances (6.160b) can be approximated by

$$\frac{1}{R_{\pm}} = \frac{1}{R}\left|1 \mp 2\frac{c}{R}\cos\theta + \frac{c^2}{R^2}\right|^{-1/2} = \frac{1}{R}\left[1 \pm \frac{c}{R}\cos\theta + O\left(\frac{c^2}{R^2}\right)\right], \quad (6.161)$$

if the separation of the singularities is small compared to the distance of the observer from the origin. The total potential is given by

$$\Phi(R,\theta) \equiv \Phi_{+}(R,\theta) + \Phi_{-}(R,\theta) = -\frac{Q}{4\pi}\left(\frac{1}{R_-} - \frac{1}{R_+}\right) = \frac{Qc}{2\pi R^2}\cos\theta + O\left(Q\frac{c^2}{R^3}\right). \quad (6.162)$$

The limit when the source and sink approach each other $c \to 0$, and the volume fluxes diverge $Q \to \infty$, so as to preserve the product, that is, the dipole moment (6.163a and b):

$$P_0 \equiv -Q: \quad P_1 = \lim_{\substack{c \to 0 \\ Q \to \infty}} -2Qc = 2P_0c, \quad \Phi_1 = -\frac{P_1}{4\pi R^2}\cos\theta, \quad (6.163a\text{--}c)$$

yields the potential of the dipole (6.163c). The dipole moment is taken from the negative to the positive monopole, hence the change of sign in (6.163a and b).

The potential of a point dipole potential can be obtained in three ways: (1) from limit of opposite source and sink (6.163a through c); (2) from a derivate impulse charge distribution (Subsection 8.1.3); and (3) from a multipolar expansion (Subsection 8.3.3). For the dipole, the stream function is obtained by the method (1), that is, a limit process similar to (6.162), but starting from (6.131a) and using

$$\Psi_1(R,\theta) = -\frac{Q}{4\pi}(\cos\theta_- - \cos\theta_+) = -\frac{2Qc}{4\pi R}\sin^2\theta = \frac{P_1}{4\pi R}\sin^2\theta, \qquad (6.164a\text{--}c)$$

in terms of the dipole moment (6.163b); in (6.164b) was used (6.149b) in spherical coordinates for large radius (6.161):

$$
\begin{aligned}
\cos\theta_\pm &= \frac{z \mp c}{R_\pm} = \frac{R\cos\theta \mp c}{R_\pm} = \left(\cos\theta \mp \frac{c}{R}\right)\left|1 \mp 2\frac{c}{R}\cos\theta + \frac{c^2}{R^2}\right|^{-1/2} \\
&= \left(\cos\theta \mp \frac{c}{R}\right)\left[1 \pm \frac{c}{R}\cos\theta + O\left(\frac{c^2}{R^2}\right)\right] \\
&= \cos\theta \mp \frac{c}{R}\left(1 - \cos^2\theta\right) + O\left(\frac{c^2}{R^2}\right) \\
&= \cos\theta \mp \frac{c}{R}\sin^2\theta + O\left(\frac{c^2}{R^2}\right);
\end{aligned}
\qquad (6.165)
$$

substitution of (6.165) in (6.164a) yields (6.164b) and hence (6.163b) also (6.164c). Thus an **axisymmetric dipole** *of moment P_1 has scalar potential (6.163c) [stream function (6.164c)] corresponding (6.72a through c) [(6.58a through c)] to*

$$v_{R1} = \frac{P_1}{2\pi R^3}\cos\theta, \quad v_{\theta 1} = \frac{P_1}{4\pi R^3}\sin\theta, \qquad (6.166a,b)$$

the spherical components of the velocity.

6.4.4 Quadrupole as a Limit of Opposite Dipoles

The axisymmetric quadrupole potential can also be obtained from: (3) a multipolar expansion (Subsection 8.3.5); (2) a second derivate impulse charge distribution (Subsection 8.2.1); (1) as the limit of opposite dipoles (Figure 6.11b), as shown next. The potential due to the dipoles (6.163c) of moments $-P_1(P_1)$ at $z = -c(z = c)$, is given by (6.167b):

$$\cos\theta_\pm = \frac{R\cos\theta \mp c}{R_\pm}: \quad 4\pi\Phi_\pm(R,\theta) = \mp P_1\frac{\cos\theta_\pm}{R_\pm^2} = \mp P_1\frac{R\cos\theta \mp c}{R_\pm^3}, \qquad (6.167a,b)$$

involving the distances (6.149a) and polar angles (6.149b) $\equiv$ (6.167a) from the singularities (Figure 6.11b). The potential (6.167b) simplifies in the far field, that is, for distances much larger than the separation of the opposite dipoles, when the asymptotic approximation (6.161) holds, to

$$
\begin{aligned}
\Phi_\pm(R,\theta) &= \mp \frac{P_1}{4\pi R^2}\left(\cos\theta \mp \frac{c}{R}\right)\left[1 \pm 3\frac{c}{R}\cos\theta + O\left(\frac{c^2}{R^2}\right)\right] \\
&= \mp \frac{P_1}{4\pi R^2}\left[\cos\theta \pm \frac{c}{R}\left(3\cos^2\theta - 1\right) + O\left(\frac{c^2}{R^2}\right)\right].
\end{aligned}
\qquad (6.168)
$$

The total potential in the same far-field approximation:

$$\Phi(R,\theta) = \Phi_+(R,\theta) + \Phi_-(R,\theta) = -\frac{P_1 c}{2\pi R^3}\left(3\cos^2\theta - 1\right) + O\left(\frac{P_1 c^2}{R^4}\right), \tag{6.169}$$

leads to

$$\Phi_2(R,\theta) = -\frac{P_2}{8\pi R^3}\left(3\cos^2\theta - 1\right), \quad P_2 \equiv \lim_{\substack{c\to 0 \\ P_1\to\infty}} 4cP_1; \tag{6.170a,b}$$

the potential (6.170a) corresponds to a longitudinal quadrupole of moment (6.170b) and spherical components (6.72a and c) of the velocity:

$$v_{R2} = \frac{3P_2}{8\pi R^4}\left(3\cos^3\theta - 1\right), \quad v_{\theta 2} = \frac{3P_2}{4\pi R^4}\cos\theta\sin\theta = \frac{3P_2}{8\pi R^4}\sin(2\theta). \tag{6.171a,b}$$

The transverse (mixed) quadrupole is the limit of two transverse dipoles (one transverse and one longitudinal dipole) and does not lead to an axisymmetric flow or field. The nonaxisymmetric quadrupoles can be obtained by the other two methods: (1) the second derivative of the Dirac impulse (Section 6.2); (2) the multipole expansion (Section 6.3). The stream function for the axisymmetric quadrupole can be obtained by a similar limit process. It is also obtained by substituting the velocity components (6.171a and b) in the exact differential (6.121b), leading to

$$\frac{8\pi}{3P_2}d\Psi_2 = \frac{\sin\theta}{R^2}\left(3\cos^2\theta - 1\right)d\theta - 2\frac{\cos\theta\sin^2\theta}{R^3}dR = d\left(\frac{\cos\theta\sin^2\theta}{R^2}\right), \tag{6.172}$$

where the following was used:

$$d\left(\cos\theta\sin^2\theta\right) = \left(-\sin^3\theta + 2\sin\theta\cos^2\theta\right)d\theta = \sin\theta\left(3\cos^2\theta - 1\right)d\theta. \tag{6.173}$$

The stream function of the axisymmetric dipole (6.164c) ≡ (6.174a) [quadrupole (6.172) ≡ (6.174b)]:

$$\Psi_1(R,\theta) = \frac{P_1}{8\pi R^2}\sin^2\theta, \quad \Psi_2(R,\theta) = \frac{3P_2}{8\pi R^2}\cos\theta\sin^2\theta, \tag{6.174a,b}$$

satisfies (6.79a and b) together with the respective Φ_1 dipole (6.163c) [quadrupole (6.170a)] potentials. The streamlines of the dipole (quadrupole) are the curves of equation (6.175a) [(6.175b)]:

$$\mathrm{const} = \frac{\sin^2\theta}{R^2}, \quad \mathrm{const} = \frac{\cos\theta\sin^2\theta}{R^2}, \tag{6.175a,b}$$

that touch the axis in two $\theta = 0, \pi$(three $\theta = 0, \pi/2, \pi$) directions.

6.4.5 Potential for an Axial Dipole and Quadrupole

The scalar potential (stream function) provides a more (less) general approach to multipoles since it is not (is) restricted to axisymmetric flow. The potential of a monopole (6.130a) ≡ (6.176a) leads by axial differentiation:

$$\Phi_0(R) = -\frac{P_0}{4\pi R}, \quad \Phi_1(R,\theta) = \frac{P_1}{4\pi}\frac{\partial}{\partial z}\left(\frac{1}{R}\right) = -\frac{P_1}{4\pi R^2}\cos\theta, \tag{6.176a,b}$$

to the potential of a dipole (6.176b) ≡ (6.163c), where the following was used:

$$\frac{\partial}{\partial z}\left(\frac{1}{R}\right) = \frac{\partial}{\partial z}\left(\left|r^2 + z^2\right|^{-1/2}\right) = -\left|r^2 + z^2\right|^{-3/2} z = -\frac{z}{R^3} = -\frac{\cos\theta}{R^2}. \tag{6.177}$$

Another axial differentiation:

$$\frac{\partial^2}{\partial z^2}\left(\frac{1}{R}\right) = -\frac{\partial}{\partial z}\left[\left|r^2 + z^2\right|^{-3/2} z\right] = 3\left|r^2 + z^2\right|^{-5/2} z^2 - \left|r^2 + z^2\right|^{-3/2}$$

$$= 3\frac{z^2}{R^5} - \frac{1}{R^3} = \frac{3\cos^2\theta - 1}{R^3}, \tag{6.178}$$

leads to the potential of a quadrupole (6.179) ≡ (6.170a):

$$\Phi_2(R,\theta) = -\frac{1}{2}\frac{P_2}{4\pi}\frac{\partial^2}{\partial z^2}\left(\frac{1}{R^2}\right) = -\frac{P_2}{8\pi R^3}\left(3\cos^2\theta - 1\right). \tag{6.179}$$

Thus the *potential of an axial multipole of order n is given by (6.180a)*:

$$\Phi_n(R,\theta) = \frac{P_n}{n!4\pi}\left(-\frac{\partial}{\partial z}\right)^n \frac{1}{R}; \quad \Phi_{n+1}(R,\theta) = -\frac{1}{n+1}\frac{P_{n+1}}{P_n}\frac{\partial}{\partial z}\left[\Phi_n(R,\theta)\right]; \tag{6.180a,b}$$

the successive multipole potentials are obtained (6.180b) by an exchange of dipole moments and axial differentiation with minus sign. This process leads from the potential of a monopole (6.176a), to an axial dipole (6.176b) and quadrupole (6.179), and can be continued to any order. The factor n! is inserted for consistency with the multipolar expansion (Section 8.3) and arises from the Taylor series in order to have the dipole moments as coefficients.

6.4.6 Axial, Transverse, and Mixed Multipoles

The same process applies to axial (6.176b) and transverse (6.181) dipole:

$$\Phi_2(R,\theta) = \frac{1}{4\pi}\left(P_z\frac{\partial}{\partial z} + P_r\frac{\partial}{\partial r}\right)\frac{1}{R} = -\frac{1}{4\pi R^2}\left(P_z\cos\theta + P_r\sin\theta\right), \tag{6.181}$$

where (6.177) was used and a similar result (6.182) was obtained:

$$\left\{\frac{\partial}{\partial z},\frac{\partial}{\partial r}\right\}\frac{1}{R} = \left\{\frac{\partial}{\partial z},\frac{\partial}{\partial r}\right\}\left|r^2 + z^2\right|^{-1/2} = \left|r^2 + z^2\right|^{-3/2}\{z,r\} = -\frac{\{z,r\}}{R^3} = -\frac{\{\cos\theta,\sin\theta\}}{R^2}. \tag{6.182}$$

Using (6.178) and two similar results:

$$\left\{\frac{\partial^2}{\partial z^2},\frac{\partial^2}{\partial r^2},\frac{\partial^2}{\partial z\partial r}\right\}\frac{1}{R} = \left\{3\frac{z^2}{R^5} - \frac{1}{R^3}, 3\frac{r^2}{R^5} - \frac{1}{R^3}, \frac{3zr}{R^5}\right\}$$

$$= R^{-3}\left\{3\cos^2\theta - 1, 3\sin^2\theta - 1, 3\sin\theta\cos\theta\right\}, \tag{6.183}$$

including

$$\frac{\partial^2}{\partial r \partial z}\left(\frac{1}{R}\right) = -\frac{\partial}{\partial r}\left(\frac{z}{R^3}\right) = \frac{3z}{R^4}\frac{\partial R}{\partial r} = \frac{3zr}{R^5} = \frac{3\cos\theta\sin\theta}{R^3} = \frac{3\sin(2\theta)}{2R^3} = \frac{\partial^2}{\partial r \partial z}\left(\frac{1}{R}\right),$$ (6.184)

leads to the potential of an axial, transverse, and mixed quadrupole:

$$\Phi_2(R,\theta) = \frac{1}{8\pi}\left(P_{zz}\frac{\partial^2}{\partial z^2} + P_{rr}\frac{\partial^2}{\partial r^2} + 2P_{zr}\frac{\partial^2}{\partial z \partial r}\right)\frac{1}{R}$$

$$= \frac{1}{8\pi R^3}\left[P_{zz}\left(3\cos^2\theta - 1\right) + P_{rr}\left(3\sin^2\theta - 1\right) + 3P_{zr}\sin(2\theta)\right].$$ (6.185)

Thus *the axial, transverse, and mixed multipoles of order n are given by*

$$\Phi_n(R,\theta) = \frac{(-)^n}{n!4\pi}\left(\sum_{m=0}^{n} P_{nm}\frac{\partial^n}{\partial z^m \partial r^{n-m}}\right)\frac{1}{R},$$ (6.186)

including: (1) axial P_z and transverse P_r dipoles (6.181); (2) axial P_{zz}, transverse P_{rr}, and mixed $P_{zr} = P_{rz}$ quadrupoles (6.185); and (3) multipoles of order n for which the number of moments is n + 1 in (6.186).

6.4.7 Potential and Velocity for Triaxial Multipoles

For triaxial multipoles in a three-dimensional space, the starting point is the inverse of the radial distance (6.187a) differentiated with regard to the Cartesian coordinates:

$$R = \left|\sum_{N=1}^{3}(x_n)^2\right|^{1/2} \quad : \quad \frac{\partial}{\partial x_n}\left(\frac{1}{R}\right) = -\frac{1}{R^2}\frac{\partial R}{\partial x_n} = -\frac{x_n}{R^3},$$ (6.187a,b)

differentiating twice (6.188) and using the identity matrix (5.278a through c) leads to

$$\frac{\partial^2}{\partial x_n \partial x_m}\left(\frac{1}{R}\right) = -\frac{\partial}{\partial x_m}\left(\frac{x_n}{R^3}\right) = -\frac{1}{R^3}\frac{\partial x_n}{\partial x_m} + \frac{3x_n}{R^4}\frac{\partial R}{\partial x_m} = \frac{1}{R^5}\left(3x_n x_m - R^2\delta_{nm}\right),$$ (6.188)

a third differentiation gives

$$\frac{\partial^3}{\partial x_n \partial x_m \partial x_k}\left(\frac{1}{R}\right) = \frac{\partial}{\partial x_k}\left[\frac{1}{R^5}\left(3x_n x_m - R^2\delta_{nm}\right)\right]$$

$$= -\frac{3}{R^5}\left(5x_n x_m x_k - x_n\delta_{mk} - x_n\delta_{mk} - x_k\delta_{nm}\right),$$ (6.189)

and the *n*th order derivative appears in (8.46). Thus *a multipole of order n has potential*:

$$\Phi_n(x_m) = \frac{(-)^{n-1}}{n!4\pi}\left(\sum_{i_1,\ldots,i_n=1}^{3} P_{i_1\ldots i_n}\frac{\partial^n}{\partial x_{i_1}\ldots\partial x_{i_n}}\right)\frac{1}{R},$$ (6.190)

for example, (1) a dipole:

$$\Phi_1(x_n) = \frac{1}{4\pi}\left(\sum_{n=1}^{3} P_n \frac{\partial}{\partial x_n}\right)\frac{1}{R} = -\frac{1}{4\pi R^3}\sum_{n=1}^{3} P_n x_n = -\frac{\vec{P}\cdot\vec{x}}{4\pi R^3}; \tag{6.191}$$

and (2) a quadrupole:

$$\Phi_2(x_n) = -\frac{1}{2}\frac{1}{4\pi}\sum_{n,m=1}^{3}\left(P_{nm}\frac{\partial^2}{\partial x_n \partial x_m}\right)\frac{1}{R}$$

$$= -\frac{1}{8\pi R^5}\left(\sum_{n,m=1}^{3} 3P_{nm}x_n x_n - R^2 \sum_{n=1}^{3} P_{nn}\right). \tag{6.192}$$

Since it is the gradient of the potential the velocity (6.193a) of the multipole of order n:

$$v_j^{(n)}(x_m) = \partial_j \Phi_n = \frac{(-)^{n-1}}{n!4\pi}\partial_j\left(\sum_{i_1,\dots,i_n=1}^{3} P_{i_1\dots i_n}\frac{\partial^n}{\partial x_{i_1}\partial x_{i_1}\dots\partial x_{i_n}}\right)\frac{1}{R}$$

$$= \frac{(-)^n}{n!4\pi}\sum_{j,i_1,\dots,i_n=1}^{3}\left(P_{ji_1\dots i_n}\frac{\partial^{n+1}}{\partial x_j\partial x_{i_1}\dots\partial x_{i_n}}\right)\frac{1}{R} = -\frac{1}{n+1}\Phi_{n+1}(x_m), \tag{6.193a,b}$$

has the same terms with reversed sign as (6.193b) the potential of the multipole of order n + 1, divided by n + 1.

6.4.8 Dipole/Quadrupole in Cartesian/Cylindrical/Spherical Coordinates

The potential of a dipole may be written explicitly in Cartesian coordinates (6.194a) ≡ (6.191) and also in cylindrical (6.194b) and spherical (6.194c) coordinates:

$$-4\pi\Phi_1(x,y,z) = \left|x^2 + y^2 + z^2\right|^{-3/2}\left(xP_x + yP_y + zP_z\right), \tag{6.194a}$$

$$-4\pi\Phi_1(r,\varphi,z) = \left|r^2 + z^2\right|^{-3/2}\left[zP_z + r\left(P_x\cos\varphi + P_y\sin\varphi\right)\right], \tag{6.194b}$$

$$-4\pi\Phi_1(R,\theta,\varphi) = R^{-2}\left[P_z\cos\theta + \sin\theta\left(P_x\cos\varphi + P_y\sin\varphi\right)\right]; \tag{6.194c}$$

the axisymmetric axial $P_1 = P_z$ dipole (6.163c) ≡ (6.176b) is the first term on the r.h.s. of (6.194c), and the nonaxisymmetric transverse P_r dipole (6.181) corresponds to the second and third terms on the r.h.s. of (6.194c).
The potential of a quadrupole is given explicitly in Cartesian coordinates by (6.192) ≡ (6.195a) and in cylindrical (spherical) coordinates by (6.195b) [(6.195c)]:

$$-8\pi\Phi_2(x,y,z) = \left|x^2 + y^2 + z^2\right|^{-5/2}$$

$$\left[P_{xx}\left(2x^2 - y^2 - z^2\right) + P_{yy}\left(2y^2 - x^2 - z^2\right) + P_{zz}\left(2z^2 - x^2 - y^2\right) - 6xyP_{xy} - 6xzP_{xz} - 6yzP_{yz}\right], \tag{6.195a}$$

$$-8\pi\Phi_2\left(r,\varphi,z\right)=\left|r^2+z^2\right|^{-5/2}\left\{P_{xx}\left[r^2\left(3\cos^2\varphi-1\right)-z^2\right]+P_{yy}\left[r^2\left(3\sin^2\varphi-1\right)-z^2\right]\right.$$

$$\left.+P_{zz}\left(2z^2-r^2\right)-3P_{xy}r^2\sin\left(2\varphi\right)-6zr\left(P_{xz}\cos\varphi+P_{yz}\sin\varphi\right)\right\},\tag{6.195b}$$

$$-8\pi\Phi_2\left(R,\theta,\varphi\right)=R^{-3}\left[P_{xx}\left(3\sin^2\theta\cos^2\varphi-1\right)+P_{yy}\left(3\sin^2\theta\sin^2\varphi-1\right)\right.$$

$$+P_{zz}\left(3\cos^2\theta-1\right)-3\sin^2\theta\sin\left(2\varphi\right)P_{xy}$$

$$\left.-3\sin\left(2\theta\right)\left(P_{xz}\cos\varphi+P_{yz}\sin\varphi\right)\right].\tag{6.195c}$$

The axisymmetric axial quadrupole (6.170a) ≡ (6.185) is $P_2 \equiv P_{zz}$, the third term on the r.h.s of (6.195c), and the nonaxisymmetric transverse P_{rr} (mixed P_{zr}) quadrupole in (6.185) corresponds to the first (last) two terms on the r.h.s. of (6.195c).

In all cases the Cartesian components of a general triaxial dipole (6.194a through c) [quadrupole (6.195a through c)] are used. The passage from Cartesian (6.194a; 6.195a) to cylindrical (6.194b; 6.195b) [spherical (6.194c; 6.195c)] coordinates uses (6.16d and e) [(6.13a, c, and d)]. *As the order (6.186a) of the multipole increases, for example, (6.196b) monopole, dipole, quadrupole, octupole, ... the number of multipole moments (6.196c) increases rapidly*:

$$n=0,2,3,4,\ldots:\quad m=2^n=1,2,4,8,16\ldots:\quad q=3^n;\tag{6.196a–c}$$

the number of independent multipole moments is

$$\bar{q}=\frac{3\times4\times5\times\cdots\times\left(2+n\right)}{n!}=\binom{n+2}{n}=\binom{n+2}{2}=\frac{\left(2+n\right)!}{n!2!}.\tag{6.196d}$$

as indicated in Table 6.1. The last results (6.196c and d) may be justified as follows: (1) a multipole of order n has n indices each taking three values in a three-dimensional space leading to 3^n moments (6.196c); (2) since the moments with the same indices in different order are equal, the number of independent moments (6.196d) must be divided by the number $n!$ of n indices; (3) the first index for a dipole can take three values; (4) the second index for a quadrupole can take also three values, but

TABLE 6.1   Number of Moments of a Three-Dimensional Multipole

			Number of Moments	
Type n	Order $m = 2^n$	Name	Total: 3^n	Independent: $\binom{n+2}{n}$
$n = 0$	1	Monopole	1	1
$n = 1$	2	Dipole	3	3
$n = 2$	4	Quadrupole	9	6
$n = 3$	8	Octopole	27	10
$n = 4$	16	Decahexapole	81	15

Note: Total number of components and number of independent components for multipoles of orders $n = 0, \ldots, 4$ in a three-dimensional space.

since a repeated index is counted twice in the permutations this is equivalent to taking four values; and (5) thus the nth index may be taken as for $n + 2$ values.

6.5 Sphere in a Stream and in a Large Cavity

The dipole limit of the Rankine body in a free stream specifies in two (three) dimensions the flow past a cylinder (Section I.28.6) [sphere (Section 6.5)] including (1) the dipole moment and added volume of fluid entrained by the body (Subsection 6.5.2); and (2) the streamlines and associated pressure distribution in the flow and on the body (Subsection 6.5.1). Both the mean flow velocity and dipole moment are reduced by the wall effect when the cylinder (Section II.8.1) [sphere (Subsection 6.5.3)] moves in a large cylindrical (spherical) cavity; the wall effect increases the added mass (Subsection 6.5.4) and the drag in accelerated motion (Subsection 6.5.5).

6.5.1 Dipole Moment and Blockage Effect

The Rankine finite body (Figure 6.10), in the limit of coincident source and sink (Figure 6.11a and b), becomes a sphere (Figure 6.12); this follows from the limit to the shape of the Rankine oval (6.155b) in the dipole limit (6.163b):

$$
\begin{aligned}
r^2 &= \lim_{\substack{c \to 0 \\ Q \to \infty}} \frac{Q}{2\pi U} \left(\frac{R\cos\theta + c}{R_-} - \frac{R\cos\theta - c}{R_+} \right) \\
&= -\lim_{c \to 0} \frac{P_1}{4\pi U} \left[\frac{1}{R_-} \left(1 + \frac{R}{c}\cos\theta \right) + \frac{1}{R_+} \left(1 - \frac{R}{c}\cos\theta \right) \right];
\end{aligned}
\tag{6.197}
$$

using (6.161) this simplifies to

$$
\begin{aligned}
1 &= -\frac{1}{r^2} \lim_{c \to 0} \frac{P_1}{4\pi UR} \left[\left(1 - \frac{c}{R}\cos\theta \right)\left(1 + \frac{R}{c}\cos\theta \right) + \left(1 + \frac{c}{R}\cos\theta \right)\left(1 - \frac{R}{c}\cos\theta \right) + O\!\left(\frac{c}{R} \right) \right] \\
&= -\frac{1}{r^2} \frac{P_1}{2\pi UR}\left(1 - \cos^2\theta \right) = -\frac{P_1}{2\pi UR}\frac{\sin^2\theta}{r^2} = -\frac{P_1}{2\pi UR^3} = -\frac{P_1}{2\pi Ua^3};
\end{aligned}
\tag{6.198}
$$

it follows from (6.198) that since the dipole moment is a constant, and the radius $R = a$ must be a constant, and the shape is a sphere. The radius a of the sphere is related to the dipole moment (6.163b) by (6.198) $\equiv$ (6.200a), as will be confirmed next by another method. This shows that whereas an arbitrary body in a

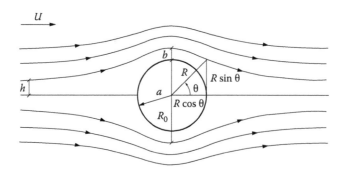

FIGURE 6.12 A rigid impermeable sphere (Figure 6.13) [cylinder without circulation (Figure I.28.4a)] in a uniform stream is represented by a point (line) dipole with dipole moment such that the flow velocity has a zero normal component on the sphere (cylinder).

stream can be represented by either (1) a multipolar series at one point or (2) a distribution of monopoles along a segment, a sphere can be represented by a point dipole (6.199a):

$$\Phi(R,\theta)=\left(UR-\frac{P_1}{4\pi R^2}\right)\cos\theta,\quad 0=v_R(a,\theta)=\lim_{R\to a}\frac{\partial\Phi(R,\theta)}{\partial R}=\left(U+\frac{P_1}{2\pi a^3}\right)\cos\theta,\qquad(6.199\text{a,b})$$

whose moment is determined by the condition of zero normal velocity at the sphere (6.199b). The dipole moment (6.199b) ≡ (6.199e) is also determined by the condition of vanishing on the sphere (6.199d) of the stream function (6.199c) corresponding to the potential (6.199a):

$$\Psi(R,\theta)=\left(\frac{1}{2}UR^2+\frac{P_1}{4\pi R}\right)\sin^2\theta:\quad \Psi(a,\theta)=0\Rightarrow P_1=-2\pi Ua^3.\qquad(6.199\text{c–e})$$

The moment of the dipole (6.198) ≡ (6.199b) ≡ (6.199e) ≡ (6.200a) is opposite to the free stream velocity because (Figure 6.10) the source is upstream and the sink downstream, that is, $P_1<0$ in (6.200a):

$$-P_1=2\pi Ua^3=\frac{3}{2}UV;\quad V\equiv\frac{4}{3}\pi a^3;\qquad(6.200\text{a,b})$$

the modulus of the dipole moment is three-halves of the volume of the sphere (6.200b) multiplied by the free stream velocity. Thus *the added volume* of a sphere in an axisymmetric potential flow is one-half of the displaced mass; the added volume of a cylinder in a plane flow (I.28.104b) is twice as large, because it equals the displaced mass. This is an instance of the property that a three-dimensional axisymmetric body has a smaller relative added mass than a two-dimensional body with the same profile, because the flow can pass more easily around it.

6.5.2 Aiming Distance and Pressure Distribution

Substituting (6.200a) into (6.199a) specifies the potential of a sphere in a stream (6.201a):

$$\Phi(R,\theta)=U\left(R+\frac{a^3}{2R^2}\right)\cos\theta;\quad \Psi(R,\theta)=\frac{U}{2}\left(R^2-\frac{a^3}{R}\right)\sin^2\theta;\qquad(6.201\text{a,b})$$

the stream function (6.201b) is that (6.199c) of the uniform stream (6.126b) plus a dipole (6.174a) with moment (6.200a). The streamlines Ψ = const can be parameterized in terms of the **aiming distance** h, defined (Figure 6.12) as the distance of the streamline from the axis at infinity (6.202a):

$$h\equiv\lim_{R\to\infty}r=\lim_{R\to\infty}R\sin\theta=\sqrt{\frac{2\Psi}{U}};\quad h^2=\left(R^2-\frac{a^3}{R}\right)\sin^2\theta;\qquad(6.202\text{a,b})$$

substitution of the limiting value (6.202a) in (6.201b) leads to the equation of the streamlines in the form (6.202b). The minimum distance d of the streamline from the sphere occurs in the transverse plane passing through the center (6.203a) and is the difference (6.203b) between the minimum radial distance R_0 along the streamline (6.203c) and the radius of the sphere:

$$\theta=\frac{\pi}{2}:\quad d+a=R_0\equiv R_{\min},\quad h^2=R_0^2-\frac{a^3}{R_0},\quad 1=\left(\frac{h}{R_0}\right)^2+\left(\frac{a}{R_0}\right)^3=\left(\frac{h}{a+d}\right)^2+\left(\frac{a}{a+d}\right)^3,\qquad(6.203\text{a–d})$$

and leads to the identity (6.203d). It is clear from (6.203c) that $R_0>h$ and no streamline touches the sphere, except for the streamline $h=0=\Psi$ leading to the stagnation points at $\theta=0,\pi$.

The potential (6.201a) and stream function (6.201b) specify the spherical components of velocity:

$$v_R(R,\theta) = U\left(1 - \frac{a^3}{R^3}\right)\cos\theta, \quad v_\theta(R,\theta) = -U\left(1 + \frac{a^3}{2R^3}\right)\sin\theta, \tag{6.204a,b}$$

and confirm that the radial velocity (6.204a) vanishes on the whole sphere (6.205a) and the tangential velocity (6.205b):

$$v_R(a,\theta) = 0, \quad v_\theta(a,\theta) = -\frac{3}{2}U\sin\theta. \tag{6.205a,b}$$

only vanishes (6.206a) at the stagnation points $\theta = 0, \pi$:

$$v_R(a,\theta) = 0 = v_\theta(a,\pi); \quad \left|v_\theta(a,\theta)\right|_{max} = \pm v_\theta\left(a,\frac{\pi}{2}\right) = \frac{3}{2}U, \tag{6.206a–c}$$

the maximum tangential velocity occurs in the transverse direction (6.206c) and is three-halves of the free stream velocity, instead of twice of the free stream velocity for the cylinder (I.28.94b), so that the coefficients in both cases are the same as for the added mass. The pressure coefficient is given in the flow by

$$C_p(R,\theta) = \frac{v_R^2 + v_\theta^2}{U^2} = \left(1 - \frac{a^3}{R^3}\right)^2\cos^2\theta + \left(1 + \frac{a^3}{2R^3}\right)^2\sin^2\theta$$

$$= 1 + \frac{a^3}{R^3}\left(3\sin^2\theta - 2\right) + \frac{a^6}{R^6}\left(1 - \frac{3}{4}\sin^2\theta\right), \tag{6.207}$$

and on the sphere by

$$C_p(a,\theta) = \frac{9}{4}\sin^2\theta \leq \left\{C_p(a,\theta)\right\}_{max} = C_p\left(a,\frac{\pi}{2}\right) = \frac{9}{4} = 2.25, \tag{6.208}$$

where the maximum occurs at the sideline positions $\theta = \pi/2$. The pressure distribution is symmetric, that is, it is unaffected by change of sign of θ, confirming that there is no drag due to the potential flow, in agreement with the symmetry of the streamlines (6.202a and b) relative to a plane orthogonal to the free stream passing through the center of the sphere.

6.5.3 Wall Effect for a Spherical Cavity

Replacing the free stream velocity U by an arbitrary constant A in the expression for the potential (6.199a) and stream function (6.199c):

$$\Phi(R,\theta) = \left(AR - \frac{P_1}{4\pi R^2}\right)\cos\theta, \quad \Psi(R,\theta) = \frac{1}{2}\left(AR^2 + \frac{P_1}{4\pi R}\right)\sin^2\theta, \tag{6.209a,b}$$

leads to the potential flow past a sphere of radius a moving at velocity U within a spherical cavity of large radius b. In this case the sphere is still represented by a dipole, but the dipole moment is no longer (6.200a), because it is changed by the wall effect of the cavity and there is no free stream at infinity; it is only when the cavity recedes to infinity that a free stream exists and the dipole moment (6.200a) is regained. The moment of the dipole representing a sphere of radius a in a large spherical cavity of radius b

(Figure 6.13) as in the case of cylinders (Section I.8.1; Figure I.8.1) must satisfy two boundary conditions, which also determine the second constant. The radial velocity is required to (1) vanish on the outer spherical boundary (6.210b); and (2) match the radial velocity of the inner sphere at its surface (6.211b):

$$b^2 \gg a^2: \quad 0 = \lim_{R \to b} \frac{\partial \Phi(R,\theta)}{\partial R} = \left(A + \frac{P_1}{2\pi b^3} \right) \cos\theta, \qquad (6.210\text{a,b})$$

$$U^2 t \ll b^2: \quad -U \cos\theta = \lim_{R \to a} \frac{\partial \Phi(R,\theta)}{\partial R} = \left(A + \frac{P_1}{2\pi a^3} \right) \cos\theta; \qquad (6.211\text{a,b})$$

in (6.211b) $-U$ appears because, if the flow moves past the sphere with velocity U, the sphere moves relative to the fluid at velocity $-U$, and the latter case must be considered for the fluid to be static at infinity (6.211a). Since the spheres are assumed to be concentric (Figure 6.13) two conditions must be met: (1) [(2)] the radius of (6.210a) [the distance traveled by (6.211a)] the inner sphere is much smaller than the radius of the outer sphere. The boundary conditions (6.210b; 6.211b):

$$-P_1 = \frac{2\pi U a^3 b^3}{b^3 - a^3}, \quad A = \frac{U a^3}{b^3 - a^3}, \quad \lim_{b \to \infty} P_1 = -2\pi U a^3, \qquad (6.212\text{a–c})$$

determine the constants (6.212a and b) and confirm that the dipole moment (6.212a) simplifies to (6.212c) ≡ (6.200a) if the cavity recedes to infinity leading to a free stream.

Thus *a sphere of radius a moving at velocity U relative to a fluid in a spherical cavity of radius b is represented as an axisymmetric flow (Figure 6.13) due to a uniform stream of velocity (6.212b), with a dipole at the origin with moment (6.212a). The constant (6.212b) is reduced by the wall effect relative to the free stream values, in proportion to the ratio (a/b)³ of the volume of the sphere and the cavity. In the limit of*

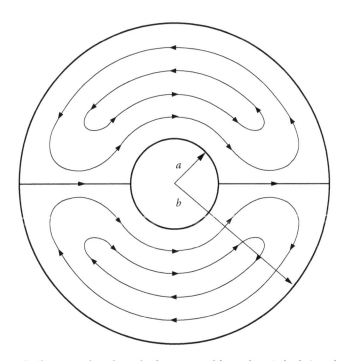

FIGURE 6.13 The wall effect on an obstacle can be demonstrated for a sphere (cylinder) in a large spherical (cylindrical) cavity without (with) circulation [Figure 6.13 (Figure II.8.1)].

a sphere in a free stream, corresponding to an infinitely large cavity $b \to \infty$, the dipole moment (6.212a) reduces to (6.212c) $\equiv$ (6.200a). The potential (6.209a) [stream function (6.209b)] is given (6.212a and b) by

$$\Phi, \Psi(R,\theta) = \frac{Ua^3}{b^3 - a^3} \left\{ \left(R + \frac{b^3}{2R^2} \right) \cos\theta, \frac{1}{2} \left(R^2 - \frac{b^3}{R} \right) \sin^2\theta \right\}; \qquad (6.213a,b)$$

the spherical components of the velocity are given by

$$v_R, v_\theta(R,\theta) = \frac{Ua^3}{b^3 - a^3} \left\{ \left(1 - \frac{b^3}{R^3} \right) \cos\theta, -\left(1 + \frac{b^3}{2R^3} \right) \sin\theta \right\}; \qquad (6.214a,b)$$

the pressure coefficient in the flow is given by

$$C_p(R,\theta) = \frac{(v_R)^2 + (v_\theta)^2}{U^2} = \frac{a^6}{\left(b^3 - a^3 \right)^2} \left\{ 1 + \frac{b^3}{R^3} \left(3\sin^2\theta - 2 \right) + \frac{b^6}{R^6} \left(1 - \frac{3}{4}\sin^2\theta \right) \right\}; \qquad (6.215)$$

the radial velocity is specified by the boundary condition (6.211b) on the inner moving sphere [(6.210b) on the outer fixed spherical cavity], and

$$v_\theta(a,\theta) = -\frac{U}{2} \frac{b^3 + 2a^3}{b^3 - a^3} \sin\theta, \quad v_\theta(b,\theta) = -\frac{3}{2} U \frac{a^3}{b^3 - a^3} \sin\theta, \qquad (6.216a,b)$$

are the corresponding azimuthal velocities (6.216a) [(6.216b)] leading to

$$C_p(a,\theta) = \frac{\left[v_R(a,\theta) \right]^2 + \left[v_\theta(a,\theta) \right]^2}{U^2} = \cos^2\theta + \left(\frac{b^3 + 2a^3}{b^3 - a^3} \frac{\sin\theta}{2} \right)^2$$

$$= \frac{4a^6 + 4a^3 b^3 \left(3\sin^2\theta - 2 \right) + b^6 \left(4 - 3\sin^2\theta \right)}{4\left(b^3 - a^3 \right)^2}, \qquad (6.217a)$$

$$C_p(b,\theta) = \left[\frac{v_\theta(b,\theta)}{U} \right]^2 = \frac{9}{4} \frac{a^6}{\left(b^3 - a^3 \right)^2} \sin^2\theta, \qquad (6.217b)$$

as the respective pressure coefficients (6.217a) [(6.217b)]. When substituting (6.214a and b) in (6.215) the following simplifications were made:

$$C_p(R,\theta) = \left(\frac{a^3}{b^3 - a^3} \right)^2 \left[\left(1 - \frac{b^3}{R^3} \right)^2 \cos^2\theta + \left(1 + \frac{b^3}{2R^3} \right)^2 \sin^2\theta \right]$$

$$= \frac{a^6}{\left(b^3 - a^3 \right)^2} \left[1 + \frac{b^3}{R^3} \left(\sin^2\theta - 2\cos^2\theta \right) + \frac{b^6}{R^6} \left(\cos^2\theta + \frac{\sin^2\theta}{4} \right) \right]$$

$$= \frac{a^6}{\left(b^3 - a^3 \right)^2} \left[1 + \frac{b^3}{R^3} \left(3\sin^2\theta - 2 \right) + \frac{b^6}{R^6} \left(1 - \frac{3}{4}\sin^2\theta \right) \right]; \qquad (6.218)$$

the pressure coefficient in the flow (6.218) $\equiv$ (6.215) specifies the pressure coefficient on the inner (6.217a) [outer (6.217b)] sphere for $R = a$ ($R = b$).

6.5.4 Kinetic Energy and Added Mass of Entrained Fluid

The kinetic energy of the flow is given by the volume integral (6.219b), which in the case of an irrotational flow (6.219a) can be expressed in terms of the potential (6.219c):

$$\vec{v} = \nabla\Phi: \quad E_c = \frac{1}{2}\int_D \rho(\vec{v}\cdot\vec{v})\mathrm{d}V = \frac{1}{2}\int_D \rho(\nabla\Phi\cdot\nabla\Phi)\mathrm{d}V. \tag{6.219a–c}$$

The first Green identity (5.227c) for $\Psi = \Phi$ leads to (6.220a) ≡ (6.220b):

$$\frac{\partial}{\partial x_i}\left(\Phi\frac{\partial\Phi}{\partial x_i}\right) = \frac{\partial\Phi}{\partial x_i}\frac{\partial\Phi}{\partial x_i} + \Phi\frac{\partial^2\Phi}{\partial x_i^2} \Leftrightarrow \nabla\cdot(\Phi\nabla\Phi) = \nabla\Phi\cdot\nabla\Phi + \Phi\nabla^2\Phi; \tag{6.220a,b}$$

in the case of a potential flow (6.221a) simplifies (6.220b) to (6.221b):

$$\nabla^2\Phi = 0: \quad \nabla\cdot(\Phi\nabla\vec{\Phi}) = \nabla\Phi\cdot\nabla\Phi; \quad \rho = \mathrm{const}: \quad E_v = \frac{\rho}{2}\int_D \nabla\cdot(\Phi\nabla\Phi)\mathrm{d}V = \frac{\rho}{2}\int_D \Phi\frac{\partial\Phi}{\partial R}\mathrm{d}S, \tag{6.221a–e}$$

since an irrotational (6.219a) and incompressible (6.221c) flow is potential the kinetic energy (6.219c) is given by the integral over the domain (6.221d), which is transformed to an integral over the boundary (6.221e), using the divergence theorem (5.163a through c). In the present case the domain D lies between the spheres of radii a and b:

$$E_v = \frac{\rho}{2}\int_{a\leq R\leq b} \nabla\cdot(\Phi\nabla\Phi)\mathrm{d}V = \frac{\rho}{2}\int_{R=a,b} \Phi\frac{\partial\Phi}{\partial R}\mathrm{d}S. \tag{6.222}$$

and the divergence theorem (5.163a through c) replaces the volume integral by a sum of integrals over the surfaces of the two spheres (6.212). The outward normal is minus (plus) unity on the inner (outer) sphere:

$$\frac{2}{\rho}E_v = -\int_{R=a} \Phi(a,\theta)\frac{\partial\Phi}{\partial R}(a,\theta)\mathrm{d}S + \int_{R=b} \Phi(b,\theta)\frac{\partial\Phi}{\partial R}(b,\theta)\mathrm{d}S. \tag{6.223}$$

The boundary conditions (6.210b) [(6.211b)] imply that the second (first) integral on the r.h.s. of (6.223) vanishes [simplifies to (6.224b)]:

$$\mathrm{d}S = h_\theta h_\varphi\,\mathrm{d}\theta\,\mathrm{d}\varphi = R^2\sin\theta\,\mathrm{d}\theta\,\mathrm{d}\varphi: \quad E_v = \frac{\rho}{2}U\int_{R=a} \Phi(a,\theta)\cos\theta\,\mathrm{d}S, \tag{6.224a,b}$$

where the area element in spherical coordinates (6.20a) ≡ (6.224a) involves the scale factors along meridians (6.14b) and parallels (6.14c). Thus the kinetic energy of the potential flow between the two spheres reduces to an integral over the surface of the moving sphere:

$$E_v = \frac{\rho}{2}Ua^2\int_0^{2\pi}\mathrm{d}\varphi\int_0^\pi \Phi(a,\theta)\sin\theta\cos\theta\,\mathrm{d}\theta, \tag{6.225}$$

involving only the potential (6.213a):

$$E_v = \rho\pi a^2 U^2 \frac{a^3}{b^3 - a^3}\left(a + \frac{b^3}{2a^2}\right)\int_0^\pi \cos^2\theta\sin\theta\,d\theta$$

$$= \frac{\rho\pi a^3 U^2}{2}\frac{b^3 + 2a^3}{b^3 - a^3}\left[-\frac{\cos^3\theta}{3}\right]_0^\pi = \rho\frac{\pi}{3}a^3 U^2\frac{b^3 + 2a^3}{b^3 - a^3}, \qquad (6.226)$$

leading to (6.226).

6.5.5 Balance of Inertia and External and Drag Forces

Thus the *kinetic energy of the potential flow due to a sphere of radius a moving with velocity U in a spherical cavity (Figure 6.13) with much larger radius b equals (6.227a) one-half of the square of the velocity of the moving sphere times (6.227b) the **added mass** of the fluid entrained by the sphere:*

$$E_v = \frac{1}{2}m_0 U^2, \quad m_0 = \frac{2\pi}{3}a^3\rho\frac{b^3 + 2a^3}{b^3 - a^3} \equiv \frac{1}{2}\rho V\mu; \qquad (6.227a,b)$$

*the added mass equals one-half the mass of the fluid displaced by the moving sphere of volume (6.200b), with a correction factor for the **wall effect** of the spherical cavity:*

$$1 < \mu \equiv \frac{b^3 + 2a^3}{b^3 - a^3} = \left(1 + 2\frac{a^3}{b^2}\right)\left[1 + \frac{a^3}{b^3} + O\left(\frac{a^6}{b^6}\right)\right] = 1 + 3\frac{a^3}{b^3} + O\left(\frac{a^6}{b^6}\right), \qquad (6.228)$$

*which (1) reduces to unity for a free stream $b \to \infty$ and (2) is always larger than unity for a cavity of finite radius $b < \infty$; and (3) the lowest-order correction is three times the ratio of the volume of the moving sphere to that of the spherical cavity. A sphere moving with constant velocity in a free stream or spherical cavity has a constant kinetic energy and experiences no drag; if the sphere is accelerated it experiences a **drag force**, whose activity or work per unit time or product by the velocity is the rate of change of the kinetic energy with time:*

$$D = \frac{1}{U}\frac{dE_v}{dt} = \frac{1}{U}\frac{d}{dt}\left(\frac{1}{2}m_0 U^2\right) = m_0\frac{dU}{dt}; \qquad (6.229)$$

*thus the drag force equals the acceleration of the sphere times the mass of fluid it displaces. If the sphere is acted by an **external force** F the latter is balanced by inertial force (6.230b) that is equal to the acceleration times the **total mass** (6.230a), that is, the mass m of the sphere plus the added mass m_0 of entrained fluid:*

$$\bar{m} = m + m_0: \quad F = \bar{m}\frac{dU}{dt} = (m + m_0)\frac{dU}{dt}, \quad m\frac{dU}{dt} = F - m_0\frac{dU}{dt} = F - D. \qquad (6.230a\text{–}c)$$

*It follows that the inertia force on the sphere (6.230c) specified by its mass times its acceleration equals the external applied force minus the drag force due to the fluid entrained by the sphere. The correction factor for the wall effect (6.228) is the same for the (1) added mass (6.227b), (2) kinetic energy (6.227a), and (3) drag force (6.229). The wall effect increases all three (1–3) due to the **blockage effect** of the cavity relative to the free stream, which forces the same volume flux through a smaller cross section; the acceleration of the flow increases the kinetic energy and also the pressure differences, leading to a higher drag.*

6.6 Spherical Vortex in a Uniform Flow

The potential (stream function) in an incompressible (irrotational) axisymmetric flow satisfies the Laplace (modified Laplace) equation, which differs only in the sign of one term (Subsection 6.6.1), if cylindrical coordinates are used; it is more convenient to use spherical coordinates to represent the flow past a sphere in a uniform stream by a dipole (Subsections 6.5.1 and 6.5.2). In an incompressible axisymmetric and rotational flow there is no potential and the stream function satisfies (Subsection 6.2.8) the modified Laplace equation forced by the vorticity; for example, a vorticity proportional to the distance from the axis leads to a spherical vortex (Subsection 6.6.2). The velocity at the surface of the spherical vortex can be matched to the velocity on the surface of a sphere in a uniform stream (Subsection 6.6.4). Thus a spherical vortex can exist in a uniform stream (Subsection 6.6.3), and corresponds to a cylinder with circulation (Section I.28.7) in two dimensions, with a difference: (1) for the cylinder the vorticity is concentrated on the axis and both a potential and stream function exist and satisfy the Laplace equation everywhere except on the axis; (2) for the spherical vortex the vorticity is distributed in the interior, so there is no potential, and the stream function satisfies a modified Poisson equation forced by the vorticity; (3) if the spherical vortex is matched to a uniform stream, then potential and stream functions satisfying the Laplace equation exist outside the sphere; and (4) the stagnation pressure is constant (not constant) in the irrotational (rotational) incompressible flow outside (inside) the spherical vortex (Subsection 6.6.5).

6.6.1 Modified Laplace and Laplace Axisymmetric Operators

In an incompressible irrotational axisymmetric flow the potential (stream function) satisfies the Laplace (modified Laplace) equation, namely,: (1) in cylindrical coordinates (6.88b) ≡ (6.231a and b) [(6.89a) ≡ (6.232a and b)]:

$$0 = \nabla^2 \Phi(r,z) = \frac{1}{r}\frac{\partial}{\partial r}\left(r\frac{\partial \Phi}{\partial r}\right) + \frac{\partial^2 \Phi}{\partial z^2} = \frac{\partial^2 \Phi}{\partial r^2} + \frac{1}{r}\frac{\partial \Phi}{\partial r} + \frac{\partial^2 \Phi}{\partial z^2}, \tag{6.231a,b}$$

$$0 = \bar{\nabla}^2 \Psi(r,z) = r\frac{\partial}{\partial r}\left(\frac{1}{r}\frac{\partial \Psi}{\partial r}\right) + \frac{\partial^2 \Psi}{\partial z^2} = \frac{\partial^2 \Psi}{\partial r^2} - \frac{1}{r}\frac{\partial \Psi}{\partial r} + \frac{\partial^2 \Psi}{\partial z^2}, \tag{6.232a,b}$$

where the difference lies only in the sign of the first-order derivative; and (2) in spherical coordinates (6.88c) ≡ (6.233a and b) [(6.89b) ≡ (6.224a and b)]:

$$0 = \nabla^2 \Phi(R,\theta) = \frac{1}{R^2}\frac{\partial}{\partial R}\left(R^2\frac{\partial \Phi}{\partial R}\right) + \frac{1}{R^2 \sin\theta}\frac{\partial}{\partial \theta}\left(\sin\theta\frac{\partial \Phi}{\partial \theta}\right)$$
$$= \frac{\partial^2 \Phi}{\partial R^2} + \frac{2}{R}\frac{\partial \Phi}{\partial R} + \frac{1}{R^2}\frac{\partial^2 \Phi}{\partial \theta^2} + \frac{\cot\theta}{R^2}\frac{\partial \Phi}{\partial \theta}, \tag{6.233a,b}$$

$$0 = \bar{\nabla}^2 \Psi(R,\theta) = \frac{\partial^2 \Psi}{\partial R^2} + \frac{\sin\theta}{R^2}\frac{\partial}{\partial \theta}\left(\frac{1}{\sin\theta}\frac{\partial \Psi}{\partial \theta}\right) = \frac{\partial^2 \Psi}{\partial R^2} + \frac{1}{R^2}\frac{\partial^2 \Psi}{\partial \theta^2} - \frac{\cot\theta}{R^2}\frac{\partial \Psi}{\partial \theta}, \tag{6.234a,b}$$

where besides the difference in the sign of the coefficient of ∂Ψ/∂θ there is an extra term in (6.233b) compared with (6.234b).

Proceeding with spherical coordinates, and seeking a potential (stream function) in the form (6.235a) [(6.236a)]:

$$\Phi(R,\theta) = \cos\theta F(R): \quad 0 = R^2 \nabla^2 \Phi = \cos\theta\left[\left(R^2 F'\right)' - 2F\right] = \cos\theta\left(R^2 F'' + 2RF' - 2F\right), \tag{6.235a,b}$$

$$\Psi(R,\theta) = \sin^2\theta\, G(R): \quad 0 = R^2 \bar{\nabla}^2 \Psi = \sin^2\theta\left(R^2 G'' - 2G\right), \tag{6.236a,b}$$

leads from (6.233a) [(6.234a)] to (6.235b) [(6.236b)]. The latter (6.235b) [(6.236b) are linear ordinary differential equations with power coefficients of Euler type, that is, with derivatives multiplied by powers with the same exponent, so that all terms are of the form $a_n r^n F^{(n)}(r)$; they have solutions as powers (6.237a) [(6.238a)] with exponents satisfying (6.237b) [(6.238b)]:

$$F(R) = R^a: \quad 0 = a(a-1) + 2a - 2 = (a-1)(a+2), \quad F(R) = A_1 R + \frac{A_2}{R^2}, \tag{6.237a-c}$$

$$G(R) = R^b: \quad 0 = b(b-1) - 2 = (b+1)(b-2), \quad G(R) = \frac{B_1}{R} + B_2 R^2; \tag{6.238a-c}$$

thus the general integral is (6.237c) [(6.238c)] where $(A_1, A_2)[(B_1, B_2)]$ are arbitrary constants. The potential (6.235a; 6.237c) $\equiv$ (6.239a) [stream function (6.236a; 6.238c) $\equiv$ 6.240a)] leads (6.79a) to the radial velocity (6.239b) [(6.240b)]:

$$\Phi(R,\theta) = \left(A_1 R + \frac{A_2}{R^2}\right)\cos\theta, \quad v_R(R,\theta) = \frac{\partial\Phi}{\partial R} = \left(A_1 - \frac{2A_2}{R^3}\right)\cos\theta, \tag{6.239a,b}$$

$$\Psi(R,\theta) = \left(\frac{B_1}{R} + B_2 R^2\right)\sin^2\theta, \quad v_R(R,\theta) = \frac{\csc\theta}{R^2}\frac{\partial\Psi}{\partial\theta} = 2\left(B_2 + \frac{B_1}{R^3}\right)\cos\theta. \tag{6.240a,b}$$

The radial velocity must (1) reduce to (6.241a) for the free stream at infinity; and (2) vanish on the surface of the sphere (6.241b):

$$\lim_{R\to\infty} v_R(R,\theta) = U\cos\theta; \quad v_R(a,\theta) = 0: \quad \left\{A_1 - U, A_1 a^3 - 2A_2\right\} = 0 = \left\{2B_2 - U, B_1 + a^3 B_2\right\}; \tag{6.241a-f}$$

the boundary conditions (6.241a and b) applied to the radial velocity (6.239b) [(6.240b)] specify the constants of integration (6.241c and d) $\equiv$ (6.242a and b) [(6.241e and f) $\equiv$ (6.242c and d)]:

$$\left\{A_1, A_2\right\} = U\left\{1, \frac{a^3}{2}\right\}, \quad \left\{B_1, B_2\right\} = \frac{U}{2}\left\{-a^3, 1\right\}; \tag{6.242a-d}$$

substitution of (6.242a and b) [(6.242c and d)] in (6.239a) [(6.239b)] leads to *potential (6.243a) [stream function (6.243b)] of a sphere of radius a in a free stream of velocity U*:

$$\Phi(R,\theta) = U\left(1 + \frac{a^3}{2R^3}\right)\cos\theta, \quad \Psi(R,\theta) = \frac{U}{2}\left(R^2 - \frac{a^3}{R}\right)\sin^2\theta, \tag{6.243a,b}$$

in agreement with (6.243a) $\equiv$ (6.201a) [(6.243b) $\equiv$ (6.201b)], leading to the velocity (6.204a and b).

6.6.2 Stream Function for a Spherical Vortex (Hill, 1894)

The simplest solution of the forced modified Laplace equation for the stream function (6.108g), excluding irrotational flow, or constant stagnation enthalpy or pressure, occurs when the latter are linear functions of the stream function (6.244a):

$$\text{const} \equiv C \equiv f(\bar{\Psi}) = -\frac{dH_0}{d\bar{\Psi}} = -\frac{1}{\rho}\frac{dp_0}{d\bar{\Psi}}, \quad \varpi = -Cr = -CR\sin\theta, \tag{6.244a,b}$$

implying that: (1) the vorticity is proportional to the distance from axis (6.244b); (2) the stagnation pressure is not constant (6.244a) but rather has a linear dependence on the stream function (6.245a):

$$\frac{p_0}{\rho} + C\bar{\Psi} = \text{const}; \quad \frac{p}{\rho} + \frac{v^2}{2} + C\bar{\Psi} = \text{const}; \tag{6.245a,b}$$

and (3) thus the pressure distribution (6.102c) is given by (6.245b). The stream function is a solution of (6.108g) ≡ (6.246):

$$CR^2 \sin^2\theta = -\varpi r = \bar{\nabla}^2\bar{\Psi} = \frac{\partial^2\bar{\Psi}}{\partial R^2} + \frac{\sin\theta}{R^2}\frac{\partial}{\partial\theta}\left(\frac{1}{\sin\theta}\frac{\partial\bar{\Psi}}{\partial\theta}\right), \tag{6.246}$$

which involves the modified Laplace operator (6.234a) with forcing (6.244b). Seeking a solution (6.247a) with the same angular dependence as for the sphere in a uniform stream (6.236a) leads to a radial dependence (6.247b) similar to (6.236b) with a forcing term on the r.h.s.:

$$\bar{\Psi}(R,\theta) = \sin^2\theta\,\bar{G}(R): \quad CR^2 = \bar{G}'' - 2\frac{\bar{G}}{R^2}. \tag{6.247a,b}$$

A particular solution of the ordinary differential equation (6.247b) with forcing is (6.248a):

$$G_0(R) = B_0 R^4: \quad CR^4 = R^2 G_0'' - 2G_0 = 10B_0 R^4, \quad B_0 = \frac{C}{10}, \tag{6.248a–c}$$

where the constant (6.248c) is determined by (6.248b). The complete integral of (6.247b) minus the particular integral (6.248a) is (6.249a), the general integral (6.238c) ≡ (6.249c) of the unforced differential equation (6.236b) ≡ (6.249b):

$$\bar{G}(R) - G_0(R) = G(R); \quad R^2\bar{G}'' - 2\bar{G} = 0: \quad \bar{G}(R) = \frac{C}{10}R^4 + \frac{B_1}{R} + B_2 R^2. \tag{6.249a–c}$$

Substituting (6.238c) and (6.248a) in (6.249a) it follows that *(6.249c) is the complete integral of the second-order differential equation (6.247b) with forcing, consisting of the sum of (1) the general integral (6.238c) of the differential equation without forcing (6.236b) involving two arbitrary constants; and (2) a particular integral (6.248a) of the forced differential equation (6.248b) without arbitrary constants (6.248c).*

The radial velocity (6.58a) corresponding to the stream function (6.247a) is

$$\bar{v}_R(R,\theta) = \frac{1}{R^2\sin\theta}\frac{\partial}{\partial\theta}\left[\sin^2\theta\,\bar{G}(R)\right] = 2\frac{\bar{G}(R)}{R^2}\cos\theta = \left(\frac{C}{5}R^2 + \frac{2B_1}{R^3} + 2B_2\right)\cos\theta. \tag{6.250}$$

Of the three constants in (6.250) the condition that the velocity be finite at the center eliminates one (6.251b), and the remaining two are related (6.241d) by the condition (6.251c) that the radial velocity is zero on a sphere of radius a:

$$v_R(0,\theta) < \infty: \quad B_1 = 0; \quad 0 = \bar{G}(a) = \frac{C}{10}a^4 + B_2 a^2, \quad B_2 = -\frac{C}{10}a^2. \tag{6.251a–d}$$

Thus are specified the constants (6.248c; 6.251b,d) in the stream function (6.247a; 6.249a) that is interpreted next (Subsection 6.6.3).

6.6.3 Stagnation Circle and Toroidal Stream Surfaces

Substituting (6.251a and c) in (6.249b) leads to (6.252a) and shows that *the stream function (6.252b):*

$$G(R) = \frac{C}{10}R^2\left(R^2 - a^2\right): \quad \bar{\Psi}(R,\theta) = \frac{C}{10}R^2\left(R^2 - a^2\right)\sin^2\theta, \tag{6.252a,b}$$

and spherical components (6.58a through c) of velocity (6.253a and b):

$$\left\{\bar{v}_R, \bar{v}_\theta, \bar{v}_\varphi\right\} = \frac{C}{5}\left\{\left(R^2 - a^2\right)\cos\theta, \left(a^2 - 2R^2\right)\sin\theta, 0\right\}, \tag{6.253a–c}$$

*correspond (Figure 6.14) to a **spherical vortex** (Hill, 1894) since: (1) the streamline* $\Psi = 0$ *consists of the circle* $R = a$ *and the axis* $\theta = 0, \pi$, *corresponding to the rotation of a half-circle around its diameter on the axis; (2) the stagnation points (6.254a and b) lie on the equatorial plane (6.254c) on a circle of radius (6.254d) corresponding to the value (6.254e) of the stream function:*

$$\bar{v}_R = 0 = \bar{v}_\theta: \quad \theta_1 = \frac{\pi}{2}, \quad R_1 = \frac{a}{\sqrt{2}}, \quad \bar{\Psi}\left(\frac{a}{\sqrt{2}}, \frac{\pi}{2}\right) = -\frac{Ca^4}{40} \equiv \bar{\Psi}_1; \tag{6.254a–e}$$

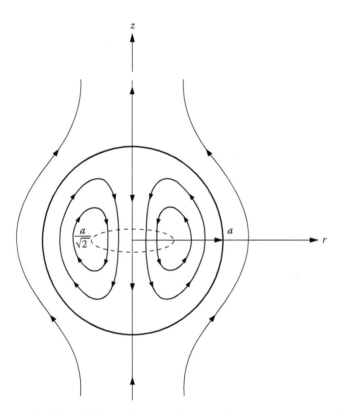

FIGURE 6.14 There is a significant difference between a line vortex that generates a rotating flow (Figure I.12.2b) and a spherical vortex (Figure 6.14) that has: (1) closed stream surfaces inside a sphere of radius a that degenerate to a stagnation ring with radius $a/\sqrt{2}$ at the equator; and (2) zero radial velocity at the surface of the sphere allowing the matching to a uniform stream with the same axis. Thus the flow is rotational inside and irrotational outside the sphere, and incompressible in all space.

(3) from (6.252b), it follows that the stream function has an extremum (6.254f,h) at the stagnation circle (6.254g,i) ≡ (6.254c–d):

$$0 = \frac{\partial \bar{\Psi}}{\partial R} = \frac{CR}{5}\left(2R^2 - a^2\right)\sin^2\theta: \quad R_1 = \frac{a}{\sqrt{2}}, \quad 0 = \frac{\partial \bar{\Psi}}{\partial \theta} = \frac{CR^2}{10}\left(R^2 - a^2\right)\sin(2\theta): \quad \theta_1 = \frac{\pi}{2};$$

$$(6.254\text{f–i})$$

(4) the extremum is a minimum (6.254j–l):

$$\lim_{\substack{R \to a/\sqrt{2} \\ \theta \to \pi/2}} \left\{ \frac{\partial^2 \bar{\Psi}}{\partial R^2}, \frac{\partial^2 \bar{\Psi}}{\partial \theta^2}, \frac{\partial^2 \bar{\Psi}}{\partial R \partial \theta} \right\}$$

$$= \lim_{\substack{R \to a/\sqrt{2} \\ \theta \to \pi/2}} \left\{ \frac{C}{5}\left(6R^2 - a^2\right)\sin^2\theta, \frac{CR^2}{5}\left(R^2 - a^2\right)\cos(2\theta), \frac{CR}{5}\left(2R^2 - a^2\right)\sin(2\theta) \right\}$$

$$= \left\{ \frac{2Ca^2}{5}, \frac{Ca^4}{10}, 0 \right\}, \tag{6.254j–l}$$

since (6.254m) is satisfied:

$$\lim_{\substack{R \to a/\sqrt{2} \\ \theta \to \pi/2}} \left[\frac{\partial^2 \bar{\Psi}}{\partial R^2} \frac{\partial^2 \bar{\Psi}}{\partial \theta^2} - \left(\frac{\partial^2 \bar{\Psi}}{\partial R \partial \theta}\right)^2 \right] = \frac{C^2 a^6}{25} > 0; \tag{6.254m}$$

(5) it can be confirmed that the difference between the stream function (6.252b) at an arbitrary point and its value (6.254e) at the stagnation circle is positive (6.254n) except at the stagnation circle (6.254c,d) ≡ (6.254g,i) where it vanishes (6.254a):

$$\Delta\bar{\Psi}(R,\theta) \equiv \bar{\Psi}(R,\theta) - \bar{\Psi}_1 = \frac{C}{10}\left[R^2\left(R^2 - a^2\right)\sin^2\theta + \frac{a^4}{4} \right]$$

$$= \frac{C}{10}\left[R^2\left(R^2 - \frac{a^2}{2}\right)^2 \sin^2\theta + \frac{a^2}{4}\cos^2\theta \right] > 0 = \Delta\bar{\Psi}\left(\frac{a}{\sqrt{2}}, \frac{\pi}{2}\right); \tag{6.254n–o}$$

(6) the stream function (6.252b) is positive outside (6.254s) [negative inside (6.254q)] the sphere, and zero on the sphere (6.254n) that is a stream surface:

$$\text{Stream surfaces} \begin{cases} \text{stagnation circle}: & \bar{\Psi}\left(\frac{a}{\sqrt{2}}, \frac{\pi}{2}\right) \equiv \bar{\Psi}_1 < 0, & (6.254\text{p}) \\[2mm] \text{toroidal}: & \bar{\Psi}_1 < \bar{\Psi}(R < a; \theta) < 0, & (6.254\text{q}) \\[2mm] \text{sphere}: & \bar{\Psi}(a, \theta) = 0, & (6.254\text{r}) \\[2mm] \text{cylindrical}: & \bar{\Psi}(R > a, \theta) > 0, & (6.254\text{s}) \end{cases}$$

thus (7) in the interior of the sphere vortex the streamlines are closed, and the stream surfaces are toroidal (6.254q), with inner limit the stagnation circle (6.254p) and the sphere (6.254r) as outer limit: (8) outside the sphere the streamlines are open in the direction of the axis since θ → 0, π implies R → ∞ for $\bar{\Psi}$ = const > 0, and the stream surfaces are cylindrical, with the corresponding velocities (6.253a through c) diverging at infinity. The constant C specifies the magnitude of the vorticity (6.244b), and is chosen next (Subsection 6.6.3) to match the spherical vortex to a uniform free stream; this also eliminates the divergence of the velocity at infinity, because (6.252b; 6.253a through c) apply only for a finite radius.

6.6.4 Matching a Spherical Vortex to a Uniform Stream

It is possible to combine a uniform stream of velocity U outside the sphere of radius a, and a spherical vortex inside, by matching the velocity on the surface (6.255a); since the velocity is tangential in both cases, only (6.253b) ≡ (6.205b) has to be matched (6.255b) on the surface of the sphere to (6.205b):

$$R = a: \quad -\frac{C}{5}a^2\sin\theta = \bar{v}_\theta(a,\theta) = v_\theta(a,\theta) = -\frac{3}{2}U\sin\theta; \quad C = \frac{15U}{2a^2}; \tag{6.255a–c}$$

this determines the remaining constant of integration (6.255c) that specifies the vorticity (6.244b). Thus for a *spherical vortex of radius a in a free stream of velocity U that has outside R > a inside [(6.256a)] the sphere: (1) the stream function (6.201b) [(6.256b) ≡ (6.252b; 6.255c)]*:

$$R < a: \quad \bar{\Psi}(R,\theta) = \frac{3}{4}UR^2\left(\frac{R^2}{a^2} - 1\right)\sin^2\theta; \tag{6.256a,b}$$

(2) the spherical components of velocity (6.204a and b) [(6.257b) ≡ (6.253a through c; 6.255c)]:

$$R < a: \quad \{\bar{v}_R, \bar{v}_\theta, \bar{v}_\varphi\} = \frac{3U}{2}\left\{\left(\frac{R^2}{a^2} - 1\right)\cos\theta, \left(1 - \frac{2R^2}{a^2}\right)\sin\theta, 0\right\}; \tag{6.257a,b}$$

(3) the reference pressure p_1 is calculated at the stagnation (6.254a and b) circle (6.254c and d) inside the spherical vortex (6.258a) where the value of the stream function (6.254e) leads (6.255c) to (6.258b):

$$R_1 = \frac{a}{\sqrt{2}} < a: \quad \bar{\Psi}_1 \equiv \bar{\Psi}\left(\frac{a}{\sqrt{2}}, \frac{\pi}{2}\right) = -\frac{5}{16}Ua^2; \tag{6.258a,b}$$

(4) the reference pressure p_1 is related to the flow pressure p by (6.245b):

$$\frac{2}{\rho}p_1 + 2C\bar{\Psi}_1 = \frac{2}{\rho}p(R,\theta) + 2C\bar{\Psi}(R,\theta) + \left[\bar{v}_R(R,\theta)\right]^2 + \left[\bar{v}_\theta(R,\theta)\right]^2; \tag{6.259}$$

and (5) the pressure coefficient is given by (6.207) [(6.260b)] outside [inside (6.260a)] the spherical vortex using the stagnation p_0 (reference p_1) pressure for the irrotational (vortical) flow:

$$R < a: \quad \bar{C}_p(R,\theta) \equiv 2\frac{p_1 - p(R,\theta)}{\rho U^2} = \frac{2C\Delta\bar{\Psi}(R,\theta) + |v(R,\theta)|^2}{U^2} \tag{6.260a,b}$$

$$|v(R,\theta)|^2 = \left[v_R(R,\theta)\right]^2 + \left[v_\theta(R,\theta)\right]^2, \tag{6.260c}$$

leading to:

$$\bar{C}_p(R,\theta) = \frac{45}{4} + \frac{9}{4}\left(1 - \frac{R^2}{a^2}\right)^2 + \frac{R^2}{a^2}\left(18\frac{R^2}{a^2} - \frac{63}{4}\right)\sin^2\theta, \tag{6.261}$$

where (6.255c; 6.256b; 6.254n; 6.257a and b) were used.

6.6.5 Pressure and Stagnation Pressure inside a Vortex

In (6.261) was made the following simplification (6.262) substituting in the pressure coefficient (6.260b) in the spherical vortex, the stream function relative to the stagnation value (6.254e; 6.255c) and the square of the modulus of the velocity (6.260a; 6.257b):

$$\bar{C}_p(R,\theta) = \frac{C^2}{5U^2}\left[R^2\left(R^2 - a^2\right)\sin^2\theta + \frac{a^4}{4}\right] + \frac{9}{4}\left[\left(\frac{R^2}{a^2} - 1\right)^2\cos^2\theta + \left(1 - \frac{2R^2}{a^2}\right)^2\sin^2\theta\right]$$

$$= \frac{45}{4a^4}\left[R^2\left(R^2 - a^2\right)\sin^2\theta + \frac{a^4}{4}\right] + \frac{9}{4} - \frac{9}{2}\frac{R^2}{a^2}\left(\cos^2\theta + 2\sin^2\theta\right) + \frac{9}{4}\frac{R^4}{a^4}\left(\cos^2\theta + 4\sin^2\theta\right)$$

$$= \frac{45}{16} + \frac{45}{4}\frac{R^2}{a^2}\left(\frac{R^2}{a^2} - 1\right)\sin^2\theta + \frac{9}{4} - \frac{9}{2}\frac{R^2}{a^2}\left(1 + \sin^2\theta\right) + \frac{9}{4}\frac{R^4}{a^4}\left(1 + 3\sin^2\theta\right) \tag{6.262}$$

The preceding expressions coincide on the sphere $R = a$ for: (1) the stream function $\Psi = 0$ in (6.201b) and (6.256b); and (2) the velocity (6.204a and b) and (6.257a and b) that is tangential (6.263a and b):

$$v_R(a,\theta) = 0, \quad v_\theta(a,\theta) = -\frac{3}{2}U\cos\theta; \quad \bar{C}_p(a,\theta) = \frac{45}{4} + \frac{9}{4}\sin^2\theta, \tag{6.263a–c}$$

the pressure coefficient inside the spherical vortex (6.261) simplifies to (6.253c) on the surface of the sphere, and is distinct from (6.208) calculated from the potential flow (6.207) outside because: (1) the pressure coefficient (6.260b) is calculated relative to the pressure p_1 at the stagnation ring (6.254c and d) inside the vortex, where the flow is rotational and the stagnation pressure is not constant; (2) the pressure coefficient (6.208) is calculated for the potential flow where the stagnation pressure p_0 is constant; and (3) the continuity of the pressure on the surface of the sphere implies that the difference of the pressure coefficients (6.143a; 6.260b):

$$\bar{C}_p(a,\theta) - C_p(a,\theta) = 2\frac{p_1 - p_0}{\rho U^2} = \frac{45}{16}, \tag{6.264}$$

is due to the difference of stagnation pressures at the stagnation ring inside the vortex and at the free stream at infinity (6.265a):

$$p_1 - p_0 = \frac{45}{8}\rho U^2 = -\frac{9}{2}\rho\frac{U}{a^2}\bar{\Psi}_1, \tag{6.265a,b}$$

(4) since the stream function vanishes at the surface of the sphere, the difference of stagnation pressures between the stagnation ring and free stream is specified (6.265b) by the stream function there (6.258b); and (5) the stagnation pressure inside the vortex is not constant and is given by

$$\bar{p}_0(R,\theta) = p(R,\theta) + \frac{1}{2}\rho\left\{\left[\bar{v}_R(R,\theta)\right]^2 + \left[\bar{v}_\theta(R,\theta)\right]^2\right\}$$

$$= p_1 - \frac{1}{2}\rho U^2\left[\bar{C}_p(R,\theta) - \frac{|v(R,\theta)|^2}{U^2}\right]. \tag{6.266}$$

The term in square brackets in (6.266) coincides with the first term on the r.h.s. of (6.262) leading to:

$$\bar{p}_0(R,\theta) = p_1 - \frac{45}{8}\rho U^2\left[\frac{R^2}{a^2}\left(\frac{R^2}{a^2} - 1\right)\sin^2\theta + \frac{1}{4}\right], \tag{6.267}$$

for *the stagnation pressure inside a spherical vortex of radius a in a free stream of velocity U and constant mass density ρ, confirming that: (1) on the stagnation circle (6.244d) inside the vortex the stagnation pressure is (6.268a):*

$$\bar{p}_0\left(\frac{a}{\sqrt{2}}, \frac{\pi}{2}\right) = p_1 ; \qquad \bar{p}_0(a,\theta) = p_1 - \frac{45}{16}\rho U^2 = p_0, \tag{6.268a,b}$$

and (2) at the surface of the vortex the stagnation pressure (6.268b) is the same (6.265a) as at infinity.

6.7 Four Axisymmetric and Nonaxisymmetric Sphere Theorems

The circle theorem concerning the introduction of a cylinder in a plane potential field (Section I.24.7) has two variants, for potential (stream function) zero on the cylinder, that is, a conductor (insulator) in an electrostatic field; the two variants are similar because both the potential and field function satisfy the Laplace equation. A second circle theorem applies to plane incompressible rotational flow, for which there is no potential, so the stream function must be used (Subsection II.2.5.3). The two circle theorems, that is, for the irrotational (rotational) plane incompressible flow, use the reciprocal point relative to the circle (Subsection I.26.8.2; Section I.35.7). The first two sphere theorems also use reciprocal point and both apply to potential fields: the first (second) sphere theorem (Subsection 6.7.1 (Subsection 6.7.2)] concerns the stream function (scalar potential) of an axisymmetric (possibly nonaxisymmetric) field, and ensures that the sphere is a stream surface (equipotential), to which the field is tangent (orthogonal); for example, the first (second) sphere theorem [Subsection 6.7.1 (Subsection 6.7.2)] applies to an insulating (conducting) sphere in an electrostatic field [Subsection 6.8.1 (Subsection 6.8.2)]. The second sphere theorem has two extensions in integral form, namely, the third (fourth) sphere theorems [Subsection 6.7.3 (Subsection 6.7.4)], both of which apply to the potential and are valid for nonaxisymmetric potential fields; the third (fourth) sphere theorem uses (does not use) the reciprocal point relative to the sphere, like (unlike) the first two sphere theorems. The third sphere theorem ensures that the sphere is a stream or field surface, like the first sphere theorem; the fourth sphere theorem does not impose a boundary condition, but is related to flows or electric currents on the surface of a sphere (Subsections 6.8.3 through 6.8.7).

6.7.1 First Sphere Theorem for the Reciprocal Stream Function (Butler, 1953)

The **reciprocal point** of $\vec{x}$ on a sphere of radius a, as for the circle (Sections I.24.7, I.26.8, and I.35.7), lies (6.269c and d) in the same direction at a distance s such that (6.269a and b) the product of the distances from the center equals the square of the radius of the sphere:

$$|\vec{x}| \equiv R, \quad \xi R = a^2, \quad \vec{\xi} = \frac{a^2}{R}\vec{e}_r = \frac{a^2}{R^2}\vec{x}, \quad \vec{x} = \frac{a^2}{\xi}\vec{e}_r = \frac{a^2}{\xi^2}\vec{\xi}; \tag{6.269a–d}$$

this definition ensures that the points $(\vec{x},\vec{\xi})$ are mutually reciprocal, that is, $\vec{\xi}$ is the reciprocal of $\vec{x}$ iff $\vec{x}$ is the reciprocal of $\vec{\xi}$. The stream function of a uniform flow with velocity U is (6.126b) in cylindrical coordinates, corresponding in spherical coordinates (6.270a) to the first term on the r.h.s. of (6.201b) and in the presence of a sphere of radius a adds (6.270b), the second term on the r.h.s. of (6.201b):

$$\Psi_0(R,\theta) = \frac{U}{2} R^2 \sin^2\theta, \quad \Psi(R,\theta) = \frac{U}{2}\left(R^2 - \frac{a^3}{R}\right)\sin^2\theta = \Psi_0(R,\theta) - \frac{R}{a}\Psi_0\left(\frac{a^2}{R},\theta\right). \qquad (6.270\text{a–c})$$

It will be shown in the sequel that the result (6.270c) holds not only for a uniform flow but also for the introduction of a sphere in any potential flow (6.270c) ≡ (6.278) that must be axisymmetric for the stream function to exist. An equivalent way to arrive at the same result is to note that the second term on the r.h.s. of (6.201b) should be the stream function of a dipole:

$$\bar{\Psi}_0(R,\theta) = -\frac{Ua^3}{2R}\sin^2\theta = -\frac{U}{2}a\xi\sin^2\theta$$

$$= -\frac{U}{2}\xi^2\frac{a}{\xi}\sin^2\theta = \frac{a}{\xi}\Psi_0(\xi,\theta) = \frac{R}{a}\Psi_0\left(\frac{a^2}{R},\theta\right). \qquad (6.271)$$

Both (6.270a through c) and (6.271) suggest the **reciprocal stream function theorem**: *if* $\Psi(R,\theta)$ *is the stream function of an axisymmetric potential flow, that is, satisfies (6.272b) the modified Laplace equation (6.234a and b), so does (6.272c) the* **reciprocal stream function** *(6.272a) for arbitrary a, and vice versa:*

$$\bar{\Psi}(R,\theta) \equiv \frac{R}{a}\Psi\left(\frac{a^2}{R},\theta\right): \quad \bar{\nabla}^2\Psi = 0 \Leftrightarrow \bar{\nabla}^2\bar{\Psi} = 0. \qquad (6.272\text{a–c})$$

The proof is made by direct calculation of (6.272c) using (6.269b):

$$\frac{\partial\bar{\Psi}}{\partial R} = \frac{\partial}{\partial R}\left[\frac{R}{a}\Psi\left(\frac{a^2}{R},\theta\right)\right] = \frac{\Psi}{a} + \frac{R}{a}\frac{\partial\Psi}{\partial\xi}\frac{\partial\xi}{\partial R} = \frac{\Psi}{a} - \frac{a}{R}\frac{\partial\Psi}{\partial\xi}, \qquad (6.273\text{a})$$

$$\frac{\partial^2\bar{\Psi}}{\partial R^2} = \frac{\partial^2}{\partial R^2}\left[\frac{R}{a}\Psi\left(\frac{a^2}{R},\theta\right)\right] = \frac{\partial}{\partial R}\left(\frac{\Psi}{a} - \frac{a}{R}\frac{\partial\Psi}{\partial\xi}\right)$$

$$= \frac{\partial\Psi}{\partial\xi}\left(\frac{1}{a}\frac{\partial\xi}{\partial R} + \frac{a}{R^2}\right) - \frac{a}{R}\frac{\partial^2\Psi}{\partial\xi^2}\frac{\partial\xi}{\partial R} = \frac{a^3}{R^3}\frac{\partial^2\Psi}{\partial\xi^2}; \qquad (6.273\text{b})$$

substituting (6.273b) in the modified Laplace operator (6.234b) leads to

$$\bar{\nabla}^2\bar{\Psi} = \left\{\frac{\partial^2}{\partial R^2} + \frac{1}{R^2}\sin\theta\frac{\partial}{\partial\theta}\left(\frac{1}{\sin\theta}\frac{\partial}{\partial\theta}\right)\right\}\left[\frac{R}{a}\Psi\left(\frac{a^2}{R},\theta\right)\right]$$

$$= \frac{a^3}{R^3}\left\{\frac{\partial^2\Psi}{\partial\xi^2} + \frac{1}{\xi^2}\sin\theta\frac{\partial}{\partial\theta}\left(\frac{1}{\sin\theta}\frac{\partial}{\partial\theta}\right)\right\}\Psi(\xi,\theta) = \frac{a^3}{R^3}\bar{\nabla}^2\Psi = 0, \qquad (6.273\text{c})$$

showing that (6.272a and b) implies (6.272c) and vice versa, as follows from

$$\bar{\nabla}^2\bar{\Psi} = \frac{a^3}{R^3}\bar{\nabla}^2\Psi = \frac{\xi^3}{a^3}\bar{\nabla}^2\Psi. \qquad (6.274)$$

Thus *the modified Laplacian (6.234b) of the reciprocal stream function (6.272a) in spherical coordinates is related by (6.274) to the modified Laplacian of the original stream function in reciprocal coordinates (6.269b).*

From (6.272a) the relation between the limits of the original (reciprocal) stream function at the origin (infinity) follows:

$$\lim_{R \to 0} \frac{\Psi(R,\theta)}{R^2} = \lim_{\xi \to \infty} \frac{\xi^2}{a^4} \Psi\left(\frac{a^2}{\xi}, \theta\right) = \lim_{\xi \to \infty} \frac{\xi^2}{a^4} \frac{a}{\xi} \bar{\Psi}(\xi, \theta) = \lim_{\xi \to \infty} \frac{\xi}{a^3} \bar{\Psi}(\xi, \theta); \qquad (6.275)$$

thus if the original stream function corresponds (6.270a) to a finite nonzero field at the origin (6.276a), then the reciprocal stream function scales (6.275) as a dipole at infinity (6.276b), and vice versa:

$$\lim_{R \to 0} \frac{\Psi(R,\theta)}{R^2} \neq 0, \infty \Leftrightarrow \lim_{R \to \infty} R\bar{\Psi}(R,\theta) \neq 0, \infty. \qquad (6.276a,b)$$

The condition (6.270b) implies (6.58a through c) that both the radial and azimuthal field decay at infinity like (6.277a and b):

$$\lim_{R \to \infty} R^3 \bar{v}_R(R,\theta) \neq 0, \infty \equiv \lim_{R \to \infty} R^3 \bar{v}_R(R,\theta), \qquad (6.277a,b)$$

$$\lim_{R \to \infty} Q = \lim_{R \to \infty} 4\pi R^2 \bar{v}_R(R,\theta) \sim \lim_{R \to \infty} \frac{4\pi}{R} = 0; \qquad (6.277c)$$

the radial velocity implies that the volume flux across the sphere of large radius (6.277c) vanishes as $R \to \infty$. This proves in general the reasoning used in the particular case (6.270a through d; 6.271) to infer that the reciprocal stream function satisfies the modified Laplace equation (6.274) and does not affect the flow in the far field (6.277a through c). The preceding results suggest the **first sphere theorem (for the reciprocal stream function)**: *if* $\Psi(R, \theta)$ *is the stream function for an axisymmetric irrotational solenoidal unbounded flow (6.272b), then the stream function (6.278):*

$$\Psi_+(R,\theta) = \Psi(R,\theta) - \bar{\Psi}(R,\theta) = \Psi(R,\theta) - \frac{R}{a} \Psi\left(\frac{a^2}{R}, \theta\right), \qquad (6.278)$$

specifies the flow (6.279a) for which: (1) the sphere $R = a$ is (6.279b) the stream surface $\Psi = 0$:

$$\bar{\nabla}^2 \Psi_+(R,\theta) = 0, \quad \Psi_+(a,\theta) = 0; \qquad (6.279a,b)$$

and (2) the far field (6.280) has the same velocity (6.58a through c):

$$\lim_{R \to \infty} \frac{1}{R \sin\theta} \left(\frac{1}{R} \frac{\partial \Psi_+}{\partial \theta}, -\frac{\partial \Psi_+}{\partial R} \right) = \lim_{R \to \infty} \frac{1}{R \sin\theta} \left(\frac{1}{R} \frac{\partial \Psi}{\partial \theta} - \frac{\partial \Psi}{\partial R} \right). \qquad (6.280)$$

The proof is as follows: (1) the stream function (6.278) for the flow, including the sphere of radius a, satisfies (6.279a) on account of (6.272b and c); (2) also (6.278) implies that the sphere is the field line $\Psi = 0$ in (6.279b); and (3) since the reciprocal stream function (6.272a) leads to zero velocity at infinity (6.277a and b) the far field (6.280) is unchanged when it is added to the base flow without sphere. If $\Psi(R, \theta)$ has no singularities inside the sphere $R < a$, then $\bar{\Psi}$ in (6.272a) has no singularities outside the sphere $a^2/R > a$, and Ψ_+ has the same singularities outside the sphere as Ψ; likewise, if $\Psi(R, \theta)$ has no singularities outside the sphere, then $\Psi_+(R, \theta)$ has the same singularities inside the sphere. Thus the sphere (circle) theorems based on the reciprocal point apply both inside and outside the cylinder (sphere). (1) The first sphere theorem specifies the stream function for the potential flow past a sphere; and (2) this corresponds to the field function of the electrostatic field (Chapters I.24 and 7) with an insulating sphere. In both cases the stream (field) function is constant on the sphere that is a stream (field) surface, that is, the field is tangent and the normal component is zero. (1) In contrast the second (first) sphere theorem [Subsection 6.7.2 (Subsection 6.7.1)] applies to the potential (stream function) and thus does not (does) require the condition of axisymmetry; and (2) the sphere is the zero equipotential (field or stream surface), and hence the field is normal (tangential), that is, has zero tangential (normal) component.

6.7.2 Second Sphere Theorem for the Reciprocal Potential (Kelvin)

In order to remove the restriction to axisymmetric potential fields and allow the consideration of nonaxisymmetric potential fields as well, the potential should be used in the sphere theorem instead of the stream function. The potential of a uniform flow is (6.126a) $\equiv$ (6.281a), that is, the first term on the r.h.s. of (6.201a); inserting a sphere of radius a corresponds to adding the second term on the r.h.s. of (6.201a) and leads to (6.281b):

$$\Phi_0(R,\theta) = UR\cos\theta, \quad \Phi(R,\theta) = U\left(R + \frac{a^3}{2R^2}\right)\cos\theta = \Phi_0(R,\theta) + \frac{a}{2R}\Phi_0\left(\frac{a^2}{R},\theta\right); \qquad (6.281\text{a–c})$$

a similar conclusion follows assuming that the second term on the r.h.s. of (6.201a) is the potential of a dipole:

$$\bar{\Phi}_0(R,\theta) = \frac{Ua^3}{2R^2}\cos\theta = \frac{U\xi^2}{2a}\cos\theta = \frac{\xi}{2a}U\xi\cos\theta = \frac{\xi}{2a}\Phi_0(\xi,\theta) = \frac{a}{2R}\Phi_0\left(\frac{a^2}{R},\theta\right). \qquad (6.282)$$

The factor $a/2R$ in (6.282) is replaced by $-a/R$ leading to a total potential (6.289) that vanishes on the sphere; in this case the latter is a field surface and the field is orthogonal to it. This suggests the **reciprocal potential theorem**: *if* $\Phi(R, \theta, \varphi)$ *is the potential of any three-dimensional irrotational solenoidal field, possibly nonaxisymmetric, that is, it is a harmonic function that satisfies the Laplace equation (6.233a and b)* $\equiv$ *(6.283b), then the* **reciprocal potential** *(6.283a) is also a harmonic function (6.283c) and vice versa:*

$$\bar{\Phi}(R,\theta,\varphi) \equiv \frac{a}{R}\Phi\left(\frac{a^2}{R},\theta,\varphi\right): \quad \nabla^2\Phi = 0 \Leftrightarrow \nabla^2\bar{\Phi} = 0. \qquad (6.283\text{a–c})$$

It will also be proven that the reciprocal potential does not change the flow at infinity and meets the boundary condition on the sphere when subtracted from the original potential, thus leading to the second sphere theorem.

The proof of (6.283a through c) is made by direct calculation of the Laplacian (6.233a), starting with the radial terms and using (6.269b):

$$R^2\frac{\partial\bar{\Phi}}{\partial R} = R^2\frac{\partial}{\partial R}\left[\frac{a}{R}\Phi\left(\frac{a^2}{R},\theta,\varphi\right)\right] = -a\Phi + aR\frac{\partial\Phi}{\partial\xi}\frac{\partial\xi}{\partial R} = -a\Phi - \frac{a^3}{R}\frac{\partial\Phi}{\partial\xi}, \qquad (6.284\text{a})$$

$$\frac{\partial}{\partial R}\left(R^2\frac{\partial\bar{\Phi}}{\partial R}\right) = -a\frac{\partial}{\partial R}\left(\Phi + \frac{a^2}{R}\frac{\partial\Phi}{\partial\xi}\right) = -a\frac{\partial\Phi}{\partial\xi}\left(\frac{\partial\xi}{\partial R} - \frac{a^2}{R^2}\right) - \frac{a^3}{R}\frac{\partial^2\Phi}{\partial\xi^2}\frac{\partial\xi}{\partial R}$$

$$= \frac{2a^3}{R^2}\frac{\partial\Phi}{\partial\xi} + \frac{a^5}{R^3}\frac{\partial^2\Phi}{\partial\xi^2} = 2\frac{\xi^2}{a}\frac{\partial\Phi}{\partial\xi} + \frac{\xi^3}{a}\frac{\partial^2\Phi}{\partial\xi^2} = \frac{\xi}{a}\frac{\partial}{\partial\xi}\left(\xi^2\frac{\partial\Phi}{\partial\xi}\right). \qquad (6.284\text{b})$$

Substitution of (6.284b) in (6.233a), the Laplace operator:

$$R^2\nabla^2\overline{\Phi} = \frac{\partial}{\partial R}\left(R^2\frac{\partial\overline{\Phi}}{\partial R}\right) + \frac{1}{\sin\theta}\frac{\partial}{\partial\theta}\left(\sin\theta\frac{\partial\overline{\Phi}}{\partial\theta}\right) + \frac{1}{\sin^2\theta}\frac{\partial^2\overline{\Phi}}{\partial\varphi^2}$$

$$= \frac{\xi}{a}\left\{\frac{\partial}{\partial\xi}\left(\xi^2\frac{\partial\Phi}{\partial\xi}\right) + \frac{1}{\sin\theta}\frac{\partial}{\partial\theta}\left(\sin\theta\frac{\partial\Phi}{\partial\theta}\right) + \frac{1}{\sin^2\theta}\frac{\partial^2\Phi}{\partial\varphi^2}\right\} = \frac{\xi^3}{a}\nabla^2\Phi = 0, \qquad (6.284c)$$

proves the equivalence of (6.283b) and (6.283c), using

$$\nabla^2\overline{\Phi} = \frac{\xi^3}{R^2 a}\nabla^2\Phi = \left(\frac{a}{R}\right)^5\nabla^2\Phi = \left(\frac{\xi}{a}\right)^5\nabla^2\Phi. \qquad (6.285)$$

Thus *the Laplacian (6.233a) of the reciprocal potential (6.283a) in spherical coordinates is related by (6.285) to the Laplacian of the original potential in reciprocal coordinates (6.269b).*

From (6.283a) the relation between the limits of the potential (reciprocal) potential at the origin (infinity) follows:

$$\lim_{R\to 0}\frac{\Phi(R,\theta,\varphi)}{R} = \lim_{R\to\infty}\frac{\xi}{a^2}\Phi\left(\frac{a^2}{\xi},\theta,\varphi\right) = \lim_{R\to\infty}\frac{\xi}{a^2}\frac{\xi}{a}\overline{\Phi}(\xi,\theta,\varphi) = \lim_{R\to\infty}\frac{\xi^2}{a^3}\overline{\Phi}(\xi,\theta,\varphi); \qquad (6.286)$$

thus if the potential scales (6.283a) as a finite nonzero field (6.281a) at the origin (6.287a) the reciprocal potential scales like a dipole at infinity (6.287b):

$$\lim_{R\to 0}\frac{\Phi(R,\theta,\varphi)}{R} \neq 0,\infty \Leftrightarrow \lim_{R\to\infty}R^2\overline{\Phi}(R,\theta,\varphi) \neq 0,\infty. \qquad (6.287a,b)$$

The latter condition (6.287b) implies (6.72a through c) that all the components of the field decay at infinity like (6.288a through c):

$$\lim_{R\to\infty}R^3\{v_R, v_\theta, v_\varphi(R,\theta,\varphi)\} = \lim_{R\to\infty}R^3\left\{\frac{\partial\Phi}{\partial R}, \frac{1}{R}\frac{\partial\Phi}{\partial\theta}, \frac{1}{R\sin\theta}\frac{\partial\Phi}{\partial\varphi}\right\} \neq, 0\infty; \qquad (6.288a\text{-}c)$$

$$\lim_{R\to\infty}Q = \lim_{R\to\infty}4\pi R^2 v_R(R,\theta,\varphi) \sim \lim_{R\to\infty}\frac{4\pi}{R} = 0. \qquad (6.288d)$$

The radial field implies (6.278d) that there is no flux through a sphere of radius R. This confirms in general the reasoning used in the particular case (6.281a through c; 6.282) to infer that the reciprocal potential satisfies the Laplace equation (6.233b) and does not affect the field at infinity (6.288a through c). The preceding results suggest the **second sphere theorem (for the reciprocal potential)**: *if* $\Phi(R, \theta, \varphi)$ *is the potential of an irrotational solenoidal unbounded field (6.283a), then the potential (6.289):*

$$\Phi_-(R,\theta,\varphi) = \Phi(R,\theta,\varphi) - \overline{\Phi}(R,\theta,\varphi) = \Phi(R,\theta,\varphi) - \frac{a}{R}\Phi\left(\frac{a^2}{R},\theta,\varphi\right), \qquad (6.289)$$

specifies the field (6.290a) for which: (1) the sphere $R = a$ is the equipotential $\Phi = 0$ in (6.290b):

$$\nabla^2\Phi_-(R,\theta,\varphi) = 0, \quad \Phi_-(a,\theta,\varphi) = 0; \qquad (6.290a,b)$$

and (2) the asymptotic potential (6.291a) [equivalently the far field (6.291b)] is the same:

$$\lim_{R\to\infty} \Phi_-(R,\theta) = \lim_{R\to\infty} \Phi(R,\theta),$$ (6.291a)

$$\lim_{R\to\infty}\left\{\frac{\partial\Phi_-}{\partial R}, \frac{1}{R}\frac{\partial\Phi_-}{\partial\theta}, \frac{1}{R\sin\theta}\frac{\partial\Phi_-}{\partial\varphi}\right\} = \lim_{R\to\infty}\left\{\frac{\partial\Phi}{\partial R}, \frac{1}{R}\frac{\partial\Phi}{\partial\theta}, \frac{1}{R\sin\theta}\frac{\partial\Phi}{\partial\varphi}\right\}.$$ (6.291b)

The proof is (1) the potential (6.289) satisfies the Laplace equation (6.290a) on account of (6.283b and c); (2) also (6.289) implies (6.290b) that the sphere is the equipotential $\Phi = 0$; and (3) the reciprocal potential (6.283a) vanishes at infinity leading the original potential unchanged asymptotically (6.291a), or equivalently the reciprocal potential leads to zero velocity at infinity (6.288a through c) and thus does not disturb the flow in the far field (6.291b). Thus the first (6.278; 6.279a and b; 6.280) [second (6.289; 6.290a and b; 6.291a and b)] sphere theorem: (1) uses the stream or field function (scalar potential) and thus is (is not) restricted to axisymmetric field; (2) ensures that the sphere is a stream or field surface (an equipotential) to which the field is tangent (orthogonal); and (3) the far field is unchanged in both cases.

6.7.3 Third Sphere Theorem for the Integral Reciprocal Potential (Weiss, 1945)

The first (second) sphere theorem is based on constructing from the original stream function (potential) a reciprocal stream function (6.272a) [reciprocal potential (6.283a)] that (1) also satisfies the modified Laplace (6.273a through c) [Laplace (6.284a through c)] equation (2) equals the original stream function (potential) on the sphere, so that the latter becomes a field surface (equipotential) for their difference; and (3) thus the difference of the original and reciprocal field function (potential) specifies a total stream function (potential) leading to a field tangent (orthogonal) to the sphere, that is, an insulating (conducting) sphere in the case of the electrostatic field. The third sphere theorem arises from the construction of a second harmonic potential, this time in integral form, that is, the **integral reciprocal potential theorem**: *if $\Phi(R, \theta, \varphi)$ is the potential of an irrotational solenoidal field, that is, a harmonic function that satisfies the Laplace equation (6.292b), then the **reciprocal integral potential** (6.292a) is also a harmonic function (6.292c) and vice versa:*

$$\tilde{\Phi}(R,\theta,\varphi) = \frac{1}{a}\int_0^{a^2/R} \eta\frac{\partial\Phi(\eta,\theta,\varphi)}{\partial\eta}d\eta: \quad \nabla^2\Phi = 0 \Leftrightarrow \nabla^2\tilde{\Phi} = 0.$$ (6.292a–c)

The proof is made by: (1) direct calculation of the Laplacian, starting with the theorem of parametric differentiation of integrals (I.13.44) applied to (6.292a):

$$\xi = \frac{a^2}{R}: \quad \frac{\partial\tilde{\Phi}}{\partial\xi} = \frac{1}{a}\frac{\partial}{\partial\xi}\int_0^{\xi}\eta\frac{\partial\Phi}{\partial\eta}d\eta = \frac{\xi}{a}\frac{\partial\Phi}{\partial\xi};$$ (6.293a,b)

and (2) use of the second sphere theorem (6.283a and c) for the derivative potential (6.294a), showing that the Laplace equation is satisfied (6.294b):

$$\Phi_\star(R,\theta,\varphi) = \frac{\partial\Phi}{\partial\xi}: \quad \nabla^2\left(\frac{\partial\tilde{\Phi}}{\partial\xi}\right) = \nabla^2\left(\frac{\xi}{a}\frac{\partial\Phi}{\partial\xi}\right) = \nabla^2\left(\frac{a}{R}\Phi_\star\right) = \nabla^2\bar{\Phi}_\star = 0.$$ (6.294a,b)

If the potential is regular at the origin (6.295a), the first integral reciprocal potential is of order (6.295b) at infinity:

$$\lim_{R \to 0} \Phi(R,\theta,\varphi) \neq 0,\infty \Leftrightarrow \lim_{R \to \infty} R^2 \tilde{\Phi}(R,\theta,\varphi) \neq 0,\infty; \tag{6.295a,b}$$

the result (6.295a and b) follows substituting (6.295a) ≡ (6.296a) in (6.292a):

$$\Phi(R,\theta,\varphi) = A_0 + A_1 R + O(R^2): \quad \tilde{\Phi} = \frac{1}{a} \int_0^\xi \left[\eta A_1 + O(\eta^2) \right] d\eta \sim O\left(\frac{\xi^2}{a} \right) = O\left(\frac{a^3}{R^2} \right), \tag{6.296a,b}$$

which leads to (6.296b) ≡ (6.295b) to the lowest order in ξ or $1/R$.

The preceding results lead to the **third sphere theorem (for the integral reciprocal potential)**: if $\Phi(R, \theta, \varphi)$ *is the potential of an irrotational solenoidal field, that is, it satisfies the Laplace equation (6.292b), so does (6.298a) the potential,*

$$\Phi_+(R,\theta,\varphi) = \Phi(R,\theta,\varphi) + \tilde{\Phi}(R,\theta,\varphi) = \Phi(R,\theta,\varphi) + \frac{1}{a} \int^{a^2/R} \xi \frac{\partial \Phi(\eta,\theta,\varphi)}{\partial \eta} d\eta, \tag{6.297}$$

that (1) scales asymptotically as (6.295b) and (2) has zero normal derivative (6.298b) on the sphere $R = a$:

$$\nabla^2 \Phi_+(R,\theta,\varphi) = 0, \quad \lim_{R \to a} \frac{\partial \Phi_+(R,\theta,\varphi)}{\partial R} = 0. \tag{6.298a,b}$$

The proof is: (1) the boundary condition (6.298b) follows from (6.297):

$$\lim_{R \to a} \frac{\partial \Phi_+}{\partial R} = \lim_{R \to a} \left(\frac{\partial \Phi}{\partial R} + \frac{d\xi}{dR} \frac{\partial \tilde{\Phi}}{\partial \xi} \right) = \lim_{R \to a} \frac{\partial \Phi}{\partial R} - \lim_{s \to a} \frac{a^2}{R^2} \frac{\partial \tilde{\Phi}}{\partial \xi}$$

$$= \lim_{R \to a} \frac{\partial \Phi}{\partial R} - \lim_{s \to a} \frac{a^2}{R^2} \frac{\xi}{a} \frac{\partial \Phi}{\partial \xi} = \lim_{R \to a} \frac{\partial \Phi}{\partial R} - \lim_{\xi \to a} \frac{\xi^3}{a^3} \frac{\partial \Phi}{\partial \xi} = 0. \tag{6.299}$$

and (2) the function Φ_+ is harmonic (6.298a) because it is the sum (6.297) of harmonic functions (6.292b and c). The third sphere theorem (6.297; 6.298a and b): (1) like the first (6.278; 6.279a and b; 6.280), [unlike the second (6.289; 6.290a and b; 6.291a and b)] specifies as potential field tangent (orthogonal) to the sphere; and (2) like the second (unlike the first) uses the potential (field function) and thus is not (is) restricted to axisymmetric fields. There is a fourth sphere theorem (Subsection 6.7.4) using a distinct integral reciprocal potential from the third (Subsection 6.7.4). The first and third (second) sphere theorems [Subsections 6.7.1 and 6.7.3 (Subsection 6.7.2)] specify potential fields both inside and outside a sphere, with zero normal (tangential) component; the fourth sphere theorem (Subsection 6.7.4) needs no boundary condition if it is applied on the surface of a sphere, for example, for surface electric currents (Section 6.8).

6.7.4 Fourth Sphere Theorem for the Integral Nonreciprocal Potential

*If an irrotational solenoidal field has a potential $\Phi(R, \theta, \varphi)$, that is, a harmonic function that satisfies the Laplace equation (6.300b), then the **integral nonreciprocal potential** (6.300a) is also a harmonic function (6.300c) and vice versa:*

$$\hat{\Phi}(R,\theta,\varphi) = \int_\infty^R \frac{\Phi(\zeta,\theta,\varphi)}{\zeta} d\zeta: \quad \nabla^2 \Phi = 0 \Leftrightarrow \nabla^2 \hat{\Phi} = 0. \tag{6.300a–c}$$

The proof follows immediately from the multipolar expansion (Subsection 8.3.6) of the potential in descending powers of the radius (6.301) ≡ (8.52):

$$\Phi(R,\theta,\varphi) = \sum_{n=1}^{\infty} R^{-n} A_n(\theta,\varphi),$$ (6.301)

leading to a similar multipolar expansion:

$$\hat{\Phi}(R,\theta,\varphi) = \sum_{n=1}^{\infty} \int_{\infty}^{R} \zeta^{-1-n} A_n(\theta,\varphi) d\zeta = -\sum_{n=1}^{\infty} R^{-n} \frac{A_n(\theta,\varphi)}{n+1},$$ (6.302)

whose coefficients do not depend on R; the dependence on latitude (Subsection 8.3.6) or longitude of the spherical harmonics $A_n(\theta, \varphi)$ is unaffected by the radial integration, with the radial decay R^{-n} corresponding to the same angular dependence $A_n(\theta, \varphi)$, and only the magnitude of the multipoles divided by $n + 1$. Thus the integral nonreciprocal potential satisfies the Laplace equation (6.300c).

6.8 Electric Charges and Currents on a Sphere

An insulating (conducting) sphere is a field surface (equipotential surface), and thus its effect on a uniform external electric field is specified [Subsection 6.8.1 (Subsection 6.8.2)] by the first (second) [Subsection 6.7.1 (Subsection 6.7.2)] sphere theorem. The electric currents on an electrically conducting medium satisfy the Ohm law (Subsection 6.8.3) leading to another analogy with a potential flow for an electrostatic field (Subsection 6.8.4). The case of electric currents on a sphere flowing from one pole to the opposite pole (Subsection 6.8.5) is considered using a current source and sink; in order to ensure that the electric current lies on the sphere (Subsection 6.8.6), that is, the normal electric field is zero, a spherical surface potential is added (Subsection 6.8.7), which is specified by the fourth sphere theorem (Subsection 6.7.4); this specifies the electric current and field at all points on the surface of the sphere, and shows that it has an inverse square root singularity at the poles (Subsection 6.8.8).

6.8.1 Insulating Sphere in an Electrostatic Field

The axisymmetric electric filed is related to the field function (6.303a and b) as the velocity from the stream function (6.58a and c) with reversed sign by convention:

$$\{E_R(R,\theta), E_\theta(R,\theta)\} = -\frac{1}{R\sin\theta}\left\{\frac{1}{R}\frac{\partial\Psi}{\partial\theta}, -\frac{\partial\Psi}{\partial R}\right\}.$$ (6.303a,b)

The stream function (6.126b) for the uniform flow with velocity U corresponds to the field function (6.304a) for a uniform electric field E_0, also with reversed sign:

$$\Psi_0(R,\theta) = -\frac{E_0}{2}R^2\sin^2\theta; \quad \Psi_+(R,\theta) = -\frac{E_0}{2}\left(R^2 - \frac{a^3}{R}\right)\sin^2\theta, \quad \Psi_+(R,\theta) = 0.$$ (6.304a–c)

Since an insulating sphere is a field surface, the first sphere theorem (6.278) is used to specify the field function (6.304b) when it is inserted in a uniform electric field E_0, confirming that it is constant on the sphere (6.304c). Substituting (6.304b) in (6.303a and b) specifies the electric field (6.305a and b):

$$E_R^+(R,\theta) = E_0\left(1 - \frac{a^3}{R^3}\right)\cos\theta, \quad E_\theta^+(R,\theta) = -E_0\left(1 + \frac{a^3}{2R^3}\right)\sin\theta,$$ (6.305a,b)

confirming that it is tangent to (6.306a and b) the sphere:

$$E_R^+(R,\theta) = 0, \quad E_\theta^+(R,\theta) = -\frac{3}{2}E_0 \sin\theta. \tag{6.306a,b}$$

The electrostatic potential is obtained substituting the electric filed (6.305a and b) in the exact differential (6.120b):

$$d\Phi_+ = -E_R^+ \, dR - E_\theta^+ R \, d\theta = -E_0\left[\left(1-\frac{a^3}{R^3}\right)\cos\theta \, dR - \left(R+\frac{a^3}{2R^2}\right)\sin\theta \, d\theta\right]$$

$$= -E_0 d\left[\left(R+\frac{a^3}{2R^2}\right)\cos\theta\right]; \tag{6.307}$$

the potential (6.307) ≡ (6.308a) simplifies to (6.308b) on the sphere:

$$\Phi_+(R,\theta) = -E_0\left(R+\frac{a^3}{2R^2}\right)\cos\theta, \quad \Phi_+(a,\theta) = -\frac{3}{2}E_0 a \cos\theta. \tag{6.308a,b}$$

Thus *the electrostatic field function (6.304b) [scalar potential (6.308a)] and electric field (6.305a and b) specify the modification of a uniform electric filed E_0 due to the insertion of an insulating sphere of radius a.* The case of an electrically conducting sphere is considered next (Subsection 6.7.2) including the distribution of surface electric charges.

6.8.2 Electric Charges on a Conducting Sphere

The first term on the r.h.s. of (6.308a) is the electrostatic potential (6.309a) of a uniform electric field E_0, and the insertion of a conducting sphere of radius a leads by the second sphere theorem (6.289) to the total potential (6.309b) that vanishes on the sphere (6.309c):

$$\Phi_0(R,\theta) = -E_0 R \cos\theta, \quad \Phi_-(R,\theta) = -E_0\left(R-\frac{a^3}{R^2}\right)\cos\theta, \quad \Phi(a,\theta) = 0. \tag{6.309a–c}$$

The minus (plus) sign in the electrostatic (6.309a) [hydrodynamic (6.281a)] potential of a uniform magnetic field E_0 (flow with velocity U) is due to different conventions (6.310a) [(6.69b)]:

$$\vec{E} = -\nabla\Phi: \quad \{E_R, E_\theta, E_\varphi\}\{R,\theta,\varphi\} = -\left\{\frac{\partial\Phi}{\partial R}, \frac{1}{R}\frac{\partial\Phi}{\partial\theta}, \frac{1}{R\sin\theta}\frac{\partial\Phi}{\partial\varphi}\right\}. \tag{6.310a,b}$$

Substituting (6.309b) in (6.310b) specifies the electric field:

$$E_R^-(R,\theta) = E_0\left(1+\frac{2a^3}{R^3}\right)\cos\theta, \quad E_\theta^-(R,\theta) = -E_0\left(1-\frac{a^3}{R^3}\right)\sin\theta, \quad E_\varphi^-(R,\theta) = 0, \tag{6.311a–c}$$

confirming that it is normal to the sphere:

$$E_\theta^-(a,\theta) = 0 = E_\varphi^-(a,\theta), \quad E_R^-(R,\theta) = 3E_0 \cos\theta = \frac{\sigma(\theta)}{\varepsilon}; \tag{6.312a–d}$$

the normal component (6.312c) specifies the density of electric change per unit area $\sigma(\theta)$ on the conducting sphere for a medium with dielectric permittivity ε. The electric field (6.301a through c) leads (6.121b) to:

$$d\Psi_- = -R\sin\theta\left(-E_\theta^- dR + E_R^- R\sin\theta \, d\theta\right)$$

$$= -E_0\left[\left(R - \frac{a^3}{R^2}\right)\sin^2\theta \, dR + \left(R^2 + \frac{2a^3}{R}\right)\sin\theta\cos\theta \, d\theta\right]$$

$$= -\frac{E_0}{2}d\left[\left(R^2 + \frac{2a^3}{R}\right)\sin^2\theta\right], \tag{6.313}$$

the field function (6.313) $\equiv$ (6.314a):

$$\Psi_-(R,\theta) = -\frac{E_0}{2}\left(R^2 + \frac{2a^3}{R}\right)\sin^2\theta, \quad \Psi_-(a,\theta) = -\frac{3}{2}E_0 a^2 \sin^2\theta, \tag{6.314a,b}$$

which takes the value (6.314b) on the sphere.

The electric charge density (6.312d) is positive $\sigma > 0$ in the direction of the external electric field $0 < \theta < \pi/2$, negative $\sigma < 0$ opposite to it $\pi/2 < \theta < \pi$, and zero $\sigma = 0$ in the transverse direction $\theta = \pi/2$. Using the spherical scale factors along meridians (6.14b) and parallels (6.14c) the area element (6.20a) on the sphere of radius a is (6.315a) and the total electric charge is zero (6.315b) because positive and negative charges balance:

$$dS = a^2 \sin\theta \, d\theta \, d\varphi: \quad e = \int_{R=a} \sigma(\theta)dS = a^2 \int_0^{2\pi} d\varphi \int_0^\pi \sigma(\theta)\sin\theta \, d\theta$$

$$= 6\pi a^2 E_0\varepsilon \int_0^\pi \cos\theta\sin\theta \, d\theta = 0. \tag{6.315a,b}$$

Since the distribution of electric charges on the sphere is symmetric (skew-symmetric) relative to the direction of (transverse to) the external electric field, the axial (transverse) dipole moment is nonzero (6.316b) [zero (6.317b)]:

$$z = a\cos\theta: \quad P_z = \int_{R=a} \sigma(\theta)z \, dS = 2\pi a^3 \int_0^\pi \sigma(\theta)\cos\theta\sin\theta \, d\theta$$

$$= 6\pi a^3 E_0\varepsilon \int_0^\pi \cos^2\theta\sin\theta \, d\theta = 4\pi a^3 E_0\varepsilon, \tag{6.316a,b}$$

$$r = a\sin\theta: \quad P_r = \int_{R=a} \sigma(\theta)r \, dS = 2\pi a^3 \int_0^\pi \sigma(\theta)\sin^2\theta \, d\theta$$

$$= 6\pi a^3 E_0\varepsilon \int_0^\pi \sin^2\theta\cos\theta \, d\theta = 0. \tag{6.317a,b}$$

Thus *a conducting sphere with radius a and uniform electric field E_0 has electrostatic potential (6.309b) [field function (6.314a)] and electric field (6.311a through c) normal to the sphere (6.312a through c).*

The normal component of the electric field specifies the electric charge density per unit area (6.312d) in a medium of dielectric permittivity ε, which leads to: (1) a zero (6.315b) total electric charge; (2) a dipole moment (6.316b; 6.317b) aligned with the external electric field (6.318c):

$$S = 4\pi a^2, \quad V = \frac{4}{3}\pi a^3: \quad \vec{P}_1 = 4\pi a^3 \vec{E}_0 \varepsilon = 3\varepsilon \vec{E}_0 V = \varepsilon \vec{E}_0 aS, \tag{6.318a–c}$$

and proportional to the dielectric permittivity times the triple of the volume of the sphere (6.318b) or the product of the area of the sphere (6.318a) by the radius. Next are considered electric currents (Subsections 6.8.3 through 6.8.8) instead of electric charges (Subsection 6.8.2) on a sphere.

6.8.3 Ohm's Law and Electrical Conductivity

Ohm's law *(6.319b) states that the electric current is proportional to the electric field, through* **electrical conductivity**, *that is, a scalar in an isotropic medium and is positive (6.319a) so that the electric current is parallel to the electric field:*

$$\sigma > 0: \quad \vec{j} = \sigma\vec{E}; \quad \vec{E} = -\nabla\Phi; \tag{6.319a–c}$$

in the electrostatic case the electric field is minus the gradient of a potential (6.310a) ≡ (6.319c). In a homogeneous medium the electrical conductivity is constant (6.320a) and the conservation of the electric current (6.320b) leads to the Laplace equation for the electrostatic potential (6.320c):

$$\sigma = \text{const}: \quad \nabla \cdot \vec{j} = 0, \quad 0 = -\frac{1}{\sigma}\nabla\vec{j} = -\nabla \cdot \vec{E} = \nabla(\nabla\Phi) = \nabla^2\Phi. \tag{6.320a–c}$$

The conservation of the electric current in steady conditions requires that the flux of the electric current through a closed regular current through a closed regular surface be zero:

$$0 = \int_{\partial D} \vec{j} \cdot d\vec{S} = \int_D (\nabla \cdot \vec{j}) dV; \tag{6.321}$$

the divergence theorem (5.163a and b) = (6.321) then implies that the divergence of the electric current be zero (6.319b) in the interior.

6.8.4 Hydrodynamic and Electric Charge/Current Monopoles

There is an analogy among the monopoles represented by: (1) the hydrodynamic (6.80b) potential (6.130a) ≡ (6.322a) due to a flow source Q > 0 or sink Q < 0 of flow rate Q; (2) the electrostatic (6.310a) potential (6.322b) due to an electric charge e in a medium of dielectric permittivity ε:

$$\Phi_v = -\frac{Q}{4\pi R}, \quad \Phi_e = \frac{e}{4\pi\varepsilon R}, \quad \Phi_j = \frac{J}{4\pi\sigma R}; \tag{6.322a–c}$$

and (3) for an electric current (6.319b) where e/ε in (6.322b) is replaced by J/σ in (6.322c). The velocity (6.323a) [electric field (6.323b and c)] of a monopole (6.322a) [(6.322b and c)] in all cases:

$$\vec{v} = \nabla\Phi_v = \vec{e}_r \frac{Q}{4\pi R^2}, \quad \vec{E} = -\nabla\Phi_e = \vec{e}_r \frac{e}{4\pi\varepsilon R^2}, \quad \vec{E} = -\nabla\Phi_j = \vec{e}_r \frac{J}{4\pi\sigma R^2}, \tag{6.323a–c}$$

is radial outward for positive (negative) Q/e/J, respectively, in (6.323a through c). In the present case the electric current due to the electric field in a conducting medium is considered (Subsection 6.8.3 through 6.8.6), not the magnetic field associated with electric currents (Chapter I.26; Sections 8.4 through 8.6).

6.8.5 Electric Currents Flowing between the Poles of a Sphere

As an example consider (Figure 6.15) a sphere of radius a with surface currents flowing from one pole to the opposite pole. The distance of an arbitrary point from the north R_+ and south R_- pole of the sphere is:

$$R_\pm = \left| x^2 + y^2 + (z \mp a)^2 \right|^{1/2} = \left| x^2 + y^2 + z^2 + a^2 \mp 2az \right|^{1/2} = \left| R^2 + a^2 \mp 2aR\cos\theta \right|^{1/2}, \qquad (6.324)$$

using spherical coordinates (6.13a through d) with axis through the poles. These distances appear in the potential:

$$\Phi(R,\theta) = \frac{J}{4\pi\sigma}\left[\frac{1}{R_+} - \frac{1}{R_-} + \tilde{\Phi}(R,\theta) \right], \qquad (6.325)$$

which consists of three terms on the r.h.s.: (1/2) the first (second) is (6.322c) a point source (sink) of current $+J(-J)$ at the north (south) pole resulting in an electric field and parallel electric current that would not be confined to the surface of the sphere; and (3) for the electric current to lie on the surface of the sphere it is necessary to add a **spherical surface potential** Φ such that the total potential has zero normal derivative on the sphere (6.326a):

$$0 = \lim_{R \to a} \frac{\partial \Phi}{\partial R}; \quad \lim_{R \to a} \frac{\partial \tilde{\Phi}}{\partial R} = \lim_{R \to \infty} \frac{\partial}{\partial R}\left(\frac{1}{R_-} - \frac{1}{R_+} \right), \qquad (6.326a,b)$$

the boundary condition (6.326a) for the total potential to ensure that electric currents flow (Figure 6.15) from the source at the north pole to the sink at the south pole only on the surface of the sphere implies the boundary condition (6.326b) for the spherical surface potential.

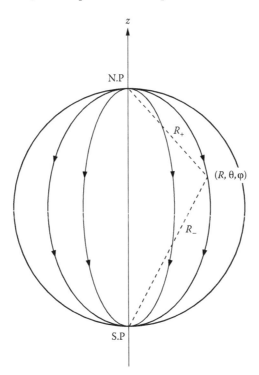

FIGURE 6.15 An example of a surface rather than a volume flow is given by the electric currents on a conducting sphere flowing from one pole to the other along the meridians.

6.8.6 Boundary Condition for Surface Electric Currents

The boundary condition that the electric currents lie on the surface of the sphere involves the derivatives of the distance (6.324) from the poles:

$$\frac{\partial R_\pm}{\partial R} = \frac{R \mp a\cos\theta}{R_\pm}, \quad \frac{\partial}{\partial R}\left(\frac{1}{R_\pm}\right) = -\frac{1}{R_\pm^2}\frac{\partial R_\pm}{\partial R} = -\frac{R \mp a\cos\theta}{R_\pm^3}, \tag{6.327a,b}$$

evaluated at an arbitrary point on the sphere:

$$\lim_{R\to a} R_\pm = a\left|2\left(1 \mp \cos\theta\right)\right|^{1/2}, \tag{6.328a}$$

$$\lim_{R\to a}\frac{\partial}{\partial R}\left(\frac{1}{R_\pm}\right) = -\lim_{R\to a}\frac{a\left(1\mp\cos\theta\right)}{2a^2\left(1\mp\cos\theta\right)}\frac{1}{R_\pm} = -\lim_{R\to a}\frac{1}{2aR_\pm}. \tag{6.328b}$$

The distance from the north (0, 0, a) [south (0, 0, −a)] pole of a sphere (Figure 6.15) is given by $R_+(R_-)$: (1) *in (6.324) for an arbitrary point, generally not on the sphere; and (2) simplifies to (6.328a) on the sphere. In particular setting* $\theta = \pi/2$ *in (6.328a) it follows that the distance from the equator to the poles is* $a\sqrt{2}$, *corresponding to third side of equilateral rectangular triangle of equal sides of length* a; *setting* $\theta = \pi(\theta = 0)$ *with the upper (lower) sign in (6.328a) gives the distance between the poles* $2a$, *that is the diameter or twice the radius. Substituting (6.328a and b) in (6.326b) leads to (6.329b).*

$$\nabla^2\tilde{\Phi} = 0, \quad \lim_{R\to a}\frac{\partial\tilde{\Phi}}{\partial R} = \frac{1}{2a}\lim_{R\to a}\left(\frac{1}{R_+} - \frac{1}{R_-}\right), \tag{6.329a,b}$$

which is equivalent to

$$\lim_{R\to a}\frac{\partial\tilde{\Phi}}{\partial R} = \frac{\left|1-\cos\theta\right|^{-1/2} - \left|1+\cos\theta\right|^{-1/2}}{\left(a\sqrt{2}\right)^3} = \frac{\left|1+\cos\theta\right|^{1/2} - \left|1-\cos\theta\right|^{1/2}}{\left(a\sqrt{2}\right)^3\sin\theta}, \tag{6.330a,b}$$

using

$$\frac{1}{\sqrt{1-\cos\theta}} - \frac{1}{\sqrt{1+\cos\theta}} = \frac{\sqrt{1+\cos\theta} - \sqrt{1-\cos\theta}}{\sqrt{1-\cos^2\theta}} = \frac{\left|1+\cos\theta\right|^{1/2} - \left|1-\cos\theta\right|^{1/2}}{\sin\theta}, \tag{6.330c}$$

The spherical surface potential is a harmonic function, that is, it satisfies the Laplace equation (6.329a), and besides meets the boundary condition (6.329b) ≡ (6.330a) ≡ (6.330b) at the surface of the sphere. In order to meet this condition an integral nonreciprocal potential is introduced next using (Subsection 6.8.7) the fourth sphere theorem (Subsection 6.7.4).

6.8.7 Spherical Surface Potential for the Electric Current

The boundary condition (6.329b) is satisfied by the function (6.331):

$$\tilde{\Phi}(R,\theta) = \frac{1}{2}\int_\infty^R\left(\frac{1}{R_+} - \frac{1}{R_-}\right)\frac{dR}{R}, \tag{6.331}$$

which is also a harmonic function, since it satisfies the Laplace equation (6.300c) by the fourth sphere theorem (6.300a). In order to determine the spherical surface potential (6.331) the primitive of the two terms in (6.331) is needed and follows (II.7.116b) from

$$\frac{d}{dR}\left[\arg\sinh\left(\frac{a\mp R\cos\theta}{R\sin\theta}\right)\right] = \left|1+\left(\frac{a\mp R\cos\theta}{R\sin\theta}\right)^2\right|^{-1/2}\frac{d}{dR}\left(\frac{a\mp R\cos\theta}{R\sin\theta}\right)$$

$$= \left|R^2\sin^2\theta+(a\mp R\cos\theta)^2\right|^{-1/2}R\sin\theta\frac{d}{dR}\left(\frac{a}{R\sin\theta}\mp\cot\theta\right)$$

$$= \left|a^2+R^2\mp 2aR\cos\theta\right|^{-1/2}\left(-\frac{a}{R}\right) = -\frac{a}{R_\pm R}. \tag{6.332a,b}$$

Thus the spherical surface potential (6.331) is $1/2a$ times the difference of (6.332b) and (6.332a) in the total potential (6.325):

$$\Phi(R,\theta) = \frac{J}{8\pi a\sigma}\left[2a\left(\frac{1}{R_+}-\frac{1}{R_-}\right)+\arg\sinh\left(\frac{a+R\cos\theta}{R\sin\theta}\right)-\arg\sinh\left(\frac{a-R\cos\theta}{R\sin\theta}\right)\right]. \tag{6.333}$$

Thus *the total potential associated with Ohmic (6.319b) electric currents flowing on the surface of a sphere from a source J at the north pole to a sink −J at the south pole (Figure 6.15) is (6.333), leading to the electric field:*

$$E_R(R,\theta) = -\frac{\partial\Phi}{\partial R} = \frac{J}{4\pi\sigma}\left[\frac{R-a\cos\theta}{R_+^3}-\frac{R+a\cos\theta}{R_-^3}+\frac{1}{2R}\left(\frac{1}{R_-}-\frac{1}{R_+}\right)\right], \tag{6.334a}$$

$$E_\theta(R,\theta) = -\frac{1}{R}\frac{\partial\Phi}{\partial\theta} = \frac{J}{8\pi\sigma}\left[2a\sin\theta\left(\frac{1}{R_+^3}-\frac{1}{R_-^3}\right)\right.$$

$$\left.+\frac{\csc\theta}{R}\left[\frac{1}{R_-}\left(\frac{R}{a}+\cos\theta\right)+\frac{1}{R_+}\left(\frac{R}{a}-\cos\theta\right)\right]\right], \tag{6.334b}$$

that is, tangent to the sphere (6.335a):

$$E_R(a,\theta)=0; \quad \vec{j}=\vec{e}_\theta j(\theta)=\vec{e}_\theta\sigma E_\theta(a,\theta)=\sigma\vec{E}, \tag{6.335a,b}$$

$$j(\theta) = \frac{J}{4\pi a^2\sqrt{2}}\frac{|1+\cos\theta|^{1/2}+|1-\cos\theta|^{1/2}}{\sin^2\theta}=j(\pi-\theta), \tag{6.335c,d}$$

the tangential component of the electric field (6.335b) shows that the electric current flows along the meridians (6.335b) with density (6.335c) that: (1) is symmetric relative to the equator (6.335d); (2) takes at the equator the finite value (6.336a):

$$j\left(\frac{\pi}{2}\right) = \frac{J}{2\pi a^2\sqrt{2}}; \quad \lim_{\theta\to 0}\theta^2 j(\theta) = \frac{J}{4\pi a^2} = \frac{j(\pi/2)}{\sqrt{2}} = \lim_{\theta\to\pi}(\theta-\pi)^2 j(\theta), \tag{6.336a,b}$$

and (3) has a singularity like θ^{-2} at the poles with coefficient (6.336b).

6.8.8 Singularities of the Electric Field at the Poles

The passage from (6.333) to (6.334a) uses (6.327b) in the first two terms and (6.332a and b) in the last term. The passage from (6.333) to (6.334b) uses in the first (last) two terms the derivatives (6.337a) [(6.327b)]:

$$-\frac{1}{R}\frac{\partial}{\partial\theta}\left(\frac{1}{R_\pm}\right) = \frac{1}{R_\pm^2 R}\frac{\partial R_\pm}{\partial\theta} = \pm\frac{a\sin\theta}{R_\pm^3}, \tag{6.337a}$$

$$-\frac{1}{R}\frac{\partial}{\partial\theta}\left[\arg\sinh\left(\frac{a\pm R\cos\theta}{R\sin\theta}\right)\right] = -\frac{1}{R}\left|1+\left(\frac{a\pm R\cos\theta}{R\sin\theta}\right)^2\right|^{-1/2}\frac{\partial}{\partial\theta}\left(\frac{a\pm R\cos\theta}{R\sin\theta}\right)$$

$$= -\frac{1}{R}\left|R^2\sin^2\theta+\left(a\pm R\cos\theta\right)^2\right|^{-1/2}R\sin\theta\frac{\mp R-a\cos\theta}{R\sin^2\theta} = \frac{\pm R+a\cos\theta}{R_\mp R\sin\theta}. \tag{6.337b}$$

Using (6.328a) it can be checked that the radial or normal electric field (6.334a) vanishes (6.325a) on the sphere:

$$\frac{4\pi\sigma}{J}E_R\left(a,\theta\right) = \lim_{R\to a}\left[\frac{1}{R_+}\left(\frac{R-a\cos\theta}{R_+^2}-\frac{1}{2R}\right)-\frac{1}{R_-}\left(\frac{R+a\cos\theta}{R_+^2}-\frac{1}{2R}\right)\right] = 0. \tag{6.338}$$

The same limit (6.338) for the tangential electric field (6.334b) specifies the distribution of the density of electric current on the sphere:

$$\frac{4\pi a^2\sqrt{2}}{J}j(\theta) = \lim_{R\to a}\frac{4\pi a^2\sigma\sqrt{2}}{J}\lim_{R\to a}E_\theta\left(R,\theta\right)$$

$$= \frac{a^2}{\sqrt{2}}\csc\theta\lim_{R\to a}\left[2a\sin^2\theta\left(\frac{1}{R_+^3}+\frac{1}{R_-^3}\right)+\frac{1+\cos\theta}{R_-a}+\frac{1-\cos\theta}{R_+a}\right]$$

$$= \frac{\csc\theta}{2}\left\{(1-\cos^2\theta)\left[|1-\cos\theta|^{-3/2}+|1+\cos\theta|^{-3/2}\right]+|1+\cos\theta|^{1/2}+|1-\cos\theta|^{1/2}\right\}$$

$$= \frac{\csc\theta}{2}\left\{(1+\cos\theta)|1-\cos\theta|^{-1/2}+(1-\cos\theta)|1+\cos\theta|^{-1/2}+|1+\cos\theta|^{1/2}+|1-\cos\theta|^{1/2}\right\}$$

$$= \csc\theta\left\{|1-\cos\theta|^{-1/2}+|1+\cos\theta|^{-1/2}\right\} = \csc\theta|1-\cos^2\theta|^{-1/2}\left\{|1+\cos\theta|^{1/2}+|1-\cos\theta|^{1/2}\right\}$$

$$= \csc^2\theta\left\{|1-\cos\theta|^{1/2}+|1+\cos\theta|^{1/2}\right\}. \tag{6.339}$$

in agreement with (6.339) ≡ (6.335c).

6.9 Two Spheres Moving Orthogonal to the Line of Centers (Stokes)

The potential flow past two cylinders (spheres) in arbitrary motion can be decomposed into the superposition of motions along [Section II.8.2 (Example 10.15)] and orthogonal [Example II.10.10 (Section 6.9)] to the line of centers. The velocity and radii of the spheres may be distinct (Subsection 6.9.1) and two cases arise for close (distant) cylinders/spheres such that the distance between the centers is not (Subsection 6.9.1) [is (Subsections 6.9.2 and 6.9.3)] large compared with the radii, when an infinite set (just a few) images are needed. The kinetic energy and added mass can be determined by the same method (Subsection 6.9.4), and the algebra is less cumbersome in the case when the distance between the centers of the spheres is large compared with the radii; in the latter case a perturbation method (Subsection 6.9.2) can be used. The case of identical cylinders/spheres moving at the same speed along (orthogonal to) the line of centers is equivalent to a single cylinder/sphere moving orthogonal (parallel) to a wall at equal distance from the two cylinders/spheres [Subsections II.8.2.6 and 7/Example 10.15.4 (Example II.10.4/Subsection 6.9.5)]. This allows a comparison of added masses (Subsection 6.9.6) and dipole moments (Subsection 6.9.7) for two- and three-dimensional bodies, namely, a cylinder (sphere): (1) in a free stream; (2) in a cylindrical (spherical) cavity; and (3) near a wall. In all cases, the blockage effect (Subsection 6.9.8) is larger for a two- than for a three-dimensional obstacle. The added mass of a cylinder moving in a cylindrical cavity is the only result not obtained before (Subsection 6.9.9).

6.9.1 Infinite Set of Pairs of Images in Two Spheres

Consider (Figure II.8.2) two spheres of radii (a,b) with centers at a distance $c > a + b$ that may or not be large compared with (a,b); the second sphere theorem for the potential (Subsection 6.7.2), which is not restricted to axisymmetric flow, leads (Figure II.8.3) to an infinite set of pairs of images (Table 6.2), as follows: (1) the first sphere, with radius a and center at the origin (6.340b), is represented for unit stream velocity (6.340a) by a dipole (6.200a) with moment (6.340c) at the origin (6.340b):

$$U = 1, R_0 = 0: \quad P_0 = -2\pi a^3; \quad R_1 = c - \frac{b^2}{c}: \quad P_1 = -P_0 \left(\frac{b}{c}\right)^3; \tag{6.340a--e}$$

(2) the image on the second sphere is located at (6.340d) and is a dipole with moment (6.340e); (3) the next image on the first sphere is located at (6.341a) and is a dipole of moment (6.341b):

$$n = 1, 2, \ldots: \quad R_{2n} = \frac{a^2}{R_{2n-1}}, \quad P_{2n} = -P_{2n-1}\left(\frac{a}{R_{2n-1}}\right)^3, \tag{6.341a,b}$$

TABLE 6.2 Images on Two Spheres

	Sphere		
First Radius a		Second Radius b	
Location of Image	Dipole Moment	Location of Image	Dipole Moment
$R_0 = 0$	$P_0 = -2\pi a^3$	$R_1 = c - \dfrac{b^2}{c}$	$P_1 = -P_0\left(\dfrac{b}{c}\right)^3$
$R_2 = \dfrac{a^2}{R_1}$	$P_2 = -P_1\left(\dfrac{a}{R_1}\right)^3$	$R_3 = c - \dfrac{b^2}{c - R_2}$	$P_3 = -P_2\left[\left(\dfrac{b}{c - R_2}\right)\right]^3$
$R_4 = \dfrac{a^2}{R_3}$	$P_4 = -P_3\left(\dfrac{a}{R_3}\right)^3$	$R_5 = c - \dfrac{b^2}{c - R_4}$	$P_5 = -P_4\left[\left(\dfrac{b}{c - R_4}\right)\right]^3$
$R_6 = \dfrac{a^2}{R_5}$	$P_6 = -P_5\left(\dfrac{a}{R_5}\right)^3$	$R_7 = c - \dfrac{b^2}{c - R_6}$	$P_7 = -P_6\left[\left(\dfrac{b}{c - R_6}\right)\right]^3$
$R_{2n} = \dfrac{a^2}{R_{2n-1}}$	$P_{2n} = -P_{2n-1}\left(\dfrac{a}{R_{2n-1}}\right)^3$	$R_{2n+1} = c - \dfrac{b^2}{c - R_{2n}}$	$P_{2n+1} = -P_{2n}\left[\left(\dfrac{b}{c - R_{2n}}\right)\right]^3$

Note: Location of images (Figure II.8.3) and magnitude of dipole moments [Table II.8.1 (Table 6.1)] for two cylinders (spheres) in a potential flow [Section II.8.2 (Section 6.9)].

and likewise for all images of even order; and (4) proceeding in the same way the odd images are located at (6.342a):

$$n = 1, 2, \ldots : \quad R_{2n+1} = c - \frac{b^2}{c - R_{2n}}, \quad P_{2n+1} = -P_{2n}\left(\frac{b}{c - R_{2n}}\right)^3.$$

(6.342a,b)

and correspond to a dipole with moment (6.342b). The exact unit potential (6.340a) is the sum of the dipole (6.176b) contributions due to all pairs of images (6.343a):

$$\Phi^{(1)}(R, \theta) = -\frac{1}{4\pi} \sum_{n=0}^{\infty} \frac{P_n}{(D_n)^2} \cos\theta, \quad \cos\theta_n = \frac{R\cos\theta - R_n}{D_n},$$

(6.343a,b)

where: (1) the distance of an arbitrary point (x, y, z) to the position $(0, 0, R_n)$ of the nth dipole image is

$$D_n = \left|x^2 + y^2 + (z - R_n)^2\right|^{1/2} = \left|x^2 + y^2 + z^2 + R_n^2 - 2zR_n\right|^{1/2} = \left|R^2 + R_n^2 - 2R_n R\cos\theta\right|^{1/2};$$

(6.344)

and (2) the distance (6.344) appears in the cosine of the angle (6.343b) of the position vector of the observer with the dipole axis that are all parallel. Substituting (6.340a; 6.341a and b; 6.342a and b) in (6.343a) specifies *the exact unit potential for a free stream of unit velocity (6.340a) of two spheres of radii a and b with centers at a distance c:*

$$\Phi^{(1)}(R, \theta) = -\frac{1}{4\pi} \sum_{n=0}^{\infty} \frac{P_n}{D_n^2} \cos\theta_n = -\frac{P_0 \cos\theta}{4\pi R^2} - \frac{1}{4\pi} \sum_{n=1}^{\infty} \frac{P_n}{D_n^2} \cos\theta_n$$

$$= \frac{a^3}{2R^2}\left\{\cos\theta + \frac{R^2}{P_0} \sum_{n=1}^{\infty}\left[\frac{P_{2n}}{(D_{2n})^2}\cos\theta_{2n} + \frac{P_{2n-1}}{(D_{2n-1})^2}\cos\theta_{2n-1}\right]\right\},$$

(6.345)

where: (1) the images of the second (first) sphere in the first (second) sphere are located (6.341a) [(6.342a)] at $(0; 0, R_{2n})[(0; 0, R_{2n+1})]$; (2) the distance from an arbitrary point is given by (6.344) and the angle with the dipole axis by (6.343b); (3) the dipole moment for even (6.341b) [odd (6.342b)] order images decay as

$$\frac{P_{2n}}{P_0} = \prod_{m=1}^{n}\left(\frac{a}{R_{2m-1}}\right)^3\left(\frac{b}{c - R_{2m-2}}\right)^3 \sim O\left(\left(\frac{ab}{c^2}\right)^{3n}\right) \sim \frac{P_{2n+1}}{P_1};$$

(6.346a,b)

thus (4) the series (6.345) for the potential converges (6.347a) with an upper bound (6.337b):

$$a + b \le c: \quad \left|\Phi^{(1)}(R, \theta)\right| \le \frac{a^3}{2R^2} \sum_{n=0}^{\infty}\left(\frac{ab}{c^2}\right)^n = \frac{a^3}{2R^2}\left[1 - \left(\frac{ab}{c^2}\right)^3\right]^{-1};$$

(6.347a,b)

(5) the convergence is more rapid the larger the distance between the centers compared with the radii of the sphere; and (6) the unit potentials of the first (6.348a) and second (6.348b) sphere are related by the transformation:

$$\left(\Phi^{(1)}, R, a, b, c\right) \leftrightarrow \left(\Phi^{(2)}, S, b, a, c\right),$$

(6.348a,b)

where

$$S = \left| x^2 + y^2 + (z-c)^2 \right|^{1/2} = \left| R^2 + c^2 - 2cR\cos\theta \right|^{1/2}; \qquad (6.349)$$

the distance of an arbitrary point from the center of the first (second) sphere is R(S). Just the first few terms of the series will be needed for distant spheres $c^2 \gg a^2, b^2$, in the **method of images**. A similar conclusion is reached by an alternative **method of perturbation potentials** used next (Subsection 6.9.2).

6.9.2 Isolated and Perturbation Potentials for Each Sphere

Consider (Figure 6.16a) two spheres of radii $a(b)$ moving with velocity $U(V)$ orthogonal to the line of centers. The potential satisfies the Laplace equation (6.233a and b) and rigid surface boundary conditions as in Subsection II.8.2.2; the total potential can be represented as a linear combination

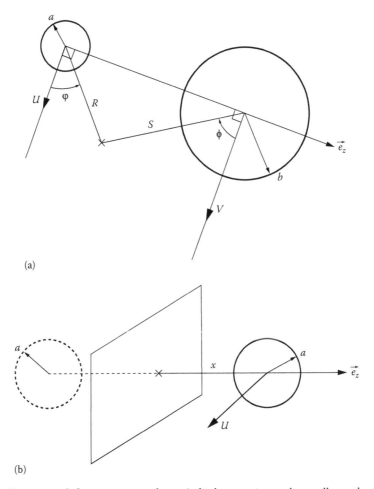

(a)

(b)

FIGURE 6.16 The potential flow past two spheres (cylinders moving orthogonally to the line of centers [Figure 6.16a (Figure II.10.1a)] includes as a particular case, when the velocities and radii are equal, the case of a sphere (cylinder) moving parallel to a wall [Figure 6.16b (Figure II.10.1b)] at equal distance from the spheres.

(6.350a) ≡ (I.8.42a) of unit potentials, each due to one sphere moving with unit velocity and the other at rest (6.350b through e) ≡ (II.8.42b through e):

$$\Phi = U\Phi^{(1)} + V\Phi^{(2)}: \quad -\frac{\partial \Phi^{(1,2)}}{\partial R}\bigg|_{R=a} = \{\cos\varphi, 0\}, \quad -\frac{\partial \Phi^{(1,2)}}{\partial S}\bigg|_{s=b} = \{0, \cos\phi\}. \tag{6.350a-e}$$

The unit potential of the first sphere consists (6.351a) of (1) the potential (6.351b) due to the first sphere in free space in the absence of the second sphere, which coincides with the second term on the r.h.s. of (6.201a), and satisfies the boundary condition (6.350b) on the first sphere; and (2) the second term is the perturbation potential due to the second sphere, which must satisfy the boundary condition (6.350d) ≡ (6.351c) on the second sphere, without disturbing to lowest order the boundary condition at the first sphere:

$$\Phi^{(1)} = \Phi_{10} + \Phi_{12}, \quad \Phi_{10} = \frac{a^3}{2R^2}\cos\varphi, \quad \frac{\partial \Phi_{12}}{\partial S}\bigg|_{S=b} = -\frac{\partial \Phi_{10}}{\partial S}\bigg|_{S=b}. \tag{6.351a-c}$$

The geometrical relation (6.352a) applies (Figure 6.16a) to the two spheres moving orthogonal to the line of centers, with the angles (φ, ϕ) measured from their velocities (U, V); the geometrical relation would be different for spheres moving along the line of centers (Example 10.15 and Figure 10.13). Using the relation (6.352a) the free potential of the first sphere (6.351b) is approximated near the second sphere by (6.352b):

$$R\cos\varphi = S\cos\phi: \quad \Phi_{10} = \frac{a^3}{2R^3}R\cos\varphi = \frac{a^3}{2R^3}S\cos\phi \sim \frac{a^3}{2c^3}S\cos\phi. \tag{6.352a,b}$$

Substitution of (6.352b) in (6.351c) leads to the boundary condition (6.353a):

$$\frac{\partial \Phi_{12}}{\partial S}\bigg|_{S=b} = -\frac{a^2}{2c^3}\cos\phi, \quad \Phi_{12} = \frac{a^3}{2c^3}\frac{b^3}{2S^2}\cos\phi, \tag{6.353a,b}$$

which is satisfied by the perturbation potential (6.353b). This specifies the complete potential of the first sphere (6.351a) ≡ (6.354a), adding to (6.351b) the perturbation due to the second sphere (6.353b); thus follows the first potential (6.354a) on the first sphere (6.354b) using the approximation (6.357a):

$$\Phi^{(1)} = \frac{a^3}{2R^2}\cos\varphi + \frac{a^3 b^3}{4c^3 S^2}\cos\phi, \quad \Phi^{(1)}(R=a) = \frac{a}{2}\cos\varphi; \tag{6.354a,b}$$

$$\Phi^{(1)} = \frac{a^3}{2c^3}S\cos\phi + \frac{a^3 b^3}{4c^3 S^2}\cos\phi, \quad \Phi^{(1)}(S=b) = \frac{3a^3 b}{4c^3}\cos\phi, \tag{6.355a,b}$$

using (6.352a) specifies the first potential near the second sphere (6.355a), and the value on its surface (6.355b).

6.9.3 Values of the Total Potentials on the Two Spheres

The second potential is obtained (6.348a and b) from the first (6.354a) interchanging (6.356a) and leading to (6.356b):

$$(a, R, \varphi) \leftrightarrow (b, S, \phi): \quad \Phi^{(2)} = \frac{b^3}{2S^2}\cos\phi + \frac{a^3 b^3}{4c^3 R^2}\cos\varphi; \tag{6.356a,b}$$

the value (6.357c) on the second sphere holds with the approximation (6.357b):

$$a^2 b^3, a^3 b^2 \ll 2c^5: \quad \Phi^{(2)}(S = b) = \frac{b}{2}\cos\phi; \tag{6.357a–c}$$

using (6.352a) the second potential (6.356b) becomes (6.358a) near the first sphere, and takes the value (6.358b) on its surface:

$$\Phi^{(2)} = \frac{b^3}{2c^3}R\cos\varphi + \frac{a^3 b^3}{4c^3 R^2}\cos\varphi, \quad \Phi^{(2)}(R = a) = \frac{3b^3 a}{4c^3}\cos\varphi. \tag{6.358a,b}$$

The passage from (6.354a) [(6.356b)] to (6.354b) [(6.357c)] involves the approximation (6.357a) [(6.357b)] that indicates the accuracy of stopping the perturbation method at the first iteration. Thus the *total potential of two spheres of radii a(b) moving with velocities U(V) orthogonal to the line of centers (Figure 6.16a) at a distance c is the linear combination (6.350a) unit potentials due to the first (6.354a) [second (6.356b)] sphere; these take the values (6.354b) [(6.358b)] on the first sphere and (6.355b) [(6.357c)] on the second sphere. These results hold with first-order accuracy (6.357a and b) in the method of perturbation potentials.*

6.9.4 Kinetic Energy and Added Mass

The kinetic energy for the potential flow past two spheres (II.8.62) $\equiv$ (6.359) is specified as for two cylinders (Subsections II.8.2.4 and II.8.2.5):

$$E_v = \frac{1}{2}\rho\left(I_1 U^2 + I_2 V^2 + 2I_3 UV\right), \tag{6.359}$$

involving the integrals (II.8.63a and b) $\equiv$ (6.360a and b) and (II.8.67a and b) $\equiv$ (6.360c):

$$I_1 = -\int_{R=a}\Phi^{(1)}\frac{\partial\Phi^{(1)}}{\partial n}dA = 2\pi a^2\int_0^\pi \Phi^{(1)}\cos\varphi\sin\varphi\,d\varphi$$

$$= \pi a^3\int_0^\pi \cos^2\varphi\sin\varphi\,d\varphi = \frac{2\pi a^3}{3}, \tag{6.360a}$$

$$I_2 = -\int_{S=b}\Phi^{(2)}\frac{\partial\Phi^{(2)}}{\partial n}dA = 2\pi b^2\int_0^\pi \Phi^{(2)}\cos\phi\sin\phi\,d\phi$$

$$= \pi b^3\int_0^\pi \cos^2\phi\sin\phi\,d\phi = \frac{2\pi b^3}{3}, \tag{6.360b}$$

$$I_3 = -\int_{R=a}\Phi^{(2)}\frac{\partial\Phi^{(1)}}{\partial n}dA = 2\pi a^2\int_0^\pi \Phi^{(2)}\cos\varphi\sin\varphi\,d\varphi$$

$$= -\int_{S=b}\Phi^{(2)}\frac{\partial\Phi^2}{\partial n}dA = 2\pi b^2\int_{S=b}\Phi^{(1)}\cos\phi\sin\phi\,d\phi$$

$$= 2\pi a^2\frac{3b^3 a}{4c^3}\int_0^\pi \cos^2\varphi\sin\varphi\,d\varphi = \frac{\pi a^3 b^3}{c^3}, \tag{6.360c}$$

where integration of an axisymmetric function over the surface of the sphere of radius a corresponds to

$$\int_{R=a} \Psi(R,\theta)dA = a^2 \int_0^{2\pi} d\phi \int_0^\pi d\theta \sin\theta \Psi(a,\theta) = 2\pi a^2 \int_0^\pi \Psi(a,\phi)\sin\phi\, d\phi. \qquad (6.360d)$$

Substituting (6.360a through c) in (6.359) it follows *the potential flow due to two spheres of radii a(b) moving at velocity U(V) orthogonal to the line of centers (Figure 6.16a) has kinetic energy:*

$$E_v = \frac{\rho}{2}\left(I_1 U^2 + I_2 V^2\right) + \rho I_3 UV = \frac{\rho\pi}{3}\left(U^2 a^3 + V^2 b^3 + 3UV\frac{a^3 b^3}{c^3}\right), \qquad (6.361)$$

that is, the sum of the kinetic energies associated with: (1) the added volume of fluid entrained by each sphere in isolation (6.360a and b), which equals one half of its volume (6.362a and b):

$$V_1 = \frac{1}{2}V_a = \frac{2\pi a^3}{3}, \quad V_2 = \frac{1}{2}V_b = \frac{2\pi b^3}{3}; \qquad (6.362a,b)$$

and (2) the added volume due to the interaction of the two spheres (6.360c) $\equiv$ (6.362c) that is proportional:

$$V_c = \frac{2\pi a^3 b^3}{c^3} = \frac{9}{8\pi c^3}V_a V_b = \frac{9}{2\pi c^3}V_1 V_2, \qquad (6.362c\text{–}e)$$

to 9/8π(9/2π) times the product of their volumes (6.362d) [displaced volumes (6.362e)] divided by the cube of the distance between the centers.

6.9.5 Sphere Moving Parallel to a Wall

The case of spheres with the same radius (6.363a) moving at the same velocity (6.363b) orthogonal to the line of centers is equivalent to images (Figure 6.16b) on a plane at half distance (6.363c); it leads to (6.363d):

$$a = b, U = V, x = \frac{c}{2}: \quad \bar{E}_v = \frac{1}{2}E_v = \frac{\rho\pi}{3}U^2 a^3\left[1 + \frac{3}{2}\left(\frac{a}{2x}\right)^3\right], \qquad (6.363a\text{–}d)$$

considering half the kinetic energy for the fluid on one side of the wall. Thus *the potential flow of a sphere of radius a moving at velocity U parallel to a wall at a distance x has a kinetic energy (6.363d) $\equiv$ (6.364a):*

$$\bar{E}_v = \frac{\bar{m}}{2}U^2, \quad \bar{m}(x) \equiv m\left(1 + \frac{3a^3}{16x^3}\right), \quad m \equiv \frac{2\pi}{3}a^3\rho = \frac{1}{2}\rho V, \qquad (6.364a\text{–}c)$$

where (6.364c) is the displaced mass and corresponds to one-half the volume of the sphere, as before (6.362a); (1) the added mass (6.364b) is increased due to the presence of the wall that is equivalent to an image sphere (6.362c) at a distance x; and (2) the added mass is maximum (6.365a) closest to the wall (6.365b):

$$x \geq a: \quad \left[\bar{m}(x)\right]_{max} = \bar{m}(a) = \frac{19}{16}m = \frac{19}{24}\rho\pi a^3 = \frac{19}{32}\rho V. \qquad (6.365a,b)$$

If the mass of the sphere is m_0, the total kinetic energy of the sphere plus entrained fluid is (6.366):

$$\text{const} = \overline{E} = \frac{1}{2}(m_0 + \overline{m})U^2 = \left[\frac{m_0}{2} + \rho\frac{\pi}{3}a^3\left(1 + \frac{3a^2}{16x^2}\right)\right]U^2, \tag{6.366}$$

corresponding to the mass m_0 of the sphere plus the added mass (6.364b and c). The total energy must be constant, and thus the velocity of the sphere moving parallel to a wall is constant; the constant is maximum (minimum) when the sphere is at infinite distance from (nearly in contact with) the wall, corresponding to the minimum (6.364c) [maximum (6.365b)] added mass:

$$\left|\frac{2\overline{E}}{m_0 + (2/3)\rho\pi a^3}\right|^{1/2} \equiv U(\infty) \geq U(x) \geq U(a) \equiv \left|\frac{2\overline{E}}{m_0 + (19/24)\rho\pi a^3}\right|^{1/2}. \tag{6.367}$$

The value of the velocity of a sphere moving close to a wall U(a) is reasonably accurate, since it was calculated by the potential perturbation to lowest order (6.357a and b) with an error for $c = 2a = 2b$ of the order $a^3b^2/2c^5 = 1/2^6 = 1/64 = 1.56\%$. In the case of a sphere moving perpendicular to a wall (Example 10.15) the motion with constant total energy is not uniform.

6.9.6 Added Mass and Fluid Entrainment by Bodies

Recalling earlier results concerning *the kinetic energy of potential flows with immersed bodies, and the* **added mass** $m_2(m_3)$ *of fluid entrained by two- (three)-dimensional bodies in the simplest case of a cylinder (sphere) of radius a is:* (1) *in isolation in a free stream of density* ρ *the added mass is (I.28.104a and b)* $\equiv$ (6.368a) [(6.364c) $\equiv$ (6.369a)] *equal to the mass density times the area (6.368b) [times one-half the volume (6.369b)] of the cylinder (sphere):*

$$m_2 = \rho\pi a^2 = \rho A, \quad A = \pi a^2, \tag{6.368a,b}$$

$$m_3 = \rho 2\frac{\pi}{3}a^3 = \frac{11}{2}\rho V, \quad V = \frac{4\pi}{3}a^2; \tag{6.369a,b}$$

(2) *when moving in a cylindrical (spherical) cavity of radius b the added mass is increased by a factor* (6.386b) $\equiv$ (6.370a) [(6.227b) $\equiv$ (6.370b)]:

$$m_{2c} = m_2\frac{b^2 + a^2}{b^2 - a^2}, \quad m_{3c} = m_3\frac{b^3 + 2a^3}{b^3 - a^3}; \tag{6.370a,b}$$

(3) *the added mass when moving at a distance x from a wall is the same for a cylinder* m_2 *(different for a sphere* m_3*) of radius a in parallel (II.8.73b)* $\equiv$ (6.371b) [(6.364b) $\equiv$ (6.372b)] *and in orthogonal (II.10.113c)* $\equiv$ (6.371c) [(6.364b) $\equiv$ (10.218b)$\equiv$] *motion:*

$$x \geq a: \quad m_{2p} = m_{2n} = m_2\left(1 + \frac{a^2}{2x^2}\right) \leq (m_{2p})_{max} = (m_{2n})_{max} = \frac{3}{2}m_2, \tag{6.371a-f}$$

$$x \geq a: \quad m_{3p} = m_3\left(1 + \frac{3a^3}{16x^3}\right) \leq (m_{3p})_{max} = \frac{19}{16}m_3, \tag{6.372a-d}$$

$$m_{3n} = m_3\left(1 + \frac{3a^3}{8x^3}\right) \leq (m_{3n})_{max} = \frac{11}{8}m_3 ; \tag{6.372e-g}$$

(4) the added mass in all cases (6.371c) [6.372b and e)] depends on the distance from the wall, and is maximum (6.371f) [6.372d and g)] for the smallest possible distance of the axis (center) of the cylinder (sphere) from the wall (6.371a) [6.372a)] that equals the radius; (5) for the cylinder m_2 that is infinitely long, the added mass (6.371a through f) is the same for motion parallel m_{2p} or orthogonal m_{2n} to the wall; (6) for the sphere m_3 the wall effect is the double $a^3/8x^3$ in (6.372e through g) in normal motion m_{3n} relative to $a^3/16x^3$ (6.372a through d) in parallel motion m_{3p}. The **added mass** equals the product of the constant mass density for an incompressible flow by the added volume; the added volume is the volume of **entrained fluid** due to the motion of the body. The added mass of the cylinder (sphere) moving in a cylindrical (spherical) cavity (6.370a) [(6.370b)] appears to diverge for equal radii $b \to a$, but in this case there is no flow; it means that for close radii $b \ge a$, the added mass is larger due to the increased wall effect, which also appears in the dipole moments (Subsection 6.9.7). The only added mass not obtained before concerns a cylinder moving in a cylindrical cavity (6.370a) ≡ (6.386b) and is calculated subsequently (Subsection 6.9.9).

6.9.7 Dipole Moment for a Cylinder and a Sphere

The cylinder (sphere) in a two- (three)-dimensional potential flow is represented by a line (point) dipole whose moment $P_1^{(2)} \left[P_1^{(3)} \right]$ is always proportional to the velocity and opposite to it, that is, equal to $-U$ times a factor that is: (1) twice (three-halves) of the area of the cross section of the cylinder (I.28.103b) ≡ (6.373a and b) [the volume of the sphere (6.200a and b) ≡ (6.374a and b)] in a free stream:

$$P_1^{(2)} = -2\pi a^2 U = -2AU, \quad A \equiv \pi a^2, \tag{6.373a,b}$$

$$P_1^{(3)} = -2\pi a^3 U = -\frac{3}{2}VU, \quad V \equiv \frac{4\pi}{3}a^3; \tag{6.374a,b}$$

and (2) moving in a cylindrical (spherical) cavity of radius b it is increased by a factor (II.8.5b) ≡ (6.375b) [(6.212a) ≡ (6.376b)]:

$$b \gg a: \quad P_{1w}^{(2)} = P_1^{(2)} \frac{b^2}{b^2 - a^2} = \frac{P_1^{(2)}}{1 - a^2/b^2} = P_1 \left[1 + \frac{a^2}{b^2} + O\left(\frac{a^4}{b^4}\right) \right], \tag{6.375a–c}$$

$$b \gg a: \quad P_{1w}^{(3)} = P_1^{(3)} \frac{b^3}{b^3 - a^3} = \frac{P_1^{(3)}}{1 - a^3/b^3} = P_1 \left[1 + \frac{a^3}{b^3} + O\left(\frac{a^6}{b^6}\right) \right]. \tag{6.376a–c}$$

*The comparison of the dipole moments is made dividing by minus the velocity times the area of the cross section (the volume) for the cylinder (sphere) both in free space (6.373a and b) [(6.374a and b)] and in a cylindrical (spherical) cavity (6.375a through c) [6.376a through c)]. The radius of the cavity should be much larger than that of the body (6.375a) [(6.376a)] and the **wall effect** increases the dipole moment more in the case of the cylinder (6.375c) than in the case of the sphere (6.376c). This is an example of the blockage effect considered next (Subsection 6.9.8).*

6.9.8 Flow Blockage in Two and Three Dimensions

The preceding comparisons show that *the coefficients are always larger for the cylinder than for the sphere for both (1) the added mass in free space (6.368a; 6.369a) ≡ (6.377a) in a cavity (6.370a and b) ≡ (6.377b and c) and for motion parallel (6.372b) ≡ (6.377d) [orthogonal (6.372e) ≡ (6.377g)] to a wall including the maximum value (6.377f) [(6.377h)]:*

$$\frac{m_2/A}{m_3/V} = 2; \quad b > a: \quad \frac{m_{2c}/m_2}{m_{3c}/m_3} = \frac{b^2 + a^2}{b^3 + 2a^3} \frac{b^3 - a^3}{b^2 - a^2} > 1, \tag{6.377a–c}$$

$$x \geq a: \quad \frac{m_{2p}/m_2}{m_{3p}/m_3} = \frac{1 + a^2/2x^2}{1 + 3a^2/16x^2} > 1, \qquad \frac{\left(m_{2p}/m_2\right)_{max}}{\left(m_{3p}/m_2\right)_{max}} = \frac{24}{19}, \qquad \text{(6.377d–f)}$$

$$\frac{m_{2n}/m_2}{m_{3n}/m_3} = \frac{1 + a^2/2x^2}{1 + 3a^2/8x^2} > 1, \qquad \frac{\left(m_{2n}/m_2\right)_{max}}{\left(m_{3n}/m_3\right)_{max}} = \frac{12}{11}, \qquad \text{(6.377g–h)}$$

and (2) the dipole moment in free space (6.373a; 6.374a) ≡ (6.378a) and in a cavity (6.375b; 6.376b):

$$\frac{P_1^{(2)}/A}{P_1^{(3)}/V} = \frac{4}{3}; \quad b > a: \quad \frac{P_{1w}^{(2)}/P_1^{(2)}}{P_{1w}^{(3)}/P_1^{(3)}} = \frac{b^3 - a^3}{b\left(b^2 - a^2\right)} > 1. \qquad \text{(6.378a–c)}$$

The implication is that a two- (three-) dimensional obstacle causes a greater (smaller) **blockage effect** *because the flow can less (more) easily pass around it. The ratios (6.377a) (6.377e and g) and (6.378a) are all larger than unity and the same applies to the other three inequalities as shown next: (1) the inequality (6.378c) follows from (6.378b) using (6.379a through d):*

$$b > a: \quad \frac{a^2}{b^2} > \frac{a^3}{b^3}, \quad 1 - \frac{a^2}{b^2} < 1 - \frac{a^3}{b^3}, \quad \frac{b^3 - a^3}{b\left(b^2 - a^2\right)} > 1; \qquad \text{(6.379a–d)}$$

(2) and the inequality (6.377d) is immediate 1/2 > 3/16; (3) the inequality (6.377c) is equivalent to

$$0 < \left(b^2 + a^2\right)\left(b^3 - a^3\right) - \left(b^2 - a^2\right)\left(b^3 + 2a^3\right)$$
$$= 2a^2b^3 - 3a^3b^2 + a^5 = 3a^2b^2\left(b - a\right) - a^2\left(b^3 - a^3\right) \equiv X; \qquad \text{(6.380)}$$

(4) the two terms on the r.h.s. of (6.380) have a common root (6.381a):

$$b^3 - a^3 = \left(b - a\right)\left(b^2 + ab + a^2\right), \qquad \text{(6.381a)}$$

leading to

$$b > a: \quad X = \left(b - a\right)\left[3a^2b^2 - a^2\left(b^2 + ab + a^2\right)\right] = a^2\left(b - a\right)\left(2b^2 - ab - a^2\right)$$
$$= a^2\left(b - a\right)\left[b\left(b - a\right) + b^2 - a^2\right] = a^2\left(b - a\right)^2\left(2b + a\right) > 0; \qquad \text{(6.381b,c)}$$

and (5) the condition (6.377b) ≡ (6.381b) implies that (6.381c) is positive, and thus proves (6.380) ≡ (6.377c). Of all the results quoted: (1) the dipole moments (6.373a and b; 6.374a and b; 6.375a through c; 6.376a through c) had all been obtained before (I.28.103b; 6.200a and b; II.8.5b; 6.212a); and (2) the added masses (6.368a and b; 6.369a and b; 6.370b; 6.371a and b; 8.372a and b) had also been obtained before (I.28.104a and b; 6.364c; 6.319b; II.8.73b ≡ II.10.113c; 10.218a) except for (6.370a) ≡ (6.386b) for a cylinder moving in a cylindrical cavity, which is obtained next (Subsection 6.9.9).

6.9.9 Added Mass of a Cylinder Moving in a Cylindrical Cavity

A cylinder with radius a moving with velocity U in a cylindrical cavity with radius b has the potential (II.8.8a) ≡ (6.382b) without circulation (6.382a):

$$\Gamma = 0: \quad \Phi\left(r, \varphi\right) = \frac{Ua^2}{b^2 - a^2}\left(r + \frac{b^2}{r}\right)\cos\varphi, \qquad \text{(6.382a,b)}$$

that (1) is a solution of the Laplace equation in polar coordinates, that is, (6.45b) without z-dependence; and (2, 3) meets the boundary condition concerning the normal velocity (6.383a) that vanishes on the fixed cavity (6.383b) of radius b and equals (6.383c) on the cylinder of radius a moving with velocity U:

$$v_r(r,\varphi) = \frac{\partial \Phi}{\partial r} = \frac{Ua^2}{b^2 - a^2}\left(1 - \frac{b^2}{r^2}\right)\cos\varphi, \quad v_r(b,\varphi) = 0, \quad v_r(a,\varphi) = -U\cos\varphi. \tag{6.383a–c}$$

The kinetic energy of the potential flow between the inner moving and outer fixed cylinder is given (Subsection 6.5.4) by the integral (6.225) over the circular perimeter of the cylinder (instead of the surface of the sphere):

$$E_v = \frac{1}{2}\rho Ua \int_0^{2\pi} \Phi(a,\varphi)\cos\varphi\, d\varphi, \tag{6.384}$$

where the potential (6.382b) may be substituted:

$$E_v = \frac{1}{2}\rho U^2 a^2 \frac{b^2 + a^2}{b^2 - a^2}\int_0^{2\pi}\cos^2\varphi\, d\varphi = \frac{\pi}{2}\rho U^2 a^2 \frac{b^2 + a^2}{b^2 - a^2}. \tag{6.385}$$

This corresponds (6.385) $\equiv$ (6.386a) to the added mass (6.386b):

$$E_v = \frac{1}{2}mU^2, \quad m = \rho U^2 \pi a^2 \frac{b^2 + a^2}{b^2 - a^2}, \tag{6.386a,b}$$

in agreement with (6.386b) $\equiv$ (6.368a; 6.370a).

NOTE 6.1: Potential Flows in Two, Two-and-Half, and Three Dimensions

The plane flow (Chapters I.12, 14, 16, 28, 34, 36, 38; II.2, 8) corresponds in three dimensions to the same flow in parallel planes, so that the flow past a closed curve or loop C becomes the flow orthogonal to the generators of a cylinder with directrix C. If instead the closed loop C is rotated around a straight line crossing it the result is a three-dimensional flow past an axisymmetric body for which the velocity lies in planes through the rotation axis (Chapter 6). The axisymmetric flow is sometimes designated two-and-a-half dimensional instead because (1) it retains rotational instead of translational symmetry of the plane flow; (2) it is a particular case of a general three-dimensional flow that may have no symmetries. The distinction between plane (2-D) and axisymmetric (2½-D) and nonaxisymmetric (3-D) flows applies to any flow such as (Classification 6.1) potential, irrotational, incompressible, or viscous. Most two-dimensional flow problems have axisymmetric analogues and three-dimensional extensions. The uniform flow is common to all cases whereas there are distinct cylindrical (axisymmetric) Rankine fairings [Section I.28.4 (Subsections 6.3.4 and 6.3.5)] and oval bodies [Section I.28.5 (Subsections 6.4.1 and 6.4.2)]. A limiting case of the Rankine body is the potential flow past a cylinder (sphere) in free space [Section I.28.6 (Subsections 6.5.1 and 6.5.2)] or inside a cylindrical (spherical) cavity [Section II.8.1 (Subsections 6.5.3 and 6.5.4)]. Some two-dimensional and three-dimensional axisymmetric and nonaxisymmetric and multidimensional fields are indicated in Classification 6.1, including fields that are irrotational or solenoidal or neither. The case of two cylinders (spheres) moving along the line of centers [Section II.8.2 (Example 10.15)] is axisymmetric in the spatial case, through not when moving perpendicular to the line of centers [Example II.10.10 (Section 6.9)]. The source sink, monopole, dipole, quadrupole, and higher-order multipoles can be considered in the plane (Chapter I.12), in axisymmetric and nonaxisymmetric configurations in space (Section 6.4; Chapter 8), and also in higher dimensions (Chapter 9). The extension of the cylindrical or line vortex (Section I.12.5) to the spherical

CLASSIFICATION 6.1 Comparison of Plane, Axisymmetric, and Nonaxisymmetric Flows

Dimensions	$N = 2$	$N = 3$: Space	$N = 3$: Space	$N \geq 4$
Flow	Plane	Axisymmetric	Nonaxisymmetric	Higher Dimensional
General				
Potential	I.12, 14, 16, 28, 34, 36, 38; II.2, 8	6	6	9
Irrotational compressible	I.14.6, 7; II.2.1	—	—	N9.3, 6, 7, 10 through 12, 16, 18, 21, 23, 25 through 52
Incompressible rotational	II.2.2 through 7; II.8.9	N.6.2, 3, 5, 7 through 10	N.6.4	N9.4, 5, 8, 10, 12, 19, 22, 24, 26, through 52
Viscous	N.II.4.11; N.6.10	N.6.5	N.II.4.4 through 10; N.6.6 and 7	N9.1, 3, 14, 15, 17, 18, 20
Multipoles				
Source/sink	I.12.4	6.3.3	—	9.7.4
Vortex	I.12.5	6.6	—	—
Monopole	I.12.6	—	—	—
Dipole	I.12.7	6.4.3 and 6.4.5	6.4.6 through 6.4.8	9.7.5
Quadrupole	I.12.8	6.4.4 and 6.4.5	6.4.6 through 6.4.8	9.7.6
Multipole	I.12.9	6.4.5	6.4.7	9.7.7 through 9.7.8
Basic				
Uniform	I.14.8	6.3.2	—	9.8.4
Fairing	I.28.4	6.3.4 and 6.3.5	—	—
Oval	I.28.5	6.4.1 and 6.4.2	—	—
Cylinder(s) or sphere(s)				
In a stream	I.28.6	6.5.1 and 6.5.2	—	—
In a cavity	II.8.1	6.5.3 and 6.5.4	—	—
Moving parallel to a wall	E.II.10.1 through 3	—	6.9.5	—
Moving orthogonally to a wall	II.8.2.6 and 7	E.10.15.4	—	—
Moving orthogonal to the line of centers	E.II.10.4	E.10.15.4	6.9.1 through 6.9.4	—
Moving along the line of centers	II.8.2.1 through 5	E.10.15.1 through 3	—	—

Each topic is indicated where the corresponding subject is addressed with the identification of book, chapter.section.subsection; N, note; E, example.

Note: Comparison on plane (spatial axisymmetric and nonaxisymmetric) potential flows, including a number of analogous problems like flow past cylinder (sphere), Rankine cylindrical (axisymmetric) fairing, and bodies and multipoles.

vortex (Section 6.6) is distinct in that it involves passing from a potential to an incompressible rotational flow. This is one aspect of the difference between irrotational and solenoidal fields when extended from two (Chapter I.12) to three (Chapters 6 and 8) and higher (Chapter 9) dimensions.

NOTE 6.2: Mathematical Methods for Potential Fields

The simplest and most effective approach to potential fields in the plane is the use of analytic functions of a complex variable (Chapters I.11 through 40; II.1 through 10), specifying both the scalar potential and field function. The use of generalized functions (Chapters 1, 3, and 5) generalizes the potential fields (Chapters 6, 8, and 9) from the plane ($N = 2$) to space ($N = 3$) and higher dimensions ($N \geq 4$). In all dimensions the irrotational field is specified by a scalar potential, namely, (1) in (Chapter I.12) the plane ($N = 2$); (2) in (Chapters 6 and 8) space ($N = 3$); and (3) in any (Chapter 9) dimension ($N \geq 4$). The solenoidal field is specified by: (1) a scalar field function in the plane (Chapters I.12 and II.2) and for axisymmetric problems in space (Chapters 6 and 8); (2) a vector potential for nonaxisymmetric problems in space (Chapters 6 and 8); and (3) solenoidal fields and potentials in higher dimension ($N \geq 4$) require multivectors (Notes 9.1 through 9.52) and lead to the tensor calculus. The plane, axisymmetric, and spatial nonaxisymmetric potentials are summarized in Table 6.3.

NOTE 6.3: Existence of Vector and Scalar Potentials

Since mostly potential flows have been considered so far, the classification of flows in Diagram 6.2 is preliminary. Two independent criteria are: (1) incompressible (compressible) flow if the dilatation, that is, divergence of velocity (6.387a) is (is not) zero:

$$\dot{D} \equiv \nabla \cdot \vec{v}: \quad \dot{D} = 0 \Rightarrow \vec{v} = \nabla \wedge \vec{A}, \quad \vec{A} = \vec{e}_z \Psi = \vec{e}_\varphi r^{-1} \Psi; \tag{6.387a–e}$$

and (2) irrotational (rotational) if the vorticity, that is, the curl of the velocity (6.388a) is (is not) zero:

$$\vec{\varpi} = \nabla \wedge \vec{v}: \quad \vec{\varpi} = 0 \Rightarrow \vec{v} = \nabla \Phi. \tag{6.388a–c}$$

The simplest case is the potential flow that is both (1) irrotational (6.388b) so that the velocity is the gradient of a scalar potential (6.388c); and (2) incompressible (6.387b) so that the velocity is the curl of a vector potential (6.387c) that reduces to a scalar stream function for plane flow (6.64a) ≡ (6.387d) and for spatial axisymmetric flow (6.60c) ≡ (6.387e). The next generalizations are: (a) irrotational compressible flow (6.388a and b) for which the velocity derives from a scalar potential (6.388c); (b) a rotational incompressible flow (6.387a and b) for which there exists a vector potential in three dimensions (6.387c) and a stream function

TABLE 6.3   Plane, Axisymmetric, and Nonaxisymmetric Potentials

Field		Irrotational	Solenoidal
Condition		$\nabla \wedge \vec{v} = 0$	$\nabla \cdot \vec{v} = 0$
Relation with potential		$\vec{v} = \nabla \Phi$	$\vec{v} = \nabla \wedge \vec{A}$
Plane: coordinates	Cartesian (x, y)	$\Phi(x, y) = \mathrm{Re}\{f(x + iy)\}$	$\Psi(x, y) = \mathrm{Im}\{f(x + iy)\}$
	Polar (r, φ)	$\Phi(r, \varphi) = \mathrm{Re}\{f(re^{i\varphi})\}$	$\Psi(r, \varphi) = \mathrm{Im}\{f(re^{i\varphi})\}$
Axisymmetric coordinates	Cylindrical (r, φ, z)	$\Phi(r, z)$	$\vec{A} = \Psi(r, z)\vec{e}_\varphi r^{-1}$
	Spherical (R, θ, φ)	$\Phi(R, \theta)$	$\vec{A} = \Psi(R, \theta)\vec{e}_\varphi R^{-1} \csc \theta$
Nonaxisymmetric	Cylindrical (r, φ, z)	$\Phi(r, \varphi, z)$	$\vec{A}(r, \varphi, z)$
	Spherical (R, θ, φ)	$\Phi(R, \theta, \varphi)$	$\vec{A}(R, \theta, \varphi)$

Note: Comparison of plane and spatial axisymmetric and nonaxisymmetric fields in terms of scalar, complex, and vector potentials and the stream function.

in the axisymmetric (6.387e) and plane (6.387d) cases. In the general case of a flow neither incompressible nor irrotational the velocity is specified by the sum (5.266a through c) ≡ (6.389b) of the gradient (curl) of a scalar (vector) potential, whose Laplacian specifies the dilatation (5.264a and b) ≡ (6.389c) [minus the vorticity (5.267b) ≡ (6.389d), imposing a gauge condition (5.267b) ≡ (6.389a) on the vector potential]:

$$\nabla \cdot \vec{A} = 0: \quad \vec{v} = \nabla \Phi + \nabla \wedge \vec{A}, \quad D = \nabla \cdot \vec{v} = \nabla^2 \Phi, \quad \vec{\varpi} = \nabla \wedge \vec{v} = -\nabla^2 \vec{A}. \tag{6.389a-d}$$

If the flow is irrotational (incompressible) the scalar (vector) potential is sufficient (6.388a through c) [(6.387a through e)] and if it is potential either can be used.

NOTE 6.4: Two Circle and Four Sphere Theorems

There are two circle theorems for a plane flow: (1) the first circle theorem for a potential flow, that is, irrotational and incompressible, when both a potential and stream function exist (Sections I.24.7 and I.35.7); and (2) the second circle theorem for an incompressible rotational flow (Section II.2.5) for the stream function, since the potential does not generally exist. There are four sphere theorems, all for three-dimensional potential flow so that the potential (stream or field function) exists in all cases axisymmetric or not (only for axisymmetric cases); (1/2) the first (second) sphere theorem for the stream function (potential) of an axisymmetric (general nonaxisymmetric) field so that the sphere is a field surface (equipotential) [Subsections 6.7.1 and 6.8.1 (Subsections 6.7.2 and 6.8.2)] corresponding to an electric insulator (conductor); and (3/4) the third (fourth) sphere theorems apply to the potential of general nonaxisymmetric flow in an integral form [Subsection 6.7.3 (Subsections 6.7.4, 6.8.3 through 6.8.8)] that is suitable for fields not constrained (constrained) to lie on a sphere. The two (four) circle (sphere) theorems appear in List 6.1.

NOTE 6.5: Classification of Flows

The classification of flows (Diagram 6.2) may start with the simplest potential flow at the bottom and end with most complex turbulent flow at the top. An initially irrotational flow will remain irrotational (Section I.14.3) for: (1) conservative external forces derived from a potential; and (2) homentropic flow, that

---

LIST 6.1 Two (Four) Circle (Sphere) Theorems

A. Circle theorems in the plane
  1. Potential and stream function in an irrotational incompressible flow: Sections I.24.7 and I.35.7
  2. Stream function in a rotational incompressible flow: Section II.2.5
B. Sphere theorems in space
  3. Stream function for an irrotational incompressible axisymmetric flow with sphere as a stream surface: Subsections 6.7.1 and 6.8.1
  4. Potential for an irrotational incompressible nonaxisymmetric flow with the sphere as an equipotential surface: Subsections 6.7.2 and 6.8.2
  5. Potential for an irrotational incompressible nonaxisymmetric flow in an integral form suitable for fields not confined to the surface of the sphere: Subsection 6.7.3
  6. Potential for an irrotational incompressible nonaxisymmetric field in an alternate integral form suitable for fields confined to the surface of a sphere: Subsections 6.7.4 and 6.8.3 through 6.8.8

Note: The two (four) circle (sphere) theorems for plane (spatial) flows including: (1) irrotational and/or incompressible flows in the plane and space; (2) axisymmetric and nonaxisymmetric flows in space.

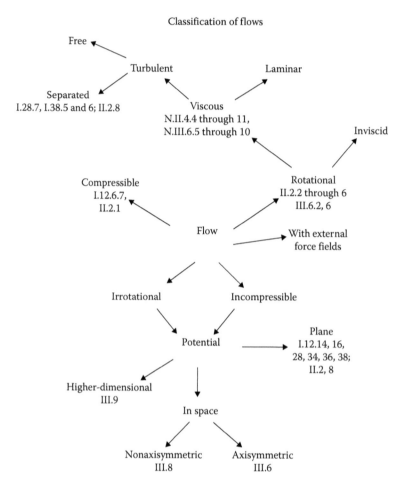

DIAGRAM 6.2 Classification of flows starting at the bottom with the simplest that have received the most attention is this and preceding books and proceeding to the more complex that will be further detailed in subsequent books of the series.

is, constant entropy or absence of heat exchanges. The entropy is not constant for (1) stratified media; and (2) dissipative processes like heat conduction and viscosity. The nonconservative force fields include magnetic forces and the analogous Coriolis force associated with rotation. Thus the rotational flows include: (1) stratified and rotating media, even inviscid and isothermal; and (2) viscosity and/or heat conduction, even in homogeneous media. The condition of incompressible flow is independent of that irrotationality, and the two combined lead to the potential flow. The potential flow, like any other flow, can be considered in: (1) two dimensions in the plane; (2) two-and-a-half dimensions if it is axisymmetric (Note 6.1); and (3) three dimensions if not axisymmetric. A viscous flow is generally rotational (Notes II.4.4 through 11) and will be laminar at low Reynolds number, but becomes turbulent above a critical Reynolds number for instability (Note II.4.10). The transition from laminar to turbulent flow can occur in the wake of a cylinder (Sections I.28.7 and II.2.8) or behind a plate (Section I.38.5) or arrow (I.38.6) or in the boundary layer of an airfoil (Section II.8.5), leading to flow separation. The turbulent flow is distinguished from the laminar flow by its random velocity fluctuations. Since the criterion of compressibility is independent of vorticity, a viscous flow, whether laminar or turbulent, can be compressible or incompressible. Thus an incompressible viscous flow has a vector potential that provides an alternative description to the vorticity (Notes 6.6 and 6.7); in the case of plane incompressible viscous flow the vector potential is replaced by the stream function (Notes II.4.10 and 11). The stream function can also be used for an incompressible viscous axisymmetric flow (Notes 6.8 through 6.10). The general classification of flows (Diagram 6.2) is detailed further for inviscid (viscous) flow in Tables 6.1 through 6.3, List 6.1, and Diagram 6.1 (Table 6.4) corresponding to Sections 6.2 through 6.9 (Notes 6.6 through 6.10).

NOTE 6.6: Vector Potential for an Incompressible Viscous Flow

The momentum equation for a Newtonian viscous fluid is the **Navier–Stokes equation** (II.4.339) ≡ (6.390a):

$$\vec{f} - \nabla p + \eta \nabla^2 \vec{v} + \left(\zeta + \frac{\eta}{3} \right) \nabla (\nabla \cdot \vec{v}) = \rho \vec{a} = \rho \left[\frac{\partial \vec{v}}{\partial t} + \nabla \left(\frac{v^2}{2} \right) + (\nabla \wedge \vec{v}) \wedge \vec{v} \right], \qquad (6.390a,b)$$

where $\vec{v}$ is the velocity, p the pressure, ρ the mass density, ζ, η the shear and bulk static viscosities, $\vec{f}$ an external applied force per unit volume; the acceleration $\vec{a}$ can be decomposed (I.14.19) ≡ (6.390b) ≡ (6.96c) into local, irrotational, and rotational parts.

For an incompressible fluid (6.391a and b):

$$\nabla \cdot \vec{v} = 0 \Leftrightarrow \rho = \text{const}: \quad \frac{\vec{f}}{\rho} - \nabla \left(\frac{p}{\rho} + \frac{1}{2} v^2 \right) + \frac{\eta}{\rho} \nabla^2 \vec{v} = \frac{\partial \vec{v}}{\partial t} + (\nabla \wedge \vec{v}) \wedge \vec{v}; \qquad (6.391a–c)$$

the bulk viscosity ζ does not appear leaving only the shear viscosity in (6.391c); the stagnation pressure (6.102c) ≡ (6.392a) can be introduced in (6.391c) ≡ (6.392b):

$$p_0 \equiv p + \frac{1}{2} \rho v^2: \quad \frac{\vec{f}}{\rho} - \nabla \left(\frac{p_0}{\rho} \right) + \frac{\eta}{\rho} \nabla^2 \vec{v} = \frac{\partial \vec{v}}{\partial t} + (\nabla \wedge \vec{v}) \wedge \vec{v}. \qquad (6.392a,b)$$

For a conservative or irrotational force field (6.393a), and constant shear viscosity (6.393b), the curl of (6.392b) leads (6.65a) to the vorticity equation (6.393c):

$$\nabla \wedge \vec{f} = 0, \eta = \text{const}: \quad \frac{\eta}{\rho} \nabla^2 \vec{\omega} = \frac{\partial \vec{\omega}}{\partial t} + \nabla \wedge (\vec{\omega} \wedge \vec{v}). \qquad (6.393a–c)$$

Since the flow is incompressible (6.391a and b) ≡ (6.387a and b) there is a vector potential (6.387c) that (6.91b) satisfies (6.393c) ≡ (6.394b):

$$\bar{\eta} = \frac{\eta}{\rho}: \quad \bar{\eta} \nabla^4 \vec{A} = \frac{\partial}{\partial t} \left(\nabla^2 \vec{A} \right) + \nabla \wedge \left[\nabla^2 \vec{A} \wedge \left(\nabla \wedge \vec{A} \right) \right], \qquad (6.394a,b)$$

where (6.394a) is the kinematic shear viscosity or **viscous diffusivity** defined as the ratio of the static shear viscosity to the mass density. For **inertial flow** (Note II.4.10) at high Reynolds number (II.4.358c) ≡ (6.395a) the nonlinear inertial forces in the last term on the r.h.s. of (6.394b) are dominant compared with the viscous forces on the l.h.s. of (6.394b) leading to (6.395b):

$$\text{Re} \equiv \frac{\rho v L}{\eta} = \frac{v L}{\bar{\eta}} \gg 1: \quad \frac{\partial}{\partial t} \left(\nabla^2 \vec{A} \right) = \nabla \wedge \left[\left(\nabla \wedge \vec{A} \right) \wedge \nabla^2 \vec{A} \right], \quad \frac{\partial \vec{\omega}}{\partial t} = \nabla \wedge (\vec{v} \wedge \vec{\omega}), \qquad (6.395a–c)$$

and the vorticity equation (6.393c) simplifies to (6.395c) that states the conservation of vorticity along streamlines (II.4.359c), corresponding to the Kelvin circulation theorem (I.14.23a through d).

NOTE 6.7: Diffusion by the Kinematic Shear Viscosity

The vector potential is not uniquely specified by its curl (6.387c) and its divergence may be chosen; the choice of zero divergence (6.396a) ≡ (6.389a) as a gauge condition leads to (6.396b):

$$\nabla \cdot \vec{A} = 0: \quad \nabla \wedge \left(\nabla \wedge \vec{A} \right) = -\nabla^2 \vec{A} + \nabla \left(\nabla \cdot \vec{A} \right) = -\nabla^2 \vec{A}, \tag{6.396a,b}$$

where (6.48a) ≡ (II.4.112) was used. The conditions (6.396a and b) are analogous to (6.397b) for an incompressible (6.397a) flow:

$$\nabla \cdot \vec{v} = 0: \quad \nabla \wedge \vec{\varpi} = \nabla \wedge \left(\nabla \wedge \vec{v} \right) = -\nabla^2 \vec{v}. \tag{6.397a,b}$$

In the case of **creeping flow** corresponding to low Reynolds number (6.398a) the equation for the vorticity (6.393c) [vector potential (6.394b)] becomes linear (6.398b) [(6.398c)]:

$$Re \ll 1: \quad \bar{\eta} \nabla^2 \vec{\varpi} = \frac{\partial \vec{\varpi}}{\partial t}, \quad \bar{\eta} \nabla^4 \vec{A} = \frac{\partial}{\partial t} \left(\nabla^2 \vec{A} \right), \tag{6.398a–c}$$

corresponding to a diffusion equation with the role of diffusivity played by the kinematic shear viscosity (6.394a). In the steady case (6.399a) the vorticity (vector potential) is a harmonic (biharmonic) function, that is, it satisfies the Laplace (6.399b) [biharmonic (6.399c)] vector equation:

$$\frac{\partial}{\partial t} = 0: \quad \nabla^2 \vec{\varpi} = 0, \quad \nabla^4 \vec{A} = 0. \tag{6.399a–c}$$

Thus *a Newtonian viscous (6.390a and b) incompressible (6.391a through c) ≡ (6.392a and b) flow under conservative external forces (6.393a) and with constant static shear viscosity (6.393b) satisfies the non-linear equation (6.393c) [(6.394b)] for the vorticity (vector potential), where the kinematic shear viscosity (6.394a) may be used. In the case of inertial flow at high Reynolds number (6.395a) the equation for the vorticity (vector potential) simplifies to (6.395c) [(6.395b)] and remains nonlinear. In the opposite case of creeping flow at flow Reynolds number (6.398a) the equation for the vorticity (vector potential) becomes linear (6.398b) [(6.398c)] and corresponds to a diffusion equation with the kinematic shear viscosity (6.394a) acting as diffusivity; the kinematic shear viscosity does not appear for steady (6.399a) incompressible (6.391a) Newtonian viscous flow under conservative external forces (6.393a) when the vorticity (vector potential) is a harmonic (6.399b) [biharmonic (6.399c)] vector function. In the case of a plane flow (6.400a) the vorticity (vector potential) reduces to a scalar (6.400b) [stream function (6.387d) ≡ (6.400c)] that is a harmonic (6.400d) [biharmonic (6.400e)] function:*

$$\vec{v} = \vec{e}_x v_x \left(x, y \right) + \vec{e}_y v_y \left(x, y \right): \quad \vec{\varpi} = \vec{e}_z \varpi, \quad \vec{A} = e_z \Psi: \quad \nabla^2 \varpi = 0 = \nabla^4 \Psi; \tag{6.400a–e}$$

the generalization to unsteady plane flow (6.401a) is the diffusion equation for the vorticity (6.401b) [vector potential (6.401c)]:

$$\vec{v} = \vec{e}_x v_x \left(x, y, t \right) + \vec{e}_y v_y \left(x, y, t \right): \quad \frac{\partial \varpi}{\partial t} - \bar{\eta} \nabla^2 \varpi = 0 = \frac{\partial}{\partial t} \left(\nabla^2 \Psi \right) - \bar{\eta} \nabla^4 \Psi, \tag{6.401a–c}$$

with the kinematic shear viscosity (6.394a) as diffusivity. In the case of a three-dimensional axisymmetric flow the Laplace (biharmonic) operator still holds for the vorticity (is replaced by a modified biharmonic operator) as shown next (Notes 6.8 and 6.9).

NOTE 6.8: Stream Function for Axisymmetric Viscous Unsteady Flow

The equation (6.108f) ≡ (6.402a) [(6.65d) ≡ (6.402b)] for the axisymmetric (plane) stream function:

$$\frac{\partial^2 \Psi}{\partial z^2} + r\frac{\partial}{\partial r}\left(r\frac{\partial \Psi}{\partial r}\right) = \overline{\nabla}^2\Psi = -r\varpi, \quad \frac{\partial^2\Psi}{\partial x^2} + \frac{\partial^2\Psi}{\partial y^2} = \nabla^2\Psi = -\varpi, \qquad (6.402a,b)$$

allows an arbitrary choice of the vorticity as a function of the stream function, and involves the modified Laplacian (Laplacian) operator in cylindrical (6.89a) ≡ (6.402a) [Cartesian (6.88a) ≡ (6.402b)] coordinates. It involves four restrictions: (1) incompressible flow (6.387a and b) in order that a vector potential exists (6.387c); (2) axisymmetric flow in order that the vector potential reduces to a stream function (6.387e); and (3/4) steady homentropic flow, thus excluding dissipative processes like viscosity. The restrictions (3) and (4) to steady, inviscid flow are not necessary, as shown in Note II.4.10, where the equations for the stream function of plane flow were obtained including unsteadiness and viscosity (II.4.356a and b) using Cartesian coordinates. In the present note are obtained the corresponding equations for the stream function of axisymmetric incompressible viscous flow using cylindrical coordinates. Using the velocity (6.57a through c) and vorticity (6.403a) for an axisymmetric flow in (6.66a) ≡ (6.403a) of an incompressible fluid (6.403b) under conservative external forces (6.403c) leads (6393c; 6.394a) to (6.403d):

$$\vec{\varpi} = \varpi\vec{e}_\varphi, \rho = \text{const}, \nabla \wedge \vec{f} = 0:$$

$$\overline{\eta}\nabla^2\left(\varpi\vec{e}_\varphi\right) = \vec{e}_\varphi\frac{\partial\varpi}{\partial t} - \nabla \wedge \left\{\frac{\varpi}{r}\vec{e}_\varphi \wedge \left[\vec{e}_r\left(\partial_z\Psi\right) - \vec{e}_z\left(\partial_r\Psi\right)\right]\right\}; \qquad (6.403a\text{--}d)$$

the nonlinear term last on the r.h.s. of (6.403d) is:

$$\nabla \wedge \left\{\frac{\varpi}{r}\left[\vec{e}_z\left(\partial_z\Psi\right) + \vec{e}_r\left(\partial_r\Psi\right)\right]\right\} = \vec{e}_\varphi\left\{\partial_z\left[\frac{\varpi}{r}\left(\partial_r\Psi\right)\right] - \partial_r\left[\frac{\varpi}{r}\left(\partial_z\Psi\right)\right]\right\}$$

$$= -\vec{e}_\varphi\left[\left(\partial_z\Psi\right)\partial_r\left(\frac{\varpi}{r}\right) - \left(\partial_r\Psi\right)\partial_z\left(\frac{\varpi}{r}\right)\right], \qquad (6.404)$$

where the curl in cylindrical coordinates (6.41b) was used. In (6.404) and (6.403d) the vorticity (6.66c) involves the modified Laplacian of the stream function, leading to (6.405):

$$\overline{\eta}\nabla^2\left(\frac{1}{r}\overline{\nabla}^2\Psi\right) = \frac{\partial}{\partial t}\left(\frac{1}{r}\overline{\nabla}^2\Psi\right) - \left(\partial_z\Psi\right)\partial_r\left(\frac{1}{r^2}\overline{\nabla}^2\Psi\right) + \left(\partial_r\Psi\right)\partial_z\left(\frac{1}{r^2}\overline{\nabla}^2\Psi\right); \qquad (6.405)$$

$$\overline{\eta}\nabla^2\Psi = \frac{\partial}{\partial t}\left(\nabla^2\Psi\right) - \left(\partial_x\Psi\right)\partial_y\left(\nabla^2\Psi\right) + \left(\partial_y\Psi\right)\partial_x\left(\nabla^2\Psi\right); \qquad (6.406)$$

this is analogous to (6.406) ≡ (II.4.3.55; II.4.352b) for the stream function in the plane. The difference between the equation for the stream function in the plane (6.406) [axisymmetric (6.405)] case lies in: (1) the geometric factors involving the distance from the axis *r* in cylindrical coordinates; and (2) the replacement of the Laplacian (6.88a through d) by the modified Laplacian (6.89a and b) operator. This leads next (Note 6.9) to the substitution of the original by the modified biharmonic operator.

NOTE 6.9: Original and Modified Biharmonic Operator

Thus *the stream function of an incompressible (6.403b) axisymmetric (6.403a) [plane (6.400b)] Newtonian viscous flow under conservative forces (6.403c) flow satisfies the fourth-order nonlinear partial differential equation (6.405) [(6.406)] in cylindrical (Cartesian) coordinates. For a creeping flow, that is, one with predominance of viscous forces, corresponding to a small Reynolds number (6.407a) [(II.4.361a)] the vorticity equation for the stream function is linearized (6.407b) [(II.4.361b) ≡ (6.407c)]:*

$$\text{Re} \equiv \frac{\rho v L}{\eta} = \frac{\rho L}{\bar{\eta}} \ll 1: \quad \frac{\partial}{\partial t}\left(\bar{\nabla}^2 \Psi\right) = \bar{\eta}\left[r\nabla^2\left(\frac{1}{r}\bar{\nabla}^2\Psi\right)\right], \quad \frac{\partial}{\partial t}\left(\nabla^2\Psi\right) = \bar{\eta}\nabla^4\Psi. \tag{6.407a–c}$$

*In the steady case (6.408a) [(II.4.361c)] it simplifies to the **modified biharmonic equation** (6.408b) [original biharmonic equation (II.4.361d) ≡ (6.408c)]:*

$$\frac{\partial\Psi}{\partial t} = 0: \quad \nabla^2\left(\frac{1}{r}\bar{\nabla}^2\Psi\right) = 0, \quad \nabla^4\Psi = 0. \tag{6.408a–c}$$

The boundary (6.409a) conditions of zero (6.409b and c) [(6.409d and e) ≡ (II.4.357b and c)] velocity (6.57a through c) [(6.63b)] at a rigid impermeable wall:

$$\partial D: \quad \frac{\partial\Psi}{\partial r}\bigg|_{\partial D} = 0 = \frac{\partial\Psi}{\partial\theta}\bigg|_{\partial D}, \quad \frac{\partial\Psi}{\partial x}\bigg|_{\partial D} = 0 = \frac{\partial\Psi}{\partial y}\bigg|_{\partial D}, \tag{6.409a–e}$$

apply in all cases, including flow that is creeping (6.407a through c), general (6.405) [(6.406)], or inertial (6.410a through c). The latter is the case (6.410b) [(6.410c)] of inertial flow at high Reynolds number (6.410a):

$$\text{Re} \gg 1: \quad \frac{1}{r}\frac{\partial}{\partial t}\left(\bar{\nabla}^2\Psi\right) = \left(\partial_z\Psi\right)\partial_r\left(\frac{1}{r^2}\bar{\nabla}^2\Psi\right) - \left(\partial_r\Psi\right)\partial_z\left(\frac{1}{r^2}\bar{\nabla}^2\Psi\right), \tag{6.410a,b}$$

$$\frac{\partial}{\partial t}\left(\nabla^2\Psi\right) = \left(\partial_x\Psi\right)\partial_y\left(\nabla^2\Psi\right) - \left(\partial_y\Psi\right)\partial_x\left(\nabla^2\Psi\right). \tag{6.410c}$$

In the steady case (6.410b) [(6.410c)] simplifies to (6.108e) ≡ (6.411a) [(II.4.363c) ≡ (6.411b)]:

$$-\bar{\varpi}r = \bar{\nabla}^2\Psi, \quad -\varpi = \nabla^2\Psi, \tag{6.411a,b}$$

where the vorticity vanishes $\varpi = 0$ *for an irrotational flow. The derivation of (6.411a) [(6.411b)] from (6.410b) [(6.410c)] uses (6.66c) [(6.65d)] leading to (6.412a) [(6.412b)]:*

$$\frac{\partial\varpi}{\partial t} = \left(\partial_z\Psi\right)\partial_z\left(\frac{\varpi}{r}\right) - \left(\partial_r\Psi\right)\partial_z\left(\frac{\varpi}{r}\right), \quad \frac{\partial\varpi}{\partial t} = \left(\partial_x\Psi\right)\left(\partial_y\varpi\right) - \left(\partial_x\varpi\right)\left(\partial_y\Psi\right); \tag{6.412a,b}$$

in the (6.413a) steady case (6.412a) [(6.412b)] becomes (6.413b) [(6.413c)]:

$$\frac{\partial\varpi}{\partial t} = 0: \quad \left(\frac{dz}{dr}\right)_\Psi = -\frac{\partial_r\Psi}{\partial_z\Psi} = -\frac{\partial_r\left(\varpi/r\right)}{\partial_z\left(\varpi/r\right)} = \left(\frac{dz}{dr}\right)_{\varpi/r}, \tag{6.413a,b}$$

$$\left(\frac{dz}{dr}\right)_\Psi = -\frac{\partial_x\Psi}{\partial_y\Psi} = -\frac{\partial_x\varpi}{\partial_y\varpi} = \left(\frac{dy}{dx}\right)_\varpi. \tag{6.413c}$$

Thus although the vorticity $\varpi(r, z)$ $[\varpi(x, y)]$ and stream function $\Psi(r, z)$ $[\Psi(x, y)]$ for an axisymmetric (plane) flow are functions of two variables, the surfaces $\Psi = $ const and $\varpi/r = $ const $(\varpi = $ const$)$ are the same (6.413b) [(6.413c)] and they must be a function of each other. It follows that $-\varpi/r = f(\Psi)$ $[-\varpi = f(\Psi)]$ where f is an arbitrary function; then (6.412a) [(6.411b)] leads to (6.116d) [(6.116c)].

NOTE 6.10: Momentum Equation for a Newtonian Viscous Fluid (Navier, 1822; Poisson, 1829b; Saint-Venant, 1843; Stokes, 1845)

The various forms of the momentum equation satisfied by the Newtonian viscous flow are summarized in Table 6.4, starting with *the Navier–Stokes equation (6.390a and b) in a domain D, with zero velocity on the boundary (6.414a):*

$$\vec{v}\big|_{\partial D} = 0: \quad \vec{f} - \nabla p - \rho \frac{\partial \vec{v}}{\partial t} = \begin{cases} -\eta \nabla^2 \vec{v} - \left(\zeta + \dfrac{\eta}{3}\right) \nabla(\nabla \cdot \vec{v}) & \text{if } \mathrm{Re} \ll 1, \\ \rho\left[\nabla\left(\dfrac{v^2}{2}\right) + (\nabla \wedge \vec{v}) \wedge \vec{v}\right] & \text{if } \mathrm{Re} \gg 1; \end{cases} \tag{6.414a–c}$$

the Navier–Stokes equation simplifies to (6.414b) [(6.414c)] for inertial (creeping) flow at high (low) Reynolds number (6.415a) when the viscous stresses are negligible (dominant) compared with the inertia force:

$$\mathrm{Re} \equiv \frac{\rho v L}{\eta} = \frac{v L}{\eta}; \frac{\partial}{\partial t} = 0: \quad \vec{f} - \nabla p = \begin{cases} -\eta \nabla^2 \vec{v} - \left(\zeta + \dfrac{\eta}{3}\right) \nabla(\nabla \cdot \vec{v}) & \text{if } \mathrm{Re} \ll 1, \\ \rho\left[\nabla\left(\dfrac{v^2}{2}\right) + (\nabla \wedge \vec{v}) \wedge \vec{v}\right] & \text{if } \mathrm{Re} \gg 1; \end{cases} \tag{6.415a–d}$$

the momentum equation simplifies further to (6.415c) [(6.415d)] for a steady flow (6.405b).

For an incompressible fluid (6.391b) $\equiv$ (6.416a), conservative external force field (6.393a) $\equiv$ (6.416b) and constant static shear (6.393b) and kinematic shear (6.394a) $\equiv$ (6.416c) viscosity the vorticity equation (6.393c) simplifies to the diffusion (6.398a and b) [conservation (6.395a and c)] equation for creeping (inertial) flow, and in the steady case (6.406d) simplifies further to (6.416e) [(6.416f)] showing that the vorticity is a harmonic vector (its outer product by the velocity is solenoidal):

$$\nabla \cdot \vec{v} = 0 = \nabla \wedge \vec{f}, \eta = \bar{\eta}\rho = \text{const}; \frac{\partial}{\partial t} = 0: \quad 0 = \begin{cases} \nabla^2 \vec{\varpi} & \text{if } \mathrm{Re} \ll 1, \\ \nabla \wedge (\vec{\varpi} \wedge \vec{v}) & \text{if } \mathrm{Re} \gg 1. \end{cases} \tag{6.416a–f}$$

TABLE 6.4 Momentum Equation for a Newtonian Viscous Fluid

Variable Flow	Velocity General	Incompressible		Stream Function	
		Vorticity	Vector Potential	Plane	Axisymmetric
General	Navier–Stokes: (6.390a and b)	Vorticity: (6.65a; 6.393a through c)	(6.394a and b)	(6.406)	(6.405)
Boundary conditions	(6.414a)	—	(6.419a)	(6.409d and e)	(6.409b and c)
Creeping	(6.414b)	(6.398a and b)	(6.398a and c)	(6.407a and c)	(6.407a and b)
Inertial	(6.414c)	(6.395a and b)	(6.395a and b)	(6.410a and c)	(6.410a and b)
Steady creeping	(6.415a and b)	(6.416a through e)	(6.417a through e)	(6.418b through d)	(6.418b and c, e)
Steady inertial	(6.415a and c)	(6.416a through d, f)	(6.417a through d, f)	(6.402b)	(6.402a)

Note: Several forms of the momentum equation and its corollaries for a viscous Newtonian fluid, starting with the Navier–Stokes equation, and including the vorticity and vector potential equations for incompressible flow and equations for the stream function in plane and axisymmetric cases.

For an incompressible flow (6.387b) the velocity is the curl of a vector potential (6.387c), and the vorticity is minus its Laplacian (6.389d) if the gauge condition (6.389a) is imposed. Thus the vorticity equation (6.393c) can be written alternatively for the vector potential (6.394b), and simplifies to (6.398c) [(6.395b)] for creeping (inertial) flow at low (6.398a) [high (6.395a)] Reynolds number, leading to (6.417e) ≡ [(6.417f)] in the steady case (6.417d):

$$\nabla \cdot \vec{v} = 0 = \nabla \wedge \vec{f}, \eta = \rho \bar{\eta} = \text{const}, \frac{\partial}{\partial t} = 0: \quad 0 = \begin{cases} \nabla^4 \vec{A} & \text{if } \mathrm{Re} \ll 1, \\ \nabla \wedge \left[\left(\nabla \wedge \vec{A} \right) \wedge \nabla^2 \vec{A} \right] & \text{if } \mathrm{Re} \gg 1, \end{cases} \tag{6.417a–f}$$

where the equation (6.417e) is biharmonic [(6.417f) shows that the outer product of the curl and Laplacian is a solentidal vector].

The boundary condition of zero velocity (6.414a) leads to (6.418a) for the vector potential, which corresponds to (6.409b and c) [(6.409d and e)] for the stream function in the case of a plane (axisymmetric) flow that specifies: (1) the velocity components (6.63d) [(6.57a through c)]; (2) the vector potential (6.387d) [(6.387e)]; and (3) the vorticity (6.65a and c) [(6.66d)]. The vorticity equation can be written in terms of the stream function for plane (6.406) [axisymmetric (6.405)] flow, simplifying to (6.407a and c) [(6.407a and b)] for creeping flow (6.407a) involving two original Laplacians (one original and one modified Laplacian). In a steady (6.418b) creeping (6.418c) flow, the stream function satisfies (6.408c) ≡ (6.418d) [(6.408b) ≡ (6.418e)] for a plane (axisymmetric) flow:

$$\nabla \wedge \vec{A}\big|_{\partial D} = 0; \frac{\partial}{\partial t} = 0, \mathrm{Re} \ll 1: \quad \nabla^4 \Psi = 0, \quad \nabla^2 \left(\frac{1}{r} \bar{\nabla}^2 \Psi \right) = 0, \tag{6.418a–e}$$

which is a biharmonic (modified biharmonic) equation, and both are linear fourth-order differential equations. In the opposite case of inertial flow (6.419b), also steady (6.419a), the stream function satisfies (6.116c) ≡ (6.411b) ≡ (6.419c) [(6.116d) ≡ (6.411a) ≡ (6.419d)] for a plane (axisymmetric) flow:

$$\frac{\partial}{\partial t} = 0; \quad \mathrm{Re} \gg 1: \quad \nabla^2 \Psi = f(\Psi), \quad \nabla^2 \Psi = r^2 f(\Psi), \tag{6.419a–d}$$

which is a nonlinear second-order partial differential equation, namely, the nonlinear Poisson (6.419c) [modified Poisson (6.419d)] equation; it simplifies to a linear Laplace (6.420d) [modified Laplace (6.420e)] equation in the absence (6.420c) of vorticity:

$$\frac{\partial}{\partial t} = 0, \mathrm{Re} \gg 1, \varpi = 0: \quad \nabla^2 \Psi = 0, \quad \bar{\nabla}^2 \Psi = 0. \tag{6.420a–e}$$

The differential equation for the stream function is of the second (fourth) order for the inertial (6.420a through e) [creeping (6.419a through e)] flow because the viscosity is neglected (taken into account) and thus only the normal (6.421a) [also the tangential (6.421b)] components of the velocity:

$$v_n = -\frac{\partial \Psi}{\partial s}, \quad v_s = \frac{\partial \Psi}{\partial n}, \tag{6.421a,b}$$

must vanish on the surface of a rigid, static, impermeable body.

6.10 Conclusion

The orthogonal curvilinear coordinates in space lead to a parallelepiped with scale factors along the axis specifying the volume (Figure 6.1a) and area (Figure 6.1b) of the faces (Figure 6.1c). The relation between Cartesian (x, y, z) and cylindrical (r, φ, z) [spherical (R, θ, φ)] coordinates (Figure 6.2) leads to: (1) the volume element (Figure 6.3a/b/c), respectively, in Cartesian/cylindrical/spherical coordinates; (2) the Cartesian (spherical) components of the velocity (Figure 6.4), for example, for [Figure 6.5a (Figure 6.5b)] uniform flow (flow due to a source or sink). For the steady axisymmetric incompressible flow the volume flux through a line joining an arbitrary point B to the axis A depends (Figure 6.6) only on the point, proving that a stream function exists. The simplest axisymmetric flows are (Figure 6.5a) the uniform stream of velocity U, and (Figure 6.5b) the flow source $Q > 0$ (sink $Q < 0$) of volume flux Q causing a radial velocity. Their superposition, that is, the insertion of a point source in a uniform stream specifies the flow (Figure 6.9) past a semi-infinite Rankine fairing of asymptotic radius b; adding an opposite sink leads (Figure 6.10) to a finite body, whose dipole limit (Figure 6.11a) is the sphere (Figure 6.12). The limit of two opposite dipoles is a quadrupole (Figure 6.11b), and so on for higher-order multipoles (Table 6.1). The sphere may move in free space (Figure 6.12) or in a cavity leading to wall effects, for example, in a large spherical cavity (Figure 6.13). An incompressible axisymmetric steady flow has stream surface obtained by rotating the streamlines around the axis, leading generally to a "bell" shape (Figure 6.7); the stagnation enthalpy and stagnation pressure are constant on the stream surfaces, both in the axisymmetric (Figure 6.7) and in the plane (Figure 6.8) case. A spherical vortex (Figure 6.14) like a sphere (Figure 6.13) can exist in a uniform stream. The electric currents can flow on a sphere (Figure 6.15) between opposite poles. A nonaxisymmetric problem concerns two dissimilar spheres moving with different velocities (Figure 6.16a) perpendicular to the line of centers; for identical spheres moving at the same velocity the image sphere may be replaced by wall (Figure 6.16b) at equal distance from the centers. This problem can be solved by using the images on the spheres (Table 6.2) or a method of perturbation potentials. The scalar (vector) potential exists for an irrotational (incompressible) flow (Table 6.3). The vector potential reduces to a stream function for a plane (axisymmetric) incompressible flow, either inviscid (Diagram 6.1) or viscous (Table 6.4). The simplest flows (Diagram 6.2) are the incompressible or rotational cases that lead to analogous problems (Classification 6.1) in the plane and axisymmetric and nonaxisymmetric in space. One such analogy is between the two circle and four sphere theorems (List 6.1).

7

Convolution, Reciprocity, and Adjointness

The fundamental solution or influence or Green function of a differential equation is defined as the solution of the equation forced by a unit impulse; the influence (Subsection 7.6.1) function depends not only on the differential equation, but also on boundary and/or initial conditions. For example, the influence function for the elastic string (bar or beam) specifies [Chapter 2 (Chapter 4)] its shape under a concentrated load; the influence function for the Poisson equation in one, two, and three dimensions (Chapters 8 and 9) specifies the potential due to respectively a plane, a line, and a point source. In all problems, such as the preceding examples, where the differential equation and boundary and/or initial conditions are linear, the principle of superposition holds (Subsection 7.6.2), specifying the solution for arbitrary forcing in terms of the influence function. The influence function specifies the field at an observation point due to a source at a forcing point; thus if it is symmetric the principle of reciprocity holds, allowing interchange of the observer and source positions (Subsection 7.6.3). The reciprocity principle holds for self-adjoint operators, and not otherwise. The linear ordinary (partial) differential operators, the adjoint operator, and related bilinear concomitant are considered [Section 7.7 (Section 7.8)] and applied to the equations of mathematical physics (Section 7.9). The superposition principle holds for all linear operators, self-adjoint or not; in the case when the influence function depends only on the relative position of source and observer, its convolution with the forcing function specifies the field, which is the solution of the linear ordinary or partial differential equation. Thus a preliminary question is the existence and properties of the convolution integral for ordinary and generalized functions (Section 7.5). This question relates to the convolution theorem (Section 7.4), which, like the Hölder (Minkowski) inequalities [Section 7.2 (Section 7.3)], is obtained for normed or metric spaces (Section 7.1).

7.1 Norm of a Function and Metric Spaces

The concepts of norm or square of the modulus for complex numbers (Section I.1.4) has analogue for vectors and functions (Subsection 7.1.1), as does the notion of distance (Subsection 7.1.2).

7.1.1 Functions with Integrable Power of the Modulus and Normed Spaces

The **normed space** $\mathcal{L}^p$ is defined as the set of functions such that the pth power of the modulus is integrable (7.1a):

$$\mathcal{L}^p\left([a,b]\right) \equiv \left\{ f(x) : \exists \int_a^b \left|f(x)\right|^p \mathrm{d}x \right\}; \quad \|f\|_p \equiv \left\{ \int_a^b \left|f(x)\right|^p \mathrm{d}x \right\}^{1/p}, \tag{7.1a,b}$$

and thus the **norm** (7.1b) defined as (7.1a) to the power $1/p$ exists; the exponents p and $1/p$ ensure that the norm (7.1b) has the same physical dimensions as the function. A normed space can be defined over any set; for example, the real interval is $a \leq x \leq b$ in (7.1a and b), or the real line $\mathcal{L}^p(|R)$ for $a = +\infty$, $b = -\infty$. There are three important particular cases. The first for $p = 2$ is the space $\mathcal{L}^2$ of **square integrable functions** (7.2a):

$$\mathcal{L}^2(a,b) \equiv \left\{ f(x): \int_a^b |f(x)|^2 \, dx < \infty \right\}; \quad \left\{ \|f\|_2 \right\}^2 \equiv \int_a^b |f(x)|^2 \, dx, \tag{7.2a,b}$$

which have the **Euclidian norm** (7.2b); the Euclidean norm (7.2b) divided by the length $|b-a|$ of a finite interval (a,b) is the root mean square value of the function. The second for $p = 1$, the space $\mathcal{L}^1$ of **absolutely integrable functions** (7.3a):

$$\mathcal{L}^1(a,b) \equiv \left\{ f(x): \int_a^b |f(x)| \, dx < \infty \right\}; \quad \|f\|_1 \equiv \int_a^b |f(x)| \, dx; \tag{7.3a,b}$$

it is a subset of the set E of integrable functions:

$$\left| \int_a^b f(x) \, dx \right| \leq \int_a^b |f(x)| \, dx, \quad \mathcal{L}^1 \subseteq \mathcal{E} \equiv \left\{ f(x): \exists \int_a^b f(x) \, dx \right\}, \tag{7.4a,b}$$

with the equality holding for functions everywhere nonnegative $f(x) \geq 0$. The absolute norm (7.3b) divided by the length $|b-a|$ of the finite interval is the mean value of the modulus of the function. Both the root mean square values and mean absolute value can be extended to infinite intervals by taking the limits $a \to -\infty$ and/or $b \to +\infty$. The third for $p = \infty$, the space $\mathcal{L}^\infty$ of **bounded integrable functions** (7.5a) with norm (7.5b):

$$\mathcal{L}^\infty \equiv \mathcal{E} \cap \mathcal{B} \equiv \left\{ f(x): \left| \int_a^b f(x) \, dx \right| < \infty \wedge \exists_{M>0} \forall_{a \leq x \leq b} |f(x)| \leq M \right\}, \quad \|f\|_\infty \leq M. \tag{7.5a,b}$$

The property (7.5b) results from

$$\|f\|_\infty \equiv \lim_{p \to \infty} \left\{ \int_a^b |f(x)|^p \, dx \right\}^{1/p} \leq \lim_{p \to \infty} \left\{ \left[|f(x)|_{\max} \right]^p (b-a) \right\}^{1/p}$$

$$= |f(x)|_{\max} \lim_{p \to \infty} (b-a)^{1/p} \leq M(b-a)^0 = M; \tag{7.6}$$

thus *the norm in the $\mathcal{L}^\infty$-space of a bounded integrable function (7.5a) is no greater than its maximum (7.5b)*. The integral of (7.1a) for an absolutely integrable function $p = 1$ gives equal weight to all values of $|f(x)|$; as p increases, the values $|f(x)| > 1$ and in particular $|f(x)|_{\max} > 1$ gain greater weight, and they dominate (7.6) the norm (7.1b) as $p \to \infty$.

7.1.2 Triangular and Projective Inequalities in a Metric Space

A **normed space** is called a **metric space** if the norm satisfies three properties:

$$\|0\| = 0, \quad \|f\| = \|-f\|, \quad \|f+g\| \leq \|f\| + \|g\|, \tag{7.7a–c}$$

which are interpreted in terms of distance between two points $PQ \equiv f(x)$ as follows: (1) if two points coincide $f = 0$, their distance is zero (7.7a); (2) the distance between two points $PQ = f$, $QP = -f$ is independent of the order (7.7b); and (3) the distance between three points $PQ = f$, $QR = g$, $PR = f + g$ satisfies the **triangular inequality** (7.7c). In a metric space, the **inner product of functions** is defined by (7.8b):

$$fg \in \mathcal{E}(a,b): \quad [f(x),g(x)] \equiv \int_a^b f(x)g(x)\mathrm{d}x, \quad |[f,g]| \le \|f\| \|g\|, \tag{7.8a–c}$$

where the product of the functions is assumed to be integrable (7.8a); if it satisfies the **projective inequality** (7.8c), then the inner product (7.8b) certainly exists. The inner product (7.8b) can be generalized (3.53a–d) to complex functions of a real variable and nonunit weighting function (Subsection II.5.7.1). The triangular inequality (7.7c) states that the length of one side of a triangle cannot exceed the sum of the lengths of the other two sides (I.3.37b); the projective inequality (7.8b) states that the inner product of two functions cannot exceed the product of their norms as is the case for vectors. The equality occurs in both cases (7.7c; 7.8c) only for constant multiples, that is, (1) *the triangular (7.9a) [projective (7.9c)] equality holds if the functions f, g are constant multiples (7.9b)*:

$$\|f + g\| = \|f\| + \|g\| \Leftrightarrow \exists_{\lambda>0} \forall_{a \le x \le b}: \quad f(x) = \lambda g(x) \Leftrightarrow |[f,g]| = \|f\| \|g\|; \tag{7.9a–c}$$

$$\|a_n + b_n\| = \|a_n\| + \|b_n\| \Leftrightarrow \exists_{\lambda>0} \forall_{n \in |N}: \quad b_n = \lambda a_n \Leftrightarrow |[a_n,b_n]| = \|a_n\| \|b_n\|; \tag{7.10a–c}$$

(2) likewise the triangular (7.10a) [projective (7.10c)] inequality holds for vectors if they are parallel (7.10b).

7.2 Schwartz and Hölder Projective Inequalities and Equalities

The projective property holds for complex numbers, stating that the real part cannot exceed the modulus, and only equals it for positive real numbers; the complex numbers are equivalent to two-dimensional vectors. The projective property applies as well to vectors in any dimension and functions (Subsection 7.2.1); it extends from square integrable functions to any normed space (Subsection 7.2.2).

7.2.1 Extension from Schwartz to Hölder Projective Inequalities

The properties (7.7a and b) obviously hold in all normed $\mathcal{L}^p$ spaces (7.1a and b). The normed spaces are metric spaces with the norm (7.1b), provided that it is shown that the triangular inequality (7.7c) [equality (7.9a)] holds for two functions that are not (are) constant multiples (7.9b). The projective inequality (7.8c) [equality (7.9c)] should then follow. The latter (Section 7.2) is proved first, followed by the former (Section 7.3). The case $p = 2$ of (7.8b; 7.9c) $\equiv$ (7.11b),

$$f,g \in \mathcal{L}^2(a,b): \quad |[f,g]|^2 = \left\{ \int_a^b f(x)g(x)\mathrm{d}x \right\}^2 \le \left\{ \int_a^b [f(x)]^2 \mathrm{d}x \right\} \left\{ \int_a^b [g(x)]^2 \mathrm{d}x \right\}$$

$$= \left(\|f\|_2 \right)^2 \left(\|g\|_2 \right)^2, \tag{7.11a,b}$$

is the **Schwartz projective inequality** (1890), *which holds for square integrable functions (7.11a) and also for vectors in an Euclidean space (7.11c):*

$$\left\{\sum_{n=1}^{N} a_n b_n\right\}^2 \leq \left\{\sum_{n=1}^{N} (a_n)^2\right\}\left\{\sum_{n=1}^{N} (b_n)^2\right\}. \tag{7.11c}$$

The Schwartz inequality can be proved directly from

$$0 \leq \frac{1}{2}\int_a^b dx \int_a^b dy\{f(x)g(y)-f(y)g(x)\}^2 = \int_a^b |f(x)|^2 dx \int_a^b |g(y)|^2 dy - \int_a^b f(x)g(x)dx \int_a^b f(y)g(y)dy. \tag{7.12}$$

The integral inequality (7.11b) extends to improper integrals of the first kind $(a = +\infty, b = -\infty)$ for square integrable function on the real line $\mathcal{L}^2(|R)$, and the discrete inequality (7.11c) applies in the limit $N \to \infty$, provided that the series of squares converge (7.13):

$$\left\{\sum_{n=1}^{\infty} a_n b_n\right\}^2 \leq \left\{\sum_{n=1}^{\infty} (a_n)^2\right\}\left\{\sum_{n=1}^{\infty} (a_n)^2\right\} < \infty. \tag{7.13}$$

If one function is bounded and integrable (7.14a) and the other function is absolutely integrable (7.14b), projective inequality (7.8b) holds (7.14c):

$$f \in \mathcal{L}^\infty \cap \mathcal{E}, g \in \mathcal{L}^1: \quad [f,g] = \left|\int_a^b f(x)g(x)dx\right| \leq \int_a^b |f(x)g(x)|dx \leq |f(x)|_{max} \int_a^b |g(x)|dx \leq \|f\|_\infty \|g\|_1. \tag{7.14a–c}$$

If $f \in \mathcal{L}^p$, $g \in \mathcal{L}^q$ then $p = 2 = q$ in (7.11a and b) and $p = \infty$, $q = 1$ in (7.14a through c), so that in both cases (7.15c) holds.

The Schwartz inequality (7.11a and b) as well as (7.14a through c) are particular cases of the **Hölder projective inequality** (1889) *in (7.15a through c), in weak form (7.15d) or strong discrete (7.15e) [continuous (7.15f)] form:*

$$1 \leq p, q < \infty, \quad 1/p + 1/q = 1: \quad [f,g] \leq \|f\|_p \|g\|_q, \tag{7.15a–d}$$

$$\sum_{n=1}^{N} |a_n b_n| \leq \left\{\sum_{n=1}^{N} |a_n|^p\right\}^{1/p} \left\{\sum_{n=1}^{N} |a_n|^q\right\}^{1/q}, \tag{7.15e}$$

$$\int_a^b |f(x)g(x)|dx \leq \left\{\int_a^b |f(x)|^p dx\right\}^{1/p} \left\{\int_a^b |g(x)|^q dx\right\}^{1/q}; \tag{7.15f}$$

the equality holds in (7.15e) or if a, b are collinear (7.10b) and in (7.15f) if the functions f, g are constant multiples (7.9b). The projective inequality (7.8c) ≡ (7.15d) is weaker than (7.15f) because

$$\left|[f,g]\right| = \left|\int_a^b f(x)g(x)\mathrm{d}x\right| \le \int_a^b \left|f(x)g(x)\right|\mathrm{d}x, \tag{7.15g}$$

$$\left|\sum_{n=1}^N a_n b_n\right| \le \sum_{n=1}^N \left|a_n b_n\right|; \tag{7.15h}$$

likewise, (7.15e) is a stronger result than the modulus of the inner product of vectors on the l.h.s. of (7.15h). In order to prove one of the equivalent formulas (7.15d through f), for example, (7.15e), note that they are homogeneous, that is, they hold for λf, μg, as well as for f, g, where λ, μ are arbitrary real constants. Thus λ, μ may be chosen such that

$$\sum_{n=1}^N \left|a_n\right|^p = 1 = \sum_{n=1}^N \left|b_n\right|^q : \quad \sum_{n=1}^N \left|a_n b_n\right| \le 1; \tag{7.16a–c}$$

so this is sufficient to prove (7.16c) for the unit vectors (7.16a and b). The case $p = 1$, $q = \infty$ is already proven (7.14a through c), so only $1 < p, q < \infty$ need be considered.

7.2.2 Proof of the Hölder Inequality in a Normed Space

Consider (Figure 7.1a) the function (7.17a):

$$y(x) = x^\alpha : \quad y(x) \le 1 + (x-1)y'(1), \tag{7.17a,b}$$

which, for (7.18a), lies below (7.17b) its tangent at the point $(x, y) = (1,1)$ so that the inequality (7.17b) ≡ (7.18b) becomes

$$0 < \alpha < 1: \quad x^\alpha \le \alpha x + 1 - \alpha, \tag{7.18a,b}$$

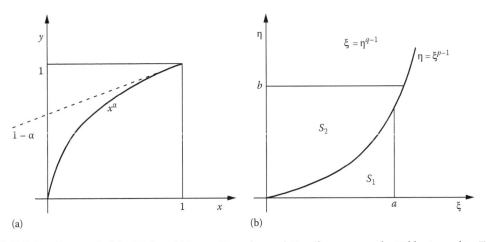

(a)   (b)

FIGURE 7.1 The proof of the Minkowski inequality and convolution theorem uses the Hölder inequality. Two alternate proofs of the latter rely on geometrical constructions: (a) the curve $y = x^\alpha$ with $0 < \alpha < 1$ lies below its tangent at $x = 1$; (b) the inverse functions $\xi = \eta^{q-1}$ and $\eta = \zeta^{p-1}$ with $1/p + 1/q = 1$ specify for $\xi = a$ and $\eta = b$ two areas respectively S_1 and S_2 whose sum is larger than the area of the parallelogram with sides (a,b).

where the equality holds for $x = 1$. Performing a change of variable (7.19a),

$$x \equiv \frac{u}{v}: \quad u^{\alpha}v^{1-\alpha} \le \alpha u + (1-\alpha)v, \tag{7.19a,b}$$

leads to the inequality (7.19b) if $u \ne v$, and equality for $u = v$, both with $u, v > 0$. Setting (7.20a and b) leads to (7.20c):

$$1 < p \equiv \frac{1}{\alpha} < \infty, \quad 1 < q \equiv \frac{1}{1-\alpha} < \infty: \quad u^{1/p}v^{1/q} \le \frac{u}{p} + \frac{v}{q}; \tag{7.20a-c}$$

choosing next (7.21a and b) leads to (7.21c and d):

$$a \equiv u^{1/p}, \quad b \equiv v^{1/q}, \quad \frac{1}{p} + \frac{1}{q} = 1: \quad ab \le \frac{a^p}{p} + \frac{a^q}{q}, \tag{7.21a-d}$$

where equality holds only for $a = b$.

Another way of obtaining the algebraic inequality (7.21d) is to consider the function (7.22b):

$$\frac{1}{p} + \frac{1}{q} = 1: \quad \eta = \xi^{p-1}, \quad \xi = \eta^{1/(p-1)} = \eta^{q-1}, \tag{7.22a-c}$$

which is represented in Figure 7.1b and whose inverse is (7.22c) on account of (7.22a) $\equiv$ (7.22d) = (7.22e) $\equiv$ (7.22f) $\equiv$ (7.22g) $\equiv$ (7.22c):

$$p+q = pq, \quad p = q(p-1), \quad q = \frac{p}{p-1}, \quad q-1 = \frac{1}{p-1}. \tag{7.22d-g}$$

The sum of the areas S_1, S_2 exceeds that of the rectangle of sides a, b:

$$ab \le \int_0^a \xi^{p-1}d\xi + \int_0^b \eta^{q-1}d\eta = \frac{a^p}{p} + \frac{b^q}{q}, \tag{7.23}$$

leading to (7.23) $\equiv$ (7.21d); the equality holds only for $a = b$. Thus it has been proven in two ways that

$$\frac{1}{p} + \frac{1}{q} = 1, \quad 1 < p, q < \infty: \quad \frac{a^p}{p} + \frac{b^q}{q} - ab \begin{cases} > 0 & \text{if } a \ne b, \\ = 0 & \text{if } a = b. \end{cases} \tag{7.24a-d}$$

Replacing a, b by a_n, b_n and summing yields:

$$\sum_{n=1}^N |a_n b_n| \le \frac{1}{p} \sum_{n=1}^N |a_n|^p + \frac{1}{q} \sum_{n=1}^N |b_n|^q = \frac{1}{p} + \frac{1}{q} = 1, \tag{7.25}$$

$$f,g \in \mathcal{L}^1(a,b): \quad \sum_{n=1}^N |a_n + b_n| \le \sum_{n=1}^N |a_n| + \sum_{n=1}^N |b_n|, \tag{7.26a,b}$$

$$\|f,g\|_1 \equiv \int_a^b |f(x)+g(x)|dx \le \int_a^b |f(x)|dx + \int_a^b |g(x)|dx = \|f\|_1 + \|g\|_1, \tag{7.26c}$$

that proves (7.16c); the equality holds in (7.25) $\equiv$ (7.16c) if $a_n = b_n$, or by relaxing the condition of unit norm if $a_n = \lambda b_n$. This completes the proof of the discrete Hölder inequality (7.15e); the integral form (7.15f) is proved similarly, by integrating (7.24a through d) with $a \equiv f$, $b \equiv g$, so that (7.16c) holds with sums replaced by integrals. The Hölder inequality (7.15a through f) can be extended from two to any number of variables (Example 10.16.1) as the generalized Hölder inequality.

7.3 Discrete/Integral Triangular and Minkowski (1896) Inequalities

The complex numbers satisfy the triangular inequality (I.3.37b), which extends to vectors in any dimension and to square integrable functions (Subsection 7.3.1); it applies also to absolutely integrable functions (Subsection 7.3.1) and extends in arbitrary normed spaces to the Minkowski inequality (Subsection 7.3.2).

7.3.1 Triangular Inequality for Absolutely or Square Integrable Functions

The proof of the triangular inequality (7.7c) is immediate for vectors (7.26b) and for absolutely integrable functions (7.26a and c) in the case of square integrable functions,

$$f, g \in \mathcal{L}^2(a,b): \quad \left\{ \sum_{n=1}^{N} (a_n + b_n)^2 \right\}^{1/2} \leq \left\{ \sum_{n=1}^{N} (a_n)^2 \right\}^{1/2} + \left\{ \sum_{n=1}^{N} (b_n)^2 \right\}^{1/2}, \tag{7.27a}$$

$$\left\{ \int_a^b |f(x) + g(x)|^2 \, dx \right\}^{1/2} \leq \left\{ \int_a^b |f(x)|^2 \, dx \right\}^{1/2} \left\{ \int_a^b |g(x)|^2 \, dx \right\}^{1/2}, \tag{7.27b}$$

the proof of the triangular inequality (7.27b) uses the Schwartz inequality (7.11b):

$$\int_a^b |f(x) + g(x)|^2 \, dx \leq \int_a^b |f(x)|^2 dx + \int_a^b |g(x)|^2 \, dx + 2 \int_a^b |f(x)g(x)| dx \leq \int_a^b |f(x)|^2 dx$$

$$+ \int_a^b |g(x)|^2 dx + 2 \left[\int_a^b |f(x)|^2 \, dx \right]^{1/2} \left[\int_a^b |g(x)|^2 \, dx \right]^{1/2} = \left\{ \left[\int_a^b |f(x)|^2 \, dx \right]^{1/2} + \left[\int_a^b |g(x)|^2 \, dx \right]^{1/2} \right\}^2; \tag{7.28}$$

an alternative proof of (7.28) using the symbolic notation (7.2b) for the $\mathcal{L}^2$ norm is:

$$\left\{ \|f + g\|_2 \right\}^2 = [f + g, f + g] = [f, f] + [g, g] + [f, g] + [g, f]$$

$$= \left\{ \|f\|_2 \right\}^2 + \left\{ \|g\|_2 \right\}^2 + 2[f, g] \leq \left\{ \|f\|_2 \right\}^2 + \left\{ \|g\|_2 \right\}^2 + 2\|f\|_2 \|g\|_2 = \left\{ \|f\|_2 + \|g\|_2 \right\}^2. \tag{7.29}$$

The Schwartz projective inequality in integral form (7.11b) is the case $p = 2 = q$ of the Hölder projective inequality (7.15f) proved before. Likewise, the triangular inequality (7.27b) is the particular case $p = 2$ of the Minkowski triangular inequality proved next.

7.3.2 Minkowski Triangular and Stronger Inequality in Normed Spaces

The triangular inequality for square (7.27a and b) [absolutely (7.26a and b)] integrable functions is the particular case $p = 2 (p = 1)$ of the **Minkowski triangular inequality** (1896):

$$1 \le p < \infty: \quad \|f + g\|_p \le \|f\|_p + \|g\|_p, \tag{7.30a,b}$$

$$\left\{ \sum_{n=1}^{N} |a_n + b_n|^p \right\}^{1/p} \le \left\{ \sum_{n=1}^{N} |a_n|^p \right\}^{1/p} + \left\{ \sum_{n=1}^{N} |b_n|^p \right\}^{1/p}, \tag{7.30c}$$

$$f, g \in \mathcal{L}^p(a,b): \quad \left\{ \int_a^b |f(x) + g(x)|^p \right\}^{1/p} \le \left\{ \int_a^b |f(x)|^p \right\}^{1/p} + \left\{ \int_a^b |g(x)|^p \right\}^{1/p}, \tag{7.30d}$$

which proves that the normed (7.1a and b) space $\mathcal{L}^p$ is (7.7c) a metric space (7.30a and b); the discrete (7.30c) [integral (7.30d)] inequalities become equalities if the vectors a, b are collinear (7.10b) [the functions f, g are constant multiples (7.9b)]. The proof of (7.30c) starts from

$$\left(|a| + |b| \right)^p = \left(|a| + |b| \right)^{p-1} |a| + \left(|a| + |b| \right)^{p-1} |b|; \tag{7.31}$$

summing over $n = 1, \ldots, N$ for the discrete case

$$\sum_{n=1}^{N} \left(|a_n| + |b_n| \right)^p = \sum_{n=1}^{N} \left(|a_n| + |b_n| \right)^{p-1} |a_n| + \sum_{n=1}^{N} \left(|a_n| + |b_n| \right)^{p-1} |b_n|; \tag{7.32}$$

applying Hölder's inequality (7.15e) to each term on the r.h.s. of (7.32) leads to

$$q = \frac{p}{p-1}: \quad \sum_{n=1}^{N} \left(|a_n| + |b_n| \right)^p \le \left\{ \sum_{n=1}^{N} \left(|a_n| + |b_n| \right)^{(p-1)q} \right\}^{1/q} \left[\left[\sum_{n=1}^{N} |a_n|^p \right]^{1/p} + \left[\sum_{n=1}^{N} |b_n|^p \right]^{1/p} \right]; \tag{7.33a,b}$$

the expression simplifies to

$$1 - \frac{1}{q} = \frac{1}{p}: \quad \left\{ \sum_{n=1}^{N} |a_n|^p \right\}^{1/p} + \left\{ \sum_{n=1}^{N} |b_n|^p \right\}^{1/p} \ge \left\{ \sum_{n=1}^{N} \left(|a_n| + |b_n| \right) \right\}^{1-1/q} = \left\{ \sum_{n=1}^{N} \left(|a_n| + |b_n| \right)^p \right\}^{1/p}. \tag{7.34a,b}$$

This is a stronger result than the Minkowski inequality in discrete form (7.30c), because

$$\left\{ \sum_{n=1}^{N} |a_n + b_n|^p \right\}^{1/p} \le \left\{ \sum_{n=1}^{N} \left(|a_n| + |b_n| \right)^p \right\}^{1/p}. \tag{7.34c}$$

The proof of (7.30d) is similar, replacing sums by integrals:

$$1 < p < \infty: \quad \left\{\sum_{n=1}^{N}|a_n|^p\right\}^{1/p} + \left\{\sum_{n=1}^{N}|b_n|^p\right\}^{1/p} \geq \left\{\sum_{n=1}^{N}\left(|a_n|+|b_n|\right)^p\right\}^{1/p} \geq \left\{\sum_{n=1}^{N}|a_n+b_n|^p\right\}^{1/p}, \qquad (7.35\text{a,b})$$

$$f,g \in \mathcal{L}^p(a,b): \quad \left\{\int_a^b |f(x)|^p\,\mathrm{d}x\right\}^{1/p} + \left\{\int_a^b |g(x)|^p\,\mathrm{d}x\right\}^{1/p}$$

$$\geq \left\{\int_a^b \left(|f(x)|+|g(x)|\right)^p\,\mathrm{d}x\right\}^{1/p} \geq \left\{\int_a^b |f(x)+g(x)|^p\,\mathrm{d}x\right\}^{1/p}. \qquad (7.35\text{c,d})$$

Thus has been proven *the* **stronger Minkowski triangular** *inequality (7.35a) in discrete (7.35b) [continuous (7.35c and d)] form, of which the original Minkowski triangular inequality (7.30a and b) is the particular case (7.30c) [(7.30d)]*. The Minkowski inequality can be extended from two (7.30a through d) to any number of variables as the generalized Minkowski inequality (Example 10.16.2).

7.4 Convolution Integral for Ordinary and Generalized Functions

The existence of the convolution integral is proved first (Subsection 7.4.1) for absolutely integrable and square integrable functions; it is then extended to all normed spaces (Subsection 7.4.2). Choosing test functions with compact support leads to the convolution of generalized functions with compact support (Subsection 7.4.3).

7.4.1 Convolution of Absolutely or Square Integrable Functions

The **convolution integral** of two functions is defined by

$$h(x) \equiv f(x) * g(x) = \int_{-\infty}^{+\infty} f(y)g(x-y)\mathrm{d}y. \qquad (7.36)$$

The existence of a convolution integral is considered first (after) for ordinary (generalized) functions [Subsections 7.4.1 and 7.4.2 (Subsection 7.4.3)]. The proof of the Hölder (Minkowski) inequality [Section 7.2 (Section 7.3)] in general normed spaces was preceded by two examples in particular spaces. Similarly, three examples in particular spaces will precede (Subsection 7.4.1) the proof for general normed spaces (Subsection 7.4.2) of the **convolution existence theorem**: *the convolution (7.37c) of two normed functions (7.37a and b) is a normed function (7.37d)*:

$$f(x) \in \mathcal{L}^p \in (|R) \wedge g(x)\mathcal{L}^q(|R) \Rightarrow f * g(x) \in \mathcal{L}^s(|R) \wedge \frac{1}{s} = \frac{1}{p} + \frac{1}{q} - 1. \qquad (7.37\text{a–d})$$

The first particular case follows applying Schwartz's inequality (7.11a) to (7.36):

$$z = x - y: \quad |h(x)| \leq \int_a^b |f(y)g(x-y)|\mathrm{d}y$$

$$\leq \left\{\int_a^b |f(x)|^2\,\mathrm{d}x\right\}^{1/2}\left\{\int_a^b |g(z)|^2\,\mathrm{d}z\right\}^{1/2} = \|f\|_2 \cdot \|g\|_2. \qquad (7.38\text{a,b})$$

Thus *the convolution of square integrable functions is a bounded function*:

$$f(x), \ g(x) \in L^2(|R) \Rightarrow f * g(x) \in L^\infty(|R) \subset B(|R); \tag{7.39}$$

in this particular case, $p = 2 = q$, $s = \infty$, in agreement with (7.37d). A second particular case is obtained integrating (7.36) over x and using the change of variable (7.38a):

$$\int_{-\infty}^{+\infty} |h(x)| dx \le \int_{-\infty}^{+\infty} dx \int_{-\infty}^{+\infty} dy |f(y)g(x-y)| \le \int_{-\infty}^{+\infty} |g(z)| dz \int_{-\infty}^{+\infty} |f(y)| dy = \|g\|_1 \cdot \|f\|_1. \tag{7.40}$$

Thus *the convolution of absolutely integrable functions is an absolutely integrable function*:

$$f(x), g(x) \in L^1(|R) \Rightarrow f * g(x) \in L^1(|R); \tag{7.41}$$

in this particular case, $p = 1 = q$ and $s = 1$, again in agreement with (7.37d). A third particular case is

$$|h(x)| \le \int_{-\infty}^{+\infty} |f(y)| |g(x-y)| dy \le |g(x-y)|_{\max} \int_{-\infty}^{+\infty} |f(y)| dy = \|g\|_\infty \|f\|_1; \tag{7.42}$$

thus *the convolution of an absolutely integrable function with a bounded function is bounded*:

$$f(x) \in L^1(|R) \wedge g(x) \in L^\infty(|R) \Rightarrow f * g(x) \in L^1(|R); \tag{7.43}$$

in this particular case, $p = 1$, $q = \infty = s$, again in agreement with (7.37d). The three particular cases (7.38a and b; 7.39), (7.40; 7.41), and (7.42; 7.43) are followed (Subsection 7.4.2) by the general proof of (7.37a through d).

7.4.2 Convolution of Ordinary Functions in Normed Spaces

To prove (7.37a through d) the functions are assumed to be normed and an upper bound for f^s norm of the convolution is obtained:

$$\left\{ \|f * g\|_s \right\}^s = \int_{-\infty}^{+\infty} |f * g(x)|^s dx = \int_{-\infty}^{+\infty} dx \left| \int_{-\infty}^{+\infty} f(y)g(x-y) dy \right|^s$$

$$\le \int_{-\infty}^{+\infty} dz \int_{-\infty}^{+\infty} dy |f(y)|^s |g(z)|^s. \tag{7.44}$$

The integrand is split into three factors (7.45c) with arbitrary exponents satisfying (7.45a and b):

$$0 < \alpha, \beta < 1: \quad |f(y)| |g(x-y)| = |f(y)|^\alpha |g(x-y)|^\beta |f(y)|^{1-\alpha} |g(x-y)|^{1-\beta}, \tag{7.45a–c}$$

and the Hölder inequality (7.15f) is used twice:

$$\left\{\left\|f \ast g\right\|_s\right\}^s \le \int\limits_{-\infty}^{+\infty} dx \left\{\int\limits_{-\infty}^{+\infty}\left|f(y)\right|^{\alpha s}\left|g(x-y)\right|^{\beta s} dy\right\}\left\{\int\limits_{-\infty}^{+\infty}\left|f(y)\right|^{(1-\alpha)r} dy\right\}^{s/r}\left\{\int\limits_{-\infty}^{+\infty}\left|g(x-y)\right|^{(1-\beta)t}\right\}^{s/t}; \quad (7.46)$$

this is equivalent to the triple Hölder inequality (Example 10.16.1), that is (10.224a, b, d, and e) with $N = 3$, and involves arbitrary parameters r, t satisfying (7.47a):

$$\frac{1}{s}+\frac{1}{r}+\frac{1}{t}=1; \quad \alpha s = p = (1-\alpha)r, \quad \beta s = q = (1-\beta)t, \quad (7.47\text{a–e})$$

since four arbitrary parameters α, β, r, t, were introduced, four equations (7.47b through e) may be imposed connecting them; the choices (7.47b through e) imply for (7.46):

$$f \in \mathcal{L}^p, g \in \mathcal{L}^q: \quad \left\{\left\|f \ast g\right\|_s\right\}^s \le \int\limits_{-\infty}^{+\infty}\left|f(y)\right|^{\alpha s} dy \int\limits_{-\infty}^{+\infty}\left|g(z)\right|^{\beta s} dz \left\{\int\limits_{-\infty}^{+\infty}\left|f(y)\right|^{(1-\alpha)r} dy\right\}^{s/r}$$

$$\left\{\int\limits_{-\infty}^{+\infty}\left|g(z)\right|^{(1-\beta)t} dz\right\}^{s/t} = \left\{\left\|f\right\|_p\right\}^{p+(1-\alpha)s}\left\{\left\|g\right\|_q\right\}^{q+(1-\beta)s} = \left\{\left\|f\right\|_p\left\|g\right\|_q\right\}^s, \quad (7.48\text{a–c})$$

where (7.47b through e) is used. This proves that the convolution of normed functions (7.48a and b) with norms (p, q) has norm s in (7.48c); it remains to relate (p, q, s) by solving (7.47a through e). The value of s in (7.37d) is calculated as follows: (1) first α, β are eliminated using (7.47b through e) and leading to (7.49a through f):

$$\frac{p}{s}=\alpha=1-\frac{p}{r}, \quad \frac{q}{s}=\beta=1-\frac{q}{t}: \quad \frac{1}{p}-\frac{1}{r}=\frac{1}{s}=\frac{1}{q}-\frac{1}{t}; \quad (7.49\text{a–f})$$

(2) then (7.49e and f) are substituted into (7.47a)

$$1=\frac{1}{r}+\frac{1}{t}+\frac{1}{s}=\frac{1}{p}-\frac{1}{s}+\frac{1}{q}-\frac{1}{s}+\frac{1}{s}=\frac{1}{p}+\frac{1}{q}-\frac{1}{s}, \quad (7.50)$$

yielding (7.50) $\equiv$ (7.37d). QED.

7.4.3 Convolution of Generalized Functions with Compact Support

The **convolution of generalized functions** (7.51a) over test functions (7.51b) with compact support is defined by (7.51c):

$$F(x), G(x) \in \mathcal{J}, \Phi(x) \in \mathcal{T}^0: \quad \left[F \ast G(x), \Phi(x)\right] \equiv \int\limits_{-\infty}^{+\infty}F \ast G(x)\Phi(x)dx$$

$$= \int\limits_{-\infty}^{+\infty}dx\int\limits_{-\infty}^{+\infty}dy\, F(y)G(x-y)\Phi(x); \quad (7.51\text{a–c})$$

performing a change of variable (7.38a) leads to

$$\left[F * G(x), \Phi(x)\right] = \int_{-\infty}^{+\infty} dy \int_{-\infty}^{+\infty} dz \, F(y)G(z)\Phi(y+z). \qquad (7.52)$$

Choosing the test function $\Phi(x)$ to have compact support, that is, to vanish outside a finite interval $x \in [a,b]$, the integration in (7.52) is performed over the strip $a \le y + z \le b$ in the (y,z)-plane (Figure 7.2a). If one of the generalized functions has compact support, for example, $F(y)$ vanishes outside the strip $c \le y \le d$, the region of integration in (7.52) is finite, namely, the parallelogram ABCD in Figure 7.2a, and the convolution integral (7.52) exists. It has been proven that *a sufficient condition for the existence (7.51c) $\equiv$ (7.52) $\equiv$ (7.53c) of the convolution integral (7.36) for generalized functions (7.53a) over test functions with compact support (7.53b) is that one of the generalized functions or its product has compact support (7.53a):*

$$F(x) \in \mathcal{J}^0 \quad \text{or} \quad G(x) \in \mathcal{J}^0 \quad \text{or} \quad F(x)G(x) \in \mathcal{J}^0 \wedge \Phi(x) \in \mathcal{T}^0 \Rightarrow F * G(x) \in \mathcal{J}, \qquad (7.53a\text{–}c)$$

where $\mathcal{J}^0$ is the set of generalized functions with compact support:

$$\mathcal{J}^0 \equiv \left\{ F(x): \quad F(x) \in \mathcal{J}^0 \wedge \exists_{a<b|R}: \quad x < a \text{ or } x > b \Rightarrow F(x) = 0 \right\}. \qquad (7.54)$$

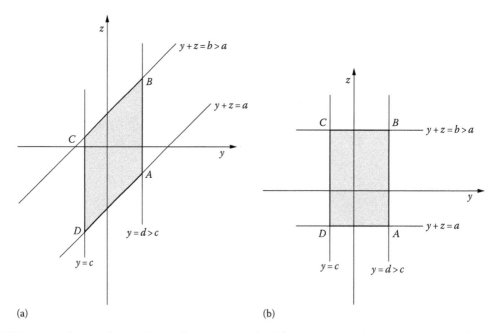

(a)   (b)

FIGURE 7.2   The convolution of two ordinary or generalized functions exists whenever the inner product by a test function is a two-dimensional integral over a compact domain, that is, not extending to infinity, for example, (a) a parallelogram if the test function has compact support and one of the functions or their product has compact support; (b) a rectangle if both functions have compact support.

An alternative sufficient condition for the existence of the convolution (7.51c) ≡ (7.52) ≡ (7.55d) of two generalized functions is that both have compact support (7.55a and b), even if the reference test does not have compact support but is integrable (7.55c):

$$F(x) \in \mathcal{J}^0 \wedge G(x) \in \mathcal{J}^0 \wedge \Phi(x) \in \mathcal{E}(|R): \quad F * G(x) \in \mathcal{J}^0. \tag{7.55a–d}$$

In this case (Figure 7.2b), the first (second) generalized function is nonzero in a vertical (horizontal) strip, and thus the integration is over their intersection in a rectangle with vertices at (A, B, C, D), and hence is finite. The unit impulse and its derivatives have compact support, namely, a point, and thus the convolution of the unit impulse or its derivatives (1) with another generalized function exists for test functions with compact support; and (2) with another unit impulse or its derivative exists for integrable test functions. Seven instances of evaluation of convolution integrals are included in Example 10.17.

7.5 Commutative, Associative, and Impulsive Properties

The convolution of ordinary functions has the commutative and associative properties (Subsection 7.5.1); since the unit impulse and its derivatives have compact support, that is, one-point, their convolution exists (Subsection 7.5.2) and leads to several simple properties (Subsection 7.5.3).

7.5.1 Commutative and Associative Properties of the Convolution of Ordinary Functions

The convolution of two functions (7.36) has the commutative (7.56a) and associative (7.56b) properties:

$$f * g = g * f, \quad f * (g * h) = (f * g) * h. \tag{7.56a,b}$$

The commutative property (7.56a) is proved (7.57) by means of a single change of variable (7.38a):

$$f * g(x) \equiv \int_{+\infty}^{+\infty} f(y)g(x-y)\,dy = \int_{-\infty}^{+\infty} g(z)f(x-z)\,dz = g * f(x). \tag{7.57}$$

Concerning the associative property (7.56b), it involves a repeated convolution:

$$f * (g * h)(x) \equiv \int_{-\infty}^{+\infty} f(y)g * h(x-y)\,dy = \int_{-\infty}^{+\infty} dy \int_{-\infty}^{+\infty} dz\ f(y)g(z)h(x-y-z); \tag{7.58}$$

performing the single change of variable (7.59a),

$$u \equiv y + z: \quad f * (g * h) \equiv \int_{-\infty}^{+\infty} du \int_{-\infty}^{+\infty} dy\ f(y)g(u-y)h(x-u)$$

$$= \int_{-\infty}^{+\infty} du\ f * g(u)h(x-u) = (f * g) * h(x), \tag{7.59a,b}$$

leads to (7.59b) that proves (7.59b) ≡ (7.56b).

7.5.2 Convolution of the Unit Impulse and Their Derivatives

The convolution of generalized functions exists if one of them has compact support, for example, the nth order derivate of the impulse has support reducing to a point; thus *the convolution with the nth derivate of the impulse: (1) exists (7.60c) for (1-a) a n-times continuously differentiable function (7.60a) over test functions with compact support (7.60b):*

$$f(x) \in C^n(|R) \wedge \Phi(x) \in \mathcal{J}^0(|R) \Rightarrow f * \delta^{(n)}(x) = (-)^n f^{(n)}(x), \qquad (7.60\text{a–c})$$

$$\{F(x) \in \mathcal{J}^0 \wedge \Phi(x) \in \mathcal{E}\} \text{ or } \{F(x) \in \mathcal{J} \wedge \Phi(x) \in \mathcal{T}^0\} \Rightarrow F * \delta^{(n)}(x) = (-)^n F^{(n)}(x); \qquad (7.61\text{a–e})$$

and (2) exists (7.61e) for (2-a) a generalized function with compact support (7.61a) over integrable test functions (7.61b) or (2-b) a generalized function (7.61c) over test functions with compact support (7.61d). In case (1) [(2)] the convolution of the ordinary (7.60c) [generalized (7.61e)] function with the nth derivative of the unit impulse is the nth derivative of the ordinary (generalized) function with plus (minus) sign if n is even (odd). The proof in the case: (1) of (7.60c) an n-times continuously differentiable ordinary function (7.62a) is the defining property (3.22a through c) $\equiv$ (7.62a and b) of the nth derivate impulse:

$$f(x) \in \mathcal{J}^n(|R): \quad f * \delta^{(n)}(x) = \int_{-\infty}^{+\infty} f(y)\delta^{(n)}(x-y)\,dy = (-)^n f^{(n)}(x); \qquad (7.62\text{a,b})$$

(2) of a generalized function (7.63a) over a test function $\Phi(x)$ with compact support (7.63b) the integral (7.51c) is used:

$$F(x) \in \mathcal{G}; \Phi \in \mathcal{T}^0: \quad \left[F * \delta^{(n)}(x), \Phi(x)\right] \equiv \int_{-\infty}^{+\infty} dx \int_{-\infty}^{+\infty} dy\, F(y)\delta^{(n)}(x-y)\Phi(x)$$

$$= \int_{-\infty}^{+\infty} dx(-)^n F^{(n)}(x)\Phi(x) = \left[(-)^n F^{(n)}(x), \Phi(x)\right]. \qquad (7.63\text{a–c})$$

In both cases, the same formula (7.60c) $\equiv$ (7.61e) is obtained.

7.5.3 Identity and Addition Properties for the Convolution of Impulses

Setting $n = 0$ in (7.60c) [(7.61e)], it follows that *the convolution of the unit impulse with (1) an ordinary continuous function (7.64a) [generalized function (7.64b)] in the conditions (7.60a and b) [(7.61a and b) or (7.61c and d)], that is, the unit impulse is the **neutral element** of the convolution, that is, it leads to an identity (7.64a):*

$$f(x) \in \mathcal{D}, \mathcal{G}: \quad f * \delta(x) = f(x), \quad \delta * \delta(x) = \delta(x). \qquad (7.64\text{a–d})$$

The property (7.64c) extends to the unit impulse itself, that is, *the impulse is unchanged by convolution with itself (7.64d)*; the proof in the latter case uses the integrals (7.51c):

$$\Phi(x) \in C(|R): \quad \left[\delta * \delta(x), \Phi(x)\right] = \int_{-\infty}^{+\infty} dx \int_{-\infty}^{+\infty} dy\, \delta(y)\delta(x-y)\Phi(x) = \int_{-\infty}^{+\infty} dy\, \delta(x)\Phi(x) = \left[\delta(x), \Phi(x)\right].$$

$$(7.65\text{a,b})$$

The property (7.64d) is generalized to derivatives, that is, *the convolution of nth and mth order derivate impulses is the $(n+m)$th order derivate impulse:*

$$n,m \in N: \quad \delta^{(n)} * \delta^{(m)}(x) = \delta^{(n+m)}(x); \tag{7.66}$$

the proof of (7.66) is similar to (7.63a through c) and (7.65a and b), that is, it uses (7.51c) again

$$\Phi(x) \in \mathcal{E}(|R): \quad \left[\delta^{(n)} * \delta^{(m)}(x), \Phi(x)\right] = \int_{-\infty}^{+\infty} dx\, \delta^{(n)} * \delta^{(m)}(x)\Phi(x)$$

$$= \int_{-\infty}^{+\infty} dx \int_{\infty}^{+\infty} dy\, \delta^{(n)}(y)\delta^{(m)}(x-y)\Phi(x)$$

$$= \int_{-\infty}^{+\infty} \delta^{(n)}(y)(-)^m \Phi^{(m)}(y)dy = (-)^{n+m} \Phi^{(n+m)}(0)$$

$$= \left[\delta^{(n+m)}(x), \Phi(x)\right]; \tag{7.67a,b}$$

also, (7.64d) is the particular case $n = 0 = m$ of (7.67b).

7.6 Principles of Superposition and Reciprocity

The influence or Green function of a differential equation is the fundamental solution obtained with forcing by a unit impulse (Subsection 7.6.1); if the differential equation is linear, with constant or variable coefficients, the solution for arbitrary forcing is obtained from the influence function using the principle of superposition (Subsection 7.6.2). If the linear differential operator is also self-adjoint, the influence function is symmetric and the principle of reciprocity also holds (Subsection 7.6.3).

7.6.1 Fundamental Solution of an Ordinary or Partial Differential Equation

A **linear ordinary (partial) differential operator** of order M is defined as (7.68) [(7.69)] polynomial of degree M of total d/dx (partial derivates $\partial/\partial \vec{x} \equiv \partial/\partial x_n$) derivates with regard to the coordinate(s) x (x_i with $i = 1, ..., N$ where N is the dimension of the space):

$$L\left(\frac{d}{dx}\right) \equiv \sum_{m=1}^{M} A_m(x)\frac{d^m}{dx^m}, \tag{7.68}$$

$$L\left(\frac{\partial}{\partial \vec{x}}\right) \equiv \sum_{m=1}^{M} A_{i_1 \dots i_m}(\vec{x})\frac{\partial^m}{\partial x_{i_1} \dots \partial x_{i_m}}, \tag{7.69}$$

with coefficients that may depend on the variable x (variables $\vec{x} \equiv x_i$), but not on the function the operator is applied to; repeated indices imply a summation over all dimensions of space, that is,

$$A_{i_1 \dots i_n}\frac{\partial^n}{\partial x_{i_1} \dots \partial x_{i_n}} \equiv \sum_{i_1=1}^{N} \dots \sum_{i_n=1}^{N} A_{i_1 \dots i_n}\frac{\partial^n}{\partial x_{i_1} \dots \partial x_{i_n}}, \tag{7.70}$$

in agreement with the Einstein (1916) **summation convention**. The **fundamental solution or influence or Green function** is defined as the solution of the ordinary (7.71a) [partial (7.71b)] differential equation forced by a one-dimensional impulse (7.71a) [impulse with the same dimension as the number of variables or dimension N of the space (7.71c)]:

$$\left\{L\left(\frac{d}{dx}\right)\right\}G(xy)=\delta(x-y);\quad \left\{L\left(\frac{\partial}{\partial\vec{x}}\right)\right\}G(\vec{x}\vec{y})=\delta(\vec{x}-\vec{y}),\quad \delta(\vec{x}-\vec{y})\equiv\prod_{n=1}^{N}\delta(x_n-y_n).\quad (7.71a\text{–}c)$$

The ordinary (partial) differential equation (7.71a) [(7.71b)] concerns the influence function in unbounded space; in the presence of boundary and/or initial conditions, these must also be satisfied by the influence function; thus the same differential equation will generally have different influence functions for distinct sets of boundary and/or initial conditions.

If $f(x)$ [$f(\vec{x})$] is an ordinary (generalized) function of one (7.72a and b) [several (7.72c and d)] variable(s),

$$f\in\mathcal{D}^M(|R)\ \text{or}\ \mathcal{G}(|R),\quad f\in\mathcal{D}^M(|R^N)\ \ \text{or}\ \ \mathcal{G}(|R^N),\qquad (7.72a\text{–}d)$$

the convolution property with the unit impulse (7.64c) and the definition of influence function (7.71a) [(7.71b)] lead to:

$$G(x;\xi)=G(x-\xi):\ \ f(x)=f*\delta(x)=f*\left\{L\frac{d}{dx}\right\}G(x)=\left\{L\frac{d}{dx}\right\}f*G(x),\qquad (7.73a,b)$$

$$G(\vec{x};\vec{\xi})=G(\vec{x}-\vec{\xi}):\ \ f(\vec{x})=f*\delta(\vec{x})=f*\left\{L\frac{\partial}{\partial\vec{x}}\right\}G(\vec{x})=\left\{L\frac{\partial}{\partial\vec{x}}\right\}f*G(\vec{x});\qquad (7.73c,d)$$

it follows that the solution of the linear ordinary (7.68) [partial (7.69; 7.70)] differential equation forced (7.73b) $\equiv$ (7.77a) [(7.73d) $\equiv$ (7.77b)] by an ordinary or generalized function (7.73b) [(7.73d)] is given by its convolution with the assumption that the influence function takes the particular form (7.73a) [(7.73c)] of a difference of variables. The latter restriction (7.73a) [(7.73c)] is not necessary, since the property (7.73b) $\equiv$ (7.74a) [(7.73d) $\equiv$ (7.74b)] depends only on the linearity of the ordinary (7.68) [partial (7.69; 7.70)] differential operator:

$$f(x)=\int_{-\infty}^{+\infty}f(y)\delta(x-y)dy=\int_{-\infty}^{+\infty}f(y)\left\{L\left(\frac{d}{dx}\right)\right\}G(x;y)dy=\left\{L\left(\frac{d}{dx}\right)\right\}\int_{-\infty}^{+\infty}f(y)G(x;y)dy,\quad (7.74a)$$

$$f(\vec{x})=\int_{-\infty}^{+\infty}f(\vec{y})\delta(\vec{x}-\vec{y})d^N\vec{y}=\int_{-\infty}^{+\infty}f(\vec{y})\left\{L\left(\frac{\partial}{\partial\vec{x}}\right)\right\}G(\vec{x};\vec{y})d^N\vec{y}=\left\{L\left(\frac{\partial}{\partial\vec{x}}\right)\right\}\int_{-\infty}^{+\infty}f(y)G(\vec{x};\vec{y})d^N\vec{y}.\quad (7.74b)$$

The interchange of the linear ordinary (7.68) [partial (7.69; 7.70)] differential operator in (7.74a) [(7.74b)] assumes that the integral is (Section I.13.8) uniformly convergent with regard to x. The same assumption is implied when the differential operator is interchanged with the convolution in (7.73a) [(7.73d)], because the integration is performed in a different variable, as can be seen by writing explicitly the integrations in (7.74a) [(7.74b)]. The only difference between (7.74a) [(7.74b)] is the use of scalar x (vector $\vec{x}$) variables, and hence integration in one (several) dimensions. The influence or Green function can be defined as the fundamental solution for any ordinary (7.77a) [partial (7.77b)] differential operator, linear

(7.68) [(7.69; 7.70)] or not. The principle of superposition and convolution property hold in the linear case (7.73a) ≡ (7.74a) [(7.73b) ≡ (7.74b)] and generally not otherwise; the linearity concerns both the differential operator and initial and/or boundary conditions.

7.6.2 Discrete and Integral Principle of Linear Superposition

The proof of (7.74a) [(7.74b)] applies to a linear *ordinary (7.68) [partial (7.69; 7.70)] differential operator of order M,*

$$\alpha_k \in |R: \quad L\left\{\sum_{k=1}^{k} \alpha_k f_k\right\} = \sum_{k=1}^{S} \alpha_k L\{f_k\}, \tag{7.75}$$

*so that the **principle of superposition** holds, in the sense that a linear combination of solutions is a solution:*

$$k = 1,\ldots,K; \quad \alpha_k \in |R: L\{f_1\} = \ldots = L\{f_K\} = 0 \Rightarrow L\left\{\sum_{k=1}^{K} \alpha_k f_k\right\} = 0; \tag{7.76}$$

if there are boundary and/or initial conditions, these must also be linear in order for the superposition principle to hold. The superposition principle can be stated in discrete (7.76), or integral form; an example of the latter is the solution of the linear ordinary (7.77a; 7.68) [partial (7.77b; 7.69; 7.70)] differential equation:

$$\left\{L\left(\frac{\mathrm{d}}{\mathrm{d}x}\right)\right\}\Phi(x) = f(x), \quad \left\{L\left(\frac{\partial}{\partial\vec{x}}\right)\right\}\Phi(\vec{x}) = f(\vec{x}), \tag{7.77a,b}$$

with arbitrary forcing, using the integral superposition in one (7.78a) [several N (7.69b)] dimensions (7.74a) ≡ (7.78a) [(7.74b) ≡ (7.78b)]:

$$\Phi(x) = \int_{-\infty}^{+\infty} f(y)G(x;y)\mathrm{d}y, \tag{7.78a}$$

$$\Phi(\vec{x}) = \int_{-\infty}^{+\infty} f(\vec{y})G(\vec{x};\vec{y})\mathrm{d}^N\vec{y}. \tag{7.78b}$$

The latter (7.78a) [(7.78b)] is a linear superposition of forcing effects weighted by the influence or Green function (7.71a) [(7.71b and c)] where the forcing function may be: (1) an ordinary M-times differentiable function (7.72a and c); and (2) a generalized function (7.72b and d). In both cases (1) and (2) it depends on one (7.78a; 7.74a; 7.73a) variable [several (7.78b; 7.74b; 7.73b) variables]. The linear superposition integral (7.78a) ≡ (7.78b) ≡ [(7.74b)] becomes a convolution integral (7.73b) ≡ (7.36) [(7.73d)] if the influence or Green function depends only on the difference of the two variables (7.73a) [(7.73c)].

7.6.3 Influence or Green (1837) Function and Self-Adjoint Operator

Besides the preceding results (Subsections 7.6.1 and 7.6.2) concerning the principle of superposition, another important property of the Green function is the **reciprocity principle**: the field at $x(\vec{x})$ due to forcing at $y(\vec{y})$ is equal to the field at $y(\vec{y})$ due to forcing at $x(\vec{x})$:

$$G(x;y) = G(y;x), \quad G(\vec{x};\vec{y}) = G(\vec{y};\vec{x}), \tag{7.79a,b}$$

if the Green function is symmetric in (7.79a) [(7.79b)]. The condition of symmetry of the Green function may be passed to the operator (7.71a):

$$G(\xi;\eta) - G(\eta;\xi) = \int_{-\infty}^{+\infty} \left[G(x;\eta)\delta(x-\xi) - G(x;\xi)\delta(x-\eta) \right] dx$$

$$= \int_{-\infty}^{+\infty} \left\{ G(x;\eta) \left\{ L\left(\frac{d}{dx} \right) \right\} G(x;\xi) - G(x;\xi) \left\{ L\left(\frac{d}{dx} \right) \right\} G(x;\eta) \right\} dx, \qquad (7.80)$$

and is equivalent to the vanishing of the integral (7.80); the implication is that the integrand is an exact differential:

$$\Psi(x) \left\{ L\left(\frac{d}{dx} \right) \right\} \Phi(x) - \Phi(x) \left\{ L\left(\frac{d}{dx} \right) \right\} \Psi(x) = \frac{dM}{dx}, \qquad (7.81)$$

with a bilinear concomitant M. Generally, *given a linear operator $L(d/dx)$, it is possible to find an **adjoint operator** $\bar{L}(d/dx)$ such that their **commutator***

$$\Psi(x) \left\{ L\left(\frac{d}{dx} \right) \right\} \Phi(x) - \Phi(x) \left\{ \bar{L}\left(\frac{d}{dx} \right) \right\} \Psi(x) = \frac{d}{dx} \left[W(\Phi,\Psi) \right] \qquad (7.82)$$

*is the exact differential of a **bilinear concomitant**; the latter is a bilinear function of (Φ, Ψ) and their derivatives up to order M – 1. In the case (7.81) the operator $L \equiv \bar{L}$ coincides with its adjoint, and thus is called a **self-adjoint operator**.* It has been proven that in one dimension (7.80; 7.81; 7.82) ≡ (7.83a through c), and can be proved similarly in several dimensions (7.84a through c), that

$$G(x;y) = G(y;x) \Leftrightarrow L\left(\frac{d}{dx} \right) \Leftrightarrow \bar{L}\left(\frac{d}{dx} \right) = \Psi L\Phi - \Phi L\Psi = \frac{dM}{dx}, \qquad (7.83a-c)$$

$$G(\vec{x};\vec{y}) = G(\vec{y};\vec{x}) \Leftrightarrow L\left(\frac{\partial}{\partial \vec{x}} \right) = \bar{L}\left(\frac{\partial}{\partial \vec{x}} \right) \Leftrightarrow \Psi L\bar{\Phi} - \Phi L\Psi = \nabla \cdot \vec{M}. \qquad (7.84a-c)$$

The following statements are equivalent in one (7.83a through c) [several (7.84a through c)] dimensions: (1) the reciprocity principle holds; (2) the influence or Green function is symmetric (7.83a) [(7.84a)]; (3) the differential operator L has a commutator that is an exact differential (7.83c) [the divergence of a vector (7.84c)] called the bilinear concomitant since it is a bilinear function of (Φ, Ψ) and its derivatives up to order M − 1; and (4) the differential operator L = $\bar{L}$ is self-adjoint (7.83b) [(7.84b)]. Two instances of unsymmetric Green functions associated with non-self-adjoint linear differential operators are presented in Example 10.19. The relation (7.82) only holds for linear operators because if the operator L (adjoint operator $\bar{L}$) is nonlinear, then (1) the first (second) term on the l.h.s. is nonlinear in $\Phi(\Psi)$; (2) since the second (first) term on the l.h.s. is linear in $\Phi(\Psi)$, their difference cannot be an exact differential, as stated on the r.h.s. of (7.82). Thus (7.82) cannot hold if either the operator L or its self-adjoint $\bar{L}$ were nonlinear. This implies that (1/2) the operator L (adjoint operator $\bar{L}$) is linear in $\Phi(\Psi)$; and (3) the concomitant is bilinear in Φ and Ψ. The adjoint linear operator and the bilinear concomitant are obtained next for any linear differential operator in three cases: (1) a linear ordinary differential operator in one variable, for example, time t or one spatial coordinate x for a one-dimensional problem (Section 7.7);

(2) a linear partial differential operator in N variables, for example, N spatial variables x_i with $i = 1, ..., N$ for an N-dimensional problem (Section 7.8); and (3) a linear space-time differential operator is a partial differential operator in $(N + 1)$ variables (t, x_i), which includes (Section 7.9) some of the most important equations of mathematical physics (Notes 7.2 through 7.17).

7.7 Operator, Adjoint, and Bilinear Concomitant

Given an ordinary linear differential operator with variable coefficients, integration by parts specifies (Subsection 7.7.1) the bilinear concomitant and the adjoint operator; the coincidence of the original and adjoint operator specifies the self-adjoint form (Subsection 7.7.2). The general linear second-order ordinary differential operator can be put into self-adjoint form by multiplying it by a suitable factor (Subsection 7.7.4). The self-adjoint (non-self-adjoint) operators correspond to conservative (dissipative) systems, for example, the linear deflection of an elastic string with nonuniform tension (Subsection 7.7.3) [a damped harmonic oscillator with variable coefficients (Subsection 7.7.5)]. The adjoint operators can be extended to differential equations of any order (Subsection 7.7.6) either with constant or variable coefficients [Subsection 7.7.7 (Subsection 7.7.8)]. An example of self-adjoint operator of second (fourth)-order is the deflection (bending) of an elastic string (bar) with nonuniform tension (stiffness) supported on springs [Subsection 7.7.3 (Subsection 7.7.9)].

7.7.1 General Second-Order Linear Ordinary Differential Operator

As an example of the procedure to find the adjoint operator and the bilinear concomitant, consider the general **linear second-order ordinary differential operator** (7.68) that is of order $N = 2$ in (7.85a) with variable coefficients (7.85b):

$$(A_0, A_1, A_2) \equiv (c, b, a): \quad L\left(\frac{d}{dt}\right) = a(t)\frac{d^2}{dt^2} + b(t)\frac{d}{dt} + c(t). \qquad (7.85a,b)$$

The adjoint operator is found starting from the first term $\Psi L \Phi$ on the l.h.s. of (7.86) and forming total differentials until in the remaining term Φ comes as a factor in $-\Phi \bar{L} \Psi$, namely:

$$\Psi\left\{L\frac{d}{dt}\right\}\Phi = \Psi(a\Phi'' + b\Phi' + c\Phi)$$

$$= (\Psi a \Phi')' - \Phi'(a\Psi)' + (\Psi b \Phi)' - \Phi(b\Psi)' + \Phi c \Psi$$

$$= \{\Psi a \Phi' - \Phi(a\Psi)' + \Psi b \Phi\}' + \Phi\{(a\Psi)'' - (b\Psi)' + c\Psi\}; \qquad (7.86)$$

the process corresponds to repeated integration by parts until all derivatives on Φ are passed over to Ψ. The last term in curly brackets on the r.h.s. of (7.76) is $\bar{L}\Psi$, and specifies the operator (7.87) adjoint to (7.85b):

$$\left\{\bar{L}\left(\frac{d}{dt}\right)\right\} \equiv \frac{d^2}{dt^2}a(t) - \frac{d}{dt}b(t) + c(t); \qquad (7.87)$$

substituting (7.85b; 7.87) into (7.86), it takes the form of a **commutation relation** (7.82), with the bilinear concomitant

$$W(\Phi, \Psi) \equiv \Psi a \Phi' - \Phi(a\Psi)' + \Psi b \Phi = a(\Phi'\Psi - \Psi'\Phi) + (b - a')\Phi\Psi. \qquad (7.88)$$

It has been proven that *the general (7.85a) ≡ (7.89a) linear ordinary second-order differential operator L has (7.87) ≡ (7.89b) adjoint $\bar{L}$*:

$$\left\{L\left(\frac{d}{dt}\right)\right\}\Phi = a(t)\Phi'' + b(t)\Phi' + c(t)\Phi, \quad \left\{\bar{L}\left(\frac{d}{dt}\right)\right\}\Psi = (a\Psi)'' - (b\Psi)' + c\Psi, \quad \text{(7.89a,b)}$$

satisfying (7.82) with the bilinear concomitant (7.88), that is, a bilinear function of (Φ, Ψ) and their first-order derivatives.

7.7.2 Second-Order Linear Self-Adjoint Operator

The difference between the operator L in (7.89a) and its adjoint $\bar{L}$ in (7.89b)

$$\left\{L\left(\frac{d}{dt}\right)\right\}\Phi - \left\{\bar{L}\left(\frac{d}{dt}\right)\right\}\Phi = a\Phi'' - (a\Phi)'' + b\Phi' + (b\Phi)'$$

$$= -a''\Phi - 2a'\Phi' + 2b\Phi' + b'\Phi = (b-a')'\Phi + 2(b-a')\Phi' \quad \text{(7.90)}$$

vanishes if

$$\left\{L\frac{d}{dt}\right\}\Phi = \left\{\bar{L}\frac{d}{dt}\right\}\Phi \Leftrightarrow b = a' \Leftrightarrow M(\Phi, \Psi) = a(\Phi'\Psi - \Psi'\Phi), \quad \text{(7.91a–c)}$$

in which case the bilinear concomitant (7.88) simplifies to the first term. It has been shown that *the most general **linear second-order self-adjoint ordinary differential operator** is the* **Sturm (1836)–Liouville (1847) operator**:

$$\left\{L\left(\frac{d}{dt}\right)\right\}\Phi = a\Phi'' + a'\Phi' + c\Phi = (a\Phi')' + c\Phi = \left\{\bar{L}\left(\frac{d}{dt}\right)\right\}\Phi; \quad \text{(7.92)}$$

its bilinear concomitant is (7.93b) ≡ (7.91c):

$$\Psi\left\{L\frac{d}{dt}\right\}\Phi - \Phi\left\{L\frac{d}{dt}\right\}\Psi = \frac{d}{dt}[W(\Phi, \Psi)], \quad W(\Phi, \Psi) = a(\Phi'\Psi - \Psi'\Phi), \quad \text{(7.93a,b)}$$

and it satisfies the commutation relation (7.93a). An example of self-adjoint linear second-order differential operator is given next (Subsection 7.7.3).

7.7.3 Elastic String under Nonuniform Tension Supported on Springs

Consider the linear deflection of an elastic string supported on linear springs (Figure 7.3a) for which the shear stress or transverse force per unit length (2.7a) is balanced by the sum of (1) minus the rate of change along the length of the vertical component of the tangential tension $T(x)$ that may be nonuniform (2.11c); and (2) the force exerted by linear springs, whose resilience $k(x)$ may vary with position:

$$f(x) = -[T(x)\zeta']' + k(x)\zeta = \left\{L\left(\frac{d}{dx}\right)\right\}\zeta(x). \quad \text{(7.94)}$$

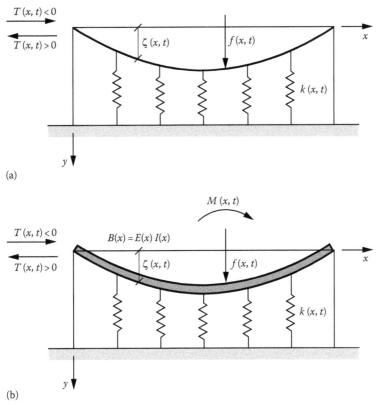

(a)

(b)

FIGURE 7.3 The weak deflection of a string (membrane) under nonuniform tension supported on linear springs with nonuniform resilience is specified by a linear self-adjoint ordinary (partial) second-order differential operator (a); the operator becomes of fourth order for (b) the weak bending of an elastic bar (plate) without or with longitudinal (in plane) tension, that is, a beam (stressed plate). The operator remains self-adjoint in the unsteady case of oscillations (vibrations) of the beam (stressed) plate for nonuniform and unsteady mass density, bending stiffness, tangential (in plane) tension, and spring resilience; the operator becomes non-self-adjoint and the principle of reciprocity fails only in the presence of damping (Figure 7.4).

The operator applied in (7.94) to the transverse deflection $\zeta(x)$ is,

$$\left\{ L\left(\frac{d}{dt}\right)\right\} = -\frac{d}{dx}T(x)\frac{d}{dx}+k(x) = \bar{L}\left(\frac{d}{dx}\right),\tag{7.95}$$

which is a linear second-order self-adjoint differential operator (7.95) ≡ (7.96a):

$$\Psi\left\{ L\frac{d}{dx}\right\}\Phi - \Phi\left\{ L\frac{d}{dx}\right\}\Psi = \frac{d}{dx}\left[W(\Phi,\Psi)\right], \quad W(\Phi,\Psi) = \Psi\Phi' - \Phi\Psi,\tag{7.96a,b}$$

with the bilinear concomitant (7.96b). It follows that *the linear deflection of an elastic string supported by linear springs (Figure 7.3a) is specified by the linear second-order differential operator (7.95) applied to the transverse displacement (7.94) with forcing. The operator is self-adjoint (7.96a) with the bilinear concomitant (7.96b). Thus are satisfied the principles of superposition and reciprocity including in the cases of nonuniform tangential tension and/or resilience of the springs.*

7.7.4 Multiplying Factor and Transformation to Self-Adjoint Form

The general ordinary linear second-order operator (7.85b) ≡ (7.97a) is not self-adjoint, that is, it does not coincide with the Sturm–Liouville operator (7.92) ≡ (7.97b):

$$a\Phi'' + b\Phi' + c\Phi = 0, \quad 0 = (p\Phi')' + q\Phi = p\Phi'' + p'\Phi + q\Phi. \tag{7.97a,b}$$

The self-adjoint (7.97b) and non-self-adjoint (7.97a) operators can be made to coincide multiplying the former by a function μ in (7.98a):

$$\mu\left(a\Phi'' + b\Phi' + c\Phi\right) = p\Phi'' + p'\Phi' + q\Phi, \quad \mu = \frac{p}{a} = \frac{p'}{b} = \frac{q}{c}, \tag{7.98a–d}$$

provided it satisfies all three conditions (7.98b through d); these are used next to determine (μ, p, q). From (7.98b and c) follows (7.99a), whose solution is (7.99b):

$$\frac{b}{a} = \frac{p'}{p} = \frac{d}{dt}(\log p), \quad p(t) = \exp\left\{ \int^{t} \frac{b(t)}{a(t)} dt \right\}, \tag{7.99a,b}$$

to within a nonessential multiplying constant which is equated to unity. Thus the function μ is given (7.98b; 7.99b) by (7.100a):

$$\mu = \frac{1}{a} \exp\left\{ \int^{b}_{a} \frac{b}{a} dt \right\}, \quad q = c\mu = \frac{c}{a} \exp\left\{ \int^{b}_{a} \frac{b}{a} dt \right\}, \tag{7.100a,b}$$

and the function q by (7.98d) ≡ (7.100b). It can be checked that the factor (7.100a) in (7.98a) transforms the non-self-adjoint (7.97a) to a self-adjoint (7.97b) differential operator:

$$\mu\left(a\Phi'' + b\Phi' + c\Phi\right) = \exp\left\{ \int^{b}_{a} \frac{b}{a} dt \right\} \left[\Phi'' + \frac{b}{a}\Phi' + \frac{c}{a}\Phi \right]$$

$$= \frac{d}{dt}\left[\Phi'\left\{ \exp \int^{b}_{a} \frac{b}{a} dt \right\} \right] + \frac{c}{a}\Phi\left\{ \exp \int^{b}_{a} \frac{b}{a} dt \right\} = (p\Phi')' + q\Phi. \tag{7.101}$$

It has been shown that *the general linear ordinary second-order differential operator (7.97a), upon multiplication by (7.100a) ≡ (7.102b), becomes a self-adjoint differential operator (7.101) ≡ (7.97b) ≡ (7.102a) with coefficients (7.102c) ≡ (7.99b) and (7.102d) ≡ (7.100b):*

$$\left\{ L\left(\frac{d}{dx}\right) \right\}\Phi \equiv (p\Phi')' + q\Phi = \left\{ \bar{L}\left(\frac{d}{dt}\right) \right\}\Phi : \quad \{\mu, p, q\} = \left\{ \frac{1}{a}, 1, \frac{c}{a} \right\} \exp\left[\int^{b}_{a} \frac{b}{a} dt \right]. \tag{7.102a–d}$$

This transformation is shown next to relate the linear harmonic oscillator without (with) damping that is specified by a linear self-adjoint (non-self-adjoint) second-order differential operator in time.

7.7.5 Undamped and Damped Harmonic Oscillators

Consider a linear harmonic oscillator, such as a mechanical (electrical) circuit [Figure 7.4a (Figure 7.4b)] consisting of a mass (self), damper (resistance), and spring (capacitor) with applied mechanical (electromotive) force. This electromechanical analogy (Section I.4) allows using next the mechanical terminology for the equation of motion that balances the external force against the sum of (1) the inertia force,

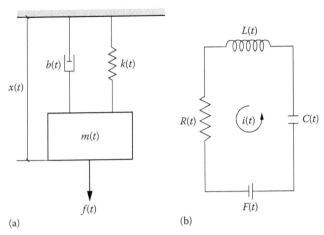

FIGURE 7.4 A harmonic oscillator without damping is specified by a linear self-adjoint second-order ordinary differential operator with time as the independent variable, for example, a mechanical circuit (Figure 7.4a) [an electrical circuit Figure 7.4b)] consisting of a mass (self) and spring (capacitor) both dependent on time with an applied (electromotive) force; adding dissipation, with a damper (resistance), either constant or time dependent, renders the operator non-self-adjoint.

that is, the time derivative of the linear momentum, specified by the product of mass and velocity; (2) the friction force of the damper that is linear the velocity with damping factor λ as the coefficient; and (3) the restoring force of the spring linear in the displacement with resilience k as the coefficient:

$$f(t) = \frac{d}{dt}\left[m(t)\frac{dx}{dt}\right] + \lambda(t)\frac{dx}{dt} + k(t)x \equiv \left\{L\left(\frac{d}{dt}\right)\right\}x(t). \qquad (7.103)$$

The coefficients may depend on time, for a body with variable mass with moving support for the damper and spring; the particular case of constant parameters is (I.4.1). The linear second-order differential operator (7.103) in time is (7.104a) $\equiv$ (7.85) and has adjoint (7.87) $\equiv$ (7.104b):

$$L\left(\frac{d}{dx}\right) = \frac{d}{dt}m\frac{d}{dt} + \lambda\frac{d}{dt} + k = m\frac{d^2}{dt^2} + (m'+\lambda)\frac{d}{dt} + k, \qquad (7.104a)$$

$$\bar{L}\left(\frac{d}{dt}\right) = \frac{d^2}{dt^2}m - \frac{d}{dt}\lambda + k = m\frac{d^2}{dt^2} + (2m'-\lambda)\frac{d}{dt} + m'' - \lambda' + k, \qquad (7.104b)$$

with (7.88) the bilinear concomitant (7.105a):

$$\Psi\left\{L\left(\frac{d}{dt}\right)\right\}\Phi - \Phi\left\{\bar{L}\left(\frac{d}{dt}\right)\right\}\Psi = \frac{d}{dt}\left[W(\Phi,\Psi)\right], \qquad (7.104c)$$

$$W(\Phi,\Psi) = \Psi m\Phi' - \Phi(m\Psi)' + \Psi\lambda\Phi = m(\Psi\Phi' - \Phi\Psi') + (\lambda - m')\Phi\Psi, \qquad (7.104d)$$

satisfying the commutation relation (7.105b). In order for the operator in (7.103) to be self-adjoint, the coefficients in (7.104a) $\equiv$ (7.104b) must coincide, leading to the two conditions (7.105a and b) that are incompatible unless the damping is zero (7.105c):

$$m' = 2\lambda, \quad m'' = \lambda' \Rightarrow \lambda = 0, \quad w(\Phi,\Psi) = m(\Psi\Phi' - \Phi\Psi') - m'\Phi\Psi, \qquad (7.105a,b)$$

Thus the operator in (7.103) is self-adjoint (7.104a) ≡ (7.104b) ≡ (7.106b) ≡ (7.106c) only in the absence of damping (7.106a):

$$b = 0: \quad L\left(\frac{d}{dt}\right) = \frac{d}{dt} m(t) \frac{d}{dt} + k(t) = \bar{L}\left(\frac{d}{dt}\right), \quad W(\Phi, \Psi) = m(\Psi\Phi' - \Phi\Psi') - m'\Phi\Psi, \quad (7.106a\text{–}d)$$

when the bilinear concomitant (7.104d) simplifies to (7.106d).

The transformation between the self-adjoint (7.106b) and non-self-adjoint (7.104a) operators is made in three steps: (1) comparing (7.104a) with the general second-order differential equation (7.104a) ≡ (7.89a) ≡ (7.107d), leading to the coefficients (7.107a through c):

$$\{a, b, c\} = \{m, m' + \lambda, k\}: \quad \left\{L\left(\frac{d}{dt}\right)\right\}\Phi = m\Phi'' + (m' + \lambda)\Phi' + k\Phi; \quad (7.107a\text{–}d)$$

(2) specifying the multiplying factor (7.100a) ≡ (7.108a):

$$\mu = \frac{1}{m}\exp\left(\int \frac{m' + \lambda}{m} dt\right) = \frac{1}{m}\exp\left(\log m + \int \frac{\lambda}{m} dt\right) = \exp\left(\int \frac{\lambda}{m} dt\right); \quad (7.108a)$$

(3) multiplying the non-self-adjoint operator (7.104a) corresponding to the damped harmonic oscillator by the factor (7.108a) leads to the self-adjoint operator (7.108b) ≡ (7.106b) corresponding to the undamped harmonic oscillator:

$$f(t)\exp\left(\int \frac{\lambda}{m} dt\right) = \exp\left(\int \frac{\lambda}{m} dt\right)\left[(mx')' + \lambda x' + kx\right]$$

$$= \left[mx'\exp\left(\int \frac{\lambda}{m} dt\right)\right]' + kx\exp\left(\int \frac{\lambda}{m} dt\right). \quad (7.108b)$$

Thus *the linear harmonic oscillator with variable coefficients (7.103) is self-adjoint (non-self-adjoint) if undamped (7.106a through d) ≡ (7.95) [damped (7.104a through d)], and thus the transformation between the two cases (7.108b) involves the addition (removal) of damping. The damping corresponds to the simplest non-self-adjoint linear differential operator involving only a first-order derivative:*

$$L\left(\frac{d}{dt}\right) = \lambda(t)\frac{d}{dt}, \quad \bar{L}\left(\frac{d}{dt}\right) = -\lambda(t)\frac{d}{dt}, \quad (7.109a,b)$$

$$\Psi\left\{L\left(\frac{d}{dt}\right)\right\}\Phi - \Phi\left\{\bar{L}\left(\frac{d}{dt}\right)\right\}\Psi = \frac{d}{dt}\left[M(\Phi, \Psi)\right], \quad M(\Phi, \Psi) = \lambda(t)\Phi\Psi, \quad (7.109c,d)$$

as follows from (7.104a through d) with $m = 0 = k$.

7.7.6 High-Order Linear Operators with Variable Coefficients (Lagrange)

The adjoint operator (7.87) ≡ (7.89b) and bilinear concomitant (7.88) can be generalized from the linear second-order differential operator (7.85b) to any order (7.68) as follows: (1) for the general term (7.110a) starting with the identity (7.110b):

$$L\left(\frac{d}{dt}\right) \equiv A_m(t)\frac{d^m}{dt^m}: \quad \Psi A_m \frac{d^m\Phi}{dt^m} = \Psi A_m \Phi^{(m)} = \left[\Psi A_m \Phi^{(m-1)}\right]' - \Phi^{(m-1)}(A_m\Psi)'$$

$$= \left[\Psi A_m \Phi^{(m-1)} - (A_m\Psi)'\Phi^{(m-2)}\right]' + \Phi^{(m-2)}(A_m\Psi)''; \quad (7.110a,b)$$

(2) continuing the integration by parts m times leads to the bilinear concomitant:

$$W(\Phi,\Psi) = \Phi^{(m-1)}A_m\Psi - \Phi^{(m-2)}(A_m\Psi)' + \cdots$$
$$+ (-)^{\ell+1}\Phi^{(m-\ell)}(A_m\Psi)^{\ell-1} + \cdots + (-)^{m-1}\Phi(A_m\Psi)^{(m-1)}; \tag{7.111}$$

(3) the last term left in (7.110b) $\equiv$ (7.112a) after (7.111) is Φ times the adjoint operator (7.112b):

$$\Psi\left\{L\left(\frac{\mathrm{d}}{\mathrm{d}t}\right)\right\}\Phi - \Phi\left\{\bar{L}\left(\frac{\mathrm{d}}{\mathrm{d}t}\right)\right\}\Psi = \frac{\mathrm{d}}{\mathrm{d}t}\big[M(\Phi,\Psi)\big], \quad \left\{\bar{L}\left(\frac{\mathrm{d}}{\mathrm{d}t}\right)\right\}\Psi = (-)^m(A_m\Psi)^{(m)}; \tag{7.112a,b}$$

(4) since the differential operator (7.68) is linear, the terms (7.110a; 7.112b; 7.111) can be added for all orders $m = 1, \ldots, M$, leading to the **adjoint operator theorem (Lagrange)**: *the **linear ordinary differential operator** of order M with variable coefficients (7.113a) has adjoint operator (7.113b):*

$$\left\{L\frac{\mathrm{d}}{\mathrm{d}t}\right\}\Phi = \sum_{m=0}^{M}A_m(t)\frac{\mathrm{d}^m\Phi}{\mathrm{d}t^m}, \quad \left\{\bar{L}\frac{\mathrm{d}}{\mathrm{d}t}\right\}\Psi = \sum_{m=0}^{\infty}(-)^m\frac{\mathrm{d}^m}{\mathrm{d}t^m}\big[A_m(t)\Psi\big]; \tag{7.113a,b}$$

they satisfy the commutation relation (7.112a):

$$W(\Phi,\Psi) = \sum_{m=0}^{M}\sum_{\ell=1}^{m}(-)^{\ell+1}\Phi^{(m-\ell)}(t)\big[A_m(t)\Psi(t)\big]^{(\ell-1)}, \tag{7.114}$$

involving the bilinear concomitant (7.111) $\equiv$ (7.114).

7.7.7 Linear Differential Operator with Constant Coefficients

In the case of constant coefficients (7.115a), it follows from (7.110a) $\equiv$ (7.115b) and (7.112b) $\equiv$ (7.115c) that a derivate of even (odd) order is (is not) self-adjoint:

$$A_m = \text{const}: \quad \left\{L\left(\frac{\mathrm{d}}{\mathrm{d}t}\right)\right\}\Phi = A_m\frac{\mathrm{d}^m\Phi}{\mathrm{d}t^m}, \quad \left\{\bar{L}\left(\frac{\mathrm{d}}{\mathrm{d}t}\right)\right\}\Psi = A_m(-)^m\frac{\mathrm{d}^m\Psi}{\mathrm{d}t^m}, \tag{7.115a–c}$$

$$W(\Phi,\Psi) = \sum_{\ell=1}^{m}(-)^{\ell+1}A_m\Phi^{(m-\ell)}(t)\Psi^{(\ell-1)}(t), \tag{7.115d}$$

Thus *a **linear ordinary differential operator with constant coefficients** (7.116a) is: (1) self-adjoint (7.116b) iff (if and only if) it constrains only derivatives of even order (7.116c):*

$$\{A_1,\ldots,A_{2M}\} = \text{const}: \quad L\left(\frac{\mathrm{d}}{\mathrm{d}t}\right) = \bar{L}\left(\frac{\mathrm{d}}{\mathrm{d}t}\right) \Leftrightarrow L\left(\frac{\mathrm{d}}{\mathrm{d}t}\right) = \sum_{m=0}^{M}A_{2m}\frac{\mathrm{d}^{2m}}{\mathrm{d}t^{2m}}; \tag{7.116a–c}$$

(2) is non-self-adjoint if it contains at least one derivative of odd order. The self-adjoint case (7.117a) corresponds to the bilinear concomitant (7.117b):

$$\Psi\left\{L\frac{\mathrm{d}}{\mathrm{d}t}\right\}\Phi - \Phi\left\{L\frac{\mathrm{d}}{\mathrm{d}t}\right\}\Psi = \frac{\mathrm{d}}{\mathrm{d}t}\big[W(\Phi,\Psi)\big], \tag{7.117a}$$

$$W(\Phi,\Psi) = \sum_{m=1}^{M}\sum_{\ell=1}^{m}(-)^{\ell+1} A_{2m}\Phi^{(2m-\ell)}\Psi^{(\ell-1)}. \tag{7.117b}$$

The non-self-adjoint case (7.118a) corresponds to the adjoint operator (7.118c):

$$\left\{L\left(\frac{d}{dt}\right)\right\} = \sum_{m=0}^{M}A_m\frac{d^m}{dt^m}, \quad \Psi\left\{L\left(\frac{d}{dt}\right)\right\}\Phi - \Phi\left\{\bar{L}\left(\frac{d}{dt}\right)\right\}\Psi = \frac{d}{dt}\big[W(\Phi,\Psi)\big], \tag{7.118a,b}$$

$$\left\{\bar{L}\left(\frac{d}{dt}\right)\right\} = \sum_{m=0}^{M}(-)^m A_m\frac{d^m}{dt^m}, \quad M(\Phi,\Psi) = \sum_{m=1}^{M}\sum_{\ell=1}^{m}(-)^{\ell+1} A_m\Phi^{(m-\ell)}\Psi^{(\ell-1)}, \tag{7.118c,d}$$

which satisfies the commutation relation (7.118b) involving the bilinear concomitant (7.118d). For example, the harmonic undamped (damped) oscillator operator (7.106b) [(7.104a)] is self-adjoint (non-self-adjoint), due to the absence (presence) of the first-order derivative (7.109a) associated with decay or amplification. A linear self-adjoint operator with variable coefficients may have derivatives of odd order, for example, the Sturm–Liouville operator (7.92), but the coefficients cannot be arbitrary, that is, satisfy the relation (7.91b).

7.7.8 High-Order Self-Adjoint Operator with Variable Coefficients

The linear second-order self-adjoint operator with constant (variable) coefficients is the undamped harmonic oscillator (7.106b) with constant mass and resilience [Sturm–Liouville operator (7.92)]. The next higher-order self-adjoint operator with variable coefficients may be expected to be (7.119a):

$$\left\{L\left(\frac{d}{dt}\right)\right\}\Phi = (A\Phi'')'' \equiv \left\{\bar{L}\left(\frac{d}{dt}\right)\right\}\Phi, \quad \Psi\left\{L\left(\frac{d}{dt}\right)\right\}\Phi - \Phi\left\{L\left(\frac{d}{dt}\right)\right\}\Psi = \frac{d}{dt}\big[W(\Phi,\Psi)\big], \tag{7.119a–c}$$

and (7.119c) confirms that it coincides with its adjoint (7.119b) and has the bilinear concomitant:

$$W(\Phi,\Psi) = (A\Phi'')'\,\Psi - (A\Psi'')'\,\Phi + A\left(\Psi''\Phi' - \Psi'\Phi''\right)$$

$$= A\left(\Phi'''\Psi - \Psi'''\Phi\right) + A\left(\Psi''\Phi' - \Psi'\Phi''\right) + A'\left(\Phi''\Psi - \Psi''\Phi\right). \tag{7.120}$$

The proof starts with multiplying the differential operator (7.119a) by Ψ and integrating by parts until Φ emerges as a factor:

$$\Psi\left(A\Phi''\right)'' = \left[\Psi\left(A\Phi''\right)'\right]' - \Psi'\left(A\Phi''\right)'$$

$$= \left[\Psi'\left(A\Phi''\right)' - \Psi'A\Phi''\right]' + \Phi''A\Psi''$$

$$= \left[\Psi\left(A\Phi''\right)' - \Psi'A\Phi'' + \Psi''A\Phi'\right]' - \Phi'\left(A\Psi''\right)'$$

$$= \left[\Psi\left(A\Phi''\right)' - \Psi'A\Phi'' + \Psi''A\Phi' - \Phi\left(A\Psi''\right)'\right]' + \Phi\left(A\Psi''\right)''. \tag{7.121}$$

The last term on the r.h.s. confirms that the operator is self-adjoint (7.119a) ≡ (7.119b), and comparing (7.121) ≡ (7.119c) specifies the bilinear concomitant (7.120) as the term in square brackets in (7.121), that is, linear in (Φ, Ψ) and have derivatives up to order 3.

The results (7.92; 7.93a and b) and (7.119a through c; 7.120) may be combined and extended to higher orders: *the linear differential operator of even order (7.122) with variable coefficients,*

$$\left\{ L\!\left(\frac{d}{dt}\right)\right\}\Phi \equiv A_0\Phi + \left(A_1\Phi'\right)' + \left(A_2\Phi''\right)'' + \left(A_3\Phi'''\right)''' + \cdots = \sum_{m=0}^{M}\left[A_m\Phi^{(m)}\right]^{(m)} \equiv \left\{ \bar{L}\!\left(\frac{d}{dt}\right)\right\}\Phi, \qquad (7.122)$$

is self-adjoint (7.119b) and has the bilinear concomitant (7.123) ≡ (7.127):

$$M(\Phi,\Psi) = \sum_{m=1}^{M}\Psi\left[A_m\,\Phi^{(m)}\right]^{(m-1)} - \Psi'\left[A_m\,\Phi^{(m)}\right]^{(m-2)} + \cdots$$

$$+ \left(-\right)^{m+1} A_m\left[\Psi^{(m-1)}\,\Phi^{(m)} - \Phi^{(m-1)}\,\Psi^{(m)}\right]$$

$$+ \cdots + \Phi'\left[A_m\,\Psi^{(m)}\right]^{(m-2)} - \Phi\left[A_m\,\Psi^{(m)}\right]^{(m-1)}. \qquad (7.123)$$

The proof consists of showing that substitution of the general term of (7.122) in (7.119c) leads to an exact differential:

$$\Psi\left[A_m\Phi^{(m)}\right]^{(m)} - \Phi\left[A_m\Psi^{(m)}\right]^{(m)} = W'. \qquad (7.124)$$

The first term on the r.h.s. of (7.124) coincides with (7.110b) exchanging (7.125a) and thus leading (7.112b; 7.114) to (7.125b):

$$\left\{A_m\Psi,\Phi\right\} \to \left\{\Psi, A_m\Phi^{(m)}\right\}: \qquad \Psi\left[A_m\,\Phi^{(m)}\right]^{(m)} = \left(-\right)^m\Psi^{(m)}\,A_m\,\Phi^{(m)}$$

$$+ \sum_{\ell=0}^{m-1}\left(-\right)^{\ell+1}\Psi^{(\ell-1)}\left[A_m\,\Phi^{(m)}\right]^{(m-\ell)}; \quad (7.125\text{a,b})$$

$$\left\{A_m\Psi,\Phi\right\} \to \left\{\Phi, A_m\Psi^{(m)}\right\}: \qquad \Phi\left[A_m\,\Psi^{(m)}\right]^{(m)} = \left(-\right)^m\Phi^{(m)}\,A_m\,\Psi^{(m)}$$

$$+ \sum_{\ell=1}^{m-1}\left(-\right)^{\ell+1}\Phi^{(\ell-1)}\left[A_m\,\Psi^{(m)}\right]^{(\ell-1)}, \quad (7.126\text{a,b})$$

the second term on the l.h.s. of (7.124) is similar to (7.126b) with the exchanges (7.126a). The difference of (7.125b) and (7.126b) leads to (7.124), where W is the difference of terms in curly brackets leading to the inner sum in

$$W(\Phi,\Psi) = \sum_{m=1}^{M}\sum_{\ell=1}^{m}(-)^{\ell+1}\left\{\Psi^{(\ell-1)}\left[A_m\Phi^{(m)}\right]^{(m-\ell)} - \Phi^{(\ell-1)}\left[A_m\Psi^{(m)}\right]^{(m-\ell)}\right\}; \qquad (7.127)$$

the outer sum adds all the terms in (7.122) leading to the bilinear concomitant (7.123) ≡ (7.127).

7.7.9 Bending of a Nonuniform Beam under Tension

The weak bending of a bar under axial tension and supported on linear springs is considered as an example of a fourth-order linear self-adjoint differential operator (Figure 7.3b). The shear stress or transverse force per unit length equals the sum of (1) the second-order derivative of the product of the curvature by the bending stiffness (4.12b); minus the (2) first-order derivative of the slope multiplied by the tangential tension (4.12b) as for the elastic string (2.11c), which is the first term on the r.h.s. of (7.94); and (3) the restoring force of a linear spring equal to the displacement times the resilience (I.4.1) as for the second term on the r.h.s. of (7.94):

$$f(x) = \frac{d^2}{dx^2}\left[E(x)I(x)\frac{d^2\zeta}{dx^2}\right] - \frac{d}{dx}\left[T(x)\frac{d\zeta}{dx}\right] + k(x)\zeta(x) \equiv \left\{L\left(\frac{d}{dx}\right)\right\}\zeta(x). \qquad (7.128)$$

In the balance equation (7.128), all coefficients can depend on position, namely, (1/2) the Young's modulus (moment of inertia of the cross section) for a bar made of an inhomogeneous material (with varying cross-section); (3) the tangential tension that may be a compression $T(x) < 0$ or traction $T(x) > 0$; and (4) the varying resilience of the springs along the length of the bar. Comparing (7.128) with the Sturm–Liouville operator (7.92; 7.93a and b) and the fourth-order self-adjoint operator (7.119a and b; 7.120) specifies the coefficients (7.129a through c) in the self-adjoint differential operator (7.129d) and bilinear concomitant (7.130):

$$\{A,a,c\} \equiv \{EI,T,k\}: \quad \bar{L}\left(\frac{d}{dx}\right) \equiv \frac{d^2}{dx^2}\left[E(x)I(x)\frac{d^2}{dx^2}\right] + \frac{d}{dx}\left[T(x)\frac{d}{dx}\right] + k(x) \equiv \bar{L}\left(\frac{d}{dx}\right), \quad (7.129a\text{–}d)$$

$$W(\Phi,\Psi) = \Psi\left(EI\Phi''\right)' - \Phi\left(EI\Psi''\right)' + EI\left(\Psi''\Phi' - \Psi'\Phi''\right) + T\left(\Phi'\Psi - \Psi'\Phi\right)$$

$$= EI\left(\Phi'''\Psi - \Psi'''\Phi\right) + EI\left(\Psi''\Phi' - \Phi''\Psi'\right) + \left(EI\right)'\left(\Phi''\Psi - \Psi''\Phi\right) + T\left(\Phi'\Psi - \Psi'\Phi\right) \qquad (7.130)$$

Thus *the weak bending of an inhomogeneous elastic bar with varying cross section under nonuniform tangential tension and supported on nonuniform springs (Figure 7.3b) is satisfied by the linear fourth-order differential operator (7.129d) applied to the transverse displacement (7.128) with forcing by the shear stress. The operator (7.129d) is self-adjoint (7.119d) with the bilinear concomitant (7.130). Thus are satisfied the principles of superposition and reciprocity, allowing for nonuniform bending stiffness, longitudinal tension, and spring resilience.*

7.8 Anisotropic, Intermediate, and Isotropic Operators

The preceding reasonings concerning adjoint, self-adjoint, and non-self-adjoint operators and their bilinear concomitants apply equally ordinary (partial) differential operators [Section 7.7 (Section 7.8)] both as concerns second order [Subsections 7.7.1 through 7.7.5 (Subsections 7.8.1 through 7.8.6)] and higher orders [Subsections 7.7.6 through 7.7.9 (Subsections 7.8.7 through 7.8.12)]. Both the ordinary (partial) differential operators, either linear or nonlinear, can have variable or constant coefficients. The linear partial differential operator is generally anisotropic, that is, it depends on direction; the particular case without directional dependence leads to the isotropic operator. There is also the intermediate operator whose coefficients are scalar but nonuniform, leading to a directional dependence specified by the derivatives of the coefficients. The linear second-order partial differential operator in general anisotropic form is considered first to obtain (1) the adjoint operator and bilinear concomitant (Subsection 7.8.1); (2) the particular self-adjoint case (Subsection 7.8.2); (3) the transformation (Subsection 7.8.3) between the preceding (1) non-self-adjoint and (2) self-adjoint cases; and (4) the intermediate and isotropic subcases (Subsection 7.8.5) of the preceding (1–3). An example of anisotropic (isotropic) self-adjoint linear second-order partial differential operator [Subsection 7.8.4 (Subsection 7.8.6)] relates to the electrostatic field of an inhomogeneous anisotropic dielectric (weak deflection of an elastic membrane under nonuniform isotropic tension). The high-order linear partial differential operator (Subsection 7.8.7) together with its adjoint and bilinear concomitants (Subsection 7.8.8) is considered in particular for (1) constant coefficients (Subsection 7.8.9); (2) self-adjoint form with constant coefficients (Subsection 7.8.10); and (3) the intermediate and isotropic subcases (Subsection 7.8.11) of the preceding (1, 2). An example of fourth-order isotropic self-adjoint linear differential operator concerns the weak bending of a nonhomogeneous elastic plate under in-plane tension and supported on springs (Subsection 7.8.12).

7.8.1 General Second-Order Linear Partial Differential Operator

*The general **linear second-order partial differential operator** is (7.131c):*

$$\nabla \equiv \partial_i \equiv \frac{\partial}{\partial x_i}, \quad \partial_{ij} \equiv \partial_i \partial_j \equiv \frac{\partial^2}{\partial x_i \partial x_j}: \quad \{L(\nabla)\}\Phi = A_{ij}(\vec{x})\partial_{ij}\Phi + B_i(\vec{x})\partial_i\Phi + C(\vec{x})\Phi, \tag{7.131a–c}$$

where the notation (7.131a and b) is used; it has an adjoint operator

$$\{\bar{L}(\nabla)\}\Psi(\vec{x}) = \partial_{ij}\left[A_{ij}(\vec{x})\Psi\right] - \partial_i\left[B_i(\vec{x})\Psi\right] + C(\vec{x})\Psi \tag{7.132}$$

and bilinear concomitant

$$W_i(\Phi, \Psi) = A_{ij}\left(\Psi\partial_j\Phi - \Phi\partial_j\Psi\right) + \left(B_i - \partial_j A_{ij}\right)\Phi\Psi, \tag{7.133}$$

satisfying the commutation relation

$$\Psi(\vec{x})\{L(\nabla)\}\Phi(\vec{x}) - \Phi(\vec{x})\{\bar{L}(\nabla)\}\Psi(\vec{x}) = \nabla \cdot \vec{W} \equiv \partial_i W_i. \tag{7.134}$$

The summation convention (7.70) is used for the independent variables in N-dimensional space, for example, the operator (7.131c) is

$$\left\{L\left(\frac{\partial}{\partial\vec{x}}\right)\right\}\Phi(\vec{x}) = \sum_{i,j=1}^{N} A_{ij}(\vec{x})\frac{\partial^2\Phi}{\partial x_i \partial x_j} + \sum_{i=1}^{N} B_i(\vec{x})\frac{\partial\Phi}{\partial x_i} + C(\vec{x})\Phi. \tag{7.135}$$

The results (7.131 through 7.134) are extensions to N-dimensions of the corresponding one-dimensional results (7.89a and b; 7.86; 7.88) and are proved similarly to (7.86) as follows:

$$\partial_i M_i + \Psi\left\{\bar{L}\left(\frac{\partial}{\partial\bar{x}}\right)\right\}\Phi \equiv \Psi\left\{L\left(\frac{\partial}{\partial\bar{x}}\right)\right\}\Phi = \Psi\left(A_{ij}\partial_i\partial_j\Phi + B_i\partial_i\Phi + C\Phi\right)$$

$$= \partial_i\left(A_{ij}\Psi\partial_j\Phi\right) - \left(\partial_j\Phi\right)\partial_i\left(A_{ij}\Psi\right) + \partial_i\left(B_i\Psi\Phi\right) - \Phi\partial_i\left(B_i\Psi\right) + C\Phi\Psi$$

$$= \partial_i\left[\Psi A_{ij}\partial_j\Phi + B_i\Phi\Psi\right] - \partial_j\left[\Phi\partial_i\left(A_{ij}\Psi\right)\right] + \Phi\left[\partial_j\partial_i\left(A_{ij}\Psi\right) - \partial_i\left(B_i\Psi\right) + C\Psi\right].$$

$$(7.136)$$

Bearing in mind the symmetry (7.137b) of the continuous second-order derivative (7.137a), the symmetric part of the matrix (7.137c),

$$\Phi\left(\bar{x}\right) \in C^2\left(\left|R^N\right.\right): \quad \partial_i\partial_j\Phi = \partial_j\partial_i\Phi, \quad 2\bar{A}_{ij} \equiv A_{ij} + A_{ji} = 2\bar{A}_{ji}, \tag{7.137a–c}$$

may be used in the first term on the r.h.s. of (7.135):

$$A_{ij}\partial_i\partial_j\Phi = \frac{1}{2}\left(A_{ij}\partial_i\partial_j\Phi + \partial_j\partial_i\Phi\right) = \frac{1}{2}\left(A_{ij} + A_{ji}\right)\partial_j\partial_i\Phi = \bar{A}_{ij}\partial_i\partial_j\Phi. \tag{7.138}$$

The symmetries (7.137b and c) allow interchange of the indices (i,j) in (7.136), and replacing $\bar{A}_{ij}$ by A_{ij} leads to

$$\partial_i M_i + \Psi\left\{\bar{L}(\nabla)\right\}\Phi = \partial_i\left[-\Phi\partial_j\left(A_{ij}\Psi\right) + \Psi A_{ij}\partial_j\Phi + B_i\Phi\Psi\right]$$

$$+ \Phi\left[\partial_i\partial_j\left(A_{ij}\Psi\right) - \partial_i\left(B_i\Psi\right) + C\Psi\right]. \tag{7.139}$$

Comparing (7.134) $\equiv$ (7.139), it follows that the terms in the first (second) square brackets on the r.h.s. of (7.139) specify the bilinear concomitant (7.133) [adjoint operator (7.132)].

7.8.2 Self-Adjoint Partial Differential Operator of the Second Order

The operator is self-adjoint (7.131) $\equiv$ (7.132) $\equiv$ (7.140b) iff the condition (7.140a) holds, simplifying the bilinear concomitant to (7.140c):

$$B_i = \frac{\partial A_{ij}}{\partial x_j} \Leftrightarrow L(\nabla) = \bar{L}(\nabla) \Leftrightarrow W_i\left(\Phi,\Psi\right) = A_{ij}\left(\Psi\partial_j\Phi - \Phi\partial_j\Psi\right). \tag{7.140a–c}$$

Thus *the most general **linear second-order self-adjoint partial differential operator** is the **partial Sturm–Liouville operator***:

$$\left\{L(\nabla)\right\}\Phi = A_{ij}\partial_{ij}\Phi + \left(\partial_j A_{ij}\right)\partial_i\Phi + C\Phi = \partial_i\left(A_{ij}\partial_j\Phi\right) + C\Phi = \left\{\bar{L}(\nabla)\right\}\Phi, \tag{7.141}$$

whose bilinear concomitant is (7.140c) $\equiv$ (7.142a):

$$W_i\left(\Phi,\Psi\right) = A_{ij}\left(\Psi\partial_j\Phi - \Phi\partial_j\Psi\right), \quad \Psi\left\{L(\nabla)\right\}\Phi - \Phi\left\{L(\nabla)\right\}\Psi = \nabla\cdot\vec{W}, \tag{7.142a,b}$$

and it satisfies the commutation relation (7.142b). The preceding results are proved as follows: (1) the difference between the adjoint (7.132) and original (7.131) operator is zero,

$$0 = \left\{\bar{L}(\nabla)\right\}\Phi - \left\{L(\nabla)\right\}\Phi = \partial_{ij}\left(A_{ij}\Phi\right) - A_{ij}\partial_{ij}\Phi - \partial_i\left(B_i\Phi\right) - B_i\partial_i\Phi$$
$$= \Phi\partial_i\left(\partial_j A_{ij} - B_i\right) + 2\left(\partial_j A_{ij} - B_i\right)\partial_i\Phi, \tag{7.143}$$

iff the condition (7.140a) is met; (2) in that case the original operator (7.131) takes the self-adjoint form (7.141); and (3) using (7.140a) the bilinear concomitant (7.133) simplifies to (7.140c) ≡ (7.142a).

7.8.3 Existence of Transformation from General to Self-Adjoint Operator

The general linear second-order ordinary differential equation (7.85b) can be transformed (7.98a) to the self-adjoint form (7.102a) by multiplying by a factor (7.102b). In order to find out whether a similar transformation applies from the general (7.131c) to the self-adjoint (7.141) linear second-order partial differential operator, consider their coincidence after multiplication of the former by a factor μ:

$$\mu\left(A_{ij}\partial_{ij}\Phi + B_j\partial_j\Phi + c\Phi\right) = \partial_i\left(P_{ij}\partial_j\Phi\right) + Q\Phi = P_{ij}\partial_{ij}\Phi + \left(\partial_i P_{ij}\right)\partial_j\Phi + Q\Phi. \tag{7.144}$$

The coincidence of the r.h.s. and l.h.s. of (7.144) leads to the conditions

$$\mu\left\{A_{ij}, B_j, c\right\} = \left\{P_{ij}, \partial_i P_{ij}, Q\right\}. \tag{7.145a–c}$$

It follows that (1) the number of conditions to be met is $N^2(N)$ components of the matrix A_{ij} (vector B_j) A_{ij} (B_j) plus one for the scalar c, for a total of $N^2 + N + 1$; (2) the number of variables is $N^2(2)$ for the matrix P_{ij} (scalar Q and μ) for a total of $N^2 + 2$; thus (3) the transformation is generally possible only if $N = 1$, that is, for an ordinary differential operator (Subsection 7.7.4); and (4) for a partial differential operator $N \geq 2$ the transformation to self-adjoint form requires $N − 1$ constraints on the coefficients (A_{ij}, B_i) that must (7.145b and c) satisfy (7.145d):

$$\mu B_j = \partial_i P_{ij} = \partial_i\left(\mu A_{ij}\right) = \mu\partial_i A_{ij} + A_{ij}\partial_i\mu; \tag{7.145d}$$

(5) this is equivalent to (7.146b) where (7.146c) the inverse matrix of A_{ij} may be used, which exists iff it has a nonzero determinant (7.146a):

$$A \equiv Det\left(A_{ij}\right) \neq 0: \quad B_j - \partial_i A_{ij} = A_{ij}\frac{1}{\mu}\partial_j\mu, \quad \partial_i\left(\log\mu\right) = A_{ij}^{-1}\left(B_j - \partial_k A_{jk}\right); \tag{7.146a–c}$$

(6) in (7.146c) appears (7.147a) leading to (7.147b) that must be an exact differential:

$$A_{ij}^{-1}\partial_k A_{jk} = \partial_i\left\{\log\left[Det\left(A_{ij}\right)\right]\right\}, \quad \partial_i\left(\log\mu\right) + \partial_i\left\{\log\left[Det\left(A_{ij}\right)\right]\right\} = A_{ij}^{-1}B_j = X_i; \tag{7.147a,b}$$

(7) therefore the transformation from non-self-adjoint to self-adjoint operator (7.144) exists iff (7.147b) is an exact differential, implying that the vector (7.147b) is the gradient of a scalar, hence irrotational (7.148a):

$$\nabla \wedge \vec{X} = 0: \quad \mu(\vec{x}) = \frac{1}{Det\left[A_{ij}(\vec{x})\right]}\exp\left(\int A_{ij}^{-1}B_j dx_i\right); \tag{7.148a,b}$$

(8) the integration of (7.147b) specifies the multiplying factor (7.148b) in (7.144; 7.145a through c). It has been shown that the general linear second-order partial differential operator (7.131c) *can be transformed to the self-adjoint form (7.141)* ≡ *(7.149)*:

$$\left\{ L(\nabla) \right\} \Phi = \left\{ \overline{L}(\nabla) \right\} \Phi = \partial_i \left(P_{ij} \partial_j \Phi \right) + Q\Phi = P_{ij}\partial_{ij}\Phi + \left(\partial_j P_{ij} \right) \partial_j \Phi + Q\Phi, \tag{7.149}$$

by multiplication by (7.148b) ≡ *(7.150a), leading to the coefficients (7.150b and c):*

$$\left\{ \mu, P_{ij}, Q \right\} = \left[Det\left(A_{ij} \right) \right]^{-1} \exp\left(\int A_{ij}^{-1} B_j dx_i \right) \left\{ 1, A_{ij}, c \right\}, \tag{7.150a–c}$$

iff the matrix A_{ij} is invertible (7.146a) and the vector (7.147b) is irrotational (7.148a); the latter condition is equivalent to the statement that the integrand in (7.148b) is an exact differential.

In the one-dimensional case, (7.149; 7.150a through c) simplifies to (7.102a through d). The general theorem (7.149; 7.150a through c) can be checked by multiplying the general second-order operator (7.131c) by μ, leading to

$$\mu\left(A_{ij}\partial_{ij}\Phi + B_j\partial_j\Phi + c\Phi \right) = \mu c\Phi + \partial_i\left(\mu A_{ij}\partial_j\Phi \right) + \left[\mu B_j - \partial_i\left(\mu A_{ij} \right) \right] \partial_j\Phi = Q\Phi + \partial_i\left(P_{ij}\partial_j\Phi \right), \tag{7.151}$$

which coincides with the self-adjoint operator (7.149), provided that the term in square brackets in (7.151) vanishes (7.152a):

$$0 = \mu\, B_j - \mu\, \partial_j A_{ij} - \partial_{ij}\, \partial_i\, \mu \Leftrightarrow A_{ij}\, \partial_j\left(\log \mu \right) = B_j - \partial_i A_{ij}\,; \tag{7.152a,b}$$

Thus it is sufficient to check (7.152b) ≡ (7.152a) ≡ (7.146b). From the expression for μ in (7.150a) follows

$$A_{ij}\, \partial_j\left(\log \mu \right) = A_{ij}\, \partial_i\left\{ -\log\left[Det\left(A_{k\ell} \right) \right] \right\} + A_{ij}\, \partial_i\left(\int A_{k\ell}^{-1}\, B_k\, dx_\ell \right)$$

$$= -\, A_{ij}\, A_{i\ell}^{-1}\, \partial_k A_{\ell k} + A_{ij}\, A_{k\ell}^{-1}\, B_k\, \partial_i x_\ell$$

$$= -\, \delta_{j\ell}\, \partial_k A_{\ell k} + A_{ij}\, A_{k\ell}^{-1}\, B_k\, \delta_{i\ell}$$

$$= -\partial_k A_{jk} + A_{ij}A_{ki}^{-1}\, B_k$$

$$= -\, \partial_i A_{ji} + \delta_{jk}\, B_k = B_j - \partial_i A_{ij}\,, \tag{7.153}$$

where were used (7.147a), and the inverse A_{ij}^{-1} and identity (5.278a–c) matrices. This confirms the coincidence of (7.153) ≡ (7.152b) ≡ (7.152a) ≡ (7.146a) ≡ (7.147b) ≡ (7.148b) and thus proves (7.151) ≡ (7.144) ≡ (7.149). Next, an example of a self-adjoint linear second-order partial differential operator (Subsection 7.8.4) is given.

7.8.4 Electrostatic Field in an Anisotropic Inhomogeneous Dielectric

The Maxwell equations for the electrostatic field (I.24.1a through c) ≡ (5.310a) state that (1) the electric field is irrotational (7.154a), hence (Equation 5.310c) an electric potential exists (7.154b):

$$\nabla \wedge \vec{E} = 0 \Leftrightarrow \vec{E} = -\nabla\Phi; \quad q = \nabla \cdot \vec{D}, \quad D_i = \varepsilon_{ij}E_j; \tag{7.154a–d}$$

(2) the electric charge density equals (7.154c) the divergence of the electric displacement. The electric displacement and field are related by (7.154d) in a medium that is (1) linear so that they are proportional;

(2) anisotropic, for example, a crystal, so that they may not be parallel; and (3) inhomogeneous so that the **dielectric permittivity tensor** ε_{ij} may be dependent on position. Substituting (7.154b) in (7.154d) and then in (7.154c) leads to

$$-q = -\nabla \cdot \vec{D} = -\partial_i D_i = -\partial_i \left(\varepsilon_{ij} E_j \right) = \partial_i \left(\varepsilon_{ij} \partial_j \Phi \right) = \varepsilon_{ij} \partial_i \partial_j \Phi + \left(\partial_j \Phi \right) \partial_i \varepsilon_{ij}. \tag{7.155}$$

It has been shown that *the electrostatic (7.156a) field in a linear, anisotropic, inhomogeneous dielectric medium with dielectric permittivity tensor ε_{ij} is specified by (7.156b) $\equiv$ (7.154b) in terms of an electric potential satisfying (7.156c):*

$$\frac{\partial}{\partial t} = 0: \quad E_i = -\partial_i \Phi, \quad q = -\partial_i \left(\varepsilon_{ij} \partial_j \Phi \right) = \left\{ L(\nabla) \right\} \Phi = \left\{ \overline{L}(\nabla) \right\} \Phi. \tag{7.156a-d}$$

The operator is self-adjoint (7.156c) $\equiv$ (7.156d) with the bilinear concomitant (7.157a) satisfying (7.157b):

$$W_i(\Phi, \Psi) = \varepsilon_{ij} \left(\Phi \partial_i \Psi - \Psi \partial_i \Phi \right), \quad \Psi \left\{ L(\nabla) \right\} \Phi - \Phi \left\{ L(\nabla) \right\} \Psi = \partial_i W_i. \tag{7.157a,b}$$

In the case of a homogeneous anisotropic dielectric medium, the permittivity tensor is constant (7.158a) and the electric potential satisfies (7.158b):

$$\varepsilon_{ij} = \text{const}: \quad q = -\varepsilon_{ij} \partial_{ij} \Phi; \quad \varepsilon_{ij} = \varepsilon \delta_{ij}: \quad -\frac{q}{\varepsilon} = \partial_{ij} \Phi = \nabla^2 \Phi. \tag{7.158a-d}$$

If in addition to homogeneous the dielectric is also isotropic, the dielectric permittivity reduces from a tensor to a scalar (7.158c) and the electric potential satisfies a Poisson equation (7.158d) $\equiv$ (I.24.5b) involving the Laplace operator. The remaining case is an inhomogeneous isotropic dielectric for which the dielectric permittivity reduces from a tensor to a scalar but depends on position (7.159a) and the electric potential satisfies (7.159b):

$$\varepsilon_{ij} = \varepsilon(\vec{x}) \delta_{ij}: \quad -q = \partial_i \left(\varepsilon \partial_i \Phi \right) = \nabla \cdot \left(\varepsilon \nabla \Phi \right) = \varepsilon \nabla^2 \Phi + \nabla \varepsilon \cdot \nabla \Phi. \tag{7.159a,b}$$

The electric potential is isotropic when it satisfies the Poisson equation (7.158d), that is, for a homogeneous (7.158a) isotropic medium (7.158c); the case of an isotropic inhomogeneous medium (7.159a) leads to the intermediate operator (7.159b) for which the electric potential has a directional dependence due to the gradient of the dielectric permittivity. The linear self-adjoint second-order isotropic (intermediate) partial differential operator leads to the homogeneous (inhomogeneous) Helmholtz operator, which includes (Subsection 7.8.5) the homogeneous (inhomogeneous) Laplace operator.

7.8.5 Homogeneous/Inhomogeneous Laplace and Helmholtz Operators

In an isotropic medium, the coefficients cannot depend on direction, implying that for the linear second-order partial differential operator (7.131c) the matrix must reduce to a scalar (7.160a) and the vector must vanish (7.160b), leading to the **isotropic linear second-order partial differential operator** (7.160c):

$$A_{ij} = A \delta_{ij}, \, B_i = 0: \quad \left\{ L(\nabla) \right\} \Phi = A(\vec{x}) \nabla^2 \Phi + C(\vec{x}) \Phi = 0, \tag{7.160a-c}$$

which is the **Helmholtz (1858) equation** *(7.160c); its particular case $C = 0$ is the* **Laplace (1825) equation**. *Its adjoint (7.161a) [bilinear concomitant (7.161b)] is*

$$\left\{ \overline{L}(\nabla) \right\} \Phi = \nabla^2 (A\Phi) + C\Phi, \quad \vec{W}(\Phi, \Psi) = A(\Psi \nabla \Phi - \Phi \nabla \Psi) - \Phi \Psi \nabla A. \tag{7.161a,b}$$

The operator is self-adjoint (7.160c) ≡ (7.161a) iff the coefficient of the Laplacian is constant (7.162a):

$$A = const: \quad \{L(\nabla)\}\Phi = \{\overline{L}(\nabla)\Phi\} = A\nabla^2\Phi + C(\vec{x})\Phi, \quad \vec{W}(\Phi,\Psi) = A(\Psi\nabla\Phi - \Phi\nabla\Psi), \quad (7.162a\text{–}d)$$

and this can be ensured by dividing (7.160c) by A; in this case C is replaced by C/A, which may depend on position through $A(\vec{x})$ and/or $C(\vec{x})$ without affecting adjointness. The bilinear concomitant simplifies to (7.162d) and satisfies (7.142b).

The **linear second-order partial differential intermediate operator** reduces the matrix to a scalar (7.160a) ≡ (7.163a), but does not exclude the vector as in (7.160b), and thus leads to (7.163b):

$$A_{ij} = A\delta_{ij}: \quad \{L(\nabla)\}\Phi = A(\vec{x})\partial_{ii}\Phi + B_i(\vec{x})\partial_i\Phi + C(\vec{x})\Phi = A\nabla^2\Phi + \vec{B}\cdot\nabla\Phi + C\Phi. \quad (7.163a,b)$$

The adjoint operator and the bilinear concomitant are, respectively:

$$\{\overline{L}(\nabla)\}\Psi = \nabla^2(A\Psi) - \nabla\cdot(\vec{B}\Psi) + C\Psi, \quad (7.163c)$$

$$\vec{W}(\Phi,\Psi) = A(\Psi\nabla\Phi - \Phi\nabla\Psi) + (\vec{B} - \nabla A)\Phi\Psi. \quad (7.163d)$$

The operator is self-adjoint (7.164b and c) if (7.164a) is met, simplifying the bilinear concomitant to (7.164d):

$$\vec{B} = \nabla A: \quad \{L(\nabla)\}\Phi = \nabla\cdot(A\nabla\Phi) + C\Phi = \{\overline{L}(\nabla)\}\Phi, \quad \vec{W} = A(\Psi\nabla\Phi - \Phi\nabla\Psi). \quad (7.164a\text{–}d)$$

Thus (7.164b) specifies the **inhomogeneous Helmholtz equation** *(7.164b) ≡ (7.164c), that includes for C = 0 the* **inhomogeneous Laplace equation**, *and applies to an isotropic inhomogeneous medium. The transformation from the intermediate non-self-adjoint operator (7.162a and b) to the intermediate self-adjoint form (7.164a through d) ≡ (7.165b through d),*

$$\nabla \wedge \left(\frac{\vec{B}}{A}\right) = 0: \quad \{L(\nabla)\}\Phi = \{\overline{L}(\nabla)\}\Phi = \nabla\cdot(P\nabla\Phi) + Q\Phi,$$

$$\{\mu,P,Q\} = A^{-N}\left(\exp\int\frac{\vec{B}\cdot d\vec{x}}{A}\right)\{1,A,C\}, \quad (7.165a\text{–}d)$$

is possible iff the vector $\vec{B}/A$ is irrotational (7.165a) so that an exact differential appears in (7.165d); the commutator (7.142b) involves the bilinear concomitant (7.161b). The proof of (7.165a through d) follows substituting (7.163a) in (7.150a through c):

$$A_{ij}^{-1} = \frac{1}{A}\delta_{ij}; \quad Det(A\delta_{ij}) = A^N, \quad A_{ij}^{-1}B_j = \frac{1}{A}B_i, \quad A_{ij}^{-1}B_j dx_i = \frac{1}{A}B_i dx_i = \frac{\vec{B}\cdot d\vec{x}}{A}; \quad (7.166a)$$

using (7.166a), the condition of existence of the transformation (7.148a and b) simplifies to (7.165a).

7.8.6 Elastic Membrane with Nonuniform Tension

The weak or linear deflection of an elastic membrane with nonuniform tension supported on springs (Figure 7.3a) is similar to the elastic string (7.94) except that: (1) the transverse displacement ζ, shear stress f, tangential tension T, and resilience of the supporting springs depend on two (x, y), instead of one (x), Cartesian coordinates; and (2) for isotropic tension the passage from the elastic string

(Chapter 2) to the elastic membrane (Sections II.6.1 and 2) corresponds to replacing the first (second) spatial derivative by the gradient (7.17a) [divergence (7.167b)]:

$$\frac{d\zeta}{dx} \to \nabla\zeta = \vec{e}_x \frac{\partial\zeta}{\partial x} + \vec{e}_y \frac{\partial\zeta}{\partial y}, \quad \frac{d^2\zeta}{dx^2} \to \nabla \cdot \nabla\zeta = \nabla^2\zeta = \frac{d^2\zeta}{dx^2} + \frac{d^2\zeta}{dy^2}. \tag{7.167a,b}$$

These substitutions lead from (7.94) to (7.168a):

$$f(x,y) = -\nabla \cdot (T\nabla\zeta) + k\zeta = -T\nabla^2\zeta + \nabla T \cdot \nabla\zeta + k\zeta = \{L(\nabla)\}\zeta = \{\bar{L}(\nabla)\}\zeta. \tag{7.168a,b}$$

Thus *the weak deflection of an elastic membrane under nonuniform isotropic tension supported on linear springs and subject to a shear stress (Figure 7.3a) is specified (7.168a and b) by an isotropic self-adjoint linear second-order partial differential operator (7.162a and b):*

$$\{L(\nabla)\}\Phi = \{\bar{L}(\nabla)\}\Phi = -\nabla \cdot (T\nabla\Phi) + k\Phi, \quad \vec{W}(\Phi,\Psi) = T(\Psi\nabla\Phi - \Phi\nabla\Psi), \tag{7.169a–c}$$

with bilinear concomitant (7.169c), corresponding to the inhomogeneous Helmholtz operator. Thus the principle of reciprocity holds for nonuniform isotropic tension T(x,y) and springs with resilience k(x,y) depending on position.

7.8.7 Self-Adjoint and Non-Self-Adjoint Differential Operators

The adjointness properties of ordinary (partial) differential operators are related by the following remark: *a partial differential operator has the adjoint and bilinear concomitant corresponding to the sum for each of its variables taken as if it was an ordinary differential operator. In particular, a partial differential operator (1) is self-adjoint if it is self-adjoint for all the variables; and (2) is not self-adjoint if it fails to be self-adjoint for at least one variable even if it is self-adjoint for all the other variables.* Some partial differential operators of fourth-order in two variables are given as examples of adjointness (nonadjointness): (1) with constant coefficients (7.170) [(7.171a through d)] because they involve only derivatives of even order (involve at least one derivative of odd order):

$$\left\{\bar{L}\left(\frac{\partial}{\partial x}, \frac{\partial}{\partial y}\right)\right\}\Phi = A\frac{\partial^4\Phi}{\partial x^4} + B\frac{\partial^4\Phi}{\partial y^4} + C\frac{\partial^4\Phi}{\partial x^2\partial y^2} + D = \left\{\bar{L}\left(\frac{\partial}{\partial x}, \frac{\partial}{\partial y}\right)\right\}\Phi, \tag{7.170}$$

$$\left\{L\left(\frac{\partial}{\partial x}, \frac{\partial}{\partial y}\right)\right\}\Phi = A\frac{\partial^4\Phi}{\partial x\partial y^3}, \quad B\frac{\partial^3\Phi}{\partial x^3}, \quad C\frac{\partial^3\Phi}{\partial x\partial y^2}, \quad D\frac{\partial\Phi}{\partial x} \neq \left\{\bar{L}\left(\frac{\partial}{\partial x}, \frac{\partial}{\partial y}\right)\right\}\Phi; \tag{7.171a–d}$$

(2) with variable coefficients (7.172) [(7.173a through d)] because all (some) terms are (are not) in the Sturm–Liouville form:

$$\left\{L\left(\frac{\partial}{\partial x}, \frac{\partial}{\partial y}\right)\right\}\Phi = \frac{\partial^2}{\partial x^2}\left\{A(x,y)\frac{\partial^2\Phi}{\partial x^2}\right\} + \frac{\partial^2}{\partial x^2}\left\{B(x,y)\frac{\partial^2\Phi}{\partial y^2}\right\}$$

$$+ \frac{\partial^2}{\partial x\partial y}\left\{C(x,y)\frac{\partial^2\Phi}{\partial x\partial y}\right\} + D(x,y)\Phi = \left\{\bar{L}\left(\frac{\partial}{\partial x}, \frac{\partial}{\partial y}\right)\right\}\Phi; \tag{7.172}$$

$$\left\{L\left(\frac{\partial}{\partial x}, \frac{\partial}{\partial y}\right)\right\}\Phi = \frac{\partial^3}{\partial x^3}\left[A(x,y)\frac{\partial\Phi}{\partial y}\right], \quad \frac{\partial^2}{\partial x^2}\left[B(x,y)\frac{\partial^2\Phi}{\partial x\partial y}\right],$$

$$\frac{\partial^2}{\partial y^2}\left[C(x,y)\frac{\partial^3\Phi}{\partial y^2\partial x}\right], \quad \frac{\partial}{\partial x}\left[D(x,y)\frac{\partial^2\Phi}{\partial x^2}\right] \neq \left\{\bar{L}\left(\frac{\partial}{\partial x}, \frac{\partial}{\partial y}\right)\right\}\Phi. \tag{7.173a–d}$$

Rather than proceeding with examples, a set of general theorems are stated concerning linear partial differential operators of any order (Subsection 7.8.8) in general (1) and particular cases that have (2) variable coefficients and are self-adjoint (Subsection 7.8.9); (3) constant coefficients and are or not self-adjoint (Subsection 7.8.10); and (4) isotropic forms (Subsection 7.8.11) of the preceding (1–3).

7.8.8 High-Order Partial Differential Operator

*In the **linear partial differential operator** (7.69) ≡ (7.174b) of order M with N variables, the notation (7.174a) is used, which is an extension of (7.131a and b):*

$$\partial_{i_1 \ldots i_m} = \frac{\partial^m}{\partial x_{i_1} \ldots \partial x_{i_m}} : \quad \left\{ L(\nabla) \right\} \Phi = \sum_{m=0}^{M} A_{i_1 \ldots i_m} \partial_{1 \ldots i_m} \Phi. \tag{7.174a,b}$$

The adjoint operator is

$$\left\{ \overline{L}(\nabla) \right\} \Psi = \sum_{m=0}^{M} (-)^m \, \partial_{i_1 \ldots i_m} \left(A_{i_1 \ldots i_m} \Psi \right), \tag{7.175}$$

and the bilinear concomitant is (7.177), satisfying:

$$\left\{ L(\nabla) \right\} \Phi - \Phi \left\{ \overline{L}(\nabla) \right\} \Psi = \partial_i \left[W_i (\Phi, \Psi) \right], \tag{7.176}$$

$$W_{i_1} (\Phi, \Psi) = \sum_{m=1}^{M} \sum_{\ell=1}^{m} (-)^{\ell+1} \left(\partial_{i_{\ell+1} \ldots i_{m-1}} \Phi \right) \partial_{i_2 \ldots i_\ell} \left(A_{i_1 \ldots i_m} \Psi \right). \tag{7.177}$$

The proof of (7.174 through 7.177) is similar to that in Subsection 7.7.6.

7.8.9 Self-Adjoint Operator with Variable Coefficients

*The **self-adjoint linear partial differential operator** must have even order:*

$$\left\{ L(\nabla) \right\} \Phi = \left\{ \overline{L}(\nabla) \right\} \Phi = \sum_{m=0}^{M} \partial_{i_1 \ldots i_m} \left(A_{i_1 \ldots i_m} \partial_{i_1 \ldots i_{2m}} \Phi \right), \tag{7.178}$$

and has the bilinear concomitant (7.180), satisfying:

$$\Psi \left\{ L(\nabla) \right\} \Phi - \Phi \left\{ \overline{L}(\nabla) \right\} \Psi = \partial_i W_i = \nabla \vec{W}, \tag{7.179}$$

$$W_{i_1} (\Phi, \Psi) = \sum_{m=1}^{M} \sum_{\ell=1}^{m} (-)^{\ell+1} \left[\left(\partial_{i_2 \ldots i_\ell} \Psi \right) \partial_{i_{\ell+1} \ldots i_{2m}} \left(A_{i_1 \ldots i_{2m}} \Phi \right) - \left(\partial_{i_2 \ldots i_\ell} \Phi \right) \partial_{i_{\ell+1} \ldots i_{2m}} \left(A_{i_1 \ldots i_{2m}} \Psi \right) \right]. \tag{7.180}$$

The proof of (7.171 through 7.173) is similar to Subsection 7.7.8.

7.8.10 High-Order Operator with Constant Coefficients

The general linear partial differential operator of order M in N variables (7.181b) and constant coefficients (7.181a) has the adjoint (7.182):

$$A_{i_1\ldots i_m} = \text{const}: \quad \{L(\nabla)\}\Phi = \sum_{m=0}^{M} A_{i_1\ldots i_m}\partial_{i_1\ldots i_m}\Phi, \tag{7.181a,b}$$

$$\{\bar{L}(\nabla)\}\Phi = \sum_{m=0}^{M}(-)^m A_{i_1\ldots i_m}\partial_{i_1\ldots i_m}\Phi, \tag{7.182}$$

and the bilinear concomitant (7.183), satisfying (7.176):

$$W_{i_1}(\Phi,\Psi) = \sum_{m=1}^{M}\sum_{\ell=1}^{m}(-)^{\ell+1} A_{i_1\ldots i_m}\left(\partial_{i_2\ldots i_\ell}\Psi\right)\left(\partial_{i_{\ell+1}\ldots i_m}\Phi\right). \tag{7.183}$$

It is not self-adjoint if there is at least one derivative of odd order. It is self-adjoint if all derivatives are of even order:

$$A_{i_1\ldots i_{2m}} = \text{const}: \quad \left\{L\left(\frac{\partial}{\partial\vec{x}}\right)\right\}\Phi = \left\{\bar{L}\left(\frac{\partial}{\partial\vec{x}}\right)\right\}\Phi = \sum_{m=0}^{M} A_{i_1\ldots i_{2m}}\partial_{i_1\ldots i_{2m}}\Phi, \tag{7.184a–c}$$

with the bilinear concomitant (7.185), satisfying (7.179):

$$W_{i_1}(\Phi,\Psi) = \sum_{m=1}^{M} A_{i_1\ldots i_m}\sum_{\ell=1}^{M}(-)^{\ell+1}\left(\partial_{i_2\ldots i_\ell}\Psi\right)\left(\partial_{i_{\ell+1}\ldots i_m}\Psi\right). \tag{7.185}$$

Concerning the proofs (7.184a and b) [(7.185)] are particular cases of (7.181a and b) ≡ (7.182) [(7.183)] taking only even terms $2m$; in turn (7.181b; 7.182; 7.183) follow, respectively, from (7.174b; 7.175; 7.177) in the case of constant coefficients (7.181a).

7.8.11 Isotropic and Intermediate High-Order Operators

In the **isotropic linear partial differential operator** *the derivatives can appear only through the Laplacian (7.186a) and thus it must be of even order (7.186b), involving multiharmonic operators (7.186c):*

$$\nabla^2 \equiv \partial_{ii} \equiv \frac{\partial^2}{\partial x_i \partial x_i}: \quad \{L(\nabla)\}\Phi = \sum_{m=0}^{M} A_m(\vec{x})\nabla^{2m}\Phi, \tag{7.186a,b}$$

$$\nabla^{2m} \equiv \partial_{i_1 i_1\ldots i_m i_m} = \frac{\partial^{2m}}{\partial x_{i_1}\partial x_{i_1}\ldots\partial x_{i_m}\partial x_{i_m}}, \tag{7.186c}$$

The adjoint operator is

$$\{\bar{L}(\nabla)\}\Psi = \sum_{m=0}^{\infty}\nabla^{2m}\left(A_m\Phi\right) \tag{7.187}$$

and the bilinear concomitant (7.188) satisfies (7.179):

$$\vec{W}(\Phi,\Psi) = \sum_{m=1}^{M}\sum_{\ell=1}^{m}(-)^{\ell+1}\left(\nabla^{2m-\ell}\Phi\right)\nabla^{\ell-1}\left(A_m\Psi\right), \tag{7.188}$$

where ∇^ℓ is the multiharmonic operator (7.186c) [gradient of the multiharmonic operator (7.189)] for
$\ell = 2m$ even $\left(\ell = 2m+1 odd\right)$:

$$\nabla^{2m+1}\Phi = \nabla\left(\nabla^{2m}\Phi\right) = \partial_i\left(\nabla^{2m}\Phi\right) = \partial_{i_1 i_1 \dots i_m i_m}\Phi. \tag{7.189}$$

The operator is self-adjoint (7.190b and c) iff the coefficients are constant (7.190a):

$$A_m = \text{const}: \quad \{L(\nabla)\}\Phi = \{\bar{L}(\nabla)\}\Phi = \sum_{m=0}^{M}A_{2m}\nabla^{2m}\Phi, \tag{7.190a–c}$$

and the bilinear concomitant simplifies to:

$$\vec{W}(\Phi,\Psi) = \sum_{m=0}^{M}A_m\sum_{\ell=1}^{2m-1}(-)^{\ell+1}\left(\nabla^{2m-\ell}\Phi\right)\left(\nabla^{\ell-1}\Psi\right) \tag{7.191}$$

and satisfies (7.179).

   The **intermediate self-adjoint linear partial differential operator** *adds to (7.164c and d) higher-order*
terms (7.192a):

$$\{L(\nabla)\}\Phi = \{\bar{L}(\nabla)\}\Phi = A_0\Phi + \nabla\cdot\left(A_1\nabla\Phi\right) + \nabla^2\left(A_2\nabla^2\Phi\right) + \nabla^3\left(A_3\nabla^3\Phi\right) + \cdots = \sum_{M=0}^{M}\nabla^m\left(A_m\nabla^m\Phi\right),$$
$$\tag{7.192a,b}$$

and it involves both multiharmonic operators (7.186c) and their gradients; the operator is self-adjoint
(7.192a) $\equiv$ (7.192b), and the bilinear concomitant (7.193) satisfies (7.179):

$$\vec{W}(\Phi,\Psi) = \sum_{m=1}^{M}\sum_{\ell=1}^{m}(-)^{\ell+1}\left[\left(\nabla^{\ell-1}\Psi\right)\nabla^{2m-\ell}\left(A_m\Phi\right)\left(\nabla^{\ell-1}\Phi\right) - \nabla^{2m-\ell}\left(A_m\Psi\right)\right]. \tag{7.193}$$

The particular case with constant coefficients (7.194a) involves (7.194b) only multiharmonic operators (7.186c):

$$A_m = \text{const}: \quad \{L(\nabla)\}\Phi = \{\bar{L}(\nabla)\}\Phi = \sum_{m=0}^{M}A_m\nabla^{2m}\Phi, \tag{7.194a–c}$$

and the bilinear concomitant simplifies to

$$\vec{W}(\Phi,\Psi) = \sum_{m=1}^{M}A_m\sum_{\ell=1}^{2m}(-)^{\ell+1}\left[\left(\nabla^{\ell-1}\Psi\right)\nabla^{m-\ell}\Phi - \left(\nabla^{\ell-1}\Phi\right)\nabla^{m-\ell}\Psi\right]. \tag{7.195}$$

The preceding results are proved as follows: (1) from (7.174b; 7.175; 7.177) selecting only Laplacian combinations of derivatives leads to (7.186b; 7.187; 7.188); (2) using constant coefficients (7.190a) leads to

(7.190b and c; 7.191); (3) the results for the intermediate operator (7.192a and b; 7.193) are proved as in Subsection 7.7.8; and (4) the case of constant coefficients (7.194a) leads to (7.194b and c; 7.195). An example of two-dimensional fourth-order intermediate linear partial differential operator is the bending of an inhomogeneous plate (Subsection 7.8.12).

7.8.12 Bending of a Plate under In-Plane Tension Supported on Springs

The two-dimensional analogue of the weak bending of an elastic bar (7.128) is the weak bending of an elastic plate with nonuniform stiffness D for which the transformations are similar (7.167a and b) to the passage from the one-dimensional string to the two-dimensional membrane:

$$f(x,y) = \nabla^2 (D\nabla^2 \zeta) - \nabla \cdot (T\nabla\zeta) + k\zeta \equiv \{L(\nabla)\}\zeta = \{\overline{L}(\nabla)\}\zeta; \tag{7.196a,b}$$

the corresponding (7.192a and b; 7.193) bilinear concomitant (7.197) satisfies (7.179):

$$\vec{W} = T(\Phi\nabla\Psi - \Psi\nabla\Phi) + \Psi\nabla^3(\nabla\Phi) - \Phi\nabla^3(\nabla\Psi) - (\nabla\Psi)\nabla^2(D\Phi) + (\nabla\Phi)\nabla^2(D\Psi). \tag{7.197}$$

Thus *the weak bending of an elastic plate with nonuniform stiffness D under in-plane compression T < 0 or traction T > 0 and supported on linear springs with resilience k is specified (Figure 7.3b) by the self-adjoint operator (7.196b) applied to the transverse displacement and forced by the shear stress. It consists of (1) the terms relative to the weak deflection of an elastic membrane (7.168a) that has no bending stiffness; and (2) the first term on the r.h.s. of (7.196a) involving the bending stiffness. The intermediate linear fourth-order partial differential operator is self-adjoint (7.196a,b) with the bilinear concomitant (7.197; 7.179) implying that both the principles of superposition and reciprocity hold with nonuniform bending stiffness, in-plane tension, and spring resilience.*

7.9 Equations of Mathematical Physics in Space-Time

A variety of physical phenomena and technological models are described by linear differential equations in space-time combining derivatives [Section 7.7 (Section 7.8)] with regard to time (position) plus cross-terms (Section 7.9). Starting with the second-order linear partial differential equation (Subsection 7.9.1), the adjoint operator and bilinear concomitants in space and time lead to the self-adjoint case (Subsection 7.9.2). The particular isotropic case includes the wave and Klein–Gordon (diffusion, telegraphy, and Schrödinger) equations for the self-adjoint (non-self-adjoint) subcases [Subsection 7.9.3 (Subsection 7.9.4)]. The corresponding intermediate self-adjoint (non-self-adjoint) operators [Subsection 7.9.5 (Subsection 7.9.6)] extend these operators to isotropic inhomogeneous media. The high-order linear partial differential operators are considered by a space-time method (Subsection 7.9.8) that may be applied in all cases: (1) constant or variable coefficients; (2) self-adjoint or non-self-adjoint; and (3) anisotropic, intermediate, or isotropic. An example of second (fourth)-order intermediate linear self-adjoint partial differential operator is given by the linear transverse vibrations of [Subsection 7.9.7 (Subsection 7.9.9)] an inhomogeneous membrane (plate).

7.9.1 Linear Second-Order Differential Operator in Space-Time

The combination of linear second-order ordinary (partial) differential operators in time (7.85b) [position (7.131c)] plus a cross-term leads to (7.198c):

$$\partial_t \equiv \frac{\partial}{\partial t}, \partial_{tt} \equiv \frac{\partial^2}{\partial t^2}: \quad \{L(\nabla, \partial_t)\}\Phi = A_{ij}(\vec{x},t)\partial_{ij}\Phi + C_i(\vec{x},t)\partial_t\partial_i\Phi$$

$$- a(\vec{x},t)\partial_{tt}\Phi + B_i(\vec{x},t)\partial_i\Phi - b(\vec{x},t)\partial_t\Phi - c(\vec{x},t)\Phi, \tag{7.198a–c}$$

where the notations (7.131a and b; 7.198a and b) are used. The adjoint operator is specified by (7.87) [(7.132)] for the temporal (spatial) part, and for the cross-term by:

$$\Psi C_i \partial_t \partial_i \Phi = \partial_t \left(\Psi C_i \partial_i \Phi \right) - \left(\partial_i \Phi \right) \partial_t \left(C_i \Psi \right)$$

$$= \partial_t \left(\Psi C_i \partial_i \Phi \right) - \partial_i \left[\Phi \partial_t \left(C_i \Psi \right) \right] + \Phi \partial_i \partial_t \left(C_i \Psi \right). \tag{7.199}$$

The last term on the r.h.s. of (7.199) appears in the adjoint operator:

$$\left\{ \overline{L} \left(\nabla, \partial_t \right) \right\} \Psi = \partial_{ij} \left(A_{ij} \Psi \right) + \partial_t \partial_i \left(C_i \Psi \right) - \partial_{tt} \left(a \Psi \right) - \partial_i \left(B_i \Psi \right) + \partial_t \left(b \Psi \right) - c \Psi. \tag{7.200}$$

The first (second) term on the r.h.s. of (7.199) goes into the temporal (7.88) [spatial (7.133)] bilinear concomitant (7.194b) [(7.194c)] in the commutator (7.194a):

$$\Psi \left\{ L \left(\nabla, \partial_t \right) \right\} \Phi - \Phi \left\{ \overline{L} \left(\nabla, \partial_t \right) \right\} \Psi = \partial_t V + \nabla \cdot \vec{W}, \tag{7.201a}$$

$$V \left(\Phi, \Psi \right) = a \left(\Phi \partial_t \Psi - \Psi \partial_t \Phi \right) - \left(b - \partial_t a \right) \Phi \Psi + \Psi C_i \partial_i \Phi, \tag{7.201b}$$

$$W_i \left(\Phi, \Psi \right) = A_{ij} \left(\Psi \partial_j \Phi - \Phi \partial_j \Psi \right) + \left(B_i - \partial_j A_{ij} \right) \Phi \Psi - \Phi \partial_t \left(C_i \Psi \right); \tag{7.201c}$$

exchanging the order of partial integrations in (7.199) exchanges the signs of the last terms in (7.194b and c).

7.9.2 Anisotropic Self-Adjoint Operator in Space-Time

It has been shown that the **linear second-order partial differential operator in space-time** (7.198c), with the notations (7.131a and b; 7.198a and b), has (1) adjoint operator (7.200); (2–4) commutator (7.201a) involving the temporal (7.201b) [spatial (7.201c)] bilinear concomitants. The self-adjoint case (7.202d and e) is (7.202a through c):

$$B_i = \partial_j A_{ij}, \quad b = \partial_t a, C_i = \text{const}, \quad \left\{ L \left(\nabla, \partial_t \right) \right\} \Phi = \left\{ \overline{L} \left(\nabla, \partial_t \right) \right\} \Phi = \partial_i \left(A_{ij} \partial_j \Phi \right) - \partial_t \left(a \partial_t \Phi \right) + C_i \partial_i \partial_t \Phi - c \Phi,$$

$$\tag{7.202a–e}$$

corresponding to the commutator (7.203) and temporal (spatial) bilinear concomitants (7.204a) [(7.204b)]:

$$\left\{ L \left(\nabla, \partial_t \right) \right\} \Phi - \Phi \left\{ L \left(\nabla, \partial_t \right) \right\} \Psi = \partial_t V + \partial_i W_i, \tag{7.203}$$

$$V \left(\Phi, \Psi \right) = a \left(\Phi \partial_t \Psi - \Psi \partial_t \Phi \right) + C_i \Psi \partial_i \Phi, \tag{7.204a}$$

$$W_i \left(\Phi, \Psi \right) = A_{ij} \left(\Psi \partial_j \Phi - \Phi \partial_j \Psi \right) - C_i \Phi \partial_t \Psi. \tag{7.204b}$$

The passage from the non-self-adjoint (7.198c) to the self-adjoint (7.202e) operators is made by multiplying the first by

$$\mu \left(\vec{x}, t \right) = \frac{1}{a \left(\vec{x}, t \right)} \frac{1}{Det \left[A_{ij} \left(\vec{x}, t \right) \right]} \exp \left\{ \int \frac{b \left(\vec{x}, t \right)}{a \left(\vec{x}, t \right)} dt + \int A_{ij}^{-1} \left(\vec{x}, t \right) B_j \left(\vec{x}, t \right) dx_i \right\}, \tag{7.205}$$

which is the product of the temporal (7.102b) [spatial (7.148b)] factors. The proof of (7.202a through e) is made subtracting the adjoint (7.200) from the original (7.198c) operator, leading to the sum of three terms: (1/2) the temporal (7.90) [spatial (7.143)] terms that vanish for (7.91b) ≡ (7.202b) [(7.140a) ≡ (7.202a)]; and (3) the remaining terms arise from the difference of the cross-terms second on the r.h.s. of (7.200) and (7.198c), leading to

$$0 = \partial_t \, \partial_i \left(C_i \Psi \right) - C_i \, \partial_t \, \partial_i \, \Psi = \left(\partial_i \, \Psi \right) \left(\partial_t C_i \right) + \left(\partial_t \, \Psi \right) \left(\partial_i C_i \right) + \Psi \, \partial_t \, \partial_i \, C_i , \tag{7.206}$$

which vanishes for (7.202c). Substitution of (7.202a through c) in (7.201b) [(7.201c)] simplifies the temporal (spatial) bilinear concomitants to (7.204b) [(7.204c)].

7.9.3 Laplace, Wave, and Klein–Gordon Equations

In the isotropic case, the matrix reduces to a scalar (7.207a) and the vectors (7.207b and c) vanish, and thus *the* **isotropic linear second-order partial differential operator in space-time** *is (7.206d):*

$$A_{ij} = A\delta_{ij}, B_i = 0 = C_i: \quad \left\{ L\left(\nabla, \partial_t \right) \right\} \Phi = A\left(\vec{x}, t \right) \nabla^2 \Phi - a\left(\vec{x}, t \right) \partial_{tt} \Phi - b\left(\vec{x}, t \right) \partial_t \Phi - c\left(\vec{x}, t \right) \Phi. \tag{7.207a–d}$$

*Omitting the first-order time derivative (7.208a) leads to the **Klein (1926)–Gordon (1926)** equation (7.208b) whose particular case (7.209a) is (7.209b) is the **wave equation (D'Alembert, 1747)**, which for (7.210a) reduces to (7.210b), the **Laplace (1820) equation**:*

$$b = 0: \quad \left\{ L\left(\nabla, \partial_t \right) \right\} \Phi = A\left(\vec{x}, t \right) \nabla^2 \Phi - a\left(\vec{x}, t \right) \partial_{tt} \Phi - c\left(\vec{x}, t \right) \Phi. \tag{7.208a,b}$$

$$b = 0 = c: \quad \left\{ L\left(\nabla, \partial_t \right) \right\} \Phi = A\left(\vec{x}, t \right) \nabla^2 \Phi - a\left(\vec{x}, t \right) \partial_t \Phi, \tag{7.209a–c}$$

$$b = c = a = 0: \quad \left\{ L\left(\nabla \right) \right\} \Phi = A\left(\vec{x} \right) \nabla^2 \Phi. \tag{7.210a–d}$$

The operator adjoint to (7.208b) is (7.211) and the temporal (spatial) bilinear concomitants (7.212a) [(7.212b)]:

$$\left\{ \overline{L}\left(\nabla, \partial_t \right) \right\} \Psi = \nabla^2 \left[A\left(\vec{x}, t \right) \Psi \right] - \partial_{tt} \Psi \left[a\left(\vec{x}, t \right) \Psi \right] - c\left(\vec{x}, t \right) \Psi, \tag{7.211}$$

$$V\left(\Phi, \Psi \right) = a\left(\Phi \partial_t \Psi - \Psi \partial_t \Phi \right) + \Phi \Psi \partial_t a, \tag{7.212a}$$

$$\vec{W}\left(\Phi, \Psi \right) = A\left(\Psi \nabla \Phi - \Phi \nabla \Psi \right) - \Phi \Psi \nabla A. \tag{7.212b}$$

The operator (7.207d) is self-adjoint (7.202a through c; 7.207a through c) if the conditions (7.213a and b) are met:

$$\nabla A = 0 = \partial_t a: \quad \left\{ L\left(\partial_t \right) \right\} \Phi = \left\{ \overline{L}\left(\partial_t \right) \right\} \Phi = A\left(t \right) \nabla^2 \Phi - a\left(\vec{x} \right) \partial_{tt} \Phi - c\left(\vec{x}, t \right) \Phi; \tag{7.213a–d}$$

the temporal (7.212a) [spatial (7.212b)] bilinear concomitants simplify to (7.214a) [(7.214b)],

$$V = a\left(\Phi \partial_t \Psi - \partial_t \Phi \right), \quad \vec{W} = A\left(\Psi \nabla \Phi - \Phi \nabla \Psi \right), \tag{7.214a,b}$$

and appear in the commutation relation (7.203).

7.9.4 Diffusion, Telegraphy, and Schrödinger Equations

*The isotropic linear second-order partial differential operator in space-time (7.207d) includes (1) for (7.215a) the **Schrödinger** (1926) **equation** (7.215b); (2) for (7.216a) the **telegraph equation** (7.216b); and (3) combining (1, 2) in (7.217a and b), the (7.217c) **diffusion equation** (Fourier, 1818):*

$$a = 0: \quad A(\vec{x},t)\nabla^2\Phi = b(\vec{x},t)\partial_t\Phi + c(\vec{x},t)\Phi, \tag{7.215a,b}$$

$$c = 0: \quad A(\vec{x},t)\nabla^2\Phi = c(\vec{x},t)\partial_{tt}\Phi + b(\vec{x},t)\partial_t\Phi, \tag{7.216a,b}$$

$$a = 0 = c: \quad A(\vec{x},t)\nabla^2\Phi = b(\vec{x},t)\partial_t\Phi. \tag{7.217a–c}$$

The operator adjoint to (7.207d) is

$$\{\bar{L}(\nabla,\partial_t)\}\Psi = \nabla^2[A(\vec{x},t)\Psi] - \partial_{tt}[a(\vec{x},t)\Psi] + \partial_t[b(\vec{x},t)\Psi] - c(\vec{x},t)\Psi, \tag{7.218}$$

and the commutator (7.203) involves the temporal (7.219) [spatial (7.212b)] bilinear concomitants

$$V(\Phi,\Psi) = a(\Phi\partial_t\Psi - \Psi\partial_t\Phi) - (b - \partial_t a)\Phi\Psi. \tag{7.219}$$

Extending the isotropic to the intermediate linear partial differential in space-time leads to the inhomogeneous Klein–Gordon and wave (diffusion, telegraphy, and Schrödinger) equations [Subsection 7.9.5 (Subsection 7.9.6)].

7.9.5 Intermediate Self-Adjoint Differential Operator

*The **intermediate second-order linear partial differential operator in space-time** follows (7.220b) from (7.198c) replacing the matrix (7.207a) ≡ (7.220a) by a scalar:*

$$A_{ij} = A\delta_{ij}: \quad \{L(\nabla,\partial_t)\}\Phi = A\nabla^2\Phi + \vec{C}\cdot\nabla(\partial_t\Phi) - a\partial_{tt}\Phi + \vec{B}\cdot\nabla\Phi - b\partial_t\Phi - c\Phi. \tag{7.220a,b}$$

Its adjoint operator is

$$\{\bar{L}(\nabla,\partial_t)\}\Psi = \nabla^2(A\Psi) + \partial_t[\nabla\cdot(\vec{C}\Psi)] - \partial_{tt}(a\Psi) - \nabla\cdot(\vec{B}\Psi) + \partial_t(b\Psi) - c\Psi. \tag{7.221}$$

The commutator (7.201a) involves the temporal (7.201b) [spatial (7.222)] bilinear concomitant:

$$\vec{W}(\Phi,\Psi) = A(\Psi\nabla\Phi - \Phi\nabla\Psi) + \Phi\Psi(\vec{B} - \nabla A) - \Phi\partial_t(\vec{C}\Psi). \tag{7.222}$$

The operator is self-adjoint (7.223d and e) if (7.223a through c) are met:

$$\vec{B} = \nabla A, b = \partial_t a, \vec{C} = \text{const}: \quad \{L(\nabla,\partial_t)\}\Phi = \{\bar{L}(\nabla,\partial_t)\}\Phi$$

$$= \nabla\cdot[A(\vec{x},t)\nabla\Phi] - \partial_t[a(\vec{x},t)\partial_t\Phi] + \vec{C}\cdot[\nabla\partial_t\Phi] - c(\vec{x},t)\Phi. \tag{7.223a–c}$$

The commutator (7.203) involves the temporal (7.224a) [spatial (7.224b)] bilinear concomitants:

$$V(\Phi,\Psi) = a(\Phi\partial_t\Psi - \Psi\partial_t\Phi) + C_i\Psi\partial_t\Phi, \quad \vec{W}(\Phi,\Psi) = A(\Psi\nabla\Phi - \Phi\nabla\Psi) - C_i\Phi\partial_t\Psi. \tag{7.224a,b}$$

*The intermediate self-adjoint linear second-order partial differential operator in space-time (7.223a) ≡ (7.223b) with (7.225a) is the **inhomogeneous Klein–Gordon equation (7.225b)**:*

$$\vec{C} = 0: \quad \left\{ L(\nabla, \partial t) \right\} \Phi = \left\{ \overline{L}(\nabla, \partial t) \right\} \Phi = \nabla \cdot (A \nabla \Phi) - \partial_t (a \partial_t \Phi) - c\Phi, \tag{7.225a,b}$$

$$\vec{C} = 0 = c: \quad \left\{ L(\nabla, \partial t) \right\} \Phi = \left\{ \overline{L}(\nabla, \partial t) \right\} \Phi = \nabla \cdot (A \nabla \Phi) - \partial_t (a \partial_t \Phi), \tag{7.226a–c}$$

$$\vec{C} = 0 = c = a: \quad \left\{ L(\nabla) \right\} \Phi = \left\{ \overline{L}(\nabla) \right\} \Phi = \nabla \cdot (A \nabla \Phi), \tag{7.227a–d}$$

*whose particular case (7.226a and b) is the **inhomogeneous unsteady wave equation (7.226c)**, which includes for (7.227a through c) the **inhomogeneous Laplace equation (7.227d)**.*

7.9.6 Inhomogeneous Diffusion, Telegraphy, and Schrödinger Equations

Adding to (7.225b) the temporal derivative leads to

$$\left\{ L(\nabla, \partial_t) \right\} \Phi = \nabla \cdot \left[A(\vec{x}, t) \nabla \Phi \right] - \partial_t \left[a(\vec{x}, t) \partial_t \Phi \right] - b(\vec{x}, t) \partial_t \Phi - c(\vec{x}, t) \Phi, \tag{7.228}$$

*which includes (1) for (7.229a) the **inhomogeneous Schrödinger equation (7.229b)**; (2) for (7.229c) the **inhomogeneous unsteady telegraph equation (7.229d)**:*

$$a = 0: \quad \nabla \cdot (A \nabla \Phi) = b \partial_t \Phi + c\Phi; \quad c = 0: \quad \nabla \cdot (A \nabla \Phi) = \partial_t (a \partial_t \Phi) + b \partial_t \Phi; \tag{7.229a–d}$$

*and (3) combining (1, 2) in (7.229a and c) ≡ (7.230a and b), the **inhomogeneous diffusion equation (7.230c)**:*

$$a = 0 = c: \quad b \partial_t \Phi = \nabla \cdot (A \nabla \Phi) = A \nabla^2 \Phi + \nabla A \cdot \nabla \Phi. \tag{7.230a–c}$$

The operator adjoint to (7.228) is:

$$\left\{ \overline{L}(\nabla, \partial_t) \right\} \Psi = \nabla \cdot \left[A(\vec{x}, t) \nabla \Psi \right] - \partial_t \left[a(\vec{x}, t) \partial_t \Psi \right] + \partial_t \left[b(\vec{x}, t) \Phi \right] - c(\vec{x}, t) \Phi; \tag{7.231}$$

*the commutator (7.201a) involves the temporal (7.219) [spatial (7.212b)] bilinear concomitant. The **Helmholtz (inhomogeneous Helmholtz) equation** (7.232c) [(7.233c)] is the particular case (7.232a and b) [≡(7.233a and b)] of the isotropic (intermediate) operator (7.207d) [(7.228)]:*

$$a = 0 = b: \quad \left\{ L(\nabla) \right\} \Phi = A(\vec{x}) \nabla^2 \Phi - c(\vec{x}) \Phi, \tag{7.232a–c}$$

$$a = 0 = b: \quad \left\{ L(\nabla) \right\} \Phi = \left\{ \overline{L}(\nabla) \right\} \Phi = \nabla \cdot (A \nabla \Phi) - c\Phi. \tag{7.233a–c}$$

The intermediate second-order linear partial differential operator in time and one (two) spatial dimensions specifies the transverse vibrations of an inhomogeneous elastic string (membrane) and is (is not) self-adjoint in the absence (presence) of damping.

7.9.7 Transverse Vibrations of an Inhomogeneous String/Membrane

The transverse vibrations of an inhomogeneous elastic string (7.94) [membrane (7.168a)] are specified by the (Figure 7.3b) balance of the transverse shear stress against the same terms as before for the spatial dependence adding (7.234) [(7.235)] as for the damped harmonic oscillator (7.103) the inertia force for variable mass density per unit length ρ and linear friction with nonuniform damping λ:

$$f(x,t) = \partial_t \left[\rho(x,t)\partial_t \zeta \right] + \lambda(x,t)\partial_t \zeta - \partial_x \left[T(x,t)\partial_x \zeta \right] + k(x,t)\zeta \equiv \left\{ L\left(\partial_x,\partial_t\right) \right\}\zeta, \quad (7.234)$$

$$f(x,y,t) = \partial_t \left[\rho(x,y,t)\partial_t \zeta \right] + \lambda(x,y,t)\partial_t \zeta - \nabla \cdot \left[T(x,y,t)\nabla \zeta \right] + k(x,y,t)\zeta \equiv \left\{ L\left(\partial_x,\partial_y,\partial_t\right) \right\}\zeta. \quad (7.235)$$

The adjoint operator is (7.236) [(7.237)]:

$$\left\{ \overline{L}\left(\nabla,\partial_t\right) \right\}\Psi = \partial_t \left(\rho\Psi\right) - \partial_t \left(\lambda\zeta\right) - \partial_x \left(T\partial_x\zeta\right) + k\zeta, \quad (7.236)$$

$$\left\{ \overline{L}\left(\nabla,\partial_t\right) \right\}\Psi = \partial_t \left(\rho\Psi\right) - \partial_t \left(\lambda\zeta\right) - \nabla \cdot \left(T\nabla\zeta\right) + k\zeta, \quad (7.237)$$

and the temporal (7.238a) [(7.238a)] and spatial (7.238b) [(7.238c)] bilinear concomitants are

$$V\left(\Phi,\Psi\right) = \rho\left(\Psi\partial_t\Phi - \Phi\partial_t\Psi\right) + \Phi\Psi\left(\lambda - \partial_t\rho\right), \quad (7.238a)$$

$$W\left(\Phi,\Psi\right) = T\left(\Phi\partial_x\Psi - \Psi\partial_x\Phi\right), \quad \vec{W}\left(\Phi,\Psi\right) = T\left(\Phi\nabla\Psi - \Psi\nabla\Phi\right). \quad (7.238b,c)$$

The superposition (reciprocity) principle holds in all cases (in the absence of damping).

7.9.8 Space-Time Method for High-Order Operators

The linear second-order partial differential operators in space-time (Subsections 7.9.1 through 7.9.7) were obtained by collecting the temporal (Subsections 7.7.1 through 7.7.5) and spatial (Subsections 7.8.1 through 7.8.6) operators and adding cross-terms. The same method could be used to form high-order linear partial differential operators based on their temporal (Subsections 7.7.6 through 7.7.9) and spatial (Subsections 7.8.7 through 7.8.12) parts, (1) requiring a more significant effort concerning cross-terms; and (2) leading to rather cumbersome expressions that will not be written here. Instead, next is illustrated a **space-time method** *for high-order operators that: (1) adds to the N spatial coordinates (7.239d) a zero or time coordinate (7.239c) so that there are (N + 1) independent variables denoted by a Greek index (7.222a) in contrast to a Latin index (7.233b) for spatial coordinates:*

$$\alpha,\beta = 0,1,\ldots,N; \quad i,j = 1,\ldots,N: \quad y_0 \equiv t, \quad y_i \equiv x_i; \quad (7.239a\text{–}d)$$

(2) the high-order linear partial differential operator take the preceding form (Section 7.8) in terms of the (N + 1) or y-coordinates (7.239a); and (3) these are split into time (7.239c) and space (7.239b and d) variables to specify the space-time operator. The space-time operator may have different orders in space and time; in this method the highest of the two orders must be used, and any unnecessary terms are assigned zero coefficients. For simplicity and as a check, the space-time method is used as alternative to derive the linear second-order partial differential operators (Subsection 7.9.1) (Subsection 7.9.8).

The linear second-order partial differential operator (7.131c) in y-coordinates (7.239a) is

$$\left\{L\left(\frac{\partial}{\partial y_\gamma}\right)\right\}\Phi = A_{\alpha\beta}\left(\vec{y}\right)\frac{\partial^2\Phi}{\partial y_\alpha \partial y_\beta} + B_\alpha\left(\vec{y}\right)\frac{\partial\Phi}{\partial y_\alpha} + C\left(\vec{y}\right)\Phi. \qquad (7.240)$$

Splitting (7.239b through d) into space, time, and cross-terms leads to:

$$\left\{L\left(\partial_i, \partial_0\right)\right\}\Phi = A_{ij}\partial_{ij}\Phi + A_{io}\partial_{io}\Phi + A_{oj}\partial_{oj}\Phi + A_{oo}\partial_{oo}\Phi + B_i\partial_i\Phi + B_0\partial_0\Phi + C\Phi. \qquad (7.241)$$

Noting (7.242a and b) and denoting the coefficients by (7.242c through f),

$$\partial_i \equiv \nabla, \partial_0 \equiv \partial_t : \quad A_{io} + A_{oi} = C_i, \quad A_{oo} = -a, \quad B_0 = -b, \quad C = -c, \qquad (7.242a\text{–}f)$$

proves the coincidence of (7.241) ≡ (7.198c). The operator adjoint to (7.131c) is (7.132) leading in y-coordinates (7.239a) to (7.244a):

$$\left\{\bar{L}\left(\frac{\partial}{\partial\bar{y}}\right)\right\}\Psi = \partial_{\alpha\beta}\left(A_{\alpha\beta}\Psi\right) - \partial_\alpha\left(B_\alpha\Psi\right) + C\Psi = \partial_{ij}\left(A_{ij}\Psi\right) + \partial_{io}\left(A_{io}\Psi\right)$$

$$+ \partial_{oj}\left(A_{jo}\Psi\right) + \partial_{oo}\left(A_{00}\Psi\right) - \partial_i\left(B_i\Psi\right) - \partial_o\left(A_{io}\Psi\right) + C\Psi = \left\{\bar{L}\left(\partial_i, \partial_0\right)\right\}\Psi; \qquad (7.243a,b)$$

the space-time decomposition (7.239b through d) leads to the adjoint second-order linear partial differential operator in space-time (7.243b) ≡ (7.200) with the same identifications (7.242a through f).

The commutator (7.134) in y-coordinates (7.244a) becomes (7.144b) in space-time (7.239b through d):

$$\Psi\left\{L\left(\frac{\partial}{\partial\bar{y}}\right)\right\}\Phi - \Phi\left\{\bar{L}\left(\frac{\partial}{\partial\bar{y}}\right)\right\}\Psi = \frac{\partial W_\alpha}{\partial y_\alpha} = \partial_0 W_0 + \partial_i W_i = \Psi\left\{L\left(\partial_i, \partial_0\right)\right\}\Phi - \Phi\left\{L\left(\partial_i, \partial_0\right)\right\}\Psi. \qquad (7.244a,b)$$

The bilinear concomitant (7.133) in y-coordinates (7.239a) is:

$$W_\alpha\left(\Phi, \Psi\right) = A_{\alpha\beta}\left(\Psi\frac{\partial\Phi}{\partial y_\beta} - \Phi\frac{\partial\Psi}{\partial y_\beta}\right) + \left(B_\alpha - \frac{\partial A_{\alpha\beta}}{\partial y_\beta}\right)\Phi\Psi; \qquad (7.245)$$

it splits into temporal (7.246) [spatial (7.247)] bilinear concomitants:

$$W_0 = A_{00}\left(\Psi\partial_0\Phi - \Phi\partial_0\Psi\right) + \left(B_0 - \partial_0 A_{00}\right)\Phi\Psi + \Psi\left(A_{i0}\partial_i\Phi + A_{0j}\partial_j\Phi\right) \equiv V_i, \qquad (7.246)$$

$$W_i = A_{ij}\left(\Psi\partial_j\Phi - \Phi\partial_j\Psi\right) + \left(B_i - \partial_j A_{ij}\right)\Phi\Psi - \Phi\partial_0\left[\left(A_{0i} + A_{i0}\right)\Psi\right]; \qquad (7.247)$$

these coincide with (7.246) ≡ (7.201b) [(7.247) ≡ (7.201c)] with the identifications (7.242a through f). The same method applies to operators of higher than the second order (Notes 7.8 and 7.9), for example, the linear vibrations of bars (plates) leads (Subsection 7.9.9) to a linear partial differential operator of the second order in time and fourth order in one (two) spatial coordinates.

7.9.9 Transverse Vibrations of Bars and Plates

The transverse vibrations of bars (plates) add to the elastic string (7.234) [membrane (7.235)] the stiffness term in (7.128) [(7.196a)] raising (keeping) the order of the operator (7.248) [(7.249)] in space to four (time at two):

$$f(x,t) = \partial_t\left[\rho(x,t)\partial_t\zeta\right] + \lambda(x,t)\partial_t\zeta + \partial_{xx}\left[B(x,t)\partial_{xx}\zeta\right]$$
$$-\partial_x\left[T(x,t)\partial_x\zeta\right] + k(x,t)\zeta \equiv \left\{L(\partial_x,\partial_t)\right\}\zeta(x,t), \tag{7.248}$$

$$f(x,y,t) = \partial_t\left[\rho(x,y,t)\partial_t\zeta\right] + \lambda(x,y,t)\partial_t\zeta + \nabla^2\left[D(x,y,t)\nabla^2\zeta\right]$$
$$-\nabla\cdot\left[T(x,y,t)\nabla\zeta\right] + k(x,y,t)\zeta \equiv \left\{L(\nabla,\partial_t)\right\}\zeta(\partial_x,\partial_y,\partial_t). \tag{7.249}$$

Thus *an inhomogeneous elastic bar (plate) with density* ρ, *nonuniform bending stiffness* $B = EI(D)$, *under tangential traction* $T > 0$ *or compression* $T < 0$, *supported on springs of resilience* k, *with linear damping* λ *subject to a shear stress* f, *has transversal vibrations* [Figure 7.3a (Figure 7.3b)] *specified by the linear partial differential operator (7.248) [(7.249)] of second order in time and fourth order in one (two) spatial variables. The adjoint operator is (7.250) [(7.251)]:*

$$\left\{\overline{L}(\partial_x,\partial_t)\right\}\Psi = \partial_t\left(\rho\partial_t\Psi\right) - \partial_t\left(\lambda\zeta\right) + \partial_{xx}\left(B\partial_{xx}\Psi\right) - \partial_x\left(T\partial_x\Psi\right) - k\Psi, \tag{7.250}$$

$$\left\{\overline{L}(\partial_x,\partial_t)\right\}\Psi = \partial_t\left(\rho\partial_t\Psi\right) - \partial_t\left(\lambda\zeta\right) + \nabla^2\left(D\nabla^2\Psi\right) - \nabla\cdot\left(T\nabla\Psi\right) + k\Psi, \tag{7.251}$$

the temporal (7.238a) and spatial [(7.169c) added to (7.197)] bilinear concomitants satisfy the commutation relation (7.201a). The principle of superposition holds in all cases because the operators are linear. The reciprocity principle holds (does not hold) in the absence (presence) of damping when the operators are (are not) self-adjoint.

NOTE 7.1: Nonlinear/Linear Green Function and the Principles of Superposition and Reciprocity

The **Green function** can be defined as the **fundamental solution** of an ordinary (partial) differential equation with suitable initial and/or boundary conditions, regardless of whether the differential operator is linear or nonlinear and self-adjoint or non-self-adjoint; the principle of superposition (reciprocity) generally holds only for linear (self-adjoint) differential operators. An example is an ordinary differential equation (7.252) for which the Green function is defined as the fundamental solution forced by a Dirac delta distribution:

$$\left\{L\left(\frac{d}{dx}\right)\right\}G(x;\xi) = \delta(x-\xi). \tag{7.252}$$

Four cases can arise: (1) if the operator is self-adjoint (7.253a), then the reciprocity principle holds (7.253b):

$$\Psi(x)\left\{L\frac{d}{dx}\right\}\Phi(x) - \Phi(x)\left\{L\frac{d}{dx}\right\}\Psi(x) = \frac{dM}{dx}: \quad G(x;\xi) = G(\xi;x), \tag{7.253a,b}$$

where M is the bilinear concomitant; (2) if the operator is linear (7.254a), the inhomogeneous differential equation (7.254b) with arbitrary forcing may be solved using the superposition principle (7.254c):

$$L\left(\frac{d}{dx}\right) \equiv \sum_{n=0}^{\infty} A_n(x)\frac{d^n}{dx^n}: \quad \left\{L\frac{d}{dx}\right\}\Phi(x) = f(x) \Rightarrow \Phi(x) = \int f(\xi)G(x;\xi)d\xi + \text{const}; \tag{7.254a–c}$$

and (3/4) if the differential operator is non-self-adjoint (nonlinear), the principle of reciprocity (superposition) does not generally hold. Several examples have been given of nonlinear (linear) Green functions, which do not (do) satisfy the principle of superposition, such as the large (small) deflection of (a) an elastic string [Sections 2.1 through 2.5 (Sections 2.6 through 2.9)]; (b) a stiff bar [Sections 4.1 through 4.6 (Sections 4.7 through 4.9)]. In both cases (a,b), the linear operator for small deflections is self-adjoint, so the principle of reciprocity also holds. The Green function is used most often for linear, self-adjoint differential operators, when both the principle of (I) superposition and (II) reciprocity hold. In this respect, it is notable that the weak bending of (transverse waves in) a bar or plate is specified by a self-adjoint operator (Subsection 7.9.9) even when the moment of inertia of the cross section, the Young's modulus of the material (bending stiffness), the longitudinal tension, the resilience of a spring support, and the mass density are all nonuniform (unsteady), that is, are functions of position (and time). Only the presence of damping causes nonadjointness as for the harmonic oscillator (Subsection 7.7.5).

NOTE 7.2: Second-Order Scalar Linear Partial Differential Equation

*The general **linear second-order partial differential equation in space-time with constant coefficients** is*

$$\left\{ L\left(\frac{\partial}{\partial x}, \frac{\partial}{\partial t} \right) \right\} \Phi = \nabla^2 \Phi - \frac{1}{c^2} \frac{\partial^2 \Phi}{\partial t^2} - \frac{1}{\alpha} \frac{\partial \Phi}{\partial t} + \beta \Phi = f(\vec{x}, t), \tag{7.255}$$

involving the following five assumptions: (a) second order, which excludes more complex coupled phenomena, for example, stiffness of a bar or plate (Note 7.3); (b) linear, which excludes large perturbations of equilibrium conditions, for example, deflection with large slope of elastic string (Sections 2.6 through 2.9) or bars (Sections 4.7 through 4.9); (c) constant coefficients excluding nonuniform (and unsteady media) for which the coefficients depend on position (time); (d) isotropic media that have at each point and instant the same properties in all directions, implying scalar coefficients in (7.255) and excluding vector coefficients like:

$$\vec{B} \cdot \nabla \left(\frac{\partial \Phi}{\partial t} \right) + \vec{D} \cdot \nabla \Phi; \tag{7.256}$$

*and (e) scalar variable Φ that could be replaced by a vector, matrix, or tensor (Note 7.5). Besides the forcing on the r.h.s. of (7.255) the terms on the l.h.s. may be interpreted (Diagram 7.3) as follows: (1) the Laplace/Poisson equation for potential fields, like potential or vortical flows (Chapters I.12, 14, 16, 28, 34, 36, 38; II.2, 8; III.6), electro(magneto)static fields [Chapter I.24 (Chapter I.26) and Sections 8.1 through 8.3 (Sections 8.4 through 8.6)], gravity field (Chapter I.18), steady heat conduction (Chapter I.32), and others (Chapter II.6); (2) the wave term specifying, for example, transverse vibrations of elastic strings and membranes (Subsection 7.9.7) and acoustic (electromagnetic) waves [Notes 7.14 through 7.17 (Notes 7.12 and 7.13)]; (3) the diffusion term applying to unsteady heat conduction, and also to diffusion of mass and electricity (Notes 7.12 and 7.13), and (4) the zero-order term analogous to the resilience of a spring of a harmonic oscillator (Subsection 7.7.5). The parameter c is the **wave speed** (7.257a) equal to the square root of the tension divided by the density for the vibrations of an elastic string (7.234) [membrane (7.235)]:*

$$c_e = \sqrt{\frac{T}{\rho}}; \quad \nabla^2 \Phi = \frac{1}{c^2} \frac{\partial^2 \Phi}{\partial t^2}; \quad \Phi(x, t) = \Psi(\vec{x}) e^{i\omega t}; \tag{7.257a–c}$$

for a wave of frequency ω in (7.257c) the wave equation (7.257b) leads to the Helmholtz equation (7.258a):

$$\nabla^2 \Phi + \beta \Phi = 0, \quad \beta = \left(\frac{\omega}{c} \right)^2 = k^2, \tag{7.258a,b}$$

corresponding to the last term in (7.255), where the coefficient is the square of the wavenumber. For a harmonic time dependence, the diffusion equation (7.259a) leads (7.257c) to (7.259b):

$$\alpha \nabla^2 \Phi = \frac{\partial \Phi}{\partial t}, \quad \nabla^2 \Psi + i\frac{\alpha}{\omega}\Psi = 0, \tag{7.259a,b}$$

which is also a Helmholtz equation (7.258a) with imaginary (7.259b) instead of real (7.258b) parameter.

NOTE 7.3: Partial Differential Equations of Mathematical Physics

Some of the most frequent **equations of mathematical physics** are (1) the Laplace (Poisson) equation (7.260a) [(7.260b)] corresponding to the first term of (7.255):

$$\nabla^2 \Phi = 0, \quad \nabla^2 \Phi = f; \tag{7.260a,b}$$

(2) the wave equation (7.257b) corresponding to the first two terms; (3) the diffusion equation (7.259a) corresponding to the first and third terms, where α is the diffusivity; (4) the Helmholtz equation (7.258a) corresponding to the first and fourth terms; (5–7) the combinations of the preceding are (5) ≡ (2, 3) the Klein–Gordon equation (7.261a), (6) ≡ (2, 4) the telegraph equation (7.261b), and (4, 5) ≡ (8) the Schrödinger equation (7.261c):

$$0 = \left\{ \nabla^2 \Phi - \frac{1}{c^2}\frac{\partial^2 \Phi}{\partial t^2} + \beta\Phi, \quad \nabla^2 \Phi - \frac{1}{c^2}\frac{\partial^2 \Phi}{\partial t^2} - \frac{1}{\alpha}\frac{\partial \Phi}{\partial t}, \quad \nabla^2 \Phi - \frac{1}{\alpha}\frac{\partial \Phi}{\partial t} + \beta\Phi \right\}; \tag{7.261a–c}$$

(8) the preceding are zero (1), one (2–4), and two (5–7) parameter subcases (Diagram 7.3) of the partial differential **equation of mathematical physics** *(7.255) whose adjoint with constant coefficients (7.262a through c) is (7.262d):*

$$c, \alpha, \beta = \text{const}: \quad \left\{ \bar{L}\left(\frac{\partial}{\partial \vec{x}}, \frac{\partial}{\partial t}\right) \right\} \Psi = \nabla^2 \Psi - \frac{1}{c^2}\frac{\partial^2 \Psi}{\partial t^2} + \frac{1}{\alpha}\frac{\partial \Psi}{\partial t} + \beta\Psi. \tag{7.262a–c}$$

Thus it is (is not) self-adjoint in the absence $\alpha = \infty$ (presence $\alpha < \infty$) of diffusion. The commutator (7.263a) involves the temporal (7.263b) [spatial (7.263c)] bilinear concomitant:

$$\Psi\left\{ L\left(\frac{\partial}{\partial \vec{x}}, \frac{\partial}{\partial t}\right) \right\}\Phi - \Phi\left\{ \bar{L}\left(\frac{\partial}{\partial \vec{x}}, \frac{\partial}{\partial t}\right) \right\}\Psi = \frac{\partial V}{\partial t} + \nabla \cdot \vec{W}, \tag{7.263a}$$

$$V(\Phi, \Psi) = \frac{1}{c^2}\left(\Phi\frac{\partial \Psi}{\partial t} - \Psi\frac{\partial \Phi}{\partial t} \right) + \frac{1}{\alpha}\Phi\Psi, \quad \vec{W}(\Phi, \Psi) = \Psi\nabla\Phi - \Phi\nabla\Psi. \tag{7.263b,c}$$

The five restrictions (a–e) on the partial differential equation of mathematical physics (7.255) have been or will be lifted one (or two) at a time in the simplest form, considering (a) an arbitrary order in space and time (Notes 7.7 and 7.8); (b) a nonlinear deflection of elastic springs (Sections 2.6 through 2.9; Examples 10.6 through 10.9) and bars (Sections 4.7 through 4.9); (c, d) inhomogeneous, unsteady, and anisotropic media (Note 7.4); and (e) the vector field as the dependent variable (Notes 7.5 and 7.6).

NOTE 7.4: Anisotropic, Nonhomogeneous, and Unsteady Media

The coefficients of the partial differential equations of mathematical physics depend on the properties of the medium (Table 7.2) that may be (1) **homogeneous (inhomogeneous)** if they do not (do) depend on position; (2) **steady (unsteady)** if they do not (do) depend on time; and (3) **isotropic (anisotropic)** if at all (some) points and instants it does not (does) depend on direction. For example, (1) an elastic plate with nonuniform mass density or bending stiffness is inhomogeneous; (2) a collapsible tube represented by an elastic string is unsteady; and (3) a membrane with different tangential stresses in distinct directions is anisotropic. The complete linear second-order partial differential operator in space-time (7.264d) adds two terms to (7.255), namely, the third and fourth on the r.h.s. of (7.264d), where the notation (7.264a through c) is used:

$$\left\{ \frac{1}{c^2}, \frac{1}{\alpha}, \beta \right\} = \frac{\{a,b,c\}}{A} : \left\{ L\left(\frac{\partial}{\partial \vec{x}}, \frac{\partial}{\partial t} \right) \right\} \Phi = A\nabla^2\Phi - a\frac{\partial^2\Phi}{\partial t^2} + \vec{C}\cdot\nabla\left(\frac{\partial\Phi}{\partial t} \right) + \vec{D}\cdot\nabla\Phi - b\frac{\partial\Phi}{\partial t} + c\Phi.$$

$$(7.264a–d)$$

Replacing the scalar A by a matrix A_{ij}, and the Laplacian (7.186a) by the matrix of second-order spatial derivates (7.131b) the operator (7.264d) becomes (7.198c). Thus there are *four main cases to consider: (1) for an inhomogeneous, steady, and isotropic medium (7.262a through c), the operator (7.255) has adjoint (7.262c) and commutator (7.263a) involving the temporal (7.263b) and spatial (7.263c) bilinear concomitants; (2) for an unsteady and inhomogeneous isotropic medium, the intermediate operator (7.220b) has self-adjoint (7.221) and commutator (7.201a) involving the temporal (7.201b) and spatial (7.222) bilinear concomitants; (3) for a steady and homogeneous anisotropic medium, the operator (7.198c) with constant coefficients (7.265a through f) has adjoint (7.265g),*

$$A_{ij}, B_i, C_i, a, b, c = \text{const} : \left\{ \overline{L}\left(\nabla, \partial t \right) \right\} \Phi = A_{ij}\partial_{ij}\Phi + C_i\partial_t\partial_i\Phi - a\partial_{tt}\Phi + B_i\partial_i\Phi - b\partial_t\Phi + c\Phi, \quad (7.265a–g)$$

and commutator (7.201a) involving the temporal (7.266a) and spatial (7.266b) bilinear concomitants:

$$V\left(\Phi, \Psi \right) = a\left(\Phi\partial_t\Psi - \Psi\partial_t\Phi \right) - b\Phi\Psi + \Psi C_i\partial_i\Phi, \quad (7.266a)$$

$$W_i\left(\Phi, \Psi \right) = A_{ij}\left(\Psi\partial_j\Phi - \Phi\partial_j\Psi \right) + B_i\Phi\Psi - \Phi C_i\partial_t\Psi; \quad (7.266b)$$

and (4) for an inhomogeneous, unsteady, and anisotropic medium, the operator (7.202e) is self-adjoint and has commutator (7.203a) involving the temporal (7.204a) and spatial (7.204b) bilinear concomitants.

NOTE 7.5: **Scalar Equation of Mathematical Physics with Vector Variable**

Replacing in (7.198c) the scalar Φ by a vector Φ_k leads to *the second-order linear scalar partial differential equation for a vector dependent variable*:

$$\left\{ L\left(\nabla, \partial_t \right) \right\} \Phi_k = A_{ijk}\, \partial_{ij}\, \Phi_k + C_{ik}\, \partial_t\, \partial_i\, \Phi_k - a_k\, \partial_{tt}\, \Phi_k$$

$$+ B_{ik}\, \partial_i\, \Phi_k - b_k\, \partial_t\, \Phi_k - C_k\, \Phi_k, \quad (7.267)$$

where the equation is scalar. The adjoint operator is

$$\left\{ \overline{L}\left(\nabla, \partial_t \right) \right\} \Psi_k = \partial_{ij}\left(A_{ijk}\Psi_k \right) + \partial_t\, \partial_i\left(C_{ik}\, \Psi_k \right) - \partial_{tt}\left(a_k\, \Psi_k \right)$$

$$- \partial_i\left(B_{ik}\, \Psi_k \right) + \partial_t\left(b_k\, \Psi_k \right) - C_k\, \Psi_k, \quad (7.268)$$

and the commutation relation

$$\Psi_k\left\{L\left(\nabla,\partial_t\right)\right\}\Phi_k - \Phi_k\left\{\overline{L}\left(\nabla,\partial_t\right)\right\}\Psi_k = \partial_t\,V_k + \partial_i\,W_{ik}\,, \tag{7.269}$$

involves the temporal (7.270a) and spatial (7.270b) bilinear concomitants:

$$V_k\left(\Phi_\ell,\Psi_\ell\right) = a_k\left(\Phi_\ell\,\partial_t\Psi_\ell - \Psi_\ell\,\partial_t\Phi_\ell\right) - \left(b_k - \partial_t\,a_k\right)\Phi_\ell\,\Psi_\ell + \Psi_\ell\,C_{ik}\,\partial_i\,\Phi_\ell, \tag{7.270a}$$

$$W_{ik}\left(\Phi_\ell,\Psi_\ell\right) = A_{ijk}\left(\Psi_\ell\,\partial_j\,\Phi_\ell - \Phi_\ell\,\partial_j\Psi_\ell\right) + \left(B_{ik} - \partial_j\,A_{ijk}\right)\Phi_\ell\,\Psi_\ell - \Phi_\ell\,\partial_t\left(C_{ik}\,\Psi_\ell\right). \tag{7.270b}$$

The self-adjoint case corresponds to the operator

$$\left\{L\left(\nabla,\partial_t\right)\right\}\Phi_k = \left\{\overline{L}\left(\nabla,\partial_t\right)\right\}\Phi_k = \partial_i\left(A_{ijk}\partial_j\Phi_k\right) - \partial_t\left(a_k\partial_t\Phi_k\right) - c_k\Phi_k, \tag{7.271}$$

and the commutation relation

$$\Psi_k\left\{L\left(\nabla,\partial_t\right)\right\}\Phi_k - \Phi_k\left\{L\left(\nabla,\partial_t\right)\right\}\Psi_k = \partial_t\,V_k + \partial_i\,W_{ik}. \tag{7.272}$$

involves the temporal (7.273a) and spatial (7.273b) bilinear concomitants:

$$V_k\left(\Phi_\ell,\Psi_\ell\right) = a_k\left(\Phi_\ell\,\partial_t\Psi_\ell - \Psi_\ell\,\partial_t\Phi_\ell\right) + C_{ik}\,\Psi_\ell\,\partial_i\,\Phi_\ell, \tag{7.273a}$$

$$W_{ik}\left(\Phi_\ell,\Psi_\ell\right) = A_{ijk}\left(\Psi_\ell\,\partial_j\,\Phi_\ell - \Phi_\ell\,\partial_j\,\Psi_\ell\right) - C_{ik}\,\Phi_\ell\,\partial_t\,\Psi_\ell, \tag{7.273b}$$

The preceding results for the non-self-adjoint (7.267; 7.268; 7.269; 7.270a and b) [self-adjoint (7.271; 7.272; 7.273a and b)] linear second-order partial differential operator in space-time with a vector dependent variable follows by analogy with the case of a scalar dependent variable (7.198c; 7.200; 7.201a through c) (7.202e; 7.203; 7.204a and b). The linear second-order partial differential operator in space-time (7.198c) leads to a scalar equation with a scalar dependent variable and has two generalizations: (1) a scalar equation with vector variable (Note 7.5); and (2) a vector equation with scalar variable would add an extra index in all coefficients:

$$\left\{L_n\left(\nabla,\partial_t\right)\right\}\Phi = A_{nij}\partial_{ij}\Phi + C_{nc}\partial_i\partial_t\Phi - a_n\partial_{tt}\Phi + B_{ni}\partial_i\Phi - b_n\partial_t\Phi - C_n\Phi. \tag{7.274}$$

The joint generalization (1) and (2) is a vector equation with a vector dependent variable (Note 7.6). Further extensions would be tensor equations with tensor dependent variable.

NOTE 7.6: Vector Equation of Mathematical Physics with Vector Variable

The combination of (7.267) and (7.274) or the double generalization of (7.198c) *is a **vector linear partial differential operator in space-time** applied to a **vector dependent variable**:*

$$\left\{L_n\left(\nabla,\partial_t\right)\right\}\Phi_k = A_{nijk}\,\partial_{ij}\,\Phi_k + C_{nki}\,\partial_t\,\partial_i\,\Phi_k - a_{nk}\,\partial_{tt}\,\Phi_k$$

$$+ B_{nik}\,\partial_i\,\Phi_k - b_{nk}\,\partial_t\,\Phi_k - C_{nk}\,\Phi_k, \tag{7.275}$$

whose adjoint is

$$\left\{\overline{L}_n\left(\nabla,\partial_t\right)\right\}\Psi_k = \partial_{ij}\left(A_{nijk}\partial_{ij}\Psi_k\right) + \partial_t\partial_i C_{nik}\Psi_k$$
$$- \partial_{tt}\left(a_{nk}\Psi_k\right) - \partial_i\left(B_{nik}\Psi_k\right) + \partial_t\left(b_{nk}\Psi_k\right) - c_{nk}\Psi_k; \tag{7.276}$$

the commutation relation

$$\Psi_k\left\{L_n\left(\nabla,\partial_t\right)\right\}\Phi_k - \Phi_k\left\{\overline{L}_n\left(\nabla,\partial_t\right)\right\}\Psi_k = \partial_t V_{nk} + \partial_i W_{nki}, \tag{7.277}$$

involves the temporal (7.278a) and spatial (7.278b) bilinear concomitants:

$$V_{nk}\left(\Phi_\ell,\Psi_\ell\right) = a_{nk}\left(\Phi_\ell\partial_t\Psi_\ell - \Psi_\ell\partial_t\Phi_\ell\right) - \left(b_{nk} - \partial_t a_{nk}\right)\Phi_\ell\Psi_\ell + \Psi_\ell C_{nki}\partial_i\Phi_\ell, \tag{7.278a}$$

$$W_{nki}\left(\Phi_\ell,\Psi_\ell\right) = A_{nijk}\left(\Phi_\ell\partial_j\Psi_\ell - \Phi_\ell\partial_j\Psi_\ell\right) + \left(B_{nik} - \partial_j A_{nijk}\right)\Phi_\ell\Psi_\ell - \Phi_\ell\partial_t\left(C_{nki}\Psi_\ell\right), \tag{7.278b}$$

The self-adjoint case corresponds to the operator:

$$\left\{L_n\left(\nabla,\partial_t\right)\right\}\Phi_k = \left\{\overline{L}_n\left(\nabla,\partial_t\right)\right\}\Phi_k = \partial_i\left(A_{nijk}\partial_j\Phi_k\right) - \partial_t\left(a_{nk}\partial_t\Phi_k\right) - c_{nk}\Phi_k, \tag{7.279}$$

which satisfies the commutation relation

$$\Psi_k\left\{L_n\left(\nabla,\partial_t\right)\right\}\Phi_k - \Phi_k\left\{L_n\left(\nabla,\partial_t\right)\right\}\Psi_k = \partial_t V_{nk} + \partial_i W_{nki}, \tag{7.280}$$

involving the temporal (7.281a) and spatial (7.281b) bilinear concomitants:

$$V_{nk}\left(\Phi_\ell,\Psi_\ell\right) = a_{nk}\left(\Phi_\ell\partial_t\Psi_\ell - \Psi_\ell\partial_t\Phi_\ell\right) + \Psi_\ell C_{nki}\partial_i\Phi_\ell, \tag{7.281a}$$

$$W_{nki}\left(\Phi_\ell,\Psi_\ell\right) = A_{nijk}\left(\Psi_\ell\partial_j\Phi_\ell - \Phi_\ell\partial_j\Psi_\ell\right) - \Phi_\ell C_{nki}\partial_t\Psi_\ell. \tag{7.281b}$$

Instead of generalizing the linear scalar second-order partial differential operator in space-time with scalar dependent variable (Note 7.4) to a vector dependent variable (Note 7.5) [vector equation (Note 7.6)], the scalar is retained for a high-order equation (Note 7.7) in an anisotropic medium, with the particular isotropic case (Note 7.8). Where the equation is scalar, a vector equation would be obtained adding one more index ℓ to all coefficients.

NOTE 7.7: High-Order Equation of Mathematical Physics

The linear scalar partial differential equation of order M in space and K in time for a scalar variable Φ in a space of dimension N allowing the medium to be anisotropic, unsteady, and nonuniform is (7.282b):

$$\partial_t^k \equiv \frac{\partial^k}{\partial t^k}: \quad \left\{L\left(\nabla,\partial_t\right)\right\}\Phi = \sum_{m=0}^{M}\sum_{k=0}^{K} {}_k A_{i_1\ldots i_m}\left(\vec{x},t\right)\partial_t^k\left[\partial_{i_1\ldots i_m}\Phi\right]; \tag{7.282a,b}$$

using the notations (7.282a; 7.174a), the adjoint operator is

$$\left\{L\left(\nabla,\partial_t\right)\right\}\Psi = \sum_{m=0}^{M}\sum_{k=0}^{K}(-)^{k+m}\partial_t^k\partial_{i_1\ldots i_m}\left(A_{i_1\ldots i_m}\Psi\right). \tag{7.282c}$$

Taking the highest (7.283a) of the temporal and spatial orders, the operator (7.282b) [its adjoint (7.282c)] may be expressed (7.283b) [(7.283c)] in space-time coordinates (7.239a through d):

$$P \equiv \sup(K,M): \quad \{L(\partial_\alpha)\}\Phi = \sum_{p=1}^{P} B_{\alpha_1 \ldots \alpha_p} \partial_{\alpha_1 \ldots \alpha_p} \Phi, \tag{7.283a,b}$$

$$\{\bar{L}(\partial_\alpha)\}\Psi = \sum_{p=1}^{P} (-)^p \, \partial_{\alpha_1 \ldots \alpha_p} \left(B_{\alpha_1 \ldots \alpha_p} \Psi \right); \tag{7.283c}$$

this gives a compact form of the commutation relation (7.284a) involving the bilinear concomitant (7.284b):

$$\Psi\{L(\partial_\alpha)\}\Phi - \Phi\{L(\partial_\alpha)\}\Psi = \partial_\alpha X_\alpha, \tag{7.284a}$$

$$X_{\alpha_1}(\Phi,\Psi) = \sum_{p=1}^{P}\sum_{\ell=1}^{P} (-)^{\ell+1} \left[(\partial_{\alpha_2 \ldots \alpha_\ell}\Psi)\partial_{\alpha_{\ell+1} \ldots \alpha_p} \left(B_{\alpha_1 \ldots \alpha_p}\Phi \right) - (\partial_{\alpha_2 \ldots \alpha_\ell}\Phi)\partial_{\alpha_{\ell+1} \ldots \alpha_p} \left(B_{\alpha_1 \ldots \alpha_p}\Psi \right) \right]. \tag{7.284b}$$

The self-adjoint case corresponds to the operator of even order (7.285a) [(7.285b)] in space-time [combined (7.239a through d)] coordinates:

$$\{L(\nabla,\partial_t)\}\Phi = \{\bar{L}(\nabla,\partial_t)\}\Phi = \sum_{m=0}^{M}\sum_{k=0}^{K} \partial_t^k \partial_{i_1 \ldots i_m} \left(_{2k}A_{i_1 \ldots i_{2m}} \partial_t^k \partial_{i_{m+1} \ldots i_{2m}}\Phi \right), \tag{7.285a}$$

$$\{L(\partial_\alpha)\}\Phi = \{\bar{L}(\partial_\alpha)\}\Phi = \sum_{p=0}^{P} \partial_{\alpha_1 \ldots \alpha_p} \left(A_{\alpha_1 \ldots \alpha_{2p}} \partial_{\alpha_{p+1} \ldots \alpha_{2p}}\Phi \right). \tag{7.285b}$$

The commutation relation (7.286a) involves the bilinear concomitant (7.286b):

$$\Psi\{L(\partial_\alpha)\}\Phi - \Phi\{L(\partial_\alpha)\}\Psi = \partial_\alpha X_\alpha, \tag{7.286a}$$

$$X_{\alpha_1} = \sum_{p=1}^{P}\sum_{\ell=1}^{P} (-)^{\ell+1} \left[(\partial_{\alpha_2 \ldots \alpha_\ell}\Psi)\partial_{\alpha_{\ell+1} \ldots \alpha_p} \left(B_{\alpha_1 \ldots \alpha_p}\Phi \right) - (\partial_{\alpha_2 \ldots \alpha_\ell}\Phi)\partial_{\alpha_{\ell+1} \ldots \alpha_p} \left(B_{\alpha_1 \ldots \alpha_p}\Psi \right) \right]. \tag{7.286b}$$

The proof of the results concerning the non-self-adjoint (7.282a through c: 7.283a through c; 7.284a and b) [self-adjoint (7.285a and b; 7.286a and b] high-order linear partial differential operator is similar to (7.174a and b; 7.175; 7.176; 7.177) [(7.178; 7.179; 7.180)]. These results apply to an inhomogeneous, unsteady, and anisotropic medium and are simplified next (Note 7.9) to the isotropic case.

NOTE 7.8: Isotropic High-Order Equation of Mathematical Physics

The spatial dependence appears only through the Laplacian in the isotropic high-order linear partial differential operator in space-time:

$$\{L(\nabla^2,\partial_t)\}\Phi = \sum_{m=0}^{M}\sum_{k=0}^{K} A_{m,k}(\vec{x},t)\partial_t^k \left(\nabla^{2m}\Phi \right). \tag{7.287a}$$

The adjoint operator is

$$\left\{\overline{L}\left(\nabla^2,\partial_t\right)\right\}\Psi = \sum_{m=0}^{M}\sum_{k=0}^{K}(-)^{k+m}\,\partial_t^k\nabla^{2m}\left(A_{m,k}\Psi\right). \tag{7.287b}$$

Using the combined space-time coordinates (7.239a through d) and the corresponding gradient (7.288a and b), the highest of the spatial and temporal orders (7.288c) is

$$\partial_\alpha^{2p} \equiv \partial_{\alpha_1\ldots\alpha_p\alpha_p}, \quad \partial_\alpha^{2p+1} = \partial_\alpha\partial_\alpha^{2p}; \quad P = \sup\left(M;\frac{K}{2}\right), \tag{7.288a–c}$$

in the operator (7.287a) [its adjoint (7.287b)] in the form (7.289a) [(7.289b)]:

$$\left\{L\left(\partial_\alpha^2\right)\right\}\Phi = \sum_{p=0}^{P}B_p\partial_\alpha^{2p}\Phi, \quad \left\{\overline{L}\left(\partial_\alpha^2\right)\right\}\Psi = \sum_{p=0}^{P}\partial_\alpha^{2p}\left(B_p\Psi\right). \tag{7.289a,b}$$

The commutation relation (7.290a) involves the bilinear concomitant (7.290b):

$$\Psi\left\{L\left(\partial_\alpha^2\right)\right\}\Phi - \Phi\left\{\overline{L}\left(\partial_\alpha^2\right)\right\}\Psi = \partial_\alpha X_\alpha, \tag{7.290a}$$

$$X_\alpha\left(\Phi,\Psi\right) = \Psi = \sum_{p=0}^{P}\sum_{\ell=1}^{P}(-)^{\ell+1}\left(\partial_\alpha^{2p-\ell}\Phi\right)\partial_\alpha^{\ell-1}\left(B_p\Psi\right). \tag{7.290b}$$

The self-adjoint case leads to the intermediate operator:

$$\left\{L\left(\partial_\alpha\right)\right\}\Phi = \left\{\overline{L}\left(\partial_\alpha\right)\right\}\Psi = \sum_{p=0}^{P}\partial_\alpha^p\left(B_p\partial_\alpha^p\Phi\right), \tag{7.291}$$

with *commutation relation (7.286a) involving the bilinear concomitant*

$$X_\alpha\left(\Phi,\Psi\right) = \sum_{p=0}^{P}\sum_{\ell=0}^{P}(-)^{\ell+1}\left[\left(\partial_\alpha^{\ell-1}\Psi\right)\partial_\alpha^{2p-\ell}\left(B_p\Phi\right)-\left(\partial_\alpha^{\ell-1}\Phi\right)\partial_\alpha^{2p-\ell}\left(B_p\Psi\right)\right]. \tag{7.292}$$

The proof of the results for the isotropic [non-self-adjoint (self-adjoint)] linear high-order partial differential operator in space-time (7.287a and b; 7.287; 7.288a through c; 7.289a and b; 7.290a and b) [(7.291; 7.287a; 7.292)] is similar to (7.186a through c; 7.187; 7.188) [(7.192a and b; 7.193)]. The adjointness properties of ordinary (partial) differential operators [Section 7.7 (Section 7.8)], and their space-time combinations (Section 7.9) including the equations of mathematical physics (Notes 7.2 through 7.4) and their generalizations (Notes 7.5 through 7.9) may be used to (1) classify the operators (Note 7.10); (2) associate them with physical problem (Note 7.11); and (3) indicate the most frequent forms (Note 7.12).

The adjoint and self-adjoint forms of (7.261b) and (7.262) and related commutators, and temporal and spatial bilinear concomitants are obtained by the methods detailed earlier (Sections 7.7 through 7.9) and summarized next with reference [Note 7.7 (Note 7.8)] to the mathematical classes (physical examples) in Table 7.1 (Table 7.2).

TABLE 7.1 Adjoint Mathematical Operators

Class	Case	Subcase	Operator	Adjoint	Concomitant(s)	Commutator
Linear ordinary differential operator	Second order	1. General	(7.85b)/(7.92)	(7.87)/(7.92)	(7.88)/(7.93b)	(7.82)/(7.93a)
		2. Conversion to self-adjoint	(7.102a through d)	(7.102a through d)	-/(7.93b)	-/(7.93a)
	High order	3. Variable coefficients	(7.113a)/(7.122)	(7.113b)/(7.122)	(7.114)/(7.123)	(7.112a)/(7.117a)
		4. Constant coefficients	(7.118a)/(7.116c)	(7.118c)/(7.116c)	(7.118d)/(7.117b)	(7.118b)/(7.117a)
Linear partial differential operator	Second order	5. Anisotropic	(7.131c)/(7.141)	(7.132)/(7.141)	(7.133)/(7.142a)	(7.134)/(7.142b)
		6. Conversion to self-adjoint	(7.149; 7.150a through c)	(7.149; 7.150a through c)	-/(7.142a)	-/(7.142b)
		7. Intermediate	(7.163b)/(7.164b)	(7.163c)/(7.164c)	(7.163d)/(7.164d)	(7.134)/(7.142b)
		8. Conversion to self-adjoint	(7.165a through d)	(7.165a and d)	-/(7.142a)	/(7.142b)
		9. Isotropic	(7.160c)/(7.162b)	(7.161a)/(7.162c)	(7.161b)/(7.162d)	(7.134)/(7.142b)
	High order	10. Variable coefficients	(7.174b)/(7.178)	(7.175)/(7.178)	(7.177)/(7.180)	(7.176)/(7.179)
		11. Constant coefficients	(7.181b)/(7.184b)	(7.182)/(7.184c)	(7.183)/(7.185)	(7.176)/(7.179)
		12. Intermediate with variable coefficients	-/(7.192b)	-/(7.192b)	-/(7.193)	-/(7.179)
		13. Intermediate with constant coefficients	-/(7.194b)	-/(7.194c)	-/(7.195)	-/(7.179)
		14. Isotropic with variable coefficients	(7.186b)/(7.190b)	(7.187)/(7.190c)	(7.188)/(7.191)	(7.176)/(7.179)
Partial differential operator in space-time	Second order	15. Anisotropic	(7.198c)/(7.202d)	(7.200)/(7.202e)	(7.201b and c)/(7.204a and b)	(7.201a)/(7.203)
		16. Conversion to self-adjoint	(7.205)	(7.205)	-/(7.204a and b)	-/(7.203)
		17. Intermediate operator	(7.220b)/(7.223b)	(7.221)/(7.223c)	(7.201b; 7.222)/(7.224a and b)	(7.201a)/(7.203)
		18. Isotropic operator	(7.207d)/(7.213c)	(7.211)/(7.213d)	(7.221a and b)/(7.214a and b)	(7.201a)/(7.203)
	High order	19. Space-time method	(7.240; 7.241)/-	(7.243a and b)/-	(7.245; 7.246, 7.247)/-	(7.201a)/-

Notes: Operator: non-self-adjoint: before slash; /self-adjoint: after slash: /...

For 19 classes of linear differential operators is indicated the adjoint operator, bilinear concomitant, and commutation relation; the nonself-adjoint (self-adjoint) cases appear before (after) the slash. The transformation from non-self-adjoint to self-adjoint is also indicated.

TABLE 7.2 Examples of Physical Operators

Class	Case	Subcase	Operator	Adjoint	Concomitant	Commutator
Linear ordinary differential operator	Second order	1. Elastic string	-/(7.94)	-/(7.95)	-/(7.96b)	-/(7.96a)
		2. Harmonic oscillator	(7.104a)/(7.106a)	(7.104b)/(7.106b)	(7.104d)/(7.106c)	(7.104c)/(7.96a)
	Fourth order	3. Elastic bar	-/(7.128)	-/(7.129d)	-/(7.130)	-/(7.96a)
Linear partial differential operator	Second order	4. Anisotropic dielectric	-/(7.156c)	-/(7.156d)	-/(7.157a)	-/(7.157b)
		5. Elastic membrane	-/(7.168a)	-/(7.168b)	-/(6.169c)	-/(7.157b)
	Fourth order	6. Elastic plate	-/(7.196a)	-/(7.196b)	-/(7.197)	-/(7.179)
Partial differential operator in space-time	Second order	7. Vibrations of string	(7.234)/-	(7.236)/-	(7.238a and b)/-	(7.201a)/-
		8. Vibrations of membrane	(7.235)/-	(7.237)/-	(7.238a and c)/-	(7.201a)/
	Fourth order	9. Vibrations of bar	(7.248)/-	(7.250)/-	(7.238a; 7.169c)/	(7.201a)/
		10. Vibrations of plate	(7.249)/-	(7.251)/-	(7.238a; 7.197)/-	(7.201a)/-

Notes: Operator: Non-self-adjoint/self-adjoint—before/after slash.

As Table 7.2 for 10 examples of physical operators including the undamped and damped harmonic oscillator, and the steady deflection (bending) and unsteady oscillations (vibrations) of strings (bars) and membranes (plates); also considered are beams (stressed plates) with elastic support and damping.

NOTE 7.9: Classification of Linear Differential Operators

The adjoint is defined only for linear differential operators that may (1) have scalar (Notes 7.7 and 7.8), vector (Notes 7.5 and 7.6), matrix, or tensor dependent variable; (2) be ordinary (partial) differential operator involving one (several) independent variables [Section 7.7 (Section 7.8)] or their combination in space-time (Section 7.9); (3) be of second-order (Subsections 7.7.1 through 7.7.5, 7.8.1 through 7.8.7, and 7.9.1 through 7.9.8) or higher (Subsections 7.7.6 through 7.7.9, 7.8.8 through 7.8.12, and 7.9.9; Notes 7.7 and 7.8) that may be different in space and time; (4) have constant or variable coefficients dependent on position (time) for inhomogeneous (unsteady) media; (5) be anisotropic (isotropic) for media with properties depending (not depending) on direction; (6) in multidimensional cases, an intermediate operator for inhomogeneous isotropic media may be defined where the gradients of the properties of the medium introduce a nonlocal isotropy; and (7) be adjoint or non-self-adjoint, affecting the commutator and bilinear concomitants. The cases (1–7) lead to a number of combinations of which 19 are listed in Table 7.1 in three levels: (1) ordinary (partial) differential operators in time (space) and their combination in space-time; (2) second- and higher-order operators; and (3) constant or variable coefficients and isotropic, anisotropic, and intermediate cases. For each of the 19 linear cases, the following are indicated in four columns: (1) the operator; (2) the adjoint operator; (3) their commutator; and (4) the bilinear concomitant(s), including temporal and spatial in space-time cases. The non-self-adjoint (self-adjoint) operators are indicated before/after the slash, and the transformations between them are indicated without slash. A blank before (after) the slash indicates that the operator is always (never) self-adjoint, as seen more frequently (Note 7.11) in Table 7.2 of physical examples. The differential operators, linear or nonlinear, may be ordinary (partial) if there is a single (are several) independent variables. The linear ordinary differential operators may have constant or variable coefficients; combined with self-adjointness or non-self-adjointness this leads to $2 \times 2 = 4$ cases.

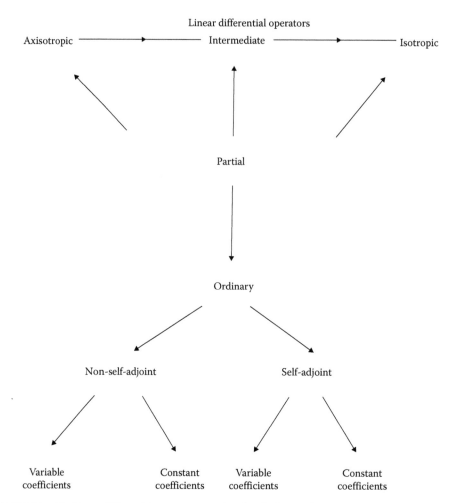

DIAGRAM 7.1 The linear differential operators may be classified according to several criteria: (1) ordinary or partial; (2) isotropic, intermediate, or anisotropic; (3) self-adjoint or non-self-adjoint; (4) variable or constant coefficients; and (5) order, for example, second or higher (Diagram 7.2).

The linear partial differential operators may be anisotropic (isotropic), and there is a third intermediate case of isotropic nonconstant coefficients (Diagram 7.1). Thus the number of cases for a linear partial differential operator is (1) with constant coefficients 4, that is, isotropic and anisotropic with or without self-adjointness; (2) with variable coefficients 6, that is, anisotropic, intermediate, or isotropic with or without self-adjointness; and (3) the total is 10 cases. The 4 (10) cases of linear ordinary (partial) differential operators (Diagram 7.1) are doubled distinguishing second- and higher-order differential operators (Table 7.1). Some examples of second- and fourth-order differential operators in mechanics and elasticity are given next (Table 7.2; Diagram 7.2).

NOTE 7.10: Second- and Fourth-Order Differential Operators

Diagram 7.2 gives some physical examples of linear differential operators: (1) the undamped (damped) harmonic oscillator [Figure 7.4a (Figure 7.4b)] is specified by a linear ordinary differential operator with time as the independent variable that is self-adjoint (non-self-adjoint); (2) the weak deflection of an elastic string (membrane) is specified by a linear ordinary (partial) differential operator with one (two)

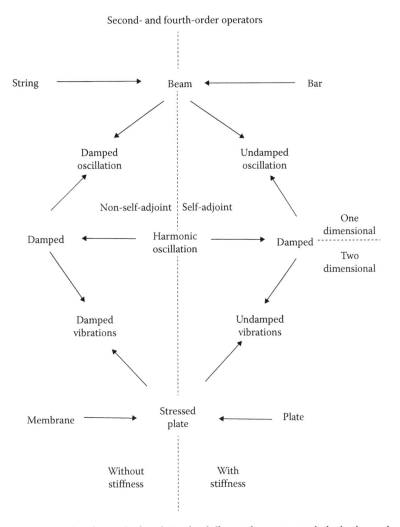

Second- and fourth-order operators

String → Beam ← Bar

Damped oscillation   Undamped oscillation

Non-self-adjoint | Self-adjoint

Damped ← Harmonic oscillation → Damped

One dimensional

Two dimensional

Damped vibrations   Undamped vibrations

Membrane → Stressed plate ← Plate

Without stiffness   With stiffness

DIAGRAM 7.2   The examples of second- (fourth-) order differential operators include the damped or undamped harmonic oscillator and deflections and oscillations of elastic strings and membranes (bending and vibrations of bars plates, beams, and stressed plates).

spatial coordinates as the variable(s) of second order due to absence of stiffness; (3) in the presence of stiffness for the weak bending of an elastic bar (plate), the linear ordinary (partial) differential operator is of order four; (4) the linear superposition of (2) and (3) leads to the weak bending of a beam (stressed plate), that is a bar (plate) with longitudinal tension (in-plane stresses); (5) all cases, an elastic (2) string (membrane), (3) bar (plate), and (4) beam (stressed plate), may be supported by linear springs [Figure 7.3a (Figure 7.3b)]; and (6) considering also the inertia force and damping leads to oscillations (vibrations) specified by a linear differential operator in space-time, with 2(3) variables, namely, time and 1(2) spatial coordinates. The only non-self-adjoint case is when damping is present; in all undamped cases, the operators are self-adjoint (Table 7.2) including nonuniform and unsteady bending stiffness, mass density, and spring resilience. The three (four)-dimensional operators in space (space-time) apply to waves and are included in the general second-order linear differential operator of mathematical physics for a scalar (vector) dependent variable [Notes 7.3 and 7.4 (Notes 7.5 and 7.6)]. The vector second-order partial differential equations in two (three) dimensions, when eliminated for one vector component, lead to a scalar fourth- (sixth-) order partial differential equation. Next are considered the main partial differential equations of mathematical physics and some generalizations.

NOTE 7.11: Physical Examples of Adjoint and Non-Self-Adjoint Operators

Table 7.1 (Table 7.2) lists 19 (10) mathematical classes (physical examples) of self-adjoint and non-self-adjoint differential operators in the same format whose relations are indicated in Diagram 7.1 (Diagram 7.2). The simplest case of the ordinary differential equation of second-order for the undamped (damped) harmonic oscillator shows that it is (is not) self-adjoint, and thus the reciprocity principle holds (fails) for nondissipative (dissipative) systems. The reciprocity principle holds even for inhomogeneous (unsteady) nondissipative systems such as the deflection (vibrations) of elastic strings, and extends to two dimensions for elastic membranes, and to higher orders for elastic bars and plates. In all cases, reciprocity is lost if damping is added, but is not affected by anisotropy. In the cases where the partial differential equations of mathematical physics have constant coefficients (Diagram 7.3), the self-adjoint (non-self-adjoint) cases are those that do not (do) involve diffusion. The linear partial differential equations of mathematical physics may be classified (Table 7.3) in four sets. The starting case (1) is the Laplace/Poisson equation for steady potential fields. The second set consists of: (2/3)

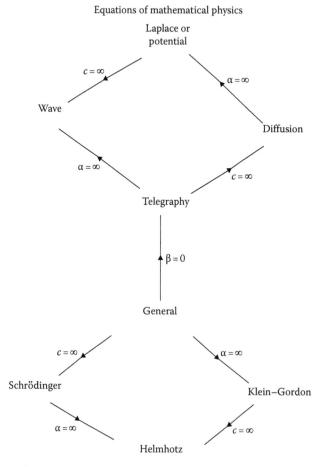

DIAGRAM 7.3 Relation between the main partial differential equations of mathematical physics and the general linear second-order partial differential operator in space-time with constant scalar coefficients for an isotropic, homogeneous, steady medium.

TABLE 7.3 Equations of Mathematical Physics: I. Constant Scalar Coefficients

Class	Name	Parameters	Equation
Simplest	1. Poisson	$c = \infty$; $\alpha, \beta = 0$	(7.260b)
Basic	2. Helmhotz	$c = \infty = \alpha$	(7.258a)
	3. Wave	$\alpha = \infty$, $\beta = 0$	(7.257b)
	4. Diffusion	$c = \infty$, $\beta = 0$	(7.259a)
Combined	5. Telegraphy	$\beta = 0$	(7.261b)
	6. Schrödinger	$c = \infty$	(7.261c)
	7. Klein–Gordon	$\alpha = \infty$	(7.261a)
Generalized	8. Isotropic	α, β, c	(7.255)
	9. Inhomogeneous	$A, \vec{B}, \vec{C}, a, b, c$	(7.264d)
	10. Anisotropic	$A_{ij}, B_i, C_i, a, b, c$	(7.198c)

Note: The four basic partial differential equations of mathematical physics can be combined in three mixed forms and lead to four generalizations.

the wave (diffusion) equation adding unsteadiness due to a finite propagation speed (finite diffusivity or dissipation); (4) the Helmholtz equation adding a term proportional to the dependent variable as for a spring with resilience coefficient. The third set consists of three mixed equations combining two effects: (5/6) wave propagation and diffusion (restitution) for the telegraph (Klein–Gordon) equation; (7) diffusion and imaginary restitution for the Schrödinger equation. The combination of the three effects of propagation, diffusion, and restitution with the Laplace operator leads to (8) the general linear second-order partial differential equation of mathematical physics for an isotropic, homogeneous, steady medium. This suggests a set of generalized partial differential of mathematical physics, (9/10) intermediate (anisotropic) instead of isotropic but still of second-order, (11/12) anisotropic (isotropic) of higher-order. These (9–12) may apply to inhomogeneous (homogeneous) and unsteady (steady) media for which the coefficients do (do not) depend respectively on position and time (Table 7.4). The equations

TABLE 7.4 Equations of Mathematical Physics: II. Generalized to Anisotropic, Inhomogeneous, and Unsteady Media

Class	Name	Nonzero Coefficients	Media Isotropic and Steady	Media Anisotropic and/or Unsteady
Simplest	1. Poisson	—	(7.260b)	(7.225d)
Basic	2. Helmholtz	A; c	(7.258a)	(7.233c)
	3. Wave	A; a	(7.257b)	(7.226c)
	4. Diffusion	A; b	(7.259a)	(7.230c)
Combined	5. Telegraph	A; a, b	(7.261a)	(7.229d)
	6. Schrödinger	A; b, c	(7.261c)	(7.229b)
	7. Klein–Gordon	A; a, c	(7.261b)	(7.225b)
Generalized	8. Isotropic	A; a, b, c	(7.207d)	—
	9. Intermediate	$A, \vec{B}, \vec{c}, a, b, c$	—	(7.220b)
	10. Anisotropic	$A_{ij}, \vec{B}, \vec{c}, a, b, c$	—	(7.198c)

Note: The partial differential of mathematical physics applies to constant scalar coefficients for homogeneous steady isotropic media (Table 7.3) and can be generalized [Table 7.4 (7.5)] to inhomogeneous and/or unsteady (anisotropic) media for which the coefficients are variable dependent on position/time (include besides scalars also vectors and a matrix).

TABLE 7.5 Equations of Mathematical Physics: III. Extensions to Vector Variables and High-Orders

Class	Name	Operator	Adjoint Operator	Commutation Relation	Bilinear Concomitant
Generalized	8. Isotropic	(7.207d)/(7.213c)	(7.211)/(7.213d)	(7.201a)/(7.203)	(7.212a and b)/ (7.214a and b)
	9. Intermediate	(7.220b)/(7.223c)	(7.221)/(7.223c)	(7.201a)/(7.203)	(7.201b; 7.222)/ (7.224a and b)
	10. Anisotropic	(7.198c)/(7.202e)	(7.200)/(7.202e)	(7.201a)/(7.203)	(7.201b and c)/ (7.204a and b)
Extended	11. Vector variable	(7.267)/(7.271)	(7.268)/(7.271)	(7.269)/(7.272)	(7.270a and b)/ (7.273a and b)
	12. Vector equation	(7.275)/(7.279)	(7.276)/(7.279)	(7.277)/(7.280)	(7.278a and b)/ (7.281a and b)
	13. High-order	(7.282b) ≡ (7.282c)/ (7.285a) ≡ (7.285b)	(7.283c)/(7.285a) ≡ (7.285b)	(7.284a)/(7.286a)	(7.284b)/(7.286b)
	14. High-order isotropic	(7.287a) ≡ (7.289a)/ (7.291)	(7.287b) ≡ (7.289b)/(7.291)	(7.290a)/(7.287a)	(7.290b)/(7.292)

Note: The partial differential equations of mathematical physics (Table 7.3) besides the generalization to anisotropic inhomogeneous and unsteady media (Table 7.4) can be extended further to high orders, vector operators, and dependent variables (Table 7.5).

of mathematical physics can also be extended to vector operators and dependent variables and higher orders in space and time (Table 7.5). The derivation of the partial differential equations of mathematical physics is illustrated next by two concluding examples: (1) electromagnetic waves in a conductor (Note 7.13) leading to the telegraph operator that is not self-adjoint due to dissipation by the Joule affect associated with electrical resistance (Note 7.13); and (2) sound in an uniformly moving medium (Note 7.14) leading to the convected wave operator (Note 7.15) that is anisotropic and self-adjoint if the velocity of the mean flow is reversed (Notes 7.16 and 7.17).

NOTE 7.12: Telegraph Equation for Electromagnetic Waves in a Conductor

The electromagnetic field is due to electric charges q (currents $\vec{j}$) and is specified by two pairs of Maxwell equations specifying the electric (7.293a) [magnetic (7.293c)] field and electric displacement (7.293b) [magnetic induction (7.293d)]:

$$\nabla \wedge \vec{E} + \frac{1}{c}\frac{\partial \vec{B}}{\partial t} = 0 = \nabla \cdot \vec{D} - q, \quad \nabla \wedge \vec{H} - \vec{j} - \frac{1}{c}\frac{\partial \vec{D}}{\partial t} = 0 = \nabla \cdot \vec{B}, \tag{7.293a–d}$$

where c is the speed of light in vacuo. The **Maxwell (1873) equations** *decouple for the electro(magneto) static (7.293a and b) [(7.293c and d)] field in the steady case [Chapter I.24 (Chapter I.26)]; they are coupled (7.293a through d) in the unsteady case, for example, for wave propagation and resistive diffusion considered together next. The* **constitutive properties** *for a linear isotropic medium state that (1/2) the electric displacement (magnetic induction) is parallel and proportional to the electric(magnetic) field through the dielectric permittivity (7.294d) [magnetic permeability (7.194e)]:*

$$\varepsilon, \mu, \sigma = \text{const}: \quad \vec{D} = \varepsilon\vec{E}, \quad \vec{B} = \mu\vec{H}; \quad \vec{j} - \vec{J} = \sigma\vec{E}; \tag{7.294a–f}$$

and (3) the electric current has a component parallel and proportional to the electric field (7.294f) through the electrical conductivity σ whose inverse is the electrical resistivity $1/\sigma$. In a homogeneous steady medium, the three physical parameters, namely, the dielectric permittivity (7.294a), the magnetic permeability (7.294b), and the electrical conductivity (7.294c) in **Ohm's law** *are all constant.*

NOTE 7.13: Electromagnetic Wave Speed and Ohmic Diffusivity

In the latter case (7.294a and b), substitution of the constitutive relations (7.294d through f) specifies the Maxwell equations (7.293a through d) in terms of the electric and magnetic field alone:

$$\nabla \wedge \vec{E} = -\frac{\mu}{c}\frac{\partial \vec{H}}{\partial t}, \quad \nabla \cdot \vec{E} = \frac{q}{\varepsilon}, \quad \nabla \cdot \vec{H} = 0, \quad \nabla \wedge \vec{H} = \vec{j} + \sigma \vec{E} + \frac{\varepsilon}{c}\frac{\partial \vec{E}}{\partial t}. \tag{7.295a–d}$$

These equations may be eliminated either for the electric (7.296a) [or magnetic (7.296b)] field using the identity (II.4.112) ≡ (6.48a):

$$\nabla^2 \vec{E} = -\nabla \wedge \left(\nabla \wedge \vec{E}\right) + \nabla\left(\nabla \cdot \vec{E}\right) = \frac{\mu}{c}\frac{\partial}{\partial t}\left(\nabla \wedge \vec{H}\right) + \frac{1}{\varepsilon}\nabla q$$
$$= \frac{\mu\varepsilon}{c^2}\frac{\partial^2 \vec{E}}{\partial t^2} + \frac{\mu\sigma}{c}\frac{\partial \vec{E}}{\partial t} + \frac{\mu}{c}\frac{\partial \vec{j}}{\partial t} + \frac{1}{\varepsilon}\nabla q, \tag{7.296a}$$

$$\nabla^2 \vec{H} = -\nabla \wedge \left(\nabla \wedge \vec{H}\right) + \nabla\left(\nabla \cdot \vec{H}\right) = -\nabla \wedge \vec{j} - \sigma\nabla \wedge \vec{E} - \frac{\varepsilon}{c}\frac{\partial}{\partial t}\left(\nabla \wedge \vec{E}\right)$$
$$= \frac{\mu\varepsilon}{c^2}\frac{\partial^2 \vec{H}}{\partial t^2} + \frac{\mu\sigma}{c}\frac{\partial \vec{H}}{\partial t} - \nabla \wedge \vec{j}. \tag{7.296b}$$

*It follows that the electric (7.296a) ≡ (7.297a) [magnetic (7.296b) ≡ (7.297b)] field in a linear homogeneous steady conductor (7.294a through f) satisfies the **equation of telegraphy** forced by the electric charges and currents:*

$$\left\{L\left(\nabla, \frac{\partial}{\partial t}\right)\right\}\vec{E}, \vec{H}(\vec{x}, t) \equiv \left\{\nabla^2 - \frac{1}{c_{en}^2}\frac{\partial^2}{\partial t^2} - \frac{1}{\alpha_e}\frac{\partial}{\partial t}\right\}\vec{E}, \vec{H}(\vec{x}, t) = \left\{\frac{1}{\varepsilon}\nabla q + \frac{\mu}{c}\frac{\partial \vec{j}}{\partial t}, -\nabla \wedge \vec{j}\right\}. \tag{7.297a–d}$$

*The **propagation speed** (7.268a) [**Ohmic diffusivity** (7.268b)] is proportional to the speed of light in vacuo and involves the dielectric permittivity (Ohmic electrical conductivity) and magnetic permeability. The adjoint operator (7.298c)*

$$c_{em} = \frac{c}{\sqrt{\mu\varepsilon}}, \quad \alpha_e = \frac{c}{\mu\sigma}: \quad \overline{L}\left(\nabla, \frac{\partial}{\partial t}\right) = \nabla^2 - \frac{1}{c_{em}^2}\frac{\partial^2}{\partial t^2} + \frac{1}{\alpha_e}\frac{\partial}{\partial t} \tag{7.298a–c}$$

shows that the reciprocity principle holds (does not hold) in the absence $\alpha_e = \infty$ (presence $\alpha_e \neq \infty$) of electrical conductivity $\sigma = 0$ $(\sigma \neq 0)$. The commutator

$$\Psi\left\{L\left(\frac{\partial}{\partial \vec{x}}, \frac{\partial}{\partial t}\right)\right\}\Phi - \Phi\left\{\overline{L}\left(\frac{\partial}{\partial \vec{x}}, \frac{\partial}{\partial t}\right)\right\}\Psi = \frac{\partial V}{\partial t} + \nabla \cdot \vec{W} \tag{7.299a}$$

involves the temporal (7.299b) [spatial (7.299c)] bilinear concomitant

$$V(\Phi, \Psi) = \frac{1}{c^2}\left(\Phi\frac{\partial \Psi}{\partial t} - \Psi\frac{\partial \Phi}{\partial t}\right) + \frac{\mu\sigma}{c^2}\Phi\Psi, \quad \vec{W}(\Phi, \Psi) = \Psi\nabla\Phi - \Phi\nabla\Psi. \tag{7.299b,c}$$

NOTE 7.14: Convected Wave Operator for Sound in a Moving Medium

The fundamental equations of fluid mechanics for a nondissipative fluid are (1) the equation of continuity or mass conservation (I.14.5) ≡ (7.300b) involving the mass density and velocity; and (2) the inviscid *momentum equation (I.14.9) $\equiv$ (7.300c) involving the pressure and the acceleration as the material deriva-tive of the velocity (I.14.6) $\equiv$ (7.300a):*

$$\frac{D}{dt} = \frac{\partial}{\partial t} + \vec{U} \cdot \nabla : \quad \frac{D\rho}{dt} + \rho \nabla \cdot \vec{U} = 0, \quad \frac{D\vec{U}}{dt} + \frac{1}{\rho} \nabla p = 0. \tag{7.300a--c}$$

The energy equation (7.301a) relating the pressure and density in a moving reference frame through the adiabatic sound speed (I.14.31c) $\equiv$ (7.301b), that is, the square root of the derivative of the pressure with regard to the density calculated at constant entropy is

$$\frac{Dp}{dt} = \left(c_s\right)^2 \frac{D\rho}{\partial t}, \quad \left(c_s\right)^2 = \left(\frac{\partial p}{\partial \rho}\right)_s. \tag{7.301a,b}$$

The total state of the fluid is assumed to consist of a uniform flow (subscripts "*o*") plus an acoustic per-turbation (primes):

$$\left\{p, \rho, \vec{U}\right\}(\vec{x}, t) = \left\{p_0, \rho_0, \vec{U}_0\right\} + \left\{p', \rho', \vec{u}\right\}(\vec{x}, t). \tag{7.302a--c}$$

For a homogeneous steady medium (7.303a through c) and perturbations of small amplitude (7.303d), the material derivative (7.300a) linearizes (7.303f):

$$p_0, \rho_0, \vec{u}_0 = \text{const}, \left|\vec{u}\right|^2 \ll \left(c_0\right)^2 = \left(\frac{\partial p_0}{\partial \rho_0}\right)_s : \quad \frac{d}{dt} = \frac{\partial}{\partial t} + \vec{u}_0 \cdot \nabla, \tag{7.303a--f}$$

and the sound speed (7.301b) is calculated for the mean state (7.303e) and is also constant.

The equations of motion (7.300b and c; 7.301a) may be linearized respectively (7.304b through d):

$$\frac{d}{dt} \nabla = \nabla \frac{d}{dt} : \quad \frac{d\rho'}{dt} = -\rho_0 \left(\nabla \cdot \vec{u}\right), \quad \frac{d\vec{u}}{dt} = -\frac{1}{\rho_0} \nabla p', \quad \frac{d\rho'}{dt} = c_0^2 \frac{d\rho'}{dt}. \tag{7.304a--d}$$

The linearized material derivative (7.303c) commutes with the gradient (7.304a) because the mean flow velocity is constant (7.303c). The equations (7.304b through d) may be eliminated for the pressure (7.305a), density (7.305b and c), or velocity (7.305d) perturbations:

$$\frac{1}{c_0^2} \frac{d^2 p'}{dt^2} = \frac{d^2 \rho'}{dt^2} = -\rho_0 \nabla \cdot \left(\frac{d\vec{u}}{dt}\right) = \rho_0 \nabla \cdot \left(\frac{1}{\rho_0} \nabla p'\right) = \nabla^2 p', \tag{7.305a}$$

$$p' = c_0^2 \rho' : \quad \frac{d^2 \rho'}{dt^2} = \frac{1}{c_0^2} \frac{d^2 p'}{dt^2} = \nabla^2 p' = \nabla^2 \left(c_0^2 \rho'\right) = c_0^2 \nabla^2 \rho', \tag{7.305b,c}$$

$$\frac{d^2 \vec{u}}{dt^2} = -\frac{1}{\rho_0} \nabla \left(\frac{dp'}{dt}\right) = -\frac{c_0^2}{\rho_0} \nabla \left(\frac{d\rho'}{dt^2}\right) = c_0^2 \nabla \left(\nabla \cdot \vec{u}\right). \tag{7.305d}$$

The proof of (7.305a) is direct and (7.305c) uses (7.305b) that follows from (7.304d) for constant sound speed (7.303e). The pressure (mass density) satisfies scalar wave equations (7.305a) [(7.305c)], whereas the velocity satisfies a vector equation (7.305d). From (7.304c) it follows (7.306a) that sound is an irrota-tional motion (7.306b) and an acoustic potential exists (7.306c):

$$0 = \nabla \wedge \nabla p' = -\rho_0 \frac{d}{dt} \left(\nabla \wedge \vec{u}\right) : \quad \nabla \wedge \vec{u} = 0, \quad \vec{u} = \nabla \Phi; \quad \dot{D} = \nabla \cdot \vec{u} = \nabla^2 \Phi. \tag{7.306a--d}$$

The dilatation (7.306d) satisfies a scalar wave equation:

$$\frac{\mathrm{d}^2 D}{\mathrm{d}t^2} = \nabla \cdot \left(\frac{\mathrm{d}^2 \vec{u}}{\mathrm{d}t^2} \right) = c_0^2 \nabla \cdot \left[\nabla (\nabla \cdot \vec{u}) \right] = c_0^2 \nabla \cdot (\nabla \dot{D}) = c_0^2 \nabla^2 \dot{D}, \tag{7.307}$$

and so does the potential. It has been shown that

$$\left\{ c_0^2 \nabla^2 - \left(\frac{\partial}{\partial t} + \vec{u}_0 \cdot \nabla \right)^2 \right\} p', \rho', \Phi, \dot{D}(\vec{x}, t) = 0; \tag{7.308a–d}$$

the pressure (7.308a), density (7.308b), and dilatation (7.306d) perturbations of the uniform flow (7.303c) of a homogeneous fluid (7.303a and b) satisfy the convected wave equation for small perturbations (7.308d), where, (1) the velocity perturbation of sound waves are irrotational (7.306a); (2) the acoustic potential (7.306c) satisfies the same convected wave equation (7.308c); and (3) the adiabatic sound speed is calculated for the mean flow (7.303e).

NOTE 7.15: Reciprocity Principle with Flow Reversal (Howe, 1975; Campos, 1978; Campos and Lau, 2012)

The convected wave operator (7.308a through d) ≡ (7.309c) in a uniform flow (7.309a and b),

$$\vec{u}_0 \equiv \vec{U} = \text{const} = c_0: \quad \left\{ L \left(\frac{\partial}{\partial \vec{x}}, \frac{\partial}{\partial t} \right) \right\} \Phi = \left\{ c_0^2 \nabla^2 - \left(\frac{\partial}{\partial t} + \vec{U} \cdot \nabla \right)^2 \right\} \Phi$$

$$= \left(c_0^2 \delta_{ij} - U_i U_j \right) \frac{\partial}{\partial x_i \partial x_j} - 2U_i \frac{\partial^2 \Phi}{\partial x_i \partial t} - \frac{\partial^2 \Phi}{\partial t^2} = \left\{ \bar{L} \left(\frac{\partial}{\partial x}, \frac{\partial}{\partial t} \right) \right\} \Phi, \tag{7.309a–d}$$

is self-adjoint (7.309d), and its commutator (7.310a) involves (7.201b) [(7.201c)] the temporal (7.310b) [spatial (7.310c)] bilinear concomitants:

$$\Psi \left\{ L \left(\frac{\partial}{\partial \vec{x}} \right) \right\} \Phi - \Phi \left\{ L \left(\frac{\partial}{\partial \vec{x}} \right) \right\} \Psi = \frac{\partial V}{\partial t} + \nabla \cdot \vec{W}, \tag{7.310a}$$

$$V(\Phi, \Psi) = \Phi \frac{\partial \Psi}{\partial t} - \Psi \frac{\partial \Phi}{\partial t} - 2U_i \Psi \frac{\partial \Phi}{\partial x_i}, \tag{7.310b}$$

$$W_i(\Phi, \Psi) = \left(c^2 \delta_{ij} - U_i U_j \right) \left(\Psi \frac{\partial \Phi}{\partial x_j} - \Phi \frac{\partial \Psi}{\partial x_j} \right) + 2U_i \Phi \frac{\partial \Psi}{\partial t}, \tag{7.310c}$$

When the reciprocity principle is applied interchanging the observer $\vec{x}$ and source $\vec{y}$ position,

$$\{\vec{x}, \vec{y}, \vec{U}\} \leftrightarrow \{\vec{y}, \vec{x}, -\vec{U}\}, \tag{7.311}$$

the direction of the mean flow must be reversed to have the same relative velocity. The proof of (7.311) follows from the following (1) the relative position vector (7.312a) implying (7.312b):

$$\vec{\ell} = \vec{x} - \vec{y}: \quad \frac{\partial}{\partial \vec{x}} = \frac{\partial}{\partial \vec{\ell}} = -\frac{\partial}{\partial \vec{y}}; \quad \frac{\mathrm{d}}{\mathrm{d}t} = \vec{U} \cdot \frac{\partial}{\partial \vec{x}} = -\vec{U} \cdot \frac{\partial}{\partial \vec{y}}; \tag{7.312a–c}$$

(2) the convected wave operator is self-adjoint (7.309b and c) and thus the material derivative (7.312c) must be the same; and (3) hence when the observer and source positions are interchanged the mean flow velocity must be reversed (7.311).

NOTE 7.16: Complex Self-Adjoint Differential Operators

The inner product may be extended from real (to complex) functions [Subsection 3.4.1 (Subsection 3.4.4)], and likewise the self-adjoint linear differential operators may be extended from real (Sections 7.7 through 7.9; Notes 7.1 through 7.16) to complex as shown next (Notes 7.16 and 7.17) in simple cases. A complex linear partial differential operator is self-adjoint if taking its complex conjugate in the commutator,

$$\Psi\{L(\nabla)\}\Phi - \Phi\{L^*(\nabla)\}\Psi = \nabla \cdot \vec{Y}, \tag{7.313}$$

leads to the divergence of a vector. Substituting in the l.h.s. of (7.313), the linear second-order partial differential operator (7.131c) with complex coefficients leads to

$$\Psi\{L(\nabla)\}\Phi - \Phi\{L^*(\nabla)\}\Psi = \Psi A_{kj}\partial_{kj}\Phi - \Phi A_{kj}^*\partial_j\Psi + \Psi B_k\partial_k\Phi - \Phi B_k^*\partial_k\Psi + \left(C - C^*\right)\Phi\Psi. \tag{7.314}$$

The r.h.s. of (7.314) is the divergence of a vector like the r.h.s. of (7.313) if (1) the scalar coefficient is a real constant (7.315a):

$$C^* = C; \quad B_k = iD_k: \quad \Psi B_k\partial_k\Phi - \Phi B_k^*\partial_k\Psi = iD_k\partial_k\left(\Phi\Psi\right); \tag{7.315a–c}$$

(2) vector coefficient is imaginary (7.315b) leading to (7.315c); and (3) the matrix coefficients is real (7.316a) leading to (7.316b):

$$A_{kj}^* = A_{kj}: \quad \Psi A_{kj}\partial_{kj}\Phi - \Phi A_{kj}^*\partial_{kj}\Psi = A_{kj}\partial_k\left(\Psi\partial_j\Phi - \Phi\partial_j\Psi\right), \tag{7.316a,b}$$

using the symmetry properties of second-order partial derivatives (7.137a through c; 7.138).

For constant coefficients (7.317a through c), the identity (7.314) ≡ (7.317d),

$$c, B_{ik}A_{kj} = \text{const}: \quad \Psi\{L(\nabla)\}\Phi - \Phi\{L^*(\nabla)\}\Psi = \partial_k\left[A_{kj}\left(\Psi\partial_j\Phi - \Phi\partial_j\Psi\right) + iD_k\Phi\Psi\right], \tag{7.317a–d}$$

specifies the complex bilinear concomitant (7.318d):

$$C, D_k, A_{kj} = \text{const} \in | R: \quad Y_k\left(\Phi, \Psi\right) = A_{kj}\left(\Psi\partial_j\Phi - \Phi\partial_j\Psi\right) + iD_k\Phi\Psi. \tag{7.318a–d}$$

The real constant coefficients (7.318a through c) appear (7.315b) in the self-adjoint (7.313) operator:

$$\{L(\nabla)\}\Phi = A_{kj}\partial_{kj}\Phi + iD_k\Phi + c\Phi. \tag{7.319}$$

It has been shown that *the second-order linear partial differential operator (7.131c) with constant complex coefficients is self-adjoint (7.313) iff the vector (scalar and matrix) coefficient(s) is imaginary (7.315b) [are real (7.315a; 7.316a)]. Thus the complex linear second-order self-adjoint differential operator (7.313) takes the form (7.319) with constant real coefficients (7.318a through c) that also appear in the bilinear concomitant (7.318d).* The extension to complex self-adjoint partial differential operators could be made in the other cases considered earlier (Sections 7.7 through 7.9; Notes 7.1 through 7.13). The present result (Note 7.16) is sufficient for application to the convected wave operator in space-time (Notes 7.14 and 7.15) and in the space-frequency domain (Note 7.17).

NOTE 7.17: Convected Wave Operator in Space-Time and Space-Frequency Domain

The convected wave operator for sound waves in a homogeneous uniform flow can be written in (1) space-time (7.309c) ≡ (7.320) as a real operator:

$$\{L(\nabla,\partial_t)\}\Phi = \left(c_0^2\delta_{kj} - U_kU_j\right)\partial_{kj}\Phi - 2U_k\partial_k\partial_t\Phi - \partial_{tt}\Phi; \tag{7.320}$$

and (2) for a wave of frequency ω in (7.321a) the space-frequency domain (7.321b) as a complex operator:

$$\Phi(\vec{x},t) = e^{-i\omega t}\Psi(\vec{x}): \quad \{L(\nabla,\partial_t)\} = \left(c_0^2\delta_{kj} - U_kU_j\right)\partial_{kj}\Psi + 2i\omega U_k\partial_k\Psi + \omega^2\Psi. \tag{7.321a,b}$$

The convected wave operator in space-time (7.320) [space-frequency domain (7.321b)] is self-adjoint (7.310a) [(7.313)] as follows from (7.198c) ≡ (7.320) [(7.319) ≡ (7.321b)] with constant coefficients (7.322a through f) [(7.323a through c)]:

$$A_{kj} = c_0^2\delta_{kj} - U_kU_j, \quad a = 1, \quad c = b = 0 = B_i = C_i, \tag{7.322a–f}$$

$$A_{kj} = c_0^2\delta_{kj} - U_kU_j, \quad D_k = 2\omega U_k, \quad c = \omega^2. \tag{7.323a–c}$$

Corresponding to the real temporal (7.310b) and spatial (7.310c) bilinear concomitants [complex bilinear concomitant (7.318d; 7.323a through c) ≡ (7.324)],

$$Y_k(\Phi,\Psi) = \left(c_0^2\delta_{kj} - U_jU_k\right)\left(\Psi\partial_i\Phi - \Phi\partial_i\Psi\right) + i2\omega U_k\Phi\Psi. \tag{7.324}$$

It can be checked that the complex linear second-order partial differential operator in space-time (7.321b) with constant coefficients is self-adjoint by substitution into the l.h.s. of (7.313):

$$\Psi\{L(\nabla)\}\Phi - \Phi\{L^*(\nabla)\}\Psi = \left(c_0^2\delta_{kj} - U_jU_k\right)\left(\Psi\partial_{kj}\Phi - \Phi\partial_{kj}\Psi\right) + i2\omega U_k\left(\Psi\partial_k\Phi + \Phi\partial_k\Psi\right)$$

$$= \partial_k\left[\left(c_0^2\delta_{kj} - U_kU_j\right)\left(\Psi\partial_j\Phi - \Phi\partial_j\Psi\right) + i2\omega U_k\Phi\Psi\right] = \partial_kY_k. \tag{7.325}$$

The r.h.s. of (7.325) ≡ (7.313) is the divergence of the bilinear concomitant (7.24) that appears in the square brackets.

7.10 Conclusion

The proof of the Minkowski inequality follows from that of the Hölder inequality, which may be stated in discrete or integral forms; these are related to an algebraic form, associated with either of the following geometrical constructions: (1) the curve $y = x^\alpha$ with $0 \leq \alpha < 1$ lies below its tangent at the point $x = 1$, which has slope α (Figure 7.1a); and (2) if $1/p + 1/q = 1$, the functions $\xi = \eta^{q-1}$ and $\eta = \xi^{p-1}$ are inverse, and the sum of the associated areas $S_2 + S_1$ is less (Figure 7.1b) than that of the rectangle (a,b). The equality occurs in Figure 7.1a (Figure 7.1b) only if $x = 1(a = b)$. Another geometric construction (Figure 7.2a) relates to the convolution of ordinary or generalized functions that involves an integration in the (y, z)-plane, over the hatched region, that is, the intersection of (1) the support $a \leq y + z \leq b$ of integrable test function with compact support, which vanishes outside a strip; and (2) the support $c \leq y \leq d$ of one of the or the product of the generalized functions that is assumed to be compact, that is, to vanish outside the interval (c,d). The resulting integral over the finite area is bounded. Another case of the existence of a convolution integral (Figure 7.2b) is for generalized functions with compact support, which vanish outside the rectangle; in this case the test function must still be integrable but need not have compact support. Figure 7.2a and b show that for the convolution of generalized functions over integrable test functions to exist, it is sufficient if two of them (out of three) have compact support. For example, the convolution of two derivatives of unit impulses exists over integrable reference functions. If the test functions are integrable and have compact support, the convolution exists for any derivative unit impulse and another generalized function. The convolution of two derivatives of the unit impulse exists for an integrable test function with no restriction to compact support. The solution of a linear ordinary or partial differential equation is specified by the convolution of the forcing term with the influence or Green function if the latter depends only on the difference of the two variables. The influence or Green function is the solution of the ordinary (partial) differential equation forced by a one-(multi)dimensional unit impulse. It can be defined for linear or nonlinear differential equations but the principle of superposition holds generally only in the linear case. Among the linear differential equations, those with a symmetric influence function satisfy the reciprocity principle; the latter requires that the linear differential operator be self-adjoint. Table 7.1 indicates the linear self-adjoint and self-adjoint differential operators in 19 mathematical classes related in Diagram 7.1; Table 7.2 indicates 10 physical examples of linear self-adjoint or non-self adjoint differential operators related in Diagram 7.2. Some of these are particular cases of the most common partial differential equations of mathematical physics listed in Table 7.3 and indicated in Diagram 7.3, which can be generalized to (1) inhomogeneous, unsteady, and anisotropic media (Table 7.4); and (2) vector and higher-order equations (Table 7.5). The harmonic oscillator without (with) damping [Figure 7.4a (Figure 7.4b)] is specified by a linear self-adjoint (non-self-adjoint) ordinary differential operator with time as the independent variable. The weak deflection (bending) of an elastic string (bar) leads [Figure 7.3a (Figure 7.3b)] to a self-adjoint second- (fourth-) order linear differential operator with the longitudinal coordinate as the independent variable; the weak deflection of an elastic membrane (plate) adds a second longitudinal coordinate as the independent variable and leads to a linear self-adjoint partial differential operator. The operator remains self-adjoint adding (1) the string (membrane) and bar (plate) to obtain a beam (stressed plate); (2) support by linear springs; and (3) the inertia force, leading to oscillations (vibrations) and a linear differential operator in space time. The self-adjointness remains if the bending stiffness, longitudinal tension, spring resilience, and mass density all depend on position and time. The self-adjointness is violated in the presence of damping, even if the damping coefficient is constant.

8

Electric/Magnetic
Multipoles and Images

A line charge in space corresponds to the generalized function two-dimensional unit impulse, and the corresponding influence or Green function for the Poisson equation specifies the potential due to the line charge. This method is distinct from, and leads to the same results as, the theory of the complex potential (Chapters I.12, 24, and 26), not only for monopoles, but also for dipoles, quadrupoles, and multiple expansions. Whereas the method using complex functions is restricted to plane problems, the use of generalized functions is valid regardless of spatial dimension; for example, the latter method extends (Section 9.2) readily from two (Chapter I.11) to three (Chapter 8) or higher (Sections 9.7 through 9.9) dimensions. The three-dimensional unit impulse represents a point charge, and the associated influence or Green function specifies the corresponding potential; the convolution integral expressing the principle of superposition then specifies the potential due to an arbitrary distribution of charges. This leads to the three-dimensional laws for the Coulombian force field which are related to plane laws (Section I.24.4), but distinct (Section 8.3) from them. In space (Section 8.1), as in the plane (Section I.12.7), a dipole is the limit of two opposite monopoles whose strength increases inversely with distance; whereas the monopole is specified by a unit impulse and is radially symmetric, a dipole involves the gradient of a unit impulse projected in the direction of the vector dipole moment and, thus, is axisymmetric (Section 8.2). A quadrupole is (Section 8.3) the limit of opposite dipoles and involves second-order derivatives of the unit impulse; since it has two axes, it is no longer generally axisymmetric, and the quadrupole moment is represented by a matrix. A three-dimensional charge distribution, like its two-dimensional counterpart, can be decomposed into a superposition of multipolar fields, with an important difference: (1) the two-dimensional multipoles have moments that are complex numbers; (2) the three-dimensional multipoles have moments of increasing complexity, namely, a scalar, a vector, a matrix, and a tensor, respectively, for mono-, di-, quadru-, and multipoles. The differences between the electrostatic (magnetostatic) field in space are greater than in the plane [Chapter I.24 (Chapter I.26)]: (1) the electrostatic field is irrotational, and is always represented by a scalar potential in two (Section I.24.2), three (Sections 8.1 through 8.3), or higher (Sections 9.7 through 9.9) dimensions; (2) the magnetostatic field is solenoidal and is represented by a pseudoscalar field function in two dimensions (Section I.26.2) and by a vector potential (Section 8.4) in three dimensions; and (3) the three-dimensional vector potential reduces to a scalar stream function not only in the plane case but also in the three-dimensional axisymmetric case (Sections 6.2 through 6.7). Both the electrostatic (magnetostatic) fields have a multipolar expansion for the scalar (vector) potential [Section 8.3 (Section 8.4)]. The lowest-order term in the vector multipolar expansion for the magnetostatic potential corresponds to a point current; it is specified by a unit impulse like the electrostatic monopole, but has a vector direction and is axisymmetric, corresponding to a magnetic dipole (Section 8.5). The limit of two opposite magnetic dipoles is a magnetic quadrupole (Section 8.6) that (1) involves derivatives of the unit impulse as an electrostatic dipole; and (2) has a quadrupole moment specified by a matrix, and its two axes imply it is generally not axisymmetric. The method of images applies to electro(magneto-)static fields and extends from the plane to space, namely, there are identical (opposite) images on an insulating (conducting) boundary, for example, (1) for a point electric charge or current near a plane (Section 8.7), that has one image; (2) for a point charge between orthogonal (parallel) planes, that has three (or an infinite number of) images (Section 8.8); and (3) for a point charge near a sphere (Section 8.9), using the axisymmetric sphere theorem (Section 6.7) involving the reciprocal point. The point charge near an insulating boundary is analogous from the point of view of images (Sections 8.6 through 8.9) to a flow source/sink near a rigid wall and leads to continuous image distributions. The continuous source distributions can be used to obtain the potential flow past bodies of arbitrary shape (Section 8.9).

8.1 Electric Dipole as the Limit of Opposite Monopoles

The simplest electric charge distribution (Figure 8.1) is a monopole specifying a radial electric field (Figure 8.1). The electric field is no longer radial (Figure 8.2) in the limit of two opposite monopoles (Figure 8.4), which specifies the electric charge distribution (Subsection 8.1.1) and electrostatic potential (Subsection 8.1.2) of a dipole (Figure 8.5); these are consistent with the influence or Green function (Subsection 8.1.3) and lead to the electrostatic field of a dipole (Subsection 8.1.4).

8.1.1 Electric Charge Density of a Monopole and a Dipole

A point electric charge P_0 at the origin (Figure 8.1) is represented by a unit impulse (8.1a) because (3.15a) the total electric charge in a domain D is P_0 (zero) if it does (does not) contain the origin (8.1b) [(8.1c)]:

$$e(\vec{x}) = P_0 \delta(\vec{x}): \quad P_0 = \int_D e(\vec{x}) dx = \begin{cases} P_0 & \text{if } \vec{O} \in D \\ 0 & \text{if } \vec{O} \notin D. \end{cases} \tag{8.1a–c}$$

Consider (Figure 8.2) the limit of two equal and opposite electric charges $\pm P_0$ at close positions $\pm \vec{y}$ in (8.2a):

$$e_\pm(\vec{x}) = \pm P_0 \delta(\vec{x} \mp \vec{y}), \quad \vec{P}_1 \equiv \lim_{P_0 \to \infty, \vec{y} \to 0} 2 P_0 \vec{y}. \tag{8.2a,b}$$

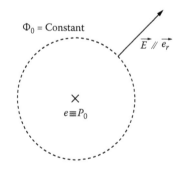

FIGURE 8.1 A point electric charge creates an isotropic electrostatic potential (radial electric field) decaying like the inverse (inverse square) of the distance.

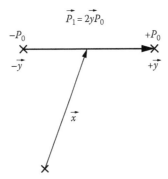

FIGURE 8.2 Two monopoles with the same strength and opposite signs as their distance decreases (strength increases) with a constant product lead to a dipole, specified by its moment and axis.

Their distance tends to zero and the charges to infinity while their product is preserved (8.2b). In order to obtain the charge distribution associated with the dipole, the three-dimensional impulse in (8.3a) is expanded in the first two terms of its Taylor series (3.66a and b) in vector form:

$$\delta(\vec{x} \pm \vec{y}) = \delta(\vec{x}) \pm (\vec{y} \cdot \nabla)\delta(\vec{x}) + O(y^2): \quad \vec{y} \cdot \nabla \equiv \vec{y} \cdot \frac{\partial}{\partial \vec{x}} \equiv \sum_{i=1}^{3} y_i \frac{\partial}{\partial x_i}, \tag{8.3a,b}$$

where (8.3b) is y times the derivative in the direction of $\vec{y}$; the total charge

$$e_+(\vec{x}) + e_-(\vec{x}) = P_0 \left\{ \delta(\vec{x} - \vec{y}) - \delta(\vec{x} + \vec{y}) \right\}$$
$$= -2P_0 (\vec{y} \cdot \nabla)\delta(\vec{x}) + O(P_0 y^2) = -(\vec{P}_1 \cdot \nabla)\delta(\vec{x}) + O(P_1 y) \tag{8.4}$$

simplifies in the limit (8.2b) to

$$e_1(\vec{x}) = \lim_{y \to \infty} \left\{ e_+(\vec{x}) + e_-(\vec{x}) \right\} = -\left\{ \vec{P}_1 \cdot \frac{\partial}{\partial \vec{x}} \right\} \delta(\vec{x}). \tag{8.5}$$

Thus *the charge distribution representing a dipole of moment (8.2b) is minus the derivate of the unit impulse along the dipole moment (8.5), which appears as the forcing term*

$$\partial_{ii}\Phi_1 \equiv \frac{\partial^2 \Phi_1}{\partial x_i^2} = \frac{\partial^2 \Phi_1}{\partial x_i \partial x_i} = \left\{ \vec{P}_1 \cdot \frac{\partial}{\partial \vec{x}} \right\} \delta(\vec{x}) = \left\{ \vec{P}_1 \cdot \nabla \right\} \delta(\vec{x}) \tag{8.6}$$

in the Poisson equation (8.6) for the dipolar potential Φ_1.

8.1.2 Electrostatic Potential of a Dipole

The electrostatic potential associated with the opposite charges (8.1a) at positions $\vec{x} \pm \vec{y}$ is (5.312c) $\equiv$ (8.7a):

$$\Phi_\pm(\vec{x}) = \pm \frac{P_0}{4\pi\varepsilon} \frac{1}{|\vec{x} \mp \vec{y}|}, \quad |\vec{x} \pm \vec{y}| = |x^2 + y^2 \pm 2\vec{x} \cdot \vec{y}|^{1/2}, \tag{8.7a,b}$$

where (8.7b) are distances from an observer at $\vec{x}$ and ε is the dielectric permittivity. If the distance between the charges $2|\vec{y}|$ is small compared with the distance from the observer (8.8a), the approximation (8.8b):

$$R \equiv |\vec{x}|: \quad \frac{1}{|\vec{x} \pm \vec{y}|} = \frac{1}{R}\left|1 \pm 2\frac{\vec{x} \cdot \vec{y}}{R^2} + \frac{y^2}{R^2}\right|^{-1/2} = \frac{1}{R} \mp \frac{\vec{x} \cdot \vec{y}}{R^3} + O\left(\frac{y^2}{R^3}\right) \tag{8.8a,b}$$

can be made in the total potential:

$$\Phi_1(\vec{x}) \equiv \Phi_+(\vec{x}) + \Phi_-(\vec{x}) = \frac{P_0}{4\pi\varepsilon}\left(\frac{1}{|\vec{x}-\vec{y}|} - \frac{1}{|\vec{x}+\vec{y}|}\right) = \frac{P_0}{2\pi\varepsilon}\left[\frac{\vec{y}\cdot\vec{x}}{R^3} + O\left(\frac{y^2}{R^3}\right)\right]. \tag{8.9}$$

The limit (8.2b) then specifies *the electrostatic potential (8.10a) due to a dipole of moment (8.2b) in a medium of dielectric permittivity* ε:

$$\Phi_1(\vec{x}) = \frac{\vec{P_1}\cdot\vec{x}}{4\varepsilon R^3} = \frac{P_1}{4\pi\varepsilon}\frac{\cos\theta}{R^2}, \quad \cos\theta = \frac{\vec{P_1}\cdot\vec{x}}{P_1 R}, \tag{8.10a,b}$$

where θ *is the angle between the dipole axis and observer direction (8.10b).*

8.1.3 Influence Function for the Electrostatic Field

The consistency of the electric charge distribution (8.5) and electrostatic potential (8.10a) of a dipole can be demonstrated by showing that they satisfy the Poisson equation (8.6). *The influence or Green function for the Poisson equation is the fundamental solution forced by a unit monopole (8.11a):*

$$\frac{\partial^2}{\partial\vec{x}^2}G(\vec{x}-\vec{y}) = \delta(\vec{x}-\vec{y}), \quad G(\vec{x}-\vec{y}) = -\frac{1}{4\pi|\vec{x}-\vec{y}|} = G(\vec{y}-\vec{x}), \tag{8.11a,b}$$

and it is (5.317c) $\equiv$ (8.11b) *the inverse distance potential (Figure 8.1)*; it is symmetric (8.11b), that is, it satisfies the reciprocity principle, because the Laplace operator in (8.10a) is self-adjoint (7.162a through c). The superposition principle holds for the Poisson equation (8.12a) because it is linear:

$$\nabla^2\Phi = -\frac{e(\vec{x})}{\varepsilon}, \quad -\varepsilon\Phi(\vec{x}) = -\frac{1}{\varepsilon}\int_D e(\vec{y})G(\vec{x}-\vec{y})\,d^3\vec{y} = e(\vec{x}) * G(\vec{x}); \tag{8.12a,b}$$

the influence function depends only on the relative position $\vec{z} = \vec{x} - \vec{y}$ leading to a convolution integral (8.12b) $\equiv$ (7.36). Thus, it follows that *the electrostatic potential (electric field) of an arbitrary electric charge distribution (Figure 8.3) in a domain D with dielectric permittivity* ε *is given by* (5.313c) $\equiv$ (8.13a) [(5.313d) $\equiv$ (8.13b)]:

$$\Phi(\vec{x}) = \frac{1}{4\pi\varepsilon}\int_D \frac{e(\vec{y})}{|\vec{x}-\vec{y}|}\,d^3\vec{y}, \tag{8.13a}$$

$$\vec{E} = -\frac{\partial\Phi}{\partial\vec{x}} = \frac{1}{4\pi\varepsilon}\int_D \frac{\vec{x}-\vec{y}}{|\vec{x}-\vec{y}|^3}e(\vec{y})\,d^3\vec{y}. \tag{8.13b}$$

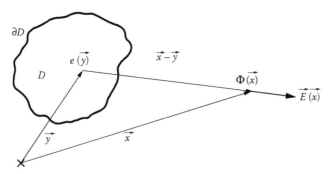

FIGURE 8.3 A distribution of sources at positions $\vec{y}$ in a domain D (e.g., electric charges) creates at an observation point $\vec{x}$ (an electrostatic) a potential that depends on the mutual distance, that is, the modulus of the relative position vector; the (electric) field is parallel (opposite) to the relative position vector for positive (negative) source [electric charges].

The passage from (8.13a) to (8.13b) uses (1) the relative position vector from the source to the observer (8.14a); (2) its modulus that is the distance between the source and the observer (8.14b); (3) the derivative of the latter (ii) with regard to the position vector of the observer (8.14c) [source (8.14d)] equals the (equals minus the) unit vector from source to observer:

$$\vec{\ell} = \vec{x} - \vec{y}, \ell \equiv |\vec{x} - \vec{y}|: \quad \vec{m} = \frac{\partial \ell}{\partial \vec{x}} = \frac{\partial}{\partial \vec{x}}\left(|\vec{x} - \vec{y}|\right) = \frac{\vec{x} - \vec{y}}{|\vec{x} - \vec{y}|} = \frac{\vec{\ell}}{|\vec{\ell}|} = -\frac{\partial \ell}{\partial \vec{y}}. \tag{8.14a–d}$$

The proof is

$$-\frac{\partial \ell}{\partial \vec{y}} = \frac{\partial \ell}{\partial \vec{x}} = \frac{1}{2\ell}\frac{\partial}{\partial \vec{x}}\left(\ell^2\right) = \frac{1}{2|\vec{x} - \vec{y}|}\frac{\partial}{\partial \vec{x}}\left[(\vec{x} - \vec{y})\cdot(\vec{x} - \vec{y})\right] = \frac{\vec{x} - \vec{y}}{|\vec{x} - \vec{y}|}, \tag{8.15}$$

which is in agreement with (5.322a through c). In the case of the monopole or point electric charge (8.1a) ≡ (8.16a), the potential is given (8.13a) by (8.16b) confirming (8.7a):

$$e(\vec{x}) = P_0\delta(\vec{x}), \quad \Phi_0(\vec{x}) = \frac{P_0}{4\pi\varepsilon}\int_{-\infty}^{+\infty}\frac{1}{|\vec{x} - \vec{y}|}\delta(\vec{y})d^3\vec{y} = \frac{P_0}{4\pi\varepsilon R}. \tag{8.16a,b}$$

The case of the electric charge distribution (8.5) due to a dipole of moment $\vec{P}_1$,

$$-\frac{1}{4\pi\varepsilon}\int_D\frac{1}{|\vec{x} - \vec{y}|}\left\{\left(\vec{P}_1\cdot\frac{\partial}{\partial\vec{y}}\right)\delta(\vec{y})\right\}d^3\vec{y} = \frac{1}{4\pi\varepsilon}\lim_{\vec{y}\to 0}\left(-\vec{P}_1\cdot\frac{\partial}{\partial\vec{y}}\right)\frac{1}{|\vec{x} - \vec{y}|}$$

$$= -\frac{1}{4\pi\varepsilon}\lim_{\vec{y}\to 0}\left(\frac{1}{|\vec{x} - \vec{y}|^2}\left(-\vec{P}_1\cdot\frac{\partial}{\partial\vec{y}}\right)|\vec{x} - \vec{y}|\right) = \frac{1}{4\pi\varepsilon}\lim_{\vec{y}\to 0}\frac{\vec{P}_1\cdot(\vec{x} - \vec{y})}{|\vec{x} - \vec{y}|^3}$$

$$= \frac{1}{4\pi\varepsilon}\frac{\vec{P}_1\cdot\vec{x}}{|\vec{x}|^3} = \frac{\vec{P}_1\cdot\vec{x}}{4\pi R^3} = \Phi_1(\vec{x}), \tag{8.17}$$

leads to the potential of the dipole (8.17) ≡ (8.10a), confirming that the Poisson (8.6) equation is satisfied.

8.1.4 Electric Field due to a Monopole and a Dipole

The electrostatic potential (8.16b) of a point charge or monopole (8.16a) ≡ (8.18a) in spherical (8.18b) [cylindrical (8.18c)] coordinates,

$$e(\vec{x}) = q\delta(\vec{x}): \quad \Phi_0(R) = \frac{P_0}{4\pi\varepsilon R} = \frac{P_0}{4\pi\varepsilon}\left|r^2+z^2\right|^{-1/2} = \Phi_0(r,z), \tag{8.18a–c}$$

specifies the electric field (8.19a and b), through the minus the gradient:

$$\vec{E}_0 = -\nabla\Phi_0 = -\vec{e}_R\frac{d\Phi}{dR} = \frac{P_0}{4\pi\varepsilon R^2}\vec{e}_R = \frac{P_0}{4\pi\varepsilon}\frac{1}{r^2+z^2}\vec{e}_r, \tag{8.19a,b}$$

and its spherical (8.20a and b) [cylindrical (8.21a and b)] components using (6.23e) [(6.23d)]:

$$\{E_{0R}, E_{0\theta}, E_{0\phi}\} = -\left\{\frac{\partial}{\partial R}, \frac{1}{R}\frac{\partial}{\partial\theta}, \frac{1}{R\sin\theta}\frac{\partial}{\partial\phi}\right\}\frac{P_0}{4\pi\varepsilon R}$$

$$= \frac{P_0}{4\pi\varepsilon R^2}\{1,0,0\} = \frac{P_0}{4\pi\varepsilon}\frac{1}{r^2+z^2}\{1,0,0\}, \tag{8.20a,b}$$

$$\{E_{0r}, E_{0z}, E_{0\varphi}\} = -\left\{\frac{\partial}{\partial r}, \frac{1}{r}\frac{\partial}{\partial z}, \frac{\partial}{\partial\varphi}\right\}\frac{P_0}{4\pi\varepsilon}\left|r^2+z^2\right|^{-1/2}$$

$$= \frac{P_0}{4\pi\varepsilon}\left|r^2+z^2\right|^{-3/2}\{r,0,z\} = \frac{P_0}{4\pi\varepsilon R^3}\{\sin\theta,0,\cos\theta\}, \tag{8.21a,b}$$

where all expressions are written both in spherical (8.18b; 8.19a; 8.20a; 8.21b) [cylindrical (8.18c; 8.19b; 8.20b; 8.21a)] coordinates. In the case of a dipole (8.5; 8.2a and b), the electrostatic potential (8.10a) leads to the electric field (8.22c) using (8.14a and b) ≡ (8.22a and b):

$$R = |\vec{x}|: \quad \frac{\partial R}{\partial\vec{x}} = \frac{\vec{x}}{R} = \vec{e}_R: \quad \vec{E}_1 = -\frac{\partial\Phi_1}{\partial\vec{x}} = -\frac{\partial}{\partial\vec{x}}\left(\frac{\vec{P}_1\cdot\vec{x}}{4\pi\varepsilon R^3}\right)$$

$$= -\frac{\vec{P}_1}{4\pi\varepsilon R^3} + \frac{3}{4\pi\varepsilon R^4}\left(\vec{P}_1\cdot\vec{x}\right)\frac{\vec{x}}{R}$$

$$= \frac{1}{4\pi\varepsilon R^5}\left[3\left(\vec{P}_1\cdot\vec{x}\right)\vec{x} - R^2\vec{P}_1\right]. \tag{8.22a–c}$$

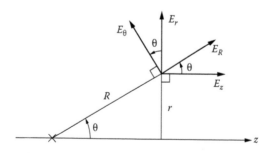

FIGURE 8.4 The cylindrical (E_z, E_r) and spherical (E_R, E_0) components of an axisymmetric vector field are related by a local rotation specified by the co-latitude, that is, the angle with the axis.

Thus *an electric dipole (8.5)* $\equiv$ *(8.23a) with moment* $\vec{P}_1$ *has an electric field (8.22c)* $\equiv$ *(8.23b):*

$$e_1(\vec{x}) = -\left(\vec{P}_1 \cdot \nabla\right)\delta(\vec{x}): \quad \vec{E}_1(\vec{x}) = \frac{1}{4\pi\varepsilon R^5}\left[3\left(\vec{P}_1 \cdot \vec{x}\right)\vec{x} - \vec{P}_1 R^2\right], \tag{8.23a,b}$$

corresponding to the electrostatic potential (8.10a) $\equiv$ *(8.24a and b):*

$$\Phi_1(R,\theta) = \frac{P_1 \cos\theta}{4\pi\varepsilon R^2} = \frac{P_1 z}{4\pi\varepsilon}\left|r^2 + z^2\right|^{-3/2} = \Phi_1(r,z), \tag{8.24a,b}$$

with the z-axis along the dipole moment (8.24b). The spherical (cylindrical) components are (8.25a through d) [(8.26a through d)]

$$\left\{E_{1R}, E_{1\theta}, E_{1\varphi}\right\} = -\left\{\frac{\partial}{\partial R}, \frac{1}{R}\frac{\partial}{\partial\theta}, \frac{1}{R\sin\varphi}\frac{\partial}{\partial\varphi}\right\}\frac{P_1 \cos\theta}{4\pi\varepsilon R^2} = \frac{P_1}{4\pi\varepsilon R^3}\left\{2\cos\theta, \sin\theta, 0\right\}$$

$$= \frac{P_1}{4\pi\varepsilon R^4}\left\{2z, r, 0\right\} = \frac{P_1}{4\pi\varepsilon}\frac{\left\{2z, r, 0\right\}}{\left(r^2 + z^2\right)^2}, \tag{8.25a–d}$$

$$\left\{E_{1r}, E_{1\varphi}, E_{1z}\right\} = -\left\{\frac{\partial}{\partial r}, \frac{1}{r}\frac{\partial}{\partial\varphi}, \frac{\partial}{\partial z}\right\}\frac{P_1 z}{4\pi\varepsilon}\left|r^2 + z^2\right|^{-3/2} = \frac{P_1}{4\pi\varepsilon}\left|r^2 + z^2\right|^{-5/2}\left\{3rz, 0, 2z^2 - r^2\right\}$$

$$= \frac{P_1}{4\pi\varepsilon R^5}\left\{3rz, 0, 2z^2 - r^2\right\} = \frac{P_1}{4\pi\varepsilon R^3}\left\{3\cos\theta\sin\theta, 0, 2\cos^2\theta - \sin^2\theta\right\}$$

$$= \frac{P_1}{4\pi\varepsilon R^3}\left\{\frac{3}{2}\sin(2\theta), 0, 3\cos^2\theta - 1\right\}. \tag{8.26a–e}$$

All expressions have been written in spherical (8.24a; 8.25a and b; 8.26d and e) [cylindrical (8.24b; 8.25d; 8.26a and b)] coordinates, plus some combination of both (8.25c; 8.26c). The spherical (8.25a through d) and cylindrical (8.26a through e) components of the axisymmetric electric field satisfy (6.51c and d). The electric field of a dipole is radial (meridional) $E_{1\theta} = 0(E_{1R} = 0)$ *in (8.25b) along* $\theta = 0, \pi$ *[transverse to* $\theta = \pi/2$*] the axis of the dipole, and has nonzero radial and meridional components in all other directions leading to a two-lobe radiation pattern (Figure 8.5).*

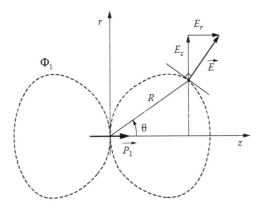

FIGURE 8.5 The directional nature of the dipole (Figure 8.3) implies that the equipotential surfaces have two lobes (Figure 8.5). Since the electric field is normal to the equipotential surfaces, it is purely radial (along meridians) along (transverse to) the axis of the dipole.

8.2 Longitudinal, Transverse, and Cross Quadrupoles

The quadrupole is the limit of two opposite dipoles (Subsection 8.2.1), leading to two axes, and hence three components (Subsection 8.2.2), namely, longitudinal, transverse, and cross components.

8.2.1 Electric Quadrupole as Limit of Opposite Dipoles

The quadrupole with moment P_{ij} is the limit of the opposite dipoles P_i at a distance y_j, as the moments tend to infinity and the distances to zero so as to preserve their product (8.27a):

$$P_{ij} = -\lim_{P_i \to \infty, y_j \to 0} 2 y_j P_i : \quad e_2(\vec{x}) = -\lim_{\vec{P} \to \infty, \vec{y} \to 0} (\vec{P} \cdot \nabla) \{ \delta(\vec{x} - \vec{y}) - \delta(\vec{x} + \vec{y}) \}$$

$$= \frac{1}{2} \left\{ P_{ij} \frac{\partial^2}{\partial x_i \partial x_j} \right\} \delta(\vec{x}), \tag{8.27a,b}$$

leading as in (8.2 through 8.5) to (8.27b); the factor (1/2) arises from the Taylor series expansion (I.23.32b) for the three-dimensional unit impulse (3.66a and b) taken one order beyond (8.3a):

$$\delta(\vec{x} \pm \vec{y}) = \delta(\vec{x}) \pm (y_i \partial_i) \delta(\vec{x}) + \frac{1}{2} (y_i y_j \partial_i \partial_j) \delta(\vec{x}) + O(y_i y_j y_k), \tag{8.28}$$

that is, to second-order, in agreement with the multipolar expansion (8.50a and b). This ensures that the definition of quadrupole moment is consistent in the electric charge distribution (8.27a and b) and in the potential (8.29; 8.30; 8.31a through c; 8.32a through d) and field (8.33a and b through 8.38a through c) obtained next. Thus *the electric charge distribution associated with a quadrupole of moment P_{ij} is one-half the double contraction with the second-order derivates of the three-dimensional impulse*:

$$\frac{\partial^2 \Phi_2}{\partial x_i \partial x_i} = \frac{1}{2} \left\{ P_{ij} \frac{\partial^2}{\partial x_i \partial x_j} \right\} \delta(\vec{x}); \tag{8.29}$$

it acts as the forcing term in the Poisson equation for quadrupole potential (8.29). Its solution specifies the quadrupole potential (8.30b):

$$\vec{e}_R = \frac{\vec{x}}{R} : \quad \Phi_2(\vec{x}) = \frac{3}{8 \pi \varepsilon R^5} P_{ij} x_i x_j - \frac{1}{8 \pi \varepsilon R^3} P_{ii} = \frac{1}{8 \pi \varepsilon R^3} (3 P_{ij} e_{Ri} e_{Rj} - P_{ii}), \tag{8.30a,b}$$

using the unit radial vector in the observer direction (8.22b) ≡ (8.30a). The electrostatic potential can be calculated from the influence or Green function (8.13a), for example, (1) with the dipole (8.2a and b) as the forcing term in (8.5):

$$4 \pi \varepsilon \Phi_1(\vec{x}) = -\left\{ P_i \frac{\partial}{\partial x_i} \right\} \frac{1}{R} = \frac{1}{R^2} P_i \frac{\partial R}{\partial x_i} = \frac{1}{R^3} P_i x_i = \frac{1}{R^3} (\vec{P}_1 \cdot \vec{x}) = P_1 \frac{\cos \theta}{R^2}, \tag{8.31a-c}$$

leading to the dipole potential in agreement with (8.10a) ≡ (8.31a through c); and (2) with the quadrupole (8.27b) as the forcing term in (8.13a):

$$8 \pi \varepsilon \Phi_2(\vec{x}) = \left\{ P_{ij} \frac{\partial^2}{\partial x_i \partial x_j} \right\} \frac{1}{R} = -P_{ij} \frac{\partial}{\partial x_i} \left(\frac{1}{R^3} x_j \right) = -P_{ij} \delta_{ij} \frac{1}{R^3} + \frac{3}{R^5} P_{ij} x_i x_j, \tag{8.32a-d}$$

leading to (8.32a through d) and proving the quadrupole potential (8.30b) ≡ (8.32d) using the identity matrix (5.278a through c).

8.2.2 Electrostatic Potential and Field of a Quadrupole

Choosing spherical (6.13a through d) [cylindrical (6.11a and b)] coordinates (Figure 6.2) along the z-axis:

$$\vec{e}_R = \vec{e}_z \cos\theta + \vec{e}_r \sin\theta, \quad \vec{e}_r = \vec{e}_x \cos\varphi + \vec{e}_y \sin\varphi, \tag{8.33a,b}$$

$$\vec{e}_R = \vec{e}_z \cos\theta + \sin\theta\left(\vec{e}_x \cos\varphi + \vec{e}_y \sin\varphi\right), \tag{8.33c}$$

the quadrupole potential (8.30b) involves one longitudinal P_{zz}, two transverse $\left(P_{xx}, P_{yy}\right)$, and three mixed $\left(P_{xy}, P_{xz}, P_{yz}\right)$ components, which appear in (8.34a) [(8.34b)]

$$8\pi\varepsilon R^3 \Phi_2\left(R,\theta,\varphi\right) = P_{zz}\left(3\cos^2\theta - 1\right) + P_{xx}\left(3\sin^2\theta\cos^2\varphi - 1\right) + P_{yy}\left(3\sin^2\theta\sin^2\varphi - 1\right)$$
$$+ 3\sin\left(2\theta\right)\left(\bar{P}_{xz}\cos\varphi + \bar{P}_{yz}\sin\varphi\right) + 3\bar{P}_{xy}\sin^2\theta\sin\left(2\varphi\right), \tag{8.34a}$$

$$8\pi\varepsilon\left|r^2 + z^2\right|^{5/2}\Phi_2\left(r,\varphi,z\right) = P_{zz}\left(2z^2 - r^2\right) + P_{xx}\left[r^2\left(3\cos^2\varphi - 1\right) - z^2\right]$$
$$+ P_{yy}\left[r^2\left(3\sin^2\varphi - 1\right) - z^2\right] + 6rz\left(\bar{P}_{xz}\cos\varphi + \bar{P}_{yz}\sin\varphi\right) + 3\bar{P}_{xy}r^2\sin\left(2\varphi\right), \tag{8.34b}$$

where the quadrupole moments are replaced by their arithmetic means (8.35a) that do not change the axial moments (8.35b):

$$2\bar{P}_{ij} \equiv P_{ij} + P_{ji} = 2\bar{P}_{ji}, \quad \bar{P}_{ii} = P_{ii}. \tag{8.35a,b}$$

Choosing the meridian (8.36a) through the observer direction (8.36b) simplifies the quadrupole potential in spherical (8.34a) [cylindrical (8.34b)] coordinates to (8.36d) [(8.36e)]

$$\varphi = 0: \quad \vec{e}_R = \vec{e}_z\cos\theta + \vec{e}_x\sin\theta; \quad 2\bar{P}_{xz} = P_{xz} + P_{zx} = 2\bar{P}_{zx}, \tag{8.36a–c}$$

$$8\pi\varepsilon R^3 \Phi_2\left(R,\theta\right) = P_{zz}\left(3\cos^2\theta - 1\right) + P_{xx}\left(3\sin^2\theta - 1\right) - P_{yy} + 3\bar{P}_{xz}\sin\left(2\theta\right), \tag{8.36d}$$

$$8\pi\varepsilon\left|r^2 + z^2\right|^{5/2}\Phi_2\left(r,z\right) = P_{zz}\left(2z^2 - r^2\right) + P_{xx}\left(2r^2 - z^2\right) - P_{yy}\left(r^2 + z^2\right) + 6rz\bar{P}_{xz}, \tag{8.36e}$$

involving one longitudinal P_{zz}, two transverse P_{xx}, P_{yy}, and one mixed $\bar{P}_{xz}$ moment (8.36c). Thus *a quadrupole (8.27b) has electrostatic potential in spherical (8.34a) [cylindrical (8.34b)] coordinates choosing the axis in the direction of the moment P_{zz} and using symmetrised quadrupole moments (8.35a,b). Measuring the azimuth (8.36a and b) from the plane through the observer and axis, the electrostatic potential simplifies to (8.36d) [(8.36e)], leading to the spherical (8.37a through c) [cylindrical (8.38a through c)] components of the electrostatic field:*

$$\left\{E_{2R}, E_{2\theta}, E_{2\varphi}\right\} = -\left\{\frac{\partial}{\partial R}, \frac{1}{R}\frac{\partial}{\partial\theta}, \frac{1}{R\sin\theta}\frac{\partial}{\partial\varphi}\right\}\Phi_2\left(R,\theta\right)$$

$$= \frac{3}{8\pi\varepsilon R^4}\left\{P_{zz}\left(3\cos^2\theta - 1\right) + P_{xx}\left(3\sin^2\theta - 1\right) - P_{yy} + \bar{P}_{xz}\sin\left(2\theta\right), \left(P_{zz} - P_{xx}\right)\sin\left(2\theta\right) - 2\bar{P}_{xz}\cos\left(2\theta\right), 0\right\}$$

$$= \frac{3}{8\pi\varepsilon}\left|r^2 + z^2\right|^{-3}\left\{P_{zz}\left(2z^2 - r^2\right) + P_{xx}\left(2r^2 - z^2\right) - P_{yy}\left(r^2 + z^2\right) + 2\bar{P}_{xz}rz, 2\left(P_{zz} - P_{xx}\right)rz - 2\bar{P}_{xz}\left(z^2 - r^2\right), 0\right\}, \tag{8.37a–c}$$

$$\{E_{2r}, E_{2\varphi}, E_{2z}\} \quad = -\left\{ \frac{\partial}{\partial r}, \frac{1}{r}\frac{\partial}{\partial \varphi}, \frac{\partial}{\partial z} \right\} \Phi_2(r,z) = \frac{3}{8\pi\varepsilon} \left| r^2 + z^2 \right|^{-7/2}$$

$$\times \Big\{ r\Big[P_{zz}\big(4z^2 - r^2\big) + P_{xx}\big(2r^2 - 3z^2\big) + P_{yy}\big(r^2 + z^2\big) \Big] + 2z\bar{P}_{xz}\big(4r^2 - z^2\big),$$

$$0, z\Big[P_{zz}\big(2z^2 - 3r^2\big) + P_{xx}\big(4r^2 - z^2\big) + P_{yy}\big(r^2 + z^2\big) \Big] + 2r\bar{P}_{xz}\big(4z^2 - r^2\big) \Big\}$$

$$= \frac{3}{8\pi\varepsilon R^4} \Big\{ \sin\theta \Big[P_{zz}\big(5\cos^2\theta - 1\big) + P_{xx}\big(5\sin^2\theta - 3\big) + P_{yy} \Big] + 2\cos\theta \bar{P}_{xz}\big(5\sin^2\theta - 1\big),$$

$$0, \cos\theta \Big[P_{zz}\big(5\cos^2\theta - 3\big) + P_{xx}\big(5\sin^2\theta - 1\big) + P_{yy} \Big] + 2\sin\theta \bar{P}_{xz}\big(5\cos^2\theta - 1\big) \Big\}$$

$$(8.38a\text{–}c)$$

using spherical (8.36d; 8.37b; 8.38c) [cylindrical (8.36e; 8.37c; 8.38b)] coordinates and involving the **longitudinal, transverse, and mixed quadrupole moments** $P_{zz}/P_{xx}, P_{yy}/\bar{P}_{xz}$ *in (8.36c).* The spherical (8.37a through c) and cylindrical (8.38a through c) components of the axisymmetric electrostatic field satisfy (6.51c and d). The longitudinal P_{zz} and transverse P_{xx} quadrupoles have a radial electric field $E_\theta = 0$ for $\sin(2\theta_1) = 0$ in three axial and transverse directions $\theta_1 = 0, \pi/2, \pi$, leading to a four-lobe radiation pattern (Figure 8.6):

$$\theta_2 = 0, \pi; r = 0, R = z: \quad \vec{E}_2 = \frac{3}{8\pi\varepsilon z^4} \Big[\vec{e}_R\big(2P_{zz} - P_{xx} - P_{yy}\big) - \vec{e}_\theta\big(P_{xz} + P_{zx}\big) \Big]$$

$$= \frac{3}{8\pi\varepsilon z^4} \Big[\mp \vec{e}_r\big(P_{xz} + P_{zx}\big) \pm \vec{e}_z\big(2P_{zz} - P_{xx} + P_{yy}\big) \Big], \qquad (8.39a\text{–}e)$$

$$\theta_3 = \frac{\pi}{2}; z = 0, R = r: \quad \vec{E}_2 = \frac{3}{8\pi\varepsilon r^4} \Big[\vec{e}_R\big(2P_{xx} - P_{zz} - P_{yy}\big) + \vec{e}_\theta\big(P_{xz} + P_{zx}\big) \Big]$$

$$= \frac{3}{8\pi\varepsilon r^4} \Big[\vec{e}_r\big(2P_{xx} - P_{zz} + P_{yy}\big) - \vec{e}_z\big(P_{xz} + P_{zx}\big) \Big]. \qquad (8.40a\text{–}e)$$

The mixed quadrupole $\bar{P}_{xz}$ is transverse $E_z = 0 \neq E_r$ in the axial $r = 0$ directions $\theta_2 = 0, \pi$, and is axial $E_r = 0 \neq E_z$ in $z = 0$ the transverse direction $\theta_3 = \pi/2$.

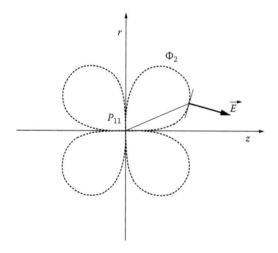

FIGURE 8.6 The use of two monopoles (Figure 8.1) to form a dipole (Figure 8.2) leads to two lobes (Figure 8.5), and the process can be continued to higher-order multipoles; next is the combination of opposite dipoles into a quadrupole whose equipotential surfaces have four lobes (Figure 8.6).

8.3 Multipolar Expansion for an Irrotational Field

The rules of spatial derivation of the inverse distance (Subsection 8.3.1) can be used in the Taylor series for the influence or Green function (Subsection 8.3.2); the latter shows that an arbitrary electric charge distribution is equivalent to a superposition of multipoles (Subsection 8.3.3) of all orders; the asymptotic spatial decay implies (Subsection 8.3.4) that the lowest-order multipole dominates at a large distance. The multipolar expansion of the influence function for the Laplace operator is equivalent to the expansion in spherical harmonics using Legendre polynomials (Subsection 8.3.6); the recurrence formula for the latter is equivalent to the rule of successive generation of multipoles (Subsection 8.3.5). The potential (stream function) for an axial multipole of arbitrary order can be obtained by differentiation (Subsection 8.3.7) of the influence function for the Laplace (modified Laplace) operator, for example, for axial dipoles, quadrupoles, and actupoles (Subsection 8.3.8).

8.3.1 Rules of Spatial Derivation of the Inverse Distance

The electrostatic field is irrotational, corresponding to the scalar potential (8.13a), which can be expanded in a multipolar series, via derivation rules similar to those used before, for example, the field (8.13b) can be derived from the potential (8.13a) using a derivation under the integral sign (Section I.13.8) with regard to the parameter $\vec{x}$ if the convergence is uniform (Section I.13.8):

$$\frac{\partial}{\partial \vec{x}} \frac{1}{|\vec{x} - \vec{y}|} = -\frac{1}{|\vec{x} - \vec{y}|^2} \frac{\partial}{\partial \vec{x}}\left(|\vec{x} - \vec{y}|\right) = -\frac{\vec{x} - \vec{y}}{|\vec{x} - \vec{y}|^3} = -\frac{\partial}{\partial \vec{y}} \frac{1}{|\vec{x} - \vec{y}|}, \tag{8.41}$$

using (8.14a through c). This involves the position vector (8.42a) of the observer $\vec{x}$ relative to the source $\vec{y}$:

$$\vec{\ell} \equiv \vec{x} - \vec{y}: \quad \ell^2 = |\vec{\ell}|^2 = \sum_{i=1}^{N} \ell_i \ell_i \equiv \ell_i \ell_i; \quad \ell \frac{\partial \ell}{\partial x_i} = \ell \frac{\partial \ell}{\partial \ell_i} = \ell_i = -\ell \frac{\partial \ell}{\partial y_i}, \tag{8.42a–c}$$

where, in (8.42b) $\equiv$ (7.70) the **summation convention**, *that a repeated index implies summation over the range of values it can take*, is used again . The formula (8.42c) states that the derivate of the modulus of a vector with regard to the vector is the unit vector in the same direction (8.43a):

$$\frac{\partial \ell}{\partial \vec{\ell}} = \frac{\vec{\ell}}{\ell} = \vec{m}: \quad \frac{\partial}{\partial \vec{x}}\left(\frac{1}{|\vec{x} - \vec{y}|}\right) = \frac{\partial}{\partial \vec{x}}\left(\frac{1}{\ell}\right) = -\frac{1}{\ell^2} \frac{\partial \ell}{\partial \vec{x}}$$

$$= -\frac{\vec{\ell}}{\ell^3} = -\frac{\vec{m}}{\ell^2} = -\frac{\partial}{\partial \vec{y}}\left(\frac{1}{\ell}\right) = -\frac{\partial}{\partial \vec{y}}\left(\frac{1}{|\vec{x} - \vec{y}|}\right), \tag{8.43a,b}$$

and provides (8.43b) an alternative proof of (8.41).

The gradient operator appears not only when passing from the potential to the field (8.41a), but also when generating higher-order multipoles; for example, using the potential, the passage from: (1) a monopole (8.18c) to a dipole (8.10a) corresponds to the application of minus the gradient (8.41); (2) a dipole (8.10a) to a quadrupole (8.32a through d) to the application of one-half of the double gradient (8.29):

$$\frac{\partial^2}{\partial x_j \partial x_i}\left(\frac{1}{|\vec{x} - \vec{y}|}\right) = \frac{\partial^2}{\partial x_j \partial x_i}\left(\frac{1}{\ell}\right) = -\frac{\partial}{\partial x_j}\left(\frac{\ell_i}{\ell^3}\right) = \frac{3}{\ell^4} \ell_i \frac{\partial \ell}{\partial x_j} - \frac{1}{\ell^3} \frac{\partial \ell_i}{\partial x_j}$$

$$= \frac{3 \ell_i \ell_j - \ell^2 \delta_{ij}}{\ell^5} = \frac{3 m_i m_j - \delta_{ij}}{\ell^3} = \frac{\partial^2}{\partial y_i \partial y_j} = \frac{\partial^2}{\partial y_i \partial y_j}\left(\frac{1}{|\vec{x} - \vec{y}|}\right), \tag{8.44}$$

where the identity matrix (5.278a through c) and the unit vector (8.43a) in the direction from the source to the observer (8.42a) are used; (3) a quadrupole (8.22c) to an octupole to the application of $(-)^3/3! = -1/6$ times the triple gradient:

$$\frac{\partial^3}{\partial x_i \partial x_j \partial x_x}\left(\frac{1}{|\vec{x}-\vec{y}|}\right) = -\frac{15}{\ell^7}\ell_i\ell_j\ell_k + \frac{3}{\ell^5}\left(\ell_i\delta_{jk} + \ell_j\delta_{ik} + \ell_k\delta_{ij}\right)$$

$$= \frac{3}{\ell^4}\left(m_i\delta_{jk} + m_j\delta_{ki} + m_k\delta_{ij} - 5m_im_jm_k\right). \tag{8.45}$$

The particular cases (8.43b; 8.44; 8.45) correspond to the **multiple gradient rule:** *(1) the n-order partial derivative of the inverse distance involves all distinct products of the identity matrix (5.278a through c) and unit vectors (8.43a) from the source to the observer; (2) the multiplying factor is* $(2n-2k-1)!!(-)^k$ *if the unit vector appears k-times; and (3) the whole is divided by the power n + 1 of distance and multiplied by* $(-)^n$:

$$\frac{\partial^n}{\partial x_{i_1}\ldots\partial x_{i_n}}\left(\frac{1}{|\vec{x}-\vec{y}|}\right) = (-)^n|\vec{x}-\vec{y}|^{-n-1}\left\{(2n-1)!!m_{i_1}\ldots m_{i_n} - (2n-3)!!\delta_{i_1i_2}m_{i_3}\ldots m_{i_n} + \cdots\right\}$$

$$= (-)^n|\vec{x}-\vec{y}|^{-n-1}\sum_{k=0}^{N}{}'\left(2N-2k-1\right)!!(-)^k\delta_{i_1i_2}\ldots\delta_{i_{2k-1}i_{2k}}m_{i_{2k+1}}\ldots m_{i_n}, \tag{8.46}$$

where a sum Σ with a prime Σ' means that all permutations of the indices are added, for example,

$$\frac{\partial^n}{\partial x_i\partial x_j\partial x_k\partial x_n}\left(\frac{1}{\ell}\right) = \frac{1}{\ell^5}\Big[105m_im_jm_km_n + 3\left(\delta_{ij}\delta_{kn} + \delta_{ik}\delta_{jn} + \delta_{in}\delta_{jk}\right)$$

$$-15\left(m_im_j\delta_{kn} + m_im_k\delta_{jn} + m_im_n\delta_{jk} + m_jm_k\delta_{in} + m_jm_n\delta_{ik} + m_km_n\delta_{ij}\right)\Big], \tag{8.47}$$

is the order following (8.45).

8.3.2 Taylor Series for the Influence Function

Using the Laurent series (I.27.23a and b), it has been shown that a singular two-dimensional potential field can be decomposed in a superposition of multipoles (I.28.21). An analogous result is obtained in three dimensions via the Taylor series (I.23.32b) of the influence or Green function (8.11b), obtained in five steps: (1) the start is the Taylor series (8.48c) of an analytic function (8.48b) of a vector variable z_i relative to a fixed point a_i in a three-dimensional space (8.48a):

$$i,j = 1,2,3; \ f \in \mathcal{A}\left(|R^3|\right): \quad f\left(\vec{\ell}\right) = f\left(\vec{a}\right) + \left(\ell_i - a_i\right)\frac{\partial f}{\partial a_i} + \frac{1}{2}\left(\ell_i - a_i\right)\left(\ell_j - a_j\right)\frac{\partial^2 f}{\partial a_i \partial a_j} + \cdots. \tag{8.48a-c}$$

where the summation convention (8.42b) is used; (2) the fixed point is chosen as the observer position (8.49a) and the variable point as the position of the observer relative to the source (8.49b), leading from (8.48c) to (8.49c):

$$\vec{a} \equiv \vec{x}, \ \vec{\ell} \equiv \vec{x} - \vec{y}: \quad f\left(\vec{x} - \vec{y}\right) = f\left(\vec{x}\right) - y_i\frac{\partial f}{\partial x_i} + \frac{1}{2}y_iy_j\frac{\partial^2 f}{\partial x_i \partial x_j} - \frac{1}{3!}y_iy_jy_k\frac{\partial^2 f}{\partial x_i \partial x_j \partial x_k} + \cdots; \tag{8.49a-c}$$

(3) the influence function (8.11b) is analytic if the observation point is outside the source region (Figure 8.3), and thus the condition (8.50a) is sufficient for the convergence of (8.50b) the Taylor series (8.49c) applied to (8.11b):

$$|\vec{x}| > |\vec{y}|: \quad -4\pi G(\vec{x} - \vec{y}) = \frac{1}{|\vec{x} - \vec{y}|} = \frac{1}{|\vec{x}|} - y_i \frac{\partial}{\partial x_i}\left(\frac{1}{|\vec{x}|}\right) + \frac{1}{2} y_i y_j \frac{\partial}{\partial x_i \partial x_j}\left(\frac{1}{|\vec{x}|}\right) + O(y_i y_j y_k); \quad (8.50a,b)$$

(4) using (8.43b) [(8.44)] with $\vec{y} = 0$ in the second (third) term on the r.h.s. of (8.50b) leads to (8.51b):

$$|\vec{y}| < |\vec{x}| \equiv R: \quad -4\pi G(\vec{x} - \vec{y}) = \frac{1}{|\vec{x} - \vec{y}|} = \frac{1}{|\vec{x}|} - \frac{x_i y_i}{|\vec{x}|^3} + \frac{x_i y_i}{2} \frac{3 x_i y_i - |\vec{x}|^2 \delta_{ij}}{|\vec{x}|^5} + O\left(\frac{y_i y_j y_k}{|\vec{x}|^4}\right)$$

$$= \frac{1}{R} + \frac{\vec{x} - \vec{y}}{R^3} + \frac{3(\vec{x} \cdot \vec{y})^2 - R^2 |\vec{y}|^2}{R^5} + O\left(\frac{|\vec{y}|^3}{R^4}\right), \quad (8.51a,b)$$

where the distance of the observer from the origin (8.51a) is larger than for any source; and (5) substituting the influence function (8.51b) in the potential (8.12b) and using the unit radial vector (8.22b) leads to

$$4\pi\varepsilon\Phi(\vec{x}) = \frac{1}{R} \int_D e(\vec{y}) d^3\vec{y} + \frac{1}{R^2} \int_D (\vec{y} \cdot \vec{e}_R) e(\vec{y}) d^3\vec{y}$$

$$+ \frac{1}{2R^3} \int_D \left\{ 3(\vec{y} \cdot \vec{e}_R)^2 - y^2 \right\} e(\vec{y}) d^3\vec{y} + O\left(\frac{1}{R^4}\right). \quad (8.52)$$

This specifies the first three terms of the multipolar expansion, which are written explicitly, and correspond to the monopole, dipole, and quadrupole considered next (Subsection 8.3.3).

8.3.3 Multipolar Expansion for an Arbitrary Electric Charge Distribution

The lowest-order term of the multipolar expansion (8.52) that dominates at a large distance,

$$\Phi_0(R) = \frac{P_0}{4\pi\varepsilon R}, \quad P_0 \equiv \int_D e(\vec{y}) d^3\vec{y}, \quad (8.53a,b)$$

is (Figure 8.1) a three-dimensional monopole (8.53a) $\equiv$ (8.18c) concentrated at the origin, with charge (8.53b) equal to the total charge in the distribution; it decays like the inverse of the distance from the origin in (8.22a). The associated electric field is radial (8.19a and b; 8.20a and b; 8.21a and b). If the total charge is zero, $P_0 = 0$, in (8.53b), the leading term in the multipolar expansion (8.52) for the electrostatic potential becomes (Figure 8.5) a three-dimensional dipole:

$$\Phi_1(\vec{x}) = \frac{\vec{P} \cdot \vec{e}_R}{4\pi\varepsilon R^2}, \quad \vec{P} \equiv \int_D \vec{y} e(\vec{y}) d^3\vec{y}; \quad (8.54a,b)$$

the dipole moment (8.54b) is the first moment of the charge distribution. The electrostatic potential of the dipole (8.54a) $\equiv$ (8.24a and b) (1) decays like R^{-2} instead of like R^{-1} for the monopole; (2) is not isotropic and leads to an axisymmetric electric field (8.22c; 8.25a through d; 8.26a through e).

If both the total electric charge (8.53b) and its dipole moment (8.54b) vanish, the multipolar expansion of the electrostatic potential (8.52) starts with the third term:

$$\Phi_2(\vec{x}) = \frac{P_{ij} e_{Ri} e_{Rj}}{4\pi\varepsilon R^3}, \quad P_{ij} \equiv \frac{1}{2} \int_D \left(3 y_i y_j - y^2 \delta_{ij}\right) e(\vec{y}) d^3\vec{y}; \tag{8.55a,b}$$

in the electrostatic potential (8.55a) of a quadrupole (Figure 8.6) appears a quadratic form involving the quadrupole tensor moment (8.55b). It decays like R^{-3}, and since it has two axes, it is generally not isotropic, as shown in (8.34a and b; 8.35a and b) [(8.36a through e)] for any observer position [observer in the plane of the quadrupole leading to the electric field (8.37a through c; 8.38a through c)]. The monopole (dipole) potential in electrostatics (8.53a) [(8.13a) ≡ (8.54a)] coincides with those of the monopole (6.130a) [dipole (6.163c)] in a potential flow, apart from the opposite sign due to the distinct convention for the electrostatic field (6.322b; 6.323b) [potential flow (6.322a; 6.323a)] that leads to the same expressions for the electric field (velocity). The potential (8.55a) of an axial quadrupole P_{zz} in (8.34a) in electrostatics [(6.170a) in a potential flow] also coincide apart from sign $P_2 = -P_{zz}$, leading to the same electric field (velocity) in terms of the quadrupole moment $P_{zz}(P_2)$.

Substituting (8.46) in the general nth order term of the Taylor series (8.50b),

$$\frac{1}{|\vec{x} - \vec{y}|} = \sum_{n=0}^{\infty} \frac{(-)^n}{n!} y_{i_1} \ldots y_{i_n} \frac{\partial^n}{\partial x_{i_1} \ldots \partial x_{i_n}} \frac{1}{|\vec{x}|}$$

$$= \sum_{n=0}^{\infty} \frac{|\vec{x}|^{-n-1}}{n!} y_{i_1} \ldots y_{i_n} \sum_{k=0}^{n} (2n - 2k - 1)!! (-)^k \delta_{i_1 i_2} \ldots \delta_{i_{2k-1} i_{2k}} e_{Ri_{2k+1}} \ldots e_{Ri_{2n}}, \tag{8.56}$$

leads (8.13a) to the electrostatic potential (8.57a):

$$\Phi(\vec{x}) = \sum_{n=0}^{\infty} \Phi_n(\vec{x}); \quad \Phi_n(\vec{x}) = \frac{|\vec{x}|^{-n-1}}{4\pi\varepsilon} P_{i_1 \ldots i_n} e_{Ri_1} \ldots e_{Ri_n}, \tag{8.57a,b}$$

consisting of a superposition of multipoles (8.57b) with moments:

$$P_{i_1 \ldots i_n} = \frac{1}{n!} \sum_{k=0}^{\leq n/2} (2n - 2k - 1)!! (-)^k \delta_{i_1 i_2} \ldots \delta_{i_{2k-1} i_{2k}} \int_D y_{i_{2k+1}} \ldots y_{i_n} e(\vec{y}) d^3\vec{y}. \tag{8.57c}$$

The electrostatic potential (8.57b) of an arbitrary multipole (8.57c) is considered next together with the monopole (8.53a and b), dipole (8.54a and b), and quadrupole (8.55a and b), which are the three lowest-order terms in the multipolar expansion (8.57a).

8.3.4 Multipole of Any Order and Asymptotic Decay

It has been shown that *an arbitrary distribution of charges* $e(\vec{y})$ *in three-dimensional space leads to a potential (8.13a), which can be represented, for an observer at any point* $(\vec{x})$, *as due to a superposition (8.57a) ≡ (8.58a) of multipoles at the origin:*

$$\Phi(\vec{x}) = \sum_{n=1}^{\infty} \Phi_n(\vec{x}), \quad \Phi_n(\vec{x}) \sim O\left(|\vec{x}|^{-n-1}\right), \tag{8.58a,b}$$

each decaying faster than the preceding (8.57b) ≡ (8.58b) so that at a large distance, only the non-vanishing multipole of lowest-order specifies the field: (i) the lowest or zero-th order is a monopole (Figure 8.1), with potential (8.53a) due to the total charge (8.53b), leading to an isotropic or radial electric field (8.19a and b; 8.20a and b; 8.21a and b); (ii) the first-order term is a dipole (Figures 8.2 and 8.4) whose potential (8.54a) involves a vector dipole moment (8.54b), which appears in the electric field (8.23a and b; 8.25a through d; 8.26a through e); (iii) the second-order term is a quadrupole (Figure 8.6) whose potential (8.55a) involves a tensor moment (8.55b) with components (8.34a and b; 8.35a and b), of which three affect the observer in the plane of the quadrupole (8.36a through d) and lead to the electric field (8.37a through c; 8.38a through c); and (iv) the m-th order term is a three-dimensional 2^n-pole that (iv-1) decays (8.57b) ≡ (8.58b) like R^{-n} (R^{-1-n}) for the electrostatic potential (electric field) is space $N = 3$ (the corresponding decay laws in the plane $N = 2$ and higher $N \geq 4$ dimensions appear in Table 9.1); (iv-2) the corresponding multipole moment is a tensor (8.57c) with n indices; (iv-3) the associated electric charge distribution is the product with contraction of (8.57c) with the n-th derivative of the Dirac unit impulse $\delta(\vec{x} - \vec{y})$ multiplied by $(-)^m/n!$ as in its Taylor series (I.23.32b) for the variable x_i:

$$e_n(\vec{y}) = \frac{(-)^n}{n!} \left\{ P_{i_1 \ldots i_n} \frac{\partial^n}{\partial x_{i_1} \ldots \partial x_{i_n}} \right\} \delta(\vec{x}); \tag{8.59}$$

(iv-4) the 2^n-pole potential satisfies the Poisson equation (8.12a) forced by (8.59):

$$-\varepsilon \frac{\partial^2 \Phi_n}{\partial x_i \partial x_i} = \frac{(-)^n}{n!} P_{i_1 \ldots i_n} \frac{\partial^n}{\partial x_{i_1} \ldots \partial x_{i_n}} \delta(\vec{x}); \tag{8.60}$$

(iv-5) the 2^n-pole potential is specified by the n-th order derivates of the influence or Green function (8.13a) concentrated at the origin with the multipole moment

$$4\pi\varepsilon\Phi_n(\vec{x}) = \frac{(-)^n}{n!} P_{i_1 \ldots i_n} \frac{\partial^n}{\partial x_{i_1} \ldots \partial x_{i_n}} \frac{1}{|\vec{x}|}; \tag{8.61}$$

(iv-6) the corresponding electric field of the 2^n-pole is given by

$$E_i^{(n)} = -\frac{\partial \Phi_n}{\partial x_i} = \frac{1}{4\pi\varepsilon} \frac{(-)^{n+1}}{n!} P_{i_1 \ldots i_n} \frac{\partial^{n+1}}{\partial x_{i_1} \ldots \partial x_{i_n} \partial x_i} \frac{1}{|\vec{x}|}; \tag{8.62}$$

and (iv-7) the axial component of the electric field of the 2^n-pole coincides with the potential of the 2^{n+1}-pole multiplied by $-(n+1)$:

$$e_{zi} = \delta_{zi}, P_{i_1 \ldots i_n i} = e_{zi} P_{i_1 \ldots i_n} : \quad E_z^{(n)} \equiv \vec{e}_z \cdot \vec{E}^{(n)} = -\frac{\partial \Phi_n}{\partial z} = -(n+1) \Phi^{(n+1)}(\vec{x}), \tag{8.63a–c}$$

*specifying the **rule of successive generation of multipoles**.* The latter also applies in the plane, for example (I.12.52a,b). The passage from (8.60) to (8.61) uses (8.13a):

$$n!4\pi\varepsilon\Phi_n(\vec{x}) = (-)^n \int_{-\infty}^{+\infty} \frac{1}{|\vec{x} - \vec{y}|} \left\{ P_{i_1 \ldots i_n} \frac{\partial^n}{\partial y_{i_1} \ldots \partial y_{i_n}} \right\} \delta(\vec{y}) d^3\vec{y}$$

$$= \lim_{\vec{y} \to 0} \left\{ P_{i_1 \ldots i_n} \frac{\partial^n}{\partial y_{i_1} \ldots \partial y_{i_n}} \right\} \frac{1}{|\vec{x} - \vec{y}|} = (-)^n \lim_{\vec{y} \to 0} \left\{ P_{i_1 \ldots i_n} \frac{\partial^n}{\partial x_{i_1} \ldots \partial x_{i_n}} \right\} \frac{1}{|\vec{x} - \vec{y}|}, \tag{8.64}$$

where the integration property of the nth derivative of the unit impulse (3.22a through c) in three-dimensional vector form was used to prove (8.64) $\equiv$ (8.62). The passage from (8.61; 8.62) to (8.63c)

$$E_z^{(n)} \equiv e_{zi} E_i^{(n)} = \frac{1}{4\pi\varepsilon}(-)^{n+1} e_{zi}\left\{P_{i_1\ldots i_n}\frac{\partial^{n+1}}{\partial x_{i_1}\ldots\partial x_{i_n}\partial x_i}\right\}\frac{1}{|\vec{x}|}$$

$$= -\frac{n+1}{4\pi\varepsilon}\frac{(-)^{n+1}}{(n+1)!}\left\{P_{i_1\ldots i_n i}\frac{\partial^{(n+1)}}{\partial x_{i_1}\ldots\partial x_{i_n}\partial x_i}\right\}\frac{1}{|\vec{x}|} = -(n+1)\Phi^{(n+1)}(\vec{x}) \qquad (8.65)$$

projects the field in the $\vec{e}_z$-direction (8.63a) and uses the component $\vec{e}_z$ of the multipole of order $n + 1$ with the remaining components corresponding (8.65) to the moment of the dipole of order n.

8.3.5 Rule of Successive Generation of Multipoles of All Orders

The last remark indicates the method of generation of the axial components of the multipoles of all orders using the derivative (8.66b) along the z-axis (6.51c) $\equiv$ (8.66a) in spherical coordinates (6.23e):

$$\vec{e}_z = \vec{e}_R\cos\theta - \vec{e}_\theta\sin\theta: \quad \frac{\partial}{\partial z} = \vec{e}_z\cdot\nabla = \cos\theta\frac{\partial}{\partial R} - \frac{\sin\theta}{R}\frac{\partial}{\partial\theta}. \qquad (8.66a,b)$$

The sequence of axial multipoles may be obtained as follows: (1) the potential of the monopole (8.53b) $\equiv$ (8.67a) leads to the electric field (8.67b)

$$\Phi_0(R) = \frac{P_0}{4\pi\varepsilon R}, \quad \vec{E}_0(R) = -\nabla\Phi_0 = -\frac{\partial\Phi}{\partial R} = \vec{e}_R\frac{P_0}{4\pi\varepsilon R^2} \qquad (8.67a,b)$$

and its axial component

$$E_{0z}(R,\theta) = \vec{e}_z\cdot\vec{E}_0 = \frac{P_0}{4\pi\varepsilon R^2}\cos\theta = -\frac{\partial\Phi_0}{\partial z}; \qquad (8.67c)$$

(2) the axial component of the monopole (8.67c) electric field coincides with the potential of the dipole (8.68b) replacing (8.68a) the monopole P_0 by the dipole moment P_1:

$$P_0 \to P_1: \quad \Phi_1(R,\theta) = \frac{P_1}{4\pi\varepsilon R^2}\cos\theta, \quad E_{1z}(R,\theta) = -\frac{\partial\Phi_1}{\partial z} = \frac{P_1}{4\pi\varepsilon R^3}(3\cos^2\theta - 1), \qquad (8.68a\text{–}c)$$

leading to the axial component of the electric field (8.68c) $\equiv$ (8.26e); (3) replacing (8.69a) in (8.68c) and (8.63c) and dividing by 2 specifies the potential of a longitudinal quadrupole (8.67b) [$\equiv$third term of (8.26e)]:

$$2P_1 \to P_2 \equiv P_{zz}: \quad \Phi_2(R,\theta) = \frac{P_2}{8\pi\varepsilon R^3}(3\cos^2\theta - 1), \qquad (8.69a,b)$$

$$E_{2z}(R,\theta) = -\frac{\partial\Phi_2}{\partial z} = \frac{P_2}{8\pi\varepsilon R^4}3\cos\theta(5\cos^2\theta - 3), \qquad (8.69c)$$

and the corresponding axial electric field (8.69c) [≡second term of (8.38c)]; and (4) replacing (8.70a) in (8.69c) and (8.63c) and dividing by 3 specifies the potential of an octupole of moment P_3:

$$3P_2 \to P_3 \equiv P_{zzz}: \quad \Phi_3(R,\theta) = \frac{P_3}{8\pi\varepsilon R^4}\cos\theta\left(5\cos^2\theta - 3\right), \tag{8.70a,b}$$

and so on. For example, in the passage (1) from (8.67a) to (8.67c) the equation used was:

$$-\frac{\partial\Phi_0}{\partial z} = -\cos\theta\frac{\partial}{\partial R}\left(\frac{P_0}{R}\right) = \frac{P_0}{R^2}\cos\theta; \tag{8.71a}$$

(2) from (8.67c) to (8.69c):

$$-\frac{\partial}{\partial z}\left(\frac{\cos\theta}{R^2}\right) = \left(-\cos\theta\frac{\partial}{\partial R} + \frac{\sin\theta}{R}\frac{\partial}{\partial\theta}\right)\frac{\cos\theta}{R^3} = \frac{2\cos^2\theta}{R^3} - \frac{\sin^2\theta}{R^3} = \frac{3\cos^2\theta - 1}{R^3}; \tag{8.71b}$$

and (3) from (8.69b) to (8.70b):

$$-\frac{\partial}{\partial z}\left(\frac{3\cos^2\theta - 1}{R^3}\right) = \left(-\cos\theta\frac{\partial}{\partial R} + \frac{\sin\theta}{R}\frac{\partial}{\partial\theta}\right)\frac{3\cos^2\theta - 1}{R^3}$$

$$= \frac{3\cos\theta}{R^4}\left(3\cos^2\theta - 1 - 2\sin^2\theta\right) = \frac{3\cos\theta}{R^4}\left(5\cos^2\theta - 3\right), \tag{8.71c}$$

with the axial differentiation operator (8.66b) appearing in all cases.

8.3.6 Spherical Harmonics and Legendre Polynomials

The influence or Green function (8.11b) has the multipolar expansion (8.51b), that is,

$$-4\pi G(\vec{x};\vec{y}) = \frac{1}{|\vec{x}-\vec{y}|} = \frac{1}{|\vec{x}|} + \frac{\vec{x}\cdot\vec{y}}{|\vec{x}|^2} + \frac{3\left(\vec{x}\cdot\vec{y}\right)^2 - |\vec{x}|^2|\vec{y}|^2}{2|\vec{x}|^5} + \frac{\left(\vec{x}\cdot\vec{y}\right)\left[5\left(\vec{x}\cdot\vec{y}\right) - 3|\vec{x}|^2|\vec{y}|^2\right]}{2|\vec{x}|^7} + O\left(\frac{|\vec{y}|^4}{|\vec{x}|^5}\right)$$

$$= \frac{1}{R} + \frac{y}{R}\cos\theta + \frac{y^2}{R^3}\frac{3\cos^2\theta - 1}{2} + \frac{y^3}{R^4}\frac{\cos\theta\left(5\cos^2\theta - 3\right)}{2} + O\left(\frac{y^4}{R^5}\right), \tag{8.72a–c}$$

where (8.72d) [(8.72e)] is the distance of the observer (source) from the origin and θ the angle between their position vectors (8.72f):

$$R = |\vec{x}|, \quad y = |\vec{y}|, \quad \vec{x}-\vec{y} = Ry\cos\theta. \tag{8.72d–f}$$

The first four terms were written explicitly in (8.72b) ≡ (8.72c) and correspond, respectively, to the monopole (8.67a), dipole (8.68b), quadrupole (8.69b), and octupole (8.70b) terms in the potential (8.12b) ≡ (8.52) ≡ (8.57a) ≡ (8.73):

$$4\pi\varepsilon\Phi(\vec{x}) = 4\pi\varepsilon\sum_{n=0}^{\infty}\Phi_n(\vec{x}) = \frac{P_0}{R} + P_1 + P_2\frac{\cos\theta}{R^3}\frac{3\cos^2\theta - 1}{2R^3} + P_3\frac{\cos\theta\left(5\cos^2\theta - 3\right)}{2R^4} + O\left(\frac{P_4}{R^5}\right). \tag{8.73}$$

The coefficients of the multipole moments correspond to the **far-field expansion in spherical harmonics**:

$$|\vec{x}| \equiv R > y \equiv |\vec{y}|: \quad |\vec{x} - \vec{y}|^{-1} = |R^2 - 2Ry\cos\theta + y^2|^{-1/2} = \sum_{m=1}^{\infty} y^n R^{-1-n} P_n(\cos\theta), \qquad (8.74a,b)$$

whose coefficients are the **Legendre** *(1785)* **polynomials** *of degree n; comparison of (8.72c) ≡ (8.73) with (8.74) identifies the first four Legendre polynomials:*

$$m = 0,1,2,3: \quad P_n(\cos\theta) = \left\{ 1, \cos\theta, \frac{3\cos^2\theta - 1}{2}, \cos\theta\frac{5\cos^2\theta - 3}{2} \right\}. \qquad (8.75a\text{–}d)$$

They can be obtained by recurrence from using the **differentiation formula**

$$(n+1)\left[P_{n+1}(\cos\theta) - \cos\theta P_n(\cos\theta) \right]$$

$$= \sin\theta\frac{d}{d\theta}\left[P_n(\cos\theta) \right] = -\sin^2\theta\frac{d\left[P_n(\cos\theta) \right]}{d(\cos\theta)}, \qquad (8.76)$$

which applies for all degrees n. The relation (8.73) between the electrostatic potential of a 2^n-pole or multipole of order n and (8.74b) the Legendre polynomial of degree n:

$$\Phi_n(\vec{x}) = \frac{n!}{4\pi\varepsilon} R^{-n-1} P_n(\cos\theta), \qquad (8.77)$$

leads to

$$(n+1)P_{n+1}(\cos\theta) = \frac{4\pi\varepsilon}{n!} R^{n+2}\Phi_{n+1}(\vec{x}) = -\frac{4\pi\varepsilon}{n!} R^{n+2}\frac{\partial}{\partial z}\left[\Phi_n(\vec{x}) \right]$$

$$= -R^{n+2}\left(\cos\theta\frac{\partial}{\partial R} - \frac{\sin\theta}{R}\frac{\partial}{\partial\theta} \right) R^{-n-1}P_n(\cos\theta)$$

$$= (n+1)\cos\theta P_n(\cos\theta) + \sin\theta\frac{d}{d\theta}\left[P_m(\cos\theta) \right] \qquad (8.78)$$

and proves (8.76), the coincidence of (1) the rule of generation of multipoles (8.63a through c); and (2) the differentiation formula (8.76) for Legendre polynomials. The Legendre polynomials are generalized to higher dimensions $N \geq 4$ as hyperspherical Legendre polynomials (Section 9.6).

8.3.7 Potentials and Field Functions for Multipoles of All Orders

The operator (8.66b) is the derivative $\partial/\partial z$ along the $\vec{e}_z$ axis in spherical coordinates, and it specifies the Legendre polynomials or spherical harmonics (8.79b):

$$\frac{\partial R}{\partial z} = \frac{z}{R} = \cos\theta, \quad P_n(\cos\theta) = (-)^n R^{n+1}\frac{\partial^n}{\partial z^n}\left(\frac{1}{R} \right), \qquad (8.79a,b)$$

where (8.79a) may be used; this implies that the potential (field function) of a 2^m-multipole, that is, a multipole of order n with moment P_n, is specified by (8.80a) [(8.80b)]:

$$\{\Phi_n, \Psi_n\}(\vec{x}) = \frac{(-)^n}{n!}\frac{P_n}{4\pi\varepsilon}\left\{ \frac{\partial^n}{\partial z^n}\left(\frac{1}{R} \right), \frac{\partial^{n+1}}{\partial z^{n+1}}(R) \right\}. \qquad (8.80a,b)$$

The potential (8.77) of the multipole of order n involves (1) the Legendre polynomial $P_n(\cos\theta)$ of degree n, in (8.74b; 8.75a through d; 8.76; 8.79b) that specifies the directivity of the potential; and (2) the constant P_n, which is the moment of the multipole of order n and specifies the magnitude of the potential, with a decay like R^{-1-n}. The potential (8.80a) [field function (8.80b)] for the axial multipole of order n corresponds to one of the moments, namely, that in the z-direction among the N^n-moments,

$$\{\Phi_n, \Psi_n\}(\vec{x}) = \frac{(-)^n}{n!} \sum_{i_1,\ldots,i_n=1}^{3} P_{i_1\ldots i_n} \frac{\partial^n}{\partial x_{i_1}\ldots\partial x_{i_n}}\left\{\frac{1}{R}, \frac{z}{R}\right\} \tag{8.81a,b}$$

in an N-dimensional space. The potential (8.81a) [field function (8.81b)] of the multipole of order n is a linear combination of partial derivatives of the monopole; the latter is the influence or Green function for the Poisson (8.82a) [modified Poisson (8.83a)] equation, which is the fundamental solution (8.82c) [(8.83c)] corresponding to the Laplace (6.88b) [modified Laplace (6.89b)] operator forced by a unit impulse:

$$\nabla^2\Phi_0(\vec{x}) = \delta(\vec{x}): \quad \nabla^2\left(\frac{1}{R}\right) = \frac{1}{R^2}\frac{\partial}{\partial R}\left[R^2\frac{\partial}{\partial R}\left(\frac{1}{R}\right)\right] = 0, \quad \Phi_0(\vec{x}) = -\frac{1}{4\pi R}, \tag{8.82a-c}$$

$$\bar{\nabla}^2\Psi_0(\vec{x}) = \delta(\vec{x}): \quad \bar{\nabla}^2(R) = \frac{\partial^2}{\partial R^2}(R) = 0, \quad \Psi_0(\vec{x}) = \frac{\partial}{\partial z}\left(-\frac{R}{4\pi}\right) = -\frac{z}{4\pi R} = -\frac{\cos\theta}{4\pi R}. \tag{8.83a-c}$$

In (8.82b) [(8.83b)] was used the Laplace (6.88b) [modified Laplace (6.89b)] operator in spherical coordinates, to obtain the influence function that coincides within a constant factor $-1/4\pi$ in (8.11b), or $P_0/(4\pi\varepsilon)\left[-Q/(4\pi)\right]$ in electrostatics (8.18b) [potential flow (6.130a)] for a monopole of moment $P_0(Q)$. The potential applies to the general (8.81a) [axial (8.80a)] multipoles that are solutions of the Laplace equation (6.88b) and specify multiaxial (uniaxial) multipoles [Subsections 6.4.6 through 6.4.8 (Subsections 8.3.5 through 8.3.8)]. The stream or field function leads to (8.81b), which are all solutions of the modified Laplace equation (6.89b); however, the field function exists only for axisymmetric fields and, thus, of all the solutions (8.81b) of the modified Laplace equation (8.89b), only the subset (8.80b) represents axial multipoles. The axial multipoles of four lowest orders are considered next in terms of both the potential and the stream/field function (Subsection 8.3.8).

8.3.8 Monopole and Axial Dipole, Quadrupole, and Octupole

The first four instances of (8.80a) [(8.80b)] are the potential (stream function) for (1) the monopole:

$$\Phi_0(R,\theta) = \frac{P_0}{4\pi\varepsilon R}, \quad \Psi_0(R,\theta) = \frac{P_0}{4\pi\varepsilon}\frac{\partial R}{\partial z} = \frac{P_0 z}{4\pi\varepsilon R} = \frac{P_0}{4\pi\varepsilon}\cos\theta \tag{8.84a,b}$$

in agreement with (8.53a) ≡ (8.67a) ≡ (8.84a) ≡ (6.130a) [(8.84b) ≡ (6.131a)], with $P_0/\varepsilon = -Q$ including a reversed sign; (2) the dipole:

$$\Phi_1(R,\theta) = -\frac{P_1}{4\pi\varepsilon}\frac{\partial}{\partial z}\left(\frac{1}{R}\right) = \frac{P_1 z}{4\pi\varepsilon R^3} = \frac{P_1}{4\pi\varepsilon R^2}\cos\theta, \tag{8.85a}$$

$$\Psi_1(R,\theta) = -\frac{P_1}{4\pi\varepsilon}\frac{\partial}{\partial z}\left(\frac{z}{R}\right) = -\frac{P_1}{4\pi\varepsilon R}\left(1 - \frac{z^2}{R^2}\right) = -\frac{P_1 r^2}{4\pi\varepsilon R^3} = -\frac{P_1}{4\pi\varepsilon R}\sin^2\theta, \tag{8.85b}$$

in agreement with (8.24a and b) ≡ (8.68b) ≡ (8.85a) ≡ (6.163c) [(8.85b) ≡ (6.164c)] with $-P_1/\varepsilon$ replaced by P_1; (3) the quadrupole:

$$\Phi_2(R,\theta) = -\frac{1}{2}\frac{P_2}{4\pi\varepsilon}\frac{\partial}{\partial z}\left(\frac{z}{R^3}\right) = -\frac{P_2}{8\pi\varepsilon R^3}\left(1-\frac{3z^2}{R^2}\right) = \frac{P_2}{8\pi\varepsilon R^3}\left(3\cos^2\theta-1\right), \tag{8.86a}$$

$$\Psi_2(R,\theta) = -\frac{1}{2}\frac{P_2}{4\pi\varepsilon}\frac{\partial}{\partial z}\left(\frac{z^2}{R^3}-\frac{1}{R}\right) = \frac{P_2}{8\pi\varepsilon}\left(-\frac{z}{R^3}-\frac{2z}{R^3}+\frac{3z^3}{R^5}\right)$$

$$= -\frac{3P_2 z}{8\pi\varepsilon R^3}\left(1-\frac{z^2}{R^2}\right) = -\frac{3P_2 z r^2}{8\pi\varepsilon R^5} = -\frac{3P_2}{8\pi\varepsilon R^2}\cos\theta\sin^2\theta, \tag{8.86b}$$

in agreement with (8.69b) ≡ (8.86a) ≡ (6.170a) [(8.86b) ≡ (6.174b)] with P_2/ε replaced by $-P_2$; and (4) the octupole:

$$\Phi_3(R,\theta) = -\frac{1}{3}\frac{P_3}{8\pi\varepsilon}\frac{\partial}{\partial z}\left(\frac{3z^2}{R^5}-\frac{1}{R^3}\right)$$

$$= -\frac{P_3}{24\pi\varepsilon}\left(\frac{6z}{R^5}-\frac{15z^3}{R^7}+\frac{3z}{R^5}\right) = \frac{P_3 z}{8\pi\varepsilon R^5}\left(\frac{5z^2}{R^2}-3\right)$$

$$= \frac{P_3}{8\pi\varepsilon R^4}\cos\theta\left(5\cos^2\theta-3\right), \tag{8.87a}$$

$$\Psi_3(R,\theta) = -\frac{1}{3}\frac{3P_3}{8\pi\varepsilon}\frac{\partial}{\partial z}\left(\frac{z^3}{R^5}-\frac{z}{R^3}\right) = \frac{P_3}{8\pi\varepsilon}\left(\frac{1}{R^3}-\frac{6z^2}{R^5}+\frac{5z^4}{R^7}\right)$$

$$= \frac{P_3}{8\pi\varepsilon R^3}\left(1-6\cos^2\theta+5\cos^4\theta\right). \tag{8.87b}$$

in agreement with (8.70b) ≡ (8.87a). The new result, namely, the stream function (8.87b) for the octupole, can be checked against the respective potential (8.87a) using (6.81a and b).

8.4 Vector/Scalar Potential and Solenoidal/Irrotational Field

The electro (magneto) static field has a scalar (vector) potential, that satisfies a scalar (vector) Poisson [Section 8.2 (Subsection 8.4.2)]; the same influence or Green function for the Laplace operator leads to the multipolar expansion for the electro (magneto) static field [Section 8.3 (Subsection 8.4.3)]. In two dimensions (Chapter I.26; Subsection 6.2.5), the vector potential reduces to a pseudo-scalar field function Ψ, that has different properties when compared with the scalar potential of an irrotational field. In three dimensions, the scalar field function does not exist generally, and when it does, namely, for axisymmetric fields (Section 6.2), it no longer satisfies the Laplace equation: it satisfies instead the modified Laplace equation. The general non-axisymmetric three-dimensional electrostatic (magnetostatic) field is specified by a scalar (vector) potential [Sections 8.1 through 8.3 (Sections 8.4 through 8.6)], leading to the scalar (vector) multipole expansion [Section 8.3 (Section 8.4)].

8.4.1 Vector Potential for a Solenoidal Field

The electrostatic field due to positive or negative charges (Chapter I.24), like the irrotational flow due to sources and sinks (Chapter I.12), the gravity field due to mass distributions (Chapter I.18), and the temperature in steady heat conduction (Chapter I.32) are all irrotational fields (8.88a) that are the gradient of a scalar potential (8.88b), satisfying the scalar Poisson equation (8.88c) with sources:

$$\nabla \wedge \vec{E} = 0, \quad \vec{E} = -\nabla \Phi, \quad \nabla^2 \Phi = -\frac{e}{\varepsilon} = -\nabla \cdot \vec{E}, \tag{8.88a–c}$$

namely, in the electrostatic (Chapter I.24) case, minus the electric charge e divided by dielectric permittivity ε of the medium. The incompressible flow due to vorticity (Chapter II.2) and the magnetic field due to electric currents (Chapter I.26) are *solenoidal fields (8.89a), which are the curl of a vector potential (8.89b). In the case of the magnetic field $\vec{H}$ in a medium of magnetic permeability* μ, *the first Maxwell equation (8.89a) implies the existence of a vector potential (8.89b):*

$$\nabla \cdot \left(\mu \vec{H} \right) = 0: \quad \mu \vec{H} = \nabla \wedge \vec{A}; \quad \frac{\vec{j}}{c} = \nabla \wedge \vec{H}: \quad \nabla^2 \vec{A} = -\frac{\mu}{c} \vec{j}, \tag{8.89a–d}$$

the second Maxwell equation (8.89c) involves the electric current $\vec{j}$ divided by the speed of light in vacuo c and for a medium with constant magnetic permeability (8.90b) leads to a Poisson equation (8.89d) for the vector potential. The latter can be obtained as follows: (1) Since the vector potential $\vec{A}$ is not uniquely defined by its curl (8.89b), the divergence can be specified, for example, it is set equal to zero by the **gauge condition** *(8.90a):*

$$\nabla \cdot \vec{A} = 0: \quad \mu = \text{const}: \quad \frac{\mu}{c} \vec{j} = \mu \nabla \wedge \vec{H} = \nabla \wedge \left(\nabla \wedge \vec{A} \right) = \left[\nabla \left(\nabla \cdot \vec{A} \right) - \nabla^2 \vec{A} \right] = -\nabla^2 \vec{A}, \tag{8.90a–c}$$

and (2) substitution of (8.89b) in the Maxwell equation (8.89c) then leads to the Poisson equation (8.90c) ≡ (8.89d) for a medium with constant magnetic permeability (8.90b), using (II.4.112) ≡ (6.48a).

8.4.2 Magnetic Field due to a Distribution of Electric Currents

In Cartesian coordinates, the vector Poisson equation (8.89d) ≡ (8.91a) is a set of three scalar equations:

$$\frac{\partial^2 A_i}{\partial x_j \partial x_j} = -\frac{\mu}{c} j_i, \vec{A}(\vec{x}) = -\frac{\mu}{c} \int_D \vec{j}(\vec{y}) G(\vec{x} - \vec{y}) \mathrm{d}^3 \vec{y}, \tag{8.91a,b}$$

for each component of the vector potential $\vec{A}$ and electric current $\vec{j}$, and thus its solution in unbounded space (8.92) ≡ (5.319b) is the vector analogue (8.91b) of (8.13a). The influence or Green function is the same (8.11b) as for Laplace operator:

$$\vec{A}(\vec{x}) = \frac{\mu}{4\pi c} \int_D \frac{\vec{j}(\vec{y})}{|\vec{x} - \vec{y}|} \mathrm{d}^3 y. \tag{8.92}$$

The curl has the property:

$$\nabla \wedge \left(\Phi \vec{U} \right) = \Phi \left(\nabla \wedge \vec{U} \right) + \nabla \Phi \wedge \vec{U}, \tag{8.93}$$

which is analogous to that of the divergence (II.6.7b), replacing $\nabla.$ by $\nabla\wedge$; it follows from the definition of curl (6.41a) Cartesian coordinates that

$$\nabla\wedge\left(\Phi\vec{U}\right)=\sum_{x,y,z}^{cycl}\vec{e}_x\left[\partial_y\left(U_z\Phi\right)-\partial_z\left(U_y\Phi\right)\right]=\Phi\sum_{x,y,z}^{cycl}\vec{e}_x\left(\partial_yU_z-\partial_zU_y\right)$$

$$+\sum_{x,y,z}^{cycl}\vec{e}_x\left(U_z\partial_y\Phi-U_y\partial_z\Phi\right)=\Phi\left(\nabla\wedge\vec{U}\right)+\nabla\Phi\wedge\vec{U},\tag{8.94}$$

using cyclic permutations of (x,y,z). The property (8.94) can be applied to the integrand in (8.92):

$$\frac{\partial}{\partial\vec{x}}\wedge\left(\frac{\vec{j}(\vec{y})}{|\vec{x}-\vec{y}|}\right)=-\vec{j}(\vec{y})\wedge\frac{\partial}{\partial\vec{x}}\left(\frac{1}{|\vec{x}-\vec{y}|}\right)=\frac{\vec{j}(\vec{y})\wedge(\vec{x}-\vec{y})}{|\vec{x}-\vec{y}|^3},\tag{8.95a}$$

where (8.42a through c; 8.43a and b) is used. Substituting (8.92) into (8.89b) and using (8.95a) specifies the magnetic field due to a distribution of electric currents:

$$\vec{H}(\vec{x})=\frac{1}{4\pi c}\frac{\partial}{\partial\vec{x}}\wedge\left(\int_D\frac{\vec{j}(\vec{y})}{|\vec{x}-\vec{y}|}d^3\vec{y}\right)$$

$$=\frac{1}{4\pi c}\int_D\left\{\frac{\partial}{\partial\vec{x}}\wedge\left[\frac{\vec{j}(\vec{y})}{|\vec{x}-\vec{y}|}\right]\right\}d^3\vec{y}$$

$$=\frac{1}{4\pi c}\int_D\frac{\vec{j}(\vec{y})\wedge(\vec{x}-\vec{y})}{|\vec{x}-\vec{y}|^3}d^3\vec{y}.\tag{8.95b}$$

The passage from (8.92) to (8.95b) can be made directly using the curl (5.185b) in index notation together with (8.43a):

$$4\pi cH_k=\frac{4\pi c}{\mu}\left(\nabla\wedge\vec{A}\right)_k=\frac{4\pi c}{\mu}e_{kmn}\partial_mA_n$$

$$=e_{kmn}\partial_m\left(\frac{j_n}{|\vec{x}-\vec{y}|}\right)=-|\vec{x}-\vec{y}|^{-3}e_{kmn}j_n\left(x_m-y_m\right)$$

$$=|\vec{x}-\vec{y}|^{-3}e_{knm}j_n\left(x_m-y_m\right)=\frac{\left[j\wedge(\vec{x}-\vec{y})\right]_k}{|\vec{x}-\vec{y}|^3}.\tag{6.95c}$$

Thus, *the vector potential (8.92) [magnetic field (8.96)]*:

$$\vec{H}(\vec{x})=\frac{1}{4\pi c}\int_D\frac{\vec{j}(\vec{y})\wedge(\vec{x}-\vec{y})}{|\vec{x}-\vec{y}|^3}d^3\vec{y},\tag{8.96}$$

is due to a distribution of electric currents $\vec{j}(\vec{y})$ over a region D with uniform magnetic permeability (8.90b) where c is the speed of light in vacuo.

8.4.3 Multipole Expansion for the Magnetostatic Field

Substituting the multipole expansion (8.50a and b) ≡ (8.72a through c) into (8.92) leads to *magnetic vector potential*:

$$|\vec{x}| > |\vec{y}|: \quad \frac{4\pi c}{\mu} \vec{A}(\vec{x}) = \frac{1}{|\vec{x}|} \int_D \vec{j}(\vec{y}) d^3\vec{y} + \frac{1}{|\vec{x}|^2} \int_D (\vec{e}_R \cdot \vec{y}) \vec{j}(\vec{y}) d^3\vec{y} + O\left(|\vec{j}| \frac{|\vec{y}|^2}{|\vec{x}|^3}\right), \quad (8.97a,b)$$

where the first (second) term represents the potential due to a magnetic dipole (quadrupole); these are the first two terms in a multipolar expansion:

$$A_i(\vec{x}) = \sum_{m=0}^{\infty} A_i^m(\vec{x}) = \frac{\mu}{4\pi c} \sum_{n=0}^{\infty} \left(P_{i_1 \dots i_n i} \frac{\partial^m}{\partial x_{i_1} \dots \partial x_{i_n}} \right) \frac{1}{|\vec{x}|}, \quad (8.98)$$

where the nth term is a multiplicity with (n + 1) indices specifying the moment of magnetic multipole of order n, that has (8.57c) as many components as an electric multipole of order n + 1:

$$P_{i_1 \dots i_n i} = \frac{1}{n!} \sum_{k=0}^{\leq n/2} {}'(2n - 2k - 1)!!(-)^k \delta_{i_1 i_2} \dots \delta_{i_{2k-1} i_{2k}} \int_D y_{i_{2k+1}} \dots y_{i_n} j_i(\vec{y}) d^3\vec{y}. \quad (8.99)$$

The two terms of lowest order (8.97a and b) in the multipolar expansion (8.98; 8.99) for the magneto-static field are the dipole and quadrupole considered next [Section 8.5 (Section 8.6)].

8.5 Point Current and Magnetic Dipole

The lowest-order term in the multipolar expansion for the magnetic vector potential is point current, for example, arising from moving electric charges (Subsection 8.5.2); it leads to a magnetic dipole (instead of electric monopole) for a moving (static) charge. The electric (Section 8.1) [magnetic (Subsection 8.5.1)] dipole has a field proportional to the projection of the vector moment along (across) the observer direction, with the same spatial decay.

8.5.1 Dipolar Vector Potential and Magnetic Field

The first term of the multipolar expansion (8.97b) for the magnetic potential

$$\vec{A}_0 = \frac{\mu \vec{J}}{4\pi c R}, \quad \vec{J} \equiv \int_D \vec{j}(\vec{y}) d^3\vec{y} \quad (8.100a,b)$$

is the vector potential (8.100a) due to the total electric current (8.100b). The magnetic field (8.89b) corresponding to the point current (8.100a) is

$$\vec{H}_0 = \frac{1}{\mu} \nabla \wedge \vec{A}_0 = \frac{1}{4\pi c} \nabla \wedge \left(\frac{\vec{J}}{R}\right) = -\frac{\vec{J}}{4\pi c} \wedge \nabla\left(\frac{1}{R}\right) = \frac{\vec{J} \wedge \vec{x}}{4\pi c R^3}, \quad (8.101)$$

where (8.93; 8.95a) were used to show that *since the vector potential of a current is parallel to the electric current (8.100a), the magnetic field (Figure 8.7) is orthogonal $\vec{N}$ to (8.102c) the plane of the electric current and position vector (8.102a and b); it decays with the square of the distance (8.102d) and is proportional to the current projected on the direction transverse to the position vector:*

$$\vec{N} \cdot \vec{J} = 0 = \vec{N} \cdot \vec{x}: \quad \vec{J} \wedge \vec{x} = JR\vec{N} \sin\theta, \quad \vec{H}_0 = \frac{J}{4\pi c R^2} \sin\theta \vec{N}; \quad (8.102a{-}d)$$

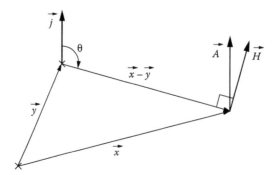

FIGURE 8.7 A point electric current $\vec{j}$ at a source position $\vec{y}$ creates at an observer position $\vec{x}$ a parallel magnetic potential $\vec{A}$; the curl of the latter specifies the magnetic induction $\vec{B}$ that is parallel to the magnetic field $\vec{H}$. The latter is orthogonal to the source current $\vec{j}$ and to the relative position vector $\vec{x} - \vec{y}$ from its location to the observer.

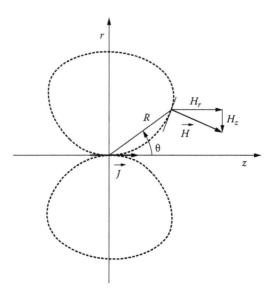

FIGURE 8.8 The two lobes for the magnetic (electric) field [Figure 8.8 (Figure 8.5)] due to a point electric current (Figure 8.7) [electric dipole (Figure 8.3)] are mutually orthogonal, implying in particular that the magnetic (electric) field is maximum transverse to (along) the axis, and zero along (transverse to) the axis.

thus the magnetic field (8.102c) is zero (maximum) (Figure 8.8) in the direction of (orthogonal to) the current $\theta = 0$ ($\theta = \pi/2$), where θ denotes the angle between the current and the position vector in (8.102c). The outer (inner) product in (8.102c) [(8.10b)] and the factor sin θ *in (8.102d) [cos θ in (8.10a)] emphasize the difference between the magnetic (Subsection 8.5.1; Figures 8.7 and 8.8) [electric (Subsection 8.1.2) and Figures 8.2 and 8.5] dipole, that is, one is maximum in the direction in which the other is zero.*

8.5.2 Electric Current due to a Moving Electric Charge

An electric charge e moving at velocity $\vec{v}$ produces an electric current (8.103a):

$$\vec{j} = e\vec{v} = \frac{e}{m}\vec{p}, \quad \vec{p} \equiv m\vec{v}, \qquad (8.103a,b)$$

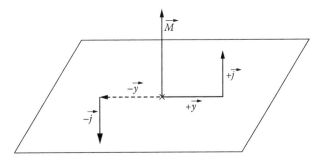

FIGURE 8.9 Two antiparallel electric currents (forces) correspond to a magnetic moment (binary moment of forces). The magnetic moment is related to the magnetic field of a quadrupole.

which is proportional to the **linear momentum** (8.103b) where m is the mass. The **angular momentum** (8.104a)

$$\vec{L} = \vec{y} \wedge \vec{p} = m\vec{y} \wedge \vec{v}, \quad \vec{M} = \frac{e}{m}\vec{L} = \frac{e}{m}\vec{y} \wedge \vec{p} = e\vec{y} \wedge \vec{v} = \vec{y} \wedge \vec{j} \tag{8.104a,b}$$

is related in the same way (Figure 8.9) to the **magnetic moment** (8.104b), which is the moment of the electric current. Next will be shown the analogy between electric current (8.103a) [magnetic moment (8.104b)] in specifying the first (second) term in the multipolar expansion for the vector potential, that is, [Section 8.5 (Section 8.6)] the magnetic dipole (quadrupole).

8.6 Quadrupolar Magnetic Potential and Field

The second term in the multipolar expansion for the vector potential specifies the magnetic quadrupole (Subsection 8.6.1) and associated magnetic field (Subsection 8.6.2).

8.6.1 Time Average of the Vector Potential

The second term of the multipolar expansion (8.97b) for the vector potential specifies a quadrupole:

$$\vec{A}_1 = \frac{\mu}{4\pi c R^3} \int_D (\vec{x} \cdot \vec{y}) \vec{j}(\vec{y}) \mathrm{d}^3\vec{y}; \tag{8.105}$$

the magnetic moment (8.104b) can be introduced in the integrand of (8.105) using

$$\vec{M} \wedge \vec{x} = (\vec{y} \wedge \vec{j}) \wedge \vec{x} = \vec{j}(\vec{y} \cdot \vec{x}) - (\vec{j} \cdot \vec{x})\vec{y}, \tag{8.106}$$

where the double outer vector product is used. This leads to the vector potential for the magnetic quadrupole:

$$\frac{8\pi c R^3}{\mu} \vec{A}_1 = 2 \int_D (\vec{x} \cdot \vec{y}) \vec{j}(\vec{y}) \mathrm{d}^3\vec{y}$$

$$= \int_D (\vec{x} \cdot \vec{y}) \vec{j}(\vec{y}) \mathrm{d}^3\vec{y} + \int_D \left[\vec{M} \wedge \vec{x} + (\vec{j} \cdot \vec{x})\vec{y}\right] \mathrm{d}^3\vec{y}$$

$$= \int_D (\vec{M} \wedge \vec{x}) \mathrm{d}^3\vec{y} + \int_D \left[\vec{j}(\vec{x} \cdot \vec{y}) + (\vec{j} \cdot \vec{x})\vec{y}\right] \mathrm{d}^3\vec{y}. \tag{8.107}$$

The electric current is given by (8.103a) $\equiv$ (8.108a) for the where dot means time derivative (8.108b):

$$\vec{j} = e\vec{v} = e\frac{d\vec{y}}{dt} = e\dot{\vec{y}}, \quad \dot{\vec{y}} \equiv \frac{d\vec{y}}{dt}. \tag{8.108a,b}$$

Using (8.108a) simplifies the last term in (8.107) to

$$\frac{1}{e}\left[\vec{j}(\vec{x}\cdot\vec{y}) + (\vec{j}\cdot\vec{x})\vec{y}\right] = \dot{\vec{y}}(\vec{y}\cdot\vec{x}) + \vec{y}(\dot{\vec{y}}\cdot\vec{x}) = \frac{d}{dt}\left[(\vec{y}\cdot\vec{x})\vec{y}\right], \tag{8.109}$$

the total derivative at the source $\vec{y}$ with regard to time of the function in curly brackets. *The* **time average** *or mean value over time* <...> *of an integrable function (8.110a) is defined by (8.110b):*

$$f(t) \in \mathcal{E}(|R): \quad \langle f(t) \rangle \equiv \lim_{T\to\infty} \frac{1}{2T} \int_{-\infty}^{+\infty} f(t)dt. \tag{8.110a,b}$$

This implies *that the mean value over time of the derivative of a function (8.108a) is zero (8.108b):*

$$f(t) \in \mathcal{D}(|R): \quad \left\langle \frac{df}{dt} \right\rangle \equiv \lim_{T\to\infty} \frac{1}{2T} \int_{-\infty}^{+\infty} \frac{df}{dt}dt = \lim_{T\to\infty} \frac{f(T)-f(-T)}{2T} = 0, \tag{8.111a,b}$$

since a function differentiable on the whole real line is bounded at infinity in both directions.

In order to consider a steady field, which is time independent, the average over time is taken in (8.107) leading by (8.111b) $\equiv$ (8.112a) to (8.112b):

$$\left\langle \left[\vec{j}(\vec{y}\cdot x) + (\vec{j}\cdot x)\vec{y}\right] \right\rangle = e\left\langle \frac{d}{dt}\left[\vec{j}(\vec{y}\cdot x)\right] \right\rangle = 0 : \frac{\mu}{8\pi cR^3} \int_D \left[\left(\langle\vec{M}\rangle\right) \wedge \vec{x}\right]d^3\vec{y} = \langle\vec{A_1}\rangle; \tag{8.112a,b}$$

the latter is equivalent to stating that

$$\langle\vec{A_1}\rangle = \frac{\mu\vec{Q}\wedge\vec{x}}{8\pi cR^3}, \quad \vec{Q} \equiv \int_d \langle\vec{M}\rangle d^3\vec{y} = \int_D \langle\vec{y}\wedge\vec{j}\rangle d^3\vec{y}; \tag{8.113a,b}$$

the time average (8.110b) of vector potential due to a magnetic quadrupole is given by (8.113a), where (8.113b) is the **total average magnetic moment** *defined as the magnetic moment (8.104b) averaged (8.110a and b) over time* <...>, *and integrated over the distribution of electric currents.*

8.6.2 Magnetic Field and Moment of a Quadrupole

The vector potential (8.113a) may be rewritten (8.114b) as

$$\nabla\left(\frac{1}{R}\right) = -\frac{\vec{x}}{R^3} : \frac{8\pi c}{\mu}\langle\vec{A_1}\rangle = \vec{Q}\wedge\frac{\vec{x}}{R^3} = -\vec{Q}\wedge\nabla\left(\frac{1}{R}\right) = \nabla\wedge\left(\frac{\vec{Q}}{R}\right), \tag{8.114a,b}$$

where particular cases are considered: (i) $\vec{y} = 0$ of (8.43a and b) $\equiv$ (8.114a); (2) $\vec{U} = \vec{Q}$ constant in (8.93). The corresponding magnetic field (8.89b) is

$$8\pi c \langle \vec{H}_1 \rangle = \frac{8\pi c}{\mu} \langle \nabla \wedge \vec{A}_1 \rangle = \nabla \wedge \left[\nabla \wedge \left(\frac{\vec{Q}}{R} \right) \right] = -\nabla^2 \left(\frac{\vec{Q}}{R} \right) + \nabla \left[\nabla \cdot \left(\frac{\vec{Q}}{R} \right) \right], \tag{8.115}$$

using the identity for the vector Laplacian (II.6.112) $\equiv$ (6.48a). The first term on the r.h.s. of (8.115) is zero (8.116):

$$\vec{Q} = \text{const}: \quad \nabla^2 \left(\frac{\vec{Q}}{R} \right) = \vec{Q} \nabla^2 \left(\frac{1}{R} \right) = 0, \tag{8.116a,b}$$

for constant $\vec{Q}$, because $1/R$ coincides within a multiplying constant with the influence or Green function (8.11a and b) for the Laplace operator, and hence using (6.46a), it follows that $1/R$ is an harmonic function (8.116c) for $R \neq 0$:

$$\nabla^2 \left(\frac{1}{R} \right) = \frac{1}{R^2} \frac{\partial}{\partial R} \left[R^2 \frac{\partial}{\partial R} \left(\frac{1}{R} \right) \right] = \frac{1}{R^2} \frac{\partial}{\partial R} (-1) = 0; \tag{8.116c}$$

the remaining term, that is, the last on the r.h.s. of (8.115), is evaluated by (8.44):

$$\vec{m} = \vec{e}_R: \quad \left\{ \nabla \left[\nabla \cdot \left(\frac{\vec{Q}}{R} \right) \right] \right\}_i = \frac{\partial}{\partial x_i} \left[\nabla \cdot \left(\frac{\vec{Q}}{R} \right) \right] = \frac{\partial^2}{\partial x_i \partial x_j} \left(\frac{Q_j}{R} \right) = Q_j \frac{\partial^2}{\partial x_i \partial x_j} \left(\frac{1}{R} \right)$$

$$= \frac{Q_j}{R^3} (3 e_{Ri} e_{Rj} - \delta_{ij}) = \frac{1}{R^3} \left[3 (\vec{Q} \cdot \vec{e}_R) e_{Ri} - Q_i \right], \tag{8.117a,b}$$

using (8.43a) $\equiv$ (8.117a) in the particular case $\vec{y} = 0$, leading to (8.117b) $\equiv$ (8.118a).

Substitution of (8.116b; 8.117b) in (8.115) yields (8.118b):

$$\frac{\partial^2}{\partial x_i \partial x_j} \left(\frac{1}{R} \right) = \frac{3 x_i x_j - R^2 \delta_{ij}}{R^2}: \quad \langle \vec{H}_1 \rangle = \frac{3 (\vec{Q} \cdot \vec{x}) \vec{x} - R^2 \vec{Q}}{8\pi c R^5}; \tag{8.118a,b}$$

choosing the z-axis along the total average magnetic moment (8.119a), the magnetic field (8.118b) becomes (8.119c):

$$\vec{N} = \vec{e}_z N; \quad \vec{e}_R = \vec{e}_z \cos\theta + \vec{e}_r \sin\theta: \quad \langle \vec{H}_1 \rangle = \frac{3 (\vec{Q} \cdot \vec{e}_R) \vec{e}_R - \vec{Q}}{8\pi c R^5} = \frac{Q}{8\pi c R^3} (3 \vec{e}_R \cos\theta - \vec{e}_z), \tag{8.119a–c}$$

where (8.119b) $\equiv$ (6.51d) relates the unit radial vector $\vec{e}_R$ to the unit vectors along $\vec{e}_z$ (perpendicular to $\vec{e}_r$) the axis. Thus *the magnetic quadrupole has time-averaged (1) vector potential (8.113a) similar to the magnetic field of a magnetic dipole (8.101), replacing total current $\vec{J}$ by total magnetic moment $\vec{Q}$ in (8.113b; 8.104b); (2) magnetic field (8.119c), whose cylindrical components are given in spherical coordinates with axis $\vec{Q}$ by (8.120a)*:

$$\{ H_{1r}, H_{1z}, H_{1\varphi} \} = \frac{Q}{8\pi c R^3} \{ 3 \sin\theta \cos\theta, 3 \cos^2\theta - 1, 0 \}, \quad P_1 = \frac{\varepsilon Q}{2c}, \tag{8.120a,b}$$

that is, by the same expression as the electric field of an longitudinal electric dipole (8.26e), with the electric dipole moment P_1 related to the magnetic quadrupole moment Q by (8.120b). The magnetic quadrupole has a directivity pattern with four lobes (Figure 8.6) because it involves two directions, namely, that of the currents $\pm\vec{j}$ and their relative positions $\pm\vec{y}$.

8.7 Image on a Conducting or Insulating Plane

The method of images extends from the plane (Chapter I.16) to space (Section 8.7) and the simplest electro(magneto)static case is a point electric charge (current) near a conducting or insulating plane [Subsection 8.7.1 (Subsection 8.7.3)]. There is an analogy between the velocity of a potential flow (electrostatic field) of a point source/sink (positive/negative electric charge) near a rigid (insulating) plane (Subsection 8.7.2). Another analogy applies to the potential flow (magnetostatic field) of a point vortex (electric current) near a rigid (insulating) plane [Subsection 8.7.4 (Subsection 8.7.3)].

8.7.1 Point Electric Charge near a Conducting or Insulating Plane

A type of axisymmetric field that appears as a dipole at a large distance is a source near a plane, if its image is opposite (Section I.16.2). In the present case, line-electric charges (Section I.24.4) are considered instead of point electric charges (Subsection 8.7.1), and the method of images can be extended to other multipoles, such as dipoles (Example 10.20). Consider (Figure 8.10) a point charge e at the position

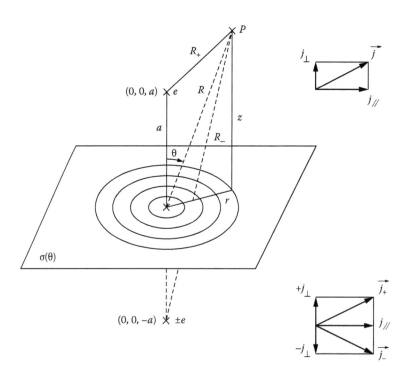

FIGURE 8.10   The effect of an insulating (conducting) plane on a nearby (1) point electric charge is equivalent to an equal (opposite) image; and (2) point electric current that has the same component parallel to the plane and the same (opposite) component normal to the plane. The insulating (conducting) plane, that is, equal (opposite) images, leads to a monopole with double strength (a dipole) in the far field, because the electric field is excluded from the lower half-space (the original electric charges causes a distribution of electric charges with opposite sign on the plane), resulting in a force of repulsion (attraction).

$z = a$ on the z-axis with image at $z = -a$, so that the position vector in cylindrical coordinates (r, φ, z) relative to the observer is (8.121a):

$$\vec{x}_\pm = \vec{e}_r r + \vec{e}_z (z \pm a), \quad R_\pm = |\vec{x}_\pm| = \left| (z \mp a)^2 + r^2 \right|^{1/2}, \quad \frac{\partial R_\pm}{\partial \vec{x}} = \frac{\vec{x}_\pm}{R_\pm}, \tag{8.121a–c}$$

and the distance (8.121b) satisfies (8.121c); the coordinate system (Figure 8.10) was chosen with (1) z-axis through the point electric charge and its image; (2) origin at equal distance from both. Thus the problem is axisymmetric and the origin lies on the plane that is the locus of the points at equal distance from the electric charge and its image. For an insulating (conducting) plane, the image point electric charge is identical $+e$ (opposite $-e$), corresponding to the upper (lower) sign in the electrostatic potential (8.117a):

$$\Phi_\pm (r, z) = \frac{e}{4\pi\varepsilon} \left(\frac{1}{R_+} \pm \frac{1}{R_-} \right), \quad \vec{E}_\pm (r, z) = \frac{e}{4\pi\varepsilon} \left(\frac{\vec{x}_+}{R_+^3} \pm \frac{\vec{x}_-}{R_-^3} \right), \tag{8.122a,b}$$

and to (8.88b; 8.121c) the electric field (8.122b) with cylindrical components:

$$\left\{ E_r^\pm, E_z^\pm, E_\varphi^\pm \right\} (r, z) = \frac{e}{4\pi\varepsilon} \left[\frac{\{r, z - a, 0\}}{R_+^3} \pm \frac{\{r, z + a, 0\}}{R_-^3} \right]. \tag{8.123}$$

Thus *a point electric charge e at a distance a from an insulating (conducting) plane has an electrostatic potential (8.122a) [electric field (8.122b) $\equiv$ (8.123)] where (8.121a) is the position vector relative to the observer and (8.121b) the distance; there is a force repulsion (attraction) normal to the plane given by*

$$\vec{F} = \pm \frac{e^2}{4\pi\varepsilon} \frac{1}{(2a)^2} \vec{e}_z = \pm \frac{e^2}{16\pi\varepsilon a^2} \vec{e}_z, \tag{8.124}$$

where $2a$ is the distance from the identical (opposite) image for an insulating (conducting) plane, corresponding to the upper (lower) signs.

A point on the plane (8.125a) is at the same distance (8.125b) from the electric charge and its image (8.121b):

$$z = 0: \quad R_\pm = \left| r^2 + a^2 \right|^{1/2} \equiv R_0; \quad \vec{E}_+ = \vec{e}_r \frac{er}{2\pi\varepsilon R_0^3}, \quad \vec{E}_- = -\vec{e}_z \frac{ae}{2\pi\varepsilon R_0^3}; \tag{8.125a–d}$$

thus the electric field due to a point charge e at a distance a from an insulating (conducting) wall is tangent (8.125c) [normal (8.125d)] to its plane. In the case of the conducting plane, the normal component of the electric field specifies the distribution of surface charges:

$$\sigma (r) = \varepsilon E_z^- (r, 0) = -\frac{ea}{2\pi} \left| r^2 + a^2 \right|^{-3/2}, \tag{8.126}$$

the total charge on the surface:

$$2\pi \int_0^\infty \sigma(r) r \, dr = -ea \int_0^\infty \left| r^2 + a^2 \right|^{-3/2} r \, dr = ea \left[\left| r^2 + a^2 \right|^{-1/2} \right]_0^\infty = -e, \tag{8.127}$$

and is equal and opposite to the original electric charge. The charge distribution, induced on the plane has an extremum (8.128a and b) at the origin, which is the point closest to the original charge:

$$\sigma(r) \geq \sigma_{min} = \sigma(0) = -\frac{e}{2\pi a^2}, \quad \sigma(r) = -\frac{ea}{2\pi r^3}\left[1 + O\left(\frac{a^2}{r^2}\right)\right], \tag{8.128a,b}$$

and the surface charge density delays like the cube inverse power (8.128b) at a large distance.

In the distance (8.121b) of the electric charge from the observer,

$$R^2 = r^2 + z^2: \quad R_\pm = \left|r^2 + z^2 + a^2 \pm 2az\right|^{1/2} = \left|R^2 \mp 2az + a^2\right|^{1/2}, \tag{8.129a,b}$$

the approximation for observer much farther from the origin than the distance of the electric charge from the plane is used:

$$R^2 = r^2 + z^2 \gg a^2: \quad \frac{1}{R_\pm} = \frac{1}{R}\left|1 \mp \frac{2az}{R^2} + \frac{a^2}{R^2}\right|^{-1/2} = \frac{1}{R} \pm \frac{az}{R^3} + \left(\frac{a^2z^2}{R^5}\right). \tag{8.129c,d}$$

Substituting (8.129d) in (8.122a) leads to *the far-field (8.130a) potential for the insulating (8.130b) [conducting (8.130c)] plane:*

$$a^2 \ll R^2: \quad \Phi_+(R,\theta) = \frac{e}{2\pi\varepsilon R}, \quad \Phi_-(R,\theta) = \frac{eaz}{2\pi R^3} = \frac{P_1\cos\theta}{4\pi\varepsilon R^2}, \quad P_1 = 2ea. \tag{8.130a–d}$$

In the case of the insulating plane, (1) the potential (8.130b) in the far field (8.130a) is double of that of a single point electric charge (8.18c with $P_0 \equiv e$) due to the presence of its identical image; and (2) the latter ensures a zero component of the electric field (8.125c) normal to the isolating plane, and thus excludes the electric field from the lower half-space $z < 0$ and doubles its strength in the upper half-space $z > 0$. In the case of a conducting plane, (1) there is an induced electric charge distribution (8.126) with opposite sign and equal total strength (8.127); (2) at a large distance (8.130a) this leads to the potential (8.130c) of a dipole whose moment (8.130d) is the product of the charge e by the distance 2a to its opposite image; and (3) the sign of the image, or induced electric charges (8.126; 8.127) is opposite to the original electric charge and cancels the monopole term (8.53b) in the multipole expansion (8.58a) for the potential, which becomes weaker since it starts with the dipole term (8.54b).

8.7.2 Flow Source/Sink near a Rigid Impermeable Plane

The case an identical image with upper signs in (8.122a and b) corresponds to an electric charge near an insulating plane, which is analogous with the substitution $e/\varepsilon \to -Q$ to (6.130a) *the potential flow due to a point source $Q > 0$ or sink $Q < 0$ of volume flux Q at a distance a from a rigid impermeable wall leading to the potential (8.131a) [velocity (8.131b)] involving the distance (8.121b) [position vector (8.121a)] relative to the observer:*

$$\Phi(r,z) = -\frac{Q}{4\pi}\left(\frac{1}{R_+} + \frac{1}{R_-}\right), \quad \vec{v}(r,z) = \frac{Q}{4\pi}\left(\frac{\vec{x}_+}{R_+^3} + \frac{\vec{x}_-}{R_-^3}\right). \tag{8.131a,b}$$

The velocity is tangential on the plane (8.132a and b):

$$v_z(r,0) = 0, \quad v_r(r,0) = \frac{Q}{4\pi}r\left|r^2 + a^2\right|^{-3/2}, \quad v_r(0,0) = 0, \tag{8.132a–c}$$

and vanishes (8.132c) at the point on the plane closest to the source/sink. The corresponding pressure distribution is specified (8.133a) by the Bernoulli law (I.14.27c) ≡ (6.102c):

$$p(r,0) = p_0 - \frac{\rho}{2}\left[v_r(r,0)\right]^2 = p_0 - \frac{\rho Q^2}{32\pi^2}\frac{r^2}{\left(r^2 + a^2\right)^3} \leq p_{max}(r,0) = p(0.0) = p_0, \qquad (8.133a,b)$$

which has a maximum (8.133b) at the closest point, that is, a stagnation point (8.132a and c). The velocity (8.134b) [dynamic pressure (8.134c)] on the surface decays at a large distance (8.134a):

$$r^2 \equiv x^2 + y^2 \gg a^2: \quad v_r(r,0) \sim \frac{Q}{4\pi r^2}, \quad p_0 - p(r,0) \sim \frac{\rho Q^2}{32\pi^2 r^4}, \qquad (8.134a\text{--}c)$$

on the second (fourth) inverse power of the distance.

For an observer at a large distance (8.129d) from the source (8.129c) ≡ (8.135a), the total potential (8.135b) is that of a source of double strength 2Q, that is, that of the original plus the identical image:

$$R^2 = x^2 + y^2 + z^2 = r^2 + z^2 \gg a^2:$$

$$\Phi = -\frac{Q}{4\pi}\left\{\left|r^2 + (z-a)^2\right|^{-1/2} + \left|r^2 + (z+a)^2\right|^{1/2}\right\} = -\frac{Q}{2\pi R}\left[1 + O\left(\frac{az}{R^2}\right)\right]. \qquad (8.135a,b)$$

The source/sink exerts on the wall a force of attraction:

$$\vec{F} = \vec{e}_z \rho \frac{Q}{2\pi}\frac{Q}{(2a)^2} = \vec{e}_z \frac{\rho Q^2}{8\pi a^2}. \qquad (8.136)$$

The image source Q at z = −a induces on the original source Q at z = a at distance 2a the vertical velocity (8.137a):

$$\dot{\vec{a}} = \frac{d\vec{a}}{dt} \equiv \vec{e}_z\frac{Q}{4\pi(2a)^2} = \vec{e}_z\frac{Q}{16\pi a^2}; \quad a(t) = \vec{e}_z\left(a_0^3 + \frac{3Qt}{16\pi}\right)^{1/3}; \qquad (8.137a,b)$$

thus if at time t_0 the source is at the position a_0, at time t it is at the position (8.137b), implying that it moves away from the wall. The result (8.137b) follows from integration of (8.137a):

$$t = \int_0^t dt = \int_{a_0}^a\frac{da}{\dot{a}} = \frac{8\pi}{Q^2}\int_{a_0}^a a^2\,da = \frac{16\pi}{3Q}\left\{\left[a(t)\right]^3 - a_0^3\right\}, \qquad (8.137c)$$

solving for a(t). The point electric charge near a conducting plane (Subsection 8.7.1) has for hydrodynamic analogue a point flow source/sink near to a plane with orthogonal velocity, for example, (1) a wall vibrating normal to its plane; and (2) a porous wall in a fluid at rest with blowing (suction) corresponding to the image source (sink).

8.7.3 Identical or Opposite Images of a Point Electric Current in a Plane

In the same geometry (Figure 8.10), with the same position vector (8.121a) and distance from observer (8.121b), the point electric charge e (current $\vec{j}$) leads to the scalar (8.122a) [vector (8.138a)] potential and electric (8.122b) [magnetic (8.138b)] field where the upper (lower) sign applies to an insulating (conducting) plane:

$$\vec{A}_\pm(x,y,z) = \frac{\mu}{4\pi c}\left(\frac{\vec{j}_+}{R_+} \pm \frac{\vec{j}_-}{R_-}\right),$$

(8.138a)

$$\vec{H}_\pm(x,y,z) = \frac{1}{4\pi c}\left(\frac{\vec{j}_+ \wedge \vec{x}_+}{R_+^3} \pm \frac{\vec{j}_- \wedge \vec{x}_-}{R_-^3}\right).$$

(8.138b)

The coordinate system (Figure 8.10) is chosen again with z-axis joining the point current to its image and origin at equal distance from them, that is where the z-axis intersects the plane; if the electric current has an horizontal component, the field is not axisymmetric and the x-axis is taken so that the electric current lies on the x,z-plane. The component of the electric current (Figure 8.10) (1) parallel to the plane is identical for the original and image; (2) the normal component is also identical (reverses sign) for the equal (opposite) image on an insulating (conducting) plane. Thus choosing the XOZ plane through the electric current, its component normal $j_\perp$ (tangent $j_{//}$) to the plane (8.139a) appears in (8.139b):

$$\vec{j}_\pm = \pm j_\perp \vec{e}_z + j_{//}\vec{e}_x: \quad \vec{j}_\pm \wedge \vec{x}_\pm = \left(\pm j_\perp \vec{e}_z + j_{//}\vec{e}_x\right) \wedge \left[(z \mp a)\vec{e}_z + \vec{e}_y y + \vec{e}_x x\right]$$

$$= \mp \vec{e}_x j_\perp y + \vec{e}_y\left[\pm j_\perp x - j_{//}(z \mp a)\right] + \vec{e}_z j_{//}y.$$

(8.139a,b)

Thus *a point current (8.139a) at a distance a from an insulating (conducting) plane has an electrostatic potential (8.138a) and magnetic field (8.138b) with upper (lower) signs, where (8.121a) is the position vector and (8.121b) the distance from the observer appearing in (8.139b).*
On the plane (8.125a) the distance is (8.125b), and using (8.139b) with $z = 0$, the magnetic field (8.138b) is given by

$$\vec{H}_+(x,y,z) = \vec{e}_z \frac{j_{//}y}{2\pi c R_0^3}, \quad \vec{H}_-(x,y,z) = \frac{\vec{e}_y(j_\perp x + j_{//}a) - \vec{e}_x j_\perp y}{2\pi c R_0^3}.$$

(8.140a,b)

Thus *the magnetic field due to a point current (8.139a) at distance a from an insulating (conducting) plane, is normal (8.140a) [tangent (8.140b)] to the plane. In the case (8.140b) of the conducting plane, the induced electric current distribution is*

$$c\{H_x^-(x,y,0), H_y^-(x,y,0)\} = \{j_x, j_y\} = \frac{1}{2\pi}\{-j_\perp y, j_\perp x + j_{//}a\}|x^2 + y^2|^{-3/2}.$$

(8.141a,b)

For an observer much farther from the origin than the distance of the electric current from the plane (8.129c) ≡ (8.142a) the vector potential (8.138a) for an insulating (conducting) plane is doubled for the component of the electric current parallel (8.142b) [orthogonal (8.142c)] to the plane:

$$R^2 \gg a^2: \quad \vec{A}_+ = \vec{e}_x j_{//}\frac{\mu}{2\pi c R}, \quad \vec{A}_- = \vec{e}_z j_\perp \frac{\mu}{2\pi c R}.$$

(8.142a–c)

If the wall is insulating (conducting) but the parallel (8.143a) [normal (8.144a)] component of the current vanishes,

$$j_{//} = 0: \quad \vec{A}_+ = \vec{P}_+ \frac{\cos\theta}{4\pi c R^2}, \quad \vec{P}_+ \equiv \vec{e}_z j_\perp 2a, \tag{8.143a–c}$$

$$j_\perp = 0: \quad \vec{A}_- = \vec{P}_- \frac{\cos\theta}{4\pi c R^2}, \quad \vec{P}_- \equiv \vec{e}_x j_{//} 2a, \tag{8.144a–c}$$

then the asymptotic potential corresponds to a quadrupole (8.143b) [(8.144b)] with vertical (8.143c) [horizontal (8.144c)] moment. The preceding results (8.142a through c; 8.143a through c; 8.144a through c) are proved substituting the asymptotic approximation of the distance (8.129c and d) in the vector potential (8.138a) due to the currents (8.139a):

$$\vec{A}_\pm = \frac{\mu}{4\pi c R}\left[\left(j_\perp\vec{e}_z + j_{//}\vec{e}_x\right)\left[1+\frac{a}{R}\cos\theta\right] \pm \left(-j_\perp\vec{e}_z + j_{//}\vec{e}_x\right)\left[1-\frac{a}{R}\cos\theta\right] + O\left(\frac{a^2}{R^2}\right)\right]. \tag{8.145}$$

To the lowest order $O(1/R)$ in (8.142a), the vector potential $\vec{A}_+(\vec{A}_-)$ is a monopole due to (8.142b) [(8.142c)] the horizontal (vertical) electric current $j_{//}(j_\perp)$, with a quadrupole $O(a^2/R^3)$ next-order correction; if the horizontal (vertical) component $j_{//}(j_\perp)$ of the electric current is zero, then the vector potential $\vec{A}_+(\vec{A}_-)$ is $O(a/R^2)$ a quadrupole (8.143a through c) [(8.144a through c)] due to the vertical (horizontal) electric current and the next-order correction is an octupole $O(a^3/R^4)$.

8.7.4 Point Vortex near a Rigid Impermeable Wall

In the case of a point vortex at a distance a from a rigid impermeable wall: (1) if the vorticity field is not axisymmetric Cartesian coordinates should be used (Subsection 8.7.3); (2) if the vorticity is orthogonal cylindrical coordinates may be used (Subsection 8.7.4) with an opposite image (8.146a) leading to the vector potential (8.146b) [velocity field (8.146c)].

$$\vec{\varpi}_\pm = \pm\vec{e}_z\varpi: \qquad \vec{A} = \frac{1}{4\pi}\left(\frac{\vec{\varpi}_+}{R_+} - \frac{\vec{\varpi}_+}{R_-}\right) = \vec{e}_z\frac{\varpi}{4\pi}\left(\frac{1}{R_+} + \frac{1}{R_-}\right),$$

$$\vec{v} = \frac{1}{4\pi}\left(\frac{\vec{\varpi}_+ \wedge \vec{x}_+}{R_+^3} - \frac{\vec{\varpi}_- \wedge \vec{x}_-}{R_-^3}\right) = \frac{\varpi}{4\pi}\vec{e}_z \wedge \left(\frac{\vec{x}_+}{R_+^3} + \frac{\vec{x}_-}{R_-^3}\right), \tag{8.146a–c}$$

similar to (8.138a) [(8.138b)] replacing $\mu\vec{j}_\pm/c$ by $\vec{\varpi}_\pm$; these substitutions follow from the *comparison of the magnetostatic field (8.89b and c) with the incompressible flow (8.147a and b):*

$$\vec{v} = \nabla \wedge \vec{A}, \quad \nabla \wedge \vec{v} = \vec{\varpi}: \quad \vec{v} \leftrightarrow \mu\vec{H}, \quad \vec{\varpi} \leftrightarrow \frac{\vec{j}}{c}, \tag{8.147a–d}$$

leading to the analogies between (1) the velocity and the magnetic induction (8.147c) that equals the magnetic field times the magnetic permeability; and (2) the vorticity and the electric current divided by the speed of light in vacuo (8.147d). The position vectors of the observer relative to the vortex and its image (8.121b) lead to:

$$\vec{e}_z \wedge \vec{x}_\pm = \vec{e}_z \wedge \left[\vec{e}_r r + (z \mp a)\vec{e}_r\right] = \vec{e}_\varphi r, \tag{8.148}$$

an azimuthal velocity (8.149a)

$$\vec{v}(r,z) = \vec{e}_\varphi \frac{\varpi r}{4\pi}\left(\frac{1}{R_+^3} + \frac{1}{R_-^3}\right), \quad \vec{v}(r,0) = \vec{e}_\varphi \frac{\varpi r}{2\pi R_0^3} = \vec{e}_\varphi \frac{\varpi r}{2\pi}\left|r^2 + a^2\right|^{-3/2}, \tag{8.149a,b}$$

which simplifies to (8.149b) on the wall. Thus *a point vortex at a distance a from a rigid impermeable wall with an orthogonal vorticity (8.146a) creates a potential flow specified by an opposite image on the wall* *leading to the vector potential (8.146b) and an azimuthal velocity (8.149a) involving the distance (8.121b) of the observer from the vortex and its image. The distances are equal on the wall (8.125a and b), leading to the azimuthal velocity (8.149b), which, by the Bernoulli law (I.14.27c) ≡ (6.102c), leads to the pressure distribution:*

$$p(r,0) = p_0 - \frac{\rho}{2}\left[v_\varphi(r,0)\right]^2 = p_0 - \frac{\rho\varpi^2 r^2}{8\pi^2}\left|r^2 + a^2\right|^{-3}. \tag{8.150}$$

The velocity (8.151b) [dynamic pressure (8.151c)] decays at a large distance (8.151a) ≡ (6.134a):

$$r^2 \equiv x^2 + y^2 \gg a^2 : \qquad v_\varphi(r,0) \sim \frac{\varpi}{2\pi r^2}, \qquad p_0 - p(r,0) \sim -\frac{\rho\varpi^2}{8\pi^2 r^4}, \tag{8.151a–c}$$

that is on the second (fourth) inverse power of distance.

At large distance (8.129d) from the vortex (8.129c) ≡ (8.152a) the vector potential (8.146b) corresponds to a monopole (8.152b) with moment (8.152c) equal to twice the vorticity:

$$R^2 \equiv r^2 + z^2 \gg a^2 : \qquad A(r,z) = \vec{e}_z \frac{P_1}{4\pi R}, \qquad P_1 = 2\varpi. \tag{8.152a–c}$$

The image induces on the vortex a velocity:

$$\frac{d\vec{a}}{dt} \equiv \dot{\vec{a}} = \vec{\varpi}_- \wedge \frac{\vec{e}_\varphi}{4\pi(2a)^2} = -\frac{\varpi}{16\pi a^2}\,\vec{e}_z \wedge \vec{e}_\varphi = \frac{\varpi}{16\pi a^2}\,\vec{e}_r, \tag{8.153}$$

which is orthogonal to the axis leading (8.137a through c) to a motion parallel to the wall (8.154a):

$$\vec{a}(t) = \vec{e}_r\left(a_0^3 + \frac{3\varpi t}{16\pi}\right)^{1/3}; \quad \vec{F} = \rho\vec{\varpi}_- \wedge \dot{\vec{a}} = -\frac{\rho\varpi^2}{16\pi a^2}\,\vec{e}_z \wedge \vec{e}_r = \vec{e}_\varphi \frac{\rho\varpi^2}{16\pi a^2} \tag{8.154a,b}$$

the vortex exerts on the wall an azimuthal force (8.154b). The case of a dipole with axis oblique to a nearby conducting or insulating plane is considered in Example 10.19.

8.8 Source/Sink Images on Perpendicular or Parallel Planes

Having compared the method of images in the simplest case of one plane (Section 8.7) [Subsection 8.7.1 (Subsection 8.7.3)] for a point electric charge (current) in electro(magneto)statics, henceforth (Sections 8.8 and 8.9) only the former will be considered, for example, for a point electric charge (Section 8.8) near orthogonal (between parallel) planes [Subsection 8.8.1 (Subsection 8.8.2)]. The conducting planes lead to induced electric charges (Subsections 8.8.1 and 8.8.2). The insulating planes lead to an electrostatic field analogous to the velocity field of the potential flow due to point sources/sinks between parallel (near orthogonal) impermeable rigid planes [Subsection 8.8.3 (Subsection 8.8.4)]. The particular case of a point electric charge at equal distance from parallel insulating (conducting) planes (Subsection 8.8.2) is simplified choosing the origin at the midpoint between the planes (Subsection 8.8.5), and it corresponds to a flow source/sink at equal distance from rigid impermeable (vibrating or porous) walls.

8.8.1 Point Charge near Orthogonal Conducting/Insulating Planes

Consider a point electric charge e at position $(0, b, a)$ relative (Figure 8.11a) two orthogonal planes $y = 0$ and $z = 0$ intersecting along the x-axis. There are three images at $(0, \pm b, \pm a)$ and the relative position vectors of the observer $\vec{x}$ are (8.155a):

$$\vec{x}_{\pm\pm} = \vec{e}_z(z \mp a) + \vec{e}_y(y \mp b) + \vec{e}_x x, \; R_{\pm\pm} \equiv |\vec{x}_{\pm\pm}| = \left|(z \mp a)^2 + (y \mp b)^2 + x^2\right|^{1/2}, \tag{8.155a,b}$$

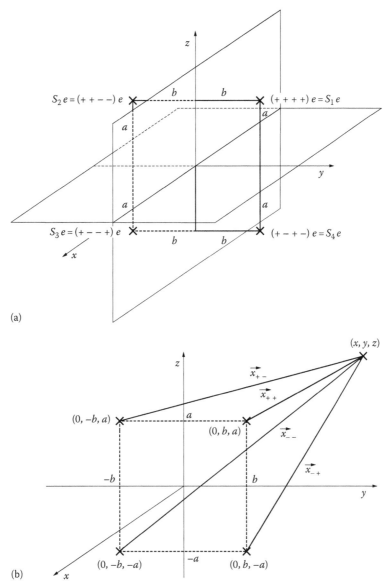

(a)

(b)

FIGURE 8.11 In the case of a point electric charge near two orthogonal planes there are (b) three images: (i/ii) one on each plane, for example, for a source in the first quadrant the image on the vertical (horizontal) plane is in the second (fourth) quadrant; (iii) one image on the line of intersection of the two planes, that is, in the third quadrant for a source in the first quadrant. There are four possible combinations (a) leading to different signs of the images; (a, b) two conducting or two insulating planes; (c, d) one conducting and one insulating plane, exchanging between horizontal and vertical. In the far-field: (1) two insulating planes, that is, all images identical, lead to a monopole with four times the strength, because the electric field is confined to a quarter of the space; (2) one insulating and one conducting plane leads to pairs of opposite electric charges, and hence a dipole perpendicular to the conducting plane; and (3) two conducting planes lead to four alternating electric charges, corresponding to two opposite dipoles or a quadrupole.

and the respective distances (Figure 8.11b) are (8.155b). *There are four possible cases (Figure 8.11a and b) for the images in the second (8.156b), third (8.156c), and fourth (8.156d) quadrants if the point electric charge in the first quadrant (8.156a)*

$$S_1 \equiv (+,+,+,+), \quad S_2 \equiv (+,+,-,-), \quad S_3 \equiv (+,-,-,+), \quad S_4 \equiv (+,-,+,-), \qquad (8.156a\text{–}d)$$

namely, (Figure 8.11a), (1/4) both planes insulating (++) [conducting (−−)], that first is (fourth or last) signs in (8.156a through d); and (2/3) horizontal plane z = 0 conducting and vertical plane y = 0 insulating (+−) [vice-versa (−+)], which is second (third) middle signs in (8.156a through d). The corresponding electrostatic potential (8.157a) and electric field (8.157b),

$$\Phi_{\pm\pm} = \frac{e}{4\pi\varepsilon}\left(\frac{S_1}{R_{++}} + \frac{S_2}{R_{+-}} + \frac{S_3}{R_{--}} + \frac{S_4}{R_{-+}}\right), \qquad (8.157a)$$

$$\vec{E}_{\pm\pm} = \frac{e}{4\pi\varepsilon}\left(S_1\frac{\vec{x}_{++}}{R_{++}^3} + S_2\frac{\vec{x}_{+-}}{R_{+-}^3} + S_3\frac{\vec{x}_{--}}{R_{--}^3} + S_4\frac{\vec{x}_{-+}}{R_{-+}^3}\right), \qquad (8.157b)$$

are not axisymmetric.

On the horizontal (8.142a) [vertical (8.142c)] plane, the distances simplify to (8.142b) [(8.142d)]:

$$z = 0: \quad R_{\pm\pm} = \left|x^2 + (y \mp b)^2 + a^2\right|^{1/2} \equiv R_{\pm}; \quad y = 0: \quad R_{\pm\pm} = \left|x^2 + b^2 + (z \mp a)^2\right|^{1/2} \equiv R^{\pm}. \qquad (8.158a\text{–}d)$$

The electric field (8.157b) on the horizontal plane (8.158a and b),

$$\vec{E}_{\pm+}(x,y,0) = \frac{e}{2\pi\varepsilon}\left[\vec{e}_x x\left(\frac{1}{R_+^3} \pm \frac{1}{R_-^3}\right) + \vec{e}_y\left(\frac{y-b}{R_+^3} \pm \frac{y+b}{R_-^3}\right)\right], \qquad (8.159a,b)$$

$$\vec{E}_{\pm-}(x,y,0) = -\frac{ea}{2\pi\varepsilon}\vec{e}_z\left(\frac{1}{R_+^3} \pm \frac{1}{R_-^3}\right), \qquad (8.159c,d)$$

is normal (tangent) if it is conducting (8.159c and d) [insulating (8.159a and b)]. Conversely, the electric field (8.157b) on (8.158c and d) the vertical plane,

$$\vec{E}_{+\pm}(x,0,z) = \frac{e}{2\pi\varepsilon}\left\{\vec{e}_x x\left[\frac{1}{(R^+)^3} \pm \frac{1}{(R^-)^3}\right] + \vec{e}_z\left[\frac{z-a}{(R^+)^3} \pm \frac{z+a}{(R^-)^3}\right]\right\}, \qquad (8.160a,b)$$

$$\vec{E}_{-\pm}(x,0,z) = -\frac{eb}{2\pi\varepsilon}\vec{e}_y\left[\frac{1}{(R^+)^3} \pm \frac{1}{(R^-)^3}\right], \qquad (8.160c,d)$$

is again normal (tangential) if it is conducting (8.160c through d) [insulating (8.160a and b)]. The normal component of the electric field specifies induced electric charges

$$\sigma_{+-}, \sigma_{--}(x,y,0) = -\frac{ea}{2\pi\varepsilon}\left\{\left|x^2 + (y-b)^2 + a^2\right|^{-3/2} \pm \left|x^2 + (y+b)^2 + a^2\right|^{-3/2}\right\} \qquad (8.161a\text{–}b)$$

$$\sigma_{-+}, \sigma_{--}(x,y,0) = -\frac{eb}{2\pi\varepsilon}\left\{\left|x^2 + b^2 + (z-a)^2\right|^{-3/2} \pm \left|x^2 + b^2 + (z+a)^2\right|^{-3/2}\right\}, \qquad (8.161c\text{,}d)$$

$$\sigma_{++}, \sigma_{-+}(x,y,0) = 0 = \sigma_{++}\,\sigma_{+-}(x,0,z), \qquad (8.161g\text{,}h)$$

four cases: (1) none if both planes are insulating (8.161e and g); (2) on (8.161a) [not on (8.161h)] the horizontal conducting (vertical insulating) plane; (3) on (8.161c) [not on (8.161f)] the vertical conducting (horizontal insulating) plane; (4) both on the vertical (8.161b) and horizontal (8.161d) conducting planes.

The distances (8.155b) of the observer from the point electric charge and its images

$$a^2 + b^2 \ll R^2 \equiv x^2 + y^2 + z^2 : \quad R_{\pm\pm} = \left| x^2 + y^2 + z^2 \pm 2\left(az + by\right) + a^2 + b^2 \right|^{-1/2}$$

$$= \left| R^2 \pm 2\left(az + by\right) + a^2 + b^2 \right|^{-1/2}, \tag{8.162a,b}$$

are approximated by (8.162d) when the distance of the observer from the origin is much larger than that of the electric charges (8.162c):

$$R^2 \gg a^2 + b^2 : \quad \frac{1}{R_{\pm\pm}} = \frac{1}{R} \left| 1 \pm 2 \frac{az + by}{R^2} + \frac{a^2 + b^2}{R^2} \right|^{-1/2}$$

$$= \frac{1}{R} - \frac{1}{R^3}\left(\pm az \pm by + \frac{a^2 + b^2}{2} \right) + \frac{3}{8R^5}\left(\pm az \pm by + \frac{a^2 + b^2}{2} \right)^2, \tag{8.162c,d}$$

leading to the asymptotic potentials

$$\Phi_{++}\left(x, y, z\right) = \frac{e}{\pi \varepsilon R} = \frac{P_0^{++}}{4 \pi \varepsilon R}; \tag{8.163a}$$

$$\left\{ \Phi_{-+}, \Phi_{+-} \right\}\left(x, y, z\right) = -\frac{e}{\pi \varepsilon R^3}\left\{ az, by \right\} = -\frac{1}{4 \pi \varepsilon R^3}\left\{ P_1^{-+}z, P_1^{+-}y \right\}, \tag{8.163b,c}$$

$$\Phi_{--}\left(x, y, z\right) = \frac{3e\,abxy}{4 \pi \varepsilon R^5} = \frac{P_2^{--}xy}{8 \pi \varepsilon R^5}, \tag{8.163d}$$

where (1) only the lowest order in (8.162c) is needed in the case of two insulating planes (8.163a) when the potential corresponds to a monopole with four times the strength (8.164a):

$$P_0^{++} = 4e; \quad \vec{P}_1^{-+} = -\vec{e}_z 4ea, \quad \vec{P}_1^{+-} = -\vec{e}_y 4eb, \quad P_1^{--} = -6eab, \tag{8.164a–d}$$

(2/3) the second-order approximation in (8.162d) is needed for one horizontal (8.163c) [vertical (8.163b)] conducting plane, leading to a vertical (8.164b) [horizontal (8.164c)] dipole of double strength; (4) the third-order approximation is needed in (8.162d) for two conducting planes corresponding to mixed or biaxial quadrupole with moment (8.154d). The moment of the biaxial quadrupole (8.164d) follows from the potential (8.163d) ≡ (8.164f) with the simplification (8.164e):

$$c \equiv \frac{a^2 + b^2}{2} : \quad \frac{8 \pi \varepsilon R^5}{3e} \Phi_{--}\left(x, y, z\right) = \left(az + by + c\right)^2 - \left(az - by + c\right)^2$$

$$+ \left(-az - by + c\right)^2 - \left(-az + by + c\right)^2 = 8abyz, \tag{8.164e,f}$$

arising from substitution of (8.162d) in (8.157a) using the last signs in (8.156a through d).

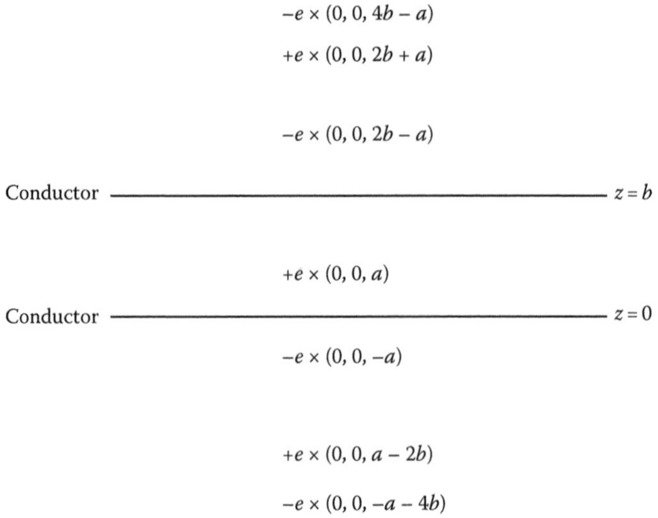

−e × (0, 0, 4b − a)
+e × (0, 0, 2b + a)
−e × (0, 0, 2b − a)
Conductor
z = b
+e × (0, 0, a)
Conductor
z = 0
−e × (0, 0, −a)
+e × (0, 0, a − 2b)
−e × (0, 0, −a − 4b)

FIGURE 8.12 In the case of a point electric charge between two parallel insulating (conducting) planes, the images are identical (alternating).

8.8.2 Identical or Alternating Images on Parallel Planes

Consider a point charge e on the axis $r = 0$, between two conducting planes $z = 0$, b at a distance $z = a$ from the lower. There are two infinite (8.165a) sets of images (Figure 8.12) at position (8.165b and c):

$$n = 0, \pm 1, \pm 2, \ldots: \quad z_m^+ = a + 2nb, \quad z_m^- = -a + 2nb; \tag{8.165a–c}$$

this can be compared with the case of a line monopole (I.40.130a through c) between parallel walls (Example I.40.12) with a different choice of origin (Figure I.40.7a). The images are identical (alternating) in the case of insulating (conducting) planes, corresponding to the upper (lower) sign in the electrostatic potential (8.166a):

$$\begin{aligned}\Phi_\pm(r,z) &= \frac{e}{4\pi\varepsilon}\sum_{n=-\infty}^{+\infty}\left[\left|r^2 + (z-a-2nb)^2\right|^{-1/2} \pm \left|r^2 + (z+a-2nb)^2\right|^{-1/2}\right] \\ &= \frac{e}{4\pi\varepsilon}\sum_{n=-\infty}^{+\infty}\left[\left|r^2 + (z-a-2nb)^2\right|^{-1/2} \pm \left|r^2 + (z+a+2nb)^2\right|^{-1/2}\right],\end{aligned} \tag{8.166a,b}$$

where n is replaced by $-n$ in the second term on the r.h.s. of (8.166b). The potential is axisymmetric like the corresponding electric field (8.1671a through c):

$$\begin{aligned}E_\varphi^\pm = 0: \quad \left\{E_r^\pm, E_z^\pm\right\} &= \frac{e}{4\pi\varepsilon}\sum_{n=-\infty}^{+\infty}\left[\{r, z-a-2nb\}\left|r^2 + (z-a-2nb)^2\right|^{-3/2}\right. \\ &\quad \left.\pm\{r, z+a+2nb\}\left|r^2 + (z+a+2nb)^2\right|^{-3/2}\right].\end{aligned} \tag{8.167a–c}$$

The electric field is vertical on the axis:

$$\vec{E}^{\pm}(0,z) = \vec{e}_z \frac{e}{4\pi\varepsilon} \sum_{n=-\infty}^{+\infty} \left[(z-a-2nb)^{-2} \pm (z+a+2nb)^{-2} \right]. \tag{8.168}$$

From (8.167a through c), it follows that the electric field is tangent to the $z = 0$, the first insulating plane (8.169a and b):

$$E_z^+(r,0) = 0, \quad E_r^+(r,0) = \frac{er}{2\pi\varepsilon} \sum_{n=-\infty}^{+\infty} \left| r^2 + (2nb+a)^2 \right|^{-3/2}; \tag{8.169a,b}$$

$$E_z^+(r,b) = 0, \quad E_r^+(r,b) = \frac{er}{2\pi\varepsilon} \sum_{n=-\infty}^{+\infty} \left| r^2 + (2nb+b+a)^2 \right|^{-3/2}. \tag{8.170a,b}$$

The electric field is also tangent to $z = b$, the second insulating plane (8.170c and d), as follows from the transformation

$$z = b: \quad n \to 1+n \Rightarrow b-a-2nb \to b-a-2(1+n)b = -(b+a+2nb) \tag{8.171a–c}$$

applied to (8.167c), leading to a tangential electric field on both planes (8.169a and b) [(8.170a and b)].

In the case of a point electric charge between conducting planes that corresponds to alternating images, the electric field is specified: (1) on the plane $z = 0$ by:

$$E_r^-(r,0) = 0: \quad \sigma_-(r) = \varepsilon E_z^-(r,0) = -\frac{e}{2\pi} \sum_{n=-\infty}^{+\infty} \left[(a+2nb) \left| r^2 + (a+2nb)^2 \right|^{-3/2} \right]; \tag{8.172a,b}$$

(2) on the plane $z = b$ by:

$$E_r^-(r,b) = \frac{er}{4\pi\varepsilon} \sum_{n=-\infty}^{+\infty} \left[\left| r^2 + (a-b-2nb)^2 \right|^{-3/2} - \left| r^2 + (b+a+2nb)^2 \right|^{-3/2} \right] = 0, \tag{8.173a}$$

$$\begin{aligned}
\sigma_+(r) &= \varepsilon E_z^-(r,b) \\
&= \frac{e}{4\pi} \sum_{n=-\infty}^{+\infty} \left[(b-a-2nb) \left| r^2 + (b-a-2nb)^2 \right|^{-3/2} - (b+a+2nb) \left| r^2 + (b+a+2nb)^2 \right|^{-3/2} \right] \\
&= -\frac{e}{2\pi} \sum_{n=-\infty}^{+\infty} (a+2nb-b) \left| r^2 + (a+2nb-b)^2 \right|^{-3/2},
\end{aligned} \tag{8.173b}$$

where the second term in (8.171a through c) was used on the r.h.s. The induced charge distribution (8.173b) follows from (8.172b) replacing $2nb$ by $2nb - b$. *An electric charge at unequal distance (Figure 8.12) from two parallel insulating (conducting) planes leads to the potential (8.166a and b) and electric field (8.167a through c) with upper (lower) signs, implying that the electric field (1) is axisymmetric (8.167a) and vertical on the axis (8.168) for both conducting and insulating walls; (2) is tangent to the two insulating walls (8.169a and b; 8.170a and b); (3) is normal (8.172a) [8.173a)] for the two conducting walls at $z = 0 (z = b)$, specifying the induced electric charge distributions (8.172b) [(8.173b)] on the two planes, which take the largest values in modulus (8.174a) [(8.174b)] at the nearest point:*

$$\sigma_-(0) = -\frac{e}{2\pi} \sum_{n=-\infty}^{+\infty} (2nb+a)^{-2}, \quad \sigma_+(0) = -\frac{e}{2\pi} \sum_{n=-\infty}^{+\infty} (2nb-b+a)^{-2}. \tag{8.174a,b}$$

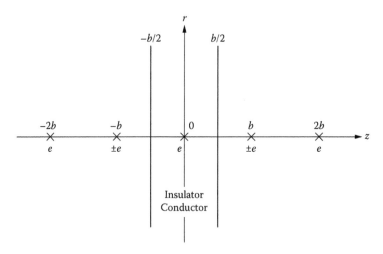

FIGURE 8.13   For a point electric charge between two parallel planes at the same (unequal) distance from both [Figure 8.13 (Figure 8.12)], the images also are (are not) at the same distance.

In the case (Figure 8.13) of a point electric charge at equal distance from the two planes (8.175a), (1) the potential (8.166a and b) [electric field (8.167a through c)] simplifies to (8.175b and c) [(8.176a through c)]:

$$b = 2a : \Phi_{\pm}(r,z) = \frac{e}{4\pi\varepsilon} \sum_{n=-\infty}^{+\infty} \left\{ \left| r^2 + \left[z - (4n+1)a \right]^2 \right|^{-1/2} \pm \left| r^2 + \left[z + (4n+1)a \right]^2 \right|^{-1/2} \right\}, \qquad (8.175\text{a–c})$$

$$E_{\varphi}^{\pm} = 0 : \quad \left\{ E_r^{\pm}, E_z^{\pm} \right\} = \frac{e}{4\pi\varepsilon} \sum_{n=-\infty}^{+\infty} \left\{ r, z - (4n+1)a \right\} \left| r^2 + \left[z - (4n+1)a \right]^2 \right|^{-3/2}$$

$$\pm \left\{ r, z + (4n+1)a \right\} \left| r^2 + \left[z + (4n+1)a \right]^2 \right|^{-3/2} ; \qquad (8.175\text{d–f})$$

(2) the image system (8.165a–c) simplifies to (8.176a,b):

$$n = 0, \pm 1, \pm 2, \ldots : \quad z_m^{\pm} = (4n \pm 1)a;$$

$$E^{\pm}(0,z) = \vec{e}_z \frac{e}{4\pi\varepsilon} \sum_{n=-\infty}^{+\infty} \left\{ \left[z - (4n+1)a \right]^{-2} \pm \left[z + (4n+1)a \right]^{-3/2} \right\}, \qquad (8.176\text{a–c})$$

(3) the electric field (8.168) remains vertical on axis (8.176c): (4) the electric field is tangential and equal (8.177a through c) on the two insulating planes (8.169a and b) ≡ (8.170a and b):

$$b = 2a : \quad E_z^+(r,0) = 0 = E_z^+(r;2a) : \quad E_z^+(r,0) = E_r^+(r;2a) = \frac{er}{2\pi\varepsilon} \sum_{n=-\infty}^{+\infty} \left| r^2 + \left[(4n+1)a \right]^2 \right|^{-3/2}$$

$$= \frac{er}{2\pi\varepsilon} \sum_{n=-\infty}^{+\infty} \left| r^2 + \left[(4n+3)a \right]^2 \right|^{-3/2} ; \qquad (8.177\text{a–e})$$

(5) the electric field is orthogonal and equals (8.178a through c) on two conducting planes (8.172a and b) ≡ (8.173a and b):

$$b = 2a : \quad E_z^-(r,0) = 0 = E_z^-(r;2a), \qquad (8.178\text{a–c})$$

$$E_r^-(r,0) = E_r^-(r;2a) = -\frac{ea}{2\pi} \sum_{n=-\infty}^{+\infty} (4n\pm1)\left|r^2 + \left[(4n\pm1)a\right]^2\right|^{-3/2}$$

$$= \sigma_\pm(r) \geq \sigma_\pm(0) = -\frac{ea}{2\pi a^2} \sum_{n=-\infty}^{+\infty} (4n\pm1)^{-2}, \tag{8.178d,e}$$

inducing the same electric charge distribution (8.178d) ≡ (8.172b) ≡ (8.173b) with the same extremum (8.178e) ≡ (8.174a and b). The two expressions (8.177d) ≡ (8.177e) [(8.178c; 8.178e) with ±signs] coincide because a translation by 2a leaves (8.176a,b) the image system unchanged (Figure 8.13) for a point charge at equal distances from two insulating or conducting planes.

8.8.3 Flow due to a Source/Sink between Parallel Planes

The case of insulating planes (8.166a and b; 8.167a through c) with +sign corresponds to the identical images also for a source/sink of flow rate $Q = -e/\varepsilon$ between parallel walls. Thus *a source/sink of flow rate Q between parallel rigid, impermeable walls at a distance a (b-a) causes a flow with potential (8.179a and b) [velocity (8.180a through c)]:*

$$\Phi(z,r) = -\frac{Q}{2\pi} \sum_{n=-\infty}^{+\infty}\left[\left|r^2 + (z-a-2nb)^2\right|^{-1/2} + \left|r^2 + (z+a-2nb)^2\right|^{-1/2}\right]$$

$$= -\frac{Q}{2\pi} \sum_{n=-\infty}^{+\infty}\left[\left|r^2 + (z+a+2nb)^2\right|^{-1/2} + \left|r^2 + (z+a+2nb)^2\right|^{-1/2}\right], \tag{8.179a,b}$$

$$v_\varphi = 0, \quad \{v_r, v_z\} = \frac{Q}{4\pi} \sum_{n=-\infty}^{\infty} \{r, z-a-2nb\}\left|r^2 + (z-a-2nb)^2\right|^{-3/2}$$

$$+ \{r, z+a+2nb\}\left|r^2 + (z+a+2nb)^2\right|^{-3/2}. \tag{8.180a–c}$$

In the second term on the r.h.s. of (8.179a) was made the substitution $n \to -n$ leading to (8.179b); the velocity field can be calculated either from (8.179a) or (8.179b), and latter was used to obtain (8.180a through c). *The velocity is tangential at the rigid impermeable walls:*

$$v_z(r,0) = 0, \quad v_r(r,0) = \frac{Qr}{2\pi} \sum_{n=-\infty}^{+\infty}\left|r^2 + (2nb+a)^2\right|^{-3/2}, \tag{8.181a,b}$$

$$v_z(r,b) = 0, \quad v_r(r,b) = \frac{Qr}{2\pi} \sum_{n=-\infty}^{+\infty}\left|r^2 + (2nb+b+a)^2\right|^{-3/2}, \tag{8.182a,b}$$

using (6.171a through c) in (8.182a and b); the velocity vanishes at the nearest point r = 0, that is, a stagnation point on both planes. The pressure distributions on the two planes,

$$p(r,0) = p_0 - \frac{\rho Q^2 r^2}{8\pi^2} \sum_{n,m=-\infty}^{+\infty} \left|r^2 + (2nb+a)^2\right|^{-3/2}\left|r^2 + (2mb+a)^2\right|^{-3/2}, \tag{8.183a}$$

$$p(r,0) = p_0 - \frac{\rho Q^2 r^2}{8\pi^2} \sum_{n,m=-\infty}^{+\infty} \left|r^2 + (2nb+b+a)^2\right|^{-3/2}\left|r^2 + (2mb+b+a)^2\right|^{-3/2}, \tag{8.183b}$$

coincide, like the velocities, in the case

$$b = 2a: \quad v_r(r,0) = \frac{Qr}{2\pi} \sum_{n=-\infty}^{+\infty} \left| r^2 + (4n+1)a^2 \right|^{-3/2}$$

$$= \frac{Qr}{2\pi} \sum_{n=-\infty}^{+\infty} \left| r^2 + (4n+3)a^2 \right|^{-3/2} = v_r(r,2a), \tag{8.184a,b}$$

$$p(r,0) = p_0 \frac{\rho Q^2 r^2}{8\pi^2} \sum_{n,m=-\infty}^{+\infty} \left| r^2 + (4n+1)^2 a^2 \right|^{-3/2} \left| r^2 + (4m+1)^2 a^2 \right|^{-3/2}$$

$$= p_0 \frac{\rho Q^2 r^2}{8\pi^2} \sum_{n,m=-\infty}^{+\infty} \left| r^2 + (4n+3)^2 a^2 \right|^{-3/2} \left| r^2 + (4m+3)^2 a^2 \right|^{-3/2} = p(r,2a) \tag{8.185}$$

of source equidistant from the two planes, because the image system (Figure 8.13) maps onto itself by a translation of 2a.

8.8.4 Source/Sink near Orthogonal Walls

The case of an electric point charge near to two insulating orthogonal planes, corresponding to the ++++ signs in (8.156a through d), is analogous to a source/sink of flow rate $Q = -e/\varepsilon$, for which the potential (8.157a) [velocity (8.157b)] is (8.186) [(8.187a through c)]:

$$\Phi(x,y,z) = -\frac{Q}{4\pi} \sum_{\pm\pm} \left| x^2 + (y \pm b)^2 + (z \pm a)^2 \right|^{-1/2}, \tag{8.186}$$

$$\{v_x, v_y, v_z\}(x,y,z) = \frac{Q}{4\pi} \sum_{\pm\pm} \{x, y \pm b, z \pm a\} \left| x^2 + (y \pm b)^2 + (z \pm a)^2 \right|^{-3/2}, \tag{8.187a-c}$$

where the sum is over all combination of pairs of $(\pm,\pm)$ signs. The velocity is tangent both to the horizontal (8.188a through d) and vertical (8.189a through d) walls,

$$z = 0: \quad v_z = 0, \quad \{v_x, v_y\} = \frac{Q}{2\pi} \sum_{\pm} \{x, y \pm b\} \left| x^2 + a^2 + (y \pm b)^2 \right|^{-3/2}, \tag{8.188a-d}$$

$$y = 0: \quad v_y = 0, \quad \{v_x, v_z\} = \frac{Q}{2\pi} \sum_{\pm} \{x, z \pm a\} \left| x^2 + a^2 + (z \pm a)^2 \right|^{-3/2}, \tag{8.189a-d}$$

$$z = 0 = y: \quad v_y = 0 = v_z; \quad v_x = Q \frac{x}{\pi} \left| x^2 + a^2 + b^2 \right|^{-3/2}, \tag{8.190a-d}$$

and lies along the intersection (8.190a through d). The pressure on the horizontal (vertical) wall is given by (8.191a) [(8.191b)]:

$$p(x,y,0) = p_0 - \frac{\rho Q^2}{8\pi^2} \sum_{\pm} \frac{x^2 + (y \pm b)^2}{\left[x^2 + a^2 + (y \pm b)^2\right]^3},$$

(8.191a)

$$p(x,0,z) = p_0 - \frac{\rho Q^2}{8\pi^2} \sum_{\pm} \frac{x^2 + (z \pm a)^2}{\left[x^2 + b^2 + (z \pm a)^2\right]^3},$$

(8.191b)

$$p(x,0,0) = p_0 - \frac{\rho Q^2}{2\pi^2} \frac{x^2}{\left(x^2 + a^2 + b^2\right)^3},$$

(8.191c)

which simplifies to (8.191c) on the intersection. The analogue two (three)-dimensional problem is a line (point) source/sink in a rectangular corner [Sections I.16.4 through I.16.7 (Subsection 8.8.5)]. The present case concerns *a point source/sink of flow near two orthogonal impermeable rigid walls leading to the potential (8.186) and velocity (8.187a through c); the velocity (corresponding pressure) simplifies to (8.188a through d; 8.189a through d) [(8.191a and b)] on the walls, leading to (8.190a through d) [(8.191c)] on their intersection.*

8.8.5 Monopole at Equal Distance from Parallel Planes

The analogous problems of the electrostatic field (hydrodynamic flow) due to a point electric charge (flow source or sink) between parallel insulating (rigid impermeable walls) or conducting (vibrating or porous blowing/suction) walls simplifies [Subsection 8.8.2 (Subsection 8.8.3)] if the distances from the two walls are equal (Figure 8.12) and the origin is chosen at the monopole (Figure 8.13). The images are located at (8.192a) and the identical (alternating) monopole moments lead to the potential (8.192b) [(8.192c)]:

$$z_n = nb: \quad \Phi^+(r,z) = \frac{e}{4\pi\varepsilon} \sum_{n=-\infty}^{+\infty} \left| r^2 + (z - nb)^2 \right|^{-1/2},$$

(8.192a,b)

$$\Phi^-(r,z) = \frac{e}{4\pi\varepsilon} \sum_{n=-\infty}^{+\infty} (-)^n \left| r^2 + (z - nb)^2 \right|^{-1/2}.$$

(8.192c)

The corresponding electric field (8.193a) [(8.193b)],

$$\{E_r^+, E_z^+\}(r,z) = \frac{e}{4\pi\varepsilon} \sum_{n=-\infty}^{+\infty} \left| r^2 + (z - nb)^2 \right|^{-3/2} \{r, z - nb\},$$

(8.193a)

$$\{E_r^-, E_z^-\}(r,z) = \frac{e}{4\pi\varepsilon} \sum_{n=-\infty}^{+\infty} (-)^n \left| r^2 + (z - nb)^2 \right|^{-3/2} \{r, z - nb\},$$

(8.193b)

is tangent (normal) to both walls (8.194) [(8.195a)]:

$$\vec{E}^+\left(r,\pm\frac{b}{2}\right)=\vec{e}_r\,\frac{er}{2\pi\varepsilon}\sum_{n=-\infty}^{+\infty}\left|r^2+\left(n\pm\frac{1}{2}\right)^2b^2\right|^{-3/2},\tag{8.194}$$

$$\vec{E}^-\left(r,\pm\frac{b}{2}\right)=-\vec{e}_z\,\frac{eb}{2\pi\varepsilon}\sum_{n=-\infty}^{+\infty}\left(n\pm\frac{1}{2}\right)\left|r^2+\left(n\pm\frac{1}{2}\right)^2b^2\right|^{-3/2}=\mp\vec{e}_z\varepsilon\sigma(r);\tag{8.195a,b}$$

the latter specifies the distribution of induced electric charges on both walls (8.195b), where the $-(+)$ sign corresponds to the downward (upward) unit normal on the upper $+b/2$ (lower $-b/2$) wall.

In the passage from (8.193a) [(8.193b)] to (8.194) [(8.195a)] (8.196a) [(8.196b)] was used:

$$\sum_{n=-\infty}^{+\infty}\left|r^2+\left(n\pm\frac{1}{2}\right)^2b^2\right|^{-3/2}\left(n\pm\frac{1}{2}\right)=0,\tag{8.196a}$$

$$\sum_{n=-\infty}^{+\infty}(-)^n\left|r^2+\left(n\pm\frac{1}{2}\right)^2b^2\right|^{-3/2}\left(n\pm\frac{1}{2}\right)=2\sum_{n=0}^{+\infty}(-)^n\left|r^2+\left(n\pm\frac{1}{2}\right)^2b^2\right|^{-3/2}\left(n\pm\frac{1}{2}\right),\tag{8.196b}$$

because the terms with n and $\pm1-n$ cancel (8.197a and b) [add (8.197c)]:

$$n\rightarrow\mp1-n:\quad n\pm\frac{1}{2}\rightarrow\mp1-n\pm\frac{1}{2}=\mp\frac{1}{2}-n=-\left(n\pm\frac{1}{2}\right),\tag{8.197a}$$

$$(-)^n\left(n\pm\frac{1}{2}\right)\rightarrow(-)^{\mp1-n}\left(\mp1-n\pm\frac{1}{2}\right)=-(-)^n\left(\mp\frac{1}{2}-n\right)=(-)^n\left(n\pm\frac{1}{2}\right).\tag{8.197b}$$

It has been shown that *a point electric charge e in a medium of dielectric permittivity ε at equal distance (8.198a) from two parallel insulating (conducting) walls (Figure 8.13) has identical (alternating) images at (8.192a), corresponding to (1) the potential (8.192b) [(8.192c)]; (2) the electric field (8.193a) [(8.193b)]; (3) the electric field is tangent (8.194) [normal (8.195a)] to the walls showing that there is no [there is [8.195b]] an induced electric charge distribution.* The relation (8.198a) for a monopole at equal distance from two insulating (conducting) walls and the translation of the origin (8.198b) proves the coincidence of the electrostatic potentials (8.192b) $\equiv$ (8.175b) $\equiv$ (8.198c) [(8.192c) $\equiv$ (8.175c) $\equiv$ (8.198d)]:

$$b=2a,\,z\rightarrow z-a:\quad\Phi^+(r,z-a)=\frac{e}{4\pi\varepsilon}\sum_{n=-\infty}^{+\infty}\left|r^2+(z-a-2na)^2\right|^{-1/2}$$

$$=\frac{e}{4\pi\varepsilon}\sum_{m=-\infty}^{+\infty}\left\{\left|r^2+\left[z-(4m+1)a\right]^2\right|^{-1/2}+\left|r^2+\left[z-(4m-1)a\right]^2\right|^{-1/2}\right\}=\Phi_+(r,z),$$

$$\tag{8.198a–c}$$

$$\Phi^-(r,z-a)=\frac{e}{4\pi\varepsilon}\sum_{n=0}^{+\infty}(-)^n\left|r^2+(z-a-2na)^2\right|^{-1/2}$$

$$=\frac{e}{4\pi\varepsilon}\sum_{n=0}^{+\infty}\left\{\left|r^2-\left[z-(4m+1)a\right]^2\right|^{-1/2}-\left|r^2+\left[z-(4m-1)a\right]^2\right|^{-1/2}\right\}=\Phi_-(r,z),\tag{8.198d}$$

and hence also of the electric fields; in (8.198c and d), the sum over all even and odd n was split into two sums, one over even $n = 2m$ and the other over odd $n = 2m - 1$ values. The method of images applies in two (three) dimensions to line (point) multipoles near: (1) a plane wall or corner [Chapter I.16; Sections I.24.4 and I.24.5, I.26.5 and I.26.6, I.36.1 and I.36.2 (Section 8.7 and Subsections 8.8.1 and 8.8.4)]; (2) between parallel walls or in rectangular wells [Sections I.36.6 through I.36.9 (Subsections 8.8.2 through 8.8.3 and 8.8.5)]; (3) a cylinder (Sections I.24.6 through I.24.8, I.26.7 and I.26.8, I.28.8 and I.29.9) [sphere (Section 8.9 next)].

8.9 Discrete or Continuous Images in Spheres

The method of images applies to a conducting (Subsection 8.9.1) sphere and specifies the force exerted upon it by a point charge due to the distribution of induced electric charges (Subsection 8.9.2). The electric charge distribution induced in the conducting sphere (Subsection 8.9.1) reduces the effect of the original electric charge in the far field (Subsection 8.9.2). The case of a point electric charge near an insulating sphere (Subsection 8.9.3) is analogous to the potential flow due to a source near a rigid impermeable sphere. The case of a rigid ≡ insulating sphere involves besides the image point source at the reciprocal point (Subsection 8.9.4) a continuous sink distribution (Subsection 8.9.5) to cancel the flow rate inside the sphere. This specifies the velocity field (Subsection 8.9.6) including on the sphere (Subsection 8.9.7). Continuous sink distributions (Subsection 8.9.8) may be used to represent the potential flow past axisymmetric bodies of different shapes such as fairings (finite) bodies [Subsection 8.9.9 (Subsection 8.9.10)] with arbitrary shape (Subsection 8.9.11). The stream function can be used to specify the flow outside (inside) a translating (rotating) body [Subsections 8.9.7 through 8.9.11 (Subsections 8.9.12 and 8.9.13)]. Whereas the method of images specifies the exact flow or electric fields of a monopole near a sphere (Subsections 8.9.1 through 8.9.9), the multipole expansion provides an approximation depending on the number of terms considered. Using the first three terms of the multipolar expansion, namely, the monopole, dipole, and quadrupole, specifies the velocity on a rigid impermeable sphere due to a nearby flow source/sink (Subsection 8.9.14); this leads to the pressure distribution and the hydrodynamic force (Subsection 8.9.15). The analogy of the potential flow (electrostatic field) allows (Subsection 8.9.16) the comparison of the force on an insulating (rigid impermeable) sphere due to a point electric charge (flow source/sink). The electric force can be compared for insulating (conducting) sphere, showing that it is smaller (larger) in modulus (Subsection 8.9.17) besides having the opposite sign, because it is repulsive (attractive). The fields and forces can also be compared (Subsection 8.9.17) between the method of images (first three terms of the multipole expansion) leading to exact formulas (leading terms only).

8.9.1 Image of a Point Charge on a Conducting Sphere

The method of images applies to (1) a circle in the plane, corresponding to a cylinder in space, using the two circle theorems (Sections I.24.6 through 8, I.26.7 and 8, I.26.7 and 8, I.28.6 through 9, I.35.7 through 9; II.2.5 and 6); (2) a sphere in space using the four sphere theorems (Sections 6.5 through 6.9 and 8.9). A point electric charge may be considered near a conducting (insulating) sphere [Subsections 8.9.1 and 8.9.2 (Subsections 8.9.3 through 8.9.7)]. Since an insulating sphere is an equipotential surface, the potential is used in a coordinate system (Figure 8.14) with the origin at the center of the sphere and z-axis passing through the point electric charge, so that the electric field is axisymmetric. The electrostatic potential due to a point charge at a distance $z = b$ on axis $r = 0$ is given by (8.199a):

$$\Phi_1(R,\theta) = \frac{e}{4\pi\varepsilon R_1}, \quad R_1 \equiv \left|r^2 + (z-b)^2\right|^{1/2} = \left|r^2 + z^2 + b^2 - 2zb\right|^{1/2}$$

$$= \left|R^2 + b^2 - 2bR\cos\theta\right|^{1/2}, \tag{8.199a–c}$$

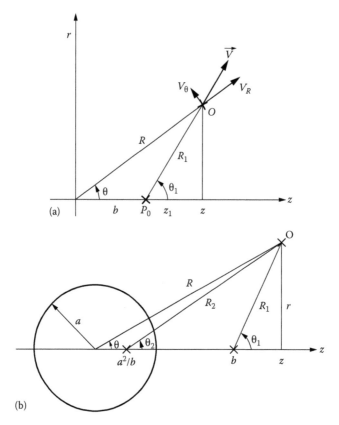

FIGURE 8.14 A point electric charge near (a) a spherical conductor (b) attracts electric charges of opposite sign more strongly to the closer (farther) side. Since the electric field is stronger (weaker) on the closer (farther) side, the latter electric charges predominate. The effect of the conducting sphere of radius a on the electric field of a point charge $\bar{e}$ at a distance b from the center is equivalent to an opposite charge $\bar{e} = -ea/b$ at the reciprocal point $c = a^2/b$. As the charge tends to infinity, $b \to \infty$, the reciprocal point goes to the origin, $c \to 0$, and the image charge to zero, $\bar{e} \to 0$. The image charge would be equal and opposite $\bar{e} = -e$ at the surface $b = a$ when the two would cancel. The far field is a monopole with the original point electric charge reduced by its image $e - \bar{e}$. The analogous hydrodynamic problem is a flow source (sink) near a porous sphere so that the flow is orthogonal to its surface, as the electric field is orthogonal to a conductor. In this case there is a single opposite image sink (source) at the reciprocal point, taking as inflow (outflow) the part of the outflow (inflow) of the original source (sink) that crosses the porous surface of the sphere. This leads to volume conservation for an incompressible fluid with no need for more than one image. The situation is quite different for a rigid impermeable surface (Figure 8.16).

where (Figure 8.14a) the distance from the observer to the source is (8.199b) [(8.199c)] in cylindrical (spherical) coordinates. By the second sphere theorem (Subsection 6.7.2), the reciprocal potential (6.283a) due to the introduction of a sphere of radius a is given by (8.200a through d):

$$\Phi_2(R,\theta) = \frac{a}{R}\Phi_1\left(\frac{a^2}{R},\theta\right) = \frac{e}{4\pi\varepsilon}\frac{a}{R}\left|\frac{a^4}{R^2} + b^2 - \frac{2ba^2}{R}\cos\theta\right|^{-1/2}$$

$$= \frac{e}{4\pi\varepsilon}\left|a^2 + \frac{b^2R^2}{a^2} - 2bR\cos\theta\right|^{-1/2} = \frac{e}{4\pi\varepsilon R_2}\frac{a}{b}, \qquad (8.200a\text{–}d)$$

where (8.201b) is the distance (Figure 8.14b) from the observer to the reciprocal point (8.201a) of the source relative to the sphere:

$$c = \frac{a^2}{b}: \quad R_2 = \left| r^2 + \left(z - \frac{a^2}{b} \right)^2 \right|^{1/2} = \left| r^2 + z^2 + \frac{a^4}{b^2} - \frac{2a^2 z}{b} \right|^{1/2}$$

$$= \left| R^2 + \frac{a^4}{b^2} - \frac{2a^2 R}{b} \cos\theta \right|^{1/2}, \tag{8.201a,b}$$

The source (8.199a and c) and perturbation (8.201b and c) potentials coincide on the conducting sphere:

$$\Phi_1(a,\theta) = \frac{e}{4\pi\varepsilon} \left| a^2 + b^2 - 2ab\cos\theta \right|^{-1/2} = \Phi_2(a,\theta), \tag{8.202}$$

showing that the total potential, that is, their difference, vanishes on the sphere, proving that it is an equipotential surface. Thus in agreement with the second sphere theorem (6.289), the difference of (8.199a and c) and (8.200d) specifies

$$\Phi_-(R,\theta) = \Phi_1(R,\theta) - \Phi_2(R,\theta) = \frac{e}{4\pi\varepsilon} \left(\frac{1}{R_1} - \frac{a}{b} \frac{1}{R_2} \right)$$

$$= \frac{e}{4\pi\varepsilon} \left[\left| R^2 + b^2 - 2bR\cos\theta \right|^{-1/2} - \left| \frac{R^2 b^2}{a^2} + a^2 - 2bR\cos\theta \right|^{-1/2} \right], \tag{8.203}$$

(1) the total potential of a point charge near a conducting sphere that consists of the potential (8.199a) of a point charge e at position b in free space (8.199b and c), as if the sphere were absent; and (2) the presence of the conducting sphere is represented (8.200d) by a point charge −ea/b with opposite sign at the conjugate point (8.201a) of the original charge (Figure 8.15a). As the charge recedes to infinity b → ∞ the image (8.201a) goes to the origin c → 0 with zero strength ea/b → 0.

8.9.2 Induced Charge Distribution and Asymptotic Field

The electrostatic potential (8.203) leads to the electric field:

$$E_\varphi^- = 0: \quad \{E_R^-, E_\theta^-\}(R,\theta) = \frac{e}{4\pi\varepsilon} \left[\left| R^2 + b^2 - 2bR\cos\theta \right|^{-3/2} \{R - b\cos\theta, b\sin\theta\} \right.$$

$$\left. - \left| \frac{R^2 b^2}{a^2} + a^2 - 2bR\cos\theta \right|^{-3/2} \left\{ \frac{b^2 R}{a^2} - b\cos\theta, b\sin\theta \right\} \right]. \tag{8.204a,b}$$

The electric field is normal to the sphere (8.205a and b):

$$E_\theta^-(a,\theta) = 0: \quad \varepsilon\vec{E}_-(a,\theta) = \vec{e}_R\sigma(\theta), \quad \sigma(\theta) = \varepsilon E_R^-(a,\theta) = -\frac{e}{4\pi} \left(\frac{b^2}{a} - a \right) \left| a^2 + b^2 - 2ab\cos\theta \right|^{-3/2}, \tag{8.205a–c}$$

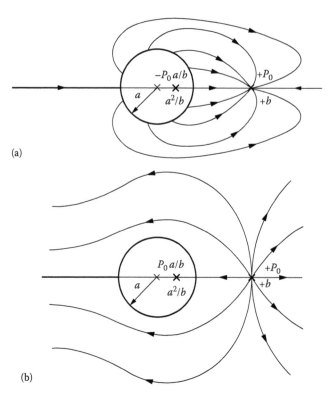

(a)

(b)

FIGURE 8.15 A point electric charge near a conducting (insulating) sphere has the field lines attracted to (repelled by) the sphere [Figure 8.15a (Figure 8.15b)]. The analogous hydrodynamic problem is a flow source/sink near a porous (impermeable) sphere so that the flow velocity is normal (tangent) to its surface. The conducting (insulating) or porous (impermeable) sphere is represented by an opposite (equal) image at the reciprocal point and does not (does) need other images [Figure 8.14 (Figure 8.16)].

and it specifies the distribution of induced electric charges (8.205b and c); the induced charges have opposite sign and are maximum (minimum) in modulus at the closest (farthest) point:

$$|\sigma(\pi)| = \frac{|e|}{4\pi a}\frac{b-a}{(b+a)^2} \leq |\sigma(\theta)| \leq \frac{|e|}{4\pi a}\frac{b+a}{(b-a)^2} = |\sigma(0)|. \qquad (8.206a\text{--}c)$$

In the far field [(8.207a) ≡ (8.199c) and (8.207b) ≡ (8.201b), the charge and its image appear to be superimposed in the potential (8.207c) ≡ (8.203):

$$R_1 = R\left[1+O\left(\frac{b}{R}\right)\right], \quad R_2 = R\left[1+O\left(\frac{a^2}{bR}\right)\right]: \qquad (8.207a,b)$$

$$\Phi_-(R,\theta) = \frac{\bar{e}}{4\pi\varepsilon R}\left[1+O\left(\frac{b}{R},\frac{a^2}{bR}\right)\right], \quad \bar{e} = e\left(1-\frac{a}{b}\right), \qquad (8.207c,d)$$

*which is due to a reduced **effective charge** (8.207d), because the induced charges on the conducting sphere oppose the original charge in sign and are smaller in modulus by a factor a/b.*

The following simplifications were made in the passage from (8.205c) to (8.206a through c):

$$|\sigma(0)| = \frac{|e|}{4\pi}\left(\frac{b^2}{a} - a\right)\left|a^2 + b^2 - 2ab\right|^{-3/2} = \frac{|e|}{4\pi a}\frac{b^2 - a^2}{(b-a)^3} = \frac{|e|}{4\pi a}\frac{b+a}{(b-a)^2},$$ (8.208a)

$$|\sigma(\pi)| = \frac{|e|}{4\pi}\left(\frac{b^2}{a} - a\right)\left|a^2 + b^2 + 2ab\right|^{-3/2} = \frac{|e|}{4\pi a}\frac{b^2 - a^2}{(b+a)^3} = \frac{|e|}{4\pi a}\frac{b-a}{(b+a)^2},$$ (8.208b)

$$\left|\frac{\sigma(0)}{\sigma(\pi)}\right| = \left|\frac{a^2 + b^2 - 2ab}{a^2 + b^2 + 2ab}\right|^{-3/2} = \left|\frac{(a+b)^2}{(b-a)^2}\right|^{3/2} = \left|\frac{b+a}{b-a}\right|^3 > 1,$$ (8.208c)

showing that the electric charge density in modulus is larger (8.208c) at the closest point (8.208a) than at the farthest point (8.208b). The electric charge creates at the reciprocal point an electric field corresponding to the first term on the r.h.s. of (8.204b):

$$\vec{E}_-\left(\frac{a^2}{b}, 0\right) = -\vec{e}_z E_R^-\left(\frac{a^2}{b}, 0\right) = -\vec{e}_z \lim_{R \to \frac{a^2}{b}} \frac{e}{4\pi\varepsilon}\frac{1}{(b-R)^2}$$

$$= -\vec{e}_z\frac{e}{4\pi\varepsilon}\left(b - \frac{a^2}{b}\right)^{-2} \equiv \vec{E}_0;$$ (8.209)

the second term on the r.h.s. is the electric field due to the image charge that does not act upon itself. The electric field (8.209) ≡ (8.210a)

$$\vec{E}_0 = -\vec{e}_z\frac{e}{4\pi\varepsilon}\left(b - \frac{a^2}{b}\right)^{-2}; \quad \vec{F}_- = -e\frac{a}{b}\vec{E}_0 = \frac{e^2}{4\pi\varepsilon}\frac{ab}{(b^2 - a^2)^2}\vec{e}_z$$ (8.210a,b)

multiplied by the image charge −e a/b specifies *the force of attraction (8.210b) exerted by the electric charge on the conducting sphere. The force of attraction is due to the induced charge distribution (8.205c) on the conducting sphere being of the opposite sign. The electric force between a charge distribution and a conductor is always attractive, because the stronger induced charges, that are closer, have the opposite sign,* for example, for a dipole near a conducting sphere (Example 10.20). The force is the same for an electric charge outside b > a (inside a < b) the sphere, that is, not affected by interchanging the position. The potential (8.213) ≡ (8.199a and c; 8.200c) is unaffected by interchange (b, R) of source and observer, that is, the reciprocity principle holds; this is consequence of the symmetry (8.11b) of the influence function for the Laplace operator, which is self-adjoint (7.162a through d). The case of an electric charge near an insulating sphere is analogous to a flow source near a rigid sphere and is considered next (Subsection 8.9.3).

8.9.3 Point Flow Source/Sink near a Rigid Sphere

A rigid impermeable sphere is a stream surface and thus the stream function is used to describe the flow due to a nearby source sink, using the same coordinate system (Figure 8.16) as before (Figure 8.14) leading to an axisymmetric flow. For a point (sink) of flow rate $Q > 0(Q < 0)$ at the origin, the stream function is (6.131a) ≡ (8.211a):

$$-\frac{4\pi}{Q}\Psi_0(R, \theta) = \cos\theta = \frac{z}{R}; \quad z_1 = R\cos\theta - b,$$ (8.211a,b)

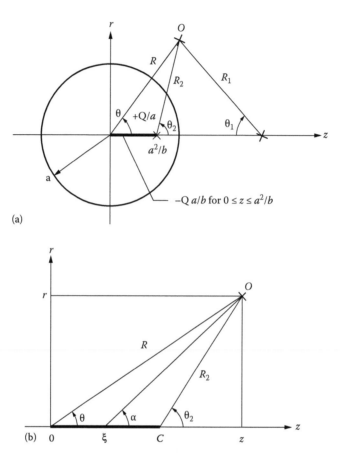

FIGURE 8.16 A positive (negative) point electric charge near an insulating sphere (Figure 8.15b) creates an electric field similar to the potential flow of a source (sink) near a rigid impermeable sphere. In order to have a tangential velocity at the surface of the sphere it is necessary to have an image source (sink) at the reciprocal point. This image source (sink) causes an outflow (inflow) of the incompressible fluid that cannot cross the impermeable surface of the sphere. Thus there must be a sink (source) with equal inflow (outflow) inside the sphere. In the case of a cylinder (Figure I.28.13), the second image is at the center. In the case of the sphere (a) the image is a continuous distribution with uniform strength extending from the center to the reciprocal point (b).

shifting to the point $z = b$ corresponds (Figure 8.14c) to the transformation $(z, R) \rightarrow (z_1, R_1)$ in (8.211b; 8.199c). This specifies the stream function for the source at $z = b$:

$$-\frac{4\pi}{Q}\Psi_1(R,\theta) = 1 + \cos\theta_1, \quad \cos\theta_1 = \frac{z_1}{R_1} = (R\cos\theta - b)\left|R^2 + b^2 - 2bR\cos\theta\right|^{-1/2}, \qquad (8.212\text{a,b})$$

where the angle (8.212b) appears in Figure 8.16a. The stream function is defined to within an added constant; the choice of constant unity in (8.212a) implies that the stream function is zero at the origin $\theta_1 = \pi$. The first sphere theorem (Subsection 6.7.1) specifies (6.272a) the reciprocal stream function (8.213a) due to the introduction of a sphere of radius a:

$$\Psi_2(R,\theta) = \frac{R}{a}\Psi_1\left(\frac{a^2}{R},\theta\right): \quad \bar{R} = \frac{R}{a}R_1\left(\frac{a^2}{R},\theta\right), \quad \cos\bar{\theta} = \frac{R}{a}\cos\theta_1\left(\frac{a^2}{R},\theta\right), \qquad (8.213\text{a–c})$$

where the same transformation applies to radius (8.213b) [angle (8.213c)] leading from (8.199c) [(8.212b)] to (8.214a) [(8.214b)]:

$$\overline{R} = \frac{R}{a}\left|\frac{a^4}{R^2} + b^2 - \frac{2ba^2}{R}\cos\theta\right|^{1/2} = \frac{R}{a}\frac{b}{R}R_2 = \frac{b}{a}R_2,$$ (8.214a)

$$\cos\overline{\theta} = \frac{R}{a}\left(\frac{a^2}{R}\cos\theta - b\right)\frac{1}{\overline{R}} = \left(a\cos\theta - \frac{b}{a}R\right)\frac{a}{R_2 b} = \left(\frac{a^2}{b}\cos\theta - R\right)\frac{1}{R_2},$$ (8.214b)

where (8.201b) was used. Thus the present problem involves, besides (8.214a and b), three more pairs (Figure 8.16b) of (distance, angle) relative to the observer: (1) relative to the origin (R,θ); (2) relative to the source (R_1,θ_1) in (8.199c; 8.212b); and (3) relative to the reciprocal point (8.201a) $\equiv$ (8.215a), that is, (R_2,θ_2) in (8.201b; 8.215b):

$$c \equiv \frac{a^2}{b}: \quad \cos\theta_2 = \frac{R\cos\theta - c}{R_2} = \left(R\cos\theta - \frac{a^2}{b}\right)\frac{1}{R_2}$$

$$= \left(R\cos\theta - \frac{a^2}{b}\right)\left|R^2 + \frac{a^4}{b^2} - 2\frac{a^2 R}{b}\cos\theta\right|^{-1/2}.$$ (8.215a,b)

The original (8.212a) and perturbation (8.213a) stream functions coincide on the sphere $R = a$:

$$\Psi_1(a,\theta) = -\frac{Q}{4\pi}\left[1 + (a\cos\theta - b)\left|a^2 + b^2 - 2ab - \cos\theta\right|^{-1/2}\right] = \Psi_2(a,\theta),$$ (8.216)

so the total stream function, that is, their difference, is zero on the sphere, proving that it is a field surface. Thus in agreement with the first sphere theorem (6.278) their difference specifies

$$\Psi_+(R,\theta) = \Psi_1(R,\theta) - \Psi_2(R,\theta) = -\frac{Q}{4\pi}\left[1 + \cos\theta_1 - \frac{R}{a}\left(1 + \cos\overline{\theta}\right)\right],$$ (8.217)

the *total stream function of the potential flow due to a source of flow rate Q at a point b near (Figure 8.15b) a rigid impermeable sphere of radius a, involving (8.212b; 8.201b; 8.214b).* It may be expected that the presence of the sphere causes an image source of flow rate Qa/b at the reciprocal point (8.215a). However, two sources $Q(Qa/b)$ at $b(a^2/b)$ do not produce a tangential velocity on sphere, because the source inside the sphere would cause an outflow across the sphere. Since the sphere is rigid and impermeable, there must be an opposite sink as another element to the image system. It is shown next that there is a second image in the sphere, namely, a continuous distribution of sinks, whose total inflow balances the outflow of the discrete source inside the sphere, so that there is no net flow across the sphere. In the analogue two-dimensional problem (Subsection I.28.8.1) of a line source/sink near a rigid impermeable cylinder, there is an equal (opposite) image at the reciprocal point (center), so that the flow rate inside the cylinder cancels. Comparing the two (three)-dimensional problem of a line (point) source/sink near a rigid impermeable cylinder (sphere), there are two images in both cases: (1) the line (point) source/sink with the same sign at the reciprocal point; and (2) the flow rate is cancelled in both cases, but differently by an opposite line sink/source at the center (a continuous distribution of sinks/sources from the reciprocal point to the center). The existence of the continuous sink/source distribution in the case of the sphere is proved next (Subsections 8.9.4 and 8.9.5).

8.9.4 Point Source and Line Sink as Images

The total stream function (8.217) can be rewritten subtracting on the r.h.s. the terms due to (1) the source in free space (8.212a); (2) the image (8.215b) of volume flux Qa/b at the reciprocal point (8.224a):

$$\frac{4\pi}{Q}\Psi_3 \equiv \frac{4\pi}{Q}\Psi_+(r,\theta) + 1 + \cos\theta_1 + \frac{a}{b}\cos\theta_2$$

$$= \frac{R}{a} + \frac{a}{b}\cos\theta_2 + \frac{R}{a}\cos\overline{\theta} \equiv D, \tag{8.218}$$

using also (8.217). The r.h.s. of (8.218) must be a second, additional image in the sphere; using (8.214b; 8.215b) it simplifies to

$$D - \frac{R}{a} = \frac{a}{b}\cos\theta_2 + \frac{R}{a}\cos\overline{\theta} = \frac{a}{R_2 b}\left(R\cos\theta - \frac{a^2}{b}\right) + \frac{R}{R_2 a}\left(\frac{a^2}{b}\cos\theta - R\right)$$

$$= -\frac{1}{R_2}\left(\frac{a^3}{b^2} + \frac{R^2}{a} - \frac{2aR}{b}\cos\theta\right) = -\frac{1}{R_2}\frac{(R_2)^2}{a} = -\frac{R_2}{a}, \tag{8.219}$$

using also (8.201b). Substituting (8.219) in (8.218), it follows that *the stream function of the potential flow due to a point source of flow rate Q at a distance b from a rigid impermeable sphere of radius a,*

$$\Psi_+(R,\theta) = -\frac{Q}{4\pi}\left(1 + \cos\theta_1 + \frac{a}{b}\cos\theta_2 + \frac{R_2 - R}{a}\right), \tag{8.220}$$

consists of three terms due (Figure 8.16a) to (1) the original source in free space (8.212a and b); (2) an image source (8.215b) of strength Qa/b [instead of −Qa/b in (8.203)] at the reciprocal point (8.215a); (3) a continuous distribution of sinks (Figure 8.16b) with uniform strength −Q/a per unit length from the origin to the reciprocal point c = a²/b. From (3), it follows that the total strength of the continuous sink distribution is −Qc/a = −Qa/b, which exactly balances the point source Qa/b, so there is no net volume flux across the sphere. It is proved next (Subsection 8.9.5) that the last term on the r.h.s. of (8.220) is indeed the stream function for the uniform continuous source distribution indicated in (3).

8.9.5 Continuous Sink along a Finite Line

The stream function (8.211a) due to a source distribution of strength $q(\xi)$ per unit length along the line (8.221a) is given by (8.221b):

$$0 \le \xi \le c = \frac{a^2}{b}: \quad -4\pi\Psi = \int_0^c q(\xi)\cos\alpha \, d\xi, \tag{8.221a,b}$$

where (8.222a and b) hold from Figure 8.16b, leading to (8.222c):

$$\xi = z - r\cot\alpha, \, d\xi = r\csc^2\alpha \, d\alpha: \quad \Psi = -\frac{r}{4\pi}\int_0^c q(z - r\cot\alpha)\cot\alpha\csc\alpha \, d\alpha, \tag{8.222a–c}$$

Thus *a distribution of sources (Figure 8.16b) along the line (8.221a) creates a potential flow with stream function (8.222c); in the case of a uniform source distribution (8.223a), the stream function simplifies to (8.223b)*:

$$q(\alpha) = \text{const}: \quad \Psi = \frac{qr}{4\pi}[\csc\alpha]_{z=0}^{z=c} = -\frac{q}{4\pi}(R - R_2). \tag{8.223a,b}$$

The stream function (8.323b) [potential (5.335)] of a uniform flow source (8.323a) [electric charge (5.331a)] distribution along a straight segment (8.221a) [(5.330a)] depends only on the distances (angles) of the observer position relative to the two ends.

The last term in (8.220) coincides with (8.223b) and corresponds (8.223b) $\equiv$ (8.224b) to a uniform sink of strength (8.224a) per unit length, from (Figure 8.16b) the origin to the reciprocal point (8.215a):

$$q = -\frac{Q}{a}: \quad \Psi_3 = \frac{Q}{4\pi a}(R - R_2) = \frac{Qr}{4\pi a}(\csc\theta - \csc\theta_2)$$

$$= \frac{Q}{4\pi a}\left[R - \left|R^2 + \frac{a^4}{b^2} - \frac{2a^2 R}{b}\cos\theta\right|^{1/2}\right]; \tag{8.224a,b}$$

the stream function (8.224b) vanishes on the median line $R_2 = R$ or $\theta_2 + \theta = \pi$ or $\alpha = \pi/2$. This completes the proof that (8.220) with (8.201b; 8.212b; 8.215b) is the stream function for a source/sink of flow rate Q at a distance b from a rigid impermeable sphere of radius a:

$$-\frac{4\pi}{Q}\Psi_+(R,\theta) = 1 - \frac{R}{a} + (R\cos\theta - b)\left|R^2 + b^2 - 2bR\cos\theta\right|^{-1/2}$$

$$+ \left|\frac{R^2}{a^2} + \frac{a^2}{b^2} - \frac{2R}{b}\cos\theta\right|^{1/2} + \left(R\cos\theta - \frac{a^2}{b}\right)\left|a^2 + \frac{b^2 R^2}{a^2} - 2bR\cos\theta\right|^{-1/2}; \tag{8.225}$$

the corresponding velocity is calculated [Subsection 8.9.6 (Subsection 8.9.7)] in the flow (on the sphere) before reconsidering the continuous sink distribution (Subsections 8.9.8 through 8.9.11).

8.9.6 Velocity Field due to a Source near a Sphere

The calculation of the flow velocity from the stream function (8.220) $\equiv$ (8.225) involves obtaining the following four: (1) the radial and azimuthal derivatives (8.226a) [(8.326b)] of the distance of the observer from the source/sink (8.199c) [its image (8.201b) in the sphere]:

$$\left\{\frac{\partial}{\partial R}, \frac{\partial}{\partial\theta}\right\} R_1 = \frac{1}{R_1}\{R - b\cos\theta, bR\sin\theta\}, \tag{8.226a}$$

$$\left\{\frac{\partial}{\partial R}, \frac{\partial}{\partial\theta}\right\} R_2 = \frac{1}{R_2}\left\{R - \frac{a^2}{b}\cos\theta, \frac{a^2 R}{b}\sin\theta\right\}; \tag{8.226b}$$

(2) the corresponding derivatives (8.227a and b) [(8.227c and d)] of the cosines of the angles with the z-axis of the position vector from the source/sink (8.212b) [its image on the sphere (8.215b)] to the observer:

$$R_1^3 \frac{\partial}{\partial R}(\cos\theta_1) = R_1^2 \cos\theta - (R\cos\theta - b)(R - b\cos\theta)$$

$$= (R_1^2 - R^2 - b^2)\cos\theta + bR\cos^2\theta + bR = bR(1 - \cos^2\theta) = bR\sin^2\theta, \tag{8.227a}$$

$$R_1^3 \frac{\partial}{\partial \theta}(\cos\theta_1) = -R_1^2 R \sin\theta - (R\cos\theta - b)bR\sin\theta$$

$$= -R\sin\theta \left(R_1^2 - b^2 + bR\cos\theta\right) = -R^2 \sin\theta (R - b\cos\theta), \qquad (8.227b)$$

$$R_2^3 \frac{\partial}{\partial R}(\cos\theta_2) = R_2^2 \cos\theta - \left(R\cos\theta - \frac{a^2}{b}\right)\left(R - \frac{a^2}{b}\cos\theta\right)$$

$$= \cos\theta \left(R_2^2 - R^2 - \frac{a^4}{b^2}\right) + \frac{a^2 R}{b} + \frac{a^2 R}{b}\cos^2\theta = \frac{a^2 R}{b} - \frac{a^2 R}{b}\cos^2\theta = \frac{a^2 R}{b}\sin^2\theta, \qquad (8.227c)$$

$$R_2^3 \frac{\partial}{\partial \theta}(\cos\theta_2) = -R_2^2 R\sin\theta - \left(R\cos\theta - \frac{a^2}{b}\right)\frac{a^2 R}{b}\sin\theta$$

$$= -R\sin\theta \left(R_2^2 + \frac{a^2 R}{b}\cos\theta - \frac{a^4}{b^2}\right) = -R^2 \sin\theta \left(R - \frac{a^2}{b}\cos\theta\right); \qquad (8.227d)$$

(3) the spherical components (6.58a through c) of the velocity are given (8.220) by

$$-\frac{4\pi}{Q} v_R (R,\theta) = \frac{1}{R^2 \sin\theta} \frac{\partial}{\partial \theta}\left(\cos\theta_1 + \frac{a}{b}\cos\theta_2 + \frac{R_2 - R}{a}\right)$$

$$= \frac{b\cos\theta - R}{R_1^3} - \frac{a}{bR_2^3}\left(R - \frac{a^2}{b}\cos\theta\right) + \frac{a}{bR_2 R}$$

$$= \frac{b\cos\theta - R}{R_1^3} + \frac{a}{bR_2^3 R}\left(R_2^2 - R^2 - \frac{a^2 R}{b}\cos\theta\right)$$

$$= \frac{b\cos\theta - R}{R_1^3} + \frac{a^3}{b^2 R_2^3 R}\left(\frac{a^2}{b} - R\cos\theta\right), \qquad (8.228a)$$

$$\frac{4\pi}{Q} v_\theta (R,\theta) = \frac{1}{R\sin\theta} \frac{\partial}{\partial R}\left(\cos\theta_1 + \frac{a}{b}\cos\theta_2 + \frac{R_2 - R}{a}\right)$$

$$= \frac{b\sin\theta}{R_1^3} + \frac{a^3 \sin\theta}{b^2 R_2^3} + \frac{1}{R_2 a\sin\theta}\left(1 - \frac{a^2}{bR}\cos\theta\right) - \frac{1}{aR\sin\theta}$$

$$= \frac{b\sin\theta}{R_1^3} + \frac{1}{R_2^3 a\sin\theta}\left[\frac{a^4}{b^2}\sin^2\theta + R_2^2\left(1 - \frac{a^2}{bR}\cos\theta\right)\right] - \frac{1}{aR\sin\theta}$$

$$= \frac{b\sin\theta}{R_1^3} + \frac{1}{R_2^3 a\sin\theta}\left[R^2 + \frac{a^4}{b^2}\left(2 + \cos^2\theta\right) - \frac{a^2 R}{b}\left(3 + \frac{a^4}{b^2 R^2}\right)\cos\theta\right] - \frac{1}{aR\sin\theta}; \qquad (8.228b)$$

(8.119c;8.201b) were used to simplify (8.227a through d) and (8.228a and b), for example,

$$\frac{a^4}{b^2}\sin^2\theta + R_2^2\left(1 - \frac{a^2}{bR}\cos\theta\right) = \frac{a^4}{b^2}\left(1 + \sin^2\theta + 2\cos^2\theta\right) + R^2 - \frac{3a^2 R}{b}\cos\theta - \frac{a^6}{b^3 R}\cos\theta$$

$$= R^2 + \frac{a^4}{b^2}\left(2 + \cos^2\theta\right) - \frac{a^2 R}{b}\left(3 + \frac{a^4}{b^2 R^2}\right)\cos\theta, \qquad (8.229)$$

in the terms in square brackets on the r.h.s. of (8.228b). Thus (8.230a and b) specify *the velocity field due to a source/sink of flow rate Q at a distance b from the center of a sphere of radius a*:

$$-\frac{4\pi}{Q}\{v_R, v_\theta\}(R,\theta) = \frac{\{b\cos\theta - R, b\sin\theta\}}{R_1^3} + \left\{0, -\frac{1}{aR\sin\theta}\right\}$$

$$+ \frac{1}{R_2^3 a}\left\{\frac{a^4}{b^2}\left(\frac{a^2}{bR} - \cos\theta\right), \frac{1}{\sin\theta}\left[R^2 + \frac{a^4}{b^2}(2+\cos^2\theta) - \frac{a^2 R}{b}\left(3+\frac{a^4}{b^2 R^2}\right)\cos\theta\right]\right\},$$

(8.230a,b)

involving the distance of the observer from the source/sink (8.199c) and its image (8.201b). The velocity is calculated next on the sphere using (8.228a and b) ≡ (8.230a and b) or (8.220) ≡ (8.225) and its approximation to $O(a/b)$ is obtained.

8.9.7 Exact and Approximate Velocity on the Sphere

The distance of a point on the sphere (8.231a) from the source/sink (8.199c) [its image (8.201b)] is (8.231b) [(8.231c)]:

$$R = a: \quad R_1 = \left|a^2 + b^2 - 2ab\cos\theta\right|^{1/2} \equiv R_0, \quad R_2 = \left|a^2 + \frac{a^4}{b^2} - \frac{2a^3}{b}\cos\theta\right|^{1/2} = \frac{a}{b}R_0. \quad (8.231a\text{–}c)$$

It follows that (1) the radial velocity (8.228a) vanishes on the sphere:

$$-\frac{4\pi}{Q}R_0^3 v_R(a,\theta) = b\cos\theta - a + \frac{b}{a}\left(\frac{a^2}{b} - a\cos\theta\right) = 0; \quad (8.232)$$

and (2) thus the velocity (8.228b) is purely tangential on the sphere:

$$\frac{4\pi}{Q}v_\theta(a,\theta) = -\frac{1}{a^2\sin\theta}\left\{1 - \frac{1}{R_0^3}\left[a^2 b\sin^2\theta + b^3 + ba^2(2+\cos^2\theta) - b^2 a\left(3+\frac{a^2}{b^2}\right)\cos\theta\right]\right\}$$

$$= -\frac{1}{a^2\sin\theta}\left[1 - \frac{1}{R_0^3}(b^3 - 3b^2 a\cos\theta + 3a^2 b - a^3\cos\theta)\right]. \quad (8.233)$$

The source/sink is outside the sphere at a distance larger than the radius (8.234a), allowing the expansion of (8.231b) in (8.234b):

$$b > a: \quad \left(\frac{b}{R_0}\right)^3 = \left|1 - \frac{2a}{b}\cos\theta + \frac{a^2}{b^2}\right|^{-3/2}$$

$$= 1 + \frac{3}{2}\left(\frac{2a}{b}\cos\theta - \frac{a^2}{b^2}\right) + \frac{15}{8}\left(\frac{2a}{b}\cos\theta - \frac{a^2}{b^2}\right)^2 + \frac{105}{6.8}\left(\frac{2a}{b}\cos\theta\right)^3$$

$$= 1 + \frac{3a}{b}\cos\theta - \frac{3a^2}{2b^2} + \frac{15a^2}{2b^2}\cos^2\theta - \frac{15a^3}{2b^3}\cos\theta + \frac{35a^3}{2b^3}\cos^3\theta; \quad (8.234a,b)$$

the expansion in (8.234b) was made up to the third order $O(a^3/b^3)$ because on substitution in (8.233) the two lowest-order terms $O(1)$ and $O(a/b)$ cancel, leading to

$$\frac{4\pi}{Q} v_\theta(a,0) = -\frac{1}{a^2 \sin\theta}\left[1 - \left(1 - \frac{3a}{b}\cos\theta + 3\frac{a^2}{b^2} - \frac{a^3}{b^3}\cos\theta\right)\right.$$

$$\left. \times\left(1 + \frac{3a}{b}\cos\theta - \frac{3a^2}{2b^2} + \frac{15a^2}{2b^2}\cos^2\theta - \frac{15a^3}{2b^3}\cos\theta + \frac{35a^3}{2b^3}\cos^3\theta\right)\right]$$

$$= \frac{1}{a^2 \sin\theta}\left[\frac{3a^2}{2b^2}\left(1 - \cos^2\theta\right) + 5\frac{a^3}{b^3}\left(\cos\theta - \cos^3\theta\right)\right]$$

$$= \frac{\sin\theta}{b^2}\left(\frac{3}{2} + 5\frac{a}{b}\cos\theta\right). \tag{8.235}$$

The azimuthal velocity on the sphere (8.235) vanishes at $\theta = 0, \pi$ that are stagnation points (8.236d,e). On axis (8.238a), but generally not on the sphere, the distance from the source/sink (8.199c) [its image (8.201b)] simplifies to (8.236b) [(8.236c)]:

$$\theta = 0, \pi: \quad R_1 = R \mp b, \quad R_2 = R \mp \frac{a^2}{b}, \quad \vec{v}(a,0) = 0 = \vec{v}(a,\pi); \tag{8.236a–e}$$

hence the azimuthal velocity (8.230b) vanishes not only on the stagnation points (8.236d and e) on the sphere, but also (8.228b) along the axis:

$$\frac{4\pi}{Q}\lim_{\theta\to0,\pi} v_\theta(R,\theta)$$

$$= \lim_{\theta\to0}\frac{1}{a\sin\theta}\left\{-\frac{1}{R} + \left(R\mp\frac{a^2}{b}\right)^{-3}\left[R^2 + \frac{3a^4}{b^2} \mp \frac{3a^2R}{b} \mp \frac{a^6}{b^3R}\right] + \frac{ab}{(R\mp b)^3}\sin^2\theta\right\}$$

$$= \lim_{\theta\to0}\frac{1}{a\theta}\left\{-\frac{1}{R} + \frac{1}{R}\left(R\mp\frac{a^2}{b}\right)^{-3}\left[\left(R\mp\frac{a^2}{b}\right)^3 + O\left(\theta^2\right)\right]\right\} = \lim_{\theta\to0}\frac{1}{a\theta R}O\left(\theta^2\right) = \lim_{\theta\to0}\frac{O(\theta)}{a R} = 0. \tag{8.237}$$

Thus the velocity is radial (8.230a) on the axis (8.226a through c):

$$\frac{4\pi}{Q} v_R(R;0,\pi) = -\frac{\pm b - R}{(R\mp b)^3} - \frac{a^2 \mp bR}{R}\frac{a^3}{b^3}\left(R\mp\frac{a^2}{b}\right)^{-3} = \frac{1}{(R\mp b)^2} + \frac{a^3}{R(bR\mp a^2)^2}. \tag{8.238}$$

It has been confirmed that *the velocity field (8.228a and b)* ≡ *(8.230a and b) of a source/sink of flow rate Q at a distance b from a sphere of radius a is (1) tangential on the sphere (8.232; 8.233); (2) radial on the axis through the source/sink and its image (8.237; 8.238); (3) the intersection of (1) and (2) is the stagnation points (8.236d and e).* The approximation (8.235) to the velocity on the sphere can be obtained by another method, namely, using the first three terms of the multipolar expansion, that is, a monopole, dipole, and quadrupole (Subsection 8.9.14). The presentation of the alternative multipole method of obtaining (8.235) is deferred to Subsection 8.9.14, after reconsidering continuous distributions of source/sinks (Subsection 8.9.5) to obtain the flow past [Subsection 8.9.8 (Subsection 8.9.9)] a semi-infinite fairing (finite body) as two examples of axisymmetric shapes (Subsection 8.9.10) in addition to [Subsection 6.3.4 (Subsection 6.4.2)] the Rankine semi-infinite fairing (finite body).

8.9.8 Line Sink in a Uniform Stream

A point (line distribution) of sinks in a uniform flow leads to a semi-infinite fairing [Subsection 6.3.4 (Subsection 8.9.8)], that is, the Rankine (a different) shape. Adding a uniform stream (6.126b) of velocity U to the uniform continuous sink distribution (8.224b) over the segment (8.221a) leads to the stream function

$$\Psi(R,\theta) = \frac{U}{2}R^2\sin^2\theta + \frac{Q}{4\pi a}\left\{R - \left|R^2 + c^2 - 2cR\cos\theta\right|^{1/2}\right\}. \tag{8.239}$$

The real axis (Figure 8.17a) belongs to the streamline (8.240):

$$\Psi(R,0) = \frac{Qc}{4\pi a} \equiv \Psi_0 = \frac{U}{2}R^2\sin^2\theta + \frac{Q}{4\pi a}\left\{R - \left|R^2 + c^2 - 2cR\cos\theta\right|^{1/2}\right\}. \tag{8.240}$$

The equation (8.248) specifies the shape of a semi-infinite body (8.241):

$$R^2\sin^2\theta = \frac{Q}{2\pi Ua}\left\{c - R + \left|R^2 + c^2 - 2cR\cos\theta\right|^{1/2}\right\}. \tag{8.241}$$

The stream function (8.239) specifies the spherical components (6.58a through c) of the velocity field using (8.224b):

$$v_R(R,\theta) = \frac{1}{R^2\sin\theta}\frac{\partial\Psi}{\partial\theta} = U\cos\theta - \frac{Qa}{4\pi R_2 Rb}, \tag{8.242a}$$

$$v_\theta(R,\theta) = -\frac{1}{R\sin\theta}\frac{\partial\Psi}{\partial R} = -U\sin\theta + \frac{Q}{4\pi a\sin\theta}\left(\frac{1}{R_2} - \frac{1}{R} - \frac{a^2}{bR_2R}\cos\theta\right), \tag{8.242b}$$

where were used (8.201a and b) and (8.226b).

By symmetry, the stagnation points can only lie on axis, that is, for $\theta = 0$ or $\theta = \pi$. For $\theta = \pi$, both terms on the r.h.s. of (8.242a) are negative, so the radial velocity cannot vanish; thus the body extends infinitely in the upstream direction (Figure 8.17a), like the Rankine fairing (Figure 6.9); it lies in the opposite direction and has a different shape (8.241) $\neq$ (6.136). In the downstream direction $\theta = 0$, the two terms on the r.h.s. of (8.242a) have opposite signs, and the velocity vanishes (8.243a) at (8.243b):

$$v_R(R,0) = 0: \quad \frac{Qa}{4\pi Ub} = RR_2 = R\left(R - \frac{a^2}{b}\right), \quad R^2 - \frac{a^2}{b}R - \frac{Qa}{4\pi Ub} = 0. \tag{8.243a–c}$$

The roots of (8.243b) $\equiv$ (8.243c) are (8.244a):

$$R_\pm = \frac{a^2}{2b}\left[1 \pm \left|1 + \frac{Qb}{\pi Ua^3}\right|^{1/2}\right]; \quad R_+ > \frac{a^2}{b} = c > 0 > R_-. \tag{8.244a,b}$$

The positive root (8.244b) is chosen in (8.244a) because it is the only one lying outside the continuous sink distribution (8.221a), and thus can correspond to a stagnation point. The velocity (8.242a and b) is continuous at the stagnation point:

$$v_\theta(R_+,0) = \lim_{\theta\to 0}\frac{Q}{4\pi aR_2\sin\theta}\left(1 - \frac{R_2}{R} - \frac{a^2}{bR}\cos\theta\right)$$

$$= \lim_{\theta\to 0}\frac{Q}{4\pi aR_2\theta}\frac{a^2}{bR}(1 - \cos\theta) = \lim_{\theta\to 0}\frac{Qa\theta}{8\pi abR_2R} = 0, \tag{8.245}$$

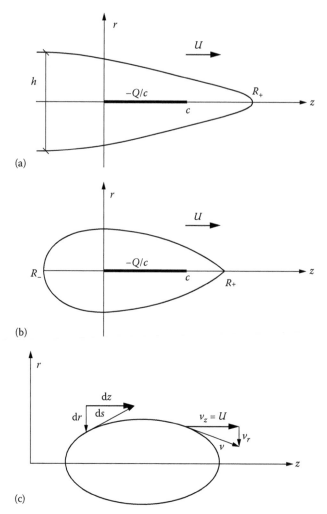

(a)

(b)

(c)

FIGURE 8.17 A line (point) sink in a uniform flow is equivalent to a Rankine (axisymmetric) fairing [Figure I.28.5 (Figure 6.5)], and a continuous distribution of sinks in a uniform stream can lead (a) to a greater variety of shapes of fairings. A pair of opposite line (point) source and sink aligned with a uniform stream is equivalent to a Rankine (axisymmetric) oval body [Figure I.28.6 (Figure 6.6)], and a continuous distribution of sinks with a point source with equal flow rate at the upstream end aligned with a uniform flow leads to a finite body (b) with a rounded nose and cusped tail. Given "a priori" any body shape, it is possible to find a source/sink distribution that reproduces this shape in a uniform stream of velocity U by ensuring (c) that the total velocity is tangential, that is, the surface of the body is a stream surface. This method also applies to a fluid in a rotating vessel (Figure 6.18).

where (8.236b) was used; thus the fairing has a smooth rounded shape at the downstream end in Figure 8.17a. It follows from the shape of the stream surface (8.241) that the fairing has a finite asymptotic width:

$$h^2 \equiv \lim_{\theta \to \pi} R^2 \sin^2 \theta = \lim_{\theta \to \pi} r^2 = \frac{Q}{2\pi Ua}(c - R + c + R) = \frac{Qc}{\pi Ua}. \tag{8.246}$$

Thus *a uniform sink distribution of strength (8.224a) along the segment (8.221a) leads (Figure 8.17a) to the stream function (8.239) of a uniform stream of velocity U past a semi-infinite fairing (8.241) of asymptotic width (8.246), with aft end at the stagnation point R_+ in (8.244a and b) facing in the downstream direction; the velocity field in the flow (8.242a and b) shows that, in particular (8.243a; 8.245), the aft end is rounded rather than cusped.*

8.9.9 Source Distribution Representing an Axisymmetric Fairing

The Rankine semi-infinite fairing due to a point sink in a uniform flow (Subsection 6.3.4) becomes a finite body adding a point source with equal flow rate upstream aligned with the flow (Subsection 6.4.2). Likewise, the infinite dimension of the fairing in Figure 8.17a is due to the net inflow $-Q$ associated with a uniform distribution of sinks of strength (8.224a) over a length (8.221a). Adding a source of strength Q at the origin cancels the inflow and should lead to a finite body. The corresponding stream function is (8.239) plus (8.211a):

$$\Psi(R,\theta) = \frac{Q}{4\pi}\left\{-\cos\theta + \frac{R}{a} - \frac{1}{a}\left|R^2 + c^2 - 2cR\cos\theta\right|^{1/2}\right\} + \frac{U}{2}R^2\sin^2\theta. \tag{8.247}$$

The streamline passing through the axis in the downstream direction (8.248) corresponds to the shape in (8.249):

$$\Psi(R,0) = \frac{Q}{4\pi}\left[-1 + \frac{R}{a} - \frac{1}{a}\left|R^2 + c^2 - 2cR\right|^{1/2}\right] = -\frac{Q}{4\pi}\left(1 - \frac{c}{a}\right), \tag{8.248}$$

$$R^2\sin^2\theta = \frac{Q}{2\pi U}\left\{\cos\theta - 1 + \frac{c-R}{a} + \frac{1}{a}\left|R^2 + c^2 - 2cR\cos\theta\right|^{1/2}\right\}; \tag{8.249}$$

the velocity adds a point source at the origin to (8.242a):

$$v_R(R,\theta) = U\cos\theta + \frac{Q}{4\pi R^2} - \frac{Qa}{4\pi R_2 Rb}, \tag{8.250}$$

leaving (8.242b) unchanged. The velocity (8.250) can vanish (8.251a) on the axis in the (8.251b) downstream (upstream) direction at a radial distance that is a root of (8.251c) with upper (lower) sign:

$$v_R(R;0,\pi) = 0: \quad 0 = R^2 R_2 \pm \frac{QR_2}{4\pi U} \mp \frac{QRa}{4\pi Ub}$$

$$= R^3 \mp \frac{a^2}{b}R^2 \pm \frac{Q}{4\pi U}\left(1 - \frac{a}{b}\right)R - \frac{Qa^2}{4\pi Ub} \equiv F_\pm(R), \tag{8.251a–c}$$

where (8.236c) is used. The cubic equation (8.251c) can have one or three real roots. The root of $F_+(F_-)$ with $R > a^2/b\,(R < 0)$ specifies the downstream (upstream) stagnation point.

8.9.10 Source Distribution Representing a Finite Body

The source at $R = 0$ ensures that there is an upstream stagnation point at $R_- < 0$; from (8.251c) follows (8.252a):

$$F_+(c) = F_+\left(\frac{a^2}{b}\right) = -\frac{Qa^3}{4\pi Ub^2} < 0 < F_+(\infty), \quad F_+(R_+) = 0, \quad \infty > R_+ > c = \frac{a^2}{b}, \tag{8.252a–c}$$

which implies that F_+ changes sign in the interval (8.252c), so it must have a root there (8.252b), that is, the downstream stagnation point in Figure 8.17b. The shape of the finite body is given by (8.249), which leads by differentiation with regard to θ to

$$2\sin^2\theta R \frac{dR}{d\theta} + 2R^2 \sin\theta\cos\theta = -\frac{Q}{2\pi U}\sin\theta + \frac{Q}{2\pi Ua}cR\sin\theta \left|R^2 + c^2 - 2cR\cos\theta\right|^{-1/2}, \quad (8.253)$$

which specifies the slope of the shape of the finite body:

$$\sin\theta \frac{dR}{d\theta} = -R\cos\theta - \frac{Q}{4\pi UR}\left[1 - R\frac{c}{a}\left|R^2 + c^2 - 2cR\cos\theta\right|^{-1/2}\right]; \quad (8.254)$$

the slope at the stagnation points satisfies

$$\lim_{\theta\to 0,\pi} \sin\theta \frac{dR}{d\theta} = \mp R - \frac{Q}{4\pi UR}\left(1 \mp \frac{c}{a}\frac{R}{R \mp c}\right) = \mp R - \frac{Q}{4\pi UR} \pm \frac{Qc}{4\pi UaR_2}$$

$$= \mp R - \frac{Q}{4\pi UR} \pm \frac{QRa}{4\pi Ub}\frac{1}{R_2 R} = \mp R - \frac{Q}{4\pi UR} + \frac{1}{R_2 R}\left(R^2 R_2 \pm \frac{QR_2}{4\pi U}\right)$$

$$= \mp R + R - \frac{Q}{4\pi UR}(1 \mp 1) = \left\{0, 2R - \frac{Q}{2\pi UR}\right\}, \quad (8.255)$$

where (8.236c; 8.251b) are used. It follows that at the downstream (upstream) stagnation point, corresponding to the upper (lower) sign in (8.255), the r.h.s. has a simple zero (is finite); since $\sin\theta$ has a simple zero in both cases, it follows that $dR/d\theta$ is finite (infinite) at the downstream (upstream) end of the body, and, therefore, is pointed (rounded), as shown in Figure 8.17b. Thus *the stream function (8.247) specifies the velocity (8.250; 8.242b) of a potential flow past a family of axisymmetric finite bodies (8.249) that are rounded (pointed) at the stagnation points upstream (downstream) specified by the roots (8.251c).* Choosing a nonuniform sink distribution (8.221b) with the same total strength $-Q$, to preserve a closed streamline, leads to other airship-like shapes besides Figure 8.17b.

8.9.11 Source Distribution for an Arbitrary Body of Revolution

On the surface of an arbitrary body of revolution (Figure 8.17c) that moves with velocity U parallel to its axis of symmetry (Figure 8.18a), the velocity component normal to the axis is (8.256a):

$$U\frac{dr}{dz} = v_r = -\frac{1}{r}\frac{\partial\Psi}{\partial z}: \quad \Psi = -\frac{1}{2}Ur^2 + \text{const}, \quad (8.256a\text{-}c)$$

using the relation (8.256b) = (6.57c) with the stream function and integrating in dr leads to (8.256c). Thus *an arbitrary source distribution on the axis leads to a stream function Ψ that represents the potential flow with free stream of velocity U past a body of revolution with the shape (8.256c)* (Rankine, 1871).

For example, choosing the stream function (8.257a) leads by (8.256c) to (8.257b):

$$\Psi = \frac{A}{R}\sin^2\theta = A\frac{r^2}{R^3}, \quad \text{const} = r^2\left(\frac{A}{R^3} + \frac{U}{2}\right). \quad (8.257a,b)$$

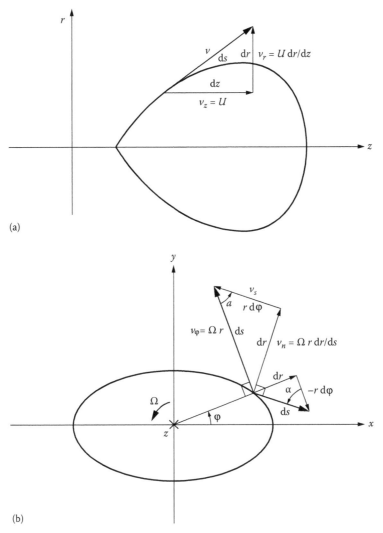

(a)

(b)

FIGURE 8.18 By adjusting the variation of the strength of the sources and sinks along a finite segment the can be obtained flow around a body of revolution with arbitrary shape (Figure 8.17c) in uniform translation (a) in terms of a stream function. The stream function also specifies (b) the potential flow inside a uniformly rotating cylinder with an arbitrary cross section.

The latter (8.257b) is satisfied on a sphere (8.258a) by (8.258b):

$$R = a: \quad A = -\frac{U}{2}a^3; \quad \Psi(R,\theta) = -\frac{Ua^3}{2R}\sin^2\theta. \tag{8.258a–c}$$

The corresponding (8.257a; 8.258b) stream function (8.258c) ≡ [second term on the r.h.s. of (6.201b)] specifies the potential flow past a sphere of radius a.

8.9.12 Flow in a Rotating Cylinder with Arbitrary Cross Section

The stream function (Subsection 8.9.11) may also be used to specify (Section II.6.9) the potential flow inside a cylinder with arbitrary cross section (Figure 8.18b) rotating around its axis with constant angular velocity Ω. At a point of distance r from the axis of rotation, the azimuthal velocity is (8.259a) implying that the velocity component normal to the boundary is (8.259b):

$$v_\varphi = \Omega r, \quad v_n = v_\varphi \frac{dr}{ds} = \Omega r \frac{dr}{ds} = \frac{\partial \Psi}{\partial s}; \quad \Psi = \frac{1}{2}\Omega r^2 + \text{const}; \qquad (8.259\text{a–d})$$

the normal velocity is the tangential derivative of the stream function (8.259c), leading the integration to (8.259d). Thus *the potential flow of a fluid in a cylindrical vessel rotating around its axis with constant angular velocity Ω is specified by the stream function (8.259d), where the shape of the vessel corresponds to a fixed value of the constant.*

As an example, consider the function in polar (8.260a) [Cartesian (8.260b)] coordinates:

$$\Psi = Cr^2 \cos(2\varphi) = Cr^2\left(\cos^2\varphi - \sin^2\varphi\right) = C\left(x^2 - y^2\right) \equiv \Psi(x,y); \qquad (8.260\text{a,b})$$

substitution in (8.259d) leads to

$$\text{const} = \frac{1}{2}\Omega\left(x^2 + y^2\right) - C\left(x^2 - y^2\right) = \left(\frac{\Omega}{2} - C\right)x^2 + \left(\frac{\Omega}{2} + C\right)y^2, \qquad (8.261)$$

which for each value of C represents a member of a family of conics, specifying the shape of the directrix of the cylinder. In the case (8.262a) of an elliptic cylinder with half-axis (a,b), the coincidence of (8.261) $\equiv$ (8.262a) implies (8.262b):

$$\frac{x^2}{a^2} + \frac{y^2}{b^2} = 1: \quad \left(\frac{\Omega}{2} - C\right)a^2 = \left(\frac{\Omega}{2} + C\right)b^2, \quad C = \frac{\Omega}{2}\frac{a^2 - b^2}{a^2 + b^2}, \qquad (8.262\text{a–c})$$

which specifies the constant (8.262c). Thus *the potential flow of a fluid inside an elliptic cylinder of cross section (8.262a) rotating with constant angular velocity Ω around its axis (Figure 8.18b) is given by the stream function (8.260b; 8.262c) $\equiv$ (8.263a) [potential (8.263b)]:*

$$\Psi(x,y) = \frac{\Omega}{2}\frac{a^2 - b^2}{a^2 + b^2}\left(x^2 - y^2\right), \quad \Phi(x,y) = -\Omega\frac{a^2 - b^2}{a^2 + b^2}xy; \qquad (8.263\text{a,b})$$

the stream function (8.263a) is the imaginary part of the complex potential (8.264):

$$\Phi(x,y) + i\Psi(x,y) = f(x+iy) = i\frac{\Omega}{2}\frac{a^2 - b^2}{a^2 + b^2}(x+iy)^2$$

$$= i\frac{\Omega}{2}\frac{a^2 - b^2}{a^2 + b^2}\left(x^2 + y^2 + 2ixy\right), \qquad (8.264)$$

and hence the real part specifies the potential (8.263b).

8.9.13 Kinetic Energy of a Flow in a Rotating Elliptic Cylinder

The kinetic energy of the incompressible fluid inside the vessel is given by

$$E_v = \frac{\rho}{2}\int v^2 \, dS = \frac{\rho}{2}\int |\nabla\Psi|^2 \, dS = \frac{\rho}{2}\int |\nabla\Phi|^2 \, dS. \qquad (8.265\text{a–c})$$

Substituting the potential (8.263c) in (8.265c) leads to

$$
E_r = \frac{\rho}{2} \iint\limits_{0 \le x^2/a^2 + y^2/b^2 \le 1} \left[\left(\frac{\partial \Phi}{\partial x} \right)^2 + \left(\frac{\partial \Phi}{\partial y} \right)^2 \right] dx \, dy
$$

$$
= \frac{\rho}{2} \Omega^2 \left(\frac{a^2 - b^2}{a^2 + b^2} \right)^2 \iint\limits_{0 \le x^2/a^2 + y^2/b^2 \le 1} \left(x^2 + y^2 \right) dx \, dy; \qquad (8.266a)
$$

the change of variable (8.266b,c) maps the ellipse (8.262a) with half-axis (a,b) to the unit circle (8.266d), so that the integration in (8.266a) reduces to (8.266e):

$$
x' \equiv \frac{x}{a}, \ y' \equiv \frac{y}{b}: \qquad 0 \le \frac{x^2}{a^2} + \frac{y^2}{b^2} = x'^2 + y'^2 \le 1,
$$

$$
E_v = \frac{\rho a^b}{2} \Omega^2 \left(\frac{a^2 - b^2}{a^2 + b^2} \right)^2 \iint\limits_{0 \le x'^2 + y'^2 \le 1} \left(a^2 x'^2 + b^2 y'^2 \right) dx' \, dy'. \qquad (8.266b\text{-}e)
$$

Introducing the radius (8.267a) in the unit circle, and noting that the integrals of x'^2 and y'^2 are idential (8.267b,c) by symmetry leads to (8.267d):

$$
r'^2 \equiv x'^2 + y'^2: \qquad \iint\limits_{0 \le r \le 1} x'^2 \, dx' \, dy' = \iint\limits_{0 \le r \le 1} y'^2 \, dx' \, dy'
$$

$$
= \frac{1}{2} \iint\limits_{0 \le r + \le 1} \left(x'^2 + y'^2 \right) dx' \, dy' = \frac{1}{2} \int_0^{2\pi} d\varphi \int_0^1 r'^3 \, dr' = \frac{\pi}{4}, \qquad (8.267a\text{-}d)
$$

where was used the area element $dx \, dy = r \, dr \, d\varphi$ in polar coordinates.

Substituting (8.267d) in (8.266e) specifies *the kinetic energy of an incompressible inviscid fluid in an elliptic cylinder (8.262a) rotating with constant angular velocity* Ω:

$$
E_v = \frac{\rho \pi a b}{8} \Omega^2 \frac{\left(a^2 - b^2 \right)^2}{a^2 + b^2} = \left(\frac{a^2 - b^2}{a^2 + b^2} \right)^2 E_0 < E_0, \quad E_0 = \frac{\rho \pi a b}{8} \Omega^2 \left(a^2 + b^2 \right), \qquad (8.268a\text{-}c)
$$

that is smaller (8.268b) than the kinetic energy for rigid body rotation (8.268c). Noting the area (8.268d) of the cross-section of the eliptic cylinder (8.262a) and the mass (8.268e) per unit length along the axis, it follows that the average velocity (8.268f) is less than for rigid body rotation (8.268g):

$$
A = \pi a b, \quad m = \rho A = \pi \rho a b:
$$

$$
\bar{v} \equiv \sqrt{\frac{2E}{m}} = \frac{\Omega}{2} \frac{\left| a^2 - b^2 \right|}{\sqrt{a^2 + b^2}} = \left| \frac{a^2 - b^2}{a^2 + b^2} \right|^{1/2} \bar{v}_0 < \frac{\Omega}{2} \sqrt{a^2 + b^2} = \sqrt{\frac{2E_0}{m}} = \bar{v}_0. \qquad (8.268d\text{-}g)
$$

The kinetic energy (8.268c) $\equiv$ (8.268j) for rigid body rotation (8.268h,i):

$$v \equiv \Omega r, \quad r^2 = x^2 + y^2: \quad E_V = \frac{\rho}{2} \iint_{0 \le x^2/a^2 + y^2/b^2 r \le 1} V^2 \, dx \, dy$$

$$= \frac{\rho \Omega^2 ab}{2} \iint_{0 \le x'^2 + y'^2 \le 1} \left(x^2 + y^2 \right) dx' \, dy' = \frac{\rho \pi ab}{8} \Omega^2 \left(a^2 + b^2 \right) \quad (8.268\text{h--j})$$

is evaluated similarly to (8.266a through e; 8.267a through d).

In the case of circular cylinder $a = b$, the kinetic energy is zero because the fluid is inviscid and does not rotate with the vessel. In a reference frame rotating with the vessel, the flow is potential and the velocity components (8.263a and b) are given by

$$\{v_x, v_y\}(x, y) = \left\{ \frac{\partial \Phi}{\partial x}, \frac{\partial \Phi}{\partial y} \right\} = \left\{ \frac{\partial \Psi}{\partial y}, -\frac{\partial \Psi}{\partial x} \right\} = \Omega \frac{b^2 - a^2}{b^2 + a^2} \{x, y\}. \quad (8.269)$$

In a reference frame at rest, the velocity of rotation:

$$\vec{\Omega} = \vec{e}_z \Omega: \quad \vec{u} \wedge \vec{x} = \Omega \vec{e}_z \wedge \left(\vec{e}_x x + \vec{e}_y y \right) = \Omega \left(\vec{e}_y x - \vec{e}_x y \right) \quad (8.270)$$

is added to (8.269), leading to the total velocity:

$$w_x = v_x + u_x = \Omega \left(\frac{b^2 - a^2}{b^2 + a^2} - 1 \right) y = -\Omega y \frac{2a^2}{a^2 + b^2} = -\omega \frac{a}{b} y, \quad (8.271\text{a})$$

$$w_y = v_y + u_y = \Omega \left(\frac{b^2 - a^2}{a^2 + b^2} + 1 \right) x = \Omega x \frac{2b^2}{b^2 + a^2} = \omega \frac{b}{a} x, \quad (8.271\text{b})$$

which is in agreement with (II.6.290a and b) $\equiv$ (8.271a and b) using (II.6.287a) $\equiv$ (8.272a):

$$\omega \equiv \Omega \frac{2ab}{b^2 + a^2}; \quad \partial_x w_y - \partial_y w_x = \omega \left(\frac{b}{a} + \frac{a}{b} \right) = \omega \frac{b^2 + a^2}{ab} = 2\Omega; \quad (8.272\text{a,b})$$

the velocity field is rotational in a reference frame at rest (8.272b) with vorticity equal to twice the angular velocity of rotation. Thus *the flow of an incompressible inviscid fluid in a rotating elliptic cylinder is irrotational with potential (8.263b) [rotational with vorticity equal to twice the angular velocity of rotation (8.272b)] in a reference frame rotating with the vessel (at rest), where the velocity is given by (8.269) [(8.271a and b)]. The kinetic energy (8.268a) is less (8.268b) for the potential flow than for rigid body rotation (8.268c)* in agreement with the Kelvin theorem (Subsection II.2.3.4) of minimum energy.

8.9.14 Pressure Distribution on a Sphere due to a Source Sink

The velocity field on a sphere due to a nearby flow point source/sink was obtained exactly (8.233; 8.231b) using the method of images (Subsections 8.9.3 through 8.9.7) involving an image point source/sink at the reciprocal point plus a continuous distribution of sinks/sources between the origin and the reciprocal point; this led to (8.235) as an approximation to the velocity on the sphere. The same result is obtained next by another method: a multipolar expansion (Subsection 8.3.2) taken to the third order, that is, consisting of monopole, dipole, and quadrupole (Subsection 8.9.14). The potential due to the source of flow rate Q at position $(b, 0)$ in the absence of the sphere is (8.199a through c) $\equiv$ (8.273):

$$\Phi_0(R,\theta) = \frac{Q}{4\pi}\left|R^2 + b^2 - 2bR\cos\theta\right|^{-1/2}$$

$$= \frac{Q}{4\pi b}\left[1 + \frac{R}{b}\cos\theta + \frac{R^2}{2b^2}\left(3\cos^2\theta - 1\right) + O\left(\frac{R^3}{b^3}\right)\right], \tag{8.273}$$

using the inverse expansion in spherical harmonics up to the third order, that is, including monopole, dipole, and quadrupole terms; this is similar to the **far-field multipole expansion** *(8.72a through c; 8.74a through c), interchanging* (y, R) *in the* **near-field multipole expansion**:

$$y \equiv |\vec{y}| > R \equiv |\vec{x}|: \quad \left|R^2 - 2y\cos\theta + y^2\right|^{-1/2} = \sum_{n=1}^{\infty} R^m y^{-1-m} P_m(\cos\theta)$$

$$= \frac{1}{y}\left[1 + \frac{R}{y}\cos\theta + \frac{R^2}{2y^2}\left(3\cos^2\theta - 1\right) + \frac{R^3}{2y^3}\cos\theta\left(5\cos^2\theta - 3\right) + O\left(\frac{R^4}{b^4}\right)\right], \tag{8.274a,b}$$

The same result (8.273) can be obtained from the first three terms (I.25.37a through c) of the binomial expansion:

$$\frac{4\pi b}{Q}\Phi_0(R,\theta) = \left|1 - 2\frac{b}{R}\cos\theta + \frac{b^2}{R^2}\right|^{-1/2}$$

$$= 1 - \frac{1}{2}\left(-2\frac{b}{R}\cos\theta + \frac{b^2}{R^2}\right) + \frac{3}{8}\left(-\frac{2b}{R}\cos\theta + \frac{b^2}{R^2}\right)^2 + O\left(\frac{b^3}{R^3}\right)$$

$$= 1 + \frac{b}{R}\cos\theta + \frac{b^2}{2R^2}\left(3\cos^2\theta - 1\right) + O\left(\frac{b^3}{R^3}\right). \tag{8.275}$$

The far-field (8.72a through c; 8.74a through c) and near-field (8.273 $\equiv$ 8.275; 8.274a and b) multipole expansions are related interchanging the observer $\vec{x}$ (source $\vec{y}$) positions with $R \equiv |\vec{x}|\left(y \equiv |\vec{x}|\right)$, in agreement with the reciprocity principle.

The point source with potential (8.273) causes a nonzero normal velocity on the sphere (8.276a):

$$\frac{\partial\Phi_0}{\partial R}\bigg|_{R=a} = \frac{Q}{4\pi b^2}\left[\cos\theta + \frac{a}{b}\left(3\cos^2\theta - 1\right) + O\left(\frac{a^2}{b^2}\right)\right] = -\frac{\partial\Phi_1}{\partial R}\bigg|_{R=a}, \tag{8.276a,b}$$

which is cancelled (8.276b) by the perturbation potential:

$$\Phi_1(R,\theta) = \frac{Q}{4\pi b^2}\left[\frac{a^3}{2R^2}\cos\theta + \frac{a^5}{3bR^3}\left(3\cos^2\theta - 1\right) + O\left(\frac{a^7}{b^2R^4}\right)\right]. \tag{8.277}$$

The total potential is the sum of the original (8.273) and perturbation (8.277) potentials, and leads to a tangential velocity on the sphere:

$$\vec{v}(a,\theta) = \vec{e}_\theta \frac{1}{a} \lim_{R \to a} \frac{\partial \Phi}{\partial \theta} = \vec{e}_\theta \frac{1}{a} \lim_{R \to a} \left(\frac{\partial \Phi_0}{\partial \theta} + \frac{\partial \Phi_1}{\partial \theta} \right)$$

$$= -\vec{e}_\theta \frac{Q}{4\pi b^2} \sin\theta \left[\frac{3}{2} + 5\frac{a}{b} \cos\theta + O\left(\frac{a^2}{b^2}\right) \right], \tag{8.278}$$

in agreement with (8.235) ≡ (8.278). The corresponding pressure distribution (6.102c) is

$$p(a,\theta) - p_0 = -\frac{\rho}{2} \left| \vec{v}(a,\theta) \right|^2$$

$$\equiv -\frac{Q^2 \rho}{32\pi^2 b^4} \sin^2\theta \left[\frac{9}{4} + 15\frac{a}{b} \cos\theta + O\left(\frac{a^2}{b^2}\right) \right]. \tag{8.279}$$

Thus *the potential flow due to a source/sink of strength Q at a distance b from a rigid sphere large compared with the radius a is specified by the total potential (8.273) + (8.277), and it leads to the tangential velocity (8.278) and pressure (8.279) on the sphere.*

8.9.15 Force on an Impermeable Sphere due to a Monopole

The pressure on the sphere corresponds to an inward radial force per unit area:

$$\frac{d\vec{F}_v}{ds} = -\vec{e}_R\, p(a,\theta) = -p(a,\theta)\left[\vec{e}_z \cos\theta + \sin\theta(\vec{e}_x \cos\varphi + \vec{e}_y \sin\varphi) \right], \tag{8.280}$$

where were used (6.13a,c and d). The area element (6.20a) of the sphere (8.281b) of radius a leads to the total force on the sphere (8.280c)

$$dS = a^2 \sin\theta\, d\theta\, d\varphi : \vec{F}_v = -a^2 \int_0^{2\pi} d\varphi \int_0^\pi d\theta \sin\theta\, p(a,\theta)\, \vec{e}_R \,, \tag{8.280b,c}$$

that is axial as should be expected by symmetry:

$$\vec{F}_v = -2\pi a^2\, \vec{e}_z \int_0^\pi p(a,\theta) \cos\theta \sin\theta\, d\theta. \tag{8.280d}$$

Substitution of the pressure (8.279) in (8.281d) leads to the integrals:

$$\int_0^\pi \sin^3\theta \cos\theta\, d\theta = \left[\frac{\sin^4\theta}{4} \right]_0^\pi = 0, \tag{8.281a}$$

$$\int_0^\pi \cos^2\theta \sin^3\theta\, d\theta = \int_0^\pi (\cos^2\theta - \cos^4\theta)\sin\theta\, d\theta = \left[\frac{\cos^5\theta}{5} - \frac{\cos^3\theta}{3} \right]_0^\pi = \frac{4}{15}. \tag{8.281b}$$

that specify the total force on the sphere:

$$\vec{F}_v = -\vec{e}_z \, 2\pi a^2 \, \frac{15 Q^2 \rho \, a}{32 \pi^2 b^5} \int_0^\pi \cos^2\theta \, \sin^2\theta \, d\theta = -\vec{e}_z \, \frac{\rho \, Q^2 \, a^3}{4\pi b^5}. \tag{8.281c}$$

If the sphere were absent, the velocity at its center (8.282a) would correspond to an acceleration (8.282b):

$$v(R) = \frac{Q}{4\pi R^2}: \quad \dot{v}(R) = \frac{dv}{dt} = \frac{dv}{dR}\frac{dR}{dt} = v\frac{dv}{dR} = -\frac{Q^2}{8\pi^2 R^5}. \tag{8.282a,b}$$

The force of attraction (8.281c) ≡ (8.283a) exerted by a source on a sphere

$$\vec{F}_v = -\vec{e}_z \rho 2\pi a^3 \dot{v}(b) = -\vec{e}_z m \dot{v}(b), \quad m = \rho 2\pi a^3 \tag{8.283a,b}$$

equals in modular the product of (1) the acceleration (8.282b) due to the source at the center of the sphere as if the latter were absent; (2) the dipole moment (6.120a) per unit velocity multiplied by the mass density (8.283b). The minus sign appears in (8.283b) because the hydrodynamic force between the source/sink and the sphere is an attraction, as explained next (Subsections 8.9.16 and 8.9.17).

8.9.16 Analogy between the Potential Flow and Electrostatics

The analogy (8.284a and b) between potential flow and electrostatics,

$$Q \leftrightarrow e, \rho \leftrightarrow \frac{1}{\varepsilon}: \quad \vec{F}_+ = -\vec{e}_z \frac{e^2 a^3}{4\pi\varepsilon b^5}, \tag{8.284a–c}$$

specifies in (8.280) the *electric force of repulsion (8.284c) exerted by a point charge at a distance b from an insulating sphere of radius a*. The result (8.284c), like (8.281c), is an approximation arising from the use of the first three terms (8.273) of the near-field multipolar expansion (8.274a and b), corresponding to a monopole, dipole, and quadrupole. The exact force can be obtained from the electric field (8.204a) ≡ (8.285a) at position $x = a^2/b$ by (8.285b):

$$\vec{E}_0(x) = -\vec{e}_z \frac{e}{4\pi\varepsilon} \frac{1}{(b-x)^2}: \quad \vec{F}_+ = \frac{a}{b}\vec{E}_0\left(\frac{a^2}{b}\right) + \int_0^{a^2/b} \left(-\frac{e}{a}\right)\vec{E}_0(x)dx, \tag{8.285a,b}$$

as the sum of two terms: (1) the image strength times the electric field of the charge at the reciprocal point; and (2) the continuous charge distribution multiplied by the electric field and integrated over its extent from the origin to the reciprocal point. The analogous result for the hydrodynamic force is the Blasius theorem (I.28.29a) that the drag force equals the velocity (8.286a) times the mass density and source strength (8.286b):

$$\vec{v}_0(x) = -\vec{e}_z \frac{Q}{4\pi} \frac{1}{(b-x)^2}, \quad F_v = -\rho Q \frac{a}{b}\vec{v}_0\left(\frac{a^2}{b}\right) - \int_0^{a^2/b} \frac{Q}{a}\vec{v}_0(x)dx. \tag{8.286a,b}$$

For comparison with the case of a conducting sphere (8.210a and b), the electric force is calculated from (8.285a and b) as

$$\vec{F}_{+} = \vec{e}_z \frac{e^2}{4\pi\varepsilon} \left\{ -\frac{a}{b}\left(b - \frac{a^2}{b}\right)^{-2} + \frac{1}{a}\left[\frac{1}{b-x}\right]_0^{a^2/b} \right\}$$

$$= \vec{e}_z \frac{e^2}{4\pi\varepsilon}\left[-\frac{ab}{\left(b^2 - a^2\right)^2} + \frac{b}{a}\frac{1}{b^2 - a^2} - \frac{1}{ab} \right]$$

$$= \vec{e}_z \frac{e^2}{4\pi\varepsilon}\left[-\frac{ab}{\left(b^2 - a^2\right)^2} + \frac{a}{b\left(b^2 - a^2\right)} \right]$$

$$= -\vec{e}_z \frac{e^2}{4\pi\varepsilon b}\frac{a^3}{\left(b^2 - a^2\right)^2} = -\vec{e}_z \frac{e^2 a^3}{4\pi\varepsilon b^5}\left(1 - \frac{a^2}{b^2}\right)^{-2}$$

$$= -\vec{e}_z \frac{e^2 a^3}{4\pi\varepsilon b^5}\left[1 + 2\frac{a^2}{b^2} + O\left(\frac{a^4}{b^4}\right) \right]; \tag{8.287}$$

this agrees with (8.284c) to lowest order on a/b and specifies the next order correction.

8.9.17 Force Exerted by a Point Charge on an Insulating/Conducting Sphere

Thus (8.287) ≡ (8.288a and b) *a point charge e at a distance b from an insulating sphere of radius a in a medium of dielectric permittivity ε exerts on the latter a repulsive force:*

$$-\frac{a^2}{b^2}\vec{F}_- = \vec{F}_+ = -\vec{e}_z \frac{e^2}{4\pi\varepsilon b}\frac{a^3}{\left(b^2 - a^2\right)^2} = -\vec{e}_z \frac{e^2 a^3}{4\pi\varepsilon b^5}\left[1 + 2\frac{a^2}{b^2} + O\left(\frac{a^4}{b^4}\right) \right], \tag{8.288a–c}$$

given by the following: (1) (8.284c), which is the term on the r.h.s. of (8.288c) to lowest order on a/b; (2) (8.288c), where the next order correction obtained from the exact expression (8.288b) appears . Comparing (8.288b) with (8.210b), it follows that the electric force exerted by the same electric charge at the same distance on an insulating (conducting) sphere is repulsive (attractive) and smaller in magnitude by a factor $(a/b)^2$ in (8.288a). The substitution of (8.286a) in (8.286b) followed by a calculation similar to (8.287), or the potential flow-electrostatics analogy (8.284a and b) applied to (8.288b), specifies

$$\vec{F}_v = -\vec{e}_z \frac{\rho Q^2 a^3}{4\pi b\left(b^2 - a^2\right)^2} = -\vec{e}_z \frac{\rho Q^2 a^3}{4\pi b^5}\left[1 + 2\frac{a^2}{b^2}O\left(\frac{a^2}{b^2}\right) \right], \tag{8.289a,b}$$

the exact hydrodynamic force of attraction (8.289a) exerted by a source/sink of flow rate Q at a distance b from a rigid impermeable sphere of radius a, in agreement with the lowest-order approximation (8.281c) in a/b, and adding the next order correction in (8.289b). The hydrodynamic force exerted by a multipole on a rigid impermeable body is always of attraction, because (1) the induced velocity is larger (smaller) on the closer (farther) side; (2) by the Bernoulli law (6.102c), since the stagnation pressure is constant, the pressure is lower (higher) on the closer (farther) side; (3) the net force on the body is toward the multipole, that is, a force of attraction. *The electric force exerted by a charge distribution on an insulating body is always repulsive,* because the dominant images have the same sign. The opposite signs apply to the force in otherwise analogous problems: (1) repulsion for the electrostatic field of a charge distribution near an insulator, because the electric field is calculated at the insulator; (2) attraction for the hydrodynamic force on a rigid impermeable body due to a multipole, because the induced velocity is calculated at the multipole. An instance of this is the electric (hydrodynamic) force exerted by a dipole (Example 10.20) on the insulator (body). In all cases the *electric (hydrodynamic) force exerted on a conductor or an insulator (a rigid impermeable body) is the same for positive and negative charge distributions (sources or sinks), and positive or negative multipoles, because their moment always appears to the square; for example, for a positive + |e| or negative − |e| electric charge (point source + |Q|), the electric (8.288 a through c) [hydrodynamic (8.289a and b)] is force the same, because it involves only* $|e|^2$ *(*$|Q|^2$*).*

NOTE 8.1: Discrete/Continuous Images in Two/Three Dimensions

The two cases of a monopole, that is, electric charge (flow source), near a conducting (insulating ≡ rigid) sphere differ in their image systems: (1) one discrete point image at the reciprocal point with strength $-ea/b(ea/b)$; (2) no more images (an additional image). The latter is a continuous image with uniform strength $-e/a$ from the origin to the reciprocal point $c = a^2/b$; its total strength $-ec/a = -ea/b$ balances the first image. A discussion of points (1) and (2) follows. The potential of a charge e at b at the closest point a on the sphere is (8.290a) and that of the image $\bar{e}$ at the reciprocal point (8.215a) is (8.290b):

$$4\pi\Phi_0 = \frac{e}{b-a}, \quad 4\pi\Phi_1 = \frac{\bar{e}}{a-a^2/b}; \tag{8.290a,b}$$

the potentials cancel (8.291a) if the image charge is (8.291b):

$$\Phi_0 + \Phi_1 = 0 \Leftrightarrow \bar{e} = -\frac{e}{b-a}\left(a - \frac{a^2}{b}\right) = -e\frac{a}{b}; \tag{8.291a,b}$$

this explains point (1) about the strength of the point image.

It must be decided whether the preceding reasoning should be applied to the potential or to the field, because the results are different. If the electric field instead of the potential had been balanced, the image strength of a monopole P_0 would be $-P_0(a/b)^2$; this is invalid since the field normal to a conductor is generally not zero. In the case of a dipole of moment P_1 near a sphere, the balancing of the electric field gives an image strength $-P_1(a/b)^3$; this is correct (Example 10.20) since the dipole electric field is tangent to the sphere at the closest and farthest points. If the dipolar potential had been balanced, the image strength would have been $-P_1(a/b)^2$, that is incorrect. Thus *the strength of the image of a monopole P_0 [dipole P_1] on a conducting sphere is* $-P_0a/b\left[-P_1(a/b)^3\right]$, *which is obtained balancing the electrostatic potential (electric field, which is tangential) at the closest and farthest points; the balancing of an electric field that is normal cannot be used to determine image strength. The preceding reasonings are valid iff (if and only if) (1) the same conclusion holds at all points on the sphere; (2) the potential (field) satisfies the scalar (vector) Poisson equation with the correct singularities as sources; and (3) the asymptotic decay is consistent.* The points on the sphere closest to and farthest from the multipole are the simplest test case, which is a necessary but not a sufficient condition for the result. The points on a circle closest to and farthest from a point b not on the circle can also be used to locate the reciprocal point $c = a^2/b$ in (I.26.63a and b); the definition of reciprocal point involves not only these two points but also all other points on the circle (Section I.35.7). Thus a reliable identification of images requires the consideration of the potential (or field/stream function) or field at all points in space, not just a few or those on the obstacles.

An example of this is the continuous image distribution for a positive monopole near a sphere that is a field surface. A positive discrete image at the reciprocal point would lead to a flux across the sphere and could not ensure a tangential velocity. A continuous negative image distribution with suitable uniform strength from the origin to the reciprocal point ensures a tangential field at the surface; this implies that there is no net flux through the sphere, so the total negative continuous image distribution must balance the discrete positive charge distribution. In the case of a sphere that is an equipotential surface, a normal field is ensured by an opposite image at the reciprocal point; there is no continuous image distribution. In both cases of a sphere that is an equipotential (field) surface, the discrete image has a smaller strength than the original. The corresponding two-dimensional problem

TABLE 8.1   Monopole Images on Cylinders (Spheres)

Case	Plane	Space
Surface of radius a	Cylinder	Sphere
Monopole	Line	Point
Strength of original at b	P	P
Images in conductor/equipotential surface		
Discrete at reciprocal point: $c = \dfrac{a^2}{b}$	$-P$	$-\dfrac{Pa}{b}$
Discrete at origin	$+P$	0
Continuous from origin to reciprocal point	0	0
Images in insulator/rigid body/field surface		
Discrete at reciprocal point	P	$\dfrac{Pa}{b}$
Discrete at origin	$-P$	0
Continuous from reciprocal point to origin	0	$-\dfrac{P}{a}$
Total strength of continuous image	0	$-P\dfrac{c}{a} = -P\dfrac{a}{b}$

Note: The corresponding problems for plane (spatial) potential fields (Classification 8.1) lead to the images of line (point) multipoles (Table 8.1) on a cylinder (sphere).

(Sections I.24.6, I.26.6, I.28.6) is a line monopole parallel to a cylinder. In this case there are two line images of equal strength in modulus to the original, namely (1) for the equipotential cylinder an opposite image at the reciprocal point and another, equal, image at the center; and (2) vice-versa if the cylinder is a field surface. Thus in the two-dimensional case, the images are always discrete. Like in the spatial case with two images, they balance each other to lead to a zero flux. The images of line (point) monopoles on cylinders (spheres) are summarized in Table 8.1.

NOTE 8.2: Plane and Spatial Potential Fields

Classification 8.1 lists 15 analogous potential problems in the plane and space. This is a just a small sample of the 150 plane potential problems listed in Classification II.8.1 and discussed in Books 1 and 2 of this series; many others also have three-dimensional extensions, beyond the few selected for illustration in Chapters 6 and 8. Some of the three-dimensional problems are (1) axisymmetric (Chapter 6), and others and (2) non-axisymmetric (Chapter 8), and there are problems with both (1, 2) variants. Many of the problems have analogies in higher dimension $N \geq 4$, often more than a variant of each problem; only a few are given as examples (Sections 9.6 through 9.9). The potential problems considered apply in two (three) dimensions to the (1) gravity field [Chapter I.18 (Note 8.3)]; (2) electrostatic field [Chapters I.24, 36 (Note 8.4)]; (3) steady heat conduction [Chapter I.32 (Note 8.5)]; (4/5) irrotational and/or incompressible flow [Chapters I.12, 14, 6, 28, 34, 36, 38; II.2, 8 (Notes 8.7 and 8.8)]; and (6) magnetostatic field [Chapters I.26, 36 (Note 8.9)]. The plane potential problems of second order involving the Laplace or Poisson equations include (1) the linear deflection of an elastic membrane [Sections II.6.1 and 2 (Note 8.6)] with variable forcing by the shear stress; and (2/3) the torsion of prisms (Sections II.6.5 through 8) and flow in a rotating vessel (Section II.6.9) for which the Poisson equation has a constant forcing term. The fourth-order potential problems specified by biharmonic equations include (1) plane elasticity (Chapter II.4); and (2) plane incompressible viscous creeping flow (Notes II.4.4 through 11).

The following notes concern the cases of the Poisson equation with nonuniform forcing, namely, (1) the gravity field due to mass distributions (Note 8.3); (2) the electrostatic field created by electric charges

CLASSIFICATION 8.1   Potential Problems in the Plane and Space

Dimension		Plane	Space
A Multipoles 4	1. Monopole	P: I.12.1 through 5	P: 6.3.3
		G: I.18.3 through 5	G: N.8.3
		E: I.24.1 through 3	E: 8.1
		M: I.26.3	U: 9.7.4
		H: I.32.4	D: II.4.7
	2. Dipole	P: I.12.7	E: 8.1
		G: I.18.9	M: 8.4
		E: I.24.4	U: 9.7.5
		M: I.26.4	D: II.4.8
	3. Quadrupole	P:I.12.8	E: 8.2
		G: I.18.9	M: 8.4
		E: I.24.4	U: 9.7.6
		M: I.26.4	
	4. Multipole	P: I.12.9	E: 8.3
		G: I.18.9	M: 8.6
		E: I.24.4	U: 9.7.1 through 7
		M: I.26.5	
B Fields 7	5. Distributions	E: I.24.3	P: 8.9, N.8.7 and 8
		G: I.18.6 through 8	E: N.5.4 and 5, N.5.7 through 9, N.5.12 and 13, N.5.17 through 21; N.8.4, 11 through 16
		M: I.26.3	M: N.5.6, N.5.10 and 11, N.5.14 through 16, N.5.22 through 27; N.8.9 and 10
			R: N.5.28 through 30
			H: N.8.5
	6. Corner uniform	P: I.16.5 through 9, 32.1 and 2	
		E: I.24.5	
		M: I.26.6	
		H: I.32.4	
	7. Fairing	P: I.28.4	P: 6.3.4 and 8.9.8 and 10
	8. Body	P: I.28.5	P: 6.4.1 and 2, 8.9.11 and 8.9.12
	9. Cylinder/sphere	P: I.28.6 through 9, II.2.8 and 9	P: 6.6 through 6.9, 8.9
		E: I.24.6 through 8	U: 9.8 and 9.9
		M: I.26.7 and 8	E: 6.6 through 6.9
		H: I.32.5 through 8	
		R: II.2.5 and 6	D: II.4.5
	10. Airfoils/wings	P: I.34, II.8.7 through 9	P: 8.9
	11. Ducts	P: I.36.4 through 9, II.2.7	P: 8.8.2 through 5
		E/M: I.36.4 through 9	
C Images 4	12. On one plane	P: I.16.1.4 through 7	E: 8.7.1
		E: I.24.4	M: 8.7.3
			P: 8.7.2 and 4
		M: I.26.5	U: 9.9.1 and 9.9.2
	13. In corner	P: I.16.5, I.32.1 through 3	E: 8.8.1
		E: I.24.5	
		M: I.26.6	
	14. Planes, plates, wells	P: I.36.3 through 9	E: 8.8.2 through 5
		E: I.36.3 through 9	
		M: I.36.3 through 9	

(continued)

CLASSIFICATION 8.1 (continued) Potential Problems in the Plane and Space

Dimension		Plane	Space
	15. On sphere	P: I.28.7 through 9; II.2.5 through 9	P: 8.9
		E: I.24.8	U: 9.8, 9.9.3 and 4
		M: I.26.8	

Note: Analogous problems for four geometries (a) plane; (b, c) spatial (b) axisymmetric and (c) non-axisymmetric; (d) higher dimensional. Seven fields are considered; (1/2) potential (rotational) flow; (3/4) electro(magneto)statics; (5) gravity; (6) elasticity; (7) steady heat conduction.

P, potential flow; E, electrostatics; G, gravity; H, heat conduction; R, rotational flow; M, magnetostatics; D, elasticity; U, higher dimension.

(Note 8.4); (3) the steady heat conduction from heat sources (Note 8.5); (4) the linear deflection of a membrane due to transverse shear stresses (Note 8.6); (5) the irrotational flow due to mass sources/sinks (Note 8.7); (6) the incompressible flow due to vorticity (8.8); and (7) the magnetostatic field due to electric currents (Note 8.9). The preceding seven cases (1–7) consist of five (two) irrotational (solenoidal) fields (1–5) [(6, 7)] with a scalar (vector) potential. They include four cases of anisotropy: (a, b) the electro(magneto)static field (2) [(7)] in a medium with anisotropic [Note 8.4 (Note 8.9)] dielectric permittivity (magnetic permeability); (c, d) steady heat conduction (linear deflection of a membrane) with anisotropic [Note 8.5 (Note 8.6)] thermal conductivity (tangential stresses). The comparison of the preceding cases [Notes 8.3, 8.4, and 8.7 (Notes 8.8 and 8.9)] includes the irrotational (solenoidal) forces (Note 8.10), that is, the gravity/electric/compression (vortical/magnetic) force. The isotropic (anisotropic) cases have been treated in Chapters 6 and 8 (which appear in Subsection 7.8.4 and are pursued further in Notes 8.11 and 8.12). The analogies among the isotropic (anisotropic) cases lead to similar solutions (Note 8.11). The anisotropic cases (Notes 8.4, 8.5, 8.6, and 8.9) involve the anisotropic Laplace operator (Note 8.12) that is self-adjoint as confirmed by its symmetric influence or Green function (Note 8.13). The latter shows that the equipotential surfaces for a monopole in an anisotropic medium are ellipsoids (Note 8.14). The multipolar expansion can be generalized to anisotropic media (Note 8.15) and specifies the forces between multipoles (Note 8.16).

NOTE 8.3: Gravity Field due to a Mass Distribution

*The acceleration of the gravity field (Chapter I.18) is an irrotational vector (8.292a) that derives from a scalar **gravity potential** (8.292b):*

$$\nabla \wedge \vec{g} = 0: \quad \vec{g} = -\nabla \Phi_g; \quad \nabla \cdot \vec{g} = -G\rho, \quad \rho \equiv \frac{dm}{dV}; \tag{8.292a–d}$$

*the gravity field is generated (8.292c) by the **mass density** per unit volume (8.292d) and involves the **gravitational constant** (8.293a) that appears in the Poisson equation (8.293b), for the gravity potential:*

$$G = 6.673 \times 10^{-11} \text{kg}^{-1} \text{m}^{-3} \text{s}^{-2}: \quad \nabla^2 \Phi_g = \nabla \cdot \nabla \Phi_g = -\nabla \cdot \vec{g} = -G\rho. \tag{8.293a,b}$$

The gravity potential (field) due to a mass distribution is given by (8.294a) [(8.294b)]:

$$\Phi_g(\vec{x}) = -\frac{G}{4\pi} \int_D \frac{\rho(\vec{y})}{|\vec{x} - \vec{y}|} d^3\vec{y}, \quad \vec{g}(\vec{x}) = -\frac{G}{4\pi} \int_D \rho(\vec{y}) \frac{\vec{x} - \vec{y}}{|\vec{x} - \vec{y}|^3} d^3\vec{y}, \tag{8.294a,b}$$

and in the case of a point mass (8.295a) simplifies to (8.295b) [(8.295c)]:

$$\rho(\vec{y}) = m\delta(\vec{y}): \quad \Phi_g = -\frac{Gm}{4\pi R}, \quad \vec{g} = -\frac{Gm}{4\pi R^3} \vec{x}. \tag{8.295a–c}$$

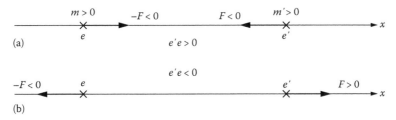

FIGURE 8.19 Newton's (Coulomb's) law of gravity (electrostatics) states that the force between two masses (electric charges) is (1) proportional to their product; (2) inversely proportional to the square of the distance; (3) parallel to their relative position; and (4) always attractive [attractive for opposite (a) charges and repulsive otherwise (b)].

*The **gravity force** exerted on another mass m' is attractive and equal to the product by the gravity field:*

$$\vec{F}_g = m'\vec{g} = \frac{Gm'm}{4\pi}\frac{\vec{x}-\vec{y}}{|\vec{x}-\vec{y}|^3},\tag{8.296a,b}$$

*leading to **Newton's (1687) law of gravity**; the force of attraction of gravity (Figure 8.19a) is proportional to the masses and inversely proportional to the square of the distance. The gravity force is generalized to*

$$\vec{F}_g = \frac{G}{4\pi}\int_{D'}d^3\vec{x}\int_D d^3\vec{y}\,\frac{\vec{x}-\vec{y}}{|\vec{x}-\vec{y}|^3}\rho'(\vec{x})\rho(\vec{y})\tag{8.297}$$

in the case of two mass distributions. The gravity force (8.297) simplifies (1) for one point mass (8.298a) to (8.298b) that leads (8.295c) to (8.298c):

$$\rho'(\vec{x}) = m'\delta(\vec{x}):\quad \vec{F}_g = \frac{m'G}{4\pi} = \int_D \frac{\vec{x}-\vec{y}}{|\vec{x}-\vec{y}|^3}\rho(\vec{y})d^3\vec{y} = m'\vec{g}(\vec{x}).\tag{8.298a–c}$$

(2) for two point masses (8.295a) then (8.298c) with (8.295a) confirms (8.296b).

NOTE 8.4: Electrostatic Field in an Anisotropic Dielectric

The electrostatic field has been considered in two (three) dimensions [Chapter I.24 (Sections I.18.3 through 9; 8.1 through 8.3, 8.7 through 8.9)] in isotropic media and also in anisotropic media (Subsection 7.8.4). The electrostatic field is irrotational (7.154a) ≡ (8.299a) and derives from an electrostatic potential (7.154b) ≡ (8.299b); the electric field is related by dielectric permittivity tensor (7.154d) ≡ (8.299c) to the electric displacement that is generated by the electric charges (7.154c) ≡ (8.299d) leading to the inhomogeneous anisotropic Poisson equation (7.155) ≡ (8.299e):

$$\nabla \wedge \vec{E} = 0;\quad \vec{E} = -\nabla\Phi_e,\quad D_i = \varepsilon_{ij}E_j,\quad -q = \nabla\cdot D,\quad D_i = \varepsilon_{ij}E_j\partial\left(\varepsilon_{ij}\partial_j\Phi_e\right) = -q.\tag{8.299a–e}$$

The operator is linear (self-adjoint) showing that the principle of superposition (reciprocity) applies with regard to the electric charge density (Section 7.8.4). In a homogeneous medium, the dielectric permittivity tensor is constant (8.300a) and the electrostatic potential satisfies an anisotropic Poisson equation (8.300b) solved in Notes 8.11 through 8.15:

$$\varepsilon_{ij} = \text{const}:\quad -q = \varepsilon_{ij}\partial_{ij}\Phi_e \equiv \sum_{i,j=1}^3 \varepsilon_{ij}\frac{\partial^2\Phi_e}{\partial x_i\partial x_j}.\tag{8.300a,b}$$

In an isotropic medium, the dielectric permittivity reduces to a scalar (8.301a) leading to the Poisson equation (8.301b):

$$\varepsilon_{ij} = \varepsilon\delta_{ij}: \quad -q = \varepsilon\delta_{ij}\frac{\partial^2\Phi_e}{\partial x_i\partial x_j} = \varepsilon\frac{\partial^2\Phi_e}{\partial x_i\partial x_j} = \varepsilon\nabla^2\Phi, \tag{8.301a,b}$$

and to the electrostatic potential (8.13a) [field (8.13b)]. The electric force on a point charge e' equals the product by the electric field (8.13b):

$$\vec{F}_e = e'\vec{E} = \frac{e'}{4\pi\varepsilon}\int_D \frac{\vec{x}-\vec{y}}{|\vec{x}-\vec{y}|^3}q(\vec{y})\mathrm{d}^3\vec{y}. \tag{8.302a,b}$$

In the case of two point electric charges (8.303a) from (8.302b) follows (8.303b):

$$q(\vec{y}) = e\delta(\vec{y}): \quad \vec{F}_e = -\frac{e'e}{4\pi\varepsilon}\frac{\vec{x}-\vec{y}}{|\vec{x}-\vec{y}|^3}, \tag{8.303a,b}$$

which is the **Coulomb law of electricity** *that is similar to Newton's law of gravity (8.296b) with one difference regarding sign: (1) since the mass is always positive the gravity force (8.286b) is always attractive; and (2) since the electric charge can be positive or negative, the electric force (8.303b) is attractive (repulsive) for charges with opposite (same) sign [Figure 8.19a (Figure 8.19b)]. Although the mass is always positive, the multipolar expansion (8.52) applies as well to the gravity field. The electric force (8.303b) extends to two electric charge distributions:*

$$\vec{F}_e = -\frac{\varepsilon}{4\pi}\int_D \mathrm{d}^3\vec{x}\int_D \mathrm{d}^3\vec{y}\,\frac{\vec{x}-\vec{y}}{|\vec{x}-\vec{y}|^3}q'(\vec{x})q(\vec{y}), \tag{8.304}$$

as for the gravity force (8.297).

NOTE 8.5: Steady Anisotropic Heat Conduction

The conservation of heat in steady conditions (Chapter I.32) states that the output (input) of heat sources (sinks) must be balanced (I.32.3) ≡ *(8.305a) by the divergence of the* **heat flux**:

$$w = \nabla\cdot\vec{G} = \partial_i G_i; \quad G_i = -k_{ij}\partial_j T; \tag{8.305a,b}$$

the **Fourier (1818) law** *states that the heat flux is proportional to the temperature gradient (8.306b) through the* **thermal conductivity tensor**. *Thus the temperature satisfies an inhomogeneous anisotropic Poisson equation (8.306a) for steady heat conduction due to heat sources or sinks:*

$$-w = \partial_i\left(k_{ij}\partial_j T\right); \quad k_{ij} = \text{const}: \quad -w = k_{ij}\partial_{ij}T, \tag{8.306a-c}$$

in an homogeneous medium, the thermal conductivity tensor is constant (8.306b) leading to an anisotropic Poisson equation (8.306c). In an isotropic medium, the thermal conductivity reduces to a scalar (8.307a) implying that (1) the heat flux (8.307b) is in the direction of decreasing temperature:

$$k_{ij} = k\delta_{ij}: \quad \vec{G} = -k\nabla T; \quad -w = k\nabla^2 T; \tag{8.307a-c}$$

and (2) the temperature satisfies a Poisson equation (8.307c) ≡ (I.32.5c). Thus a distribution of heat source/sinks creates a temperature (8.308a) [heat flux (8.308b)] in steady conditions:

$$T(\vec{x}) = \frac{1}{4\pi k} \int_D \frac{w(\vec{y})}{|\vec{x} - \vec{y}|} d^3\vec{y}, \quad \vec{G}(\vec{x}) = \frac{1}{4\pi} \int_D \frac{\vec{x} - \vec{y}}{|\vec{x} - \vec{y}|^3} w(\vec{y}) d^3\vec{y}. \tag{8.308a,b}$$

The superposition and reciprocity principles (7.156c and d) are satisfied by (7.306a).

NOTE 8.6: Deflection of a Membrane under Anisotropic Stresses

The deflection (8.309a) of an elastic membrane (Sections II.6.1 and 2) is linear if the slope is small everywhere (II.6.12a) ≡ (8.309a and b) in which case it satisfies an inhomogeneous Poisson equation (II.6.12b) ≡ (8.309c):

$$z = \zeta(x, y), |\nabla\zeta|^2 \ll 1: \quad -f(x, y) = \nabla \cdot (T\nabla\zeta), \quad f = \frac{dF}{dxdy}, \tag{8.309a–d}$$

where (1) the forcing term is the **shear stress** or transverse force per unit area (8.309d); (2) the tangential tension T is assumed to be isotropic. Otherwise (8.310a), the inhomogeneous Poisson equation (8.309c) becomes anisotropic as well (8.310b):

$$T_{ij} \neq T\delta_{ij}: \quad -f = \partial_i (T_{ij}\partial_i\zeta). \tag{8.310a,b}$$

In the case of uniform anisotropic (8.311a) [isotropic (8.311c)] tension, the transverse displacement satisfies an anisotropic (isotropic) Poisson equation (8.311b) [(8.311d)]:

$$T_{ij} = \text{const}: \quad -f = T_{ij}\partial_{ij}\zeta; \quad T = \text{const}: \quad \nabla^2\zeta = -\frac{f}{T}. \tag{8.311a–d}$$

In the case of isotropic (8.312a) uniform (8.312b) tension, the linear (8.312c) deflection of an elastic membrane by a distribution of transverse shear stress is given by (8.312d):

$$T_{ij} = T\delta_{ij}, T = \text{const}, |\nabla\zeta|^2 \ll 1: \quad \zeta(x) = \frac{1}{4\pi T} \int_D \frac{f(\vec{y})}{|\vec{x} - \vec{y}|} d^3\vec{y}; \tag{8.312a–d}$$

The Cauchy principal value,

$$D' = D - \{|\vec{x} - \vec{y}| < \varepsilon\}: \quad _C\!\!\int_D \frac{f(\vec{y})}{|\vec{x} - \vec{y}|} d^3\vec{y} = \lim_{\varepsilon \to 0} \int_{D'} \frac{f(\vec{y})}{|\vec{x} - \vec{y}|} d^3\vec{y}, \tag{8.312e–f}$$

of the integral (8.312f) excludes the neighborhood of the observer (8.312e) in the case when it lies inside the source distribution. *If the displacement is measured at a point $\vec{x}$ inside the region where there is $\vec{y}$ a shear stress $f(\vec{y})$, the singularity $|\vec{x} - \vec{y}| = 0$ in (8.312d) is avoided by taking the Cauchy principal value of the integral (Section I.17.9)* that (1) has been demonstrated for the gravity inside the mass (Sections I.18.6 and 7); and (2) applies as well in similar cases such as the electric field (8.13a and b) inside a distribution of electric charges.

NOTE 8.7: Irrotational Flow due to Mass Sources/Sinks

The fifth and last irrotational field (Notes 8.3 through 8.7) is the irrotational flow (8.313a) for which the velocity derives from a scalar hydrodynamic potential (8.313b):

$$\vec{\varpi} \equiv \nabla \wedge \vec{v} = 0: \quad \vec{v} = \nabla \Phi_v; \quad \nabla \cdot \vec{v} = \dot{D}, \quad \dot{D} = \nabla \cdot \nabla \Phi_v = \nabla^2 \Phi_v; \tag{8.313a–d}$$

*the flow **sources (sinks)** specify a positive (negative) **dilatation** (8.313c) that forces the Poisson equation for the potential (8.313d). Thus a distribution of volume sources/sinks creates a flow with potential (8.314a) [velocity (8.314b)]:*

$$\Phi_v(\vec{x}) = -\frac{1}{4\pi} \int_D \frac{\dot{D}(\vec{y})}{|\vec{x} - \vec{y}|} d^3\vec{y}, \quad \vec{v}(\vec{x}) = \frac{1}{4\pi} \int_D \frac{\vec{x} - \vec{y}}{|\vec{x} - \vec{y}|^3} \dot{D}(\vec{y}) d^3\vec{y}. \tag{8.314a,b}$$

*If the flow is **homentropic**, that is, if there are no heat exchanges or entropy gradients (6.315a), the pressure p and density ρ are related by the **Bernoulli (1738) equation** (I.14.25c) ≡ (6.315c):*

$$S = \text{const}; \quad \vec{F}_a = -\nabla \Phi_a: \quad \frac{\partial \Phi}{\partial t} + \frac{1}{2}(\nabla \Phi)^2 + \int \frac{dp}{\rho} + \Phi_a = \text{const}, \tag{8.315a–c}$$

*assuming that all the external **applied forces** are **conservative** (8.315b), that is, derivative from a scalar potential. For a steady (8.316a), incompressible (8.316b) flow in the absence of external applied forces (8.316c), the Bernoulli equation (8.315c) takes the simplest form (8.316d) ≡ (6.102c):*

$$\frac{\partial}{\partial t} = 0, \rho = \text{const}, \vec{F}_a = 0: \quad p + \frac{1}{2}\rho v^2 = p_0 = \text{const}, \tag{8.316a–d}$$

where p_0 is the stagnation pressure.

NOTE 8.8: Incompressible Flow due to Vorticity

The passage from irrotational (to solenoidal) fields [Notes 8.3 through 8.7 (Notes 8.8 and 8.9)] is illustrated by the example of the *irrotational (8.313a) [incompressible (8.317a)]* flow for which the velocity derives from a scalar *(vector)* potential (8.313b) [(8.317b)]:

$$\nabla \cdot \vec{v} = 0: \quad \vec{v} = \nabla \wedge \vec{A}; \quad \vec{\varpi} = \nabla \wedge \vec{v} = \nabla \wedge (\nabla \wedge \vec{A}) = -\nabla^2 \vec{A}; \tag{8.317a–d}$$

*the dilatation (8.313) [**vorticity** (8.317c)] forces the Poisson equation for the scalar (8.313d) [vector (8.317d)] potential. In the derivation of the latter was imposed the **gauge condition** (8.318b) on the vector potential:*

$$\nabla \cdot \vec{A} = 0; \quad \vec{A} = \vec{e}_z \Psi(x,y), \quad \vec{A} = \vec{e}_\varphi \frac{\Psi(r,z)}{r} = \vec{e}_\varphi \frac{\Psi(R,\theta)}{R\sin\theta}; \tag{8.318a–d}$$

*the vector potential reduces to a scalar **stream function** in the case of plane flow (8.318b) ≡ (6.64a) [three-dimensional axisymmetric flow in cylindrical (8.318c) ≡ (6.60c) or spherical (8.318d) ≡ (6.60d) coordinates)]. In a steady (8.319a), axisymmetric (8.319b), incompressible (8.319c), and homentropic (8.319d) flow, the vorticity depends only on the stream function (6.106h) and is related to the pressure and velocity by (8.319e):*

$$\frac{\partial}{\partial t} = 0 = \frac{\partial}{\partial \varphi}; \quad \rho, S = \text{const}: \quad \frac{1}{\rho}dp + \vec{v} \cdot d\vec{v} = -\frac{\varpi(\Psi)}{r}d\Psi. \tag{8.319a–e}$$

$$\frac{\partial}{\partial t} = 0 = \frac{\partial}{\partial z}; \quad \rho, S = \text{const}: \quad \frac{1}{\rho} dp + \vec{v} \cdot d\vec{v} = -\varpi(\Psi) d\Psi. \tag{8.320a–e}$$

The corresponding (8.320a, c, and d) plane flow (8.320b), the vorticity velocity, the stream function, and the pressure are related by (8.320e). The Bernoulli equation (8.316d) corresponds to zero vorticity on the r.h.s. of (8.319e) [(8.320e)], and is generalized to (8.321a) [(8.321b)]:

$$\frac{P}{\rho} + \frac{v^2}{2} + \frac{1}{r} \int \varpi(\Psi) d\Psi = \text{const}, \quad \frac{P}{\rho} + \frac{v^2}{2} + \int \varpi(\Psi) d\Psi = \text{const}, \tag{8.321a,b}$$

in the case of axisymmetric (8.319b) [plane (8.320b)] steady [(8.319a) ≡ (8.320a)], incompressible [(8.319c) ≡ (8.320c)], homentropic [(8.319d) ≡ (8.320d)] flow.

NOTE 8.9: Magnetic Field in an Anisotropic Medium

Another example of the passage from an irrotational (to a solenoidal) field is the *steady electro(magneto) static field for which the electric field (8.299a) [magnetic induction (8.322a)] is irrotational (solenoidal) and thus derives (8.299b) [(8.322b)] from a scalar (vector) electric (**magnetic**) potential*:

$$\nabla \cdot \vec{B} = 0: \quad \vec{B} = \nabla \wedge \vec{A}; \quad \frac{\vec{j}}{c} = \nabla \wedge \vec{H}, \quad B_i = \mu_{ij} H_j, \tag{8.322a–d}$$

*the electric (magnetic) field is generated (8.299c) [(8.322c)] by electric charges (currents) that specify the electric displacement $\vec{D}$ (magnetic field $\vec{H}$). The electric displacement $\vec{D}$ (magnetic induction $\vec{B}$) is related to the electric $\vec{E}$ (magnetic $\vec{H}$) field through (7.299d) [(8.322d)] the dielectric permittivity (**magnetic permeability**) tensor. The* **curl** *(8.323b) like the* **outer product** *(8.323c) of vectors in three dimensions (8.323a) uses the* **permutation symbol** *(5.182a through c):*

$$i, j, k = 1, 2, 3: \quad \left(\vec{A} \wedge \vec{B}\right)_i = e_{ijk} A_j B_k, \quad \left(\nabla \wedge \vec{H}\right)_i = e_{ijk} \partial_j H_k. \tag{8.323a–c}$$

Using the inverse μ_{ij}^{-1} of the magnetic permeability tensor (8.324a), the system of equations (8.322c, d, and b) may be eliminated:

$$\mu_{ij} \mu_{jk}^{-1} = \delta_{ik}: \quad \frac{j_i}{c} = e_{ijk} \partial_j H_k = e_{ijk} \partial_j \left(\mu_{k\ell}^{-1} B_\ell\right) = e_{ijk} \partial_j \left(\mu_{k\ell}^{-1} e_{\ell mn} \partial_m A_n\right), \tag{8.324a,b}$$

for the vector potential (8.324b).

Thus *a distribution of electric currents (8.322c) in a medium with inverse magnetic permeability tensor (8.322d; 8.324a) creates a vector magnetic potential (8.322b) that satisfies a vector inhomogeneous anisotropic Poisson equation*:

$$j_i = c e_{ijk} e_{\ell mn} \partial_j \left[\mu_{k\ell}^{-1} \left(\partial_m A_n\right)\right]. \tag{8.325}$$

If the medium is homogeneous, the magnetic permeability tensor is constant (8.326a) and the magnetic potential satisfies the vector anisotropic Poisson equation (8.326b):

$$\mu_{ij} = \text{const}: \quad j_i = c \mu_{k\ell}^{-1} e_{ijk} e_{\ell mn} \partial_{jm} A_n. \tag{8.326a,b}$$

If the medium is inhomogeneous but isotropic (8.327a), the vector potential satisfies the vector inhomogeneous Poisson equation (8.327b):

$$\mu_{ij} = \mu(\vec{x}) \delta_{ij}: \quad \vec{j} = c \nabla \wedge \vec{H} = \nabla \wedge \left(\mu \vec{B}\right) = \nabla \wedge \left[\mu \left(\nabla \wedge \vec{A}\right)\right]. \tag{8.327a,b}$$

If the medium is isotropic (8.327a) ≡ (8.328a) and homogeneous (8.326a) ≡ (8.328b), the vector potential satisfies a vector Poisson equation (8.328d) ≡ (8.90c) using the gauge condition (8.90a) ≡ (8.328c):

$$\mu_{ij} = \mu\delta_{ij}, \mu = const; \quad \nabla \cdot \vec{A} = 0: \quad \frac{\mu}{c}\vec{j} = \nabla \wedge \left(\nabla \wedge \vec{A}\right) = -\nabla^2\vec{A}. \tag{8.328a-d}$$

The vector magnetic (velocity) potential reduces to a scalar field (stream) function in the plane (8.318b) [axisymmetric (8.318c and d)] cases.

NOTE 8.10: Gravity/Electric/Compression and Vortical/Magnetic Forces

*The irrotational fields have forces equal to the product of the field by the source, namely, the mass/electric charge/dilatation times the gravity field/the electric field/minus the mass flux specify the gravity (8.329a) ≡ (8.296a) electric (8.329b) ≡ (8.302a)/**compression force** (8.329c) ≡ (I.28.49d):*

$$\vec{F}_g = m'\vec{g}, \quad \vec{F}_e = e'\vec{E}, \quad \vec{F}_b = -\rho\vec{v}\dot{D}' = -\rho\vec{v}\left(\nabla\cdot\vec{v}'\right). \tag{8.329a-c}$$

In the case of two sources, that is, two masses/two electric charges/two sources-sinks, this leads to the Newton (8.296b)/Coulomb (8.303b)/analogue compression force (8.330),

$$\vec{F}_b = \rho\vec{v}\dot{D}' = -\frac{\rho\dot{D}'\dot{D}}{4\pi}\frac{\vec{x} - \vec{y}}{\left|\vec{x} - \vec{y}\right|^3}, \tag{8.330}$$

*between two source/sinks $\dot{D}', \dot{D}$ at positions $\vec{x}, \vec{y}$, showing that [Figure 8.19a (Figure 8.19b)] they attract (repel) each other if they have opposite (equal) signs, that is, for a source and sink pair (two sources or two sinks). The solenoidal field has forces equal to the outer vector product of the force by the field, namely, the **magnetic (vortical) force** is equal (8.331b) [(8.332b)] to the vector product of the electric current (vorticity) by the magnetic induction (velocity) times a constant 1/c (the mass density ρ):*

$$\vec{j} = c\nabla \wedge \vec{H}: \quad \vec{F}_m = \frac{1}{c}\vec{j} \wedge \vec{B} = \frac{1}{c}\left(\nabla \wedge \vec{H}\right) \wedge \vec{B}, \tag{8.331a-c}$$

$$\vec{\omega} = \nabla \wedge \vec{v}: \quad \vec{F}_\ell = \rho\vec{v} \wedge \vec{\omega} = \rho\vec{v} \wedge \left(\nabla \wedge \vec{v}\right); \tag{8.332a-c}$$

the magnetic (vortical) force can be written alternatively (8.331c) [(8.332c)] expressing the electric current (8.331a) [vorticity (8.332a)] in terms of the magnetic field (velocity).

In the case of an isotropic medium (8.333a) ≡ (8.322d; 8.327a), the magnetic force (8.331c) becomes (8.333b):

$$\vec{B} = \mu\vec{H}: \quad \vec{F}_m = \frac{\mu}{c}\vec{j} \wedge \vec{H} = \frac{\mu}{c}\left(\nabla \wedge \vec{H}\right) \wedge \vec{H}. \tag{8.333a,b}$$

Using the magnetic field due to a point current (8.101) leads to the magnetic force (8.334a) and analogous vortical force (8.334b):

$$\vec{F}_m = \frac{\mu}{4\pi c}\frac{\vec{J}' \wedge \left[\vec{J} \wedge \left(\vec{x} - \vec{y}\right)\right]}{\left|\vec{x} - \vec{y}\right|^3}, \quad \vec{F}_\ell = \frac{1}{4\pi}\frac{\vec{\omega}' \wedge \left[\vec{\omega} \wedge \left(\vec{x} - \vec{y}\right)\right]}{\left|\vec{x} - \vec{y}\right|^3}. \tag{8.334a,b}$$

Thus has been obtained the **Biot–Savart law** *(8.334a) [analogous vortical force law (8.334b)] specifying the force between two point electric currents* $\left(\vec{J},\vec{J}'\right)$ *[vortices* $\left(\vec{\varpi};\vec{\varpi}'\right)$*] at positions* $\left(\vec{y},\vec{x}\right)$*. The magnetic force (8.334a) may be written as*

$$\vec{F}_m = \frac{\mu}{4\pi c}\,\frac{\vec{J}\left[\left(\vec{x}-\vec{y}\right)\cdot\vec{J}'\right]-\left(\vec{x}-\vec{y}\right)\left(\vec{J}'\cdot\vec{J}\right)}{\left|\vec{x}-\vec{y}\right|^3}. \tag{8.335}$$

If (Figure 8.20a) the electric currents are parallel (8.336a) and orthogonal to the relative position vector (8.336b), the magnetic force (8.327c),

$$\vec{J}\wedge\vec{J}'=0,\ \vec{J}'\cdot\left(\vec{x}-\vec{y}\right)=0:\quad \vec{F}_m = -\frac{\mu J'J}{4\pi c}\,\frac{\vec{x}-\vec{y}}{\left|\vec{x}-\vec{y}\right|^2}, \tag{8.336a–c}$$

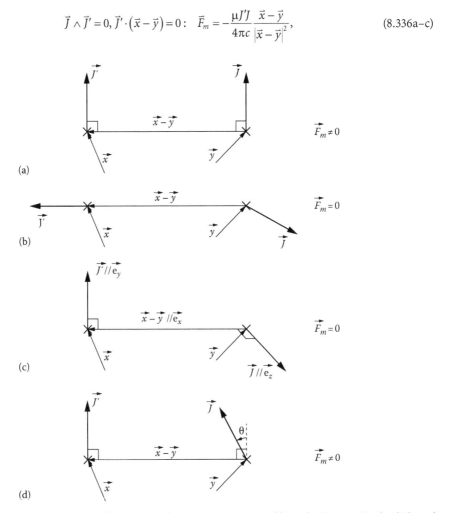

FIGURE 8.20 The Biot–Savart law for the magnetic force is (Figure 8.19a and b) to the Newton (Coulomb) law of gravity (electrostatics) in the case (a) of parallel electric currents orthogonal to their relative position. Otherwise the magnetic force is usually different from the gravity or electric forces. The latter are never zero, whereas the magnetic force can vanish, for example, if (1) one electric current is parallel to the relative position (b); and (2) the two electric currents are orthogonal to the relative position and to each other (c). The case (a) [(c)] of magnetic force like the electric force (zero instead) is the parallel (orthogonal) extreme of (d) two electric currents perpendicular to their relative position and at an arbitrary angle between them. These differences apply to all irrotational (gravity, electric, compression) and solenoidal (vortical, magnetic) forces.

resembles Newton's (Coulomb's) law for the gravity (8.296b) [electric (8.303b)] force with the opposite (same) sign. If (Figure 8.20b) one current lies along the relative position vector (8.337a), the magnetic force is zero (8.337b):

$$\left(\vec{x}-\vec{y}\right)\wedge\vec{J}=0 \Rightarrow \vec{F}_m=0 \Leftarrow \vec{J}\cdot\vec{J}'=0=\vec{J}'\cdot\left(\vec{x}-\vec{y}\right), \qquad (8.337a\text{–}d)$$

If (Figure 8.20c) the currents and relative position vector form an orthogonal triad (8.337c and d), the magnetic force is also zero (8.337b). If (Figure 8.20d) one current is orthogonal to the position vector (8.338a) and makes an angle θ with the other current (8.338b), the magnetic force is given by (8.338c)

$$\vec{J}'\cdot\left(\vec{x}-\vec{y}\right)=0; \quad \left(\vec{J}'\cdot\vec{J}\right)=\vec{J}'\vec{J}\cos\theta: \quad \vec{F}_m=-\frac{\mu\vec{J}'\vec{J}}{4\pi c}\frac{\vec{x}-\vec{y}}{\left|\vec{x}-\vec{y}\right|^3}\cos\theta, \qquad (8.338a\text{–}c)$$

which includes (8.336c) [(8.337b)] for $\theta=0\left(\theta=\pi/2\right)$.

NOTE 8.11: Potential, Sources, and Constitutive Tensor

The preceding irrotational (solenoidal) fields [Notes 8.3 through 8.7 (Notes 8.8 and 8.9)] involve (1) a scalar (vector) potential satisfying a Poisson equation; (2) the sources of the field as the forcing term; (3) scalar (tensor) coefficients representing the properties of the isotropic (anisotropic) medium; and (4) constant coefficients (coefficients depending on position) correspond to a homogenous (inhomogeneous) medium. Thus for homogeneous media, four sets of analogue cases arise, namely, anisotropic/isotropic irrotational/solenoidal fields. In the case of *irrotational anisotropic fields, the scalar potential in a homogeneous anisotropic medium satisfies the anisotropic Poisson equation (8.339a):*

$$C_{ij}\partial_{ij}\Phi=\Lambda: \quad \Phi\equiv\left\{\Phi_e,T,\zeta\right\}, \quad C_{ij}=\left\{\varepsilon_{ij},k_{ij},T_{ij}\right\}, \quad \Lambda\equiv-\left\{q,w,f\right\}, \qquad (8.339a\text{–}d)$$

where for the electrostatic field (8.299e)/steady heat conduction (8.306c)/linear deflection of a membrane (8.311b), (1) the potential (8.339b) is the electrostatic potential/temperature/transverse deflection; (2) the properties of the medium are represented (8.339c) by the tensors dielectric permittivity/thermal conductivity/tangential stresses; and (3) the source or forcing term (8.339d) is minus the density of electric charge/heat source or sink/shear stress or force per unit area. The particular isotropic cases of (8.339a through d) are joined by the gravity field (8.293b) [irrotational flow (8.313d)] in the Poisson equation (8.340a):

$$C\nabla^2\Phi=\Lambda: \quad \Phi\equiv\left\{\Phi_e,T,\zeta,\Phi_g,\Phi_v\right\}, \quad C=\left\{\varepsilon,k,T,\frac{1}{G},1\right\}, \quad \Lambda=-\left\{q,w,f,-\rho,-\dot{D}\right\}, \qquad (8.340a\text{–}d)$$

by (1) adding the gravity (electrostatic) potential (8.340b); (2) adding the inverse of the gravitational constant (8.293a) [unity] representing the medium; and (3) adding the mass density (subtracting the dilatation) as the source. The isotropic solenoidal fields, namely, incompressible flow (8.317d) [magnetostatic field (8.328d)], satisfy a vector Poisson equation (8.341a):

$$C\nabla^2\vec{A}=\vec{\Lambda}: \quad C=\left\{1,\frac{c}{\mu}\right\}, \quad \vec{\Lambda}=-\left\{\vec{\varpi},\vec{j}\right\}, \quad c=3\times10^8\,\text{ms}^{-1}, \qquad (8.341a\text{–}d)$$

involving the vector potential and (1) unity [the ratio of the speed of light in vacuo (8.341d) to the magnetic permeability] representing the medium (8.341b); and (2) minus the vorticity (electric current) as the vector source (8.341c). The vector Poisson equation (8.341a) in Cartesian coordinates consists of three scalar Poisson equations (8.340a), and the latter will be considered in the sequel (Notes 8.12 through 8.15) as a particular case of the anisotropic form (8.339a).

NOTE 8.12: Principal Directions, Eigenvalues and Isotropy

For a potential with continuous second-order derivatives (8.342a), (1) the derivatives are symmetric (8.342b); and (2) the **constitutive tensor** can be replaced (8.342a through c) by its symmetric part:

$$\Phi(\vec{x}) \in C^2\left(\left|R^3\right|\right): \quad \partial_{ij}\Phi = \partial_{ij}\Phi, \quad 2\bar{C}_{ij} \equiv C_{ij} + C_{ji} = 2\bar{C}_{ji}. \tag{8.342a–c}$$

The symmetric constitutive tensor can be (Figure 8.21a) diagonalized (8.343b and c) in its principal reference frame with eigenvalues C_i in

$$\xi_i \equiv \frac{\eta_i}{\sqrt{C_i}}: \quad \bar{C}_{ij} = C_i \delta_{ij} = \begin{cases} C_i & \text{if } i = j, \\ 0 & \text{if } i \neq j; \end{cases} \tag{8.343a–c}$$

the coordinates in the principal reference frame are rescaled (8.343a) using the eigenvalues, so that the anisotropic Laplace operator,

$$\sum_{i,j=1}^{3} C_{ij} \frac{\partial^2}{\partial x_i \partial x_j} = \sum_{i=1}^{3} C_i \frac{\partial^2}{\partial \eta_i^2} = \sum_{i=1}^{3} \frac{\partial^2}{\partial \xi_i \partial \xi_j}, \tag{8.344}$$

becomes the isotropic Laplace operator (8.344) for which the influence function is known (8.11b). It has been shown that *the homogeneous anisotropic Laplace operator (8.339a) for a potential with continuous second-order derivatives (8.342a) can be transformed to the homogeneous isotropic Laplace operator (8.344) by (1) taking the symmetric part (8.342c) of the constitutive tensor; (2) choosing principal axis in which it is diagonal (8.343b and c); and (3) using the eigenvalues to rescale the coordinates (8.343a).*

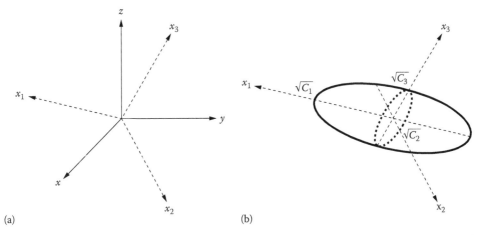

(a) (b)

FIGURE 8.21 The potential field in an anisotropic medium is affected by its constitutive tensor C_{ij}, specifying how the physical properties vary with direction. Since the tensor is symmetric, it can be diagonalized by choosing a set of orthogonal principal axis (a), with the anisotropy of the medium described by the three eigenvalues leading to equipotential surfaces of a monopole that are ellipsoids (b).

In the isotropic case (8.345a), all the eigenvalues are equal and (8.339a) ≡ (8.345b) reduces to (8.340a) ≡ (8.345c) in any orthogonal frame:

$$C_{ij} = C\delta_{ij}: \quad C_{ij}\partial_{ij}\Phi = \Lambda \leftrightarrow C\nabla^2\Phi = \Lambda. \tag{8.345a–c}$$

Thus the influence function (8.11a) for the homogeneous isotropic Laplace operator (8.345c) specifies that for the homogeneous anisotropic Laplace operator (8.345b) via the transformation (8.343a through c).

NOTE 8.13: Influence Function for an Anisotropic Medium

It has been shown that *the influence or Green function for the anisotropic Laplace operator (8.345b) ≡ (8.346b) with constant coefficients (8.346a),*

$$C_{ij} = \text{const}: \quad \left\{C_{ij}\frac{\partial^2}{\partial x_i \partial x_j}\right\}G(\vec{x};\vec{y}) = \delta(\vec{x};\vec{y}), \tag{8.346a,b}$$

is given (8.11b) by –1/4π divided (8.346a) by the distance (8.346d),

$$G(\vec{x};\vec{y}) = -\frac{1}{4\pi|\vec{z}|} = G(\vec{y};\vec{x}), \quad z_i = \frac{x_i - y_i}{\sqrt{C_i}}, \quad |\vec{z}| = \left|\sum_{i=1}^{3}\frac{(x_i - y_i)^2}{C_i}\right|^{-1/2}, \tag{8.347a–d}$$

calculated for (1) the relative position $\vec{x} - \vec{y}$ measured in the principal reference frame (Figure 8.21a) of the constitutive tensor where it is diagonal (8.343b and c); and (2) its eigenvalues (8.343a) are used to rescale the coordinates (8.347c). Since the influence function (8.347a) is symmetric (8.347b) and depends only on the difference of coordinates, the principle of superposition leads (7.78b) to a convolution integral specifying the potential for an arbitrary source (8.345b) ≡ (8.339a):

$$\Phi(x_i) = \Lambda * G = \int_{-\infty}^{+\infty}\Lambda(\vec{\xi})G(\vec{x};\vec{y})\mathrm{d}^3\vec{\xi}$$

$$= -\frac{1}{4\pi}\int_{-\infty}^{+\infty}\Lambda\left(\frac{y_i}{\sqrt{C_i}}\right)\left|\sum_{i=1}^{3}\frac{(x_i - y_i)^2}{C_i}\right|^{-1/2}\prod_{i=1}^{3}\frac{\mathrm{d}y_i}{\sqrt{C_i}}. \tag{8.348}$$

The rescaling (8.343a) of the coordinates along the principal directions using the eigenvalues of the constitutive matrix (8.343b and c) applies to (1) the source term (8.339d); (2) the modified distance (8.347c and d) from the source to the observer; and (3) to the volume element in (8.348).

NOTE 8.14: Spherical, Spheroidal, and Ellipsoidal Equipotential Surfaces

In the case of an isotropic medium (8.345a) ≡ (8.349a) or (8.343b and c) equal eigenvalues (8.349b), the potential (8.348) simplifies (8.349c) to the isotropic form like (8.288a):

$$C_{ij} = C\delta_{ij}; \quad C_i \equiv C: \quad \Phi(x_i) = -\frac{1}{4\pi C}\int_{-\infty}^{+\infty}\frac{\Lambda(\vec{x})}{|\vec{x} - \vec{y}|}\mathrm{d}^3\vec{y}. \tag{8.349a–c}$$

The potential (8.348) of an arbitrary source (8.345b) in an anisotropic medium with constitutive tensor (8.342c; 8.343b and c) for a monopole (8.350a) simplifies to (8.350b):

$$S(\vec{x}) = P_0\delta(\vec{x} - \vec{y}): \quad \Phi_0(x_i) = -\frac{P_0}{4\pi|\vec{z}|}, \quad \text{const} = \sum_{i=1}^{3}\frac{(x_i - y_i)^2}{C_i}, \tag{8.350a–c}$$

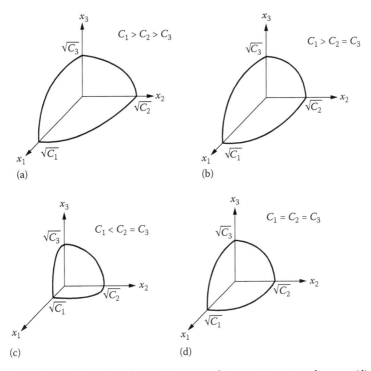

FIGURE 8.22 The equipotential surfaces for a point monopole in an isotropic medium are (d) spheres (Figure 8.2) because the properties of the medium are the same in all directions (Figure 8.20d), as implied by the three identical eigenvalues. For a uniaxial medium with two equal eigenvalues, there is axisymmetry around the axis of the third distinct eigenvalue: the spherical equipotential surfaces are stretched (squashed) into a prolate (b) [oblate(c)] spheroid [Figure 8.20b (Figure 8.20c)] like a rugby ball (a flying saucer) if the distinct eigenvalue is larger (smaller) than the other two. For a biaxial medium with (a) three distinct eigenvalues (Figure 8.20a), the equipotential surface is an ellipsoid, whose section by all three coordinate planes are distinct ellipses, and thus is not a surface of revolution.

*showing that the equipotential surfaces are ellipsoids (8.350c) (Figure 8.21b) with (1) axis along (Figure 8.21a) the principal directions of the constitutive tensor (8.343b and c); and (2) semiaxis proportional to the square root of the eigenvalues. The equipotential surface (Table 8.2) is (1) a **ellipsoid** (Figure 8.21b ≡ 8.22a) for a **biaxial** medium with distinct eigenvalues in all three directions; (2) a **prolate (oblate)** spheroid in a **uniaxial** medium with two equal eigenvalues that are smaller (larger) than the other [Figure 8.22b (Figure 8.22c)]; and (3) a sphere (Figure 8.22d) in the isotropic case (8.349c) corresponding to equal eigenvalues (8.349b).*

TABLE 8.2 Equipotential Surfaces in an Anisotropic Medium

Medium	Equipotentials	Eigenvalues of Constitutive Tensor	Figure
Biaxial	Ellipsoid	$C_1 > C_2 > C_3$	8.22a
Uniaxial	Prolate spheroid	$C_1 > C_2 = C_3$	8.22b
	Oblate spheroid	$C_1 < C_2 = C_3$	8.22c
Isotropic	Sphere	$C_1 = C_2 = C_3$	8.22d

Note: The sphere is the equipotential surface of a monopole in an isotropic medium, and is deformed into a spheroid (ellipsoid) in a uniaxial (biaxial) anisotropic medium.

NOTE 8.15: Multipolar Expansion in an Anisotropic Medium

The potential (8.348) for an arbitrary source in an anisotropic medium leads to the multipolar expansion (8.351a and b),

$$|\vec{x}| > |\vec{y}|: \quad \Phi(\vec{x}) = \sum_{n=0}^{\infty} \Phi_n(\vec{x}), \quad -4\pi\Phi_n(\vec{x}) = P_{i_1 \ldots i_n} \frac{\partial^m}{\partial \xi_{i_1} \ldots \partial \xi_{i_n}} \frac{1}{|\vec{z}|}, \tag{8.351a-c}$$

$$E_i^{(n)} \equiv \frac{\partial \Phi_n}{\partial x_i} = -\frac{1}{4\pi} P_{i_1 \ldots i_n} \frac{\partial^{n+1}}{\partial \xi_{i_1} \ldots \partial \xi_{i_n} \partial x_i} \frac{1}{|\vec{z}|}, \tag{8.351d}$$

with (1) general multipole moment (8.57c) → (8.352),

$$P_{i_1 \ldots i_n} = \frac{1}{n!} \sum_{k=0}^{\leq n/2} (2n - 2k - 1)!!(-)^k \delta_{i_1 i_2} \ldots \delta_{i_{2k-1} i_{2k}}$$

$$\int_D \frac{y_{i_{2k+1}} \ldots y_{i_n}}{\sqrt{C_{i_{2k+1}} \ldots C_{i_n}}} \Lambda\left(\frac{y_k}{\sqrt{C_k}}\right) \prod_{j=1}^{3} \frac{dy_j}{\sqrt{C_j}}, \tag{8.352}$$

in the potential (8.351c) and field (8.351d); and (2) differentiation (8.351d) [integration (8.352)] with regard to rescaled coordinates (8.343a) and relative position (8.347c). The lowest-order term is a monopole with moment (8.353a), potential (8.353b) ≡ (8.350b), and field (8.353c):

$$P_0 = \int_D \Lambda\left(\frac{y_k}{\sqrt{C_k}}\right) \prod_{j=1}^{3} \frac{dy_j}{\sqrt{C_j}}: \quad \Phi_0(\vec{x}) = -\frac{P_0}{4\pi|\vec{z}|}, \quad E_{oi} = \frac{P_0}{4\pi|\vec{z}|^3} \frac{x_i}{C_i}. \tag{8.353a-c}$$

The second-order term is a dipole with moment (8.354a), potential (8.354b), and field (8.354c):

$$P_i = \int_D \Lambda\left(\frac{y_k}{\sqrt{C_k}}\right) \frac{y_i}{\sqrt{C_i}} \prod_{j=1}^{3} \frac{dy_j}{\sqrt{C_j}}: \quad \Phi_1(x_i) = -\frac{1}{4\pi|\vec{z}|} \sum_{i=1}^{3} \frac{P_i x_i}{C_i}, \tag{8.354a,b}$$

$$E_{1i} = \frac{1}{4\pi|\vec{z}|^5} \sum_{j=1}^{3} \left(\frac{P_j x_j x_i}{\sqrt{C_j C_i}} - |\vec{z}|^2 P_i\right). \tag{8.354c}$$

The product of the vector field by the scalar source specifies the force vector for irrotational fields (Note 8.16).

NOTE 8.16: Force between Multipoles in an Anisotropic Medium

The potential (8.348) in an anisotropic medium leads to the field (8.355):

$$E_i = -\frac{\partial \Phi}{\partial x_i} = \frac{1}{4\pi} \int_D \Lambda\left(\frac{y_k}{\sqrt{C_k}}\right) \frac{x_i - y_i}{\sqrt{C_i}} \left|\sum_{\ell=1}^{3} \frac{(x_\ell - y_\ell)^2}{C_i}\right|^{-3/2} \prod_{j=1}^{3} \frac{dy_j}{\sqrt{C_j}}, \tag{8.355}$$

and the latter to the force between two source distributions:

$$4\pi F = \int_{D'} E_i \left(x_\ell - y_\ell \right) \Lambda' \left(\frac{y_m}{\sqrt{C_m}} \right) \prod_{n=1}^{3} \frac{dy_n}{\sqrt{C_n}}$$

$$= \int_{D'} \prod_{n=1}^{3} \frac{dy_n}{\sqrt{C_n}} \int_{D} \prod_{\ell=1}^{3} \frac{dy_j}{\sqrt{C_j}} \Lambda' \left(\frac{y_m}{\sqrt{C_m}} \right) \Lambda \left(\frac{y_k}{\sqrt{C_k}} \right) \frac{z_i}{|\vec{z}|^3}.$$ (8.356a,b)

If one source is a multipole (8.357b) of order n, the force is (8.357c) involves derivatives with regard to (8.357a):

$$\xi_i = \frac{x_i}{\sqrt{C_i}}; \quad \Lambda'(\xi_i) = Q_{i_1 \ldots i_n} \frac{\partial^n}{\partial \xi_{i_1} \ldots \partial \xi_{i_n}} : 4\pi F_i = Q_{i_1 \ldots i_n} \frac{\partial^n}{\partial \xi_{i_1} \ldots \partial \xi_{i_n}} \int_D \Lambda \left(\frac{y_k}{\sqrt{C_k}} \right) \frac{z_i}{|\vec{z}|^3} \prod_{j=1}^{3} \frac{dy_j}{\sqrt{C_j}}.$$ (8.357a–c)

If the second source is a multipole (8.358b) of order m, the force (8.358c) also involves derivatives with regard to (8.358a):

$$\eta_j = \frac{y_j}{\sqrt{C_j}}; \quad \Lambda(\eta_j) = P_{j_1 \ldots j_m} \frac{\partial^n}{\partial \eta_{j_1} \ldots \partial \eta_{j_m}} : \quad 4\pi F_i = Q_{i_1 \ldots i_n} P_{j_1 \ldots j_m} \frac{\partial^{n+m}}{\partial \xi_{i_1} \ldots \partial \xi_{i_n} \partial \eta_{j_1} \ldots \partial \eta_{j_m}} \frac{z_i}{|\vec{z}|^3}.$$ (8.358a–c)

The simplest case is the force (8.359c) between two monopoles (8.359a and b) in an anisotropic medium:

$$n = 0 = m: \quad F_i = \frac{Q_0 P_0}{4\pi |\vec{z}|^3} \frac{z_i}{\sqrt{C_i}};$$ (8.359a–c)

Another example is the force (8.360c) between two dipoles (8.360a and b) in an anisotropic medium:

$$n = 1 = m: \quad 4\pi \vec{F} = \left(\vec{Q} \cdot \frac{\partial}{\partial \vec{x}} \right) \left(\vec{P} \cdot \frac{\partial}{\partial \vec{y}} \right) \frac{\vec{z}}{|\vec{z}|^3}.$$ (8.360a–c)

The force (8.358c) between multipole of orders n and m decays with distance R like R^{-2-n-m} and thus is a short-range effect for high orders.

8.10 Conclusion

The field or potential due to a charge distribution (Figure 8.3) may be represented by a superposition of multipoles, namely, by order of faster decay: (1) a point charge (Figure 8.1) creates an isotropic field, represented (Figure 8.4) in spherical (R,θ) or cylindrical (z,r) coordinates; (2) a dipole creates an anisotropic field, because it is a limit of opposite charges (Figure 8.2) which introduces (Figure 8.5) a preferred direction, leading to a directivity with two lobes; (3) a quadrupole, which has two preferred directions and four lobes (Figure 8.6); and (4) higher-order multipoles have more lobes. The vector potential due to an electric current (Figure 8.7) is parallel to it, and the magnetic field is orthogonal to the plane of the current and position vector from the current to the observer; it vanishes in the direction of the current and is maximum in the transverse direction leading to a pair of lobes (Figure 8.8) for a magnetic dipole that are rotated by $\pi/2$ relative to an electric dipole (Figure 8.5). The association of opposite currents at a short distance with constant product leads to a magnetic quadrupole, whose moment (Figure 8.9) is analogous to the angular momentum of a pair of particles moving under a torque; since it involves two directions, it leads to a magnetic field reassembling (Figure 8.6) a quadrupole electric field.

The presence of obstacles near mono-, di-, or multipoles, can be accounted for by the method of images, illustrated by four examples: (1) a single equal (opposite) image in one insulating (conducting) plane (Figure 8.10); (2) three images on two orthogonal planes (Figure 8.11a) with identical (opposite) images on insulating (conducting) planes (Figure 8.11b); (3) an infinite set of images in two parallel planes leading to identical (alternating) images for two parallel insulating (conducting) planes (Figure 8.12); the preceding cases such as (1) simplify for a monopole at equal distance from the two planes (Figure 8.13); (4) (Table 8.1) the image on a sphere (Figure 8.14a), which is specified by the second sphere theorem (Figure 8.14b), leading to the electrostatic field of a point charge near a conducting sphere (Figure 8.15a). The case of a point electric charge near an insulating sphere (Figure 8.15b) is analogous to the potential flow of a source near a rigid sphere; it involves a continuous distribution of sinks (Figure 8.16a). The potential flow past a body of revolution can be represented by a suitable distribution of sources along the axis (Figure 8.16b), for example, (1) an infinite fairing with a rounded tail (Figure 8.17a); (2) a "bubble"-shaped airship with a rounded nose and pointed tail (Figure 8.17b); and (3) a body of arbitrary shape (Figure 8.18a) specified by a stream function. The latter also applies (Figure 8.18a) to a fluid in a uniformly rotating cylindrical vessel with arbitrary cross section (Figure 8.18b).

The analogous (Classification 8.1) irrotational (solenoidal) potential fields, namely, gravity, electrostatic field, and irrotational flow (incompressible flow and magnetic field) lead to forces. The forces for solenoidal fields, that is, the magnetic (vortical) force(s) between two parallel electric currents (vortices) orthogonal to the relative position velocity, are similar (Figure 8.20a) to Newton's (Coulomb's) law for the gravity (electric) field (Figure 8.19a and b): (1) it is proportional to the product of source strengths divided by the square of the distance and has the direction of the relative position vector; and (2) the direction is the same as (opposite to) the electric (gravity) force. For the irrotational forces, the source is a monopole without directional effect (Figure 8.19a and b). For the rotational forces, for example, the magnetic force, different directions of the electric currents can lead to zero (Figure 8.20b and c) or nonzero (Figure 8.20d) magnetic force. The anisotropic media are relevant for the dielectric permittivity (magnetic permeability) in electro(magneto) statics, and thermal conductivity (tangential stresses) for steady heat condition (linear deflection of an elastic membrane). Choosing principal axis (Table 8.2) for which the constitutive tensor is diagonalized (Figure 8.21a), the equipotentials for a monopole are (1) generally an ellipsoid (Figure 8.21b) for a biaxial medium with distinct eigenvalues in three directions (Figure 8.22a); (2/3) a prolate (oblate) spheroid [Figure 8.22b (Figure 8.22c)] for a uniaxial medium with two coincident eigenvalues and the third larger (smaller); and (4) a sphere (Figure 8.22d) for three identical eigenvalues corresponding to an isotropic medium.

9

Multidimensional Harmonic Potentials

A harmonic potential or function is a solution of the Laplace equation, and it has a singularity at a point multipole, where Poisson's equation is satisfied. The analytic (generalized) functions, used as solutions of the Laplace equation in the plane (in space), are thus particular harmonic functions in two (three) dimensions. There are general properties of harmonic potentials that are common to both plane and spatial cases and extend to any higher dimension, for example, the boundary conditions (Section 9.1) that render the solution of the Laplace and Poisson equation unique (1) in a compact region, that is, in a region of finite extent; and (2) if the region is noncompact, that is, includes the point-at-infinity, an additional asymptotic condition is needed for the unicity of solution of the Laplace and Poisson equations. The proof of (1) and (2) uses the Green integral identities that can also be applied to (3) the field (Section 9.2) due to a distribution of sources in a bounded region. The field due to a distribution of sources in a unbounded region can be obtained via the influence or Green function, which is a ramp function/logarithmic potential/inverse distance potential/or inverse power distance potential in, respectively, one/two/three/or more dimensions $N = 1/2/3/\geq 4$; this includes the logarithmic (Newtonian) potential in two (three) dimensions (Section 9.4). The properties of harmonic functions that apply in any dimension (Section 9.3) include (4) the theorem of the extrema that a function harmonic in a region takes the maximum and the minimum value on the boundary; for example, a function harmonic (analytic) in a three- (two-) dimensional region has the maximum (minimum) modulus on the boundary surface (curve); (5) the mean value theorem that a function harmonic on a hypersphere has at the center the mean of the values on the boundary; for example, if a function is harmonic (analytic) in a sphere (disk), the value at the center is the mean over the spherical surface (circular boundary); (6) the theorem of constancy (nullity) that at a function harmonic in a region and constant (zero) on the boundary is also constant (zero) in the interior; for example, if a function is harmonic (analytic) in a three- (two-) dimensional region and is constant (zero) on the boundary, then it is constant (zero) in the interior as well. All these properties can be interpreted in terms of (a) the temperature in steady heat conduction, the scalar potential for irrotational, incompressible flow, the gravity field and the electrostatic field; and (b) the vector potential for the magnetostatic field and rotational flow. They lead to the Newton/Coulomb (Biot–Savart) law for the force [Section 9.4 (Section 9.5)] in the gravity/electric (magnetic) field that is irrotational (solenoidal). The vector potential exists in three dimensions (Section 9.5) and reduces to a scalar field function in two dimensions, and its extension beyond three dimensions requires multivectors (Notes 9.20 through 9.52) that are considered in the context of tensor calculus (Notes 9.6 through 9.52) in order to address questions of unicity and invariance. The scalar potential applies (Section 9.4) in one, two, and three dimensions, and extends to any higher dimension, for example, as concerns (7) the multipolar expansion that leads to generalized or hyperspherical Legendre polynomials (Section 9.6); (8) the specification of the potential and the field of multidimensional multipoles in terms of hyperspherical and hypercylindrical coordinates (Section 9.7); (9) the hypersphere theorem (Section 9.8) that generalizes the circle (sphere) theorem [Section I.24.6 (Section 6.6)] to a hypersphere in a space of any dimension; and (10) the method of images also extends (Section 9.9) from straight lines (Chapter I.16)/planes (Sections 8.7 and 8.8) [circles (Section I.24.7)/spheres (Section 8.9)] to hyperplanes (hyperspheres).

9.1 Boundary and Asymptotic Conditions for Unicity

The first Green identity (Subsection 9.1.1) is used to prove the unicity of the solution of the Poisson and Laplace equations in a bounded region, that is, a compact domain (Subsection 9.1.3) for three kinds of boundary conditions (Subsection 9.1.2); in an unbounded region or a noncompact domain (Subsection 9.1.5), an additional asymptotic condition must be imposed to retain unicity (Subsection 9.1.4).

9.1.1 Divergence Theorem, the Two Green Identities, and the Kinetic Energy

The two Green (1828) identities (Subsection 5.7.9) arise from the divergence theorem, that is, one of the integral identities applying to the invariant differential operators (Section 5.5); next, only concepts directly relevant to the proof of unicity of the solutions of the Laplace and Poisson equations are recalled. A **regular surface** S is a surface without edges or cups such that there is a unique unit normal vector $\vec{N}$ at each point, and in each neighborhood of area dS, a unique area vector $d\vec{S} = \vec{N}\,dS$ exists. A closed surface ∂D divides the three-dimensional space $|R^3$ into an inner D and an outer $|R^3-D$ region, and $\vec{N}$ is taken to be the outward normal to ∂D that points from D to $|R^3-D$. These concepts appear (Subsection 5.7.1) in the **divergence theorem**: *if $\vec{A}(\vec{x})$ is a vector, continuously differentiable (9.1a) in a region D with a closed regular boundary ∂D, then the volume integral dV of divergence of the vector equals its flux across the surface (9.1b):*

$$\vec{A} \in C^1(D): \int_D (\nabla \cdot \vec{A})dV = \int_{\partial D} (\vec{A} \cdot \vec{N})dS = \int_{\partial D} \vec{A} \cdot d\vec{S}. \qquad (9.1a,b)$$

The divergence operator can be applied to the product of a scalar Φ and a vector $\vec{U}$:

$$\nabla \cdot (\Phi \vec{U}) = \Phi(\nabla \cdot \vec{U}) + \vec{U} \cdot \nabla \Phi; \qquad (9.2a)$$

the property (9.2a) is proved as follows:

$$\frac{\partial(\Phi U_i)}{\partial x_i} = \Phi \frac{\partial U_i}{\partial x_i} + U_1 \frac{\partial \Phi}{\partial x_i}, \qquad (9.2b)$$

where the repeated index i implies summation over N dimensions $i = 1,2,3,...N$, as in (5.153c).

If Φ is a scalar function with (9.3a) continuous second-order derivates,

$$\Phi \in C^2(D): \quad \vec{A} = \Phi\nabla\Phi \in C^1(D), \quad \nabla \cdot \vec{A} = (\nabla\Phi)^2 + \Phi\nabla^2\Phi, \qquad (9.3a\text{–}c)$$

the vector (9.3b) has continuous first-order derivatives (9.3b), and divergence is continuous and given by (9.3c), where the following relation was used (9.2b):

$$\nabla \cdot (\Phi\nabla\Phi) - \nabla\Phi \cdot \nabla\Phi = \Phi\nabla \cdot (\nabla\Phi) = \Phi\nabla \cdot \frac{\partial\Phi}{\partial\vec{x}} = \Phi\frac{\partial^2\Phi}{\partial x_i^2} = \Phi\partial_{ii}\Phi = \Phi\nabla^2\Phi. \qquad (9.4)$$

Substituting (9.3b and c; 9.4) in (9.1b) leads to the **first Green identity** (9.5a and b) [≡ (5.230a through c) with $\Phi = \Psi$]:

$$\Phi \in C^2(D): \quad \int_D (\Phi \nabla^2 \Phi + \nabla \Phi \cdot \nabla \Phi) dV = \int_{\partial D} \Phi \frac{\partial \Phi}{\partial N} dS, \qquad (9.5a,b)$$

where the **derivate along the normal** is denoted by

$$\Phi \in C^1(D): \quad \frac{\partial \Phi}{\partial N} = \vec{N} \cdot \nabla \Phi = \vec{N} \cdot \frac{\partial \Phi}{\partial \vec{x}} = N_i \frac{\partial \Phi}{\partial x_i}. \qquad (9.6a,b)$$

A particular case of (9.6a) is the energy theorem for a potential flow:

$$\nabla^2 \Phi = 0, \quad \vec{v} = \nabla \Phi: \quad \frac{2}{\rho} E_v = \int_D |\vec{v}|^2 \, dV = \int_D (\nabla \Phi \cdot \nabla \Phi) dV = \int_{\partial D} \Phi \frac{\partial \Phi}{\partial N} dS, \qquad (9.7a\text{--}d)$$

the kinetic energy (9.7c) of a potential flow (9.7a) with velocity (9.7b) and mass density ρ in a domain D may be evaluated as an integral (9.7d) over the boundary ∂D.

Replacing one Φ in (9.3b) with Ψ, and then interchanging Φ, Ψ and subtracting leads to the vector (9.8b):

$$\Phi, \Psi \in C^2(|R^3): \quad \vec{A} = \Psi \nabla \Phi - \Phi \nabla \Psi, \quad \nabla \cdot \vec{A} = \Psi \nabla^2 \Phi - \Phi \nabla^2 \Psi, \qquad (9.8a\text{--}c)$$

whose divergence is given by (9.8c), where (9.3b and c) or (9.2a) was applied to the functions (9.8a) with a continuous second-order derivate. Substituting (9.8b and c) in (9.1a and b) yield the **second Green identity** (9.9a and b) ≡ (5.231a through c):

$$\Phi, \Psi \in C^2(D): \quad \int_D (\Psi \nabla^2 \Phi - \Phi \nabla^2 \Psi) dV = \int_{\partial D} \left(\Psi \frac{\partial \Phi}{\partial N} - \Phi \frac{\partial \Psi}{\partial N} \right) dS, \qquad (9.9a,b)$$

where (9.6a and b) was used. Thus, *if the function Φ has (functions Φ, Ψ have) a continuous second-order derivate(s) (9.5a) [(9.9a)] in a region D with a closed regular boundary ∂D, then they satisfy the first (9.5b) [second (9.9b)] Green identity where (9.6a and b) is the derivative along the unit outward normal.* This theorem was used in Subsections I.28.1.1 and II.8.2.4.

9.1.2 Cauchy–Dirichlet, Neumann, and Robin Problems

The Laplace equation is the unforced form of the Poisson equation (9.10a), in the case when there are no sources:

$$D: \quad \nabla^2 \Phi = \Lambda, \quad \partial D: \quad \lambda \Phi + \frac{\partial \Phi}{\partial N} = g; \qquad (9.10a,b)$$

for example, the Poisson equation (9.10a) applies to temperature $\Phi \equiv T$ for steady heat conduction (I.32.5b and c) ≡ (8.307c) in a region D, with $\Lambda \equiv -w/k$ sources of heat of output w per unit volume in a medium with thermal conductivity k; the most general boundary condition (I.32.7c) is of a convective type (9.10b), indicating that the difference of temperature $\Phi - \Phi_0$ between the body D and the surrounding fluid, multiplied by the surface thermal conductivity h, equals the heat flux (9.11b) $\equiv$ (8.307b):

$$h(\Phi - \Phi_0) = J_n = -k\frac{\partial \Phi}{\partial N}; \quad \lambda = \frac{h}{k}, \quad g = \frac{h}{k}\Phi_0 = \lambda \Phi_0; \qquad \text{(9.11a–c)}$$

comparing (9.11a) $\equiv$ (9.10b) leads to (9.11b and c). The parameter λ in (9.11b) is positive (9.12a) so that if the body is hotter (cooler) than the fluid (9.12b), then (9.12c) implies that the heat flux (9.12d) is toward the fluid (body):

$$\lambda = \frac{h}{k} > 0: \quad \Phi >< \Phi_0 = \frac{kg}{h} = \frac{g}{\lambda} \Leftrightarrow \frac{\partial \Phi}{\partial N} <> 0 \Leftrightarrow J_n >< 0; \qquad \text{(9.12a–d)}$$

in both cases, the heat flows from the hotter to the cooler region. The solution of the the Poisson (9.10a) or Laplace equation in a region D, with a convective-type boundary condition (9.10b), is the **mixed or Robin problem** (1886) with parameter $0 < \lambda < \infty$. The extreme values of the parameter specify two simpler problems, namely, (1) for $\lambda = 0(\lambda = \infty, g/\lambda = finite \to \Phi_0)$, it leads to the **Neumann (1961) (Cauchy–Dirichlet, 1850) problem** of finding the solution of the Poisson (9.10a) or Laplace equation in a region with a given normal derivative (9.13a) [value (9.13b)] on the boundary:

$$\partial D: \quad \frac{\partial \Phi}{\partial N} = g, \quad \Phi = \Phi_0. \qquad \text{(9.13a,b)}$$

In the case of steady heat conduction in a body, this corresponds to giving the heat sources w in the interior (9.10a); and (2) specifying the heat flux $J_n = -k\partial\Phi/\partial n = -kg$ in (9.13a) [temperature (9.13b)] at the surface. This is necessary and sufficient to determine the temperature at all points in a compact region, that is, a region that does not extend to infinity (Subsection 9.1.3); in the case of a noncompact region, that is, one extending to infinity, an additional asymptotic condition (Subsection 9.1.4) is needed.

Concerning the Cauchy–Dirichlet (9.10a; 9.13a), Neumann (9.10a; 9.13b), and Robin (9.10a and b) problems or other boundary value problems, three questions arise: (1) **existence**: does the problem have a solution, that is, is there any function Φ satisfying the conditions stated?; (2) **unicity**: assuming at least one solution exists, is it unique? Or do several distinct functions satisfy the required conditions?; and (3) **determination**: if it is known that the solution exists and is unique, can it be calculated? If yes, then by which method? The approach to these three different aspects of the same problem can be very different. For example, many solutions of Cauchy–Dirichlet and Neumann problems were obtained before (Chapters I.12, 14, 16, 18, 24, 26, 28, 32, 34, 36, 38; II.2, 8; 6, 8) and also for some of the mixed problems, using complex or generalized functions, but all these particular cases prove neither existence nor unicity in general. The general proof of existence may be quite different from the proof of unicity; both may fail to give any hint of a method to find the explicit solution in a particular case. In this section, the question of unicity only is addressed. The boundary condition of Neumann type (9.13a) is not affected by adding a constant to the potential, but the Cauchy–Dirichlet (9.13d) and the mixed (9.10b) conditions are affected. Thus, the following **inner unicity theorem** may be expected to be valid: *the Poisson (9.10a) or Laplace ($\Lambda = 0$) equation in a compact region D, that is, the interior of a closed regular boundary ∂D, has for solution a potential Φ with continuous second-order derivates that has given values (9.13b), or a normal derivate (9.13a), or a linear combination of them (9.10b) on the boundary; the solution is unique for the Cauchy–Dirichlet (9.10a; 9.13b) and mixed (9.10a and b) problems, and is determined with an added constant $\Phi + const$ for the Neumann (9.10a; 9.13a) problem.*

9.1.3 Inner Unicity Problem for a Compact Region

First, the unicity theorem is proved for the interior D of a closed regular surface, namely, the **inner problem** for a **compact region**, that is, a region that does not extend to infinity. Consider two functions Φ_1 and Φ_2, both satisfying the Poisson $\Lambda \neq 0$ or Laplace $\Lambda = 0$ equation (9.10a) in D, together with one of the boundary conditions (9.10b), (9.13a), or (9.13b). The difference of two functions, $\Phi_0 \equiv \Phi_1 - \Phi_2$, satisfies the same equations because they are all linear and in unforced form, that is, with zero on their r.h.s.:

$$\nabla^2\Phi_0\big|_D = 0 = \left[\Phi_0, \frac{\partial\Phi_0}{\partial N}, \Phi_0\lambda + \frac{\partial\Phi_0}{\partial N}\right]_{\partial D}; \qquad (9.14\text{a--d})$$

thus, it satisfies the Laplace equation (9.14a) in the region D and meets on the boundary ∂D one of the three conditions (9.14b through d) in unforced form. Applying the first Green identity (9.5b) to Φ_0 and using (9.14a) leads to:

$$\int_D (\nabla\Phi_0)^2\,dV = \int_{\partial D} \Phi_0 \frac{\partial\Phi_0}{\partial N}\,dS = 0, \qquad (9.15)$$

where the r.h.s. is (1/2) zero for the first two boundary conditions (9.13b) [(9.13a)] corresponding to the Cauchy–Dirichlet (Neumann) problem in (9.14b) [(9.14c)]; (3) the case of the mixed problem (9.10b), that is, the last (9.14d) boundary condition leads to

$$\lambda > 0: \quad 0 \le \int_D (\nabla\Phi_0)^2\,dV = -\int \lambda(\Phi_0)^2\,dS \le 0, \qquad (9.16\text{a,b})$$

and since in (9.16b) the r.h.s. is negative and the l.h.s. is positive by (9.12a) $\equiv$ (9.16a), both must be zero. Thus, for any of the three boundary conditions, it follows that:

$$0 = \int_D (\nabla\Phi_0)^2\,dV \Rightarrow \Phi_0 \equiv \Phi_1 - \Phi_2 = \text{const}, \qquad (9.17)$$

which implies that the potential is unique to within a constant. The constant can be calculated on the boundary (9.14b through d), showing that, for the Cauchy–Dirichlet (9.14b) and Robin (9.14d) [Neumann (9.14c)] problems, the solution is unique, $\Phi_2 = \Phi_1$ (unique to within an added constant, $\Phi_2 = \Phi_1 + \text{const}$). QED.

9.1.4 Asymptotic Condition for an Unbounded Region

The outer problem considers the solution of Poisson $\Lambda \ne 0$ or Laplace $\Lambda = 0$ equation (9.10a) outside a regular surface ∂D, where one of the boundary conditions holds. Whereas the case of the inner problem concerns a compact region, that is, not including the point at infinity, the outer problem concerns a **noncompact region,** and the point at infinity must be excluded by adding to the inner boundary ∂D an outer boundary, namely, the surface at infinity, S_∞, so that (9.15) is replaced by

$$\int_D (\nabla\Phi_0)^2\,dV = \int_{\partial D + S\infty} \Phi_0 \frac{\partial\Phi_0}{\partial N}\,dS; \qquad (9.18)$$

the preceding proof of unicity (9.15; 9.16a and b; 9.17) will remain valid from the inner to the outer problem provided in the latter case, the surface at infinity makes no contribution to the integral:

$$\int_{S\infty} \Phi \frac{\partial\Phi}{\partial N}\,dS = 0. \qquad (9.19)$$

The latter equation (9.19) can be met by imposing a suitable asymptotic condition on the potential, for example, by assuming that the potential decays asymptotically as (9.20a) an inverse power k of the distance:

$$\Phi(\vec{x}) = O\left(|\vec{x}|^{-k}\right), \quad \frac{\partial \Phi}{\partial N} = O\left(|\vec{x}|^{-k-1}\right), \tag{9.20a,b}$$

such that (9.20b) the integral (9.19) vanishes.

The integral (9.19) may be evaluated in one/two/three/or more dimensions for a straight line (9.21a through c)/a circle in a plane (9.22a through c)/a sphere in space (9.23a through c)/or a hypersphere in an N-dimensional space (9.24a through d):

$$N = 1, \quad x \in |R: \quad \int \Phi \frac{d\Phi}{dx} dx \sim O\left(|x|^{-2k}\right), \tag{9.21a-c}$$

$$N = 2, \quad \vec{x} \in |R^2; \quad \int_{|\vec{x}|=r} \Phi \frac{\partial \Phi}{\partial N} ds = O\left(|\vec{x}|^{-2k}\right), \tag{9.22a-c}$$

$$N = 3, \quad \vec{x} \in |R^3: \quad \int_{|\vec{x}|=R} \Phi \frac{\partial \Phi}{\partial N} dS = O\left(|\vec{x}|^{1-2k}\right), \tag{9.23a-c}$$

$$N \geq 4, \quad \vec{x} \in |R^N: \quad \int_{|\vec{x}|=R} \Phi \frac{\partial \Phi}{\partial N} dS = O\left(|\vec{x}|^{N-2k-2}\right), \tag{9.24a-c}$$

because the perimeter/area/hyper-area of the circle/sphere/hypersphere scales as $\sim O\left(|\vec{x}|\right)/O\left(|\vec{x}|^2\right)/O\left(|\vec{x}|^{N-1}\right)$. The condition (9.19) is met (1) on the line (9.21a through c) if $k > 0$; (2) in the plane (9.22a through c) if $k > 0$, for example, not for a monopole $\Phi \sim \log r$ but for a dipole or higher order multipoles; (3) in space (9.23a through c) for $k > 1/2$, for example, for monopoles $\Phi \sim O(R^{-1})$, as well as for dipoles and higher order multipoles; and (4) in N-dimensional space for all multipoles. Statement (4) can be proved as follows: (4-a) the decay of (9.24c) at infinity requires (9.25b):

$$N \geq 3: \quad k > \frac{N}{2} - 1; \quad \Phi(\vec{x}) \sim O\left(|\vec{x}|^{2-N-n}\right); \tag{9.25a-c}$$

(4-b) the potential of a multipole of order m decays as $|\vec{x}|^{-1-n}$ in a three-dimensional space $N = 3$ and as (9.25c) in an N-dimensional space (9.25a); (4-c) the condition (9.20a) $\equiv$ (9.25a) leads to (9.26a):

$$k = N + n - 2: \quad N + n - 2 = k > \frac{N}{2} - 1, \quad n > 1 - \frac{N}{2}, \tag{9.26a-c}$$

(4-d) substituting (9.26a) in (9.25b) yields (9.26b), implying (9.26c); and (4-e) the latter (9.26c) is satisfied by any multipole $n = 0,1,\ldots,3$ or in three or more dimensions (9.25a) in agreement with cases (3) and (4). The asymptotic condition (9.26c) also applies $N = 2$ in the plane (9.17) for all multipoles except the monopole, in agreement with case (2). QED.

9.1.5 Outer Unicity Problem for a Noncompact Region

Thus, the following **outer unicity theorem** has been proved: *the solution of the Poisson $\Lambda \neq 0$ or Laplace $\Lambda = 0$ equation (9.10a) in a noncompact region, for example, the exterior of a closed, regular surface* *(curve) in space (in the plane), is unique for the Dirichlet (9.14b) [Robin (9.10b)] boundary condition and is determined to within an added constant for the Neumann boundary condition (9.14c), assuming that the asymptotic condition (9.19) is also is met. A sufficient condition for the latter is that the potential at infinity,*

$$N = 1,\ x \in| R:\quad \Phi(x) = o(1) \Leftrightarrow \lim_{x \to \infty} \Phi(\vec{x}) = 0, \tag{9.27a–d}$$

$$N = 2,\ x \in| R^2:\quad \Phi(\vec{x}) = o(1) \Leftrightarrow \lim_{|\vec{x}| \to \infty} \Phi(\vec{x}) = 0, \tag{9.28a–d}$$

$$N = 3,\ \vec{x}G \,|\, R^3:\quad \Phi(\vec{x}) = o\left(\left|\vec{x}\right|^{-1/2}\right) \Leftrightarrow \lim_{|\vec{x}| \to \infty} \left|\vec{x}\right|^{1/2} \Phi(\vec{x}) = 0, \tag{9.29a–d}$$

$$N \geq 1,\ \vec{x} \in| R^N:\quad \Phi(\vec{x}) = o\left(\left|\vec{x}\right|^{1-N/2}\right) \Leftrightarrow \lim_{|\vec{x}| \to \infty} \left|\vec{x}\right|^{1-N/2} \Phi(\vec{x}) = 0. \tag{9.30a–d}$$

(1) decays (9.27a through d) in one dimension; (2) decays in the plane (9.28a through d); (3) decays faster than an inverse square root in space (9.29a through d); (4) decays faster than $R^{1-N/2}$ in an N-dimensional space. The asymptotic condition applies to multipoles of all orders in a space of dimension three or more; in the plane, it fails only for monopoles, that is, line sources, but not for dipoles or higher order multipoles. The asymptotic decay laws for multipoles (Table 9.1) that were used before are proved next by obtaining the influence or Green function for the Laplace operator in any dimension.

TABLE 9.1 Spatial Dependence of Potential Fields

Dimension		Plane	Space	Higher
N		$N = 2$	$N = 3$	$N \geq 4$
Distance		$r \equiv \left\|x^2 + y^2\right\|^{1/2}$	$R \equiv \left\|x^2 + y^2 + z^2\right\|^{1/2}$	$R \equiv \left\|\sum\limits_{i=1}^{N} (x_i)^2\right\|^{1/2}$
Potential	Monopole	$\log r$	R^{-1}	R^{2-N}
	Dipole	r^{-1}	R^{-2}	R^{1-N}
	Quadrupole	r^{-2}	R^{-3}	R^{-N}
	2^n-pole	r^{-n} (*)	R^{-n-1}	R^{2-N-n}
Field	Monopole	r^{-1}	R^{-2}	R^{1-N}
	Dipole	r^{-2}	R^{-3}	R^{-N}
	Quadrupole	r^{-3}	R^{-4}	R^{-1-N}
	2^n-pole	r^{-1-n}	R^{-2-n}	R^{1-N-n}
Force	Two monopoles	r^{-1}	R^{-2}	R^{1-N}
	Monopole + dipole	r^{-2}	R^{-3}	R^{-N}
	Two dipoles	r^{-3}	R^{-4}	R^{-1-N}
	Monopole + quadrupole	r^{-3}	R^{-4}	R^{-1-N}
	Dipole + quadrupole	r^{-4}	R^{-5}	R^{-2-N}
	Two quadrupoles	r^{-5}	R^{-6}	R^{-3-N}
	Monopole + 2^n-pole	r^{-1-n}	R^{-2-n}	R^{1-N-n}
	Dipole + 2^n-pole	r^{-2-n}	R^{-3-n}	R^{-N-n}
	Quadrupole + 2^n-pole	r^{-3-n}	R^{-4-n}	R^{-1-N-n}
	2^n-pole + 2^n-pole	r^{-1-n-m}	R^{-2-n-m}	$R^{1-N-n-m}$

Note: Spatial dependence on the distance from the origin in the plane ($N = 2$), space ($N = 3$), and hyperspace ($N \geq 4$) of (1/2) the potential and the field of monopoles ($n = 0$, $2^0 = 1$), dipoles ($n = 1$, $2^1 = 1$), quadrupoles ($n = 2$, $2^2 = 4$), and 2^n multipoles or arbitrary-order multipoles; and (3) force between the preceding multipoles.

[a] Does not apply to a monopole $n \neq 0$; is replaced by the logarithmic potential for $n = 0$.

9.2 Influence Function and Source Distributions

The influence or Green function (Subsection 9.2.2) is the fundamental solution for the Laplace operator with hyperspherical symmetry (Subsection 9.2.1). It can be used to specify the potential due to an arbitrary source distribution in free space (Subsection 9.2.3). It is extended to a bounded region by the (Subsection 9.2.4) second Green identity (Subsection 9.1.1), leading to source distributions on the boundary (Subsection 9.2.5).

9.2.1 Laplace Operator with Hyperspherical Symmetry

Consider an N-dimensional space (9.31a) with coordinates (9.31b) for which the radial distance from the origin is given by (9.31c):

$$i = 1,\ldots,N; \quad \vec{x} \equiv (x_1, x_2, \ldots, x_N) \in |R: \quad R = \sum_{i=1}^{N} (x_i)^2 \equiv x_i x_i, \tag{9.31a–c}$$

where a repeated index (9.31c) means summation over (9.31a). The derivatives of first (second) order of the radial distance are (9.32a) [(9.32b)], the latter involving the identity matrix (5.278a through c):

$$\frac{\partial R}{\partial x_i} = \frac{x_i}{R}, \quad \frac{\partial^2 R}{\partial x_i \partial x_j} = \frac{\delta_{ij}}{R} - \frac{x_i x_j}{R^3}. \tag{9.32a,b}$$

They specify the first-order (9.33b) [second-order (9.33c)] derivatives of a scalar function with **hyperspherical symmetry** (9.33a):

$$\Phi(x_1, \ldots, x_N) = \Phi(R): \quad \frac{\partial \Phi}{\partial x_i} = \frac{d\Phi}{dR} \frac{\partial R}{\partial x_i} = \frac{d\Phi}{dR} \frac{x_i}{R}, \tag{9.33a,b}$$

$$\frac{\partial^2 \Phi}{\partial x_i \partial x_j} = \left(\frac{\delta_{ij}}{R} - \frac{x_i x_j}{R^3} \right) \frac{d\Phi}{dR} + \frac{x_i x_j}{R^2} \frac{d^2\Phi}{dR^2}; \tag{9.33c}$$

summing over $i, j = 1, \ldots, N$ in (9.33c) and noting (9.34a) leads to (9.34b):

$$\delta_{ii} = N: \quad \frac{\partial^2 \Phi}{\partial x_i \partial x_i} = \frac{d^2\Phi}{dR^2} + \frac{N-1}{R} \frac{d\Phi}{dR}. \tag{9.34a,b}$$

This is the **Laplacian with hyperspherical symmetry** in N dimensions,

$$R \equiv \left| \sum_{i=1}^{N} (x_i)^2 \right|^{1/2} : \quad \nabla^2 \equiv \sum_{i=1}^{N} \frac{\partial^2 \Phi}{\partial x_i^2} = \frac{d^2\Phi}{dR^2} + \frac{N-1}{R} \frac{d\Phi}{dR} = \frac{1}{R^{N-1}} \frac{d}{dR} \left(R^{N-1} \frac{d\Phi}{dR} \right), \tag{9.35a,b}$$

which includes as particular cases in one/two/three dimensions the line (9.36a through c)/cylindrical (9.37a through c)/spherical (9.38a through d) Laplacian:

$$N = 1: \quad x \equiv x_1: \quad \nabla^2 \equiv \frac{d^2}{dx^2}, \tag{9.36a–c}$$

$$N = 2: \quad r \equiv \left|x^2 + y^2\right|^{1/2}: \quad \nabla^2 = \frac{1}{r}\frac{d}{dr}\left(r\frac{d\Phi}{dr}\right) = \frac{d^2\Phi}{dr^2} + \frac{1}{r}\frac{d\Phi}{dr}, \tag{9.37a–c}$$

$$N = 3: \quad R \equiv \left|x^2 + y^2\right|^{1/2}: \quad \nabla^2 \equiv \frac{1}{R^2}\frac{d}{dR}\left(R^2\frac{d\Phi}{dR}\right) = \frac{d^2\Phi}{dR^2} + \frac{2}{R}\frac{d\Phi}{dR} = \frac{1}{R}\frac{d^2}{dR^2}(R\Phi). \tag{9.38a–d}$$

The Laplacian with line (9.36c)/cylindrical (9.37c)/spherical (9.38c) symmetry is the first term of the Laplacian in Cartesian (6.44)/cylindrical (6.45a and b)/spherical (6.46a and b) coordinates.

9.2.2 Green Function for the Laplace Operator in Any Dimension

The influence or Green function for the N-dimensional Laplacian is the fundamental solution of the Poisson equation forced by N-dimensional Dirac impulse distribution:

$$\sum_{i=1}^{N}\frac{\partial^2}{\partial x_i^2}\left[G_N(x_n; y_n)\right] = \frac{\partial^2}{\partial \vec{x}^2}\left[G_N(\vec{x}; \vec{y})\right] = \delta(\vec{x} - \vec{y}) \equiv \prod_{i=1}^{N}\delta(x_i - y_i). \tag{9.39}$$

The latter has hyperspherical symmetry (5.306) and involves the area σ_N of the unit hypersphere in N dimensions, and thus forces the Laplacian with hyperspherical symmetry (9.35b) whose solution is the influence function:

$$R^{1-N}\frac{d}{dR}\left(R^{N-1}\frac{dG_N}{dR}\right) = \sigma_N^{-1}R^{1-N}\delta(R). \tag{9.40}$$

A first integration (9.40) involves the Heaviside unit jump (9.41b):

$$R > 0: \quad R^{N-1}\frac{dG_N}{dR} = \sigma_N^{-1}H(R) = \frac{1}{\sigma_N}, \quad G_N(R) = \frac{1}{\sigma_N}\int R^{1-N}dR, \tag{9.41a–c}$$

leading to (9.41c) by a second integration; an arbitrary constant was omitted since the potential can be defined to within a constant. The integral (9.41c) is specified by (9.42a through d) [(9.43a and b)] for the plane $N = 2$ (other dimensions $N \neq 2$):

$$R \equiv \left|\vec{x} - \vec{y}\right|, N = 2, \sigma_2 = 2\pi: \quad G_2(\vec{x}; \vec{y}) = \frac{1}{\sigma_2}\log R = \frac{1}{2\pi}\log\left|\vec{x} - \vec{y}\right|, \tag{9.42a–d}$$

$$2 \neq N = 1, 3, 4, \ldots: \quad G_N(\vec{x}; \vec{y}) = \left[\sigma_N(2 - N)\right]^{-1}\left|\vec{x} - \vec{y}\right|^{2-N} = G_N(\vec{x} - \vec{y}). \tag{9.43a–c}$$

Thus the **influence or Green function for the Laplace operator** *(9.39)* ≡ *(9.40) in N dimensions (9.41c) has been obtained, which includes (2) the logarithmic potential (9.42d) in the plane (9.42b) involving the circumference of the unit circle (9.42c)* ≡ *(5.301a) and (1/3) the one- (three-) dimensional Green function (9.44a) [(9.45a)] that is a ramp function (9.44a–c) [the inverse distance (9.45a through c)],*

$$N = 1: \quad G_1(x - y) = \begin{cases} 0 & \text{if } x \leq y, \\ x - y & \text{if } x \geq y, \end{cases} \tag{9.44a–c}$$

$$N = 3, \sigma_3 = 4\pi: \quad G_3(\vec{x} - \vec{y}) = -\left\{\sigma_3\left|\vec{x} - \vec{y}\right|\right\}^{-1} = -\left\{4\pi\left|\vec{x} - \vec{y}\right|\right\}^{-1}, \tag{9.45a–c}$$

with the latter (9.45c) involving the area of the unit sphere (9.45b); and (4) both are particular cases of the area of the unit hypersphere (5.301a and b) that appears in the Green function (9.43b) in higher $N \geq 4$ dimensions (9.43a). In all cases, the influence function depends only on the relative distance (9.42a), implying that (a) it is symmetric (7.79a and b); (b) the reciprocity principle allowing interchange of observer and source holds; and (c) this agrees with the Laplace operator being self-adjoint (7.213a–d). Since the Poisson equation (9.10a) is linear, the principle of superposition is used next (Subsection 9.2.3) to obtain the solution (Subsection 9.2.3) with arbitrary forcing.

9.2.3 Arbitrary Source Distribution in Free Space

The solution of the Poisson equation (9.46a) ≡ (9.10a) forced by an arbitrary source distribution is given (7.78a and b) by the convolution with the symmetric influence or Green function (9.46b):

$$\nabla^2 \Phi = \Lambda(\vec{x}), \quad \Phi(\vec{x}) = \Lambda * G(\vec{x}) \equiv \int_{-\infty}^{+\infty} \Lambda(\vec{y}) G(\vec{x} - \vec{y}) d\vec{y}; \qquad (9.46a,b)$$

substituting, respectively, (9.44a through c)/(9.42a through d)/(9.45a through c)/(9.43a through c), it follows that *the potential as a solution of the Poisson equation (9.46a) forced by an arbitrary source distribution in free space, that is, unbounded space is specified in one (9.47a)/two (9.48a)/three (9.49a)/higher (9.50a) dimensions,*

$$N = 1, \ \sigma_1 = 1: \quad \Phi(\vec{x}) = \int_{-\infty}^{\infty} (x - y) \Lambda(y) dy, \qquad (9.47a\text{–}c)$$

$$N = 2, \ \sigma_2 = 2\pi: \quad 2\pi \Phi(\vec{x}) = \int \Lambda(\vec{y}) \log|\vec{x} - \vec{y}| d^2\vec{y}, \qquad (9.48a\text{–}c)$$

$$N = 3, \ \sigma_3 = 4\pi: \quad -4\pi \Phi(\vec{x}) = \int \frac{\Lambda(\vec{y})}{|\vec{x} - \vec{y}|} d^2\vec{y}, \qquad (9.49a\text{–}c)$$

$$2 \neq N = 1,3,4: \quad \sigma_N(2 - N)\Phi(\vec{x}) = \int \Lambda(\vec{y})|\vec{x} - \vec{y}|^{2-N} d^N\vec{y}, \qquad (9.50a,b)$$

as the convolution of the source with the influence or Green function (9.44a through c)/(9.42a through d)/ (9.45a through c)/(9.43a through c) involving the perimeter of the unit circle (9.48b)/area of the unit sphere (9.49b)/hypersphere (5.301a and b), respectively, in the **Poisson integral** *(9.47c)/(9.48c)/(9.49c)/(9.50b).* The case of a source distribution in a bounded region leads to additional sources on the boundary as shown next (Subsection 9.2.4).

9.2.4 Poisson Equation in a Compact versus Noncompact Region

The first Green identity of (9.5a and b) has been used to prove unicity theorems (Section 9.1) for the inner (outer) problems of Cauchy–Dirichlet, Neumann, and Robin type [Subsections 9.1.2 and 9.1.3 (Subsections 9.1.4 and 9.1.5)]; next, the second Green identity (9.9a and b) is applied to the Poisson equation generalizing the Poisson integral from free space (Subsection 9.2.3) to a compact (Subsection 9.2.4) region. Choosing (1) the function Ψ to be the influence function (9.51a) for the Poisson equation

(9.51b) ≡ (9.39) in any number of dimensions, that is, the potential due to a point monopole of unit magnitude and (2) the potential Φ due to an arbitrary source distribution (9.51c) ≡ (9.46a):

$$\Psi \equiv G: \quad \nabla^2 \Psi = \delta(\vec{x} - \vec{y}), \quad \nabla^2 \Phi = \Lambda(\vec{x}); \tag{9.51a-c}$$

the second Green identity (9.9a and b) leads to (9.52a) where (9.52b) the substitution property (1.68) of the N-dimensional Dirac unit impulse (5.147a and b) is used:

$$\int_{\partial D} \left(G \frac{\partial \Phi}{\partial N} - \Phi \frac{\partial G}{\partial N} \right) dS - \int_D G \Lambda dV = - \int_D \Phi(\vec{y}) \delta(\vec{x} - \vec{y}) d^N \vec{y} = -\Phi(\vec{x}). \tag{9.52a,b}$$

The preceding result may be restated: *the solution of the Poisson equation (9.10a) ≡ (9.51c) ≡ (9.53a) in a domain D satisfies the integral identity (9.52b) ≡ (9.53b):*

$$D: \quad \nabla^2 \Phi = \Lambda; \quad \Phi(\vec{x}) = \int_D G(\vec{x}; \vec{y}) \Lambda(\vec{y}) d^N \vec{y} + \int_{\partial D} \left\{ \Phi(\vec{y}) \frac{\partial}{\partial \vec{y}} [G(\vec{x}; \vec{y})] \cdot d\vec{S} - G(\vec{x}; \vec{y}) \frac{\partial \Phi(\vec{y})}{\partial \vec{y}} \cdot d\vec{S} \right\}, \tag{9.53a,b}$$

*where (1) the volume integral represents the real source distribution $\Lambda(\vec{y})$ in the domain D and (2) the surface integral represents an **equivalent source distribution** on the boundary consisting of monopole (dipole) terms involving the influence or Green function (its normal derivative). In particular, in one (9.54a through c)/two (9.55a through c)/three (9.56a through c)/more (9.57a through c) dimensions, the **volume and surface sources** (1) + (2) lead to the total potential:*

$$N = 1, \sigma_1 = 1: \quad \Phi(x) = \int_y^\infty (x - y) \Lambda(y) dy + \Phi(y), \tag{9.54a-c}$$

$$N = 2, \sigma_2 = 2\pi: \quad 2\pi \Phi(\vec{x}) = \int_D \Lambda(\vec{y}) \log|\vec{x} - \vec{y}| d^2 \vec{y} + \int_{\partial D} \left[\Phi(\vec{y}) \frac{(\vec{x} - \vec{y}) \cdot \vec{N}}{|\vec{x} - \vec{y}|^2} - \left(\vec{N} \cdot \frac{\partial \Phi}{\partial \vec{y}} \right) \log|\vec{x} - \vec{y}| \right] dS,$$
$$\tag{9.55a-c}$$

$$N = 3, \sigma_3 = 4\pi: \quad 4\pi \Phi(\vec{x}) = -\int_D \frac{\Lambda(\vec{y})}{|\vec{x} - \vec{y}|} d^3 \vec{y} + \int_{\partial D} \left[\left(\vec{N} \cdot \frac{\partial \Phi}{\partial \vec{y}} \right) - \Phi(\vec{y}) \frac{(\vec{x} - \vec{y}) \cdot \vec{N}}{|\vec{x} - \vec{y}|^2} \right] \frac{dS}{|\vec{x} - \vec{y}|}, \tag{9.56a-c}$$

$$2 \neq N = 3, 4, \ldots: \quad \sigma_N (2 - N) \Phi(\vec{x}) = \int_D \Lambda(\vec{y}) |\vec{x} - \vec{y}|^{2-N} d^N \vec{y}$$

$$+ \int_{\partial D} \left[\Phi(\vec{y}) \frac{(\vec{x} - \vec{y}) \cdot \vec{N}}{|\vec{x} - \vec{y}|^2} (2 - N) - \left(\vec{N} \cdot \frac{\partial \Phi}{\partial \vec{y}} \right) \right] |\vec{x} - \vec{y}|^{2-N} dS_N, \tag{9.57a-c}$$

where $\vec{N}$ is the unit outer normal to the boundary and the Green functions (9.44a through c), (9.42a through d), (9.45a through c), and (9.43a through c) were used.

9.2.5 Volume and Surface Source Distributions

The preceding result (9.53b) ≡ (9.54c; 9.55c; 9.56c; 9.57c) is not the solution of the Poisson equation (9.53a) in the region D because it is not possible to specify independently the monopole strength Φ and the dipole moment $-\partial\Phi/\partial N$ on the boundary. According to the unicity theorem (Subsection 9.1.2), only one of Φ and $\partial\Phi/\partial N$ can be specified on the boundary ∂D, and the other is then determined by the solution of the problem. Thus, if both Φ and $\partial\Phi/\partial N$ are specified, the two conditions will be incompatible (redundant) if the solutions of the Cauchy–Dirichlet and Neumann problems with a given Φ and a given $(\partial\Phi/\partial N)$ are distinct (coincide). It follows that the integral identity (9.53b) ≡ (9.54c; 9.55c; 9.56c; 9.57c) is equivalent to the Poisson equation (9.53a) in the region D and not a solution of it; the region D can be compact or noncompact if the asymptotic condition (9.19) is met (9.21a through c)/(9.22a through c)/ (9.23a through c)/(9.24a through c). The integral equation (9.53b) for the potential becomes a solution of the Poisson equation (9.53a) if the surface integrals vanish, for example, if $\partial D \equiv S_\infty$ is the surface at infinity, and the asymptotic condition (9.19) is met by the potential. Omitting the boundary integral that is the last term on the r.h.s. of (9.53b), it follows that the solution of the Poisson equation (9.53a) ≡ (9.46a) ≡ (9.10a) in an unbounded medium is the first term of (9.53b) that coincides with (9.46b) ≡ (9.47a through c)/(9.48a through c)/(9.49a through c)/(9.50a and b); the latter was obtained using the convolution property of generalized functions instead of the second Green theorem for the Poisson equation, both being linear processes. Thus, *the potential as a solution of the Poisson equation (9.53a) with sources (1) satisfies in one/two/three/more dimensions (9.54a through c/9.55a through c/9.56a through c/9.57a through c) in a compact domain; (2) applies as well in a noncompact domain if it satisfies the asymptotic conditions (9.21a through c/9.22a through c/9.23a through c/9.24a through c); and (3) is specified by (9.47a through c/9.48a through c/9.49a through c/9.50a through c) in free space. The integral identity (9.53b) for the potential is equivalent to the Poisson equation (9.53a).* Also, (1) and (2) are not solutions of the Poisson equation, unless the monopole Φ and the dipole $\partial\Phi/\partial N$ potentials can be specified consistently on the boundary; since there are no boundary terms in (3), the latter is a solution of the Poisson equation in free space. The solution of the Laplace equation in the plane with given values on a circle also leads to Poisson-type integrals (Sections I.37.8.9 and I.39.1.2).

9.3 Mean Value, Extrema, and Constancy Theorems

The first (second) Green identity has been used [Section 9.1 (Section 9.2)] to prove the unicity theorem (obtain the Poisson integral) for the Laplace operator and to lead to further properties of harmonic functions that are solutions of the Laplace equation. The second Green theorem can also be used (Subsection 9.3.1) to prove that a harmonic function in a hypersphere takes at the center the mean of the surface values (Subsection 9.3.2), which implies the theorem of the extrema that a nonconstant harmonic function in an arbitrary domain takes the maximum value on the boundary (Subsection 9.3.3); if it is nonzero, it also takes the minimum value on the boundary. The exception is a harmonic function in a domain that is constant (zero) on the boundary and hence takes the same constant (zero) value in the interior. It follows (Subsection 9.3.4) that two harmonic functions in a domain that are equal (have an equal normal derivative) on the boundary coincide (differ by a constant).

9.3.1 Volume Sources and Surface Flux

A simpler form of the second Green identity (9.9a and b) results considering the auxiliary function $\Psi = 1$ to be unity is as follows:

$$\Phi \in C^2(D): \quad \int_D \nabla^2\Phi\, dV = \int_{\partial D} \frac{\partial\Phi}{\partial N}\, dS. \qquad (9.58\text{a,b})$$

This states that for *the Poisson equation (9.46a)* ≡ *(9.59a), the total source strength (9.59c) in a domain D, with a closed regular boundary, is equal to the normal derivative of the potential integrated over the surface,*

$$\nabla^2\Phi = \Lambda, \quad \vec{U} = \nabla\Phi: \quad \int_D \Lambda dV = \int_{\partial D}(\vec{U}\cdot\vec{N})dS = \int_{\partial D}\vec{U}\cdot d\vec{S}, \tag{9.59a–c}$$

which is equal to the total flux of the field (9.59b) across the boundary ∂D. The result (9.59c) can also be proved from:

$$\Lambda = \nabla^2\Phi = \nabla\cdot(\nabla\Phi) = \nabla\cdot\vec{U}, \tag{9.60}$$

which states that *the source strength is the divergence of the field*; applying the divergence theorem (9.1a and b) leads to (9.61) ≡ (9.59c):

$$\int_D \Lambda dV = \int_D (\nabla\cdot\vec{U})dV = \int_{\partial D}\vec{U}\cdot d\vec{S}. \tag{9.61}$$

This result can be applied to any irrotational field.

In the case of the gravity (9.62a) [electrostatic (9.63a)] field, the source is the mass (9.62b) [electric charge (9.63b)] density:

$$\vec{g} = -\nabla\Phi_g \equiv \vec{U}; \quad \nabla\cdot\vec{g} = -G\rho \equiv \Lambda: \quad m \equiv \int_D \rho dV = -\frac{1}{G}\int_D(\nabla\cdot\vec{g})dV = -\frac{1}{G}\int_{\partial D}\vec{g}\cdot d\vec{S}, \tag{9.62a–c}$$

$$\vec{E} = -\nabla\Phi_e = -\vec{U}, \quad \nabla\cdot\vec{E} = \frac{q}{\varepsilon} \equiv \Lambda: \quad e = \int_D q dV = \varepsilon\int_D(\nabla\cdot\vec{E})dV = \varepsilon\int_{\partial D}\vec{E}\cdot d\vec{S}, \tag{9.63a–c}$$

the total mass m (electric charge e) contained in a region D is specified by the ratio of the outflux of the gravity (electric) field through the boundary ∂D divided by the minus gravitational constant G (by the dielectric permittivity of the medium ε). Likewise, for steady heat conduction:

$$\vec{G} = -k\nabla T, \quad \nabla\cdot\vec{G} = w \equiv \Lambda:$$

$$W = \int_D w dV = \int_D(\nabla\cdot\vec{G})dV = \int_D\vec{G}\cdot d\vec{S} = -k\int_D\frac{\partial T}{\partial N}dS, \tag{9.64a–c}$$

the total output of the heat sources/sinks (9.64b) in a domain (9.64c) equals the heat flux (9.64a) through the boundary in the direction of decreasing temperature for a medium with thermal conductivity k.

In the absence of sources Λ = 0, the Poisson equation (9.59a) is replaced by the Laplace equation (9.65) whose solutions form the set $\mathcal{H}$ of **harmonic functions**:

$$\mathcal{H}(D) \equiv \left\{\Phi(\vec{x}): \quad \Phi \in C^2(D) \wedge \forall_{\vec{x}\in D} : \nabla^2\Phi = 0\right\}; \tag{9.65}$$

the harmonic functions are functions with continuous second-order derivatives that satisfy the Laplace equation. For a function harmonic (9.65) ≡ (9.66a) in a region *D*, the l.h.s. of (9.58b) vanishes (9.66b), so that it simplifies to (9.66c):

$$\Phi \in \mathcal{H}(D): \quad \nabla^2\Phi = 0 \Rightarrow \int_{\partial D}\frac{\partial\Phi}{\partial N}dS = 0. \tag{9.66a–c}$$

Thus, *if a function is harmonic (9.66a) ≡ (9.65) in a region D with a closed regular boundary ∂D, the integral of the normal derivative over the surface is zero*; the interpretation is that if there are no sources in the region D, there is no net flux across the boundary ∂D, that is, an influx over a part $E \subset \partial D$ of the boundary is compensated for by an equal and opposite outflow over the remainder, $\partial D - E$. The analogous result for two-dimensional harmonic functions, that is, complex differentiable or analytic functions,

$$\mathcal{D}(|C) \equiv \{f(z): z \in C \wedge \exists f'(z)\} \equiv \mathcal{A}(D), \tag{9.67}$$

is that (9.67) ≡ (9.68a) the integral of the derivative over a closed regular curve or loop is zero (9.68b),

$$f \in \mathcal{A}(D): \quad \int_{\partial D} f'(z)\,\mathrm{d}z = \int_{\partial D} \frac{\mathrm{d}f}{\mathrm{d}z}\,\mathrm{d}z = \int_{\partial D} \mathrm{d}f = 0. \tag{9.68a,b}$$

because the function is single-valued; this is the Cauchy first integral theorem (I.15.1).

9.3.2 Linear, Circular, Spherical, and Hyperspherical Mean Value

An analytic function satisfies the mean value theorem on a circle (Section I.23.1), which can be proved from (9.55c) or (9.68b), together with a similar theorem for a sphere from (9.56c) or (9.66c). The Laplace equation (9.69a) has no source term $\Lambda = 0$ in (9.53a), so that the first integral on the r.h.s. of (9.53b) vanishes, leading to (9.69b):

$$\nabla^2 \Phi = 0: \quad \Phi(\vec{x}) - \int_{\partial D} \Phi(\vec{y}) \left\{ \vec{N} \cdot \frac{\partial}{\partial \vec{y}} G(\vec{x} - \vec{y}) \right\} \mathrm{d}S = -\int_{\partial D} \left\{ G(\vec{x}; \vec{y}) \vec{N} \cdot \frac{\partial \Phi}{\partial \vec{y}} \right\} \mathrm{d}S. \tag{9.69a,b}$$

The Green function (9.43c) is constant on a hypersphere and thus the r.h.s. of (9.69b) vanishes on account of (9.66c):

$$\int_{|\vec{x}-\vec{y}|=R} G(\vec{x} - \vec{y}) \frac{\partial \Phi}{\partial N} \mathrm{d}S = G(R) \int_{|\vec{x}-\vec{y}|=R} \frac{\partial \Phi}{\partial N} \mathrm{d}S = 0; \tag{9.70a}$$

substitution of (9.70a) and the influence function (9.43c) simplify (9.69b) to

$$\Phi(\vec{x}) = \int_{|\vec{x}-\vec{y}|=R} \Phi(\vec{y}) \frac{\mathrm{d}}{\mathrm{d}R} \left[\frac{R^{2-N}}{\sigma_N(2-N)} \right] \mathrm{d}S = \frac{1}{\sigma_N R^{N-1}} \int_{|\vec{x}-\vec{y}|=R} \Phi(\vec{y}) \mathrm{d}S, \tag{9.70b}$$

where the area of the hypersphere of radius R is (9.71b) ≡ (5.291a):

$$\Phi(\vec{x}) \in \mathcal{H}\left(|\vec{x}-\vec{y}| \le R\right): \quad S_N = \sigma_N R^{N-1}, \quad \Phi(\vec{x}) = \frac{1}{S_N} \int_{|\vec{x}-\vec{y}|=R} \Phi(\vec{y}) \mathrm{d}S. \tag{9.71a-c}$$

Thus, the **hyperspherical mean value theorem** has been proved: *a harmonic function (9.71a) in a hypersphere of radius R takes on the surface (9.71b) a mean value (9.71c) equal to the value at the center. In particular, in one (9.72a through d)/two (9.73a through d)/three (9.74a through d)/more (9.71a through c) dimensions,*

$$w = 1; \quad \Phi \in \mathcal{H}(x-y, x+y), \quad S_1 = 2(x-y): \quad \Phi(x) = \frac{\Phi(x+y) + \Phi(x-y)}{2}, \tag{9.72a-d}$$

$$N = 2; \quad \Phi \in \mathcal{H}\left(|\vec{x} - \vec{y}| \le r\right), S_2 = 4\pi r: \quad \Phi(\vec{x}) = \frac{1}{2\pi r} \int\limits_{|\vec{x} - \vec{y}| = r} \Phi(\vec{x}) ds, \qquad \text{(9.73a–d)}$$

$$N = 3; \quad \Phi \in \mathcal{H}\left(|\vec{x} - \vec{y}| \le R\right), S_3 = 4\pi R^2: \quad \Phi(\vec{x}) = \frac{1}{4\pi R^3} \int\limits_{|\vec{x} - \vec{y}| = R} \Phi(\vec{x}) dS, \qquad \text{(9.74a–d)}$$

namely (1) in the one-dimensional case, a harmonic function is a linear function and has the arithmetic mean property (9.72a); and (2/3) in the plane (space) a function harmonic in a disk (sphere), takes on circle (surface) a mean value equal to the value in the center (9.73d) [(9.74d)]. The plane case (2) coincides with the Cauchy second integral theorem (Section I.15.3).

9.3.3 Lemma of the Maximum and the Minimum on the Boundary

A harmonic function in the plane, that is, an analytic function, satisfies the lemma of the maximum modulus (Section I.23.2), for which there is an analogous result in any dimension, namely the **theorem of the extrema**: *if a nonconstant (9.75b) function Φ is harmonic (9.75a) in a region D, its maximum (9.75d) and minimum (9.75c) lie on the boundary ∂D:*

$$\Phi \in \mathcal{H}(D), \Phi(\vec{x}) \ne \text{const}: \quad \{\Phi(\partial D)\}_{\min} < \Phi(D - \partial D) < \{\Phi(\partial D)\}_{\max}. \qquad \text{(9.75a–d)}$$

To prove the theorem, let us consider $P \in D - \partial D$ to be an interior point in the region D and $\Phi(P) = A$ the value of the harmonic function at P. If ε is the distance from the point P to the boundary, there is an R such that $0 < R < \varepsilon$ and the sphere E with center at P and radius R is wholly contained in D; thus, the mean value of Φ over the sphere E is A. Since the function Φ is not a constant, there are points on E where it takes values larger and smaller than A. Hence, an interior point P cannot be either a maximum or a minimum of the function Φ, that is, the extrema must be on the boundary. QED. As an example, consider the electrostatic field in a hollow sphere; since the singularity at the center is excluded, the electrostatic potential is harmonic inside the hollow sphere. Since it decays in modulus with distance, the maximum (minimum) is on the inner (outer) surface; if the monopole potential is negative, the maximum (minimum) is on the outer (inner) surface, and vice versa if it is positive; in both cases, the extrema are on the boundary. The latter statement extends to a hollow body of arbitrary shape, provided there are no singularities of the electrostatic field in it. If there are singularities in the body, the potential is infinite at the singular point and the extremal theorem could not possibly hold.

9.3.4 Identity, Constancy, and Nullity Theorems

The extremal theorem has excluded the "trivial" case of a harmonic function reducing to a constant. The latter is covered in the proof of the unicity theorem (Subsection 9.1.3) for the Cauchy–Dirichlet (Subsection 9.1.2) and also for the Neumann problems, which can be re-stated as the **theorem of constancy**: *a function harmonic in a region (9.76a) is a constant (9.76b) iff it is constant on the boundary (9.76c) or iff it has zero normal derivatives on the boundary (9.76d):*

$$\Phi \in \mathcal{H}(D): \quad \Phi|_D = \text{const} \Leftrightarrow \Phi|_{\partial D} = \text{const} \Leftrightarrow \partial\Phi/\partial n|_{\partial D} = 0. \qquad \text{(9.76a–d)}$$

The case when the constant is zero, $c = 0$, leads to the **theorem of nullity**: *a function harmonic in a region (9.77a) and zero on the boundary (9.77b) is zero in the interior (9.77c):*

$$\Phi \in \mathcal{H}(D): \quad \Phi|_D = 0 \Leftrightarrow \Phi|_{\partial D} = 0. \qquad \text{(9.77a–c)}$$

Both theorems can be re-stated,

$$\Phi, \Psi \in \mathcal{H}(D): \quad \Phi\big|_D = \Psi\big|_{\partial D} \Leftrightarrow \Phi\big|_{\partial D} = \Psi\big|_{\partial D}, \tag{9.78a--c}$$

$$\Phi, \Psi \in \mathcal{H}(D): \quad (\Phi - \Psi)\big|_D = \text{const} \Leftrightarrow \partial\Phi/\partial n\big|_{\partial D} = \partial\Psi/\partial n\big|_{\partial D}. \tag{9.79a--c}$$

as the **identity theorems**: *two functions* Φ, Ψ *harmonic in a region (9.78a) [(9.79a)] coincide (9.78b) [coincide to within an added constant (9.79b)] iff their values (their normal derivatives (9.79a)] coincide on the boundary.* As an example, consider the electrostatic field due to a conductor, that is, an equipotential $\Phi = \text{const}$, such that the tangential electric field is zero, $E_s = \partial\Phi/\partial s = 0$; if $\partial\Phi/\partial n = 0$ on the conductor, the potential is constant and the field is zero everywhere; thus, a conductor $\Phi = \text{const}$ creates an electrostatic field only if $0 \neq \partial\Phi/\partial n = \sigma/\varepsilon$, that is, there are surface electric charges. In the case of a potential flow (9.80a), if the normal (9.80b) and the tangential (9.80c) components of velocity both vanish at a boundary, the fluid is at rest everywhere (9.80d):

$$\nabla^2\Phi\big|_D = 0 \wedge \partial\Phi/\partial s\big|_{\partial D} = 0 = \partial\Phi/\partial n\big|_{\partial D} \Rightarrow \Phi\big|_D = 0. \tag{9.80a--d}$$

Thus, a potential flow cannot represent a viscous fluid because the latter would have zero velocity at a wall (Notes 1.6 and 2.6). For a potential flow, at least one component of the velocity at a boundary must be nonzero, for example, the tangential (normal) component for a rigid wall at rest (a moving or pulsating wall). For a plane elastic medium, there is no deformation unless there is a compression or shear or both at a boundary.

9.4 Irrotational Fields and the Newton/Coulomb Laws

An important class of harmonic functions is the scalar potential of irrotational fields (Subsection 9.4.1), for example, in one (Subsection 9.4.2), two (Subsection 9.4.3), and three (Subsection 9.4.4) dimensions; it specifies the electrostatic (gravity) field leading to Newton (Coulomb) law of gravity (electric) forces between monopoles (Subsection 9.4.5). The forces between multipoles were considered (Note 8.16) for the analogous irrotational fields in three dimensions (Notes 8.3 through 8.8). Next, the fields and forces in one, two, and three dimensions are considered, taking as an example the electrostatic field.

9.4.1 Electrostatics in One/Two/Three Dimensions

The electrostatic field $\vec{E}$ due to a distribution of electric charges with density $e(\vec{x})$ is specified (9.81a and b) by the negative of the gradient of the scalar electric potential $\Phi(\vec{x})$:

$$\vec{E} = -\nabla\Phi \equiv -\frac{\partial\Phi}{\partial\vec{x}}, \quad -\frac{q(\vec{x})}{\varepsilon} = \nabla^2\Phi = \sum_{i=1}^{N} \frac{\partial^2\Phi}{\partial x_i \partial x_i}, \tag{9.81a,b}$$

that satisfies the Laplace (9.69a) [Poisson (9.81b)] outside $q = 0$ (inside $q \neq 0$) the charges, where ε is the dielectric permittivity of the medium and $\vec{x}$ is the position vector in N-dimensional space:

$$\vec{x} = \sum_{i=1}^{N} x_i \vec{e}_i = \vec{e}_x x, \vec{e}_x x + \vec{e}_y y, \vec{e}_x x + \vec{e}_y y + \vec{e}_z z; \tag{9.81c--f}$$

the formulation (9.81a and b) applies to charges in free space of any dimension (9.81c), and the three lowest, namely, one- (9.98e)/two- (9.81f)/three- (9.81g) dimensional cases are considered next.

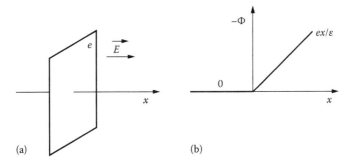

(a)　　　　　　　　　　　　(b)

FIGURE 9.1　A one-dimensional electric field corresponds to a uniform electric charge on a plane; since the flux is independent of the distance from any parallel plane, the field is (a) constant, and the potential is (b) proportional to the distance from the plane. The common factor is the electric charge density per unit area divided by the dielectric permittivity of the medium.

If the charge distribution is an impulse (9.82a), then its integral over a domain D is zero (9.82b) [nonzero (9.82c)] if it does not (does) include the origin:

$$q(\vec{x}) = e\delta(\vec{x}), \quad \int_D q(\vec{x})\mathrm{d}\vec{x} = \begin{cases} e & \text{if } \vec{0} \in D, \\ 0 & \text{if } \vec{0} \in D. \end{cases} \tag{9.82a–c}$$

Thus, (9.82a) represents (1) in one dimension, a plane $x = 0$ with uniform surface charge density (Figure 9.1a); (2) in two dimensions, a line charge $x = 0 = y$ with uniform charge per unit length (Figure 9.2a); and (3) in three dimensions, a point charge at the origin $x = y = z = 0$ (Figure 9.3a). In all three cases, the potential due to the impulse charge is specified by the influence or Green function (9.83a) that is the solution of the Poisson equation forced by an impulse:

$$\Phi_0(\vec{x}) = -\frac{e}{\varepsilon}G(\vec{x}): \quad \frac{\mathrm{d}^2\Phi_0}{\mathrm{d}x^2} = -\frac{e}{\varepsilon}\delta(x), \tag{9.83a,b}$$

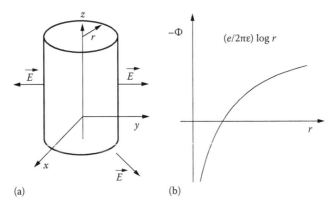

(a)　　　　　　　　　　　　(b)

FIGURE 9.2　A line charge uniform along its length leads to a constant flux across a cylinder with the same axis independent of the radius. Hence, the electric field (a) decays as the inverse of the distance from the axis and the potential (b) varies as the logarithm. In the factor e/ε in the one-dimensional case (Figure 9.1), e now means the electric charge per unit length and is divided by the perimeter of the unit circle, 2π.

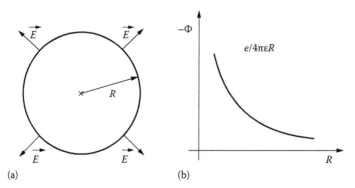

(a) (b)

FIGURE 9.3 A point electric charge in space leads to a constant flux across any sphere with the same center independent of the radius. Hence, the electric field (a) decays as the inverse square of the radial distance from the origin and the potential (b) decays as the inverse of the distance. In the factor $e/2\pi$ in the plane case (Figure 9.2), e now means the total point electric charge and the area of the unit sphere, 4π, is used in $e/4\pi$. In a hyperspace $N \geq 4$ of any dimension, the area of a hypersphere scales as R^{N-1}. Hence, the constant flux of a point source across a sphere is independent of the radius if the field decays as R^{1-N}. Thus, the potential decays as R^{2-N}, excluding the case $N = 2$ of the logarithmic potential (Figure 9b). Together with the area σ_N of the unit hypersphere, this specifies the influence or Green function that is the fundamental solution for the hyperspherical Laplace operator.

$$\frac{1}{r}\frac{d}{dr}\left(r\frac{d\Phi_0}{dr}\right) = -\frac{e}{\varepsilon}\delta(x)\delta(y) = \frac{e}{2\pi\varepsilon r}\delta(r), \tag{9.83c}$$

$$\frac{1}{R^2}\frac{d}{dR}\left(R^2\frac{d\Phi_0}{dR}\right) = -\frac{e}{\varepsilon}\delta(x)\delta(y)\delta(z) = -\frac{e}{4\pi\varepsilon R^2}\delta(R), \tag{9.83d}$$

where (1) the Laplacian was written in Cartesian/polar/spherical (9.83b/c/d) $\equiv$ (6.44/6.45b/6.46a) coordinates, depending on only $x/r/R$, because of the plane/cylindrical radial symmetry (9.36a through c/9.37a through c/9.38a through d); (2) likewise for the Dirac impulse distribution in one dimension (9.83b) and with cylindrical (9.83c) $\equiv$ (5.60a and b) [spherical (9.83d) $\equiv$ (5.307a and b)] symmetry. *The plane (9.83b), cylindrical (9.83c), and spherical (9.83d) Laplacian are, respectively, the cases $N = 1$, 2, and 3 of the Laplacian with hyperspherical symmetry in N dimensions (9.39) $\equiv$ (9.40).* Next, the irrotational fields in one (Subsection 9.4.1), two (Subsection 9.4.3), and three (Subsection 9.4.4) dimensions will be considered successively, deferring the higher dimensions to Sections 9.6 through 9.9.

9.4.2 Ramp Potential due to a Charged Plane

Starting with the one-dimensional case, the solution of (9.83b) is the jump in the field (9.84a and b),

$$\vec{E}_0(x) = -\vec{e}_x\frac{d\Phi_0}{dx} = \vec{e}_x\frac{e}{\varepsilon}H(x) = \begin{cases} 0 & \text{if } x < 0, \\ \dfrac{e}{\varepsilon} & \text{if } x > 0, \end{cases} \tag{9.84a) and (9.84b}$$

$$\Phi_0(x) = -\frac{e}{\varepsilon}G(x) = \frac{e}{\varepsilon}H^{(-1)}(x) = \begin{cases} 0 & \text{if } x < 0, \\ -\dfrac{ex}{\varepsilon} & \text{if } x > 0, \end{cases} \tag{9.85a) and (9.85b}$$

and the unit ramp (3.68a,b) for the potential (9.85a and b). Thus, the potential is (1) a constant before the charge $x \leq 0$ and the constant may be set to zero, because the potential is defined to within an arbitrary additive constant and (2) linear, $-e\, x/\varepsilon$, for $x \geq 0$, which corresponds to passing a plane $x = 0$ in three dimensions, with uniform charge. The result is *that a uniformly charged e plane x = 0 causes the potential to change from constant to linear, $-e\, x/\varepsilon$, and the field to jump from zero to a constant, e/ε, normal to the plane,* corresponding (Subsection I.36.5.1) to an infinite flat-plate condenser. *The potential (electric field), due to a distribution e(x) of charges uniform (9.87b) on planes x = const [Figure 9.1b (Figure 9.1a)], is given by (9.86) [(9.87a)]:*

$$\Phi(x) = -\frac{1}{\varepsilon}\int_{-\infty}^{+\infty} G(x-\xi)q(\xi)\,\mathrm{d}\xi = \frac{1}{\varepsilon}\int_{-\infty}^{x}(\xi-x)q(\xi)\,\mathrm{d}\xi; \qquad (9.86)$$

differentiating once (twice) with regard to x:

$$E(x) = -\frac{\mathrm{d}\Phi}{\mathrm{d}x} = \frac{1}{\varepsilon}\int_{-\infty}^{x} q(\xi)\,\mathrm{d}\xi, \quad \frac{\mathrm{d}^2\Phi}{\mathrm{d}x^2} = -\frac{q(x)}{\varepsilon}, \qquad (9.87a,b)$$

it is confirmed that the electric field is given by (9.87a) [the potential satisfies the Poisson equation (9.87b) in one dimension]. The potential (9.85a and b) with $-e/\varepsilon = 1$ coincides with the influence function (9.44a through c) with $y = 0$ and the one-dimensional degenerate Poisson integral (9.47a through c) coincides with (9.86) for the source $\Lambda = -q/\varepsilon$. The linear potential satisfies the mean value property (9.72a through d). The passage from (9.86) to (9.87a)

$$q(-\infty) = 0: \quad \varepsilon\frac{\mathrm{d}\Phi}{\mathrm{d}x} = \int_{-\infty}^{x}\frac{\mathrm{d}(\xi-x)}{\mathrm{d}x}q(\xi)\,\mathrm{d}\xi + \left[(\xi-x)q(\xi)\right]_{-\infty}^{x} = -\int_{-\infty}^{x} q(\xi)\,\mathrm{d}\xi, \qquad (9.88a,b)$$

uses the (9.88b) rule of parametric differentiation of integrals (I.13.46a and b) as well as (9.88a). The theory of the linear potential shows that real functions provide a suitable approach to the one-dimensional Poisson and Laplace equations.

9.4.3 Logarithmic Potential due to a Line Charge

Proceeding to (9.89a) the two-dimensional case (9.83c) replaces the Cartesian impulse by a polar one (9.89b):

$$r \equiv |x^2 + y^2|^{1/2}: \quad \frac{\mathrm{d}}{\mathrm{d}r}\left(r\frac{\mathrm{d}\Phi_0}{\mathrm{d}r}\right) = -\frac{e}{2\pi\varepsilon}\delta(r); \qquad (9.89a,b)$$

the cylindrical influence or Green function is (Figure 9.2a) the potential due to a line charge of uniform density e per unit length along the oz-axis $x = 0 = y$ or $r = 0$; the factor $1/2\pi$ arises (5.60a and b) from the perimeter of the unit circle centered at the origin or the area per unit length of the cylinder of unit radius with the line charge as the axis. Integrating (9.89b) once leads to (9.90b):

$$r > 0: \quad r\frac{\mathrm{d}\Phi_0}{\mathrm{d}r} = -\frac{e}{2\pi\varepsilon}H(r) = -\frac{e}{2\pi\varepsilon}, \qquad (9.90a,b)$$

where the Heaviside unit jump may be replaced by unity for (9.90a) outside the line charge. Integrating (9.91a) once more leads the logarithmic potential (9.91d):

$$\frac{\mathrm{d}\Phi_0}{\mathrm{d}r} = -\frac{e}{2\pi\varepsilon r}, \quad \Phi_0(r) = -\frac{e}{2\pi\varepsilon}\log r = -\frac{e}{\varepsilon}G(r), \qquad (9.91a,b)$$

where r denotes the distance from the observer to the source. Thus, *the field due to a line charge (Figure 9.2a) decays as the inverse of the distance (9.92a) and the potential (Figure 9.2b) varies as the negative of the logarithm of the distance (9.92b):*

$$\vec{E}_0 = \frac{e}{2\pi\varepsilon |\vec{x}|}\vec{e}_r, \quad \Phi_0(\vec{x}) = -\frac{e}{2\pi\varepsilon}\log|\vec{x}|; \tag{9.92a,b}$$

the common factor is the charge per unit length e, divided by the dielectric permeability ε, and the geometric factor 2π; in the present two-dimensional case, the latter is the perimeter of the unit circle or the area per unit length of the cylinder of unit radius. In (9.92a and b), the observer is at r and the source is at the origin; for a source at $\vec{y}$, the influence or Green function is given by (9.91b), with r replaced by the relative distance:

$$\vec{x} = \vec{e}_x x + e_y y: \quad -\frac{\varepsilon}{e}\Phi_0(\vec{x} - \vec{y}) \equiv G(\vec{x} - \vec{y}) = \frac{1}{2\pi}\log|\vec{x} - \vec{y}| = G(\vec{y} - \vec{x}). \tag{9.93a,b}$$

The cylindrical Green function (9.93b) is symmetric in $(\vec{x}, \vec{y})$, showing that the position of the observer and the source can be interchanged without altering the potential, that is, the reciprocity principle holds, because the Laplace operator is self-adjoint (7.162a through d).

The Laplace operator is also linear so that the superposition principle holds: *the electrostatic potential (9.94) [field (9.95)] due to an electric charge distribution q(ȳ) in a medium of dielectric permittivity ε is given by*

$$\Phi(\vec{x}) = -\frac{1}{2\pi\varepsilon}\iint_D q(\vec{y})\log|\vec{x} - \vec{y}|\,d^2\vec{y}, \tag{9.94}$$

$$\vec{E}(\vec{x}) = \frac{1}{2\pi\varepsilon}\iint_D q(\vec{y})\frac{\vec{x} - \vec{y}}{|\vec{x} - \vec{y}|^3}\,d^2\vec{y}, \tag{9.95}$$

where (1) the potential (9.94) arises from the convolution of the charge distribution $q(\vec{y})$ with the influence function (9.93b):

$$\Phi(\vec{x}) = -\frac{1}{\varepsilon}\iint q(\vec{y})G(\vec{x} - \vec{y})d^2\vec{y} = -\frac{q}{\varepsilon} * G(x), \tag{9.96}$$

a result obtained before (2.59a and b) for an elastic string and proven generally in the sequel (7.78a and b); and (2) the field (9.95) can be obtained from a similar superposition principle applied to (9.92a) as minus the gradient (9.81a) to (9.94) and using the formula to differentiate inside the integral modulus of a vector (8.14a–d). The potential (9.94) and the field (9.95) involve an integration over the region D occupied by the charges in the case (9.89a and b) of a two-dimensional domain. If the charges are concentrated along a curve L, then (9.94; 9.95) become line integrals: this agrees with the potential at a point (x, y) being the real part of the complex potential (I.24.14a):

$$\Phi(x + iy) = -\frac{1}{2\pi\varepsilon}\mathrm{Re}\left\{\int_L q(\xi)\log(\xi - x - iy)d\xi\right\}$$

$$= -\frac{1}{2\pi\varepsilon}\int_L q(\xi)\log|\xi - x - iy|d\xi, \tag{9.97}$$

where the charge density $q(\xi)$ is assumed to be real on the curve L in the (x, y) plane. The two formulas obtained via two-dimensional potential theory (complex analytic functions) coincide (9.94) ≡ (9.97), for example, for electrostatic potential due to a charge distribution in a domain:

$$\vec{y} = \vec{e}_x \xi + \vec{e}_y \eta: \quad \Phi(x, y) = -\frac{1}{4\pi\varepsilon} \iint_D q(\xi, \eta) \log\left|(x - \xi)^2 + (y - \eta)^2\right| d\xi \, d\eta. \tag{9.98}$$

Both on the potential due to a charge distribution in a domain (9.94) ≡ (9.98) or along a curve (9.97), the extension from two to three dimensions can be made by assuming uniformity in the direction perpendicular to the (x, y) plane.

9.4.4 Inverse-Distance Potential due to a Point Charge

Dropping the assumption mentioned last in Subsection 9.4.3 leads to a three-dimensional electric charge distribution. The corresponding Green function satisfies (9.83d), where the three-dimensional (9.99a) impulse distribution has been replaced (9.99b) by a spherically symmetric one:

$$R \equiv \left|x^2 + y^2 + z^2\right|^{1/2}: \quad \frac{d}{dR}\left(R^2 \frac{d\Phi_0}{dR}\right) = -\frac{e}{4\pi\varepsilon} \delta(R), \tag{9.99a,b}$$

and the factor 4π arises (5.307a and b) from the area of the unit sphere. Integrating (9.99b) once leads to

$$R > 0: \quad -\frac{d\Phi_0}{dR} = \frac{e}{4\pi\varepsilon R^2} H(R) = \frac{e}{4\pi\varepsilon R^2}; \tag{9.100a,b}$$

it follows that *the field due to a point charge is radial and decays as the inverse square of the distance:*

$$\vec{E}_0(R) = \frac{e}{4\pi\varepsilon R^2} \vec{e}_R, \quad \Phi_0(R) = \frac{e}{4\pi\varepsilon R}, \tag{9.101a,b}$$

and the potential varies with the inverse of the distance (Figure 9.3b). The factor common to (9.101a and b) is the electric charge e divided by the dielectric permittivity ε, as in the one-dimensional case (9.84a and b; 9.85a and b); the geometrical factor is 4π in the three-dimensional, instead of 2π in the two-dimensional (9.92a and b) case, because the area of the unit sphere replaces the perimeter of the unit circle.
The potential at $\vec{x}$ due to a point charge at $\vec{y}$ is given by

$$\left\{\sum_{i=1}^{3} \frac{\partial^2}{\partial x_i \partial x_i}\right\} \Phi_0(\vec{x}; \vec{y}) = -\frac{e}{\varepsilon} \delta(\vec{x} - \vec{y}), \tag{9.102}$$

by $-(e/\varepsilon)$ times the Green function for the three-dimensional Poisson equation:

$$-\frac{e}{\varepsilon} \Phi_0(\vec{x}, \vec{y}) \equiv G(\vec{x} - \vec{y}) = -\frac{1}{4\pi|\vec{x} - \vec{y}|} = G(\vec{x} - \vec{y}); \tag{9.103}$$

it is symmetric, so that, as in two dimensions (9.93b), it satisfies the **reciprocity principle**: *the potential at a point $\vec{x}$ due to a point charge at $\vec{y}$ is equal to the potential at $\vec{y}$ due to a similar point charge at $\vec{x}$.* The principle of superposition, in the form of a convolution integral, three-dimensional instead of two-dimensional (9.96), may be used,

$$\Phi(\vec{x}) = -\frac{e}{\varepsilon} * G(\vec{x}) = -\frac{1}{\varepsilon} \int q(\vec{y}) G(\vec{x}; \vec{y}) d^3\vec{y}, \tag{9.104}$$

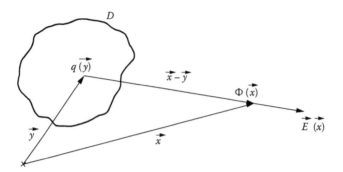

FIGURE 9.4 The electrostatic field is specified by a Poisson equation that is linear and thus the principle of superposition holds: the field or potential at an observer position $\vec{x}$ due to a distribution of electric charge $e(\vec{y})$ in a domain D is specified by the convolution integral with the influence function that depends on the relative distance, $|\vec{x} - \vec{y}|$.

to obtain from (9.103) *the potential (9.105) [electric field (9.106)] due to a charge distribution in a three-dimensional space [Figure 9.3b (Figure 9.3a)]*:

$$\Phi(\vec{x}) = -\frac{1}{4\pi\varepsilon}\int_D \frac{q(\vec{y})}{|\vec{x}-\vec{y}|}d^3\vec{y}, \tag{9.105}$$

$$\vec{E}(\vec{x}) = \frac{1}{4\pi\varepsilon}\int_D q(\vec{y})\frac{\vec{x}-\vec{y}}{|\vec{x}-\vec{y}|^3}d^3\vec{y}; \tag{9.106}$$

the formula (9.106) follows from (9.101a), as (9.105) does from (9.101b), that is, replacing the position vector $\vec{x}$ relative to the origin by the position vector $\vec{x} - \vec{y}$ relative to the charge $\vec{y}$ that is no longer at the origin, so that an integration over the charge distribution $q(\vec{y})$ is performed. Also, (9.106) can be derived from (9.105) using (9.81a). If the charges occupy a three-dimensional region (Figure 9.4), then (9.105; 9.106) are volume integrals, which reduce to surface (line) integrals for charge distributions in two (one) dimensions.

9.4.5 Gravity and Electric (Coulomb) Forces

A charge e' in a field $\vec{E}$ is subject (I.24.15a) ≡ (8.302a) to a force $e'\vec{E}$; considering a charge distribution $q'(\vec{y})$ in a field $\vec{E}(\vec{y})$, the total force on it is given by

$$N = 1,2,3: \quad \vec{F}_e = \int q'(\vec{y})\vec{E}(\vec{y})d^N\vec{y}, \tag{9.107a,b}$$

where (9.107b) applies the one-/two-/three-dimensional cases (9.107a). Thus, *the force between two charge distributions is given (1) in one dimension (charge uniform on plane x = const) by*

$$\vec{F}_e = \frac{1}{\varepsilon}\vec{e}_x \int_{-\infty}^{+\infty} dx \int_{-\infty}^{+\infty} dy\, q(y)q'(x), \tag{9.108}$$

the product of the charges divided by the dielectric permittivity, integrated over the intervals of the real axis occupied by the charges and (2) in two dimensions [charge uniform on lines (x, y) = const] by

$$\vec{\ell} \equiv \frac{\vec{x}-\vec{y}}{|\vec{x}-\vec{y}|}: \quad \vec{F}_e = \frac{1}{2\pi\varepsilon}\iint_{D'} d^2\vec{x} \iint_D d^2\vec{y}\, q(\vec{y})\frac{\vec{x}-\vec{y}}{|\vec{x}-\vec{y}|^2}q'(\vec{x}), \tag{9.109a,b}$$

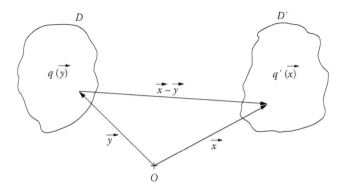

FIGURE 9.5 A point charge e' in an electric field $\vec{E}(\vec{x})$ is acted upon by an electric force $e'\vec{E}(\vec{x})$. The principle of superposition extends this to the electric force between two electric charge distributions $q(\vec{x})$ and $q'(\vec{y})$ in two domains, respectively, $\vec{x} \in D$ and $\vec{y} \in D'$.

where the factor $\left\{2\pi|\vec{x} - \vec{y}|\right\}^{-1}$ is due to the area per unit length of the cylinder of radius $|\vec{x} - \vec{y}|$, the integrand in (9.109b) also involves the unit vector (9.109a) joining the charge distributions, and the integrations are performed over the areas D,D' occupied by the charges; whereas (3) in three dimensions, the charge need not be uniform in any direction:

$$\vec{F}_e = \frac{1}{4\pi\varepsilon} \iiint_D d^3\vec{y} \iiint_{D'} d^3\vec{x}\, q(\vec{y}) \frac{\vec{x} - \vec{y}}{|\vec{x} - \vec{y}|^3} q'(\vec{x}), \tag{9.110}$$

there is a factor $\left\{4\pi|\vec{x} - \vec{y}|^2\right\}^{-1}$ due to the area of the sphere of radius $|\vec{x} - \vec{y}|$, the integrand involves the unit vector (9.109a) between charges, and the integrations are performed over the volumes they occupy (Figure 9.5).

In the one-dimensional case (9.108), the force is calculated by evaluating integrals over the axis; in the two-dimensional case (9.109), there may be area (line) integrals for charges over two-dimensional regions (along curves); and in the three-dimensional case (9.110), there may be volume, surface, and line integrals, respectively, for charges in domains, on surfaces, and along curves. All integrations disappear for the forces between a point charge e' at $\vec{x} = \vec{a}$ and (1) a charged plane e at $x = b$:

$$e'(x) = e'\delta(\vec{x} - \vec{a}), \quad e(y) = e\delta(y - b): \quad \vec{F} = \frac{e'e}{\varepsilon}\operatorname{sgn}(b - a)\vec{e}_x; \tag{9.111a-c}$$

(2) a line charge e at $\vec{x} = \vec{b}$:

$$e'(\vec{x}) = e'\delta(\vec{x} - \vec{a}), \quad e(\vec{y}) = e\delta(\vec{y} - \vec{b}): \quad \vec{F} = \frac{e'e}{2\pi\varepsilon}\frac{\vec{b} - \vec{a}}{|\vec{b} - \vec{a}|^2}; \tag{9.112a-c}$$

and (3) another point charge e at $\vec{x} = \vec{b}$:

$$e'(\vec{x}) = e\delta(\vec{x} - \vec{a}), \quad e(\vec{y}) = e\delta(\vec{y} - \vec{b}): \quad \vec{F} = \frac{e'e}{4\pi\varepsilon}\frac{\vec{b} - \vec{a}}{|\vec{b} - \vec{a}|^3}. \tag{9.113a-c}$$

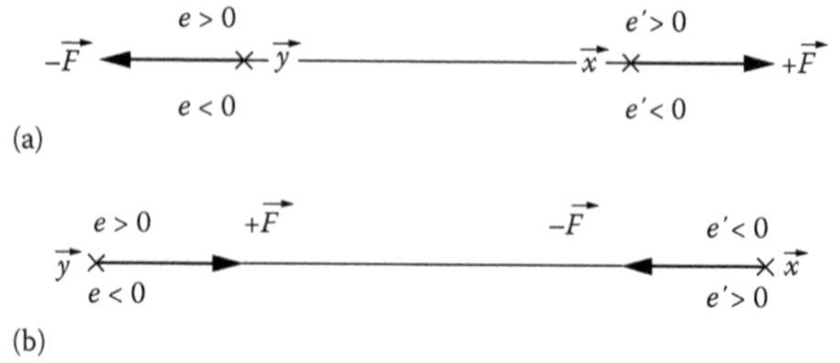

(a)

(b)

FIGURE 9.6   A particular case of electric force (Figure 9.5) concerns two point electric charges (this figure) that (a) repel [(b) attract] each other if they have the same (opposite) sign. The Coulomb law states that the modulus of the force is (1) proportional to the product of the charges, due to the linearity or superposition principle; (2) inversely proportional to the square of the distance as the electric field of a point source in space (Figure 9.3b); and (3) the inverse of the area of the unit sphere appears as a factor inherited from (2).

Thus, the **Coulomb law of forces**, *in one, two, and three dimensions,* has been obtained:

$$\vec{F}_e = \vec{e}_r \frac{e'e}{\varepsilon} \left\{ \mathrm{sgn}(x - y), \vec{e}_r \frac{1}{2\pi} \frac{\vec{x} - \vec{y}}{\left|\vec{x} - \vec{y}\right|^2}, \vec{e}_R \frac{1}{4\pi} \frac{\vec{x} - \vec{y}}{\left|\vec{x} - \vec{y}\right|^3} \right\}, \qquad (9.114\text{a–c})$$

stating that the force between (1) a point charge and a charged plane (9.114a) is proportional to the charges and independent of the distance; (2) a point and a line charge (9.114b) are proportional to their product divided by the area per unit length of the cylinder with axis along the line charge and radius equal to their distances; and (3) two point charges (9.114c) are proportional to their product divided by the area of the sphere with center at one point charge and radius equal to their distance. The force has the direction of the line joining the point charges (3) [orthogonal to line charge (2) or the conducting plane (1)] and is in all cases repulsive (attractive) for charges of same (opposite) signs, that is, both charges positive or negative (Figure 9.6a) [one positive and one negative (Figure 9.6b)]. The same laws apply (Figure 8.19) to the Newtonian force between masses (m'm), with the difference that it is always attractive because m'm > 0 and the inverse of the dielectric permeability 1/ε is replaced by minus the gravitational constant, −G (Newton, 1687). The three-dimensional Coulomb law (9.114c) ≡ (8.303b) is an example of irrotational force for several analogous fields (Notes 8.3 through 8.7) and may also be considered for multipoles and anisotropic fields (Notes 8.11 through 8.17). The forces (Note 8.10) associated with irrotational (solenoidal) fields include in particular the Coulomb (Biot–Savart) force for the electro(magneto)static field [Section 9.4 (Section 9.5)] up to three dimensions. Potentials and fields in higher dimensions are deferred to Sections 9.6 through 9.8 and Notes 9.3 through 9.52.

9.5 Solenoidal Fields and Biot–Savart Force

The irrotational (solenoidal) field has a [Section 9.4 (Section 9.5)] scalar (vector) potential [Subsection 9.4.1 (Subsection 9.5.1)], leading to a Coulomb/Newton (Biot–Savart) force [Subsection 9.4.5 (Subsection 9.5.2)]; these forces apply to a distribution of sources, for example, a point (Subsection 9.5.3) or a continuous (Subsection 9.5.2) current electric distribution in space. For line currents or current distributions normal to a plane, the vector potential reduces to a scalar field function, and the theory of complex analytic functions can be used (Subsection 9.5.5). In three dimensions, the vector potential is not unique and a gauge condition may be imposed (Subsection 9.5.4).

9.5.1 Vector/Scalar Potential and Irrotational/Solenoidal Field

The irrotational flow due to sources and sinks (Chapter I.12 and Note 8.7), the gravity field due to masses (Chapter I.18 and Note 8.3), the electrostatic field due to positive or negative charges (Chapter I.24.6 and Note 8.4), and the temperature in steady heat conduction (Chapter I.32 and Note 8.5) are all **irrotational fields**.

Taking the electrostatic field (9.115a) as an example: it is the gradient of a scalar potential (9.115b) satisfying the scalar Poisson equation (9.115c) with source:

$$\nabla \wedge \vec{E} = 0, \quad \vec{E} = -\nabla \Phi, \quad \nabla^2 \Phi = -\frac{q}{\varepsilon}, \tag{9.115a–c}$$

that is, in the electrostatic (Chapter I.24 and Note 8.4) case, the source is minus the electric charge density q divided by the dielectric permittivity ε of the medium. The incompressible flows due to vorticity (Chapter 6), or the magnetic field due to electric currents (Chapter 8), are **solenoidal fields**. *Taking as an example the magnetostatic field (9.116a), it is the curl of a vector potential (9.116b), satisfying the vector Poisson equation (9.116c), with source:*

$$\nabla \cdot \vec{H} = 0, \quad \mu \vec{H} = \nabla \wedge \vec{A}, \quad \nabla^2 \vec{A} = -\frac{\mu}{c}\vec{j}, \tag{9.116a–c}$$

namely, in the magnetostatic case, the source is (Chapter I.26 and Note 8.10) minus the electric current $\vec{j}$ divided by the speed of light in vacuo c, multiplied by the magnetic permeability μ of the medium. In two dimensions (Chapter I.26), the vector potential reduces to a scalar field function Ψ that is not quite analogous to the scalar potential Φ because it arises from a curl (9.116b) rather than from a gradient (9.115b). In three dimensions, the scalar field function does not exist generally and when it does, namely, for axisymmetric fields (Chapter 6), it (1) satisfies a modified (6.87a and b) instead of the original (6.85a and b) Laplace equation (Subsection (6.2.6)) and (2) it is related to but not coincident with one component of the vector potential (6.60a through d). In order to consider the incompressible flow due to vorticity, or the magnetostatic field due to currents in three dimensions, the theory of the scalar potential (Section 9.4) is extended to the vector potential (Section 9.5), including the passage from scalar (Section 8.3) to vector (Section 8.6) multipole expansions.

9.5.2 Magnetic Field and Biot–Savart Force

The vector Poisson equation (9.116c) is in Cartesian coordinates x_i a set of three scalar equations,

$$\frac{\partial^2 A_i}{\partial x_j \partial x_j} = -\frac{\mu}{c}j_i, \tag{9.117}$$

for each component of the vector potential A_i and electric current j_i, and thus its solution in unbounded space is

$$\vec{A}(\vec{y}) = \frac{\mu}{4\pi c} \int_D \frac{\vec{j}(\vec{y})}{|\vec{x} - \vec{y}|} d^3\vec{y}; \tag{9.118}$$

the curl has the property (8.93) that is analogous to that of the divergence (9.2a), replacing $\nabla\cdot$ by $\nabla\wedge$; for example, (8.93) substituted in (9.118) specifies the magnetic field (9.116b) due to an electric current distribution over a region D:

$$\vec{H}(\vec{x}) = \frac{1}{4\pi c}\frac{\partial}{\partial \vec{x}} \wedge \left(\int_D \frac{\vec{j}(\vec{y})}{|\vec{x} - \vec{y}|} d^3\vec{y} \right) = \frac{1}{4\pi c} \int_D \frac{\vec{j}(\vec{y}) \wedge (\vec{x} - \vec{y})}{|\vec{x} - \vec{y}|^3} d^3\vec{y}. \tag{9.119}$$

In the presence of another current distribution $\vec{j}'(\vec{x})$ over a region D', there is a **Biot–Savart force**:

$$\vec{F}_m = \frac{\mu}{c} \int_{D'} \left\{ \vec{j}'(\vec{x}) \wedge \vec{H}(\vec{x}) \right\} d^3\vec{x}$$

$$= \frac{\mu}{4\pi c^2} \int_{D'} d^3\vec{x} \int_{D} d^3\vec{y} \left\{ \vec{j}'(\vec{x}) \wedge \left[\vec{j}(\vec{y}) \wedge (\vec{x} - \vec{y}) \right] \right\} |\vec{x} - \vec{y}|^{-3}. \tag{9.120}$$

Thus, has been obtained *the vector potential (9.118) and the magnetic field (9.119) due to a distribution of electric currents $\vec{j}(\vec{y})$ over a region D, and the magnetic or Biot–Savart force (9.120) between this and another electric current distribution $\vec{j}'(\vec{x})$ over a region D'.*

9.5.3 Continuous and Discrete Electric Current Distributions

In the case of two point electric currents (9.121a and b),

$$\vec{j}(\vec{y}) = \vec{J}\delta(\vec{y} - \vec{a}), \quad \vec{j}'(\vec{x}) = \vec{J}'\delta(\vec{x} - \vec{b}): \quad \vec{A}(\vec{x}) = \frac{\mu}{4\pi c} \frac{\vec{J}}{|\vec{x} - \vec{y}|}, \tag{9.121a–c}$$

the (1) vector potential (9.119) is parallel to the current (9.121c), (2) the magnetic field (9.119) is orthogonal (9.122) to the current and the position vector relative to the observer,

$$\vec{H}(\vec{x}) = \frac{1}{\mu} \nabla \wedge \vec{A} = \frac{1}{4\pi c} \frac{\vec{J} \wedge (\vec{x} - \vec{y})}{|\vec{x} - \vec{y}|^3}, \tag{9.122}$$

and (3) the magnetic force is orthogonal to the second current and the magnetic field of the first current, and it has components along the first current (position vector) if the second current is not orthogonal to the position vector (first current):

$$\vec{F}_m = \frac{\mu}{c} \vec{J}' \wedge \vec{H} = \frac{\mu}{4\pi c^2} \frac{\vec{J}' \wedge \left[\vec{J} \wedge (\vec{x} - \vec{y}) \right]}{|\vec{x} - \vec{y}|^3}$$

$$= \frac{\mu}{4\pi c^2} |\vec{x} - \vec{y}|^{-3} \left\{ \vec{J} \left[\vec{J}'.(\vec{x} - \vec{y}) \right] - (\vec{x} - \vec{y})(\vec{J} \cdot \vec{J}') \right\}; \tag{9.123}$$

there is no magnetic force if the second current is orthogonal to the plane of the first current and the position vector.

9.5.4 Gauge Condition for the Vector Potential

In all cases, the vector potential is not uniquely specified by (9.116b) because the Poisson equation (9.116c) $\equiv$ (9.124a),

$$-\frac{\mu}{c}\vec{j} = -\mu\nabla \wedge \vec{H} = -\nabla \wedge (\nabla \wedge \vec{A}) = \nabla^2\vec{A} - \nabla(\nabla \cdot \vec{A}), \quad \nabla \cdot \vec{A} = 0, \tag{9.124a,b}$$

is obtained only if $\nabla.\vec{A}$ is a constant; this constant may be set to zero (9.124b), specifying the **gauge condition** for the vector potential. *The magnetic field is unchanged (9.125b) upon adding the gradient of a scalar potential* χ *to the vector potential (9.125a)*,

$$\vec{A}' = \vec{A} + \nabla\chi, \quad \nabla\wedge\vec{A}' = \vec{H}, \quad 0 = \nabla\cdot\vec{A}' - \nabla\cdot\vec{A} = \nabla^2\chi, \tag{9.125a–c}$$

provided (9.124b) $\equiv$ *(9.125c) the scalar potential is harmonic.* For a plane magnetostatic field, the electric currents are normal to the plane (9.126a):

$$\vec{j} = \vec{e}_z J, \quad \vec{A} = \vec{e}_z\Psi: \quad \mu\vec{H} = \nabla\wedge(\vec{e}_z\Psi) = \vec{e}_x\frac{\partial\Psi}{\partial y} - \vec{e}_y\frac{\partial\Psi}{\partial x}, \tag{9.126a–c}$$

and so is the vector potential (9.126b) that reduces to a field function, specifying the magnetic field (9.126a through c). The case of plane magnetostatics can also be considered using complex functions (Chapter I.26), as shown next.

9.5.5 Field Function for Plane Magnetostatics

A line current J in plane magnetostatics (9.127a) corresponds to the stream function (9.127b) similar to (9.118), replacing (9.48a through c; 9.49a through c) the area of the unit sphere 4π by the perimeter of the unit circle 2π and the inverse distance by the logarithmic potential:

$$\vec{j} = \vec{e}_z J: \quad \Psi = \frac{\mu J}{2\pi c}\log|\vec{x}| = \frac{\mu J}{4\pi c}\log|x^2 + y^2|; \tag{9.127a,b}$$

the latter corresponds to the magnetic field (9.128):

$$\{H_x, H_y\} = \frac{1}{\mu}\left\{\frac{\partial\Psi}{\partial y}, -\frac{\partial\Psi}{\partial x}\right\} = \frac{J}{2\pi c}\frac{\{y, -x\}}{x^2 + y^2}. \tag{9.128}$$

Using the theory of functions of a complex variable, the complex potential (I.26.13a) due to a line current is (9.129a):

$$f(z) = -\frac{iJ}{2\pi c}\log z, \quad H^*(z) = -\frac{df}{dz} = \frac{iJ}{2\pi cz}, \tag{9.129a,b}$$

which leads (I.26.13b) to the same magnetic field (9.129b) in agreement with (9.128) $\equiv$ (9.130):

$$H_x - iH_y = \frac{iJ}{2\pi c}\frac{1}{x + iy} = \frac{iJ}{2\pi c}\frac{x - iy}{x^2 + y^2} = \frac{J}{2\pi c}\frac{y + ix}{x^2 + y^2}. \tag{9.130}$$

For plane magnetostatics, the line currents (9.131a and b) are orthogonal to the plane and hence parallel to each other and orthogonal to the position vector (9.131c); the magnetic force is given by (9.123), replacing (9.131d) the area of the sphere by the unit length of cylinder of the same radius:

$$\vec{j} = J\vec{e}_3, \quad \vec{j}' = J'\vec{e}_3, \quad \vec{j} / / \vec{j}' \perp (\vec{x} - \vec{y}), \quad 4\pi|\vec{x} - \vec{y}|^2 \to 2\pi|\vec{x} - \vec{y}|; \tag{9.131a–d}$$

this leads to the Biot–Savart force:

$$\vec{F} = \frac{\mu J'J}{2\pi c^2} \frac{\vec{y}}{|\vec{y}|^2} = \frac{\mu}{2\pi c^2} \frac{J'J}{x^2+y^2}(\vec{e}_x x + \vec{e}_y y),$$

(9.132)

In the context of complex function theory, the force between line currents (I.26.15) is specified by:

$$F_x - iF_y \equiv F_m^* = -i\frac{\mu J'}{c} H^*(z) = \frac{\mu J'J}{2\pi c^2}\frac{1}{z} = \frac{\mu}{2\pi c^2}\frac{J'J}{x^2+y^2}(x-iy),$$

(9.133)

in agreement with (9.132) ≡ (9.133). Having checked the consistency of the various approaches to potential field theory on line, plane, and space, that is, in one, two, and three dimensions, higher dimensions are considered next.

9.6 Hyperspherical or Generalized Legendre Polynomials (Campos and Cunha, 2012)

The irrotational fields are specified by a scalar potential in any dimension, leading to a straightforward generalization to dimensions higher than the third, which requires a modest extension of ordinary concepts; a starting point can be the use of the Poisson integral (Subsection 9.6.1) obtained from the influence or Green function (Subsection 9.6.2) for the Laplace operator with hyperspherical symmetry: it specifies the multipolar expansion in any dimension (Subsection 9.6.2), which involves the hyperspherical Legendre polynomials. The properties of hyperspherical multipoles of arbitrary order are related to those of hyperspherical Legendre polynomials: (1) the free-space influence function is the generating function (Subsection 6.9.3); (2) it specifies the values of hyperspherical Legendre polynomials at particular points (Subsection 9.6.4); (3) the values at all points follow from explicit formulas for all the coefficients of the polynomial (Subsection 6.9.5); (4) these formulas involve powers, and if cos θ is used as an argument, they can be written as linear combinations of cosines of multiple angles (Subsection 6.9.6); (5) the hyperspherical Legendre polynomials of higher degree can also be obtained from those of lower degree using a recurrence formula (Subsection 6.9.7); (6) the recurrence formula can be combined with differentiation formulas, for example, in the rule of generation of higher order multidimensional multipoles (Subsection 6.9.8); (7) the differentiation formulas lead to the differential equation satisfied by the hyperspherical Legendre polynomials (Subsection 6.9.9); and (8) the hyperspherical Legendre functions are a generalization allowing all parameters, namely, degree n and order α, to be complex (Subsection 6.9.10) and relate to differential (integral) representations like the Rodrigues (Schlaffi) formulas.

9.6.1 Multidimensional Potential, Field, and Force

*The influence or Green function (9.43a through c) for the Laplace operator (9.39) ≡ (9.40) applies in all dimensions except for the logarithmic potential in two dimensions (9.42a through d). It leads (9.50a through c) to the **irrotational field**,*

$$N = 1,2,\dots: \quad \vec{U} \equiv \nabla\Phi = \int_D \Lambda(\vec{y})|\vec{x}-\vec{y}|^{1-N} \, d^N\vec{y},$$

(9.134)

*for an arbitrary source (9.46a), which applies in any dimension because (9.134) follows from (9.47a through c; 9.49a through c; 9.50a and b) [(9.48a through c)] for $N \neq 2(N = 2)$. The **force** between two source distributions is*

$$\vec{F} = \frac{1}{\sigma_N} \int_{D'} \Lambda'(\vec{x}) \vec{U}(\vec{x}) d^N \vec{x} = \frac{1}{\sigma_N^2} \int_{D'} d^N \vec{x} \int_D d^N \vec{y} |\vec{x} - \vec{y}|^{1-N} \Lambda'(\vec{x}) \Lambda(\vec{y}). \tag{9.135}$$

The case of point sources (9.136a and b),

$$\Lambda(\vec{y}) = \lambda \delta(\vec{x} - \vec{y}), \quad \Lambda'(\vec{x}) = \lambda' \delta(\vec{x} - \vec{a}): \tag{9.136a,b}$$

$$N \neq 2: \quad \Phi(\vec{x}) = \left\{ \frac{\lambda}{\sigma_N(2-N)} \right\} |\vec{x} - \vec{y}|^{2-N}; \quad N = 2: \quad \Phi(\vec{x}) = \frac{\lambda}{2\pi} \log|\vec{x} - \vec{y}|; \tag{9.137a-d}$$

$$N = 1,2,\ldots: \quad \vec{U} \equiv \nabla \Phi(\vec{x}) = \frac{\lambda}{\sigma_N} |\vec{x} - \vec{y}|^{1-N}, \quad \vec{F} = \frac{\lambda' \lambda}{\sigma_N} |\vec{x} - \vec{y}|^{1-N}, \tag{9.138a,b}$$

leads to (1) the potential (9.137c and d) [(9.137a and b)] in two (other) dimensions and (2/3) the field (9.138a) and the force (9.138b) in both cases.

9.6.2 Definition of Generalized or Hyperspherical Legendre Polynomials

The influence or Green function corresponds to the potential of a unit point source $\lambda = 1$ in two (9.137c and d) $\equiv$ (9.139c and d) [other (9.137a and b) $\equiv$ (9.140c and d)] dimensions and depends only on the modulus (9.139a) and the angle (9.140a) of the position vectors of the source and the observer:

$$|\vec{x}| = R > |\vec{y}| = y, N = 2: \quad 2\pi G(\vec{x}; \vec{y}) = \log \left| R^2 + y^2 - 2Ry \cos \theta \right|$$

$$= \log R + \log \left| 1 - 2 \frac{y}{R} \cos \theta + \frac{y^2}{R^2} \right|, \tag{9.139a-d}$$

$$\vec{x} \cdot \vec{y} = Ry \cos \theta; \quad N \neq 2: \quad (N-2)\sigma_N G(\vec{x} - \vec{y}) = \left| R^2 + y^2 - 2Ry \cos \theta \right|^{1-N/2}$$

$$= R^{1-N/2} \left| 1 - 2 \frac{y}{R} \cos \theta + \frac{y^2}{R^2} \right|^{1-N/2}. \tag{9.140a-d}$$

The nonplane case (9.140b) leads to the **hyperspherical Legendre polynomials** of degree m and order α that are defined by (9.141c):

$$\alpha = N - 3; \quad a < 1: \quad \left| 1 - 2a \cos \theta + a^2 \right|^{-1/2 - \alpha/2} = \sum_{n=0}^{\infty} a^n P_{n,\alpha}(\cos \theta), \tag{9.141a-c}$$

so that they coincide with the original (8.74a and b) Legendre polynomials for $\alpha = 0$, that is, in three dimensions, $N = 3$ in (9.141a); the relation (9.141a) arises from the equality of exponents in (9.140d) and (9.141c). Substitution of (9.141b) in (9.140d) with the parameters (9.141a; 9.142a),

$$a \equiv \frac{y}{R} = \frac{|\vec{y}|}{|\vec{x}|} < 1: \quad (N-2)\sigma_N G(\vec{x}; \vec{y}) = \sum_{n=0}^{\infty} |\vec{x}|^{1-N/2-n} |\vec{y}|^n P_{n,N-3}(\cos \theta), \tag{9.142a,b}$$

leads to the near-field (9.142a) multipolar expansion (9.142b). Expanding (9.141b) by the binomial theorem (I.25.37a through c) to $O(a^4)$ leads to (9.143):

$$|1 - 2a\cos\theta + a^2|^{-1/2-\alpha/2} = 1 + \frac{1+\alpha}{2}a(2\cos\theta - a) + \frac{1+\alpha}{2}\frac{3+\alpha}{2}\frac{a^2}{2}(2\cos\theta - a)^2$$

$$+ \frac{1+\alpha}{2}\frac{3+\alpha}{2}\frac{5+\alpha}{2}\frac{a^3}{6}(2\cos\theta - a)^3 + O(a^4)$$

$$= 1 + a(1+\alpha)\cos\theta + a^2\frac{1+\alpha}{2}[(3+\alpha)\cos^2\theta - 1]$$

$$+ a^3\frac{1+\alpha}{2}(3+\alpha)\cos\theta\left[\frac{5+\alpha}{3}\cos^2\theta - 1\right] + O(a^4). \tag{9.143}$$

Comparison of (9.143) $\equiv$ (9.141c) specifies the hyperspherical Legendre polynomials of four lowest degrees and arbitrary order:

$$P_{0,\alpha}(\cos\theta) = 1, \quad P_{1,\alpha}(\cos\theta) = (1+\alpha)\cos\theta, \tag{9.144a,b}$$

$$P_{2,\alpha}(\cos\theta) = \frac{1+\alpha}{2}[(3+\alpha)\cos^2\theta - 1], \tag{9.144c}$$

$$P_{3,\alpha}(\cos\theta) = \frac{1+\alpha}{2}\left(1+\frac{\alpha}{3}\right)\cos\theta[(5+\alpha)\cos^2\theta - 3]; \tag{9.144d}$$

the first four hyperspherical (9.144a through d) reduce to the original Legendre polynomials (8.75a through d) for $\alpha = 0$ in (9.141a) or in three dimensions, $N = 3$. The generating function (9.141c) for the hyperspherical Legendre polynomials has been used in (9.142b) in the far field $a < 1$ or $y < R$ in (9.141b) $\equiv$ (9.142a); it can also be applied in the near field (9.145a), leading to (9.145b):

$$a > 1: \quad a^{-1-\alpha}\left|1 - \frac{2}{a}\cos\theta + \frac{1}{a^2}\right|^{-1/2-\alpha/2} = \sum_{n=0}^{\infty} a^{-1-\alpha-n}P_{n,\alpha}(\cos\theta); \tag{9.145a,b}$$

the corresponding near-field (9.146a) multipolar expansion for the N-dimensional influence or Green function (9.140d) is (9.146b):

$$\frac{1}{a} \equiv \frac{R}{y} = \frac{|\vec{x}|}{|\vec{y}|} < 1: \quad (N-2)\sigma_N G(\vec{x} - \vec{y}) = \sum_{n=0}^{\infty} |\vec{x}|^{-1+N/2+n}|\vec{y}|^{2-N-n}P_{m,\alpha}(\cos\theta). \tag{9.146a,b}$$

Thus, *the influence or Green function (9.140a through d) for the unit monopole in $N \neq 2$ dimensions (9.43a through c) has multipolar expansion (9.142b) [(9.146b)] for the observer in the far field (9.142a) [near field (9.146a)]. Both multipolar expansions involve the hyperspherical or generalized Legendre polynomials (9.141a through c), of which the four of the lowest degree are (9.144a through d) with an arbitrary order. The hyperspherical Legendre polynomials of all degrees appear in the* **far- (near-) field multipolar expansion** *(9.141a through c) [(9.145a and b)] that is valid for a(1/a) satisfying (9.150a through d).* The latter is one of the several properties of the hyperspherical Legendre polynomials of arbitrary degree that relate to hyperspherical harmonics of any order and are obtained next (Subsections 9.6.3 through 9.6.10). It is possible to proceed directly to Section 9.7.1, where the first three terms of the multidimensional multipolar expansion are obtained from a Taylor series expansion of the influence function; this is sufficient for the three hyperspherical harmonics of lowest order, namely, the multidimensional monopole, dipole, and quadrupole considered in the sequel (Sections 9.7 through 9.9).

9.6.3 Hyperspherical Legendre Generating Function

The **generating function** for hyperspherical Legendre polynomials (9.141c) is specified by (9.147b) [(9.148b)] in the far (9.147a) [near (9.148a)] field:

$$|a| < 1: \quad L(a,z;\alpha) \equiv \left|1 - 2az + a^2\right|^{-1/2-\alpha/2} = \sum_{n=0}^{\infty} a^n P_{n,\alpha}(z); \tag{9.147a,b}$$

$$|a| > 1: \quad L(a,z;\alpha) = a^{-1-\alpha} \left|1 - 2\frac{z}{a} + \frac{1}{a^2}\right|^{-1/2-\alpha/2} = \sum_{n=0}^{\infty} a^{-1-\alpha-n} P_{n,\alpha}(z). \tag{9.148a,b}$$

The series expansion (9.147b) [(9.148b)] applies (Section I.25.9) in a circle with center at the origin $a = 0$ and radius so as to exclude any singularities of the function. For absolute convergence (Section I.21.3), this requires the expression in modulus to be nonzero (9.149a and b):

$$z \equiv \cos\theta: \quad 1 - 2az + a^2 = (1-a)^2 + 2a(1-\cos\theta) > 0, \quad \{a, \cos\theta\} \ne \{1,1\}. \tag{9.149a–c}$$

which holds if a and $\cos\theta$ are not both unity (9.149c); thus, absolute convergence holds (1) for (9.147b) [(9.148b)] inside (outside) the unit disk (9.147a) [(9.148a)] and (2) also on the unit disk $|a| = 1$ for both (9.147b) and (9.148b), except if $a = 1$ and $\cos\theta = 1$. For uniform convergence (Section I.21.6) of (9.147b) [(9.148b)], a neighborhood of (9.149c) must be excluded, implying (9.150a, b, and d) [(9.150a, c, and d)]:

$$|1 - z| = |1 - \cos\theta| \ge \varepsilon > 0: \quad |a| \le 1 - \delta; \quad |a| \ge 1 + \delta, \quad 0 < \delta < 1. \tag{9.150a–d}$$

Since (9.150a through d) include (9.149c), the conditions (9.150a, b, and d) [(9.150a, c, and d)] ensure total convergence (Section I.21.7) that is absolute and uniform convergence of (9.147b) [(9.148b)]. Thus, *the far- (9.147b) [near (9.148b)] field expansion in spherical harmonics is (1) absolutely convergent for (9.149c) and (2) totally (that is, absolutely and uniformly) convergent for (9.150a, b, and d) [(9.150a, c, and d)].* In the latter case, it can be differentiated term-by-term, integrated, or deranged, and limits may be taken under the integral sign (Section I.21.6).

9.6.4 Particular Values of the Hyperspherical Legendre Polynomials

The generating function (9.147b) specifies:

$$z = 1: \quad L(a,1;\alpha) = |1 - a|^{-1-\alpha} = \sum_{n=0}^{\infty} (-)^n \binom{-1-\alpha}{n} a^n, \tag{9.151a,b}$$

$$z = -1: \quad L(a,-1;\alpha) = |1 + a|^{-1-\alpha} = \sum_{n=0}^{\infty} \binom{-1-\alpha}{n} a^n, \tag{9.152a,b}$$

$$z = 0: \quad L(a,0;\alpha) = |1 + a^2|^{-1/2-\alpha/2} = \sum_{n=0}^{\infty} \binom{-1/2-\alpha/2}{n} a^{2n}, \tag{9.153a,b}$$

the *values of the hyperspherical Legendre polynomials at points* $z = \pm 1, 0$:

$$z = 1, \theta = 0: \quad P_{n,\alpha}(1) = (-)^n \binom{-1-\alpha}{n} = \frac{(\alpha+1)(\alpha+2)...(\alpha+n)}{n!}, \qquad (9.154a\text{-}c)$$

$$z = -1; \theta = \pi: \quad P_{n,\alpha}(-1) = \binom{-1-\alpha}{n} = \frac{(-)^n}{n!}(\alpha+1)(\alpha+2)...(\alpha+n) = (-)^n P_{n,\alpha}(-1), \qquad (9.155a\text{-}d)$$

$$z = 0, \theta = \pm\frac{\pi}{2}: \quad P_{2m+1,\alpha}(0) = 0, \quad P_{2m,\alpha}(0) = \binom{-1/2-\alpha/2}{m} = \frac{(-)^m}{2^m}\frac{(\alpha+1)(\alpha+3)...(\alpha+2m-1)}{m!}, \qquad (9.156a\text{-}d)$$

The values at the origin of the hyperspherical Legendre polynomials have been separated for odd $n = 2m + 1$ *(even* $n = 2m$) *degree in* (9.156c) [(9.156d)]. *The values* (9.154a through c; 9.155a through c; 9.156a through c) *can be checked for the first four hyperspherical Legendre polynomials* (9.144a through d). *They simplify* (9.143a) $\equiv$ (9.157a) *to*

$$P_n(z) \equiv P_{n,0}(z): \quad P_n(1) = 1, \quad P_n(-1) = (-)^n, \quad P_{2m+1}(0) = 0, \quad P_{2m}(0) = (-)^m \frac{(2m-1)!!}{(2m)!!}, \qquad (9.157a\text{-}e)$$

for the original Legendre polynomials. The values of the hyperspherical Legendre polynomials at the origin are given by (9.156d) $\equiv$ (9.158a):

$$P_{2n}(0) = \binom{-1/2-\alpha/2}{m} = \frac{(-)^m}{m!} \frac{1+\alpha}{2} \frac{3+\alpha}{2} ... \frac{\alpha+2m-1}{2}$$

$$= \frac{(-)^m}{2^m} \frac{(\alpha+1)(\alpha+3)...(\alpha+2m-1)}{m!}, \qquad (9.158a)$$

which simplifies to (9.158c) $\equiv$ (9.157e) for the original Legendre polynomials:

$$\alpha = 0: \quad P_{2m}(0) = P_{2m,0}(0) = \frac{(-)^m}{m!} \frac{(2m-1)!!}{2^m} = \frac{(-)^m}{m!2^m} \frac{(2m)!}{m!2^m} = \frac{(-)^m}{2^{2m}} \frac{(2m)!}{(m!)^2}, \qquad (9.158b,c)$$

using either double or single factorials.

9.6.5 Explicit Coefficients for the Generalized Legendre Polynomials

Applying the binomial series (I.25.37a through c) to the generating function (9.147b) twice leads to the double sum (9.159b):

$$n = j + k: \quad |1 - 2az + a^2|^{-1/2-\alpha/2} = |1 + a(a - az)|^{-1/2-\alpha/2}$$

$$= \sum_{j=0}^{\infty} \binom{-1/2-\alpha/2}{j} a^j (a - 2z)^j$$

$$= \sum_{j=0}^{\infty} \sum_{k=0}^{j} \binom{-1/2-\alpha/2}{j} \binom{j}{k} a^{j+k} (-2z)^{j-k}$$

$$= \sum_{j=0}^{\infty} \sum_{k=0}^{j} \frac{\alpha+1}{2} \frac{\alpha+3}{2} ... \frac{\alpha+2j-1}{2} \frac{(-)^{2j-k} a^{j+k} (2z)^{j-k}}{k!(j-k)!}$$

$$= \sum_{n=0}^{\infty} a^n \sum_{k=0}^{\leq n/2} \frac{(-)^k 2^{-k} z^{j-k}}{k!(2n-2k)!} (\alpha+1)(\alpha+3)...(\alpha+2n-2k-1), \qquad (9.159a,b)$$

where a change of index of summation was made (9.159a) and terms with negative factorial $(n - 2k)! = \infty$ for $2k > n$ are omitted. Comparing (9.159b) and (9.147b) gives *the explicit formula for the coefficients of the hyperspherical Legendre polynomials*:

$$P_{n,\alpha}(z) = \sum_{k=0}^{\leq n/2} \frac{(-)^k 2^{-k} z^{n-2k}}{k!(n-2k)!}(\alpha+1)(\alpha+3)...(\alpha+2n-2k-1).$$ (9.160)

The case (9.161a) $\equiv$ (9.161c) of (9.160) leads (9.143a) to the original Legendre polynomials:

$$\alpha = 0: \quad P_n(z) = \sum_{k=0}^{\leq n/2} (-)^k 2^{-k} z^{n-2k} \frac{(2n-2k-1)!!}{k!(n-2k)!},$$ (9.161a,b)

$$P_n(z) = P_{n,0}(z): \quad P_n(z) = 2^{-n} \sum_{k=0}^{\leq n/2} (-)^k z^{n-2k} \frac{(2n-2k)!}{k!(n-k)!(n-2k)!},$$ (9.161c,d)

where the following simplification was made:

$$(2n-2k-1)!! \equiv (2n-2k-1)(2n-2k-3)...3.1 = \frac{(2n-2k)!}{(n-k)!2^{n-k}}.$$ (9.162)

The general formula (9.160) can be checked against the hyperspherical Legendre polynomials of degrees up to three (9.144a through d) and apply as well to higher degrees.

9.6.6 Explicit Expressions in terms of Cosines of Angles and Multiple Angles

The explicit expression (9.160) for hyperspherical harmonics is a polynomial of degree n of $\cos\theta$:

$$P_{n,\alpha}(\cos\theta) = \sum_{k=0}^{\leq n/2} \frac{(-)^k 2^{-k} \cos^{n-2k}\theta}{k!(n-2k)!}(\alpha+1)(\alpha+3)...(\alpha+2n-2k-1).$$ (9.163)

The powers of the cosines can be expressed as linear combinations of cosines of multiple angles (II.5.78a and b). Thus, the hyperspherical Legendre polynomials can be expressed, as an alternative to (9.163), by a linear combination of cosines of angles that are multiples of θ, up to $n\theta$. This can be obtained most simply by rewriting the generating function (9.147b; 9.149a) in the form:

$$L(a,\cos\theta;\alpha) = |1 - a(e^{i\theta} + e^{-i\theta}) + a^2|^{-1/2-\alpha} = |1-ae^{i\theta}|^{-1/2-\alpha}|1-ae^{-i\theta}|^{-1/2-\alpha};$$ (9.164)

using the binomial series again leads (9.159a) to the double sum:

$$\sum_{n=0}^{\infty} a^n P_{n,\alpha}(\cos\theta) = \sum_{j,k=0}^{\infty} (-)^{j+k} a^{j+k} \binom{-1/2-\alpha/2}{j} e^{ij\theta} \binom{-1/2-\alpha/2}{k} e^{-ik\theta}$$

$$= \sum_{j,k=0}^{\infty} \frac{\alpha+1}{2} \frac{\alpha+3}{2}...\frac{\alpha+2j-1}{2} \frac{\alpha+1}{2} \frac{\alpha+3}{2}...\frac{\alpha+2k-1}{2} a^{j+k} \frac{\exp[i(j-k)\theta]}{j!k!}$$

$$= \sum_{n=0}^{\infty} a^n \sum_{k=0}^{\leq n/2} \frac{\exp[i(j-k)\theta]}{k!(2n-k)!}(\alpha+1)(\alpha+3)...(\alpha+2k-1)\,\alpha(\alpha+3)...(\alpha+2n-2k-1),$$

(9.165)

where the change of variable of summation (9.159a) was used, thus limiting the inner sum in (9.165) to $0 = k \leq n/2$. For even $n = 2m$ (odd $n = 2m + 1$) degree the hyperspherical Legendre polynomials (9.165) are given by (9.166a) [(9.166b)]:

$$P_{2n,\alpha}(\cos\theta) = \frac{2^{-2m}}{(m!)^2}\left[(\alpha+1)(\alpha+3)...(\alpha+2m-1)\right]^2$$

$$+ 2^{1-2m}\sum_{k=0}^{m-1}\frac{\cos\left[2(m-k)\theta\right]}{k!(2m-k)!}(\alpha+1)(\alpha+3)...(\alpha+2k-1)(\alpha+1)(\alpha+3)...(\alpha+4m-2k-1),$$

$$(9.166a)$$

$$P_{2m+1,\alpha}(\cos\theta) = 2^{-2m}\sum_{k=0}^{m}\frac{\cos\left[(2m-2k+1)\theta\right]}{k!(2m-k+1)!}(\alpha+1)(\alpha+3)...(\alpha+2k-1)(\alpha+1)(\alpha+3)...(\alpha+4m-2k+1).$$

$$(9.166b)$$

These are *the explicit formulas (9.166a and, b) for the hyperspherical Legendre polynomials in terms of cosines of angles that are multiples* θ, *instead of powers of* $\cos\theta$ *in (9.163). From (9.166a) [(9.166b)] follow alternative expressions for the hyperspherical Legendre polynomials of degree two (9.167a)* $\equiv$ *(9.144c) [three (9.167b)* $\equiv$ *(9.144d)]*:

$$P_{2,\alpha}(\cos\theta) = \frac{1+\alpha}{2}\left[\frac{3+\alpha}{2}\cos(2\theta) + \frac{1+\alpha}{2}\right],$$

$$(9.167a)$$

$$P_{3,\alpha}(\cos\theta) = \frac{1+\alpha}{2}\frac{3+\alpha}{2}\left[\frac{5+\alpha}{2}\frac{\cos(3\theta)}{3} + \frac{1+\alpha}{2}\cos\theta\right].$$

$$(9.167b)$$

The particular case (9.168a) leads from (9.163; 9.162) to (9.168b and c),

$$\alpha = 0: \quad P_{n,0}(\cos\theta) \equiv P_n(\cos\theta) = \sum_{k=0}^{\leq n/2}(-)^k\frac{(2n-2k-1)!!(-)^k 2^{-k}}{k!(n-2k)!}\cos^{n-2k}\theta$$

$$= 2^{-n}\sum_{k=0}^{\leq n/2}\frac{(2n-2k)!(-)^k}{k!(n-k)!(n-2k)!}\cos^{n-2k}\theta,$$

$$(9.168a-c)$$

in agreement with (9.161b and d); the same particular case (9.168a) of (9.166a) [(9.166b)] leads to the original Legendre polynomials of even $n = 2m$ *(9.169a and b) [odd* $n = 2m + 1$ *(9.170a and b)] degree:*

$$P_{2m}(\cos\theta) = \left[\frac{(2m-1)!!\,2^{-m}}{m!}\right]^2 + 2^{1-2m}\sum_{k=0}^{m-1}\frac{\cos\left[2(m-k)\theta\right]}{k!(2m-k)!}(2k-1)!!(4m-2k-1)!!,$$

$$= 2^{-4m}\left[\frac{(2m)!}{m!}\right]^2 + 2^{1-4m}\sum_{k=0}^{m-1}\frac{(2k)!(4m-4k)!}{\left[k!(2m-k)!\right]^2}\cos\left[2(m-k)\theta\right],$$

$$(9.169a,b)$$

$$P_{2,m+1}(\cos\theta) = 2^{-2m}\sum_{k=0}^{m}\frac{\cos\left[(2m-2k+1)\theta\right]}{k!(2m-k+1)!}(2k-1)!!(4m-2k+1)!!$$

$$= 2^{-4m}\sum_{k=0}^{m}\frac{(2k)!(4m-2k+2)!}{\left[k!(2m-k+1)!\right]^2}\cos\left[(2m-2k+1)\theta\right],$$

$$(9.170a,b)$$

The passage from (9.168b; 9.169a; 9.170a) to (9.168c; 9.169b; 9.170b) involves a change from double to single factorials similar to (9.162). The proof of coincidence of the hyperspherical Legendre polynomials of degree two (9.167a) ≡ (9.144c) [three (9.167b) ≡ (9.144d)] uses the trigonometric identities (I.3.31b) ≡ (9.171a) [(I.3.33b) ≡ (9.171b)],

$$\cos(2\theta) = 2\cos^2\theta - 1, \quad \cos(3\theta) = \cos\theta(4\cos^2\theta - 3),$$

(9.171a,b)

in (9.167a) [(9.167b)], leading to (9.172a) [(9.172b)]:

$$P_{2,\alpha}(\cos\theta) = \frac{1+\alpha}{2}\left[\frac{3+\alpha}{2}(2\cos^2\theta - 1) + \frac{1+\alpha}{2}\right]$$

$$= \frac{1+\alpha}{2}[(3+\alpha)\cos^2\theta - 1],$$

(9.172a)

$$P_{2,\alpha}(\cos\theta) = \frac{1+\alpha}{2}\frac{3+\alpha}{2}\cos\theta\left[\frac{5+\alpha}{2}\frac{4\cos^2\theta - 3}{3} + \frac{1+\alpha}{2}\right]$$

$$= \frac{1+\alpha}{2}\left(1 + \frac{2}{3}\alpha\right)\cos\theta[(5+\alpha)\cos^2\theta - 3].$$

(9.172b)

These are three equivalent expressions for the hyperspherical Legendre polynomials of degree two (9.144c) ≡ (9.172a) ≡ (9.167a) [three (9.144d) ≡ (9.172b) ≡ (9.167b)].

9.6.7 Recurrence Formula for the Hyperspherical Harmonics

The hyperspherical Legendre polynomials of higher degree may be obtained from those of lower degree via a recurrence formula. The latter can be obtained by differentiating with regard to the parameter a the uniformly convergent (9.150a, b, and d) series expansion for the generating function (9.147b), leading to

$$\sum_{n=1}^{\infty} na^{n-1}P_{n,\alpha}(z) = \frac{d}{da}\{|1 - 2az + a^2|^{-1/2-\alpha/2}\} = (1+\alpha)(z-a)|1 - 2az + a^2|^{-3/2-\alpha};$$

(9.173)

this is equivalent to

$$(1 - 2az + a^2)\sum_{n=0}^{\infty}(n+1)a^n P_{n+1,\alpha}(z) = (1+\alpha)(z-a)\sum_{n=0}^{\infty}a^n P_{n,\alpha}(z).$$

(9.174)

Equating the coefficients of powers of a in (9.174) leads to

$$(n+1)P_{n+1,\alpha}(z) - 2nzP_{n,\alpha}(z) + (n-1)P_{n-1,\alpha}(z) = (1+\alpha)[zP_{n,\alpha}(z) - P_{n-1,\alpha}(z)].$$

(9.175)

This simplifies to *the recurrence formula (9.176) for hyperspherical Legendre polynomials*:

$$(n+1)P_{n+1,\alpha}(z) = (1 + 2n + \alpha)zP_{n,\alpha}(z) - (n+\alpha)P_{n-1,\alpha}(z).$$

(9.176)

The particular case (9.177a),

$$\alpha = 0: \quad (n+1)P_{n+1}(z) = (1+2n)zP_n(z) - nP_{n-1}(z),$$

(9.177a,b)

is the recurrence formula (9.177b) for the original Legendre polynomials. This more general formula (9.176) can be used to obtain the hyperspherical Legendre polynomials of higher degree from those of lower degree, for example, starting with (9.144a) ≡ (9.178a),

$$P_{0,\alpha}(z) = 1: \quad P_{1,\alpha}(z) = (1+\alpha)zP_{0,\alpha}(z) = (1+\alpha)z, \tag{9.178a,b}$$

$$2P_{2,\alpha}(z) = (3+\alpha)zP_{1,\alpha}(z) - (1+\alpha)P_{0,\alpha}(z) = (1+\alpha)[(3+\alpha)z^2 - 1], \tag{9.178c}$$

$$3P_{3,\alpha}(z) = (5+\alpha)zP_{2,\alpha}(z) - (2+\alpha)P_{1,\alpha}(z)$$

$$= \frac{1+\alpha}{2}z\{(5+\alpha)[(3+\alpha)z^2 - 1] - 2(2+\alpha)\}$$

$$= \frac{1+\alpha}{2}z(3+\alpha)[(5+\alpha)z^2 - 3], \tag{9.178d}$$

and using (9.176) with *m* = 1, 2, 3 leads, respectively, to (9.178b) ≡ (9.144b), (9.178c) ≡ (9.144c), and (9.178d) ≡ (9.144d).

9.6.8 Four Differentiation Formulas for the Hyperspherical Legendre Polynomials

The series for the generating function (9.147b) is uniformly convergent with regard to *a(z)* in (9.150b) [(9.150a)], and thus (I.21.45c) can be differentiated term-by-term (9.173) [(9.179b)] with regard to *a(z)*:

$$P'_{n,\alpha}(z) \equiv \frac{d}{dz}[P_{n,\alpha}(z)]: \quad \sum_{n=0}^{\infty} a^n P'_{n,\alpha}(z) = \frac{d}{dz}\{|1 - 2az + a^2|^{-1/2-\alpha/2}\}$$

$$= (1+\alpha)a|1 - 2az + a^2|^{-3/2-\alpha/2}. \tag{9.179a,b}$$

Using in (9.179b) first (9.147b) and after (9.174) leads to (9.180):

$$(z-a)\sum_{n=0}^{\infty} a^n P'_{n,\alpha}(z) = a(1+\alpha)\frac{z-a}{1-2az+a^2}\sum_{n=0}^{\infty} a^n P_{n,\alpha}(z)$$

$$= \sum_{n=0}^{\infty}(n+1)a^{n+1}P_{n+1,\alpha}(z). \tag{9.180}$$

Equating the coefficients of powers of *a* leads to

$$nP_{n,\alpha}(z) = zP'_{n,\alpha}(z) - P'_{n-1,\alpha}(z); \quad \alpha = 0: \quad nP_n(z) = zP'_n(z) - P'_{n-1}(z), \tag{9.181a–c}$$

which is *the first differentiation formula for the hyperspherical (9.181a) [original (9.181c)] Legendre polynomials, and depends only on the degree n and not on the order α.*
    Differentiation of the recurrence formula (9.176) with regard to *z* leads to

$$(n+1)P'_{n+1,\alpha}(z) = (1+2n+\alpha)zP'_{n,\alpha}(z)$$

$$+ (1+2n+\alpha)P_{n,\alpha}(z) - (n+\alpha)P'_{n-1,\alpha}(z). \tag{9.182}$$

Substituting $P'_{n,\alpha}(z)$ from (9.181a) leads to

$$(n+1)P'_{n+1,\alpha}(z) = (1+2n+\alpha)[nP_{n,\alpha}(z) + P'_{n-1,\alpha}(z)]$$

$$+(1+2n+\alpha)P_{n,\alpha}(z) - (n+\alpha)P'_{n-1,\alpha}(z)$$

$$= (n+1)[(1+2n+2\alpha)P_{n,\alpha}(z) + P'_{n-1,\alpha}(z)]; \tag{9.183}$$

this is *the second differentiation formula for the hyperspherical (9.184a) [original (9.184b and c)] Legendre polynomials*:

$$P'_{n+1,\alpha}(z) - P'_{n-1,\alpha}(z) = (1+2n+\alpha)P_{n,\alpha}(z), \tag{9.184a}$$

$$\alpha = 0: \quad P'_{n+1}(z) - P'_{n-1}(z) = (1+2n)P_n(z). \tag{9.184b,c}$$

Adding (9.184a) to (9.181a) yields *the third differentiation formula for the hyperspherical (9.185a) [original (9.185b and c)] Legendre polynomials*:

$$P'_{n+1,\alpha}(z) - zP'_{n,\alpha}(z) = (1+n+\alpha)P_{n,\alpha}(z), \tag{9.185a}$$

$$\alpha = 0: \quad P'_{n+1}(z) - zP'_n(z) = (1+n)P_n(\alpha). \tag{9.185b,c}$$

There is one recurrence (9.176) [(9.177a and b)] and there are four differentiation (9.181a; 9.184a; 9.185a; 9.188) [(9.181b and c; 9.184b and c; 9.185b and c; 9.186b)] formulas for the hyperspherical (original) Legendre polynomials. The last differentiation formula is obtained next and leads to the differential equation for the hyperspherical Legendre polynomials (Subsection 9.6.9).

9.6.9 Hyperspherical Legendre Differential Equation

The rule for generation of multipoles in three dimensions corresponds to the differentiation formula for the Legendre polynomials:

$$z \equiv \cos\theta: \quad (n+1)[P_{n+1}(z) - zP_n(z)] = (z^2-1)P'_n(z), \tag{9.186a,b}$$

as follows from (8.76) $\equiv$ (9.186b) using the variable (9.186a). Substitution of (9.177b) in (9.186b) leads to:

$$(z^2-1)P'_n(z) = n[zP_n(z) - P_{n-1}(z)]. \tag{9.186c}$$

The extension of (9.186c) to the hyperspherical Legendre polynomials is obtained by multiplying (9.181a) by z and adding (9.185a) to n replaced by $n-1$:

$$nzP_{n,\alpha}(z) + P'_{n,\alpha}(z) = z^2 P'_{n,\alpha}(z) + (n+\alpha)P_{n-1,\alpha}(z). \tag{9.187}$$

This may be rearranged as:

$$(z^2-1)P'_n(z) = nzP_{n,\alpha}(z) - (n+\alpha)P_{n-1,\alpha}(z), \tag{9.188}$$

and is *the fourth differentiation formula for the hyperspherical (9.188) [original (9.186c)] Legendre polynomials. The differentiation formula (9.188) for the hyperspherical Legendre polynomials corresponds to the rule of generation of multidimensional multipoles and it reduces to (9.186b) $\equiv$ (9.186c) for $\alpha = 0$.*

In order to obtain the differential equation satisfied by the hyperspherical Legendre polynomials, the starting point is a further differentiation with regard to z of the fourth differentiation formula (9.188), leading to

$$(z^2 - 1)P''_{n,\alpha}(z) + 2zP'_{n,\alpha}(z) - nP_{n,\alpha}(z) = nzP'_{n,\alpha}(z) - (n+\alpha)P'_{n-1,\alpha}(z); \tag{9.189}$$

substitution of (9.181a) in the last term on the r.h.s. of (9.189) leads to

$$\left(z^2 - 1\right)P''_{n,\alpha}\left(z\right) + \left(2 - n\right)z\,P'_{n,\alpha}\left(z\right) = n\,P_{n,\alpha}\left(z\right) + \left(n+\alpha\right)\left[\,n\,P'_{n,\alpha}\left(z\right) - z\,P_{n,\alpha}\left(z\right)\,\right]$$

$$= n\left(n+\alpha+1\right)P'_{n,\alpha}\left(z\right) - \left(n+\alpha\right)z\,P'_{n,\alpha}\left(z\right). \tag{9.190}$$

Thus, *the linear second-order differential equation satisfied by the hyperspherical Legendre polynomials has been obtained:*

$$(1-z^2)P''_{n,\alpha}(z) - (2+\alpha)zP'_{n,\alpha}(z) + n(n+1+\alpha)P_{n,\alpha} = 0. \tag{9.191}$$

The **hyperspherical Legendre differential** *equation (9.191) contains as a particular case (9.191a) the original Legendre differential equation,*

$$\alpha = 0: \quad (1-z^2)P''_n(z) - 2zP'_n(z) + n(n+1)P_n(z) = 0, \tag{9.191a,b}$$

satisfied by the original Legendre polynomials.

9.6.10 Rodrigues and Schlaffi Integrals

Since (9.147b) is the Stirling–Maclaurin series (I.23.34a and b) expansion of the generating function in powers of a, the coefficient is given by

$$P_{n,\alpha}(z) = \frac{1}{n!}\lim_{a\to 0}\frac{\partial^n}{\partial a^n}\left\{[1-2az+a^2]^{-1/2-\alpha/2}\right\}. \tag{9.192}$$

Cauchy third theorem (I.15.13) may be used in (9.192a), leading to a loop integral moving clockwise around the point $\zeta = a$:

$$\frac{1}{n!}\frac{\partial^n}{\partial a^n}\left[(1-2az+a^2)^{-1/2-\alpha/2}\right] = \frac{1}{2\pi i}\int\limits^{(a+)}(1-2\zeta z+\zeta^2)^{-1/2-\alpha/2}(\zeta-a)^{-n-1}d\zeta; \tag{9.193}$$

substitution of (9.193) in (9.192) specifies *the hyperspherical Legendre polynomial as a complex loop integral moving clockwise around the origin:*

$$P_{n,\alpha}(z) = \frac{1}{2\pi i}\int\limits^{(0+)}(1-2\zeta z+\zeta^2)^{-1/2-\alpha/2}\zeta^{-n}d\zeta. \tag{9.194}$$

The expression (9.194) holds for complex α, not necessarily an integer, and may be used to define the hyperspherical Legendre functions; likewise, n may also be complex in (9.194), leading to fractional derivatives. The case of nonintegral n or α leads to branch-points and branch-cuts (Chapter I.7) in the integrand (9.194) and affects the path of integration. Differentiating term-by-term after using the binomial theorem (I.25.37a through c) leads to

$$\frac{d^n}{dz^n}[(z^2-1)^n] = \frac{d^n}{dz^n}\left\{\sum_{k=0}^{\infty} \frac{n!(-)^k}{k!(n-k)!} z^{2n-2k}\right\}$$

$$= \sum_{k=0}^{\leq n/2} (-)^k \frac{n!(2n-2k)!(-)^k}{k!(n-k)!(n-2k)!} z^{n-2k}. \qquad (9.195)$$

Comparison with (9.161d) specifies the **Rodrigues formula** (1813) *for the Legendre polynomials (9.196a):*

$$P_n(z) = \frac{2^{-n}}{n!} \frac{d^n}{dz^n}[(z^2-1)^n] = \frac{2^{-n}}{2\pi i} \int^{(z+)} (\zeta^2-1)^n (\zeta-z)^{-n} d\zeta, \qquad (9.196a,b)$$

corresponding to (9.196b) the **Schlaffi integral** *(1881),* on account of the Cauchy third theorem (I.15.13). The extension of (9.196b) to nonintegral n involves differintegration deformation of the path of integration around the branch-points $\zeta = \pm 1, z$ and the branch-cuts associated with them. The preceding account is a sample of the properties of the hyperspherical and the original Legendre polynomials that can be treated more fully together with the associated Legendre functions in the theory of special functions.

9.7 Multipoles in Hyperspherical and Hypercylindrical Coordinates

The multidimensional multipolar expansion is specified (1) exactly for hyperspherical harmonics of all orders in terms of the hyperspherical Legendre polynomials (Section 9.6); and (2) using the Taylor series for the influence or Green function (Subsection 9.7.1), for the hyperspherical harmonics of four lowest orders, namely, multidimensional monopoles/dipoles/quadrupoles and octupoles (respectively, Subsections 9.7.4 through 9.7.7). Their potentials and fields in any dimension can be expressed in terms of hyperspherical (or hypercylindrical) coordinates [Subsection 9.7.2 (Subsection 9.7.3)]. The multipolar expansion (Subsection 9.7.1) specifies axisymmetric multipoles (Subsections 9.7.4 through 9.7.7) that are a particular case of the nonaxisymmetric multipoles (Subsection 9.7.8) obtained by differentiating the influence function with regard to the coordinates. The axisymmetric (nonaxisymmetric) multipolar expansions [Subsections 9.7.4 through 9.7.7 (Subsection 9.7.8)] derive from the influence function (Subsection 9.7.1) and can be expressed in hyperspherical (or hypercylindrical) coordinates [Subsection 9.7.2 (Subsection 9.7.3)].

9.7.1 Multidimensional Multipolar Expansion

The multipolar (Section 9.6) expansion may also be obtained in integral form, using derivatives of the influence function (9.43a through c) of the Laplace operator:

$$\frac{\partial}{\partial x_i}[\sigma_N G(\vec{x}, y)] = \frac{1}{2-N} \frac{\partial}{\partial x_i}\left[|\vec{x}-\vec{y}|^{2-N}\right] = |\vec{x}-\vec{y}|^{-N}(x_i-y_i), \qquad (9.197)$$

$$\frac{\partial^2}{\partial x_i \partial x_j}[\sigma_N G(\vec{x};\vec{y})] = -N(x_i-y_i)(x_j-y_j)|\vec{x}-\vec{y}|^{-N-2} + |\vec{x}-\vec{y}|^{-N}\delta_{ij}, \qquad (9.198)$$

where (2.198) involves the identity matrix (5.278a through c). This specifies the first three terms of the Taylor series:

$$e_{ri} = \frac{x_i}{|\vec{x}|}: \quad \sigma_N G(\vec{x}; \vec{y}) = \frac{|\vec{x}|^{2-N}}{2-N} - x_i y_i |\vec{x}|^{-N}$$

$$- \frac{1}{2} y_i y_j (\delta_{ij} R^2 - N x_i x_i) |\vec{x}|^{-N-2} + O(x_i x_j x_k |\vec{x}|^{-N-4}). \tag{9.199a,b}$$

Substituting in the generalized Poisson integral (9.46b) specifies as in (8.51b):

$$\sigma_N \Phi(\vec{x}) = |\vec{x}|^{2-N} \left\{ \frac{P_0}{2-N} - |\vec{x}|^{-2} P_i x_i - \frac{|\vec{x}|^{-4}}{2} P_{ij} (N x_i x_j - |\vec{x}|^2 \delta_{ij}) + O(|\vec{x}|^{-3}) \right\}, \tag{9.200}$$

the N-dimensional multipole expansion for the potential of an arbitrary source distribution:

$$P_0 \equiv \int_D \Lambda(\vec{y}) d^N \vec{y}, \quad \vec{P}_i \equiv \int_D y_i \Lambda(\vec{y}) d^N \vec{y}, \quad P_{ij} = \int_D y_i y_j \Lambda(\vec{y}) d^N \vec{y}, \tag{9.201a–c}$$

with moments for the monopole (9.201a), dipole (9.201b), and quadrupole (9.201c) terms. The potentials and the fields of the multipoles in the plane (space) are represented conveniently [Chapters I.12, 16 (Chapters 6, 8)] using polar (spherical or cylindrical) coordinates [Section I.11.4 (Section 6.1)]. Their extension to hyperspherical (hypercylindrical) coordinates [Subsection 9.7.2 (Subsection 9.7.3)] likewise specifies the potential and the fields of the multidimensional monopoles, dipoles, quadrupoles, octupoles, and higher order multipoles (Subsections 9.7.4 through 9.7.7).

9.7.2 Radial, Longitudinal, and Multicolatitude Coordinates

The polar (spherical) coordinates [Figure 9.7a (Figure 9.7b)] are specified by (9.202a through d) [(9.203a through f)]:

$$0 \leq r < \infty, \quad 0 \leq \varphi < 2\pi: \quad x_1 = r \cos\varphi, \quad x_2 = r \sin\varphi, \tag{9.202a–d}$$

$$0 \leq R < \infty, \quad 0 \leq \theta \leq \pi, \quad 0 \leq \varphi < 2\pi:$$
$$x_1 = R \cos\theta, \quad x_2 = R \sin\theta \cos\varphi, \quad x_3 = R \sin\theta \sin\varphi, \tag{9.203a–f}$$

where $R(r)$ is the distance from the origin (axis) and $\varphi(\theta)$ is the longitude (colatitude from the north pole). The **four-dimensional spherical coordinates** have two colatitudes:

$$0 \leq R < \infty, \quad 0 \leq \theta_1, \theta_2 \leq \pi, \quad 0 \leq \varphi < 2\pi: \quad x_1 = R \cos\theta_1,$$
$$x_2 = R \sin\theta_1 \cos\theta_2, \quad x_3 = R \sin\theta_1 \sin\theta_2 \cos\varphi, \quad x_4 = R \sin\theta_1 \sin\theta_2 \sin\varphi. \tag{9.204a–g}$$

The **hyperspherical coordinates,**

$$0 \leq R < \infty, \quad 0 \leq \theta_1, ..., \theta_{N-2} \leq \pi, \quad 0 \leq \varphi < 2\pi: \quad x_1 = R \cos\theta_1, \tag{9.205a–d}$$

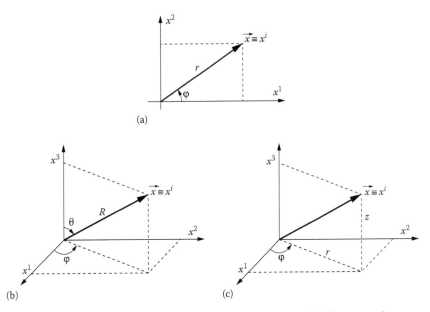

(a)

(b)   (c)

FIGURE 9.7 The polar coordinates (a) may be used to study plane potential fields. A two-dimensional source is (a) in three dimensions a line source and the cylindrical coordinates correspond to the translation of the polar coordinates (a) along an axis passing through the origin and perpendicular to their plane. The point source in space is (b) represented most simply by spherical coordinates that reduce to polar coordinates in (a) the equatorial plane $\theta = \pi/2$. The spherical (b) [cylindrical] coordinates in the space $N = 3$ may (Figure 6.2) be extended (c) to hyperspherical (hypercylindrical) coordinates in the hyperspace $N > 4$ by adding $N - 2(N - 3)$ colatitude angles as additional coordinates. They may be used in multipolar expansions to identify the multipoles of all orders and force between them, including asymptotic scalings (Table 9.1).

$$n = 2,\ldots,N-2: \quad x_n = R\sin\theta_1\sin\theta_2\ldots\sin\theta_{n-1}\cos\theta_n, \tag{9.205e,f}$$

$$\{x_{N-1},x_N\} = R\sin\theta_1\sin\theta_2\ldots\sin\theta_{N-3}\sin\theta_{N-2}\{\cos\varphi,\sin\varphi\}, \tag{9.205g,h}$$

have one longitude, (N − 2), colatitudes, and one radial distance,

$$R \equiv \left\{\left|(x_1)^2+\cdots+(x_N)^2\right|^{1/2} : \left|(x_1)^2+(x_2)^2+(x_3)^2+(x_4)^2\right|^{1/2},\right.$$
$$\left.\left|(x_1)^2+(x_2)^2+(x_3)^2\right|^{1/2}, \left|(x_1)+(x_2)^2\right|^{1/2}\right\}, \tag{9.206a–d}$$

which applies in two (9.202a through d; 9.206d), three (9.203a through f; 9.206c), four (9.204a through g; 9.206h), and N dimensions (9.205a through h; 9.206a). *The last N − M − 1 dimensions of the hyperspherical coordinates (9.207a and b),*

$$0 \le \varphi < 2\pi, \quad 0 \le \theta_M,\theta_{M+1},\ldots,\theta_{N-2} \le \pi: \quad R_M = R\cos\theta_1\cos\theta_2\ldots\cos\theta_{M-1}, \tag{9.207a–c}$$

lie on a hypersphere of dimension N − M − 1 with radius (9.207c).

9.7.3 Hyperspherical and Hypercylindrical Coordinates

The cylindrical coordinates in space (Figure 9.7c) add a Cartesian coordinate z orthogonal to the plane (9.202a through d) of polar coordinates (9.208a through f):

$$-\infty < z < +\infty, \quad 0 \leq r < \infty, \quad 0 \leq \varphi < 2\pi: \quad x_1 = z, \quad x_2 = R\cos\varphi, \quad x_3 = R\sin\varphi. \qquad (9.208\text{a--f})$$

The **four-dimensional cylindrical coordinates** have one colatitude (9.209a through h):

$$-\infty < z < +\infty, \quad 0 \leq r < \infty, \quad 0 \leq \varphi < 2\pi, \quad 0 \leq \theta \leq \pi:$$
$$x_1 = z, \quad x_2 = r\cos\theta, \quad x_3 = r\sin\theta\cos\varphi, \quad x_4 = r\sin\theta\sin\varphi. \qquad (9.209\text{a--h})$$

The hypercylindrical coordinates,

$$-\infty < z < \infty, \quad 0 \leq r < \infty, \quad 0 \leq \theta_1, \quad \theta_2,...,\theta_{N-3} \leq \pi, \quad 0 \leq \varphi < 2\pi: \quad x_1 = z, \quad x_2 = r\cos\theta_1, \qquad (9.210\text{a--f})$$

$$n = 2,...,N-3: \quad x_{n+1} = r\sin\theta_1 \sin\theta_2 ... \sin\theta_{n-1}\cos\theta_n, \qquad (9.210\text{g,h})$$

$$\{x_{N-1},x_N\} = r\sin\theta_1 \sin\theta_2,...,\sin\theta_{N-3}\{\cos\varphi,\sin\varphi\}, \qquad (9.210\text{i,j})$$

have one Cartesian coordinate, one longitude, $N-3$ colatitudes, and one radial distance from the axis:

$$r \equiv \left\{ |(x_2)^2 + \cdots + (x_N)^2|^{1/2}: \quad |(x_2)^2 + (x_3)^2 + (x_4)^2|^{1/2}, \quad |(x_2)^2 + (x_3)^2|^{1/2} \right\}, \qquad (9.211\text{a--c})$$

which applies in three [(9.208a through f); (9.211c)], four (9.209a through h; 9.211b), and N dimensions (9.210a through c; 9.211a). The hyperspherical (9.205a through h; 9.206a) [hypercylindrical (9.210a through j; 9.211a)] coordinates have (1) the same longitude φ and $(N-3)$ colatitudes $\theta_1, ..., \theta_{N-3}$; (2) they differ only in the distance from the origin (9.206a) [from the axis (9.211a)]; and (3) the last latitude θ_{N-2} in the hyperspherical coordinates is replaced by the Cartesian coordinate z for the hypercylindrical coordinates. These are related by the first colatitude as for spherical and cylindrical coordinates in space:

$$R^2 = r^2 + z^2, \quad z = R\cos\theta_1, \quad r = R\sin\theta_1. \qquad (9.212\text{a--c})$$

The last $N-M-2$ hypercylindrical coordinates (9.213a and b),

$$0 \leq \varphi < 2\pi, 0 \leq \theta_M, \theta_{M+1},...,\theta_{N-3} \leq \pi: \quad r_M = r\sin\theta_1 \sin\theta_2 ... \sin\theta_{M-1}, \qquad (9.213\text{c})$$

lie on a hypersphere of dimension $N-M-2$ with radius (9.213c).

9.7.4 Potential and Field of a Multidimensional Monopole

*The **multidimensional monopole** has a potential corresponding to the first term (9.201a) of the multipolar expansion (9.200) involving (9.144a) the hyperspherical Legendre polynomial of degree zero, that is unity, and thus need not appear explicit,*

$$(2-N)\Phi_0(\vec{x}) = P_0 |\vec{x}|^{2-N} = P_0 R^{2-N} = P_0 R^{2-N} P_{0,N-3}(\cos\theta_1); \qquad (9.214\text{a--c})$$

the corresponding to the field is (9.215a and b) [(9.216a through c)] in the hyperspherical (hypercylindrical) coordinates:

$$\vec{U}_0 = \nabla\Phi_0 = \frac{P_0}{\sigma_N} x_i \,|\,\vec{x}\,|^{-N} = \frac{P_0}{\sigma_N} R^{1-N} \vec{e}_R, \tag{9.215a,b}$$

$$\vec{U}_0 = \frac{P_0}{\sigma_N} R^{1-N} (\vec{e}_z \cos\theta_1 + \vec{e}_r \sin\theta_1) = \frac{P_0}{\sigma_N} \,|\, r^2 + z^2 \,|^{1/2 - N/2} (\vec{e}_z z + \vec{e}_r r); \tag{9.216a-c}$$

the multidimensional monopole:

$$S_N = \sigma_N R^{N-1}: \qquad \int_{|\vec{x}|=R} \frac{\vec{U}_0 . \vec{x}}{R} \,dS = \frac{P_0}{\sigma_N} R^{1-N} S_N = P_0, \tag{9.217a,b}$$

has flux (9.217b) across a hypersphere of area (9.217a) that is a constant independent of the radius. The potential (9.214a through c) and the field (9.215a and b) reduce for $N = 3$ to the point-monopole in space (8.67a and b). The line monopole in the plane (I.12.27a) corresponds to the logarithmic potential (9.42a through d) $\equiv$ (9.218b) and is excluded (9.218a) from the potential (9.214) that applies to all other dimensions $N \neq 2$:

$$N = 2: \quad \Phi_0(r) = \frac{P_0}{2\pi} \log r; \quad \vec{U}_0 = \nabla\Phi_0 = \frac{P_0}{2\pi r} \vec{e}_r, \tag{9.218a-c}$$

the field (9.215a and b) applies to all dimensions, including $N = 2$ in (9.218c).

9.7.5 Potential and Field of a Multidimensional Dipole

The **multidimensional dipole** has potential specified by the second term (9.219b) of the multipolar expansion (9.200):

$$\sigma_N \Phi_1(\vec{x}) = -P_i x_i \,|\,\vec{x}\,|^{-N} = -P_1 R^{1-N} \cos\theta_1$$

$$= -\frac{P_1}{N-2} R^{1-N} P_{1,N-3}(\cos\theta_1), \tag{9.219a-c}$$

and (1) is specified in hyperspherical coordinates with axis along the dipole moment by (9.219a) $\equiv$ (9.219b); (2) involves (9.219c) the hyperspherical Legendre polynomial of first degree (9.144b); and (3) corresponds (9.219c) to the monopole field (9.216c) component in the axial direction. The corresponding dipolar field,

$$\vec{U}_1(\vec{x}) = \nabla\Phi_1 = \frac{1}{\sigma_N} \frac{\partial}{\partial\vec{x}}\left(\frac{\vec{P}_1 . \vec{x}}{|\,\vec{x}\,|^N}\right) = -\frac{R^{-N}}{\sigma_N}\left[\vec{P}_1 - N\frac{\vec{x}}{R^2}(\vec{P}_1 . \vec{x})\right], \tag{9.220}$$

has components parallel (167a) [transverse (167b)] to the axis,

$$U_{1z} = \frac{P_1}{\sigma_N}(N\cos^2\theta_1 - 1)R^{-N}, \quad U_{1r} = \frac{P_1}{\sigma_N} NR^{-N}\cos\theta_1\sin\theta_1, \tag{9.221a,b}$$

corresponding to hypercylindrical components. The particular case $N = 3$ corresponds to a dipole in (8.26d). The dipole in the plane (9.222a) has complex potential (9.222b) $\equiv$ (I.12.43a) and conjugate velocity (9.222c):

$$N = 2: \quad f_1(z) = -\frac{P_1}{2\pi z}, \quad v_1^{\star}(z) = \frac{df}{dz} = \frac{P_1}{2\pi z^2}. \tag{9.222a-c}$$

The Cartesian components of the velocity (9.223b) in polar coordinates (9.223a),

$$z = re^{i\varphi}: \quad v_1^* \equiv v_{1x} - iv_{1y} = \frac{P_1}{2\pi r^2} e^{-i2\varphi} = \frac{P_1}{2\pi r^2} [\cos(2\varphi) - i\sin(2\varphi)], \qquad (9.223a,b)$$

are given by

$$\{v_{1x}, v_{1y}\} = \frac{P_1}{2\pi r^2} \{\cos(2\varphi), \sin(2\varphi)\} = \frac{P_1}{2\pi r^2} \{2\cos^2\varphi - 1, 2\sin\varphi\cos\varphi\}, \qquad (9.224a,b)$$

which coincide with (9.221a and b) $\equiv$ (9.224a and b) for $\theta_1 = \varphi$, $R = r$, $N = 2$, $\sigma_2 = 2\pi$.

9.7.6 Potential and Field of a Multidimensional Quadrupole

*The potential (9.225) of a **multidimensional quadrupole** (9.201c) is given by*

$$-2\sigma_N \Phi_2(\vec{x}) = (NP_{ij}x_i x_j - P_{ii}R^2)R^{-2-N}$$

$$= \left\{ N[P_{11}(x_1)^2 + 2P_{12}x_1 x_2 + P_{22}(x_2)^2] - (P_{11} + P_{22})R^2 \right\} R^{-2-N}, \qquad (9.225)$$

choosing the (x_1, x_2) plane so as to contain the two quadrupole axes. The case of a longitudinal quadrupole (9.226a and b) simplifies potential (9.225) to (9.226c):

$$P_{11} \equiv P_2, \quad P_{12} = 0 = P_{22}: \quad -2\sigma_N \Phi_2(\vec{x}) = P_2 R^{-N}(N\cos^2\theta_1 - 1)$$

$$= \frac{P_2}{N-1} R^{-N} P_{2,N-3}(\cos\theta_1), \qquad (9.226a\text{–}c)$$

involving (9.142b) the hyperspherical Legendre polynomial of second degree (9.144c). The multidimensional quadrupole potential corresponds to (1) the third term in the multipolar expansion (2.200); (2) the component along the axis of the dipolar field (9.221a), replacing P_1 by $-P_2$; and (3) the hyperspherical Legendre polynomial (9.144c) of degree two. The corresponding field has hyperspherical (hypercylindrical) components (9.227a) [(9.227b)]:

$$\vec{U}_2 = \nabla \Phi_2 = \frac{N}{2} \frac{P_2}{\sigma_N} R^{-N-1} \left[\vec{e}_R(N\cos^2\theta_1 - 1) + \vec{e}_\theta \sin(2\theta_1) \right], \qquad (9.227a)$$

$$= \frac{N}{2} \frac{P_2}{\sigma_N} R^{-N-1} \left\{ \vec{e}_z \cos\theta_1 \left[(N+2)\cos^2\theta_1 - 3 \right] - \vec{e}_r \sin\theta_1 \left[(N+2)\cos^2\theta_1 - 1 \right] \right\}. \qquad (9.227b)$$

The passage from (9.227a) to (9.227b) uses the relations,

$$\vec{e}_R = \vec{e}_z \cos\theta_1 + \vec{e}_r \sin\theta_1, \quad \vec{e}_\theta = -\vec{e}_z \sin\theta_1 + \vec{e}_r \cos\theta_1, \qquad (9.228a,b)$$

between the first two (9.212a through c) hypercylindrical and hyperspherical coordinates; they coincide with the relation between the cylindrical and the spherical coordinates (6.51d) $\equiv$ (9.228a and b) because the remaining coordinates are not involved. Substitution of (9.228a and b) in (9.227a) gives

$$\frac{2\sigma_N}{P_2}\frac{R^{1+N}}{N}\vec{U}_z = \vec{e}_z\left[\cos\theta_1(N\cos^2\theta_1 - 1) - \sin\theta_1\sin(2\theta_1)\right]$$

$$+\vec{e}_r\left[\sin\theta_1(N\cos^2\theta_1 - 1) + \cos\theta_1\sin(2\theta_1)\right]$$

$$= \vec{e}_z\cos\theta_1(N\cos^2\theta_1 - 1 - 2\sin^2\theta_1)$$

$$+\vec{e}_r\sin\theta_1(N\cos^2\theta_1 - 1 + 2\cos^2\theta_1)$$

$$= \vec{e}_z\cos\theta_1\left[(N+2)\cos^2\theta_1 - 3\right] + \vec{e}_r\sin\theta_1\left[(N+2)\cos^2\theta_1 - 1\right]. \tag{9.229}$$

in agreement with (9.227b) $\equiv$ (9.229).

The particular case $N = 3$ corresponds to an axial quadrupole in space (8.38c). A quadrupole in the plane (9.230a) has complex potential (9.230b) $\equiv$ (I.12.55a) and conjugate velocity (9.230c):

$$N = 2: \quad f_2(z) = -\frac{P_2}{2\pi z^2}, \quad v_2^*(z) = \frac{df_z}{dz} = \frac{P_2}{\pi z^3}. \tag{9.230a-c}$$

The Cartesian components of the velocity are given by

$$v_2^* = v_{2x} - iv_{2y} = \frac{P_2}{\pi r^3}e^{-3i\varphi} = \frac{P_2}{\pi r^3}\left[\cos(3\varphi) - i\sin(3\varphi)\right]. \tag{9.231}$$

Using (9.171b) leads to:

$$v_{2x} = \frac{P_2}{\pi r^3}\cos(3\varphi) = \frac{P_2}{\pi r^3}\cos\varphi(4\cos^3\varphi - 3\cos\varphi), \tag{9.232}$$

which coincides with the $\vec{e}_z$ component of (9.227b). Using the relation (I.3.33c) $\equiv$ (9.233):

$$\sin(3\varphi) = \sin\varphi(3 - 4\sin^2\varphi) = \sin\varphi(4\cos^2\varphi - 1), \tag{9.233}$$

leads to:

$$v_{2y} = \frac{P_2}{\pi r^3}\sin(3\varphi) = \frac{P_2}{\pi r^3}\sin\varphi(4\cos^3\varphi - 1), \tag{9.234}$$

which coincides with the $\vec{e}_r$ component of (9.227b) $\equiv$ (9.234) for $N = 2, \sigma_2 = 2\pi, \varphi = \theta_1, R = r$. The potential of an axial multidimensional octupole follows from the axial component.

9.7.7 Potential of an Octupole and Other Multipoles

The potential of an axial octupolar field follows from the axial component of the quadrupolar field (9.227b) replacing P_2 by $-P_3$:

$$\Phi_3(\vec{x}) = -\frac{N}{2}\frac{P_3}{\sigma_N}R^{-N-1}\cos\theta_1\left[(N+2)\cos^2\theta_1 - 3\right], \tag{9.235}$$

in agreement with the spatial case $N = 3$ in (8.70b). In the plane (9.236a; 9.237a), the potential of the axial dipole (9.219b) [quadrupole (9.226b)] is (9.236b) [(9.237b)]:

$$N = 2: \quad -\Phi_1(r, \varphi) = \frac{P_1}{2\pi r} \cos\varphi = \text{Re}\left(\frac{P_1}{2\pi} r^{-1} e^{-i\varphi}\right) = \text{Re}\left(\frac{P_1}{2\pi z}\right), \tag{9.236a,b}$$

$$\sigma_2 = 2\pi: \quad -\Phi_2(r, \varphi) = \frac{P_2}{2\pi r^2}(2\cos^2\varphi - 1) = \frac{P_2}{2\pi r^2}\cos(2\varphi)$$

$$= \text{Re}\left(\frac{P_2}{2\pi} r^{-2} e^{-2i\varphi}\right) = \text{Re}\left(\frac{P_2}{2\pi z^2}\right), \tag{9.237a,b}$$

in agreement with (9.222b) $\equiv$ (I.12.43a) [(9.130b) $\equiv$ (I.12.55a)]. The signs in (I.12.43a; I.12.55a) are the same as in (9.236b; 9.237b) because they concern the hydrodynamic potential $\vec{v} = \nabla\Phi$, and would be reversed for the electric potential $\vec{E} = -\nabla\Phi$; the signs in the fields are the same in both cases. The octupole in the plane has complex potential (9.238a) $\equiv$ (12.59b with $n = 3$) corresponding to the real potential (9.238b):

$$f_3(z) = -\frac{P_3}{2\pi z^3}; \quad -\Phi_3(r, \varphi) = \text{Re}\left(\frac{P_3}{2\pi z^3} e^{-i3\varphi}\right) = \frac{P_3}{2\pi r^3}\cos(3\varphi)$$

$$= \frac{P_3}{2\pi r^3}\cos\varphi(4\cos^2\varphi - 3), \tag{9.238a,b}$$

where (9.171b) was used to prove the coincidence of (9.238b) $\equiv$ (9.235) for $N = 2$. The complex potentials for the dipole (9.222b)/quadrupole (9.232)/octupole (9.238a) coincide, respectively, with (I.12.43a/55a/59b). It has been shown that the **multidimensional octupole** *has potential (9.235), corresponding in three (two) dimensions to (8.70b) [(9.238b)].* The multidimensional multipolar expansion (Subsection 9.7.1) leads to axisymmetric multidimensional multipoles (Subsections 9.7.3 through 9.7.7). These are a particular case of the general, nonaxisymmetric multidimensional multipoles (Subsection 9.7.8) that can be obtained by differentiating the influence function with regard to the coordinates.

9.7.8 Multidimensional Nonaxisymmetric Multipoles

The N-dimensional influence or Green function for the Laplace operator (9.39; 9.43a through c) specifies by differentiation with regard to the coordinates the potential of a **multidimensional multipole:**

$$(2 - N)\sigma_N \Phi_n(\vec{x}) = P_{i_1 \cdots i_n} \frac{\partial^n}{\partial x_{i_1} \cdots \partial x_{i_m}} |\vec{x}|^{2-N}, \tag{9.239}$$

where σ_N is the area of the N-dimensional unit hypersphere (5.301a and b). The three lowest orders are (1) the multidimensional monopole (9.214a through c); (2) using (9.197), the multidimensional dipole:

$$-\sigma_N \Phi_1(\vec{x}) = \frac{P_i}{2 - N} \frac{\partial}{\partial x_i} |\vec{x}|^{2-N} = P_i x_i |\vec{x}|^{-N}; \tag{9.240}$$

and (3) using (2.198), the multidimensional quadrupole:

$$\sigma_N \Phi_2(\vec{x}) = \frac{P_{ij}}{2 - N} \frac{\partial}{\partial x_i \partial x_j} |\vec{x}|^{2-N} = \left(P_{ii} - \frac{N}{|\vec{x}|^2} P_{ij} x_i x_j\right) |\vec{x}|^{-N}; \tag{9.241}$$

the axisymmetric multipole is the first component $P_{1...1}$ in (9.239). For example, the multidimensional dipole (9.240) is given by

$$-|r^2 + z^2|^{N/2} \sigma_N \Phi_1(r, \theta_1, \dots, \theta_{N-3}, \varphi, z)$$
$$= P_1 z + P_2 r \cos\theta_1 + r \sum_{n=2}^{N-3} P_{N+1} \sin\theta_1 \dots \sin\theta_{n-1} \cos\theta_n$$
$$+ r \sin\theta_1 \dots \sin\theta_{N-3}(P_{N-1}\cos\varphi + P_N \sin\varphi), \tag{9.242}$$

$$-R^N \sigma_N \Phi_1(R, \theta_1, \dots, \theta_{N-2}, \varphi)$$
$$= P_1 R \cos\theta_1 + R \sin\theta_1 \sum_{n=2}^{N-3} P_{N+1} \sin\theta_1 \dots \sin\theta_{n-1} \cos\theta_n$$
$$+ R \sin\theta_1 \dots \sin\theta_{n-2}(P_{n-1}\cos\varphi + P_N \sin\varphi), \tag{9.243}$$

using hypercylindrical (9.242) [hyperspherical (9.243)] coordinates (9.212a through c). The potential of the axisymmetric multidimensional multipole is the first term on the r.h.s. of (9.243) ≡ (9.242) ≡ (9.219b).

9.8 Hypersphere Theorem and Insertion in a Uniform Field

The insertion of a hypersphere in a uniform field is represented by a dipole, with distinct dipole moments in the cases of a zero normal (tangential) field [Subsection 9.8.1 (Subsection 9.8.4)]. This suggests the hypersphere theorem (Subsection 9.8.2) as the multidimensional extension of the circle (sphere) theorem in the plane, involving a reciprocal hyperpotential (Subsection 9.8.3).

9.8.1 Insertion of a Hypersphere in a Uniform Field

A uniform field $\vec{U}$ corresponds to the first term in the potential:

$$\Phi(\vec{x}) = \vec{U}.\vec{x} - (\vec{P}_1.\vec{x})\frac{R^{-N}}{\sigma_N} = \cos\theta_1 \left[UR - \frac{P_1}{\sigma_N} R^{1-N} \right], \tag{9.244}$$

while the second term corresponds to a dipole (9.219b) of moment P_1. The normal component of the field on a hypersphere radius a vanishes (9.245a),

$$0 = \lim_{R \to a} \frac{\partial \Phi_+}{\partial R} = \cos\theta_1 \left[U + \frac{N-1}{\sigma_N} P_1^- a^{-N} \right], \quad P_1^+ \equiv -\frac{\sigma_N}{N-1} U a^N, \tag{9.245a,b}$$

for a dipole moment (9.245b) that simplifies for a cylinder (9.246a through c) [sphere (9.246d through f)]:

$$N = 2, \quad \sigma_2 = 2\pi: \quad P_1^+ = -2\pi U a^2; \quad N = 3, \quad \sigma_3 = 4\pi: \quad P_1^+ = -2\pi U a^3. \tag{9.246a–f}$$

in agreement with (I.28.89a) [(6.199e)].

Comparing with the volume of the hypersphere (9.247a) ≡ (5.291b) of radius a:

$$V_N = N^{-1}\sigma_N a^N: \quad P_1^+ = -f_N V_N U, \quad f_N = \frac{N}{N-1} = \left(1 - \frac{1}{N}\right)^{-1}, \tag{9.247a–c}$$

the dipole moment (9.247b) ≡ (9.245b) has a factor (9.247c) that is (1) $f_2 = 2$ for a cylinder (9.246c); (2) $f_3 = 3/2$ for a sphere (9.246f); (3) $f_4 = 4/3$ in four dimensions $N = 4$; and (4) reduces with increasing dimension tending to a minimum $f_N \to 1$ as $N \to \infty$. Thus, *a hypersphere of radius a in a uniform field $\vec{U}$ corresponds to a dipole of moment (9.245b) ≡ (9.247b), related to the volume (9.247a) of the hypersphere by the factor (9.247c); hence, it equals the product of the volume and the uniform field U, and a factor decreasing with the dimension of the space. The superposition of the dipole with moment (9.245b) on the uniform field U leads (9.244) to a potential (9.248) [field (9.249a and b)]:*

$$\Phi_+(R,\theta_1) = U\left[R + \frac{a^N}{N-1}R^{1-N} \right]\cos\theta_1, \tag{9.248}$$

$$U_R^+(R,\theta_1) = U\left[1 - \left(\frac{a}{R}\right)^N \right]\cos\theta_1, \quad U_\theta^+(R,\theta_1) = -U\left[1 + \frac{1}{N-1}\left(\frac{a}{R}\right)^N \right]\sin\theta_1; \tag{9.249a,b}$$

the radial component vanishes on the hypersphere (9.250a):

$$U_R'^+(a,\theta_1) = 0, \quad U_\theta^+(a,\theta_1) = -\frac{N}{N-1}U\sin\theta_1, \tag{9.250a,b}$$

where the tangential component simplifies to (9.250b). This agrees (9.250b) for $N = 2(N = 3)$ with the case of the cylinder (I.28.94b) [sphere (6.205b)]. In the case $N = 3$ of an insulating sphere, (9.248) [(9.249a and b)] reduces to (6.308a) [(6.305a and b)] with $U = -E_0(U = E_0)$ due to the different convention on the sign of the potential $v = \nabla\Phi = -\vec{E}$.

9.8.2 Hypersphere Theorem for the Potential

The inverse point (9.251a) ≡ (6.269b) ≡ (I.24.52a) with regard (Figure 8.14b) to the hypersphere of radius a transforms the potential of a uniform flow (9.251b) into (9.251c):

$$\xi = \frac{a^2}{R}: \quad \Phi_0(R,\theta_1) \equiv UR\cos\theta_1 = U\frac{a^2}{\xi}\cos\theta_1 \equiv \Phi_0\left(\frac{a^2}{\xi},\theta_1\right); \tag{9.251a–c}$$

the latter should be the potential of a dipole, that is, the second term in (9.248) ≡ (9.252c):

$$R \equiv |\vec{x}|, \xi \equiv |\vec{y}|: \quad (N-1)\Phi_1(R,\theta_1) = a^N U R^{1-N}\cos\theta_1$$

$$= \left(\frac{a}{R}\right)^{N-2}\frac{a^2}{R}U\cos\theta = \left(\frac{R}{a}\right)^{2-N}\Phi_0\left(\frac{a^2}{R},\theta_1\right). \tag{9.252a–c}$$

The reciprocal hyperpotential coincides in the plane $N = 2$ (space $N = 3$) with the first circle (second sphere) theorem (I.24.47) [(6.283a)]. This suggests **the theorem of the reciprocal hyperpotential:**

(9.253b) is a potential that satisfies the N-dimensional Laplace equation iff the **reciprocal hyperpotential** *(9.253a) is also a harmonic function (9.253c):*

$$\bar{\Phi}(R,\theta_1,\ldots,\theta_{N-2},\varphi) \equiv \left(\frac{R}{a}\right)^{2-N} \Phi\left(\frac{a^2}{R},\theta_1,\ldots,\theta_{N-2},\varphi\right) : \nabla^2\Phi = 0 \Leftrightarrow \nabla^2\bar{\Phi} = 0. \qquad (9.253\text{a--c})$$

From this follows immediately the **hypersphere theorem**: *if Φ is the harmonic (9.254b) potential of a field in free space, then (9.254a) is the harmonic potential (9.254c) for which the corresponding field is tangent (9.254d) to the hypersphere of radius a:*

$$\Phi_-(R,\theta_1,\ldots,\theta_{N-2},\varphi) \equiv \Phi(R,\theta_1,\ldots,\theta_{N-2},\varphi) - \left(\frac{R}{a}\right)^{2-N} \Phi\left(\frac{a^2}{R},\theta_1,\ldots,\theta_N,\varphi\right), \qquad (9.254\text{a})$$

$$\nabla^2\Phi = 0 = \nabla^2\Phi_-, \quad \lim_{R\to a}\frac{\partial\Phi_-}{\partial R} = 0. \qquad (9.254\text{b--d})$$

If the scaling near the origin (9.255a) corresponds to a finite nonzero field, then the reciprocal potential (9.255b) corresponds to a dipole at infinity:

$$\lim_{R\to 0} R^{-1}\Phi(R) \neq 0, \quad \lim_{R\to\infty} R^{N-1}\bar{\Phi}(R) = \lim_{R\to\infty} R\Phi\left(\frac{a^2}{R}\right) \neq 0; \qquad (9.255\text{a,b})$$

the latter scaling (9.255b) $\equiv$ (9.256b) implies a decay (9.256a) of the field (9.256c) leading to a zero flux (9.256d) across a hypersphere of infinite radius:

$$R \to \infty: \quad \Phi(R) \sim R^{1-N}, \quad \vec{U}(R) \sim \nabla\Phi \sim \vec{e}_R R^{-N}, \quad U(R)S_N \sim R^{-1}. \qquad (9.256\text{a--d})$$

This agrees with and generalizes the particular case (9.251a through c; 9.252a through c) used to infer the reciprocal hyperpotential. In space $N = 3$ then, (9.255a and b) reduces to (6.287a and b). All of the preceding results follow from the proof that is next made that the reciprocal hyperpotential is a harmonic function.

9.8.3 Reciprocal Hyperpotential as a Harmonic Function

The proof that the reciprocal hyperpotential (9.253a) satisfies the N-dimensional Laplace equation (9.253c) is an extension of the second sphere theorem (6.284a through c) and is made similarly in five steps: (1) The radial derivative of the reciprocal hyperpotential (9.253a) involves:

$$\frac{\partial\bar{\Phi}}{\partial R} - \frac{2-N}{a}\left(\frac{R}{a}\right)^{1-N}\Phi = \left(\frac{R}{a}\right)^{2-N}\frac{\partial}{\partial R}\left[\Phi\left(\frac{a^2}{R}\right)\right]$$

$$= \left(\frac{R}{a}\right)^{2-N}\frac{d\xi}{dR}\frac{\partial\Phi}{\partial s} = -\frac{a^2}{R^2}\left(\frac{R}{a}\right)^{2-N}\frac{\partial\Phi}{\partial\xi} = -\left(\frac{R}{a}\right)^{-N}\frac{\partial\Phi}{\partial\xi}, \qquad (9.257\text{a})$$

where the variables $(\theta_1, ..., \theta_{N-2}, \varphi)$ in the N-dimensional Laplacian that are not differentiated have been omitted. (2) The radial part of the N-dimensional Laplacian (9.35b) involves:

$$\frac{\partial}{\partial R}\left(R^{N-1}\frac{\partial\bar{\Phi}}{\partial R}\right) = \frac{\partial}{\partial R}\left\{R^{N-1}\frac{\partial}{\partial R}\left[\left(\frac{R}{a}\right)^{2-N}\Phi\left(\frac{a^2}{R}\right)\right]\right\}$$

$$= \frac{\partial}{\partial R}\left\{R^{N-1}\left[\frac{2-N}{a}\left(\frac{R}{a}\right)^{1-N}\Phi - \left(\frac{R}{a}\right)^{-N}\frac{\partial\Phi}{\partial\xi}\right]\right\}$$

$$= \frac{\partial}{\partial R}\left[(2-N)a^{N-2}\Phi - \frac{a^N}{R}\frac{\partial\Phi}{\partial\xi}\right]$$

$$= \frac{\partial\Phi}{\partial\xi}\left[(2-N)a^{N-2}\frac{d\xi}{dR} + \frac{a^N}{R^2}\right] - \frac{a^N}{R}\frac{\partial^2\Phi}{\partial\xi^2}\frac{d\xi}{dR}$$

$$= \frac{\partial\Phi}{\partial\xi}\left[\frac{a^N}{R^2} - (2-N)a^{N-2}\left(\frac{a}{R}\right)^2\right] + \frac{a^N}{R}\left(\frac{a}{R}\right)^2\frac{\partial^2\Phi}{\partial\xi^2}$$

$$= a^N\left(\frac{N-1}{R^2}\frac{\partial\Phi}{\partial\xi} + \frac{a^2}{R^3}\frac{\partial^2\Phi}{\partial\xi^2}\right)$$

$$= a^{N-4}\left[(N-1)\xi^2\frac{\partial\Phi}{\partial\xi} + \xi^3\frac{\partial^2\Phi}{\partial\xi^2}\right]$$

$$= a^{N-4}\xi^3\left[\xi^{1-N}\frac{\partial}{\partial\xi}\left(\xi^{N-1}\frac{\partial\Phi}{\partial\xi}\right)\right]. \tag{9.257b}$$

(3) It follows that if $\bar{\Phi}$ satisfies the radial part (9.35b) of the N-dimensional Laplacian (9.257b) if Φ does, and vice versa:

$$R^{1-N}\frac{\partial}{\partial R}\left(R^{N-1}\frac{\partial\bar{\Phi}}{\partial R}\right) = \left(\frac{a}{R}\right)^{N-1}\left(\frac{\xi}{a}\right)^3\left[\xi^{1-N}\frac{\partial}{\partial\xi}\left(\xi^{N-1}\frac{\partial\Phi}{\partial\xi}\right)\right]; \tag{9.258}$$

the nonradial part of the Laplacian is the same for Φ and $\bar{\Phi}$; (4) The Laplacian in hyperspherical coordinates adds to (9.258) a term $O(R^{-2})$ involving derivatives with regard to other variables (9.205a through c):

$$\nabla^2\bar{\Phi} = R^{1-N}\frac{\partial}{\partial R}\left(R^{N-1}\frac{\partial\bar{\Phi}}{\partial R}\right) + \frac{1}{R^2}L\left(\frac{\partial}{\partial\theta_1}, ..., \frac{\partial}{\partial\theta_{N-2}}, \frac{\partial}{\partial\varphi}\right)\bar{\Phi}; \tag{9.259}$$

(5) The operator L does not involve derivatives with regard to R, and thus applies only to the second factor in (9.253a):

$$\nabla^2\bar{\Phi} = \left(\frac{R}{a}\right)^{1-N}\left(\frac{\xi}{a}\right)^3\xi^{1-N}\frac{\partial}{\partial\xi}\left(\xi^{N-1}\frac{\partial\Phi}{\partial\xi}\right) + \frac{1}{R^2}\left(\frac{R}{a}\right)^{2-N}L\Phi$$

$$= \left(\frac{R}{a}\right)^{1-N}\left(\frac{\xi}{a}\right)^3\left[\xi^{1-N}\frac{\partial}{\partial\xi}\left(\xi^{N-1}\frac{\partial\Phi}{\partial\xi}\right) + \frac{1}{aR}\left(\frac{a}{\xi}\right)^3 L\Phi\right]$$

$$= \left(\frac{R}{a}\right)^{-2-N}\left[\xi^{1-N}\frac{\partial}{\partial\xi}\left(\xi^{N-1}\frac{\partial\Phi}{\partial\xi}\right) + \frac{1}{\xi^2}L\Phi\right] = \left(\frac{R}{a}\right)^{-2-N}\nabla^2\Phi; \tag{9.260}$$

(6) Thus, if the original hyperpotential is a harmonic function (9.253b), the reciprocal hyperpotential (9.253a) is also (9.261) a harmonic (9.253c) function as follows from (9.260) ≡ (9.261):

$$\nabla^2\bar{\Phi} = \left(\frac{R}{a}\right)^{-2-N}\nabla^2\Phi = \left(\frac{\xi}{a}\right)^{N+2}\nabla^2\Phi. \tag{9.261}$$

Thus, *the Laplacian of the reciprocal potential (9.253a) in hyperspherical coordinates (9.259) is related by (9.261) to the Laplacian of the original potential in reciprocal coordinates (9.251a)*. In the three-dimensional $N = 3$ case, (9.261) simplifies to (6.285).

9.8.4 Equipotential Hypersphere in a Uniform Field

The introduction of a hypersphere of radius a in a uniform field is represented by the potential (9.244) corresponding to adding a dipole whose moment is given (1) by (9.245b) in the case (9.245a) of a zero normal component of the field; and (2) in the case of a zero tangential of the field, the hypersphere is an equipotential (9.262a):

$$0 = \Phi_-(a,\theta_1) = \cos\theta_1\left[U_0 a - \frac{P_1^+}{\sigma_N}a^{1-N}\right], \quad P_1^- = \sigma_N a^N U_0, \tag{9.262a,b}$$

leading to the dipole moment (9.262b). Substituting (9.262b) in (9.244) specifies *the potential (9.263) [field (9.264a and b)] for a hypersphere of radius a that is the equipotential $\Phi_+ = 0$ in a uniform external field:*

$$\Phi_-(R,\theta_1) = U_0 R\cos\theta_1\left[1 - \left(\frac{a}{R}\right)^N\right], \tag{9.263}$$

$$U_R^-(R,\theta_1) = U_0\cos\theta_1\left[1 + (N-1)\left(\frac{a}{R}\right)^N\right], \quad U_\theta^-(R,\theta_1) = -U_0\sin\theta_1\left[1 - \left(\frac{a}{R}\right)^N\right]. \tag{9.264a,b}$$

The field is orthogonal to the hypersphere (9.265a):

$$U_\theta^-(a,\theta_1) = 0, \quad U_R^-(a,\theta) = NU_0\cos\theta_1 \equiv \sigma(\theta_1), \tag{9.265a,b}$$

corresponding to a surface source distribution (9.265b). The potential of a uniform stream (9.250a) ≡ (9.266a) leads by the hypersphere theorem (9.254a) to the potential for a uniform field with the hypersphere as the equipotential surface (9.266b):

$$\Phi_0(R,\theta) = UR\cos\theta_1 : \quad \Phi_-(R,\theta) = UR\cos\theta_1 - \left(\frac{R}{a}\right)^{2-N}U\frac{a^2}{R}\cos\theta_1$$

$$= UR\cos\theta_1\left[1 - \left(\frac{a}{R}\right)^N\right], \tag{9.266a,b}$$

in agreement with (9.266b) ≡ (9.263). In the three-dimensional case, (9.263; 9.264a and b) reduce, respectively, to (6.309b; 6.311a and b) with $U = -E_0(U = E_0)$ due to the different convention and the sign of the potential $\vec{U} = \nabla\Phi = -\vec{E}$.

9.9 Images on Hyperplanes and Hyperspheres

The method of images extends from: (1) lines (planes) in two (three) dimensions [Section I.16.1 (Section 8.8)] to hyperplanes in any dimension (Subsections 9.9.1 and 9.9.2) and (2) cylinders (spheres) [Sections I.24.8, I.26.8, and I.28.8 (Section 6.9)] to hyperspheres (Subsections 9.9.3 and 9.9.4). The image on a hyperplane (hypersphere) may be used in two ways: (1) replacing the obstacle to determine its effect on the potential and the field [Subsection 9.9.1 (Subsection 9.9.3)], including the asymptotic forms at large distances and (2) to specify the force on the obstacle as the force exerted by the original on its image [Subsection 9.9.2 (Subsection 9.9.4)].

9.9.1 Equal or Opposite Images on a Hyperplane

Consider a monopole P_0 at a distance a from the hyperplane $x_1 = 0$ in an N-dimensional space, and add to the potential (9.267a) its equal Φ_+ (opposite Φ_-) image:

$$\Phi_\pm(\vec{x}) = \frac{P_0}{\sigma_N}(R_-^{2-N} \pm R_+^{2-N}), \quad R_\pm = \left|(x_1 \pm a)^2 + \sum_{m=2}^{N}(x_m)^2\right|^{1/2}, \quad (9.267\text{a–c})$$

where $R_-(R_+)$ is the distance of the observer from the monopole (9.267b) [its image (9.267c)]. The field corresponding to the potentials (9.267a),

$$U_1^\pm = \frac{\partial \Phi_\pm}{\partial x_1} = \frac{P_0}{\sigma_N}(2-N)\left[R_-^{-N}(x_1 - a) \pm R_+^{-N}(x_1 + a)\right], \quad (9.268\text{a})$$

$$m = 2,\ldots,N: \quad U_m^\pm = \frac{\partial \Phi_\pm}{\partial x_m} = \frac{P_0}{\sigma_N}(2-N)x_m(R_-^{-N} \pm R_+^{-N}), \quad (9.268\text{b,c})$$

has normal (9.268a) [tangential (9.268b and c)] components. The upper sign corresponds to an insulating hyperplane (9.269a and d) where the normal component of the field vanishes (9.269b) but not the tangential component (9.269c):

$$x_1 = 0: \quad U_1^+ = 0, \quad U_m^+ = \frac{2P_0}{\sigma_N}(2-N)x_m R_0^{-N}, \quad R_0 = \left|a^2 + \sum_{m=2}^{N}(x_m)^2\right|^{1/2}. \quad (9.269\text{a–d})$$

The lower sign corresponds to a conducting hyperplane (9.270a) where the tangential components of the field vanish (9.270b), but not the normal component (9.270c):

$$x_1 = 0: \quad U_m^- = 0, \quad U_1^- = -2a\frac{P_0}{\sigma_N}(N-2)R_0^{-N}. \quad (9.270\text{a–c})$$

The distance of the monopole and its image from a point on the hyperplane is the same for both (9.269d). Thus, *a monopole of moment P_0 at a distance a from a hyperplane has potential (9.267a) and field (9.268a through c), where (1) the upper (lower) sign corresponds to a field tangent (9.269a through c) [normal (9.270a through c)] to the hyperplane and (2) the distance(s) of a point on the hyperplane (an arbitrary point not necessarily on the hyperplane) from the dipole and its image is (9.269d) [are (9.267b and c)].*

9.9.2 Far Field of a Monopole near a Hyperplane

The distance of the observer from the monopole R_- (image R_+) is given (9.267c) by

$$R_\pm = \left| \sum_{n=1}^{N} (x_n)^2 \pm 2x_1 a + a^2 \right|^{1/2} = |R^2 \pm 2aR\cos\theta_1 + a^2|^{1/2}; \tag{9.271}$$

in the far field (9.272a), that is, for an observer much farther from the origin than the monopole from the hyperplane, (9.271) simplifies to (9.272b):

$$R^2 \gg a^2 : \quad R_\pm = R\left[1 \pm \frac{a}{R}\cos\theta_1\right], \quad R_\pm^{2-N} = R^{2-N}\left[1 \mp (N-2)\frac{a}{R}\cos\theta_1\right]. \tag{9.272a–c}$$

The leading terms in the distance from the observer (9.272c), substituted in (9.267a), specify a far field potential:

$$\Phi_+(\vec{x}) \sim 2\frac{P_0}{\sigma_N}R^{2-N}, \quad \Phi_-(\vec{x}) \sim 2\frac{P_0}{\sigma_N}(N-2)aR^{1-N}\cos\theta_1, \tag{9.273a,b}$$

that is, (1) the double for equal images (9.273a) and (2) for opposite images (9.273b) $\equiv$ (9.274a and b):

$$\Phi_-(\vec{x}) \sim \left(\frac{P_1}{\sigma_N}\right)R^{1-N}\cos\theta_1, \quad P_1 \equiv 2P_0 a(N-2), \tag{9.274a,b}$$

a dipole (9.274a) of moment (9.274b) with the same sign as P_0, because the original P_0 (image $-P_0$) is at a $(-a)$. The force between the monopole and its equal (opposite) image is:

$$\vec{F} = \pm\vec{e}_1 (P_0)^2 \frac{N-2}{\sigma_N}(2a)^{-N}, \tag{9.275}$$

and is a repulsive (attractive) force on the insulating (conducting) hyperplane. *A monopole of moment P_0 at a distance a from an insulating (conducting) hyperplane where the field is tangential (normal) is represented by an identical (opposite) image, implying that (1) the force on the hyperplane, that equals the force on the image (9.275), is repulsive (attractive) and (2) the equal (opposite) image causes a far field with double strength (9.273a) [that decays faster (9.273b) as a dipole (9.274a) of moment (9.274b)].*

9.9.3 Reciprocal Point and Image on Hypersphere

A monopole of moment P_0 at $x_1 = b$ creates a potential (9.276a and b):

$$\Phi_0(\vec{x}) = \frac{P_0}{\sigma_N}\left|(x_1 - b)^2 + \sum_{m=2}^{N}(x_m)^2\right|^{1-N/2} = \frac{P}{\sigma_N}\left|\sum_{m=2}^{N}(x_m)^2 - 2x_1 b + b^2\right|^{1-N/2}$$

$$= \frac{P_0}{\sigma_N}\left|R^2 + b^2 - 2bR\cos\theta_1\right|^{1-N/2} = \Phi_0(R,\theta_1), \tag{9.276a,b}$$

using hyperspherical coordinates (9.205a and b), with the axis joining the monopole to the origin:

$$x_1 = R\cos\theta_1, \quad \sum_{m=2}^{N}(x_m)^2 = R^2 - (x_1)^2 = R^2\sin^2\theta_1. \tag{9.277a,b}$$

A conducting hypersphere of radius a is an equipotential and corresponds to subtracting from (9.276b) its reciprocal potential (9.253a), leading to

$$\Phi_-(R,\theta_1) = \Phi_0(R,\theta_1) - \left(\frac{R}{a}\right)^{2-N}\Phi_0\left(\frac{a^2}{R},\theta_1\right)$$

$$= \frac{P_0}{\sigma_N}\left[R_+^{2-N} - \left(\frac{R}{a}\right)^{2-N}R_-^{2-N}\right], \tag{9.278}$$

where the distance from the monopole (its image on the hypersphere) is (9.279a) [(9.279b)]:

$$R_+ \equiv \left|R^2 + b^2 - 2bR\cos\theta_1\right|^{1/2}, \quad R_- \equiv \left|\frac{a^4}{R^2} + b^2 - 2\frac{a^2 b}{R}\cos\theta_1\right|^{1/2}. \tag{9.279a,b}$$

The distances (9.279a and b) coincide (9.280b) on the hypersphere (9.280a):

$$R = a: \quad R_\pm = \left|a^2 + b^2 - 2ab\cos\theta_1\right|^{1/2} \equiv R_0, \quad \Phi_-(a,\theta_1) = 0, \tag{9.280a–c}$$

implying that it is an equipotential (9.280c) in the case of opposite images.
   From (9.279a and b) follow

$$\left\{\frac{\partial}{\partial R},\frac{1}{R}\frac{\partial}{\partial\theta_1}\right\}R_+ = \frac{1}{R_+}\{R - b\cos\theta_1, b\sin\theta_1\}, \tag{9.281a,b}$$

$$\left\{\frac{\partial}{\partial R},\frac{1}{R}\frac{\partial}{\partial\theta_1}\right\}R_- = \frac{1}{R_-}\left(\frac{a}{R}\right)^2\left\{b\cos\theta_1 - \frac{a^2}{R}, b\sin\theta_1\right\}; \tag{9.282a,b}$$

using (9.281a and b; 9.282a and b) in (9.278) specifies the radial (9.283a) [tangential (9.283b)] components of the field associated with the potential (9.278):

$$U_R^-(R,\theta) = \frac{\partial\Phi_-}{\partial R} = \frac{P_0}{\sigma_N}(2-N)\left\{R_+^{-N}R + \left(\frac{a}{R}\right)^N R_-^{-N}\frac{a^2}{R}\right.$$

$$\left. - b\cos\theta_1\left[R_+^{-N} + R_-^{-N}\left(\frac{a}{R}\right)^N\right] - \frac{1}{a}\left(\frac{a}{R}\right)^{N-1}R_-^{2-N}\right\}, \tag{9.283a}$$

$$U_\theta^-(R,\theta) = \frac{1}{R}\frac{\partial\Phi_-}{\partial\theta} = \frac{P_0}{\sigma_N}(2-N)\left[R_+^{-N} - \left(\frac{a}{R}\right)^N R_-^{-N}\right]b\sin\theta_1. \tag{9.283b}$$

Thus, *a monopole of moment P_0 at a distance b from the center of a sphere of radius a that is an equipotential surface has potential (9.278) and field (2.283a and b) involving the distance of the observer from the monopole (9.279a) [its image (9.284b) at the reciprocal point (9.284a)]:*

$$c = \frac{a^2}{b}: \quad \bar{R} = |R^2 + c^2 - 2cR\cos\theta|^{1/2} = \left|R^2 + \frac{a^4}{b^2} - 2\frac{a^2 R}{b}\cos\theta\right|^{1/2} = \frac{R}{b}R_-; \qquad (9.284a,b)$$

in (9.284c) we used (9.279b).

9.9.4 Force Exerted by a Source on a Hypersphere

The potential (9.278) implies that the hypersphere is an equipotential (9.280c) and hence the tangential component of the field vanishes (9.285a):

$$U_\theta^-(a,\theta_1) = 0, \quad U_R^+(a,\theta_1) = -(N-2)\frac{P_0}{\sigma_N}R_0^{-N}\left(2a - 2b\cos\theta_1 - \frac{R_0^2}{a}\right), \qquad (9.285a,b)$$

and the radial component (9.285b) specifies the surface source distribution. The force between the monopole and its opposite image is

$$\vec{F} = \vec{e}_1(P_0)^2 \sigma_N\left(b - \frac{a^2}{b}\right)^{1-N}, \qquad (9.286)$$

and specifies the attractive force exerted by the monopole on the conducting hypersphere. For an observer in the far field, that is, much farther than the distance of the monopole from the center of the sphere (9.287a), the distances (9.279a and b) simplify to (9.287b and c):

$$R^2 \gg b^2: \quad R_+^{2-N} = R^{2-N}\left|1 - 2\frac{b}{R}\cos\theta_1 + \frac{b^2}{R^2}\right|^{1-N/2}$$

$$= R^{2-N}\left\{1 + (N-2)\frac{b}{R}\cos\theta_1 + O\left(\frac{b^2}{R^2}\right)\right\}, \qquad (9.287a,b)$$

$$R_-^{2-N} = b^{2-N}\left|1 - 2\frac{a^2}{bR}\cos\theta_1 + \frac{a^4}{b^4 R^2}\right|^{1-N/2} = b^{2-N}\left[1 + (N-2)\frac{a^2}{bR}\cos\theta_1 + O\left(\frac{a^4}{b^2 R^2}\right)\right]. \qquad (9.287c)$$

Substituting (9.287b and c) in (9.278) leads to the potential in the far field:

$$\Phi_-(R,\theta_1) = \frac{P_0}{\sigma_N}R^{2-N}\left\{1 - \left(\frac{a}{b}\right)^{N-2} + (N-2)\frac{b}{R}\cos\theta_1\left[1 - \left(\frac{a}{b}\right)^N\right]\right\}. \qquad (9.288)$$

Thus, *a monopole of moment P_0 at a distance b from the center of a hypersphere of radius a that is an equipotential surface induces a surface distribution (9.285a and b) with reverse sign, leading to an attractive* *force (9.286). The potential (9.288) ≡ (9.289a) in the far field (9.287a) consists of (1) a leading monopole term with moment (9.289b) because the original monopole is stronger than the distribution it induces on the hypersphere:*

$$\Phi_-(R,\theta_1) = \frac{\bar{P}_0}{\sigma_N} R^{2-N} - \frac{\bar{P}_0}{\sigma_N} R^{1-N} \cos\theta_1 : \quad \bar{P}_0 = P_0\left[1 - \left(\frac{a}{b}\right)^{N-2}\right],$$

$$\bar{P}_1 = (2-N)P_0 b\left[1 - \left(\frac{a}{b}\right)^N\right]; \qquad (9.289a–c)$$

(2) the next order term is a dipole with moment (9.289c). The case of an insulating hypersphere would correspond to a tangential field; the hypersphere would be a field surface. The existence of a field surface relates to the properties of solenoidal fields and multivectors in higher dimensions (Notes 9.3 through 9.52) that are considered after reviewing: (1) the asymptotic scaling of multipoles (Note 9.1) and (2) the properties of irrotational fields (Note 9.2).

NOTE 9.1: Asymptotic Decay of Multipoles in All Dimensions

Table 9.1 indicates the decay laws in the plane ($N = 2$), space ($N = 3$), and higher dimension ($N \geq 4$) for monopoles ($n = 0$), dipoles ($n = 1$), quadrupoles ($n = 2$), and arbitrary multipoles ($n \geq 3$) as concerns (1) the scalar potential, (2) the irrotational fields, and (3) the forces between multipoles. The starting point is $n = 0$, the monopole, whose potential is specified by the influence or Green function (9.39) in two (9.42a through d) [higher (9.43a through c)] dimensions. The $n = 1$, $2^n = 2$ dipole decays faster by a factor $1/R$ and a similar extra factor applies for $n = 2$, $2^n = 4$, the quadrupole, $n = 3$, $2^n = 8$, the octupole, and so on so that the extra factor is R^{-n} for the 2^n multipole. The field has an extra factor $1/R$ relative to the potential and the general power decay law applies in all dimensions. The force exerted by a multipole of order n on a monopole decays as the field in N dimensions, R^{1-N-n}; if the multipole of order m acts on a multipole of order n, the force has another factor R^{-m} leading a decay $R^{1-N-n-m}$.

NOTE 9.2: Thirty Properties of the Harmonic Potential for an Irrotational Vector Field

Table 9.1 is an example of one of at least 30 properties of irrotational fields that have analogues for any space dimension: (1) the existence of a scalar potential; (2) the Laplace operator with hyperspherical symmetry; (3) the corresponding fundamental solution corresponding to forcing by a unit impulse and leading to the influence or Green function; (4) the Poisson integral for the field of an arbitrary source distribution; (5) the representation of boundaries by hypersurface monopole and dipole source distributions; (6–8) the boundary conditions of three kinds (Cauchy–Dirichlet, Neumann, Robin) for the unicity of solution of the Laplace and the Poisson equation in compact domains; (9) the asymptotic condition for the extension to unbounded domains; (10 and 11) the properties of harmonic functions that are solutions of Laplace equation, as concerns (10) the mean value on a hypersphere and (11) the extrema, (12) constancy, (13) nullity, and (14) unicity in an arbitrary domain; (15 and 16) the flux balance (conservation) in the (15) presence [(16) absence] of sources; (17 and 18) the outer (inner) multipolar expansion for a large (small) radius; (19) the hyperspherical (original) Legendre polynomials representing axisymmetric multipoles in three (higher) dimensions; (20) the general nonaxisymmetric multipoles represented by spatial derivatives of the influence function; (21–27) the use of (21) hyperspherical and (22) hypercylindrical coordinates to specify the potential and the fields of (23) monopoles, (24) dipoles, (25) quadrupoles, (26) octupoles, and (27) higher order multipoles; (28) the hypersphere theorem; (29 and 30) the method of images on (29) hyperplanes and (30) hyperspheres. The preliminary algebraic examples (Notes 9.1 through 9.4) precede a fuller treatment of the subject based on tensor calculus (Notes 9.5 through 9.52); the tensor calculus underlies the theory of the potential, and thus the following brief account is a fitting conclusion of the first three books of this series.

NOTE 9.3: Comparison of Irrotational and Solenoidal Vector Fields

Many of the preceding properties of irrotational vector fields have analogies and differences for solenoidal fields in two and three dimensions. For solenoidal fields, there is already a significant difference between (a) the scalar stream function [(b) vector potential] in the plane (space), which satisfy the scalar (vector) Laplace/Poisson equation without/with sources; (c) the axisymmetric stream function in space satisfies a modified Laplace equation, with a distinct operator and a Poisson nonlinear forcing in the rotational case. This suggests that the extension to more than three dimensions is less straightforward for solenoidal than for irrotational vector fields. An irrotational or curl-free vector field (9.290a) is the gradient of a scalar potential (9.290b) in any dimension:

$$\partial_i U_j = \partial_j U_i \Leftrightarrow U_i = \partial_i \Phi. \tag{9.290a,b}$$

The proof of (1) sufficiency is immediate (9.291b):

$$\Phi \varepsilon C^2(\mid R^N \mid): \quad \partial_i U_j - \partial_j U_i = \partial_{ij}\Phi - \partial_{ji}\Phi = 0, \tag{9.291a,b}$$

from the symmetry of the partial derivatives of a potential with continuous second-order derivatives (9.291a) and (2) necessity uses the curl integral theorem (Subsection 5.8.5).

A solenoidal or divergence-free vector field (9.292a) derives from a potential (9.292b) that is a skew-symmetric matrix or a **bivector** (9.292c):

$$0 = \partial_i U_i, \quad U_i = \partial_j C_{ij}, \quad C_{ij} = -C_{ji}, \quad C_{ii} = 0; \tag{9.292a–d}$$

hence, its diagonal components (9.292d) are zero. The proof of necessity as mentioned in Subsection 5.8.6, is less straightforward, and is deferred to note 9.30, as a particular case of a more general theorem on multivectors; the proof of sufficiency,

$$\partial_i U_i = \partial_{ij} C_{ij} \equiv \sum_{i,j=1}^{N} \partial_{ij} C_{ij} = \sum_{i=1}^{N} \partial_{ii} C_{ii} + \sum_{i=1}^{N}\sum_{j=1}^{i-1} (\partial_{ij} C_{ij} + \partial_{ji} C_{ji})$$

$$= \sum_{i=1}^{N}\sum_{j=1}^{i-1} (\partial_{ij} C_{ij} - \partial_{ji} C_{ji}) = 0, \tag{9.293}$$

uses the symmetry (skew symmetry) of the partial derivatives (9.290b) [bivector potential (9.292c and d)], implying that the sum of products or contraction is zero (9.293). In two dimensions (9.294a), the bivector or skew-symmetric matrix has only one distinct component, namely, the stream function (9.294b):

$$N = 2: \quad \Psi_{ij} = \begin{bmatrix} 0 & \Psi \\ -\Psi & 0 \end{bmatrix}, \quad \vec{A} = \vec{e}_z \Psi, \tag{9.294a–c}$$

which may be taken normal to the plane (9.294c) in the curl:

$$\vec{U} = \nabla \wedge \vec{A} = \nabla \wedge (\vec{e}_z \Psi) = \begin{bmatrix} \vec{e}_x & \vec{e}_y & \vec{e}_z \\ \partial_x & \partial_y & \partial_z \\ 0 & 0 & \Psi \end{bmatrix} = \vec{e}_x \partial_y \Psi - \vec{e}_y \partial_x \Psi, \tag{9.294d}$$

leading to the field (9.294c and d) ≡ (6.63d). In three dimensions (9.295a), the dual of the bivector is the vector potential (9.295b) ≡ (9.295c),

$$N = 3: \quad A_k = e_{kij} \Psi_{ij}, \quad \Psi_{ij} = e_{ijk} A_k, \tag{9.295a–c}$$

and the field is the curl of the vector potential (9.296a) ≡ (6.59b):

$$U_i = \partial_j A_{ij} = e_{ijk}\partial_j A_k \iff \vec{U} = \nabla \wedge \vec{A}. \tag{9.296a,b}$$

Whereas the scalar potential of an irrotational vector field exists in any dimension, the potential representation of a solenoidal vector field depends on the space dimension; this is demonstrated most simply for solenoidal vector fields in the plane and space (Note 9.4).

NOTE 9.4: Solenoidal Vector Fields in the Plane, Space, and Higher Dimensions

In the plane (9.297a) ≡ (9.294a), a 2×2 skew-symmetric matrix has only one independent component that may be identified with the stream or the field function (9.297b and c) ≡ (9.294b):

$$N = 2: \quad C_{11} = 0 = C_{22}, \quad C_{12} \equiv \Psi = -C_{21}; \tag{9.297a–c}$$

the two components of the vector field are specified by the field function through:

$$U_1 = \partial_j C_{1j} = \partial_1 C_{11} + \partial_2 C_{12} = \partial_2\Psi, \quad U_2 = \partial_j C_{2j} = \partial_i C_{21} = -\partial_1\Psi. \tag{9.298a,b}$$

in agreement with (9.298a and b) ≡ (9.294d) ≡ (6.63d) ≡ (I.12.14).

In space (9.299a) ≡ (9.295a), a skew-symmetric matrix has three independent components that specify a vector potential (9.299b through e) ≡ (6.295b):

$$N = 3: \quad C_{11} = C_{22} = C_{33} = 0, \quad C_{12} = A_3 = -C_{21},$$

$$C_{31} = A_2 = -C_{13}, \quad C_{23} = A_1 = -C_{32}, \tag{9.299b–c}$$

corresponding in matrix notation to

$$\begin{bmatrix} C_{11} & C_{12} & C_{13} \\ C_{21} & C_{22} & C_{23} \\ C_{31} & C_{32} & C_{33} \end{bmatrix} = \begin{bmatrix} 0 & C_{12} & C_{13} \\ -C_{12} & 0 & C_{23} \\ -C_{13} & -C_{23} & 0 \end{bmatrix} = \begin{bmatrix} 0 & A_3 & -A_2 \\ -A_3 & 0 & A_1 \\ A_2 & -A_1 & 0 \end{bmatrix} \tag{9.299d}$$

The solenoidal vector field (9.292b) ≡ (9.300a through c) is the curl of the vector potential:

$$U_1 = \partial_j C_{1j} = \partial_1 C_{11} + \partial_2 C_{12} + \partial_3 C_{13} = \partial_2 A_3 - \partial_3 A_2, \tag{9.300a}$$

$$U_2 = \partial_j C_{2j} = \partial_1 C_{21} + \partial_2 C_{22} + \partial_3 C_{23} = \partial_3 A_1 - \partial_1 A_3, \tag{9.300b}$$

$$U_2 = \partial_j C_{3j} = \partial_1 C_{31} + \partial_2 C_{32} + \partial_3 C_{33} = \partial_1 A_2 - \partial_2 A_1, \tag{9.300c}$$

or in the notation of determinants:

$$\vec{U} = \begin{bmatrix} \vec{e}_1 & \vec{e}_2 & \vec{e}_3 \\ \partial_1 & \partial_2 & \partial_3 \\ A_1 & A_2 & A_3 \end{bmatrix}$$

$$= \vec{e}_1(\partial_2 A_3 - \partial_3 A_2) + \vec{e}_2(\partial_3 A_1 - \partial_1 A_3) + \vec{e}_3(\partial_1 A_2 - \partial_2 A_1), \tag{9.301}$$

in agreement with (9.300a through c) ≡ (9.301) ≡ (9.296a) ≡ (9.286b) ≡ (6.59b).

In any dimension N, the vector field has N components (9.302a):

$$\#U_i = N, \quad \#C_{ij} = N^2, \quad *\#C_{ij} = \frac{(N^2 - N)}{2} = \frac{N(N-1)}{2}, \tag{9.302a–c}$$

and the bivector potential has (9.302b) components; since it is a skew-symmetric matrix, the N diagonal components are zero (9.292d) and the off-diagonal components differ in sign (9.292c) by pairs, and so the number of independent components of the bivector potential is (9.302c). Thus, *the bivector potential (9.292b) has a smaller/the same/a larger number of components than the solenoidal field (9.292a) in the plane/space/higher dimension:*

$$N = 2,3,4,\ldots: \quad Q_+ = \frac{*\#C_{ij}}{\#U_i} = \frac{N-1}{2} = \left\{\frac{1}{2}, 1, \frac{3}{2}, \ldots\right\}, \tag{9.302d,e}$$

where #(*#) denotes the number of components (9.302a and b) [number of independent components (9.299c)]. It follows that for a space dimension $N \geq 4$, the bivector potential is more complicated than the vector solenoidal field as concerns the number of components and thus there is less benefit in using it. The preceding examples suggest that the extension to any dimension of potential theory, including both irrotational and solenoidal fields, and the operators gradient, divergence, and curl, involves multivectors (Notes 9.5 through 9.52).

NOTE 9.5: Unicity and Invariance of Differential Operators

The extension of potential theory to higher dimensions is best done in the context of tensor calculus, to address issues of unicity and invariance that also apply in the plane and space. The concept of **invariance** with regard to reference frames may be explained as follows: (1) in order to describe a physical phenomenon or a geometrical property, a reference frame is needed; (2) however, the choice of reference frame is arbitrary and should not affect the final result; and (3) thus, the geometric and the physical equations should be frame-invariant, that is, hold for any reference frame in a given set or group. This applies to tensor equations, leading to a tensor algebra (Notes 9.6 through 9.15) that is frame-invariant; the invariance is then extended to tensor analysis (Notes 9.16 through 9.18) including differentiation and integration. This leads to the unique properties of the invariant differential operators, namely, the gradient curl and divergence that are used in the potential theory of multivectors in any dimension (Notes 9.18 through 9.52). The following account includes the bare minimum of tensor calculus to consider multivector fields and potentials in any dimension; besides extending potential theory to any dimension, it also provides further insight into plane and spatial potential theory.

For example, although curl in the plane (9.303a) has appeared many times:

$$\Omega = \partial_x A_y - \partial_y A_x, \quad \Omega_1 = \partial_x A_y - 2\,\partial_y A_x, \quad \Omega_2 = \partial_x \left(A_x + A_y\right), \tag{9.303a–c}$$

seldom if ever has appeared (9.303b) or (9.303c). Why? It will be shown in the sequel that (9.303a is invariant whereas (9.303b), (9.303c) or other combinations are not invariant for general curvilinear coordinate transformations. *The only linear combinations with constant coefficients of first-order partial derivatives of a vector in the plane that are invariant are: (1) the curl (9.303a); (2) the divergence (9.303d):*

$$\Delta \equiv \partial_x A_x + \partial_y A_y: \quad \Delta_1 = \partial_x A_x - \partial_y A_y, \quad \Delta_2 = 2\,\partial_x A_x - \partial_y A_y, \tag{9.303d–f}$$

for example (9.303e), (9.303f) and other forms distinct from (9.303a and 9.303d) are not invariant. In order to prove the preceding statement related to (9.303a through f) and address invariance in general in any space dimension it is necessary to start with coordinate transformations (Note 9.6).

NOTE 9.6: Curvilinear and Rectilinear Coordinate Transformations

A set of points in an N-dimensional space may be referenced by N coordinates (9.304a) using a **reference frame** x^i. Using another reference frame $x^{i'}$ instead, the coordinates must be related by **coordinate**

transformation that is a one-to-one or bijective function (Section I.9.1) that is assumed to have continuous first-order derivatives (9.304b):

$$i, i' = 1, \ldots, N : \quad x^{i'} = x^{i'}(x^i) \in C^1(\,|\, R^N). \tag{9.304a,b}$$

Differentiating (9.304b) it follows that **the infinitesimal displacements** *on the two reference frames are related (9.305a) by the* **direct transformation matrix** *(9.305b):*

$$dx^{i'} = X_i^{i'} dx^i, \quad X_i^{i'} \equiv \frac{dx^{i'}}{\partial x^i}. \tag{9.305a,b}$$

that is, the $N \times N$ matrix of partial derivatives.

Since the function (9.304b) is one-to-one, it has an inverse (9.306a) that is also one-to-one and continuously differentiable, leading to (9.306b and c):

$$x^i = x^i(x^{i'}) \in C^1(\,|\, R^N) : \quad dx^i = X_{i'}^i dx^{i'}, \quad X_{i'}^i \equiv \frac{\partial x^i}{\partial x^{i'}}, \tag{9.306a–c}$$

where (9.306c) is the **inverse transformation matrix** because its product by the direct transformation matrix is the identity matrix (5.278a through c):

$$X_{i'}^i X_j^{i'} = \frac{\partial x^i}{\partial x^{i'}} \frac{\partial x^{i'}}{\partial x_j} = \frac{\partial x^i}{\partial x^j} = \delta_j^i. \tag{9.307}$$

Applying determinants to (9.307) it follows that

$$1 = Det(\delta_j^i) = Det(X_{i'}^i X_j^{i'}) = Det(X_{i'}^i) Det(X_j^{i'}) = X Det(X_{i'}^i), \tag{9.308}$$

the determinants are algebraically inverse:

$$X_{i'}^i X_j^{i'} = \delta_j^i : \quad Det(X_i^{i'}) \equiv X = \frac{1}{Det(X_{i'}^i)}, \tag{9.309a,b}$$

where X is the **Jacobian** of the transformation.

The linear homogeneous system of N equations (9.305a) is invertible if the Jacobian is not zero (9.310a):

$$X \equiv Det(X_i^{i'}) \neq 0; \quad 0 \neq Det(X_{i'}^i) = \frac{1}{X} \Rightarrow X \neq \infty, \tag{9.310a–c}$$

the same condition (9.310b) applied to (9.306b) implies it is not infinite (9.310c). All the preceding properties hold in a **curvilinear geometry** defined as the set of one-to-one coordinate transformations with continuous first-order derivatives (9.304b) and a Jacobian that is neither zero (9.310a) nor infinity (9.310c):

$$\mathbb{X}_N \equiv \left\{ i, i' = 1, \ldots, N : x^{i'}(x^i) \in C^1(\,|\, R) \wedge X \equiv Det\left(\frac{\partial x^{i'}}{\partial x} \right) \neq 0, \infty \right\}. \tag{9.311}$$

A particular case is the **rectilinear geometry** using linear coordinate transformations:

$$\mathbb{A}_N \equiv \left\{ i, i' = 1, \ldots, N : x^{i'} = A_i^{i'} x^i + B^{i'} \wedge A \equiv Det(A_i^{i'}) \neq 0 \right\}. \tag{9.312}$$

Starting from a Cartesian reference frame, a curvilinear (rectilinear) geometry allows curved (only straight) coordinate axes [Figure 9.8a (Figure 9.8b)] generally oblique. After the definition of coordinate transformations (Note 9.6), the transformation of other quantities between reference frames is considered (Note 9.7).

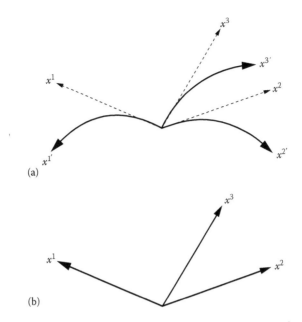

FIGURE 9.8 The Cartesian (Figure 6.2) [polar, cylindrical, and spherical (Figure 9.7)] coordinates are orthogonal and rectilinear (curvilinear). The curvilinear (rectilinear) coordinates are oblique [a (b)] if at least one pair of coordinate axes does not intersect at right angles. The rectilinear oblique (b) coordinates x^i were chosen to be locally tangent to the curvilinear oblique (a) coordinates $x^{i'}$. In a curvilinear geometry, the transformation of coordinates $x^{i'}(x^i)$ is nonlinear and has continuous first-order derivatives whose determinant, that is, the Jacobian, is neither zero nor infinity.

NOTE 9.7: Scalars and Contravariant/Covariant Vectors

The simplest quantity is a **scalar** that has the same value that is specified by the same number in every reference frame; for example, *in classical, that is, nonrelativistic, physics,* **time** *is a scalar because it is the same (9.313a) in every reference frame:*

$$t' = t: \quad \frac{dx^{i'}}{dt'} = \frac{\partial x^{i'}}{\partial x^i} \frac{dx^i}{dt'} = X_i^{i'} \frac{dx^i}{dt}. \tag{9.313}$$

It follows from the infinitesimal displacement (9.305a) that (9.313b) **velocity** *defined (9.314a) as the derivative of position with regard to time is a* **contravariant vector** (Figure 9.9a) *since it transforms (9.314b) with the direct transformation matrix (9.305b):*

$$v^i \equiv \frac{dx^i}{dt}: \quad v^{i'} = X_i^{i'} v^i. \tag{9.314a,b}$$

Another example is **potential** *that is a scalar (9.315a); its* **gradient** *or partial derivatives with regard to position (9.315b),*

$$\Phi' = \Phi: \quad \frac{\partial \Phi'}{\partial x^{i'}} = \frac{\partial \Phi}{\partial x^i} \frac{\partial x^i}{\partial x^{i'}} = X_{i'}^i \frac{\partial \Phi}{\partial x^i}, \tag{9.315a,b}$$

is a covariant vector (9.316a):

$$F_i = \partial_i \Phi \equiv \frac{\partial \Phi}{\partial x^i}: \quad F_{i'} = X_{i'}^i F_i, \tag{9.316a,b}$$

because (Figure 9.9b) it transforms (9.316b) with the inverse transformation matrix (9.306c). The transformation law for a contravariant (9.314b) [covariant (9.316b)] vector is generalized next to tensors (Note 9.8).

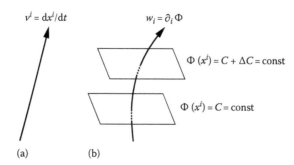

(a)   (b)

FIGURE 9.9 The velocity (a) is a contravariant vector represented by an arrow (Figure 5.12a) or an inner 1-direction; it is a contravariant vector because it is larger for smaller units of measurement; that is, a larger velocity is a bigger arrow. A covariant vector (b), such as the gradient of a scalar, is orthogonal to the equipotential hypersurfaces and specifies an outer 1-direction: it is a covariant vector because it is larger for larger units of measurement, that is, an irrotational field is stronger for closer equipotential surfaces. These true contravariant and covariant vectors apply in any space dimension, in contrast to with the "false" vectors that exist only in a three-dimensional space, such as the contravariant bivector, specifying an inner 2-direction (Figure 5.12b) whose dual is a covariant vector capacity (Figure 9.10d) and specifies an outer 1-direction.

NOTE 9.8: Transformation Law for Absolute Tensors

Consider the simplest coordinate transformation corresponding to the same change of scale along all axes (9.317a) leading to the direct (9.305b) ≡ (9.317b) [inverse (9.306c) ≡ (9.317c)] transformation matrix:

$$X^{i'} = \lambda \delta_i^{i'} x^{i'}: \quad X_i^{i'} \equiv \frac{\partial x^{i'}}{\partial x^i} = \lambda \delta_i^{i'}, \quad X_{i'}^{i} = \frac{1}{\lambda} \delta_{i'}^{i}. \tag{9.317a–c}$$

The components of a contravariant (9.314b) [covariant (9.316b)] vector are multiplied (divided) by the scale in (9.318a) [(9.318b)]:

$$v^{i'} = \lambda \delta_i^{i'} v^i, \quad F_{i'} = \frac{1}{\lambda} \delta_{i'}^{i} F_i. \tag{9.318a,b}$$

If the unit of measurement is increased, then λ is smaller and the contravariant (covariant) vector has smaller (larger) components. Hence, the name (1) **contravariant** vector for a vector like the velocity, whose magnitude varies in opposition to the size of the measuring units, for example, the same velocity $v = 1$ m/s = 100 cm/s has smaller magnitude 1 in larger units (meters) and larger magnitude 100 in smaller units (centimeters) and (2) the reverse is the case for a **covariant** vector that has larger magnitude in larger units, for example, $F = 100$ m^{-1} = 1 cm^{-1}. The **direct product** of a contravariant (9.314b) and a covariant (9.316b) vector is a **mixed matrix** (9.319a) whose transformation law (9.319b) involves a direct (9.305b) [inverse (9.306c)] transformation matrix for the contravariant (covariant) index placed above (below):

$$T_j^i = v^i F_j: \quad T_{j'}^{i'} = X_i^{i'} X_{j'}^{j} T_j^i. \tag{9.319a,b}$$

This may be generalized to the **transformation law of an absolute tensor** *with p upper (q lower) indices specifying the **contravariance (covariance)** and **order** p + q:*

$$\mathcal{X}_N: \quad T_{j_1 \ldots j_q}^{i_1 \ldots i_p} = X_{i_1}^{i_1'} \ldots X_{i_p}^{i_p'} X_{j_1'}^{j_1} \ldots X_{j_q'}^{j_q} T_{j_1 \ldots j_q}^{i_1 \ldots i_p}, \tag{9.320}$$

that involves one direct (9.305b) [inverse (9.306c)] transformation matrix for each contravariant (covariant) index. The transformation law can be considered for both physical (mathematical) quantities [Note 9.7 (Note 9.9)].

NOTE 9.9: Permutation and Identity Symbols of Order N

The permutation symbol has been defined in two (5.254a and b) [three (5.182a through c)] dimensions, corresponding to (9.321a) [(9.321b)]:

$$e_{ij} \to e_{i_1 i_2}, \quad e_{ijk} \to e_{i_1 i_2 i_3}; \tag{9.321a,b}$$

In N dimensions in covariant or contravariant form:

$$e_{i_1 \ldots i_N} = e^{i_1 \ldots i_N} = \begin{cases} 0 & \text{if there are repeated indices} \\ +1 & \text{if } (i_1, \ldots, i_N) \text{ is even permutation} \\ -1 & \text{if } (i_1, \ldots, i_N) \text{ is odd permutation.} \end{cases} \tag{9.322a–c}$$

The permutation symbol is (1) zero for repeated indices and (2/3) if all N indices take distinct values, it is +1 (−1) if $(i_1 \ldots i_N)$ is an even (odd) permutation of $(1,2,\ldots,N)$. The product of the covariant and the contravariant permutation symbols (9.322a through c) of order N is the **identity symbol** of order N,

$$\delta_{j_1 \ldots j_N}^{i_1 \ldots i_N} = e^{i_1 \ldots i_N} e_{j_1 \ldots j_N} = \begin{cases} 0 & \text{if there are repeated indices} \\ +1 & \text{equal parity} \\ -1 & \text{opposite parity,} \end{cases} \tag{9.323a–c}$$

and is (1) zero if there are repeated indices in either $(i_1 \ldots i_N)$ or $(j_1 \ldots j_N)$ and (2/3) if neither $(i_1 \ldots i_N)$ nor $(j_1 \ldots j_N)$ has repeated indices, then it is +1(−1) if they have the same (opposite) parity, that is, transform into each other by an even (odd) number of permutations. The permutation symbols can be used to define the determinant of a matrix (Note 9.10).

NOTE 9.10: Matrix, Cofactors, Determinant, and Inverse

A **matrix** or a tensor of order $p + q = 2$ can be (1) mixed (9.324b) if $p = 1 = q$ and (2/3) contravariant (9.324a) [covariant (9.324c)] if $p = 2, q = 0 (p = 0, q = 2)$:

$$A^{i'j'} = X_i^{i'} X_j^{j'} A^{ij}, \quad A_{j'}^{i'} = X_i^{i'} X_{j'}^{j} A_j^{i}, \quad A_{i'j'} = X_{i'}^{i} X_{j'}^{j} A_{ij}. \tag{9.324a–c}$$

The case of a mixed matrix (9.324b) is considered next for the definition of **determinant**:

$$Det(A_j^i) = \frac{1}{N!} e_{i_1 \ldots i_N} e^{j_1 \ldots j_N} A_{j_1}^{i_1} \ldots A_{j_N}^{i_N}, \tag{9.325}$$

that is, a sum where (1) all terms with repeated indices either $(i_1, \ldots, i_N)$ or $(j_1, \ldots, j_N)$ vanish; (2) thus, the products of N terms of the matrix have one factor from each line and column; (3) the sign is +(−) iff $(i_1, \ldots, i_N)$ and $(j_1, \ldots, j_N)$ are even (odd) permutations of each other; and (4) each set of N distinct indices appears $N!$ times, hence the division by $N!$. *The determinant (9.325) is a linear function of the matrix elements, whose coefficients form (9.326a) the **matrix of cofactors**:*

$$Det(A_j^i) = A_j^i \overset{c}{A}_i^j; \quad k \neq i: \quad A_j^i \overset{c}{A}_k^j = 0, \tag{9.326a–c}$$

*if the elements of a **row [line or (column)]** are multiplied by the cofactors of another row (9.326b), the result is zero (9.326c) because the product involves repeated indices in all terms. The two cases (9.326a) and (9.326b and c) are included in (9.327a) involving the identity matrix:*

$$A_j^i \overset{c}{A}_k^j = [Det(A_\ell^i)]\delta_k^i; \quad A_j^i \overset{I}{A}_k^j = \delta_k^i. \tag{9.327a,b}$$

The **inverse matrix**, if it exists, is such that, if multiplied by the original matrix, it leads to an identity matrix (9.327b). Comparing (9.327a and b), it follows that,

$$Det(A_\ell^k) \neq 0: \quad \overset{I}{A_j^i} = [Det(A_\ell^k)]^{-1} \overset{c}{A_j^i}, \tag{9.328a,b}$$

*a **nonsingular matrix**, that is, with a nonzero determinant (9.328a), has an inverse matrix (9.328b) specified by the cofactors divided by the determinant. The cofactors (9.326a) are obtained from the determinant (9.325) by deleting the corresponding line and column. The use of the Jacobian (its algebraic inverse) as the determinant of the direct (9.310a) [inverse (9.310b)] transformation matrix leads to the transformation law for the contravariant (covariant) permutation symbol (Note 9.11).*

NOTE 9.11: Jacobian and Tensor Densities/Capacities

The Jacobian is defined as the determinant (9.325) of the direct transformation matrix (9.310a):

$$X \equiv Det(X_i^{i'}) = \frac{1}{N!} e^{i_1 \ldots i_N} e_{j_1 \ldots j_N} X_{i_1}^{j_1} \ldots X_{i_N}^{j_N}. \tag{9.329}$$

Multiplying by the contravariant permutation symbol (9.322a through c) introduces the identity symbol (9.323a through c) with the same order:

$$X e^{i_1 \ldots i_N} = \frac{1}{N!} e^{i_1 \ldots i_N} \delta_{j_1 \ldots j_N}^{i_1 \ldots i_N} X_{i_1}^{j_1} \ldots X_{i_N}^{j_N}. \tag{9.330}$$

The identity symbol owes its name to the property:

$$\delta_{j_1 \ldots j_N}^{i_1 \ldots i_N} A_{i_1 \ldots i_N} = N! A_{j_1 \ldots j_N}. \tag{9.331}$$

Using (9.331) in (9.330) leads to:

$$e^{i_1 \ldots i_N} = \frac{1}{X} X_{i_1}^{i_1} \ldots X_{i_n}^{i_N} e^{i_1 \ldots i_N}, \tag{9.332}$$

showing that *the contravariant permutation symbol is a contravariant tensor (9.320a) of order N = p, q = 0, with the inverse of the Jacobian as an extra factor, and hence is a pseudotensor **density**.*

A similar proof applies to the determinant (9.325) of the inverse transformation matrix (9.310c):

$$\frac{1}{X} \equiv Det(X_{i'}^{i}) = \frac{1}{N!} e_{i_1 \ldots i_N} e^{j_1 \ldots j_N} X_{j_1}^{i_1} \ldots X_{j_N}^{i_N}; \tag{9.333}$$

multiplying by the covariant permutation symbol and using the properties (9.323a through c; 9.331) of the identity symbol,

$$\frac{1}{X} e'_{i_1 \ldots i_N} = \frac{1}{N!} e_{i_1 \ldots i_N} \delta_{j_1 \ldots j_N}^{i_1 \ldots i_N} X_{i_1}^{j_1} \ldots X_{i_N}^{j_N} = e_{i_1 \ldots i_N} X_{i_1}^{i_1} \ldots X_{i_N}^{i_N}, \tag{9.334}$$

shows that *the covariant permutation symbol is a covariant tensor (9.320) of order N = q, p = 0, with the Jacobian (9.310a) as an extra factor, and hence is a pseudotensor **capacity**. The contravariant (covariant) permutation symbol of order N is a contravariant (covariant) pseudotensor density (capacity) of order N* *only in a space of dimension N*, and not in any other space dimension; for example, in the plane (space), only the permutation symbol with two (5.254a and b) [three (5.182a through c)] indices is a pseudotensor of order 2(3). The reason for the designation capacity (density) arises from the volume element (Note 9.13) [other related quantities (Note 9.14)] that first requires the consideration of some tensor algebra (Note 9.12).

NOTE 9.12: Transpose, Sum, Mixing, and Alternation

The **transpose** of a tensor is obtained by interchanging indices; for example, (1) *a matrix or a tensor of order two has only one transpose (9.335a), obtained by exchanging the lines and the columns:*

$$A_{ij} \to A_{ji}: \quad A_{ijk} \to A_{jki}, A_{kij}, A_{ikj}, A_{jik}, A_{kji}; \tag{9.335a,b}$$

*(2) a tensor of order three has five transposes (9.335b); and (3) a tensor of order N has N! − 1 transposes or **isomers**. The tensor transformation law (9.325) is preserved iff the transposition applies only to indices with the same variance; for example,*

$$A_{k\ell}^{ij} \to A_{k\ell}^{ji}, A_{\ell k}^{ij}, A_{\ell k}^{ji}. \tag{9.336}$$

*The sum of two tensors with the same contravariance and covariance (p, q) is another tensor with the same contravariance and covariance (p, q). The **mixing** is the sum of the tensor and all transposes divided by their number and preserves the tensor character; for example, for covariant matrices (9.337a) [tensor of order three (9.337b)]:*

$$A_{(ij)} \equiv \frac{1}{2}(A_{ij} + A_{ji}), \quad A_{(ijk)} = \frac{1}{6}(A_{ijk} + A_{jki} + A_{kij} + A_{ikj} + A_{jik} + A_{kji}), \tag{9.337a,b}$$

$$A_{[ij]} = \frac{1}{2}(A_{ij} - A_{ji}), \quad A_{[ijk]} = \frac{1}{6}(A_{ijk} + A_{jki} + A_{kij} - A_{ikj} - A_{jik} - A_{kji}). \tag{9.338a,b}$$

*The **alternation** also sums the tensor and all transposes and divides by their number, but changes sign for odd permutations of indices: for example, for a covariant matrix (9.338a) [tensor or order three (9.338b)]. The mixing (alternation) differs only in the positive (negative) sign of odd permutations of indices and leads to **symmetric (skew-symmetric)** tensors. For example,*

$$A_{ij} = A_{(ij)} + A_{[ij]}, \quad A_{(ij)} = \frac{A_{ij} + A_{ji}}{2} = A_{(ji)}, \quad A_{[ij]} = \frac{A_{ij} + A_{ji}}{2} = A_{[ji]}, \tag{9.339a–e}$$

a covariant matrix has a unique decomposition (9.339a) into the sum of a symmetric (9.339c) [skew-symmetric (9.339e)] part specified by its mixing (9.339b) [alternation (9.339d)]. The result applies to any pair of indices of the same variance. The permutation (9.322a through c) [and identity (9.333a through c)] symbols are skew-symmetric tensors of order N ($N + N$) and, together with the alternation, relate to the volume, surface and subspace elements (Note 9.13).

NOTE 9.13: Volume, Surface, and Subspace Elements

The **infinitesimal volume element** is the product of the infinitesimal displacements along all coordinate axes:

$$i = 1, \ldots, N: \quad dV = \prod_{i=1}^{N} dx^i = dx^1 dx^2 \ldots dx^N. \tag{9.340a,b}$$

The skew-symmetric contravariant tensor of order N is defined by the alternated product of N infinitesimal displacements:

$$dV^{i_1\ldots i_N} = dx^{[i_1}\ldots dx^{i_N]},\qquad(9.341)$$

and (1) has N^N components; (2) only $N!$ components are not zero, namely, for distinct values of the indices $(i_1,\ldots i_N)$; and (3) only one component is independent, since all components equal the volume element (9.340a and b) with +(−) sign if $(i_1,\ldots i_N)$ is an even (odd) permutation of $(1, 2, \ldots, N)$. The inner product by the covariant permutation symbol (9.322a through c) specifies the **volume element** *(9.342a) that is a scalar capacity since its transformation law involves (9.342b) the product by the Jacobian:*

$$dV = e_{i_1\ldots i_N}dx^{i_1}\ldots dx^{i_N},\qquad dV' = X\,dV.\qquad(9.342a,b)$$

The N-vector or skew-symmetric tensor of (9.341) of order N and its dual (9.342a) are equivalent, but the latter is a simpler pseudoscalar representation; the volume element is a pseudoscalar because it is a number that depends on the reference frame, unless the Jacobian of the transformation is unity.

The **infinitesimal hypersurface element** *is a contravariant $(N-1)$-vector specified by the alternated product of $N-1$ infinitesimal displacements (9.343a):*

$$dS^{i_2\ldots i_N} = dx^{[i_2}\ldots dx^{i_N]},\qquad dS_{i_1} = e_{i_1\ldots i_N}dS^{i_2\ldots i_N},\qquad dS_{i'} = XX_{i'}^i\,dS_i,\qquad(9.343a\text{–}c)$$

whose dual (9.343b) is a covariant vector capacity (9.343c). The infinitesimal volume (9.341; 9.342a and b) [surface (9.342a through c)] element is the particular case $M = N(M = N - 1)$ of the **infinitesimal element of an M-dimensional sub-space** *(9.344a) specified by the alternated product of M infinitesimal displacements (9.344b):*

$$0 \le M \le N:\quad dV^{i_1\ldots i_M} = dx^{[i_1}\ldots dx^{i_M]};\quad dV_{i_{M+1}\ldots i_N} = e_{i_1\ldots i_N}dV^{i_1\ldots i_M},\qquad(9.344a\text{–}c)$$

The dual of the contravariant M-vector (9.344a; 9.345) is a covariant $(N-M)$ vector capacity (9.344c; 9.346):

$$dV^{i'_1\ldots i'_M} = X_{i_1}^{i'_1}\ldots X_{i_M}^{i'_M}dV^{i_1\ldots i_M},\qquad(9.345)$$

$$dV_{i'_{M+1}\ldots i'_N} = XX_{i'_{M+1}}^{i_{M+1}}\ldots X_{i'_N}^{i_N}dV_{i_{M+1}\ldots i_N}.\qquad(9.346)$$

The pseudotensor capacity is related to volume (9.341; 9.342a and b), surface (9.343a through c), and subspace (9.344a through c; 9.345; 9.346) elements, and the density relates to pseudotensors or relative tensors (Note 9.14).

NOTE 9.14: Weight and Pseudotensor or Relative Tensor Transformation Law

Since mass is a scalar (9.347a) in classical physics and the volume element (Figure 9.10a) a scalar capacity (9.342b), the **mass density** *(9.347b) is (Figure 9.10b) a scalar density (9.347c):*

$$dm' = dm:\quad \rho \equiv \frac{dm}{dV},\quad \rho' = \frac{1}{X}\rho.\qquad(9.347a\text{–}c)$$

Multiplying by the velocity (Figure 5.10a) that is a contravariant vector (9.314b), it leads to (Figure 9.10c) a contravariant vector density (9.348c),

$$J^i = \rho v^i = v^i\frac{dm}{dV},\quad J^{i'} = \frac{1}{X}X_i^{i'}J^i,\qquad(9.348a\text{–}c)$$

that is (1) the momentum per unit volume (9.348b), that is, mass times velocity per unit volume and (2) equivalently the **mass flux** *(9.348a) or mass that crosses the unit area normal to the velocity per*

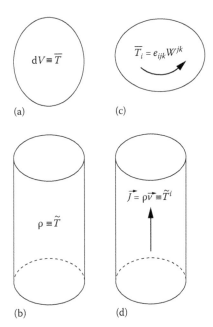

FIGURE 9.10 The volume element (a) is a pseudoscalar since it is a number that changes in a coordinate transformation by multiplication by the Jacobian and hence the volume element is a scalar capacity. Since the mass is an absolute scalar in classical or nonrelativistic physics, the mass density, that is, mass per unit volume, is a scalar density (b) that changes in coordinate transformations by division by the Jacobian. The mass flux or momentum per unit volume, which is the product of the mass density and the velocity (Figure 5.12a ≡ Figure 9.9a), is a contravariant vector density (d). Starting with a contravariant bivector in a three-dimensional space its dual is a covariant vector capacity (c), that combines the capacity (a) with the covariant vector (Figure 9.9b). The capacities (densities) are identified by an overbar (tilde), namely the scalar capacity $\overline{T}$ (density $\widetilde{T}$) in (a) [(b)], and the covariant (contravariant) vector capacity $\overline{T}_i$ (density $\widetilde{T}^i$) in (c) [(d)].

unit time. A further multiplication by the velocity (9.349a) leads to the **momentum flux** *(9.349b) that is a contravariant matrix density (9.349c):*

$$T^{ij} = J^i v^j = \rho v^i v^j, \quad T^{i'j'} = \frac{1}{X} X_i^{i'} X_j^{j'} T^{ij}. \tag{9.349a–c}$$

This suggests the generalization of the tensor transformation law (9.320) to the **pseudotensor or relative tensor transformation law**,

$$\mathbf{X}_N: \; \overset{(\vartheta)\,i_1'\ldots i_p'}{T}{}_{j_1'\ldots j_q'} = X^{\vartheta} X_{i_1}^{i_1'}\ldots X_{i_q}^{i_q'} X_{j_1'}^{j_1}\ldots X_{j_q'}^{j_q} \overset{(\vartheta)\,i_1\ldots i_p}{T}{}_{j_1\ldots j_q}, \tag{9.350}$$

that differs from the tensor transformation law (9.320) by multiplication by the Jacobian (9.310a) with the **weight** ϑ *as exponent, including the capacity (density) for* $\vartheta = 1(\vartheta = -1)$*. The capacity (density) corresponds to N alternated contravariant (9.341) ≡ (9.342a and b) (covariant) indices, and thus the weight is a shorthand for not writing the kN indices. The general pseudotensor transformation law is the basis for an invariant algebra involving the operations (1) sum, transposition, alternation, and mixing (Note 9.12) and (2) product, contraction, and transvection (Note 9.15).*

NOTE 9.15: Product, Contraction, Transposition, and Invariance

Momentum flux (9.349a) is an example of **direct product** of a contravariant vector (the velocity) and a contravariant vector density (the mass flux). The *direct product of tensors adds the* *contravariance, covariance, and weight.* Another direct product is that of a contravariant vector (the velocity) by a covariant vector (the force) in (9.351a):

$$F_i v^j = D_i^j; \quad D_i^i = F_i v^i = \vec{F} . \vec{v} = \frac{\vec{F} . d\vec{x}}{dt} = \frac{dW}{dt} \equiv A, \tag{9.351a–e}$$

since one index is contravariant and the other covariant, the sum over all values or **contraction** (9.351b) leads to a scalar (9.352):

$$A' = F_{i'} v^{i'} = X_{i'}^i F_i X_j^{i'} v^j = \delta_j^i F_i v^j = F_i v^i = A. \tag{9.352}$$

The product followed by a contraction is a **transvection**; *the simplest transvection, of a covariant vector (the force) by a contravariant vector (the velocity), leads to a scalar, namely, the* **activity** *or* **work** *per unit time.* Another example is a contravariant bivector (9.352b) whose dual in a three-dimensional space is a covariant vector capacity:

$$\bar{T}_i = e_{ijk} W^{jk}, \quad W^{jk} = \tilde{e}^{jki} \bar{T}_i; \tag{9.352a,b}$$

its geometric representation (Figure 9.10c) includes the area due to the weight and the inner two-direction of rotation as a bivector. *In (9.352a and b) and henceforth, a multivector density (capacity) is denoted by a tilda (overbar) to distinguish it from other absolute tensors or pseudotensors.*

Combining the preceding operations, it follows that *the transvection of two pseudotensors with con-travariance $p_1(p_2)$, covariance $q_1(q_2)$, and weight ϑ_1 (ϑ_2), with (9.353a) contractions is a pseudotensor with weight (9.353b), contravariance (9.353c), and covariance (9.353d):*

$$s \le p_1, p_2, q_1, q_2: \quad \vartheta = \vartheta_1 + \vartheta_2, \quad p = p_1 + p_2 - s, \quad q = q_1 + q_2 - s. \tag{9.353a–d}$$

Combining the preceding results leads to the **invariance property of tensor equations**: *a tensor equation involving only invariant algebraic operations (sum, transposition, mixing, alternation, product, contraction, and transvection) has the same form in every reference frame.* The proof is made in three steps: (1) performing all the operations, the tensor equation can be put into the form (9.354a) of a tensor equal to zero:

$$T^{i_1 \dots i_p}_{j_1 \dots j_q} = 0 \Rightarrow T^{i'_1 \dots i'_p}_{j'_1 \dots j'_q} = 0; \tag{9.354a,b}$$

(2) by the pseudotensor transformation law (9.350), if a tensor is zero in one reference frame (9.354a), it is also zero in every other reference frame (9.354b); and (3) hence, the tensor equation has the same form in every reference frame, proving invariance. QED. The invariant tensor algebra (Notes 9.6 through 9.15) is next extended (Notes 9.16 through 9.24) to tensor analysis including differentiation and integration. The combination of tensor algebra and analysis establishes the unique properties of the invariant differential operators divergence and curl in the tensor calculus that are the basis of a higher dimensional potential theory (Notes 9.25 through 9.52). Before starting with the differentiation of tensors (Notes 9.17 through 9.19), a preliminary result is obtained concerning the derivative of a determinant (Note 9.16), and in particular of the Jacobian.

NOTE 9.16: Derivative of a Determinant and Differential of the Jacobian

The determinant (9.325) is a linear function of the matrix elements (9.326a), with the matrix of cofactors as the coefficients; thus, *the derivative of the determinant with regard to the matrix is the matrix of cofactors (9.355a), that is, (9.328b) the inverse matrix multiplied by the determinant (9.355b):*

$$\frac{\partial}{\partial A_j^i}[Det(A_\ell^k)] = \overset{c}{A}_i^j = \overset{I}{A}_i^j [Det(A_\ell^k)]; \tag{9.355a,b}$$

$$\overset{I}{A_i^j} = \frac{1}{Det(A_\ell^k)} \frac{\partial}{\partial A_i^j}[Det(A_\ell^k)] = \frac{\partial}{\partial A_i^j}\{\log[Det(A_\ell^k)]\}, \tag{9.355c}$$

equivalently, the inverse matrix is (9.355c) the transpose of the derivative with regard to the matrix of the logarithm of the determinant. The preceding result can be applied to the Jacobian (9.310a), that is, the determinant of the direct transformation matrix; the latter is the inverse (9.309a) of the inverse transformation matrix (9.306c) in:

$$\frac{\partial X}{\partial X_i^{i'}} = X_{i'}^i X, \quad X_{i'}^i = \frac{\partial}{\partial X_i^{i'}}(\log X). \tag{9.356a,b}$$

From (9.356a and b) follows the derivative of *the Jacobian with regard to the first (9.357a and f) and the second (9.358a through f) coordinate system:*

$$\partial_i X \equiv \frac{\partial X}{\partial x^i} \equiv \frac{\partial}{\partial x^i}\Big[Det(X_j^{j'})\Big] = \frac{\partial}{\partial x^i}\left[Det\left(\frac{\partial x^{j'}}{\partial x^j}\right)\right]$$

$$= \frac{\partial X_j^{j'}}{\partial x^i} \frac{\partial}{\partial X_j^{j'}}\Big[Det(X_j^{j'})\Big] = X_j^{j'} X \frac{\partial X_j^{j'}}{\partial x^i} = X \frac{\partial x^j}{\partial x^{i'}} \frac{\partial^i x^{j'}}{\partial x^i x^j}, \tag{9.357a–f}$$

$$\partial_{i'} X \equiv \frac{\partial X}{\partial x^{i'}} \equiv \frac{\partial}{\partial x^{i'}}\Big[Det(X_j^{j'})\Big] = \frac{\partial x^i}{\partial x^{i'}} \frac{\partial}{\partial x^i}\Big[Det(X_j^{j'})\Big]$$

$$= X_{i'}^i X_{j'}^j X \frac{\partial X_j^{j'}}{\partial x^i} = X \frac{\partial x^i}{\partial x^{i'}} \frac{\partial x^j}{\partial x^{j'}} \frac{\partial^i x^{j'}}{\partial x^i x^j}. \tag{9.358a–f}$$

In the passage from (9.357a through f) to (9.358a through f), was used (9.359a):

$$\partial_{i'} \equiv \frac{\partial}{\partial x^{i'}} = \frac{\partial x^i}{\partial x^{i'}} \frac{\partial}{\partial x^i} = X_{i'}^i \partial_i, \quad \partial_i = \frac{\partial}{\partial x^i} = \frac{\partial x^{i'}}{\partial x^i} \frac{\partial}{\partial x^{i'}} = X_i^{i'} \partial_{i'}, \tag{9.359a,b}$$

which will appear together with its inverse (9.359b) in the differentiation of tensors (Note 9.17).

NOTE 9.17: Differentiation of a Tensor with regard to the Coordinates

A pseudotensor with weight ϑ and contravariance and covariance unity $p = 1 = q$ has (9.350) the transformation law:

$$\overset{(\vartheta)}{T_{j'}^{i'}} = X^\vartheta X_i^{i'} X_{j'}^j \overset{(\vartheta)}{T_j^i}; \tag{9.360}$$

the derivative with regard to the coordinates is given by

$$\partial_{k'} \overset{(\vartheta)}{T_{j'}^{i'}} = X^\vartheta X_{k'}^k X_i^{i'} X_{j'}^j \partial_k \overset{(\vartheta)}{T_j^i} + \vartheta X^{\vartheta-1}(\partial_{k'} X) X_i^{i'} X_{j'}^j \overset{(\vartheta)}{T_j^i}$$

$$+ X^\vartheta (\partial_{k'} X_i^{i'}) X_{j'}^j \overset{(\vartheta)}{T_j^i} + X^\alpha X_i^{i'} X_k^{k'} (\partial_k X_{j'}^{j'}) \overset{(\vartheta)}{T_j^i}, \tag{9.361}$$

where (1) (9.359a) was used on the r.h.s. of (9.361) and (2) the coordinate transformations are assumed to have continuous second-order derivatives:

$$\mathfrak{X}_N^2 \equiv \left\{ i, i' = 1, \ldots, N : x^{i'}(x^i) \in C^2(|R^N) \wedge X = Det\left(\frac{\partial x^{i'}}{\partial x^i}\right) \neq 0, \infty \right\}, \tag{9.362}$$

leading to a **curvilinear geometry with curvature** that is a continuous derivative one order higher than for the basic curvilinear geometry (9.311).

The transformation law (9.361) is that of a pseudotensor (9.350) iff only the first term on the r.h.s. is present; this will be the case for a constant transformation matrix, that is, for a rectilinear geometry (9.312), and this applies with any number of covariant and contravariant indices. It has been shown that, *in a rectilinear geometry (9.312) ≡ (9.363a) consisting of linear coordinate transformations, the partial derivative with regard to the coordinates of a pseudotensor (9.350) of weight ϑ, contravariance p, and covariance q,*

$$\mathfrak{A}_N : \quad \overset{(\vartheta)i'_1 \ldots i'_p}{T_{j'_1 \ldots j'_q}} = A^\vartheta A_{i_1}^{i'_1} \ldots A_{i_p}^{i'_p} A_{j'_1}^{j_1} \ldots A_{j'_q}^{j_q} \overset{(\vartheta)i_1 \ldots i_p}{T_{j_1 \ldots j_q}}, \tag{9.363}$$

is a pseudotensor with the same weight ϑ, contravariance p, and covariance increased by one unit q + 1:

$$\mathfrak{A}_N : \quad \partial_{k'} \overset{(\vartheta)i'_1 \ldots i'_p}{T_{j'_1 \ldots j'_q}} = A^\vartheta A_{k'}^k A_{i_1}^{i'_1} \ldots A_{i_p}^{i'_p} A_{j'_1}^{j_1} \ldots A_{j'_q}^{j_q} \partial_k \overset{(\vartheta)i_1 \ldots i_p}{T_{j_1 \ldots j_q}}, \tag{9.364}$$

because the only nonconstant factor to be differentiated (9.359a) is the tensor.

NOTE 9.18: Invariant Differentiation/Integration in a Rectilinear/Curvilinear Geometry

The preceding result may be explained as follows: *(1) In a rectilinear geometry (9.312), the transformation matrices are constant; (2) hence, the pseudotensor transformation law (9.363) is the same at all points; (3) thus, a tensor can be summed at different points; (4) in particular, the* **partial derivative**, *defined as the limit of the* **incremental ratio**,

$$\partial_k T_{j_1 \ldots j_q}^{i_1 \ldots i_p} = \lim_{\Delta x_k \to 0} \frac{1}{\Delta x_k}\left[T_{j_1 \ldots j_q}^{i_1 \ldots i_p}(x_k + \Delta x_k) - T_{j_1 \ldots j_q}^{i_1 \ldots i_p}(x_k) \right], \tag{9.365}$$

is the difference of two tensors at infinitesimally close points divided by the infinitesimal displacement, and is thus a pseudotensor with covariance increased by unity; and (5) the integral of a tensor is a sum at different points of a domain and is a tensor with the same weight, covariance, and contravariance:

$$\mathfrak{A}_N : \quad U_{j_1 \ldots j_q}^{i_1 \ldots i_p} = \int T_{j_1 \ldots j_q}^{i_1 \ldots i_p} dV, \tag{9.366}$$

and is invariant in a rectilinear geometry. The preceding results do not extend generally to a curvilinear geometry (9.311) because (1) the transformation matrices (9.305b; 9.306c) are a function of position and (2) the sum of a tensor at different points is not a tensor. Thus, unless additional structures are imposed: *(3) the partial derivative with regard to the coordinates of a curvilinear pseudotensor (9.320) is a tensor only in the exceptional cases of the* **invariant differential operators***, gradient, curl, and divergence (Notes 9.19 through 9.22) and (4) the integration is possible only for a scalar that has the same transformation law at all points, leading (Notes 9.23 through 9.30) to the curl (divergence) integral theorems of Stokes (Gauss).*

NOTE 9.19: Gradient of a Scalar and Curl/Divergence of a Vector

The invariant differential operators will be considered initially in the simplest case of scalars and vectors, leading to three cases (I–III). The first, Case I, is the gradient (9.316a) or partial derivative of a scalar (9.315a) that is (9.315b) a covariant vector (9.316b). The gradient or partial derivative of any tensor other than a scalar is not a tensor in a curvilinear geometry because the l.h.s. of (9.361) does not reduce to the first term. The second, Case II, concerns a covariant vector (9.316b) whose partial derivative (9.359a) has the transformation law:

$$\partial_{k'} F_{i'} = X_{i'}^i X_{k'}^k \partial_k F_i + F_i \partial_{k'} X_{i'}^i. \tag{9.367}$$

It follows that (1) in a rectilinear geometry (9.312), the transformation matrices are constant, the second term on the r.h.s. of (9.367) vanishes, and the first term on the r.h.s. of (9.367) specifies the transformation law of a covariant bivector, in agreement with (9.363; 9.364); (2) the transformation law of a covariant bivector also applies in (9.367) in a curvilinear geometry (9.311) iff the second-term on the r.h.s. is made to vanish; and (3) in a curvilinear geometry with curvature (9.362) $\equiv$ (9.368a), the double partial derivatives are symmetric (5.207b) $\equiv$ (9.368b) and the alternation (9.338a) is zero (9.368c):

$$x^{i'}(x^i) \in C^2(|\,R^N) : \quad \frac{\partial^2}{\partial x_i \partial x_j} = \frac{\partial^2}{\partial x_j \partial x_i} \Leftrightarrow 2\partial_{[ij]} = \partial_{ij} - \partial_{ji} = 0. \tag{9.368a–c}$$

It has been shown that *the alternated partial derivative or the **curl** (9.369c) of a continuously differentiable (9.369b) covariant vector (9.316b) in a curvilinear geometry with curvature (9.362) $\equiv$ (9.369a) is a covariant bivector (9.369d):*

$$\mathfrak{X}_N^2; \quad F_i \in C^1(R^N): \quad G_{ki} = 2\partial_{[k} F_{i]} = \partial_k F_i - \partial_i F_k, \quad G_{k'i'} = X_{k'}^k X_{i'}^i G_{ki}; \tag{9.369a–d}$$

this coincides with the curl of a vector (5.177a through c) $\equiv$ (9.369a through d). The tensor invariance would not be possible for a covariant pseudovector because the weight would add a nonvanishing term on the r.h.s. of (9.367), like the second term on the r.h.s. of (9.361).

The last, Case III, is a contravariant vector with weight ϑ corresponding (9.350) to the transformation law:

$$\overset{(\vartheta)}{A^{i'}} = X^\vartheta X_i^{i'} \overset{(\vartheta)}{A^i}, \tag{9.370}$$

whose contracted partial derivative is:

$$\partial_{i'} \overset{(\vartheta)}{A^{i'}} = X^\vartheta X_i^{i'} \partial_{i'} \overset{(\vartheta)}{A^i} + X_i^{i'} \overset{(\vartheta)}{A^i} \vartheta X^{\vartheta-1} \partial_{i'} X + X^\vartheta \overset{(\vartheta)}{A^i} \partial_{i'} X_i^{i'}; \tag{9.371}$$

using (9.359a) and (9.358e) simplifies (9.371) to (9.372):

$$\partial_{i'} \overset{(\vartheta)}{A^{i'}} = X^\vartheta \partial_i \overset{(\vartheta)}{A^i} + \overset{(\vartheta)}{A^i} X^\vartheta \left(\vartheta X_i^{i'} X_{j'}^k X_i^j \partial_k X_{j'}^{j'} + \partial_{j'} X_i^{j'} \right)$$

$$= X^\vartheta \partial_i \overset{(\vartheta)}{A^i} + \overset{(\vartheta)}{A^i} X^\vartheta \left(\vartheta \delta_i^k \partial_{j'} X_k^{j'} + \partial_{j'} X_i^{j'} \right)$$

$$= X^\vartheta \partial_i \overset{(\vartheta)}{A^i} + (\vartheta+1) X^\vartheta \overset{(\vartheta)}{A^i} \partial_{j'} X_i^{j'}, \tag{9.372}$$

The transformation law of a pseudoscalar is obtained in (9.372) iff the second term on the r.h.s. vanishes, that is, $\vartheta = -1$ for a density. It has been shown that *in a curvilinear geometry with curvature (9.362) ≡ (9.373a), the contracted partial derivative or **divergence** (9.373d) of a continuously differentiable (9.373b) contravariant vector density (9.374c) is a scalar density (9.374e):*

$$\mathcal{X}_N^2; \quad \overset{(-1)}{A^i}, \quad \tilde{B}^i \in C^1(|R^N): \quad \tilde{B}^{i'} = X^{-1}X_i^{i'}\tilde{B}^{i'}, \quad \tilde{B} = \partial_i\tilde{B}^i, \quad \tilde{B}' = X^{-1}\tilde{B}, \tag{9.373a–e}$$

where tilda $\sim$ denotes pseudotensor density or weight $\vartheta = -1$. This corresponds to the divergence of a vector (9.373a through e) $\equiv$ (5.153a through c). The curl (9.369a through d) [divergence (9.373a through c)] may be generalized from vectors (Note 9.19) to multivectors (Note 9.20).

NOTE 9.20: Curl and Divergence of Multidimensional Multivectors

The gradient (9.316a) may be considered as a particular case of the curl (9.369c) with one index and hence no alternation. The general case is a continuously differentiable (9.374a) covariant tensor of order M with transformation law (9.374b):

$$A_{i_1...i_M} \in C^1(|R^N): \quad A_{i_1'...i_M'} = X_{i_1'}^{i_1}...X_{i_M'}^{i_M}A_{i_1...i_M}. \tag{9.374a,b}$$

Its partial derivative has transformation law:

$$\partial_{i_{M+1}'}A_{i_1'...i_M'} = X_{i_1'}^{i_1}...X_{i_M'}^{i_M}\partial_{i_{M+1}'}A_{i_1...i_M} + A_{i_1...i_M}\sum_{m=1}^{M}X_{i_1'}^{i_1}...X_{i_{m-1}'}^{i_{m-1}}X_{i_{m+1}'}^{i_m}...X_{i_M'}^{i_M}\partial_{i_{M+1}'}X_{i_M'}^{i_M}, \tag{9.375}$$

that is, (1) the transformation law (9.320) of a covariant tensor of order $M + 1$ for a rectilinear geometry (9.312) when the r.h.s. reduces to the first term and (2) the second term on the r.h.s. also vanishes in a curvilinear geometry with curvature (9.362) using the alternation (9.368a through c), leading to (9.376b):

$$1 \leq M \leq N-1: \quad \partial_{[i_{M+1}'}A_{i_1'...i_M']} = X_{i_1'}^{i_1}...X_{i_{M+1}'}^{i_{M+1}}\partial_{[i_{M+1}}A_{i_1...i_M]}, \tag{9.376a,b}$$

where (1) (9.359a) is used on the r.h.s. of (9.368b) and (2) the multivector has order $M + 1 \leq N$ in (9.368a); otherwise for $M + 1 > N$, there would always be repeated indices and it would vanish by skew symmetry. It has been shown that *in a curvilinear geometry with curvature (9.362) ≡ (9.377a), a continuously differentiable (9.377d) covariant M-vector (9.377c) of order (9.377b) has for an alternated partial derivative or a **curl** a covariant (M + 1)-vector (9.377e):*

$$\mathcal{X}_N^2; \quad 1 \leq M \leq N-1: \quad A_{[i_1'...i_M']} = X_{i_1'}^{i_1}...X_{i_M'}^{i_M}A_{[i_1...i_M]}, \tag{9.377a–c}$$

$$A_{i_1...i_M} \in C^1(|R^N): \quad \partial_{[i_{M+1}'}A_{i_1'...i_M']} = X_{i_1'}^{i_1}...X_{i_{M+1}'}^{i_{M+1}}\partial_{[i_{M+1}}A_{i_1...i_M]}. \tag{9.377d,e}$$

The particular case $M = 0(M = 1)$ is the gradient of a scalar (9.316a) [curl of a covariant vector (9.369a through d)]. The tensor invariance would not apply to a covariant pseudo-M-vector because the weight would add to the r.h.s. of (9.375) a nonvanishing term like the second term on the r.h.s. of (9.361).

Considering a continuously differentiable (9.378a) contravariant pseudotensor of order M with transformation law (9.378b):

$$\overset{(\vartheta)i_1...i_M}{A} \in C^1(|R^N): \quad \overset{(\vartheta)i_1'...i_M'}{A} = X^\vartheta X_{i_1}^{i_1'}...X_{i_M}^{i_M'}\overset{(\vartheta)i_1...i_M}{A}, \tag{9.378a,b}$$

its contracted partial derivatives has transformation law:

$$\partial_{i_M'} \overset{(9)^{i_1'\ldots i_M'}}{A} = X^9 X_{i_1}^{i_1'} \ldots X_{i_M}^{i_M'} X_{i_M'}^k \partial_k \overset{(9)^{i_1\ldots i_M}}{A} + 9 X^{9-1}(\partial_{i_M'} X) X_{i_1}^{i_1'} \ldots X_{i_M}^{i_M'} \overset{(9)^{i_1\ldots i_M}}{A}$$

$$+ \overset{(9)^{i_1\ldots i_M}}{A} X^9 \sum_{m=1}^{M} X_{i_1}^{i_1'} \ldots X_{i_{m-1}}^{i_{m-1}'} X_{i_{m+1}}^{i_{m+1}'} \ldots X_{i_M}^{i_M'} X_{i_m'}^k \partial_k X_{i_m}^{i_m'}, \tag{9.379}$$

where (9.359a) was used. In a curvilinear geometry with curvature (9.362), all terms on the r.h.s. of (9.379) vanish by alternation (9.368a through c) except three:

$$\partial_{i_M'} \overset{(9)}{A}{}^{[i_1'\ldots i_M']} = X^9 X_{i_1}^{i_1'} \ldots X_{i_{M-1}}^{i_{M-1}'} \delta_{i_M}^k \partial_k \overset{(9)}{A}{}^{[i_1\ldots i_M]}$$

$$+ X^9 X_{i_1}^{i_1'} \ldots X_{i_{M-1}}^{i_{M-1}'} \overset{(9)}{A}{}^{[i_1\ldots i_M]} \left(\frac{9}{X} X_{i_M}^{i_M'} \partial_{i_M'} X + X_{i_M'}^k \partial_k X_{i_M}^{i_M'} \right); \tag{9.380a}$$

the term in parentheses on the r.h.s. of (9.380a) vanishes (9.380c) for a density (9.380b), using (9.358c; 9.359a) as in (9.372):

$$9 = -1: \quad X_{i_M'}^k \partial_k X_{i_M}^{i_M'} - \frac{1}{X} X_{i_M}^{i_M'} \partial_{i_M'} X = \partial_{i_M'} X_{i_M}^{i_M'} - X_{i_M}^{i_M'} X_{i_M'}^i X_{j'}^j \partial_j X_i^{j'}$$

$$= \partial_{i'} X_{i_M}^{i'} - \delta_{i_M}^i X_{j'}^j \partial_j X_i^{j'} = \partial_{i'} X_{i_M}^{i'} - \partial_{j'} X_{i_M}^{j'} = 0. \tag{9.380b,c}$$

Substitution of (9.380b,c) reduces the r.h.s. of (9.380a) to the first term, leading to the transformation law of a contravariant $(M-1)$-vector density:

$$1 \le M \le N: \quad \partial_{i_M'} \tilde{A}^{[i_1'\ldots i_M']} = \frac{1}{X} X_{i_1}^{i_1'} \ldots X_{i_{M-1}}^{i_{M-1}'} \partial_{i_M} \tilde{A}^{[i_1\ldots i_M]}. \tag{9.381}$$

It has been shown that *in a curvilinear geometry with curvature (9.362) ≡ (9.382a), a continuously differentiable (9.382c) contravariant M-vector (9.382b) density (9.382d) has for alternated and contracted partial derivative or **divergence** a contravariant $(M-1)$-vector density (9.382e):*

$$\mathcal{X}_N^2; \quad 1 \le M \le N: \quad \tilde{A}^{[i_1\ldots i_M]} = \frac{1}{X} X_{i_1}^{i_1'} \ldots X_{i_M}^{i_M'} \tilde{A}^{[i_1'\ldots i_M']}, \tag{9.382a-c}$$

$$\tilde{A}^{i_1\ldots i_M} \in C^1(|R^N): \quad \partial_{i_M'} \tilde{A}^{[i_1'\ldots i_M']} = \frac{1}{X} X_{i_1}^{i_1'} \ldots X_{i_{M-1}}^{i_{M-1}'} \partial_{i_M} \tilde{A}^{[i_1\ldots i_M]}. \tag{9.382d,e}$$

The curl (9.377a through e) and the divergence (9.382a through e) of multivectors are considered next together with their duals (Note 9.21) in the plane and space (Note 9.22).

NOTE 9.21: Dual Multivector and Invariant Differential Operators

The preceding results may be summarized as the **theorem on invariant differential operators**: *in a curvilinear geometry with curvature (9.362), there are only four cases (I–IV) where a linear combination with constant coefficients of the partial derivatives of a pseudotensor forms a pseudotensor, namely, the three invariant differential operators (cases I–III) or their product (Case IV). The first case is the partial derivative or **gradient** (9.316a) of a scalar (9.315a), that is, a covariant vector (9.316b), for which the dual is not* *needed [the gradient of a tensor is not a tensor in (9.361)]. The second (third) case is the **curl (divergence)** of a continuously differentiable covariant M-vector (9.383d) [contravariant M-vector density (9.384d)] in a curvilinear geometry with curvature (9.362) $\equiv$ (9.383a) $\equiv$ (9.384a) whose alternated (and contracted) partial derivative is a covariant (M + 1)-vector (9.383c) [contravariant (M−1)-vector density (9.384c)]:*

$$\mathfrak{X}_N^2: \quad 1 \le M \le N-1: \quad (CurlA_{i_1\ldots i_M})_{im+1} \equiv \partial_{[i_{M+1}} A_{i_1\ldots i_M]}, \tag{9.383a-c}$$

$$A_{i_1\ldots i_M} \in C^1\left(|R^N\right): \quad \tilde{A}^{i_{M+1}\ldots i_N} = e^{i_1\ldots i_N}\partial_{i_{M+1}} A_{i_1\ldots i_M}, \tag{9.383d,e}$$

$$\mathfrak{X}_N^2: \quad 1 \le M \le N-1: \quad \left(Div\tilde{A}^{i_1\ldots i_M}\right)_{im} \equiv \partial_{i_M} \tilde{A}^{[i_1\ldots i_M]}, \tag{9.384a-c}$$

$$\tilde{A}^{i_1\ldots i_M} \in C^1\left(|R^N\right): \quad A_{i_M\ldots i_N} = e_{i_1\ldots i_N}\partial_k \tilde{A}^{[i_1\ldots i_{M-1}k]}, \tag{9.384d,e}$$

whose dual is a contravariant (N − M − 1)-vector density (9.383e) [covariant (N − M)-vector (9.384e)] obtained by transvection with the contravariant (covariant) permutation symbol (9.322a through c). The tensor or multivector invariant differential operators curl (9.383a through e) and divergence (9.384a through e) simplify in the plane (space), that is, in two (three) dimensions, using also the duals (Note 9.22).

NOTE 9.22: Divergence and Curl in the Plane and Space

In the plane $N = 2$, scalars $M = 0$, vectors $M = 1$ and bivectors $M = 2$, and permutation symbols (9.322a through c) with two indices (5.254a and b) may be considered. A scalar leads in all dimensions to the gradient (9.315a and b; 9.316a and b). A continuously differentiable (9.385d) covariant vector (9.385c) in the plane (9.385a and b) has a curl that is a bivector (9.385e) whose dual is a scalar density (9.385f) $\equiv$ (5.256a,b):

$$\mathfrak{X}_2^2, \quad N=2, \quad M=1, \quad A_i \in C^1\left(|R^2\right): \quad B_{ij} = 2\partial_{[i}A_{j]}, \quad \tilde{B} = e^{ij}\partial_i A_j. \tag{9.385a-f}$$

There are no more cases of the curl because $M \le N - 1 = 1$. The divergence of a contravariant vector density is a scalar density (9.386a) and coincides with (5.153a through c):

$$\partial_i\tilde{A}^i = \tilde{B}; \quad \partial_i\tilde{A}^{ij} = \tilde{B}^i, \quad \tilde{A}^{ij} = e^{ij}\tilde{A}, \quad \tilde{B}^i = e^{ij}\partial_i\tilde{A}, \tag{9.386a-d}$$

the divergence of a contravariant bivector density is a contravariant vector density (9.386b) and the dual leads to the gradient (9.386d) of a scalar density (9.386c). There are no more cases of the divergence since $M \le N = 2$.

In a three-dimensional space $N = 3$, scalars $M = 0$, vectors $M = 1$, bivectors $M = 2$, and trivectors $M = 3$ together with the permutation symbols with three indices (5.182a through c) may be considered. The curl of a continuously differentiable (9.387d) covariant vector (9.387c) in a three-dimensional space (9.387a and b) is a covariant bivector (9.387e) whose dual is a contravariant vector density (9.387f) and coincides with (5.177a through c) $\equiv$ (5.185a and b):

$$\mathfrak{X}_3^2; \quad N=3, \quad M=1, \quad A_i \in C^1\left(|R^3\right): \quad B_{ij} = 2\partial_{[i}A_{j]}, \quad \tilde{B}^k = e^{kij}B_{ij} = e^{kij}\partial_i A_j; \tag{9.387a-f}$$

$$A_{jk} = e_{jk\ell}\tilde{A}^\ell, \quad B_{ijk} = 3!\partial_{[i}A_{jk]} = e_{jk\ell}\partial_i\tilde{A}^\ell, \quad \tilde{B} \equiv e^{ijk}B_{ijk} = e^{ijk}\partial_i A_{jk} = \partial_i\tilde{A}^i. \tag{9.388a-e}$$

The curl of a covariant bivector is a covariant trivector (9.388b) whose dual (9.388d) is a scalar density (9.388e), namely, the divergence of the dual (9.388a) contravariant vector density. There are no more cases of curl since $M \le N - 1 = 2$.

The divergence in a three-dimensional space, (9.389a and b) may be applied to a continuously differentiable (9.389d) contravariant vector (9.389c) density, leading to a scalar density (9.389e), in agreement with (5.153a through c):

$$\mathbf{X}_3^2; \quad N = 3, \quad M = 1; \quad A^i \in C^1\left(|R^3\right): \quad \tilde{B} = \partial_i \tilde{A}^i. \tag{9.389a–e}$$

A contravariant bivector (9.390a) [trivector (9.390c)] density has for divergence a contravariant vector (bivector) density:

$$\tilde{B}^i = \partial_j \tilde{A}^{ij}; \quad \tilde{A}^{ijk} = e^{ijk} A, \quad \tilde{B}^{ij} = \partial_k \tilde{A}^{ijk} = e^{ijk} \partial_k A, \quad B_k = e_{kij} \tilde{B}^{ij} = \partial_k A; \tag{9.390a–f}$$

the dual of the latter (9.390e) is the gradient of a scalar (9.390f) that is the dual of the trivector (9.390b). The use of dual multivectors allows the reduction of the divergence in the plane $N = 2$ and space $N = 3$ always to scalars $M = 0$ or vectors $M = 1$, sometimes interchanging the gradient, curl, and divergence. It follows that the double application of the tensor curl and the divergence can mix an ordinary vector curl and divergence (Note 9.23).

NOTE 9.23: Double Application of the Curl and Divergence

The symmetry (9.368b) or zero alternation (9.368c) property of the continuous second-order derivatives implies,

$$Curl\ Curl = 0: \quad N \geq M \geq 0, \quad A_{i_1 \ldots i_M} \in C^2\left(|R^N\right), \quad \partial_{[i_{M+2}} \partial_{i_{M+1}} A_{i_1 \ldots i_M]} = 0, \tag{9.391a–d}$$

$$Div\ Div = 0: \quad N \geq M \geq 2, \quad \tilde{A}^{i_1 \ldots i_M} \in C^2\left(|R^N\right), \quad \partial_{i_M} \partial_{i_{M-1}} \tilde{A}^{[i_1 \ldots i_M]} = 0, \tag{9.392a–d}$$

that *the double curl (9.391a) [divergence (9.392a)] of a twice continuously differentiable (9.391c) [(9.392c)] covariant M-vector (9.391b) [contravariant M-vector density (9.392b)] is zero (9.391d) [(9.392d)]*. The simplest case $M = 0$ of the double-curl property (9.391a through d) states that *the curl of the gradient of a twice continuously differentiable scalar is zero*:

$$\mathbf{X}_N^2: \quad \Phi \in C^2\left(|R^N\right): \quad \partial_{[i} \partial_{j]} \Phi = 0 \Leftrightarrow curl\ grad = 0, \tag{9.393a–d}$$

in agreement with (5.243a and b). The simplest case $M = 2$ of the double-divergence property (9.391a through d) states (9.394a through c):

$$\mathbf{X}_N^2; \quad \tilde{A}^{ij} \in C^2\left(|R^N\right): \quad 0 = \partial_i \partial_j \tilde{A}^{[ij]}; \tag{9.394a–c}$$

in a three-dimensional space (9.395a), the dual covariant vector (9.395c) satisfies (9.395d) stating that the divergence of the curl of a twice continuously differentiable vector (9.395b) is zero (9.395e):

$$\mathbf{X}_3^2; \quad \tilde{A}^{ij} \in C^2\left(|R^3\right): \quad \tilde{A}^{ij} = e^{ijk} A_k: \quad 0 = \partial_i (e^{ijk} \partial_j A_k) \Leftrightarrow div\ curl = 0, \tag{9.395a–e}$$

in agreement with (5.249a and d). The tensor analysis includes the differentiation (integration) of tensors [Notes 9.16 through 9.23 (Notes 9.24 through 9.29)] that demonstrate the unique properties of both the curl and the divergence in a curvilinear geometry.

NOTE 9.24: Interior and Exterior Representations of a Subspace

A regular **subspace of dimension** M *of an N-dimensional space has two equivalent representations:* (1) the **interior representation** *by the N space coordinates (9.396a) as continuously differentiable functions (9.396c) of M* **subspace coordinates** *(9.396b), with the matrix of partial derivatives of maximum possible rank (9.396d) to ensure that there are M linearly independent continuous* **tangent vectors** *(9.396e):*

$$n = 1,\ldots,N; m = 1,\ldots,M \le N; \quad x^n(u^m) \in C^1\left(|R^M\right): \quad Ra\left(\frac{\partial x^n}{\partial x^m}\right) = M, \quad T_m^n \equiv \frac{\partial x^n}{\partial x^n}, \qquad (9.396a\text{-}d)$$

$$\alpha = m+1,\ldots,N; \quad v^\alpha(x^n) \in C^1\left(|R^N\right): \quad Ra\left(\frac{\partial v^\alpha}{\partial x^n}\right) = N - M, \quad N_n^\alpha = \frac{\partial v^\alpha}{\partial x^n}, \qquad (9.397a\text{-}d)$$

(2) eliminating the M **integral coordinates** *leads to N − M* **external coordinates** *(9.397a) that are continuously differentiable functions of the space coordinates (9.397b), with the matrix of partial derivatives of maximum possible rank (9.397c) to ensure that there are N − M linearly independent continuous* **normal vectors** *(9.397d) as its* **exterior representation***, in agreement with (5.114a through d). The interior (exterior) representation of an M-dimensional sub-space of an N-dimensional space (9.344a) leads to the contravariant M-vector (9.344b) [dual covariant (N − M)-vector capacity (9.344c)]. The exterior (interior) representation of a subspace leads to the curl (divergence) theorem for a multivector [Note 9.25 (Note 9.26)] that generalizes (Note 9.27) the Stokes (Gauss) theorem in the plane and space (Note 9.28).*

NOTE 9.25: Stokes Theorem for a Covariant Multivector

Using the interior representation by tangent vectors, the integral over an M-dimensional subspace (9.344a) with a contravariant M-vector volume element (9.344b) is invariant in a curvilinear geometry (9.311) iff the integrand is a scalar; thus, there must be a transvection with a covariant M-vector (9.398a):

$$I_1 = \int\limits_{D_M \equiv \partial D_{M+1}} A_{i_1\ldots i_M}\, dV^{i_1\ldots i_M}; \quad I_2 = \int\limits_{D_M} \partial_{[i_{M+1}} A_{i_1\ldots i_M]}\, dV^{i_1\ldots i_M} = I_1, \qquad (9.398a\text{-}c)$$

if the M-dimensional subspace is closed, it is the boundary of an $(M + 1)$-dimensional subspace in its interior whose volume element is a contravariant $(M + 1)$-vector; the integrand becomes a scalar by transvection with a covariant $(M + 1)$-vector, such as (9.377a through e) the curl of the M-vector (9.398b). The two invariant integrals (9.338a and b) involve the same covariant M-vector and their equality (9.338c) leads to

$$\mathcal{X}_N^2; \quad 1 \le M \le N - 1; \quad A_{i_1\ldots i_M} \in C^1(D_{M+1}):$$

$$\int\limits_{D_M \equiv \partial D_{M+1}} A_{i_1\ldots i_M}\, dV^{i_1\ldots i_M} = \int\limits_{D_{M+1}} \partial_{i_{M+1}} A_{i_1\ldots i_M}\, dV^{i_1\ldots i_{M+1}}. \qquad (9.399a\text{-}d)$$

the **curl or Stokes theorem for multivectors**: *consider a curvilinear N-dimensional geometry with curvature (9.362)* ≡ *(9.399a) with a covariant M-vector (9.339b) continuously differentiable in a sub-space*

(9.339c) of dimension M + 1, whose boundary is a closed regular subspace of dimension M; then the integral of the covariant M-vector over the boundary (9.398a) equals (9.398c) ≡ (9.339d) the integral of its curl over the interior (9.338b).

NOTE 9.26: Gauss Theorem for a Contravariant Multivector Density

Using the dual exterior representation by normal vectors, a sub-space of dimension $N - M$ has for a volume element a covariant M-vector capacity; an integral invariant in a curvilinear geometry must have a scalar integrand, implying a transvection by a contravariant M-vector density (9.400a):

$$I_3 = \int\limits_{D_{N-M} \equiv \partial D_{N-M+1}} \tilde{A}^{i_1 \ldots i_M} d\bar{V}_{i_1 \ldots i_M}; \quad I_4 = \int\limits_{D_{N-M+1}} \partial_{i_M} \tilde{A}^{i_1 \ldots i_M} d\bar{V}_{i_1 \ldots i_{M-1}} = I_3, \qquad (9.400\text{a–c})$$

the divergence (9.382a through e) of the contravariant M-vector density is a contravariant $(M - 1)$-vector density and leads to a scalar integrand (9.400b) when transvected with a covariant $(M - 1)$-vector density, that is, the volume element of a subspace of dimension $N - (M - 1) = N - M + 1$ that could be the interior of the sub-space of dimension $N - M$ if it is closed. The invariant integrals (9.400a and b) involve the same contravariant M-vector density and their equality leads to:

$$\mathbf{X}_N^2; \quad 1 \le M \le N; \quad \tilde{A}^{i_1 \ldots i_M} \in C^1(D_{N-M+1}):$$

$$\int\limits_{D_{N-M} \equiv \partial D_{N-M+1}} \tilde{A}^{i_1 \ldots i_M} d\bar{V}_{i_1 \ldots i_M} = \int\limits_{D_{N-M+1}} \left(\partial_{i_M} \tilde{A}^{i_1 \ldots i_M} \right) d\bar{V}_{i_1 \ldots i_{M-1}}, \qquad (9.401\text{a–d})$$

the **divergence or Gauss theorem for multivector densities**: *consider an N-dimensional curvilinear geometry with curvature (9.362) ≡ (9.401a) with a contravariant M-vector density (9.400b) continuously differentiable (9.400c) in a sub-space of dimensional N − M + 1, whose boundary is a closed regular subspace of dimension N − M. Then the integral of the contravariant M-vector density over the boundary (9.400a) equals (9.400c) ≡ (9.401d) the integral of its divergence in the interior.* The Stokes (Gauss) or curl (divergence) theorems for multivectors [Note 9.25 (Note 9.26)] were inferred from invariance considerations, and a formal proof follows [Note 9.27 (Note 9.28)].

NOTE 9.27: Proof of Multivector Stokes Theorem

The two proofs are similar and that of the curl or Stokes multivector theorem is taken first. The proof uses an $(M + 1)$ coordinate f such that

$$x_{i_{M+1}} \equiv f \begin{cases} > 0 & \text{in the interior of } D_{M+1}, \\ = 0 & \text{on the boundary } \partial D_{M+1} \equiv D_M, \\ < 0 & \text{in the exterior of } D_{M+1}, \end{cases} \qquad (9.402\text{a,b,c})$$

it is positive (negative) in the interior (9.402a) [exterior (9.402c)] and zero on the boundary (9.402b) of the subspace of dimension $M + 1$. The integral on the r.h.s. of (9.399d; 9.344b) may be extended to all values of f using the unit jump:

$$\int\limits_{D_{M+1}} \partial_{[i_{M+1}} A_{i_1 \ldots i_M]} dx^{i_1} \ldots dx^{i_{M+1}} = \int\limits_{-\infty}^{+\infty} dx^{i_{M+1}} \int\limits_{D_M} dx^{i_1} \ldots dx^{i_M} H(f) \partial_{[i_{M+1}} A_{i_1 \ldots i_M]}. \qquad (9.403)$$

Performing an integration by parts on the r.h.s. of (9.403) leads to

$$\int_{-\infty}^{+\infty} dx^{i_{M+1}} \int_{D_M} dx^{i_1} \dots dx^{i_M} H(f) \delta_{[i_{M+1}} A_{i_1 \dots i_M]}$$

$$= \int_{-\infty}^{+\infty} dx^{i_{M+1}} \int_{D_M} dx^{i_1} \dots dx^{i_M} \partial_{[i_{M+1}} \left\{ A_{i_1 \dots i_M]} H(f) \right\}$$

$$- \int_{-\infty}^{+\infty} dx^{i_{M+1}} \frac{\partial f}{\partial x_{[i_{M+1}}} \delta(f) \int_{D_M} dx^{i_1} \dots dx^{i_M} A_{i_1 \dots i_M]}, \tag{9.404}$$

where (1) the first term on the r.h.s. of (9.404) is evaluated at infinity where the unit jump vanishes:

$$\int_{-\infty}^{+\infty} dx^{i_{M+1}} \int_{D_M} dx^{i_1} \dots dx^{i_M} \partial_{[i_{M+1}} \left\{ A_{i_1 \dots i_M]} H(f) \right\} = \left[\int_{D_M} dx^{i_1} \dots dx^{i_M} A_{i_1 \dots i_M} H(f) \right]_{f=\infty} = 0; \tag{9.405}$$

and (2) the second term on the r.h.s. of (9.404) has a factor minus unity taking the unit outward normal in (5.149c) instead of inward in (9.402a through c):

$$- \int_{-\infty}^{+\infty} dx_{i_{M+1}} \frac{\partial f}{\partial x_{i_{M+1}}} \delta(f) = \int_{-\infty}^{+\infty} df \, \delta(f) = 1. \tag{9.406}$$

Substituting (9.405; 9.406) in (9.404), it reduces to the l.h.s. of (9.399d), proving the theorem.

NOTE 9.28: Proof of the Multivector Gauss Theorem

The generalized Stokes or curl (9.399a through d) [Gauss or divergence (9.401a through d)] theorems apply to a covariant M-vector (contravariant M-vector density) and similar proofs [Note 9.27 (Note 9.28)] apply. Thus, the proof of the multivector divergence or the Gauss theorem is similar to that of the multivector curl or Stokes theorem (9.403 through 9.406) using the coordinate (9.402a through c) normal to the boundary, in the integral on the r.h.s. of (9.401d):

$$\int_{D_{N-M+1}} \left(\partial_{i_M} \tilde{A}^{i_1 \dots i_M} \right) d\bar{V}_{i_1 \dots i_{M-1}} = \int_{-\infty}^{+\infty} H(f) \left(\partial_{i_M} \tilde{A}^{i_1 \dots i_M} \right) d\bar{V}_{i_1 \dots i_{M-1}}, \tag{9.407}$$

where the unit jump restricts the integration from all space to the domain. An integration by parts gives

$$\int_{D_{N-M+1}} \left(\partial_{i_M} \tilde{A}^{i_1 \dots i_M} \right) d\bar{V}_{i_1 \dots i_{M-1}} = \int_{-\infty}^{+\infty} \partial_{i_M} \left[H(f) \tilde{A}^{i_1 \dots i_M} \right] d\bar{V}_{i_1 \dots i_{M-1}} - \int_{-\infty}^{+\infty} \tilde{A}^{i_1 \dots i_M} \delta(f) \frac{\partial f}{\partial x_{i_M}} d\bar{V}_{i_1 \dots i_{M-1}}, \tag{9.408}$$

where (1) the first term on the r.h.s. of (9.408) is evaluated at infinity where the unit jump vanishes:

$$\int_{-\infty}^{+\infty} \partial_{i_M} \left[H(f) \tilde{A}^{i_1 \dots i_M} \right] d\bar{V}_{i_1 \dots M-1} = \left[\int_{D_{N-M}} H(f) \tilde{A}^{i_1 \dots i_M} d\bar{V}_{i_1 \dots i_{M-1}} \right]_{f=0} = 0; \tag{9.409a}$$

(2) in the second term on the r.h.s. of (9.408), the unit impulse limits the integration to the boundary with unit outward normal (5.149c):

$$-\int_{-\infty}^{+\infty} \tilde{A}^{i_1\ldots i_M} \delta(f) \frac{\partial f}{\partial x_{i_M}} d\bar{V}_{i_1\ldots i_{M-1}} = \int_{-\infty}^{+\infty} \tilde{A}^{i_1\ldots i_M} \delta(f) N_{i_M} d\bar{V}_{i_1\ldots i_{M-1}}$$

$$= \int_{-\infty}^{+\infty} \tilde{A}^{i_1\ldots i_M} \delta(f) d\bar{V}_{i_1\ldots i_M}$$

$$= \int_{D_{N-M}} \tilde{A}^{i_1\ldots i_M} d\bar{V}_{i_1\ldots i_{M-1}}. \tag{9.409b}$$

Substitution of (9.409a and b) in (9.408) proves the generalized Gauss theorem (9.401a through d). The curl (divergence) of a covariant M-vector (9.377a through e) [contravariant M-vector density (9.382a through e)] and the corresponding multivector Stokes (9.399a through d) [Gauss (9.401a through d)] integral theorems include as particular cases the vector Gauss (5.163a through c) [Stokes (5.192a through c) $\equiv$ (5.210a and b)] theorem, as shown next (Note 9.29), in a three-dimensional space.

NOTE 9.29: Three-Dimensional Gauss and Stokes Theorems

The Stokes theorem (5.192a through c) $\equiv$ (5.210a and b) is the particular case of the generalized Stokes theorem (9.401a through d) for a covariant vector integrated over a closed regular loop that is the boundary of a regular surface:

$$\int_{D_1 \equiv \partial D_2} A_i \, dx^i = \int_{D_2} \partial_{[i} A_{j]} dx^i \, dx^j. \tag{9.410}$$

Using the axial vector dual curl (9.411a) [exterior area element (9.411b)],

$$\partial_{[i} A_{j]} = e_{ijk} (\nabla \wedge \vec{A})_k, \quad dS_k = e_{ijk} dx^i \, dx^j, \tag{9.411a,b}$$

in (9.410) leads in vector notation to

$$\int_L \vec{A}.d\vec{x} = \int_S e_{ijk} \left(\nabla \wedge \vec{A}\right)_k dx^i dx^j = \int_S \left(\nabla \wedge \vec{A}\right)_k dS_k = \left(\nabla \wedge \vec{A}\right).d\vec{S}, \tag{9.412}$$

which coincides with (9.412) $\equiv$ (5.210a and b).

The divergence theorem for a vector in N dimensions follows from the multivector Gauss theorem (9.401a through d) in the simplest case, $M = 1$:

$$\mathbf{X}_N^2: \quad \tilde{A}^i \in C^1(D_N): \quad \int_{D_{N-1} \equiv \partial D_N} \tilde{A}^i d\bar{S}_i = \int_{D_N} (\partial_i \tilde{A}^i) dV, \tag{9.413a–c}$$

in agreement with (5.163a through c). The multivector curl or Stokes (9.399a through d) [divergence or Gauss (9.401a through d)] theorems extend the original vector curl or Stokes (9.412) $\equiv$ (5.210a and b) [divergence or Gauss (9.413b through d) ≡ (5.163a through c)] theorems, and are all invariant in a curvilinear geometry with curvature (9.362) since the integrands are absolute scalars with zero weight. *The three-dimensional gradient (5.170a through c) [curl (5.180a through c)] theorems have vector integrands, and thus are invariant in a rectilinear geometry (9.312) but not in a curvilinear geometry (9.311) with curvature (9.362).* The scalar (vector) potential of an irrotational (5.240a through d) [solenoidal (5.247a through d)] vector field involves operators that are invariant in a curvilinear geometry, and thus extend to multivectors [Note 9.30 (Note 9.31)].

NOTE 9.30: Irrotational Multivector Field and Covariant Potential

A continuously differentiable (9.414b) covariant M-vector field (9.414a) is **irrotational** if its curl is zero (9.414c):

$$1 \le M \le N-1; \quad U_{i_1 \ldots i_M} \in C^1\left(|R^N\right): \quad \partial_{[i_{M+1}} U_{i_1 \ldots i_M]} = 0; \tag{9.414a–c}$$

$$S \subset \mathfrak{X}_N^2: \quad \Phi_{i_1 \ldots i_{M-1}} \in C^2\left(|R^N\right): \quad U_{i_1 \ldots i_M} = \partial_{[i_M} \Phi_{i_1 \ldots i_{M-1}]}. \tag{9.415a–c}$$

The condition (9.414c) is met by a covariant M-vector field that is the curl (9.415c) of a covariant $(M-1)$-vector potential with continuous second-order derivatives (9.415b), requiring (9.415a) a curvilinear geometry with curvature (9.362) to prove that:

$$\partial_{[i_M} U_{i_1 \ldots i_M]} = \partial_{[i_{M+1}} \partial_{i_M} \Phi_{i_1 \ldots i_{M-1}]} = 0. \tag{9.416}$$

This corresponds to the **theorem on irrotational multivector fields**: *a continuously differentiable (9.414b) M-vector (9.414a) field is irrotational (9.414c) iff it is the curl (9.415c) of a covariant (M − 1)-vector potential that has continuous second-order derivatives (9.415b) in a star-shaped region in a curvilinear geometry with curvature (9.415a) ≡ (9.362).* The proof of the sufficient condition (9.416) is an immediate consequence of the double-curl property (9.391a through d). The proof of the necessary condition is less elementary (Note 9.33) and leads to the requirement for a star-shaped region (9.415a). The preceding result (corresponding theorem) for irrotational (solenoidal) multivectors applies [Note 9.30 (Note 9.31)] to fields and potentials that are absolute covariant (contravariant densities).

NOTE 9.31: Solenoidal Multivector Field and Contravariant Density Potential

A continuously differentiable (9.417a) contravariant M-vector density (9.417b) is **solenoidal** iff its divergence (9.417c) is zero:

$$1 \le M \le N-1; \quad \tilde{U}^{i_1 \ldots i_M} \in C^1\left(|R^N\right): \quad \partial_{i_M} \tilde{U}^{i_1 \ldots i_M} = 0; \tag{9.417a–c}$$

$$S \subset \mathfrak{X}_N^2: \quad \tilde{A}^{i_1 \ldots i_{M+1}} \in C^2\left(|R^N\right): \quad \tilde{U}^{i_1 \ldots i_M} = \partial_{i_{M+1}} \tilde{A}^{i_1 \ldots i_{M+1}}, \tag{9.418a–c}$$

The condition (9.417c) is met by a contravariant M-vector density that is the divergence of a contravariant $(M+1)$-vector potential (9.418c) that must have continuous second-order derivatives (9.318b) requiring a curvilinear geometry with curvature (9.418a) to prove:

$$\partial_{i_M} \tilde{U}^{i_1 \ldots M} = \partial_{i_{M+1}} \partial_{i_M} A^{[i_1 \ldots M-1]} = 0. \tag{9.419}$$

This corresponds to the **theorem on multivector solenoidal fields**: *a continuously differentiable (9.417b) contravariant M-vector (9.417a) density is solenoidal (9.417c) iff it is the divergence of a contravariant*

(M + 1)-vector density potential (9.418c) that has continuous second-order derivatives (9.418b) in a star-shaped region (9.418a) in a differentiable geometry with curvature (9.362). The proof of the sufficient condition (9.418a through c) ⇒ (9.417a through c) is an immediate consequence (9.419) of the double-divergence property (9.392a through d); the proof of the necessary condition is less elementary and is similar for irrotational and next (Note 9.32) for solenoidal multivector fields.

NOTE 9.32: Differential Identity for Multivector Fields

A **star-shaped region** with center at $\vec{x}_0$ is defined (Figure 9.11) as the set of points lying on straight lines through $\vec{x}_0$ in all directions $\vec{a}$ for a parameter t in the unit interval (9.420a):

$$E \equiv \left\{ \vec{x} : \vec{x} = \vec{x}_0 + \vec{a}t, \vec{a} \in \left| R^n, 0 \le t \le 1 \right\}; \quad U_{i_1 \cdots i_M} \in C^1(E), \quad (9.420a,b)$$

considering an M-vector with continuous first-order derivatives (9.420b) in a star-shaped region (9.420a), the linear operator defined by (9.421a),

$$L\{U_{i_1 \cdots i_M}\} \equiv \sum_{m=1}^{M} (-)^{m-1} x_{i_m} \int_0^1 t^{m-1} U_{i_1 \cdots i_M} (x_k t) dt$$

$$= \sum_{m=1}^{M} x_{i_1} \int_0^1 t^{m-1} U_{i_1 \cdots i_M} (x_k t) dt \equiv (-)^{M-1} \Phi_{i_2 \cdots i_M} \in C^1(E), \quad (9.421a\text{–}c)$$

is (1) rewritten (9.421b), moving the summed or the contracted index i_m to the first position i_1, with $(m - 1)$ changes of sign due to skew symmetry and (2) it specifies an $(M - 1)$-vector (9.421c) with

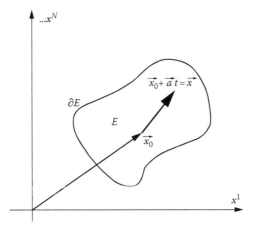

FIGURE 9.11 A star-shaped region D consists of points issued along straight lines from a "center" point; there may be several or a subregion of "center points." From each center point, the whole boundary ∂D of a star-shaped domain D is visible. The star-shaped domain is used to (1) prove the existence of a covariant $(M - 1)$-vector [contravariant $(M + 1)$-vector density] potential of a covariant M-vector (contravariant M-vector density) field that is irrotational (solenoidal), that is, has a zero curl (divergence) and (2) also leads to an explicit formula for the potential in terms of the field.

continuous first-order derivatives in the star-shaped region (9.420a). Thus, its curl exists and is calculated by

$$
\partial_{[i_{M+1}}\Phi_{i_2}\cdots_{i_M]} = \partial_{[i_{M+1}}L\{U_{i_1}\cdots_{i_M]}\}
$$

$$
= \sum_{m=1}^{M}(-)^{m-1}x_{i_m}\int_0^1 t^{m-1}\partial_{[i_{M+1}}(x_k t)\partial_k U_{[i_2}\cdots_{i_M]}\,dt
$$

$$
+ \sum_{m=1}^{M}(-)^{m-1}\partial_{[i_{M+1}}x_{i_{|m|}}\int_0^1 t^{m-1}U_{i_2}\cdots_{i_M]}(x_k t)\,dt
$$

$$
= \sum_{m=1}^{M}(-)^{m-1}x_{i_m}\int_0^1 t^{m}\partial_k U_{[i_2}\cdots_{i_M]}\,\delta_{k]i_{M+1}}\,dt
$$

$$
+ \sum_{m=1}^{M}(-)^{m-1}\delta_{i_m[i_{M+1}}\int_0^1 t^{m-1}U_{i_2}\cdots_{i_M]}\,dt
$$

$$
= \sum_{m=1}^{M}(-)^{m-1}x_{i_m}\int_0^1 t^{m}\partial_{[i_{M+1}}U_{i_2}\cdots_{i_M]}\,dt
$$

$$
+ M(-)^{M-1}\int_0^1 t^{M-1}U_{i_2}\cdots_{i_M}(tx^t)\,dt,
\tag{9.422}
$$

where (1) in both terms the identity matrix was used and (2) in the second-term on the r.h.s. of (9.422), the contracted index was moved to the first position. The r.h.s. of (9.368) involves the M-vector (its curl) in the second (first) term.

NOTE 9.33: Multivector Potential of a Multivector Field

The linear operator (9.421a) may also be applied to the curl of the M-vector, that is, an $(M + 1)$-vector:

$$
L\{\partial_{[i_{M+1}}U_{i_1}\cdots_{i_M]}\} \equiv \sum_{m=1}^{M+1}(-)^{m-1}x_{i_m}\int_0^1 t^{m-1}\partial_{[i_{M+1}}U_{i_1}\cdots_{i_M]}(x_k t)\,dt
$$

$$
= \sum_{m=1}^{M}(-)^{m-1}x_{i_m}\int_0^1 t^{m-1}\partial_{[i_{M+1}}U_{i_1}\cdots_{i_M]}(x_k t)\,dt + (-)^{M}x_{i_{M+1}}\int_0^1 t^{M}\partial_{[i_{M+1}}U_{i_1}\cdots_{i_M]}(x_k t)\,dt,
\tag{9.423}
$$

where the first $m = 1,\ldots,M$ terms (last $m = M + 1$ term) of the sum were separated. The first terms on the r.h.s. of (9.422) and (9.423) are equal and cancel by subtraction:

$$
L\{\partial_{[i_{M+1}}U_{i_1}\cdots_{i_M]}\} - \partial_{[i_{M+1}}\left[L\{U_{i_1}\cdots_{i_M]}\}\right]
$$

$$
= (-)^{M}\int_0^1\left(t^{M}x_{i_{M+1}}\partial_{[i_{M+1}}U_{i_1}\cdots_{i_M]} + Mt^{M-1}U_{i_1}\cdots_{i_N}\right)dt
$$

$$
= (-)^{M}\int_0^1 \frac{d}{dt}\left[t^{M}U_{i_1}\cdots_{i_M}\left(x_{i_{M+1}}t\right)\right]dt = (-)^{M}\left[t^{M}U_{i_1}\cdots_{i_M]}(x_x t)\right]_0^t = (-)^{M}U_{i_1}\cdots_{i_M},
\tag{9.424}
$$

where the integrand is an exact differential evaluated at the ends of the interval. Thus, has been proved the **differential multivector identity**: *an M-vector with continuous first-order derivatives in a star-shaped region (9.420a) satisfies the differential identity:*

$$
(-)^{M}U_{i_1}\cdots_{i_M} = L\{\partial_{[i_{M+1}}U_{i_1}\cdots_{i_M]}\} - \partial_{[i_{M+1}}\left[L\{U_{i_1}\cdots_{i_M}\}\right],
\tag{9.425}
$$

involving the linear differential operator (9.421a) applied to its curl (divergence) in the first (second) term on the r.h.s. of (9.425). In the preceding identity (9.425), the first (second) term on the r.h.s. is zero for an irrotational M-vector field (9.414c) ≡ (9.426a) [involves (9.426b) an $(M-1)$-vector potential (9.421c)]:

$$\partial_{[i_{M+1}} U_{i_1 \cdots i_M]} = 0 \Rightarrow U_{i_1 \cdots i_M} = \partial_{[i_1} \Phi_{i_2 \cdots i_M]}. \tag{9.426a,b}$$

This *proves the existence of a covariant $(M-1)$-vector potential (9.426a and b) of an irrotational (9.414c) M-vector (9.414b) field in a star-shaped region (9.420a), and also shows how it can be calculated (9.421a through c).* The particular constant irrotational vector field (9.427a) leads to the scalar potential (9.427b) in agreement with (5.245a through d):

$$U_i = \text{const}: \quad \Phi = U_i x_i \int_0^1 dt = U_i x_i = \vec{U}.\vec{x}; \tag{9.427a,b}$$

$$U_i = \text{const}: \quad A_i = x_j (U_{ji} - U_{ij}) \int_0^1 t\, dt = -\frac{1}{2} e_{ijk} x_j \tilde{U}_k, \quad A = \frac{1}{2}\vec{U} \wedge \vec{x}. \tag{9.428a–c}$$

The particular case $M = 2$ of (9.421a through c) for a constant bivector field (9.428a) leads to the vector potential (9.428b), corresponding to (9.428c), in agreement with (5.252a through d).

NOTE 9.34: Three-Dimensional Scalar and Vector Potentials

The simplest case $M = 1$ of (9.426a) is an irrotational covariant vector field (9.429a) that is the gradient of a scalar potential (9.429b):

$$\partial_i U_j = \partial_j U_i \Rightarrow U_i = \partial_i \Phi; \quad U_i = e_{ijk} \tilde{U}_k; \tag{9.429a–c}$$

in a three-dimensional space, the bivector (9.429a) has a dual axial vector or contravariant vector density (9.429c); this result coincides with (5.240a through d). The next case $M = 2$ is a solenoidal bivector (9.430a) that derives from a vector potential (9.430b):

$$\partial_{[i} U_{jk]} = 0 \Rightarrow U_{jk} = \partial_{[j} A_{k]}; \tag{9.430a,b}$$

in a three-dimensional space, the permutation symbol with three indices is used (1) in (9.430a), which has only one independent component and corresponds to the divergence (9.431b) of the dual vector or contravariant vector density (9.431a):

$$\tilde{U}^i = e^{ijk} U_{jk}: \quad 0 = e^{ijk} \partial_i U_{jk} = \partial_i \tilde{U}^i; \tag{9.431a,b}$$

and (2) in (9.430b), which involves the curl of the vector potential:

$$\tilde{U}^i = e^{ijk} U_{jk} = e^{ijk} \partial_j A_k = \left(\nabla \wedge \vec{A} \right)^i; \tag{9.432}$$

the implication from (9.430a) ≡ (9.431b) to (9.430b) ≡ (9.431) corresponds to the theorem (5.247a through d) of existence of a vector potential of a solenoidal field. The tensor algebra (Notes 9.6 through 9.15) and analysis (Notes 9.16 through 9.34) have been applied to multivectors in a curvilinear geometry (9.311) with curvature (9.362) and no other structure, implying the strictest or most demanding invariance conditions; for example, the notions of metric, arc length, distance, or modulus of a vector were never used before (Notes 9.6 through 9.34). They are introduced next (Notes 9.35), leading to additional properties (Notes 9.36 through 9.52).

NOTE 9.35: Changing the Contravariance, Covariance, and Weight

The transformation law of pseudotensors (9.350) in a curvilinear geometry (9.311) does not allow change of contravariance q, covariance p, or weight ϑ. A covariant vector corresponds to a contravariant vector by **lowering the index** via a transvection (9.433a) with a **covariant metric tensor**:

$$A_i = g_{ij}A^j; \quad A^i = g^{ij}A_j, \quad g_{ij}g^{jk} = \delta_i^k, \tag{9.433a–c}$$

the inverse process of **raising an index** by assigning a contravariant vector to a covariant vector uses the **contravariant metric tensor** (9.433b). Successive application of (9.433a and b) leads to (9.434a):

$$A_i = g_{ij}A^j = g_{ij}g^{jk}A_k; \quad A_i = \delta_i^k A_k, \tag{9.434b}$$

comparison with (9.434a) ≡ (9.434b) shows that *the covariant (9.433a) and the contravariant (9.433b) metric tensors are inverse matrices (9.433c)*. The transformation law (9.320) for the covariant metric tensor is (9.435a):

$$g_{i'j'} = X_{i'}^i X_{j'}^j g_{ij}, \quad Det(g_{i'j'}) = \left[Det(X_{i'}^i) \right]^2 Det(g_{ij}), \tag{9.435a,b}$$

and taking determinants leads to (9.435b). Using (9.310c), it follows that,

$$g \equiv Det(g_{ij}): \quad g' = X^{-2}g, \quad \sqrt{g'} = X^{-1}\sqrt{g}, \tag{9.436a–c}$$

the determinant of the covariant metric tensor (9.436a) is (9.436b) a pseudoscalar (9.350) with weight $\vartheta = -2$ and can be used (9.436c) to change the weight of the tensors.

NOTE 9.36: Absolute and Pseudotensors in a Metric Geometry

A metric geometry is defined as a curvilinear geometry (9.311) with an additional structure, namely, a covariant metric tensor with a nonzero determinant,

$$\mathcal{M}_N = \left\{ X_N : g_{ij} \in X_N \wedge g \equiv Det(g_{ij}) \neq 0 \right\}, \tag{9.437}$$

so that there is an inverse, namely, the contravariant metric tensor. *In a metric geometry, it is possible to change the contravariance q, covariance p, and weight ϑ of a pseudotensor (9.350), leading to an absolute tensor (9.320) or vice versa. For example, a contravariant matrix with weight ϑ is transformed to a covariant absolute matrix by:*

$$B_{ij} = g^{\vartheta/2} g_{ik} g_{j\ell} \overset{(\vartheta)}{A^{k\ell}}. \tag{9.438a}$$

For example, *the contravariant (covariant) permutation symbol (9.322a through c) is a contravariant (covariant) M-vector density (9.332) [capacity (9.334)] and leads to an* **absolute contravariant (covariant) permutation tensor** *(9.438b) [(9.438c)]:*

$$\varepsilon^{i_1 \ldots i_N} = \frac{1}{\sqrt{g}} \tilde{e}^{i_1 \ldots i_N}, \quad \varepsilon_{i_1 \ldots i_N} = \sqrt{g}\, \tilde{e}_{i_1 \ldots i_N}, \tag{9.438b,c}$$

that is an absolute contravariant (covariant) M-vector dividing (multiplying) by the square root of the determinant of the covariant metric tensor (9.436a and b), which is a scalar density (9.436c).

NOTE 9.37: Inner Product, Modulus, and Angle of Vectors

The transformation law for a contravariant (9.314b) and a covariant (9.316b) vector shows that

$$\boldsymbol{\mathfrak{X}}_N: \quad A' \equiv F_{i'}v^{i'} = X_{i'}^i F_i X_j^{i'} v^j = \delta_j^i F_i v^j = F_i v^i = A, \qquad (9.439\text{a,b})$$

the **inner *product*** (9.439b) *of a vector is invariant, that is, a scalar in a curvilinear geometry (9.311) = (9.439a), iff one vector is covariant and the other contravariant. The **norm** (9.440b) or the square of the modulus (9.440c) of a vector is the inner product by itself and is invariant (9.440d) ≡ (9.440e) only in a metric (9.437) curvilinear geometry (9.440a):*

$$\boldsymbol{\mathfrak{M}}_N: \quad \left\| \vec{A} \right\| \equiv \left(\vec{A}.\vec{A} \right) = A_i A^i = g_{ij} A^i A^j = g^{ij} A_i A_j. \qquad (9.440\text{a–e})$$

In a metric space, (9.441a) is defined by the **angle between two vectors** (9.441b) *as the inner product divided by the moduli:*

$$\boldsymbol{\mathfrak{M}}_N: \quad \cos^2 \theta = \frac{\left(\vec{A}.\vec{B} \right)}{\left| \vec{A} \right|^2 \left| \vec{B} \right|^2} = \frac{g_{ij} g_{k\ell} A^i B^j A^k B^\ell}{g_{ij} g_{k\ell} A^i A^j B^k B^\ell}. \qquad (9.441\text{a,b})$$

When squaring the inner product or a contraction, different indices must be used,

$$(A_i B^i)^2 = A_i B^i A_j B^j = \sum_{i,j=1}^N A_i B^i A_j B^j = \left(\sum_{i=1}^N A_i B^i \right)^2 \neq \sum_{i=1}^N (A_i)^2 (B^i)^2 = A_i A_i B^i B^i, \qquad (9.442\text{a,b})$$

to indicate that there is a double sum (9.442a) not a single sum (9.442b).

NOTE 9.38: Contravariant and Covariant Base Vectors

The **contravariant (covariant) base** vectors have one component, unity, along the coordinate axes and all others are zero, and thus coincide (9.443a) [(9.443b)] with the identity matrix in their reference frame of origin:

$$e^i_{\ j} \overset{*}{=} \delta^i_j \overset{*}{=} e^i_{\ j}: \quad e^{i'}_{\ j} = X_i^{i'} e^i_{\ j} \overset{*}{=} X_i^{i'} \delta^i_j \overset{*}{=} X_j^{i'}, \quad e^i_{\ j'} = X_{j'}^j e^i_{\ j} \overset{*}{=} X_{j'}^j \delta^i_j \overset{*}{=} X_{j'}^i; \qquad (9.443\text{a–d})$$

In (9.443a) [(9.443b)] should be distinguished: (1) the **internal index** $\vec{e}_i \left(\vec{e}^{\,j} \right)$ that identifies the *i*th contravariant (*j*th covariant) base vector, and belongs to the kernel letter "*e*"; (2) the free index $e_i^j \left(e_i^{\,j} \right)$ that take the running values $j = 1, \dots, N (i = 1, \dots, N)$ and represents a component of the base vector. The different transformation law for contravariant (9.314b) [covariant (9.316b)] vectors implies that *in another reference frame, the contravariant (covariant) base vectors coincide with the direct (9.443c) ≡ (9.305b) [inverse (9.443d) ≡ (9.306c)] transformation matrix. The relations (9.443a through d) are not invariant in a curvilinear geometry (9.311), that is, they depend on the reference frame, and this is indicated by the asterisk over the equality sign* $\overset{*}{=}$. *The contravariant and covariant base vectors are orthogonal in any reference frame,*

$$\boldsymbol{\mathfrak{X}}_N: \quad e^i_{\ k'} e^{k'}_{\ j} \overset{*}{=} X_k^i X_j^{k'} = \delta^i_j, \quad e^i_{\ k} e^k_{\ j} = \delta^i_k \delta^k_j = \delta^i_j, \quad \vec{e}^{\,i}.\vec{e}_j = \delta^i_j, \qquad (9.444\text{a–d})$$

and hence this is an invariant property (9.444b and c) ≡ (9.444d) in a curvilinear geometry (9.311) ≡ (9.444a).

NOTE 9.39: Oblique, Orthogonal, and Orthonormal Reference Frames

The inner product of the contravariant (covariant) base vectors specifies the covariant (contravariant) metric tensor (9.445b) [(9.446b)]:

$$g_{i'j'} = \delta_{i'j'}: \quad \underset{i}{e^{i'}}\,\overset{*}{\underset{j}{e^{j'}}} = X^{i'}_i X^{i'}_j = X^{i'}_i X^{j'}_j \delta_{i'j'} = g_{ij}, \tag{9.445a,b}$$

$$g^{i'j'} = \delta^{i'j'}: \quad \underset{i'}{e^i}\,\overset{*}{\underset{i'}{e^j}} = X^i_{i'} X^j_{i'} = X^i_{i'} X^j_{j'} \delta^{i'j'} = g^{ij}, \tag{9.446a,b}$$

*relative to a **Cartesian reference frame** where the metric tensor coincides with the identity matrix (9.445a) [(9.446a)]. Using either representation in a metric geometry (9.437) ≡ (9.447a), the covariant (contravariant) metric tensor, that is, the inner product of the contravariant (9.445b) ≡ (9.447a) [covariant (9.446b) ≡ (9.447b)] base vectors,*

$$g_{ij} = \underset{i}{e^k}\,\underset{j}{e_k} = \left(\underset{i}{\vec{e}}\cdot\underset{j}{\vec{e}}\right), \quad g^{ij} = \underset{k}{e^i}\,\overset{j}{e^k} = \left(\overset{i}{\vec{e}}\cdot\overset{j}{\vec{e}}\right), \tag{9.447a,b}$$

*can be used to classify **the reference frames** into the following: (1) an **oblique** frame if the metric tensor is not diagonal, with the diagonal elements specifying the norm or the square of the modulus of the contravariant (covariant) base vectors (9.448a) [(9.449a)] and the nondiagonal elements specifying their angles (9.448b) [(9.449b)]:*

$$g_{ii} = \left(\underset{i}{\vec{e}}\cdot\underset{i}{\vec{e}}\right) = \left\|\underset{i}{\vec{e}}\right\|, \quad g_{ij} = \left|\underset{i}{\vec{e}}\right|\left|\underset{j}{\vec{e}}\right|\cos(\theta_{ij}), \tag{9.448a,b}$$

$$g^{ii} = \left(\overset{i}{\vec{e}}\cdot\overset{i}{\vec{e}}\right) = \left\|\overset{i}{\vec{e}}\right\|, \quad g^{ij} = \left|\overset{i}{\vec{e}}\right|\left|\overset{j}{\vec{e}}\right|\cos(\theta_{ij}), \tag{9.449a,b}$$

*the angles being the same because the contravariant and covariant base vectors are mutually orthogonal (9.444d); (2) an **orthogonal** frame if all axes are mutually perpendicular (9.450a), implying that the contravariant (covariant) metric tensor is diagonal (9.450b) [(9.450c)], with the diagonal elements specified by the **scale factors** or the moduli of the base vectors (9.450d) [(9.450e)],*

$$\theta_{ij} = \frac{\pi}{2}: \quad g_{ij} = (h_i)^2 \delta_{ij}, \quad g^{ij} = (h_i)^{-2}\delta_{ij}, \tag{9.450a–c}$$

$$h_i = \left|\underset{i}{\vec{e}}\right| = \left(\left|\overset{i}{\vec{e}}\right|\right)^{-1}, \quad \sqrt{g} = h_1 \ldots h_N = \prod_{i=1}^{N} h_i, \tag{9.450d–f}$$

*which also specify the determinant of the metric tensor (9.450f); and (3) an **orthonormal** frame if the base vectors have unit length (9.451a) so that the scale factors can be omitted,*

$$h_i = \left|\underset{i}{\vec{e}}\right| = 1: \quad g_{ij} = \delta_{ij}, \quad g^{ij} = \delta^{ij}, \quad g = 1, \tag{9.451a–d}$$

and the metric tensor reduces to the identity matrix (9.451b and c), so its determinant is unity (9.451d).

NOTE 9.40: Arc Length, Distance, and Geodesic

The **arc length** is defined as the modulus (9.452a) of the infinitesimal displacement (9.305a), which is a contravariant vector:

$$\mathcal{M}_N: \quad (\mathrm{d}s)^2 = |\mathrm{d}x^i| = g_{ij}\mathrm{d}x^i\mathrm{d}x^j = g^{ij}\mathrm{d}x_i\mathrm{d}x_j. \tag{9.452a–d}$$

The arc length is (i) a quadratic form (9.452b through d) involving the metric tensor, hence the designation metric geometry (9.452a) ≡ (9.437) for general oblique coordinates (9.448a and b; 9.449a and b). In (ii) an **orthogonal metric geometry** *(9.452a):*

$$\mathcal{O}_N \equiv \left\{ \mathcal{M}_N : g_{ij} = (h_i)^2 \delta_{ij} \right\}: \quad (\mathrm{d}s)^2 = \sum_{i=1}^{N}(h_i)^2(\mathrm{d}x^i)^2 = \sum_{i=1}^{N}(h_i)^{-2}(\mathrm{d}x_i)^2, \tag{9.453a–c}$$

the arc length is (9.450a through f) a sum of squares, with the scale factors as coefficients (9.453b and c); and (iii) in (9.454a) an **orthonormal or Cartesian geometry** *(Descartes 1637), the arc length is (9.451a through d) the sum of the squares of the differentials of the coordinates:*

$$\mathcal{N}_N = \left\{ \mathcal{M}_N : g_{ij} = \delta_{ij} \right\}: \quad (\mathrm{d}s)^2 = \sum_{i=1}^{N}(\mathrm{d}x^i)^2 = \sum_{i=1}^{N}(\mathrm{d}x_i)^2, \tag{9.454a–c}$$

where the contravariant and the covariant infinitesimal displacements coincide (9.454b and c). Since the arc length is a scalar in all cases, it can be integrated along a curve to specify its **length** *(9.455a):*

$$L = \int_A^B \mathrm{d}s: \quad \mathrm{d}L = 0 < \mathrm{d}^2 L; \quad \mathcal{A}\,\mathcal{M}_N: \quad d(P,Q) \equiv \left| \underset{P}{\vec{x}} - \underset{Q}{\vec{x}} \right| = \left| g_{ij}\left(\underset{P}{x^i} - \underset{Q}{x^i} \right)\left(\underset{P}{x^j} - \underset{Q}{x^j} \right) \right|^{1/2}, \tag{9.455a–d}$$

the shortest curve (9.455b and c) between two points is the **geodesic** *and may be unique or not. The length of all geodesics between two points is the same and specifies the distance between the points. In a rectilinear metric geometry (9.312; 9.437) ≡ (9.455d) the geodesic is a straight line and the* **distance** *between two points is given by (9.455d).*

NOTE 9.41: Contravariant/Covariant or Parallel/Orthogonal Projections

The preceding results (Notes 9.35 through 9.40) have the following geometrical interpretation, visualized most readily in two dimensions (Figure 9.12a through d); *(1) the contravariant (covariant) base vectors are (Figure 9.12a) tangent (orthogonal) to the coordinate curves (coordinate hyperplanes containing all other coordinate curves); (2) the contravariant (9.456a and b) [covariant (9.456c and d)] components of a vector are obtained by projection on the contravariant (9.443a) [covariant (9.443b)] base vectors:*

$$v^i = v^j \delta_j^i \overset{*}{=} \underset{j}{v}\,e^i, \quad v_i = v_j \delta_i^j \overset{*}{=} \underset{j}{v}\,e_i; \tag{9.456a–d}$$

hence, the projection (Figure 9.12b) is parallel (orthogonal) to the coordinate axes in an oblique metric geometry (9.437); (3) in an orthogonal metric geometry (9.453a), the contravariant and covariant base vectors are parallel and differ only in modulus if the scale factors are not unity; and (4) in an orthonormal metric geometry (9.454a), all scale factors are unity, and hence the contravariant and covariant base vectors coincide (Figure 9.12c), and the contravariant and covariant components are equal for a vector (Figure 9.12d) or any absolute (9.320) or pseudo- (9.350) tensor. The last interpretation (4) is the simplest case of a Cartesian geometry.

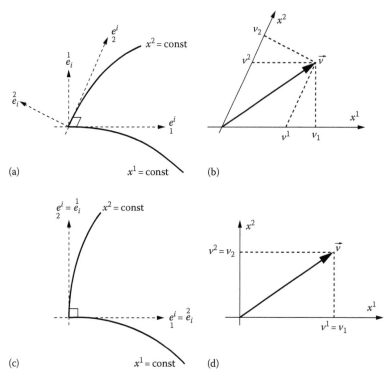

FIGURE 9.12 For a curvilinear oblique coordinate system (a), the contravariant (covariant) base vectors that are tangent (orthogonal) to the coordinate curves (hyperplanes) (1) are orthogonal in pairs and (2) form two oblique systems. The contravariant (covariant) components of a vector obtained (b) by projection parallel (orthogonal) to the coordinate axes are also distinct. In an orthogonal coordinate system (c), the contravariant and covariant base vectors are parallel and thus coincide with scale vectors. If the scale factors are unity, that is, a Cartesian local reference frame, the covariant and the contravariant components of a vector coincide (d).

NOTE 9.42: Physical Components and Dimensional Consistency

In an orthogonal, curvilinear (5.311), metric (5.437) geometry (9.453a), taking as an example the velocity (9.314a), which is a contravariant vector (9.314b), the scale factors can be used to introduce a third type of components, in addition to the contravariant and covariant components, namely, the physical components: (1/2) the contravariant (covariant) components of the velocity are (9.314a) $\equiv$ (9.457a) [(9.457b)]:

$$v^i = \frac{dx^i}{dt}, \quad v_i = g_{ij}v_j = (h_i)^2 \delta_{ij}v_j = \left(h_{\underline{i}}\right)^2 v_i; \quad v_{(i)} = h_{\underline{i}}v^i, \qquad (9.457\text{a–c})$$

(3) the **physical components** (9.457c) are intermediate in the sense they use one scale factor (the underlined symbol $h_{\underline{i}}$ means no use of the sum convention) in (9.457c), instead of none (two) in the contravariant (9.457a) [covariant (9.457b)] components. Considering, for example, spherical coordinates with scale factors (6.14a through c) and using a dot to denote the time derivative, it follows

$$v^i = \dot{x}^i = \{\dot{R}, \dot{\theta}, \dot{\phi}\}, \quad v_i = (h_i)^2 \dot{x}^i = \{\dot{R}, R^2\dot{\theta}, R^2 \sin^2\theta\dot{\phi}\}, \qquad (9.458\text{a,b})$$

$$v_{(i)} \equiv h_{\underline{i}}\dot{x}^i = \{\dot{R}, R\dot{\theta}, R\sin\theta\dot{\phi}\}, \qquad (9.458\text{c})$$

that (1) the contravariant components (9.458a) have different dimensions, namely, length per unit time for v^1 (inverse time for v^2, v^3) as a linear (angular) velocity; (2) the covariant components (9.458c) also do not have the same dimensions, namely, length (square of length) per unit time for v_1 (v_2, v_3); and (3) the physical components (9.458c) all have the same physical dimensions of length per unit time. The dimensional consistency of the physical components of the velocity is ensured in every reference frame because the arc length (9.453b and c) per unit time (9.459b) is a scalar in an orthogonal geometry (9.453a) $\equiv$ (9.459a):

$$\oplus_N : \quad \left(\frac{\mathrm{d}s}{\mathrm{d}t}\right)^2 = \sum_{i=1}^{N}(h_i)^2\left(\frac{\mathrm{d}x^i}{\mathrm{d}t}\right)^2 = \sum_{i=1}^{N}(h_iv^i) = \sum_{i=1}^{N}\left[v_{(i)}\right]^2; \tag{9.459a,b}$$

also, the use of physical components incorporates the scale factors, leading from an orthogonal (9.453a) to an orthonormal (9.454a) geometry. Thus, *the physical components of a pseudotensor (9.350),*

$$\oplus_N : \quad T^{(i_1)...(i_p)}_{(j_1)...(j_q)} = \left(\prod_{n=1}^{N}h_n\right)^{g/2} h_{i_1}...h_{i_p}\left(h_{j_1}...h_{j_p}\right)^{-1}{}^{(g)^{h}...i_p} T^{h...i_p}_{j_1...j_q}, \tag{9.458a,b}$$

are obtained as follows: (1/2) for each contravariant (covariant) index, there is multiplication (division) by a scale factor, without summation and (3) the weight is removed (9.438) by using the determinant of the metric tensor (9.450f). The physical components exist in an orthogonal curvilinear metric geometry (9.453a) $\equiv$ (9.458a); all have the same physical dimension and can be treated as Cartesian tensors.

NOTE 9.43: Physical Components of the Invariant Differential Operators

The gradient (9.316a) of a scalar (9.315a) is a covariant vector (9.316b) whose physical components are obtained dividing by the scale factors:

$$\oplus_N : \quad (\nabla\Phi)_i = \frac{1}{h_i}\frac{\partial\Phi}{\partial x_i} = (h_i)^{-1}\partial_i\Phi, \tag{9.459a,b}$$

in agreement with (6.23a and b). For the curl of a vector, the physical components,

$$\oplus_N : \quad \left(\nabla\wedge\vec{A}\right)_{(ij)} = (h_ih_j)^{-1}\partial_i\left[h_jA_{(j)}\right], \tag{9.460a,b}$$

can be obtained as follows: (1) the physical components of the vector are made covariant by multiplying by the scale factor; (2) the curl is applied as an alternated partial derivative; and (3) the curl is a covariant bivector and its physical components are obtained by dividing by the scale factors. The result (9.460a,b) coincides with (6.40a and b) in three dimensions and simplifies to (6.42) in two dimensions, suppressing one coordinate and one scale factor. The result (9.460) may be extended,

$$\oplus_N : \quad \left[curl\, A_{(i_1)...(i_M)}\right]_{(i_{M+1})} = \left(h_{i_1}...h_{i_{M+1}}\right)^{-1}\partial_{[i_{M+1}}\left[h_{i_1}...h_{i_M}A_{(i_1)...(i_M)]}\right], \tag{9.461a,b}$$

to the physical dimensions of the curl of an *M*-vector.

The physical components of the divergence of a vector are obtained,

$$\oplus_N : \quad \nabla.\vec{A} = \frac{1}{\sqrt{g}}\partial_i\left[\frac{\sqrt{g}}{h_i}A_{(i)}\right] = \left(\prod_{k=1}^{N}h_k\right)^{-1}\partial_i\left[\left(\prod_{\substack{\ell=1 \\ \ell\neq i}}^{N}h_\ell\right)A_i\right]. \tag{9.462a-c}$$

as follows: (1) the physical components are made contravariant by dividing by the scale factor; (2) the density or weight $\vartheta = -1$ is introduced by multiplying by the square root of the determinant of the metric tensor (9.450f); (3) the divergence is the contracted partial derivative and leads to a scalar density; and (4) dividing by (2) restores an absolute scalar. In the three-(two-) dimensional case, (9.462c) simplifies to (6.29) [(6.33)]. The result (9.462b) is extended to the divergence of an M-vector:

$$\textbf{9}_N: \quad \partial_{(i_M)} A^{(i_1)...(i_M)} = \frac{h_{i_1}...h_{i_{M-1}}}{\sqrt{g}} \partial_{i_M} \left[\frac{\sqrt{g}}{h_{i_1}...h_{i_M}} A^{(i_1)...(i_M)} \right]. \tag{9.463a,b}$$

Thus, *the physical components of the invariant differential operators, namely, (1) the gradient of a scalar (9.459a and b)* $\equiv$ *(6.23a and b); (2) the curl of a vector (9.460), in particular in a two- (6.42) [three- (6.40a and b)] dimensional space, and the generalization (9.461a and b) to an M-vector in an N-dimensional space; and (3) the divergence of a vector (9.462a through c), in particular in a two- (6.33) [three- (6.29)] dimensional space, and its generalization (9.463a and b) to an M-vector in an N-dimensional space have been obtained in an orthogonal (9.453a), metric (9.437), curvilinear (9.311) geometry.* The absolute (physical) components of the invariant differential operators exist in a curvilinear (9.311), metric (9.347), oblique [orthogonal (9.453a)] geometry [Note 9.43 (Note 9.44)]. The absolute invariant differential operators, that is, without weight are considered next (Note 9.44).

NOTE 9.44: Absolute Invariant Differential Operators

In an oblique metric (9.437), curvilinear (9.311) geometry (9.464a), the gradient of a scalar (9.315a) has no weight and is an absolute covariant (9.316a) $\equiv$ (9.464b) [contravariant (9.464c)] vector:

$$\textbf{9}_N: \quad (\nabla \Phi)_i = \partial_i \Phi, \quad (\nabla \Phi)^i = g^{ij} \partial_j \Phi. \tag{9.464a–c}$$

The curl of an M-vector is made absolute using the absolute permutation symbol (9.438b):

$$\textbf{9}_N: \quad B^{i_{M+2}...i_N} = \varepsilon^{i_1...i_N} \partial_{i_{M+1}} A_{i_1...i_M} = \frac{1}{\sqrt{g}} e^{i_1...i_N} \partial_{i_{M+1}} A_{i_1...i_M}. \tag{9.465a,b}$$

The divergence of a covariant vector does not exist in a curvilinear geometry (9.311) without metric. In a metric geometry (9.357), the divergence of a covariant vector can be made into an absolute scalar:

$$\textbf{9}_N: \quad \nabla.\vec{A} = \frac{1}{\sqrt{g}} \partial_j \left(\sqrt{g} g^{ij} A_j \right), \tag{9.466a,b}$$

which is obtained as follows: (1) the covariant index is raised using the contravariant metric tensor (9.433b); (2) the density or weight $\vartheta = -1$ is obtained by multiplying (9.436c) by the square root of the metric tensor; (3) the divergence is the contracted partial derivative; and (4) it leads to a scalar density made into an absolute scalar by dividing by the square root of the covariant metric tensor. This can be generalized to the divergence of a covariant M-vector field:

$$\textbf{9}_N: \quad C^{i_1...i_{M-1}} = \frac{1}{\sqrt{g}} \partial_{i_M} \left(\sqrt{g} g^{i_1 j_1} ....g^{i_M j_M} A_{[j_1...j_M]} \right), \tag{9.467a,b}$$

which is a contravariant $(M - 1)$-vector.

The Laplacian of a scalar (9.468a) is the divergence (9.466b) of the gradient that is a covariant vector (9.466b), leading to:

$$\mathfrak{M}_N: \quad \nabla^2 \Phi = \nabla.(\nabla \Phi) = \frac{1}{\sqrt{g}} \partial_j \left(\sqrt{g} \, g^{ij} \partial_j \Phi \right).$$

(9.468a,b)

Thus, we have obtained in *a metric (9.437), curvilinear (9.311) geometry the absolute invariant differential operators, that is, with zero weight, namely, (1) the covariant (9.464b) [contravariant (9.464c) components of the gradient of a scalar; (2/3) the curl (9.465b) [divergence (9.467b)] of a covariant M-vector as a contravariant (N − M − 1)-vector [(M − 1)-vector]; (4) the divergence of a covariant vector (9.466b) as a scalar; and (5) the Laplacian of a scalar (9.468a and b) as a scalar that in an orthogonal geometry (9.454a) can be written in terms of scalar factors as:*

$$\mathfrak{G}_N: \quad \nabla^2 \Phi = \frac{1}{\sqrt{g}} \partial_i \left[\frac{\sqrt{g}}{(h_i)^2} \partial_i \Phi \right] = \left(\prod_{k=1}^{N} h_k \right)^{-1} \partial_i \left[\frac{1}{h_i} \left(\prod_{\substack{\ell=1 \\ \ell \neq i}}^{N} h_\ell \right) \partial_i \Phi. \right]$$

(9.469a,b)

In a two- (three-) dimensional space, the Laplacian of a scalar in physical components simplifies to (6.47) [(6.43)]. The tensor algebra (Notes 9.6 through 9.15) and analysis (Notes 9.16 through 9.32) in curvilinear and metric (Notes 9.35 through 9.44) geometries are sufficient to extend the potential theory to multivectors in any dimension (Notes 9.45 through 9.52).

NOTE 9.45: Potential of Irrotational Multivector Fields

In a curvilinear geometry (9.311) with curvature (9.362), a covariant M-vector field that is irrotational, that is, has zero curl (6.414a through c), is the curl of a covariant (M − 1)-vector potential (6.415a through c). The covariant M-vector field may be generated by a source forcing its divergence (9.470b) ≡ (9.467b) in a metric geometry (9.437) ≡ (9.470a):

$$\mathfrak{M}_N^2: \quad \tilde{\Lambda}^{j_1 \ldots j_{M-1}} = \partial_{j_M} \left(\sqrt{g} \, g^{i_1 j_1} \ldots g^{i_M j_M} \, U_{[i_1 \ldots i_M]} \right),$$

(9.470a,b)

leading to a **contravariant multivector Poisson equation**:

$$\mathfrak{M}_N^2: \quad \tilde{\Lambda}^{j_1 \ldots j_{M-1}} = \partial_{j_M} \left(\sqrt{g} \, g^{i_1 j_1} \ldots g^{i_M j_M} \partial_{[i_M} \Phi_{i_1 \ldots i_{M-1}]} \right).$$

(9.471a,b)

In an orthogonal geometry (9.453a) ≡ (9.472a), the generalized Poisson equation (9.471b) involves only scale factors (9.472b):

$$\mathfrak{G}_N^2: \quad \tilde{\Lambda}^{j_1 \ldots j_{M-1}} = \partial_k \left(\frac{h_{i_{M+1}} \ldots h_{i_N}}{h_{i_1} \ldots h_{i_M}} \partial_{[k} \Phi_{i_1 \ldots i_{M-1}]} \right).$$

(9.472a,b)

In an orthonormal geometry (9.454a) ≡ (9.473a), it simplifies further to (9.473b):

$$\mathfrak{N}_N^2: \quad \tilde{\Lambda}^{i_1 \ldots i_{M-1}} = \partial_{kk} \Phi_{i_1 \ldots i_{M-1}};$$

(9.473a,b)

this is a set of Cartesian Poisson equations whose solution,

$$\mathfrak{N}_N^2: \quad \Phi_{i_1 \ldots i_{M-1}}(\vec{x}) = G_N * \Lambda_{i_1 \ldots i_{M-1}}(\vec{x}) = \frac{\sigma_N^{-1}}{(N-2)} \int_D |\vec{x} - \vec{y}|^{2-N} \Lambda_{i_1 \ldots i_{M-1}}(\vec{y}) \mathrm{d}^N \vec{y},$$

(9.474a,b)

is the convolution (7.36) with the N-dimensional influence function (9.43a through c) for the Cartesian Laplace operator (6.44) in free space. It is modified in two dimensions (9.42a through d) and in compact regions (9.53a and b).

NOTE 9.46: Number of Independent Components of a Multivector

Since each index of a tensor in an N-dimensional geometry takes N values, the M-vector field [$(M-1)$-vector potential] has a larger (smaller) number of components:

$$\#U_{i_1\ldots i_M} = N^M > \#\Phi_{i_1\ldots i_{M-1}} = N^{M-1}; \tag{9.475a,b}$$

however, since both are skew-symmetric, the number of independent components is less than indicated in (9.475a and b), as shown next. For the M-vector field (9.475a), (1) the second index i_2 cannot repeat the first, otherwise it vanishes; hence, it takes $N-1$ values; (2) the third index i_3 takes only $N-2$ values; and (3) and so on, the Mth index i_M takes $N-M+1$ values:

$$N_0 \equiv {}^*\#U_{i_1\ldots i_M} = \frac{N(N-1)\ldots(N-M+1)}{N!} = \frac{N!}{M!(N-M)!} = \binom{N}{M}; \tag{9.476}$$

the division by $M!$ arises because the components of the M-vector with the same indices in any order are equal in modulus and differ in sign only for odd permutations. Likewise, for the $(M-1)$-vector irrotational potential, the number of independent components is:

$$N_- \equiv {}^*\#\Phi_{i_1\ldots i_{M-1}} = \binom{N}{M-1} = \frac{N!}{(M-1)!(N-M+1)!}. \tag{9.477}$$

Thus, *an irrotational covariant M-vector field (9.414a through c) [covariant $(M-1)$-vector irrotational potential (9.415a through c)] has (9.475a) [(9.475b)] components, of which (9.476) [(9.477)] are independent. The ratio of (9.476) and (9.477) is (9.478a):*

$$M_- \equiv \frac{N_0}{N_-} = \frac{(M-1)!(N-M+1)!}{M!(N-M)!} = \frac{N-M+1}{M}, \tag{9.478}$$

showing that,

$$M_- \equiv \frac{N_0}{N_-} = \frac{N-M+1}{M} \begin{cases} <1 & \text{if } M > \dfrac{N+1}{2}, \\[2mm] =1 & \text{if } M = \dfrac{N+1}{2}, \\[2mm] >1 & \text{if } M < \dfrac{N+1}{2}, \end{cases} \tag{9.479a,b,c}$$

an $(M-1)$-vector potential has a smaller (9.479c)/equal (9.479b)/larger (9.349a) number of independent components than the irrotational M-vector field in an N-dimensional space for values of M, respectively, less/equal/larger than $(N+1)/2$. For example, an irrotational vector field $M=1$ always has more components N than the potential, except in a one-dimensional space $N=1$ where both are scalars. An irrotational bivector field $M=2$ has more independent components than its vector potential in a space of dimension up to $N>3$, an equal number for $N=3$, and a larger number for $N<3$. The latter value is next compared with solenoidal M-vectors and their $(M+1)$-vector potentials (Notes 9.46 and 9.47).

NOTE 9.47: Potential of Multivector Solenoidal Fields

In a curvilinear geometry (9.311) with curvature (9.362), a contravariant M-vector density is solenoidal iff it has zero divergence (9.417a through c), in which case it is the divergence of a contravariant (M + 1)-vector density potential (9.418a through c). The contravariant (M + 1)-vector field may be generated by forcing its curl (9.480b) ≡ (9.465b) in a metric geometry (9.437) ≡ (9.480a):

$$\mathfrak{M}_N^2: \quad \Omega_{i_1\dots i_{M+1}} = \partial_{[i_{M+1}}\left(\frac{1}{\sqrt{g}}\, g_{|i_1|j_1}\cdots g_{i_M|j_M]}\tilde{U}^{j_1\dots j_M}\right); \tag{9.480a,b}$$

leading to a **covariant multivector Poisson equation**:

$$\mathfrak{M}_N^2: \quad \Omega_{i_1\dots i_{M+1}} = \partial_{[i_{M+1}}\left(\frac{1}{\sqrt{g}}\, g_{|i_1|j_1}\, g_{i_M|j_M]}\partial_k\tilde{A}^{j_1\dots j_M k}\right). \tag{9.481a,b}$$

In an orthogonal geometry (9.453a) ≡ (9.482a), the metric tensor is replaced by scale factors in (9.482b):

$$\mathfrak{O}_N^2: \quad \Omega_{i_1\dots i_{M+1}} = \partial_{[i_{M+1}}\left(\frac{h_1\dots h_{i_M}}{h_{i_{M+1}}\dots h_{i_N}}\partial_k\tilde{A}^{i_1\dots i_M k}\right). \tag{9.482a,b}$$

In an orthonormal geometry (9.453a) ≡ (9.483a), it simplifies further to a **modified multivector Poisson equation** *(9.483b):*

$$\mathfrak{M}_N^2: \quad \Omega_{i_1\dots i_{M+1}} = \partial_k\left(\partial_{[i_{M+1}}A_{i_1\dots i_M]k}\right), \tag{9.483a,b}$$

that is distinct from (9.473a and b). This can be seen in the simplest case (9.484b) ≡ (9.484c) of a vector (9.484a),

$$M = 0: \quad \Lambda_i = \partial_i(\partial_k A_k) \Leftrightarrow \vec{\Lambda} = \nabla(\nabla.\vec{A}), \tag{9.484a-c}$$

that involves a grad div rather than a Laplacian in (9.473a and b). The number of components is considered [Note 9.46 (Note 9.48)] for both irrotational (Note 9.45) [solenoidal (9.47)] multivector fields and potentials.

NOTE 9.48: Independent Components of Multivector Fields and Potentials

The number of components of a solenoidal M-vector (9.485a) ≡ (9.475a) is less than that (9.485b) of its $(M + 1)$-vector potential:

$$\#U_{i_1\dots i_{M+1}} = N^M < N^{M+1} = \#A_{i_1\dots i_{M+1}}. \tag{9.485a,b}$$

The number of independent components is (9.476) for the M-vector, solenoidal or not, and (9.486) for the $(M + 1)$-vector potential is

$$N_+ \equiv *\#A_{i_1\dots i_{M+1}} = \binom{N}{M+1} = \frac{N!}{(M+1)!(N-M-1)!}. \tag{9.486}$$

Thus, *a solenoidal M-vector field (9.312b) [(M + 1)-vector potential (9.418a through c)] has (9.485a) [(9.485b)] components, of which (9.476) [(9.486)] are independent. The ratio of (9.486) and (9.476),*

$$M_+ = \frac{N_0}{N_+} = \frac{(M+1)!(N-M-1)!}{M!(N-M)!} = \frac{M+1}{N-M}, \tag{9.487}$$

shows that,

$$M_+ \equiv \frac{N_0}{N_+} = \frac{M+1}{N-M} \begin{cases} <1 & \text{if } M < \dfrac{N-1}{2}, \\[2mm] =1 & \text{if } M = \dfrac{N-1}{2}, \\[2mm] >1 & \text{if } M > \dfrac{N-1}{2}, \end{cases} \tag{9.488a,b,c}$$

the (M + 1)-vector potential has a smaller (9.488c)/equal (9.488b)/larger (9.488a) number of independent components than the solenoidal M-vector field in an N-dimensional space for values of M, respectively, larger/equal/smaller than (N − 1)/2. For example, for a solenoidal vector field M = 1, the solenoidal potential has a smaller/equal/larger number of independent components in the plane (9.297a through c; 9.298a and b)/space (9.299a through d; 9.300a through c; 9.301)/higher dimension (9.302a through e). Thus, the solenoidal potential is a complication as regards the number of independent components compared with the field for dimensions more than three.

NOTE 9.49: Nonirrotational, Nonsolenoidal Multivector Fields

In the general case of a nonirrotational (9.489c), nonsolenoidal (9.489d), continuously differentiable (9.489b) M-vector field, a covariant representation may be used in a curvilinear (9.311), metric (9.437) geometry (9.489a):

$$\mathcal{M}_N: \quad U_{i_1 \ldots i_M} \in C^1 \left(| R^N \right): \quad \partial_{[i_{M+1}} U_{i_1 \ldots i_M]} = \Omega_{i_1 \ldots i_{M+1}},$$

$$\partial_{i_M} \left(\sqrt{g}\, g^{i_1 j_1} \ldots g^{i_M j_M} U_{[j_1 \ldots j_M]} \right) = \Lambda^{i_1 \ldots i_{M-1}}. \tag{9.489a–d}$$

In a metric geometry with curvature (9.362) ≡ (9.490a), a representation of the M-vector as the sum of an irrotational (and a solenoidal) part (9.490b) may be sought:

$$\mathcal{M}_N^2: \quad U_{i_1 \ldots i_M} = \partial_{[i_M} \Phi_{i_1 \ldots i_{M-1}]} + \frac{1}{\sqrt{g}} g_{[i_1|j_1|} \ldots g_{i_M|j_M|} \partial_{j_{M+1}]} \tilde{A}^{j_1 \ldots j_{M+1}}; \tag{9.490a,b}$$

the (M − 1)-vector covariant irrotational [(M + 1)-vector contravariant density solenoidal] potential satisfies a multivector (9.491) [modified multivector (9.492)] Poisson equation,

$$\Lambda^{i_1 \ldots i_{M-1}} = \partial_{i_M} \left(\sqrt{g}\, g^{i_1 j_1} \ldots g^{i_M j_M} \partial_{[j_M} \Phi_{j_1 \ldots j_{M-1}]} \right), \tag{9.491}$$

$$\Lambda_{i_1 \ldots i_{M+1}} = \partial_{[i_{M+1}} \left[\frac{1}{\sqrt{g}} g_{[i_1|j_1|} \ldots g_{i_M|j_M|} \partial_{j_{M+1}} \tilde{A}^{j_1 \ldots j_{M+1}} \right], \tag{9.492}$$

that can be simplified (1) to (9.472a and b) [(9.482a and b)] in an orthogonal geometry(9.453a) and (2) to (9.473a and b) [(9.483a and b)] in an orthonormal geometry (9.453b). A necessary condition for the existence of the decomposition (9.490a and b) is (9.493a),

$$N^2 > 1 + 3M(N-M): \quad M_0 = \frac{N!}{(M+1)!(N-M+1)!}\left[N^2 + 3M(M-N)-1\right], \qquad (9.493a,b)$$

*in which case (9.493b) is the number of **gauge conditions**, that is, the number of arbitrary components of the potentials, or equivalently, the number of independent conditions that can be imposed on the potentials. The proof follows (Note 9.50).*

NOTE 9.50: Number of Gauge Conditions for the Potentials

Since (9.491) [(9.492)] follow immediately from (9.490a and b), only (9.493a and b) needs to be proved. The number of components of the $(M-1)$-vector potential plus the $(M+1)$-vector potential always exceeds the number of components of the M-vector field:

$$\#\Phi_{i_1\ldots i_{M-1}} + \#A_{i_1\ldots i_{M+1}} - U_{i_1\ldots i_M} = N^{M-1} + N^{M+1} - N^M = N^M\left(N + \frac{1}{N} - 1\right) > 0, \qquad (9.494)$$

but they are not all independent. The number of gauge conditions is the sum of the number of independent components of the $(M-1)$-vector irrotational (9.477) [plus $(M+1)$-vector solenoidal (9.487)] potentials minus the number of independent components of the M-vector field (9.476):

$$
\begin{aligned}
^*\#\Phi_{i_1\ldots i_{M-1}} + {}^*\#A_{i_1\ldots i_{M+1}} - {}^*\#U_{i_1\ldots i_M} &= N_+ + N_- - N_0 = \binom{N}{M-1} + \binom{N}{M+1} - \binom{N}{M} \\
&= N!\left[\frac{1}{(M-1)!(N-M+1)!} + \frac{1}{(M+1)!(N-M-1)!} - \frac{1}{M!(N-M)!}\right] \\
&= \frac{N!}{(M+1)!(N-M+1)!}[M(M+1) + (N-M)(N-M+1) - (M+1)(N-M+1)] \\
&= \frac{N!}{(M+1)!(N-M+1)!}\left[N^2 + 3M(M-N)-1\right] = M_0.
\end{aligned}
\qquad (9.495)
$$

This proves the number of gauge conditions (9.493b) ≡ (9.495) that must be positive (9.493a) for the decomposition to be possible. The number of gauge conditions is indicated in Table 9.2 for subspaces of dimension $M = 1 - 7$ of a space of dimension $N = 1 - 10$. In the case of a three-dimensional $N = 3$ vector field $M = 1$, the decomposition is possible by imposing $Q_0 \equiv 1$ one gauge condition as shown next.

NOTE 9.51: Table of Multivector Irrotational and Solenoidal Fields and Potentials

Comparing the covariant $(M-1)$–vector (9.415a through c) [contravariant $(M+1)$-vector density (9.418a through c)] potential of a covariant (contravariant density) irrotational (9.414a through c) [solenoidal (9.417a through c)] M-vector field, it follows (9.479a through c) [(9.488a through c)] that (1) the number of independent components is the same for (9.479b) [(9.488b)]; (2) the potential is simpler than the field in the sense that it has a smaller number of independent components (9.479c) [(9.488c)] decreasing (increasing) the dimension of the M vector in the N-dimensional space; and (3) in the opposite case, the potential has a larger number of independent components than the field, and its use may be a complication or hindrance rather than a benefit. Table 9.2 compares for spaces of dimension up to ten

TABLE 9.2   Number of Independent Components of an M-Vector Field and of Its Irrotational (Solenoidal) (M − 1)[(M + 1)]-Vector Potentials

Number of independent components of An M-vector field: −/N₀/−//−; (M − 1)-vector irrotational potential: N_-/−/−//−; (M + 1)-vector solenoidal potential: −/−/N_+//−; Number of independent gauge conditions: −/−/−//M₀

N_- (9.477)/N₀(9.476)/N_+ (9.486)/M₀ = N_- + N_+ − N₀ (9.493b)

Dimension of space N	M = 1	M = 2	M = 3	M = 4	M = 5	M = 6	M = 7
Vector field: N₀	M = 1	M = 2	M = 3	M = 4	M = 5	M = 6	M = 7
Irrotational: N_-	M − 1 = 0	M − 1 = 1	M − 1 = 2	M − 1 = 3	M − 1 = 4	M − 1 = 5	M − 1 = 6
Solenoidal: N_+	M + 1 = 2	M + 1 = 3	M + 1 = 4	M + 1 = 5	M + 1 = 6	M + 1 = 7	M + 1 = 8
N = 2	1/2/1/0	2/1/0//1	−/0/−//−	−/0/−//−	−/0/−//−	−/0/−//−	−/0/−//−
N = 3	1/3/3/1	3/3/1//1	3/1/0//2	−/0/−//−	−/0/−//−	−/0/−//−	−/0/−//−
N = 4	1/4/6/3	4/6/4//2	6/4/1//3	4/1/0//3	−/0/−//−	−/0/−//−	−/0/−//−
N = 5	1/5/10/6	5/10/10//5	10/10/5//5	10/5/1//6	5/1/0//4	−/0/−//−	−/0/−//−
N = 6	1/6/15/10	6/15/20//11	15/20/15//10	20/15/6//11	15/6/1//10	6/1/0//5	−/0/−//−
N = 7	1/7/21/15	7/21/35//21	21/35/35//21	35/35/21//21	35/21/7//21	21/7/1//15	7/1/0//6
N = 8	1/8/28/21	8/28/56//36	28/56/70//42	56/70/56//42	70/56/28//42	56/28/8//36	28/8/1//21
N = 9	1/9/36/28	9/36/84//57	36/84/126//78	84/126/126//84	126/126/84//84	126/84/36//78	84/36/9//57
N = 10	1/10/45/36	10/45/120//85	45/120/210//135	120/210/252//162	210/252/210//168	252/210/120//162	210/120/45//135

Notes: N₀: Number of independent components of an M-vector field in a space with N dimensions (9.476).

N_-: Number of independent components of the (M − 1)-vector potential of an irrotational M-vector field (9.477).

N_+: Number of independent components of the (M + 1)-vector potential of a solenoidal M-vector field (9.486).

−: An M-vector in an N-dimensional space with M > N is null.

N₀, $N_\pm$ = 0: The field or potential does not exist—if the field does not exist, the potential is not considered.

M₀: Number of independent gauge conditions (9.493b) in the representation of an M-vector field by (M−1) irrotational and (M + 1) solenoidal potential (Note 9.49).

The difference M₀ = N_+ + N_- − N₀ specifies the number of independent gauge conditions that may be imposed on the potentials. An M-vector in an N-dimensional space with M > N is null and the representation of the M-vector as the sum of the curl (divergence) of an (M − 1)-vector [(M + 1)-vector] potential is possible only if M₀ ≥ 0. The table indicates N_-/N₀/N_+/M₀ for all space dimensions N = 1,…,10 and all M-vectors M = 1,…,7, with blanks in all null or impossible cases.

Number of independent components (N₀) of an M-vector in an N-dimensional space, and also of the (M − 1)-vector irrotational [(M + 1)-vector solenoidal] potential N_-(N_+) representing its irrotational (solenoidal) part.

and M up to seven the number of independent components of (1) an M-vector field (9.476) and (2–3) an $(M-1)$-vector potential (9.477) [$(M+1)$-vector potential (9.486)] of an irrotational (solenoidal) M-vector field. It shows the cases in which the potential is simpler, comparable, or more complex than the field as regards the number of independent components; hence, it indicates when the use of potentials is a benefit as a representation of the fields. That is always the case for irrotational vector fields, but not always for vector solenoidal fields; for multivector irrotational or solenoidal multivector fields, the $(M-1)$-vector (9.479a through c) [$(M+1)$-vector (9.488a through c)] potential is generally simpler than the M-vector irrotational (solenoidal) field when M is far from (close to) the dimension of the space as seen in Table 9.2. A space of dimension N has nonzero M-vectors of dimension $M \leq N$; thus, an M-vector irrotational (solenoidal) field can have a nonzero $(M-1)$-vector [$(M+1)$-vector] potential only if $M \leq N(M \leq N-1)$. Table 9.2 also indicates the number of gauge conditions (9.493b) that apply to the potentials of M-vectors that are neither irrotational nor solenoidal (Notes 9.49 and 9.50).

NOTE 9.52: Invariant Potential Theory of Multivectors

The condensed and selective account of tensor calculus (Notes 9.6 through 9.52) is aimed at potential theory; it is not a replacement for a more gradual and thorough presentation of the tensor calculus, with precise definitions and more illustrative examples, that would extend over one or more books. The index notation was used for multivectors and invariant differential operators, rather than the approach of differential forms, to make the invariance properties such as covariance and contravariance and weight of tensors explicit. The tensor calculus underlies potential theory even in two (three) dimensions in Books 1 and 2 (Book 3). For example, the two-dimensional curl has the form (9.496a) with parameters (9.496b and c) because it is the only one invariant in a curvilinear geometry:

$$\nabla \wedge \vec{A} = \vec{e}_z(\lambda \partial_x A_y + \mu \partial_y A_x), \quad \lambda = 1 = -\mu; \tag{9.496a–c}$$

nevertheless, the tensor calculus can be omitted when presenting potential theory in the plane or space, but it becomes unavoidable when considering multivectors in higher dimensional spaces. The present account has addressed (1) the tensor algebra (Notes 9.6 through 9.15) and analysis (Notes 9.16 through 9.34) in curvilinear geometries and (2) the tensor calculus in metric geometries (Notes 9.35 through 9.52). The topics selected were those more relevant to (1) demonstrate how tensor invariance underlies plane and space potential theory and (2) outline the extension to the invariant potential theory of multivectors in any dimension (Notes 9.46 through 9.52). The tensor calculus also underlies most mathematical and all physical theories, with a scope far beyond the present summary account.

9.10 Conclusion

The influence or Green function, that specifies the potential due to an impulse distribution: (Figure 9.1) in one dimension is a linear ramp function, that is, the potential is constant before the conducting plane $x = 0$, and linear after; (Figure 9.2) in two dimensions is a logarithmic function of distance from the line charge; and (Figure 9.3) in three dimensions, varies with the inverse of the distance from a point charge. The potential and the field due to an arbitrary distribution of charges (Figure 9.4) may be obtained by integrating over it; in the presence of another charge distribution (Figure 9.5), there is generally a mutual force (Figure 9.6) that is (a) repulsive [(b) attractive] for point charges of the same (opposite) signs. The potentials (Figures 9.1b, 9.2b, and 9.3b) and the fields (Figures 9.1a and b and 9.3b) of an electric charge distribution (Figure 9.4) lead to multipoles whose decay is indicated in Table 9.1. The latter includes higher dimensional multipoles that are described by hyperspherical (hypercylindrical) coordinates, as extensions of the spherical (cylindrical) coordinates [Figure 9.7b (Figure 9.7c)] whose common particular case $\theta = \pi/2(z = 0)$ is polar coordinates (Figure 9.7a). A further extension is to consider M-vector fields in an N-dimensional space, and its representation by an $(M-1)$-vector irrotational and an $(M + 1)$-vector solenoidal potential, leaving free M_0 gauge conditions (Table 9.2). The existence theorems for the multivector potentials of irrotational or solenoidal multivector fields hold in a star-shaped region (Figure 9.11) with an explicit formula for the potentials. These results are obtained in the context of the tensor calculus in a curvilinear geometry (Figure 9.8a) that includes as particular case the rectilinear geometry (Figure 9.8b). Both are represented in oblique coordinates (Figure 9.8a), implying that (1) the contravariant (covariant) base vectors tangent (normal) to the coordinate curves (hypersurfaces) are distinct (Figure 9.12a) and (2) the corresponding contravariant (covariant) components of a vector obtained by projection parallel (orthogonal) to the coordinate axes are also distinct (Figure 9.12b). The contravariant and covariant base vectors (components of a vector) coincide [Figure 9.12c (Figure 9.12d)] to within scale factors for orthogonal curvilinear coordinates. The polar vector (Figures 5.12a, 9.9a) and the covariant vector (Figure 9.9b) are the simplest examples of absolute tensors, besides the absolute scalar. The volume element is a scalar capacity (Figure 9.10a), implying that the mass density (flux) is a scalar (contravariant vector) density [Figure 9.10b (Figure 9.10d)]; the covariant (contravariant) bivector in a three-dimensional space has for dual a contravariant (covariant) vector density (Figure 5.12b) [capacity (Figure 9.10c)]. The covariant bivector, through its dual contravariant vector density, with weight removed in a metric geometry, becomes an absolute vector, corresponding to an axial vector.

Twenty Examples

10.1 Examples 10.1 through 10.20

EXAMPLE 10.1: DERIVATIVES OF DISCONTINUOUS FUNCTIONS

Use the corresponding generalized functions to obtain the derivatives of all orders of the following ordinary functions that (1) are discontinuous; (2) have angular points, that is, discontinuous derivative; or (3) vanish outside a finite or infinite interval, with a discontinuity of the function and/or of its derivative at one (two) end point(s). First, consider the functions' negative exponential (hyperbolic sine) in the positive real half-axis, continued symmetrically to the negative real half-axis in Figure 10.1a (b), and obtain their derivates of all orders. Second, consider the functions' circular cosine (sine) in Figure 10.1c (d) vanishing outside one period and obtain their derivatives of all orders. Third, consider the functions' hyperbolic cosine (sine) in Figure 10.1e (f) vanishing on the negative real axis and obtain their derivatives of all orders.

E10.1.1 Angular Points and Points of Discontinuity

The two functions can be represented by the generalized functions indicated at the top of Table 10.1, and their derivatives of all orders follow from (1.108a and b). For example, for the first function,

$$F(x) = e^{-|x|} = \begin{cases} e^x \equiv f_1(x) = f_1^{(n)}(x) & \text{if } x \leq 0, \\ e^{-x} = f_2(x) = (-)^n f_2^{(n)}(x) & \text{if } x \geq 0. \end{cases} \tag{10.1}$$

The jumps of the derivatives at the origin are

$$F^{(n)}(-0) \equiv f_1^{(n)}(0) = 1, \quad F^{(n)}(+0) \equiv f_2^{(n)}(0) = (-)^n: \quad \Delta f^{(2n)}(0) = 0, \quad \Delta f^{(2n+1)}(0) = -2. \tag{10.2a-d}$$

Substitution of (10.2c,d) in (1.108d) leads to (1) for derivatives of even $2N$ (odd $2N + 1$) order of the function (10.1) to coefficients involving the jumps at the origin of derivatives of order $2N - m - 1$ ($2N - m$); (2) these jumps are zero (10.2a) except for odd order $2n - 1$ in (10.2d) implying $2n - 1 = 2N - m - 1$ ($2N - m$), that is, $m = 2N - 2n$ ($m = 2N - 2n - 1$); (3) since the jumps (10.2d) are all equal to -2 in the last sum on the r.h.s. of (1.108b), $2N - 2n$ ($2N - 2n - 1$) may be replaced by $2n$ ($2n + 1$) in the corresponding last term on the r.h.s. of (10.3a) [(10.3b)]:

$$F^{(2N)}(x) = e^x H(-x) + e^{-x} H(-x) - 2 \sum_{n=0}^{N-1} \delta^{(2n)}(x), \tag{10.3a}$$

$$F^{(2N+1)}(x) = e^x H(-x) - e^{-x} H(-x) - 2 \sum_{n=0}^{N-1} \delta^{(2n+1)}(x), \tag{10.3b}$$

which is in agreement with the first column of Table 10.1. An alternative method is presented using the sign function (1.86a–c):

$$F(x) = |\sinh(x)| = \operatorname{sgn} x \sinh x = \begin{cases} -\sinh x & \text{if } x \leq 0, \\ +\sinh x & \text{if } x \geq 0. \end{cases} \tag{10.4a}$$

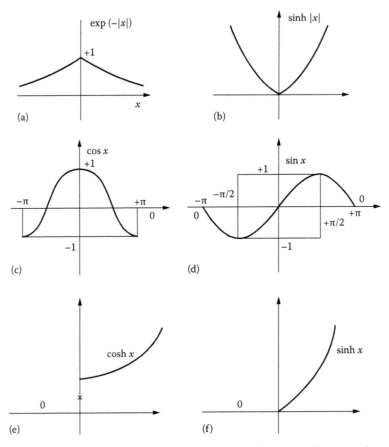

FIGURE 10.1 Examples of functions with finite discontinuities (c and e) or angular points where the derivative is discontinuous (a,b,d,f). In all cases the left- and right-hand derivatives exist and the corresponding generalized functions are infinitely differentiable, with the derivatives of all orders appearing in Tables 10.1/10.2/10.3, respectively, for the functions in a,b/c,d/e,f.

TABLE 10.1   Derivatives of Functions Involving the Modulus

$f(x)$	$\exp(-\lvert x \rvert)$	$\lvert \sinh x \rvert$
$F(x)$	$e^x H(-x) + e^{-x} H(x)$	$\sinh x \, \mathrm{sgn}\, x$
$F'(x)$	$e^x H(-x) - e^{-x} H(x)$	$\cosh x \, \mathrm{sgn}\, x$
$F''(x)$	$e^x H(-x) + e^{-x} H(x) - 2\delta(x)$	$\sinh x \, \mathrm{sgn}\, x + 2\delta(x)$
$F^{(2N)}(x)$	$e^x H(-x) + e^{-x} H(x) - 2\sum_{n=0}^{N-1} \delta^{(2n)}(x)$	$\sinh x \, \mathrm{sgn}\, x + 2\sum_{n=0}^{N-1} \delta^{(2n)}(x)$
$F^{(2N+1)}(x)$	$e^x H(-x) - e^{-x} H(x) - 2\sum_{n=0}^{N-1} \delta^{(2n+1)}(x)$	$\cosh x \, \mathrm{sgn}\, x + 2\sum_{n=0}^{N-1} \delta^{(2n+1)}(x)$

Note: The exponential of minus (hyperbolic sine of) the modulus [Figure 10.1a (b)] have an angular point at the origin, where the right- and left-hand derivatives exist and are different; the corresponding generalized functions are infinitely differentiable and their derivatives of all orders are indicated.

The first derivative is

$$F'(x) - \text{sgn}\, x \cosh x = \sinh x\, 2\delta(x) = 2\sinh 0\, \delta(x) = 0, \tag{10.5a}$$

using the derivative of the sign function (1.88a) and the substitution property (1.78b). A second differentiation yields

$$F''(x) - \text{sgn}\, x \sinh x = \cosh x\, 2\delta(x) = 2\cosh 0\, \delta(x) = 2\delta(x). \tag{10.5b}$$

Using

$$\frac{d^{2n}}{dx^{2n}}(\sinh x) = \sinh x, \qquad \frac{d^{2n+1}}{dx^{2n+1}}(\sinh x) = \cosh x, \tag{10.6a,b}$$

and continuing by iteration yields the result in the second column of Table 10.1. Both methods can be used to obtain all results.

E10.1.2 Derivatives of Circular Functions Vanishing outside One Period

The derivatives in Table 10.2 are obtained as in Table 10.1, replacing the ordinary functions by generalized functions and using (1.108b). The derivate of order $(N + 1)$ in the second column coincides with the derivate of order N in the first column.

E10.1.3 Derivatives of Hyperbolic Functions Vanishing in the Negative Real Axis

The derivatives in Table 10.3 are obtained as in Tables 10.1 and 10.2, replacing the ordinary functions by generalized functions (1.95a and b) and using (1.108a and b). The derivate of order N in the second column coincides with the derivate of order $(N - 1)$ in the first column.

TABLE 10.2 Derivatives of Circular Functions Vanishing outside One Period

$\lvert x \rvert < \pi$	$\cos x$	$\sin x$
$\lvert x \rvert > \pi$	0	0
$F(x)$	$\cos x\{H(x+\pi) - H(x-\pi)\}$	$\sin x\{H(x+\pi) - H(x-\pi)\}$
$F'(x)$	$-\sin x\{H(x+\pi) - H(x-\pi)\} - \delta(x+\pi) + \delta(x-\pi)$	$\cos x\{H(x+\pi) - H(x-\pi)\}$
$F''(x)$	$-\cos x\{H(x+\pi) - H(x-\pi)\} - \delta'(x+\pi) + \delta'(x-\pi)$	$-\sin x\{H(x+\pi) - H(x-\pi)\} - \delta(x+\pi) + \delta(x-\pi)$
$F^{(2N-1)}(x)$	$(-)^N \sin x\{H(x+\pi) - H(x-\pi)\} +$ $\displaystyle\sum_{n=0}^{N-1}(-)^n\{\delta^{(2n)}(x-\pi) - \delta^{(2n)}(x-\pi)\}$	$(-)^{N-1}\cos x\{H(x+\pi) - H(x-\pi)\} +$ $\displaystyle\sum_{n=0}^{N-2}(-)^n\{\delta^{(2n+1)}(x-\pi) - \delta^{(2n+1)}(x-\pi)\}$
$F^{(2N)}(x)$	$(-)^N \cos x\{H(x+\pi) - H(x-\pi)\} +$ $\displaystyle\sum_{n=0}^{N-1}(-)^n\{\delta^{(2n+1)}(x-\pi) - \delta^{(2n+1)}(x+\pi)\}$	$(-)^N \sin x\{H(x+\pi) - H(x-\pi)\} +$ $\displaystyle\sum_{n=0}^{N-1}(-)^n\{\delta^{(2n)}(x-\pi) - \delta^{(2n)}(x+\pi)\}$

Note: The function equal to the circular cosine (sine) for one period centered at the origin and zero otherwise [Figure 10.1c (d)] has finite discontinuities (angular) points at $\pm\pi$; the derivatives of all orders are indicated for the corresponding generalized functions.

TABLE 10.3   Derivatives of Hyperbolic Functions Vanishing on the Negative Real Axis

$x > 0$	$\cosh x$	$\sinh x$
$x < 0$	0	0
$F(x)$	$\cosh x H(x)$	$\sinh x H(x)$
$F'(x)$	$\sinh x H(x) + \delta(x)$	$\cosh x H(x)$
$F''(x)$	$\cosh x H(x) + \delta'(x)$	$\sinh x H(x) + \delta(x)$
$F^{(2N-1)}(x)$	$\sinh x H(x) + \sum_{n=0}^{N-1} \delta^{(2n)}(x)$	$\cosh x H(x) + \sum_{n=0}^{N-2} \delta^{(2n+1)}(x)$
$F^{(2N)}(x)$	$\cosh x H(x) + \sum_{n=0}^{N-1} \delta^{(2n+1)}(x)$	$\sinh x H(x) + \sum_{n=0}^{N-1} \delta^{(2n)}(x)$

Note: The function equal to the hyperbolic cosine (sine) for positive variable and zero for negative variable [Figure 10.1e (f)] has a finite discontinuity (angular point) at the origin, and as a generalized function is infinitely differentiable with the derivatives of all orders indicated.

EXAMPLE 10.2: DEFLECTION OF A HEAVY STRING WITH DISCONTINUOUS MASS DENSITY

A string of length L has different mass densities on each half. Consider the weak deflection of the string under its own weight, with uniform tension and supports at the same height.

The weak deflection of the string is specified by (2.69a) with distinct mass density $\rho_1(\rho_2)$ in the first (second) half:

$$-T\frac{d^2\zeta}{dx^2} = \begin{cases} \rho_1 g & \text{for } 0 \le x < L/2, & (10.7a) \\ \rho_2 g & \text{for } L/2 < x \le L. & (10.7b) \end{cases}$$

Integrating twice with the boundary condition (10.8a) [(10.8c)] leads to (10.8b) [(10.8d)]

$$\begin{aligned} \xi(0) &= 0: \\ \xi(L) &= 0: \end{aligned} \qquad -\frac{2T}{g}\zeta(x) = \begin{cases} (\rho_1 x + A)x & \text{if } 0 \le x \le \frac{L}{2}, & (10.8a,b) \\ [\rho_2(x-L)+B](x-L) & \text{if } \frac{L}{2} \le x \le L. & (10.8c,d) \end{cases}$$

The constants of integration are determined (10.9b) [(10.9d)] by the conditions of continuity of the displacement (10.9a) [slope (10.9c)]:

$$\zeta\left(\frac{L}{2}-0\right) = \zeta\left(\frac{L}{2}+0\right): \quad \rho_1 L + 2A = \rho_2 L - 2B; \qquad (10.9a,b)$$

$$\zeta'\left(\frac{L}{2}-0\right) = \zeta'\left(\frac{L}{2}+0\right) \quad \rho_1 L + A = B - \rho_2 L. \qquad (10.9c,d)$$

The solution of (10.9b and d) is

$$4A = -(\rho_2 + 3\rho_1)L, \qquad 4B = (\rho_1 + 3\rho_2)L. \qquad (10.10a,b)$$

Substituting (10.10a and b) in (10.8b and d) specifies *the weak deflection of a string under its own weight under tension T with two halves of different densities (10.7a and b) and supports at the same height (10.8a and c):*

$$\zeta(x) = -\frac{g}{8T} \times \begin{cases} x\left[4\rho_1 x - (\rho_2 + 3\rho_1)L\right] & \text{if } 0 \le x \le \frac{L}{2}, \quad (10.11a) \\ (x-L)\left[4\rho_2 x + (\rho_1 - \rho_2)L\right] & \text{if } \frac{L}{2} \le x \le L. \quad (10.11b) \end{cases}$$

The shape consists of two half-parabolas, and thus the slope is linear:

$$\zeta'(x) = -\frac{g}{8T} \times \begin{cases} 8\rho_1 x - (\rho_2 + 3\rho_1)L & \text{if } 0 \le x \le \frac{L}{2}, \quad (10.12a) \\ 8\rho_2 x + (\rho_1 - 5\rho_2)L & \text{if } \frac{L}{2} \le x \le L. \quad (10.12b) \end{cases}$$

The deflection at the midpoint is the same (10.13a) ≡ (2.65a) as for a homogeneous string with density equal to the arithmetic mean of densities (10.13b):

$$\zeta\left(\frac{L}{2}\right) = (\rho_1 + \rho_2)\frac{gL^2}{16T} = \frac{\bar{\rho}gL^2}{8T}, \quad 2\bar{\rho} \equiv \rho_1 + \rho_2. \quad (10.13a,b)$$

For unequal densities (10.14a), (1) the slope (10.12a and b) is not zero at the midpoint (10.14b):

$$\rho_1 \ge \rho_2 : \zeta'\left(\frac{L}{2}\right) = (\rho_2 - \rho_1)\frac{gL}{8T}; \quad \zeta'(a) = 0: \quad a = \left(3 + \frac{\rho_2}{\rho_1}\right)\frac{L}{8} \le \frac{L}{2}; \quad (10.14a\text{–}d)$$

and (2) it is zero (6.14c) on the side (Figure 10.2) with larger density (10.14d). The maximum deflection is (10.15a)

$$\delta \equiv \zeta_{max} = \zeta(a) = \frac{\rho_1 gL^2}{128T}\left(3 + \frac{\rho_2}{\rho_1}\right)^2 = \frac{(3 + \rho_2/\rho_1)^2}{8(1 + \rho_2/\rho_1)}\zeta\left(\frac{L}{2}\right); \quad (10.15a,b)$$

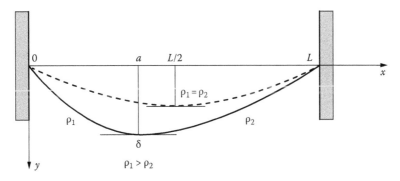

FIGURE 10.2 The weak deflection of an elastic string under its own weight under constant tension with suspension points at the same height and one half with higher uniform mass density than the other half leads to an unsymmetric shape with the maximum deflection on the heavier side.

the ratio of the maximum deflection to the deflection at midpoint (10.15b) is unity only for the homogeneous string. The sum of the reactions at the supports (10.16a and b) is minus the total mass of the string (10.16c):

$$R_- \equiv F(0) = -T\zeta'(0) = -\frac{L}{8}(\rho_2 + 3\rho_1)g, \tag{10.16a}$$

$$R_+ \equiv -F(L) = T\zeta'(L) = -\frac{L}{8}(\rho_1 + 3\rho_2)g, \quad F_- + F_+ = -\frac{L}{2}(\rho_1 + \rho_2)g. \tag{10.16b,c}$$

The slopes and reaction forces are larger in modulus at the support corresponding to the half-length of the string with larger mass density:

$$\rho_1 \geq \rho_2: \quad \zeta'(0) = \frac{|R_-|}{T} = (\rho_2 + 3\rho_1)\frac{gL}{8T} \geq (\rho_1 + 3\rho_2)\frac{gL}{8T} = \frac{|R_+|}{T} = -\zeta'(L). \tag{10.17a,b}$$

In the case $\rho_1 = \rho_2 = \rho$ of a homogeneous string, (10.11a and b; 10.12a and b; 10.13a $\equiv$ 10.15a; 10.16a and b) reduce, respectively, to (2.64f; 2.65b; 2.65a; 2.66a).

EXAMPLE 10.3: DEFLECTION OF A NONUNIFORM HEAVY STRING WITH CONTINUOUS MASS DENSITY

Consider the weak deflection of an elastic string with varying cross-section:

$$0 < \lambda < 1; \quad 0 \leq x \leq L: \quad \rho(x) = \rho_0\left[1 - \lambda\left(\frac{2x}{L} - 1\right)^2\right], \tag{10.18a–c}$$

whose mass per unit length is given by (10.18a through c).

The inhomogeneous string has a symmetric mass distribution (10.19a), with maximum in the middle and minimum at the ends (10.19b):

$$\rho(x) = \rho(L - x): \quad \rho_{max} = \rho\left(\frac{L}{2}\right) = \rho_0 \geq \rho(x) \geq \rho(1 - \lambda) = \rho(0) = \rho(L) = \rho_{min}. \tag{10.19a,b}$$

The weak deflection of a string under tension T in the gravity field g is specified (2.69a) by

$$-T\frac{d^2\zeta}{dx^2} = g\rho(x) = \rho_0 g\left(1 - \lambda + 4\lambda\frac{x}{L} - 4\lambda\frac{x^2}{L^2}\right), \tag{10.20a}$$

where the mass density (10.18c) was substituted. Integrating twice leads to

$$-T\zeta(x) = \rho_0 g\left(\frac{1-\lambda}{2}x^2 + \frac{2\lambda}{3}\frac{x^3}{L} - \frac{\lambda}{3}\frac{x^4}{L^2} + Ax + B\right); \tag{10.21}$$

the constants of integration are determined (10.22c and d) by the boundary conditions (10.22a and b)

$$\zeta(0) = 0 = \zeta(L): \quad B = 0, \quad \frac{A}{L} = \frac{\lambda}{6} - \frac{1}{2}. \tag{10.22a–d}$$

Substituting (10.22c and d) in (10.21) specifies the *weak deflection (10.23) of a string with non-uniform density (10.18c), by its own weight under the uniform tension T and with supports at the same height (10.22a and b)*:

$$\zeta(x) = \frac{\rho_0 g x}{6T}\left[(3-\lambda)L - 3(1-\lambda)x - 4\lambda\frac{x^2}{L} + 2\lambda\frac{x^3}{L^2}\right]; \tag{10.23}$$

the deflection (10.23) corresponds to the slope:

$$\zeta'(x) = \frac{\rho_0 g}{6T}\left[(3-\lambda)L - 6(1-\lambda)x - 12\lambda\frac{x^2}{L} + 8\lambda\frac{x^3}{L^2}\right]. \tag{10.24}$$

Since the distribution of mass density (10.18c) is symmetric (10.19a) relative to the mid-position, (1) the slope is zero (10.25a), corresponding to the maximum deflection (10.25b):

$$\zeta'\left(\frac{L}{2}\right) = 0: \quad \delta = \zeta_{\max} = \zeta\left(\frac{L}{2}\right) = \frac{\rho_0 g L^2}{8T}\left(1 - \frac{\lambda}{6}\right); \tag{10.25a,b}$$

and (2) the reactions at the supports are equal to half the weight:

$$R_\pm = -T\zeta'(0) = T\zeta'(L) = -\frac{\rho_0 g L}{2}\left(1 - \frac{\lambda}{3}\right) = g\int_0^{L/2}\rho(x)\,dx = -\frac{P}{2}. \tag{10.26}$$

The maximum deflection (10.25b) and reactions at the supports (10.26) are smaller than for a homogeneous string of density ρ_0 because (Figure 10.3) the density (10.18c) is less than ρ_0 everywhere except at the midpoint where it takes the value ρ_0. In the case $\lambda = 0$ of a homogeneous string, (10.23; 10.24; 10.25b; 10.26) reduce, respectively, to (2.64f; 2.65b; 2.65a; 2.66b).

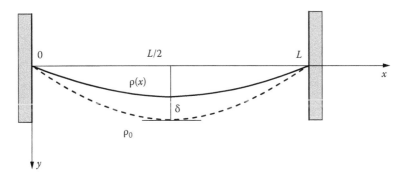

FIGURE 10.3 The weak deflection of an elastic string under its own weight under constant tension with supports at the same height leads to a symmetric shape if the mass density per unit length is nonuniform but symmetric: if the nonuniform mass density $\rho(x)$ has a maximum $\rho_0 \geq \rho(x)$ at the mid-position, the maximum deflection is less than for a uniform string of mass density ρ_0.

EXAMPLE 10.4: SHAPE OF A HEAVY STRING SUSPENDED FROM UNEQUAL HEIGHTS

Consider the weak deflection of a homogeneous elastic string under uniform tension with the two suspension points at unequal heights.

A homogeneous elastic string, of mass density ρ, under tension T, subject to its own weight in the gravity field g, has a shape specified by the differential equation (2.69a) $\equiv$ (10.27c):

$$\zeta(0)=0, \quad \zeta(L)=h: \quad -T\frac{d^2\zeta}{dx^2}=\rho g, \tag{10.27a–c}$$

with boundary conditions (10.27a and b) specified by the suspension points at different heights. Integrating (10.27c) twice yields (10.28a)

$$-\frac{2T}{\rho g}\zeta(x)=x^2+Ax+B: \quad B=0, \quad A=-L-\frac{2Th}{\rho gL}, \tag{10.28a–c}$$

where the constants of integration (10.28b and c) are determined from the boundary conditions (10.27a and b). Substituting (10.28b and c) in (10.28a) specifies *the weak deflection of a homogeneous string of density ρ, by its own weight in a uniform gravity field g under uniform tension T, suspended at unequal heights* (10.27a and b):

$$\zeta(x)=x\left[\frac{\rho g}{2T}(L-x)+\frac{h}{L}\right], \quad \zeta'(x)=\frac{\rho g}{T}\left(\frac{L}{2}-x\right)+\frac{h}{L}, \tag{10.29a,b}$$

The shape (10.29a) (Figure 10.4a) is a parabola unsymmetric about the mid-position corresponding to adding (1) the symmetric parabola for supports at the same height (2.64f); and (2) the straight

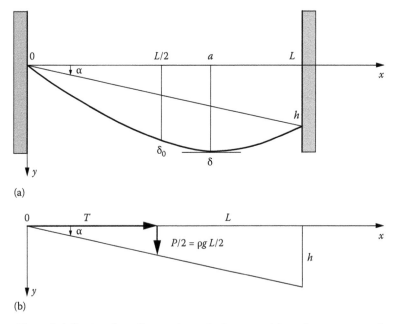

(a)

(b)

FIGURE 10.4 The weak deflection of a uniform string under its own weight under constant tension with suspension points at unequal heights leads to an unsymmetric shape with maximum deflection closer to lower support (a); the ratio of the horizontal distance to the height difference between the supports equals (b) the ratio of the tension (i.e., horizontal) to half the total weight (i.e., vertical).

line *h x/L* joining the two supports. The slope (10.29b) is not zero (10.30a) at the mid-position, which does not correspond to the maximum deflection (10.30b):

$$\zeta'\left(\frac{L}{2}\right) = \frac{h}{L}, \quad \zeta\left(\frac{L}{2}\right) = \frac{h}{2} + \frac{\rho g L^2}{8T} < \delta = \zeta_{max} = \zeta(a). \tag{10.30a,b}$$

The maximum deflection (10.31c) ≡ (10.32c) (Figure 10.4b) occurs closer to the lower suspension point (10.31b):

$$\zeta'(a) = 0: \quad a = \frac{L}{2} + \frac{hT}{\rho g L} = \frac{L}{2}(1 + \mu) > \frac{L}{2}, \quad \mu \equiv \frac{2hT}{\rho g L^2}. \tag{10.31a–d}$$

The dimensionless parameter (10.31d) ≡ (10.32c) (Figure 10.4b) compares (1) the weight of half the string (10.32a), that would be the reaction at the supports if they were at the same height; and (2) the horizontal tension *T* in the undeflected string converted to the vertical direction through the angle (10.32b) ≡ (2.32d) with the horizontal of the line joining the suspension points at different height:

$$P = \rho g L, \quad \alpha = \arctan\left(\frac{h}{L}\right): \quad \mu \equiv \frac{2T}{P} \tan \alpha. \tag{10.32a–c}$$

The parameter (10.31d) ≡ (10.32d) appears in the maximum deflection (10.29a, 10.31a) ≡ (10.33a):

$$\delta = \zeta(a) = \zeta_{max} = \frac{\rho g L^2}{8T}(1 + \mu)^2 = \frac{\rho g L^2}{8T}\left(1 + \frac{4hT}{\rho g L^2} + \frac{4h^2 T^2}{\rho^2 g^2 L^4}\right) > \frac{\rho g L^2}{8T}\left(1 + \frac{4hT}{\rho g L^2}\right)$$

$$= \zeta\left(\frac{L}{2}\right) = \delta_0 > \frac{\rho g L^2}{8T}, \tag{10.33a–c}$$

the maximum deflection (10.33a) (1) occurs closer to the lower support (10.31c); (2) is larger than the deflection (10.33b) at mid-position; (3) both are larger than the deflection at mid-position for supports at the same height (10.33c) ≡ (2.65a); and (4) is larger than the height of the lowest support (10.34):

$$\delta - h = \frac{\rho g L^2}{8T}\left[(1 + \mu^2) - 4\mu\right] = \frac{\rho g L^2}{8T}(1 - \mu)^2 > 0. \tag{10.34}$$

The maximum slope is at the origin and the linear approximation (2.3a) requires that it be small:

$$1 \gg |\zeta'(0)|^2 = \left(\frac{\rho g L}{2T} + \frac{h}{L}\right)^2 = \frac{\rho g L}{2T}(1 + \mu)^2 = \frac{P}{2T}(1 + \mu)^2, \tag{10.35}$$

where (2.67a) ≡ (10.32a) is the weight of the string. Thus the linear approximation requires that (1) the total weight of the string be small compared with twice the tension; (2) the height difference between the supports be small compared with the horizontal distance, which, to this level of approximation, coincides with the length of the string; and (3) the slope of the straight line joining the suspension points be small, implying that the slope is small everywhere along the string. The reaction force in modulus is larger at the higher (10.36a) than at the lower (10.36b) support and their sum is minus the weight of the string (10.36c):

$$-R_\pm = \{T\zeta'(0), -T\zeta'(L)\} = \frac{\rho g L}{2} \pm \frac{h}{L}T = \frac{\rho g L}{2}(1 \pm \mu) = \frac{P}{2}(1 \pm \mu), \quad R_- + R_+ = -P. \tag{10.36a,b}$$

In the case *h* = 0 of suspension points at the same height, (10.29a; 10.29b; 10.33a through c; 10.36a and b) reduce, respectively, to (2.64f; 2.65b; 2.65a; 2.66b).

EXAMPLE 10.5: LINEAR AND NONLINEAR DEFLECTION OF A STRING BY TWO CONCENTRATED FORCES

Consider the nonlinear (linear) deflection of an elastic string under constant tension by two concentrated transverse forces (Figure 10.5) and show that the principle of superposition does not (does) apply.

E10.5.1 Reactions at the Supports for Nonlinear Deflection

The nonlinear deflection of a string under constant tension is specified by (2.6) with the shear stress corresponding to two transverse forces $P_\pm$ at positions $x = x_\pm$:

$$T\left\{\left|1+\zeta'^2\right|^{-1/2}\zeta'\right\}' = -P_-\delta(x-x_-) - P_+\delta(x-x_+).\tag{10.37}$$

A first integration,

$$T\left|1+\zeta'^2\right|^{-1/2}\zeta' = A - P_-\,H(x-x_-) - P_+\,H(x-x_+),\tag{10.38}$$

leads to the transverse force

$$F(x) = -T\zeta'\left|1+\zeta'^2\right|^{-1/2} = \begin{cases} -A = R_- & \text{if } 0 \le x < x_-, & (10.39a)\\ P_- - A & \text{if } x_- < x < x_+, & (10.39b)\\ P_+ + P_- - A = -R_+ & \text{if } x_+ < x \le L, & (10.39c) \end{cases}$$

specifying the reactions at the supports and involving the constant of integration A. Solving (10.39a through c) for the constant slopes between the supports and the points of application of the forces,

$$\zeta'(x) = \begin{cases} a \equiv \left|\dfrac{T^2}{A^2}-1\right|^{-1/2} & \text{if } 0 \le x < x_-, & (10.40a)\\[3ex] b \equiv \left|\dfrac{T^2}{(A-P_-)^2}-1\right|^{-1/2} & \text{if } x_- < x < x_+, & (10.40b)\\[3ex] c \equiv \left|\dfrac{T^2}{(A-P_--P_+)^2}-1\right|^{-1/2} & \text{if } x_+ < x \le L, & (10.40c) \end{cases}$$

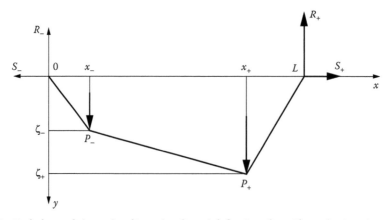

FIGURE 10.5 Both the weak (strong) or linear (nonlinear) deflection of a uniform elastic string under constant tension with supports at the same height by two concentrated forces at arbitrary positions consists of straight segments between (1/2) each support and the nearest concentrated force; and (3) the two concentrated forces.

shows that the shape of the string (Figure 10.5) consists of straight segments:

$$\zeta(x) = \begin{cases} ax & \text{if } 0 \le x < x_-, \quad (10.41\text{a}) \\ x_- a + b(x - x_-) & \text{if } x_- \le x \le x_+, \quad (10.41\text{b}) \\ c(x - L) & \text{if } x_+ \le x \le L, \quad (10.41\text{c}) \end{cases}$$

namely, (1/2) a straight line from the first (second) support (10.41b) [(10.41c)] at $x = 0$ ($x = L$) with slope (10.40a) [(10.40c)] up to the first (10.41a) [second (10.41c)] concentrated force; and (3) between the two forces the slope is

$$\frac{\zeta - \zeta_-}{x - x_-} = b = \frac{\zeta_+ - \zeta_-}{x_+ - x_-}, \quad \zeta_- \equiv \zeta(x_-) = x_- a, \quad \zeta_+ \equiv \zeta(x_+) = c(x_+ - L), \quad (10.42\text{a–c})$$

leading to (10.41b).

The condition of continuity at the point of application of first force (10.43a) is met by (10.42a through c) and the condition of continuity at the point of application of the second force (10.43b) is met by (10.43c):

$$\zeta(x_- - 0) = \zeta(x_- + 0), \quad 0 = \zeta(x_+ - 0) - \zeta(x_+ + 0) = x_- a + b(x_+ - x_-) - c(x_+ - L). \quad (10.43\text{a–c})$$

Substituting (10.40a) in (10.43b) leads to a polynomial of degree six:

$$0 = \frac{x_- A}{\sqrt{T^2 - A^2}} + \frac{(x_+ - x_-)(A - P_-)}{\sqrt{T^2 - (A - P_-)^2}} - \frac{(L - x_+)(A - P_- - P_+)}{\sqrt{T^2 - (A - P_- - P_+)^2}}, \quad (10.44)$$

whose roots specify the constant of integration A. A real root specifies the reaction at the first (second) support (10.39a) [(10.39b)]. In the case when the two concentrated forces add (10.45a) at the same points (10.45b), the polynomial of degree six (10.44) reduces to degree four (10.45c):

$$P = P_+ + P_-, \quad x_- = \xi = x_+: \quad A^2 \xi^2 \left[T^2 - (A - P)^2 \right] = (A - P)^2 (L - \xi)^2 (T^2 - A^2), \quad (10.45\text{a–c})$$

in agreement with (10.45c) $\equiv$ (2.118). Thus *the nonlinear deflection of an elastic string under constant tension by two forces concentrated at arbitrary positions (Figure 10.5) leads to (1) the transverse force (10.39a through c) and reactions at the first R_- (second R_+) support where A is a real root of (10.44); and (2) the same constant of integration specifying the slopes (10.40a through c) and the trapezoidal shape (10.41a through c) of the string.* The shape of the string (10.40a through c) is not the sum of two influence functions (2.117a and b), showing that the superposition principle does not hold for nonlinear deflections.

E10.5.2 Two Arbitrary Concentrated Forces and Points of Application

In the linear case when the applied forces are much smaller than the longitudinal tension (10.46a), the slopes (10.40a through c) simplify to (10.46b through d)

$$(P_\pm)^2 \ll T^2: \quad -T\zeta'(x) = \begin{cases} A = -R_- & \text{if } 0 \le x < x_-, \quad (10.46\text{b}) \\ A - P_- & \text{if } x_- < x < x_+, \quad (10.46\text{a–c}) \\ A - P_- - P_+ = R_+ & \text{if } x_- < x < x_+. \quad (10.46\text{d}) \end{cases}$$

Integration with the boundary conditions (2.39b and c) leads to the shape of the string:

$$
T\zeta(x) = \begin{cases} Ax & \text{if } 0 \le x \le x_-, & (10.47a) \\ Ax_- + (A - P_-)(x - x_-) & \text{if } x_- \le x \le x_+, & (10.47b) \\ (A - P_- - P_+)(x - L) & \text{if } x_+ \le x \le L. & (10.47c) \end{cases}
$$

The condition at the first (10.43a) [second (10.43b)] support is met by (10.47a and b) [leads by (10.47b and c) to (10.48)]:

$$
0 = Ax_- + (A - P_-)(x_+ - x_-) - (A - P_- - P_+)(x_+ - L) = AL + P_-(x_- - L) + P_+(x_+ - L). \quad (10.48)
$$

The latter (10.48) is the balance of moments relative to the second support and specifies the reaction force at the first support (10.49a):

$$
R_- = -A = -P_-\left(1 - \frac{x_-}{L}\right) - P_+\left(1 - \frac{x_+}{L}\right); \quad R_+ = -P_+ - P_- - R_- = -\frac{P_- x_- + P_+ x_+}{L}. \quad (10.49a,b)
$$

The reaction force at the second support (10.49b) follows either from the balance of forces (10.46d) or from the balance of moments around the first support.

Substitution of (10.49a and b) in (10.47a through c) specifies the shape of the string

$$
T\zeta(x) = \begin{cases} x\left[P_-\left(1 - \frac{x_-}{L}\right) + P_+\left(1 - \frac{x_+}{L}\right) \right] & \text{if } 0 \le x < x_-, & (10.50a) \\ x\left[P_+\left(1 - \frac{x_+}{L}\right) - P_-\frac{x_-}{L} \right] + P_- x_- & \text{if } x_- \le x \le x_+, & (10.50b) \\ \left(1 - \frac{x}{L}\right)(P_- x_- + P_+ x_+), & \text{if } x_+ \le x \le L. & (10.50c) \end{cases}
$$

and the slopes

$$
LT\zeta'(x) = \begin{cases} P_-(L - x_-) + P_+(L - x_+) & \text{if } 0 \le x < x_-, & (10.51a) \\ P_+(L - x_+) - P_- x_- & \text{if } x_- < x < x_+, & (10.51b) \\ -P_- x_- - P_+ x_+ & \text{if } x_+ < x \le L. & (10.51c) \end{cases}
$$

Thus *the linear deflection of an elastic string under constant tension by two transverse concentrated forces at arbitrary positions (Figure 10.5) leads to (1) the reaction forces (10.49a and b); (2) the slopes (10.51a through c); and (3) the deflections (10.50a through c) that coincide (10.50a through c) $\equiv$ (10.52a) with the sum of the influence functions (2.41b and c) $\equiv$ (10.52b and c) for a unit load multiplied by the respective concentrated forces:*

$$
\zeta(x) = P_- G(x; x_-) + P_+ G(x; x_+): \quad TG(x; x_\pm) = \begin{cases} x\left(1 - \frac{x_\pm}{L}\right) & \text{if } x \le x_\pm, \\ x_\pm\left(1 - \frac{x}{L}\right) & \text{if } x \ge x_\pm. \end{cases} \quad (10.52a\text{–}c)
$$

In the case of forces applied at equal distance from the supports (10.53a and b),

$$x_- = \frac{L}{3}, \quad x_+ = \frac{2L}{3}: \quad P_\pm = P, \quad P_\pm = \pm P, \tag{10.53a–d}$$

for equal (10.53c) [opposite (10.53d)] force (10.51a through c) simplifies to (2.79a through c) [(2.87a through c)].

EXAMPLE 10.6: LINEAR AND NONLINEAR DEFLECTION BY A UNIFORM PLUS A CONCENTRATED LOAD

Consider the nonlinear (linear) deflection of an elastic string under uniform tension with supports at the same height subject to a concentrated force at an arbitrary position and its own weight for a constant total mass and confirm that the principle of superposition is not (is) satisfied.

E10.6.1 Nonlinear Deflection by Mixed Load

The nonlinear deflection of an elastic string under constant tension is specified by (2.9b), where the shear stress is taken as the superposition of (1) a concentrated force at arbitrary position (2.113b); and (2) the weight of the string considered in the case (2.171a through d) of constant mass

$$-\rho g - P\delta(x-\xi) = T\left\{\zeta' \left|1+\zeta'^2\right|^{-1/2}\right\}'. \tag{10.54}$$

A first integration gives

$$-F(x) = T\zeta' \left|1+\zeta'^2\right|^{-1/2} = -\rho gx - PH(x-\xi) + A = \begin{cases} A - \rho gx & \text{if } 0 \le x < \xi & (10.55a) \\ A - P - \rho gx & \text{if } \xi < x \le L & (10.55b) \end{cases}$$

where the constant of integration A is related to the reactions at the supports (10.56a and b)

$$R_- = F(0) = -A, \quad R_+ = -F(L) = -A - P - \rho gL, \quad R_+ + R_- = -P - \rho gL, \tag{10.56a–c}$$

which satisfies the balance of transverse forces (10.56c). Solving (10.55a) [(10.55b)] for the slopes leads to (10.57a) [(10.57b)]:

$$\zeta'(x) = \begin{cases} (A-\rho gx)\left|T^2 - (A-\rho gx)^2\right|^{-1/2} & \text{if } 0 \le x < \xi & (10.57a) \\ (A-P-\rho gx)\left|T^2 - (A-P-\rho gx)^2\right|^{-1/2} & \text{if } \xi < x \le L. & (10.57b) \end{cases}$$

Integrating the slope before (10.57a) [after (10.57b)] the point of application of the concentrated force with the boundary condition at the first (2.64b) [second (2.64c)] support leads to (10.58a) [(10.58b)]

$$\rho g\zeta(x) = \begin{cases} \left|T^2 - (A-\rho gx)^2\right|^{1/2} - \left|T^2 - A^2\right|^{1/2} & \text{if } 0 \le x \le \xi & (10.58a) \\ \left|T^2 - (A-P-\rho gx)^2\right|^{1/2} - \left|T^2 - (A-P-\rho gL)^2\right|^{1/2} & \text{if } \xi \le x \le L. & (10.58b) \end{cases}$$

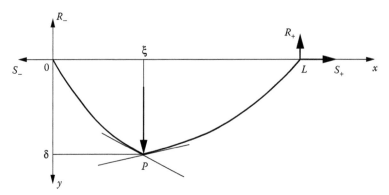

FIGURE 10.6 The deflection of a uniform elastic string under constant tension with supports at the same height by the combination of (1) a concentrated force at an arbitrary position plus (2) a uniform load leads to a curved shape with an angular point where the concentrated force is applied; the principle of superposition does (does not) apply for weak (strong) deflection corresponding to (1) small (large) maximum slope; and (2) total load smaller (not much smaller) than the tangential tension.

The constant A is determined by the continuity of displacement (10.58) at the point of application of the force:

$$0 = \rho g\left[\zeta(\xi+0)-\zeta(\xi-0)\right] = \left|T^2-(A-P-\rho g\xi)^2\right|^{1/2} - \left|T^2-(A-P-\rho gL)^2\right|^{1/2}$$
$$- \left|T^2-(A-\rho g\xi)^2\right|^{1/2} + \left|T^2-A^2\right|^{1/2}. \tag{10.59}$$

Thus *the nonlinear deflection of an elastic string under constant tension due (Figure 10.6) to a transverse concentrated force at an arbitrary position (10.54) plus its own weight in the case of constant mass (2.171a and b) leads to the reaction forces at the supports (10.56a through c) where A is a positive real root of (10.59), which also appears in the slope (10.57a and b) and shape (10.58a and b) of the string.* These expressions can be expanded in terms of a lowest-order linear approximation plus nonlinear corrections of all higher orders as in Sections 2.6 and 2.7. In the sequel (E10.6.2), only the lowest-order linear approximation is considered as a confirmation that the principle of superposition (1) does not hold in the nonlinear case, for example, (10.58a and b) is not the sum of (2.117a and b) and (2.176); and (2) does apply in the linear case when the sum of (2.41b and c) and (2.64f) is obtained using a different method from Sections 2.3 and 2.4.

E10.6.2 Lowest-Order or Linear Approximation

The lowest two orders in

$$\left|T^2-(A-P-\rho g\xi)^2\right|^{1/2} = T - \frac{1}{2T}(A-P-\rho g\xi)^2 + O\left(\frac{(A-P-\rho g\xi)^4}{T^2}\right), \tag{10.60}$$

correspond to the linear approximation in (10.59)

$$0 = (A-P-\rho gL)^2 - (A-P-\rho g\xi)^2 + (A-\rho g\xi)^2 - A^2$$
$$= \rho g\left[\rho gL^2 + P(L-\xi) - 2AL\right]. \tag{10.61}$$

This specifies (10.56a and b) the reactions at the supports

$$R_- = -A = -\frac{\rho gL}{2} - P\left(1-\frac{\xi}{2L}\right), \quad R_+ = -\rho gL - P - R_- = -\frac{\rho gL}{2} - P\frac{\xi}{L}, \tag{10.62a,b}$$

which satisfy (10.56c) and are the sum of (1) the reactions due to the concentrated force (2.39d and e); and (2) half the weight of the string (2.66b). To the lowest order (10.60), the displacement before (10.58a) [after (10.58b)] the point of application of the force is (10.63a) [(10.63b)]:

$$\zeta\left(x<\xi\right)=\frac{1}{2\rho gT}\left[A^2-\left(A-\rho gx\right)^2\right]=-\frac{\rho g}{2T}x^2+\frac{A}{T}x=\frac{\rho g}{2T}x\left(L-x\right)+\frac{P}{T}\left(1-\frac{\xi}{L}\right)x, \quad (10.63a)$$

$$\zeta\left(x>\xi\right)=\frac{1}{2\rho gT}\left[\left(A-P-\rho gL\right)^2-\left(A-P-\rho gx\right)^2\right]=\frac{\rho g}{2T}\left(L^2-x^2\right)-\frac{L-x}{T}\left(A-P\right)$$

$$=\frac{\rho g}{2T}\left(L^2-x^2\right)-\frac{L-x}{T}\left(\frac{\rho gL}{2}-P\frac{\xi}{L}\right)=\frac{\rho g}{2T}x\left(L-x\right)+\frac{P}{T}\left(1-\frac{x}{L}\right)\xi. \quad (10.63b)$$

The shape of the string (10.63a and b) $\equiv$ (10.64a and b)

$$\zeta\left(x\right)=\frac{\rho g}{2T}x\left(L-x\right)+\frac{P}{T}\times\begin{cases}\left(1-\dfrac{\xi}{L}\right)x & \text{if } 0\leq x\leq\xi \quad (10.64a)\\[2ex]\left(1-\dfrac{\xi}{L}\right)\xi & \text{if } \xi\leq x\leq L, \quad (10.64b)\end{cases}$$

is the sum of the deflections due to (1) the weight (2.64f); and (2) the concentrated force (2.41b and c). The principle of superposition applies to the deflections and slopes and their extreme values, as it did to the reaction forces at the supports (2.61a and b). The method used in the present example gives the same results as Sections 2.3, 2.4, 2.6, and 2.7 together, in a combined way.

EXAMPLE 10.7: LINEAR AND NONLINEAR DEFLECTION OF A STRING WITH NONUNIFORM TENSION

Consider the linear and nonlinear deflection of an elastic string by a concentrated force at arbitrary position, corresponding to the influence or Green function, in the case of nonuniform tangential tension varying linearly with the coordinate between the supports (Figure 10.7a and b). Consider the particular case of weakly nonuniform tension.

E10.7.1 Linear Deflection with Nonuniform Tension

The nonuniform tension varying linearly with the distance between the supports is given by (10.65a) implying a tension (10.65b) [(10.65c)] at the first (second) support (Figure 10.7a):

$$T\left(x\right)=T_0\left(1+\lambda\frac{x}{L}\right):\quad T\left(0\right)=T_0,\quad T\left(L\right)=T_0\left(1+\lambda\right). \quad (10.65a\text{-}c)$$

The linearly nonuniform tension could arise for an elastic string hanging vertically in the gravity field (Figure 4.17) for an elastic displacement (4.428c) $\equiv$ (10.65d) with Young modulus E (and zero inelastic modulus $\overline{E}=0$), where the applied force per unit area or tension is the weight of the string up to the end at $x=L$:

$$E\frac{du}{dx}=-T\left(x\right)=-\rho g\left(L-x\right);\quad T_0=\rho gL,\quad \lambda=-1, \quad (10.65d\text{-}f)$$

the tension (10.65d) $\equiv$ (10.65a) corresponds to the values (10.65e) for the tension at the support, and slope (10.65f). In this case the two supports would lie in the same vertical line, and the transverse force would be horizontal, corresponding to a rotation of Figure 10.7b counterclockwise by ninety degrees. The deflection by a transverse concentrated force specifies the influence or Green function; bearing in mind that the tension is nonuniform (2.11c), it follows that (2.29c) must be replaced by

$$\frac{d}{dx}\left\{T\left(x\right)\frac{d}{dx}\left[G\left(x;\xi\right)\right]\right\}=-P\delta\left(x-\xi\right). \quad (10.66)$$

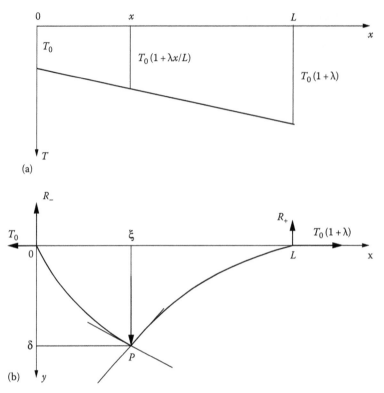

FIGURE 10.7 The strong (weak) deflection of a uniform string with suspension points at the same height in the case of nonuniform longitudinal tension (a) with a concentrated transverse force at an arbitrary position (b) specifies nonlinear (linear) the influence or Green function leading to a shape that is not a triangle but rather a "curved triangle" with concave and convex sides joined at the angular point where the concentrated force is applied.

A first integration yields

$$T_0\left(1+\lambda\frac{x}{L}\right)\frac{d}{dx}\Big[G(x;\xi)\Big]=A-PH(x-\xi)=\begin{cases}A=-R_- & \text{if } 0\le x<\xi,\\ A-P=R_+ & \text{if } \xi<x\le L.\end{cases}\qquad\begin{array}{c}(10.67a)\\(10.67b)\end{array}$$

The integration of the slope before (10.67a) [after (10.67b)] the point of application of the force is made from the boundary condition at first (10.68a) [second (10.69a)] support, leading to (10.68b) [(10.69b)]:

$$G(0;\xi)=0:\quad G(x\le\xi;\xi)=\frac{A}{T_0}\int_0^x\frac{d\xi}{1+\lambda\xi/L}=\frac{AL}{T_0\lambda}\log\left(1+\lambda\frac{x}{L}\right),\qquad(10.68a,b)$$

$$G(L;\xi)=0:\quad G(x\ge\xi;\xi)=\frac{A-P}{T_0}\int_L^x\frac{d\xi}{1+\lambda\xi/L}=\frac{(A-P)L}{T_0\lambda}\left[\log\left(1+\lambda\frac{x}{L}\right)-\log(1+\lambda)\right].\qquad(10.69a,b)$$

The continuity of the displacement at the point of application of the concentrated force

$$0=\frac{T_0\lambda}{L}\Big[G(\xi-0;\xi)-G(\xi+0;\xi)\Big]=P\log\left(1+\lambda\frac{\xi}{L}\right)+(A-P)\log(1+\lambda)\qquad(10.70a,b)$$

specifies the constant A.

Thus *the linear deflection of a string nonuniform tension (10.65a) by a concentrated force at an arbitrary point leads (10.66) to (1) the reactions (10.67a and b) at the supports (10.71a and b) that satisfy the balance of forces (10.71c):*

$$-T_0\left(1+\lambda\frac{x}{L}\right)G'\left(x<\xi;\xi\right)=-A=R_-=-P\left[1-\frac{\log\left(1+\lambda\xi/L\right)}{\log\left(1+\lambda\right)}\right], \tag{10.71a}$$

$$T_0\left(1+\lambda\frac{x}{L}\right)G'\left(x>\xi;\xi\right)=A-P=R_+=-P\frac{\log\left(1+\lambda\xi/L\right)}{\log\left(1+\lambda\right)}=-P-R_-; \tag{10.71b,c}$$

(2) the corresponding slopes, which have a jump at the point of application of the force:

$$G'\left(\xi+0;\xi\right)-G'\left(\xi-0;\xi\right)=-\frac{P}{T\left(\xi\right)}=\frac{P}{T_0}\frac{L}{L+\lambda\xi}; \tag{10.72}$$

and (3) the shape of the string before (10.68b) [after (10.69b)] the point of application of the concentrated force (10.73a) [(10.73b)]:

$$G\left(x;\xi\right)=G\left(\xi;x\right)=\frac{PL}{T_0\lambda}\times\begin{cases}\left[1-\dfrac{\log\left(1+\lambda\xi/L\right)}{\log\left(1+\lambda\right)}\right]\ \log\left(1+\lambda\dfrac{x}{L}\right)\ \text{if}\ 0\leq x\leq\xi, & (10.73a)\\[4mm]\left[1-\dfrac{\log\left(1+\lambda x/L\right)}{\log\left(1+\lambda\right)}\right]\ \log\left(1+\lambda\dfrac{\xi}{L}\right)\ \text{if}\ \xi\leq x\leq L, & (10.73b)\end{cases}$$

which is symmetric in (x,ξ), confirming that the reciprocity principle applies with nonuniform tension (Subsection 7.7.9). In the case of increasing tension (10.74a, 10.75a), the slope increases in modulus (decreases) from the upper (10.74b) [lower (10.75b)] support to the point of application of the concentrated force (10.74c) [(10.75c)]:

$$\lambda>0:\quad -G'\left(L;\xi\right)=\frac{P}{T_0}\frac{\eta}{1+\lambda}<\frac{P}{T_0}\frac{\eta}{1+\lambda\xi/L}=-G'\left(\xi+0;\xi\right), \tag{10.74a–c}$$

$$\eta\equiv\frac{\log\left(1+\lambda\xi/L\right)}{\log\left(1+\lambda\right)}:\quad G'\left(0;\xi\right)=\frac{P}{T_0}\left(1-\eta\right)>\frac{P}{T_0}\frac{1-\eta}{1+\lambda\xi/L}=G'\left(\xi-0;\xi\right), \tag{10.75a–c}$$

implying that the shape of the string (Figure 10.7b) is convex (concave) before (after) the point of application of the concentrated force.

E10.7.2 Case of Weakly Nonuniform Tension

In the case of weakly nonlinear tension, only the lowest-order terms in the series for the logarithm (I.21.64b) ≡ (II.3.47a and b) ≡ (10.74a) are used:

$$|\lambda|<1:\quad \log\left(1+\lambda\frac{x}{L}\right)=\lambda\frac{x}{L}-\frac{1}{2}\left(\lambda\frac{x}{L}\right)^2+O\left(\left(\lambda\frac{x}{L}\right)^3\right). \tag{10.76a,b}$$

Substitution in (10.70b) leads to

$$0=P\frac{\xi}{L}\left(1-\frac{\lambda\xi}{2L}\right)+\left(A-P\right)\left(1-\frac{\lambda}{2}\right),\quad A=P-P\frac{\xi}{L}\frac{1-\lambda\xi/2L}{1-\lambda/2}. \tag{10.77a,b}$$

To the same order of approximation, the reactions at the supports are given by

$$R_- = -A = -P + P\frac{\xi}{L}\left[1 + \frac{\lambda}{2}\left(1 - \frac{\xi}{L}\right)\right], \tag{10.78a}$$

$$R_+ = -P - R_- = -P\frac{\xi}{L}\left[1 + \frac{\lambda}{2}\left(1 - \frac{\xi}{L}\right)\right], \tag{10.78b}$$

which (1) coincide with (2.39d and e) for uniform tension $\lambda = 0$; and (2) add the lowest-order correction for nonuniform tension. The leading terms of the shape of the string (10.73a and b) are:

$$G\left(x \le \xi; \xi\right) = \frac{PL}{T_0\lambda}\left(1 - \frac{\xi}{L}\right)\lambda\frac{x}{L} = \frac{P}{T_0}\left(1 - \frac{\xi}{L}\right)x, \tag{10.79a}$$

$$G\left(x \le \xi; \xi\right) = \frac{PL}{T_0\lambda}\left(1 - \frac{x}{L}\right)\lambda\frac{\xi}{L} = \frac{P}{T_0}\left(1 - \frac{x}{L}\right)\xi, \tag{10.79b}$$

which coincide with the influence functions (2.41b and c) in the case of uniform tension.

E10.7.3 Nonlinear Deflection with Nonuniform Tension

The nonuniform tension (10.65a) is substituted in (2.6) for the case of the nonlinear deflection of an elastic string by a concentrated force (2.113b) at an arbitrary position:

$$\zeta(x) \equiv G(x;\xi): \quad \left\{T(x)\zeta'\left|1 + \zeta'^2\right|^{-1/2}\right\}' = -P\delta(x - \xi). \tag{10.80a,b}$$

A first integration yields

$$T_0\left(1 + \lambda\frac{x}{L}\right)\zeta'\left|1 + \zeta'^2\right|^{-1/2} = A - PH(x - \xi) = \begin{cases} A = R_- & \text{if } 0 \le x < \xi, \\ A - P = R_+ & \text{if } \xi < x \le L. \end{cases} \tag{10.81a} \atop \tag{10.81b}$$

Solving for the slopes leads to

$$G'(0;\xi) = \begin{cases} \left|\left[\frac{T_0}{A}\left(1 + \lambda\frac{x}{L}\right)\right]^2 - 1\right|^{-1/2} & \text{if } 0 \le x < \xi, \\ \left|\left[\frac{T_0}{A - P}\left(1 + \lambda\frac{x}{L}\right)\right]^2 - 1\right|^{-1/2} & \text{if } \xi \le x < L. \end{cases} \tag{10.82a} \atop \tag{10.82b}$$

The integration of (10.82a) leads (I.7.33) $\equiv$ (I.7.123b) to an inverse hyperbolic cosine:

$$\int \left|\left[\frac{T_0}{A}\left(1 + \lambda\frac{x}{L}\right)\right]^2 - 1\right|^{-1/2} dx = \frac{AL}{T_0\lambda}\arg\cosh\left[\frac{T_0}{A}\left(1 + \lambda\frac{x}{L}\right)\right]. \tag{10.83}$$

Using (10.83) in (10.82a [(10.82b)]) to obtain the shape of the string before (10.84a) [after (10.84b)] the point of application of the force by integration from the boundary condition at the first (10.68a) [second (01.69a)] support leads to

$$G(x;\xi) = \begin{cases} \dfrac{AL}{T_0\lambda}\left\{\arg\cosh\left[\frac{T_0}{A}\left(1 + \lambda\frac{x}{L}\right)\right] - \arg\cosh\left(\frac{T_0}{A}\right)\right\} & \text{if } 0 \le x \le \xi, \\ \dfrac{(A - P)L}{T_0\lambda}\left\{\arg\cosh\left[\frac{T_0}{A - P}\left(1 + \lambda\frac{x}{L}\right)\right] - \arg\cosh\left(T_0\frac{1 + \lambda}{A - P}\right)\right\} & \text{if } \xi \le x \le L. \end{cases} \tag{10.84a,b}$$

The continuity at the point of application of the force specifies the constant A:

$$0 = \frac{T_0 \lambda}{L}\left[G(\xi - 0; \xi) - G(\xi + 0; \xi)\right] = A\left\{\operatorname{arg\,cosh}\left[\frac{T_0}{A}\left(1 + \lambda\frac{\xi}{L}\right)\right] - \operatorname{arg\,cosh}\left(\frac{T_0}{A}\right)\right\}$$

$$-(A - P)\left\{\operatorname{arg\,cosh}\left[\frac{T_0}{A - P}\left(1 + \lambda\frac{\xi}{L}\right)\right] - \operatorname{arg\,cosh}\left[\frac{T_0}{A - P}(1 + \lambda)\right]\right\}. \tag{10.85}$$

Thus *the nonlinear deflection of an elastic string under nonuniform tension (10.65a) due to a concentrated force at an arbitrary position (10.80a and b) leads to (1) reaction at the supports (10.78a and b) where A is a positive real root of (10.85); (2) slopes (10.82a and b); and (3) shapes (10.84a and b) specifying the nonlinear Green or influence function.* The verification of the consistency of the linear (E10.7.1) and nonlinear (E10.7.3) shapes requires (E10.7.5) the power series for inverse hyperbolic cosine in (10.84a and b) for a variable larger than unity that is obtained next (E10.7.4).

E10.7.4 Power Series for the Inverse Hyperbolic Cosine

The method in Section II.7.4 can be applied starting with the derivative of the inverse hyperbolic cosine (II.7.123b) ≡ (10.86b) that can be expanded in binomial series (I.25.37) ≡ (10.86c) with coefficients (II.6.55b) ≡ (2.159a and b), valid outside the unit disk (10.86a):

$$|z| > 1: \frac{d}{dz}(\operatorname{arg\,cosh} z) = \frac{1}{\sqrt{z^2 - 1}} = \frac{1}{z}\left(1 - \frac{1}{z^2}\right)^{-1/2}$$

$$= \frac{1}{z}\sum_{n=0}^{\infty}\binom{-1/2}{n}\left(-\frac{1}{z^2}\right)^n = \frac{1}{z} + \sum_{n=1}^{\infty} a_n z^{-2n-1}. \tag{10.86a–c}$$

Integration leads to

$$\operatorname{arg\,cosh} z - \log z = C + \sum_{n=1}^{\infty} a_n \frac{z^{-2n}}{2n}, \tag{10.87}$$

where C is a constant, which can be determined in the limit $z \to \infty$ when the other terms on the r.h.s. vanish. The definition of hyperbolic cosine (10.88a) for large variables leads to (10.88b) ≡ (10.88c) specifying the value of the constant (10.88d):

$$z \equiv \cosh\zeta = \frac{e^\zeta + e^{-\zeta}}{2}, z = \lim_{\zeta\to\infty}\cosh\zeta = \frac{e^\zeta}{2}: \quad \lim_{z\to\infty}\operatorname{arg\,cosh} z = \lim_{\zeta\to\infty}\zeta = \lim_{\zeta\to\infty}\log(2z)$$

$$= \log z + \log 2, C = \log 2. \tag{10.88a–d}$$

Substituting (10.88d) and (2.159b) in (10.87) leads to the power series for the inverse hyperbolic cosine (10.89b):

$$|z| > 1: \quad \operatorname{arg\,cosh} z = \log 2 + \log z + \sum_{n=1}^{\infty}\frac{(2n-1)!!}{(2n)!!}\frac{z^{-2n}}{2n}, \tag{10.89a,b}$$

which is valid outside the unit disk (10.89a). The coefficients (10.90a) of the series (10.89b) have the ratio (10.90b):

$$u_n(z) \equiv \frac{(2n-1)!!}{(2n)!!}\frac{z^{-2n}}{2n}: \quad \frac{u_{n+1}(z)}{u_n(z)} = \frac{1}{z^2}\frac{2n+1}{2n+2}\frac{2n}{2n+2} = \frac{1}{z^2}\frac{1 + 1/2n}{(1 + 1/n)^2}$$

$$= \frac{1}{z^2}\left(1 + \frac{1}{2n}\right)\left[1 - \frac{1}{2n} + O\left(\frac{1}{n^2}\right)\right] = \frac{1}{z^2}\left[1 + O\left(\frac{1}{n^2}\right)\right], \tag{10.90a,b}$$

implying by the combined convergence test (Subsection I.29.9.1) that the *power series for the inverse hyperbolic cosine (10.89b) is (1/2) absolutely (and uniformly) convergent outside the unit disk* $|z| > 1$ *[outside a larger disk* $(z) \geq 1 + \varepsilon$ *with* $\varepsilon > 0$)]; *(3) divergent inside the unit disk; and (4/5) on the circle of convergence* $|z| = 1$, *oscillating at all points except at* $z = \pm 1$ *where it diverges* (Table 10.4).

E10.7.5 Linear Approximation and Nonlinear Corrections of All Orders

Substitution of the leading term of the power series for the inverse hyperbolic cosine (10.89b) in (10.84a and b) leads to

$$G(x \leq \xi; \xi) = \frac{AL}{T_0\lambda}\left\{\log\left[\frac{T_0}{A}\left(1+\lambda\frac{x}{L}\right)\right] - \log\left(\frac{T_0}{A}\right)\right\} = \frac{AL}{T_0\lambda}\log\left(1+\lambda\frac{x}{L}\right), \quad (10.91a)$$

$$G(x \geq \xi; \xi) = \frac{(A-P)L}{T_0\lambda}\left\{\log\left[\frac{T_0}{A-P}\left(1+\lambda\frac{x}{L}\right)\right] - \log\left(\frac{T_0}{A-P}(1+\lambda)\right)\right\}$$
$$= \frac{(A-P)L}{T_0\lambda}\left[\log\left(1+\lambda\frac{x}{L}\right) - \log(1+\lambda)\right], \quad (10.91b)$$

which coincides with the linear solution (10.68b) ≡ (10.91a) [(10.69b) ≡ (10.91b)]. The remaining terms of the series (10.89a and b) specify the nonlinear corrections of all orders:

$$G(x \leq \xi; \xi) = \frac{AL}{T_0\lambda}\log\left(1+\lambda\frac{x}{L}\right) + \sum_{n=1}^{\infty}\frac{(2n-1)!!}{(2n)!!}\frac{(T_0/A)^{2n}}{2n}\left[\left(1+\lambda\frac{x}{L}\right)^{2n}-1\right], \quad (10.92a)$$

$$G(x \geq \xi; \xi) = \frac{(A-P)L}{T_0\lambda}\left\{\log\left(1+\lambda\frac{x}{L}\right) - \log(1+\lambda)\right.$$
$$\left. + \sum_{n=1}^{\infty}\frac{(2n-1)!!}{(2n)!!}\frac{\left[T_0/(A-P)\right]^{2n}}{2n}\left[\left(1+\lambda\frac{x}{L}\right)^{2n}-(1+\lambda)^{2n}\right]\right\} \quad (10.92b)$$

In the power series (10.89b) for the inverse hyperbolic cosine outside the unit disk (10.89a) appear the coefficients in (10.87) ≡ (10.93) where (2.159b) is used:

$$\frac{1}{2n}\binom{-1/2}{n} = \frac{a_n}{2n} = \frac{(2n-1)!!}{(2n)!!\,2n}. \quad (10.93)$$

Thus *the influence function for the nonlinear deflection of an elastic string under nonuniform tension (10.65a) by a concentrated transverse force (10.80a and b) is given exactly by (10.84a and b) consisting of (1) the linear approximation (10.68b; 10.69b); and (2) the nonlinear corrections of all orders (10.92a and b). The constant* A *is given by (10.77b) in the linear case and is a positive real root of (10.85) in the nonlinear case*:

E10.7.6 Power Series for Inverse Hyperbolic and Circular Functions

Using the relations (II.7.185a) ≡ (10.94a), (II.7.186a) ≡ (10.94b), and (II.7.186b) ≡ (10.94c)

$$\operatorname{arg\,sech} z = \operatorname{arg\,cosh}\left(\frac{1}{z}\right), \operatorname{arc\,cos} z = \mp i \operatorname{arg\,cosh}(iz). \quad (10.94a,b)$$

$$\operatorname{arg\,sec} z = \mp i \operatorname{arg\,sech}(-iz) = \mp i \operatorname{arg\,cosh}\left(\frac{i}{z}\right), \quad (10.94c)$$

leads from (10.89a and b), respectively, to the series

$$|z| < 1: \quad \operatorname{arg\,sech} z = \log 2 - \log z + \sum_{n=1}^{\infty} \frac{(2n-1)!!}{(2n)!!} \frac{z^{2n}}{2n}, \tag{10.95a,b}$$

$$|z| > 1: \quad \pm i \operatorname{arc\,cos} z = \log 2 + i\frac{\pi}{2} + \log z + \sum_{n=1}^{\infty} \frac{(2n-1)!!}{(2n)} \frac{(-)^n}{2n} z^{-2n}, \tag{10.96a,b}$$

$$|z| < 1: \quad \pm i \operatorname{arc\,sec} z = \log 2 + i\frac{\pi}{2} - \log z + \sum_{n=1}^{\infty} \frac{(2n-1)!!}{(2n)!} \frac{(-)^n}{2n} z^{2n}. \tag{10.97a,b}$$

Applying the relations (II.7.189a) $\equiv$ (10.98a), (II.7.189b) $\equiv$ (10.98b), (II.7.190a) $\equiv$ (10.98c), and (II.7.190b) $\equiv$ (10.98d)

$$\operatorname{arc\,cos} z + \operatorname{arc\,sin} z = \frac{\pi}{2} = \operatorname{arc\,sec} z + \operatorname{arc\,csc} z, \tag{10.98a,b}$$

$$\pm \operatorname{arg\,cosh} z + \operatorname{arg\,sinh}(iz) = i\frac{\pi}{2} = \pm \operatorname{arg\,sech} z - i \operatorname{arg\,csch}(iz), \tag{10.98c,d}$$

respectively, to (10.96a and b), (10.97a and b), (10.89a and b), and (10.95a and b) leads to

$$|z| > 1: \quad \operatorname{arc\,sin} z = \frac{\pi}{2} \pm i \log 2 \mp \frac{\pi}{2} \pm i \log z \pm i \sum_{n=1}^{\infty} \frac{(2n-1)!!}{(2n)!!} \frac{(-)^n}{2n} z^{-2n}, \tag{10.99a,b}$$

$$|z| < 1: \quad \operatorname{arc\,csc} z = \frac{\pi}{2} \pm i \log 2 \mp \frac{\pi}{2} \pm i \log z \pm i \sum_{n=1}^{\infty} \frac{(2n-1)!!}{(2n)!!} \frac{(-)^n}{2n} z^{2n}, \tag{10.100a,b}$$

$$|z| > 1: \quad \operatorname{arg\,sinh} z = i\frac{\pi}{2} \mp i\frac{\pi}{2} \mp \log 2 \mp \log z \mp \sum_{n=1}^{\infty} \frac{(2n-1)!!}{(2n)!!} \frac{(-)^n}{2n} z^{-2n}, \tag{10.101a,b}$$

$$|z| < 1: \quad \operatorname{arg\,csch} z = -\frac{\pi}{2} \mp i \log 2 \pm \log z \pm \frac{\pi}{2} \mp i \sum_{n=1}^{\infty} \frac{(2n-1)!!}{(2n)!!} \frac{(-)^n}{2n} z^{2n}. \tag{10.102a,b}$$

Using the combined convergence test (I.29.1.1) as in (10.90a and b) follows *the convergence at all points of the complex plane (Table 10.4) of the 8 power series for inverse circular and hyperbolic functions (10.89a and b), (10.95a and b), (10.96a and b), (10.97a and b), (10.99a and b), (10.100a and b), (10.101a and b), (10.102a and b). Together with the 16 other power series in Subsection II.7.4.1 this completes a set of 24 power series expansions for all inverse circular and hyperbolic (2) functions cosine, sine, secant, cosecant, tangent, and cotangent (6) valid inside or outside (2) the unit disk. These $2 \times 6 \times 2 = 24 = 3 \times 8$ power series were proved in 3 sets of 8 by transformations of only 3 power series, namely, (1) for the inverse hyperbolic cosine (10.89a and b); and (2/3) for the inverse circular sine (II.7.167a through c) and the tangent (II.7.169a and b).*

TABLE 10.4   Power Series for Inverse Circular and Hyperbolic Functions

Function	Series	D.	D.	O.	A.C.	T.C.
arg cosh	(10.89b)	$\|z\| < 1$	$z = \pm 1$	$\|z\| = 1 \neq \pm z$	$\|z\| > 1$	$\|z\| \geq 1 + \varepsilon$
arg sinh	(10.101b)					
arg sech	(10.95b)	$\|z\| > 1$	$z = \pm 1$	$\|z\| = 1 \neq \pm z$	$\|z\| < 1$	$\|z\| \leq 1 - \delta$
arg csch	(10.102b)					
arc cos	(10.96b)	$\|z\| < 1$	$z = \pm i$	$\|z\| = 1 \neq \pm iz$	$\|z\| > 1$	$\|z\| \geq 1 + \varepsilon$
arc sin	(10.99b)					
arc sec	(10.97b)	$\|z\| > 1$	$z = \pm i$	$\|z\| = 1 \neq \pm iz$	$\|z\| < 1$	$\|z\| \leq 1 - \delta$
arc csc	(10.100b)					

Note: $\varepsilon > 0 < \delta < 1$. The convergence is indicated at all points of the complex plane for the power series of the inverse circular and hyperbolic cosine, sine, secant, and cosecant. Table II.7.4 gives the same data for the remaining inverse circular and hyperbolic functions. Thus 24 series for cyclometric functions are considered 16(8) in Table II.7.4 (10.4).

EXAMPLE 10.8 PRODUCTS AND INTEGRALS INVOLVING DERIVATES OF IMPULSES

Use the derivative of the generalized function Dirac delta and unit impulse to (1) evaluate the integrals

$$\{A_{m,n}; B_{m,n}; C_{m,n}; D_{m,n}\} \equiv \int\limits_{-\infty}^{+\infty} x^m \{\cos x; \sin x; \cosh x; \sinh x\} \delta^{(n)}(x)\,dx, \qquad (10.103a\text{-}d)$$

for all integer values of m, n; (2) simplify the products of analytic functions

$$f(x) = e^x, \cos x, \sin x, \sinh x, \cosh x, (x+a)^k \qquad (10.104a\text{-}f)$$

by the nth derivative of the unit impulse.

E10.8.1 Evaluation of Integrals Involving Derivative Impulses

The integrals are evaluated using (3.79a through c) followed by (3.22a through c):

$$m \leq n: A_{m,n} \equiv \int\limits_{-\infty}^{+\infty} x^m \cos x\, \delta^{(n)}(x)\,dx = (-)^m \left[\frac{n!}{(n-m)!}\right]^{+\infty}_{-\infty}\!\!\!\int \cos x\, \delta^{(n-m)}(x)\,dx$$

$$= (-)^n \frac{n!}{(n-m)!!} \lim_{x \to 0} \frac{d^{n-m}}{dx^{n-m}}(\cos x). \qquad (10.105)$$

Bearing in mind that

$$\cos^{(2q)}(0) = (-)^q, \quad \cos^{(2q+1)}(0) = 0, \qquad (10.106a,b)$$

the expression (10.105) vanishes if either n or m is odd and the other even; if both are even (10.107a) [odd (10.107c)] then (10.105) leads to (10.107b) [(10.107d)]:

$$m = 2p \leq n = 2q: \quad A_{2p,2q} = (-)^{q-p}\frac{(2q)!}{(2q-2p)!}, \qquad (10.107a,b)$$

$$m = 2p+1 \leq n = 2q+1: \quad A_{2p+1,2q+1} = (-)^{q-p+1}\frac{(2q+1)!}{(2q-2p)!}. \qquad (10.107c,d)$$

These results are indicated in Table 10.5 together with other cases that are treated similarly.

TABLE 10.5 Integrals Involving Derivative Impulses

Integral	$A_{m,n}$	$B_{m,n}$	$C_{m,n}$	$D_{m,n}$
Equation	(10.103a)	(10.103b)	(10.103c)	(10.103d)
$m = 2p \leq n = 2q$	$\dfrac{(-)^{q-p}(2q)!}{(2q-2p)!}$	0	$\dfrac{(2q)!}{(2q-2p)!}$	0
$m = 2p \leq n = 2q+1$	0	$\dfrac{(-)^{q-p+1}(2q+1)!}{(2q-2p+1)!}$	0	$\dfrac{(2q+1)!}{(2q-2p+1)!}$
$m = 2p+1 \leq n = 2q$	0	$\dfrac{(-)^{q-p+1}(2q)!}{(2q-2p-1)!}$	0	$\dfrac{(2q)!}{(2q-2p-1)!}$
$m = 2p+1 \leq n = 2q+1$	$\dfrac{(-)^{q-p+1}(2q+1)!}{(2q-2p)!}$	0	$\dfrac{(2q+1)!}{(2q-2p)!}$	0

Note: Evaluation for all pairs of even and odd values of (m, n) of the integrals (10.103a through d) involving derivatives of order n of the unit impulse times the power with exponent m times a circular or hyperbolic cosine or sine.

E10.8.2 Products of Derivative Impulses by Analytic Functions

The following expansions result from the Stirling–Maclaurin series for the analytic function (3.64), followed by multiplication term-by-term by derivative impulses (3.79a through c):

$$e^x \delta^{(n)}(x) = \sum_{m=0}^{n} (-)^m \binom{n}{m} \delta^{(n-m)}(x), \tag{10.108}$$

$$\cos x \, \delta^{(n)}(x) = \sum_{m=0}^{\leq n/2} (-)^m \binom{n}{2m} \delta^{(n-2m)}(x), \tag{10.109}$$

$$\sin x \, \delta^{(n)}(x) = -\sum_{m=0}^{\leq (n-1)/2} (-)^m \binom{n}{2m+1} \delta^{(n-2m-1)}(x), \tag{10.110}$$

$$\sinh x \, \delta^{(n)}(x) = -\sum_{m=0}^{\leq (n-1)/2} \binom{n}{2m+1} \delta^{(n-2m-1)}(x), \tag{10.111}$$

$$\cosh x \, \delta^{(n)}(x) = \sum_{m=0}^{\leq n/2} \binom{n}{2m} \delta^{(n-2m)}(x), \tag{10.112}$$

$$(x+a)^k \delta^{(n)}(x) = \sum_{m=0}^{\leq n,k} (-)^m \frac{k!\,n!}{m!(k-m)!(n-m)!} a^{k-m} \delta^{(n-m)}(x), \tag{10.113}$$

For example,

$$e^x \delta^{(n)}(x) = \sum_{m=0}^{\infty} \frac{x^m}{m!} \delta^{(n)}(x) = \sum_{m=0}^{\infty} (-)^m \frac{n!}{m!(n-m)!} \delta^{(n-m)}(x) \tag{10.114}$$

coincides with (10.108) ≡ (10.114). Another method to arrive at the same result is to use the definition of derivative impulse (3.22a through c) involving a test function:

$$\left[e^x\delta^{(n)}(x),\Phi(x)\right] \equiv \int_{-\infty}^{+\infty} e^x\Phi(x)\delta^{(n)}(x)dx = (-)^n \lim_{x\to 0}\frac{d^n}{dx^n}\left[e^x\Phi(x)\right]$$

$$=(-)^n\sum_{m=0}^{n}\binom{n}{m}\Phi^{(n-m)}(0)=\left[(-)^{2n-m}\sum_{m=0}^{n}\binom{n}{m}\delta^{(n-m)}(x),\Phi(x)\right], \quad (10.115)$$

where the Leibnitz chain rule (I.13.31) is used. The two results (10.114) ≡ (10.115) ≡ (10.108) are equivalent. The same two methods apply to (10.109 through 10.113).

EXAMPLE 10.9: INFLUENCE FUNCTION FOR A FOURTH-ORDER DIFFERENTIAL EQUATION

Obtain the influence or Green function for all real values of the coefficients (10.116a and b) in the following linear fourth-order ordinary differential equation with constant coefficients (l.o.d.e.c.c.):

$$a,b\in|R, a\neq 0: \quad a\frac{d^4}{dx^4}\left[G(x;\xi)\right]+bG(x;\xi)=\delta(x-\xi). \quad (10.116a-c)$$

Compare with the particular case of linear bending of a bar clamped horizontally at one end and free at the other end with a concentrated force applied at an arbitrary position.

The Fourier transform (3.153c) of the l.o.d.e.c.c. (10.116c) is (10.117a):

$$\left(ak^4+b\right)\tilde{G}(x;\xi)=\frac{e^{-ik\xi}}{2\pi}, \quad G(x;\xi)=\frac{1}{2\pi}\int_{-\infty}^{+\infty}\frac{e^{ik(x-\xi)}}{ak^4+b}, \quad (10.117a-c)$$

leading by inversion (3.153c) to the influence function (10.117b). There are three cases (I, II, III) and four subcases (IA, IB, II, III) for the evaluation of the integral.

E10.9.1 Real Symmetric and Imaginary Conjugate Roots

In case I, the roots of the characteristic polynomial (10.118b) for (10.118a) are a real symmetric and an imaginary conjugate pair (10.118c) with the same modulus (10.118d):

$$A:\frac{b}{a}<0: \quad P_4(k)=ak^4+b=a\prod_{n=1}^{4}(k-k_n), \quad (10.118a,b)$$

$$k_{1-4}=\sqrt[4]{-\frac{b}{a}}=\sqrt[4]{\left|\frac{b}{a}\right|}=c\sqrt[4]{1}=c\sqrt[2]{\pm 1}=\pm c, \pm ic, \quad c\equiv\left|\frac{b}{a}\right|^{1/4}, \quad (10.118c,d)$$

leading to the factorization:

$$P_4(k)=ak^4+b=a(k-c)(k+c)(k-ic)(k+ic). \quad (10.118c)$$

For $x > \xi$ ($x < \xi$), the path of integration is closed in the upper (lower) half k-complex plane [Figure 10.8a (b)] with the simple pole at $+ic$ ($-ic$) in the interior. Concerning the poles at $\pm c$ on the real axis, a choice of indentation must be made, for example, indenting both poles $\pm c$

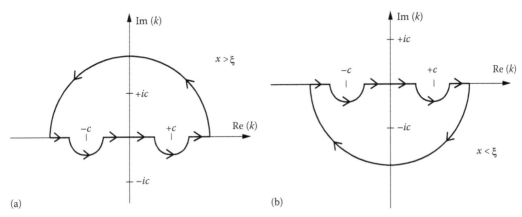

(a)   (b)

FIGURE 10.8 The influence or Green function of an ordinary linear differential operator with constant coefficients of order N can be calculated by evaluating a Fourier integral in the complex plane with poles whose order adds to N. Thus, in the case of a fourth-order operator, there are four poles counted by order of multiplicity. If besides the leading fourth-order derivative there is only one more term linear in the function, and the two coefficients are real, four subcases arise. In the subcase IA, the poles lie symmetrically on the positive and negative real and imaginary axes. Closing the path of integration along the real axis by a large half-circle in the upper (lower) complex plane [Figure 10.8a (b)], (1) the pole in the positive (negative) imaginary axis lies inside; and (2) the poles on the real axis are indented (excluded).

in the upper loop [Figure 10.8a (b). Then the influence function (10.117b) is evaluated as $\pm 2\pi$ i ($\pm \pi i$) times the residue at the pole in the interior (on the boundary) leading, for $x < \xi$ ($x > \xi$), to the following:

$$G\left(x<\xi,\xi\right)=-\frac{i}{a}\lim_{k\to -ic}\frac{e^{ik(x-\xi)}}{\left(k-ic\right)\left(k^2-c^2\right)}=-\frac{e^{c(x-\xi)}}{4c^3a},\qquad (10.119a)$$

$$\begin{aligned}
G\left(x>\xi,\xi\right) &=\frac{i}{a}\lim_{k\to ic}\frac{e^{ik(x-\xi)}}{\left(k+ic\right)\left(k^2-c^2\right)}+\frac{i}{2a}\lim_{k\to c}\frac{e^{ik(x-\xi)}}{\left(k+c\right)\left(k^2+c^2\right)} \\
&\quad +\frac{i}{2a}\lim_{k\to -c}\frac{e^{ik(x-\xi)}}{\left(k-c\right)\left(k^2+c^2\right)} \\
&=-\frac{e^{-c(x-\xi)}}{4c^3a}+\frac{i}{8c^3a}\left[e^{ic(x-\xi)}-e^{-ic(x-\xi)}\right] \\
&=-\frac{e^{-c(x-\xi)}}{4c^3a}-\frac{1}{4c^3a}\sin\left[c\left(x-\xi\right)\right].
\end{aligned}\qquad (10.119b)$$

Thus *the l.o.d.e.c.c. (10.116c) has in the case (10.118a through e) the influence function*

$$IA:\ 0<\frac{b}{a}\equiv c^4:\ G\left(x;\xi\right)=-\frac{1}{4c^3a}\times\begin{cases}\exp\left[c\left(x-\xi\right)\right] & \text{if } x<\xi, \\[2mm] \exp\left[c\left(\xi-x\right)\right]+\sin\left[c\left(x-\xi\right)\right] & \text{if } x>\xi.\end{cases}\qquad (10.120a\text{-}c)$$

The influence function is symmetric, $G(x;\xi) = G(\xi;x)$, for the first terms in (10.120b and c) because the differential operator in (10.116c) is self-adjoint; the second term on the r.h.s. of (10.120c) arises from an asymmetric indentation of the path integration.

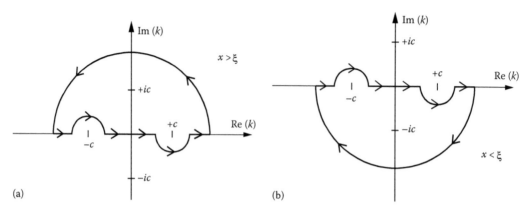

FIGURE 10.9 As in Figure 10.8 for the subcase IA, with a single difference in subcase IB concerning item (2): the indentation includes one pole (the opposite pole), for example, [Figure 10.9a (b)], the path of integration along the real axis is deformed to indent the pole on the positive (negative) real axis.

E10.9.2 Symmetric and Unsymmetric Indentation of the Paths of Integration

In the case of symmetric indentation of the paths of integration, leaving the pole at $+c$ $(-c)$ in the upper (lower) loop [Figure 10.9a (b)], the influence functions are given by

$$G(x><\xi,\xi)=\pm\frac{i}{a}\lim_{k\to\pm ic}\frac{e^{ik(x-\xi)}}{\left(k\pm ic\right)\left(k^{2}-c^{2}\right)}\pm\frac{i}{2a}\lim_{k\to\pm c}\frac{e^{ik(x-\xi)}}{\left(k\pm c\right)\left(k^{2}+c^{2}\right)}$$

$$=-\frac{e^{\mp c(x-\xi)}}{4c^{3}a}+\frac{i}{8c^{3}a}e^{\pm ic(x-\xi)}. \tag{10.121}$$

Thus *the influence function for the l.o.d.e.c.c. (10.116c) in the case (10.118a through e) is given in the case I, subcase IA (IB) by (10.120b and c) [(10.122b and c)] for an unsymmetric (symmetric) indentation [Figures 10.8a and b (Figures 10.9a and b)] of the poles on the real axis:*

$$IB:0<\frac{b}{a}\equiv c^{4}:\ \ G(x;\xi)=-\frac{1}{4c^{3}a}\times\begin{cases}\exp\left[c(x-\xi)\right]-\dfrac{i}{2}\exp\left[ic(\xi-x)\right] & \text{if } x<\xi\,,\\[2mm]\exp\left[c(\xi-x)\right]-\dfrac{i}{2}\exp\left[ic(x-\xi)\right] & \text{if } x>\xi,\end{cases} \tag{10.122a-c}$$

showing that the influence function is (1) symmetric $G(x;\xi)\equiv(\xi;x)$ [unsymmetric $G(x;\xi)\neq G(\xi;x)$]; and (2) real (complex). Thus an unsymmetric indentation can lead to a complex influence function for a real ordinary differential equation. Both the symmetric (Figure 10.9a and b) and unsymmetric (Figure 10.8a and b) indentations could be made in reverse, and the calculation would be similar.

E10.9.3 Two Pairs of Complex Conjugate Roots

In the case II, the characteristic polynomial (10.118b) for (10.123a) has two pairs of complex conjugate roots (10.123b) lying on the diagonals of the four quadrants at a distance (10.118d) from the origin (Figure 10.10a):

$$\frac{b}{a}>0:\ \ k_{1-4}=\sqrt[4]{-\frac{b}{a}}=\sqrt[4]{-c^{4}}=c\sqrt[4]{-1}=c\sqrt{\pm i}=c\,\frac{\pm1\pm i}{\sqrt{2}}=\pm\lambda\pm i\lambda,\ \ \lambda=\frac{c}{\sqrt{2}}=\frac{1}{\sqrt{2}}\left|\frac{b}{a}\right|^{1/4}$$

$$\tag{10.123a,b}$$

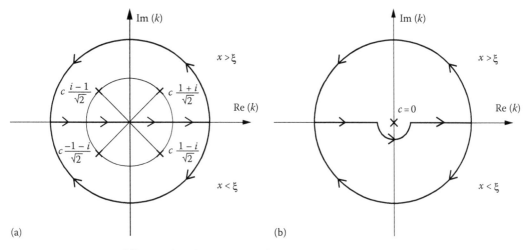

FIGURE 10.10 Case II following the subcases IA (IB) of the case I in Figure 10.8 (10.9) involves again four simple poles at the same distance from the origin, this time not on the real and imaginary axes (Figures 10.8 and 10.9) but rather on the diagonals of quadrants (a). In this case, closing the path of integration by a large half-circle in the upper or lower half-complex plane leaves two poles inside in both instances. The remaining case III with all poles coincident at the origin as a pole of order four leads to an indentation either for the upper (b) or the lower path of integration.

leading to the factorization:

$$P_4(k) = a\,k^4 + b = a\left(k - \lambda - i\,\lambda\right)\left(k - \lambda + i\,\lambda\right)\left(k + \lambda - i\,\lambda\right)\left(k + \lambda + i\,\lambda\right) \qquad (10.123e)$$

Thus the influence function (10.117b) for $x > \xi$ ($x < \xi$) equals $+2\pi i$ ($-2\pi i$) times the residues at the poles in the upper (lower) complex k-plane leading to

$$
\begin{aligned}
G\left(x \gtrless \xi\right) = {}& \pm \frac{i}{a} \lim_{k \to \lambda \pm i\lambda} \frac{e^{ik(x-\xi)}}{\left(k - \lambda \pm i\,\lambda\right)\left(k + \lambda + i\,\lambda\right)\left(k + \lambda - i\,\lambda\right)} \\
& \pm \frac{i}{a} \lim_{k \to -\lambda \pm i\lambda} \frac{e^{ik(x-\xi)}}{\left(k + \lambda \pm i\,\lambda\right)\left(k - \lambda - i\,\lambda\right)\left(k - \lambda + i\,\lambda\right)} \\
= {}& \frac{1}{8\lambda^3 a}\,\frac{1}{1 \pm i}\,\exp\!\left[\,i\lambda\left(x-\xi\right) \pm \lambda\left(x-\xi\right)\,\right] \\
& + \frac{1}{8\lambda^3 a}\,\frac{1}{1 \mp i}\,\exp\!\left[\,-i\lambda\left(x-\xi\right) \mp \lambda\left(x-\xi\right)\,\right] \\
= {}& \frac{1}{c^3\,a\,4\sqrt{2}}\,\exp\!\left[\,\mp \lambda\left(x-\xi\right)\,\right] \\
& \times \left\{\left(1 \mp i\right)\exp\!\left[\,i\lambda\left(x-\xi\right)\,\right] - \left(1 \pm i\right)\exp\!\left[\,-i\lambda\left(x-\xi\right)\,\right]\right\}. \qquad (10.124)
\end{aligned}
$$

Thus *the influence function for the l.o.d.e.c.c. (10.116c) in the case (10.123a through c) is*

$$\text{II}: \quad 0 < \frac{b}{a} = c^4 = 4\lambda^2 :$$

$$
G(x;\xi) = \frac{1}{8\lambda^3 a} \times
\begin{cases}
\exp\!\left[\lambda\left(x-\xi\right)\right]\left\{\cos\!\left[\lambda\left(x-\xi\right)\right] + \sin\!\left[\lambda\left(x-\xi\right)\right]\right\} & \text{if } x \le \xi, \\[2mm]
\exp\!\left[\lambda\left(\xi-x\right)\right]\left\{\cos\!\left[\lambda\left(x-\xi\right)\right] + \sin\!\left[\lambda\left(\xi-x\right)\right]\right\} & \text{if } x \ge \xi,
\end{cases}
\qquad (10.125a\text{-}c)
$$

which is symmetric because the operator is self-adjoint.

E10.9.4 Quadruple Root or Pole of Order Four

The remaining case III (10.126a) is a quadruple root at the origin (10.126b) (Figure 10.10b) that is indented for the upper loop, so that (1) the integral for $x < \xi$ vanishes; and (2) the integral for $x > \xi$ is evaluated (10.126c) as πi times the residue (I.15.33b) at the pole of order four:

$$\text{III}: b = 0, k_{1-4} = 0: \quad G(x;\xi) = \frac{1}{2\pi} \int\limits_{-\infty}^{+\infty} \frac{e^{ik(x-\xi)}}{ak^4} dk$$

$$= H(x-\xi)\frac{i}{2a}\frac{1}{3!}\lim_{k\to 0}\frac{d^3}{dk^3}\left[e^{ik(x-\xi)}\right]$$

$$= \frac{(x-\xi)^3}{12a} H(x-\xi). \tag{10.126a–c}$$

Thus *the influence function for the l.o.d.e.c.c. (10.125a) is (10.127b and c):*

$$a\frac{d^4}{d\lambda^4}\left[G(x;\xi)\right] = \delta(x-\xi): \quad G(x;\xi) = \begin{cases} 0 & \text{if } x < \xi, \\ \dfrac{(x-\xi)^3}{12a} & \text{if } x > \xi. \end{cases} \tag{10.127a–c}$$

This is applied next to the linear bending of a beam.

E10.9.5 Linear Bending of a Clamped-Free Beam

The differential equation for the linear bending of a beam (4.89) $\equiv$ (10.128a) is of the form (10.127a) with parameter (10.128b):

$$EI\frac{d^4\zeta}{d\lambda^4} = P\delta(x-\xi), \quad a = \frac{EI}{P}: \zeta(x) = A + Bx + Cx^2 + Dx^3 + \frac{P}{12EI}(x-\xi)^3 H(x-\xi); \tag{10.128a–c}$$

to the influence function (10.127c) a polynomial of third degree was added in (10.128c), which is the solution of the unforced l.o.d.e.c.c. (10.128a) with zero on the r.h.s. The boundary conditions for a bar clamped horizontally at one end (10.127a and b) $\equiv$ (4.90b and c) and free at the other end (10.127c and d) $\equiv$ (4.90d and e):

$$\zeta(0) = 0 = \zeta'(L), \quad \zeta''(L) = 0 = \zeta'''(L):$$

$$A = 0 = B, \quad D = -\frac{P}{12EI}, \quad C = -3DL - \frac{P}{4EI}(L-\xi) = \frac{P\xi}{4EI}, \tag{10.129a–h}$$

determine the four constants of integration (10.129e through h). Substitution of (10.129e through h) in (10.128c) specifies the shape of the elastic of the bent bar:

$$\zeta(x) = \frac{P}{12EI}\left[x^2(3\xi - x) + (x-\xi)^3 H(x-\xi)\right], \tag{10.130}$$

which coincides with (4.96a and b) $\equiv$ (10.131a and b):

$$\zeta(x) = \frac{P}{12EI} \times \begin{cases} x^2(3\xi - x) & \text{if } 0 \le x \le \xi. & (10.131a) \\ \xi^2(3x-\xi) & \text{if } \xi \le x \le L. & (10.131b) \end{cases}$$

This was a particular case of the influence function for the l.o.d.e.c.c. (10.116b); all cases are summarized in Table 10.6.

TABLE 10.6 Influence Function for the Fourth-Order l.o.d.e.c.c. (10.116b)

Case		Subcase		Influence	
Number	Parameter	Letter	Identation	Function	Figure
I	$\dfrac{b}{a} < 0$	IA	Unsymmetric	(10.120a through c)	10.8a and b
		IB	Symmetric	(10.122a and b)	10.9a and b
II	$\dfrac{b}{a} > 0$	—	—	(10.125a through c)	10.10a
III	$b = 0$	—	—	(10.126a through c)	10.10b

Note: The influence or Green function for a linear ordinary differential equation with constant coefficients can be obtained evaluating a Fourier integral in the complex plane. In the case of a fourth-order equation having besides the leading fourth-order derivative only a term involving the function there are three cases and four subcases indicated in Table 10.6.

EXAMPLE 10.10: CLAMPED HEAVY BAR SUPPORTED BY A STRAIGHT SPRING AT THE FREE END

Consider the weak bending of a uniform bar under its own weight, clamped at one end and supported at the other end by a straight spring exerting a force proportional to the displacement from the undeflected position.

The bar of length L, with constant cross section of moment of inertia I, homogeneous of mass density ρ, and the Young's modulus E, weak bending under its own weight in the gravity field g, has a shape specified by (4.55a) ≡ (10.132a):

$$\text{EI}\,\zeta''''(x) = \rho g; \quad \zeta(0) = 0 = \zeta'(0). \tag{10.132a–c}$$

The boundary conditions (10.132b and c) correspond to clamping at one endpoint $x = 0$. The other $x = L$ endpoint is supported by a straight spring (Figure 10.11), exerting no torque, but only a force proportional to the displacement *and opposite to it through the resilience p of the spring*:

$$\zeta''(L) = 0, \quad F(L) = EI\zeta'''(L) = p\zeta(L); \tag{10.133a,b}$$

The solution of (10.132a) satisfying the boundary conditions (10.132b and c) is

$$\zeta(x) = \frac{\rho g}{24\text{EI}} x^2 \left(x^2 + AxL + BL^2 \right). \tag{10.134}$$

The constants of integration are determined (10.133a and b) from the boundary conditions (10.131a and b):

$$b \equiv \frac{\text{EI}}{pL^3}: \quad 6 + 3A + B = 0 = 1 + A + B - b(24 + 6A), \tag{10.135a–c}$$

involving the dimensionless parameter (10.135a) that compares the resilience of the linear spring p with the bending stiffness of the bar. Solving (10.135b and c) specifies the constants of integration (10.136a and b)

$$2(1 + 3b)\{A, B\} = \{-24b - 5, 3(1 + 12b)\}, \tag{10.136a,b}$$

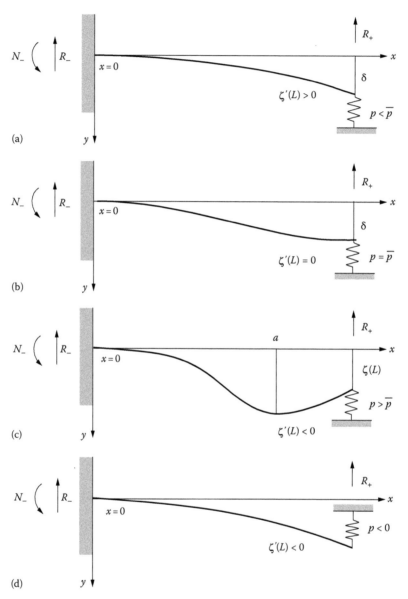

FIGURE 10.11 The linear deflection of a uniform bar horizontally clamped at one end by its own weight leads to a positive slope at the free end (a) that may be countered by using a linear spring as support. If the spring has the critical resilience, it causes the bar to be horizontal at the far end (b); if the resilience of the linear spring is smaller (larger) than the critical value, the slope at the far end [Figure 10.11a (c)] remains positive (is reversed into negative). In a through c, the spring is positioned below the bar, opposing its deflection, and insuring elastic stability. If the position of the linear spring is reversed to above the bar (d) it increases the deflection and for sufficiently high resilience causes an unbounded deflection corresponding to linear elastic instability.

appearing in

$$\zeta(x) = \frac{\rho g}{24EI} x^2 \left[x^2 + AL(x - 3L) - 6L^2 \right], \tag{10.137}$$

the shape of the elastica.

Thus *the weak bending of a bar under its own weight (10.132a), clamped at one end (10.132b and c) and held at the other end by a straight spring (10.133a and b) leads (10.135a through c, 10.136a and b) to the shape of the elastica (10.134) $\equiv$ (10.137) and slope:*

$$\zeta'(x) = \frac{\rho g}{24EI} x \left[4x^2 + 3AL(x - 2L) - 12L^2 \right], \tag{10.138}$$

$$\frac{24EI}{\rho g L^3} \zeta'(L) = -3A - 8 = \frac{24b - 1}{2 + 6b}. \tag{10.139}$$

The slope at the far end (10.139) vanishes (10.140a) for (10.140b) $\equiv$ (10.140c) a linear spring of **critical resilience** *(10.140d):*

$$\zeta'(L) = 0: \quad b = \frac{1}{24}, \quad A = -\frac{8}{3}, \quad \bar{p} \equiv \frac{24EI}{L^3}. \tag{10.140a–d}$$

Thus, three cases arise: (1/2) if the resilience of the spring is (10.141a) less than (equal to) the critical value (10.140d), the slope is positive (zero) at the far end [Figure 10.11a (b)], where the maximum deflection (10.141b) occurs:

$$p \le \bar{p}: \quad \delta \equiv \zeta(L) = -\frac{\rho g L^4}{24EI}(2A + 5) = \frac{\rho g L^4}{8EI} \frac{3b}{1 + 3b}; \tag{10.141a,b}$$

and (3) if the resilience of the spring exceeds (10.142a) the critical value (10.140d), the slope is negative (Figure 10.11c) at the far end and the maximum deflection (10.142d) occurs before (10.142b) at (10.142c):

$$p > \bar{p}: \quad \zeta'(a) = 0, \quad \frac{a}{L} = \frac{3A}{8} \left(-1 + \left| 1 + \frac{32(A + 2)}{3} \right|^{1/2} \right), \quad \zeta(a) > \zeta(L). \tag{10.142a–d}$$

The bending moment (10.143a) specifies the value at the clamped support (10.143b):

$$-M(x) = EI\zeta''(x) = \frac{\rho g}{4} \left[2x^2 + AL(x - L) - 2L^2 \right], \tag{10.143a}$$

$$N_- = -M(0) = -\frac{\rho g L^2}{4}(A + 2) = \frac{\rho g L^2}{8} \frac{1 + 12b}{1 + 3b}. \tag{10.143b}$$

The transverse force (10.144a) specifies the following: (1) the reaction at the end of the bar (10.144b), which corresponds to the force exerted by the spring (10.133b) to cause the deflection (10.141b):

$$F(x) = EI\zeta'''(x) = \rho g \left(x + \frac{AL}{4} \right), \tag{10.144a}$$

$$R_+ = F(L) = \rho g L \left(1 + \frac{A}{4}\right) = \frac{\rho g L}{8} \frac{3}{1+3b} = \frac{EI}{L^3 b} \zeta(L) = p\zeta(L), \qquad (10.144b)$$

$$R_- = -F(0) = -\frac{\rho g A L}{4} = \frac{\rho g L}{8} \frac{5+24b}{1+3b}, \quad R_+ + R_- = \rho g L; \qquad (10.144c,d)$$

and (2) the sum with the reaction force (10.144c) at the clamped support equals the total weight of the bar (10.144d). The preceding analysis assumes that the linear spring is placed below the end of the bar (Figure 10.11a through c) so as to oppose the bending, corresponding to p > 0 in (10.133b) and b > 0 in (10.135a). If the spring is placed above the bar to increase the bending (Figure 10.11d), then b < 0; there would (10.145a) be divergence (10.137) in (10.136a and b):

$$\zeta(L) = \infty: \quad b = -3, \quad p = -EI/3L^3, \qquad (10.145a–c)$$

for (10.145b) corresponding to the minimum spring resilience (10.145c) for instability. In the case p = 0, the linear spring is absent, leading to b = ∞ in (10.135a) and A = −4 in (10.136a); then, the deflection (10.137) ≡ (4.78a) coincides with that for a clamped-free bar. The preceding results for a linear spring are collected in Table 10.7 together with those for a rotary spring (Example 10.11).

EXAMPLE 10.11: CANTILEVER BAR SUPPORTED ON A ROTARY SPRING AT THE FREE END

Consider the weak bending of a uniform bar under its own weight, clamped at one end and supported at the other end on a rotary spring that exerts a torque proportional to the slope.

The equation of the elastica (10.132a) and boundary conditions (10.132b and c) at the clamped end still hold; the replacement of a straight (Figure 10.11a through d) by a rotary (Figure 10.12a and b) spring at the other end leads, instead of (10.133a and b), to the boundary conditions (10.146a and b), stating that the rotary spring causes no transverse force but only a bending moment proportional to the slope (10.146b):

$$\zeta'''(L) = 0, \quad M(L) = -EI\zeta''(L) = q\zeta'(L). \qquad (10.146a,b)$$

The method of solution is similar for a linear (rotary) spring in Example 10.10 (10.11), and the results are listed together in Table 10.7. The boundary conditions (10.146a and b) determine (10.147b and c) constants of integration in (10.134):

$$b \equiv \frac{qL}{EI}: \quad A = -4, \quad B = 2\frac{3+2b}{1+b}, \qquad (10.147a–c)$$

in terms of the parameter (10.147a) comparing the resilience of the rotary spring q with the bending stiffness of the bar. Substituting (10.147b) in (10.134), it follows that the weak bending of a bar under its own weight (10.132a), clamped at one end (10.132b and c) and supported at the other end on a rotary spring (10.146a and b) leads (10.134; 10.147a through c) to the shape of the elastica (10.148a) and the slope (10.148b):

$$\{\zeta(x), \zeta'(x)\} = \frac{\rho g}{24EI}\{x^2(x^2 - 4xL + BL^2), 2x(2x^2 - 6xL + BL^2)\}, \qquad (10.148a,b)$$

The slope at the far end is positive (10.149a) if the condition (10.149b) is met (10.149c):

$$\zeta'(L) = \frac{\rho g L^3}{12EI}(B-4) = \frac{\rho g}{6EI}\frac{L^3}{1+b} > 0: \quad 0 < B - 4 = \frac{2}{1+b}, \quad b > -1; \qquad (10.149a–c)$$

thus a rotary spring at the end of a clamped bar cannot change its curvature if b > 0, unlike the linear spring (Example 10.10). The maximum deflection is always at the far end:

$$\delta = \zeta_{max} = \zeta(L) = \frac{\rho g L^4}{24EI}(B-3) = \frac{\rho g L^4}{24EI}\frac{3+b}{1+b}. \tag{10.150}$$

The same conclusion can be obtained from the slope (10.148b) whose roots (10.151a) are (10.151b):

$$\zeta'(a) = 0: \quad \frac{a}{L} = \frac{3}{2}\left[1 \pm \sqrt{1 - \frac{2B}{9}}\right] = \frac{3}{2}\left[1 \pm \sqrt{\frac{b-3}{9(1+b)}}\right] \tag{10.151a,b}$$

leading to two cases: (1) if $b < 3$ the roots (10.151b) are complex; (2) if $b \geq 3$ one root is negative and the other exceeds $3L/2$. In both cases there are no roots in the interval $0 \leq a \leq L$, so the slope never vanishes. *The bending moment (10.152a) specifies the value at (1) the clamped support (10.152b):*

$$-M(x) = EI\zeta''(x) = \frac{\rho g}{12}\left(6x^2 - 12xL + BL^2\right), \tag{10.152a}$$

$$N_- = -M(0) = -\frac{\rho g BL^2}{12} = -\frac{\rho g L^2}{6}\frac{3+2b}{1+b}, \tag{10.152b}$$

$$N_+ = M(L) = -\frac{\rho g L^2}{12}(B-6) = \frac{\rho g L^2}{6}\frac{b}{1+b} = \frac{bEI}{L}\zeta'(L) = q\zeta'(L), \tag{10.152c,d}$$

$$N_+ + N_- = -\frac{\rho g L^2}{2}, \tag{10.152e}$$

and (2) far end (10.152c), which equals (10.152d) the bending moment due to the rotary spring (10.146b) for the slope (10.149b). The sum of the two moments is (10.152d). The transverse force (10.153a) confirms that the reaction at the support is the weight of the bar (10.153b):

$$F(x) = EI\zeta'''(x) = \rho g(L-x), \quad R_- = F(0) = \rho g L = P. \tag{10.153a,b}$$

In the preceding analysis, it is assumed that the rotary spring is curled so as to oppose the bending (Figure 10.12a) corresponding to q > 0 in (10.146b) and b > 0 in (10.147a). If the rotary spring is curled to increase the bending (Figure 10.12b) corresponding to b < 0, there is instability (10.154a and b) in (10.148a and b) if (10.154c) holds (10.147c):

$$\zeta(L) = \infty = \zeta'(L): \quad B = \infty, \quad b = -1, \quad q = -EI/L, \tag{10.154a–e}$$

which corresponds to the minimum rotary spring stiffness for instability (10.154e). The case $q = 0$ when there is no rotary spring, leads to $b = 0$ in (10.147a) and $B = 6$ in (10.147c); then, the deflection (10.148b) $\equiv$ (4.78a) is that of a clamped-free bar. The main difference (Table 10.7) between an elastic bar with a clamped end and a straight (rotary) spring at the opposite end is that the curvature can (cannot) be reversed [Figure 10.11 (10.12)] if $b > 0$.

EXAMPLE 10.12: PROPERTIES OF IMPULSE AS LIMIT(S) OF FAMILIES OF FUNCTIONS

Use the limit(s) of a family of analytic functions to obtain: (1) the Fourier transform of the unit impulse; (2) the unit impulse with cylindrical symmetry.

TABLE 10.7   Weak Bending of Heavy Clamped-Sprung Uniform Bar

One End Clamped Plus Other with Spring		Example 10.10 Linear	Example 10.11 Rotary
Parameters	b	$b \equiv EI/pL^3$	$b \equiv EI/qL$
	A	$A = -\dfrac{5+24b}{2(1+3b)}$	$A = -4$
	B	$B = \dfrac{3}{2}\dfrac{1+12b}{1+3b}$	$B = 2\dfrac{3+2b}{1+b}$
Elastica: shape $\zeta(x) = \dfrac{\rho g}{24EI} \times \ldots$		$x^2[x^2 + AL(x-3L) - 6L^2]$	$x^2(x^2 - 4xL + BL^2)$
Slope: $\zeta'(x) = \dfrac{\rho g L^3}{24EI} \times \ldots$		$x[4x^2 + 3AL(x-2L) - 12L^2]$	$2x(2x^2 - 6xL + BL^2)$
Slope at far end: $\zeta'(L) = \dfrac{\rho g L^3}{24EI} \times \ldots$		$-3A - 8 = \dfrac{24b-1}{2(1+3b)}$	$2B - 8 = \dfrac{4}{1+b}$
Critical stiffness for $\zeta'(L) = 0$		$\bar{p} = \dfrac{24EI}{L^3}$	$\bar{q} = \infty$
Maximum deflection: $p = \bar{p}$		1	—
$\delta = \zeta(L) = \dfrac{\rho g L^3}{24EI} \times \ldots$		$\dfrac{1}{9}$	
Maximum deflection: $p < \bar{p}$		$-2A - 5 = \dfrac{3b}{1+3b}$	$B - 3 = \dfrac{3+b}{1+b}$
$\delta = \zeta(L) = \dfrac{\rho g L^3}{24EI} \times \ldots$			
Maximum deflection: $p < \bar{p}$		$\dfrac{3A}{8}\left(-1 + \left\lvert 1 + \dfrac{32(A+2)}{3A^2}\right\rvert^{1/2}\right)$	—
$\delta = \zeta(a) \, at \, \dfrac{a}{L} = x \cdots$			
Bending moment: $-M(x) = \dfrac{\rho g}{4}$		$2x^2 + AL(x-L) - 2L^2$	$2x^2 - 4xL + \dfrac{B}{3}L^2$
$N_- = -M(0) = \rho g L^2 \times \ldots$		$-\dfrac{1}{2} - \dfrac{A}{4} = \dfrac{1+12b}{8+24b}$	$\dfrac{B}{12} = \dfrac{3+2b}{6+6b}$
At supports: $N_+ = M(L) = \rho g L^2 \times \ldots$		0	$-\dfrac{B}{12} + \dfrac{1}{2} = \dfrac{b}{6+8b}$
Transverse force: $F(x) = \rho g \times \ldots$		$x + \dfrac{AL}{4}$	$x - L$
At supports: $R_- = -F(0) = \rho g L \times \ldots$		$-\dfrac{A}{4} = \dfrac{5+24b}{8+24b}$	1
$R_+ = F(L) = \rho g L \times \ldots$		$1 + \dfrac{A}{4} = \dfrac{3}{8+24b}$	0
Negative spring stiffness for instability		$\underline{p} = -\dfrac{EI}{3L^3}$	$\underline{q} = -\dfrac{EI}{L}$

Note: Comparison of the weak bending of a uniform bar under its own weight, clamped horizontally at one end when supported at the opposite by a straight (rotary) spring [Figure 10.11 (10.12)], which may be positioned to (i) oppose bending [Figure 10.11a through c (10.12a)] ensuring elastic stability; and (ii) increase bending [Figure 10.11d (10.12b)] possibly leading to elastic instability.

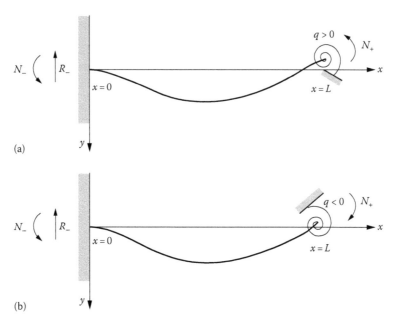

FIGURE 10.12 The weak deflection of a uniform bar under its own weight with one end clamped horizontally and the other supported on a straight (rotary) spring [Figure 10.11 (10.12)] leads to linear (1) elastic stability if the spring is positioned below the bar [Figure 10.11a through c (10.12a)] to oppose bending (twist); and (2) elastic instability if the spring has sufficiently high resilience and is placed above [Figure 10.11d (10.22a)] the bar to increase bending (twist).

E10.12.1 Fourier Transform of the Unit Impulse

The proof that the Fourier transform of the unit impulse is

$$2\pi\tilde{\delta}(k) = \int_{-\infty}^{+\infty}\delta(x)e^{-ikx}dx = 1 \tag{10.155}$$

can be made using the definition of the unit impulse as the limit of a family of functions (1.37) [(1.39a)]. The Fourier transform (3.153c) is calculated for the family of functions (1.37) using the Gaussian integral (1.13):

$$2\pi\tilde{\delta}_\sigma(k) = \int_{-\infty}^{+\infty}\delta_\sigma(x)e^{-ikx}dx = \frac{1}{\sqrt{\pi\sigma}}\int_{-\infty}^{+\infty}\exp\left(-ikx - \frac{x^2}{\sigma}\right)dx$$

$$= \frac{1}{\sqrt{\pi\sigma}}\int_{-\infty}^{+\infty}\exp\left(-\frac{x^2}{\sigma}\right)\cos(kx)dx = \exp\left(-\frac{\sigma}{4}k^2\right), \quad (10.156)$$

which is a cosine transform. The Fourier transform of the family of functions (1.39a) is calculated by the method of residues (Section I.17.5):

$$\text{Im}(k)<>0: \quad 2\pi\tilde{\Delta}_\sigma(k) = \int_{-\infty}^{+\infty}\Delta_\sigma(x)e^{-ikx}dx = \frac{\sigma}{\pi}\int_{-\infty}^{+\infty}\frac{e^{-ikx}}{(x-i\sigma)(x+i\sigma)}dx$$

$$= \pm 2\pi i\frac{\sigma}{\pi}\lim_{x\to\pm i\sigma}(x\mp i\sigma)\frac{e^{-ikx}}{(x-i\sigma)(x+i\sigma)}$$

$$= \pm 2i\sigma\lim_{x\to\pm i\sigma}\frac{e^{-ikx}}{x\pm i\sigma} = \exp(\pm k\sigma), \tag{10.157a,b}$$

where: (1) the path of integration along the real x-axis is closed by a half-circle with center at the origin and large radius in the upper (lower) half x-plane for $\mathrm{Im}(k) > 0$ [$\mathrm{Im}(k) < 0$] in (10.157a); and (2) since the loop is taken in the positive (negative) direction, that is, counterclockwise (clockwise), the integral equals $+2\pi i$ ($-2\pi i$) the residue at the pole $x = +i\,\sigma$ ($x = -i\sigma$) in the interior. Although the Fourier transforms of the two families of functions (10.156; 10.157b) are different (10.158a),

$$\tilde{\delta}(k) \neq \tilde{\Delta}_\sigma(k); \quad \lim_{\sigma \to 0} \tilde{\delta}_\sigma(k) = \lim_{\sigma \to 0} \tilde{\Delta}_\sigma(k) = \frac{1}{2\pi} = \tilde{\delta}(k), \tag{10.158a,b}$$

they have the same limit (10.158b), proving (10.155).

E10.12.2 Cylindrical and Two-Dimensional Impulse

The family of functions (10.159b) converges to the unit impulse (10.159c) with cylindrical symmetry (10.159a):

$$r = \left| x^2 + y^2 \right|^{1/2} : \quad \gamma_\sigma(r) = \frac{1}{\pi}\left(\frac{\sigma}{r^2 + \sigma^2} \right)^2, \quad \lim_{\sigma \to 0} \gamma_\sigma(r) = \delta(r), \tag{10.159a–c}$$

which corresponds to the two-dimensional unit impulse. The proof of the main result follows from two statements: (1) the limit function is zero everywhere (10.160a and b) except at the origin where it is infinite (10.160c and d):

$$r \neq 0 : \lim_{\sigma \to 0} \gamma_\sigma(r) = 0; \quad r = 0 : \lim_{\sigma \to 0} \gamma_\sigma(0) = \lim_{\sigma \to 0} \frac{\pi}{\sigma} = \infty; \tag{10.160a–d}$$

and (2) the integral over a domain including the origin is unity:

$$\int_0^{2\pi} d\varphi \int_0^\infty dr\, r \frac{1}{\pi}\left(\frac{\sigma}{r^2 + \sigma^2} \right)^2 = \int_0^\infty dr \frac{2\sigma^2 r}{\left(r^2 + \sigma^2 \right)^2} = -\left[\frac{\sigma^2}{r^2 + \sigma^2} \right]_0^\infty = 1. \tag{10.161}$$

Comparing with (10.162a), (10.162b) follows:

$$\int_{-\infty}^{+\infty} dx \int_{-\infty}^{+\infty} dy\, \delta(x)\delta(y) = 1, \quad 2\pi r \delta(r) = \delta(x)\delta(y); \tag{10.162a,b}$$

this result is equivalent to

$$\delta(x)\delta(y) = \lim_{\sigma \to 0} \Delta_\sigma(x,y), \quad \Delta_\sigma(x,y) = 2\pi r \gamma_\sigma(r) = \frac{2\sigma^2 \left| x^2 + y^2 \right|^{1/2}}{\left(\sigma^2 + x^2 + y^2 \right)^2}. \tag{10.163a,b}$$

Thus are obtained two examples of the *nonuniform limits of a family of analytic functions (10.163b) [(10.159b)] specifying the two-dimensional Cartesian (10.163a) [cylindrical (10.159c)] unit impulses that are related by (10.162b) $\equiv$ (5.60b).*

EXAMPLE 10.13: DECOMPOSITION OF IMPULSES
WHOSE ARGUMENT IS A FUNCTION

Perform the following decompositions of (1) unit impulses (10.164 through 10.168) whose argument is a real function with simple zeros:

$$\delta\big(\sinh(ax)\big) = \delta(ax) = \frac{1}{a}\delta(x), \tag{10.164}$$

$$\delta\big(\sinh(x^2 - a^2)\big) = \frac{1}{2a}\big\{\delta(x-a) - \delta(x+a)\big\}, \tag{10.165}$$

$$\delta\big(e^x(x^2 - a^2)\big) = \frac{1}{2a}\big\{e^a\delta(x-a) - e^{-a}\delta(x+a)\big\}, \tag{10.166}$$

$$\delta(\cos x) = \sum_{n=-\infty}^{+\infty} (-)^{n+1}\,\delta\left(x - n\pi - \frac{\pi}{2}\right); \tag{10.167}$$

$$\delta(\sin x) = \sum_{n=-\infty}^{+\infty} (-)^n\,\delta(x - n\pi); \tag{10.168}$$

(2) derivative impulses (10.169 through 10.171) whose arguments are functions with simple zeros:

$$\delta'(x^2 - a^2) = \frac{1}{2a}\big[\delta'(x-a) - \delta'(x+a)\big] + \frac{1}{2a^2}\big[\delta(x-a) + \delta(x+a)\big], \tag{10.169}$$

$$\delta'\big((x-a)\log x\big) = \frac{1}{1-a}\delta'(x-1) + \frac{1}{\log a}\delta'(x-a)$$
$$+ \frac{1+a}{(1-a)^2}\delta(x-1) + \frac{2}{a\log^2 a}\delta(x-a), \tag{10.170}$$

$$\delta'(\sin x) = \sum_{n=-\infty}^{+\infty} (-)^n\,\delta'(x - n\pi); \tag{10.171}$$

and (3) products of derivative impulses (10.172 through 10.175) whose arguments are real functions of two real variables with common single roots:

$$\delta(\sin x)\delta(\sin y) = \sum_{n,m=-\infty}^{+\infty} (-)^{m+n}\,\delta(x - n\pi)\delta(y - m\pi), \tag{10.172}$$

$$\delta(x-y)\delta(x^2 - a^2) = \frac{1}{2a}\big\{\delta(x-a)\delta(y-a) - \delta(x+a)\delta(y+a)\big\}, \tag{10.173}$$

$$\delta(xy + a^2)\delta(x^2 - a^2) = -\frac{1}{2a^2}\big\{\delta(x-a)\delta(y+a) + \delta(x+a)\delta(y-a)\big\}, \tag{10.174}$$

$$\delta(xy - ax)\delta(x^2 - y^2) = -\frac{1}{2a^2}\big\{\delta(x-a) + \delta(x+a)\big\}\delta(y-a). \tag{10.175}$$

The decompositions are made into simple (10.164 through 10.171) [products of simple (10.172 through 10.175)] unit impulses, whose arguments are linear in the variable itself.

E10.13.1 Unit Impulse of Function of One Variable

The formulas (10.164 through 10.168) are particular cases of (5.11a through c) ≡ [(10.176a through c) in

$$f(x_n)=0 \neq f'(x_n): \quad \delta(f(x))=\sum_{n=0}^{N}\frac{\delta(x-x_n)}{f'(x_n)}, \qquad (10.176a\text{-}c)$$

where the functions (10.177a through d)

$$f(x)\equiv\left\{\sinh(ax),\sinh(x^2-a^2),e^x(x^2-a^2),\cos x,\sin x\right\} \qquad (10.177a\text{-}e)$$

have simple roots:

$$f(x_n)=0:x_n\equiv\left\{x_1=0;x_{1,2}=\pm a;x_{1,2}=\pm a;x_n=n\pi+\frac{\pi}{2};x_n=n\pi\right\}; \qquad (10.178a\text{-}f)$$

the slopes

$$f'(x)=\left\{a\cosh(ax),2x\cosh(x^2-a^2),e^x(x^2-a^2+2x),-\sin x,\cos x\right\} \qquad (10.179a\text{-}c)$$

at the zeros are:

$$f'(x_n)=\left\{a,\pm 2a,\pm 2ae^{\pm a},(-)^{n+1},(-)^n\right\} \qquad (10.180a\text{-}e)$$

The zeros (10.178b through f) and corresponding slopes (10.180a through e) appear in (10.176c) leading, respectively, to (10.164) through (10.168).

E10.13.2 Derivative Unit Impulse of a Function of One Variable

The formulas (10.169) through (10.171) are particular cases of (5.16a through d) ≡ (10.181):

$$\delta'(f(x))=\sum_{n=1}^{N}\left\{-\left[f'(x_n)\right]^{-1}\delta'(x-x_n)+f''(x_n)\left[f'(x_n)\right]^{-2}\delta(x-x_n)\right\}, \qquad (10.181)$$

where the functions (10.182a through c) have (1) simple roots (10.182d through f):

$$f(x)=\left\{x^2-a^2,(x-a)\log x,\sin x\right\}: \quad x_n\equiv\{x_{1,2}=\pm a;x_{1,2}=1,a;x_n=n\pi\}; \qquad (10.182a\text{-}f)$$

and (2/3) first (second) derivatives (10.183a through c) [(10.183d through f)]:

$$f'(x)=\left\{2x,1-\frac{a}{x}+\log x,\cos x\right\}, \quad f''(x)=\left\{2,\frac{a+x}{x^2},-\sin x\right\}, \qquad (10.183a\text{-}f)$$

$$f'(x_n)=\left\{\pm 2a;1-a,\log a;(-)^n\right\}, \quad f''(x_n)\equiv\left\{2;1+a,\frac{2}{a};0\right\}, \qquad (10.184a\text{-}f)$$

with slopes (10.184a through c) [curvatures (10.184d through f)]. Substituting the zeros (10.182d through f) of the functions (10.182a through c), together with the corresponding slopes (10.184a through c) and curvature (10.184d through f) in (10.181) leads, respectively, to (10.169) through (10.171).

E10.13.3 Product of Unit Impulses of Functions of Two Variables

The formulas (10.172 through 10.176) are particular cases of (5.77a and b) ≡ (10.185):

$$\delta\big(f(x,y)\big)\delta\big(g(x,y)\big) = \sum_{n=1}^{N}\big[\Delta(x_n,y_n)\big]^{-1}\delta(x-x_n)\delta(y-y_n), \qquad (10.185)$$

where (1) the pair of functions (10.186a through d)

$$\{f(x,y),g(x,y)\} = \{\sin x, \sin y; x-y, x^2-a^2; xy+a^2, x^2-a^2; xy-ax, x^2-y^2\}, \qquad (10.186a\text{-d})$$

$$\{x_n,y_n\} = \{x_x = n\pi, y_m = m\pi; x_{1,2} = \pm a = y_{1,2}; x_{1,2} = \pm a = -y_{1,2}; x_{1,2} = \pm a, y_{1,2} = a\}, \qquad (10.187a\text{-d})$$

have common roots (10.187a through d); (2) the Jacobian functions (10.188a through d) have values (10.189a through d) at the roots:

$$\Delta(x,y) = \frac{\partial f}{\partial x}\frac{\partial g}{\partial y} - \frac{\partial f}{\partial y}\frac{\partial g}{\partial x} = \{\cos x \cos y, 2x, -2x^2, 2y(a-y)-2x^2\}, \qquad (10.188a\text{-d})$$

$$\Delta(x_n,y_m) = \{(-)^{n+m}, \pm 2a, -2a^2, -2a^2\}, \qquad (10.189a\text{-d})$$

Substitution of the Jacobian (10.189a through d) at the common zeros (10.187a through d) of the pairs of functions of two variables (10.186a through d) in (10.185) leads, respectively, to (10.172) through (10.176).

The product of (10.168) with distinct variables x and y leads to (10.172). The attempt to apply (5.13e) ≡ (10.190)

$$2a\delta(x^2-a^2) = \delta(x-a) - \delta(x+a), \qquad (10.190)$$

to (10.173 through 10.175)

$$
\begin{aligned}
2a\delta(x-y)\delta(x^2-a^2) &= \delta(x-y)\big[\delta(x-a)-\delta(x+a)\big] \\
&= \delta(a-y)\delta(x-a) - \delta(-a-y)\delta(x+a) \\
&= \delta(x-a)\delta(y-a) - \delta(x+a)\delta(y+a),
\end{aligned} \qquad (C)
$$

$$
\begin{aligned}
2a^2\,\delta(xy-ax)\delta(x-y^2) &= \frac{2a^2}{x}\delta(y-a)\frac{\delta(x-y)-\delta(x+y)}{2x} \\
&= \frac{a^2}{x^2}\delta(y-a)\big[\delta(x-a)-\delta(x+a)\big] = \delta(y-a)\big[\delta(x-a)-\delta(x+a)\big] \quad (D)
\end{aligned}
$$

$$
\begin{aligned}
2a^2\,\delta(xy-ax)\delta(x^2-y^2) &= \frac{2a^2}{x}\delta(y-a)\frac{\delta(x-y)-\delta(x+y)}{2x} \\
&= \frac{a^2}{x^2}\delta(y-a)\big[\delta(x-a)-\delta(x+a)\big] = \delta(y-a)\big[\delta(x-a)-\delta(x+a)\big] \quad (E)
\end{aligned}
$$

uses (1) first the linear property [(5.12d) that is valid; and (2) second the substitution property (1.78a and b) that is valid (not valid) for the product of an impulse and an ordinary function (product of two impulses). Thus (D) $\neq$ (10.174) is in error, although fortuitously (c) = (10.173) and (E) = (10.175); the reliable method to consider the product of two impulses is to use (10.185) $\equiv$ (5.77a and b).

EXAMPLE 10.14: EVALUATION OF DOUBLE INTEGRALS OF ORDINARY AND GENERALIZED FUNCTIONS

Evaluate the following double integrals involving products of generalized functions:

$$I_1 \equiv \int\limits_{-\infty}^{+\infty}\int \delta^{(n)}(x)\delta^{(m)}(y)e^{ax+by}\,dxdy = (-)^{n+m}a^n b^m, \tag{10.191}$$

$$I_2 \equiv \int\limits_{-\infty}^{+\infty}\int \delta^{(n)}(x)H(-y)e^{ax+by}\,dxdy = \frac{(-a)^n}{b}, \tag{10.192}$$

$$I_3 \equiv \int\limits_{-\infty}^{+\infty}\int H(-x)H(y)e^{ax-by}\,dxdy = \frac{1}{ab}, \tag{10.193}$$

and the following single integrals involving Gaussian and exponential functions:

$$I_4 \equiv \int\limits_{0}^{+\infty}\left(x^2+a^2\right)^n e^{-x^2}\,dx = \sqrt{\pi}\sum_{k=0}^{n}\binom{n}{k}a^{2n-2k}(2k-1)!!2^{-k-1}, \tag{10.194}$$

$$I_5 \equiv \int\limits_{-\infty}^{+\infty}x\left(x^2+a^2\right)^n e^{-x^2}\,dx = \frac{1}{2}\sum_{k=0}^{n}\binom{n}{k}a^{2n-2k}k!, \tag{10.195}$$

$$I_6 \equiv \int\limits_{-\infty}^{+\infty}\left(x^2+a^2\right)^n e^{-|x|}\,dx = 2\sum_{k=0}^{n}\binom{n}{k}a^{2n-2k}(2k)!, \tag{10.196}$$

$$I_7 \equiv \int\limits_{-\infty}^{+\infty}x\left(x^2+a^2\right)^n e^{-|x|}\,dx = 2\sum_{k=0}^{n}\binom{n}{k}a^{2n-2k}(2k+1)!, \tag{10.197}$$

assuming a, b real and positive.

E10.14.1 Double Integrals with Generalized Functions

The evaluation of the integrals (10.191 through 10.193) uses the properties of generalized functions

$$I_1 = (-)^{n+m}\lim_{x,y\to 0}\frac{\partial^{n+m}}{\partial x^n\partial y^m}e^{ax+by} = (-)^{n+m}a^n b^m, \tag{10.198}$$

$$I_2 = (-)^n \lim_{x \to 0} \frac{d^n}{dx^n} \int_{-\infty}^{+\infty} H(-y)e^{ax+by}dy = (-)^n a^n \int_{-\infty}^{0} e^{by}dy = \frac{(-a)^n}{b}, \tag{10.199}$$

$$I_3 = \int_{-\infty}^{0} e^{ax}dx \int_{0}^{\infty} e^{-by}dy = \frac{1}{ab} \tag{10.200}$$

associated with inner products by (1) derivative impulses (3.21a and b) for (10.191) ≡ (10.198) and (10.192) ≡ (10.199); and (2) the unit jump (3.1) for (10.192) ≡ (10.199) and (10.193) ≡ (10.200).

E10.14.2 Integrals with Gaussians or Exponentials

The Gaussian integrals (10.194) ≡ (10.201a) [(10.195) ≡ (10.201b)] are evaluated using the binomial theorem (I.25.37a through c)

$$I_4 = \sum_{k=0}^{n} \binom{n}{k} a^{2n-2k} \int_{-\infty}^{+\infty} x^{2k} e^{-x^2} dx, \quad I_5 = \sum_{k=0}^{n} \binom{n}{k} a^{2n-2k} \int_{0}^{+\infty} x^{2k+1} e^{-x^2} dx, \tag{10.201a,b}$$

and substituting (5.298) [(5.300)]. The last two integrals (10.196) ≡ (10.202a) [(10.197) ≡ (10.202b)] are similar to (10.201a) [(10.201b)]:

$$I_6 = 2 \sum_{k=0}^{n} \binom{n}{k} a^{2n-2k} \int_{0}^{+\infty} x^{2k} e^{-x} dx, \quad I_7 = 2 \sum_{k=0}^{n} \binom{n}{k} a^{2n-2k} \int_{0}^{\infty} x^{2k+1} e^{-x} dx, \tag{10.202a,b}$$

involving exponentials:

$$\int_{0}^{\infty} x^m e^{-x} = \lim_{a \to 1} (-)^m \frac{\partial^m}{\partial a^m} \int_{0}^{\infty} e^{-ax} dx$$

$$= \lim_{a \to 1} (-)^m \frac{\partial^m}{\partial a^m} \frac{1}{a} = m! \lim_{a \to 1} a^{-m-1} = m!; \tag{10.203}$$

substitution of (10.2003), which is a particular case of the Gamma function (1.149a and b; 1.156a and b) in (10.202a) [(10.202b)], yields (10.196) [(10.197)].

EXAMPLE 10.15: POTENTIAL FLOW PAST TWO SPHERES MOVING ALONG THE LINE OF CENTERS

Consider the potential flow past two spheres with distinct radii moving with different velocities along the line of centers; use the particular case of spheres with the same radius and velocity to consider the potential flow of a sphere moving in a direction normal two wall. Assume that the radii are small compared with the distance between spheres or from the wall.

E10.15.1 Perturbation of the Potential of One Sphere by Another Sphere

Consider two spheres with radii a and b moving with velocity U and V, respectively, along the line of centers, which are at a mutual distance c (Figure 10.13a). The total potential (6.350a) is the linear combination of unit potentials due to each sphere satisfying the boundary conditions (6.350b through e). The unit potential of the first sphere in isolation is given by (6.351b) ≡ (10.204b), and using the geometric relation (6.204a) leads to (6.204c) near the second sphere:

$$R\cos\varphi + S\cos\phi = c: \quad \Phi_{10} = \frac{a^3}{2R^2}\cos\varphi = \frac{a^3}{2R^3}(c - S\cos\phi). \tag{10.204a-c}$$

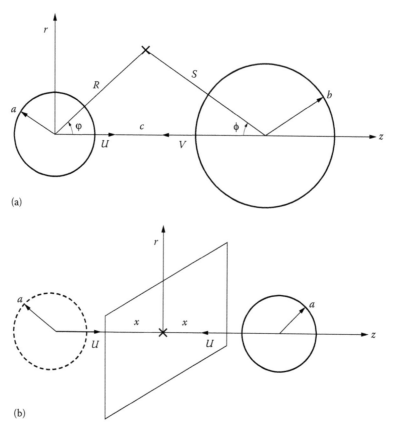

(a)

(b)

FIGURE 10.13 The potential flow due to two spheres with different radii and different velocities moving along (perpendicular to) the line of centers [Figure 10.13a (6.16a)] may be obtained by the method of perturbation potentials, as for cylinders [Figure II.8.4a (II.10.1a)]. The case of equal radii and velocities leads to the potential flow of a sphere moving perpendicular (parallel) to a wall [Figure 10.13b (6.16b)], as for cylinders [Figure II.8.4b (II.10.1b)].

The distances R and S from the centers of the first and second sphere (Figure 10.13a) are related by (10.205a)

$$R = \left|c^2 + S^2 - 2cS\cos\phi\right|^{1/2} = c\left[1 - \frac{S}{c}\cos\phi + O\left(\frac{S^2}{c^2}\right)\right], \qquad (10.205a,b)$$

which leads to the asymptotic approximation (10.205b) near the second sphere. Substitution of (10.205b) in (10.204c) specifies the free potential of the first sphere near the second:

$$\Phi_{10} = \frac{a^3}{2c^2}\left[1 - \frac{S}{c}\cos\phi\right]\left[1 + 3\frac{S}{c}\cos\phi + O\left(\frac{S^2}{c^2}\right)\right]$$

$$= \frac{a^3}{2c^2} + \frac{a^3}{c^3}S\cos\phi + O\left(\frac{a^3S^2}{c^5}\right). \qquad (10.206)$$

The free potential of the first sphere (10.206) creates at the second sphere a normal velocity (10.207a)

$$\left.\frac{\partial\Phi_{10}}{\partial S}\right|_{S=b} = \frac{a^3}{c^3}\cos\phi = -\left.\frac{\partial\Phi_{12}}{\partial S}\right|_{S=b}, \qquad \Phi_{12} = \frac{a^3}{c^3}\frac{b^3}{2S^2}\cos\phi, \qquad (10.207a,b)$$

which would violate the rigid wall boundary condition; the latter is restored by adding a the perturbation potential (10.207b). The sum of (10.207b) and (10.204b) specifies the total unit potential of the first sphere (10.208b), including the effect of the second sphere to order (10.208a):

$$a^3 b^2 \ll c^5: \quad \Phi^{(1)} = \Phi_{10} + \Phi_{12} = \frac{a^3}{2R^2}\cos\varphi + \frac{a^3 b^3}{2c^3 S^2}\cos\phi; \qquad (10.208\text{a,b})$$

this specifies the value of the first potential on the first sphere (10.209b):

$$\left(a,U,R,\phi\right) \leftrightarrow \left(b,V,S,\varphi\right): \quad \left.\Phi^{(1)}\right|_{R=a} = \frac{a}{2}\cos\varphi, \; \left.\Phi^{(2)}\right|_{S=b} = \frac{b}{2}\cos\phi, \qquad (10.209\text{a–c})$$

and by interchange of variables (10.209a) the second potential (10.209c) on the second sphere is obtained.

E10.15.2 Value of the Unit Potentials on Each Sphere

To calculate the kinetic energy and added mass (E10.15.3), the two unit potentials are needed on both spheres. The interchange of variables (10.209a) specifies the second potential (10.210b) from the first (10.208b) to order (10.210a):

$$a^2 b^3 \ll c^5: \quad \Phi^{(2)} = \frac{b^3}{2S^2}\cos\phi + \frac{a^3 b^3}{2c^3 R^2}\cos\varphi. \qquad (10.210\text{a,b})$$

By interchange of variables (10.209a) or directly, the expansion (10.205b) gives

$$S = \left|c^2 + R^2 - 2cR\cos\varphi\right|^{1/2} = c\left[1 - \frac{R}{c}\cos\varphi + O\left(\frac{R^2}{c^2}\right)\right]; \qquad (10.211)$$

substitution of (10.211) together with the geometric relation (10.204a) in the first term of the second potential (10.210b) yields

$$\Phi_{20} = \frac{b^3}{2S^3}\left(c - R\cos\varphi\right) = \frac{b^3}{2c^2}\left[1 - \frac{R}{c}\cos\varphi\right]\left[1 + 3\frac{R}{c}\cos\varphi + O\left(\frac{R^2}{c^2}\right)\right]$$

$$= \frac{b^3}{2c^2} + \frac{b^3 R}{c^3}\cos\varphi + O\left(\frac{b^3 R^2}{c^5}\right). \qquad (10.212)$$

Adding the second term in (10.210b), the second potential near the first sphere takes the value (10.213a), and using (10.209a) in (10.213a), the first potential takes the value (10.213b) on the second sphere:

$$\left.\Phi^{(2)}\right|_{R=a} = \frac{b^3}{2c^2} + \frac{3b^3 a}{2c^3}\cos\varphi; \quad \left.\Phi^{(1)}\right|_{S=b} = \frac{a^3}{2c^2} + \frac{3a^3 b}{2c^3}\cos\phi, \qquad (10.213\text{a,b})$$

Thus *two spheres of radii a and b moving with velocity U and V along the line of centers (Figure 10.13a) at a distance c have a total potential (10.214a) that is a linear combination of the unit potentials (10.208b) [(10.210b)]; these take the values (10.209b) [(10.213a)] on the first sphere and (10.213b) [(10.209c)] on the second sphere. The first iteration of the method of perturbation potentials is valid with an accuracy (10.208a; 10.210a).*

E10.15.3 Kinetic Energy and Added Mass

The total potential (6.350a) ≡ (10.214a) corresponds to the kinetic energy (10.214b) ≡ (6.359):

$$\Phi = U\Phi^{(1)} + U\Phi^{(2)}: \quad \frac{2}{\rho}E_v = I_1 U^2 + I_2 V^2 + 2I_3 UV; \qquad (10.214a,b)$$

the coefficients are specified by the integrals (6.360a through c) ≡ (10.215a through c):

$$I_1 = -\int_{R=a} \Phi^{(1)} \frac{\partial \Phi^{(1)}}{\partial x} dA = 2\pi a^2 \int_0^\pi \Phi^{(1)} \cos\varphi \sin\varphi \, d\varphi$$

$$= \pi a^3 \int_0^\pi \cos^2\varphi \sin\varphi \, d\varphi = \frac{2}{3}\pi a^3, \qquad (10.215a)$$

$$I_2 = -\int_{S=a} \Phi^{(2)} \frac{\partial \Phi^{(2)}}{\partial x} dA = 2\pi b^2 \int_0^\pi \Phi^{(2)} \cos\phi \sin\phi \, d\phi$$

$$= \pi b^3 \int_0^\pi \cos^2\phi \sin\phi \, d\phi = \frac{2}{3}\pi b^3, \qquad (10.215b)$$

$$I_3 = -\int_{R=a} \Phi^{(2)} \frac{\partial \Phi^{(1)}}{\partial x} dA = 2\pi a^2 \int_0^\pi \Phi^{(2)} \cos\varphi \sin\varphi \, d\varphi$$

$$= \frac{3\pi a^3 b^3}{c^3} \int_0^\pi \cos^2\varphi \sin\varphi \, d\varphi = \frac{2\pi a^3 b^3}{c^3}$$

$$= -\int_{S=b} \Phi^{(1)} \frac{\partial \Phi^{(2)}}{\partial x} dA = 2\pi b^2 \int_0^\pi \Phi^{(1)} \cos\phi \sin\phi \, d\phi, \qquad (10.215c)$$

where the following were used in the evaluation of the integrals (10.215a through c): (1) the area element (6.360d) on the sphere of radius *a*; and (2) the values of the unit potentials on the spheres (10.209b and c; 213a and b). Substituting (10.215a through c) in (10.214b) specifies

$$E_v = \frac{\pi}{3}\rho\left(a^3 U^3 + b^3 U^3\right) + 2\pi\rho\frac{a^3 b^3}{c^3} UV, \qquad (10.216)$$

which is *the kinetic energy of the potential flow of two spheres of radii a and b moving with veloci-ties U and V orthogonal to the line of centers (Figure 3.13a) at a distance c.*

E10.15.4 Sphere Moving Perpendicular to a Wall

The case of spheres with the same radius (10.217a) and velocity (10.217b) corresponds (Figure 10.13b) to images on a wall at half-distance (10.217c); since the flow occupies a half-space, one half of (10.216) is taken as *the kinetic energy (10.217d) of a sphere of radius a moving with velocity U perpendicular to a wall (Figure 10.13b) at a distance z from the center:*

$$a = b, \quad U = V, \quad c = 2x: \quad E = \frac{E_v}{2} = \pi\rho U^2 a^3\left(\frac{1}{3} + \frac{a^3}{8x^3}\right) = \frac{\overline{m}}{2}U^2, \qquad (10.217a\text{–}d)$$

The added mass (10.218a) is the mass of displaced fluid (10.218b)

$$\bar{m} = \frac{2\pi}{3}\rho a^3 \left[1 + 3\left(\frac{a}{2x}\right)^3 \right] = \frac{m}{2}\left(1 + \frac{3a^3}{8x^3} \right), \quad m = \frac{4}{3}\pi\rho a^3, \qquad (10.218\text{a,b})$$

with a factor in (10.218a) depending on the ratio of the radius to the distance from the wall. If the sphere has mass m_0, the total kinetic energy is:

$$\bar{E} = \left(\frac{m_0}{2} + \frac{\pi}{3}\rho a^3 + \frac{\pi}{8}\rho\frac{a^6}{x^3} \right) U^2. \qquad (10.219)$$

The velocity of the sphere is minimum (maximum) at infinity (when it touches the wall):

$$\left| \frac{6\bar{E}}{3m_0 + 2\pi\rho a^3} \right|^{1/2} \equiv U(\infty) \le U(x) \le U(a) = \left| \frac{24\bar{E}}{12m_0 + 11\pi\rho a^3} \right|^{1/2}. \qquad (10.220)$$

The velocity at an arbitrary distance from the wall is given by

$$\frac{dx}{dt} = U(x) = U_\infty \left| 1 + \mu\frac{a^3}{x^3} \right|^{-1/2}, \quad \frac{1}{\mu} \equiv \frac{8}{3} + \frac{4m_0}{\pi\rho a^3}; \qquad (10.221\text{a,b})$$

and the trajectory is specified by:

$$U_\infty t = x - a + \frac{\mu a}{4}\left(1 - \frac{a^2}{x^2} \right) + a\sum_{n=2}^{\infty}\frac{(2n-3)!!}{n!}\frac{(-)^{n-1}}{3n-1}\left(\frac{\mu}{2}\right)^n\left[1 - \left(\frac{a}{x}\right)^{3n-1} \right]. \qquad (10.222)$$

The result (10.222) follows from (10.221a) integrating from time zero $t = 0$ at the wall $x = a$, and using the binomial series (I.25.37a through c):

$$U_\infty t = \int_a^x \frac{U_\infty}{U(\xi)}d\xi = \int_a^x \left| 1 + \mu\frac{a^3}{\xi^3} \right|^{1/2} d\xi$$

$$= \sum_{n=0}^{\infty}\binom{1/2}{n}\left(\mu a^3\right)^n\int_a^x \xi^{1-3n}\,d\xi$$

$$= \int_a^x\left(1 + \frac{\mu a^3}{2\xi^3} \right)d\xi + \sum_{n=2}^{\infty}\left(\frac{1}{2}\right)\left(-\frac{1}{2}\right)\left(-\frac{3}{2}\right)\cdots\left(\frac{1}{2} - n + 1\right)\frac{\left(\mu a^3\right)^n}{n!}\int_a^x \xi^{-3n}d\xi$$

$$= \left[\xi\right]_a^x - \left[\frac{\mu a^3}{4\xi^2}\right]_a^x - \sum_{n=2}^{\infty}\frac{(2n-3)(2n-1)\cdots 3.1}{n!\,2^n}\frac{\left(-\mu a^3\right)^n}{1-3n}\left[\xi^{1-3n}\right]_a^x; \qquad (10.223)$$

the series (10.223) (10.222) converges faster for larger distance from the wall.

EXAMPLE 10.16: GENERALIZED HÖLDER AND MINKOWSKI IDENTITIES

Prove the extension of the Hölder (7.15a through f) [Minkowski (7.30a through d)] inequalities from $N = 2$ functions to an arbitrary integer number.

E10.16.1 Generalized Hölder Inequality in Discrete (Integral) Form

The generalized projective or Hölder inequality is (10.224c) [(10.224e)] in discrete (integral) form:

$$i = 1, \dots, m; \quad 1 \le p_i < \infty: \quad \sum_{n=1}^{N} \left| \prod_{i=1}^{m} a_{i,n} \right| \le \prod_{i=1}^{m} \left\{ \sum_{n=1}^{N} \left| a_{i,n} \right|^{p_i} \right\}^{1/p_i}, \qquad (10.224\text{a--c})$$

$$\sum_{i=1}^{m} \frac{1}{p_i} = 1: \quad \int_a^b \left| \prod_{i=1}^{m} f_i(x) \right| dx \le \prod_{i=1}^{m} \left\{ \int_a^b \left| f_i(x) \right|^{p_i} dx \right\}^{1/p_i}, \qquad (10.224\text{d,e})$$

where the parameters satisfy (10.224a,b,d), and the equalities only hold if all functions f_i (vectors a_i) in the integral (10.224e) [sum (10.224c)] are constant multiples (10.225b) [10.225a)]:

$$a_{2,k} = \lambda_2 a_{1,k}, \quad a_{3,k} = \lambda_3 a_{1,k}, \dots, a_{m,k} = \lambda_m a_{1,k}. \qquad (10.225\text{a})$$

$$f_2(x) = \lambda_2 f_1(x), \quad f_3(x) = \lambda_3 f_1(x), \dots, f_m(x) = \lambda_m f_1(x), \qquad (10.225\text{b})$$

The result holds for all: (1) integer m in the discrete (continuous) case (10.224c) [(10.224e)]; and (2) integer N in the discrete case (10.224c).

It suffices to prove one of the inequalities, for example, (10.224e). The proof is by induction: (1) first (10.224e) holds for $m = 2$, when it coincides with the Hölder inequality (7.15f); and (2) if (10.224e) holds for $m \ge 2$, it also holds for $m + 1$ as follows:

$$\int_a^b \left| \prod_{i=1}^{m+1} f_i(x) \right| dx = \int_a^b \left| \prod_{i=1}^{m} f_i(x) \right| \left| f_{m+1}(x) \right| dx$$

$$\le \left\{ \int_a^b \left| \prod_{i=1}^{m} f_i(x) \right|^{p_i} dx \right\}^{1/p_i} \left\{ \int_a^b \left| f_{m+1}(x) \right|^{p_{m+1}} dx \right\}^{1/p_{m+1}} \le \prod_{i=1}^{m+1} \left\{ \int_a^b \left| f_i(x) \right|^{p_i} dx \right\}^{1/p_i},$$

$$(10.226)$$

where

$$\frac{1}{p_0} \equiv \sum_{i=1}^{m} \frac{1}{p_i}, \quad 1 = \frac{1}{p_0} + \frac{1}{p_{m+1}} = \sum_{i=1}^{m+1} \frac{1}{p_i}; \qquad (10.227\text{a,b})$$

the Hölder inequality is applied to p_0 in (10.227a) and $1/p_{m+1}$ in (10.227b). The triple Hölder inequality is used (7.46) to show that the convolution of normed functions is a normed function.

E10.16.2 Generalized Minkowski Inequality in Discrete (Integral) Form

The generalized triangular Minkowski or inequality is (10.228b) (10.228c)] in discrete (integral) form:

$$1 \le p < \infty: \quad \left\{ \sum_{n=1}^{N} \left| \sum_{i=1}^{m} a_{i,n} \right|^p \right\}^{1/p} \le \sum_{i=1}^{m} \left\{ \sum_{n=1}^{N} \left| a_{i,n} \right|^p \right\}^{1/p}. \qquad (10.228\text{a,b})$$

$$\left\{\int_a^b \left| \sum_{i=1}^m f_i(x) \right|^p dx \right\}^{1/p} \le \sum_{i=1}^m \left\{ \int_a^b \left| f_i(x) \right|^p \right\}^{1/p}, \tag{10.228c}$$

where the parameter satisfies (10.228a), and the equality in (10.228b) [(10.229c)] applies in the case (10.225a) [(10.225b)].

The proof is by induction as for the generalized Hölder inequality (10.224e) $\equiv$ (10.226; 10.227a and b) and is done next for the discrete (10.228b) instead of the integral form: (1) first (10.228b) holds for $m = 2$ when it reduces to the Minkowski inequality (7.30c); and (2) second if it holds for $m \ge 2$, it also holds for $m + 1$, by

$$\left\{ \sum_{n=1}^N \left| \sum_{i=1}^{m+1} a_{i,n} \right|^p \right\}^{1/p} \le \left\{ \sum_{n=1}^N \left[\left| \sum_{i=1}^m a_{i,n} \right|^p + \left| a_{m+1,n} \right|^p \right] \right\}^{1/p}$$

$$\le \left\{ \sum_{i=1}^m \left| \sum_{n=1}^N a_{i,n} \right|^p \right\}^{1/p} + \left\{ \sum_{n=1}^N \left| a_{m+i,n} \right|^p \right\}^{1/p} \le \sum_{i=1}^{m+1} \left\{ \sum_{n=1}^N \left| a_{i,m} \right|^p \right\}^{1/p}, \tag{10.229}$$

where the Minkowski inequality was applied to $a_{m+1,n}$ and to the aggregate $i = 1, 2, \ldots, m$ of all preceding terms $a_{i,n}$.

EXAMPLE 10.17: EVALUATION OF CONVOLUTION INTEGRALS

Evaluate the following convolution integrals:

$$\exp(-|x|) * x = 2x \int_0^\infty e^{-\xi} d\xi = 2x. \tag{10.230}$$

$$\exp(-|x|) * x^2 = 2 \int_0^\infty e^{-\xi} (x^2 + \xi^2) d\xi = 2(2 + x^2), \tag{10.231}$$

$$\exp(-|x|) * \exp(ix) = 2e^{ix} \int_0^\infty e^{-\xi} \cos\xi \; d\xi = e^{ix}, \tag{10.232}$$

$$\exp(-x^2) * x \equiv \int_{-\infty}^{+\infty} e^{-\xi^2} (x - \xi) d\xi = x\sqrt{\pi}, \tag{10.233}$$

$$\exp(-x^2) * x^2 \equiv \int_{-\infty}^{+\infty} e^{-\xi^2} (x^2 - 2x\xi + \xi^2) d\xi = \left(x^2 + \frac{1}{4} \right) \sqrt{\pi}. \tag{10.234}$$

$$\exp(-x^2) * \exp(x) \equiv \int_{-\infty}^{+\infty} e^{-\xi^2 - i\xi} d\zeta = e^{ix} e^{-1/4} \sqrt{\pi}. \tag{10.235}$$

$$\exp(-x^2) * \exp(x) \equiv \int_{-\infty}^{+\infty} e^{-\xi^2} e^{\xi - x} \; d\xi = \sqrt{\pi} \; e^{1/4} \cosh x. \tag{10.236}$$

involving exponential (10.230 through 10.232) [Gaussian (10.233 through 10.236)] functions.

The first three convolution integrals (10.230 through 10.232) involve exponential functions:

$$e^{-|x|} * x = \int_{-\infty}^{0} e^{\xi}(x-\xi)d\xi + \int_{0}^{\infty} e^{-\xi}(x-\xi)d\xi$$

$$= \int_{0}^{\infty} e^{-\xi}\left[(x-\xi)+(x+\xi)\right]d\xi = 2x \int_{0}^{\infty} e^{-\xi}d\xi = 2x, \tag{10.237}$$

$$e^{-|x|} * x^2 = \int_{-\infty}^{0} e^{\xi}(x-\xi)^2 + \int_{0}^{\infty} e^{-\xi}(x-\xi)^2 d\xi = \int_{0}^{\infty} e^{-\xi}\left[(x-\xi)^2 + (x+\xi)^2\right]d\xi$$

$$= 2 \int_{0}^{\infty} e^{-\xi}(x^2+\xi^2)d\xi = 2(2+x^2), \tag{10.238}$$

$$e^{-|x|} * e^{ix} = \int_{-\infty}^{0} e^{\xi}e^{i(x-\xi)}d\xi + \int_{0}^{\infty} e^{-\xi}e^{i(x-\xi)}d\xi = \int_{0}^{\infty} e^{-\xi}\left[e^{i(x-\xi)} + e^{i(x+\xi)}\right]d\xi$$

$$= e^{ix} \int_{0}^{\infty} \left[e^{-\xi(1+i)} + e^{\xi(i-1)}\right]d\xi = e^{ix}\left[\frac{1}{i+1} - \frac{1}{i-1}\right] = e^{ix}; \tag{10.239}$$

(10.203) was used in (10.238). The integrals (10.233 through 10.237) involve Gaussian functions:

$$e^{-x^2} * x^2 = \int_{-\infty}^{+\infty} e^{-\xi^2}(x-\xi)d\xi = x\sqrt{\pi}, \tag{10.240}$$

$$e^{-x^2} * x^2 = \int_{-\infty}^{+\infty} e^{-\xi^2}(x-\xi)^2 d\zeta = \int_{-\infty}^{+\infty}(x^2+\xi^2)e^{-\xi^2}d\xi = \left(x^2 + \frac{1}{2}\right)\sqrt{\pi}, \tag{10.241}$$

$$e^{-x^2} * e^{ix} = \int_{-\infty}^{+\infty} e^{-\xi^2}e^{i(x-\xi)}d\xi = e^{ix} \int_{-\infty}^{+\infty} e^{-i\xi-\xi^2}d\xi = \sqrt{\pi}e^{-1/4}e^{ix} \tag{10.242}$$

$$\eta \equiv \xi - \frac{1}{2}: \quad e^{-x^2} * e^{-x} = e^x \int_{-\infty}^{+\infty} e^{-\xi^2-\xi} \, d\xi = e^{x+1/4} \int_{-\infty}^{+\infty} e^{-\eta^2} \, d\eta = \sqrt{\pi} \, e^{1/4} \, e^x. \tag{10.243a,b}$$

(1.1a), (1.17a), and (1.13) were used, respectively, in (10.240), (10.241) and (10.242). In (10.243b) was used the change of variable (10.243a) and (1.1a).

EXAMPLE 10.18: UNSYMMETRIC INFLUENCE FUNCTION
AND NONRECIPROCAL FORCING

Obtain the influence or Green function for the two simplest linear ordinary differential equations with constant coefficients of the lowest odd order, which is one (three). Show that it does not satisfy the reciprocity principle. Use the superposition principle to obtain the solution for arbitrary forcing. Consider the case of forcing by a derivative impulse for the third-order equation.

E10.18.1 Green Function for the First-Order Differential Equation

A linear differential operator of odd order is never self-adjoint (Subsection 7.7.7) and thus does not satisfy the reciprocity principle. The simplest examples is a first (third)-order ordinary differential equation [E.10.18.1 (E.10.18.2–4)] consisting only of the highest order derivative with coefficient unity (10.244a) [(10.247a)]. The simplest first-order differential operator with constant coefficients (10.244a) has fundamental solution or Green or influence function (10.244b):

$$\frac{\mathrm{d}}{\mathrm{d}x}\big[G(x;\xi)\big]=\delta(x-\xi), \quad G(x;\xi)=H(x-\xi)+A, \tag{10.244a,b}$$

where the constant of the integration is determined by the boundary condition (10.245a):

$$0=G(0;\xi)=A: \quad G(x;\xi)=\begin{cases}0 & \text{if } x<\xi,\\ 1 & \text{if } x>\xi;\end{cases} \tag{10.245a–c}$$

the Green function (10.245b and c) is discontinuous at $x=\xi$. The solution (10.246d) of the equation (10.246c) with the initial condition (10.246a) and forcing by an integrable function (10.246b)

$$y(0)=0: \quad y'=f(x), \quad y(x)=\int_0^x f(\xi)\,\mathrm{d}\xi=\int_{-\infty}^{+\infty}f(\xi)G(x;\xi)\,\mathrm{d}\xi \tag{10.246a–e}$$

coincides with the principle of superposition (10.243e) applied to the influence function (10.245b and c); the Green function (10.245b and c) is not symmetric because the operator (10.244a) is not self-adjoint.

E10.18.2 Green Function for the Third-Order Differential Equation

The simplest third-order differential operator (10.247a) with constant coefficients has a fundamental solution:

$$\frac{\mathrm{d}^3}{\mathrm{d}x^3}\big[G(x;\xi)\big]=\delta(x-\xi), \quad \frac{\mathrm{d}^2}{\mathrm{d}x^2}\big[G(x;\xi)\big]=H(x-\xi)+A, \tag{10.247a,b}$$

$$G'(x;\xi-0)=G'(x;\xi+0): \quad \frac{\mathrm{d}^2}{\mathrm{d}x}\big[G(x;\xi)\big]=\begin{cases}Ax+B & \text{if } x\le\xi,\\ (1+A)x-\xi+B & \text{if } x\ge\xi,\end{cases} \tag{10.248a–c}$$

$$G(x;\xi-0)=G(x;\xi+0): \quad G(x;\xi)=\begin{cases}A\dfrac{x^2}{2}+Bx+C & \text{if } x\le\xi,\\[2mm] (1+A)\dfrac{x^2}{2}+(B-\xi)x+\dfrac{\xi^2}{2}+C & \text{if } x\ge\xi,\end{cases} \tag{10.249a–c}$$

which is specified by three successive integrations (10.247b)/(10.248b and c)/(10.249b and c); in each of the last two integrations (10.248b and c) [(10.249b and c)] one constant was eliminated using a continuity condition (10.248a) [(10.249a)]. The remaining three constants can be determined (10.250d through f) from the three boundary conditions (10.250a through c):

$$G(0;\xi)=G'(0;\xi)=0=G(L;\xi): \quad B=0=C, \quad A=-1+2\frac{\xi}{L}-\frac{\xi^2}{L^2}; \tag{10.250a–f}$$

substituting (10.250d through f) in (10.249b and c) specifies *the influence or Green function (10.251a and b) for the third-order differential equation (10.247a) with boundary conditions (10.250a through c):*

$$
G(x;\xi) = \begin{cases}
x^2\left(-\dfrac{1}{2}+\dfrac{\xi}{2}-\dfrac{\xi^2}{2L^2}\right) & \text{if } 0 \le x \le \xi, \qquad (10.251a) \\[4mm]
\xi\left[\dfrac{\xi}{2}-x+\dfrac{x^2}{L}\left(1-\dfrac{\xi}{2L}\right)\right] & \text{if } \xi \le x \le L. \qquad (10.251b)
\end{cases}
$$

that is unsymmetric in (x, ξ) and does not satisfy the reciprocity principle because the differential operator is not self-adjoint.

E10.18.3 Arbitrary Forcing of a Third-Order Differential Equation

The inhomogeneous differential equation of third-order (10.252b):

$$
f(x) \in \mathcal{E}(|R): \quad y'''(x) = f(x), \quad y(0) = y'(0) = 0 = y(L), \qquad (10.252a\text{–}e)
$$

with boundary conditions (10.252c through e) and forcing by an integrable function (10.252a), has a particular solution specified by convolution with the Green function (10.251a and b):

$$
y(x) = \int_0^x \xi\left[\frac{\xi}{2} - x + \frac{x^2}{L}\left(1 - \frac{\xi}{2L}\right)\right] f(\xi)\,d\xi
$$

$$
+ x^2 \int_x^L \left(-\frac{1}{2} + \frac{\xi}{L} - \frac{\xi^2}{2L^2}\right) f(\xi)\,d\xi. \qquad (10.253)
$$

Thus is obtained *the solution (10.253) of the third-order differential equation (10.252b) with boundary conditions (10.252c through e) and forcing by an integrable function (10.252a).*

E10.18.4 Forcing of a Third-Order Differential Equation by a Derivative Impulse

The solution of (10.254a)

$$
y'''(x) = \delta'(x-\eta), \quad y''(x) = \delta(x-\eta) + A, \quad y'(x) = H(x-\eta) + Ax + B \qquad (10.254a\text{–}c)
$$

can be obtained by (1) using the Green function in (10.253), which automatically satisfies the boundary conditions (10.252c through e) in the convolution integrals:

$$
0 \le x \le \eta: \quad y(x;\eta) = x^2 \int_x^L \left(-\frac{1}{2} + \frac{\xi}{L} - \frac{\xi^2}{2L^2}\right)\delta'(\xi-\eta)\,d\xi
$$

$$
= -x^2 \lim_{\xi \to \eta} \frac{d}{d\xi}\left(-\frac{1}{2} + \frac{\xi}{L} - \frac{\xi^2}{2L^2}\right) = -\frac{x^2}{L} + \frac{x^2\eta}{L^2}, \qquad (10.255a,b)
$$

$$
\eta \le x \le L: \quad y(x;\eta) = \int_0^x \xi\left[\frac{\xi}{2} - x + \frac{x^2}{L}\left(1 - \frac{\xi}{2L}\right)\right]\delta'(\xi-\eta)\,d\xi
$$

$$
= -\lim_{\xi \to \eta} \frac{d}{d\xi}\left[\frac{\xi^2}{2} - x\xi + \frac{x^2}{L}\left(\xi - \frac{\xi^2}{2L}\right)\right]
$$

$$
= -\eta + x - \frac{x^2}{L}\left(1 - \frac{\eta}{L}\right); \qquad (10.256a,b)
$$

and (2) by successive integration of (10.254a through c), leading to (10.257a and b)

$$y(x) = \begin{cases} \dfrac{A}{2}x^2 + Bx + C & \text{if } 0 \le x \le \eta, & (10.257a) \\[3mm] \dfrac{A}{2}x^2 + (B+1)x + C - \eta & \text{if } \eta \le x \le L, & (10.257b) \end{cases}$$

where one constant of integration was eliminated by continuity at $x = \eta$. The three remaining constants of integration are determined (10.258a through c) from the boundary conditions (10.252c through e)

$$B = 0 = C; \quad A = \frac{2}{L}\left(\frac{\eta}{L} - 1\right) : y(x; \eta) = \begin{cases} \dfrac{x^2}{L}\left(\dfrac{\eta}{L} - 1\right) & \text{if } 0 \le x \le \eta, \\[3mm] x - \eta + \dfrac{x^2}{L}\left(\dfrac{\eta}{L} - 1\right) & \text{if } \eta \le x \le L; \end{cases} \quad (10.258\text{a--e})$$

substitution of (10.258a through c) in (10.257a and b) specifies *the solution (10.258d and e) ≡ (10.255a and b; 10.256a and b) of the third-order differential equation (10.254a) with forcing by a derivative impulse and boundary conditions (10.252c through e). The Green function for a differential equation of order N has continuous derivatives up to order N – 2, for example, the first (10.248a through c) in (10.247a); if the forcing is by a derivative impulse, the solution has continuous derivatives up to order N – 3, for example, none (10.254c) in (10.254a). If an ordinary differential equation of order N is forced by the kth derivative of a unit impulse, the solution has continuous derivatives up to order N-k-2.*

EXAMPLE 10.19: DIPOLE OBLIQUE TO AN INSULATING/CONDUCTING PLANE

Consider the electrostatic field due to a dipole whose moment is oblique to a conducting/insulating plane. Determine the electric force exerted on the plane, and also the scaling of the electrostatic potential in the far-field.

E10.19.1 Electrostatic Potential due to the Dipole and Its Image

For the electric (magnetic) field of a point electric current (dipole with moment oblique) to a nearby conducting/insulating plane [Figure 8.10 (10.14)], (1) the z-axis is chosen through the current (dipole) with origin on the plane, so that its coordinates are $(0, 0, \pm a)$ for the dipole and its image on the plane; and (2) the position vector relative to the observer at (x, y, z) is (8.121a) [(10.259a)] and its modulus specifies the distance (8.121b) [10.259b)] using cylindrical (Cartesian) coordinates because the problem is (is not) axisymmetric:

$$\vec{x}_\pm = \vec{x} \mp \vec{e}_z a = \vec{e}_x x + \vec{e}_y y + \vec{e}_z(z \mp a), \quad (10.259a)$$

$$R_\pm = |\vec{x}_\pm| = \left|(z \mp a)^2 + x^2 + y^2\right| = \left|x^2 + y^2 + z^2 + a^2 \mp 2az\right|^{1/2}. \quad (10.259b)$$

In addition, (3) also the x-axis is chosen so that the dipole moment (10.260a) lies in the (x, z)-plane:

$$\vec{P}_1 = \vec{e}_x P_{\backslash\backslash} + \vec{e}_z P_\perp; \quad \vec{P}_1^+ = \vec{e}_x P_{\backslash\backslash} - \vec{e}_z P_\perp = -\vec{P}_1^-; \quad (10.260\text{a--c})$$

(4) for the image in an insulating plane (10.260b) the parallel (normal) component is the same (opposite); and (5) the image in a conducting plane (10.260c) is opposite. The total electrostatic potential (8.10a) of the dipole (10.260a) and its image (10.260b) is

$$4\pi\varepsilon\,\Phi_\pm(x, y, z) = \frac{\vec{P}_1 \cdot \vec{x}_+}{R_+^3} + \frac{\vec{P}_1^\pm \cdot \vec{x}_-}{R_-^3} = \frac{P_{\backslash\backslash}x + P_\perp(z-a)}{R_+^3} \pm \frac{P_{\backslash\backslash}x - P_\perp(z+a)}{R_-^3}, \quad (10.261\text{a,b})$$

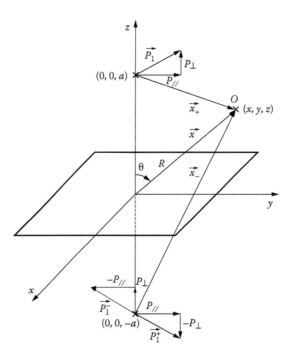

FIGURE 10.14 A point electric dipole near a plane has an image on the vertical to the plane and at the same distance. In the case of an insulating plane the image dipole has an equal horizontal (opposite vertical) component for the dipole moment. The dipole moment is reversed for a conducting plane, so that the dipole moment of the image has an equal vertical (opposite horizontal) component relative to the original dipole.

Thus *a point dipole at a distance a from an insulating (conducting) plane (Figure 10.14), with moment (10.260a) oblique to the plane, in a medium of dielectric permittivity ε has an electrostatic potential (10.261a) [(10.261b)] involving the position (10.259a) and distance (10.259b) of the observer relative to the dipole and its image (10.260b) [(10.260c)].*

E10.19.2 Electric Field of a Dipole near a Conducting/Insulating Plane

From (10.259b) follows

$$\left\{\frac{\partial}{\partial x},\frac{\partial}{\partial y},\frac{\partial}{\partial z}\right\}R_{\pm} = \frac{\{x,y,z\mp a\}}{R_{\pm}}, \quad \left\{\frac{\partial}{\partial x},\frac{\partial}{\partial y},\frac{\partial}{\partial tz}\right\}\frac{1}{R_{\pm}^3} = -3\frac{\{x,y,z\mp a\}}{R_{\pm}^5}. \qquad (10.262\text{a,b})$$

Using (10.262a) in (10.261a and b) specifies the Cartesian components of the electric field:

$$4\pi\varepsilon E_x^{\pm}(x,y,z) = -\frac{\partial\Phi_{\pm}}{\partial x} = \frac{P_\|(3x^2-R_+^2)+3P_\perp x(z-a)}{R_+^5} \pm \frac{P_\|(3x^2-R_-^2)-3P_\perp x(z+a)}{R_-^5}, \quad (10.263\text{a})$$

$$4\pi\varepsilon E_y^{\pm}(x,y,z) = -\frac{\partial\Phi_{\pm}}{\partial y} = 3y\left[\frac{P_\| x+P_\perp(z-a)}{R_+^5} \pm \frac{P_\| x-P_\perp(z+a)}{R_-^5}\right], \qquad (10.263\text{b})$$

$$4\pi\varepsilon E_z^{\pm}(x,y,z) = -\frac{\partial\Phi_{\pm}}{\partial z} = \frac{3P_\| x(z-a)+P_\perp\left[3(z-a)^2-R_+^2\right]}{R_+^5}$$

$$\pm\frac{3P_\| x(z+a)-P_\perp\left[3(z+a)^2-R_-^2\right]}{R_-^5}. \qquad (10.263\text{c})$$

On the plane (8.119a) ≡ (10.264a), the dipole and its image are at the same distance (8.125b) ≡ (10.264b):

$$z = 0: \quad R_\pm(x, y, 0) = \left| x^2 + y^2 + a^2 \right|^{1/2} = R_0, \tag{10.264a,b}$$

showing that the electric field (10.263a through c) is (1) tangent to an insulating plane (10.265):

$$2\pi\varepsilon R_0^5 \vec{E}_+(x, y, 0) = \vec{e}_x \left[P_{\backslash\backslash}\left(2x^2 - y^2 - a^2\right) - 3P_\perp ax \right] + \vec{e}_y 3y \left(P_{\backslash\backslash} x - P_\perp a\right); \tag{10.265a–c}$$

and (2) orthogonal to a conducting plane (10.266b):

$$\Phi_-(x, y, 0) = 0: \quad \varepsilon \vec{E}_-(x, y, 0) = -\vec{e}_z \frac{3P_{\backslash\backslash}\,ax - P_\perp\left(2a^2 - x^2 - y^2\right)}{2\pi R_0^5} = \vec{e}_z\,\sigma_-(x, y), \tag{10.266a–c}$$

and specifies the induced electric charge distribution (10.266c) on the plane that is (10.261b) an equipotential surface (10.266a). Thus *a point dipole with moment (10.260a) oblique to an insulating (conducting) plane (Figure 10.14) creates an electric field (10.263a through c) that is corresponding to the potential (10.261a) [(10.261b)]. The case of the conducting plane corresponds to an equipotential surface (10.266a) with an induced electric charge distribution (10.266c).*

E10.19.3 Force Exerted by a Dipole on a Conducting/Insulating Plane

A dipole of moment $\vec{P}_1$ corresponds to an electric charge density (8.5) ≡ (10.267a) and when placed in an electric field $\vec{E}$ is subject to an electric force (10.267b)

$$e_\pm(\vec{x}) = -\left\{ \vec{P}_1^\pm \cdot \frac{\partial}{\partial \vec{x}} \right\} \delta(\vec{x} - \vec{y}): \quad \vec{F}_\pm = -\int \vec{E}(\vec{y}) \left\{ \vec{P}_1^\pm \cdot \frac{\partial}{\partial \vec{y}} \right\} \delta(\vec{x} - \vec{y})\, d^3\vec{y} = \left\{ \vec{P}_1^\pm \cdot \frac{\partial}{\partial \vec{x}} \right\} \vec{E}(\vec{x}), \tag{10.267a,b}$$

where was used (3.18a through c). In the case of two dipoles (10.267b) leads to (10.360c). The force exerted by the dipole (10.260a) on the insulating (conducting) plane equals the force exerted on its image:

$$\vec{F}_\pm = \pm \lim_{x, y \to 0} \left(P_{\backslash\backslash} \frac{\partial}{\partial x} - P_\perp \frac{\partial}{\partial z} \right) \frac{1}{4\pi\varepsilon R_+^5} \left\{ \vec{e}_x P_{\backslash\backslash} \left[\left(3x^2 - R_+^2\right) + 3P_\perp x(z - a) \right] \right.$$
$$\left. + \vec{e}_y 3y \left[P_{\backslash\backslash} x + P_\perp(z - a) \right] + \vec{e}_z \left[3P_{\backslash\backslash} x(z - a) + P_\perp \left(3(z - a)^2 - R_+^2\right) \right] \right\}, \tag{10.268}$$

where only the first term on the r.h.s. of (10.263a through c) has been retained, because the dipole acts on its image and the image exerts no force upon itself. From (10.259b) and (10.262a) follows

$$\lim_{\substack{x, y \to 0 \\ z \to -a}} \left\{ 1, \frac{\partial}{\partial x}, \frac{\partial}{\partial y}, \frac{\partial}{\partial z} \right\} R_+ = \{2a, 0, 0, -1\}. \tag{10.269a–d}$$

Using (10.269a through d) in (10.268) leads to

$$\mp 4\pi\varepsilon(2a)^4 \vec{F}_\pm = 3P_{\backslash\backslash}\left(P_\perp \vec{e}_x + P_{\backslash\backslash}\vec{e}_z \right) + 5P_\perp \left(2P_\perp \vec{e}_z - P_{\backslash\backslash}\vec{e}_x \right) + P_\perp \left(2P_{\backslash\backslash}\vec{e}_x - 4P_\perp \vec{e}_z \right). \tag{10.270}$$

The electric force exerted by a dipole on an insulating (conducing) wall at a distance a has always the vertical direction:

$$\mp 64\pi\varepsilon a^4 \vec{F}_{\pm} = 3\left(P_{\parallel}^2 + 2P_{\perp}^2\right)\vec{e}_z, \tag{10.271}$$

The component of the dipole moment orthogonal to the plane exerts the double of the force relative to an horizontal component of the same magnitude.

E10.19.4 Dipole with Double Strength or Quadrupole in the Far-Field

Using spherical coordinates (6.13a through d) ≡ (10.272a through c)

$$R = \left|x^2 + y^2 + z^2\right|^{1/2}, \quad z = R\cos\theta, \quad x = R\sin\theta\cos\phi, \tag{10.272a–c}$$

the distance from the observer to the dipole and its image (10.259b) scales in the far-field as

$$R_{\pm} = \left|R^2 \mp 2aR\cos\theta + a^2\right|^{1/2} = R\left[1 \mp \frac{a}{R}\cos\theta + O\left(\frac{a^2}{R^2}\right)\right], \tag{10.273a}$$

$$\frac{1}{R_{\pm}^3} = \frac{1}{R^3}\left[1 \pm 3\frac{a}{R}\cos\theta + O\left(\frac{a^2}{R^2}\right)\right], \tag{10.273b}$$

implying (10.273b), and hence (10.273c through e):

$$a^2 \ll R^2: \qquad \frac{1}{R_+^3} + \frac{1}{R_-^3} = \frac{2}{R^3}, \quad \frac{1}{R_+^3} + \frac{1}{R_-^3} = 6\frac{a}{R}\cos\theta. \tag{10.273c–e}$$

Rewriting the electrostatic potential (10.261b) in the form,

$$4\pi\varepsilon\Phi_{\pm}(x,y,z) = \left(P_{\parallel}x - P_{\perp}a\right)\left(\frac{1}{R_+^3} \pm \frac{1}{R_-^3}\right) + P_{\perp}z\left(\frac{1}{R_+^3} \mp \frac{1}{R_-^3}\right) \tag{10.274}$$

and substituting (10.272b and c; 10.273), it follows (10.273d and e) that *the far-field (10.273a) of an electric dipole is given (1) for an insulating plane:*

$$4\pi\varepsilon R^3\Phi_+(x,y,z) = 2\left(P_{\parallel}x - P_{\perp}a\right) + 6P_{\perp}z\frac{a}{R}\cos\theta = 2P_{\parallel}x + 2P_{\perp}a\left(3\cos^2\theta - 1\right), \tag{10.275}$$

corresponding to the electrostatic potential

$$\Phi_+(R,\theta,\phi) = \frac{\vec{P}_{1\parallel}}{4\pi R^2}\sin\theta\cos\varphi + \frac{\vec{P}_{2\perp}}{8\pi\varepsilon R^3}\left(3\cos^2\theta - 1\right) \tag{10.276}$$

of the sum of a horizontal dipole $P_{1\parallel}$ in the x-direction (6.194c) [vertical quadrupole P_{zz} in the z-direction) (6.195c)] with moment (10.277a) [(10.277b)]

$$\vec{P}_{1\parallel} = \vec{e}_x 2P_{\parallel}, \quad \vec{P}_{2\perp} = \vec{e}_z 4P_{\perp}a; \tag{10.277a,b}$$

and (2) for a conducting plane:

$$4\pi\varepsilon R^3 \Phi_-\left(x,y,z\right) = 2P_\perp z + 6\left(P_\parallel x - P_\perp a\right)\frac{a}{R}\cos\theta, \qquad (10.278)$$

corresponding to the potential

$$\Phi_-\left(R,\theta,\phi\right) = \frac{P_{1\perp}}{4\pi\varepsilon R^2}\cos\theta + \frac{P_\parallel}{4\pi\varepsilon R^3}3\cos\theta\sin\theta\cos\phi \qquad (10.279)$$

of the sum of a vertical dipole P_z in the z-direction (6.194c) [cross-quadrupole P_{zx} in the (x, z)-plane (6.195c)] with moment (10.280a) [(10.281b)]

$$\vec{P}_{1\perp} = \vec{e}_z 2P_{1\perp}, \vec{P}_{2\parallel} = \vec{e}_x 4P_\parallel a. \qquad (10.280\text{a,b})$$

The dipoles have twice the moment of the original dipole (10.277a) [(10.280a)], and the moment of the quadrupoles (10.277b) [(10.280b)] equals the moment of the dipole multiplied by 4a, that is twice the distance 2a from its image. The third and last terms on the r.h.s. of (10.278) were omitted when passing to (10.279) because they are of lower order $O(R^{-4})$; thus, they would be part of an octupole, and other octupole terms were omitted in (10.273).

EXAMPLE 10.20: ELECTROSTATIC FIELD OF A DIPOLE ORTHOGONAL TO A CONDUCTING/INSULATING SPHERE

By using the appropriate sphere theorem (Figure 10.16), consider the electrostatic field of a dipole with axis orthogonal to a sphere (Figure 10.15) in the conducting (insulating) cases [Figure 10.17a (b)] and calculate the force (Figure 10.18a and b) exerted on the sphere.

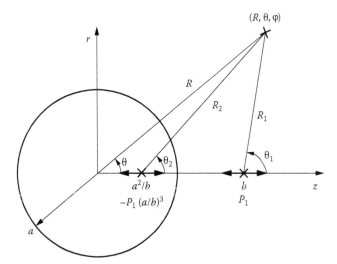

FIGURE 10.15 In the case of a dipole with moment orthogonal to the sphere, the image dipole lies at the reciprocal point and its moment equals in modulus that of the original dipole multiplied by the cube of the ratio of the radius of the sphere to the distance of the original dipole from the center. The original dipole may be outside the sphere and the image inside or vice versa.

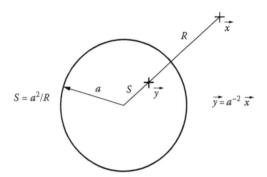

FIGURE 10.16   The image on a plane (sphere) [Figure 10.14 (10.15)] lies at the reciprocal point on the normal to the plane (on the radial line through the center) at the same distance (at a distance that when multiplied by the distance of the original from the center equals the square of the radius).

E10.20.1 Induced Charge Distribution on a Conducting Sphere

The electrostatic potential of a dipole of moment P_1 at the position $R = b$, $\theta = 0$ with axis $\theta = 0$ (Figure 10.15) in an infinite medium of dielectric permeability ε is (8.10a) $\equiv$ (10.278)

$$\frac{4\pi\varepsilon}{P_1}\Phi_1(R,\theta) = \frac{\cos\theta_1}{R_1^2} = \frac{z_1}{R_1^3} = (R\cos\theta - b)\left|R^2 + b^2 - 2bR\cos\theta\right|^{-3/2},\tag{10.281}$$

where R_1 is the distance of the observer from the dipole (8.199b) and θ_1 is the angle of the relative position vector with the axis (8.212b).

By the second sphere theorem (6.289), the insertion of a conducting sphere of radius a (Figure 10.16) involves the potential at the reciprocal point (6.169b) leading to the total potential

$$\Phi_-(R,\theta) = \Phi_1(R,\theta) - \frac{a}{R}\Phi_1\left(\frac{a^2}{R},\theta\right) = \frac{P_1}{4\pi\varepsilon}\left\{(R\cos\theta - b)\left|R^2 + b^2 - 2bR\cos\theta\right|^{-3/2}\right.$$

$$\left. - \frac{a}{R}\left(\frac{a^2}{R}\cos\theta - b\right)\left|\frac{a^4}{R^2} + b^2 - \frac{2a^2 b}{R}\cos\theta\right|^{-3/2}\right\},\tag{10.282a}$$

which can be written in the alternative form

$$\frac{4\pi\varepsilon}{P_1}\Phi_-(R,\theta) = (R\cos\theta - b)\left|R^2 + b^2 - 2bR\cos\theta\right|^{-3/2}$$

$$- \left(R\cos\theta - \frac{bR^2}{a^2}\right)\left|a^2 + \frac{b^2 R^2}{a^2} - 2bR\cos\theta\right|^{-3/2}.\tag{10.282b}$$

The total potential (10.282a) $\equiv$ (10.282b) vanishes on the sphere; it is the electrostatic potential of an axial dipole (Figure 10.17a) at a distance b from the center of a conducting sphere of radius a. The electric field is normal to the sphere:

$$E_\theta^-(a,\theta) = 0,\quad \varepsilon E_R^-(a,\theta) = \frac{P_1}{4\pi a}\left|a^2 + b^2 - 2ab\cos\theta\right|^{-5/2}$$

$$\times\left(b^3 + ab^2\cos\theta - 5ba^2 + 3a^3\cos\theta\right) = \sigma(\theta),\tag{10.283a,b}$$

it specifies the distribution of electric charges induced on the sphere, whose density is largest (smallest) in modulus at the closest (farthest) point:

$$\left|\sigma(0)\right| = \frac{|P_1|}{4\pi a} \frac{\left|b^3 + ab^2 - 5ba^2 + 3a^3\right|}{(a-b)^5} \le \left|\sigma(\theta)\right|$$

$$\le \frac{|P_1|}{4\pi a} \frac{\left|b^3 - ab^2 - 5ba^2 - 3a^3\right|}{(a+b)^5} = \sigma(\pi). \tag{10.284}$$

The passage from (10.282b) to (10.283a and b) corresponds to the calculation of the radial electric field on the sphere:

$$\frac{4\pi\varepsilon}{P_1} E_{\bar{R}}(a,\theta) = -\frac{4\pi\varepsilon}{P_1} \lim_{R\to a} \frac{\partial \Phi_-}{\partial R}$$

$$= -\frac{2b}{a} \left|a^2 + b^2 - 2ab\cos\theta\right|^{-3/2}$$

$$+ 3(a\cos\theta - b)\left(a - \frac{b^2}{a}\right)\left|a^2 + b^2 - 2ab\cos\theta\right|^{-5/2}$$

$$= \left|a^2 + b^2 - 2ab\cos\theta\right|^{-5/2}\left[3(a\cos\theta - b)\left(a - \frac{b^2}{a}\right) - \frac{2b}{a}\left(a^2 + b^2 - 2ab\cos\theta\right)\right]$$

$$= \left|a^2 + b^2 - 2ab\cos\theta\right|^{-5/2}\left(\frac{b^3}{a} + b^2\cos\theta - 5ab + 3a^2\cos\theta\right); \tag{10.285a}$$

this specifies the density of induced electric charges on the sphere:

$$\left|a^2 + b^2 - 2ab\cos\theta\right|^{5/2}\frac{4\pi a}{P_1}\sigma(\theta) = b^3 + b^2 a\cos\theta - 5a^2 b + 3a^3\cos\theta, \tag{10.285b}$$

which is in agreement with (10.283b).

E10.20.2 Dipole Image on the Sphere and in the Far Field

The exact potential (10.282b) can be written in the form

$$\Phi_-(R,\theta) = \frac{P_1}{4\pi\varepsilon}\left\{(R\cos\theta - b)\left|R^2 + b^2 - 2bR\cos\theta\right|^{-3/2}\right.$$

$$\left. -\frac{a^3}{b^3}\left(R\cos\theta - \frac{R^2 b}{a^2}\right)\left|R^2 + \frac{a^4}{b^2} - \frac{2a^2 R}{b}\cos\theta\right|^{-3/2}\right\}$$

$$= \frac{P_1}{4\pi\varepsilon}\left[\frac{\cos\theta_1}{R_1^2} - \left(\frac{a}{b}\right)^3\frac{\cos\theta_3}{R_2^2}\right]. \tag{10.286}$$

Thus *an axial dipole with dipole moment P_1 at a distance b from the center of a conducting sphere of radius a (Figure 10.17a) has an image dipole with moment (10.287a), as shown by the exact electrostatic potential (10.286) where (1) $R_1(R_2)$ is the distance (8.199b) [(8.201b)] from the observer to the dipole at a (its image on the sphere at a^2/b); (2) of the two $\theta_1(\theta_2)$ angles (8.212b) [(8.215b)], only the first appears in (10.286); (3) the angle (8.214b) $\equiv$ (10.287b) is replaced in (10.286) by (10.287c):*

$$\bar{P}_1 = -P_1\left(\frac{a}{b}\right)^3: \quad R_2\cos\theta_2 = \left(R\cos\theta - \frac{a^2}{b}\right), \quad R_2\cos\theta_3 = R\cos\theta - b\frac{R^2}{a^2}; \tag{10.287a-c}$$

and (4) the two angles (10.287b and c) do not coincide on the sphere $R \equiv a$. In the far field, the dipole and its image on the sphere lead to an asymptotic potential (10.288):

$$
\begin{aligned}
\Phi_-(R,\theta) &= \frac{P_1}{4\pi\varepsilon}\left[\frac{a}{b^2 R} + \left(1 + 2\frac{a^3}{b^3}\right)\frac{\cos\theta}{R^2} + O\left(\frac{b^2}{R^4},\frac{a^2}{bR^3}\right)\right] \\
&= \frac{\tilde{P}_0}{4\pi\varepsilon R} + \frac{\tilde{P}}{4\pi\varepsilon R^2}\cos\theta + O\left(\frac{P_1 b}{R^4},\frac{P_1 a^2}{bR^3}\right),
\end{aligned}
\tag{10.288}
$$

where (1) the first term is a monopole of moment (10.289a); (2) the second term is a dipole whose moment (10.289b) adds to the original dipole twice the image (10.289c); and (3) the remaining terms on the r.h.s. of (10.288) are quadrupoles and higher-order multipoles.

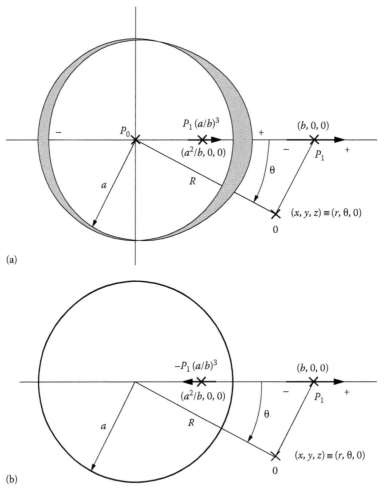

(a)

(b)

FIGURE 10.17 A dipole with moment orthogonal to a conducting (insulating) sphere [Figure 10.17a (b)] creates an electrostatic field corresponding to an equal (opposite) image dipole (Figure 10.16) at the reciprocal point (Figure 10.15). The insulating sphere has a tangential electric field and no surface electric charges (Figure 10.17b). On the conducting sphere a distribution of electric charges is induced, with maximum (minimum) density in modulus at the point closest to (farthest from) the dipole, and with opposite (equal) sign to the closest electric charge of the dipole. The dipole exerts an electric force (Figure 10.18) of attraction (repulsion) on the conducting (insulating) sphere [Figure 10.17a (b)].

$$\tilde{P_0} = P_1 \frac{a}{b^2}, \quad \tilde{P_1} = P_1 \left(1 + 2\frac{a^3}{b^3}\right) = P_1 + 2\bar{P_1}, \quad \bar{P_1} = \frac{a^3}{b^3}P_1. \tag{10.289a–c}$$

The physical explanation for the appearance of a monopole term and the derivation of (10.288) from (10.286) are considered next.

E10.20.3 Appearance of a Monopole Term and Asymptotic Scaling

The appearance of a monopole term shows that the presence of a conducting sphere leads to an enhanced far-field effect larger $O(R^{-1})$ than the original dipole $O(R^{-2})$ in free space; in addition, the image dipole has the same direction and appears with its strength $P_1(a/b)^3$ doubled. The reason (Figure 10.17a) is that (1) a positive dipole has a negative (positive) charge closer to (farther from) the conducting sphere; hence (2) stronger positive (weaker negative) charges accumulate on the side of the conducting sphere closest to (farthest from) an image dipole; (3) this corresponds to an image dipole with the same sign; and (4) the imbalance in the charge distribution on the conducting sphere leads to the appearance of a positive monopole term. The latter violates the condition of zero flux at infinity; it implies that the conducting sphere at zero potential is "earthed," and collects electric charges from the "ground." In the plane case, a line charge near a conducting cylinder (Section I.24.8) has images both at the reciprocal point and at the center. The passage from the exact potential (10.286) ≡ (10.290)

$$\frac{4\pi\varepsilon R^2}{P_1}\Phi_-(R,\theta) = \left(\cos\theta - \frac{b}{R}\right)\left|1 - \frac{2b}{R}\cos\theta + \frac{b^2}{R^2}\right|^{-3/2}$$

$$- \frac{a^3}{b^3}\left(\cos\theta - \frac{bR}{a^2}\right)\left|1 - \frac{2a^2}{bR}\cos\theta + \frac{a^4}{b^2R^2}\right|^{-3/2} \tag{10.290}$$

to the asymptotic approximation (10.288) ≡ (10.291) includes the two lowest orders

$$\frac{4\pi\varepsilon R^2}{P_1}\Phi_-(R,\theta) = \left(\cos\theta - \frac{b}{R}\right)\left[1 + \frac{3b}{R}\cos\theta + O\left(\frac{b^2}{R^2}\right)\right]$$

$$- \frac{a^3}{b^3}\left(\cos\theta - \frac{bR}{a^2}\right)\left[1 + \frac{3a^2}{bR}\cos\theta + O\left(\frac{a^4}{b^2R^2}\right)\right] \tag{10.291}$$

to retain (1) the term $O(R^{-2})$ corresponding to the original dipole; and (2) all images of the same or lower order. The monopole moment (10.289a) corresponds to the total electric charge on the sphere, as shown next (E10.20.6) by a method that applies to axial multipole moments of any order (E10.20.5).

E10.20.4 Total Electric Charge Induced on the Conducting Sphere

Integrating the induced electric charge density over the sphere of radius a (6.360d) specifies the total electric charge:

$$\tilde{P_0} = \int_{R=a}\sigma(\theta)\,dA = 2\pi a^2\int_0^\pi\sigma(\theta)\sin\theta\,d\theta. \tag{10.292}$$

For (10.283b), this corresponds to a monopole with moment

$$\tilde{P_0} = \frac{P_1 a}{2}\int_0^\pi\left[b\left(b^2 - 5a^2\right) + a\left(b^2 + 3a^2\right)\cos\theta\right]\left|a^2 + b^2 - 2ab\cos\theta\right|^{-5/2}\sin\theta\,d\theta. \tag{10.293}$$

The integral is of the type

$$I = \sum_{n=0}^{N} c_n \int_0^{\pi} \cos^n \theta \left| a^2 + b^2 - 2ab\cos\theta \right|^{-m/2} \sin\theta\, d\theta, \tag{10.294}$$

which can be evaluated using the exact differential

$$\frac{d}{d\theta}\left\{ \left| a^2 + b^2 - 2ab\cos\theta \right|^{-m/2+1} \right\} = \left| a^2 + b^2 - 2ab\cos\theta \right|^{-m/2} (2-m)ab\cos\theta \tag{10.295}$$

in n successive integrations by parts.

The method is illustrated next in the case ($n = 1$, $m = 5$) for (10.293) involving one integration by parts:

$$\tilde{P}_0 = -\frac{P_1}{6b} \int_0^{\pi} \left[b\left(b^2 - 5a^2\right) + a\left(b^2 + 3a^2\right)\cos\theta \right] \frac{d}{d\theta}\left\{ \left| a^2 + b^2 - 2ab\cos\theta \right|^{-3/2} \right\} d\theta$$

$$= -\frac{P_1}{6b} \left\{ \left[b\left(b^2 - 5a^2\right) + a\left(b^2 - 3a^2\right)\cos\theta \right] \left| a^2 + b^2 - 2ab\cos\theta \right|^{-5/2} \right\}_0^{\pi}$$

$$- P_1 a \frac{b^2 + 3a^2}{6b} \int_0^{\pi} \left| a^2 + b^2 - 2ab\cos\theta \right|^{-3/2} \sin\theta\, d\theta. \tag{10.296}$$

The second term on the r.h.s. of (10.296) is again an exact differential:

$$\tilde{P}_0 + \frac{P_1}{6b}\left[\frac{b\left(b^2 - 5a^2\right) - a\left(b^2 + 3a^2\right)}{\left(b+a\right)^3} - \frac{b\left(b^2 - 5a^2\right) + a\left(b^2 + 3a^2\right)}{\left(b-a\right)^3} \right]$$

$$= P_1 \frac{b^2 + 3a^2}{6b^2} \left\{ \left| a^2 + b^2 - 2ab\cos\theta \right|^{-1/2} \right\}_0^{\pi}$$

$$= P_1 \frac{b^2 + 3a^2}{6b^2}\left(\frac{1}{b+a} - \frac{1}{b-a} \right) = -\frac{P_1 a}{3b^2}\frac{b^2 + 3a^2}{b^2 - a^2}, \tag{10.297}$$

completing the evaluation of the integral (10.293).

E10.20.5 Method for the Axial Multipole Moment of All Orders

In the case of the monopole moment, (10.297) is simplified:

$$\tilde{P}_0\left(b^2 - a^2\right) + P_1 a \frac{b^2 + 3a^2}{3b^2} = \frac{P_1}{6b}\left(b^2 - a^2\right)\left[\frac{b\left(b^2 - 5a^2\right) + a\left(b^2 + 3a^2\right)}{\left(b-a\right)^3} - \frac{b\left(b^2 - 5a^2\right) - a\left(b^2 + 3a^2\right)}{\left(b+a\right)^3} \right]$$

$$= \frac{P_1}{6b}\frac{1}{\left(b^2 - a^2\right)^2}\left\{ \left(b+a\right)^3\left[b\left(b^2 - 5a^2\right) + a\left(b^2 + 3a^2\right) \right] \right.$$

$$\left. - \left(b-a\right)^3\left[b\left(b^2 - 5a^2\right) - a\left(b^2 + 3a^2\right) \right] \right\}$$

$$= \frac{P_1 a}{3}\frac{\left(b^2 + 3a^2\right)^2 + \left(a^2 + 3b^2\right)\left(b^2 - 5a^2\right)}{\left(b^2 - a^2\right)^2}$$

$$= \frac{4}{3}P_1 a \frac{b^4 + a^4 - 2a^2 b^2}{\left(b^2 - a^2\right)^2} = \frac{4}{3}P_1 a. \tag{10.298}$$

The monopole moment corresponding to the induced electric charge distribution on the sphere is

$$\tilde{P}_0 = \frac{P_1 a}{b^2 - a^2}\left(\frac{4}{3} - \frac{b^2 + 3a^2}{3b^2}\right) = P_1\frac{a^2}{b}, \tag{10.299}$$

which is in agreement with (10.289a).

Thus *the total electric charge induced on a conducting sphere of radius a by a dipole with moment P_1 orthogonal to the sphere and at a distance b from its center (10.299) agrees with the moment (10.289a) of the leading monopole term of the electrostatic potential in the far field (10.288). The axial multipole moments of all orders,*

$$\tilde{P}_n = \int_{R=a} z^n \sigma(\theta)\,dA = 2\pi a^{n+2}\int_0^\pi \cos^n\theta\,\sigma(\theta)\sin\theta\,d\theta, \tag{10.300}$$

can be calculated for the induced surface electric change distribution (10.283b) by

$$P_n = \frac{P_1}{2}a^{n+1}\int_0^\pi \cos^n\theta\Big[b\big(b^2 - 5a^2\big) + a\big(b^2 - 3a^2\big)\cos\theta\Big]\sin\theta\,\big|a^2 - b^2 - 2ab\cos\theta\big|^{-5/2}\,d\theta; \tag{10.301}$$

this integral is of the type (10.294) and can be evaluated by the preceding methods (10.295 through 10.297).

E10.20.6 Moments of the Dipole Image on the Sphere

The field function of an axial dipole at position $(R = b, \theta = 0)$ in space is given by (6.164c) $\equiv$ (10.302)

$$\frac{4\pi\varepsilon}{P_1}\Psi_1(R,\theta) = -\frac{\sin^2\theta_1}{R_1} = -\frac{r^2}{R_1^3} = -R^2\sin^2\theta\big|R^2 + b^2 - 2bR\cos\theta\big|^{-3/2}, \tag{10.302}$$

where the hydrodynamic dipole moment P_1 is replaced by the electric dipole moment $-P_1/\varepsilon$. The field (10.302) [stream (6.164c)] function for an electrostatic (hydrodynamic) dipole has opposite signs due to the convention of opposite signs in relation with the electric field (10.303a through c) [velocity (6.58a through c)]:

$$\{E_R, E_\theta, E_\varphi\} = \{-R^{-2}\csc\theta\,\partial_\theta\Psi,\ R\csc\theta\,\partial_R\Psi,\ 0\}. \tag{10.303a-c}$$

The insertion of an insulating sphere (Figure 10.17b) of radius a is specified by the first sphere theorem (6.278), leading to the total field function:

$$\Psi_+(R,\theta) = \Psi_1(R,\theta) - \frac{R}{a}\Psi_1\left(\frac{a^2}{R},\theta\right) = -\frac{P_1}{4\pi\varepsilon}\sin^2\theta$$

$$\left\{R^2\big|R^2 + b^2 - 2bR\cos\theta\big|^{-3/2} - \frac{a^3}{R}\left|\frac{a^4}{R^2} + b^2 - \frac{2a^2 b}{R}\cos\theta\right|^{-3/2}\right\}. \tag{10.304}$$

Rewriting the exact field function in the form

$$\Psi_+(R,\theta) = -\frac{P_1}{4\pi\varepsilon}R^2\sin^2\theta\left\{\big|R^2 + b^2 - 2bR\cos\theta\big|^{-3/2} - \frac{a^3}{b^3}\left|R^2 + \frac{a^4}{b^2} - \frac{2a^2 R}{b}\cos\theta\right|^{-3(2}\right\}$$

$$= -\frac{P_1}{4\pi\varepsilon}R^2\sin^2\theta\left(\frac{1}{R_1^3} - \frac{a^3}{b^3}\frac{1}{R_2^3}\right), \tag{10.305}$$

it follows that *the total field function (10.304) ≡ (10.305) (Figure 10.16) is due to the real dipole of moment P_1 at $R = b$ plus its image with moment (10.287a) at the reciprocal point a^2/b on the sphere of radius a, the respective distances from the observer being (8.199c) and (8.201b). The image dipole (10.287a) has moment with opposite sign. In the case of a dipole with positive moment $P_1 > 0$ orthogonal to the sphere (Figure 10.17b), the negative (positive) charge is closer (farther) from the sphere; the opposite image has the negative (positive) charge closer to the sphere. Thus the closest electric charges for the dipole and its image on an insulating sphere always have the same sign, leading to a force of repulsion (Section E.10.20.8).*

E10.20.7 Insulating Sphere as an Electrostatic Field Surface

In the far field, the two dipoles appear superimposed, with the dipole moments (10.287a) added together (10.306b) in the asymptotic field function (10.306a)

$$\Psi_+(R,\theta) = -\frac{\hat{P}_1}{4\pi\varepsilon R}\sin^2\theta\left[1+O\left(\frac{b}{R},\frac{a^2}{bR}\right)\right], \quad \hat{P}_1 = P_1\left(1-\frac{a^3}{b^3}\right) = P_1 + \overline{P}_1. \quad (10.306a,b)$$

The total dipole moment (10.306b) consists of: (1) the original dipole moment; and (2) the image dipole moment (10.287a), with opposite sign, and the modulus is reduced relative to the original dipole by a factor $(a/b)^3$; as the dipole moves away from the sphere $b \to \infty$ the image dipole at the reciprocal point goes to the center $a^2/b \to 0$ with a vanishing moment $\overline{P}_1 \to 0$. The electrostatic field function of an axial dipole of moment P_1 at a distance b from an insulating sphere of radius a (10.304) ≡ (10.305) ≡ (10.307) is

$$\Psi_+(R,\theta) = -\frac{P_1}{4\pi\varepsilon}R^2\sin^2\theta\left\{\left|R^2+b^2-2bR\cos\theta\right|^{-3/2} - \frac{a^3}{R^3}\left|\frac{a^4}{R^2}+b^2-\frac{2a^2R}{b}\cos\theta\right|^{-3/2}\right\}, \quad (10.307)$$

which leads to a tangential electric field on the sphere:

$$E_R^+(a,\theta) = 0, \quad E_\theta^+(a,\theta) = -\frac{3P_1}{4\pi\varepsilon}\sin\theta\left(b^2-a^2\right)\left|a^2+b^2-2ab\cos\theta\right|^{-5/2}. \quad (10.308a,b)$$

The electric field vanishes at the points of the sphere closest to and farthest from the dipole (10.309a and b)

$$\vec{E}_+(a,0) = 0 = \vec{E}_+(a,\pi), \quad \vec{E}_+\left(a,\pm\frac{\pi}{2}\right) = \pm\vec{e}_\theta\frac{3P_1}{4\pi\varepsilon}\left(b^2-a^2\right)\left|b^2+a^2\right|^{-5/2}, \quad (10.309a\text{-}c)$$

and takes the extreme values (10.309c) at the side positions. The electric field on the sphere (10.308a and b) follows from the field function (10.304) using (10.303a through c), for example,

$$\frac{4\pi\varepsilon}{P_1}E_\theta^+(a\theta) = -\frac{4\pi\varepsilon}{P_1}\lim_{R\to a}\frac{1}{R\sin\theta}\frac{\partial\Psi_+}{\partial R}$$

$$= -\sin\theta\left\{3\left|a^2+b^2-2ab\cos\theta\right|^{-3/2}+6a\left(b\cos\theta-a\right)\left|a^2+b^2-2ab\cos\theta\right|^{-5/2}\right\}; \quad (10.310)$$

this simplifies to

$$-\frac{4\pi\varepsilon}{3P_1}\left|a^2+b^2-2ab\cos\theta\right|^{5/2}E_\theta^+(a,\theta)$$

$$= \sin\theta\left[a^2+b^2-2ab\cos\theta-2a\left(a-b\cos\theta\right)\right] = \left(b^2-a^2\right)\sin\theta, \quad (10.311)$$

which coincides with (10.308b) ≡ (10.311).

E10.20.8 Electric Force Exerted by the Dipole on an Insulating/Conducting Sphere

In the case of a dipole with moment orthogonal to an insulating (conducting) sphere the image [Figure 10.17b(a)] has a dipole moment smaller by a^3/b^3 in modulus (10.287a) with the same (opposite) sign (10.312a). Thus the electric force exerted by a dipole with moment orthogonal to an insulating (conducting) sphere may be calculated in four steps: (1) the dipole moment (10.312b) of the image is used in the dipole operator (10.312c):

$$P_1^\pm = \pm \bar{P}_1 = \mp P_1 \frac{a^3}{b^3} : \quad \vec{P}_1^\pm = \vec{e}_z \, P_1^\pm = \mp \vec{e}_z \, P_1 \frac{a^3}{b^3}, \quad \vec{P}_1^\pm . \nabla = P_1^\pm \frac{\partial}{\partial z} = \mp P_1 \frac{a^3}{b^3} \frac{\partial}{\partial z}; \quad (10.312\text{a–c})$$

(2) the axial electric field is that (8.26e) of the original dipole in free space:

$$E_{0z}(z,\theta) = \lim_{\theta \to 0} \frac{P_1}{4\pi\varepsilon} \frac{3\cos^2\theta - 1}{(z-b)^3} = \frac{P_1}{2\pi\varepsilon} (z-b)^3 ; \quad (10.313)$$

(3) applying the dipole operator (10.312c) to the electric field (10.313) specifies (10.267b) the electric force (10.314b):

$$z = \frac{a^2}{b} : \quad \vec{F}_\pm = \left\{ \vec{P}_1^\pm . \nabla \right\} \vec{E}_0(\vec{x}) = \vec{e}_z \, P_1^\pm \frac{\partial}{\partial z} \left[E_{0z}(z,0) \right]$$

$$= \mp \vec{e}_z \frac{P_1 \bar{P}_1}{2\pi\varepsilon} \frac{\partial}{\partial z} \left[(z-b)^{-3} \right] = \pm \vec{e}_z \frac{3\bar{P} P_1}{2\pi\varepsilon} (z-b)^{-4} ; \quad (10.314\text{a,b})$$

and (4) the force (10.314b) is evaluated (10.315) at the position (10.304a) of the image. Thus a dipole of moment P_1 *at a distance b from the center of an insulating (conducting) sphere of radius a exerts a repulsive (attractive) axial force:*

$$\vec{F}_\pm = \pm \vec{e}_z \frac{3\bar{P}_1 P_1}{2\pi\varepsilon} \left(b - \frac{a^2}{b} \right)^{-4} = \mp \vec{e}_z \frac{3P_1^2}{2\pi\varepsilon} \frac{a^3}{b^3} \left(\frac{b}{b^2 - a^2} \right)^4 . \quad (10.315)$$

The electric force exerted by a dipole on a conducting sphere is analogous to the hydrodynamic force on a rigid impermeable sphere, both being attractive, whereas the electric force on a conducting sphere is repulsive. Both electric and hydrodynamic forces are considered next.

E10.20.9 Hydrodynamic and Electric Force Exerted by a Dipole on a Sphere

Thus for a hydrodynamic dipole near a rigid impermeable sphere the transformation (10.316a,b) leads to a force of attraction (10.316c):

$$\frac{1}{\varepsilon} \leftrightarrow \rho, \quad \vec{F}_- \leftrightarrow \vec{F}_v : \quad \vec{F}_v = \vec{e}_z \frac{3\rho}{2\pi} P_1^2 \frac{a^3 b}{(b^2 - a^2)^4} , \quad (10.316\text{a,b})$$

which is similar to the electric force of attraction exerted by a dipole on a conducting sphere. In the case of an insulating (conducting) sphere, the image dipole moment with the same (opposite) sign as the original leads to a force of repulsion (attraction). The force of attraction exerted by a dipole on a nearby conducting sphere (Figure 10.17a) may be explained as follows: (1) the dipole induces on the sphere a distribution of electric charges with higher density in modulus on the closest side; (2) the electric charges on the sphere on the closest side have opposite sign to that of

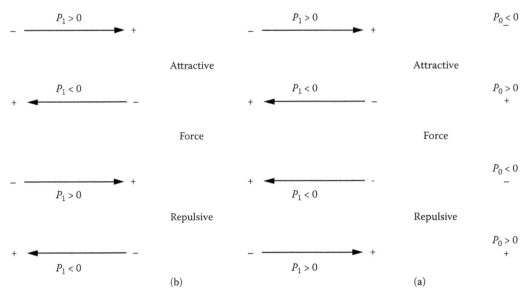

FIGURE 10.18 The Coulomb law states that there is an electric force repulsion (attraction) between [Figure 9.6a (b)] point electric charges with the same (opposite) signs. It follows that (1) there is repulsion (attraction) between a point dipole and a point electric charge (a) if the closest electric charge of the dipole is of the same (opposite) sign to the point electric charge; and (2) there is repulsion (attraction) between two dipoles (b) if the closest electric charges have the same (opposite) signs, that is, the dipole moments are antiparallel (parallel).

the closest electric charge of the dipole; and (3) hence the net electric force is an attraction of the sphere toward the dipole. Figure 10.18 illustrates the electric forces between dipoles and monopoles: (1) two dipoles with opposite (same) sign have charges with opposite (equal) sign that are closer and thus attract (repel) each other; and (2) a dipole is attracted (repelled) by a point charge if it is of the opposite (same) sign as the electric charge on the closest side of the dipole. For the rigid impermeable sphere in the hydrodynamic case, the dipole exerts a force of attraction because at the closer (farther) side the: (1) induced velocity is larger (smaller); hence (2) the pressure is smaller (larger). Thus the net force on the sphere is an attraction toward the dipole. A point flow source (sink) near a rigid impermeable sphere also exerted a force of attraction (Subsection 8.9.15) and has two images (Subsection 8.9.4), namely, a point source (sink) at the reciprocal point whose flow rate is absorbed by a continuous sink (sources) with the same total strength distributed uniformly from the reciprocal point to the center of the sphere. In the case of a point dipole near a rigid impermeable sphere, there is only one dipole image at the reciprocal point, because the dipole has no net flow: the outflow of the source is the inflow of the sink.

NOTE 10.1: Theory of Functions and Potential Fields

The first three books of the present Mathematics and Physics for Science and Technology series form a first subset or "Course on the Theory of Functions and Potential Fields." The theory of functions concerns mostly functions of a complex complex variable [the transcendental functions (elementary and higher) and the generalized functions] that were applied [Books 1 and 2(3)] to potential fields in the plane (space and higher dimensions). The complex functions have three types of general infinite representations: (1) series; (2) infinite products; and (3) continued fractions. The series may be power series (series of fractions) for analytic (meromorphic) functions, which include most of the functions found as solutions of specific problems (Note 10.2). The Laurent series of an analytic function in the neighborhood of an isolated singularities consists, for its principal part, of a multipolar expansion; the multipoles are extended from the plane to space and higher dimensions using generalized functions (Note 10.3).

The simplest multipoles are monopoles and the force between two monopoles (Note 10.4) is specified by: (1) the Newton (Coulomb) law for the gravity (electrostatic) field created by mass (electric charge) distributions; and (2) the Biot–Savart law for the incompressible flow (magnetostatic field) created by vortices (electric currents). The vector fields may be (1) irrotational, (2) solenoidal, or and (3) neither and can be extended to multivectors in multidimensional spaces (Note 10.5). The influence or Green function (Note 10.6) is an example of the close relation (Note 10.7) between (1) complex and generalized functions; (2) vector and tensor geometry; and (3) specified by ordinary and partial differential equations, which form the next subset of the present series as a "Course on Boundary and Initial Value Problems".

NOTE 10.2: Classification of Functions and Function Spaces

The function spaces have been introduced as needed along the first three books, and all are denoted by capital calligraphic letters. A few useful classes of functions may be mentioned: (1) the harmonic functions that are solutions of the Laplace equation and are relevant to irrotational and/or incompressible flows (Chapters I.12, 14, 16, 28, 32, 36, 38; (II.2, 8; III.6), the gravity field (Chapter I.18), electro- and magnetostatic fields (Chapters I.24, 26, 36; III.6), capillarity, elastic membranes, and torsion of prisms (Chapter II.6); (2) the multiharmonic functions, such as the biharmonic functions appearing in plane elasticity and viscous flows (Chapters II.4 and III.6); (3) the orthogonal systems of functions related to Fourier series (Chapter II.5) and integrals (Chapters I.17 and III.5); and (4) the test functions used to define generalized functions (Chapters III.1, 3, 5).

A different classification of functions (List 10.1) concern those occurring most commonly by an arguably increasing order of complexity: (1) the simplest are the rational functions specified by the ratio of polynomials (Chapter I.31); (2) the simplest irrational or elementary transcendental functions are the exponential and the logarithm functions (Chapter II.3) and the circular and the hyperbolic functions and inverses (Chapters II.5, 7); (3) the auxiliary functions include the Gamma function (Notes III.1.8 through III.1.12) and the digamma functions (Subsection I.29.5.2); and (4) the specific functions that are solutions of nonlinear ordinary differential equations include the elliptic functions (Sections I.39.8 and I.39.9); (5) the special functions that are solutions of linear ordinary differential equations with variable coefficients include the Gaussian (Section I.29.9), confluent (Section II.1.9), and generalized (Example I.30.20 and Note II.3.3) hypergeometric functions and the Hermite (Section III.1 and Notes III.1.13 and III.1.14) and the Legendre (Section III.8.3) polynomials; (6) the extended functions are generalizations of the classical special functions such as the hyperspherical Legendre polynomials (Section III.9.6) that extend the Legendre polynomials and multipolar expansion in a three-dimensional space to higher dimensions.

NOTE 10.3: Axisymmetric and Nonaxisymmetric Multipolar Expansions

An analytic function with an isolated singularity has in its neighborhood a Laurent series of ascending and descending powers (Chapter I.25). When applied to the complex conjugate velocity of a plane potential flow (Section I.28.2), the ascending (descending) powers correspond to corner flows (Sections I.14.8 and I.14.9) [multipoles of all orders (Chapter I.12)]. Thus the principal part (Chapter I.27) of the Laurent series represents a multipolar expansion in the plane. The axisymmetric multipolar expansion is extended to the three (higher)-dimensional space using (List 10.2) the original (hyperspherical) Legendre polynomials [Sections III.8.3 and III.8.4 (III.9.6)]. In addition to the axisymmetric multipoles, there are non-axisymmetric multipoles (List 10.3). The non-axisymmetric (axisymmetric) multipoles have only (not only) a potential and no (but also a) stream or field function, and can be extended to anisotropic media (Notes III.8.11 through III.8.15).

NOTE 10.4: Newton, Coulomb, and Biot–Savart Forces

The multipolar expansion can be used to specify the forces and moments in a plane flow (Section I.28.2 and I.28.3 and Examples I.30.14 and I.30.15), for a variety of body shapes (Chapters I.28, 34, 36, 38; III.2, 8; III.6). There are (List 10.4) two kinds of forces (Notes III.8.1 through III.8.16), namely, (1) those associated with irrotational fields, such as the Newton (Coulomb) force [(Chapter I.18 (I.24)] due to the gravity (electric) field of masses (electric charges), including as analogues the compression force between flow source/sinks (Sections I.28.2 and I.28.3); and (2) the solenoidal fields, such as the Biot–Savart force between [(Chapter I.28 (I.26)] vortices (electric currents). These forces between monopoles (Note III.8.10) extend to multipoles (Note III.8.16) in anisotropic media (Notes III.8.11 through III.8.15).

NOTE 10.5: Multivector Irrotational and Solenoidal Fields

Multipoles (List 10.5) can be classified based on six criteria: (1) The axial multipoles that appear in the multipolar expansion in terms of the original (Section III.8.3 and III.8.4) or hyperspherical (Section III.9.6) Legendre polynomials correspond to unidirectional derivatives (Subsection III.6.4.5) and are a subset of the multiaxial multipoles arising from all derivatives of all orders of the monopole with regard to all directions (Subsections III.6.4.6 through 6.4.8); (2) the medium may be isotropic or anisotropic (Notes III.8.1 through III.8.16); (3) the multipoles may lie in the plane (Chapters I.12, 24, 26), in space (Chapters III.6, 8), or in higher dimension (Chapter III.9); (4) the multipole in the plane is a line multipole in space, distinct from a point multipole, and in an N-dimensional space there are multipole distributions of any subspace dimension $0 \leq M \leq N$, such as point multipoles $M = 0$, line multipoles $M = 1$, and multipoles on surfaces $M = 2$, hypersurfaces $M = N - 1$, or domains $M = N$; (5) the potential fields are forced by steady multipoles, because the time dependence is irrelevant for instantaneous propagation, whereas for wave propagation at a finite speed, the unsteady or time-dependent multipoles imply different times of emission by the source and reception by the observer (Note III.10.8); and (6) the vector fields considered before (1) to (5) can be extended to M-vector fields in N-dimensional spaces using tensor calculus (algebra and analysis) and the related differential geometry. The distinct geometries or boundary conditions lead to different free space potentials (Note 10.6), that is, without sources; the forced solutions may be obtained using the influence or Green function (Note 10.7).

NOTE 10.6: Scalar Potential and Flows/Fields in Several Dimensions

The Classification II.8.1 lists 150 distinct potential flow and field problems considered in the first two books of this series. Most of those 150 plane potential problems have three-dimensional analogues, both axisymmetric (Chapter III.6) and non-axisymmetric (Chapter III.8). From these were selected 20 potential flow/field problems discussed in the present book and included in Classification III.9.1; other spatial potential problems arise is electromagnetism, fluid dynamics, solid mechanics, and gravitational and quantum fields. The 20 spatial potential flow/field problems considered in Chapters 6 and 8 of this book are listed in Classification 9.1, together with the analogue plane potential problems; Classification III.9.1 also lists 10 potential problems in any dimension, including more than 3 (Chapter III.9) that are analogous to 10 of 20 spatial potential problems (Chapters III.6 and III.8). This reduced sample is not due to the any impossibility to find analogues in more than three dimensions to all 20 spatial potential problems, or even of the 150 plane problems in Classification III.9.1. Since the application is less immediate, a smaller sample is sufficient to show that the methods of potential theory apply in any dimension, with suitable modifications.

NOTE 10.7: Multidimensional Influence or Green Functions

The main modification for scalar potential fields in any dimension is the influence or Green function in free space, namely, (1) the logarithmic potential in the plane; (2) the inverse distance potential in space; and (3) the inverse power distance potential in higher dimension. With the choice of the appropriate free space influence function, the potential problems in the plane (Chapters I.12, 14, 16, 18, 24, 26, 28, 32, 34, 36, 38; II.2, 8), in space (Chapters III.6, 8), and in higher dimensions (Chapter III.9) share similar methods and properties, such as, (1) the monopoles, dipoles, quadrupoles, multipoles, and multipolar expansions; (2) the images on straight lines/planes/ hyperplanes and the images on cylinders/spheres/ hyperspheres associated with the circle/sphere/hypersphere theorems; (3) unicity, constancy, and asymptotic theorems in the plane, space, or higher dimension valid for arbitrary regions; and (4) the maxima and minima (mean value) theorems in arbitrary domains (circles/spheres/hyperspheres). Thus there is an analogy between (1) the analytic functions in the plane (Chapters I.11, 13, 15, 17, 19, 21, 23, 25, 27, 29, 31, 33, 35, 37, 39); (2) harmonic functions in space (Chapters III.6 and III.8); (3) hyperharmonic functions in arbitrary dimensions (Chapter III.9); and (4) multiharmonic functions (Chapter II.4). All these properties associated with the Laplace operator may be modified (or lost) in more general time-dependent operators; for example, the wave (diffusion) equations have (do not have) an unsteady multipolar expansion. Thus the extension of analytic, generalized, and harmonic functions beyond the Laplace and associated operators leads to other differential equations considered in subsequent books in this series.

NOTE 10.8: Relation of Potential Theory with Differential Equations

There is a close relation between generalized functions, multivectors, and differential equations, for example, (1) the condition that a covariant (contra variant density) M-vector is irrotational (9.414a through c) [solenoidal (9.417a through c)] is a set of (10.316) [(10.317)] simultaneous independent partial differential equations with p (q) dependent variable and the N-dimensional space coordinates as independent variables:

$$M_+ \equiv \overset{*}{\#} \partial_{[i_{M+1}} U_{i_1 \cdots i_M]} = \binom{N}{M+1} = \frac{N!}{(M+1)!(N-M-1)!}, \tag{10.316}$$

$$M_- \equiv \overset{*}{\#} \partial_{i_M} \tilde{U}^{i_1 \cdots i_M} = \binom{N}{M-1} = \frac{N!}{(M-1)!(N-M+1)!}; \tag{10.317}$$

(2) the implication (9.414a through c) $\Rightarrow$ (9.415a through c) [(9.417a through c $\Rightarrow$ (9.418a through c)] corresponds to an existence theorem stating that the solution exists in the form of the curl (divergence) of a covariant $(M-1)$ – vector (10.318a) [contravariant $(M+1)$ – vector density (10.319a)] potential whose number of independent components is (10.318b) [(10.319b)]:

$$U_{i_1 \cdots i_M} = \partial_{[i_M} \Phi_{i_1 \cdots i_{M\,m-1}]} : \quad \overset{*}{\#} \Phi_{i_1 \cdots i_{M-1}} = \binom{N}{M-1} = M_-, \tag{10.318a,b}$$

$$\tilde{U}^{i_1 \cdots i_M} = \partial_{i_{M+1}} \tilde{A}^{i_1 \cdots i_{M+1}} : \quad \overset{*}{\#} A^{i_1 \cdots i_{M+1}} = \binom{N}{M-1} = M_+; \tag{10.319a,b}$$

(3) thus M_+ (M_-) is the number of conditions to be satisfied and M_- (M_+) is the number of functions used to satisfy these conditions; (4) the two numbers generally do not coincide:

$$M_- - M_+ = \binom{N}{M-1} - \binom{N}{M+1} = \frac{N!}{(M+1)!(N-M+1)!} \left[M(M+1) - (N-M+1)(N-M) \right]$$

$$= \frac{N!(2M-N)(N+1)}{(M+1)!(N-M+1)!}; \tag{10.320}$$

(5) the number of conditions can equal the number of unknowns only in an even-dimensional space $N = 2M$ for an M-vector, that is, a multivector with number of indices equal to half the dimension of the space; and (6) in all other cases there are either too many conditions (or too many free functions) showing that the conditions are not independent (more gauge conditions can be imposed on the potential). The preceding is an example of the existence and unicity of solution of a single or a system of ordinary

or partial differential equations; it may also be possible to prove other general properties of the solution, For example, for harmonic functions that are solutions of the Laplace equation (Section III.9.2). Books 1–3 are concerned, respectively, with the theory of complex, transcendental, and generalized functions, with applications mainly to the Laplace operator, in the form of Laplace and Poisson, and also biharmonic, equations: (1) with a variety of boundary conditions, for example, potential flows past different body shapes; and (2) in various physical contexts, such as potential or rotational flow, electro or magnetostatic field, gravity field, elasticity, and steady heat conduction. A natural sequel is to consider (1) more general classes of ordinary and partial differential equations whose properties and solutions are enabled by the preceding theories of functions; and (2) a wider range of physical phenomena than steady potential fields, for example, combining with time-dependent and dissipative phenomena, to lead not only to trajectories, but also to vibration, wave propagation, and diffusion processes.

10.2 Conclusion

The examples of functions with jump discontinuities (angular points), that is, finite discontinuities of the function (its first-order derivative) that can be differentiated indefinitely by replacing them by generalized functions include (1) the symmetrized exponential (hyperbolic sine) in Figure 10.1a (b) and Table 10.1; (2) the one-period circular cosine (sine) in Figure 10.1c (d) and Table 10.2; and (3) the hyperbolic cosine (sine) on the positive real axis in Figure 10.1e (f) and Table 10.3. The cases of linear deflection of an elastic string include (1) a discontinuous mass density with distinct values on each half (Figure 10.2); (2) an inhomogeneous string with continuously varying mass density (Figure 10.3); and (3) a homogeneous string with suspension points at unequal height (Figure 10.4), for which the shape is unsymmetric (Figure 10.4a) due to the difference of height between the supports (Figure 10.4b). The cases of nonlinear deflection of an elastic string include: (1) two concentrated forces at arbitrary positions (Figure 10.5); (2) a uniform plus a concentrated load (Figure 10.6); and (3) a concentrated load (Figure 10.7b) with nonuniform tension (Figure 10.7a). The latter solution (3) involves inverse hyperbolic functions in the nonlinear case, with the linear approximation corresponding to the lowest-order nonzero terms in their power series expansion (Table 10.4). Another infinite process reducing to a single term is the evaluation of integrals by derivatives of the unit impulse (Table 10.5).

The influence or Green function for a simple fourth-order linear ordinary differential equation with constant coefficients can be calculated using Fourier integrals (Table 10.6) whose evaluation (Figures 10.8a,b, 10.9a,b and 10.10a,b) depends on: (1) the location of four poles on the complex plane; (2) the path of integration that closes the real axis; and (3) choice of indenting or not the poles on the real axis. Concerning the weak bending of an elastic bar under its own weight, with one end clamped, and the other held by a straight spring (Figure 10.11), there are three possibilities for opposing the deflection: (1) the critical spring stiffness leads to a horizontal slope at the far end (Figure 10.11a); and (2/3) a smaller (larger) value does not (does) change the curvature of the bar, and [Figure 10.11b (c)] the point of maximum deflection remains at the far end (is moved toward the inside). If the spring is in the reverse position (Figure 10.11a through c), increasing the bending in, (Figure 10.11d), it can lead to instability or failure. Replacing the straight spring (Figure 10.11) by a rotary spring (Figure 10.12) does not change the curvature of the bar in any case; it can decrease bending (Figure 10.12a), or in reverse position increase bending (Figure 10.12b), possibly leading to divergence. The cantilever bars with free end supported on a linear or rotary spring can be compared (Table 10.7).

Two spheres moving along the line joining the centers (Figure 10.13a) create a potential flow that can be calculated by a perturbation method if the distance between the centers is large compared with the radii; the case of spheres with the same radius moving at the same velocity corresponds to images on a plane at equal distance from each other (Figure 10.13b) and specify the potential flow due to a sphere moving parallel to a wall. The method of images can be applied to a dipole [Figure(s) 10.14 (10.15 through 10.17)] near a conducting or insulating plane (sphere), and specifies the electrical (Figure 10.18) force and the analogous hydrodynamic force. The dipole moment is allowed (assumed) to be oblique to the plane (Figure 10.14 [orthogonal to the sphere (Figure 10.16)]; the image [Figure 10.14 (10.15)] on the plane (sphere) is at equal distance (at the reciprocal point). In the case of a dipole with axis orthogonal to a conducting sphere, the image at the reciprocal point (Figure 10.17a) induces surface charges that have axial symmetry and enhance the far field relative to an isolated dipole. If the sphere is insulating (Figure 10.17b) with the dipole axis still pointing away from the center, the electrostatic field is tangent instead of normal to the sphere, that is, the latter changes from an equipotential to a field surface, and the far field is weakened by the opposite image dipole. The force between a dipole and a monopole (Figure 10.18) is repulsive (attractive) if the two closest charges have the same (opposite) signs (Figure 10.18a); the force between two collinear dipoles is repulsive (attractive) if they are parallel (antiparallel), that is, the closest two charges have the same (opposite) signs (Figure 10.18b). The first three books of the series concern the theory of real, complex, and generalized functions, illustrated by applications to plane, spatial, and higher-dimensional fields, and the contents are briefly recalled in five (Lists 10.1 through 10.5).

LIST 10.1 Classification of Functions

  A. *Function spaces*
    1. Capital calligraphic alphabet
  B. *Useful classes of functions:*
    2. Harmonic: I.12, 14, 16, 18, 24, 26, 28, 32, 34, 36, 38; II.2, 8; III.6, 8, 9
    3. Multiharmonic: II.4, III.6
    4. Orthogonal: II.5, III.5
    5. Test functions: III.3
    6. Generalized functions: III.1, 3, 5
  C. *Frequently used functions:*
    7. Rational: I.31
    8. Elementary transcendental: II.5, 7, 9
    9. Auxiliary: I.29–5.2, N.III.1.8–12
   10. Specific: I.39.8–9
   11. Special: I.1.29.9, E.I.10.30, II.1.9, II.3.3; N.III.1.13–14; III.8.3
   12. Extended: III.9.6.

LIST 10.2 Multipolar Expansions

  A. *In the plane:*
    1. Gravity field: I.18.9
    2. Potential flow: I.28.2.2; E.I.30.14–15
  B. *In space:*
    3. Potential flow
    4. Electrostatics: 8.3
    5. Magnetostatics: 8.6
    6. Anisotropic medium: 8.14
  C. *In hyperspace:*
    7. Irrotational field: 9.5 and 9.6

LIST 10.3 Nonaxisymmetric Multipoles

  A. *In space:*
    1. Isotropic medium: 6.4.6
    2. Anisotropic medium: N.8.15
  C. *In hyperspace:*
    3. Isotropic medium: 9.7.8

LIST 10.4 Forces

A. *Two-dimensional:*
1. Gravity: I.18.5, I.28.3.4
2. Electric: I.24.3, I.28.3.2
3. Magnetic: I.26.3, I.28.3.2
4. Compression: I.28.3.2
5. Vortical: I.28.3.2; II.2.2.9
6. Stagnation: I.28.3.4

B. *Three-dimensional:*
7. Gravity: N.8.3
8. Electric: N.8.4
9. Magnetic: N.8.10
10. Compression: N.8.10
11. Vortical: N.8.10

B. *Higher dimensional:*
12. Asymptotic scaling: T.9.1

LIST 10.5 Classification of Multipoles

A. *Axis:*
1. Axis: 6.4.3 through 6.4.5
2. Multiaxial: 6.4.6 through 6.4.8

B. *Isotropic:*
3. Isotropic: 6.4.3 through 6.4.8
4. Anisotropic: N.8.11 through N.8.16

C. *Dimension:*
5. One-dimensional: 9.4.2
6. Plane: 9.4.3
7. Space: 5.6.3; 9.4.4 through 9.4.5
8. Higher dimensional: 9.1 through 9.3

D. *Distribution:*
9. Point-multipole: 6.4.3 through 6.4.8
10. Line-multipole: 9.4.3
11. Surface distribution: 5.2.5
12. Subspace distribution: N.9.25 and N.9.26
13. Spatial distribution: N.9.13

E. *Time dependence:*
14. Steady forcing of potential fields
15. Unsteady wave sources

F. *Vector fields:*
16. Irrotational vector field: scalar potential: 5.8.1
17. Solenoidal vector field: vector potential: 5.8.2
18. General vector field: 5.8.4

G. *Tensor fields:*
19. Irrotational covariant M-vector field: covariant $(M - 1)$ – vector potential: N.9.30
20. Solenoidal contravariant density M-vector field: contravariant $(M + 1)$ – vector density potential: N.9.31
21. General tensor: N.9.11

Bibliography

The bibliography of Book 3 mostly complements those of Books 1 and 2, whose relevance is unchanged. The bibliography on general mathematics (Section 1 of Book 1) and theoretical physics (Section 1 of Book 2) is complemented by engineering technology (Section 1 of Book 3). The bibliography on real and complex functions (Sections 2 and 3 of Book 1 expanded in Sections 2 and 3 of Book 2) is complemented by generalized functions (Section 2 of Book 3). The extension of the potential theory in Books 1 and 2 to higher dimensional spaces and multivectors uses the tensor calculus (Section 3 of Book 3). The hydrodynamics and aerodynamics (Sections 4 and 5 of Book 1) are complemented by other aspects of fluid dynamics (Section 4 of Book 3). The elasticity and plasticity relate to structures (Sections 4 and 5 of Book 2) and also to materials (Section 5 of Book 3). The electricity and magnetism (Section 6 of Book 1) rather than thermodynamics and heat (Section 6 of Book 2) are more closely related to optics and electronics (Section 6 of Book 3). In each of the three books, the six sections of the bibliography include different approaches to the subject; those that have most influenced the contents of the present book are indicated by one, two, or three asterisks.

1. Engineering Technology

Attenborough, M. *Engineering Mathematics Exposed*. McGraw-Hill, New York, 1994.

*Avalone, E. A. and Baumeister III, T. (ed.) *Standard Handbook for Mechanical Engineers*. McGraw-Hill, New York, 1916, 10th edn., 1986.

Bronwell, A. *Advanced Mathematics in Physics and Engineering*. McGraw-Hill, New York, 1953.

Dettman, J. W. *Mathematical Methods in Physics and Engineering*. McGraw-Hill, 1962, reprinted Dover, New York, 1988.

*Harris, C. M. (ed.) *Shock and Vibration Handbook*. McGraw-Hill, New York, 1995.

Harris, C. M. *Handbook of Acoustical Measurements and Noise Control*, 3rd edn. McGraw-Hill, reprinted Acoustical Society of America, New York, 1998.

*Hicks, T. G. (ed.) *Standard Handbook of Engineering Calculations*. McGraw-Hill, 1972, New York, 3rd edn., 1995.

Johnson, R. W. (ed.) *The Handbook of Fluid Mechanics*. CRC Press, Boca Raton, FL, 1998.

Kreysig, E. *Advanced Engineering Mathematics*, 7th edn. John Wiley, New York, 1993.

Lanczos, C. *Applied Analysis*. Prentice-Hall, 1956, reprinted Dover, New York, 1988.

**Lide, D. R. (ed.) *Handbook of Chemistry and Physics*. CRC Press, Boca Raton, FL, 1957, 82nd edn., 2001.

*Menzel, D. H. (ed.) *The Fundamental Formulas of Physics*. Dover, New York, 1960.

O'Neil, P. V. *Advanced Engineering Mathematics*. PWS-Kent Publishing, Boston, MA, 1995.

Orszag, S. A. and Bender, C. M. *Advanced Mathematical Methods for Scientists and Engineers*. McGraw-Hill, New York, 1978.

Parker, S. P. (ed.) *Encyclopedia of Physics*. McGraw-Hill, 1991, New York, 2nd edn., 1993.

*Pelegrin, M. and Hollister, W. M. (eds.) *Aeronautics and Space Systems*. Pergamon Press, Oxford, 1993.

Pipes, L. A. *Applied Mathematics for Engineers and Physicists*. McGraw-Hill, New York, 1946, 2nd edn., 1958.

Poularikas, A. D. (ed.) *The Transforms and Applications*. CRC Press, Boca Raton, FL, 1999.

Sokolnikoff, I. S. and Redheffer, R. M. *Mathematics of Physics and Modern Engineering*. McGraw-Hill, New York, 1958, 2nd edn., 1966.

Stroud, K. A. *Engineering Mathematics*. McMillan, New York, 1970, 4th edn., 1995.

Tucker, A. B. (ed.) *The Computer Science and Engineering Handbook*. CRC Press, Boca Raton, FL, 1996.

*Woan, G. *Handbook of Physics Formulas*. Cambridge University Press, Cambridge, 2000.

Wylie, C. R. *Advanced Engineering Mathematics*. McGraw-Hill, New York, 1951.

Wylie, C. R. and Barrett, L. C. *Advanced Engineering Mathematics*. McGraw-Hill, New York, 1960, 5th edn., 1982.

2. Generalized Functions

Antosik, P., Mikusinski, J., and Sikorski, R. *Generalized Functions: The Sequential Approach*. Elsevier, New York, 1973.

Caratheodory, C. *Measure and Integration*. Birkhauser Verlag, 1956, Basel, Chelsea Publications, New York, 1963.

Critescu, R. and Marinescu, G. *Applications of the Theory of Distributions*. John Wiley, New York, 1973.

*Dirac, P. A. M. *Principles of Quantum Mechanics*. Oxford University Press, Oxford, 1st edn., 1930, 3rd edn., 1947.

Duvaut, G. and Lyons, J. L. *Les inéquations en mécanique et physique*. Dunod, Paris, 1972.

Dwight, H. B. *Tables of Integrals*. MacMillan, New York, 1934, 3rd edn., reprinted Dover, New York, 1957.

*Edwards, G. *Integral Calculus*. McMillan, New York, 1900, Chelsea Publishing, New York, 1960, 2 Vols.

Ferreira, J. C. *Teoria das distribuições,* Fundação Calouste Gulbenkian, Lisbon, 1990.

**Guelfand, I. M., Chilov G. E., Graev, M. I., and Vilenkin, N. J. *Les distributions*. Dunod, Paris, or Academic Press, New York, 5 Vols.

Gillespie, R. P. *Integration*. Oliver & Boyd, London, 1939, 6th edn., 1955.

*Heaviside, O. *Electromagnetic Theory*. MacMillan, 1876, London, Chelsea Publishers, New York, 1956, 3 Vols.

*Jones, D. S. *Generalized Functions*. McGraw-Hill, New York, 1966.

**Kellogg, O. D. *Foundations of Potential Theory*. Dover, New York, 1953.

Lattés, R. and Lyons, J. L. *Méthode de quasi-reversibilité et applications*. Dunod, Paris, 1967.

Lebesgue, L. *Leçons sur l'integration*. 1928, reprinted Chelsea, New York, 1973.

Lederman, W. *Multiple Integrals*. Butler & Tanner, London, 1966.

**Lighthill, M. J. *Fourier Analysis and Generalized Functions*. Cambridge University Press, Cambridge, 1958.

Lyons, J. L. and Magènes, E. *Problémes aux limites non-homogénes*. Dunod, Paris, 1970, 3 Vols.

*MacMillan, W. D. *The Theory of the Potential*. Dover, New York, 1958.

Saks, S. *Theory of the Integral*. Warsaw, 1937.

**Schwartz, L. *Les Distributions*. Hermann, Paris, 1950, 2nd edn., 1966.

Taylor, S. R. *Measure and Integration*. Cambridge University Press, Cambridge, 1966.

Vladimirov, V. *Les distributions en physique mathematique*. Editions Mir, Moscow, 1979.

3. Tensor Calculus

Birkhoff, G. and Beatley, R. *Basic Geometry*. Scott & Foresman, 1940, 2nd edn., 1944, reprinted Chelsea, New York, 1972.

Bold, B. *Famous Problems in Geometry*. Dover, New York, 1969.

Bonola, R. *Non-Euclidean Geometry*. Open Court Publishing, 1912, reprinted Dover, New York, 1955.

Bowen, R. M. and Wang, C. C. *Introduction to Vectors and Tensors*. Plenum Press, London, 1976, 2 Vols.

*Brillouin, L. *Les tenseurs en mécanique et elasticité*. Masson, Paris, 1936, 2éme edition, 1948.

Chasles, C. *Traité des sections coniques*. Gauthier-Villars, Paris, 1865.

Coolidge, J. L. *A Treatise on the Circle and the Sphere*. Cambridge, 1916, reprinted Chelsea Publications, New York, 1971.

Graustein, W. C. *Higher Geometry*. MacMillan, New York, 1940.

Hay, G. E. *Vector and Tensor Analysis*. Dover, New York, 1953.

Hilbert, D. and Cohn-Vossen, H. *Geometry and the Imagination*. Gottingen, 1932, reprinted Chelsea Publications, New York, 1952.

Jeffreys, H. *Cartesian Tensors*. Cambridge University Press, Cambridge, 1931.

Klein, F. *Famous Problems in Geometry*. Gottingen, 1895, translation Beman, W. W. and Smith, D. E., 1897, reprinted Chelsea Publications, New York, 1962.

Klein, F. *Nicht-Euklidische geometrie*. Hannover, 1927, reprinted Chelsea Publications, New York, 1972.

*Lamé, G. *Leçons sur les coordonées curvilignes et leus divers applications*. Gauthier-Villars, Paris, 1859.

Lass, H. *Vector and Tensor Analysis*. McGraw-Hills, New York, 1950.

**Levi-Civita, T. *The Absolute Differential Calculus*. Blackie, London, 1926, reprinted Dover, New York, 1977, 1972.

*Lichnerowicz, A. *Élements de calcul tensoriel*, Armand Colin, Paris, 1950.

Lie, S. *Vorlesungen ueber continuerliche Gruppen*. Leipzig, 1893, reprinted Chelsea Publications, New York, 1971.

*McConnell, A. J. *Applications of Tensor Analysis*. Blakie, London, 1931, reprinted Dover, New York, 1957.

Newell, H. E. *Vector Analysis*. McGraw-Hill, New York, 1955.

Salmon, G. *Treatise on Conic Sections*, 6th edn. Chelsea Publication, New York, 1972.

*Schouten, J. A. *Der Ricci-Kalkul*. Springer, Berlin, 1956.

**Schouten, J. A. *Ricci Calculus*. Springer, Berlin, 1956.

***Schouten, J. A. *Tensor Analysis for Physicists*. Oxford University Press, 1954, Oxford, 2nd edn., 1956.

*Schouten, J. A. and Kulk, W. V. D. *Pfaff s Problem and Its Generalizations*. Oxford University Press, Oxford, 1949.

Schouten, J. A. and Struik, D. J. *Einfuhrung in die Differential Geometrie*. P. Noordhoff, Netherlands, 1935, 2 Vols.

*Simons, S. *Vector Analysis*. Pergamon Press, Oxford, 1964.

Sommerville, D. M. Y. *Bibliography of Non-Euclidean Geometry*. London, U.K., 1911, reprinted Chelsea Publications, New York, 1970.

*Spain, B. *Tensor Calculus*. Oliver & Boyd, London, 1953, 3rd edn., 1960.

Stove, M. H. *Linear Transformations and Hilbert Space*. American Mathematical Society, Providence, RI, 1932.

Synge, J. L. and Schild, A. *Tensor Calculus*. University of Toronto Press, Toronto, 1949.

Tovar de Lemos, A. F. *Tensores cartesianos*. Lisbon, 1977.

Wills, P. *Vector Analysis with an Introduction to the Tensor Analysis*. 1931, Dover, New York, 1958.

Wrede, R. C. *Vector and Tensor Analysis*. Wiley, 1963, reprinted Dover, New York, 1972.

Yano, K. *Lie Derivatives and Its Applications*. North-Holland, the Netherlands, 1955.

4. Fluid Dynamics

Bear, J. *Dynamics of Fluids in Porous Media*. Elsevier, 1972, reprinted Dover, New York, 1988.

Boon, J. P. and Yip, S. *Molecular Hydrodynamics*. McGraw-Hill, 1980, reprinted Dover, New York, 1991.

*Chandrasekhar, S. *Ellipsoidal Figures of Equilibrium*. Yale University Press, Yale, 1969, reprinted Dover, New York, 1987.

Dutton, J. A. *Dynamics of Atmospheric Motion*. McGraw-Hill, 1976, reprinted Dover, New York, 1986.

*Eckart, C. *Hydrodynamics of Oceans and Atmospheres*. Pergamon Press, Oxford, 1960.

Friendlander, S. *Geophysical Fluid Dynamics*. North-Holland, the Netherlands, 1980.

Gennes, P. G. de and Prost, J. *Physics of Liquid Crystals*, Clarendon, Oxford, 1993.

*Greenspan, H. P. *Rotating Fluids*. Cambridge University Press, Cambridge, 1968.

Hamrock, B. J. *Fluid Film Lubrication*. McGraw-Hill, New York, 1993.
Happel, J. and Brenner, H. *Low Reynolds Number Hydrodynamics*. Kluwer, the Netherlands, 1991.
Hines, C. O. *The Upper Atmosphere in Motion*. American Geophysical Union, Washington, DC, 1982.
Isenberg, C. *Soap Films and Soap Bubbles*. Tieto, 1978, reprinted Dover, New York, 1992.
Kuo, C. Y. *Engineering Hydrology*. American Society of Civil Engineers, Reston, VA, 1993.
Lencastre, A. *Hidráulica geral*. Author edition, Lisbon, 1996.
*Lighthill, M. J. *Mathematical Biofluiddynamics*. Society for Industrial Mathematics, Philadelphia, PA, 1970.
Lockhurst,G. R. and Gray, G. W. *The Molecular Physics of Liquid Crystals*. Academic Press, New York, 1979.
Newman, J. N. *Marine Hydrodynamics*. MIT Press, Cambridge, MA, 1978.
Pedley, T. J. *Fluid Dynamics of Large Blood Vessels*. Cambridge University Press, Cambridge, 1974.
Pedlosky, J. *Geophysical Fluid Dynamics*. Springer, Berlin, 1979.
Philips, O. M. *The Dynamics of the Upper Ocean*. Cambridge University Press, Cambridge, 1966, 2nd edn., 1977.
Quintela, A. C. *Hidráulica*. Fundação Calouste Gulbenkian, Lisbon, 1981.
Saucier, W. J. *Meteorological Analysis*. University of Chicago Press, Chicago, OH, 1955, reprinted Dover, New York, 1983.
*Turner, J. S. *Buoyancy Effects in Fluids*. Cambridge University Press, Cambridge, 1973.
Wilks, D. S. *Statistical Methods in the Atmospheric Sciences*. Academic Press, New York, 1995.

5. Materials

Atkins, A. G. and Mai, Y-W. *Elastic and Plastic Fracture*. Wiley, New York, 1985.
Atkinson, J. *Mechanics of Soils and Foundations*. McGraw-Hill, New York, 1993.
Boldyrev, A. K. *Cristalografia*. Editorial Labor, Barcelona, 1934.
Borges, F. S. *Elementos de cristalografia*. Fundação Calouste Gulbenkian, Lisbon, 1980.
Branco, C. A. G. *Mecânica dos materiais*. Fundação Calouste Gulbenkian, Lisbon, 1998.
Broek, D. *Elementary Engineering Fracture Mechanics*. Martinus Nijhoff, 1974, the Netherlands, 4th edn., 1986.
Broek, D. *Practical Use of Fracture Mechanics*. Kluwer, the Netherlands, 1988.
Chen, W. F. and Liu, X. L. *Limit Analysis in Soil Mechanics*. Elsevier, New York, 1990.
Colangelo, V. J. and Heiser, F. A. *Analysis of Metallurgical Failures*. Wiley, New York, 1979.
Collins, J. A. *Failure of Materials in Mechanical Design*. Wiley, New York, 1981.
Cowin, S. C. *Bone Mechanics*. CRC Press, Boca Raton, FL, 2001.
*Cottrell, A. H. *The Mechanical Properties of Matter*. Wiley, New York, 1964.
Cowin, S. C. *Bone Mechanics*. CRC Press, Boca Raton, FL, 2001.
Craig, R. F. *Soil Mechanics*. Span, 1974, 7th edn., 2004.
Das, B. M. *Geotechnical Engineering*. Brooks-Cole, Belmont, CA, 2002.
Davis, H. E., Troxell, G. E., and Hanck, G. F. W. *The Testing of Engineering Materials*. McGraw-Hill, New York, 1955, 4th edn., 1982.
Davison, B. and Owens, G. W. *Steel Designers Manual*. Blackwell, London, 1992.
Fang, H. Y. *Foundation Engineering Handbook*. Chapman & Hall, London, 1991.
Flint, E. *Principes de cristallographie*. Editions Mir, Moscow, 1981.
Gittus, J. *Creep, Vicoelasticity and Creep Fracture in Solids*. Applied Science Publishers, London, 1975.
Guy, A. G. *Essentials of Materials Science*. McGraw-Hill, New York, 1976.
Hogarth, C. A. and Blitz, J. *Techniques of Non-Destructive Testing*. Butterworths, London, 1960.
Jaswon, M. A. and Rose, M. A. *Chrystal Symmetry*. Wiley, New York, 1983.
Juran, J. M., Gryna, F. M., and Bingham, R. S. *Quality Control Handbook*. McGraw-Hill, New York, 1951, 3rd edn., 1979.
Kalpakjian, S. *Tool and Die Failures Source Book*. American Society for Metals, Materials Park, OH, 1982.
Kennett, B. L. N. *The Seismic Wavefield*. Cambridge University Press, Cambridge, 2002.
Kinloch, A. J. and Young, R. J. *Fracture Behaviour of Polymers*. Elsevier, New York, 1983.

Kobayashi, A. S. *Handbook on Experimental Mechanics*. Prentice-Hall, New York, 1987.

Kraus, H. *Creep Analysis*. Wiley, New York, 1980.

Lambe, W. T. and Whitman, R. V. *Soil Mechanics*. Wiley, New York, 1969, 2nd edn., 1979.

Lee, W. H. K., Kanamiri, H., Jennings, P. C., and Kisslinger, C. *International Handbook of Earthquake and Engineering Seismology*. Elsevier, New York, 2002, 2 Vols.

Lewis, W. J. *Treatise on Crystallography*. Cambridge University Press, Cambridge, 1899.

Liebowitz, H. *Fracture*. Academic Press, New York, 1968–1972, 7 Vols.

Little, R. E. *Tables for Estimating Median Fatigue Limits*. American Society for Testing and Materials, West Conshohocken, Pennsylvania, 1981.

Mallet, J. L. *Geomodelling*. Oxford University Press, Oxford, 2003.

McMaster, R. C. *Non-Destructive Testing Handbook*. American Society Metals, Materials Park, OH, 1982, 2 Vols.

Milles, A. K. *Unified Constitutive Equations for Creep and Plasticity*. Elsevier, New York, 1987.

Mitchell, J. K. *Fundamentals of Soil Behaviour*. Wiley, New York, 1993.

Montgomery, D. C. *Design and Analysis of Experiments*. Wiley, New York, 1997, 5th edn., 2001.

Nicolaevskiy, V. N. *Geomechanics and Fluid Dynamics*. Kluwer, the Netherlands, 1996.

Niebel, B. W. and Draper, A. B. *Product Design and Process Engineering*. McGraw-Hill, New York, 1974.

Nielsen, M. P. *Limit Analysis and Concrete Plasticity*. CRC Press, Boca Raton, FL, 1999.

Ozkaya, N. and Nordin, M. *Fundamentals of Biomechanics*. Springer, Berlin, 1998, 3 Vols.

Paz, M. *International Handbook of Earthquake Engineering*. Chapman & Hall, London, 1994.

Pearson, J. R. A. *Mechanics of Polymer Processing*. Elsevier, the Netherlands, 1985.

Philips, F. C. *An Introduction to Crystallography*. Oliver & Boyd, London, 1946, 4th edn., 1971.

Ramsay, J. G. and Lisle, R. J. *Modern Structural Geology*. Academic Press, New York, 2000.

Rabinowicz, E. *Friction and Wear of Materials*. Wiley, New York, 1965.

Rice, R. C. *Fatigue Design Handbook*. Society of Automotive Engineers, Washington, 1981.

Sarkar, A. D. *Friction and Wear*. Academic Press, New York, 1980.

Salvendy, G. *Handbook of Industrial Engineering*. Wiley, New York, 2001.

Seider, W. D, Seade, J. D., and Lewis, D. R. *Product and Process Design Principles*. Wiley, New York, 1999.

Shearer, P. M. *Introduction to Seismology*. Cambridge University Press, Cambridge, 1999.

Sirotine, Y. and Chaskolskaya, M. *Physique des cristaux*. Nauka, Moscow, 1979, Editions Mir, 1984.

Skinner, B. J. and Porter, S. C. *Physical Geology*. Wiley, New York, 1997.

Waddell, J. J. and Dobrowolski, J. A. *Concrete Construction Handbook*. McGraw-Hill, New York, 1993.

Weibull, W. *Fatigue and Analysis of Results*. Pergamon Press, Oxford, 1961.

Winter, D. A. *Biomechanics and Motor Control of Human Movement*. Wiley, New York, 2003.

6. Optics and Electronics

Bartee, T. C. *Digital Computer Fundamentals*. McGraw-Hill, New York, 1960, 2nd edn., 1966.

Boldea, I. and Nasar, S. A. *The Induction Machine Handbook*. CRC Press, Boca Raton, FL, 2002.

Brennan K. F. *The Physics of Semiconductors*. Cambridge University Press, Cambridge, 1999.

Buchdahl, H. A. *Hamiltonian Optics*. Cambridge University Press, Cambridge, 1993.

Chow, W. W and Koch, S. W. *Semiconductor-Laser Fundamentals*. Springer, Berlin, 1999.

Chuang, S. L. *Physics of Optoelectronic Devices*. Wiley, New York, 1995.

Conrady, A. E. *Applied Optics and Optical Design*. Oxford University Press, Oxford, 1929.

*Copson, E. T. *Geometrical Optics*. Cambridge University Press, Cambridge, 1942.

Cormen, T. H., Leiserson, C. E., and Rivest, R. L. *Introduction to Algorithms*. MIT Press, Cambridge, MA, 2000.

Ditchburn, R. W. *Light*. reprinted Dover, New York, 1991.

Dmitriev, V. and Tarassov, L. *Optique Non-Lineaire Appliquée*. Nauka, Moscow, 1982.

Grob, B. *Basic Television and Video Systems*. McGraw-Hill, New York, 1949, 5th edn., 1984.

Hayt, W. H. *Engineering Circuit Analysis*. McGraw-Hill, New York, 1962, 4th edn., 1987.

Hecht, E. *Óptica*. Addison-Wesley, Fundação Calouste Gulbenkian, Lisbon, 1991.

*Jackson, J. D. *Electromagnetic Theory*. Addison-Wesley, New York, 1967.

Kuo, F. *Network Analysis and Synthesis*. Wiley, New York, 1962, 2nd edn., 1966.

*Lorentz, H. A. *Problems of Modern Physics*. Ginn, London, 1927, reprinted Dover, New York, 1967.

Maissel, L. I. and Glang, R. *Handbook of Thin Film Technology*. McGraw-Hill, New York, 1970.

Mandel, L. and Wolf, E. *Optical, Coherence and Quantum Optics*. Cambridge University Press, Cambridge, 1995.

*Moon, P. and Spencer, D. A. *Foundations of Electrodynamics*, Van Nostrand, New York, 1960.

Palik, E. D. *Handbook of Optical Constants of Solids*. Academic Press, New York, 1985, 2 Vols.

Ralston, A. and Meek, C. L. *Encyclopedia of Computer Science*. Van Nostrand, New York, 1976.

Ramo, S., Whinnery, J. R., and Duzer, T. *Fields and Waves in Communications Electronics*. Wiley, New York, 1965, 3rd edn., 1994.

Richter, R. *Elektrishe machinen*. Birkhauser Verlag, Basel, 1954, 4 Vols.

Sholnik, M. I. *Introduction to Radar Systems*. McGraw-Hill, New York, 1962, 2nd edn., 1980.

Saleh, B. E. A. and Teich, M. C. *Fundamentals of Photonics*. McGraw-Hill, New York, 1962, 2nd edn., 1980.

Sippl, C. J. *Computer Dictionary and Handbook*. Howard Sams, Indianapolis, IN, 1966.

Snyder, A. W and Love, J. D. *Optical Waveguide Theory*. Chapman & Hall, London, 1983.

Sommerville, I. *Software Engineering*. Addison-Wesley, New York, 1982, 3rd edn., 1989.

Sullivan, R. J. *Microwave Radar*. Artech House, Norwood, MA, 2000.

Svelto, O. *Principles of Lasers*. Plenum Press, London, 1989.

Terman, F. E. *Electronic and Radio Engineering*. McGraw-Hill, New York, 1932, 4th edn., 1955.

Wohlfarth, E. P. *Ferro-Magnetic Materials*. North-Holland, the Netherlands, 1980, 5 Vols.

Yariv, A. *Quantum Electronics*. Wiley, New York, 1975, 3rd edn., 1989.

Yacoub, M. D. *Wireless Technology*. CRC Press, Boca Raton, FL, 2002.

Yu, P. and Cardona, M. *Fundamentals of Semi-Conductors*. Cambridge University Press, Cambridge, 1999.

References

(To authors mentioned in the text, in chronological order, indicating: date, name, publication)

1678   Hooke, R. *De Potentia Restitutiva*. London, U.K.

1687   Newton, I. *Philosophiae naturalis principia mathematica*. Cambridge University Press, reprinted University of California Press, 1934, 2 Vols.; *Exposition des découvertes philosophiques de M. le chevalier Newton*, C. MacLaurin, Royal Society of London, traduction Acatémic des Sciences de Paris, 1746.

1729   Euler, L. Letter to Ch. Goldbach

1738   Bernoulli, D. *Hydrodinamica*. Argentorati.

1744   Bernoulli, D. Véritable hypothése de la résistance des solides, avec demonstration de la courbure des corps qui font ressort, Geneva, Switzerland *(Gesammelte Werke, 2)*.

1744   Euler, L. De curvis elasticis, in *Methodus inveniendi lineas curvas maximi and minimi proprietate gaudentes*. Lausanne.

1747   D'Alembert, J. R. Essai sur les vibrations des cordes. *Opuscules Mathématiques* 1,1–47 Paris, France.

1752   Euler, L. Principles géneraux du mouvement des fluides. Histoire de l'Académie de Berlin 1755, in Truesdell, C. A. *Rational Fluid Mechanics 1687–1765*.

1759   Euler, L. De Principis motus fluidorum. *Novi Commentari Academia Petropolitana*. 14,1: *Opera Omnia*, Basel Akademie.

1785   Legendre, A. M. *Mémoires de Mathématique et Physique presentées à l'Academie des Sciences par divers savants* 10, 411–434.

1807   Young, T. *A Course of Lectures on Natural Philosophy and the Mechanical Arts*. London, U.K., reprinted Kelland, 1845.

1809   Gauss, C. F. Theorie motus corporus coelestium. Gottingen, *Werke* 7.

1813   Euler, L. Commentatio in fractionem continuam qua illustris La Grange potestates binomials expressif. *Mémoires de l'Académie Imperiale des Sciences de St. Petersburg*, 6.

1813   Poisson, S. D. Remarques sur une équation qui se présent dans la théorie des attractions spheroides. *Nouveau Bulletin de la Société Philomatique de Paris* 3, 388–392.

1814   Rodrigues, O. *Correspondence de l'École Polytechnique* 3, 361–385

1818   Fourier, J. B. J. *Theorie analytique de la chaleur*. Paris, France, reprinted Dover, 1955.

1820   Laplace, P. S. *Mécanique Celeste*. Gaulthier-Villars, reprinted Chelsea, 1950, 5 Vols.

1822   Navier, C. L. M. H. Mémoire sur les lois du mouvement des fluides. *Mémoires de l'Academie des Sciences de Paris* 6, 389

1825   Laplace, P. S. *Mécanique Céleste*. Bachelier, Paris, France, reprinted Chelsea, 1982, 5 Vols.

1827   Jacobi, C. G. J. *Astronomische Nachrichten* 6, 123; *Fundamenta nova theoriae functionum ellipticarum*, Konigsberg; *Gesammelte Werke* 1, 49–239, reprinted Chelsea, 1969.

1828   Green, G. Essay on electricity and magnetism, *Mathematical Papers*. Cambridge University Press, Nottingham, 1871, p. 3.

1828   Ostrogradski, M. V. *On heat theory* (in Russian).

1829a   Poisson, S. D. Mémoire sur l'équilibre et le mouvement des corps élastiques. *Memöires de l'Académies des Siences de Paris*, 8.

1829b   Poisson, S. D. Mémoire sur les équations generales de l'équilibre et mouvement des corps élastiques et des fluides. *Journal de l'Ecole Polytechnique 8, 1*.

1836   Sturm, J. C. F. *Journal de Mathematiques*. 1, 106.

1837   Green, G. On the laws of reflection and refraction of light at the common surface of two non-crystallized media. *Cambridge Philosophical Transactions (Mathematical Papers, 245)*.

1843   Saint-Venant, B. *Comptes Rendus de l 'Academie des Sciences* 13, 1240.

1845   Stokes, G. On the theories of internal friction of fluids in motion. *Philosophical Transactions of the Cambridge Philosophical Society* 8, 187 (*Papers* 1, 75).

1847   Liouville, J. *Leçons* and Borchardt, C. W. *Journal fur reine and angewandte Mathematik* 277–310 (1880).

1850   Dirichlet, P. G. J. *Abhandlungen der Königliche Preussiche Akademie der Wissenschaften* 99–116.

1854   Stokes, G.G. *The Correspondence between Sir George Gabriel Stokes and Sir William Thomson, Baron Kelvin of Largs*, Volume 1: 1846–1869. Ed. B. Wilson, Cambridge University Press, 1990.

1856   Saint-Venant, B. *Journal de Mathématiques 2,1*.

1858   Rouché, E. *Journal de l'École Polytechnique, 21*.

1858   Helmholtz, H. L. *Crelle* 40.

1861   Hankel, H. Zun allgeneime theorie des Bewegung des Flussigkeiten *Gottingen*.

1864   Hermite, Ch. Sur un nouveau development em série de functions. *Comptes Rendus de l'Academic des Sciences Paris* 58, 93–100 (*Ouevres* 2, 293–308).

1869   Thomson, J. J. On vortex motion. *Transaction of the Edinburg Society* 25 *Papers* 413.

1871   Rankine, W. J. M. On the mathematical theory of stream lines, especially those with four foci and upwards. *Philosophical Transactions of the Royal Society*, 167, 267–306.

1873   Maxwell, J. C. *Treatise of Electricity and Magnetism*. Clarendon Press, Oxford 3rd edn., 1891, 2 Vols. reprinted Dover, 1954.

1876   Heaviside, O. *Electromagnetic Theory*. Reprinted Chelsea, 1956, 3 Vols.

1886   Robin, G. 1886 Sur la distribution d'électricité à la surface des conductteurs fermés et des conducteurs ouverts. *Annales Scientifiques de l'École Normale Supérieure* 3, 31–358.

1889   Hölder, O. Ueber ein Mitteelwerthsatz. *Nachrichten der Gesellscghaft der Wissenschaften Göttingen*, 38–47.

1890   Schwartz, H. A. *Abhandlungen*. Teubner, 1890 Berlin, Germany, reprinted Chelsea, 1972.

1894   Hill, M. J. M. On a spherical vortex. *Transactions of the Royal Society* 135.

1896   Minkowski, H. *Geometrie des Zahlen*. Leipzig, 1, 115–117, reprinted Chelsea, 1953.

1916   Einstein, A. Die Grundlage der allgemeinen relativitatstheorie. *Annalen der Physik* 49, 769 (*Sitzberichte der Preussichen Akademie der Wissenschaften Berlin* 778–786 1915).

1921   Timoshenko, S. On the corrections for shear of the differential equation for the transverse vibration of prismatic bars. *Philosophical Magazine* 41, 744–746.

1926   Schrodinger, E. *Annalen der Physik* 79, 361.

1926   Klein, O. *Zeitschrift fur Physik* 37, 895.

1926   Gordon, W. *Zeitschrift fur Physik* 40, 121.

1930   Dirac, P. A. M. *Quantum Mechanics*. Cambridge University Press, 3rd edn., 1947.

1945   Weiss, P. *Proceedings of the Cambridge Philosophical Society* 40.

1953   Butler, S. F. J. *Proceedings of the Cambridge Philosophical Society* 49, 169–174.

1961   Neumann, J. v. *Collected Works*. Pergamon Press, 1961, 4 Vols..

1961   Vlasov, B. Z. *Thin Walled Elastic Beams*. (Israel program of scientific translations, Jerusalem, Israel).

1975   Howe, M. S. Contribution to the theory of aerodynamic sound with applications to excess jet noise and the theory of the flute. *Journal of Fluid Mechanics* 72, 625–673.

1978   Campos, L. M. B. C. On the emission of sound by an ionized inhomogeneity. *Proceedings of the Royal Society A* 359, 65–91.

1992   Campos, L. and Viaño, J. M. Mathematical modelling of rods. In *Handbook of Numerical Analysis*. eds. Ciarlet, P. G. and Lyons, J. L. Elsevier.

1995   Johnson, N. L., Kotz, S., and Balakrishnan, N. *Continuous Univariate Probability Distributions*. Wiley

2004   Campos, L. M. B. C. and Marques, J. M. G. On a combination of gamma and generalized error distributions with application to aircraft flight path deviations. *Communications in Statistics*, 10, 2307–2332.

2011   Campos, L. M. B. C. and Lau, F. J. P. On sound generation by moving surfaces and convected sources in a flow. *International Journal of Aeroacoustics*, 11,103–136.

2012   Campos, L. M. B. C. and Cunha, F. S. R. P. On hyperspherical Legendre polynomials and higher dimensional multipole expansions. *Journal of Inequalities and Special Functions*, 3, 1–28.

Index